Artificial Intelligence: Foundations, Theory, and Algorithms

More information about this series at http://www.springer.com/series/13900

Fabio Cuzzolin

The Geometry of Uncertainty

The Geometry of Imprecise Probabilities

Fabio Cuzzolin
Department of Computing & Communication
Oxford Brookes University
Oxford, UK

ISSN 2365-3051 ISSN 2365-306X (electronic)
Artificial Intelligence: Foundations, Theory, and Algorithms
ISBN 978-3-030-63155-0 ISBN 978-3-030-63153-6 (eBook)
https://doi.org/10.1007/978-3-030-63153-6

This Springer imprint is published by the registered company Springer Nature Switzerland AG
The registered company address is: Gewerbestrasse 11, 6330 Cham, Switzerland

To my dearest parents Elsa and Albino, whom I owe so much to, my beloved wife Natalia and my beautiful son Leonardo

Preface

Uncertainty

Uncertainty is of paramount importance in artificial intelligence, applied science, and many other areas of human endeavour. Whilst each and every one of us possesses some intuitive grasp of what uncertainty is, providing a formal definition can prove elusive. Uncertainty can be understood as a lack of information about an issue of interest for a certain agent (e.g., a human decision maker or a machine), a condition of limited knowledge in which it is impossible to exactly describe the state of the world or its future evolution.

According to Dennis Lindley [1175]:

" *There are some things that you know to be true, and others that you know to be false; yet, despite this extensive knowledge that you have, there remain many things whose truth or falsity is not known to you. We say that you are uncertain about them. You are uncertain, to varying degrees, about everything in the future; much of the past is hidden from you; and there is a lot of the present about which you do not have full information. Uncertainty is everywhere and you cannot escape from it* ".

What is sometimes less clear to scientists themselves is the existence of a hiatus between two fundamentally distinct forms of uncertainty. The first level consists of somewhat '*predictable*' variations, which are typically encoded as probability distributions. For instance, if a person plays a fair roulette wheel they will not, by any means, know the outcome in advance, but they will nevertheless be able to predict the frequency with which each outcome manifests itself (1/36), at least in the long run. The second level is about '*unpredictable*' variations, which reflect a more fundamental uncertainty about the laws themselves which govern the outcome. Continuing with our example, suppose that the player is presented with ten different doors, which lead to rooms each containing a roulette wheel modelled by a different probability distribution. They will then be uncertain about the very game they are supposed to play. How will this affect their betting behaviour, for instance?

Lack of knowledge of the second kind is often called *Knightian* uncertainty [1007, 831], from the Chicago economist Frank Knight. He would famously distinguish 'risk' from 'uncertainty':

"*Uncertainty must be taken in a sense radically distinct from the familiar notion of risk, from which it has never been properly separated ... The essential fact is that 'risk' means in some cases a quantity susceptible of measurement, while at other times it is something distinctly not of this character; and there are far-reaching and crucial differences in the bearings of the phenomena depending on which of the two*

is really present and operating ... It will appear that a measurable uncertainty, or 'risk' proper, as we shall use the term, is so far different from an unmeasurable one that it is not in effect an uncertainty at all."
In Knight's terms, 'risk' is what people normally call *probability* or *chance*, while the term 'uncertainty' is reserved for second-order uncertainty. The latter has a measurable consequence on human behaviour: people are demonstrably averse to unpredictable variations (as highlighted by *Ellsberg's paradox* [569]).

This difference between predictable and unpredictable variation is one of the fundamental issues in the philosophy of probability, and is sometimes referred to as the distinction between *common cause* and *special cause* [1739]. Different interpretations of probability treat these two aspects of uncertainty in different ways, as debated by economists such as John Maynard Keynes [961] and G. L. S. Shackle.

Probability

Measure-theoretical probability, due to the Russian mathematician Andrey Kolmogorov [1030], is the mainstream mathematical theory of (first-order) uncertainty. In Kolmogorov's mathematical approach probability is simply an application of measure theory, and uncertainty is modelled using additive measures.

A number of authors, however, have argued that measure-theoretical probability theory is not quite up to the task when it comes to encoding second-order uncertainty. In particular, as we discuss in the Introduction, additive probability measures cannot properly model missing data or data that comes in the form of *sets*. Probability theory's frequentist interpretation is utterly incapable of modelling 'pure' data (without 'designing' the experiment which generates it). In a way, it cannot even properly model continuous data (owing to the fact that, under measure-theoretical probability, every point of a continuous domain has zero probability), and has to resort to the 'tail event' contraption to assess its own hypotheses. Scarce data can only be effectively modelled asymptotically.

Bayesian reasoning is also plagued by many serious limitations: (i) it just cannot model ignorance (absence of data); (ii) it cannot model pure data (without artificially introducing a prior, even when there is no justification for doing so); (iii) it cannot model 'uncertain' data, i.e., information not in the form of propositions of the kind 'A is true'; and (iv) again, it is able to model scarce data only asymptotically, thanks to the Bernstein–von Mises theorem [1841].

Beyond probability

Similar considerations have led a number of scientists to recognise the need for a coherent mathematical theory of uncertainty able to properly tackle all these issues. Both alternatives to and extensions of classical probability theory have been proposed, starting from de Finetti's pioneering work on subjective probability [403]. Formalisms include possibility-fuzzy set theory [2084, 533], probability intervals

[784], credal sets [1141, 1086], monotone capacities [1911], random sets [1344] and imprecise probability theory [1874]. New original foundations of subjective probability in behavioural terms [1877] or by means of game theory [1615] have been put forward. The following table presents a sketchy timeline of the various existing approaches to the mathematics of uncertainty.

Imprecise-probabilistic theories: a timeline

Approach	Proposer(s)	Seminal paper	Year
Interval probabilities	John Maynard Keynes	A treatise on probability	1921
Subjective probability	Bruno de Finetti	Sul significato soggettivo della probabilità	1931
Theory of previsions	Bruno de Finetti	La prévision: ses lois logiques, ses sources subjectives	1937
Theory of capacities	Gustave Choquet	Theory of capacities	1953
Fuzzy theory	Lotfi Zadeh, Dieter Klaua	Fuzzy sets	1965
Theory of evidence	Arthur Dempster, Glenn Shafer	Upper and lower probabilities induced by a multivalued mapping; A mathematical theory of evidence	1967, 1976
Fuzzy measures	Michio Sugeno	Theory of fuzzy integrals and its applications	1974
Credal sets	Isaac Levi	The enterprise of knowledge	1980
Possibility theory	Didier Dubois, Henri Prade	Théorie des possibilités	1985
Imprecise probability	Peter Walley	Statistical reasoning with imprecise probabilities	1991
Game-theoretical probability	Glenn Shafer, Vladimir Vovk	Probability and finance: It's only a game!	2001

Sometimes collectively referred to as *imprecise probabilities* (as most of them comprise classical probabilities as a special case), these theories in fact form, as we will see in more detail in Chapter 6, an entire hierarchy of encapsulated formalisms.

Belief functions

One of the most popular formalisms for a mathematics of uncertainty, the *theory of evidence* [1583] was introduced in the 1970s by Glenn Shafer as a way of representing epistemic knowledge, starting from a sequence of seminal papers [415, 417, 418] by Arthur Dempster [416]. In this formalism, the best representation of chance is a *belief function* rather than a traditional probability distribution. Belief functions assign probability values to *sets* of outcomes, rather than single events. In this sense, belief functions are closely related to *random sets* [1268, 1857, 826]. Important work on the mathematics of random set theory has been conducted in recent years by Ilya Molchanov [1302, 1304].

In its original formulation by Dempster and Shafer, the formalism provides a simple method for merging the evidence carried by a number of distinct sources (called *Dempster's rule of combination* [773]), with no need for any prior distributions [1949]. The existence of different levels of granularity in knowledge representation is formalised via the concept of a 'family of compatible frames'.

The reason for the wide interest the theory of belief functions has attracted over the years is that it addresses most of the issues probability theory has with the handling of second-order uncertainty. It starts from the assumption that observations are indeed set-valued and that evidence is, in general, in support of propositions rather than single outcomes. It can model ignorance by simply assigning mass to the whole sample space or 'frame of discernment'. It copes with missing data in the most natural of ways, and can coherently represent evidence on different but compatible sample spaces. It does not 'need' priors but can make good use of them whenever there is actual prior knowledge to exploit. As a direct generalisation of classical probability, the theory's rationale is relatively easier to grasp. Last but not least, the formalism does not require us to entirely abandon the notion of an event, as is the case for Walley's imprecise probability theory [1874]. In addition, the theory of evidence exhibits links to most other theories of uncertainty, as it includes fuzzy and possibility theory as a special case and it relates to the theory of credal sets and imprecise probability theory (as belief functions can be seen as a special case of convex sets of probability measures). Belief functions are infinitely-monotone capacities, and have natural interpretations in the framework of probabilistic logic, and modal logic in particular.

Since its inception, the formalism has expanded to address issues such as inference (how to map data to a belief function), conditioning, and the generalisation of the notion of entropy and of classical results from probability theory to the more complex case of belief functions. The question of what combination rule is most appropriate under what circumstances has been hotly debated, together with that of mitigating the computational complexity of working with sets of hypotheses. Graphical models, machine learning approaches and decision making frameworks based on belief theory have also been developed.

A number of questions still remain open, for instance on what is the correct epistemic interpretation of belief functions, whether we should actually manipulate intervals of belief functions rather than single quantities, and how to formulate an effective general theory for the case of continuous sample spaces.

Aim(s) of the book

The principal aim of this book is to introduce to the widest possible audience an original view of belief calculus and uncertainty theory which I first developed during my doctoral term in Padua. In this *geometric approach to uncertainty*, uncertainty measures can be seen as points of a suitably complex geometric space, and there manipulated (e.g. combined, conditioned and so on).

The idea first sprang to my mind just after I had been introduced to non-additive probabilities. Where did such objects live, I wondered, when compared to classical, additive probabilities defined on the same sample space? How is their greater complexity reflected in the geometry of their space? Is the latter an expression of the greater degree of freedom these more complex objects can provide?

For the reasons mentioned above, my attention was first drawn to belief functions and their combination rule which, from the point of view of an engineer, appeared to provide a possible principled solution to the sensor fusion problems one encounters in computer vision when making predictions or decisions based on multiple measurements or 'features'. Using the intuition gathered in the simplest case study of a binary domain, I then proceeded to describe the geometry of belief functions and their combination in fully general terms and to extend, in part, this geometric analysis to other classes of uncertainty measures.

This programme of work is still far from reaching its conclusion – nevertheless, I thought that the idea of consolidating my twenty-year work on the geometry of uncertainty in a monograph had some merit, especially in order to disseminate the notion and encourage a new generation of scientists to develop it further. This is the purpose of the core of the book, Parts II, III and IV.

In the years that it took for this project to materialise, I realised that the manuscript could serve the wider purpose of illustrating the rationale for moving away from probability theory to non-experts and interested practitioners, of which there are many. Incidentally, this forced me to reconsider from the very foundations the reasons for modelling uncertainty in a non-standard way. These reasons, as understood by myself, can be found in the Introduction, which is an extended version of the tutorial I gave on the topic at IJCAI 2016, the International Joint Conference on Artificial Intelligence, and the talk I was invited to give at Harvard University in the same year.

The apparent lack of a comprehensive treatise on belief calculus in its current, modern form (and, from a wider perspective, of uncertainty theory) motivated me to make use of this book to provide what turned out to be probably the most complete summary (to the best of my knowledge) of the theory of belief functions. The entire first part of the book is devoted to this purpose. Part I is not quite a 'manual' on belief calculus, with easy recipes the interested practitioner can just follow, but does strive to make a serious effort in that direction. Furthermore, the first part of the book concludes with what I believe to be the most complete compendium of the various approaches to uncertainty theory, with a specific focus on how do they relate to the theory of evidence. All major formalisms are described in quite some detail, but an effort was really made to cover, albeit briefly, all published approaches to a mathematics of uncertainty and variations thereof.

Finally, the last chapter of the book advances a tentative research agenda for the future of the field, inspired by my own reflections and ideas on this. As will become clearer in the remainder of this work, my intuition brings me to favour a random-set view of uncertainty theory, driven by an analysis of the actual issues with data that expose the limitations of probability theory. As a result, the research problems I propose tend to point in this direction. Importantly, I strongly believe that, to break the near-monopoly of probability theory in science, uncertainty theory needs to measure itself with the really challenging issues of our time (climate change, robust artificial intelligence), compete with mainstream approaches and demonstrate its superior expressive power on their own grounds.

Last but not least, the book provides, again to the best of my knowledge, the largest existing bibliography on belief and uncertainty theory.

Structure and topics

Accordingly, as explained, this book is divided into five Parts.

Part I, 'Theories of uncertainty', is a rather extensive recapitulation of the current state of the art in the mathematics of uncertainty, with a focus on belief theory. The Introduction provided in Chapter 1 motivates in more detail the need to go beyond classical probability in order to model realistic, second-order uncertainty, introduces the most significant approaches to the mathematics of uncertainty and presents the main principles of the theory of belief functions. Chapter 2 provides a succinct summary of the basic notions of the theory of belief functions as formulated by Shafer. Chapter 3 digs deeper by recalling the multiple semantics of belief functions, discussing the genesis of the approach and the subsequent debate, and illustrating the various original frameworks proposed by a number of authors which use belief theory as a basis, while developing it further in original ways. Chapter 4 can be thought of as a manual for the working scientist keen on applying belief theory. It illustrates in detail all the elements of the evidential reasoning chain, delving into all its aspects, including inference, conditioning and combination, efficient computation, decision making and continuous formulations. Notable advances in the mathematics of belief functions are also briefly described. Chapter 5 surveys the existing array of classification, clustering, regression and estimation tools based on belief function theory. Finally, Chapter 6 is designed to provide the reader with a bigger picture of the whole field of uncertainty theory, by reviewing all major formalisms (the most significant of which are arguably Walley's imprecise probability, the theory of capacities and fuzzy/possibility theory), with special attention paid to their relationship with belief and random set theory.

Part II, 'The geometry of uncertainty', is the core of this book, as it introduces the author's own geometric approach to uncertainty theory, starting with the geometry of belief functions. First, Chapter 7 studies the geometry of the space of belief functions, or *belief space*, both in terms of a simplex (a higher-dimensional triangle) and in terms of its recursive bundle structure. Chapter 8 extends the analysis to Dempster's rule of combination, introducing the notion of a conditional subspace and outlining a simple geometric construction for Dempster's sum. Chapter 9 delves into the combinatorial properties of plausibility and commonality functions, as equivalent representations of the evidence carried by a belief function. It shows that the corresponding spaces also behave like simplices, which are congruent to the belief space. The remaining Chapter 10 starts extending the applicability of the geometric approach to other uncertainty measures, focusing in particular on possibility measures (consonant belief functions) and the related notion of a consistent belief function.

Part III, 'Geometric interplays', is concerned with the interplay of uncertainty measures of different kinds, and the geometry of their relationship. Chapters 11 and 12 study the problem of transforming a belief function into a classical probability measure. In particular, Chapter 11 introduces the *affine family* of probability transformations, those which commute with affine combination in the belief space. Chapter 12 focuses instead on the *epistemic* family of transforms, namely 'relative belief' and 'relative plausibility', studies their dual properties with respect to Dempster's sum, and describes their geometry on both the probability simplex and the belief space. Chapter 13 extends the analysis to the consonant approximation problem, the problem of finding the possibility measure which best approximates a given belief function. In particular, approximations induced by classical Minkowski norms are derived, and compared with classical outer consonant approximations. Chapter 14 concludes Part III by describing Minkowski consistent approximations of belief functions in both the mass and the belief space representations.

Part IV, 'Geometric reasoning', examines the application of the geometric approach to the various elements of the reasoning chain illustrated in Chapter 4. Chapter 15 tackles the conditioning problem from a geometric point of view. Conditional belief functions are defined as those which minimise an appropriate distance between the original belief function and the 'conditioning simplex' associated with the conditioning event. Analytical expressions are derived for both the belief and the mass space representations, in the case of classical Minkowski distances. Chapter 16 provides a semantics for the main probability transforms in terms of credal sets, i.e., convex sets of probabilities. Based on this interpretation, decision-making apparatuses similar to Smets's transferable belief model are outlined.

Part V, 'The future of uncertainty', consisting of Chapter 17, concludes the book by outlining an agenda for the future of the discipline. A comprehensive statistical theory of random sets is proposed as the natural destination of this evolving field. A number of open challenges in the current formulation of belief theory are highlighted, and a research programme for the completion of our geometric approach to uncertainty is proposed. Finally, very high-impact applications in fields such as climate change, rare event estimation, machine learning and statistical learning theory are singled out as potential triggers of a much larger diffusion of these techniques and of uncertainty theory in general.

Acknowledgements

This book would not have come to existence had I not, during my doctoral term at the University of Padua, been advised by my then supervisor Ruggero Frezza to attend a seminar by Claudio Sossai at Consiglio Nazionale delle Ricerche (CNR). At the time, as every PhD student in their first year, I was struggling to identify a topic of research to focus on, and felt quite overwhelmed by how much I did not know about almost everything. After Claudio's seminar I was completely taken by the idea of non-additive probability, and ended up dedicating my entire doctorate to the study of the mathematics of these beautiful objects. Incidentally, I need to thank Nicola Zingirian and Riccardo Bernardini for enduring my ramblings during our coffee breaks back then.

I would like to acknowledge the encouragement I received from Glenn Shafer over the years, which has greatly helped me in my determination to contribute to belief theory in terms of both understanding and impact. I would also like to thank Art Dempster for concretely supporting me in my career and for the insights he shared with me in our epistolary exchanges (not to mention the amazing hospitality displayed during my visit to Harvard in 2016).

Running the risk of forgetting somebody important, I would like to thank Teddy Seidenfeld, Gert de Cooman, Matthias Troffaes, Robin Gong, Xiao-Li Meng, Frank Coolen, Thierry Denoeux, Paul Snow, Jonathan Lawry, Arnaud Martin, Johan Schubert, Anne-Laure Jousselme, Sebastien Destercke, Milan Daniel, Jim Hall, Alberto Bernardini, Thomas Burger and Alessandro Antonucci for the fascinating conversations that all deeply influenced my work. I am also thankful to Ilya Molchanov for his help with better understanding the links between belief function theory and random set theory.

Finally, I am grateful to my Springer editor Ronan Nugent for his patience and support throughout all those years that passed from when we started this project to the moment we could actually cheer its successful outcome.

Oxford, United Kingdom *Fabio Cuzzolin*

September 2020

Table of Contents

Part II The geometry of uncertainty

Part IV Geometric reasoning

Part V The future of uncertainty

Introduction

1.1 Mathematical probability

The mainstream mathematical theory of uncertainty is measure-theoretical probability, and is mainly due to the Russian mathematician Andrey Kolmogorov [1030]. As most readers will know, in Kolmogorov's mathematical approach[1] probability is simply an application of *measure theory* [783], the theory of assigning numbers to sets. In particular, Kolmogorov's probability measures are *additive* measures, i.e., the real value assigned to a set of outcomes is the sum of the values assigned to its constituent elements. The collection Ω of possible outcomes (of a random experiment or of a decision problem) is called the *sample space*, or universe of discourse. Any (measurable) subset A of the universe Ω is called an *event*, and is assigned a real number between 0 and 1.

Formally [1030], let Ω be the sample space, and let 2^Ω represent its power set $2^\Omega \doteq \{A \subset \Omega\}$. The power set is also sometimes denoted by $\mathcal{P}(\Theta)$.

Definition 1. *A collection $\mathcal{F}$ of subsets of the sample space, $\mathcal{F} \subset 2^\Omega$, is called a* σ-algebra *or* σ-field *if it satisfies the following three properties:*

- *$\mathcal{F}$ is non-empty: there is at least one $A \subset \Omega$ in $\mathcal{F}$;*
- *$\mathcal{F}$ is closed under complementation: if A is in $\mathcal{F}$, then so is its complement, $\overline{A} = \{\omega \in \Omega, \omega \notin A\} \in \mathcal{F}$;*
- *$\mathcal{F}$ is closed under countable union: if $A_1, A_2, A_3, \cdots$ are in $\mathcal{F}$, then so is $A = A_1 \cup A_2 \cup A_3 \cup \cdots$,*

[1] A recent study of the origins of Kolmogorov's work has been done by Shafer and Vovk: http://www.probabilityandfinance.com/articles/04.pdf.

F. Cuzzolin, *The Geometry of Uncertainty*, Artificial Intelligence: Foundations, Theory, and Algorithms, https://doi.org/10.1007/978-3-030-63153-6_1

where $\cup$ *denotes the usual set-theoretical union.*

Any subset of Ω which belongs to such a σ-algebra is said to be *measurable*. From the above properties, it follows that any σ-algebra $\mathcal{F}$ is closed under *countable intersection* as well by De Morgan's laws: $\overline{A \cup B} = \overline{A} \cap \overline{B}$, $\overline{A \cap B} = \overline{A} \cup \overline{B}$.

Definition 2. *A* probability measure *over a σ-algebra $\mathcal{F} \subset 2^{\Omega}$, associated with a sample space Ω, is a function $P : \mathcal{F} \to [0, 1]$ such that:*

- $P(\emptyset) = 0$;
- $P(\Omega) = 1$;
- *if* $A \cap B = \emptyset,\ A, B \in \mathcal{F}$ *then* $P(A \cup B) = P(A) + P(B)$ (additivity).

A simple example of a probability measure associated with a spinning wheel is shown in Fig. 1.1.

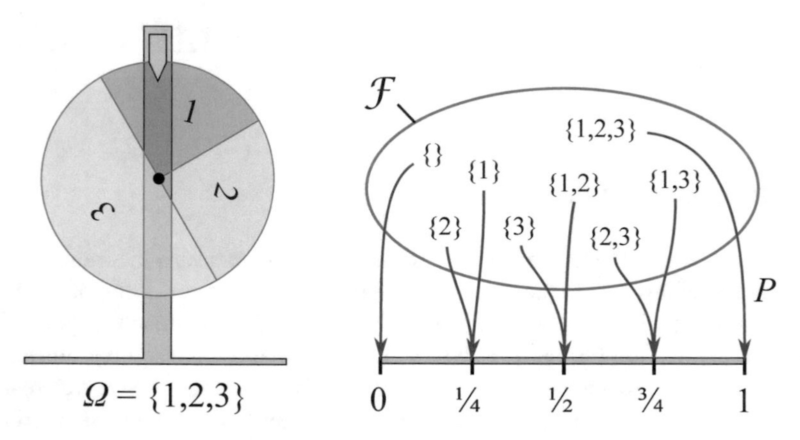

Fig. 1.1: A spinning wheel is a physical mechanism whose outcomes are associated with a (discrete) probability measure (adapted from original work by Ziggystar, `https://commons.wikimedia.org/wiki/File:Probability-measure.svg`).

A sample space Ω together with a σ-algebra $\mathcal{F}$ of its subsets and a probability measure P on $\mathcal{F}$ forms a *probability space*, namely the triplet $(\Omega, \mathcal{F}, P)$. Based on the notion of a probability space, one can define that of a *random variable*. A random variable is a quantity whose value is subject to random variations, i.e., to 'chance' (although, as we know, what chance is is itself subject to debate). Mathematically, it is a function X from a sample space Ω (endowed with a probability space) to a measurable space E (usually the real line $\mathbb{R}$)[2]. Figure 1.4 (left) illustrates the random variable associated with a die.

[2]However, the notion of a random variable can be generalised to include mappings from a sample space to a more structured domain, such as an algebraic structure. These functions are called *random elements* [643].

To be a random variable, a function $X : \Omega \to \mathbb{R}$ must be *measurable*: each measurable set in E must have a pre-image $X^{-1}(E)$ which belongs to the σ-algebra $\mathcal{F}$, and therefore can be assigned a probability value. In this way, a random variable becomes a means to assign probability values to sets of real numbers.

1.2 Interpretations of probability

1.2.1 Does probability exist at all?

When one thinks of classical examples of probability distributions (e.g. a spinning wheel, a roulette wheel or a rolling die), the suspicion that 'probability' is simply a fig leaf for our ignorance and lack of understanding of nature phenomena arises.

Assuming a view of the physical world that follows the laws of classical Newtonian mechanics, it is theoretically conceivable that perfect knowledge of the initial conditions of say, a roulette wheel, and of the impulse applied to it by the croupier would allow the player to know exactly what number would come out. In other words, with sufficient information, any phenomenon would be predictable in a completely deterministic way. This is a position supported by Einstein himself, as he was famously quoted as saying that 'God does not play dice with the universe'. In Doc Smith's Lensman series [1736], the ancient race of the Arisians have such mental powers that they compete with each other over foreseeing events far away in the future to the tiniest detail.

A first objection to this argument is that 'infinite accuracy' is an abstraction, and any actual measurements are bound to be affected by a degree of imprecision. As soon as initial states are not precisely known, the nonlinear nature of most phenomena inexcapably generates a chaotic behaviour that effectively prevents any accurate prediction of future events. More profoundly, the principles of quantum mechanics seem to suggest that probability is not just a figment of our mathematical imagination, or a representation of our ignorance: the workings of the physical world seem to be inherently probabilistic [119]. However, the question arises of why the finest structure of the physical world should be described by *additive* measures, rather than more general ones (or *capacities*: see Chapter 6).

Finally, as soon as we introduce the human element into the picture, any hope of being able to predict the future deterministically disappears. One may say that this is just another manifestation of our inability to understanding the internal workings of a system as complex as a human mind. Fair enough. Nevertheless, we still need to be able to make useful predictions about human behaviour, and 'probability' in a wide sense, is a useful means to that end.

1.2.2 Competing interpretations

Even assuming that (some form of mathematical) probability is inherent in the physical world, people cannot agree on what it is. Quoting Savage [45]:

"It is unanimously agreed that statistics depends somehow on probability. But, as to what probability is and how it is connected with statistics, there has seldom been such complete disagreement and breakdown of communication since the Tower of Babel. Doubtless, much of the disagreement is merely terminological and would disappear under sufficiently sharp analysis".

As a result, probability has multiple competing interpretations: (1) as an objective description of frequencies of events (meaning 'things that happen') at a certain persistent rate, or 'relative frequency' – this is the so-called *frequentist* interpretation, mainly due to Fisher and Pearson; (2) as a degree of belief in events (interpreted as statements/propositions on the state of the world), regardless of any random process – the *Bayesian* or *evidential* interpretation, first proposed by de Finetti and Savage; and (3) as the propensity of an agent to act (or gamble or decide) if the event happens – an approach called *behavioural* probability [1874].

Note that neither the frequentist nor the Bayesian approach is in contradiction with the classical mathematical definition of probability due to Kolmogorov: as we will see in this book, however, other approaches to the mathematics of uncertainty do require us to introduce different classes of mathematical objects.

1.2.3 Frequentist probability

In the frequentist interpretation, the (aleatory) probability of an event is its relative frequency in time. When one is tossing a fair coin, for instance, frequentists say that the probability of getting a head is 1/2, not because there are two equally likely outcomes (due to the structure of the object being tossed), but because repeated series of large numbers of trials (a *random experiment*) demonstrate that the empirical frequency converges to the limit 1/2 as the number of trials goes to infinity.

Clearly, it is impossible to actually complete the infinite series of repetitions which constitutes a random experiment. Guidelines on the design of 'practical' random experiments are nevertheless provided, via either *statistical hypothesis testing* or *confidence interval analysis*.

Statistical hypothesis testing A *statistical hypothesis* is a conjecture on the state of the world which is testable on the basis of observing a phenomenon modelled by a set of random variables.

In hypothesis testing, a dataset obtained by sampling is compared with data generated by an idealised model. A hypothesis about the statistical relationship between the two sets of data is proposed, and compared with an idealised 'null' hypothesis which rejects any relationship between the two datasets. The comparison is considered statistically significant if the relationship between the datasets would be an unlikely realisation of the null hypothesis according to a threshold probability, called the *significance level*. Statistical hypothesis testing is thus a form of confirmatory data analysis.

The steps to be followed in hypothesis testing are:[3]

[3] https://en.wikipedia.org/wiki/Statistical_hypothesis_testing.

1. State the null hypothesis H_0 and the alternative hypothesis H_1.
2. State the statistical assumptions being made about the sample, e.g. assumptions about the statistical independence or the distributions of the observations.
3. State the relevant *test statistic* T (i.e., a quantity derived from the sample).
4. Derive from the assumptions the distribution of the test statistic under the null hypothesis.
5. Set a significance level (α), i.e., a probability threshold below which the null hypothesis is rejected.
6. Compute from the observations the observed value t_{obs} of the test statistic T.
7. Calculate the *p-value*, the probability (under the null hypothesis) of sampling a test statistic 'at least as extreme' as the observed value.
8. Reject the null hypothesis, in favour of the alternative one, if and only if the p-value is less than the significance level threshold.

In hypothesis testing, false positives (i.e., rejecting a valid hypothesis) are called 'type I' errors; false negatives (not rejecting a false hypothesis) are called 'type II' errors. Note that if the p-value is above α, the result of the test is inconclusive: the evidence is insufficient to support a conclusion.

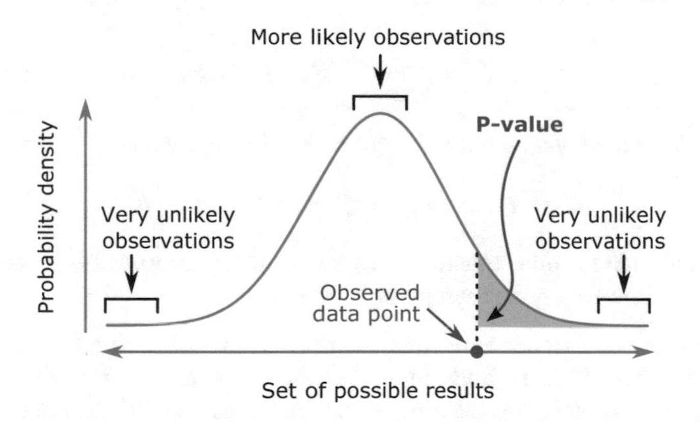

Fig. 1.2: Notion of p-value (adapted from https://upload.wikimedia.org/wikipedia/commons/3/3a/P-value_in_statistical_significance_testing.svg).

P-values The notion of a p-value is crucial in hypothesis testing. It is the probability, under the assumption of hypothesis H, of obtaining a result equal to or more extreme than what was actually observed, namely $P(X \geq x|H)$, where x is the observed value (see Fig. 1.2).

The reason for not simply considering $P(X = x|H)$ when assessing the null hypothesis is that, for any continuous random variable, such a conditional probability is equal to zero. As a result we need to consider, depending on the situation, a right-tail event $p = \mathbb{P}(X \geq x|H)$, a left-tail event $p = \mathbb{P}(X \leq x|H)$, or a double-tailed event: the 'smaller' of $\{X \leq x\}$ and $\{X \geq x\}$.

Note that the p-value is not the probability that the null hypothesis is true or the probability that the alternative hypothesis is false: frequentist statistics does not and cannot (by design) attach probabilities to hypotheses.

Maximum likelihood estimation A popular tool for estimating the parameters of a probability distribution which best fits a given set of observations is *maximum likelihood estimation* (MLE). The term *likelihood* was coined by Ronald Fisher in 1922 [620]. He argued against the use of 'inverse' (Bayesian) probability as a basis for statistical inferences, proposing instead inferences based on likelihood functions.

Indeed, MLE is based on the *likelihood principle*: all of the evidence in a sample relevant to model parameters is contained in the likelihood function. Some widely used statistical methods, for example many significance tests, are not consistent with the likelihood principle. The validity of such an assumption is still debated.

Definition 3. *Given a* parametric model $\{f(.|\theta), \theta \in \Theta\}$, *a family of conditional probability distributions of the data given a (vector) parameter* θ, *the maximum likelihood estimate of* θ *is defined as*

$$\hat{\theta}_{\text{MLE}} \subseteq \left\{ \arg\max_{\theta \in \Theta} \mathcal{L}(\theta\,;\, x_1, \ldots, x_n) \right\},$$

where the likelihood of the parameter given the observed data $x_1, \ldots, x_n$ *is*

$$\mathcal{L}(\theta\,;\, x_1, \ldots, x_n) = f(x_1, x_2, \ldots, x_n \mid \theta).$$

Maximum likelihood estimators have no optimal properties for finite samples: they do have, however, good asymptotic properties:

- *consistency*: the sequence of MLEs converges in probability, for a sufficiently large number of observations, to the (actual) value being estimated;
- *asymptotic normality*: as the sample size increases, the distribution of the MLE tends to a Gaussian distribution with mean equal to the true parameter (under a number of conditions[4]);
- *efficiency*: MLE achieves the *Cramer–Rao lower bound* [1841] when the sample size tends to infinity, i.e., no consistent estimator has a lower asymptotic mean squared error than MLE.

1.2.4 Propensity

The propensity theory of probability [1415], in opposition, thinks of probability as a *physical* propensity or tendency of a physical system to deliver a certain outcome. In a way, propensity is an attempt to explain why the relative frequencies of a random experiment turn out to be what they are. The law of large numbers is interpreted

[4] http://sites.stat.psu.edu/~dhunter/asymp/fall2003/lectures/pages76to79.pdf.

as evidence of the existence of invariant single-run probabilities (as opposed to the relative frequencies of the frequentist interpretation), which do emerge in quantum mechanics, for instance, and to which relative frequencies tend at infinity.

What propensity exactly means remains an open issue. Popper, for instance, has proposed a theory of propensity, which is, however, plagued by the use of relative frequencies in its own definition [1439].

1.2.5 Subjective and Bayesian probability

In *epistemic* or *subjective* probability, probabilities are degrees of belief assigned to events by an individual assessing the state of the world, whereas in frequentist inference a hypothesis is typically tested without being assigned a probability.

The most popular theory of subjective probability is perhaps the *Bayesian* framework [419], due to the English clergyman Thomas Bayes (1702–1761). There, all degrees of belief are encoded by additive mathematical probabilities (in Kolmogorov's sense). It is a special case of *evidential* probability, in which some *prior* probability is updated to a *posterior* probability in the light of new evidence (data). In the Bayesian framework, *Bayes' rule* is used sequentially to compute a posterior distribution when more data become available, namely whenever we learn that a certain proposition A is true:

$$P(B|A) = \frac{P(B \cap A)}{P(A)}. \tag{1.1}$$

Considered as an operator, Bayes' rule is inextricably related to the notion of *conditional* probability $P(B|A)$ [1143].

Bayes proved a special case of what is now called Bayes' theorem (1.1) in a paper entitled 'An essay towards solving a problem in the doctrine of chances'. Pierre-Simon Laplace (1749–1827) later introduced a general version of the theorem. Jeffreys' 'Theory of Probability' [894] (1939) played an important role in the revival of the Bayesian view of probability, followed by publications by Abraham Wald [1870] (1950) and Leonard J. Savage [1537] (1954).

The statistician Bruno de Finetti produced a justification for the Bayesian framework based on the notion of a *Dutch book* [404]. A Dutch book is made when a clever gambler places a set of bets that guarantee a profit, no matter what the outcome of the bets themselves. If a bookmaker follows the rules of Bayesian calculus, de Finetti argued, a Dutch book cannot be made. It follows that subjective beliefs must follow the laws of (Kolmogorov's) probability if they are to be coherent.

However, Dutch book arguments leave open the possibility that non-Bayesian updating rules could avoid Dutch books – one of the purposes of this book is indeed to show that this is the case. Justification by axiomatisation has been tried, but with no great success. Moreover, evidence casts doubt on the assumption that humans maintain coherent beliefs or behave rationally at all. Daniel Kahneman[5] won a Nobel Prize for supporting the exact opposite thesis, in collaboration with Amos

[5] https://en.wikipedia.org/wiki/Daniel_Kahneman.

Tversky. People consistently pursue courses of action which are bound to damage them, as they do not understand the full consequences of their actions.

For all its faults (as we will discuss later), the Bayesian framework is rather intuitive and easy to use, and capable of providing a number of 'off-the-shelf' tools to make inferences or compute estimates from time series.

Bayesian inference In Bayesian inference, the prior distribution is the distribution of the parameter(s) before any data are observed, i.e. $p(\theta|\alpha)$, a function of a vector of hyperparameters α. The likelihood is the distribution of the observed data $\mathbf{X} = \{x_1, \cdots, x_n\}$ conditional on its parameters, i.e., $p(\mathbf{X}|\theta)$. The distribution of the observed data marginalised over the parameter(s) is termed the *marginal likelihood* or *evidence*, namely

$$p(\mathbf{X}|\alpha) = \int_\theta p(\mathbf{X}|\theta) p(\theta|\alpha) \, \mathrm{d}\theta.$$

The *posterior distribution* is then the distribution of the parameter(s) after taking into account the observed data, as determined by Bayes' rule (1.1):

$$p(\theta|\mathbf{X}, \alpha) = \frac{p(\mathbf{X}|\theta) p(\theta|\alpha)}{p(\mathbf{X}|\alpha)} \propto p(\mathbf{X}|\theta) p(\theta|\alpha). \tag{1.2}$$

The *posterior predictive distribution* is the distribution of a new data point x', marginalised over the posterior:

$$p(x'|\mathbf{X}, \alpha) = \int_\theta p(x'|\theta) p(\theta|\mathbf{X}, \alpha) \, \mathrm{d}\theta,$$

amounting to a distribution over possible new data values. The *prior predictive distribution*, instead, is the distribution of a new data point marginalised over the prior:

$$p(x'|\alpha) = \int_\theta p(x'|\theta) p(\theta|\alpha) \, \mathrm{d}\theta.$$

By comparison, prediction in frequentist statistics often involves finding an optimum point estimate of the parameter(s) (e.g., by maximum likelihood), not accounting for any uncertainty in the value of the parameter. In opposition, (1.2) provides as output an entire probability distribution over the parameter space.

Maximum a posteriori estimation *Maximum a posteriori (MAP) estimation* estimates a single value θ for the parameter as the mode of the posterior distribution (1.2):

$$\hat{\theta}_{\mathrm{MAP}}(x) \doteq \arg\max_\theta \frac{p(x|\theta) \, p(\theta)}{\displaystyle\int_\vartheta p(x|\vartheta) \, p(\vartheta) \, \mathrm{d}\vartheta} = \arg\max_\theta p(x|\theta) \, p(\theta).$$

MAP estimation is not very representative of Bayesian methods, as the latter are characterised by the use of distributions over parameters to draw inferences.

1.2.6 Bayesian versus frequentist inference

Summarising, in frequentist inference unknown parameters are often, but not always, treated as having fixed but unknown values that are not capable of being treated as random variates. Bayesian inference, instead, allows probabilities to be associated with unknown parameters. The frequentist approach does not depend on a subjective prior that may vary from one investigator to another. However, Bayesian inference (e.g. Bayes' rule) can be used by frequentists.[6]

Lindley's paradox is a counter-intuitive situation which occurs when the Bayesian and frequentist approaches to a hypothesis-testing problem give different results for certain choices of the prior distribution.

More specifically, Lindley's paradox[7] occurs when:

- the result x is 'significant' by a frequentist test of H_0, indicating sufficient evidence to reject H_0 say, at the 5% level, while at the same time
- the posterior probability of H_0 given x is high, indicating strong evidence that H_0 is in better agreement with x than H_1.

This can happen when H_0 is very specific and H_1 less so, and the prior distribution does not strongly favour one or the other.

It is not really a paradox, but merely a consequence of the fact that the two approaches answer fundamentally different questions. The outcome of Bayesian inference is typically a probability distribution on the parameters, given the results of the experiment. The result of frequentist inference is either a 'true or false' (binary) conclusion from a significance test, or a conclusion in the form that a given confidence interval, derived from the sample, covers the true value.

Glenn Shafer commented on the topic in [1590].

1.3 Beyond probability

A long series of students have argued that a number of serious issues arise whenever uncertainty is handled via Kolmogorov's measure-theoretical probability theory. On top of that, one can argue that something is wrong with both mainstream approaches to probability interpretation. Before we move on to introduce the mathematics of belief functions and other alternative theories of uncertainty, we think it best to briefly summarise our own take on these issues here.

1.3.1 Something is wrong with probability

Flaws of the frequentistic setting The setting of frequentist hypothesis testing is rather arguable. First of all, its scope is quite narrow: rejecting or not rejecting

---

[6] See for instance www.stat.ufl.edu/~casella/Talks/BayesRefresher.pdf.

[7] See onlinelibrary.wiley.com/doi/10.1002/0470011815.b2a15076/pdf.

a hypothesis (although confidence intervals can also be provided). The criterion according to which this decision is made is arbitrary: who decides what an 'extreme' realisation is? In other words, who decides what is the right choice of the value of α? What is the deal with 'magical' numbers such as 0.05 and 0.01? In fact, the whole 'tail event' idea derives from the fact that, under measure theory, the conditional probability (p-value) of a point outcome is zero – clearly, the framework seems to be trying to patch up what is instead a fundamental problem with the way probability is mathematically defined. Last but not least, hypothesis testing cannot cope with pure data, without making additional assumptions about the process (experiment) which generates them.

The issues with Bayesian reasoning Bayesian reasoning is also flawed in a number of ways. It is extremely bad at representing ignorance: Jeffreys' uninformative priors [895] (e.g., in finite settings, uniform probability distributions over the set of outcomes), the common way of handling ignorance in a Bayesian setting, lead to different results for different reparameterisations of the universe of discourse. Bayes' rule assumes the new evidence comes in the form of certainty, 'A is true': in the real world, this is not often the case. As pointed out by Klir, a precise probabilistic model defined only on some class of events determines only interval probabilities for events outside that class (as we will discuss in Section 3.1.3).

Finally, model selection is troublesome in Bayesian statistics: whilst one is forced by the mathematical formalism to pick a prior distribution, there is no clear-cut criterion for how to actually do that.

In the author's view, this is the result of a fundamental confusion between the original Bayesian description of a person's subjective system of beliefs and the way it is updated, and the 'objectivist' view of Bayesian reasoning as a rigorous procedure for updating probabilities when presented with new information.

1.3.2 Pure data: Beware of the prior

Indeed, Bayesian reasoning requires modelling the data *and* a prior. Human beings do have 'priors', which is just a word for denoting what they have learned (or think they have learned) about the world during their existence. In particular, they have well-sedimented beliefs about the likelihood of various (if not all) events. There is no need to 'pick' a prior, for prior (accumulated) knowledge is indeed there. As soon as we idealise this mechanism to, say, allow a machine to reason in this way, we find ourselves forced to 'pick' a prior for an entity (an algorithm) which does not have any past experience, and has not sedimented any beliefs as a result. Nevertheless, Bayesians content themselves by claiming that all will be fine in the end, as, asymptotically, the choice of the prior does not matter, as proven by the Bernstein–von Mises theorem [1841].

1.3.3 Pure data: Designing the universe?

The frequentist approach, on its side, is inherently unable to describe pure data without having to make additional assumptions about the data-generating process.

Unfortunately, in nature one cannot 'design' the process which produces the data: data simply come our way. In the frequentist terminology, in most applications we cannot set the 'stopping rules' (think of driverless cars, for instance). Once again, the frequentist setting brings to the mind the image of a nineteenth-century scientist 'analysing' (from the Greek elements *ana* and *lysis*, breaking up) a specific aspect of the world within the cosy confines of their own laboratory.

Even more strikingly, it is well known that the same data can lead to opposite conclusions when analysed in a frequentist way. The reason is that different random experiments can lead to the same data, whereas the parametric model employed (the family of probability distributions $f(.|\theta)$ which is assumed to produce the data) is linked to a specific experiment.[8]

Apparently, however, frequentists are just fine with this [2131].

1.3.4 No data: Modelling ignorance

The modelling of ignorance (absence of data) is a major weakness of Bayesian reasoning. The typical solution is to pick a so-called 'uninformative' prior distribution, in particular *Jeffreys' prior*, the Gramian of the Fisher information matrix $\mathcal{I}$ [895]:

$$p(\theta) \propto \sqrt{\det \mathcal{I}(\theta)}, \quad \mathcal{I}(\theta) \doteq E\left[\left(\frac{\partial}{\partial\theta}\log f(\mathbb{X}|\theta)\right)^2 \middle| \theta\right]. \tag{1.3}$$

Unfortunately, Jeffreys' priors can be improper (unnormalised). Most importantly, they violate the strong version of the likelihood principle: when using Jeffreys' prior, inferences about a parameter θ depend not just on the probability of the observed data as a function of θ, *but also on the universe* Ω of all possible experimental outcomes. The reason is that the Fisher information matrix $\mathcal{I}(\theta)$ is computed from an expectation (see (1.3)) over the chosen universe of discourse.

This flaw was pointed out by Glenn Shafer in his landmark book [1583], where he noted how the Bayesian formalism cannot handle multiple hypothesis spaces ('families of compatible frames', in Shafer's terminology: see Section 2.5.2) in a consistent way.

In Bayesian statistics, to be fair, one can prove that the asymptotic distribution of the posterior mode depends only on the Fisher information and not on the prior – the so-called *Bernstein–von Mises theorem*. The only issue is that the amount of information supplied must be large enough. The result is also subject to the caveat [644] that the Bernstein–von Mises theorem does not hold almost surely if the random variable considered has a countably infinite probability space.
As A. W. F. Edwards put it [564]:

"It is sometimes said, in defence of the Bayesian concept, that the choice of prior distribution is unimportant in practice, because it hardly influences the posterior distribution at all when there are moderate amounts of data. The less said

[8] http://ocw.mit.edu/courses/mathematics/18-05-introduction-to-probability-\and-statistics-spring-2014/readings/MIT18_05S14_Reading20.pdf

about this 'defence' the better."

In actual fact, 'uninformative' priors can be dangerous, i.e., they may bias the reasoning process so badly that it can recover only asymptotically.[9]

As we will see in this book, instead, reasoning with belief functions does not require any prior. Belief functions encoding the available evidence are simply combined as they are, whereas ignorance is naturally represented by the 'vacuous' belief function, which assigns a mass equal to 1 to the whole hypothesis space.

1.3.5 Set-valued observations: The cloaked die

A die (Fig. 1.3) provides a simple example of a (discrete) random variable. Its probability space is defined on the sample space $\Omega = \{\text{face1}, \text{face } 2, \ldots, \text{face } 6\}$, whose elements are mapped to the real numbers $1, 2, \cdots, 6$, respectively (no need to consider measurability here).

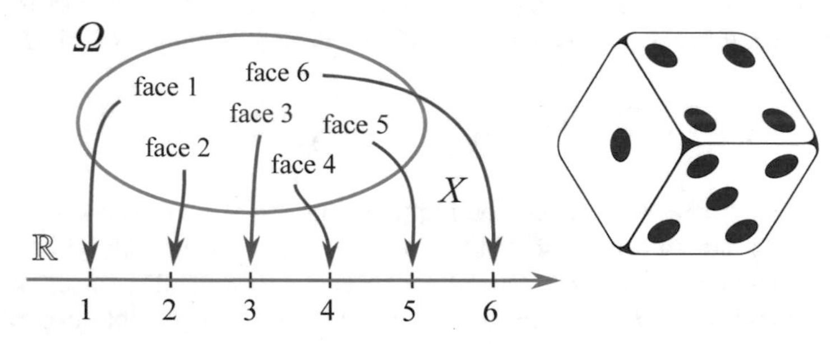

Fig. 1.3: The random variable X associated with a die.

Now, imagine that faces 1 and 2 are cloaked, and we roll the die. How do we model this new experiment, mathematically? Actually, the probability space has not changed (as the physical die has not been altered, its faces still have the same probabilities). What has changes is the mapping: since we cannot observe the outcome when a cloaked face is shown (we assume that only the top face is observable), both face 1 and face 2 (as elements of Ω) are mapped to the set of possible values $\{1, 2\}$ on the real line $\mathbb{R}$ (Fig. 1.4). Mathematically, this is called a *random set* [1268, 950, 1344, 1304], i.e., a set-valued random variable.

A more realistic scenario is that in which we roll, say, four dice in such a way that for some the top face is occluded, but some of the side faces are still visible, providing information about the outcome. For instance, I might be able to see the top face of the Red die as ⚄, the Green die as ⚂ and the Purple die as ⚃ but, say, not the outcome of the Blue die. Still, if I happen to observe the side faces ⚂ and ⚀ of Blue, I can deduce that the outcome of Blue is in the set $\{2, 4, 5, 6\}$.

[9] http://andrewgelman.com/2013/11/21/hidden-dangers-noninformative-priors/.

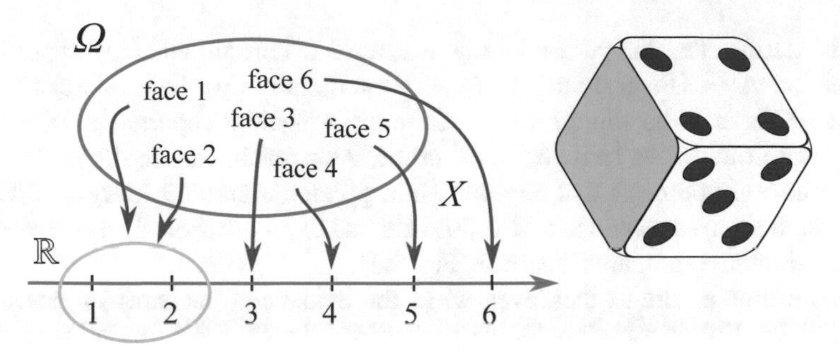

Fig. 1.4: The random set (set-valued random variable) associated with the cloaked die in which faces 1 and 2 are not visible.

This is just an example of a very common situation called *missing data*: for part of the sample that I need to observe in order to make my inference, the data are partly or totally missing. Missing data appear (or disappear?) everywhere in science and engineering. In computer vision, for instance, this phenomenon is typically called 'occlusion' and is one of the main nuisance factors in estimation.

The bottom line is, whenever data are missing, observations are inherently set-valued. Mathematically, we are sampling not a (scalar) random variable but a set-valued random variable – a random set. My outcomes are sets? My probability distribution has to be defined on sets.

In opposition, traditional statistical approaches deal with missing data either by deletion (discarding any case that has a missing value, which may introduce bias or affect the representativeness of the results); by single imputation (replacing a missing value with another one, e.g. from a randomly selected similar record in the same dataset, with the mean of that variable for all other cases, or by using a stochastic regression model); or multiple imputation (averaging the outcomes across multiple imputed datasets using, for instance, stochastic regression). Multiple imputation involves drawing values of the parameters from a posterior distribution, therefore simulating both the process generating the data and the uncertainty associated with the parameters of the probability distribution of the data.

When using random sets, there is no need for imputation or deletion whatsoever. All observations are set-valued: some of them just happen to be pointwise. Indeed, when part of the data used to estimate a probability distribution is missing, it has been shown that what we obtain instead is a convex set of probabilities or *credal set* [1141] (see Section 3.1.4), of the type associated with a *belief function* [401].

1.3.6 Propositional data

Just as measurements are naturally set-valued, in various scenarios evidence is directly supportive of propositions. Consider the following classical example [1607].

Suppose there is a murder, and three people are on trial for it: Peter, John and Mary. Our hypothesis space is therefore $\Theta = \{\text{Peter}, \text{John}, \text{Mary}\}$. There is a witness: he testifies that the person he saw was a man. This amounts to supporting the proposition $A = \{\text{Peter}, \text{John}\} \subset \Theta$. However, should we take this testimony at face value? In fact, the witness was tested and the machine reported an 80% chance that he was drunk when he reported the crime. As a result, we should partly support the (vacuous) hypothesis that any one among Peter, John and Mary could be the murderer. It seems sensible to assign 80% chance to proposition A, and 20% chance to proposition Θ (compare Chapter 2, Fig. 2.1).

This example tells us that, even when the evidence (our data) supports whole *propositions*, Kolmogorov's additive probability theory forces us to specify support for *individual outcomes*. This is unreasonable – an artificial constraint due to a mathematical model that is not general enough. In the example, we have no elements to assign this 80% probability to either Peter or John, nor information on how to distribute it among them. The cause is the additivity constraint that probability measures are subject to.

Kolmogorov's probability measures, however, are not the only or the most general type of measure available for sets. Under a minimal requirement of *monotonicity*, any measure can potentially be suitable for describing probabilities of events: the resulting mathematical objects are called *capacities* (see Fig. 1.5). We will study capacities in more detail in Chapter 6. For the moment, it suffices to note that random sets are capacities, those for which the numbers assigned to events are given by a probability distribution. Considered as capacities (and random sets in particular), belief functions therefore naturally allow us to assign mass directly to propositions.

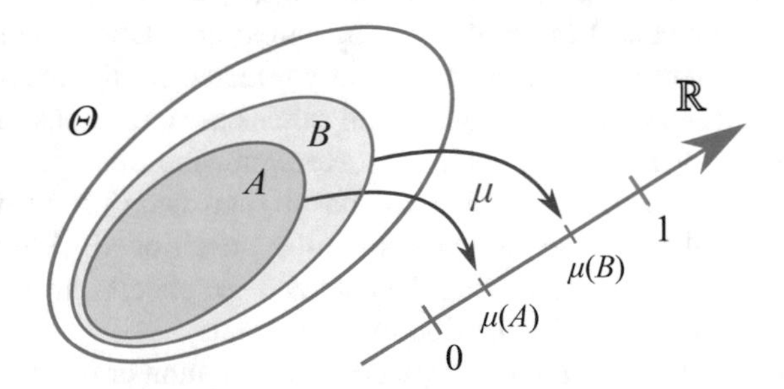

Fig. 1.5: A capacity μ is a mapping from 2^{Θ} to $[0, 1]$, such that if $A \subset B$, then $\mu(A) \leq \mu(B)$.

1.3.7 Scarce data: Beware the size of the sample

The current debate on the likelihood of biological life in the universe is an extreme example of inference from very scarce data. How likely is it for a planet to give birth to life forms? Modern analysis of planetary habitability is largely an extrapolation of conditions on Earth and the characteristics of the Solar System: a weak form of the old anthropic principle, so to speak. What people seem to do is model perfectly the (presumed) causes of the emergence of life on Earth: the planet needs to circle a G-class star, in the right galactic neighbourhood, it needs to be in a certain habitable zone around a star, have a large moon to deflect hazardous impact events, . . . The question arises: how much can one learn from a single example? More, how much can one be sure about what they have learned from very few examples?

Another example is provided by the field of *machine learning*, the subfield of computer science which is about designing algorithms that can learn from what they observe. The main issue there is that machine learning models are typically trained on ridiculously small amount of data, compared with the wealth of information truly contained in the real world. Action recognition tools, for instance, are trained (and tested) on benchmark datasets that contain, at best, a few tens of thousands of videos – compare that with the billions of videos one can access on YouTube. How can we make sure that they learn the right lesson? Should they not aim to work with sets of models rather than precise models?

As we will see in Chapter 17, random set theory can provide more robust foundations for machine learning 'in the wild'.[10] Statistical learning theory [1849, 1851, 1850] derives generalisation bounds on the error committed by trained models on new test data by assuming that the training and test distributions are the same. In opposition, assuming that both distributions, while distinct, belong to a given random set allows us to compute bounds which are more robust to real-world situations – this concept is illustrated in Fig. 1.6.

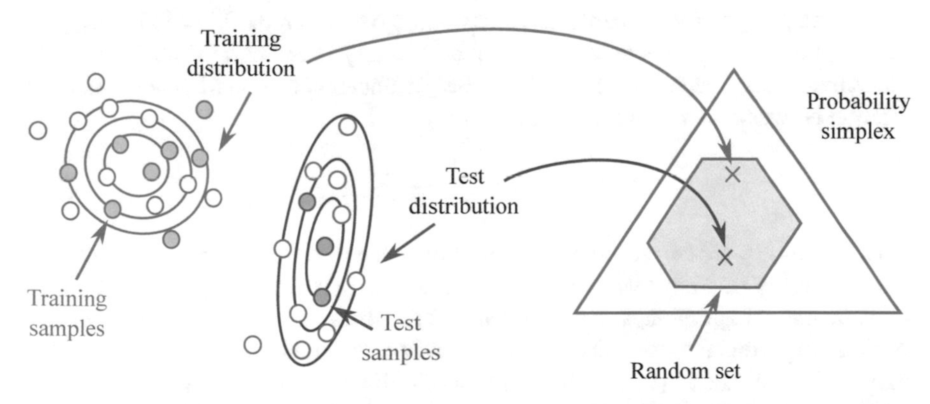

Fig. 1.6: A random-set generalisation of statistical learning theory, as proposed in Chapter 17.

Constraints on 'true' distributions From a statistical point of view, one can object that, even assuming that the natural description of the variability of phenomena is a probability distribution, under the law of large numbers probability distributions

[10] http://lcfi.ac.uk/news-events/events/reliable-machine-learning-wild/.

are the outcome of an infinite process of evidence accumulation, drawn from an infinite series of samples. In all practical cases, then, the available evidence may only provide some sort of constraint on the unknown, 'true' probability governing the process [1589]. Klir [988], among others, has indeed argued that 'imprecision of probabilities is needed to reflect the amount of information on which they are based. [This] imprecision should decrease with the amount of [available] statistical information.'

Unfortunately, those who believe probabilities to be limits of relative frequencies (the frequentists) never really 'estimate' a probability from the data – they merely assume ('design') probability distributions for their p-values, and test their hypotheses on them. In opposition, those who do estimate probability distributions from the data (the Bayesians) do not think of probabilities as infinite accumulations of evidence but as degrees of belief, and content themselves with being able to model the likelihood function of the data.

Both frequentists and Bayesians, though, seem to be happy with solving their problems 'asymptotically', thanks to the limit properties of maximum likelihood estimation, and the Bernstein–von Mises theorem's guarantees on the limit behaviour of posterior distributions. This hardly fits with current artificial intelligence applications, for instance, in which machines need to make decisions on the spot to the best of their abilities.

Logistic regression In fact, frequentists do estimate probabilities from scarce data when performing stochastic regression.

Logistic regression, in particular, allows us, given a sample $Y = \{Y_1, \ldots, Y_n\}$, $X = \{x_1, \ldots, x_n\}$, where $Y_i \in \{0, 1\}$ is a binary outcome at time i and x_i is the corresponding measurement, to learn the parameters of a conditional probability relation between the two, of the form

$$P(Y = 1|x) = \frac{1}{1 + e^{-(\beta_0 + \beta_1 x)}}, \tag{1.4}$$

where β_0 and β_1 are two scalar parameters. Given a new observation x, (1.4) delivers the probability of a positive outcome $Y = 1$. Logistic regression generalises deterministic linear regression, as it is a function of the linear combination $\beta_0 + \beta_1 x$. The n trials are assumed independent but not equally distributed, for $\pi_i = P(Y_i = 1|x_i)$ varies with the index i (i.e., the time instant of collection): see Fig. 1.7.

The parameters β_0, β_1 of the logistic function (1.4) are estimated by the maximum likelihood of the sample, where the likelihood is given by

$$L(\beta|Y) = \prod_{i=1}^{n} \pi_i^{Y_i} (1 - \pi_i)^{Y_i}.$$

Unfortunately, logistic regression appears inadequate when the number of samples is insufficient or when there are too few positive outcomes (1s) [970]. Also, inference by logistic regression tends to underestimate the probability of a positive outcome (see Section 1.3.8).

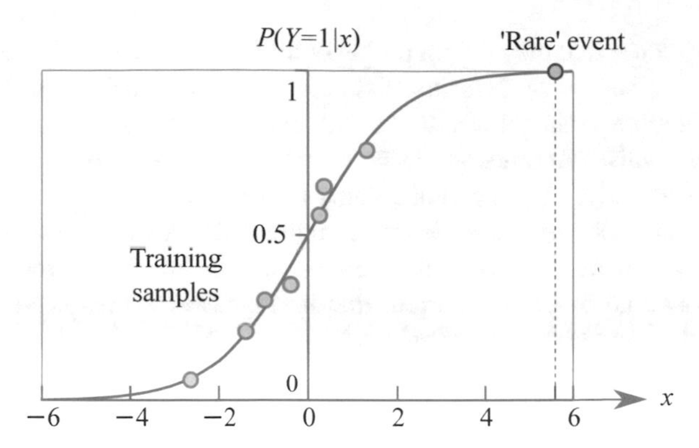

Fig. 1.7: Logistic regression and the notion of a 'rare' event.

Confidence intervals A major tool by which frequentists deal with the size of the sample is *confidence intervals.*

Let X be a sample from a probability $P(.|\theta, \phi)$ where θ is the parameter to be estimated and ϕ a nuisance parameter. A confidence interval for the parameter θ, with confidence level γ, is an interval $[u(X), v(X)]$ determined by the pair of random variables $u(X)$ and $v(X)$, with the property

$$\mathbb{P}(u(X) < \theta < v(X)|\theta, \phi) = \gamma \quad \forall(\theta, \phi). \tag{1.5}$$

For instance, suppose we observe the weight of 25 cups of tea, and we assume it is normally distributed with mean μ. Since the (normalised) sample mean Z is also normally distributed, we can ask, for instance, what values of the mean are such that $P(-z \leq Z \leq z) = 0.95$. Since $Z = (\overline{X} - \mu)/(\sigma/\sqrt{n})$, this yields a (confidence) interval for μ, namely

$$P(\overline{X} - 0.98 \leq \mu \leq \overline{X} + 0.98).$$

Confidence intervals are a form of interval estimate. Their correct interpretation is about 'sampling samples': if we keep extracting new sample sets, 95% (say) of the time the confidence interval (which will differ for every new sample set) will cover the true value of the parameter. Alternatively, there is a 95% probability that the calculated confidence interval from some future experiment will encompass the true value of the parameter. One cannot claim, instead, that a specific confidence interval is such that it contains the value of the parameter with 95% probability.

A Bayesian version of confidence intervals also exists, under the name of *credible*[11] intervals.

1.3.8 Unusual data: Rare events

While the term 'scarce data' denotes situations in which data are of insufficient *quantity*, *rare events* [587] denotes cases in which the training data are of insufficient

[11] https://www.coursera.org/learn/bayesian/lecture/hWn0t/credible-intervals.

quality, in the sense that they do not properly reflect the underlying distribution. An equivalent term, coined by Nassim Nicholas Taleb, is 'black swans'. This refers to an unpredictable event (also called a 'tail risk') which, once it has occurred, is (wrongly) rationalised in hindsight as being predictable/describable by the existing risk models. Basically, Knightian uncertainty is presumed not to exist, typically with extremely serious consequences. Examples include financial crises and plagues, but also unexpected scientific or societal developments. In the most extreme cases, these events may have never even occurred: this is the case for the question 'will your vote will be decisive in the next presidential election?' posed by Gelman and King in [673].

What does consitute a 'rare' event? We can say that an event is 'rare' when it covers a region of the hypothesis space which is seldom sampled. Although such events hardly ever take place when a single system is considered, they become a tangible possibility when very many systems are assembled together (as is the case in the real world). Given the rarity of samples of extreme behaviours (tsunamis, power station meltdowns, etc.), scientists are forced to infer probability distributions for the behaviour of these systems using information captured in 'normal' times (e.g. while a nuclear power plant is working just fine). Using these distributions to extrapolate results at the 'tail' of the curve via popular statistical procedures (e.g. logistic regression, Section 1.3.7) may then lead to sharply underestimating the probability of rare events [970] (see Fig. 1.7 again for an illustration). In response, Harvard's G. King [970] proposed corrections to logistic regression based on oversampling rare events (represented by 1s) with respect to normal events (0s). Other people choose to drop generative probabilistic models entirely, in favour of discriminative ones [74]. Once again, the root cause of the problem is that uncertainty affects our very models of uncertainty.

Possibly the most straightforward way of explictly modelling second-order (Knightian) uncertainties is to consider sets of probability distributions, objects which go under the name of *credal sets*. In Chapter 17 we will show instead how belief functions allow us to model this model-level uncertainty, specifically in the case of logistic regression.

1.3.9 Uncertain data

When discussing how different mathematical frameworks cope with scarce or unusual data, we have implicitly assumed, so far, that information comes in the form of certainty: for example, I measure a vector quantity x, so that my conditioning event is $A = \{x\}$ and I can apply Bayes' rule to update my belief about the state of the world. Indeed this is the way Bayes' rule is used by Bayesians to reason (in time) when new evidence becomes available. Frequentists, on the other hand, use it to condition a parametric distribution on the gathered (certain) measurements and generate p-values (recall Section 1.2.3).

In many situations this is quite reasonable: in science and engineering, measurements, which are assumed to be accurate, flow in as a form of 'certain' (or so it is often assumed) evidence. Thus, one can apply Bayes' rule to condition a parametric model given a sample of such measurements $x_1, \ldots, x_T$ to construct likelihood functions (or p-values, if you are a frequentist).

Qualitative data In many real-world problems, though, the information provided cannot be put in a similar form. For instance, concepts themselves may be not well defined, for example 'this object is dark' or 'it is somewhat round': in the literature, this is referred to as *qualitative* data. Qualitative data are common in decision making, in which expert surveys act as sources of evidence, but can hardly be put into the form of measurements equal to sharp values.

As we will see in Section 6.5, *fuzzy theory* [2083, 973, 531] is able to account for not-well-defined concepts via the notion of *graded membership* of a set (e.g. by assigning every element of the sample space a certain degree of membership in any given set).

Unreliable data Thinking of measurements produced by sensor equipment as 'certain' pieces of information is also an idealisation. Sensors are not perfect but come with a certain degree of reliability. Unreliable sensors can then generate faulty (outlier) measurements: can we still treat these data as 'certain'? They should rather be assimilated to false statements issued with apparent confidence.

It then seems to be more sensible to attach a degree of reliability to any measurements, based on the past track record of the data-generating process producing them. The question is: can we still update our knowledge state using partly reliable data in the same way as we do with certain propositions, i.e., by conditioning probabilities via Bayes' rule?

Likelihood data Last but not least, evidence is often provided directly in the form of whole probability distributions. For instance, 'experts' (e.g. medical doctors) tend to express themselves directly in terms of chances of an event happening (e.g. 'diagnosis A is most likely given the symptoms, otherwise it is either A or B', or 'there is an 80% chance this is a bacterial infection'). If the doctors were frequentists, provided with the same data, they would probably apply logistic regression and come up with the same prediction about the conditional probability $P(\text{disease}|\text{symptoms})$: unfortunately, doctors are not statisticians.

In addition, sensors may also provide as output a probability density function (PDF) on the same sample space: think of two separate Kalman filters, one based on colour, the other on motion (optical flow), providing a Gaussian predictive PDF on the location of a target in an image.

Jeffrey's rule of conditioning Jeffrey's rule of conditioning [1690, 1227] is a step forward from certainty and Bayes' rule towards being able to cope with uncertain data, in particular when the latter comes in the form of another probability distribution. According to this rule, an initial probability P 'stands corrected' by a second probability P', defined only on a certain number of events.

Namely, suppose that P is defined on a σ-algebra $\mathcal{A}$, and that there is a new probability measure P' on a subalgebra $\mathcal{B}$ of $\mathcal{A}$.
If we require that the updated probability P''

1. has the probability values specified by P' for events in $\mathcal{B}$, and
2. is such that $\forall B \in \mathcal{B}$, $X, Y \subset B$, $X, Y \in \mathcal{A}$,

$$\frac{P''(X)}{P''(Y)} = \begin{cases} \frac{P(X)}{P(Y)} & \text{if } P(Y) > 0, \\ 0 & \text{if } P(Y) = 0, \end{cases}$$

then the problem has a unique solution, given by

$$P''(A) = \sum_{B \in \mathcal{B}} P(A|B)P'(B). \tag{1.6}$$

Equation (1.6) is sometimes also called the *law of total probability*, and obviously generalises Bayesian conditioning (obtained when $P'(B) = 1$ for some B).

Beyond Jeffrey's rule What if the new probability P' is defined on the same σ-algebra $\mathcal{A}$? Jeffrey's rule cannot be applied. As we have pointed out, however, this does happen when multiple sensors provide predictive PDFs on the same sample space.

Belief functions deal with uncertain evidence by moving away from the concept of *conditioning* (e.g., via Bayes' rule) to that of *combining* pieces of evidence simultaneously supporting multiple propositions to various degrees. While conditioning is an inherently asymmetric operation, in which the current state of the world and the new evidence are represented by a probability distribution and a single event, respectively, combination in belief function reasoning is completely symmetric, as both the current beliefs about the state of the world and the new evidence are represented by a belief function.

Belief functions are naturally capable of encoding uncertain evidence of the kinds discussed above (vague concepts, unreliable data, likelihoods), as well as of representing traditional 'certain' events. Qualitative concepts, for instance, are represented in the formalism by *consonant* belief functions (Section 2.8), in which the supported events are nested – unreliable measurements can be naturally portrayed as 'discounted' probabilities (see Section 4.3.6).

1.3.10 Knightian uncertainty

Second-order uncertainty is real, as demonstrated by its effect on human behaviour, especially when it comes to decision making. A classical example of how Knightian uncertainty empirically affects human decision making is provided by *Ellsberg's paradox* [565].

Ellsberg's paradox A *decision* problem can be formalised by defining:

- a set Ω of states of the world;
- a set $\mathcal{X}$ of consequences;
- a set $\mathcal{F}$ of acts, where an act is a function $f : \Omega \to \mathcal{X}$.

Let $\succcurlyeq$ be a *preference relation* on $\mathcal{F}$, such that $f \succcurlyeq g$ means that f is at least as desirable as g. Given $f, h \in \mathcal{F}$ and $E \subseteq \Omega$, let fEh denote the act defined by

$$(fEh)(\omega) = \begin{cases} f(\omega) & \text{if } \omega \in E, \\ h(\omega) & \text{if } \omega \notin E. \end{cases} \quad (1.7)$$

Savage's *sure-thing principle* [1411] states that $\forall E, \forall f, g, h, h'$,

$$fEh \succcurlyeq gEh \Rightarrow fEh' \succcurlyeq gEh'.$$

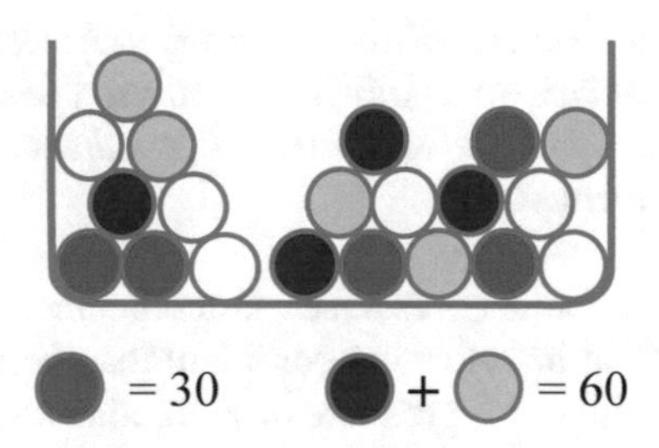

Fig. 1.8: Ellsberg's paradox.

Now, suppose you have an urn containing 30 red balls and 60 balls which are either black or yellow (see Fig. 1.8). Then, consider the following gambles:

- f_1: you receive 100 euros if you draw a red (R) ball;
- f_2: you receive 100 euros if you draw a black (B) ball;
- f_3: you receive 100 euros if you draw a red or a yellow (Y) ball;
- f_4: you receive 100 euros if you draw a black or a yellow ball.

In this example, $\Omega = \{R, B, Y\}$, $f_i : \Omega \to \mathbb{R}$ and $\mathcal{X} = \mathbb{R}$ (consequences are measured in terms of monetary returns). The four acts correspond to the mappings in the following table:

	R	B	Y
f_1	100	0	0
f_2	0	100	0
f_3	100	0	100
f_4	0	100	100

Empirically, it is observed that most people strictly prefer f_1 to f_2, while strictly preferring f_4 to f_3. Now, pick $E = \{R, B\}$. By the definition (1.7),

$$f_1\{R,B\}0 = f_1, \quad f_2\{R,B\}0 = f_2, \quad f_1\{R,B\}100 = f_3, \quad f_2\{R,B\}100 = f_4.$$

Since $f_1 \succcurlyeq f_2$, i.e., $f_1\{R,B\}0 \succcurlyeq f_2\{R,B\}0$, the sure-thing principle would imply that $f_1\{R,B\}100 \succcurlyeq f_2\{R,B\}100$, i.e., $f_3 \succcurlyeq f_4$. This empirical violation of the sure-thing principle is what constitutes the so-called Ellsberg paradox.

Aversion to 'uncertainty' The argument above has been widely studied in economics and decision making,[12] and has to do with people's instinctive aversion to (second-order) uncertainty. They favour f_1 over f_2 because the former ensures a guaranteed $\frac{1}{3}$ chance of winning, while the latter is associated with a (balanced) interval of chances between 0 and $\frac{2}{3}$. Although the average probability of success is still $\frac{1}{3}$, the lower bound is 0 – people tend to find that unacceptable.

Investors, for instance, are known to favour 'certainty' over 'uncertainty'. This was apparent, for instance, from their reaction to the UK referendum on leaving the European Union:

> "*In New York, a recent meeting of S&P Investment Advisory Services five-strong investment committee decided to ignore the portfolio changes that its computer-driven investment models were advising. Instead, members decided not to make any big changes ahead of the vote.*"[13]

Does certainty, in this context, mean a certain outcome of an investor's gamble? Certainly not. It means that investors are confident that their models can fit the observed patterns of variation. In the presence of Knightian uncertainty, human beings assume a more cautious, conservative behaviour.

Climate change An emblematic application in which second-order uncertainty is paramount is climate change modelling. Admittedly, this constitutes an extremely challenging decision-making problem, where policy makers need to decide whether to invest billions of dollars/euros/pounds in expensive engineering projects to mitigate the effects of climate change, knowing that the outcomes of their decision will be apparent only in twenty to thirty years' time.

Rather surprisingly, the mainstream in climate change modelling is *not* about explicitly modelling uncertainty at all: the onus is really on developing ever more complex dynamical models of the environment and validating their predictions. This is all the more surprising as it is well known that even deterministic (but nonlinear) models tend to display chaotic behaviour, which induces uncertainty in predictions of their future state whenever initial conditions are not known with certainty. Climate change, in particular, requires making predictions very far off in the future: as dynamical models are obviously much simplified versions of the world, they become more and more inaccurate as time passes.

What are the challenges of modelling statistical uncertainty explicitly, in this context? First of all, the lack of priors (ouch, Bayesians!) for the climate space,

[12] http://www.econ.ucla.edu/workingpapers/wp362.pdf.

[13] http://www.wsj.com/articles/global-investors-wake-up-to-brexit-threat\-1466080015.

whose points are very long vectors whose components are linked by complex dependencies. Data are also relatively scarce, especially as we go back in time: as we just saw, scarcity is a source of Knightian uncertainty as it puts constraints on our ability to estimate probability distributions. Finally, hypothesis testing cannot really be used either (too bad, frequentists!): this is clearly not a designed experiment where one can make sensible assumptions about the underlying data-generating mechanism.

1.4 Mathematics (plural) of uncertainty

It is fair to summarise the situation by concluding that something is wrong with both Kolmogorov's mathematical probability and its most common interpretations.

As discussed in our Preface, this realisation has led many authors to recognise the need for a mathematical theory of uncertainty capable of overcoming the limitations of classical mathematical probability. While the most significant approaches have been briefly recalled there, Chapter 6 is devoted to a more in-depth overview of the various strands of uncertainty theory.

1.4.1 Debate on uncertainty theory

Authors from a variety of disciplines, including, statistics, philosophy, cognitive science, business and computer science, have fuelled a lively debate [890] on the nature and relationships of the various approaches to uncertainty quantification, a debate which to a large extent is still open.

Back in 1982, a seminal paper by psychologists Kahneman and Tversky [943] proposed, in stark contrast to formal theories of judgement and decision, a formalisation of uncertainty which contemplates different variants associated with frequencies, propensities, the strength of arguments or direct experiences of confidence.

Bayesian probability: Detractors and supporters A number of authors have provided arguments for and against (Bayesian) probability as the method of choice for representing uncertainty.

Cheeseman, for instance, has argued that probability theory, when used correctly, is sufficient for the task of reasoning under uncertainty [255] ('In defense of probability'), while advocating the interpretation of probability as a measure of belief rather than a frequency ratio. In opposition, in his 'The limitation of Bayesianism' [1893], Wang has pointed out a conceptual and notational confusion between the explicit and the implicit condition of a probability evaluation, which leads to seriously underestimating the limitations of Bayesianism. The same author [1892] had previously argued that psychological evidence shows that probability theory is not a proper descriptive model of intuitive human judgement, and has limitations even as a normative model. A new normative model of judgement under uncertainty was then designed under the assumption that the system's knowledge and resources are insufficient with respect to the questions that the system needs to answer.

In his response to Shafer's brief note [1590] on Lindley's paradox (see Section 1.2.6), Lindley himself [426] argued that the Bayesian approach comes out of Shafer's criticism quite well. Indeed, Lindley's paradox has been the subject of a long list of analyses [828, 706, 1828].

More recently (2006), Forster [635] has considered from a philosophical perspective what he called the likelihood theory of evidence [87], which claims that all the information relevant to the bearing of data on hypotheses is contained in the likelihoods, showing that there exist counter-examples in which one can tell which of two hypotheses is true from the full data, but not from the likelihoods alone. These examples suggest that some forms of scientific reasoning, such as the consilience of inductions [1619], cannot be represented within the Bayesian and likelihoodist philosophies of science.

Relations between uncertainty calculi A number of papers have presented attempts to understand the relationships between apparently different uncertainty representations. In an interesting 1986 essay, for instance, Horvitz [841] explored the logical relationship between a small number of intuitive properties for belief measures and the axioms of probability theory, and discussed its relevance to research on reasoning under uncertainty.

In [822], Henkind analysed four of (what were then) the more prominent uncertainty calculi: Dempster–Shafer theory, fuzzy set theory, and the early expert systems MYCIN [200, 778] and EMYCIN. His conclusion was that there does not seem to be one calculus that is the best for all situations. Other investigators, including Zimmerman, have supported an application-oriented view of modelling uncertainty, according to which the choice of the appropriate modelling method is context dependent. In [2132], he suggested an approach to selecting a suitable method to model uncertainty as a function of the context. Pearl's 1988 survey of evidential reasoning under uncertainty [1401] highlighted a number of selected issues and trends, contrasting what he called *extensional* approaches (based on rule-based systems, in the tradition of classical logic) with *intensional* frameworks (which focus on 'states of the world'), and focusing on the computational aspects of the latter methods and of belief networks of both the Bayesian and the Dempster–Shafer type.

Dubois and Prade [540] pointed out some difficulties faced by non-classical probabilistic methods, due to their relative lack of maturity. A comparison between the mathematical models of expert opinion pooling offered by Bayesian probabilities, belief functions and possibility theory was carried out, proving that the Bayesian approach suffers from the same numerical stability problems as possibilistic and evidential rules of combination in the presence of strongly conflicting information. It was also suggested that possibility and evidence theories may offer a more flexible framework for representing and combining subjective uncertain judgements than the framework of subjective probability alone.

Kyburg [1088] also explored the relations between different uncertainty formalisms, advocating that they should all be thought of as special cases of sets of probability functions defined on an algebra of statements. Thus, interval probabilities should be construed as maximum and minimum probabilities within a set of distributions, belief functions should be construed as lower probabilities, etc.

Philippe Smets [1702], on his side, surveyed various forms of imperfect data, classified into either imprecision, inconsistency or uncertainty. He argued that the greatest danger in approximate reasoning is the use of inappropriate, unjustified models, and made a case against adopting a single model and using it in all contexts (or, worse, using all models in a somewhat random way) [1684]. The reason is that ignorance, uncertainty and vagueness are really different notions which require different approaches. He advised that, before using a quantified model, we should:

1. Provide canonical examples justifying the origin of the numbers used.
2. Justify the fundamental axioms of the model and their consequences, via 'natural' requirements.
3. Study the consequence of the derived models in practical contexts to check their validity and appropriateness.

A common error, he insisted, consists in accepting a model because it 'worked' nicely in the past, as empirical results can only falsify a model, not prove that it is correct.

Approximate reasoning Several papers have dealt with the issue of uncertainty in the context of artificial intelligence, expert systems or, as it is sometimes referred to, *approximate reasoning*.

In a 1987 work, Shafer [1597] discussed the challenges arising from the interaction of artificial intelligence and probability, identifying in particular the issue of building systems that can design probability arguments. Thompson [1816], after acknowledging that there was no general consensus on how best to attack evidential reasoning, proposed a general paradigm robust enough to be of practical use, and used it to formulate classical Bayes, convex Bayes, Dempster–Shafer, Kyburg and possibility approaches in a parallel fashion in order to identify key assumptions, similarities and differences. Ruspini [1512] argued in 1991 that approximate reasoning methods are sound techniques that describe the properties of a set of conceivable states of a real-world system, using a common framework based on the logical notion of 'possible worlds'. In his 1993 book, Krause [1067] supported the view that an eclectic approach is required to represent and reason under the many facets of uncertainty. Rather than the technical aspects, that book focuses on the foundations and intuitions behind the various schools. Chapter 4 of it, 'Epistemic probability: The Dempster–Shafer theory of evidence', is devoted entirely to belief theory.

The recent debate A more recent publication by Castelfranchi et al. [235], has focused on the central role of 'expectations' in mental life and in purposive action, reducing them in terms of more elementary ingredients, such as beliefs and goals. There, the authors allow the possibility that beliefs in a proposition and its negation do not add up to one, as in the belief function framework.

Gelman [672] has pointed out the difficulties with both robust Bayes and belief function approaches, using a simple example involving a coin flip and a boxing/wrestling match. His conclusions are that robust Bayes approaches allow ignorance to spread too broadly, whereas belief functions inappropriately collapse to simple Bayesian models.

Keppens [956] has argued that the use of subjective probabilities in evidential reasoning (in particular for crime investigation) is inevitable for several reasons, including lack of data, non-specificity of phenomena and fuzziness of concepts in this domain. His paper argues that different approaches to subjective probability are really concerned with different aspects of vagueness.

1.4.2 Belief, evidence and probability

As recalled in the Preface, this book mostly focuses on the theory of belief functions, one of the most widely adopted formalisms for a mathematics of uncertainty, its geometric interpretation and its links with other uncertainty theories.

Belief functions as random sets The notion of a belief function originally derives from a series of seminal publications [415, 417, 418] by Arthur Dempster on upper and lower probabilities induced by multivalued mappings. Given a probability distribution p on a certain sample space, and a one-to-many map from such a sample space to another domain, p induces a probability distribution (a *mass assignment*) on the power set of the latter [415], i.e., a *random set* [1268, 1344]. A very simple example of such a mapping was given in the cloaked die example (Figure 1.4).

The term *belief function* was coined by Glenn Shafer [1583], who proposed to adopt these mathematical objects to represent evidence in the framework of subjective probability, and gave an axiomatic definition of them as non-additive (indeed, superadditive) probability measures. As mentioned when recalling Jeffrey's rule (Section 1.3.9), in belief theory conditioning (with respect to an event) is replaced by combination (of pieces of evidence, represented by belief functions).

As a result, as we will see in more detail in Part I, the theory of belief functions addresses all the issues with the handling of uncertainty we discussed in this Introduction. It does not assume an infinite amount of evidence to model imprecision, but uses all the available partial evidence, coping with missing data in the most natural of ways. It properly represents ignorance by assigning mass to the whole sample space or 'frame of discernment', and can coherently represent evidence on different but compatible domains. Furthermore, as a straightforward generalisation of probability theory, its rationale is rather neat and does not require us to entirely abandon the notion of an event (as opposed to Walley's imprecise probability theory [1874]), although it can be extended to assign basic probabilities to real-valued functions [153, 1078] rather than events. Finally, it contains as special cases both fuzzy set theory and possibility theory.

Belief theory as evidential probability Shafer called his 1976 proposal [1583] 'A mathematical theory of evidence', whereas the mathematical objects it deals with are termed 'belief functions'. Where do these names come from, and what interpretation of probability (in its wider sense) do they entail?

In fact, belief theory is a theory of epistemic probability: it is about probabilities as a mathematical representation of knowledge (never mind whether it is a human's knowledge or a machine's). *Belief* is often defined as a state of mind in which a person thinks something to be the case, with or without there being empirical evidence in support. *Knowledge* is a rather more controversial notion, for it is regarded by some as the part of belief that is true, while others consider it as that part of belief which is *justified* to be true. *Epistemology* is the branch of philosophy concerned with the theory of knowledge. *Epistemic probability* (Fig. 1.9) is the study of probability as a representation of knowledge.

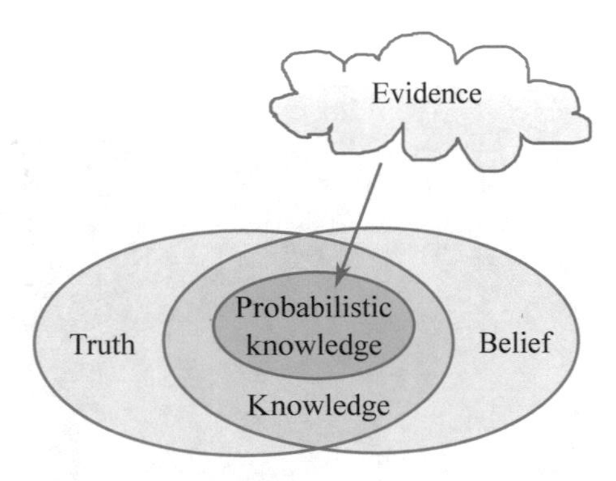

Fig. 1.9: Belief function theory is an instance of evidential, epistemic probability.

The theory of evidence is also, as the name itself suggests, a theory of *evidential* probability: one in which the probabilities representing one's knowledge are induced ('elicited') by the evidence at hand. In probabilistic logic [586, 205], statements such as 'hypothesis H is probably true' are interpreted to mean that the empirical evidence E supports hypothesis H to a high degree – this degree of support is called the *epistemic probability* of H given E. As a matter of fact, Pearl and others [1405, 1682, 137] have supported a view of belief functions as probabilities on the logical causes of a certain proposition (the so-called *probability of provability* interpretation), closely related to modal logic [800].

The rationale for belief function theory can thus be summarised as follows: there exists evidence in the form of probabilities, which supports degrees of belief on the matter at hand. The space where the (probabilistic) evidence lives is different from the hypothesis space (where belief measures are defined). The two spaces are linked by a one-to-many map, yielding a mathematical object known as a random set [415]. In Chapter 2 we will recall the basic elements of the theory of evidence and the related mathematical definitions.

Part I

Theories of uncertainty

2

Belief functions

The *theory of evidence* [1583] was introduced in the 1970s by Glenn Shafer as a way of representing epistemic knowledge, starting from a sequence of seminal papers [415, 417, 418] by Arthur Dempster [424]. In this formalism, the best representation of chance is a *belief function* rather than a traditional probability distribution (Definition 2). Belief functions assign probability values to *sets* of outcomes, rather than single events: their appeal rests on their ability to naturally encode evidence in favour of propositions (see Section 1.3.6). The theory embraces the familiar idea of assigning numbers between 0 and 1 to measure degrees of support but, rather than focusing on *how* these numbers are determined, it concerns itself with the mechanisms driving the *combination* of degrees of belief.

The formalism indeed provides a simple method for merging the evidence carried by a number of distinct sources (called *Dempster's rule of combination* [773]), with no need for any prior distributions [1949]. In this sense, according to Shafer, it can be seen as a theory of probable reasoning. The existence of different levels of granularity in knowledge representation is formalised via the concept of a *family of compatible frames*.

As we recall in this chapter, Bayesian reasoning (see Section 1.2.5) is actually contained in the theory of evidence as a special case, since:

1. Bayesian functions form a special class of belief functions.
2. Bayes' rule is a special case of Dempster's rule of combination.

In the following we will overlook most of the emphasis Shafer put in his essay on the notion of 'weight of evidence', as in our view it is not strictly necessary to the comprehension of what follows.

F. Cuzzolin, *The Geometry of Uncertainty*, Artificial Intelligence: Foundations, Theory, and Algorithms, https://doi.org/10.1007/978-3-030-63153-6_2

Chapter outline

We start by describing Arthur Dempster's original setting for the notions of lower and upper probabilities (Section 2.1), based on the idea of mapping a probability measure on a source domain onto something more general (indeed, a belief function) on the domain of interest, thanks to (multivalued) functions that associate elements of the former with subsets of the latter. The rationale for Dempster's rule as a means to combine belief functions induced by separate independent sources is also recalled.

Shafer's own definition of belief functions as set functions induced by a basic probability assignment on the power set of the universe of discourse (the 'frame of discernment') is introduced in Section 2.2, together the dual notion of a plausibility function. Belief and plausibility functions correspond to Dempster's lower and upper probabilities.

Dempster's rule of combination is explicitly introduced in Section 2.3 with the help of simple examples, and the distinction between combination and conditioning when manipulating belief measures is highlighted.

Simple support functions, as belief functions induced by a single elementary piece of evidence, are recalled in Section 2.4, together with the notion of separable support functions (Dempster sums of simple support functions), and that of internal conflict of a separable support belief function.

The issue of belief functions defined over distinct, but related, frames of discernment is discussed in Section 2.5. Based on the notion of a refining of a given disjoint partition, the concept of a family of compatible frames is proposed as the natural framework for dealing with belief functions on different representations. The consistency of belief functions defined on different frames, and the restriction, marginalisation and vacuous extension operators are also discussed.

Support functions (i.e., restrictions of separable support functions) and the notion of evidence interaction are presented in Section 2.6.

Quasi-support functions are then defined as limits of series of separable support functions (Section 2.7). The Bayesian formalism is criticised there because of the infinite amount of evidence carried by a Bayesian belief function (a probability measure), and the fact (already hinted at in the Introduction, Section 1.3.4) that uniform priors are incompatible when defined on different but related frames of discernment.

Finally, consonant belief functions, as belief functions with nested focal elements, are defined and their main properties recalled (Section 2.8).

2.1 Arthur Dempster's original setting

Let us first introduce the main notions of the formalism with the help of Shafer's murder trial example (Section 1.3.6), formalised by the diagram in Fig. 2.1.

Here, Ω is the space where the evidence lives, in a form of the probability distribution P such that $P(\text{Drunk}) = 0.2$, $P(\text{Not drunk}) = 0.8$. The actual sample

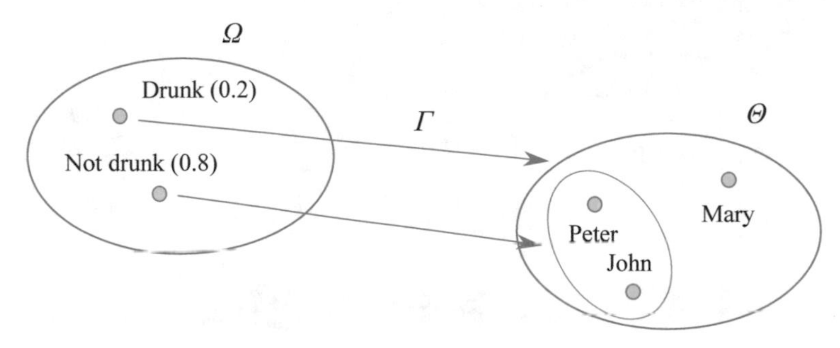

Fig. 2.1: The multivalued mapping $\Gamma : \Omega \rightarrow 2^{\Theta}$ encoding the murder trial example.

(hypothesis) space is, however, the set Θ of outcomes of the murder trial. The crucial fact is that the *elements* of the (evidence) sample space Ω are mapped to *subsets* of Θ by the constraints of the problem: for example, if the witness was not drunk, it follows that the culprit would be either Peter or John (a man). As we discussed informally in the Introduction, this mapping amounts to a random set, i.e, a set-valued random variable (recall Section 1.3.5). As a consequence, the probability distribution P on Ω induces a *mass assignment* $m : 2^{\Theta} \rightarrow [0, 1]$ on the subsets of Θ, via the multivalued (one-to-many) mapping $\Gamma : \Omega \rightarrow 2^{\Theta}$. In the example, Γ maps not drunk $\in \Omega$ to $\{\text{Peter, John}\} \subset \Theta$; the corresponding mass function is $m(\{\text{Peter, John}\}) = 0.8$, $m(\Theta) = 0.2$.

What happens if we have a number of separate bodies of evidence, each represented by such a random set, and we wish to take all of them into account prior to drawing inferences on the state of the world, or making a decision?

Imagine that, in our example, a new piece of evidence is made available: a blond hair is found on the crime scene. Suppose that we also know that there is a probability 0.6 that the room had been cleaned before the murder. This second body of evidence can then be encoded by the mass assignment $m_2(\{\text{John, Mary}\}) = 0.6$, $m_2(\Theta) = 0.4$. Once again, sources of evidence are given to us in the form of probability distributions in some space relevant to (but not coinciding with) the problem.

Dempster's original proposal for combining these two pieces of evidence rests on the random set interpretation of belief functions (Fig. 2.2). Selecting the 'codes'[14] $\omega_1 \in \Omega_1$ and $\omega_2 \in \Omega_2$ implies that the correct answer θ to the question of whose set of outcomes is Θ (e.g., in the murder example, 'who is the murderer?') must belong to $\Gamma_1(\omega_1) \cap \Gamma_2(\omega_2) \subset \Theta$. If the codes are selected *independently*, then the probability that the pair (ω_1, ω_2) is selected is just the product of the individual probabilities, $P_1(\{\omega_1\}) \cdot P_2(\{\omega_2\})$. Whenever $\Gamma_1(\omega_1) \cap \Gamma_2(\omega_2) = \emptyset$, the pair (ω_1, ω_2) cannot be selected, and hence the joint distribution on $\Omega_1 \times \Omega_2$ must be conditioned to eliminate such pairs.

The resulting rule of combination is called *Dempster's rule*: its behaviour and properties are described in Section 2.3.

---

[14] A terminology introduced by Glenn Shafer in his canonical examples [1607].

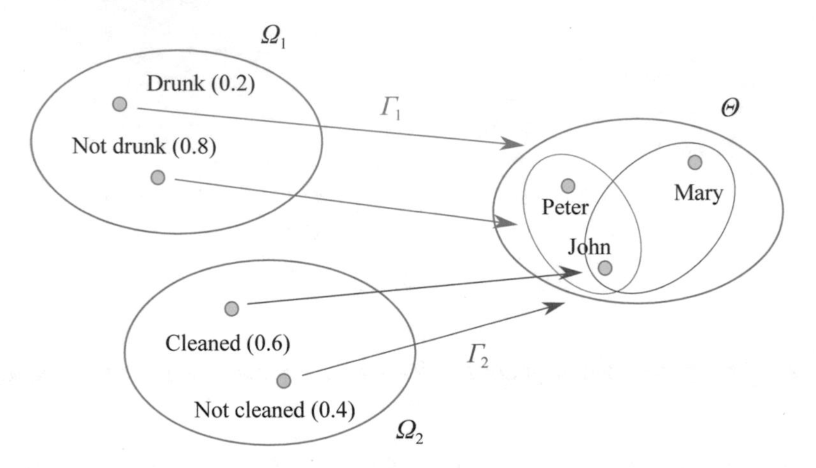

Fig. 2.2: Dempster's combination of mass assignments associated with two multivalued mappings, Γ_1 and Γ_2, acting on independent sources in the murder trial example.

2.2 Belief functions as set functions

In his 1976 essay [1583], Glenn Shafer reinterpreted Dempster's work as the foundation for a mathematical theory of evidential probability, centred around the notion of a belief function. In his approach, the role of the probability space (Ω, P) inducing a belief function is neglected, and all the focus is on the domain Θ on whose power set the mass assignment is defined.

Following [1583], we call the latter a *frame*[15] *of discernment* (FoD).

2.2.1 Basic definitions

Basic probability assignments

Definition 4. *A* basic probability assignment *(BPA) [66] over an FoD Θ is a set function [527, 442] $m : 2^\Theta \rightarrow [0, 1]$ defined on the collection 2^Θ of all subsets of Θ such that*

$$m(\emptyset) = 0, \quad \sum_{A \subseteq \Theta} m(A) = 1.$$

The quantity $m(A)$ is called the *basic probability number* or 'mass' [1081, 1079] assigned to a subset (*event*) $A \in 2^\Theta$, and measures the belief committed exactly to A. The elements of the power set 2^Θ associated with non-zero values of m are called the *focal elements* of m, and their set-theoretical union is called its *core*,

[15]For a note about the intuitionistic origin of this terminology see Rosenthal, *Quantales and Their Applications* [1501].

$$\mathcal{C}_m \doteq \bigcup_{A \subseteq \Theta : m(A) \neq 0} A. \tag{2.1}$$

Belief functions Now suppose that the available empirical evidence induces a basic probability assignment on a given FoD Θ.

Definition 5. *The* belief function *(BF) associated with a basic probability assignment $m : 2^\Theta \to [0,1]$ is the set function $Bel : 2^\Theta \to [0,1]$ defined as*

$$Bel(A) = \sum_{B \subseteq A} m(B). \tag{2.2}$$

The frame of discernment Θ is typically interpreted as the set of possible answers to a given problem, exactly one of which is correct. For each event $A \subset \Theta$, the quantity $Bel(A)$ then takes on the meaning of a *degree of belief* that the true answer lies in A, and represents the total belief committed to a set of possible outcomes A by the available evidence m.

Example 1: the Ming vase. *A simple example (taken from [1583]) may clarify the notion of degree of belief. We are looking at a vase that is represented as a product of the Ming dynasty, and we are wondering whether the vase is genuine. If we denote by θ_1 the possibility that the vase is original, and by θ_2 the possibility that it is counterfeit, then*

$$\Theta = \{\theta_1, \theta_2\}$$

is the set of possible outcomes, and

$$\mathcal{P}(\Theta) = 2^\Theta = \{\emptyset, \Theta, \{\theta_1\}, \{\theta_2\}\}$$

is the (power) set of all its subsets. A belief function Bel on Θ will represent the degree of belief that the vase is genuine as $Bel(\{\theta_1\}) = m(\{\theta_1\})$ (by (2.2)), and the degree of belief that the vase is a fake as $Bel(\{\theta_2\}) = m(\{\theta_2\})$ (note that the belief function assumes values on the subsets *$\{\theta_1\}$ and $\{\theta_2\}$ rather than on the* elements *θ_1, θ_2 of Θ).*

Definition 4 imposes a simple constraint on these degrees of belief, namely

$$Bel(\{\theta_1\}) + Bel(\{\theta_2\}) \leq 1,$$

since $m(\{\theta_1\}) + m(\{\theta_2\}) + m(\Theta) = 1$ and $m(A) \geq 0$ for all subsets A. The mass value $m(\Theta)$ of the whole outcome space Θ encodes, instead, evidence which does not specifically support either of the two precise answers θ_1 and θ_2, and is therefore an indication of the level of uncertainty about the problem.

As Example 1 illustrates, belief functions readily lend themselves to the representation of ignorance, in the form of the mass assigned to the whole FoD. Indeed, the simplest possible belief function is the one which assigns all the basic probability to Θ: we call it the *vacuous* belief function.

Bayesian reasoning, in comparison, has trouble with the whole idea of encoding ignorance, for it cannot distinguish between 'lack of belief' in a certain event A ($1 - Bel(A)$ in our notation) and 'disbelief' (the belief $Bel(\bar{A})$ in the negated event $\bar{A} = \Theta \setminus A$), owing to the additivity constraint: $P(A) + P(\bar{A}) = 1$ (see Definition 2). As discussed in the Introduction, Bayesian reasoning typically represents the complete absence of evidence by assigning an equal degree of belief to every outcome in Θ. As we will see in Section 2.7.4, this generates incompatible results when we consider distinct descriptions of the same problem at different levels of granularity.

Moebius transform Given a belief function Bel, there exists a unique basic probability assignment inducing it via (2.2). The latter can be recovered by means of the *Moebius inversion formula*,[16]

$$m(A) = \sum_{B \subseteq A} (-1)^{|A \setminus B|} Bel(B), \tag{2.3}$$

so that there is a 1–1 correspondence between the two set functions m and Bel [720].

2.2.2 Plausibility and commonality functions

Other mathematical representations of the evidence generating a belief function Bel are the *degree of doubt* or *disbelief* $Dis(A) \doteq Bel(\bar{A})$ in an event A and, most importantly, its *upper probability* or *plausibility*,

$$Pl(A) \doteq 1 - Dis(A) = 1 - Bel(\bar{A}), \tag{2.4}$$

as opposed to its *lower probability* or belief value $Bel(A)$. The quantity $Pl(A)$ expresses the amount of evidence *not against* an event (or proposition) A [347]. The *plausibility function* $Pl : 2^\Theta \to [0,1]$, $A \mapsto Pl(A)$ conveys the same information as Bel, and can be expressed as a function of the basic plausibility assignment m as follows:

$$Pl(A) = \sum_{B \cap A \neq \emptyset} m(B) \geq Bel(A).$$

Finally, the *commonality function* $Q : 2^\Theta \to [0,1]$, $A \mapsto Q(A)$ maps each subset to a *commonality number* $Q(A)$, which can be interpreted as the amount of mass which can move freely through the entire event A:

$$Q(A) \doteq \sum_{B \supseteq A} m(B). \tag{2.5}$$

Example 2: a belief function. *Let us consider a belief function Bel on a frame of size 3,* $\Theta = \{x, y, z\}$, *with basic probability assignment* $m(\{x\}) = 1/3$,

[16] See [1762] for an explanation in terms of the theory of monotone functions over partially ordered sets.

$m(\Theta) = 2/3$. Its belief values on all the possible events of Θ are, according to (2.2), $Bel(\{x\}) = m(\{x\}) = 1/3$, $Bel(\{y\}) = 0$, $Bel(\{z\}) = 0$, $Bel(\{x, y\}) = m(\{x\}) = 1/3$, $Bel(\{x, z\}) = m(\{x\}) = 1/3$, $Bel(\{y, z\}) = 0$ and $Bel(\Theta) = m(\{x\}) + m(\Theta) = 1$.

To appreciate the difference between belief, plausibility and commonality values let us consider in particular the event $A = \{x, y\}$. Its belief value

$$Bel(\{x, y\}) = \sum_{B \subseteq \{x,y\}} m(B) = m(\{x\}) = \frac{1}{3}$$

represents the amount of evidence which surely supports *$\{x, y\}$ – the total mass $Bel(A)$ is guaranteed to involve* only *elements of $A = \{x, y\}$. On the other hand,*

$$Pl(\{x, y\}) = 1 - Bel(\{x, y\}^c) = \sum_{B \cap \{x,y\} \neq \emptyset} m(B) = 1 - Bel(\{z\}) = 1$$

measures the evidence not surely against *it – the plausibility value $Pl(A)$ accounts for the mass that* might *be assigned to some element of A.*

Finally, the commonality number

$$Q(\{x, y\}) = \sum_{A \supseteq \{x,y\}} m(A) = m(\Theta) = \frac{2}{3}$$

summarises the amount of evidence which can (possibly) equally support *each of the outcomes in $A = \{x, y\}$ (i.e., x and y).*

2.2.3 Bayesian belief functions

In the theory of evidence, a (finite) probability function is simply a belief function satisfying the additivity rule for disjoint sets.

Definition 6. *A* Bayesian *belief function $Bel : 2^\Theta \rightarrow [0, 1]$ satisfies the additivity condition*

$$Bel(A) + Bel(\bar{A}) = 1$$

whenever $A \subseteq \Theta$.

Further, the following proposition can be proved [1583].

Proposition 1. *A belief function $Bel : 2^\Theta \rightarrow [0, 1]$ is Bayesian if and only if $\exists p : \Theta \rightarrow [0, 1]$ such that $\sum_{\theta \in \Theta} p(\theta) = 1$ and*

$$Bel(A) = \sum_{\theta \in A} p(\theta) \quad \forall A \subseteq \Theta.$$

In other words, Bayesian belief functions are in 1–1 correspondence with finite probability distributions.

2.3 Dempster's rule of combination

Belief functions representing distinct bodies of evidence can be combined by means of *Dempster's rule of combination* [526], also called the *orthogonal sum.*

2.3.1 Definition

Definition 7. *The* orthogonal sum $Bel_1 \oplus Bel_2 : 2^\Theta \rightarrow [0,1]$ *of two belief functions* $Bel_1 : 2^\Theta \rightarrow [0,1]$, $Bel_2 : 2^\Theta \rightarrow [0,1]$ *defined on the same FoD* Θ *is the unique belief function on* Θ *whose focal elements are all the possible intersections* $A = B \cap C$ *of focal elements* B *and* C *of* Bel_1 *and* Bel_2*, respectively, and whose basic probability assignment is given by*

$$m_{Bel_1 \oplus Bel_2}(A) = \frac{\displaystyle\sum_{B \cap C = A} m_1(B) m_2(C)}{1 - \displaystyle\sum_{B \cap C = \emptyset} m_1(B) m_2(C)}, \tag{2.6}$$

where m_i *denotes the BPA of the input belief function* Bel_i.

Figure 2.3 pictorially expresses Dempster's algorithm for computing the basic probability assignment of the combination $Bel_1 \oplus Bel_2$ of two belief functions. Let a unit square represent the total, unitary probability mass that one can assign to subsets of Θ, and associate horizontal and vertical strips with the focal elements $B_1, \ldots, B_k$ and $C_1, \ldots, C_l$ of Bel_1 and Bel_2, respectively. If their width is equal to the mass value of the corresponding focal element, then their area is equal to the mass $m(B_i)$ ($m(C_j)$). The area of the intersection of the strips related to any two focal elements B_i and C_j is then equal to the product $m_1(B_i) \cdot m_2(C_j)$, and is committed to the intersection event $A = B_i \cap C_j$. As more than one such rectangle can end up being assigned to the same subset A (as different pairs of focal elements can have the same intersection), we need to sum all these contributions, obtaining

$$m_{Bel_1 \oplus Bel_2}(A) \propto \sum_{i,j: B_i \cap C_j = A} m_1(B_i) m_2(C_j).$$

Finally, as some of these intersections may be empty, we need to discard the quantity

$$\sum_{i,j: B_i \cap C_j = \emptyset} m_1(B_i) m_2(C_j).$$

We do so by normalising the resulting BPA, obtaining (2.6).

Note that, by Definition 7, *not all pairs of belief functions admit an orthogonal sum* – two belief functions are combinable if and only if their cores (2.1) are not disjoint, i.e., $\mathcal{C}_1 \cap \mathcal{C}_2 \neq \emptyset$ or, equivalently, iff there exist a focal element of Bel_1 and a focal element of Bel_2 whose intersection is non-empty.

Proposition 2. *[1583] If $Bel_1, Bel_2 : 2^\Theta \to [0,1]$ are two belief functions defined on the same frame Θ, then the following conditions are equivalent:*

- *their Dempster combination $Bel_1 \oplus Bel_2$ does not exist;*
- *their cores (2.1) are disjoint, i.e., $\mathcal{C}_{Bel_1} \cap \mathcal{C}_{Bel_2} = \emptyset$;*
- *$\exists A \subset \Theta$ s.t. $Bel_1(A) = Bel_2(\bar{A}) = 1$.*

Example 3: Dempster combination. *Consider a frame of discernment containing only four elements, $\Theta = \{\theta_1, \theta_2, \theta_3, \theta_4\}$.*

We can define a belief function Bel_1 there, with basic probability assignment

$$m_1(\{\theta_1\}) = 0.7, \quad m_1(\{\theta_1, \theta_2\}) = 0.3.$$

Such a belief function then has just two focal elements, $A_1 = \{\theta_1\}$ and $A_2 = \{\theta_1, \theta_2\}$. As an example, its belief values on the events $\{\theta_4\}$, $\{\theta_1, \theta_3\}$, $\{\theta_1, \theta_2, \theta_3\}$ are respectively $Bel_1(\{\theta_4\}) = m_1(\{\theta_4\}) = 0$, $Bel_1(\{\theta_1, \theta_3\}) = m_1(\{\theta_1\}) + m_1(\{\theta_3\}) + m_1(\{\theta_1, \theta_3\}) = 0.7 + 0 + 0 = 0.7$ and $Bel_1(\{\theta_1, \theta_2, \theta_3\}) = m_1(\{\theta_1\}) + m_1(\{\theta_1, \theta_2\}) = 0.7 + 0.3 = 1$ (so that the core of Bel_1 is $\{\theta_1, \theta_2\}$).

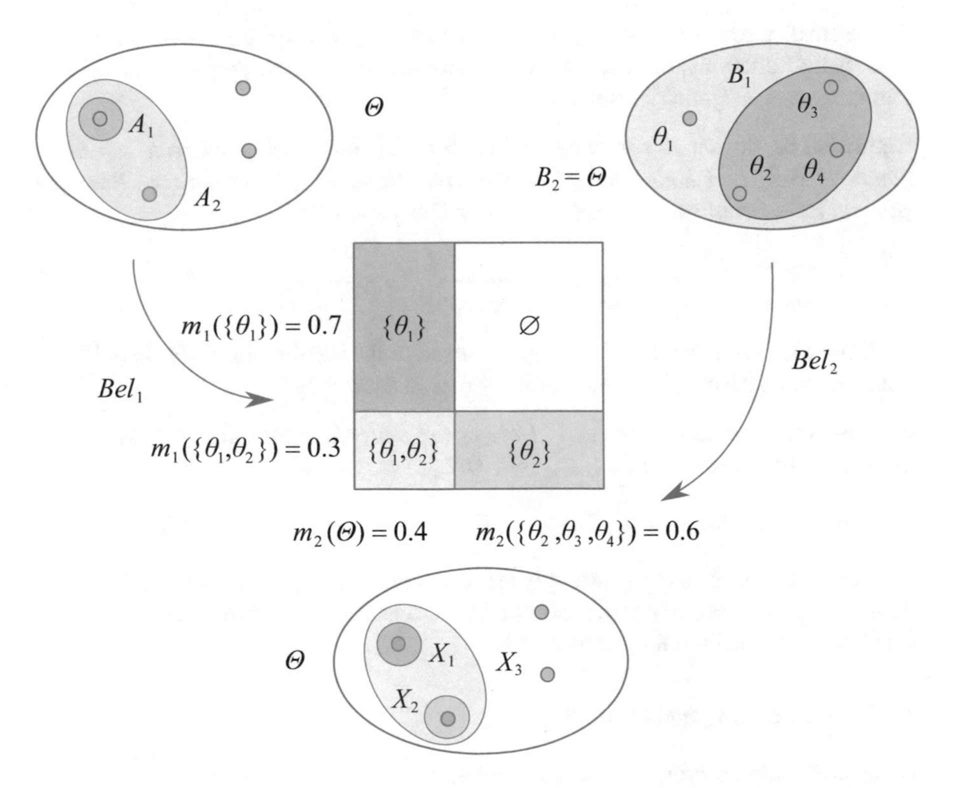

Fig. 2.3: Example of Dempster combination, and its graphical representation.

Now, let us introduce another belief function Bel_2 on the same FoD, with focal elements $B_1 = \{\theta_2, \theta_3, \theta_4\}$, $B_2 = \Theta$, and with BPA

$$m_2(\{\theta_2, \theta_3, \theta_4\}) = 0.6, \quad m_2(\Theta) = 0.4.$$

This pair of belief functions (Fig. 2.3, top) are combinable, as their cores $\mathcal{C}_1 = \{\theta_1, \theta_2\}$ *and* $\mathcal{C}_2 = \{\theta_1, \theta_2, \theta_3, \theta_4\}$ *are trivially not disjoint.*

Dempster combination ((2.6), represented in Fig. 2.3, middle) yields a new belief function on the same FoD, with focal elements (Fig. 2.3, bottom) $X_1 = \{\theta_1\} = A_1 \cap B_2$, $X_2 = \{\theta_2\} = A_2 \cap B_1$ *and* $X_3 = \{\theta_1, \theta_2\} = A_2 \cap B_2$, *and with BPA*

$$m(\{\theta_1\}) = \frac{0.7 \cdot 0.4}{1 - 0.42} = 0.48, \quad m(\{\theta_2\}) = \frac{0.3 \cdot 0.6}{1 - 0.42} = 0.31,$$
$$m(\{\theta_1, \theta_2\}) = \frac{0.3 \cdot 0.4}{1 - 0.42} = 0.21,$$

where the mass $0.6 \cdot 0.2 = 0.42$ *originally assigned to the empty set is used to normalise the result.*

2.3.2 Weight of conflict

The normalisation constant in (2.6) measures the *level of conflict* between the two input belief functions, for it represents the amount of evidence they attribute to contradictory (i.e., disjoint) subsets.

Definition 8. *We define the* weight of conflict $\mathcal{K}(Bel_1, Bel_2)$ *between two belief functions* Bel_1 *and* Bel_2, *defined on the same frame of discernment, as the logarithm of the normalisation constant in their Dempster combination,*

$$\mathcal{K} = \log \frac{1}{1 - \sum_{B \cap C = \emptyset} m_1(B) m_2(C)}. \tag{2.7}$$

Dempster's rule can be trivially generalised to the combination of n belief functions, in which case weights of conflict combine additively.

Proposition 3. *Suppose* $Bel_1, \ldots, Bel_{n+1}$ *are belief functions defined on the same frame* Θ, *and assume that* $Bel_1 \oplus \cdots \oplus Bel_{n+1}$ *exists. Then*

$$\mathcal{K}(Bel_1, \ldots, Bel_{n+1}) = \mathcal{K}(Bel_1, \ldots, Bel_n) + \mathcal{K}(Bel_1 \oplus \cdots \oplus Bel_n, Bel_{n+1}).$$

Dempster's rule has been strongly criticised in the past, for a number of reasons. As a result, numerous alternative combination rules have been proposed. The matter is thoroughly considered in Section 4.3.

2.3.3 Conditioning belief functions

Dempster's rule describes the way in which the assimilation of new evidence Bel' changes our beliefs previously encoded by a belief function Bel, determining new belief values given by $Bel \oplus Bel'(A)$ for all events A. As we have already highlighted, a new body of evidence is not constrained to be in the form of a single proposition A known with certainty, as happens in Bayesian reasoning.

Yet, the incorporation of new certainties is permitted in the theory of evidence, as a special case. In fact, this special kind of evidence is represented by belief functions of the form

$$Bel'(A) = \begin{cases} 1 & \text{if } B \subset A, \\ \\ 0 & \text{if } B \not\subset A, \end{cases}$$

where Bel is the proposition known with certainty. Such a belief function is combinable with the original BF Bel as long as $Bel(\bar{B}) < 1$, and the result has the form

$$Bel(A|B) \doteq Bel \oplus Bel' = \frac{Bel(A \cup \bar{B}) - Bel(\bar{B})}{1 - Bel(\bar{B})}$$

or, expressing the result in terms of plausibilities (2.4),

$$Pl(A|B) = \frac{Pl(A \cap B)}{Pl(B)}. \tag{2.8}$$

Equation (2.8) strongly reminds us of Bayes' rule of conditioning (1.1) – Shafer calls it *Dempster's rule of conditioning*. Whereas Dempster's rule (2.6) is symmetric in the role assigned to the two pieces of evidence Bel and Bel', owing to the commutativity of set-theoretical intersection, in Bayesian reasoning such symmetry is lost.

Conditioning in belief theory is surveyed in much more detail in Chapter 4, Section 4.5.

2.4 Simple and separable support functions

In the theory of evidence, a body of evidence (a belief function) usually supports more than one proposition (subset) of a frame of discernment. The simplest situation, however, is that in which the evidence points to a single non-empty subset $A \subset \Theta$.

Assume that $0 \leq \sigma \leq 1$ is the degree of support for A. Then, the degree of support for a generic subset $B \subset \Theta$ of the frame is given by

$$Bel(B) = \begin{cases} 0 & \text{if } B \not\supset A, \\ \sigma & \text{if } B \supset A,\ B \neq \Theta, \\ 1 & \text{if } B = \Theta. \end{cases} \tag{2.9}$$

Definition 9. *The belief function $Bel : 2^\Theta \rightarrow [0,1]$ defined by (2.9) is called a* simple support function *focused on A. Its basic probability assignment is $m(A) = \sigma$, $m(\Theta) = 1 - \sigma$ and $m(B) = 0$ for every other $B \subset \Theta$.*

2.4.1 Heterogeneous and conflicting evidence

We often need to combine evidence pointing towards different subsets, A and B, of a frame of discernment. When $A \cap B \neq \emptyset$ these two propositions are said to be 'compatible', and we say that the associated belief functions represent *heterogeneous* evidence. In this case, if σ_1 and σ_2 are the masses committed to A and B, respectively, by two simple support functions Bel_1 and Bel_2, we have as a result that their Dempster combination has BPA

$$m(A \cap B) = \sigma_1\sigma_2, \quad m(A) = \sigma_1(1-\sigma_2), \quad m(B) = \sigma_2(1-\sigma_1),$$
$$m(\Theta) = (1-\sigma_1)(1-\sigma_2).$$

Therefore, the belief values of $Bel = Bel_1 \oplus Bel_2$ are as follows:

$$Bel(X) = Bel_1 \oplus Bel_2(X) = \begin{cases} 0 & X \not\supset A \cap B, \\ \sigma_1\sigma_2 & X \supset A \cap B,\ X \not\supset A, B, \\ \sigma_1 & X \supset A,\ X \not\supset B, \\ \sigma_2 & X \supset B,\ X \not\supset A, \\ 1-(1-\sigma_1)(1-\sigma_2) & X \supset A, B,\ X \neq \Theta, \\ 1 & X = \Theta. \end{cases}$$

As our intuition would suggest, the combined evidence supports $A \cap B$ with a degree $\sigma_1\sigma_2$ equal to the product of the individual degrees of support for A and B.

When the two propositions have empty intersection, $A \cap B = \emptyset$, instead, we talk about *conflicting* evidence. The following example is once again taken from [1583].

Example 4: the alibi. *A criminal defendant has an alibi: a close friend swears that the defendant was visiting his house at the time of the crime. This friend has a good reputation: suppose this commits a degree of support of* $1/10$ *to the innocence of the defendant (*I*). On the other hand, there is a strong, actual body of evidence providing a degree of support of* $9/10$ *for his guilt (*G*).*

To formalise this case, we can build a frame of discernment $\Theta = \{G, I\}$*, so that the defendant's friend provides a simple support function focused on* $\{I\}$ *with* $Bel_I(\{I\}) = m_I(\{I\}) = 1/10$*, while the hard piece of evidence corresponds to another simple support function,* Bel_G*, focused on* $\{G\}$ *with* $Bel_G(\{G\}) = m_G(\{G\}) = 9/10$*. Their orthogonal sum* $Bel = Bel_I \oplus Bel_G$ *then yields*

$$Bel(\{I\}) = \frac{1}{91}, \quad Bel(\{G\}) = \frac{81}{91}.$$

The effect of the testimony has mildly eroded the force of the circumstantial evidence.

2.4.2 Separable support functions

In general, belief functions can support more than one proposition at a time. The next simplest class of BFs is that of 'separable support functions'.

Definition 10. *A* separable support function *Bel is a belief function that is either simple or equal to the orthogonal sum of two or more simple support functions, namely*

$$Bel = Bel_1 \oplus \cdots \oplus Bel_n,$$

where $n \geq 1$ and Bel_i is simple $\forall i = 1, \ldots, n$.

A separable support function Bel can be decomposed into simple support functions in different ways. More precisely, given one such decomposition $Bel = Bel_1 \oplus \cdots \oplus Bel_n$ with foci $A_1, \ldots, A_n$ and denoting by $\mathcal{C}$ the core of Bel, each of the following is a valid decomposition of Bel in terms of simple support functions:

- $Bel = Bel_1 \oplus \cdots \oplus Bel_n \oplus Bel_{n+1}$ whenever Bel_{n+1} is the vacuous belief function on the same frame;
- $Bel = (Bel_1 \oplus Bel_2) \oplus \cdots \oplus Bel_n$ whenever $A_1 = A_2$;
- $Bel = Bel'_1 \oplus \cdots \oplus Bel'_n$ whenever Bel'_i is the simple support function focused on $A'_i \doteq A_i \cap \mathcal{C}$ such that $Bel'_i(A'_i) = Bel_i(A_i)$, if $A_i \cap \mathcal{C} \neq \emptyset$ for all i.

However, the following proposition is true.

Proposition 4. *If Bel is a non-vacuous, separable support function with core $\mathcal{C}$, then there exists a unique collection $Bel_1, \ldots, Bel_n$ of non-vacuous simple support functions which satisfy the following conditions:*

1. $n \geq 1$.
2. $Bel = Bel_1$ *if* $n = 1$, *and* $Bel = Bel_1 \oplus \cdots \oplus Bel_n$ *if* $n \geq 1$.
3. $\mathcal{C}_{Bel_i} \subset \mathcal{C}$.
4. $\mathcal{C}_{Bel_i} \neq \mathcal{C}_{Bel_j}$ *if* $i \neq j$.

This unique decomposition is called the *canonical decomposition* (see Section 2.4.4) of the separable support function Bel.

Some intuition on what a separable support function may represent is provided by the following result [1583].

Proposition 5. *If Bel is a separable belief function, and A and B are two of its focal elements with $A \cap B \neq \emptyset$, then $A \cap B$ is also a focal element of Bel.*

The set of focal elements of a separable support function is closed under set-theoretical intersection. Such a belief function Bel is coherent in the sense that if it supports two propositions, then it must support the proposition 'naturally' implied by them, i.e., their intersection. Proposition 5 gives us a simple method to check whether a given belief function is indeed a separable support function.

2.4.3 Internal conflict

Since a separable support function can support pairs of disjoint subsets, it flags the existence of what we can call 'internal' conflict.

Definition 11. *The* weight of internal conflict $\mathcal{W}$ *for a separable support function* Bel *is defined as:*

- $\mathcal{W} = 0$ *if* Bel *is a simple support function;*
- $\mathcal{W} = \inf \mathcal{K}(Bel_1, \ldots, Bel_n)$ *for the various possible decompositions of* Bel *into simple support functions* $Bel = Bel_1 \oplus \cdots \oplus Bel_n$ *if* Bel *is not simple.*

It is easy to see (see [1583] again) that $\mathcal{W} = \mathcal{K}(Bel_1, \ldots, Bel_n)$, where $Bel_1 \oplus \cdots \oplus Bel_n$ is the canonical decomposition of Bel.

2.4.4 Inverting Dempster's rule: The canonical decomposition

The question of how to define an inverse operation to Dempster's combination rule naturally arises. If Dempster's rule reflects a modification of one's system of degrees of belief when the subject in question becomes familiar with the degrees of belief of another subject and accepts the arguments on which these degrees are based, the inverse operation should enable one to erase the impact of this modification and to return to one's original degrees of belief, supposing that the reliability of the second subject is put into doubt.

Namely, a belief function can be decomposed into a Dempster sum of simple support functions,[17]

$$m = \bigoplus_{A \subsetneq \Theta} m_A^w, \tag{2.10}$$

where m_A^w denotes the simple belief function such that

$$m_A^w(B) = \begin{cases} 1 - w & B = A, \\ w & B = \Theta, \\ 0 & \forall B \neq A, \Theta, \end{cases}$$

and the weights $w(A)$ satisfy $w(A) \in [0, +\infty)$ for all $A \subsetneq \Theta$. Note that, as a result, the mass that m_A^w assigns to some subsets of Θ may be negative: in the literature, this is called a *pseudo-belief function*.

The inversion problem was solved via algebraic manipulations by Smets in [1698]. Following Smets, we term *dogmatic* a belief function for which the FoD Θ is not a focal element. A belief function is *categorical* if it is both simple and dogmatic [451].

Proposition 6. *Any non-dogmatic belief function* $Bel : 2^\Theta \rightarrow [0,1]$ *admits a unique canonical decomposition of the form (2.10), where*

$$w(A) = \prod_{B : A \subseteq B \subseteq \Theta} Q(B)^{(-1)^{|B|-|A|+1}}$$

and $Q(B)$ *is the commonality value (2.5) of* $B \subseteq \Theta$ *according to* Bel.

[17] In fact, the decomposition is formally defined for pseudo-belief functions, in which case the conjunctive combination $\textcircled{\cap}$ applies: see Section 4.3.2.

Importantly, for separable support functions, the decomposition is such that $w(A) \in [0,1]$ for all A (i.e., separable support functions decompose into proper belief functions).

The weight representation of the Dempster sum also applies to the combination of two or more belief functions. Namely, if $Bel_1 = \oplus m_A^{w_A^1}$, $Bel_2 = \oplus m_A^{w_A^2}$, then

$$Bel_1 \oplus Bel_2 = \bigoplus_{A \subseteq \Theta} m_A^{w_A^1 \cdot w_A^2},$$

i.e., Dempster's rule amounts to a simple multiplication of the weights of the simple components.

Kramosil also proposed a solution to the inversion problem, based on his measure-theoretic approach [1064]. As we will see in Chapter 4, other combination rules such as the conjunctive and disjunctive combinations admit simple inverses.

2.5 Families of compatible frames of discernment

One appealing idea in the theory of evidence is the sensible claim that our knowledge of any given problem is inherently imperfect and imprecise. As a consequence, new evidence may allow us to make decisions on more detailed decision spaces (represented by frames of discernment). All these frames need to be 'compatible' with each other, in a sense that we will make precise in the following.

2.5.1 Refinings

A frame can certainly be assumed to be compatible with another one if it can be obtained by introducing new distinctions, i.e., by splitting some of its possible outcomes into finer ones. This idea is embodied by the notion of *refining*.

Definition 12. *Given two frames of discernment Θ and Ω, a map $\rho : 2^\Theta \rightarrow 2^\Omega$ is said to be a* refining *if it satisfies the following conditions:*

1. $\rho(\{\theta\}) \neq \emptyset \; \forall \theta \in \Theta$.
2. $\rho(\{\theta\}) \cap \rho(\{\theta'\}) = \emptyset$ *if* $\theta \neq \theta'$.
3. $\cup_{\theta \in \Theta} \rho(\{\theta\}) = \Omega$.

In other words, a refining maps the coarser frame Θ to a disjoint partition of the finer one Ω (see Fig. 2.4).

The finer frame is called a *refinement* of the first one, whereas Θ is termed a *coarsening* of Ω. Both frames represent sets of admissible answers to a given decision problem (see Chapter 4 as well) – the finer one is nevertheless a more detailed description, obtained by splitting each possible answer $\theta \in \Theta$ in the original frame of discernment. The image $\rho(A)$ of a subset A of Θ consists of all the outcomes in Ω that are obtained by splitting an element of A.

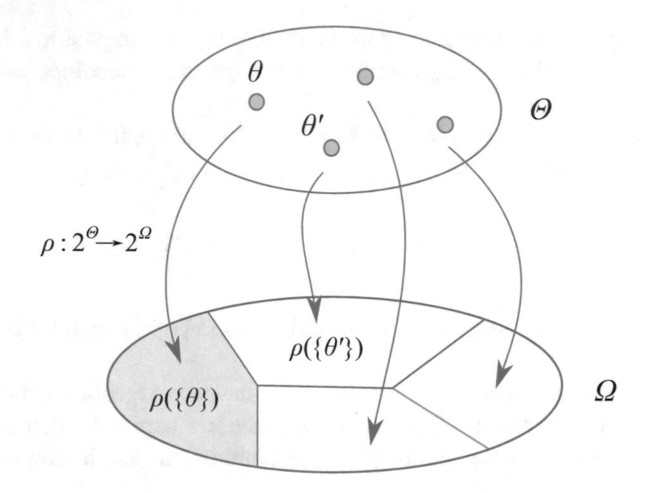

Fig. 2.4: A refining between two frames of discernment Θ and Ω.

A refining $\rho : 2^\Theta \rightarrow 2^\Omega$ is not, in general, onto; in other words, there are subsets $B \subset \Omega$ that are not images of subsets A of Θ. Nevertheless, we can define two different ways of associating each subset of the more refined frame Ω with a subset of the coarser one Θ.

Definition 13. *The* inner reduction *associated with a refining* $\rho : 2^\Theta \rightarrow 2^\Omega$ *is the map* $\underline{\rho} : 2^\Omega \rightarrow 2^\Theta$ *defined as*

$$\underline{\rho}(A) = \left\{ \theta \in \Theta \middle| \rho(\{\theta\}) \subseteq A \right\}. \tag{2.11}$$

The outer reduction *associated with* ρ *is the map* $\bar{\rho} : 2^\Omega \rightarrow 2^\Theta$ *given by*

$$\bar{\rho}(A) = \left\{ \theta \in \Theta \middle| \rho(\{\theta\}) \cap A \neq \emptyset \right\}. \tag{2.12}$$

Roughly speaking, $\underline{\rho}(A)$ is the largest subset of Θ that implies $A \subset \Omega$, while $\bar{\rho}(A)$ is the smallest subset of Θ that is implied by A. As a matter of fact [1583], the following proposition is true.

Proposition 7. *Suppose* $\rho : 2^\Theta \rightarrow 2^\Omega$ *is a refining,* $A \subset \Omega$ *and* $B \subset \Theta$*. Let* $\bar{\rho}$ *and* $\underline{\rho}$ *be the related outer and inner reductions. Then* $\rho(B) \subset A$ *iff* $B \subset \underline{\rho}(A)$*, and* $A \subset \rho(B)$ *iff* $\bar{\rho}(A) \subset B$*.*

2.5.2 Families of frames

The existence of distinct admissible descriptions at different levels of granularity of the same phenomenon is encoded in the theory of evidence by the concept of a *family of compatible frames* (see [1583], pages 121–125), whose building block is the notion of a refining (Definition 12).

Definition 14. *A non-empty collection of finite non-empty sets $\mathcal{F}$ is a* family of compatible frames of discernment *with refinings $\mathcal{R}$, where $\mathcal{R}$ is a non-empty collection of refinings between pairs of frames in $\mathcal{F}$, if $\mathcal{F}$ and $\mathcal{R}$ satisfy the following requirements:*

1. Composition of refinings*: if $\rho_1 : 2^{\Theta_1} \rightarrow 2^{\Theta_2}$ and $\rho_2 : 2^{\Theta_2} \rightarrow 2^{\Theta_3}$ are in $\mathcal{R}$, then $\rho_2 \circ \rho_1 : 2^{\Theta_1} \rightarrow 2^{\Theta_3}$ is in $\mathcal{R}$.*
2. Identity of coarsenings*: if $\rho_1 : 2^{\Theta_1} \rightarrow 2^{\Omega}$ and $\rho_2 : 2^{\Theta_2} \rightarrow 2^{\Omega}$ are in $\mathcal{R}$ and $\forall \theta_1 \in \Theta_1 \; \exists \theta_2 \in \Theta_2$ such that $\rho_1(\{\theta_1\}) = \rho_2(\{\theta_2\})$, then $\Theta_1 = \Theta_2$ and $\rho_1 = \rho_2$.*
3. Identity of refinings*: if $\rho_1 : 2^{\Theta} \rightarrow 2^{\Omega}$ and $\rho_2 : 2^{\Theta} \rightarrow 2^{\Omega}$ are in $\mathcal{R}$, then $\rho_1 = \rho_2$.*
4. Existence of coarsenings*: if $\Omega \in \mathcal{F}$ and $A_1, \ldots, A_n$ is a disjoint partition of Ω, then there is a coarsening in $\mathcal{F}$ which corresponds to this partition.*
5. Existence of refinings*: if $\theta \in \Theta \in \mathcal{F}$ and $n \in \mathbb{N}$, then there exists a refining $\rho : 2^{\Theta} \rightarrow 2^{\Omega}$ in $\mathcal{R}$ and $\Omega \in \mathcal{F}$ such that $\rho(\{\theta\})$ has n elements.*
6. Existence of common refinements*: every pair of elements in $\mathcal{F}$ has a* common refinement *in $\mathcal{F}$.*

Roughly speaking, two frames are compatible if and only if they concern propositions which can both be expressed in terms of propositions of a common, finer frame. The algebraic properties of families of compatible frames in terms of lattices and order relations were extensively studied in [369, 329, 351].

By property 6, each collection of compatible frames has many common refinements. One of these is particularly simple.

Theorem 1. *If $\Theta_1, \ldots, \Theta_n$ are elements of a family of compatible frames $\mathcal{F}$, then there exists a unique frame $\Theta \in \mathcal{F}$ such that:*

1. *There exists a refining $\rho_i : 2^{\Theta_i} \rightarrow 2^{\Theta}$ for all $i = 1, \ldots, n$.*
2. *$\forall \theta \in \Theta \; \exists \theta_i \in \Theta_i$ for $i = 1, \ldots, n$ such that*

$$\{\theta\} = \rho_1(\{\theta_1\}) \cap \cdots \cap \rho_n(\{\theta_n\}).$$

This unique frame is called the *minimal refinement* $\Theta_1 \otimes \cdots \otimes \Theta_n$ of the collection $\Theta_1, \ldots, \Theta_n$, and is the simplest space in which we can compare propositions pertaining to different compatible frames. Furthermore, the following proposition is true.

Proposition 8. *If Ω is a common refinement of $\Theta_1, \ldots, \Theta_n$, then $\Theta_1 \otimes \cdots \otimes \Theta_n$ is a coarsening of Ω. Furthermore, $\Theta_1 \otimes \cdots \otimes \Theta_n$ is the only common refinement of $\Theta_1, \ldots, \Theta_n$ that is a coarsening of every other common refinement.*

Example 5: number systems. *Figure 2.5 illustrates a simple example of a family of compatible frames. A real number r between 0 and 1 can be expressed, for instance, using either binary or base-5 digits. Furthermore, even within one number system of choice (e.g., the binary system), a real number can be represented with different* *degrees of approximation, using for instance one or two digits. Each of these quantised versions of r is associated with an interval of $[0, 1]$ (purple rectangles) and can be expressed in a common frame (their common refinement, Definition 14, property 6), for example by selecting a two-digit base-10 approximation.*

Refining maps between coarser and finer frames are easily interpreted, and are depicted in Fig. 2.5.

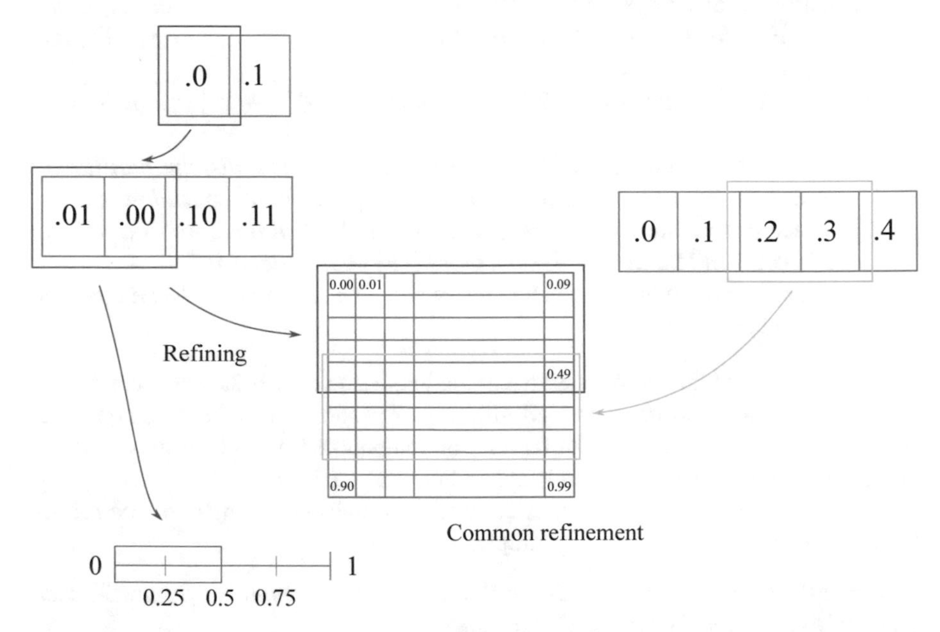

Fig. 2.5: The different digital representations of the same real number $r \in [0, 1]$ constitute a simple example of a family of compatible frames.

2.5.3 Consistent and marginal belief functions

If Θ_1 and Θ_2 are two compatible frames, then two belief functions $Bel_1 : 2^{\Theta_1} \rightarrow [0, 1]$, $Bel_2 : 2^{\Theta_2} \rightarrow [0, 1]$ can potentially be expressions of the same evidence.

Definition 15. *Two belief functions Bel_1 and Bel_2 defined over two compatible frames Θ_1 and Θ_2 are said to be* consistent *if*

$$Bel_1(A_1) = Bel_2(A_2)$$

whenever

$$A_1 \subset \Theta_1, \; A_2 \subset \Theta_2 \text{ and } \rho_1(A_1) = \rho_2(A_2), \; \rho_i : 2^{\Theta_i} \rightarrow 2^{\Theta_1 \otimes \Theta_2},$$

where ρ_i is the refining between Θ_i and the minimal refinement $\Theta_1 \otimes \Theta_2$ of Θ_1, Θ_2.

In other words, two belief functions are consistent if they agree on (attach the same degree of belief to) propositions which are equivalent in the minimal refinement of their FoDs.

A special case is that in which the two belief functions are defined on frames connected by a refining $\rho : 2^{\Theta_1} \rightarrow 2^{\Theta_2}$ (i.e., Θ_2 is a refinement of Θ_1). In this case Bel_1 and Bel_2 are consistent iff

$$Bel_1(A) = Bel_2(\rho(A)), \quad \forall A \subseteq \Theta_1.$$

The belief function Bel_1 is called the *restriction* or *marginal* of Bel_2 to Θ_1, and their mass values have the following relation:

$$m_1(A) = \sum_{A=\bar{\rho}(B)} m_2(B), \tag{2.13}$$

where $A \subset \Theta_1$ and $B \subset \Theta_2$, and $\bar{\rho}(B) \subset \Theta_1$ is the inner reduction (2.11) of Bel.

2.5.4 Independent frames

Two compatible frames of discernment are *independent* if no proposition discerned by one of them trivially implies a proposition discerned by the other. Obviously, we need to refer to a common frame: by Proposition 8, what common refinement we choose is immaterial.

Definition 16. *Let $\Theta_1, \ldots, \Theta_n$ be compatible frames, and $\rho_i : 2^{\Theta_i} \rightarrow 2^{\Theta_1 \otimes \cdots \otimes \Theta_n}$ the corresponding refinings to their minimal refinement. The frames $\Theta_1, \ldots, \Theta_n$ are said to be* independent *if*

$$\rho_1(A_1) \cap \cdots \cap \rho_n(A_n) \neq \emptyset \tag{2.14}$$

whenever $\emptyset \neq A_i \subset \Theta_i$ for $i = 1, \ldots, n$.

The notion of independence of frames is illustrated in Fig. 2.6.

In particular, it is easy to see that if $\exists j \in [1, \ldots, n]$ such that Θ_j is a coarsening of some other frame Θ_i, $|\Theta_j| > 1$, then $\{\Theta_1, \ldots, \Theta_n\}$ are *not* independent. Mathematically, families of compatible frames are collections of Boolean subalgebras of their common refinement [1650], as (2.14) is nothing but the independence condition for the associated Boolean subalgebras. [18]

[18] The following material is adapted from [1650].

Definition 17. *A* Boolean algebra *is a non-empty set $\mathcal{U}$ provided with the internal operations*

$$\begin{array}{rlrrlrrl} \cap : \mathcal{U} \times \mathcal{U} & \longrightarrow & \mathcal{U}, & \cup : \mathcal{U} \times \mathcal{U} & \longrightarrow & \mathcal{U}, & \neg : \mathcal{U} & \longrightarrow \mathcal{U}, \\ A, B & \mapsto & A \cap B, & A, B & \mapsto & A \cup B, & A & \mapsto \neg A \end{array}$$

called respectively meet, join *and* complement, *characterised by the following properties:*

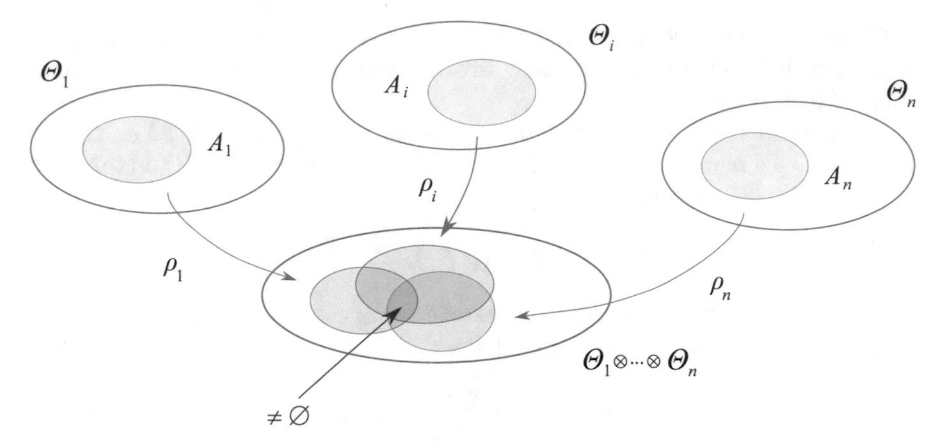

Fig. 2.6: Independence of frames.

2.5.5 Vacuous extension

There are occasions on which the impact of a body of evidence on a frame Θ is fully discerned by one of its coarsenings Ω, i.e., no proposition discerned by Θ receives greater support than what is implied by propositions discerned by Ω.

Definition 20. *A belief function $Bel : 2^\Theta \to [0,1]$ on Θ is the* vacuous extension *of a second belief function $Bel_0 : 2^\Omega \to [0,1]$, where Ω is a coarsening of Θ, whenever*

$$Bel(A) = \max_{B \subset \Omega,\ \rho(B) \subseteq A} Bel_0(B) \quad \forall A \subseteq \Theta.$$

$$\begin{array}{cc} A \cup B = B \cup A, & A \cap B = B \cap A, \\ A \cup (B \cup C) = (A \cup B) \cup C, & A \cap (B \cap C) = (A \cap B) \cap C, \\ (A \cap B) \cup B = B, & (A \cup B) \cap B = B, \\ A \cap (B \cup C) = (A \cap B) \cup (A \cap C), & A \cup (B \cap C) = (A \cup B) \cap (A \cup C), \\ (A \cap \neg A) \cup B = B, & (A \cup \neg A) \cap B = B. \end{array}$$

As a special case, the collection $(2^S, \subset)$ of all the subsets of a given set S is a Boolean algebra.

Definition 18. *$\mathcal{U}'$ is a subalgebra of a Boolean algebra $\mathcal{U}$ iff whenever $A, B \in \mathcal{U}'$ it follows that $A \cup B, A \cap B$ and $\neg A$ are all in $\mathcal{U}'$.*

The 'zero' of a Boolean algebra $\mathcal{U}$ is defined as $0 = \cap_{A \in \mathcal{U}} A$.

Definition 19. *A collection $\{\mathcal{U}_t\}_{t \in T}$ of subalgebras of a Boolean algebra $\mathcal{U}$ is said to be* independent *if*

$$A_1 \cap \cdots \cap A_n \neq 0 \tag{2.15}$$

whenever $0 \neq A_j \in \mathcal{U}_{t_j}$, $t_j \neq t_k$ for $j \neq k$.

Compare (2.15) and (2.14).

We say that Bel is 'carried' by the coarsening Ω.

This is a strong feature of belief theory: the vacuous belief function (our representation of ignorance) is left unchanged when moving from one hypothesis set to another, unlike what happens with Fisher's uninformative priors in Bayesian reasoning (see Section 1.3.1).

2.6 Support functions

The class of separable belief functions is not sufficiently large to describe the impact of a body of evidence on any frame of a family of compatible frames, as not all belief functions are separable ones.

Let us consider a body of evidence inducing a separable belief function Bel' over a certain frame Ω of a family $\mathcal{F}$: the 'impact' of this evidence on a coarsening Θ of Ω is naturally described by the restriction $Bel = Bel'|2^\Theta$ of Bel' (2.13) to Θ.

Definition 21. *A belief function $Bel : 2^\Theta \rightarrow [0,1]$ is a* support function *if there exists a refinement Ω of Θ and a separable support function $Bel' : 2^\Omega \rightarrow [0,1]$ such that $Bel = Bel'|2^\Theta$.*

In other words, a support function [1015] is the restriction of some separable support function.

As can be expected, not all support functions are separable support functions. The following proposition gives us a simple equivalent condition.

Proposition 9. *Suppose Bel is a belief function, and $\mathcal{C}$ its core. The following conditions are equivalent:*

- *Bel is a support function;*
- *$\mathcal{C}$ has a positive basic probability number, $m(\mathcal{C}) > 0$.*

Since there exist belief functions whose core has mass zero, Proposition 9 tells us that not all belief functions are support functions (see Section 2.7).

In [2108], Zhang discussed Shafer's weight-of-conflict conjecture, which implies that the notions of weight of evidence and weight of internal conflict, originally introduced for separable support functions, can be extended in a natural way to all support functions. He showed that such an extension to support functions can be carried out whether or not the weight-of-conflict conjecture is true.

2.6.1 Families of compatible support functions in the evidential language

In its 1976 essay [1583], Glenn Shafer distinguished between a 'subjective' and an 'evidential' vocabulary, maintaining a distinction between objects with the same mathematical description but different philosophical interpretations. In this brief summary, we have mainly stressed the subjective view of belief functions – nevertheless, they do possess an evidential interpretation, according to which a BF is induced by a *body of evidence* $\mathcal{E}$ living in a completely separate domain.

Each body of evidence $\mathcal{E}$ supporting a belief function Bel (see [1583]) simultaneously affects the whole family $\mathcal{F}$ of compatible frames of discernment that the domain of Bel belongs to, determining a support function over every element of $\mathcal{F}$. We say that $\mathcal{E}$ determines a *family of compatible support functions* $\{s_{\mathcal{E}}^{\Theta}\}_{\Theta \in \mathcal{F}}$.

The complexity of this family depends on the following property.

Definition 22. *The evidence $\mathcal{E}$* affects $\mathcal{F}$ sharply *if there exists a frame $\Omega \in \mathcal{F}$ that carries $s_{\mathcal{E}}^{\Theta}$ for every $\Theta \in \mathcal{F}$ that is a refinement of Ω. Such a frame Ω is said to* exhaust the impact *of $\mathcal{E}$ on $\mathcal{F}$.*

Whenever Ω exhausts the impact of $\mathcal{E}$ on $\mathcal{F}$, $s_{\mathcal{E}}^{\Omega}$ determines the whole family $\{s_{\mathcal{E}}^{\Theta}\}_{\Theta \in \mathcal{F}}$, for any support function over any given frame $\Theta \in \mathcal{F}$ is the restriction to Θ of $s_{\mathcal{E}}^{\Omega}$'s vacuous extension (Definition 20) to $\Theta \otimes \Omega$.

A typical example in which the evidence affects the family sharply is statistical inference, in which case both the frames and the evidence are highly idealised [1583].

2.6.2 Discerning the relevant interaction of bodies of evidence

Let us consider a frame Θ, its coarsening Ω, and a pair of support functions s_1, s_2 on Θ determined by two distinct bodies of evidence $\mathcal{E}_1, \mathcal{E}_2$. Applying Dempster's rule directly on Θ yields the following support function on its coarsening Ω:

$$(s_1 \oplus s_2)|2^{\Omega},$$

while its application on the coarser frame Θ after computing the restrictions of s_1 and s_2 to it yields

$$(s_1|2^{\Omega}) \oplus (s_2|2^{\Omega}).$$

In general, the outcomes of these two combination strategies will be different. Nevertheless, a condition on the refining linking Ω to Θ can be imposed which guarantees their equivalence.

Proposition 10. *Assume that s_1 and s_2 are support functions over a frame Θ, their Dempster combination $s_1 \oplus s_2$ exists, $\bar{\rho} : 2^{\Theta} \rightarrow 2^{\Omega}$ is an outer reduction, and*

$$\bar{\rho}(A \cap B) = \bar{\rho}(A) \cap \bar{\rho}(B) \tag{2.16}$$

holds wherever A is a focal element of s_1 and B is a focal element of s_2. Then

$$(s_1 \oplus s_2)|2^{\Omega} = (s_1|2^{\Omega}) \oplus (s_2|2^{\Omega}).$$

In this case Ω is said to *discern the relevant interaction* of s_1 and s_2. Of course, if s_1 and s_2 are carried by a coarsening of Θ, then that latter frame discerns their relevant interaction.

The above definition generalises to entire bodies of evidence.

Definition 23. *Suppose $\mathcal{F}$ is a family of compatible frames, $\{s^{\Theta}_{\mathcal{E}_1}\}_{\Theta\in\mathcal{F}}$ is the family of support functions determined by a body of evidence $\mathcal{E}_1$, and $\{s^{\Theta}_{\mathcal{E}_2}\}_{\Theta\in\mathcal{F}}$ is the family of support functions determined by a second body of evidence $\mathcal{E}_2$. Then, a frame $\Omega \in \mathcal{F}$ is said to discern the relevant interaction of $\mathcal{E}_1$ and $\mathcal{E}_2$ if*

$$\bar{\rho}(A \cap B) = \bar{\rho}(A) \cap \bar{\rho}(B)$$

whenever Θ is a refinement of Ω, where $\bar{\rho} : 2^{\Theta} \to 2^{\Omega}$ is the associated outer reduction, A is a focal element of $s^{\Theta}_{\mathcal{E}_1}$ and Bel is a focal element of $s^{\Theta}_{\mathcal{E}_2}$.

2.7 Quasi-support functions

The class of support functions still does not cover all possible belief functions.

2.7.1 Limits of separable support functions

Let us consider a finite power set 2^{Θ}. A sequence $f_1, f_2, \ldots$ of set functions on 2^{Θ} is said to tend to a limit function f if

$$\lim_{i\to\infty} f_i(A) = f(A) \quad \forall A \subset \Theta. \tag{2.17}$$

The following proposition can be proved [1583].

Proposition 11. *If a sequence of belief functions has a limit, then that limit is itself a belief function.*

In other words, the class of belief functions is closed with respect to the limit operator (2.17). The latter provides us with an insight into the nature of non-support functions.

Proposition 12. *If a belief function $Bel : 2^{\Theta} \to [0,1]$ is not a support function, then there exists a refinement Ω of Θ and a sequence $s_1, s_2, \ldots$ of separable support functions over Ω such that*

$$Bel = \left(\lim_{i\to\infty} s_i\right)\Big|2^{\Theta}.$$

Definition 24. *We call belief functions of this class* quasi-support functions.

It should be noted that

$$\left(\lim_{i\to\infty} s_i\right)\Big|2^{\Theta} = \lim_{i\to\infty} (s_i|2^{\Theta}),$$

so that we can also say that s is a limit of a sequence of support functions.

2.7.2 Bayesian belief functions as quasi-support functions

The following proposition investigates some of the properties of quasi-support functions.

Proposition 13. *Suppose $Bel : 2^\Theta \rightarrow [0,1]$ is a belief function over Θ, and $A \subset \Theta$ is a subset of Θ. If $Bel(A) > 0$ and $Bel(\bar{A}) > 0$, with $Bel(A) + Bel(\bar{A}) = 1$, then Bel is a quasi-support function.*

It easily follows that Bayesian belief functions (the images of discrete probability measures in the space of belief functions: see Section 2.2.3) are quasi-support functions, unless they commit all their probability mass to a single element of the frame. Furthermore, it is easy to see that vacuous extensions of Bayesian belief functions are also quasi-support functions.

As Shafer remarks, people used to thinking of beliefs as chances can be disappointed to see them relegated to a peripheral role, as beliefs that cannot arise from actual, finite evidence. On the other hand, statistical inference already teaches us that chances can be evaluated only after infinitely many repetitions of independent random experiments.[19]

2.7.3 Bayesian case: Bayes' theorem

Indeed, as it commits an infinite amount of evidence in favour of each possible element of a frame of discernment, a Bayesian belief function tends to obscure much of the evidence additional belief functions may carry with them.

Definition 25. *A function $pl_s : \Theta \rightarrow [0, \infty)$ is said to express the* relative plausibilities *of singletons under a support function $s : 2^\Theta \rightarrow [0,1]$ if*

$$pl_s(\theta) = c \cdot Pl_s(\{\theta\}) \tag{2.18}$$

for all $\theta \in \Theta$, where Pl_s is the plausibility function for s and the constant c does not depend on θ.

The quantity (2.18) is often also called the *contour function*.

Proposition 14. Bayes' theorem. *Suppose Bel_0 and s are a Bayesian belief function and a support function, respectively, on the same frame Θ. Suppose $pl_s : \Theta \rightarrow [0, \infty)$ expresses the relative plausibilities of singletons under s. Suppose also that their Dempster sum $Bel' = s \oplus Bel_0$ exists. Then Bel' is Bayesian, and*

$$Bel'(\{\theta\}) = K \cdot Bel_0(\{\theta\}) pl_s(\theta) \quad \forall \theta \in \Theta,$$

where

$$K = \left(\sum_{\theta \in \Theta} Bel_0(\{\theta\}) pl_s(\theta) \right)^{-1}.$$

[19]Using the notion of *weight of evidence*, Shafer gave a formal explanation of this intuitive observation by showing that a Bayesian belief function indicates an *infinite* amount of evidence in favour of *each* possibility in its core [1583].

This implies that the combination of a Bayesian belief function with a support function requires nothing more than the latter's relative plausibilities of singletons.

It is interesting to note that the latter functions behave multiplicatively under combination, as expressed in the following proposition.

Proposition 15. *If $s_1, \ldots, s_n$ are combinable support functions, and pl_i represents the relative plausibilities of singletons under s_i for $i = 1, \ldots, n$, then $pl_1 \cdot pl_2 \cdot \ldots \cdot pl_n$ expresses the relative plausibilities of singletons under $s_1 \oplus \cdots \oplus s_n$.*

This provides a simple algorithm for combining any number of support functions with a Bayesian BF.

2.7.4 Bayesian case: Incompatible priors

Having an established convention for how to set a Bayesian prior would be useful, as it would prevent us from making arbitrary and possibly unsupported choices that could eventually affect the final result of our inference process. Unfortunately, the only natural such convention (a uniform prior: see Section 1.3.4) is strongly dependent on the frame of discernment at hand, and is sensitive to both the refining and the coarsening operators.

More precisely, on a frame Θ with n elements, it is natural to represent our ignorance by adopting an uninformative uniform prior assigning a mass $1/n$ to every outcome $\theta \in \Theta$. However, the same convention applied to a different compatible frame Ω of the same family may yield a prior that is incompatible with the first one. As a result, the combination of a given body of evidence with one arbitrary such prior can yield almost any possible result [1583].

Example 6: Sirius's planets. (From [1583]). *A team of scientists wonder whether there is life around Sirius. Since they do not have any evidence concerning this question, they adopt a vacuous belief function to represent their ignorance on the frame of discernment*

$$\Theta = \{\theta_1, \theta_2\},$$

where θ_1, θ_2 are the answers 'there is life' and 'there is no life'. They can also consider the question in the context of a more refined set of possibilities. For example, our scientists may raise the question of whether there even exist planets around Sirius. In this case the set of possibilities becomes

$$\Omega = \{\zeta_1, \zeta_2, \zeta_3\},$$

where $\zeta_1, \zeta_2, \zeta_3$ are, respectively, the possibility that there is life around Sirius, that there are planets but no life, and that there are no planets at all. Obviously, in an evidential set-up our ignorance still needs to be represented by a vacuous belief function, which is exactly the vacuous extension of the vacuous BF previously defined on Θ.

From a Bayesian point of view, instead, it is difficult to assign consistent degrees of belief over Ω and Θ, both symbolising the lack of evidence. Indeed, on Θ a *uniform prior yields* $p(\{\theta_1\}) = p(\{\theta_1\}) = 1/2$*, while on* Ω *the same choice will yield*

$$p'(\{\zeta_1\}) = p'(\{\zeta_2\}) = p'(\{\zeta_3\}) = \frac{1}{3}.$$

The frames Ω *and* Θ *are obviously compatible (as the former is a refinement of the latter). The vacuous extension of* p *onto* Ω *generates a Bayesian distribution*

$$p(\{\zeta_1\}) = \frac{1}{3}, \quad p(\{\zeta_1, \zeta_2\}) = \frac{2}{3},$$

which is obviously inconsistent with p'.

2.8 Consonant belief functions

To conclude this brief review of evidence theory, we wish to recall a class of belief functions which is, in some sense, opposed to that of quasi-support functions – that of *consonant* belief functions (Fig. 2.7).

Definition 26. *A belief function is said to be* consonant *if its focal elements* $A_1, \ldots, A_m$ *are nested:* $A_1 \subset A_2 \subset \cdots \subset A_m$.

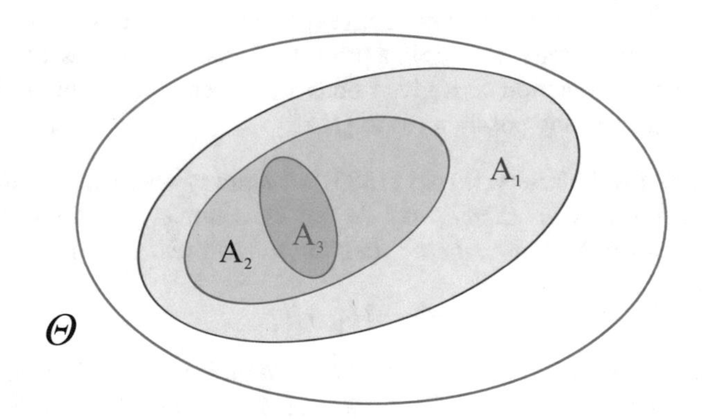

Fig. 2.7: Consonant belief functions have nested focal elements.

The following proposition illustrates some of their properties [1583].

Proposition 16. *If* Bel *is a belief function with plausibility function* Pl*, then the following conditions are equivalent:*

1. Bel *is consonant.*
2. $Bel(A \cap B) = \min(Bel(A), Bel(B))$ *for every* $A, B \subset \Theta$.
3. $Pl(A \cup B) = \max(Pl(A), Pl(B))$ *for every* $A, B \subset \Theta$.
4. $Pl(A) = \max_{\theta \in A} Pl(\{\theta\})$ *for all non-empty* $A \subset \Theta$.

5. *There exists a positive integer n and a collection of simple support functions $s_1, \ldots, s_n$ such that $Bel = s_1 \oplus \cdots \oplus s_n$ and the focus of s_i is contained in the focus of s_j whenever $i < j$.*

Consonant belief functions represent collections of pieces of evidence *all pointing in the same direction*. Moreover, the following is true.

Proposition 17. *Suppose $s_1, \ldots, s_n$ are non-vacuous simple support functions with foci $C_1, \ldots, C_n$, respectively, and $Bel = s_1 \oplus \cdots \oplus s_n$ is consonant. If C denotes the core of Bel, then all the sets $C_i \cap C$, $i = 1, \ldots, n$, are nested.*

By condition 2 of Proposition 16, we have that

$$0 = Bel(\emptyset) = Bel(A \cap \bar{A}) = \min(Bel(A), Bel(\bar{A})),$$

i.e., either $Bel(A) = 0$ or $Bel(\bar{A}) = 0$ for every $A \subset \Theta$. Comparing this result with Proposition 13 explains, in part, why consonant and quasi-support functions can be considered as representing diametrically opposite subclasses of belief functions.

3

Understanding belief functions

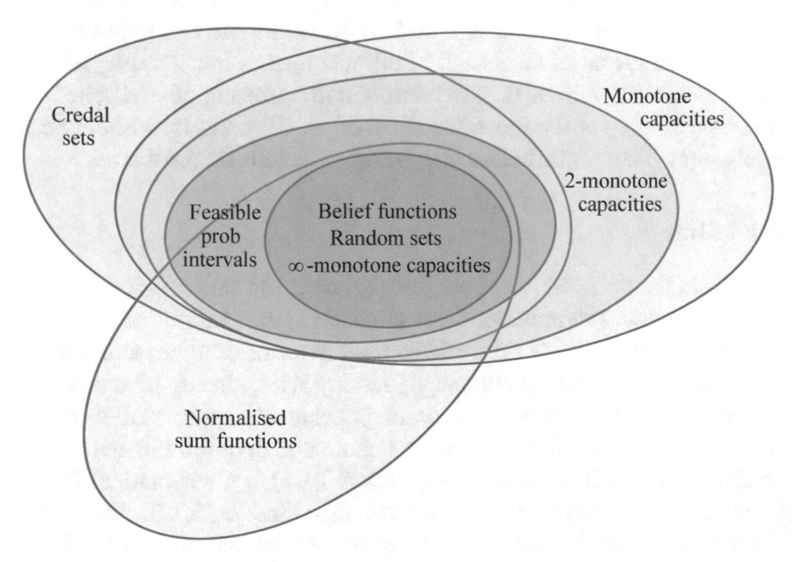

Fig. 3.1: The notions of a belief function, a (finite) random set and an ∞-monotone capacity are mathematically equivalent. Their relationship with a number of more general uncertainty measures is depicted here – the list is not exhaustive.

From the previous chapter's summary of the basic notions of the theory of evidence, it is clear that belief functions are rather complex objects. Figure 3.1 gives

F. Cuzzolin, *The Geometry of Uncertainty*, Artificial Intelligence: Foundations, Theory, and Algorithms, https://doi.org/10.1007/978-3-030-63153-6_3

a pictorial representation of the relation between the theory of belief functions and other mathematical theories of uncertainty. As a consequence, multiple, equivalent mathematical definitions of what a 'belief function' is (in addition to that given in Definition 5) can be provided. Furthermore, belief functions can be given distinct and sometimes even conflicting interpretations, in terms of what kind of (Knightian, or second-order) uncertainty is modelled by these mathematical objects.

The fact that belief functions (and random sets in general) are more complex than standard probability measures has, in fact, often hindered their adoption. Their multiple facets as mathematical entities has caused arguments about their 'true' nature, and about what is the 'correct' way of manipulating them [1363]. Most of the debate has taken place at the semantic level, for different interpretations imply the use of entirely different operators to construct a theory of reasoning with belief functions. The purpose of this chapter is to provide an overview of the manifold mathematical definitions and epistemic interpretations of belief measures, *en passant* pointing out the most common misunderstandings, and to succinctly recall the long-standing debate on their nature and on that of uncertainty itself.

To add a further layer of complexity, in the almost fifty years since its inception the theory of evidence has obviously evolved much, thanks to the efforts of several scientists, to the point that this terminology now subsumes a number of different takes on the common mathematical concept of a (finite) random set. Several students have proposed their own framework or formulation for evidential reasoning, partly in response to early strong criticisms made by scholars such as Judea Pearl [1406, 1408]. This is, for instance, the case for Philippe Smets's transferable belief model. We will recall here the various frameworks centred on the notion of a belief measure, reserving more space for the most impactful of such proposals, without neglecting less widely known but nevertheless interesting variations on the theme.

Chapter outline

Accordingly, in the first part of the chapter (Section 3.1) we provide an overview of the main mathematical formulations and semantic interpretations of the theory.

We start from Dempster's original proposal in terms of upper and lower probabilities induced by multivalued mappings (Section 3.1.1, already hinted at in Section 2.1), also termed *compatibility relations*, and explain Dempster's rule's rationale in that context. We then recall the alternative axiomatic definition of belief functions as generalised, non-additive measures (Section 3.1.2), a mathematical formulation closely related to the notion of an *inner measure* (Section 3.1.3). In a robust statistical perspective, belief functions can also be interpreted as (a specific class of) convex sets of probability measures (*credal sets*, Section 3.1.4). The most general mathematical framework upon which to rest the notion of a belief measure, however, is that of random set theory (Section 3.1.5). Most notably, the latter allows us to naturally extend the concept of a belief function to continuous domains. A number of behavioural (betting) interpretations have also been proposed (Section 3.1.6), which are inspired by de Finetti's betting view of probability measures and have strong links with Walley's imprecise probability theory (see Section 6.1). Section

3.1.7 concludes the first part of the chapter by briefly dispelling some common misconceptions about the nature of belief functions, namely their supposed equivalence to credal sets, families of probability distributions, confidence intervals (see Section 1.3.7) or second-order distributions.

Section 3.2 recalls in quite some detail the long-standing debate on the epistemic nature of belief functions and their 'correct' interpretation. After a brief description of Shafer's evolving position on the matter, and of the early support received by his mathematical theory of evidence (Section 3.2.1), we focus in particular on the debate on the relationship between evidential and Bayesian reasoning (Section 3.2.3). After summarising Pearl's and others' criticisms of belief theory (Section 3.2.4), we argue that most of the misunderstandings are due to confusion between the various interpretations of these objects (Section 3.2.5), and summarise the main rebuttals and formal justifications which have been advanced in response (Section 3.2.6).

Finally, in the last part of the chapter (Section 3.3) we review the variety of frameworks based on (at least the most fundamental notions of) the theory of belief functions which have been formulated in the last fifty years, and the various generalisation and extensions put forward by a number of authors. We start with the approaches which share the original rationale in terms of multivalued mappings/compatibility relations (Section 3.3.1), including Kohlas and Monney's theory of hints, Kramosil's probabilistic analysis and Hummel and Landy's statistics of experts' opinions. We then move on to what is arguably the most widely adopted approach, Smets's transferable belief model (Section 3.3.2), to later recall the main notions of Dezert–Smandarache theory (Section 3.3.3) and of Dempster and Liu's Gaussian belief functions (Section 3.3.4). Extensions of the notion of a belief function to more general domains, such as lattices and Boolean algebras, are considered in Section 3.3.5. Qualitative models are described in Section 3.3.6, whereas Section 3.3.7 groups together various interval or credal extensions of the notion of a basic probability assignment, including Denoeux's imprecise belief structures and Augustin's generalised BPAs. Finally, a rather comprehensive review of less well-known (but in some cases fascinating) approaches, including, among others, Zadeh's 'simple view' of belief functions as necessity measures associated with second-order relations and Lowrance's evidential reasoning, concludes the chapter (Section 3.3.8).

3.1 The multiple semantics of belief functions

Belief functions were originally introduced by Dempster as a simple application of probability theory to multiple domains linked by a multivalued map (compatibility relation) [414]. Later, as we saw, Shafer interpreted them as mathematical (set) functions encoding evidence in support of propositions [1583].

As we discuss here, further alternative interpretations of these objects can also be given in terms of generalised or non-additive probabilities (Section 3.1.2), inner measures ([1511, 583]: see Section 3.1.3), convex sets of probability measures (credal sets, Section 3.1.4), or random sets ([1344, 826]: Section 3.1.5). Zadeh's 'simple view' of the theory of evidence is also discussed in Section 3.3.8.

Finally, by considering situations where uncertainty is partially resolved, Jaffray [876] proved that there exists a one-to-one correspondence between belief functions and coherent betting systems *à la* de Finetti. This result was extended further by Flaminio and Godo in [625].

3.1.1 Dempster's multivalued mappings, compatibility relations

As we hinted in Section 2.1, the notion of a belief function [1607, 1597] originally derives from a series of publications by Dempster on upper and lower probabilities induced by multivalued mappings, in particular [415], [417] and [418].

Multivalued mappings The idea that intervals, rather than probability values, should be used to model degrees of belief had already been suggested and investigated by many, including Fishburn, Good and Koopman [619, 705, 1036, 1037, 1735].

Dempster's original contribution was to define upper and lower probability values in terms of statistics of set-valued functions defined over a measure space. In [415], for instance, he gave examples of the use of upper and lower probabilities in terms of finite populations with discrete univariate observable characteristics. Shafer later reformulated Dempster's work by identifying his upper and lower probabilities with epistemic probabilities or 'degrees of belief', i.e., the quantitative assessments of one's belief in a given fact or proposition. The following sketch of the nature of belief functions is taken from [1598]: another analysis of the relation between belief functions and upper and lower probabilities was developed in [1673].

Let us consider a problem in which we have probabilities (coming from arbitrary sources, for instance subjective judgement or objective measurements) for a question Q_1 and we want to derive degrees of belief for a related question Q_2. For example, as in the murder trial scenario of Section 2.1, Q_1 could be a judgement about the reliability of a witness, and Q_2 a decision about the truth of a reported fact. In general, each question will have a number of possible answers, only one of them being correct.

Let us denote by Ω and Θ the sets of possible answers to Q_1 and Q_2, respectively. We are given a probability measure P on Ω, and we want to derive a degree of belief $Bel(A)$ that $A \subset \Theta$ contains the correct response to Q_2 (see Fig. 3.2).

If we call $\Gamma(\omega)$ the subset of answers to Q_2 compatible with $\omega \in \Omega$, each element ω tells us that the answer to Q_2 is somewhere in A whenever $\Gamma(\omega) \subset A$. The degree of belief (2.2) $Bel(A)$ of an event $A \subset \Theta$ is then the total probability (in Ω) of all the answers ω to Q_1 that satisfy the above condition, namely

$$Bel(A) = P(\{\omega | \Gamma(\omega) \subset A\}). \tag{3.1}$$

Analogously, the degree of plausibility (2.4) of $A \subset \Theta$ is defined as

$$Pl(A) = P(\{\omega | \Gamma(\omega) \cap A \neq \emptyset\}). \tag{3.2}$$

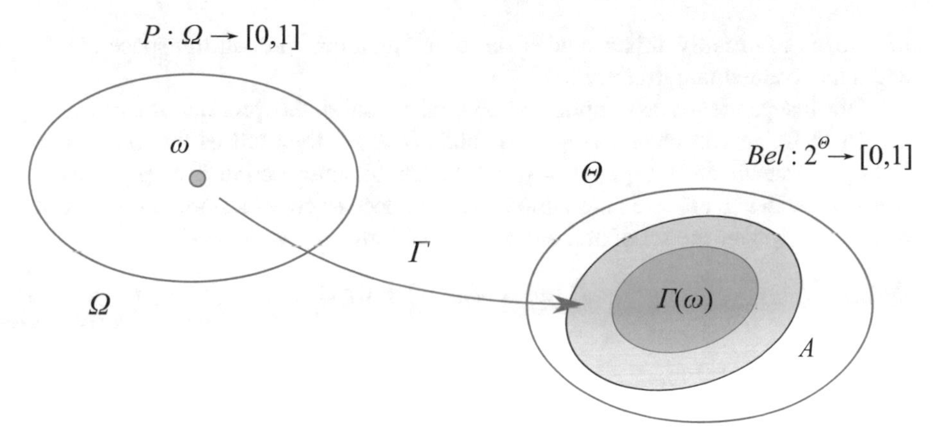

Fig. 3.2: Compatibility relations and multivalued mappings. A probability measure P on Ω induces a belief function Bel on Θ whose values on the events A of Θ are given by (3.1).

The map $\Gamma : \Omega \rightarrow 2^{\Theta}$ (where 2^{Θ} denotes, as usual, the collection of subsets of Θ) is called a *multivalued mapping* from Ω to Θ. Such a mapping Γ, together with a probability measure P on Ω, induces a *belief function* on Θ,

$$\begin{aligned} Bel : 2^{\Theta} & \rightarrow [0,1] \\ A \subset \Theta \mapsto Bel(A) & \doteq \sum_{\omega \in \Omega : \Gamma(\omega) \subset A} P(\omega). \end{aligned}$$

Compatibility relations Obviously, a multivalued mapping is equivalent to a *relation*, i.e., a subset C of $\Omega \times \Theta$. The *compatibility relation* associated with Γ,

$$\mathcal{C} = \Big\{ (\omega, \theta) | \theta \in \Gamma(\omega) \Big\}, \tag{3.3}$$

describes the subset of answers θ in Θ compatible with a given $\omega \in \Omega$.

As Shafer himself admits in [1598], compatibility relations are just a different name for multivalued mappings. Nevertheless, several authors (among whom are Shafer [1596], Shafer and Srivastava [1612], Lowrance [1214] and Yager [2021]) chose this approach to build the mathematics of belief functions.

As we will see in Section 3.3.1, a number of frameworks based on belief function theory rest on generalisations of the concept of a compatibility relation.

Dempster combination Consider now two multivalued mappings Γ_1, Γ_2 inducing two belief functions over the same frame Θ, where Ω_1 and Ω_2 are their domains and P_1, P_2 are the associated probability measures over Ω_1 and Ω_2, respectively.

Under the assumption that the items of evidence generating P_1 and P_2 are independent, we wish to find the belief function resulting from pooling the two pieces of evidence. Formally, this is equivalent to finding a new probability space (Ω, P) and a multivalued map from Ω to Θ.

The independence assumption allows us to build the product space $(P_1 \times P_2, \Omega_1 \times \Omega_2)$: two outcomes $\omega_1 \in \Omega_1$ and $\omega_2 \in \Omega_2$ then tell us that the answer to Q_2 is somewhere in $\Gamma_1(\omega_1) \cap \Gamma_2(\omega_2)$. When this intersection is empty, the two pieces of evidence are in contradiction. We then need to condition the product measure $P_1 \times P_2$ over the set of non-empty intersections

$$\Gamma_1(\omega_1) \cap \Gamma_2(\omega_2) \neq \emptyset, \tag{3.4}$$

obtaining

$$\begin{gathered} \Omega = \left\{ (\omega_1, \omega_2) \in \Omega_1 \times \Omega_2 \middle| \Gamma_1(\omega_1) \cap \Gamma_2(\omega_2) \neq \emptyset \right\}, \\ P = P_1 \times P_2|_{\Omega}, \quad \Gamma(\omega_1, \omega_2) = \Gamma_1(\omega_1) \cap \Gamma_2(\omega_2). \end{gathered} \tag{3.5}$$

It is easy to see that the new belief function Bel is linked to the pair of belief functions being combined by Dempster's rule, as defined in (2.6).

The combination of compatibility relations defined by (3.4) can be termed 'optimistic', as the beliefs (rather than the doubts) of the two sources are combined. In other words, the two sources are assumed to agree on the intersection of the propositions they support. Indeed, this is a hidden assumption of Dempster's combination rule (3.5), as important as the assumption of independence of sources.

An alternative approach supported by, among others, Kramosil [1056, 1049] is based on the dual idea that doubts should be shared instead. As a result, an outcome $\theta \in \Theta$ is incompatible with $\omega \in \Omega$ if it is considered incompatible by all the sources separately, namely whenever

$$\Gamma_1(\omega) \cup \Gamma_2(\omega) = \emptyset.$$

This results in what Smets calls the *disjunctive rule of combination*, which we will discuss in Section 4.3.2.

3.1.2 Belief functions as generalised (non-additive) probabilities

A third major interpretation of belief functions is as non-additive probabilities.

Let us go back to Kolmogorov's classical definition of a probability measure (Chapter 1, Definition 2). If we relax the third constraint

$$\text{if } A \cap B = \emptyset, \ A, B \in \mathcal{F}, \text{ then } P(A \cup B) = P(A) + P(B),$$

to allow the function $P : \mathcal{F} \to [0,1]$ to satisfy additivity only as a lower bound, and restrict ourselves to finite sets, we obtain Shafer's second definition of a belief function [1583].

Definition 27. *Suppose Θ is a finite set, and let $2^\Theta = \{A \subseteq \Theta\}$ denote the set of all subsets of Θ. A belief function on Θ is a function $Bel : 2^\Theta \to [0,1]$ from the power set 2^Θ to the real interval $[0,1]$ such that:*

- $Bel(\emptyset) = 0$;
- $Bel(\Theta) = 1$;
- *for every positive integer* n *and for every collection* $A_1, \ldots, A_n \in 2^\Theta$,

$$Bel(A_1 \cup \ldots \cup A_n) \geq \sum_i Bel(A_i) - \sum_{i<j} Bel(A_i \cap A_j) + \ldots + (-1)^{n+1} Bel(A_1 \cap \ldots \cap A_n). \quad (3.6)$$

The following proposition can be proven [1583].

Proposition 18. *Definitions 27 and 5 are equivalent formulations of the notion of a belief function.*

The condition (3.6), called *superadditivity*, obviously generalises Kolmogorov's additivity (Definition 2). Belief functions can then be seen as generalisations of the familiar notion of a (discrete) probability measure, and the theory of evidence as a generalised probability theory [364].

Objects which satisfy the conditions of Definition 27 are called *infinitely monotone capacities*: we will come back to this point in Chapter 6, Section 6.2.

3.1.3 Belief functions as inner measures

Belief functions can also be assimilated into *inner measures*.

Definition 28. *Given a probability measure* P *defined over a* σ*-field of subsets* $\mathcal{F}$ *of a finite set* $\mathcal{X}$*, the* inner probability *of* P *is the function* P_* *defined by*

$$P_*(A) = \max \left\{ P(B) \middle| B \subset A, \ B \in \mathcal{F} \right\}, \quad A \subset \mathcal{X} \quad (3.7)$$

for each *subset* A *of* $\mathcal{X}$*, not necessarily in* $\mathcal{F}$.

The inner probability value $P_*(A)$ represents the degree to which the available probability values of P suggest that we should believe in an event A for which a probability value is not directly available.

Now, let us define as the domain $\mathcal{X}$ of the inner probability function P_* the compatibility relation $\mathcal{C}$ (3.3) associated with a multivalued mapping Γ, and choose as the σ-field $\mathcal{F}$ on $\mathcal{C}$ the collection

$$\mathcal{F} = \left\{ \mathcal{C} \cap (E \times \Theta) \middle| E \subset \Omega \right\}. \quad (3.8)$$

Each element $\mathcal{C} \cap (E \times \Theta)$ of $\mathcal{F}$ is the set of compatibility pairs $C = (\omega, \theta)$ such that $\omega \in E$. It is then natural to define a probability measure Q over the σ-field (3.8) which depends on the original measure P on Ω:

$$Q : \mathcal{F} \to [0, 1], \quad \mathcal{C} \cap (E \times \Theta) \mapsto P(E).$$

The inner probability measure associated with Q is then the following function on $2^{\mathcal{C}}$:

$$Q_* : 2^{\mathcal{C}} \to [0,1],$$
$$\mathcal{A} \subset \mathcal{C} \mapsto Q_*(\mathcal{A}) = \max\Big\{P(E)\Big| E \subset \Omega,\ \mathcal{C} \cap (E \times \Theta) \subset \mathcal{A}\Big\}.$$

We can then compute the inner probability of the subset $\mathcal{A} = \mathcal{C} \cap (\Omega \times A)$ of $\mathcal{C}$ which corresponds to a subset A of Θ as

$$\begin{aligned} Q_*(\mathcal{C} \cap (\Omega \times A)) &= \max\{P(E)|E \subset \Omega,\ \mathcal{C} \cap (E \times \Theta) \subset \mathcal{C} \cap (\Omega \times A)\} \\ &= \max\{P(E)|E \subset \Omega,\ \omega \in E \wedge (\omega, \theta) \in \mathcal{C} \Rightarrow \theta \in A\} \\ &= P(\{\omega|(\omega,\theta) \in \mathcal{C} \Rightarrow \theta \in A\}), \end{aligned}$$

which, by the definition of a compatibility relation, becomes

$$P(\{\omega|\Gamma(\omega) \subset A\}) = Bel(A),$$

i.e., the classical definition (3.1) of the belief value of A induced by a multivalued mapping Γ. This connection between inner measures and belief functions appeared in the literature in the second half of the 1980s [1511, 1515, 583].

As we will see in Section 4.9.4, inner measures have been used by Kramosil to propose four possible generalisations of belief measures to arbitrary spaces.

3.1.4 Belief functions as credal sets

Credal sets and lower probabilities Belief functions admit the following order relation:

$$Bel \leq Bel' \equiv Bel(A) \leq Bel'(A) \ \ \forall A \subset \Theta, \tag{3.9}$$

called *weak inclusion*. A belief function Bel is weakly included in Bel' whenever its belief values are dominated by those of Bel' for all the events of Θ.

A probability distribution P which weakly includes a belief function Bel, $P(A) \geq Bel(A)\ \forall A \subset \Theta$, is said to be *consistent* with Bel [1086]. Each belief function Bel thus uniquely identifies a set of probabilities consistent with it,

$$\mathcal{P}[Bel] = \Big\{P \in \mathcal{P}\Big| P(A) \geq Bel(A)\Big\}, \tag{3.10}$$

(where $\mathcal{P}$ is the set of all probabilities one can define on Θ), of which it is its lower envelope (see Fig. 3.3). Accordingly, the theory of evidence is seen by some authors as a special case of robust statistics [1572].

The fact that belief functions are mathematically equivalent to lower envelopes $\underline{P}$ of sets of probability measures explains the terminology *lower probability* which is sometimes use to denote these objects: $Bel(A) = \underline{P}(A)$. Dually, the corresponding plausibility measure is the upper envelope $\overline{P}$ of the set (3.10): $Pl(A) = \overline{P}(A)$. A more formal definition of lower probabilities will be given in Section 6.1.1.

Convex sets of probabilities are often called *credal sets* [1141, 2094, 344, 48]. Given an arbitrary credal set Cr, its lower and upper probabilities are defined as

$$\underline{P}(A) \doteq \inf_{P \in Cr} P(A), \quad \overline{P}(A) \doteq \inf_{P \in Cr} P(A). \tag{3.11}$$

A number of scholars, as it turns out, have argued in favour of representing beliefs (see Section 1.4.2) in terms of convex sets of probabilities, including Koopman [1037], Good [705] and Smith [1734, 1735].

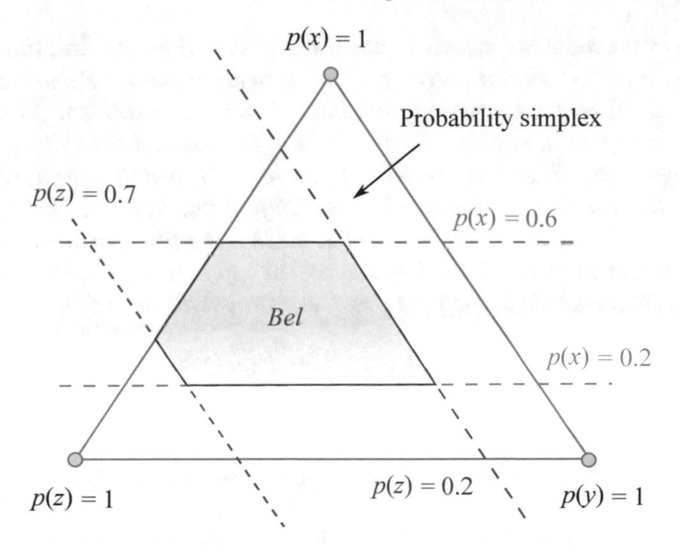

Fig. 3.3: A belief function is a credal set with boundaries determined by lower and upper bounds (3.10) on probability values.

Extremal probabilities of credal sets associated with belief functions Although the set $\mathcal{P}[Bel]$ (3.10) is a polytope in the simplex $\mathcal{P}$ of all probabilities we can define on Θ, not all credal sets there 'are' belief functions. Credal sets associated with BFs have vertices of a very specific type (see Fig. 3.3). The latter are all the distributions P^π induced by a permutation $\pi = \{x_{\pi(1)}, \ldots, x_{\pi(|\Theta|)}\}$ of the singletons of $\Theta = \{x_1, \ldots, x_n\}$ of the form [245, 339]

$$P^\pi[Bel](x_{\pi(i)}) = \sum_{A \ni x_{\pi(i)};\ A \not\ni x_{\pi(j)}\ \forall j<i} m(A). \tag{3.12}$$

Such an extremal probability (3.12) assigns to a singleton element put in position $\pi(i)$ by the permutation π the mass of all the focal elements containing it, but not containing any elements preceding it in the permutation order [1879].

A characterisation of the extremal points of credal sets generated by 2-alternating capacities, which generalises (3.12), was given by Miranda et al. in [1293]. Some results on the set of k-*additive* belief functions (whose focal elements have cardinality up to k, rather than up to 1 as in Bayesian BFs) which dominate a given belief function Bel were provided in [1296].

Belief functions as special credal sets A landmark work by Kyburg [1086] established a number of results relating belief updating for belief functions, and for credal sets (Section 3.1.4).

Proposition 19. *Closed convex sets of classical probability functions include Shafer's probability mass functions as a special case.*

Example 7: a credal set which is not a belief function. (From [1086]). *As an example of a credal set not associated with a belief function, Kyburg considered a compound experiment consisting of either (i) tossing a fair coin twice, or (ii) drawing a coin from a bag containing 40% two-headed and 60% two-tailed coins and tossing it twice. The two parts (i) and (ii) are performed in some unknown ratio* p*, so that, for example, the probability that the first toss lands heads is* $p * 1/2 + (1-p) * 0.4$*,* $0 < p < 1$*. Let* A *be the event that the first toss lands heads, and* B *the event that the second toss lands tails. The representation by a convex set of probability functions yields the following lower probabilities (3.11):*

$$\underline{P}(A \cup B) = 0.75 < 0.9 = \underline{P}(A) + \underline{P}(B) - \underline{P}(A \cap B) = 0.4 + 0.5 - 0. \quad (3.13)$$

By Theorem 2.1 of [1583], instead,

$$Bel(A \cup B) \geq Bel(A) + Bel(B) - Bel(A \cap B), \quad (3.14)$$

and therefore $\underline{P}$ *in (3.13) is not a belief function. It is still possible to compute a mass function, but the masses assigned to the union of any three atoms will be negative.*

Belief functions with negative masses are widely discussed later in this Book.

Credal and belief updating Kyburg also showed that the impact of 'uncertain evidence' (represented either as a simple support function, Definition 9, or as a probability 'shift' of the kind associated with Jeffrey's rule, Section 4.6.2) can be represented by Dempster conditioning (see Section 2.3.3) in Shafer's framework, whereas it is represented in the framework of convex sets of classical probabilities by classical conditioning. As a credal set is a convex set of probabilities, it suffices to condition its vertices to obtain the vertices of the updated credal set.

The following result relates updating in belief and robust Bayesian theory.

Proposition 20. ([1086], Theorem 4). *The probability intervals resulting from Dempster updating are included in (and may be properly included in) the intervals that result from applying Bayesian updating to the associated credal set.*

Whether this is a good or a bad thing, Kyburg argues, depends on the situation. Examples in which one of the two operators leads to more 'appealing' results are provided.

Paul Black [166] emphasised the importance of Kyburg's result by looking at simple examples involving Bernoulli trials, and showing that many convex sets of probability distributions generate the same belief function.

3.1.5 Belief functions as random sets

Given a multivalued mapping $\Gamma : \Omega \to 2^{\Theta}$, a straightforward conceptual step is to consider the probability value $P(\omega)$ as directly attached to the subset $\Gamma(\omega) \subset \Theta$, rather than to the element ω of Ω. What we obtain is a *random set* in Θ, i.e., a probability measure on a collection of subsets (see [717, 716, 1268] for the most complete introductions to the matter). The degree of belief $Bel(A)$ of an event A then becomes the total probability that the random set is contained in A.

Note the subtle difference between the two interpretations: whereas in Dempster's setting the masses of the subsets of Θ are induced by source probabilities in a classical probability space, one can directly define a random set on Θ making no reference to external probability measures. Nevertheless, as informally anticipated in the Introduction, a random set can also be thought of as a set-valued random variable $X : \Omega \to 2^\Theta$ which maps elements of a sample space to sets of possible outcomes (recall the cloaked-die example in Fig. 1.4), in accordance with Dempster's multivalued set-up. This approach has been emphasised in particular by Nguyen [708], [1353, 1344] and Hestir [826], and continued in [1611].

Random set theory first appeared in the context of stochastic-geometry theory, thanks to the independent work of Kendall [950] and Matheron [1268], to be later developed by Dempster, Nguyen, Molchanov [1304] and others into a theory of imprecise probability.

Formally, consider a multivalued mapping $\Gamma : \Omega \to 2^\Theta$. The *lower inverse* of Γ is defined as [1344]

$$\begin{aligned} \Gamma_* : 2^\Theta &\to 2^\Omega, \\ A &\mapsto \Gamma_*(A) \doteq \{\omega \in \Omega : \Gamma(\omega) \subset A, \Gamma(\omega) \neq \emptyset\}, \end{aligned} \tag{3.15}$$

while its *upper inverse* is

$$\begin{aligned} \Gamma^* : 2^\Theta &\to 2^\Omega, \\ A &\mapsto \Gamma^*(A) \doteq \{\omega \in \Omega : \Gamma(\omega) \cap A \neq \emptyset\}. \end{aligned} \tag{3.16}$$

Given two σ-fields (see Chapter 1, Definition 1) $\mathcal{A}, \mathcal{B}$ on Ω, Θ, respectively, Γ is said to be *strongly measurable* iff $\forall B \in \mathcal{B}$, $\Gamma^*(B) \in \mathcal{A}$. The *lower probability* measure on $\mathcal{B}$ is defined as $P_*(B) \doteq P(\Gamma_*(B))$ for all $B \in \mathcal{B}$. By (3.1) the latter is nothing but a belief function.

Nguyen proved that, if Γ is strongly measurable (i.e., the upper inverse of any element of the σ-field on Θ belongs to the σ-field on Ω), the (cumulative) probability distribution $\hat{P}$ of the random set [1344] coincides with the lower probability measure

$$\hat{P}[I(B)] = P_*(B) \quad \forall B \in \mathcal{B},$$

where $I(B)$ denotes the interval $\{C \in \mathcal{B}, C \subset B\}$. In the finite case, the probability distribution of the random set Γ is precisely the basic probability assignment (Definition 4) associated with the lower probability or belief function P_*.

An extensive analysis of the relations between Smets's transferable belief model and the theory of random sets can be found in [1687].

3.1.6 Behavioural interpretations

Smets's transferable belief model [1683], as we will see later in this chapter, is centred around a betting interpretation in which, whenever decisions are made, belief measures representing states of belief are converted by 'pignistic' probabilities (from the Latin *pignus*, wager), after which standard expected utility is adopted. We will elaborate further on this later.

Ville's interpretation A proper betting interpretation of degrees of belief has been recently proposed by Shafer, in the context of his game-theoretical approach to probability (see Chapter 6, Section 6.13).

In de Finetti's work [403], we make a probability judgement $P(A) = p$ by saying that p is the price at which we are willing to buy or sell tickets that pay 1 if A happens. According to Ville, instead, we make a probability judgement $P(A) = p$ by saying that if we do offer such bets on A, and on a sequence of similar events in similar but independent circumstances, then an opponent will not succeed in multiplying the capital they risk in betting against us by a large factor (we will return to this in Section 6.13). Based on the latter, a Ville argument for the classical conditional probability $P(B|A) = P(A \cap B)/P(A)$ being the price for a B ticket in the new situation where we have learned that A has happened (and nothing more) can be constructed. The argument works by showing that, if there were a strategy for multiplying capital by exploiting $P(B|A) = P(A \cap B)/P(A)$, a similar strategy would exist for $P(A)$ and $P(A \cap B)$, against the hypothesis. The construction of such a strategy depends crucially on the assumption that nothing else is known except that A has happened – this is termed *judgement of irrelevance*.

Shafer claims that various operations in the Dempster–Shafer calculus also require Ville-like judgements of irrelevance. More precisely, we can give degrees of belief $Bel(A) = \{P(\omega) : \Gamma(\omega) \subset A\}$ (3.1) a Ville interpretation under the following conditions:

1. The probability distribution P has a Ville interpretation: no betting strategy will beat the probabilities it gives for the variable X with values in Ω.
2. The multivalued mapping Γ has the following meaning:
$$\text{if } X = \omega, \text{ then } \theta \in \Gamma(\omega). \tag{3.17}$$
3. Learning the relationship (3.17) between X and θ does not affect the impossibility of beating the probabilities for X (this is the irrelevance judgement).

In other words, learning that the probability P is the source of a multivalued mapping does not affect its betting interpretation.

The Ville interpretation that derives from these conditions is then the following: a strategy that buys for $Bel(A)$ tickets that pay 1 if $\theta \in A$ (and makes similar bets on the strength of similar evidence) will not multiply the capital it risks by a large factor. This judgements-of-irrelevance argument extends to conditioning as well.

Kerkvliet and Meester's B-consistency An intriguing behavioural interpretation of belief functions has been recently advanced in [957].

Definition 29. *A* betting function $R : \mathcal{L} \to [0, 1]$ *is a binary function on the set* $\mathcal{L}$ *of gambles such that* $\forall X \in \mathcal{L}\ \exists \alpha_X \in \mathbb{R}$ *such that* $R(X + \alpha) = 0$ *for* $\alpha < \alpha_X$, *while* $R(X + \alpha) = 1$ *for* $\alpha \geq \alpha_X$.

Given a betting function,

$$Buy_R(X) \doteq \max\{\alpha \in \mathbb{R} : R(X - \alpha) = 1\}$$

is the maximum price an agent is willing to pay for the gamble X. In Walley's terminology, the betting function determines the set of desirable gambles $\mathcal{D} = \{X : R(X) = 1\}$, whereas $Buy_R(X)$ coincides with the lower prevision $\underline{P}(X)$ of X.

A *belief valuation* $\mathcal{B}$ is a belief function $Bel_\mathcal{B}$ such that $Bel_\mathcal{B}(A) = 1$ if $A \supseteq B$, 0 otherwise. For any belief valuation $\mathcal{B}$, the *guaranteed revenue* $G_\mathcal{B} : \mathcal{L} \to \mathbb{R}$ is defined as:

$$G_\mathcal{B}(X) \doteq \max_{A \supseteq B} \min_{\theta \in A} X(\theta).$$

We can then define the following notion.

Definition 30. *A betting function R is* B-consistent *if, for all $X_1, \ldots, X_N \in \mathcal{L}$ and $Y_1, \ldots, Y_M \in \mathcal{L}$ such that $\sum_{i=1}^N G_\mathcal{B}(X_i) \leq \sum_{j=1}^M G_\mathcal{B}(X_j)$ for every belief valuation $\mathcal{B}$, we have that*

$$\sum_{i=1}^N Buy_R(X_i) \leq \sum_{j=1}^M Buy_R(X_j).$$

The authors of [957] proved the following results.

Proposition 21. *Bel is a belief function if and only if there exists a coherent (in Walley's sense) and B-consistent R such that $Bel(A) = Buy_R(1_A)$ [957].*

Proposition 22. *If R is a coherent betting function, then R is B-consistent if and only if there is a BPA m such that*

$$Buy_R(X) = \sum_{A \subseteq \Theta} m(A) \min_{\theta \in A} X(\theta)$$

for all $X \in \mathcal{L}$.

In other words, after adding B-consistency to imprecise probabilities' rationality axioms, lower previsions reduce to belief functions.

3.1.7 Common misconceptions

Belief functions are not (arbitrary) credal sets We already know that, although any belief function is in 1–1 correspondence with a convex set of probability distributions (a credal set), belief functions are a very special class of credal sets, those induced by a random set mapping.

This is examplified in Fig. 3.3. A belief function corresponds to a set of probabilities determined by lower and upper bounds on probability values of events. For instance, a belief function on $\Theta = \{x, y, z\}$ may be associated with the bounds $0.2 \leq P(x) \leq 0.6$ and $0.2 \leq P(z) \leq 0.7$, among others. As a result, each face of its credal set is parallel to one of the level sets $P(A) = \text{const}$, for some event $A \subset \Theta$.

An arbitrary credal set (Fig. 3.4 (a)), instead, will be a polytope in the same probability simplex, but its boundaries will not, in general, be parallel to those level sets.

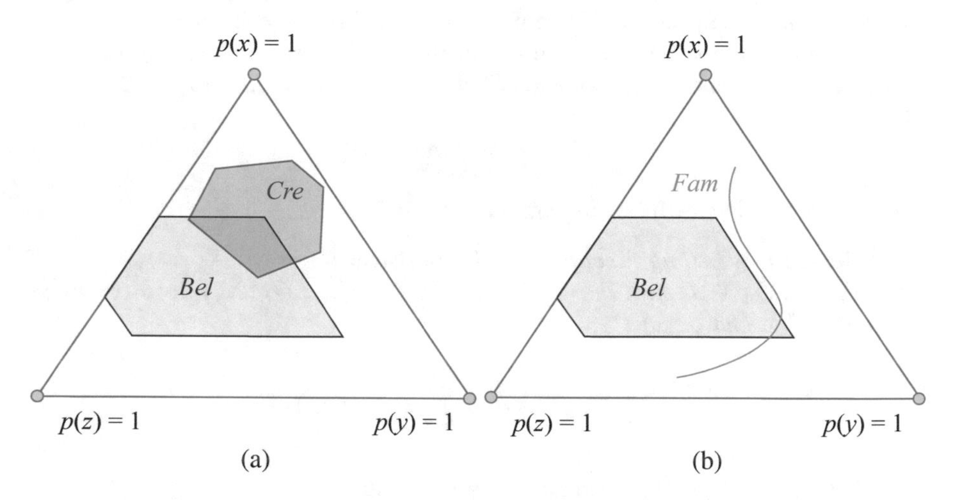

Fig. 3.4: (a) The boundaries of an arbitrary credal set are not necessarily parallel to the level sets $P(A) = \text{const}$. (b) Although a credal set (and thus a belief function) can be seen as a parameterised family of probability distributions, not all such families (e.g., the dark green curve 'Fam' plotted here) correspond to a belief function.

Belief functions are not parameterised families of distributions Similarly, a parameterised family of distributions on Θ is a subset of the set of all possible distributions (just like belief functions and credal sets). However, once again, not all families of probability distributions correspond to a belief function (Fig. 3.4(b)). For instance, the family of Gaussian PDFs on a given sample space with zero mean and arbitrary variance, $\{\mathcal{N}(0, \sigma), \sigma \in \mathbb{R}^+\}$, is not a belief function.

Belief functions are not confidence intervals As we saw in Section 1.3.7, confidence intervals are intervals of probability distributions, each associated with a value of a certain parameter θ. However, confidence intervals (1.5),

$$\mathbb{P}(u(X) < \theta < v(X)|\theta, \phi) = \gamma \quad \forall(\theta, \phi),$$

are always one-dimensional, as they are associated with a single scalar parameter. As a result, only in the case of a binary hypothesis set $\Theta = \{x, y\}$, with $\theta = P(x)$, can they even formally be put into correspondence with a belief function – in all other cases, even a formal correspondence cannot be enforced. At any rate, their interpretation is entirely different: confidence intervals are a form of interval estimate, with no probability attached.

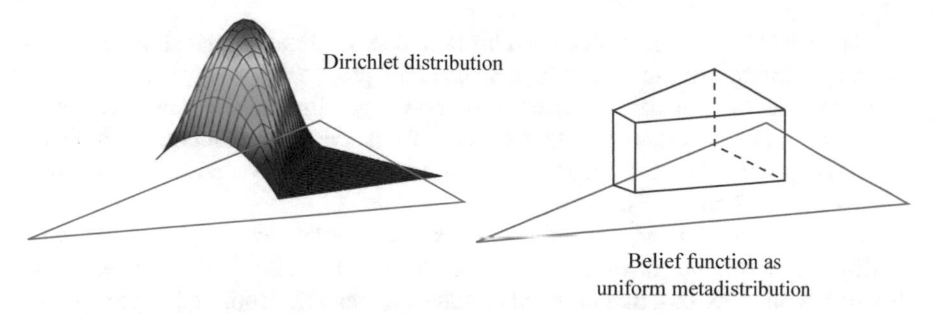

Fig. 3.5: Second-order distributions such as the Dirichlet distribution (left) are the result of general Bayesian inference. Belief functions (right) can be seen as 'indicator' second-order distributions when defined on the sample space Ω, or as convex sets of second-order distributions when defined on the parameter space Θ.

Belief functions are not second-order distributions Unlike hypothesis testing, general Bayesian inference leads, as we saw in the Introduction, to probability distributions over the space of parameters. The latter can then be thought of as *second-order* probabilities (e.g. the Dirichlet distribution, Fig. 3.5, left), i.e., probability distributions on hypotheses which are themselves probabilities.

Belief functions can also be defined either on the hypothesis space Ω for the problem at hand or on the parameter space Θ, i.e., the space of possible values of the parameter. When defined on Ω, belief functions are sets of PDFs and can then be seen as 'indicator' second-order distributions (see Fig. 3.5, right). When defined on the parameter space Θ, instead, they amount to families of second-order distributions.

As we will see in more detail in the next chapter, in these two cases inference with belief functions generalises MLE/MAP and general Bayesian inference, respectively.

3.2 Genesis and debate

The axiomatic imprint that Shafer originally gave to his work, and which we recalled in part in Chapter 2, might appear arbitrary at first glance [1632, 974].

For example, Dempster's rule is not given any convincing justification in his seminal thesis [1583], stimulating the question of whether an entirely different rule of combination could be chosen. This question has indeed been addressed by several authors ([1860, 2089, 1595] among others), the majority of whom have tried to provide solid axiomatic support for Dempster and Shafer's choice of this specific mechanism for combining evidence.

3.2.1 Early support

As a matter of fact, Shafer's own position has also shifted over time.

In 1976 [1584], he first illustrated his rationale for a mathematical theory of evidence, claiming that the impact of evidence on a proposition may either support it to various degrees, or cast doubt on it to various possible degrees. Indeed [1594], belief function theory has even older antecedents than Bayesian theory, as similar arguments appeared in the work of George Hooper (1640–1723) and James Bernoulli (1654–1705), as emphasised in [1591].

Shafer's proposal for a new theory of statistical evidence and epistemic probability received rather favourable attention [616, 1938]. Fine [616] commented in his review that the fact that probability takes its meaning from and is used to describe phenomena as diverse as propensities for actual behaviour (the behavioural interpretation), propositional attitudes of belief (subjective degrees of belief) and experimental outcomes under prescribed conditions of unlinked repetitions (the frequentist view) has long been the source of much controversy, resulting in a dualistic conception of probability as being jointly epistemic (oriented towards 'subjective' belief assessment) and aleatory (focused on the 'objective' description of the outcomes of 'random' experiments), with most of the present-day emphasis on the latter.
We have ourselves ventured into this discussion in the introduction to this book.

Lowrance and Garvey [1216], in particular, were early supporters of the use of Shafer's mathematical theory of evidence as a framework for 'evidential reasoning' in expert systems. At the time, reasoning with knowledge-based expert systems was a very hot topic, and the debate much less biased than today towards Bayesian graphical models. Gordon and Shortliffe [711] argued that the advantage of belief function theory over other approaches is its ability to model, in expert reasoning, the narrowing of the hypothesis set with the accumulation of evidence. Strat and Lowrance [1773] focused, rather, on the difficulty of generating explanations for the conclusions drawn by evidential reasoning systems based on belief functions, and presented a methodology for augmenting an evidential-reasoning system with a versatile explanation facility.

Interestingly, Curley and Golden [323] constructed in 1994 a very interesting experiment related to legal situations, in which a belief assessor was interested in judging the degree of support or justification that the evidence afforded hypotheses, to determine (a) if subjects could be trained in the meanings of belief function responses, and (b) once trained, how they used those belief functions in a legal setting. They found that subjects could use belief functions, identified limits to their descriptive representativeness, and discovered patterns in the way subjects use belief functions which inform our understanding of their uses of evidence.

3.2.2 Constructive probability and Shafer's canonical examples

An important evolution of Shafer's thought is marked by [1607], where he formulated a 'constructive' interpretation of probability in which probability judgements are made by comparing the situation at hand with abstract *canonical examples*, in which the uncertainties are governed by known chances.

These can, for instance, be games played repeatedly for which the limit relative frequencies of the possible outcomes are known (e.g. think of designing a random experiment in hypothesis testing, in such a way that the probability describing the outcomes can be assumed to be known). He claimed at the time that when one does not have sufficient information to compare the situation at hand with a classical, probability-governed experiment, other kinds of examples may be judged to be appropriate and used to give rise to a mathematical description of thc uncertainty involved in terms of belief functions.

The canonical examples from which belief functions are to be constructed are based on 'coded messages' $c_1, \ldots, c_n$, which form the values of a random process with prior probabilities $p_1, \ldots, p_n$ [1589]. Each message c_i has an associated subset A_i of outcomes, and carries the message that the true outcome is in A_i. The masses representing the current state are simply the probabilities (with respect to this random process) of receiving a message associated with a subset A. Shafer then offers a set of measures on belief functions to assist in fitting parameters of the coded-message example to instances of subjective notions of belief.

Later, however, Hummel and Landy [858] argued that Shafer's constructive probability techniques disregard the statistical-theoretical foundations from which the theory was derived, exactly because they are based on fitting problems to scales of canonical examples. As a matter of fact, Shafer's presentation before the Royal Statistical Society [1589] was heavily criticised by several discussants, who asked for a closer connection to be made between the canonical examples and the interpretation of belief values. As reported in [858], 'Prof. Barnard, for example, states that "the connections between the logical structure of the ... example and the story of the uncertain codes is not at all clear". Prof. Williams desired "a deeper justification of the method and a further treatment of unrelated bodies of evidence", while Prof. Krantz simply stated that "[a] comparison of evidence to a probabilistically coded message seems strained". Prof. Fine summarised the problem by stating that "the coded message interpretation is ignored when actually constructing belief functions, calling into question the relevance of the canonical scales" '.

3.2.3 Bayesian versus belief reasoning

Much analysis has obviously been directed towards understanding or highlighting the fundamental difference between classical Bayesian and belief function reasoning [1673, 419, 1168]. Lindley [1174], for instance, put forward arguments, based on the work of Savage and de Finetti, aiming to show that probability theory is the unique correct description of uncertainty. Other authors have instead argued that the additivity axiom is too restrictive when one has to deal with uncertainty deriving from partial ignorance [1086].

In a chapter on Cognitive Science, Shafer and Tversky [1613] described and compared the semantics and syntax of the Bayesian and belief function languages, and investigated designs for probability judgement afforded by the two languages. Wakker [1868] argued that a 'principle of complete ignorance' plays a central role in decisions based on Dempster belief functions, in that whereas the Bayesian approach requires that uncertainty about the true state of nature be probabilised, belief functions assume complete ignorance, permiting strict adherence to the available data. Laskey [1097] examined Shafer's canonical examples in which beliefs over the hypotheses of interest are derived from a probability model for a set of auxiliary hypotheses, noting that belief reasoning differs from Bayesian reasoning in that one does not condition on those parts of the evidence for which no probabilities are specified. The significance of this difference in conditioning assumptions is illustrated with two examples giving rise to identical belief functions, but different Bayesian probability distributions.

Scozzafava [1570] argued that if the Bayesian approach is set up taking into account all the relevant aspects, i.e., if there is no beforehand-given structure on the set of events, and if probability is interpreted as degree of belief, then the Bayesian approach leads to very general results that fit (in particular) with those obtained from the belief function model. Nevertheless, Klopotek and Wierzcho [1006] maintained that previous attempts to interpret belief functions in terms of probabilities [1086, 785, 582] had failed to produce a fully compatible interpretation, and proposed in reponse three models: a 'marginally correct approximation', a 'qualitative' model and a 'quantitative' model.

In 'Perspectives on the theory and practice of belief functions' [1598], Shafer reviewed in 1992 the work conducted until then on the interpretation, implementation and mathematical foundations of the theory, placing belief theory within the broader topic of probability, and that of probability itself within artificial intelligence. This work stimulated an important discussion of the foundations and applicability of belief function theory.

Although agreeing with Shafer that there are situations where belief functions are appropriate, Wasserman [1916] raised a number of questions motivated by statistical considerations. He argued that the betting paradigm has a status in the foundations of probability of a different nature than that of the canonical examples that belief function theory is built on. In addition to using the betting metaphor to make judgements, we use this analogy at a higher level to judge the theory as a whole. Wasserman's thesis was that a similar argument would make belief functions easier to understand. Wasserman also questioned the separation of belief from frequency, arguing that it is a virtue of subjective probability that it contains frequency probability as a special case, by way of de Finetti's theory of exchangeability, and hinted at the notion of asymptotics in belief functions.

3.2.4 Pearl's criticism

Judea Pearl [1399, 1406, 1408, 1407] contributed very forcefully to the debate in the early 1990s. He claimed that belief functions have difficulties in representing incomplete knowledge, in particular knowledge expressed in conditional sentences, and that encoding if–then rules as belief function expressions (a standard practice at the time) leads to counter-intuitive conclusions. He found that the updating process for belief functions violates what he called 'basic patterns of plausibility', so that the resulting beliefs cannot serve as a basis for rational decisions. As for evidence pooling, although belief functions offer in his view a rich language for describing the evidence, the available combination operators cannot exploit this richness and are challenged by simpler methods based on likelihood functions. In [1404], he argued that a causal-network formulation of probabilities facilitates a representation of confidence measures that does not require the use of higher-order probabilities, and contrasted this idea with the way Dempster–Shafer 'intervals' (*sic*) represent confidence about probabilities.

Detailed answers to several of the criticisms raised by Pearl were provided by Smets in 1992 [1686], within his transferable belief model interpretation (see Section 3.3.2). In the same year, Dubois and Prade tried to clarify some aspects of the theory of belief functions, addressing most of the questions raised by Pearl in [1399, 1408]. They pointed out that their mathematical model could be useful beyond a theory of evidence, for the purpose of handling imperfect statistical knowledge. They compared Dempster's rule of conditioning with upper and lower conditional probabilities, concluding that Dempster's rule is a form of updating, whereas the second operation expresses 'focusing' (see Section 4.5). Finally, they argued that the concept of focusing models the meaning of uncertain statements in a more natural way than updating.

Wilson [1942] also responded to Pearl's criticisms, in a careful and precise manner. He noted that Pearl criticised belief functions for not obeying the laws of Bayesian belief, whereas these laws lead to well-known problems in the face of ignorance, and seem unreasonably restrictive. He argued that it is not reasonable to expect a measure of belief to obey Pearl's 'sandwich' principle, whereas the standard representation of 'if–then' rules in Dempster–Shafer theory, criticised by Pearl, is in his view justified and compares favourably with a conditional probability representation.
Shafer himself [1600] addressed Pearl's remarks by arguing that the interpretation of belief functions is controversial because the interpretation of probability is controversial. He summarised his constructive interpretation of probability, probability bounds and belief functions, and explained how this interpretation bears on many of the issues raised.

A further rebuttal by Pearl [1409] responded to Shafer's, Wilson's and Dubois's comments and discussed the degree to which his earlier conclusions affected the applicability of belief functions in automated reasoning tasks. In this author's view, most of the logical arguments Pearl put forward in [1409] are pretty shaky [1443], and in most cases completely unrelated to the issues originally raised in [1408].

3.2.5 Issues with multiple interpretations

The fact that belief functions possess multiple interpretations (Section 3.1), all of which seem sensible from a certain point of view, ignited, especially in the early 1990s, a debate to which many scholars have contributed.

In a landmark paper, Halpern and Fagin [785, 785] underlined two different views of belief functions: (i) as generalised probabilities (see Sections 3.1.2 and

3.1.3) and (ii) as a way of representing evidence, which, in turn, can be understood as a mapping from probability functions to probability functions. Under the first interpretation, they argue, it makes sense to think of updating a belief function because of its nature of a generalised probability. If we think of belief functions as a mathematical representation of evidence instead, using combination rules to merge two belief functions is the natural thing to do. Their claim is that many problems about the use of belief functions can be explained as a consequence of a confusion of these two interpretations. As an example, they cited comments by Pearl [1406, 1408] and others that belief theory leads to incorrect or counter-intuitive answers in a number of situations.

A comment on this dual interpretation of belief theory was provided by Lingras and Wong [1176]. These authors argued that the 'compatibility' interpretation (Section 3.1.1) is useful when limited information regarding the relationship between two frames of discernment is available, while the 'probability allocation' view (in which probability mass is assigned to propositions, based on some body of evidence) comes in when the evidence cannot be explicitly expressed in terms of propositions.

The many interpretations of Dempster–Shafer theory were considered by Smets [1683, 1696]. He reviewed there both the knowledge representation and the belief update aspects of the various mathematical models: within classical probability, as a theory of upper and lower probabilities, Dempster's model, his own transferable belief model (TBM), an 'evidentiary value' model, and the provability of modal propositions or 'necessity' model. He argued that none of these models is the best, for each has its own domain of application. If a probability measure exists and can be identified, he maintained, the Bayesian model is to be preferred. If a probability measure exists but its values cannot be precisely estimated, upper and lower probability models (of which Dempster's original proposal is a special case) are adequate. Finally, if a probability measure is not known to exist, the TBM should be used, as it formalises an agent's belief without reference to an underlying unknown probability.

In [1693], the same author gave an axiomatic justification for the use of belief functions to quantify partial beliefs, whereas in [1686] he rebutted Pearl's criticisms [1406] by accurately distinguishing the different epistemic interpretations of the theory of evidence (echoing Halpern et al. [785]), focusing in particular on his TBM (Section 3.3.2).

The transferable belief model interpretation of belief functions is reviewed in more detail later in the chapter.

3.2.6 Rebuttals and justifications

A number of authors have proposed rebuttals to Pearl's and others' criticisms [1948], sometimes based (as in Smets's case) on an axiomatic justification that cuts all bridges with the original statistical rationale.

A very comprehensive rebuttal of criticisms of evidential reasoning based on Dempster–Shafer calculus was presented by Ruspini, Lowrance and Strat in 1992 [1515], addressing theoretical soundness, decision support, evidence combination, complexity and a detailed analysis of the so-called 'paradoxes' generated by the application of the theory. They showed that evidential reasoning can be interpreted in terms of classical probability theory, and belief function theory considered as a generalisation of probabilistic reasoning based on the representation of ignorance by intervals of possible values, without resorting to non-probabilistic or subjectivist explanations, in clear opposition to Smets's position. They pointed out a confusion between the (then) current state of development of the theory (especially in what was then the situation for decision making) and its potential usefulness. They also considered methodological criticisms of the approach, focusing primarily on the alleged counter-intuitive nature of Dempster's combination formula, showing that such results are the result of its misapplication.

Once again, Smets [1703] wrote extensively on the axiomatic justification for the use of belief functions [1693]. Essentially, he postulated that degrees of belief are quantified by a function in $[0, 1]$ that gives the same degrees of belief to subsets that represent the same propositions according to an agent's evidential corpus. He derived the impact of coarsening and refining a frame of discernment, and of the conditioning process. Finally, he proposed a closure axiom, according to which any measure of belief can be derived from other measures of belief defined on less specific frames. In [1713], he argued that it is easy to show that any mathematical function used to represent quantified beliefs should be a 2-monotone Choquet capacity (or *probability interval*: see Chapter 6, Section 6.3). In order to show that it must be monotone of infinite order (see Definition 27), and thus a belief function, several extra rationality requirements need to be introduced. One of them is based on the negation of a belief function, a concept introduced by Dubois and Prade [527].

Shafer himself [1601] produced a very extensive, and somewhat frustrated, document addressing the various contributions to the discussion of the interpretation of belief functions. The main argument is that the disagreement over the 'correct' meaning of belief functions boils down, fundamentally, to a lack of consensus on how to interpret probability, as belief functions are built on probability. In response, he illustrated his own constructive interpretation of probability, probability bounds and belief functions and related it to the views and concerns of Pearl, Smets, Ruspini and others, making use of the main canonical examples for belief functions, namely the partially reliable witness and its generalisation, the randomly coded message.

Neapolitan [1340] elaborated further on Shafer's defence in two ways: (1) by showing that belief functions, as Shafer intends them to be interpreted, use probability theory in the same way as the traditional statistical tool, significance testing; and (2) by describing a problem for which the application of belief functions yields a meaningful solution, while a Bayesian analysis does not.

3.3 Frameworks

Starting from Dempster and Shafer's mathematical foundations, a number of researchers have proposed their own variations on the theme of belief functions, with various degrees of success. Arguably the frameworks with the highest impact, to date, are Smets's transferable belief model (Section 3.3.2), by far the most widely adopted; Kohlas and Monnet's theory of hints (Section 3.3.1); and the so-called Dezert–Smandarache theory (DSmT, Section 3.3.3).

In this last part of the chapter we review these frameworks, starting from those which extend the original idea of a multivalued mapping. We cover the TBM and DSmT quite extensively, together with Dempster and Liu's Gaussian belief functions. We then discuss the various extensions of the notion of a belief function to more general domains, qualitative models, and interval or credal extensions of the notion of a basic probability assignment.

We conclude the section by surveying in somewhat less detail other significant but less well-known frameworks.

3.3.1 Frameworks based on multivalued mappings

A number of frameworks are indeed elaborations of the notion of a multivalued mapping illustrated in Section 3.1.1. This is the case, for instance, for the theory of hints and Kramosil's measure-theoretical analysis, but also for Hummel and Landy's statistical formulation in terms of opinions of experts, Yen's conditional multivalued mappings and Lin's probabilistic multivalued random variables, among others.

Kohlas and Monney's mathematical theory of hints [1016, 1025, 1018, 1305] is a vehicle for introducing the notion of a belief function on real numbers, inspired by Dempster's multivalued setting. Indeed, some of the relations introduced in [1714] and [1027, 1016] were already present in [417]. An introduction to the topic can be found in [1016], whereas for a more detailed illustration the reader can refer to [1027]. The following summary of the theory of hints is taken from [471].

Functional models In [1027], Kohlas and Monney describe the process by which a data point x is generated from a parameter θ and some random element ω:

$$f : \Theta \times \Omega \rightarrow X, \quad x = f(\theta, \omega), \tag{3.18}$$

where X denotes the set of possible values of x, Θ denotes the domain of the parameter θ, and Ω denotes that of the random element ω. The function f together with a probability measure $P : \Omega \rightarrow [0, 1]$ constitutes a *functional model* for a statistical experiment $\mathcal{E}$.

A functional model induces a parametric family of probability distributions (*statistical specifications*) on the sample space X, via

$$P_\theta(x) = \sum_{\omega : x = f(\theta, \omega)} P(\omega). \tag{3.19}$$

Note that different functional models may induce the same statistical specifications, i.e., the former (3.18) contain more information than the latter (3.19).

Assumption-based reasoning Consider an experiment $\mathcal{E}$ represented by a functional model $x = f(\theta, \omega)$ with probabilities $P(\omega)$. Suppose that the observed outcome of the experiment is x. What can be inferred about the value of the unknown parameter θ? As discussed in the Introduction, this inference problem is central in statistics.

The basic idea of assumption-based reasoning is to assume that a random element ω generated the data, and then to determine the consequences of this assumption for the parameter. The observation x induces an event in Ω, namely

$$v_x \doteq \{\omega \in \Omega | \exists \theta \in \Theta : x = f(\theta, \omega)\}. \tag{3.20}$$

Once we know that $v_x \subset \Omega$ has happened, in a Bayesian setting we need to condition the initial probabilities $P(\omega)$ with respect to v_x, obtaining $P'(\omega) = P(\omega)/P(v_x)$, whose probability measure is trivially $P'(A) = \sum_{\omega \in A} P'(\omega)$, $A \subset \Omega$.

Note that it is unknown what element $\omega \in v_x$ has actually generated the observation. Assuming ω was the cause, the possible values for the parameter θ are obviously restricted to the set

$$T_x(\omega) = \{\theta \in \Theta | x = f(\theta, \omega)\}.$$

Summarising, an observation x in a functional model (f, P) generates a structure

$$\mathcal{H}_x = (v_x, P', T_x, \Theta), \tag{3.21}$$

which Kohlas and Monney call a *hint*.

A theory of hints In general, if Θ denotes the set of possible answers to a question of interest, then a hint on Θ is a quadruple of the form $\mathcal{H} = (\Omega, P, \Gamma, \Theta)$, where Ω is a set of assumptions, P is a probability measure on Ω reflecting the probability of the different assumptions, and Γ is a mapping between the assumptions and the power set of Θ, $\Gamma : \Omega \rightarrow 2^{\Theta}$. If assumption ω is correct, then the answer is certainly within $\Gamma(\omega)$.

It is easy to see that the mathematical setting of hints is identical to Dempster's multivalued mapping framework described in Section 3.1.1. Degrees of support and plausibility can then be computed as in (3.1), (3.2). What is (arguably) different is their interpretation. While Dempster interprets them as lower and upper bounds on the amount of probability assigned to $A \subset \Theta$, Kohlas and Monney do not assume the existence of an unknown true probability of A, but rather adopt Pearl's probability of provability interpretation [1405] (see Section 6.6.8), related to the classical artificial intelligence paradigm called 'truth-maintenance systems' [1100]. These systems contain a symbolic mechanism for identifying the set of assumptions needed to create a proof of a hypothesis A, so that when probabilities are assigned to the assumptions support and plausibility functions can be obtained. In the theory of hints, these assumptions ω form an argument for the hypothesis $A \subset \Theta$, and their probability is the weight $P(\omega)$ assigned to each argument.

Functional models and hints Given a hint (3.21) induced by a functional model, we can assess any hypothesis $H \subseteq \Theta$ about the correct value of the parameter with respect to it. The arguments for the validity of H are the elements of the set $u_x(H) = \{\omega \in v_x : T_x(\omega) \subseteq H\}$, with degree of support $P'(u_x(H))$: those compatible with H are $v_x(H) = \{\omega \in v_x : T_x(\omega) \cap H \neq \emptyset\}$, with degree of plausibility $P'(v_x(H))$.

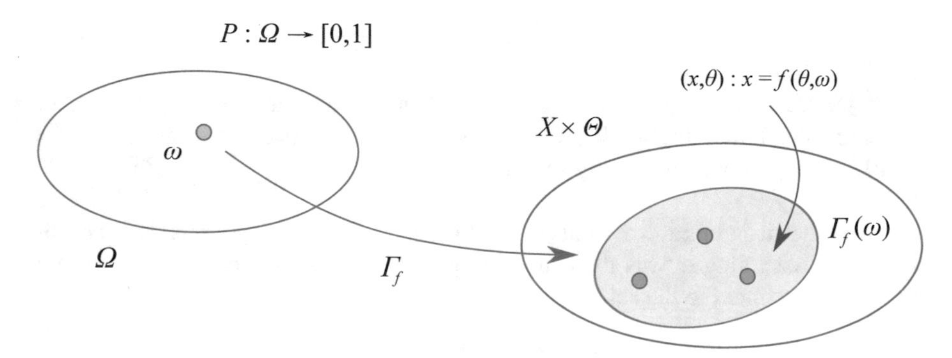

Fig. 3.6: A functional (parametric) model $x = f(\theta, \omega)$ can be represented as a hint $\mathcal{H}_f$ (3.22).

Importantly, not only the result of the inference but also the experiment that is used to make the inference can be expressed in terms of hints. Indeed, a functional model (f, P) can be represented by the hint (Fig. 3.6)

$$\mathcal{H}_f = (\Omega, P, \Gamma_f, X \times \Theta), \tag{3.22}$$

where

$$\Gamma_f(\omega) = \{(x, \theta) \in X \times \Theta | x = f(\theta, \omega)\},$$

while an observation x can be represented by the hint

$$\mathcal{O}_x = (\{v_x\}, P, \Gamma, X \times \Theta),$$

where v_x is the assumption (3.20) stating that x has been observed, which is true with probability $P(\{v_x\}) = 1$, and $\Gamma(v_x) = \{x\} \times \Theta$.

This equation is justified by the fact that no restriction can be imposed on Θ when the observed value x is the only piece of information that is being considered. The hints $\mathcal{H}_f$ and $\mathcal{O}_x$ represent two pieces of information that can be put together in order to determine the information about the parameter that can be derived from the model and the data, resulting in the combined hint $\mathcal{H}_f \oplus \mathcal{O}_x$. By marginalising on Θ, we obtain the desired information about the value of θ: it is easy to show that the result is (3.21).

By extension, the hint derived by a series of n repeated experiments $\mathcal{E}_i$ under functional models f_i, $i = 1, \ldots, n$, whose outcomes are $x_1, \ldots, x_n$, can be written as

$$\mathcal{H}_{x_1,\ldots,x_n} = \mathcal{H}_{x_1} \oplus \cdots \oplus \mathcal{H}_{x_n}.$$

Probabilistic assumptions-based reasoning is then a natural way to conduct statistical inference [471, 1306].

Kramosil's probabilistic analysis The theory of evidence can be developed in an axiomatic way quite independent of probability theory (Section 3.1.2), by way of axioms which come from a number of intuitive requirements that a sensible uncertainty calculus should satisfy. On the other side, we have learned that belief theory can be seen as a sophisticated application of probability theory in the random set context (Section 3.1.1 and 3.1.5).

Starting from this point of view, Ivan Kramosil [1056, 1055, 1057, 1052, 1047, 1058, 1062] published a number of papers in which he exploited measure theory[20] to expand the theory of belief functions beyond its classical scope. The topics of his investigation varied from Boolean and non-standard-valued belief functions [1063, 1058], with application to expert systems [1047], to the extension of belief functions to countable sets (see Chapter 4, Section 4.9.4 [1061]) and the introduction of a strong law of large numbers for random sets [1065].

Unfortunately, Kramosil's work does not seem to have received sufficient recognition. A complete analysis of Kramosil's random-set approach is obviously beyond the scope of this chapter. An extensive review of Kramosil's work can be found in a series of technical reports produced by the Academy of Sciences of the Czech Republic [1049, 1051, 1050]. Here we will just briefly mention a few interesting contributions.

Belief functions induced by partial compatibility relations These are discussed in [1051], Chapter 8. Given a standard compatibility relation $\mathcal{C} \subset \Omega \times \Theta$ (3.3), the *total generalised compatibility relation* induced on $\mathcal{P}(\Omega) \times \mathcal{P}(\Theta)$ by $\mathcal{C}$ is the relation $\mathcal{C}^* \subset \mathcal{P}(\Omega) \times \mathcal{P}(\Theta)$ such that

$$(X, Y) \in \mathcal{C}^* \text{ iff } \exists \omega \in X, \theta \in Y \text{ such that } (\omega, \theta) \in \mathcal{C}.$$

A *partial* generalised compatibility relation is then a relation defined on a proper subset D of $\mathcal{P}(\Omega) \times \mathcal{P}(\Theta)$, such that it can be extended to a proper total relation (i.e., it is the restriction to D of some total generalised compatibility relation).

The main result of Chapter 8 of [1051] is that any belief function defined by a fragment of the total generalised compatibility relation induced by an original compatibility relation $\mathcal{C}$ is a lower approximation to the original belief function associated with $\mathcal{C}$.

[20] A similar measure-theoretical analysis of the relationship between Dempster–Shafer and Bayesian theory was later conducted by Peri [1418].

Theorem 2. *Let $\mathcal{C}_1, \mathcal{C}_2 : \mathcal{P}(\Omega) \times \mathcal{P}(\Theta) \rightarrow [0,1]$ be two partial generalised compatibility relations such that $\mathcal{C}_i = \mathcal{C}^* \downarrow Dom(\mathcal{C}_i)$ for both $i = 1,2$, where $\mathcal{C}^*$ is the total generalised compatibility relation associated with a compatibility relation $\mathcal{C}$ on $\Omega \times \Theta$. Let $Dom(\mathcal{C}_1) \subset Dom(\mathcal{C}_2) \subset \mathcal{P}(\Omega) \times \mathcal{P}(\Theta)$ hold, and define*

$$\bar{\mathcal{C}}_i(\omega, \theta) = \min \left\{ \mathcal{C}_i(X,Y) \middle| (X,Y) \in Dom(\mathcal{C}_i), \omega \in X, \theta \in Y \right\}$$

whenever this value exists, and $\bar{\mathcal{C}}_i(\omega,\theta) = 1$ otherwise, for both $i = 1,2$. Then the following inequalities hold for each $A \subset \Theta$:

$$Bel_{\bar{\mathcal{C}}_1}(A) \leq Bel_{\bar{\mathcal{C}}_2}(A), \quad Pl_{\bar{\mathcal{C}}_1}(A) \leq Pl_{\bar{\mathcal{C}}_2}(A),$$

where $Bel_{\bar{\mathcal{C}}_i} : 2^\Theta \rightarrow [0,1]$ is the belief function induced by the partial compatibility relation $\mathcal{C}_i$.

Signed belief functions A note is due on the notion of a *signed* belief function [1048], in which the domain of classical belief functions is replaced by a measurable space equipped with a signed measure, i.e., a σ-additive set function which can also take values outside the unit interval, including negative and infinite values.

Definition 31. *A* signed measure *is a mapping $\mu : \mathcal{F} \rightarrow \mathbb{R}^* = (-\infty, +\infty) \cup \{-\infty\} \cup \{+\infty\}$ which is σ-additive, namely, for any sequence $A_1, A_2, \ldots, A_n, \ldots$ of pairwise disjoint sets in $\mathcal{F}$*

$$\mu\left(\bigcup_{n=1}^{\infty} A_n\right) = \sum_{n=1}^{\infty} \mu(A_n), \tag{3.23}$$

$\mu(\emptyset) = 0$, and it takes at most one of the values $-\infty$, $+\infty$ (in order to avoid expressions such as $(\infty - \infty)$).

Signed measures can then be used to generalise random variables, and by extension set-valued random variables. For instance, one can define a 'basic signed measure assignment' and a corresponding 'signed belief function', and define a Dempster sum for them [1055, 1059], under a straightforward extension of statistical independence.

An assertion analogous to the Jordan decomposition theorem for signed measures can be stated and proved [1062] (see also [1050], Chapter 11), according to which each signed belief function, when restricted to its finite values, can be defined by a linear combination of two classical probabilistic belief functions, under the assumption that the frame of discernment is finite.

Hummel and Landy's statistical view In [858], Hummel and Landy provided an interpretation of evidence combination in belief theory in terms of statistics of expert opinions. This approach is closely related to Kyburg's [1086] explanation of belief theory within a lower-probability framework, in which beliefs are viewed as extrema of opinions of experts.

In this interpretation, evidence combination relates to statistics of experts in a space of 'expert opinions', who combine information in a Bayesian fashion. As a result, Dempster's rule of combination, rather than an extension of Bayes' rule for combining probabilities, reduces to the Bayesian updating of Boolean assertions, while tracking multiple opinions. A more general formulation was suggested in which opinions are allowed to be probabilistic, as opposed to the Boolean opinions that are implicit in the Dempster formula.

Probabilistic opinions of experts Formally, consider a set of experts $\mathcal{E} = \Omega$, where each expert $\omega \in \mathcal{E}$ is attributed a certain weight $\mu(\omega)$, and maintains a set of possible outcomes (called 'labels' in [858]) $\Gamma(\omega)$ for a question with universe (sets of possible answers) $\Lambda = \Theta$. Crucially, we also assume that each expert has a probabilistic opinion p_ω on Λ, representing expert ω's assessment of the probability of occurrence of the labels, which satisfies the following for all ω:

$$p_\omega(\lambda) \geq 0 \; \forall \lambda \in \Lambda, \quad p_\omega(\lambda) > 0 \text{ iff } \lambda \in \Gamma(\omega),$$

and

$$\text{either} \sum_{\lambda \in \Lambda} p_\omega(\lambda) = 1 \text{ or } p_\omega(\lambda) = 0 \; \forall \lambda, \tag{3.24}$$

with the last constraint describing the case in which expert ω has no opinion on the matter.

This setting generalises Dempster's multivalued setting, which corresponds to the special case in which only *Boolean opinions* of the experts

$$x_\omega(\lambda) = \begin{cases} 1 & p_\omega(\lambda) > 0, \\ 0 & p_\omega(\lambda) = 0 \end{cases} \tag{3.25}$$

are used (as $\Gamma(\omega) = \{\lambda \in \Lambda : x_\omega(\lambda) = 1\}$). If we regard the space of experts $\mathcal{E} = \Omega$ as a sample space, then each $x_\omega(\lambda)$ can be thought of as a sample of a random (Boolean) variable $x(\lambda)$. In a similar way, the $p_\omega(\lambda)$s can be seen as samples of a random variable $p(\lambda)$. The state of the system is then fully determined by statistics on the set of random variables $\{x(\lambda)\}_{\lambda \in \Lambda}$, computed in the space of experts using the weights $\mu(\omega)$ of the individual experts, via the counting measure $\mu(B) = \sum_{\omega \in B} \mu(\omega)$, $B \subset \mathcal{E}$.

Space of probabilistic opinions of experts

Definition 32. *Let $K = \{k_\lambda\}$ be a set of positive constants indexed over the label set Λ. The* space of probabilistic opinions of experts *$(\mathcal{N}, K, \otimes)$ is defined by*

$$\mathcal{N} = \Big\{ (\mathcal{E}, \mu, P) \Big| \mu \text{ measure on } \mathcal{E}, \; P = \{p_\omega\}_{\omega \in \mathcal{E}} \Big\},$$

where the probabilistic opinions p_ω satisfy the constraint (3.24), and the following binary operation is defined:

$$(\mathcal{E}, \mu, P) = (\mathcal{E}_1, \mu_1, P_1) \otimes (\mathcal{E}_2, \mu_2, P_2) \tag{3.26}$$

such that $\mathcal{E} = \mathcal{E}_1 \times \mathcal{E}_2$, $\mu(\{(\omega_1, \omega_2)\}) = \mu_1(\{\omega_1\}) \cdot \mu_2(\{\omega_2\})$, *and*

$$p_{(\omega_1,\omega_2)}(\lambda) = \frac{p_{\omega_1}(\lambda) p_{\omega_2}(\lambda) k_\lambda^{-1}}{\sum_{\lambda'} p_{\omega_1}(\lambda') p_{\omega_2}(\lambda') k_{\lambda'}^{-1}}$$

whenever the denominator is non-zero, and $p_{(\omega_1,\omega_2)}(\lambda) = 0$ *otherwise.*

The combination operation (3.26) can be interpreted as a Bayesian combination, where k_λ is the prior probability of λ, and expresses the generation of a consensus between two sets of experts $\mathcal{E}_1$ and $\mathcal{E}_2$, obtained by pairing one expert from $\mathcal{E}_1$ with one expert from $\mathcal{E}_2$.

The authors of [858] proved that this space maps homomorphically onto a belief space. Similarly, a *space of Boolean opinions of experts* can be formed by replacing the probabilistic opinions p_ω with the indicator functions (3.25) in Definition 32.

Belief measures from statistics of experts Let X then be the set of Boolean opinions of experts. We can define

$$\tilde{m}(A) = \frac{\mu(\omega \in \mathcal{E} | x_\omega = x_A)}{\mu(\mathcal{E})} = \frac{\mu(\omega \in \mathcal{E} | \Gamma(\{\omega\}) = A)}{\mu(\mathcal{E})},$$

where x_A is the indicator function of the event $A \subset \Lambda$. If we view the experts as endowed with prior probabilities $\mu(\omega)/\mu(\mathcal{E})$, and say that

$$\text{Prob}_{\mathcal{E}}(\text{event}) = \frac{\mu(\{\omega \in \mathcal{E} | \text{event is true for } \omega\})}{\mu(\mathcal{E})},$$

we find that $\tilde{m}(A) = \text{Prob}_{\mathcal{E}}(x_\omega = x_A)$. Under this interpretation, the belief on a set A is the joint probability

$$Bel(A) = \sum_{B \subseteq A} m(B) = \text{Prob}_{\mathcal{E}'}(x(\lambda) = 0 \text{ for } \lambda \notin A),$$

where $\mathcal{E}'$ is the set of experts expressing an opinion.

For further details of this framework, see [858].

Context models In [670], Gebhardt and Kruse developed a *context model* of vagueness and uncertainty to provide a formal semantic foundation for several different uncertainty theories, and belief theory in particular.

In their framework, *valuated vague characteristics* show a formal analogy to the concept of a random set developed by Matheron and Nguyen (see Section 3.1.5), in the sense that each *context* $c \in C$[21] is mapped by a vague characteristic γ to a subset $\gamma(c) \subset \Theta$ of the frame of discernment Θ. The (somewhat subtle) difference is that vague characteristics should not be interpreted as indivisible set-valued observations, but as imperfect descriptions of the actual state, under a certain 'context'.

[21] We use the original notation, rather than the usual Ω, for the context space to stress the different interpretations of the two spaces.

As in Dempster–Shafer theory, the inherent imprecision of γ only allows us to compute lower and upper bounds of 'acceptance degrees' of events $A \subset \Theta$. Operations such as the *context conditioning* of a vague characteristic by another and the combination of a number of vague characteristics (see Definition 12 in [670]) were also introduced.

Definition 33. *Let $\gamma_1, \gamma_2 : C \to 2^\Theta$ be two vague characteristics. We call the following the* context conditioning *of γ_1 by γ_2:*

$$\gamma_1|\gamma_2 : C \to 2^\Theta, \quad (\gamma_1|\gamma_2)(c) \doteq \begin{cases} \gamma_1(c) & \gamma_1(c) \subseteq \gamma_2(c), \\ \emptyset & \textit{otherwise}. \end{cases} \tag{3.27}$$

Conditional and evidential multivalued mappings In [2060], Yen extended the original multivalued mapping of Dempster's formulation (Section 3.1.1) to a probabilistic setting which uses conditional probabilities to express uncertainty in the mapping itself between the set Ω where the probabilistic evidence lives and the frame of discernment Θ where focal elements reside (see Fig. 3.7).

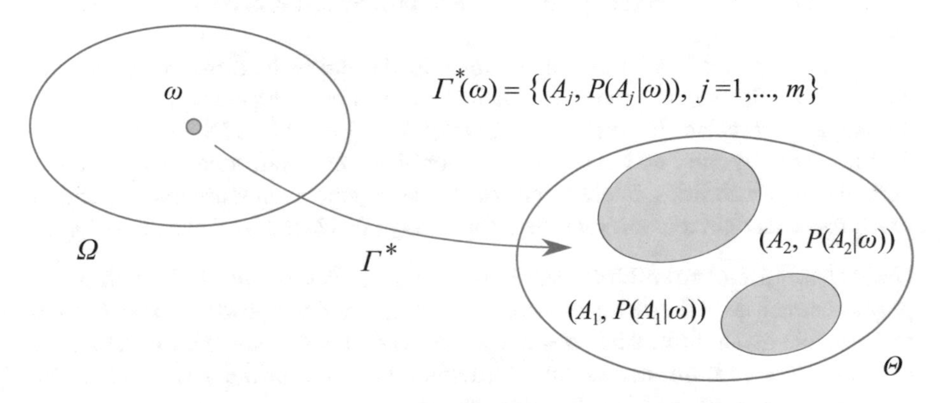

Fig. 3.7: Notion of conditional multivalued mapping (compare Fig. 3.2).

Going back to Fig. 3.2, if the mapping Γ is known with certainty to associate $\omega \in \Omega$ with $A \subset \Theta$, then $P(A|\omega) = 1$, whereas $P(A^c|\omega) = 0$. We can say that the deterministic multivalued setting of Dempster is associated with binary conditional probabilities on the mapping.

Definition 34. *A* probabilistic multivalued mapping *from a space Ω to a space Θ is a function*

$$\Gamma^* : \Omega \to 2^{2^\Theta \times [0,1]}$$

in which the image of an element ω of Ω is a collection of subset–probability pairs of the form

$$\Gamma^*(\omega) = \Big\{ (A_{i_1}, P(A_{i_1}|\omega)), \ldots, (A_{i_m}, P(A_{i_m}|\omega)) \Big\}$$

subject to the following conditions:

1. $A_{i_j} \neq \emptyset, \;\; j = 1, \ldots, m$;
2. $A_{i_j} \cap A_{i_k} = \emptyset$ *whenever* $j \neq k$;
3. $P(A_{i_j}|\omega) > 0, \;\; j = 1, \ldots, m$;
4. $\sum_{j=1}^{m} P(A_{i_j}|\omega) = 1$.

Each A_{i_j} is called a *granule*, and the collection $\{A_{i_j}, j\}$ is the *granule set* associated with ω. Roughly speaking, each focal element of standard belief theory is broken down into a collection of disjoint granules, each with an attached (conditional) probability. The mass of each focal element $A \subset \Theta$, in this extended multivalued framework, can then be expressed in a way which recalls the law of total probability:

$$m(A) = \sum_{\omega \in \Omega} P(A|\omega) P(\omega).$$

In this context, Dempster's rule is used to combine belief updates rather than absolute beliefs, obtaining results consistent with Bayes' theorem. Guan later presented in [741] a method for combining evidence from different evidential sources based on Yen's conditional compatibility relations, and proved that it gives the same results as Yen's.

A further generalisation of Yen's probabilistic multivalued mapping was presented in [1203] in which uncertain relations $\omega \to A$ between elements $\omega \in \Omega$ and subsets A of Θ replace the disjoint partitions $\{A_{i_j}, j\}$ of Θ (see Definition 34, item 2) considered by Yen, and mass functions on these uncertain relations generalise conditional probabilities. Interestingly, evidential mappings that use mass functions to express uncertain relationships were separately introduced by Guan et al. [740].

Non-monotonic compatibility relations Compatibility relations (Section 3.1.1) play a central role in the theory of evidence, as they provide knowledge about values of a variable given information about a second variable. A compatibility relation is called *monotonic* if an increase in information about the primary variable cannot result in a loss of information about the secondary variable.

Yager [2011] has investigated an extension of the notion of a compatibility relation which allows for non-monotonic relations. A belief function Bel_1 is said to be 'more specific' than another one, Bel_2, whenever

$$[Bel_1(A), Pl_1(A)] \subseteq [Bel_2(A), Pl_2(A)]$$

for all events $A \subseteq \Theta$.

Definition 35. *A type I compatibility relation $\mathcal{C}$ on $\Omega \times \Theta$ is such that:*
(i) for each $\omega \in \Omega$, there exists at least one $\theta \in \Theta$ such that $\mathcal{C}(\theta, \omega) = 1$;
(ii) for each $\theta \in \Theta$, there exists at least one $\omega \in \Omega$ such that $\mathcal{C}(\theta, \omega) = 1$.

Yager showed that the usual (type I) compatibility relations are always monotonic, i.e., if $\mathcal{C}$ is a type I compatibility relation between Ω and Θ and $Bel_1 \subset Bel_2$ are two belief functions on Ω, then $Bel_1^* \subset Bel_2^*$, where Bel_i^* is the belief function induced on Θ by the compatibility relation $\mathcal{C}$.

Type II compatibility relations were then introduced as follows. Let $\mathcal{X} = 2^{\Omega} \setminus \{\emptyset\}$.

Definition 36. *A type II compatibility relation C on $\mathcal{X} \times \Theta$ is such that for each $X \in \mathcal{X}$, there exists at least one $\theta \in \Theta$ such that $C(X, \theta) = 1$.*

It can be shown [2011] that a special class of type II relations, called 'irregular', are needed to represent non-monotonic relations between variables.

Belief functions based on probabilistic multivalued random variables A probabilistic multivalued random variable (PMRV) [1169] generalises the concept of a random variable, as a mapping from a sample space (endowed with a probability measure) to a target space.

Definition 37. *A* probabilistic multivalued random variable *from Ω to Θ is a function $\mu : \Omega \times \Theta \rightarrow [0, 1]$ such that, for all $\omega \in \Omega$,*

$$\sum_{\theta \in \Theta} \mu(\omega, \theta) = 1.$$

It is easy to see that, whereas compatibility relations are equal to 1 for a selection of pairs (ω, θ), a PMRV associates with each ω a probability distribution on its set of compatible outcomes Θ. If μ is a PMRV, we can define an 'inverse' mapping as

$$\mu^{-1}(\theta) = \left\{ \omega \in \Omega \middle| \mu(\omega, \theta) \neq 0 \right\}. \tag{3.28}$$

Note that μ^{-1} is really just the multivalued mapping $\Theta \rightarrow 2^{\Omega}$ induced by the compatibility relation $\mathcal{C}(\theta, \omega) = 1$ iff $\mu(\omega, \theta) \neq 0$ (this time, the direction of the multivalued mapping is reversed).

If (Ω, P_Ω) is a probability space, a PMRV $\mu : \Omega \times \Theta \rightarrow [0, 1]$ induces a probability P_Θ on Θ as follows:

$$P_\Theta(\theta) = \sum_{\omega \in \mu^{-1}(\theta)} P_\Omega(\omega) \mu(\omega, \theta). \tag{3.29}$$

If only the induced probability measure (3.29) and the inverse mapping (3.28) are know, the PMRV induces a basic probability assignment on Ω as follows:

$$m(A) = \sum_{\mu^{-1}(\theta) = A} P_\Theta(\theta) = \sum_{\mu^{-1}(\theta) = A} \sum_{\omega \in A} P_\Omega(\omega) \mu(\omega, \theta).$$

By such an interpretation, belief and plausibility measures are respectively the lower and upper estimations of the probability P_Ω on the sample space Ω. As the authors of [1169] note, in Shafer's definition the probability distribution on Ω is arbitrary and unrelated to the compatibility relation, whereas in theirs the probability distribution on Θ is induced by the PMRV, which also defines the multivalued mapping μ^{-1}. This framework exhibits significant similarities with the weak-belief approach of 'smearing' the probability measure on a pivot variable via a multivalued mapping (see Chapter 4, Section 4.1.1).

A significant difference emerges when deriving Dempster's combination rule. Consider two PMRVs μ_1, μ_2 from Ω to Θ_1, Θ_2, respectively. Since $\mu_1^{-1}(\theta_1) \cap \mu_2^{-1}(\theta_2) = \emptyset$ if and only if $\mu_1^{-1}(\theta_1) = \emptyset$ and $\mu_2^{-1}(\theta_2) = \emptyset$, it follows that the independence assumption behind Dempster's rule implies

$$\sum_{A \cap B = \emptyset} m_1(A) m_2(B) = 0,$$

and no normalisation is necessary, because conflict does not arise (under the assumption of independence).

3.3.2 Smets's transferable belief model

In his 1990s seminal work [1677], Philippe Smets introduced the *transferable belief model* as a framework based on the mathematics of belief functions for the quantification of a rational agent's degrees of belief [1701]. The TBM is by far the approach to the theory of evidence which has achieved the largest diffusion and impact. Its philosophy, however, is very different from Dempster's (and Shafer's) original probabilistic semantics for belief functions, as it cuts all ties with the notion of an underlying probability measure to employ belief functions directly to represent an agent's belief.

As usual here, we have room only for a brief introduction to the principles of the TBM. Aspects of the framework which concern evidence combination ([1689], Section 4.3.2), probability transformation (Section 4.7.2), conditioning (Section 4.6.1) and decision making (Section 4.8.1) are discussed in more detail in Chapter 4. In [1730, 1697] (and also [1707, 1732]) the reader will find an extensive explanation of the features of the transferable belief model. An interesting criticism of the TBM in terms of Dutch books was presented by Snow in [1740].

As far as applications are concerned, the transferable belief model has been employed to solve a variety of problems, including data fusion [1723], diagnostics [1704] and reliability issues [1688], and the assessment of the value of a candidate [522].

Credal and pignistic levels In the TBM, beliefs are represented at two distinct levels (Fig. 3.8):

1. A *credal* level (from the Latin word *credo* for 'I believe'), where the agent's degrees of belief in a phenomenon are maintained as belief functions.
2. A *pignistic* level (from the Latin *pignus*, 'betting'), where decisions are made through an appropriate probability function, called the *pignistic function*,

$$BetP(A) = \sum_{B \subseteq A} \frac{m(B)}{|B|}. \tag{3.30}$$

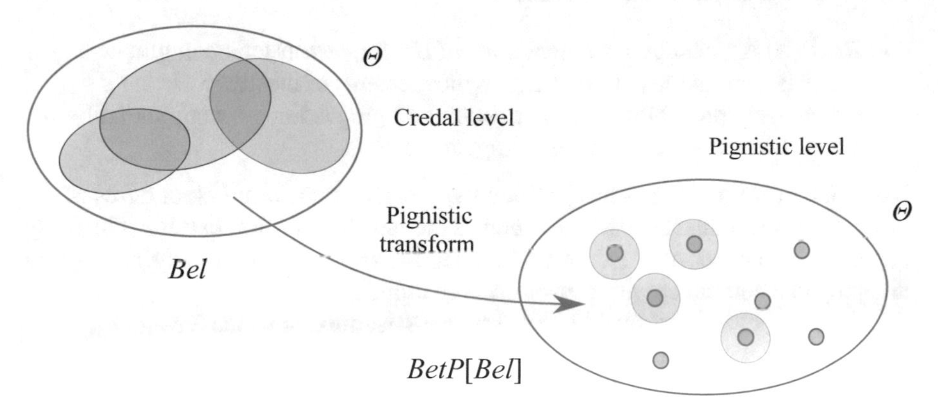

Fig. 3.8: Credal and pignistic levels in the transferable belief model.

The credal level At the credal level, each agent is characterised by an 'evidential corpus', a collection of pieces of evidence that they have collected in their past. This evidential corpus has an effect on the frame of discernment associated with a certain problem (e.g., who is the culprit in a certain murder). As in logic-based approaches to belief theory (compare Chapter 6, Section 6.6), a frame of discernment Θ is a collection of possible worlds ('interpretations', in the logical language), determined by the problem at hand, and a (logical) proposition is mapped the subset of possible worlds in which it holds true.

The basic assumption postulated in the TBM is that the impact of a piece of evidence on an agent's degrees of belief consists in an allocation of parts of an initial unitary amount of belief among all the propositions in the frame of discernment. The mass $m(A)$ is the part of the agent's belief that supports A, i.e. that the 'actual world' $\theta \in \Theta$ is in $A \subset \Theta$ and that, owing to lack of information, does not support any strict subproposition of A.

As in Shafer's work, then, each piece of evidence directly supports a *proposition*. To underline the lack of any underlying probabilistic model, Smets uses the terminology 'basic belief assignment' (BBA), rather then Shafer's 'basic probability assignment', to refer to mass assignments. Note that Smets does not claim that every form of uncertainty or imprecision can be represented by a belief function, but that uncertain evidence *induces* a belief function in an agent's belief system [1732]. Belief is *transferred* in the model whenever sharp information is available (of the form 'B is true') via Dempster's (unnormalised) rule of conditioning, so justifying the model's name.

For a discussion of evidence combination in the TBM, we refer the reader to Chapter 4, Section 4.3.2.

The pignistic level The pignistic function was axiomatically derived by Smets in [1678] as the only transformation which meets a number of rationality requirements:

- the probability value $BetP(x)$ of x depends only on the mass of propositions containing x;

- $BetP(x)$ is a continuous function of $m(B)$, for each B containing x;
- $BetP$ is invariant to permutations of the elements of the frame Θ;
- the pignistic probability does not depend on propositions not supported by the belief function which encodes the agent's beliefs.

Note that, over time, Smets has justified the pignistic transform first in terms of the principle of insufficient reason, and then as the only transform satisfying a linearity constraint. However, as we show in Chapter 11, $BetP$ is *not* the only probability mapping that commutes with convex combination.

Finally, decisions are made in a utility theory setting, in which Savage's axioms hold [1537] (see Section 4.8.1).

An interesting point Smets makes is that, although beliefs are necessary ingredients for our decisions, that does not mean that beliefs cannot be entertained without being manifested in terms of actual behaviour [1737]. In his example, one may have some beliefs about the status of a traffic light on a particular road in Brussels even though one is not in Brussels at the moment and does not intend to make a decision based on this belief. This is in stark constrast with behavioural interpretations of probability, of which Walley's imprecise probability theory is a significant example (as we will see in Section 6.1).

We will also see in Chapter 4, Section 4.7.2 that the pignistic transform is only one of many possible probability transforms, each coming with a different rationale.

Open-world assumption and unnormalised belief functions Within the TBM, positive basic belief values can be assigned to the empty set itself, generating *unnormalised* belief functions (UBFs) $m : 2^\Theta \rightarrow [0,1]$, $m(\emptyset) \neq 0$ [1685].

The belief values of unnormalised belief functions are computed in the usual way,

$$Bel(A) = \sum_{\emptyset \neq B \subseteq A} m(B),$$

whereas the function which takes into account the mass of the empty set is called the *implicability function*, and is typically denoted by $b : 2^\Theta \rightarrow [0,1]$:

$$b(A) = \sum_{\emptyset \subseteq B \subseteq A} m(B) = Bel(A) + m(\emptyset). \tag{3.31}$$

Implicability and commonality (recall (2.5)) are linked by the following relation:

$$\bar{b}(A) = Q(\bar{A}),$$

where $\bar{b}$ is the implicability function associated with the BBA $\bar{m}(A) = m(\bar{A})$.

As a matter of fact, UBFs and their space were first introduced by Hummel and Landy [858], where they were called 'generalised belief functions'. The structure of the Abelian monoid formed by their space, endowed with a (generalised) Dempster sum, was also studied there, and recalled in [1859].

Unnormalised belief functions are indeed a sensible representation under an 'open-world assumption' that the hypothesis set (frame of discernment) itself is not known with certainty (in opposition to the 'closed-world' situation, in which all alternative hypotheses are perfectly known). Note that the implicability function of a UBF is no longer a capacity, since it does not meet the requirement $\mu(\emptyset) = 0$ (see Definition 95).

Unnormalised belief functions and also pseudo-belief functions with negative mass assignments will arise in the geometric approach described in this book.

TBM versus other interpretations of belief theory As we have mentioned, the transferable belief model should not be considered as a generalised probability model – in the TBM, no links are assumed between belief functions and any underlying probability space (although they may exist). It is, rather, a normative model for a 'rational' agent's subjective beliefs about the external world. In [1683], Smets explictly stressed this point when comparing his creature with other interpretations of Dempster–Shafer theory, such as Dempster's upper and lower probability model, Pearl's probability of provability (basically amounting to a modal logic interpretation of belief theory: see Section 6.6.7) and Gardenfors's *evidentiary value* model [666], among others. The final message is a strong criticism of the careless use of Dempster combination when mixed with the interpretation of belief functions as lower bounds on an unknown probability distribution.

3.3.3 Dezert–Smarandache theory (DSmT)

The basis of DSmT [490, 491, 1665, 1155] is the rejection of the principle of the excluded middle, since for many problems (especially in sensor fusion) the nature of hypotheses themselves (i.e., the elements of the frame of discernment) is known only vaguely. As a result, a 'precise' description of the set of possible outcomes is difficult to obtain, so that the 'exclusive' hypotheses θ cannot be properly identified or separated.

Note that Cholvy [282], by studying the relation between evidence theory and DSmT using propositional logic, managed to show that DSmT can be reformulated in the classical framework of Dempster–Shafer theory, and that any DSmT combination rule corresponds to a rule in the classical framework. The conclusion was that the interest of DSmT lies in its compactness rather than its expressive power.

Hyperpower sets, free and hybrid models A cornerstone of DSmT is the notion of a *hyperpower set* [492], or 'Dedekind lattice'.

Definition 38. *[1665] Let $\Theta = \{\theta_1, \ldots, \theta_n\}$ be a frame of discernment. The* hyperpower set D^Θ *is defined as the set of all composite propositions built from elements of Θ using the $\cup$ and $\cap$ operators such that:*

1. *$\emptyset, \theta_1, \ldots, \theta_n \in D^\Theta$.*
2. *If $A, B \in D^\Theta$, then $A \cup B$ and $A \cap B$ belong to D^Θ.*
3. *No other elements belong to D^Θ, except those constructed using rules 1 and 2.*

In the above definition, there seems to be an implicit assumption that the elements of the set belong to a Boolean algebra. Indeed, as the authors of [1665] state, 'The generation of hyper-power set ... is closely related with the famous Dedekind's problem ... on enumerating the set of isotone Boolean functions.' An upper bound on the cardinality of D^Θ is obviously $2^{2^{|\Theta|}}$.

For example, when $\Theta = \{\theta_1, \theta_2, \theta_3\}$ has cardinality 3, the hyperpower set has as elements

$$\emptyset, \theta_1, \ldots, \theta_3, \theta_1 \cup \theta_2, \ldots, \theta_2 \cup \theta_3, \ldots, \theta_i \cap \theta_j, \ldots, \theta_1 \cup \theta_2 \cup \theta_3,$$

and therefore has cardinality 19 (see Fig. 3.9). In general, it will follow the sequence of Dedekind numbers[22] 1, 2, 5, 19, 167, 7580, ... Note that the classical complement is not contemplated, as DSmT rejects the law of the excluded middle.

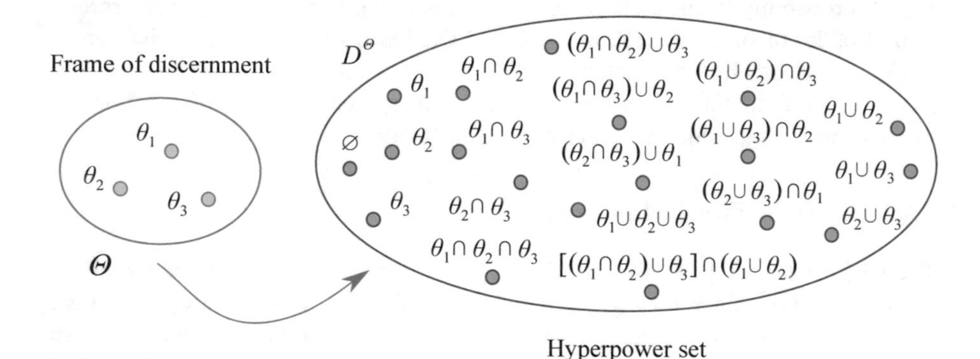

Fig. 3.9: Hyperpower set (free model, right) D^Θ associated with the frame of discernment $\Theta = \{\theta_1, \theta_2, \theta_3\}$ (left).

A hyperpower set is also called a *free model* in DSmT. If the problem at hand allows some constraints to be enforced on the hypotheses which form the frame (e.g. the non-existence or disjointness of some elements of D^Θ), we obtain a so-called *hybrid* model. The most restrictive hybrid model is the usual frame of discernment of Shafer's formulation, in which all hypotheses are disjoint.

Several issues about the ordering or partial ordering of the elements of the hyperpower sets involved in DSmT were examined by Dezert and Smarandache in [494] .

Generalised belief functions

Definition 39. *Let* Θ *be a frame of discernment. A* generalised basic belief assignment *is a map* $m : D^\Theta \to [0,1]$ *such that* $m(\emptyset) = 0$ *and* $\sum_{A \in D^\Theta} m(A) = 1$.

---

[22] N. J. A. Sloane, *The On-line Encyclopedia of Integer Sequences*, 2003 (sequence no. A014466): see `https://oeis.org/`.

The *generalised belief and plausibility functions* are then, trivially,

$$Bel(A) = \sum_{B \subseteq A, B \in D^\Theta} m(B), \quad Pl(A) = \sum_{B \cap A \neq \emptyset, B \in D^\Theta} m(B).$$

For all $A \in D^\Theta$, it still holds that $Bel(A) \leq Pl(A)$ – however, in the free model (whole hyperpower set), we have $Pl(A) = 1$ for all $A \in D^\Theta$, $A \neq \emptyset$.

Rules of combination Under the free DSmT model, the rule of combination is

$$m(C) = \sum_{A,B \in D^\Theta, A \cap B = C} m_1(A) m_2(B) \quad \forall C \in D^\Theta,$$

where this time A and B are elements of the hyperpower set (i.e., conjunctions and disjunctions of elements of the initial frame). Just like the TBM's disjunctive rule, this rule of combination is commutative and associative.

Obviously, this formalism is potentially very computationally expensive. Dezert and Smarandache, however, note that in most practical applications bodies of evidence allocate a basic belief assignment only to a few elements of the hyperpower set.

In the case of a hybrid model, things get much more complicated. The rule of combination assumes the form

$$m(A) = \phi(A) \Big[S_1(A) + S_2(A) + S_3(A) \Big],$$

where $\phi(A)$ is the 'characteristic non-emptiness function' of a set A, which is 1 for all sets A which have been forced to be empty under the constraints of the model, and 0 otherwise, and

$$S_1(A) = \sum_{X_1, \ldots, X_k \in D^\Theta, (X_1 \cap \ldots \cap X_k) = A} \prod_{i=1}^{k} m_i(X_i)$$

(S_2 and S_3 have similar expressions, which depend on the list of sets forced to be empty by the constraints of the problem).

A generalisation of the classical combination rules to DSmT hyperpower sets has also been proposed by Daniel in [385]. Daniel has also contributed to the development of DSmT in [383]. Interestingly, an extension of the theory of evidence to include non-exclusive elementary propositions was separately proposed by Horiuchi in 1996 [839].

3.3.4 Gaussian (linear) belief functions

The notion of a *Gaussian belief function* [1189] was proposed by Dempster [421, 1599] and was formalised by Liu in 1996 [1191].

Technically, a Gaussian belief function is a Gaussian distribution over the members of the parallel partition of a hyperplane (Fig. 3.10). The idea is to encode each proposition (event) as a linear equation, so that all parallel subhyperplanes of a given hyperplane are possible focal elements and a Gaussian belief function is a Gaussian distribution over these subhyperplanes. For instance, we can write $3X + 5Y \sim \mathcal{N}(2, 10)$ to mean that the frame of discernment is $\mathbb{R}^2$, the focal elements are the hyperplanes parallel to $3X + 5Y = 0$, and the mass value for each hyperplane $s(w) = \{(x, y)|3x + 5y = w\}$ is the density value $f(w) \sim \mathcal{N}(2, 10)$.

As focal elements (hyperplanes) cannot intersect, the framework is less general than standard belief theory, where the focal elements normally have non-empty intersection. However, it is more general than Shafer's original finite formulation, as focal elements form a continuous domain.

A nice matrix calculus can be derived for linear belief functions. A linear belief function on $X = (X_1, \ldots, X_n)$ (with $\mathbb{R}^n$ as the frame of discernment) can be more conveniently expressed as an $(n + 1) \times n$ matrix as follows [1194]:

$$M(X) = \begin{bmatrix} \mu \\ \Sigma \end{bmatrix}, \tag{3.32}$$

where μ and Σ are the mean vector and the covariance matrix, respectively, of the associated multivariate Gaussian distribution.

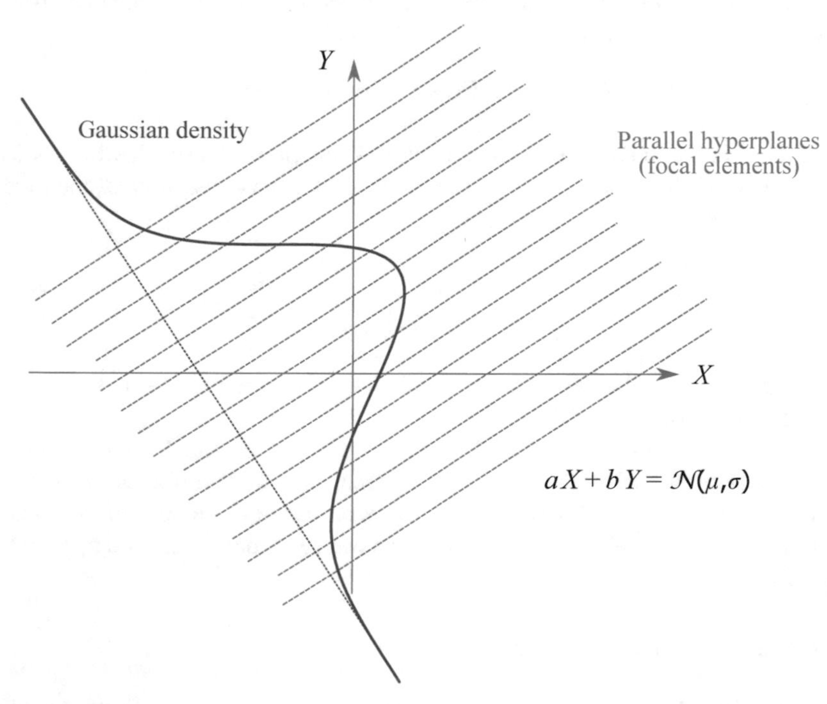

Fig. 3.10: Graphical representation of the concept of a Gaussian belief function.

By adapting Dempster's rule to the continuous case, Liu also derived a rule of combination and proved its equivalence to Dempster's geometrical description [421]. In [1192], Liu proposed a join-tree computation scheme for expert systems using Gaussian belief functions, after he proved that their rule of combination satisfies the axioms of Shenoy and Shafer [1632].

This framework was applied to portfolio evaluation in [1195].

3.3.5 Belief functions on generalised domains

A number of frameworks are centred around the idea of extending the domain of belief measures from the original power sets of finite frames of discernment to more general spaces.

Belief functions on lattices This approach, proposed by Michel Grabisch [721], extends belief functions from the Boolean algebra of subsets (which is in fact a lattice) to *any* lattice. This can be useful, for instance, in cases in which some events are not meaningful, in non-classical logic, or when considering coalitions in multi-agent games.

A *lattice* is a partially ordered set for which inf (the greatest lower bound) and sup (the least upper bound) exist for each pair of elements. A Boolean algebra is a special case of a lattice. A lattice is *De Morgan* when it admits negation, and *complete* if all collections of elements (not just pairs) admit inf and sup. In particular, a complete lattice has an initial element $\wedge = \inf L$ and a final element $\vee = \sup L$.

A *capacity* can then be defined as a function v on L such that (1) $v(\wedge) = 0$, (2) $v(\vee) = 1$ and (3) $x \leq y$ implies $v(x) \leq v(y)$ (see Chapter 6, Definition 95). The Moebius transform $m : f(x) = \sum_{y \leq x} m(y)$ of any function f on a lattice $(L, \leq)$ can also be defined. In particular, a belief function can be seen as a function on a lattice L such that $Bel(\wedge) = 0$, $Bel(\vee) = 1$ and its Moebius transform is non-negative. The conjunctive combination of two belief functions Bel_1, Bel_2 assumes the form

$$m_1 \oplus m_2(x) = \sum_{y_1 \wedge y_2 = x} m_1(y_1) m_2(y_2), \quad x, y_1, y_2 \in L.$$

Commonality, possibility and necessity measures with the usual properties can also be defined. Interestingly, any capacity is a belief function iff L is linear (in other words, a total order) [95].

A special case of belief functions on lattices was also developed in [476], where a formalism was proposed for representing uncertain information on set-valued variables using BFs. A set-valued variable X on a domain Ω is a variable taking zero, one or several values in Ω. As defining mass functions on the frame 2^{2^Ω} is not feasible because of the double-exponential complexity involved, the authors of [476] proposed to work on a restricted family of subsets of 2^Ω endowed with a lattice structure.

In [472], Denoeux and Masson noted that when Θ is linearly ordered, working with just intervals drastically reduces the complexity of calculations. They showed that this trick can be extrapolated to frames endowed with an arbitrary lattice structure, not necessarily a linear order. This principle makes it possible to apply the Dempster–Shafer framework to very large frames such as the power set, the set of partitions or the set of preorders of a finite set.

Distributive lattices The above approach was recently extended by Zhou [2114, 2115] to distributive lattices. This more general theory was applied there to a simple epistemic logic, the first-degree-entailment fragment of relevance logic, providing a sound and complete axiomatisation for reasoning about belief functions for this logic. Zhou showed that the complexity of the satisfiability problem of a belief formula with respect to the class of the corresponding Dempster–Shafer structures is NP-complete.

Kohlas's uncertain information in graded semilattices In [1019, 1022], Kohlas showed that functions monotone to order ∞, such as Choquet capacities, can be defined on semilattices, and studied random variables with values in some kinds of graded semilattices. It turns out that random variables in this algebra themselves form an algebra of the same kind. Their probability distributions correspond to belief functions. Kohlas therefore proposed a natural generalisation of evidence theory to a general structure related to probabilistic argumentation systems.

Belief functions on Boolean algebras A few authors [737, 756] have explored extensions of belief theory to general Boolean algebras rather than power sets, including spaces of propositions and infinite frames.

Indeed, such a generalisation can be trivially achieved by replacing the set-theoretical intersection $\cap$ and $\cup$ with the meet and join operators of an arbitrary Boolean algebra $\langle \mathcal{U}, \vee, \wedge, 1, 0 \rangle$ (see footnote 18). Guan and Bell [737], in particular, produced generalisations of a number of the main results of [1583] in this context. Previously [756], Guth had considered how mass assignments on Boolean algebras can propagate through a system of Boolean equations, as a basis for rule-based expert systems and fault trees.

3.3.6 Qualitative models

Preference relations and qualitative belief structures In [1954] Wong, Lingras and Yao argued that preference relations can provide a more realistic model of random phenomena than quantitative probability or belief functions.

The rationale is that if propositions in the source space Ω are described by a preference relation (compare Section 1.3.10) rather than a probability measure, a multivalued mapping can still be applied. The result is a pair of *lower and upper preference relations*.

Definition 40. *Let Ω and Θ be two domains linked by a multivalued mapping $\Gamma : \Omega \to 2^\Theta$. Given a preference relation $\succ$ on Ω, the lower $\succ_*$ and upper $\succ^*$ preference relation on 2^Θ are defined as*

$$A \succ_* B \Leftrightarrow \Gamma_*(A) \succ \Gamma_*(B), \quad A \succ^* B \Leftrightarrow \Gamma^*(A) \succ \Gamma^*(B), \tag{3.33}$$

where Γ_*, Γ^* denote the lower inverse (3.15) and the upper inverse (3.16), respectively, of the multivalued mapping Γ.

Pessimistic and optimistic propagations of preference relations of the form (3.33) can also be applied to those relations which are induced by a belief (or, respectively, plausibility) measure in the source frame Ω. Namely, given a belief function $Bel : 2^\Omega \rightarrow [0,1]$, one can define

$$A \succ_{Bel} B \Leftrightarrow Bel(A) > Bel(B), \quad A \succ_{Pl} B \Leftrightarrow B^c \succ_{Bel} A^c. \tag{3.34}$$

Conversely, Wong et al. [1958] have proven that, given a preference relation on 2^Ω, there exists a belief function consistent with $\succ$ in the sense of (3.34) if and only if $\succ$ satisfies the following axioms:

1. *Asymmetry*: $A \succ B$ implies $\neg(B \succ A)$.
2. *Negative transitivity*: $\neg(A \succ B)$ and $\neg(B \succ C)$ imply $\neg(A \succ C)$.
3. *Dominance*: for all $A, B \in 2^\Omega$, $A \supseteq B$ implies $A \succ B$ or $A \sim B$.
4. *Partial monotonicity*: for all $A, B, C \in 2^\Omega$, if $A \supset B$ and $A \cap C \neq \emptyset$ then $A \succ B$ implies $(A \cup C) \succ (B \cup C)$.
5. *Non-triviality*: $\Omega \succ \emptyset$.

Here, $A \sim B$ denotes the absence of strict preference.

In order to actually use preference relations for reasoning under uncertainty, it is necessary to perform sequential and parallel combinations of propagated information in a qualitative inference network; this is discussed in [1954].

Relation-based evidential reasoning In closely related work [42], An et al. argued that the difficult and ill-understood task of estimating numerical degrees of belief for the propositions to be used in evidential reasoning (an issue referred to as 'inference': see Chapter 4, Section 4.1) can be avoided by replacing estimations of absolute values with more defensible assignments of relations.

This leads to a framework based on representing arguments such as 'evidence e supports alternative set A' and their relative strengths, as in 'e_1 supports A_1 better than e_2 supports A_2'. An et al. proved that belief functions (in a precise sense) are equivalent to a special case of the proposed method, in which all arguments are based on only one piece of evidence.

Qualitative Dempster–Shafer theory Parsons [1391] separately introduced the idea of using the theory of evidence with qualitative values, when numerical values are not available. To cope with this lack of numbers, Parsons proposed to use qualitative, semiqualitative and linguistic values, and to apply a form of order-of-magnitude reasoning.

Suppose that in the classical trial example (Section 2.1) the witness says that she is sure it was a man, but is unwilling to commit herself to a specific numerical value. This piece of information ($m(\{Peter, John\})$ = something) can be converted to a 'qualitative' value via a mapping

$$[\cdot] : \mathbb{R} \to \{+, 0\}$$

so that any mass known to be non-zero is represented by the qualitative value '+'. This enables us to manipulate mass values using well-established methods of qualitative arithmetic [175]. Mass assignments are still combined by Dempster's rule, where addition and multiplication of real numbers are now replaced by qualitative addition and multiplication (Table 3.1):

$\oplus$	+	0
+	+	+
0	+	0

$\otimes$	+	0
+	+	0
0	0	0

Table 3.1: Qualitative addition (left) and multiplication (right) operators.

This qualitative approach can be further refined by using linguistic values, such as 'None', 'Little', etc., associated with intervals of quantitative values. Interval arithmetic can then be used to implement Dempster's rule on such linguistic mass assignments. The idea clearly resembles that of fuzzy assignments (Section 6.5).

Non-numeric belief structures An interesting, albeit dated, publication by Wong, Wang and Yao [1956] introduced *non-numeric belief* as the lower envelope of a family of incidence mappings, which can be thought of as non-numeric counterparts of probability functions. Likewise, non-numeric conditional belief can be defined as the lower envelope of a family of conditional incidence mappings. Such definitions are consistent with the corresponding definitions for belief functions and come in closed-form expressions.

Consider a situation in which the set of possible worlds is described by W, and an *incidence mapping* $i : 2^\Theta \to 2^W$ exists such that if $w \in i(A)$ then A is true, while it is false otherwise. The mapping associates each proposition A with the set of worlds (interpretations) in which it is true. When the evidence is not sufficient to specify i completely, it may be possible to specify lower and upper bounds

$$\underline{F}(A) \subseteq i(A) \subseteq \overline{F}(A)$$

on the true incidence sets. A set of lower bounds is called an *interval structure* [1956] if it satisfies the following axioms:

1. $\underline{F}(\emptyset) = \emptyset$.
2. $\underline{F}(\Theta) = W$.
3. $\underline{F}(A \cap B) = \underline{F}(A) \cap \underline{F}(B)$.
4. $\underline{F}(A \cup B) \supseteq \underline{F}(A) \cup \underline{F}(B)$.

After observing the close relationships between the above qualitative axioms and the quantitative axioms of belief functions, we may regard the lower bound of an interval structure as a non-numeric belief and the upper bound as the corresponding non-numeric plausibility.

3.3.7 Intervals and sets of belief measures

Imprecise belief structures *Imprecise belief structures* (IBSs), introduced by Denoeux [445, 446], are sets of belief structures whose masses on focal elements $\{A_i, i\}$ satisfy interval-valued constraints $\mathbf{m} = \{m : a_i \leq m(A_i) \leq b_i\}$ (see Fig. 3.11), and express imprecision in the belief of a rational agent within the transferable belief model. Note that, however, since

$$m(A_i) \leq \min\left\{b_i, 1 - \sum_{j \neq i} a_j\right\},$$

the intervals $[a_i, b_i]$ specifying an IBS are not unique. Upper and lower bounds on m determine interval ranges for belief and plausibility functions, and also for pignistic probabilities.

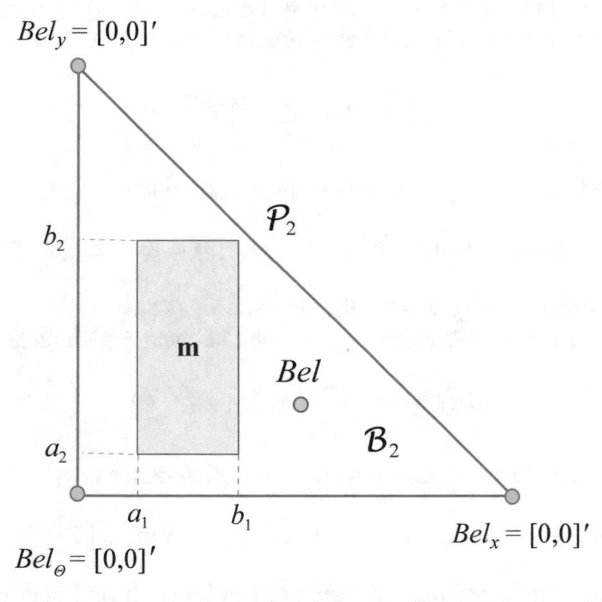

Fig. 3.11: Imprecise belief structures $\mathbf{m}$ versus belief measures Bel in the binary case. $\mathcal{B}_2$ denotes the set of all belief functions on $\Theta = \{x, y\}$, while $\mathcal{P}_2$ denotes the set of all probability measures there.

The combination of two IBSs $\mathbf{m}_1, \mathbf{m}_2$ can be defined either as

$$\mathbf{m}' = \left\{m = m_1 \circledast m_2 \middle| m_1 \in \mathbf{m}_1, m_2 \in \mathbf{m}_2\right\} \tag{3.35}$$

or as the IBS $\mathbf{m} = \mathbf{m}_1 \circledast \mathbf{m}_2$ with bounds

$$m^-(A) = \min_{(m_1,m_2) \in \mathbf{m}_1 \times \mathbf{m}_2} m_1 \circledast m_2(A), \; m^+(A) = \max_{(m_1,m_2) \in \mathbf{m}_1 \times \mathbf{m}_2} m_1 \circledast m_2(A), \tag{3.36}$$

for all $A \subset \Theta$, where $\circledast$ denotes any combination operator for individual mass functions. It is easy to see that $\mathbf{m}' \subset \mathbf{m}$.

The definition (3.36) results in a quadratic programming problem. An iterative algorithm for its solution was proposed in [445], Section 4.1.2.

Interval-valued focal weights Yager [2020] considered a similar situation in which the masses of the focal elements lie in some known interval, allowing us to model more realistically situations in which the basic probability assignments cannot be precisely identified. He pointed out in [2020] that this amounts to uncertainty in the actual belief structure of the possibilistic type.

Interval Dempster–Shafer approaches A different formal setting, based on Yager's fuzzy connectivity operators, was proposed by Lee in [1112].

An *interval basic probability assignment* is defined as a function $M : 2^\Theta \rightarrow [0,1] \times [0,1]$, mapping (in a similar way to Denoeux's IBSs) each subset to a pair of upper and lower mass values. A belief measure can then be defined as

$$Bel(A) = \sum_{B \subseteq A} M(B),$$

where this time, however, $\sum$ denotes the interval summation

$$[a,b] + [c,d] = [u(a,c), u(c,d)], \quad u(a,b) \doteq \min[1, \|[a,b]'\|_{L_p}],$$

where $p \in [0,\infty]$ and $\|\cdot\|_p$ denotes the classical L_p norm.

Two interval-valued BPAs can be combined in a way which recalls conjunctive combination:

$$M(C) = \sum_{A \cap B = C} M(A) \star M(B),$$

where $\star$ denotes an 'interval multiplication' operator, defined as

$$[a,b] \star [c,d] = [i(a,c), i(c,d)], \quad i(a,b) \doteq 1 - \min[1, \|\mathbf{1} - [a,b]'\|_{L_p}].$$

Interval summation and multiplication in fact follow from Yager's popular proposals on fuzzy connectivity operators [1999]. Normalisation is not necessary in this setting and can be ignored. This interval approach duly reduces to classical belief theory when $p = 1$ and point intervals are considered.

Generalised basic probability assignments Augustin [68] extended further the idea of imprecise belief structures by considering *sets* of basic probability assignments, rather than intervals of mass values.

Definition 41. [68] *Let $(\Omega, \mathcal{P}(\Omega))$ be a finite measurable space, and denote by $\mathcal{M}(\Omega, \mathcal{P}(\Omega))$ the set of all basic probability assignments on $(\Omega, \mathcal{P}(\Omega))$. Every non-empty, closed subset $S \subseteq \mathcal{M}(\Omega, \mathcal{P}(\Omega))$ is called a* generalised basic probability assignment *on $(\Omega, \mathcal{P}(\Omega))$.*

Augustin proved the following proposition.[23]

Proposition 23. ([68], generalised belief accumulation) *For every generalised basic probability assignment S, the set function $\mathbb{L} : \mathcal{P}(\Omega) \rightarrow [0,1]$ with*

$$\mathbb{L}(A) \doteq \min_{m \in S} \sum_{\emptyset \neq B \subseteq A} m(B), \quad \forall A \in \mathcal{P}(\Omega),$$

is well defined, and is a lower probability.

Generalised belief accumulation offers an alternative to the combination of individual belief functions: the more the assignments differ from each other, the wider the intervals of the resulting lower probability (called by some authors, such as Weichselberger, the 'F-probability' [1921, 1922]).

The opposite can be proven as well: every convex set of probability measures can be obtained by generalised belief accumulation [68] (although the correspondence is not one-to-one). Indeed, given a credal set, one just needs to take the convex closure of the Bayesian belief functions associated with the vertex probability distributions.

Sets of basic probability assignments thus constitute an appealing constructive approach to imprecise probability, which allows a very flexible modelling of uncertain knowledge.

3.3.8 Other frameworks

Many other frameworks have been proposed – here we list them, roughly in order of their impact in terms of citations (as of July 2017) [1180, 1463, 1969]. For some of them, not many details are available, including Zarley's evidential reasoning system [2095], Fister and Mitchell's 'entropy-based belief body compression' [622], Peterson's 'local Dempster–Shafer theory' [1422] and Mahler's customisation of belief theory via a priori evidence [1232].

For all the others, we provide a brief description in the following.

Zadeh's 'simple view' In Zadeh's 'simple view' [2089], Dempster–Shafer theory is viewed in the context of relational databases, as an instance of inference/retrieval techniques for second-order relations. In the terminology of relational databases, a first-order relation is a relation whose elements are atomic rather than set-valued, while a second-order relation associates entries i with sets D_i of possible values (e.g., person i has an age in the range $[22, 26]$).

Given a query set Q, the possibility of certain entries satisfying the query can be measured (e.g., whether person i has an age in the range $[20, 25]$). Namely, for each entry, (i) Q is possible if the query set intersects the set of possible values for that entry (which Zadeh calls the 'possibility distribution'); (ii) Q is certain (necessary) if $D_i \subset Q$; and (iii) Q is not possible if the two do not intersect.

[23] The terminology has been changed to that of the present book.

We can then answer questions of the kind 'what fraction of entries satisfy the query?' (e.g. what percentage of employees are between 20 and 25 years of age) in terms of necessity and possibility measures defined as follows:

$$N(Q) = \sum_{D \subset Q} p_D, \quad \Pi(Q) = \sum_{D \cap Q \neq \emptyset} p_D,$$

where p_D is the fraction of entries whose set of possible values is exactly D. As such, $N(Q)$ and $\Pi(Q)$ can be computed from a mere histogram (count) of entries with the same values, rather than the entire table.

From this perspective, belief and plausibility measures are, respectively, the certainty (or necessity) and possibility of the query set Q in the context of retrieval from a second-order relation in which the data entries are possibility distributions.

Lowrance and Strat's framework Garvey, Lowrance, Fischler and Strat's early evidential reasoning framework [668, 1218] uses Shafer's theory of belief functions in its original form for encoding evidence.

Their original contribution in [668] was a set of inference rules for computing belief/plausibility intervals of dependent propositions, from the mass assigned to the focal elements. Their framework for evidential reasoning systems [1218] instead focuses more on the issue of specifying a set of distinct frames of discernment, each of which defines a set of possible world situations, and their interrelationships, and on establishing paths for the bodies of evidence to move on through distinct frames by means of evidential operations, eventually converging on spaces where the target questions can be answered. As such, their work is quite closely related to the author of this book's algebraic analysis of families of compatible frames [326, 329, 364], and his belief-modelling regression approach to pose estimation in computer vision [365, 366, 370, 704].

This is done through a compatibility relation, a subset $\Theta_{A,B} \subset \Theta_A \times \Theta_B$ of the Cartesian product of the two related frames (their common refinement in Shafer's terminology). A compatibility mapping taking statements A_k in Θ_A to obtain statements of Θ_B can then be defined as

$$C_{A \mapsto B}(A_k) = \left\{ b_j \in \Theta_B \middle| (a_i, b_j) \in \Theta_{A,B}, a_i \in A_k \right\}.$$

Interestingly, in dynamic environments compatibility relations can be used to reason over time, in which case a compatibility relation represents a set of possible state transitions (see [364], Chapter 7).

Given evidence encoded as a belief function on Θ_A, we can obtain a projected belief function on Θ_B via

$$m_B(B_j) = \sum_{C_{A \mapsto B}(A_i) = B_j} m_A(A_i).$$

The framework was implemented using Grasper II,[24] a programming language extension of LISP which uses graphs, encoding families (or, in Lowrance's terminology, 'galleries') of frames, as primitive data types.

'Improved' evidence theory In [589], Fan and Zuo 'improved' standard evidence theory by introducing a fuzzy membership function, an importance index and a conflict factor to address the issues of scarce and conflicting evidence, together with new decision rules.

Josang's subjective evidential reasoning Josang devised in [916] a framework based on belief theory for combining and assessing subjective evidence from different sources. The central notion of this framework is a new rule called the *consensus operator*, based on statistical inference, originally developed in [918].

Josang's approach makes use of an alternative representation of uncertain probabilities by probability density functions over a (probability) variable of interest (in other words, second-order probability distributions: compare Section 6.4), obtained by generalising the beta family:

$$f(\alpha, \beta) = \frac{\Gamma(\alpha+\beta)}{\Gamma(\alpha)\Gamma(\beta)} p^{\alpha-1}(1-p)^{\beta-1},$$

where α, β are the parameters specifying the density function, and Γ denotes the gamma distribution, for frames of discernment of arbitrary atomicity in a three-dimensional representation with parameters r, s and a. A mapping between this three-dimensional representation and belief functions is then applied as follows:

$$Bel(A) = \frac{r}{r+s+2}, \quad Dis(A) = Bel(A^c) = \frac{s}{r+s+2}.$$

After this mapping, Dempster and consensual combination can be compared [918].

In further work [917], Josang extended the notion of a 'base rate', used in standard probability for default and conditional reasoning, to the case of belief functions.

Connectionist evidential reasoning *Connectionist evidential reasoning* [96] is an interesting approach in which a multilayer perceptron neural network is implemented to calculate, for each source of information, posterior probabilities for all the classes $c_1, \ldots, c_n$ in a decision problem. Then, the estimated posterior probabilities $p(c_i|X)$ given the data X are mapped to the masses of a consonant belief function on the frame $\Theta = \{c_1, \ldots, c_n\}$ of all classes, via the formula

$$m(\{c_1, \ldots, c_i\}) = \frac{p(c_i|X) - p(c_{i+1}|X)}{p(c_1|X)}, \quad i = 1, \ldots, n.$$

The most interesting contribution of [96] was a network realisation of Dempster–Shafer evidential reasoning, via a five-layer neural network called the 'DSETnet',

[24] `http://www.softwarepreservation.org/projects/LISP/massachusetts/Lowrance-Grasper_1.0.pdf`

in which the first layer contains the input masses of the two belief functions to be combined, layer 2 contains multiplication nodes in which the masses are multiplied, layer 3 implements the summation of masses associated with the same focal element, layer 4 is for normalisation, and layer 5 computes the posterior of each class by a pignistic transform. In order to tackle the issue of dependence between sources, the DSETnet was tuned for optimal performance via a supervised learning process.

Hybrid approach based on assumptions In [372], a hybrid reasoning scheme that combines symbolic and numerical methods for uncertainty management was presented by D'Ambrosio. The hybrid was based on symbolic techniques adapted from *assumption-based truth maintenance systems* (ATMSs) [405], combined with Dempster–Shafer theory, as extended in Baldwin's support logic programming system [76].

ATMSs are symbolic approaches to recording and maintaining dependencies among statements or beliefs. The central idea of such an approach is that for a derived statement, a set of more primitive ones supporting that statement are produced. Each primitive statement, which is called an *assumption*, is assumed to be true if there is no conflict. Assumptions can be interpreted as the primitive data from which all other statements (or beliefs or data) can be derived.

In [372], hybridisation was achieved by viewing an ATMS as a symbolic algebra system for uncertainty calculations. D'Ambrosio pointed out several advantages over conventional methods for performing inference with numerical certainty estimates, in addition to the ability to dynamically determine hypothesis spaces. These advantages include improved management of dependent and partially independent evidence, faster run-time evaluation of propositional certainties, and the ability to query the certainty value of a proposition from multiple perspectives.

Belief with minimum commitment In [846], an approach to reasoning with belief functions, fundamentally unrelated to probabilities and consistent with Shafer and Tversky's canonical examples, was proposed by Hsia.

Basically, the idea is to treat all available partial information, in the form of marginal or conditional beliefs, as constraints that the overall belief function needs to satisfy. The principle of minimum commitment, a variant of the principle of minimum specificity [529], then prescribes the least committed such belief function (in the usual weak inclusion order (3.9)).

This reasoning setting is very much related to the total belief problem (compare Chapter 17, Section 17.1.3), i.e., the generalisation of the law of total probability to belief measures. Unfortunately, in his paper Hsia does not go beyond stating the formal principle of this reasoning framework, and does not provide a methodology for actually computing the least committed belief function, which is the output of the framework itself.

A theory of mass assignments Baldwin [81, 80] proposed a theory of mass assignments for evidential reasoning, which correspond mathematically to Shafer's basic probability assignments, but are treated using a different rule of combination. The latter uses an assignment algorithm subject to constraints derived from operations research.

In [80], an algebra for mass assignments was given, and a conditioning process was proposed which generalises Bayesian updating to the case of updating prior mass assignments, with uncertain evidence expressed as another mass assignment.

Evidence theory of exponential possibility distributions Tanaka studied in [1796] a form of evidence theory which uses 'exponential' possibility distributions, of the form

$$\Pi_A(x) = \exp -(x-a)^T D_A (x-a),$$

where A denotes a fuzzy set, a is a mean vector and D_A is a symmetric positive-definite matrix. A rule of combination similar to Dempster's rule can then be defined for such exponential possibility distributions, as follows:

$$\Pi_{A_1 \oplus A_2}(x) = \kappa \Pi_{A_1}(x) \cdot \Pi_{A_2}(x),$$

where κ is a normalisation factor such that $\max_x \Pi_{A_1 \oplus A_2}(x) = 1$.

Marginal and conditional possibilities were also discussed in [1796], and a posterior possibility derived from the prior one as in Bayes' formula.

A set-theoretic framework Lu and Stephanou [1220] proposed in 1984 a set-theoretic framework based on belief theory for uncertain knowledge processing in which (1) the user enters input observations with an attached degree of belief ('certainty', in these authors' terminology); (2) each piece of evidence receiving non-zero certainty activates a mapping to an output space (a multivalued mapping) in which its certainty is multiplied by that of the mapping, and is thus propagated to a proposition in the output space; and (3) the consensus among all propositions with non-zero certainties is computed by Dempster's rule, and a degree of support is associated with each conclusion.

Interestingly, the inverse of the rule of combination, which these authors call the 'rule of decomposition', was derived in [1220] for separable support belief functions (Section 2.4).

Temporal evidential reasoning Fall [588] described a system which uses frame-like objects called 'models' that propagate the effects of a piece of evidence through time, and uses Gordon and Shortliffe's theory of propagation on diagnostic trees ([710]: see Section 4.7.4) to combine the effects of the active models.

Conditioned Dempster–Shafer theory Mahler [1233] described a conditioned Dempster–Shafer (CDS) theory (also called 'modified Dempster–Shafer' (MDS) in [623]) in which the composition operator can be 'conditioned' to reflect the influence of any kind of a priori knowledge which can be modelled as a belief measure. CDS is firmly grounded in probability theory via the theory of random sets. It is also a generalisation of Bayesian theory to the case when both evidence and a priori knowledge are ambiguous.

Mahler first provides an alternative, equivalent definition of Dempster's rule as

$$m_{1\oplus 2}(A \cap B) \propto m_1(A) m_2(B) \alpha_\oplus(A, B), \tag{3.37}$$

where

$$\alpha_\oplus(A, B) = \frac{\rho(A \cap B)}{\rho(A)\rho(B)}$$

is an *agreement function* such that $\rho(X) = 1$ if $X \neq \emptyset$, and 0 otherwise. An MDS agreement function is then obtained by replacing ρ as above with an a priori probability distribution q, which induces a combination rule parameterised by q, thus generalising Dempster's rule in its form (3.37).

MDS can be given a random-set justification as follows. Given a probability space $(\Omega, \mathcal{F}, P)$, let $\Gamma : \Omega \to 2^\Theta$ be a random set (multivalued mapping), and $X : \Omega \to \Theta$ a classical random variable on Θ. X induces a probability measure $P'(A) = P(\omega : X(\omega) \in A)$, $A \subset \Theta$, there. This probability measure can be extended to any random set Γ via

$$P'(\Gamma) = P(\omega \in \Omega : X(\omega \in \Gamma(\omega))).$$

Mahler proceeds to show that the MDS agreement function is a probabilistic function based on this special kind of random variable: a random (sub)set.

In [623], Fixsen and Mahler argued that MDS is also closely related to Smets's pignistic probabilities, which in the MDS framework become true posterior distributions.

Self-conditional probabilities In [305] Cooke and Smets presented an interesting interpretation of belief functions as normalised self-conditional expected probabilities, and studied their mathematical properties in that framework. The self-conditional interpretation considers surplus belief in an event emerging from a future observation, conditional on the event occurring, whereas Dempster's original interpretation involves partial knowledge of a belief state.

Plausible reasoning Guan and Bell [739] described the mathematical foundations of an original knowledge representation and evidence combination framework, and related it to Dempster–Shafer theory.

Definition 42. *Given an observation space $\mathcal{X}$ and a frame of discernment Θ, a* pl-function *is a function $pl(\theta|x) : \mathcal{X} \to [0, 1]$ defined on the observation space for each value $\theta \in \Theta$, which represents the extent to which we do not wish to rule θ out after observing $x \in \mathcal{X}$.*

Guan and Bell's representations, called *pl-functions*, endowed with a simple multiplicative combination rule, were shown to be equivalent to a subclass of the family of mass functions as described by Shafer with Dempster's rule as the combination function. Their simpler combination rule, however, has a complexity which is linear in the cardinality of the frame of discernment.

4

Reasoning with belief functions

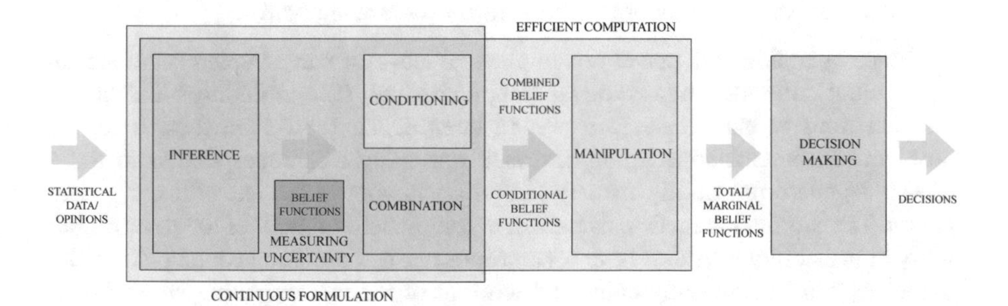

Fig. 4.1: When reasoning with uncertainty in the framework of belief theory, the first step is inferring a belief function from the available (uncertain) data (Section 4.1). Various ways of measuring the degree of uncertainty encoded by a belief function exist (Section 4.2). The resulting belief function(s) are subsequently combined (rather than conditioned: see Section 4.3), although conditioning, as a special case, can be applied when the new evidence is hard (Section 4.5). Continuous formulations of the notions of a belief function, inference and combination have been proposed (Section 4.9). Conditional belief functions can be combined to achieve a single total function, propagated in a graphical model or manipulated in the framework of the generalised Bayes theorem (Section 4.6). Efficient implementations of all reasoning steps are available (Section 4.7). Finally, decisions can be made based on the resulting belief function description (Section 4.8). Regression, classification and estimation are covered in Chapter 5.

F. Cuzzolin, *The Geometry of Uncertainty*, Artificial Intelligence: Foundations, Theory, and Algorithms, https://doi.org/10.1007/978-3-030-63153-6_4

Belief functions are fascinating mathematical objects, but most of all they are useful tools designed to achieve a more natural, proper description of one's uncertainty concerning the state of the external world. Reasoning with belief functions involves a number of steps which closely recall those of Bayesian reasoning (see Fig. 4.1):

1. *Inference*: building a belief function from (uncertain) data, which can be either statistical or qualitative in nature.
2. Measuring the *uncertainty* associated with a belief measure.
3. *Reasoning*: updating a belief measure when new data are available. This can happen either:
   - by *combination* with another belief function, encoding evidence which does not naturally come in the form of a conditioning event; or
   - by *conditioning* the original belief function(s) with respect to a new event A, in the special case in which hard evidence is available.
4. *Manipulating* conditional and multivariate belief functions, via either:
   - the generalised Bayes theorem (GBT), formulated by Philippe Smets;
   - their propagation within graphical models similar to classical Bayesian networks;
   - the generalisation of the total probability theorem, to combine the input conditional belief functions into a single total function.
5. *Applying* the resulting belief function(s) for decision making, regression, classification, estimation or optimisation purposes, among others.

Since operating with power sets implies, at least for naive implementations, an exponential rather than a polynomial complexity, the issue of finding efficient implementations of belief calculus has long been debated.[25] As we show here, the problem can be tackled using approximate reasoning techniques similar to those used in standard probability, namely Monte Carlo approaches and efficient propagation. Transformation methods based on mapping belief functions to leaner probability or possibility measures have been proposed. In addition, maximal plausibility outcomes can be efficiently computed without explicitly computing whole belief functions.

Last but not least, in order to address most real-world problems, the original, discrete formulation of evidence theory needs to be generalised to continuous frames of discernment. Several proposals have been made to this end: among them, we argue here, the most promising is the mathematical formalism provided by random set theory.

Chapter outline

Inference is dealt with in Section 4.1, whereas Section 4.2 discusses various proposals for measuring the 'entropy' of a belief function, and the main principles of

[25] A good survey of algorithms for operating with belief functions is provided by a recent book by Reineking [1479].

uncertainty. The mathematics of evidence combination and conditioning is illustrated in Sections 4.3 and 4.5, respectively. Belief function and Bayesian reasoning are compared in a data fusion example in Section 4.4.

The manipulation of conditional and multivariate belief functions is described in Section 4.6, while Section 4.7 is devoted to the efficient implementation of belief calculus. Strategies for making decisions under uncertainty represented by belief functions are recalled in Section 4.8.

The last part of the chapter is devoted to the various mathematical formulations for defining and manipulating belief functions on arbitrary, continuous domains (including the real line, Section 4.9), and to some interesting facets of belief functions in terms of their geometry and algebra (Section 4.10).

4.1 Inference

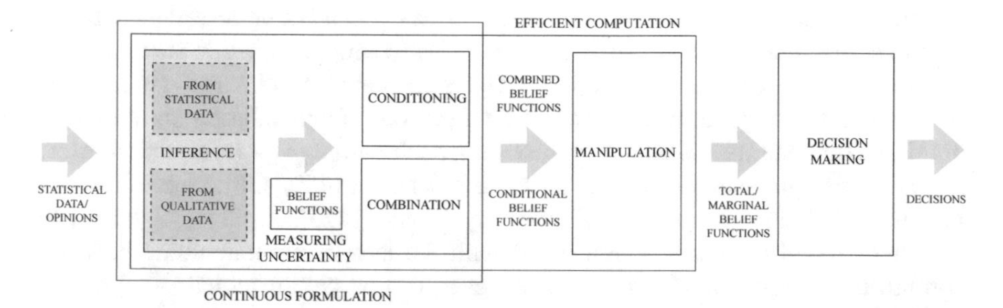

Inference is the first step in any estimation or decision problem. In the context of belief theory, inference means constructing a belief function from the available evidence [1001, 1002, 872].

As we argued in the Introduction, belief functions (as opposed to Kolmogorov's additive probability measures) can represent a wide range of what we have called 'uncertain' data, from classical statistical samples [1571] to qualitative expert judgements, sometimes expressed in terms of mere preferences of one outcome over another. A number of different approaches have been proposed: a very general exposition by Chateauneuf and Vergnaud can be found in [250]. Another reference document on inference with belief functions is [604], which summarises a variety of the most useful and commonly applied methods for obtaining belief measures, and their mathematical kin probability boxes. The topic has been the subject of patents as well (`https://www.google.ie/patents/US6125339`).

Here we briefly survey the main proposals on both inference from statistical data (Section 4.1.1), with a focus on the likelihood-based approach and Dempster's auxiliary-variable idea, and inference from qualitative evidence (Section 4.1.2). Inference with partial data is recalled in Section 4.1.3. A comparison of the results of Bayesian, frequentist and belief function inference in the case of a series of coin tosses is provided in Section 4.1.4.

4.1.1 From statistical data

As far as inference from statistical data is concerned, the two dominant approaches are Dempster's proposal based on an auxiliary variable, and Wasserman and Shafer's principle based on the likelihood function.

The problem can be posed as follows. Consider a parametric model (see Definition 3), i.e., a family of conditional probability distributions $f(x|\theta)$ of the data given a parameter θ,

$$\Big\{f(x|\theta), x \in \mathbb{X}, \theta \in \Theta\Big\}, \tag{4.1}$$

where $\mathbb{X}$ is the observation space and Θ the parameter space. Having observed x, how do we quantify the uncertainty about the parameter θ, without specifying a prior probability distribution?

Likelihood-based approach Given a parametric model (4.1), we want to identify (or compute the support for) parameter values which better describe the available data. An initial proposal for a likelihood-based support function was made by Shafer [1583], and immediately supported by Seidenfeld [1571]. Later [1913, 1914] Wasserman also argued in favour of belief function inference based on the likelihood function and its application to problems with partial prior information. Most recently, this approach has been endorsed by Denoeux [459, 458], whereas its axiomatic foundations have been criticised by Moral [1313]. A major objection to likelihood-based inference is associated with the lack of commutativity between combination and conditioning – this, however, can be circumvented [518] by assuming that the set of hypotheses or parameter values is rich enough.

Consider the following requirements [460]. Given the parametric model (4.1), we have:

- *Likelihood principle*: the desired belief function $Bel_\Theta(\cdot|x)$ on the space Θ of parameter values should be based on the likelihood function $L(\theta; x) \doteq f(x|\theta)$ only.
- *Compatibility with Bayesian inference*: when a Bayesian prior P_0 on Θ is available, combining it with $Bel_\Theta(\cdot|x)$ using Dempster's rule should yield the Bayesian posterior:
$$Bel_\Theta(\cdot|x) \oplus P_0 = P(\cdot|x).$$
- *Principle of minimum commitment*: among all the belief functions which meet the previous two requirements, $Bel_\Theta(\cdot|x)$ should be the least committed (see Section 4.2.3).

These constraints lead to uniquely identifying $Bel_\Theta(\cdot|x)$ as the consonant belief function (see Section 2.8) whose contour function (2.18) $pl(\theta|x)$ is equal to the normalised likelihood:

$$pl(\theta|x) = \frac{L(\theta; x)}{\sup_{\theta' \in \Theta} L(\theta'; x)}. \tag{4.2}$$

The associated plausibility function is

$$Pl_\Theta(A|x) = \sup_{\theta \in A} pl(\theta|x) = \frac{\sup_{\theta \in A} L(\theta; x)}{\sup_{\theta \in \Theta} L(\theta; x)} \quad \forall A \subseteq \Theta,$$

while the corresponding multivalued mapping $\Gamma_x : \Omega \to 2^\Theta$ is:

$$\Gamma_x(\omega) = \left\{ \theta \in \Theta \middle| pl(\theta|x) \geq \omega \right\},$$

where $\Omega = [0, 1]$ and the source probability there is the uniform one.

Denoeux [460] has argued that the method can be extended to handle low-quality data (i.e., observations that are only partially relevant to the population of interest, and data acquired through an imperfect observation process). Indeed, an extension of the expectation-maximization (EM) algorithm, called the evidential EM (E2M) algorithm, has been proposed which maximises a generalised likelihood function which can handle uncertain data [458]. Zribi and Benjelloun [2137] have proposed an iterative algorithm to estimate belief functions based on maximum likelihood estimators in this generalised EM framework. An interesting discussion of the connection between belief functions and the likelihood principle can be found in [17].

Example 8: Bernoulli sample. *Let $x = (x_1, \ldots, x_n)$ consist of independent Bernoulli observations and $\theta \in \Theta = [0, 1]$ be the probability of success. A* Bernoulli trial *(or binomial trial) is a random experiment with exactly two possible outcomes, 'success' and 'failure', in which the probability of success is the same every time the experiment is conducted (repeated trials are independent). The contour function (4.2) of the belief function obtained by likelihood-based inference is*

$$pl(\theta|x) = \frac{\theta^y (1-\theta)^{n-y}}{\hat{\theta}^y (1-\hat{\theta})^{n-y}},$$

where $y = \sum_{i=1}^n x_i$ and $\hat{\theta}$ is the maximum likelihood estimate.

Robust Bayesian inference Wasserman [1915] noted that the mathematical structure of belief functions makes them suitable for generating classes of prior distributions to be used in robust Bayesian inference. In particular, the upper and lower bounds of the posterior probability of a (measurable) subset of the parameter space may be calculated directly in terms of upper and lower expectations.

Assume a parametric model $\{f(x|\theta), \theta \in \Theta\}$ (4.1), and that we want to apply Bayesian inference to derive a posterior distribution on Θ, given a prior P there. If a prior cannot be accurately specified, we might consider a credal set (an 'envelope', in Wasserman's terminology) Π and update each probability in the envelope using Bayes' rule, obtaining a new envelope Π_x conditioned on $x \in X$. We denote by $P_*(A) \doteq \inf_{P \in \Pi} P(A)$ and $P^*(A) \doteq \sup_{P \in \Pi} P(A)$ the lower and upper bounds on the prior induced by the envelope Π, and define $L_A(\theta) = L(\theta) I_A(\theta)$, where $L(\theta) = f(x|\theta)$ is the likelihood and I_A is the indicator function on A.

Proposition 24. *Let Π be the credal set associated with a belief function on Θ induced by a source probability space $(\Omega, \mathcal{B}(\Omega), \mu)$ via a multivalued mapping Γ. If $L(\theta)$ is bounded, then for any $A \in \mathcal{B}(\Theta)$*

$$\inf_{P_x \in \Pi_x} P_x(A) = \frac{E_*(L_A)}{E_*(L_A) + E^*(L_{A^c})}, \quad \sup_{P_x \in \Pi_x} P_x(A) = \frac{E^*(L_A)}{E^*(L_A) + E_*(L_{A^c})},$$

where $E^*(f) = \int f^*(\omega)\mu(d\omega)$, $E_*(f) = \int f_*(\omega)\mu(d\omega)$ *for any real function* f *on* Θ, *and where* $f^*(\omega) \doteq \sup_{\theta \in \Gamma(\omega)} f(\theta)$ *and* $f_*(\omega) \doteq \inf_{\theta \in \Gamma(\omega)} f(\theta)$.

The integral representation of probability measures consistent with a belief function given by Dempster [414] can also be extended to infinite sets, as shown by Wasserman.

Proposition 25. ([1915], Theorem 2.1) *A probability measure* P *is consistent with a belief function* Bel *induced on an arbitrary frame of discernment* Θ *by a source probability space* $(\Omega, \mathcal{B}(\Omega), \mu)$ *via a multivalued mapping* Γ *if and only if there exists, for* μ*-almost all* $\omega \in \Omega$, *a probability measure* P_ω *on* $\mathcal{B}(\Theta)$ *supported by* $\Gamma(\omega)$ *such that*

$$P(A) = \int_\Omega P_\omega(A)\mu(d\omega)$$

for each $A \in \mathcal{B}(\Theta)$, *where* $\mathcal{B}(.)$ *is a* σ*-algebra on the argument domain.*

Dempster's auxiliary variable In Dempster's approach to inference, the parametric (sampling) model (4.1) is supplemented by the assumption that an *a-equation* holds, of the form

$$X = a(\Theta, U),$$

where U is an (unobserved) *auxiliary variable* with a known probability distribution μ independent of θ. This representation is quite common in the context of sampling and data generation. For instance, in order to generate a continuous random variable X with cumulative distribution function[26] F_θ, one might draw U from the uniform distribution $\mathcal{U}([0,1])$ on the interval $[0,1]$, and set

$$X = F_\theta^{-1}(U).$$

The equation $X = a(\Theta, U)$ defines a multivalued mapping (a.k.a. 'compatibility relation': compare Section 3.1.1) $\Gamma : U \to 2^{\mathbb{X} \times \Theta}$ as follows:

$$\Gamma : u \mapsto \Gamma(u) = \Big\{ (x, \theta) \in \mathbb{X} \times \Theta \Big| x = a(\theta, u) \Big\} \subset \mathbb{X} \times \Theta.$$

Under the usual measurability conditions (see Section 3.1.5), the probability space $(U, \mathcal{B}(U), \mu)$ and the multivalued mapping Γ induce a belief function $Bel_{\Theta \times \mathbb{X}}$ on $\mathbb{X} \times \Theta$.

Conditioning (by Dempster's rule) $Bel_{\Theta \times \mathbb{X}}$ on θ yields the desired (belief) sample distribution on $\mathbb{X}$, namely

$$Bel_{\mathbb{X}}(A|\theta) = \mu\{u : a(\theta, u) \in A\}, \quad A \subseteq \mathbb{X}.$$

[26]The *cumulative distribution function* (CDF) of a random variable X is defined as $F(x) \doteq P(X \leq x)$.

Conditioning it on $X = x$ yields instead a belief measure $Bel_\Theta(\cdot|x)$ on Θ (see Fig. 4.2)

$$Bel_\Theta(B|x) = \frac{\mu\{u : M_x(u) \subseteq B\}}{\mu\{u : M_x(u) \neq \emptyset\}}, \quad B \subseteq \Theta, \tag{4.3}$$

where $M_x(u) \doteq \{\theta : x = a(\theta, u)\}$.

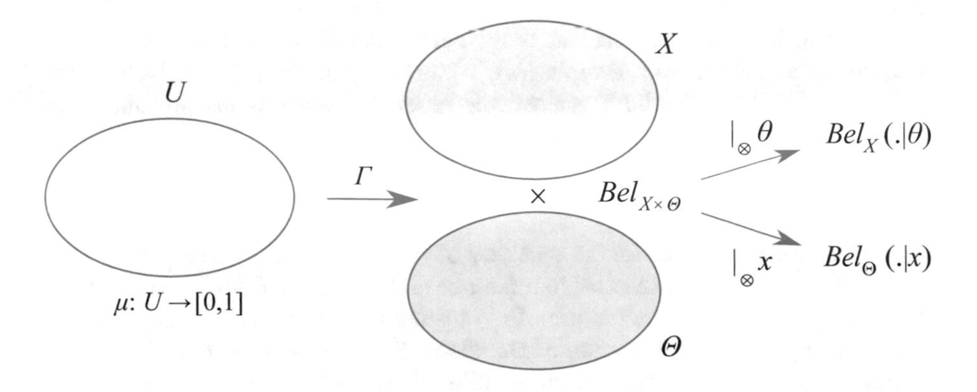

Fig. 4.2: Inference *à la Dempster* via an auxiliary variable U.

Example 9: Bernoulli sample. *In the same situation as Example 8, consider the sampling model*

$$X_i = \begin{cases} 1 & \text{if } U_i \leq \theta, \\ 0 & \text{otherwise,} \end{cases}$$

where $U = (U_1, \ldots, U_n)$ *has pivotal measure* $\mu = \mathcal{U}([0,1]^n)$*. After the number of successes* $y = \sum_{i=1}^n x_i$ *has been observed, the belief function* $Bel_\Theta(\cdot|x)$ *is induced by a random closed interval (see also Section 4.9.2)*

$$[U_{(y)}, U_{(y+1)}],$$

where $U_{(i)}$ *denotes the i-th-order statistics from* $U_1, \ldots, U_n$*. Quantities such as* $Bel_\Theta([a,b]|x)$ *or* $Pl_\Theta([a,b]|x)$ *can then be readily calculated.*

Dempster's model has several nice features: it allows us to quantify the uncertainty in Θ after observing the data *without having to specify a prior distribution* on Θ. In addition, whenever a Bayesian prior P_0 is available, combining it with $Bel_\Theta(\cdot|x)$ using Dempster's rule yields the Bayesian posterior: $Bel_\Theta(\cdot|x) \oplus P_0 = P(\cdot|x)$ (it is compatible with Bayesian inference). However, it often leads to cumbersome or even intractable calculations except for very simple models, which results in the need for the use of Monte Carlo simulations (see Section 4.7.3). More fundamentally, the analysis depends on the a-equation $X = a(\Theta, U)$ and the auxiliary variable U, which are not observable and not uniquely determined for a given statistical model $\{f(\cdot;\theta), \theta \in \Theta\}$.

An original, albeit rather neglected, publication by Almond [29] explored the issues involved in constructing belief functions via Dempster's model [29], with a focus on belief function models for Bernoulli and Poisson processes.

Dempster's (p, q, r) interpretation Dempster has recently come back to a statistical perspective on belief function theory [422], and statistical inference with these objects.

In [468], he proposed a semantics for belief functions whereby every formal assertion is associated with a triple (p, q, r), where p is the probability 'for' the assertion, q is the probability 'against' the assertion, and r is the probability of 'don't know'. In terms of belief values

$$p = Bel(A), \quad q = Bel(A^c), \quad r = 1 - p - q.$$

Arguments were presented for the necessity of $r = 1 - Bel(A) - Bel(A^c)$.

The methodology was applied to inference and prediction from Poisson counts, including an introduction to the use of a join-tree model structure to simplify and shorten computation. The relation of Dempster–Shafer theory to statistical significance testing was elaborated on, introducing along the way the concept of a *dull* null hypothesis, i.e., a hypothesis which assigns mass 1 to an interval of values $L_1 \leq L \leq L_2$. Poisson's a-probabilities and values for p, q and r can then be readily derived.

Fiducial inference Inference based on a-equations is an example of *fiducial inference*, introduced by Fisher in 1935 [621]. There, using 'pivotal' variables, Fisher moves from logical statements about restrictions of parameter spaces to probabilistic statements about parameters, subject to a number of caveats: for example, the pivotal variables must be sufficient statistics, and must be continuous.

Let Θ be the parameter of interest, X a sufficient statistic[27] rather than the observed data, and U the auxiliary variable (*pivotal quantity*). The crucial assumption underlying the fiducial argument is that each one of (X, Θ, U) is uniquely determined by the a-equation given the other two. The pivotal quantity U is assumed to have an a priori distribution μ, independent of Θ. Prior to the experiment, X has a sampling distribution that depends on Θ; after the experiment, however, X is no longer a random variable (as it is measured).

To produce a posterior distribution for Θ, the variability in X prior to the experiment must somehow be transferred, after the experiment, to Θ. This happens through the so-called *continue to believe* assumption. If we 'continue to believe' that U is distributed according to μ even after X has been observed, we can derive a distribution for Θ which we call the *fiducial distribution*.

Example 10: fiducial inference. *As an example, consider the problem of estimating the unknown mean of an $N(0, 1)$ population based on a single observation X. The*

[27] A statistic $t = T(S)$ is a function of the observed sample S. The statistic is sufficient if $P(s|t, \theta) = P(s|t)$.

a-equation is in this case $X = \Theta + \Psi^{-1}(U)$, *where* $\Psi(.)$ *is the CDF of the* $N(0,1)$ *distribution. Assume the pivotal quantity has prior distribution* $\mu = \mathcal{U}(0,1)$. *The crucial point is that, for a fixed* θ, *the events* $\{\Theta \leq \theta\}$ *and* $\{U \geq \Psi(X-\theta)\}$ *are the same. Hence, their probabilities need to be the same. If we 'continue to believe', then, the* fiducial probability *of* $\{\Theta \leq \theta\}$ *is* $\Psi(\theta - X)$. *In other words, the* fiducial distribution of Θ given X *is*

$$\Theta \sim N(X, 1).$$

We have then obtained a posterior for Θ, *without requiring a prior for it.*

The foundations of Dempster's inference approach within the fiducial inference framework were discussed by Almond in [30]. Dempster himself later revisited these ideas, claiming that the theory of belief functions moves between logical statements and probabilistic statements through subjective statements about the state of knowledge in the form of upper and lower probabilities.

Inferential models and weak belief In [1258], Martin and Liu presented a new framework for probabilistic statistical inference without priors, as an alternative to Fisher's fiducial inference, standard belief function theory and Bayesian inference with default priors, based on the notion of an *inferential model* (IM), and on ideas first published in [1259, 2104].

The IM approach attempts to accurately predict the value u^* of the auxiliary variable before conditioning on $X = x$. The benefit of focusing on u^* rather than on θ is that more information is available about the former: indeed, all that is known about θ is that it sits in Θ, while u^* is known to be a realisation of a draw U from an a priori distribution (in the above example, $N(0,1)$) that is fully specified by the postulated sampling model. For better prediction of u^*, Martin and Liu adopted a so-called 'predictive random set', which amounts to a sort of 'smearing' of this distribution for U. When combined with the multivalued mappings linking observed data, parameters and auxiliary variables, the overall random set produces prior-free, data-dependent probabilistic assessments of the uncertainty about θ.

Namely, given $Bel_\Theta(.|x)$, Dempster's posterior on Θ given the observable x (4.3), we say that another belief function Bel^* on Θ specifies an 'inferential model' there if

$$Bel^*(A) \leq Bel_\Theta(A) \quad \forall A \subseteq \Theta.$$

Weak belief (WB) is then a method for specifying a suitable belief function within an inferential model. This can be done, for example, by modifying the multivalued mapping Γ or the pivot distribution μ.

Whereas both the fiducial and the Dempster inference approaches uphold a 'continue to believe' assumption which states that u^* can be predicted by taking draws from the pivotal measure μ, WB weakens this assumption by replacing the draw $u \sim \mu$ with a set $S(u)$ containing u, which is in fact equivalent to replacing μ with a belief function. Given a pivotal measure μ, WB constructs a belief function on U by choosing a set-valued mapping $\mathcal{S} : U \to 2^U$ that satisfies $u \in \mathcal{S}(u)$. As this is a multivalued mapping from the domain of the auxiliary variable onto itself, the pivot distribution μ induces a belief function on its own domain.

The quantity $\mathcal{S}$ is called a *predictive random set*, a kind of 'smearing' of the pivot distribution μ for u which amounts to allowing uncertainty in the pivot variable itself. At this point we have three domains, $\mathbb{X}, U$ and Θ, and two multivalued mappings, $\Gamma : U \rightarrow 2^{\mathbb{X}\times\Theta}$ and $\mathcal{S} : U \rightarrow 2^U$. The two belief functions on U and $\mathbb{X}\times\Theta$ can be extended to $U\times\mathbb{X}\times\Theta$. By combining them there, and marginalising over U, we obtain a belief function whose random set is

$$\Gamma_{\mathcal{S}}(u) = \bigcup_{u'\in\mathcal{S}(u)} \Gamma(u'),$$

which is, by construction, dominated by the original Dempster's posterior: in Martin and Liu's terminology, it is a valid inferential model.

In [1259], various methods for building the mapping $\mathcal{S}$ were discussed, including 'maximal belief'. The authors of [2104] illustrated their weak belief approach with two examples: (i) inference about a binomial proportion, and (ii) inference about the number of outliers ($\mu_i \neq 0$) based on the observed data $x_1, \ldots, x_n$ under the model $X_i \sim \mathcal{N}(\mu_i, 1)$. Further extensions of the WB idea have been proposed by Martin [1259] and Ermini [1108], under the name 'elastic belief'.

Frequentist interpretations A number of attempts to reconcile belief theory with the frequentist interpretation of probability need also to be mentioned.

Walley and Fine's frequentist theory of lower and upper probability In an interesting even if not very recent paper, predating Walley's general theory of imprecise probability, Walley and Fine [1878] attempted to formulate a frequentist theory of upper and lower probability, considering models for independent repetitions of experiments described by interval probabilities, and suggesting generalisations of the usual concepts of independence and asymptotic behaviour.

Formally, the problem is to estimate a lower probability $\underline{P}$ from a series of observations $\epsilon_1, \ldots, \epsilon_n$. Walley and Fine consider the following estimator for $\underline{P}$:

$$\underline{r}_n(A) = \min\Big\{ r_j(A) : k(n) \leq j \leq n \Big\}, \quad k(n) \rightarrow \infty,$$

where $r_j(A)$ is the relative frequency of event A after ϵ_j. Let $\underline{P}^\infty$ describe the independent and identically distributed (i.i.d.) repetitions $\epsilon_1, \ldots$. One can prove that this estimation process succeeds, in the sense that

$$\lim_{n\rightarrow\infty} \frac{\underline{P}^\infty(G^c_{n,\delta})}{\underline{P}^\infty(G_{n,\delta})} = 0 \quad \forall\delta > 0,$$

where $G_{n,\delta}$ is the event $|\underline{r}_n(A) - \underline{P}(A)| < \delta$.

This result parallels Bernoulli's law of large numbers: the confidence that $\underline{r}_n(A)$ is close to $\underline{P}(A)$ grows with the sample's size. The opposite view was supported by Lemmers [1132], who used the traditional semantic model of probability theory to show that, if belief functions are based on reasonably accurate sampling or observation of a sample space, then the beliefs and upper probabilities as computed according to Dempster–Shafer theory cannot be interpreted as frequency ratios.

In a subsequent work by Fierens and Fine [612], credal sets, with their associated lower and upper envelopes, were assumed to represent the appropriate imprecise probabilistic description, without any reference to a 'true', unknown measure. These authors then asked themselves how envelopes might manifest themselves in the data, and proposed inference methods for such data to obtain credal models. They identified, in particular, long finite sequences that exhibit observably persistent oscillations of relative frequencies, occurring in a highly complex pattern of transition times, closely visiting given measures and no others.

A theory of confidence structures In recent work, Perrussel et al. [1420] attempted to reconcile belief function theory with frequentist statistics, using conditional random sets.

Formally, given a statistical model expressed as a family of conditional probability distributions $P_{x|\theta} : \mathcal{F}_x \to [0,1]$ of the values of an observable random variable $x \in \mathcal{X}$, given a parameter $\theta \in \Theta$, we wish to construct a belief function $Bel_{\theta|x} : 2^{\Theta} \to [0,1]$. An *observation-conditional random set* on the parameter θ is a mapping $\Gamma(\omega, x) \subset \Theta$ which depends also on the value of the observable x. The associated belief value is

$$Bel_{\theta|x}(A) = P_{\omega}(\{\omega \in \Omega : \Gamma(\omega, x) \subseteq A\})$$

for all $x \in \mathcal{X}$, $A \subset \Theta$. The problem then reduces to that of constructing an appropriate observation-conditional random set from the given statistical model.

A *confidence set* is a set-estimator for the parameter. It is a function of the observable

$$C : \mathcal{X} \to \mathcal{P}(\Theta), \quad C(x) \subset \Theta, \tag{4.4}$$

designed to cover the true parameter value with a specified regularity. Confidence intervals (see Section 1.3.7) are the most well-known type of confidence set. The *Neyman–Pearson confidence* associated with a confidence set is then the frequentist probability of drawing x such that $C(x)$ covers the true parameter value. A *confidence structure* is then an observation-conditional random set whose source probabilities P_{ω} are commensurate with the Neyman–Pearson confidence, namely

$$P_{x|\theta}\left(\left\{x \in \mathcal{X} : \theta \in \bigcup_{\omega \in A} \Gamma(\omega, x)\right\}\right) \geq P_{\omega}(A), \quad \forall A \subset \Omega, \theta \in \Theta. \tag{4.5}$$

In other words, a confidence structure is a means for constructing confidence sets of the following form:

$$C(A; x) = \bigcup_{\omega \in A} \Gamma(\omega, x), \quad A \subset \Omega,$$

subject to the condition that the coverage probability of $C(A; x)$ is greater than or equal to the source probability, $P_{\omega}(A)$ (4.5).

Since every $B \subset \Theta$ can be put in the form $B = C(A; x)$ for some $A \subset \Omega$, $x \in \mathcal{X}$, all sets of parameter values are confidence intervals. The belief value of such a set measures how much confidence is associated with that set, as it returns a lower bound on the coverage probability of the confidence set associated with $B \subset \Theta$ for a given observation x.

In [1420], various (inference) methods for constructing confidence structures from confidence distributions (Section 3.1), pivots (Section 3.2) and p-values (Section 3.3) were illustrated. Confidence structures on a group of input variables can be propagated through a function to obtain a valid confidence structure on the output of that function. In [1420], this method was compared and contrasted with other statistical inference methods.

From confidence intervals As we mentioned, confidence intervals can also be exploited to quantify beliefs about the realisation of a discrete random variable X with unknown probability distribution P_X. In [452], a solution which is less committed than P_X, and converges towards the latter in probability as the size of the sample tends to infinity, was proposed. Namely, each confidence interval can be thought of as a family of probability measures, specifically a set of (reachable) probability intervals (see Section 6.3). One can then apply the formulae (6.11) to obtain the lower and upper probabilities associated with a confidence interval.

Shafer's discussion of belief functions and parametric models Shafer himself [1589] illustrated back in 1982 three different ways of doing statistical inference in the belief framework (i.e., deriving a belief function on $\mathbb{X} \times \Theta$ which has the prior BF Bel_Θ as its marginal on Θ and $f(.|\theta)$ as its conditional given θ), according to the nature of the evidence which induces the available parametric model.

These are: (i) distinct and independent observations for each $f(.|\theta)$ (e.g. the symptoms x which arise from a medical condition θ); (ii) a single empirical frequency distribution (e.g. the parametric models induced by error distributions; see our discussion above on fiducial inference); and (iii) no evidence except the conviction that the phenomenon is purely random. As we have seen when discussing Dempster's inference, the above constraints do not uniquely identify a belief function on $\mathbb{X} \times \Theta$.

In case (i), Shafer argued in support of Smets's 'conditional embedding' approach of extending each $P_\theta = f(.|\theta)$ on $\mathbb{X} \times \Theta$ and combining them by Dempster's rule (see Section 4.6.1). Case (ii), since the parametric model arises from the probability distribution of the error, can be treated using fiducial inference. Case (iii) can draw inspiration from the Bayesian treatment given there, in which the random variable X is thought of as one of a sequence $\mathbf{X} = (X_1, X_2, \ldots)$ of unknown quantities taking values in $\mathbb{X} = \{1, \ldots, k\}$, and the Bayesian beliefs on $\mathbf{X}$ are expressed by a countably additive and symmetric probability distribution P. Additivity and symmetry imply that limiting frequencies exist, $P(\lim_{n\rightarrow\infty} f(x,n) \text{ exists}) = 1$, where $f(x,n)$ is the proportion of the quantities $X_1, \ldots, X_n$ that equal $x \in \mathbb{X}$, and that the probability distribution $P_\theta, \theta \in \Theta$ of the parameter can be recovered by conditioning P on these limiting frequencies.

Shafer thus argued in favour of constructing a belief function for the sequence $\mathbf{X} = (X_1, X_2, \ldots)$ that satisfies:

1. $Bel(\lim_{n\to\infty} f(x,n) \text{ exists}) = 1$ for all $x \in \mathbb{X}$.
2. $Bel(X_1 = x_1, \ldots, X_n = x_n | \lim_{n\to\infty}[f(1,n), \ldots, f(k,n)] = \theta)$ is equal to $P_\theta(x_1), \ldots, P_\theta(x_n)$ for all $x_1, \ldots, x_n \in \mathbb{X}$.
3. $Bel(\lim_{n\to\infty}[f(1,n), \ldots, f(k,n)] \in A) = 0 \; \forall A \subsetneq \Theta$.

Such belief functions exist, and can be constructed. Interestingly, in the case $\mathbb{X} = \{0,1\}$, the solution is unique and its marginal for (θ, X_1) is a BF on $\mathbb{X} \times \Theta$ which corresponds to Dempster's generalised Bayesian solution. In the general case $k > 2$, at least some solutions to the above constraints also have marginals on $\mathbb{X} \times \Theta$ which correspond to Dempster's solution, but it seems that others do not. No proofs of the above arguments were given in [1589].

Shafer concluded by pointing out how the strength of belief calculus is really about allowing inference under partial knowledge or ignorance, when simple parametric models are not available.

Other statistical approaches Other people have worked on a statistical approach to belief function inference [1840].

An early publication by Fua [652], for instance, proposed a combination of the methods of Strat and Shafer to exploit continuous statistical information in the Dempster–Shafer framework, based on deriving continuous possibility and mass functions from PDFs. Among other relevant work, a belief function generalisation of Gibbs ensembles was proposed by Kong [1034]. Srivastava and Shafer [1759] discussed the integration of statistical evidence from attribute sampling with non-statistical evidence within the belief function framework. They also showed how to determine the sample size required in attribute sampling to obtain a desired level of belief that the true attribute occurrence rate of the population lies in a given interval, and what level of belief is obtained for a specified interval given the sample result. Edlefsen et al. [563] presented a Dempster–Shafer approach to estimating limits from Poisson counting data with nuisance parameters, by deriving a posterior belief function for the 'Banff upper limits challenge' three-Poisson model. Novelty detection, i.e., the problem of testing whether an observation may be deemed to correspond to a given model, was studied in the belief function framework by Aregui and Denoeux [54].

4.1.2 From qualitative data

A number of studies have been published on inference from qualitative and uncertain data [678] as well. Among them are Wong and Lingras's perceptron idea, Bryson et al.'s qualitative discrimination process and Ben Yaghlane's constrained optimisation framework. Other proposals include Bryson et al.'s [1355, 197] approach to the generation of quantitative belief functions, which includes linguistic quantifiers to avoid the premature use of numeric measures.

Wong and Lingras's perceptron model In particular, Wong and Lingras [1958, 1953] proposed a method for generating belief functions from a body of qualitative *preference relations* between propositions. Preferences are not needed for all pairs of propositions.

In this setting expert opinions are expressed through two binary relations: preference, $\cdot >$, and indifference, $\sim$. The goal is to build a belief function Bel such that $A \cdot > B$ iff $Bel(A) > Bel(B)$ and $A \sim B$ iff $Bel(A) = Bel(B)$. Wong and Lingras proved that such a belief function exists if $\cdot >$ is a 'weak order' and $\sim$ is an equivalence relation. Their algorithm can be summarised as shown in Algorithm 1.

Algorithm 1 Wong and Lingras

---

procedure BFFROMPREFERENCES($\cdot >, \sim$)

Consider all propositions that appear in the preference relations as potential focal elements;

elimination step: if $A \sim B$ for some $B \subset A$ then A is not a focal element;

a perceptron algorithm is used to generate the mass m by solving the system of remaining equalities and disequalities.

end procedure

A negative feature of this approach is that it arbitrarily selects one solution out of the many admissible ones. Also, it does not address possible inconsistencies in the given body of expert preferences.

Qualitative discrimination processes Osei-Bryson and co-authors [198, 1355, 197, 1370] later proposed a different approach which implicitly (through qualitative assignments) and explicitly (through vague interval pairwise comparisons) provides for different levels of preference relationships. This allows an expert to use linguistic quantifiers, avoiding the premature use of numeric measures, and to identify input data that are inconsistent with the theory of belief functions.

There, the expert assigns propositions first to a 'broad' category bucket, then to a corresponding 'intermediate' bucket, and finally to a 'narrow' category bucket. Such a qualitative scoring table is used to identify and remove non-focal propositions by determining if the expert is indifferent regarding any propositions and their subsets in the same or a lower narrow bucket (as in Wong and Lingras's elimination step). Next ('imprecise pairwise comparisons'), the expert is required to provide numeric intervals to express their beliefs about the relative truthfulness of the propositions. The consistency of the above information is checked, and a mass interval is provided for every focal element. Finally, the expert re-examines the results, and restarts the process if they think this is appropriate.

Ben Yaghlane's constrained optimisation To address the issues in Wong and Lingras's method, Ben Yaghlane et al. proposed an approach which uses preference and indifference relations which obey the same axioms, but converts them into a constrained optimisation problem [111].

The objective is to maximise the entropy/uncertainty of the belief function to be generated (in order to select the least informative one), under constraints derived from the input preferences/indifferences, in the following way:

$$A \cdot > B \leftrightarrow Bel(A) - Bel(B) \geq \epsilon, \quad A \sim B \leftrightarrow |Bel(A) - Bel(B)| \leq \epsilon.$$

Here, ϵ is a constant specified by the expert. Various uncertainty measures can be plugged into this framework (see Section 4.2). Ben Yaghlane et al. proposed various mono- and multi-objective optimisation problems based on this principle.

Other work In a variation on the theme of inference from preferences, Ennaceur et al. [575] supported an approach which constructs appropriate quantitative information from preference relations, despite incompleteness and incomparability in the preference orderings. Compared with existing methods based on strict preferences and indifferences (see above), those authors' model was claimed to be able to provide additional interpretation values.

Yang [2048] claimed that existing methods have paid little attention to the bounded rationality of the expert from which a belief function is built, and introduced a confidence belief function generated from a set of BFs sampled from an expert by interacting with a number of reliable information providers in a valid time interval. The method involves three steps: (1) dividing the set of belief functions into a number of non-conflicting or consistent subgroups, (2) forming the confidence-interval-valued belief structures (IBSs; see Section 3.3.7) of the groups and (3) integrating these IBSs into a confidence IBS for the entire set.

4.1.3 From partial knowledge

A series of interesting papers have investigated a slightly different problem, namely that of constructing a belief function from *partial knowledge* about the probabilities involved, in particular in the transferable-belief-model framework (recall Section 3.3.2).

Inference from partial knowledge in the TBM In [1695], Smets assumed that we know that a random process is governed by a distribution which belongs to a set $\mathcal{P}$ of probabilities, and asked what belief values follow from this knowledge. This setting is very similar to Wasserman's robust Bayesian inference.

In [55], Aregui and Denoeux proposed as a solution the pl-most committed BPA (see Definition 43) in the set of BPAs that are less committed than all probability measures in the given set $\mathcal{P}$. In [58], instead, the solution is the lower envelope of a set of pignistic probabilities $\mathcal{P}$, in the case in which the latter set is a belief function. This is the case, in particular, when the set $\mathcal{P}$ is a p-box (see Section 6.8), or when it is constructed from a multinomial confidence region. The most committed consonant belief function possessing this property is finally selected. In [56], this work was extended to the case of continuous random variables, based on step or continuous confidence bands. Caron et al. [230] considered the case in which partial knowledge is encoded by an n-dimensional Gaussian betting density function, and provided an explicit formulation of the least committed basic belief density, again in the TBM framework. Finally, Doré et al. [508] reviewed various principles for eliciting belief functions from probabilities, in particular the pignistic principle, and evaluated their impact on pattern recognition.

Building complete belief functions from partial ones An interesting, related question is how we obtain a full belief function from a partially specified one.

Hsia [846] proposed in response to adopt the principle of minimum commitment (a stronger version of the minimum-specificity principle): he did not, however, consider the issue of how to effectively calculate the resulting belief function. Lemmer and Kyburg [1134] studied the situation in which what we know are the belief-plausibility intervals for singletons, and proposed an algorithm whose result, however, is not the least informative belief function in general. In a more recent publication, Moral and Campos [1314] presented a procedure based on the principle of minimum commitment, and a new axiom which they called the 'focusing' principle. The resulting procedure was compared with other methods based on the minimum-specificity [529] and least-commitment principles.

Inference in machine learning A number of papers have been devoted to the learning of belief functions for machine learning purposes, especially clustering and classification.

A method based on the distance between the sample data under test and the model of attributes of species was proposed in [900], with a focus on fuzzy environments [901]. Bentabet et al. [127] proposed a fuzzy c-means clustering approach (see Section 5.1.1) to mass function generation from image grey levels in an image-processing context. Matsuyama [1270] suggested a way of computing a basic probability assignment from similarity measures between observed data and object categories for classification.

Finally, a very original inference approach based on regressing belief functions via a support vector machine (SVM) in the context of statistical learning theory has been proposed in [1089]. Fuzzy membership functions and belief functions were treated there as constraints of the SVM convex optimisation problem, creating a fuzzy SVM and a belief SVM classifier.

4.1.4 A coin toss example

To understand the difference between Bayesian, frequentist and belief function statistical inference, let us consider a simple coin toss experiment.

We toss the coin $n = 10$ times, obtaining the sample

$$X = \{H, H, T, H, T, H, T, H, H, H\}$$

with $k = 7$ successes (heads, H) and $n - k = 3$ failures (tails, T). The parameter of interest which governs the process is the probability $\theta = p = P(H)$ of heads in a single toss. The inference problem then consists in gathering information about the value of p, in the form of either a point estimate, the acceptability of certain guesses about its value or a probability distribution of the possible values of p.

General Bayesian inference In the Bayesian setting, the trials are typically assumed to be independent (as they are obviously equally distributed). The likelihood of the sample X thus follows a binomial law,[28]

$$P(X|p) = p^k(1-p)^{n-k}. \tag{4.6}$$

We can then apply Bayes' rule to get the following posterior (see Fig. 4.3(a)):

$$P(p|X) = \frac{P(X|p)P(p)}{P(X)} \sim P(X|p), \tag{4.7}$$

the last part holding whenever we do not have a priori information about the prior distributions of p and X. The maximum likelihood estimate for $\theta = p$ is simply the peak of the likelihood function (4.7).

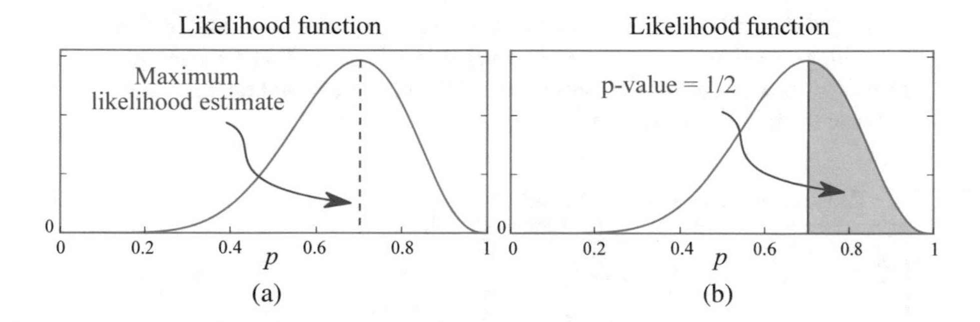

Fig. 4.3: (a) Bayesian inference applied to the coin toss experiment yields a posterior distribution (4.7) which tracks the likelihood of the sample, in the absence of prior information. The MLE of the parameter p is shown. (b) Frequentist hypothesis testing compares the area under the conditional distribution function to the right of the hypothesised value (in this case $p = 0.7$) with the assumed confidence level.

Frequentist hypothesis testing What would a frequentist do? Well, it seems reasonable that the value of p should be k/n, i.e., the fraction of successes observed for this particular sample. We can then test this hypothesis in the classical frequentist setting (Chapter 1, Section 1.2.3). Once again, this implies assuming independent and equally distributed trials, so that the conditional distribution of the sample is the binomial distribution (4.6). We can then compute the p-value for, say, a confidence level of $\alpha = 0.05$.

As depicted in Fig. 4.3(b), the right-tail p-value for the hypothesis $p = k/n$ (the integral area in pink) is equal to $\frac{1}{2}$, which is much greater than $\alpha = 0.05$. Hence, the hypothesis cannot be rejected.

[28] https://en.wikipedia.org/wiki/Binomial_distribution.

Likelihood-based belief function inference Applying likelihood-based belief function inference (see Section 4.1.1) to the problem yields the following belief measure, conditioned on the observed sample X, over the frame $\Theta = [0,1]$ of possible values for the parameter $\theta = p$:

$$Pl_\Theta(A|X) = \sup_{p \in A} \hat{L}(p|X), \quad Bel_\Theta(A|X) = 1 - Pl_\Theta(A|X),$$

where $\hat{L}(p|X)$ is the normalised version of (4.6), for every $A \subseteq \Theta$.

The random set associated with this belief measure, the one induced by the likelihood (4.6), is depicted in Fig. 4.4(a). Every normalised likelihood value $\omega \in \Omega = [0,1]$ is mapped to the following subset of $\Theta = [0,1]$:

$$\Gamma_X(\omega) = \left\{ \theta \in \Theta \middle| Pl_\Theta(\{\theta\}|X) \geq \omega \right\} = \left\{ \theta \in \Theta \middle| \hat{L}(p|X) \geq \omega \right\},$$

which is an interval centred around the MLE of p (purple in the figure). As we know, this random set determines an entire convex envelope of PDFs on the parameter space $\Theta = [0,1]$, a description more general than that provided by general Bayesian inference (a single PDF over Θ).

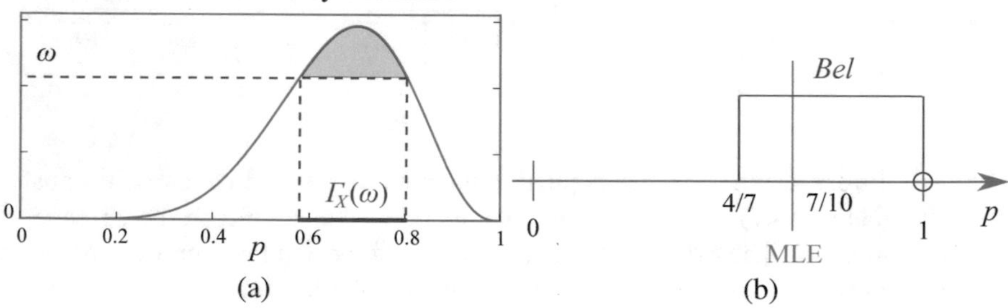

Fig. 4.4: (a) The random set induced by the binomial likelihood function in the coin toss example. (b) Applying the procedure to the normalised counts produces instead a belief function on $\Omega = \{H,T\}$, which corresponds to the interval of parameter values $1/3 \leq p < 1$ depicted.

The same procedure can applied to the normalised empirical counts $\hat{f}(H) = \frac{7}{7} = 1$, $\hat{f}(T) = \frac{3}{7}$, rather than to the normalised likelihood function. We can do that by imposing the condition $Pl_\Omega(H) = 1$, $Pl_\Omega(T) = \frac{3}{7}$ on $\Omega = \{H,T\}$, and looking for the least committed belief function there with these plausibility values. In this second case, we get the mass assignment

$$m(H) = \frac{4}{7}, \quad m(T) = 0, \quad m(\Omega) = \frac{3}{7},$$

which corresponds to the credal set $Bel = \{\frac{4}{7} \leq p < 1\}$ depicted in Fig. 4.4. Note that $p = 1$ needs to be excluded, as the available sample evidence reports that we have had $n(T) = 3$ counts already, so that $1 - p \neq 0$. This outcome (a belief function on $\Omega = \{H, T\}$) 'robustifies' the classical MLE.

In conclusion, this example shows that general Bayesian inference produces a continuous PDF on the parameter space $\Theta = [0, 1]$ (i.e., a second-order distribution), while MLE/MAP estimation delivers a single parameter value, corresponding to a single PDF on $\Omega = \{H, T\}$. In opposition, generalised maximum likelihood outputs a belief function on Ω (i.e., a convex set of PDFs there), which generalises the MAP/maximum likelihood estimates. Both likelihood-based and Dempster-based belief function inference, however, produce a BF on the parameter space Θ – this amounts to an entire convex set of second-order distributions, in an approach which is inherently more cautious than general Bayesian inference.

In Chapter 17, we will propose an entirely different lower- and upper- likelihood approach to statistical inference, which generates an interval of belief functions rather than a single belief measure, when presented with a series of repeated trials.

4.2 Measuring uncertainty

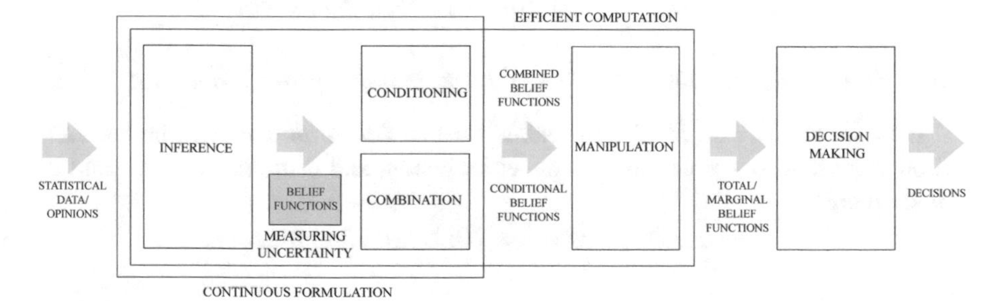

The question of how to measure the level of uncertainty associated with a belief function is a tricky one, as authors such as Yager [2001] and Klir [980] have argued that there are several characteristics of uncertainty such as conflict (or discord or dissonance) and non-specificity (also called vagueness, ambiguity or imprecision). Different proposals thus differ in the way they tackle these various aspects. Two main approaches can nevertheless be identified: one based on establishing partial order relations between belief measures, in terms of the information they contain, and another based on generalising to belief functions the classical notion of entropy.

4.2.1 Order relations

Let m_1 and m_2 be two mass functions on Θ.

Definition 43. *The BPA* m_1 *is* pl-more committed *than* m_2 *(denoted by* $m_1 \sqsubseteq_{pl} m_2$*) if* $Pl_1(A) \leq Pl_2(A)$*,* $\forall A \subseteq \Theta$*.*

Weak inclusion (3.9), which we saw in Section 3.1.4 and can be denoted by $\sqsubseteq_{bel}$, is strictly related to pl-inclusion, as

$$Pl_1(A) \leq Pl_2(A) \; \forall A \subset \Theta \; \equiv \; Bel_2(A) \leq Bel_1(A) \; \forall A \subset \Theta,$$

i.e., $m_2 \sqsubseteq_{bel} m_1$.

Definition 44. *The BPA m_1 is* q-more committed *than m_2 (denoted by $m_1 \sqsubseteq_q m_2$) if $Q_1(A) \leq Q_2(A)$, $\forall A \subseteq \Theta$.*

Both pl- and q-orderings are extensions of set inclusion, as for categorical belief functions:

$$m_A \sqsubseteq_{pl} m \Leftrightarrow m_A \sqsubseteq_q m \Leftrightarrow A \subseteq B.$$

In both cases, the greatest element is the vacuous mass function on Θ.

A more general notion of partial order in the belief space is that of 'specialisation', which amounts to mass redistribution to subsets of the original focal elements.

Definition 45. *We say that m_1 is a specialisation of m_2, $m_1 \sqsubseteq_S m_2$, if m_1 can be obtained from m_2 by distributing the mass $m_2(B)$ of each focal element to arbitrary subsets of B:*

$$m_1(A) = \sum_{B \subseteq \Theta} S(A, B) m_2(B), \quad \forall A \subseteq \Theta,$$

where $S(A, B)$ denotes the proportion of the mass $m_2(B)$ transferred to $A \subseteq B$.

$S = [S(A, B) : A, B \subseteq \Theta]$ is called the *specialisation matrix*. The specialisation order is also a generalisation of set inclusion, and it implies both pl- and q-commitment:

$$m_1 \sqsubseteq_S m_2 \Rightarrow (m_1 \sqsubseteq_{pl} m_2) \wedge (m_1 \sqsubseteq_q m_2).$$

4.2.2 Measures of entropy

A rather complete review of most of the uncertainty measures for belief functions [1438] can be found in a very recent paper by Jirousek and Shenoy [909, 908]. An older review of the topic was presented by Pal et al. [1378].

The problem was tackled in the more general framework of discrete Choquet capacities by Marichal in [1242]. The case of credal sets was investigated by Abellán and Moral [7, 6, 5] and Abellán and Gomez [8]. A general discussion of the different facets of uncertainty, randomness, non-specificity and fuzziness, and the correct way to quantify them was given by Pal [1377]. A formal taxonomy of hybrid uncertainty representations was proposed by Joslyn and Rocha [931], whereas an original concept of 'credibility' of evidence was proposed by Yager [2015, 2012]. The problem of the uniqueness of the information measure in the theory of evidence was also investigated by Ramer [1458].

Generalisations of classical entropy Some of the proposed measures of uncertainty for belief functions are directly inspired by Shannon's entropy of probability measures [1616],

$$H_s[P] = -\sum_{x\in\Theta} P(x)\log P(x). \tag{4.8}$$

Nguyen's measure is a direct generalisation of the Shannon entropy in which probability values are replaced by mass values [1346],

$$H_n[m] = -\sum_{A\in\mathcal{F}} m(A)\log m(A), \tag{4.9}$$

where $\mathcal{F}$ is, as usual, the list of focal elements of m. In Yager's entropy [2001], probabilities are (partly) replaced by plausibilities:

$$H_y[m] = -\sum_{A\in\mathcal{F}} m(A)\log Pl(A). \tag{4.10}$$

The latter quantity is equal to 0 for consonant or consistent belief functions (such that $A_i \cap A_j \neq \emptyset$ for all focal elements), while it is maximal for belief functions whose focal elements are disjoint and have equal mass (including Bayesian ones).

Hohle's *measure of confusion* [832], supported by Dubois and Ramer [549], is the dual measure in which belief measures replace probabilities:

$$H_o[m] = -\sum_{A\in\mathcal{F}} m(A)\log Bel(A). \tag{4.11}$$

All these quantities, however, capture only the 'conflict' portion of uncertainty.

In a rather neglected publication, the notions of the *upper entropy* and *lower entropy* of a belief function, as a generalisation of the Shannon entropy, were introduced by Chau, Lingras and Wong in [251]. Whereas the upper entropy measures the amount of information conveyed by the evidence currently available, the lower entropy measures the maximum possible amount of information that could be obtained if further evidence became available. Maluf [1240], instead, examined an entropy measure as a monotonically decreasing function, symmetrical to the measure of dissonance.

Measures of specificity A number of measures have also been proposed to capture the *specificity* of belief measures, i.e., the degree of concentration of the mass assigned to focal elements.

A first measure of non-specificity was presented by Klir, and later extended by Dubois and Prade [543]:

$$H_d[m] = \sum_{A\in\mathcal{F}} m(A)\log|A|. \tag{4.12}$$

This can be considered as a generalisation of Hartley's entropy ($H = \log(|\Theta|)$[807]) to belief functions. This measure is also discussed in Chapter 7 of [1051] (where it is called $W = \sum m(A)|A|/|\Theta|$).

A more sophisticated such measure was proposed by Pal [1379]:

$$H_a[m] = \sum_{A \in \mathcal{F}} \frac{m(A)}{|A|},$$

which assesses the dispersion of the pieces of evidence generating a belief function, and shows clear connections to the pignistic function. Non-specificity was also investigated by Kramosil [1053]. A separate proposal, based on commonality, was made by Smets [1672]:

$$H_t = \sum_{A \in \mathcal{F}} \log \left(\frac{1}{Q(A)} \right). \quad (4.13)$$

More recently, [1667] Smarandache et al. reinterpreted some measures of uncertainty in the theory of belief functions and introduced a new measure of contradiction.

Composite measures *Composite* measures, such as Lamata and Moral's $H_l[m] = H_y[m] + H_d[m]$ [1090], are designed to capture both entropy and specificity.

Yager's entropy $H_y[m]$, however, was criticised by Klir and Ramer for expressing conflict as $A \cap B = \emptyset$ rather than $B \not\subseteq A$, while $H_d[m]$ (they argued) does not measure to what extent two focal elements disagree (i.e., the size of $A \cap B$). In response [994, 996, 989, 983], Klir and Ramer proposed a *global uncertainty measure*, defined as

$$H_k[m] = D[m] + H_d[m], \quad (4.14)$$

where

$$D(m) = - \sum_{A \in \mathcal{F}} m(A) \log \left[\sum_{B \in \mathcal{F}} m(B) \frac{|A \cap B|}{|B|} \right],$$

further modified by Klir and Parviz in [995]. Ramer and Klir focused in a later work [1460] on the mathematics of the above measure of discord, and of the total uncertainty measure defined as the sum of discord and non-specificity. A measure of strife was introduced in [1854].

Vejnarova [1853] argued, however, that neither the measure of discord nor that of total uncertainty proposed by Klir satisfies the subadditivity requirement, generally taken as necessary for a meaningful measure of information. Pal et al. [1379] also argued that none of these composite measures is really satisfactory, as they do not admit a unique maximum. There is no sound rationale for simply adding conflict and non-specificity measures together, and, finally, some are computationally very expensive. They proposed instead the following:

$$H_p = \sum_{A \in \mathcal{F}} m(A) \log \left(\frac{|A|}{m(A)} \right). \quad (4.15)$$

Abellán and Masegosa [2] also recently analysed and extended the requirements for total uncertainty measures in belief theory. Abellán [1] later proposed a total uncertainty measure which combines two different measures of non-specificity, as it possesses a number of desirable properties.

Credal measures In the credal interpretation of belief functions, Harmanec and Klir's *aggregated uncertainty* (AU) [801, 794] is defined as the maximum Shannon entropy of all the probabilities consistent with the given belief function:

$$H_h[m] = \max_{P \in \mathcal{P}[Bel]} \{H_s(P)\}. \tag{4.16}$$

As these authors proved [801], $H_h[m]$ is the minimal measure meeting a set of rationality requirements, which include symmetry, continuity, expansibility, subadditivity, additivity, monotonicity and normalisation. An algorithm for computing the AU was presented in [805]. An improved algorithm was given by Wierman in [1932]. The AU (4.16) itself was, once again, criticised by Klir and Smith for being insensitive to arguably significant changes in evidence, and replaced by a linear combination of AU and non-specificity $I(m)$.

Maeda and Ichihashi [1231] proposed a composite measure

$$H_i[m] = H_h[m] + H_d[m], \tag{4.17}$$

which incorporates non-specificity; the first component consists of the maximum entropy of the set of probability distributions consistent with m, and the second component is the generalised Hartley entropy (4.12) defined by Dubois and Prade [543].

Both H_h and H_i are still characterised by high computational complexity: in response, Jousselme et al. [936] put forward their *ambiguity measure* (AM), as the classical entropy of the pignistic function:

$$H_j[m] = H_s[BetP[m]]. \tag{4.18}$$

Note that in [992], however, Klir argued that the attempt to demonstrate that the AM qualifies as a measure of total aggregated uncertainty was flawed, owing to a particular error in the proof of one of the principal theorems in the paper.

A summary In 2016, Jirousek and Shenoy [909, 908] analysed all these proposals, assessing them versus the following desirable properties:

1. *Non-negativity*: $H[m] \geq 0$, with equality iff $m(x) = 1$ for some $x \in \Theta$.
2. *Maximum entropy* [3]: $H[m] \leq H[m_\Theta]$, where m_Θ is the vacuous mass.
3. *Monotonicity*: if $|\Theta_1| \leq |\Theta_2|$, then $H[m_{\Theta_1}] \leq H[m_{\Theta_2}]$ (similarly to Shannon's Axiom 2).
4. *Probability consistency*: if $m = p$ is Bayesian, then $H[m] = H_s[p]$ (the entropy we use for assessing belief measures should be a generalisation of the Shannon entropy for probability measures).
5. *Additivity*: $H[m_1 \oplus m_2] = H[m_1] + H[m_2]$, where $\oplus$ denotes the Dempster sum.

The last assumption (similar to the compound axiom for Shannon's entropy) is, in this author's view, rather arguable. Nevertheless, it is satisfied by all the above proposals. According to [908], however, only the Maeda–Ichihashi proposal (4.17) possesses all these properties.

Fei et al. [594] argued that their 'Deng entropy',

$$H_e[m] = \sum_{A \in \mathcal{F}} m(A) \log \left(\frac{2^{|A|} - 1}{m(A)} \right), \tag{4.19}$$

also a generalisation of the Shannon entropy, has a more rational empirical behaviour.

We will need to resort to a generalised measure of entropy for belief functions when discussing our proposals on random-set random forest classifiers (Chapter 17) and generalised maximum entropy models (Section 17.3.2).

4.2.3 Principles of uncertainty

The *maximum entropy principle* (MEP) [888], due to Jaynes, has an important role in probability theory, as it allows one to express ignorance there by selecting as the probability distribution which best represents the current state of knowledge the one with the largest entropy. The maximum entropy principle generalises Laplace's *principle of insufficient reason* (PIR) [1096], which states that:

> "If we know that one of k possible alternatives will actually take place, but we have no idea whatsoever whether any of these alternatives is more likely than any other one, then we should assign each of the alternatives probability $1/k$."

The principle of insufficient reason is the origin of the argument about the correct way of representing ignorance in probability that we covered in the Introduction.

A number of principles of uncertainty, some of them related to maximum entropy, apply to belief measures, and are discussed below. An interesting paper by Klir [984] reviewed and discussed the meaning and utility of various principles of uncertainty, which apply to several uncertainty theories.

Least commitment The *least commitment principle* has been extensively employed for inference purposes, and rests on the notion of partial (informational) ordering between belief measures (Section 4.2.1).

Definition 46. *The* least commitment principle *states that, when several belief functions are compatible with a set of constraints, the least informative (according to some informational ordering) should be selected, if such a belief function exists.*

Both maximum entropy and least commitment are examples of *principles of maximum uncertainty* [984].

Minimum specificity The principle of *minimum specificity* was proposed by Dubois and Prade [529] as a basis for evidential reasoning, in order to select, from among bodies of evidence, the one which suitably represents the available information.

The imprecision of a focal element A is expressed by its cardinality $|A|$ or, more generally, by any increasing function $f(|A|)$ of it. If the function f is decreasing, $f(|A|)$ can be called a *measure of specificity* of the focal element. The latter extends to whole belief measures as

$$f(Bel) = \sum_{A \subseteq \Theta} m(A) f(|A|). \tag{4.20}$$

Minimum specificity then prescribes that one should select, from among a given collection of belief measures, the one such that (4.20) is minimal.

This principle leads to a rule of combination which does not presuppose any independence assumption [529].

Minimum description length The maximum entropy principle is closely related to the *minimum description length* (MDL) principle [733], i.e., the notion of describing the data using fewer symbols than needed to encode it literally. The MDL principle is a formalisation of Occam's razor in which the best hypothesis (a model and its parameters) for a given set of data is the one that leads to the best compression of the data, and was introduced by Jorma Rissanen in 1978 [1489].

4.3 Combination

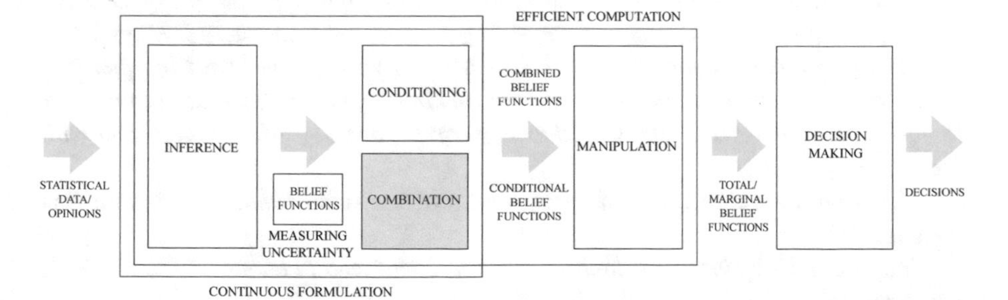

The question of how to update or revise the state of belief represented by a belief function when new evidence becomes available is central in the theory of evidence. In Bayesian reasoning, this role is performed by Bayes' rule, as evidence comes in the form of a single conditioning event. In the theory of belief functions, instead, as we have seen, new uncertain evidence is assumed to be encoded by an additional belief function, rather than by a simple event. The issue of combining the belief function representing our current knowledge state with the new one encoding the new evidence arises.

After an initial proposal by Arthur Dempster (Definition 7), several other aggregation operators have been proposed, based on different assumptions about the nature and properties of the sources of evidence which determine the belief functions to be combined, leaving the matter still far from settled. In particular, a hierarchy of operators, from the least to the most conservative, can be identified.

4.3.1 Dempster's rule under fire

Dempster's rule [696] is not really given a convincing justification in Shafer's seminal book [1583], leaving the reader wondering whether a different rule of combination could be chosen instead [1606, 2089, 584, 371, 1780, 1718]. This question has been posed by several authors (e.g. [1860, 2089, 1595] and [1941], among others), most of whom tried to provide axiomatic support for the choice of this mechanism for combining evidence. Smets, for instance, tried [1719] to formalise the concept of distinct evidence that is combined by an orthogonal sum. Walley [1873], instead, characterised the classes of belief and commonality functions for which independent statistical observations can be combined by Dempster's rule, and those for which the latter is consistent with Bayes' rule. The two were compared by Cinicioglu and Shenoy in [287].

Early on, Seidenfeld [1571] objected to the rule of combination in the context of statistical evidence, suggesting it was inferior to conditioning.

Example 11: Zadeh's counter-example. *Most famously, Zadeh [2085] formulated an annoying example for which Dempster's rule seems to produce counter-intuitive results. Since then, many authors have used Zadeh's example either to criticise Dempster–Shafer theory as a whole or as a motivation for constructing alternative combination rules. In the literature, Zadeh's example appears in different but essentially equivalent versions of disagreeing experts: we report here Haenni's version [766]. Suppose a doctor uses* $\Theta = \{M, C, T\}$ *to reason about the possible condition of a patient (where* M *stands for meningitis,* C *for concussion and* T *for tumour). The doctor consults two other experts* E_1 *and* E_2*, who provide him with the following answers:*

E_1: *'I am 99% sure it's meningitis, but there is a small chance of 1% that it is concussion'.*

E_2: *'I am 99% sure it's a tumour, but there is a small chance of 1% that it is concussion'.*

These two statements can be encoded by the mass functions (see Fig. 4.5)

$$m_1(A) = \begin{cases} 0.99 & A = \{M\}, \\ 0.01 & A = \{C\}, \\ 0 & \textit{otherwise}, \end{cases} \qquad m_2(A) = \begin{cases} 0.99 & A = \{T\}, \\ 0.01 & A = \{C\}, \\ 0 & \textit{otherwise}, \end{cases} \tag{4.21}$$

whose (unnormalised) Dempster combination is

$$m(A) = \begin{cases} 0.9999 & A = \{\emptyset\}, \\ 0.0001 & A = \{C\}, \\ 0 & \textit{otherwise}. \end{cases}$$

As the two masses are highly conflicting, normalisation yields a categorical be-

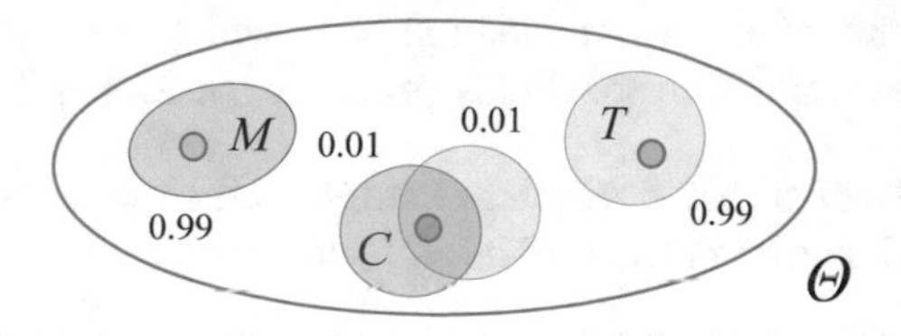

Fig. 4.5: The two belief functions m_1 (in light purple) and m_2 (in light blue) in Zadeh's classical 'paradox'.

lief function focused on $\{C\}$ – a strong statement that it is definitively concussion, although both experts had left it as only a fringe possibility.

Zadeh's dilemma was discussed in 1983 by Yager [2002], who suggested a solution based on the inclusion of a 'hedging' element. In [766] Haenni showed, however, that the counter-intuition in Zadeh's example is not a problem with Dempster's rule, but a problem with Zadeh's own model, which Haenni claimed does not correspond to reality.

First of all, the mass functions in (4.21) are Bayesian (i.e., probability measures): thus, Bayesian reasoning leads to the very same conclusions. The example would then lead us to reject Bayes' rule as well. Secondly, diseases are never exclusive, so that it may be argued that Zadeh's choice of a frame of discernment is misleading and is the root of the apparent 'paradox'.

Finally, experts are never fully reliable. In the example, they disagree so much that any person would conclude that one of the them is just wrong. This can be addressed by introducing two product frames

$$\Theta_1 = \{R_1, U_1\} \times \Theta, \quad \Theta_2 = \{R_2, U_2\} \times \Theta$$

and by *discounting* the reliability of each expert prior to combining their views (i.e., by assigning a certain probability $P(R_i)$ to their being reliable). The result of such a combination, followed by a marginalisation on the original frame Θ, adequately follows intuition.

A number of other authors have reasoned about similar discounting techniques, as we will see in Section 4.3.6.

Dubois and Prade's analysis In 1985, Dubois and Prade [525] had already pointed out that, when analysing the behaviour of Dempster's rule of combination, assessing a zero value or a very small value may lead to very different results. Theirs was also a criticism of the idea of having 'certain' evidence ('highly improbable is not impossible'), which led them to something similar to discounting.

In a 1986 note [526], the same authors proved the uniqueness of Dempster's rule under a certain independence assumption, while stressing the existence of alternative rules corresponding to different assumptions or different types of combination.

Eventually, in 1988, Dubois and Prade came to the conclusion that the justification for the pooling of evidence by Dempster's rule was problematic [532]. As a response, they proposed a new combination rule (which we will review in Section 4.3.2) based on their minimum specificity principle (Section 4.2.3).

Lemmer's counter-example A counter-example to the use of Dempster's rule was proposed in [1133], and can be described as follows.

Example 12: Lemmer's counter-example. *Imagine balls in an urn which have a single 'true' label. The set Θ of these labels (or, rather, the set of propositions expressing that a ball has a particular label from the set of these labels) functions as the frame of discernment. Belief functions are formed empirically on the basis of evidence acquired from observation processes, which are called 'sensors'. These sensors attribute to the balls labels which are subsets of Θ. The labelling of each sensor is assumed to be accurate in the sense that the frame label of a particular ball is consistent with the attributed label. Each sensor s gives rise to a BPA m_s in which $m_s(A)$, $A \subset \Theta$, is the fraction of balls labelled A by sensor s. Then, owing to the assumed accurateness of the sensors, $Bel(A)$ is the minimum fraction of balls with frame label $\theta \in A$, and $Pl(A)$ the maximum fraction of balls which could have as frame label an element of A.*

Lemmer's example shows that the Dempster combination of belief functions which derive from accurate labelling processes does not necessarily yield a belief function which assigns 'accurate' probability ranges to each proposition. Voorbraak [1860] countered that Dempster–Shafer theory is a generalisation of Bayesian probability theory, whereas Lemmer's sample space interpretation is a generalisation of the interpretation of classical probability theory.

Voorbraak's re-examination In [1860], Voorbraak also analysed randomly coded messages, Shafer's canonical examples for Dempster–Shafer theory, in order to clarify the requirements for using Dempster's rule. His conclusions were that the range of applicability of Dempster–Shafer theory was rather limited, and that in addition these requirements did not guarantee the validity of the rule, calling for some additional conditions. Nevertheless, the analysis was conducted under Shafer's constructive probability interpretation. Voorbraak provided his own 'counter-intuitive' result on the behaviour of Dempster's rule. Let $\Theta = \{a, b, c\}$, and let m, m' be two mass assignments such that $m(\{a\}) = m(\{b, c\}) = m'(\{a, b\}) = m'(\{c\}) = 0.5$. Therefore

$$m \oplus m'(\{a\}) = m \oplus m'(\{b\}) = m \oplus m'(\{c\}) = \frac{1}{3},$$

a result which is deemed counter-intuitive, for the evidence given to $\{a\}$ or $\{c\}$ is precisely assigned to it by at least one of the two belief functions, while that assigned to $\{b\}$ was never precisely given to it in the first place.

To the author of this book, this sounds like a confused argument in favour of taking into account the entire structure of the filter[29] of focal elements with a given intersection when assigning the mass to the focal elements of the combined belief function.

Axiomatic justifications In 1992, Klawonn and Schwecke [974] presented a set of axioms that uniquely determine Dempster's rule, and which reflect the intuitive idea of partially movable evidence masses. In a related paper [975], these authors examined the notion of a specialisation matrix (e.g., the matrix encoding the redistribution of mass from events to singletons; see Definition 45), and showed that Dempster's rule of conditioning corresponds essentially to the least committed specialisation, whereas Dempster's rule of combination results from commutativity requirements. A dual concept of generalisation was also described.

Liu [1199] supported a number of claims: (i) that the condition for combination in Dempster's original combination framework is stricter than that required by Dempster's combination rule in the standard form of the theory of evidence; (ii) that some counter-intuitive results of using Dempster's combination rule pointed out in some papers are caused by overlooking different independence conditions required by Dempster's original combination framework, and (iii) that in Dempster's combination rule, combinations are performed at the target information level.

A nice paper by Wilson [1943] took an axiomatic approach to the combination of belief functions similar to Klawonn's. The following requirements were formulated there:[30]

Definition 47. *A combination rule* $\pi : s \mapsto P^s : \Omega^s \to [0,1]$ *mapping a collection* $s = \{(\Omega_i, P_i, \Gamma_i), i \in I\}$ *of random sets to a probability distribution* P^s *on* $\Omega^s \doteq \times_{i \in I} \Omega_i$ *is said to* respect contradictions *if, for any finite such collection* s *of combinable random sets and* $\omega \in \Omega^s$, *whenever* $\Gamma(\omega) \doteq \bigcap_{i \in I} \Gamma_i(\omega_i) = \emptyset$ *we have that* $P^s(\omega) = 0$.

It is easy to see that any sensible combination rule must respect contradictions.

Definition 48. *A combination rule* π *is said to* respect zero probabilities *if, for any combinable multiple-source structure* s *and* $\omega = (\omega_i, i \in I) \in \Omega^s$, *if* $P_i(\omega_i) = 0$ *for some* $i \in I$, *then* $P^s(\omega) = 0$.

If $P_i(\omega_i) = 0$ for some i, then ω_i is considered impossible (since frames are finite). Therefore, since ω is the conjunction of the propositions ω_i, ω should clearly have zero probability.

In [1943], Dempster's rule was claimed to be the only combination rule respecting contradictions and zero probabilities, as long as two additional assumptions were valid. Unfortunately, no proof was given. A benefit of this approach, however, is that it makes the independence or irrelevance assumptions explicit.

---

[29] An *ultrafilter* is a maximal subset of a partially ordered set. A *filter* $\mathcal{F}$ is such that if $A \in \mathcal{F}$ and $A \subset B$, then $B \in \mathcal{F}$. An example is the collection $\{A \supseteq M\}$, the *principal filter* generated by M.

[30] Once again, the original statements have been translated into the more standard terminology used in this book.

Absorptive behaviour The issue is still open to debate, and some interesting points about the behaviour of Dempster's rule have recently been made. In 2012, Dezert, Tchamova et al. [498] challenged the validity of Dempster–Shafer theory by using an example derived from Zadeh's classical 'paradox' to show that Dempster's rule produces counter-intuitive results.

Example 13: Dezert and Tchamova's paradox. *This time, the two doctors generate the following mass assignments over $\Theta = \{M, C, T\}$:*

$$m_1(A) = \begin{cases} a & A = \{M\}, \\ 1-a & A = \{M, C\}, \\ 0 & \text{otherwise}, \end{cases} \quad m_2(A) = \begin{cases} b_1 & A = \{M, C\}, \\ b_2 & A = \Theta, \\ 1 - b_1 - b_2 & A = \{T\}. \end{cases} \tag{4.22}$$

Assuming equal reliability of the two doctors, Dempster combination yields $m_1 \oplus$

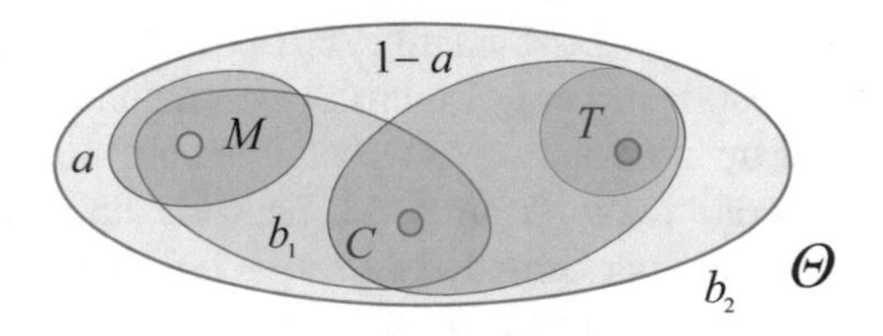

Fig. 4.6: The two belief functions in Dezert and Tchamova's example.

$m_2 = m_1$, i.e., Doctor 2's diagnosis is completely absorbed by that of Doctor 1 ('does not matter at all', in the authors' terminology) [1802] (see Fig. 4.6).

The interesting feature of this example is that the 'paradoxical' behaviour is not a consequence of conflict (as in other counter-examples), but of the fact that, in Dempster combination, every source of evidence has a 'veto' power over the hypotheses it does not believe to be possible – in other words, evidence is combined only for 'compatible' events/propositions (see Chapter 3, Section 3.1.1). Mathematically, this translates into an 'absorptive' behaviour, whose theoretical extent needs to be better analysed in the future.

Criticisms by other authors We conclude this section by briefly summarising other contributions to the debate on Dempster's rule.

In 2012, Josang and Pope [923] analysed Dempster's rule from a statistical and frequentist perspective and proved, with the help of simple examples based on coloured balls, that Dempster's rule in fact represents a method for the serial combination of stochastic constraints rather than a method for cumulative fusion of belief functions, under the assumption that subjective beliefs are an extension of frequentist beliefs. Wang [1891] argued that considering probability functions as special cases of belief functions while using Dempster's rule for combining belief functions leads to an inconsistency. As a result, he rejected some fundamental postulates of the theory, and introduced a new approach for uncertainty management that shares many intuitive ideas with Dempster–Shafer theory while avoiding this problem.

Smets [1728] claimed that the reason which led some authors to reject Dempster–Shafer theory was an inappropriate use of Dempster's rule of combination. That paper discussed the roots of this mismanagement, two types of errors, and the correct solutions for both types within the transferable-belief-model interpretation.

Following a similar line of reasoning, Liu and Hong [1201] argued that Dempster's original idea of evidence combination is, in fact, richer than what has been formulated in the rule. They concluded that, by strictly following what Dempster suggested, no counter-intuitive results should arise when combining evidence. Bhattacharya [148] analysed instead the 'non-hierarchical' aggregation of belief functions, showing that the values of certain functions defined on a family of belief structures decrease when the latter are combined by Dempster's rule, including the width of the belief–plausibility interval.

A method of dispelling the 'absurdities' (probably paradoxes) of Dempster's rule of combination was proposed in [2072], based on having all experts make their decision on the same focused collection. In [809], Hau et al. demonstrated that the orthogonal sum is not robust when one is combining highly conflicting belief functions. It was also shown that Shafer's discounted belief functions suffer from this lack of robustness with respect to small perturbations in the discount factor. A modified version of Dempster's rule was proposed to remedy this difficulty. In [779], a concrete example of the use of the rule presented by Weichselberger and Pohlmann was discussed, showing how their approach has to be modified to yield an intuitively adequate result. Finally, the authors of [2065] described a model in which masses are represented as conditional granular distributions. By comparing this model with Zadeh's relational model, they showed how Zadeh's conjecture on combinability does not affect the applicability of Dempster's rule.

4.3.2 Alternative combination rules

Yager's proposals In [2009], Yager highlighted some concerns with Dempster's orthogonal sum inherent in the normalisation due to conflict, and introduced both a practical recipe for using Dempster's rule and an alternative combination rule.

The latter is based on the view that conflict is generated by non-reliable information sources. In response, the conflicting mass (denoted here by $m_\cap(\emptyset)$) is reassigned to the whole frame of discernment Θ:

$$m_Y(A) = \begin{cases} m_\cap(A) & \emptyset \neq A \subsetneq \Theta, \\ m_\cap(\Theta) + m_\cap(\emptyset) & A = \Theta, \end{cases} \tag{4.23}$$

where $m_\cap(A) \doteq \sum_{B \cap C = A} m_1(B) m_2(C)$ and m_1, m_2 are the masses to be combined. Note that earlier (in 1984) [2003] Yager had already proposed alternatives to Dempster's rule based on interpreting plausibility and belief as a special case of the compatibility of a linguistically quantified statement within a database consisting of an expert's fragmented opinion.

Separately [2006], Yager discussed a rule of inference called the 'entailment principle', and extended it to situations in which the knowledge is a mixture of possibilistic and probabilistic information, which he called Dempster–Shafer granules. He discussed the conjunction of these Dempster–Shafer granules and showed that Dempster's rule of combination is a special application of conjunction followed by a particular implementation of the entailment principle.

Dubois and Prade's minimum specificity rule The combination operator proposed by Dubois and Prade[31] [532] comes from applying the minimum specificity principle (Section 4.2.3) to cases in which the focal elements B, C of two input belief functions do not intersect.

This results in assigning their product mass to $B \cup C$, namely

$$m_D(A) = m_\cap(A) + \sum_{B\cup C=A, B\cap C=\emptyset} m_1(B) m_2(C). \tag{4.24}$$

Obviously, the resulting belief function dominates that generated by Yager's rule.

Smets's combination rules in the TBM Just as Dempster does, Smets also assumes that all sources to be combined are reliable – conflict is merely the result of an incorrectly defined frame of discernment.

Conjunctive rule Rather than normalising (as in Dempster's rule) or reassigning the conflicting mass $m_\cap(\emptyset)$ to other non-empty subsets (as in Yager's and Dubois's proposals), Smets's *conjunctive rule* leaves the conflicting mass with the empty set,

$$m_{\textcircled{\cap}}(A) = \begin{cases} m_\cap(A) & \emptyset \neq A \subseteq \Theta, \\ m_\cap(\emptyset) & A = \emptyset, \end{cases} \tag{4.25}$$

and thus is applicable to *unnormalised* belief functions (recall Section 3.3.2). As Lefevre et al. noted [1118], a similar idea was already present in [2009], in which a new hypothesis was introduced in the existing frame to receive the conflicting mass. The conjunctive rule of combination (under a different name) also appeared around the same time in a publication by Hau and Kashyap [808]. This rule was called 'combination by exclusion' by Yamada [2034].

Using the conjunctive rule (4.25) amounts to adhering to an 'open-world assumption' (see 3.3.2 again) in which the current frame of discernment only approximately describes the true set of possible outcomes (hypotheses).

Disjunctive rule In Dempster's original random-set idea, consensus between two sources is expressed by the intersection of the supported events. When the *union* of the supported propositions is taken to represent such a consensus instead, we obtain what Smets called the *disjunctive* rule of combination,

$$m_{\textcircled{\cup}}(A) = \sum_{B\cup C=A} m_1(B) m_2(C). \tag{4.26}$$

[31] We follow here the same notation as in [1118].

It is interesting to note that under disjunctive combination, $Bel_1 \circledcup Bel_2(A) = Bel_1(A) * Bel_2(A)$, i.e., the belief values of the input belief functions are simply multiplied. Such a 'dual' combination rule was mentioned by Kramosil as well in [1056].

The disjunctive rule was termed 'combination by union' by Yamada [2034]. A mixed conjunctive/disjunctive rule was proposed by Martin and Osswald in [1254].

Denoeux's cautious and bold rules

Cautious rule Another major alternative to Dempster's rule is the so-called *cautious rule* of combination [455, 451], based on Smets's canonical decomposition [1698] of non-dogmatic (i.e., such as $m(\Theta) \neq 0$) belief functions into (generalised) simple belief functions (see Section 2.4.4).

Definition 49. *Let m_1 and m_2 be two non-dogmatic basic probability assignments. Their combination using the* cautious *(conjunctive) rule is denoted as $m_1 \owedge m_2$, and is defined as the mass assignment with the following weight function:*

$$w_{1 \owedge 2}(A) = w_1(A) \wedge w_2(A) \doteq \min\{w_1(A), w_2(A)\}, \quad A \in 2^\Theta \setminus \{\Theta\}. \tag{4.27}$$

In [455, 451], Denoeux proved (Proposition 1) that the cautious combination of two belief functions is the w-least committed BF in the intersection $S_w(m_1) \cap S_w(m_2)$, where $S_x(m)$ is the set of belief functions x-less committed than m (where $x \in \{pl, q, s, w\}$ denotes the plausibility/belief, commonality, specialisation and weight-based orderings, respectively). Notably, the cautious operator is commutative, associative and *idempotent* – this latter property makes it suitable for combining belief functions induced by reliable but possibly overlapping bodies of evidence.

A normalised version of the cautious rule can also be defined by replacing the conjunctive combination with Dempster's rule [455]. A cautious conjunctive rule which differs from Denoeux's was supported by Destercke et al. [487]. These authors argued that when the sources of information are not independent, one must identify a cautious merging rule that adds a minimal amount of information to the inputs, according to the principle of minimum commitment. The resulting cautious merging rule is based on maximising the expected cardinality of the resulting belief function.

Bold rule A dual operator, the *bold disjunctive rule*, can also be introduced once we note that if m is an unnormalised BPA, its complement $\overline{m}$ (which satisfies $\overline{m}(A) = m(A^c)$ for all $A \subset \Theta$) is non-dogmatic and can thus be (canonically) decomposed as $\overline{m} = \circledcap_{A \subsetneq \Theta} m_A^{\overline{w}}$ (compare Section 2.4.4).

Let us introduce the notation $v(A^c) \doteq \overline{w}(A)$. We can then prove the following proposition.

Proposition 26. ([455], Proposition 10) *Any unnormalised belief function can be uniquely decomposed as the following disjunctive combination:*

$$m = \circledcup_{A \neq \emptyset} m_{A, v(A)}, \tag{4.28}$$

where $m_{A,v(A)}$ is the UBF assigning mass $v(A)$ to $\emptyset$ and $1 - v(A)$ to A.

Denoeux calls (4.28) the *canonical disjunctive decomposition* of m.

Let $\mathcal{G}_x(m)$ then be the set of basic probability assignments x-more committed than m. The bold combination corresponds to the most committed element in the intersection set $\mathcal{G}_v(m_1) \cap \mathcal{G}_v(m_2)$.

Definition 50. *Let m_1 and m_2 be two unnormalised*[32] *basic probability assignments. The v-most committed element in $\mathcal{G}_v(m_1) \cap \mathcal{G}_v(m_2)$ exists and is unique. It is defined by the following disjunctive weight function:*

$$v_{1 \circledvee 2}(A) = v_1(A) \wedge v_2(A), \quad A \in 2^\Theta \setminus \{\emptyset\}.$$

The bold combination of m_1 and m_2 is defined as

$$m_1 \circledvee m_2 = \textcircled{\cup}_{A \neq \emptyset} m_{A, v_1(A) \wedge v_2(A)}. \tag{4.29}$$

The fact that the bold disjunctive rule is only applicable to unnormalised belief functions is, however, a severe restriction, as admitted by Denoeux.

Pichon and Denoeux [1428] pointed out that the cautious and unnormalised Dempster's rules can be seen as the least committed members of families of combination rules based on triangular norms and uninorms, respectively. Yet another combination rule, called the cautious-adaptive rule, was proposed in [600], based on a generalised discounting procedure defined for separable BPAs, to be applied to the source correlation derived from the cautious rule.

Consensus operator From a completely different perspective, Josang introduced in [915] a *consensus* operator, and showed how it can be applied to dogmatic conflicting opinions, i.e., when the degree of conflict is very high, overcoming the shortcomings of Dempster's and other existing rules.

Let the *relative atomicity* of $A \subset \Theta$ with respect to $B \subset \Theta$ be defined as

$$a(A/B) = \frac{|A \cap B|}{|B|},$$

as a measure of how much of Bel is overlapped by A. Josang represents an agent's degrees of belief in an event A as a tuple (which he calls an *opinion*), defined as $o = (b(A) \doteq Bel(A), d(A) \doteq 1 - Pl(A), u(A) \doteq Pl(A) - Bel(A), a(A) \doteq a(A/\Theta))$.

Definition 51. *Let*

$$o_1 = (b_1(A), d_1(A), u_1(A), a_1(A)), \quad o_2 = (b_2(A), d_2(A), u_2(A), a_2(A))$$

be opinions held by two agents about the same proposition/event A. The consensus combination *$o_1 \oplus o_2$ is defined as*

[32] 'Subnormal', in Denoeux's terminology.

$$o_1 \oplus o_2 = \begin{cases} \left(\frac{b_1 u_2 + b_2 u_1}{\kappa}, \frac{d_1 u_2 + d_2 u_1}{\kappa}, \frac{u_1 u_2}{\kappa}, \frac{a_1 u_2 + a_2 u_1 - (a_1 + a_2) u_1 u_2}{u_1 + u_2 - 2 u_1 u_2} \right) & \kappa \neq 0, \\ \left(\frac{\gamma b_1 + b_2}{\gamma + 1}, \frac{\gamma d_1 + d_2}{\gamma + 1}, 0, \frac{\gamma a_1 + a_2}{\gamma + 1} \right) & \kappa = 0, \end{cases} \tag{4.30}$$

where $\kappa = u_1 + u_2 - u_1 u_2$ *and* $\gamma = u_2/u_1$.

The consensus operator is derived from the posterior combination of beta distributions, and is commutative and associative, besides satisfying

$$b_{o_1 \oplus o_2} + d_{o_1 \oplus o_2} + u_{o_1 \oplus o_2} = 1.$$

Clearly, (4.30) is a combination operator which acts on lower and upper probabilities, rather than on belief functions per se.

A 'cumulative' rule and an 'averaging' rule of belief fusion were also presented by Josang et al. in [920]. These represent generalisations of the consensus operator to independent and dependent opinions, respectively, and are applicable to the combination of general basic probability assignments. Josang et al. argued that these rules can be derived directly from classical statistical theory, and produce results in line with human intuition. In particular, the cumulative rule is equivalent to a posteriori updating of Dirichlet distributions, while the averaging rule is equivalent to averaging the evidence provided by the latter. Both are based on a bijective mapping between Dirichlet distributions and belief functions, described in [920].

Averaging and distance-based methods A number of proposals for evidence combination are based on some way of computing the 'mean' of the input mass functions. Such an averaging method was proposed by Murphy [1328], who suggested that if all the pieces of evidence are available at the same time, one can average their masses, and calculate the combined masses by combining the average values multiple times.

Deng and co-authors [2071, 1159, 438] proposed a modified averaging method to combine belief functions based on a measure of evidence distance, in which the weight (importance) of each body of evidence is taken into account. Namely, the degree of credibility $Crd(m_i)$ of the i-th body of evidence is computed as

$$Crd(m_i) = \frac{Sup(m_i)}{\sum_j Sup(m_j)}, \quad Sup(m_i) \doteq \sum_{j \neq i} 1 - d(m_i, m_j),$$

where $d(.,.)$ is a distance between mass functions. The latter can be used to compute a weighted average of the input masses as

$$\tilde{m} = \sum_i Crd(m_i) \cdot m_i.$$

As in Murphy's approach, one can then use Dempster's rule to combine the resulting weighted average n times, when n is the number of input masses.

A number of similar approaches can be found in the literature. Du et al. [516], for instance, proposed to use evidence distance to extract the intrinsic characteristics of the existing evidence sources in the case of high conflict. A similar combination method based on introducing weight factors was proposed in [2059]. In [1974], a similarity matrix was calculated and a credit vector was derived. As in Murphy's approach, the available evidence is finally averaged using the normalised credit vector and combined $n-1$ times by Dempster's rule. In an original variation on this theme, in [656] the mutual degree of support of several pieces of evidences was calculated by use of evidence distance; the eigenvector associated with the maximum eigenvalue of the evidence support degree matrix so calculated can be then employed as a weight vector. An evidence discount coefficient can be computed and used to modify each piece of evidence prior to Dempster combination. In [655], the amount of conflict between pieces of evidence was first evaluated by use of both evidence distance and conflicting belief, and every piece of evidence was given a weight coefficient according to its degree of conflict with the others. Belief functions were modified based on their weight coefficient, and finally combined by Dempster's rule.

Albeit rather empirical, these methods try to address the crucial issue with Dempster combination (already pointed out in Section 4.3.1), namely that each piece of evidence has 'veto' powers on the possible consensus outcomes. If any of them gets it wrong, the combined belief function will never give support to the 'correct' hypothesis.

Other proposals Many other combination rules have been proposed over the years [247, 1920, 2082, 104, 267, 1821, 27, 218].

Surveys Some authors have tried to provide a survey of this 'jungle' of combination strategies.

As an example, in [919] Josang and Daniel discussed and compared various strategies for dealing with 'dogmatic' beliefs, including Lefevre's weighting operator (see Section 4.3.3), Josang's own consensus operator (Section 4.3.2) and Daniel's MinC approach. Wang et al. [1903] reviewed and critically analysed existing combination approaches. In these authors' view, the existing approaches either ignore normalisation or separate it from the combination process, leading to irrational or suboptimal interval-valued belief structures. In response, in [1903] a new 'logically correct' approach was developed, in which combination and normalisation are optimised together rather than separately. Yamada [2034], on his side, proposed a 'combination by compromise' model, based on the idea of sharing the mass $m_1(X) \cdot m_2(Y)$ among the subsets of $X \cup Y$ which 'qualify', namely $X \cap Y = C$, $X \cap Y^c = X_Y$, $X^c \cap Y = Y_X$, $X = C \cup X_Y$, $Y = C \cup Y_X$, $(X \cup Y)(X \cap Y) = X_Y \cup Y_X$ and $X \cup Y = C \cup X_Y \cup Y_X$. Several alternative strategies were proposed.

Prioritised belief structures The focus of [2016], again due to Yager, was to provide a procedure for aggregating 'prioritised' belief structures. Yager argued that a belief measure m_1 has priority over another measure m_2 if our confidence in m_1 is such that we are not willing to sacrifice any information from m_1. In other words, m_2 can only enhance the information coming from m_1, and not alter it. This translates into assigning the mass $m_1(A) \cdot m_2(B)$ to the subset with characteristic function

$$D(x) = \max\Big\{\min\{A(x), Bel(x)\}, 1 - Poss[B/A]\Big\},$$

where

$$Poss[B/A] = \max_{x \in \Theta} \min\{A(x), Bel(x)\} = \begin{cases} 1 \text{ if } A \cap B \neq \emptyset, \\ 0 \text{ if } A \cap B = \emptyset, \end{cases}$$

and $A(x)$ is the characteristic function of the set A. In [2016], an alternative to the normalisation step used in Dempster's rule was also suggested, inspired by non-monotonic logics.

Non-exhaustive frames The combination process on non-exhaustive frames of discernment was analysed by Janez and Appriou in an original series of papers [885]. In a previous work (in French), [884] these authors had already presented methods based on a technique called 'deconditioning' which allows the combination of such sources. For instance, the principle of minimum commitment can be employed: if a belief function is defined on a proper subset $\Theta' \subset \Theta$ of the whole frame of discernment Θ, the mass of each of its focal elements $A \subseteq \Theta'$ is transferred to the union $A \cup \Theta \setminus \Theta'$. When dealing with two belief functions defined on $\Theta_1, \Theta_2 \subset \Theta$, they are first both deconditioned to $\Theta_1 \cup \Theta_2$, where they can eventually be combined.

Weighted belief distributions In [2044], the concept of a *weighted belief distribution* (WBD) was proposed to characterise evidence, as opposed to classical basic probability assignments. The application of the orthogonal sum to WBDs leads to a new combination rule, of which Dempster's rule is a special case which applies when each piece of evidence is fully reliable.

ACR and PCR Florea et al. [631] presented the class of *adaptive combination rules* (ACRs), defined as

$$m_{\text{ACR}}(A) = \alpha(\kappa) m_{\circledcup}(A) + \beta(\kappa) m_{\circledcap}(A), \tag{4.31}$$

where $\kappa = m_{\circledcap}(\emptyset)$ is the usual Dempster conflict, and α, β are functions from $[0, 1]$ to $[0, +\infty)$ required to satisfy a number of sensible conditions. The idea behind the *proportional conflict redistribution* (PCR) rule [631], instead, is to transfer conflicting masses proportionally to non-empty sets involved in the model (compare Yamada's approach). The latter can happen in different ways, leading to distinct versions of the PCR rule. Both rules then allow one to deal with highly conflicting sources.

In related work, Leung et al. [1140] incorporated ideas from group decision making into the theory of evidence to propose an integrated approach to automatically identifying and discounting unreliable evidence. An adaptive robust combination rule that incorporates the information contained in the consistent focal elements was then constructed to combine such evidence. This rule adjusts the weights of the conjunctive and disjunctive rules according to a function of the consistency of focal elements, in a fashion similar to the ACR approach.

Idempotent fusion rules Destercke and Dubois [484, 483] noted that, when the dependencies between sources are ill known, it is sensible to require idempotence from a combination rule, as this property captures the possible redundancy of dependent sources. In [483], they thus studied the feasibility of extending the idempotent fusion rule of possibility theory (the 'minimum') to belief functions. They reached, however, the conclusion that, unless we accept the idea that the result of the fusion process can be a family of belief functions, such an extension is not always possible.

Joshi's rule Yet another combination rule was proposed by Joshi in [925], to address the asymmetrical treatment of evidence and some counter-intuitive results of Yager, Zhang [2106] and disjunctive combination. Namely,

$$m(A) = \begin{cases} (1+\kappa) \cdot \sum\limits_{B \cap C = A} m_1(B) \cdot m_2(C) & A \neq \Theta, \\ 1 - \sum\limits_{X \subseteq \Theta} m(X) & A = \Theta, \end{cases} \tag{4.32}$$

where κ is the classical Dempster conflict $\kappa = \sum_{B \cap C = \emptyset} m_1(B) m_2(C)$. The underlying assumption of this rule is that our level of ignorance is directly proportional to the degree of conflict. In the absence of conflict, this combination rule reduces to Dempster's.

Logic-based belief revision Ma et al. [1227, 1226] argued that in a belief revision process, in which the prior knowledge of an agent is altered by upcoming information, combination is inherently asymmetric. Most combination rules proposed so far in the theory of evidence, however, especially Dempster's rule, are symmetric. Assuming the input information is reliable, it should be retained, whilst the prior information should be changed minimally to that effect. A revision rule which best fits the revision framework was then proposed, which reduces to Dempster's rule when the input is strongly consistent with the prior belief function, and generalises Jeffrey's rule of updating, Dempster's rule of conditioning and a form of logic-based AGM (Alchourrón, Gärdenfors and Makinson) revision [26].

Hierarchical combination schemes Rather than introducing yet another combination rule, Klein et al. [977] proposed in 2010 to use existing ones as part of a hierarchical and conditional combination scheme. In each subproblem, the number of constraints is reduced so that an appropriate rule can be selected and applied.

Others We conclude by citing other relevant publications, in rough chronological order and without much detail.

In an early work, Hummel et al. [859] built on their framework of statistics of experts' opinions (Section 3.3.1) to address the combination of bodies of dependent information, giving a model for parameterising their degree of dependence. In 1990 [1080], Kruse and Schwecke argued that the concept of specialisation generalises Dempster's rule. This is founded on the fact that modelling uncertain phenomena always entails a simplifying coarsening which arises from the renunciation of a description to the depth that a perfect image would require. A generalised evidence combination formula relaxing the requirement of evidence independence was proposed by Wu in 1996 [1970]. Around the same time, Bell and Guan [104] supported a view of the orthogonal sum as a means of reducing inconsistency in data arising from different sources. Also in 1996, a rather empirical alternative rule of combination, which adapts the belief-updating process based on a contextual weighting parameter for combining evidence over time, was developed by Murphy in [1330]. In the same year, an algorithmic approach to combining belief functions was proposed by Tonn [1820] which is capable of handling inconsistent evidence. A modification of the Dempster–Shafer theory was employed by Zeigler in 1998 [2097] to formulate a computationally feasible approach to evidence accumulation, whose properties were formulated axiomatically. An 'absorptive' combination rule was proposed by Zhang et al. in 2000 [2110]. An original method for combining several belief functions in a common frame based on the concept of an unnormalised mass function was presented by Monney in [1308], which relates to the notion of the Dempster specialisation matrix [1707]. Some authors proposed in 2004 to adjust the input BPAs prior to applying Dempster's rule, as in the discounting approach (compare the 'disturbance of ignorance' technique in [1173]). Also in 2004, based on the theoretical grounding of Dempster's combination rule on random set theory, Zhou [2130] analysed all possible combination rules, to finally propose an optimal Bayes combination rule. More recently, in 2008, a thesis by Wang [1896] put forward an improved combination method based on the reliability and correlation between pieces of evidence. Finally, in 2010 Valin et al. [1839] introduced a 'ThresholdedDS' strategy, which prevents a belief measure becoming too certain of its decision upon accumulation of supporting evidence.

4.3.3 Families of combination rules

A separate line of research has investigated the formulation of whole families of combination rules, as a more flexible approach to evidence combination.

Lefevre's combination rules parameterised by weighting factors Lefevre et al.'s analysis [1118] starts from observing that, as conflict in Dempster combination increases with the number of information sources, a strategy for reassigning the conflicting mass becomes essential.

The family of combination rules they propose in response redistributes the conflicting mass to each proposition A of a set of subsets $\mathcal{P} = \{A\}$ according to a weighting factor $w(A, \mathbf{m})$, where $\mathbf{m} = \{m_1, \ldots, m_j, \ldots, m_J\}$. Namely,

$$m(A) = m_{\cap}(A) + m^c(A), \tag{4.33}$$

where[33]

$$m^c(A) = \begin{cases} w(A, \mathbf{m}) \cdot m_{\cap}(\emptyset) & A \in \mathcal{P}, \\ 0 & \text{otherwise}, \end{cases}$$

under the constraint that the weights are normalised: $\sum_{A \in \mathcal{P}} w(A, \mathbf{m}) = 1$.

This (weighted) family subsumes Smets's and Yager's rules when $\mathcal{P} = \{\emptyset\}$ and $\mathcal{P} = \{\Theta\}$, respectively. We get Dempster's rule when $\mathcal{P} = 2^\Theta \setminus \{\emptyset\}$ with weights

$$w(A, \mathbf{m}) = \frac{m_{\cap}(A)}{1 - m_{\cap}(\emptyset)} \quad \forall A \in 2^\Theta \setminus \{\emptyset\}.$$

Dubois and Prade's operator can also be obtained by setting suitable weight factors. The authors of [1118] proposed to learn the most appropriate weights for a specific problem.

Similar conflict redistribution strategies have been proposed by other authors. A similar idea, for instance, was presented in [1147], where the conflicting mass is distributed to every proposition according to its average supported degree. In [435], a global conflict index is first calculated as the weighted average of the local conflicts. Then, a validity coefficient is defined to show the effect of conflicting evidence on the results of the combination. Han et al. [790] suggested a modified combination rule which is based on the ambiguity measure, a recently proposed uncertainty measure for belief functions. Weight factors based on the AM of the bodies of evidence are used to reallocate conflicting mass assignments.

However, Lefevre's proposal of a parameterised combination rule was criticised by Haenni [763]. Lefevre et al. replied to Haenni in [1119], from the point of view of the transferable belief model, as opposed to the probabilistic argumentation systems proposed by Haenni.

Denoeux's families induced by t-norms and conorms Cautious and bold rules are themselves, as shown in [455], specific members of infinite families of conjunctive and disjunctive combination rules, based on triangular norms and conorms.

A *t-norm* is a commutative and associative binary operator $\top$ on the unit interval satisfying the monotonicity property

$$y \leq z \Rightarrow x \top y \leq x \top z, \quad \forall x, y, z \in [0, 1],$$

and the boundary condition $x \top 1 = x, \forall x \in [0, 1]$. A *t-conorm* $\perp$ possesses the same three basic properties (commutativity, associativity and monotonicity), and differs only in the boundary condition $x \perp 0 = x$. T-norms and t-conorms are usually interpreted as generalised conjunction and disjunction operators, respectively, in fuzzy logic. Denoeux noted that the conjunctive combination is such that

$$w^c_{1 \circledcirc 2}(A) = w^c_1(A) \cdot w^c_2(A), \tag{4.34}$$

[33] Note that Lefevre et al. [1118] mistakenly wrote $A \subseteq \mathcal{P}$ in their definition (17), rather than $A \in \mathcal{P}$ (since $\mathcal{P}$ is a collection of subsets).

where $w_1^c(A) = 1 \wedge w(A)$. New rules for combining non-dogmatic belief functions can then be defined by replacing the minimum $\wedge$ in (4.34) by a positive t-norm:[34]

$$m_1 \circledast_{\top,\perp} m_2 = \textcircled{\cap}_{A \subset \Theta} m_A^{w_1(A) *_{\top,\perp} w_2(A)}, \tag{4.35}$$

where $*_{\top,\perp}$ is the following operator in $(0, +\infty)$:

$$x *_{\top,\perp} y = \begin{cases} x \top y & x \vee y \leq 1, \\ x \wedge y & x \vee y > 1 \text{ and } x \wedge y \leq 1, \\ (1/x \perp 1/y)^{-1} & \text{otherwise}, \end{cases}$$

where $\top$ is a positive t-norm and $\perp$ a t-cornorm, for all $x, y > 0$.

The cautious rule corresponds to $\circledast_{\wedge,\vee}$.

α-junctions The family of combination rules termed *α-junctions* was proposed by Smets [1720] as the associative, commutative and linear operators for belief functions with a neutral element. α-junction operators depend on a single parameter α, and generalise conjunction, disjunction and exclusive disjunction.

Namely, let m_1 and m_2 be two BPAs on Θ, represented as column vectors $\mathbf{m}_i = [m_i(A), A \subseteq \Theta]'$, $i = 1, 2$ (see Section 4.7.1). Suppose that we wish to build a BPA $m_{12} = f(m_1, m_2)$. Then all such operators f which satisfy the axioms of linearity ($f(pm_1 + qm_2) = pf(m_1) + qf(m_2)$), commutativity, associativity, existence of a neutral element $\mathbf{m}_{\text{vac}}$, anonymity and context preservation (if $Pl_1(A) = 0$ and $Pl_2(A) = 0$ for some $A \subseteq \Theta$, then $Pl_{12}(A) = 0$) are stochastic matrices of the form

$$\mathbf{m}_{12} = K_{m_1} \mathbf{m}_2, \tag{4.36}$$

where

$$K_{m_1} = \sum_{A \subseteq \Theta} m_1(A) K_A.$$

Smets [1720] proved that the $2^{|\Theta|} \times 2^{|\Theta|}$ matrices K_A depend only on $\mathbf{m}_{\text{vac}}$ and one parameter $\alpha \in [0, 1]$. Furthermore, he showed that there are only two solutions for $\mathbf{m}_{\text{vac}}$: either $\mathbf{m}_{\text{vac}} = \mathbf{m}_\Theta$ or $\mathbf{m}_{\text{vac}} = \mathbf{m}_\emptyset$. Hence, there are only two sets of solutions [1429].

In [1429], Pichon showed that α-junctions correspond to a particular form of knowledge about the truthfulness of the sources generating the belief functions to be combined.

Other families Families of update and combination rules were also considered by several other authors [1372].

In [2010], quasi-associative operators were proposed by Yager for representing a class of evidence combination operators, including averaging and Dempster–Shafer ones. Gilboa and Schmeidler [691] put forward a family of update rules for

[34] A t-conorm can replace the combination of the diffidence components of the weights [455].

non-additive probabilities, of which Bayes' and Dempster's rules are extreme cases [439]. A family of update rules even more general than that of Gilboa and Schmeidler was introduced later by Young et al. [2074].

More recently, Wen [1925] presented a unified formulation of combination rules on the basis of random set theory which encompasses all the classical combination rules, including the Dempster sum. Such a formulation has the potential, according to Wen, to provide original ideas for constructing more applicable and effective combination rules. Indeed, a new combination rule was constructed in [1925] which overcomes a class of counter-intuitive phenomena described by Zadeh.

4.3.4 Combination of dependent evidence

We know that independence of sources is a crucial assumption behind Dempster combination. What to do when the sources of evidence are *not* independent is an interesting, albeit rather surprisingly unexplored, question.

Smets [1719] explored the concept of distinct pieces of evidence, to try and provide a formal definition of distinctness and compare it with that of stochastic independence. Kramosil [1054] showed that, for a specific but large enough class of probability measures, an analogue of Dempster combination can be obtained which uses some non-standard and Boolean-like structures over the unit interval of real numbers, and does not rely on the assumption of statistical independence. In 2003, Cattaneo [236] suggested using the least specific combination which minimises conflict, among those induced by a simple generalisation of Dempster's rule. This increases the monotonicity of the reasoning and helps us to manage situations of dependence. In a later paper [1309], the same author introduced two combination rules with no assumptions about the dependence of the information sources, based on cautious combinations of plausibility and commonality functions, respectively. The properties of these rules and their connection with Dempster's rule and the minimum rule of possibility theory were studied in [1309].

More recently (2008), Yager [2023] suggested an approach to the aggregation of non-independent belief measures via weighted aggregation, where the weights are related to the degree of dependence. As the aggregation is non-commutative, however, one needs to consider the problem of how best to sequence the evidence. Fu and Yang [649] investigated the conjunctive combination of belief functions from dependent sources based on a *cautious conjunctive rule* (CCR). Here, weight functions in the canonical decomposition of a belief function (see Section 2.4.4) are divided into positive and negative ones. The positive and negative weight functions of two belief functions are used to construct a new partial ordering, which is then used to derive a new rule for combining BFs from dependent sources. In 2016, Su et al. [1775] proposed a method for combining dependent bodies of evidence which takes the significance of the common information sources into consideration.

4.3.5 Combination of conflicting evidence

As we learned from the debate on Dempster's rule (Section 4.3.1), most criticisms of the original combination operator focus on its behaviour in situations in which the various pieces of evidence are highly conflicting.

In Section 4.3.2, we have already met a number of combination rules which have been proposed to address this issue: the TBM solution (4.3.2), where masses are not renormalised and conflict is stored in the mass given to the empty set; Yager's rule (4.23), according to which conflict is transferred to the universe; and Dubois and Prades solution (4.24), in which the masses resulting from pairs of conflicting focal elements are transferred to the union of these subsets. The most relevant work on the issue of conflict is probably [1328], where it was argued that averaging (Section 4.3.2) best solves the issue of normalisation, but it does not offer convergence towards certainty, nor a probabilistic basis. The 'jungle' of combination rules that have been proposed as a result of the conflict problem was discussed by Smets in [1716].

Measures of conflict First of all, one needs to define an appropriate measure of conflict.[35]

Liu's measure In a significant piece of work, Liu [1196] provided a formal definition of when two basic probability assignments are in conflict, using a combination of both the combined mass assigned to the empty set before normalisation, and the distance between betting commitments of beliefs, namely the following.

Definition 52. ([1196], Definition 10) *Let m_1 and m_2 be two BPAs. Let*

$$cf(m_1, m_2) = \langle m_\cap(\emptyset), difBetP_{m_1}^{m_2} \rangle \tag{4.37}$$

be a two-dimensional measure in which $m_\cap(\emptyset)$ is (as usual) the mass of uncommitted belief when m_1 and m_2 are combined by Dempster's rule, and $difBetP_{m_1}^{m_2}$ is the distance between the betting commitments associated with m_1 and m_2 (measured by the pignistic function), namely

$$difBetP_{m_1}^{m_2} = \max_{A \subset \Theta} \Big(|BetP[m_1](A) - BetP[m_2](A)| \Big).$$

The mass functions m_1 and m_2 are said to be in conflict *iff both $difBetP > \epsilon$ and $m_\cap(\emptyset) > \epsilon$ hold, where $\epsilon \in [0, 1]$ is a threshold of conflict tolerance.*

The author of [1196] argued that only when both measures are high is it safe to say that the evidence is in conflict. The idea was applied further to possibility measures in [1197].

[35] In a different context, a definition of consistency applicable to the theory of evidence was introduced in [1008].

Conflict as distance A radically different interpretation of conflict views it as a function of the 'distance' between the bodies of evidence involved (see Section 4.10.1).

George and Pal [677], for instance, formulated a set of axioms a metric distance should satisfy in order to quantify such conflict. A unique expression for the conflict between propositions was so derived. The average conflict between propositions would then give a measure of total conflict for a given body of evidence.

Martin et al. [1251], more recently, proposed some alternative measures of conflict via distances, which can be used further for an a posteriori estimation of the relative reliability of sources of information which does not need any training or prior knowledge. In a later paper [1249], the same authors argued against using $m_{\cap}(\emptyset)$ as a measure of conflict. After showing some counter-intuitive results associated with some proposed measures of conflict, they defined a new distance-based measure which satisfies a number of desirable properties, weighted by a degree of inclusion.

Internal and external conflict An important distinction is that between *internal* conflict (between focal elements of the same belief function) and *external* conflict (between belief functions). Daniel has conducted much work in this area.

In [384, 388], three new different approaches to the problem (combinational, plausibility and comparative) were presented and compared with Liu's interpretation of conflict. The notions of non-conflicting and conflicting parts of a belief function were formally introduced in [386]. Several statements about the algebra of belief functions on a general finite frame of discernment were also discussed, and the non-conflicting part of a BF uniquely identified. Schubert [1561], for his part, defined and derived the internal conflict of a belief function by decomposing it into a set of generalised simple support functions (GSSFs). The internal conflict of a belief function is then the conflict in the Dempster combination of all such GSSFs (except the one supporting the empty set).

Other work Other significant work in this area was conducted by Destercke and Burger [481, 482]. They proposed to study the notion of conflict from a different perspective, starting by examining consistency and conflict on sets, and extracting from this setting basic properties that measures of consistency and conflict should have. This basic scheme was then extended to belief functions (which generalise the notion of a set) in different ways, without making assumptions about (in)dependence of sources. An analysis of conflict functions in a generalised power space was also recently conducted by Lifang Hu et al. [852]. Finally, an alternative measure of evidence conflict based on the pignistic transform was defined in [851]. There, an 'improved' Dempster–Shafer algorithm would first weigh the 'importance' of each piece of evidence, and then combine the preprocessed evidence using Dempster's rule.

Redistribution approaches A number of approaches to dealing with conflict are based on the idea of redistributing mass among focal elements, similarly to the rationale of several major combination rules (see Section 4.3.2).

One such framework, for instance, was proposed by Lefevre et al. [1120], based on combination operators which allow an arbitrary redistribution of the conflicting mass. In [374, 373, 376], two alternative ways of distributing the contradiction among non-empty subsets of the frame of discernment were studied. In particular, a so called minC (associative) combination operator [381] was presented. Daniel also conducted an analysis of various combination operators with respect to their commutativity with coarsening and refinement [375].

Martin et al. presented in [1253] a generalised proportional conflict redistribution rule within the Dezert–Smarandache extension of the Dempster–Shafer theory, after studying and comparing different combination rules. Finally, Xu et al. [1981] proposed a new conflict reassignment approach based on the 'grey relational analysis' notion. The proposed approach can automatically evaluate the reliability of the information sources and distinguish the unreliable information.

Other conflict resolution methods Besides redistribution methods, a number of other contributions on the topic exist [1552, 1919, 675, 747]. As usual, we summarise some of them here in rough chronological order.

An early interesting empirical comparison between the results of applying Dempster's rule of combination and conditional probabilities to a problem involving conflicting evidence was conducted in [488]. The normalisation constant was shown to be tied to a prior probability of the hypothesis if equality was to occur, forcing further relationships between the conditional probabilities and the prior. Ways of incorporating prior information into the belief function framework were also explored.

In a series of papers [1543, 1546, 1547, 1549], Schubert defined instead a criterion function called a *metaconflict function*, which allows us to partition a collection of several pieces of evidence into subsets. Separately [1557, 1310], he proposed that each piece of evidence be discounted in proportion to the degree to which it contributes to the conflict, in a sequence of incremental steps.

In an original paper from 1996 [1465], Ray and Krantz reduced conflict among belief functions to conflict among *schemata*, and argued in favour of weighting schemata to compute common-ground inferences. In [2109], Zhang proposed two concrete algorithms to address the problem of high conflict, with and without a prior.

In [217], Campos and Cavalcante presented a modified combination rule that allows the combination of highly conflicting pieces of evidence, avoiding the tendency to reward low-probability but common possible outcomes of otherwise disjoint hypotheses (as in Zadeh's paradox). Their combination rule reads as

$$m_1 \Psi m_2(A) = \frac{Z \cdot \sum_{B \cap C = A} m_1(B) m_2(C)}{1 + \log(1/\kappa)}, \tag{4.38}$$

where κ is the classical Dempster conflict $\sum_{B \cap C = \emptyset} m_1(B) m_2(C)$, and Z is a normalisation factor.

More recently, Guan [748] developed a method for dealing with seriously conflicting evidence which uses an 'additive' strategy to modify the properties of the conflicting evidence. On their side, Lefevre and Elouedi [1123, 1122] discussed how to preserve the role of conflict as an alarm bell announcing that there is some kind of disagreement between sources. They proposed a process based on dissimilarity measures which allows one to preserve some conflict after the combination, by keeping only the part of it which reflects the disagreement between belief functions.

Finally, in [2046] evidence conflict was investigated based on the 'degree of coherence' between two sources of evidence, and methods for dealing with evidence conflict were analysed. A new 'paradoxical' combination algorithm based on an 'absolute' difference factor and a 'relative' difference factor of two pieces of evidence was proposed.

4.3.6 Combination of (un)reliable sources of evidence: Discounting

Discounting *Discounting*, introduced by Shafer in [1583], is a powerful tool for representing the *degree of reliability* of a source of evidence, i.e., to what degree we can trust the source and the numerical values it attaches to events.

Very simply, a belief function Bel is 'discounted' by reserving some mass to (typically) the whole frame of discernment, and normalising the mass assignment of the original focal elements:

$$m^\alpha(\Theta) = \alpha + (1-\alpha)m(\Theta), \quad m^\alpha(A) = (1-\alpha)m(A) \; \forall A \subsetneq \Theta. \tag{4.39}$$

Note that the original mass of Θ is also subject to discounting, prior to being incremented by α. A more formal definition which makes use of the conjunctive combination of the original mass function on Θ with a second one defined on the binary frame $\mathcal{R} = \{\text{source } S \text{ is reliable, source } S \text{ is unreliable}\}$ in the product frame $\Theta \times \mathcal{R}$ can also be provided [1282]. In particular, Bayesian belief functions (the belief-theoretical counterparts of probability measures) can be discounted to allow for less than perfect reliability of the generating process. Note that Guan et al. [742] showed back in 1993 that discounting does not commute with the orthogonal sum.

At any rate, the question of how to assess source reliability arises [1433, 1250].

Assessing reliabilities In response, Elouedi et al. [572] developed a method for evaluation of the reliability of a sensor, based on finding the discounting factor which minimises the distance between the pignistic probabilities computed from the discounted beliefs and the original ones. Similarly, the discounting factors for several sensors that are supposed to work jointly are computed by minimising the distance between the pignistic probabilities computed from the combined discounted belief functions and the actual combination. Liu et al. [658] proposed to characterise the dissimilarity of different sources of evidence using a distance metric and a conflict coefficient based on the pignistic transform, and presented a method for estimating weighting factors for the available pieces of evidence which uses their dissimilarity measure so calculated. Klein [976], instead, proposed a method that computes discounting rates using a *dissent* measure, which compares each source with the average mass function across all sources. Delmotte et al. [410] proposed to estimate discounting factors from the conflict arising between sources, and from past knowledge about the qualities of those sources. Under the assumption that conflict is generated by defective sources, an algorithm was outlined for detecting them and mitigating the problem.

Generalisations of Shafer's discounting

De-discounting The discounting operation only allows one to weaken a source, whereas it is sometimes useful to strengthen it when it is deemed to be too cautious. For that purpose, the *de-discounting* operation was introduced as the inverse of discounting by Denoeux and Smets [475].

Suppose we have a mass function $m^\alpha : 2^\Theta \rightarrow [0,1]$ generated by a source S, and we know that the belief function has been discounted with a rate α. We can then recompute the original, non-discounted mass assignment simply by using [1282]

$$m = \frac{m^\alpha - m_\Theta}{1-\alpha}, \tag{4.40}$$

where m_Θ is, as usual, the vacuous belief function on Θ.

Contextual discounting Mercier et al. [1286, 1287] proposed an extension of the discounting operation which allows us to use more detailed information regarding the reliability of the source in different contexts, i.e., conditionally on different hypotheses regarding the variable of interest. The resulting *contextual discounting* operation is parameterised by a discount rate *vector*, rather than a single scalar value α. Later [1284], these authors provided new interpretations of discounting, de-discounting and extended and contextual discounting, and introduced two families of correction mechanisms which generalise non-contextual and contextual discounting, respectively. In [1285], contextual discounting was extended further, and its specific link to canonical disjunctive decomposition was highlighted.

Contextual discounting was used in [1283] to provide an objective way of assessing the reliability of a sensor, generalising the approach proposed by Elouedi, Mellouli and Smets [572].

Other approaches In [1252], Pichon et al. proposed a general approach to information correction and fusion for belief functions, in which not only may information items be irrelevant, but sources may lie as well. A new correction scheme was then introduced which takes into account uncertain metaknowledge about a source's relevance and truthfulness, generalising Shafer's discounting (4.39).

Smandarache et al. [1666] presented an approach to combining sources of evidences with different *importances*, as distinguished from Shafer's 'reliabilities'. Haenni proposed in [768] a general model of partially reliable sources which includes several previously known results as special cases. Finally, Zhu and Basir [2123] extended the conventional discounting scheme by allowing the discount rate to be outside the conventional range $[0,1]$. The resulting operation can perform both discounting and de-discounting of a belief function.

4.4 Belief versus Bayesian reasoning: A data fusion example

After our review of inference and combination in the theory of evidence, we are ready to see a concrete example of the difference between the fusion processes in belief and Bayesian theory.

Suppose we want to estimate the class of an object (e.g. 'chair', 'table') appearing in an image, based on salient ('feature') measurements extracted from the image itself. One way to do this is, for instance, to feed the image pixel intensity values as a vector of inputs to a convolutional neural network,[36] which is trained to generate a discriminative feature vector in response. We may thus proceed to capture a training set of images, complete with annotated object labels. Assuming that the feature data are distributed according to a PDF of a specified family (e.g. a mixture of Gaussians), we can learn from the training data the likelihood function $p(y|x)$ of the object class y given the image feature vector x.

Suppose as well that n different 'sensors' (e.g., n convolutional nets) extract n features x_i from each image, yielding the sample $x_1, \ldots, x_n$. In order to infer the correct object class, the information coming from the n feature measurements needs to be merged.

Let us now compare how such a data fusion process works under the Bayesian and the belief function paradigms.

4.4.1 Two fusion pipelines

Bayesian fusion The likelihoods of the individual features are computed using the n likelihood functions learned during training, $p(x_i|y)$, for all $i = 1, \ldots, n$. Measurements are typically assumed to be conditionally independent, yielding the product likelihood: $p(\mathbf{x}|y) = \prod_i p(x_i|y)$. Bayesian inference is then applied, typically after assuming uniform priors (for there is no reason to assume otherwise), yielding the following conditional posterior distribution of the object's class (Fig. 4.7):

$$p(y|\mathbf{x}) \sim p(\mathbf{x}|y) = \prod_i p(x_i|y).$$

Belief function fusion In belief theory, for each feature type i a separate belief function is learned from each individual likelihood $p(x_i|y)$, for example via the likelihood-based approach supported by Shafer and Denoeux (Section 4.1.1). This yields n belief functions $Bel(Y|x_i)$, all of them defined on subsets Y of the range $\mathcal{Y}$ of possible object classes.

A combination rule $\odot$ (e.g. $\textcircled{\cap}$, $\oplus$, $\textcircled{\cup}$: Section 4.3) is then applied to compute an overall belief function, obtaining (see Fig. 4.8):

$$Bel(Y|\mathbf{x}) = Bel(Y|x_1) \odot \ldots \odot Bel(Y|x_n), \quad Y \subseteq \mathcal{Y}.$$

[36] http://ufldl.stanford.edu/tutorial/supervised/ConvolutionalNeuralNetwork/.

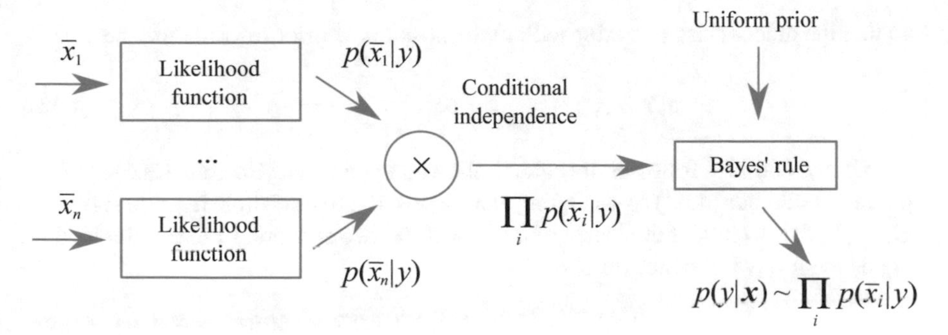

Fig. 4.7: The typical fusion process in the Bayesian framework.

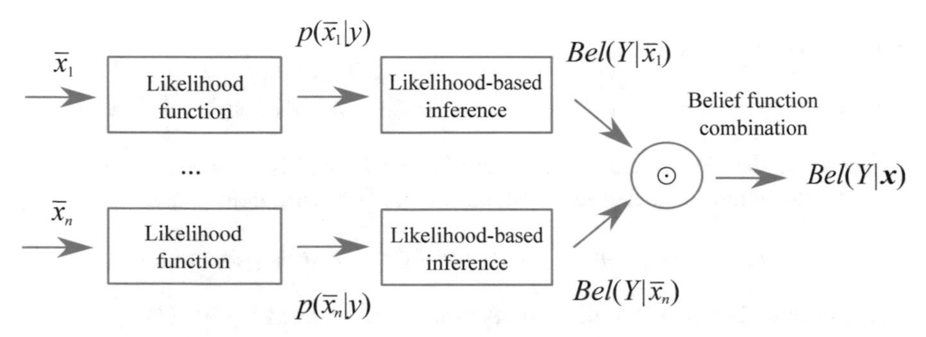

Fig. 4.8: The typical fusion process in the belief theory framework.

4.4.2 Inference under partially reliable data

So far, in our toy fusion example, we have assumed that the data are measured correctly, and can be perfectly trusted. What if the data-generating process is not completely reliable? We have covered this aspect in Section 4.3.6.

Suppose we simply wish to detect an object of interest (e.g., a car) within the given image: this is a binary decision, yes (Y) or no (N). Imagine that two sensors produce image features x_1 and x_2, but that we have learned from the training data that both sensors are reliable only 20% of the time. At test time, we measure x_1 and x_2 from the available test image, and apply the learnt likelihood functions, obtaining the following normalised likelihoods:

$$p(x_1|Y) = 0.9, \; p(x_1|N) = 0.1; \quad p(x_2|Y) = 0.1, \; p(x_2|N) = 0.9. \tag{4.41}$$

Unluckily, however, sensor 2 gets it wrong: although $p(x_2|Y) \ll p(x_2|N)$, the object is actually there. How do the two fusion pipelines cope with such an outcome?

The Bayesian scholar, having assumed that the two sensors/processes are conditionally independent, will multiply the likelihoods obtained, yielding

$$p(x_1, x_2|Y) = 0.9 * 0.1 = 0.09, \quad p(x_1, x_2|N) = 0.1 * 0.9 = 0.09,$$

so that the object class posterior will, in the absence of prior information, be

$$p(Y|x_1, x_2) = \frac{1}{2}, \quad p(N|x_1, x_2) = \frac{1}{2}. \tag{4.42}$$

Shafer's faithful follower, instead, is allowed to discount (Section 4.3.6) the empirical likelihoods (4.41) by assigning a mass $\alpha = 0.2$ to the whole hypothesis space $\Theta = \{Y, N\}$, to reflect the learnt reliability of the two sensors. In the example, this yields the following belief functions:

$$\begin{array}{lll} m(Y|x_1) = 0.9 * 0.8 = 0.72, & m(N|x_1) = 0.1 * 0.8 = 0.08, & m(\Theta|x_1) = 0.2; \\ m(Y|x_2) = 0.1 * 0.8 = 0.08, & m(N|x_2) = 0.9 * 0.8 = 0.72, & m(\Theta|x_2) = 0.2. \end{array}$$

Combining them by Dempster's rule yields the following belief function Bel on the binary decision frame $\{Y, N\}$:

$$m(Y|x_1, x_2) = 0.458, \quad m(N|x_1, x_2) = 0.458, \quad m(\Theta|x_1, x_2) = 0.084. \tag{4.43}$$

When using the disjunctive rule for combination (which is, as we know, the least committed combination operator) instead, we get Bel' with mass assignment

$$m'(Y|x_1, x_2) = 0.09, \quad m'(N|x_1, x_2) = 0.09, \quad m'(\Theta|x_1, x_2) = 0.82. \tag{4.44}$$

The corresponding (credal) sets of probabilities are depicted in Fig. 4.9.

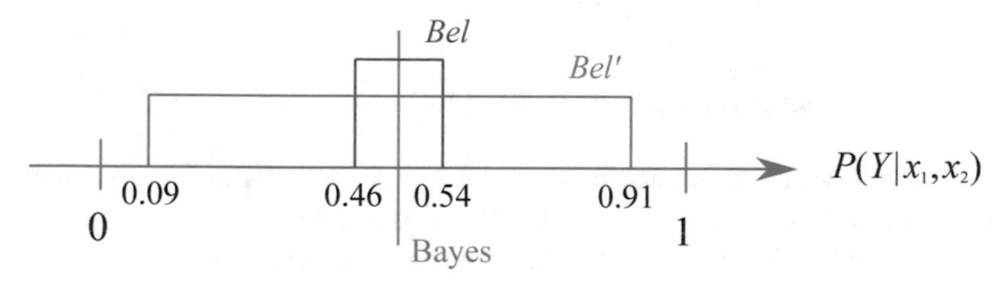

Fig. 4.9: Credal sets associated with the belief functions Bel (4.43) and Bel' (4.44) in the example of fusion of unreliable data, versus the Bayesian estimate (4.42).

As can be observed, the credal interval implied by Bel is quite narrow. This is explained by the fact that the sensor reliability was estimated to be 80%, while we were unlucky to get a faulty measurement among two (50%). The disjunctive rule is much more cautious about the correct inference: as a result, the associated credal set is much wider.

4.5 Conditioning

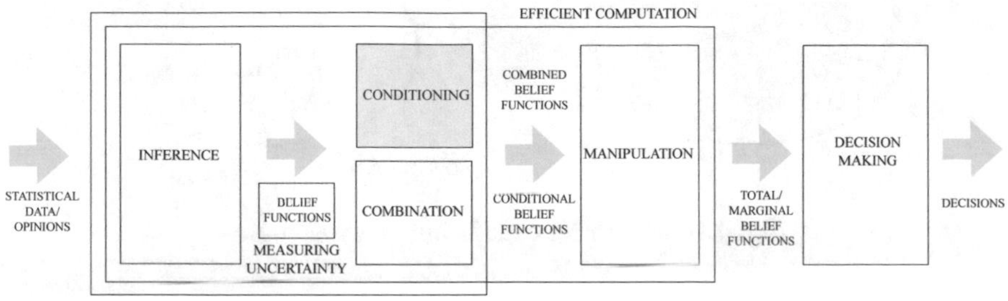

In Bayesian reasoning, where all evidence comes in the form of a proposition A being true, conditioning (as we know) is performed via Bayes' rule. Belief functions can also be conditioned, rather than combined, whenever they need to be updated based on similar hard evidence. However, just as in the case of combination rules, a variety of conditioning operators can be defined for belief functions [245, 584, 880, 691, 439, 2080, 873], many of them generalisations of Bayes' rule itself.

Here we review some of the most significant proposals: the original Dempster conditioning, Fagin and Halpern's lower and upper envelopes of conditional probabilities, Suppes's geometric conditioning, Smets's unnormalised conditioning, and Spies's formulation based on sets of equivalent events under multivalued mappings. Conditioning within the framework of the geometric approach supported by this book will be discussed in detail in Chapter 15.

Finally, an overview of the existing major conditioning approaches is provided, which shows that they form a nested family, supporting an argument in favour of working with pairs of belief functions rather than individual ones.

4.5.1 Dempster conditioning

Dempster's rule of combination (2.6) naturally induces a conditioning operator.

Given a conditioning event $B \subset \Theta$, the 'logical' (or 'categorical', in Smets's terminology) belief function Bel_B such that $m(B) = 1$ is combined via Dempster's rule with the a priori BF Bel. The resulting measure $Bel \oplus Bel_B$ is the conditional belief function given B *à la Dempster*, which we denote by $Bel_\oplus(A|B)$ (see Fig. 4.10 for a simple example).

Quite significantly, in terms of belief and plausibility values, Dempster conditioning yields

$$\begin{aligned} Bel_\oplus(A|B) &= \frac{Bel(A \cup \bar{B}) - Bel(\bar{B})}{1 - Bel(\bar{B})} = \frac{Pl(B) - Pl(B \setminus A)}{Pl(B)}, \\ Pl_\oplus(A|B) &= \frac{Pl(A \cap B)}{Pl(B)}. \end{aligned} \tag{4.45}$$

Namely, it can be thought of as the outcome of Bayes' rule when applied to plausibility rather than probability measures.

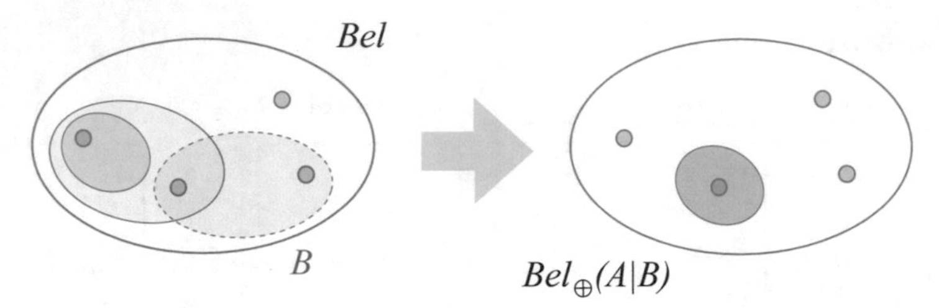

Fig. 4.10: Dempster conditioning: an example.

4.5.2 Lower and upper conditional envelopes

Fagin and Halpern [584], in opposition, proposed an approach based on the credal (robust Bayesian) interpretation of belief functions, as lower envelopes of a family of probability distributions (recall Section 3.1.4):

$$Bel(A) = \inf_{P \in \mathcal{P}[Bel]} P(A).$$

They defined the conditional belief function associated with Bel as the *lower envelope* (that is, the infimum) *of the family of conditional probability functions* $P(A|B)$, where P is consistent with Bel:

$$Bel_{\text{Cr}}(A|B) \doteq \inf_{P \in \mathcal{P}[Bel]} P(A|B), \quad Pl_{\text{Cr}}(A|B) \doteq \sup_{P \in \mathcal{P}[Bel]} P(A|B). \tag{4.46}$$

It is straightforward to see that this definition reduces to that of the conditional probability whenever the initial belief function is Bayesian, i.e., its credal set consists of a single probability measure.

This notion, quite related to the concept of the inner measure (see Section 3.1.3), was considered by other authors too, including Dempster, Walley and Kyburg [1086], who arrived at the conclusion that the probability intervals generated by Dempster updating were included in those generated by Bayesian updating. Indeed, the following closed-form expressions for the conditional belief and plausibility in the credal approach can be proven:

$$\begin{aligned} Bel_{\text{Cr}}(A|B) &= \frac{Bel(A \cap B)}{Bel(A \cap B) + Pl(\bar{A} \cap B)}, \\ Pl_{\text{Cr}}(A|B) &= \frac{Pl(A \cap B)}{Pl(A \cap B) + Bel(\bar{A} \cap B)}. \end{aligned} \tag{4.47}$$

Note that lower and upper envelopes of *arbitrary* sets of probabilities are not in general belief functions, but these actually are, as Fagin and Halpern proved. A comparison with the results of Dempster conditioning showed that they provide a more conservative estimate, as the associated probability interval is included in that resulting from Dempster conditioning:

$$Bel_{\text{Cr}}(A|B) \leq Bel_{\oplus}(A|B) \leq Pl_{\oplus}(A|B) \leq Pl_{\text{Cr}}(A|B).$$

Fagin and Halpern argued that Dempster conditioning behaves unreasonably in the context of the classical 'three prisoners' example [584], originally discussed by Diaconis [499].

4.5.3 Suppes and Zanotti's geometric conditioning

In [1690], Philippe Smets pointed out the distinction between *revision* and *focusing* in the conditional process, following previous work by Dubois and Prade. Although these are conceptually different operations, in probability theory they are both expressed by Bayes' rule: as a consequence, this distinction is not given as much attention as in other theories of uncertainty. However, in belief theory these principles do lead to distinct conditioning rules. The application of revision and focusing to belief theory has been explored by Smets in his transferable belief model – thus no random set generating Bel nor any underlying convex sets of probabilities are assumed to exist.

In *focusing* no new information is introduced, as we merely focus on a specific subset of the original hypothesis space. When applied to belief functions, this yields Suppes and Zanotti's *geometric conditioning* [1788]:

$$Bel_{\mathrm{G}}(A|B) = \frac{Bel(A \cap B)}{Bel(B)}, \quad Pl_{\mathrm{G}}(A|B) = \frac{Bel(B) - Bel(B \setminus A)}{Bel(B)}. \tag{4.48}$$

This was proved by Smets using Pearl's 'probability of provability' [1405] interpretation of belief functions (Section 6.6.8).

It is interesting to note that geometric conditioning is somewhat dual to Dempster conditioning, as it amounts to replacing probability values with belief values in Bayes' rule:

$$Pl_{\oplus}(A|B) = \frac{Pl(A \cap B)}{Pl(B)} \leftrightarrow Bel_{\mathrm{G}}(A|B) = \frac{Bel(A \cap B)}{Bel(B)}.$$

Unlike Dempster and conjunctive conditioning (see the next paragraph), geometric conditioning does not seem to be linked to an underlying combination operator, although the issue remains open.

4.5.4 Smets's conjunctive rule of conditioning

As opposed to focusing, in *belief revision* [691, 1417], a state of belief is modified to take into account a new piece of information.

When applied to belief functions, this results in what Smets called the *unnormalised*[37] conditioning operator, whose output is the belief function with mass assignment[38]

[37] 'Unnormalized' in the original American-inspired spelling.

[38] This author's notation. As this conditioning operator is induced by the conjunctive rule of combination, $m_{\textcircled{\cap}}(A|B) = m \textcircled{\cap} m$, we consider the notation $m_{\textcircled{\cap}}$ more coherent and sensible than Smets's original one.

$$m_{\textcircled{\cap}}(A|B) = \begin{cases} \sum_{X \subseteq B^c} m(A \cup X) & A \subseteq B, \\ 0 & A \not\subset B. \end{cases} \tag{4.49}$$

Remember that in the transferable belief model (see Section 3.3.2) belief functions which assign mass to $\emptyset$ do exist, under the 'open-world' assumption.

The belief and plausibility values of the resulting belief function are, respectively,

$$Bel_{\textcircled{\cap}}(A|B) = \begin{cases} Bel(A \cup \bar{B}) & A \cap B \neq \emptyset, \\ 0 & A \cap B = \emptyset, \end{cases}$$

$$Pl_{\textcircled{\cap}}(A|B) = \begin{cases} Pl(A \cap B) & A \not\supset B, \\ 1 & A \supset B. \end{cases}$$

As it turns out, conjunctive (unnormalised) conditioning is more committal than Dempster's rule, namely

$$Bel_{\oplus}(A|B) \leq Bel_{\textcircled{\cap}}(A|B) \leq Pl_{\textcircled{\cap}}(A|B) \leq Pl_{\oplus}(A|B).$$

According to Klawonn, $Bel_{\textcircled{\cap}}(.|B)$ is also the minimum-commitment specialisation of Bel such that $Pl(B^c|B) = 0$ [975].

4.5.5 Disjunctive rule of conditioning

We can define a conditioning operator dual to conjunctive conditioning, as the one induced by the disjunctive rule of combination: $m_{\textcircled{\cup}}(A|B) = m \textcircled{\cup} m$.

The mass assignment of the resulting conditional belief function is

$$m_{\textcircled{\cup}}(A|B) = \sum_{X \subseteq B} m(A \setminus B \cup X) \quad A \supseteq B,$$

while $m_{\textcircled{\cup}}(A|B) = 0$ for all $A \not\supset B$. Clearly, disjunctive conditioning assigns mass only to subsets containing the conditioning event Bel. The associated belief and plausibility values are

$$Bel_{\textcircled{\cup}}(A|B) = \begin{cases} Bel(A) & A \supset B, \\ 0 & A \not\supset B, \end{cases} \quad Pl_{\textcircled{\cup}}(A|B) = \begin{cases} Pl(A) & A \cap B = \emptyset, \\ 1 & A \cap B \neq \emptyset, \end{cases}$$

i.e., disjunctive conditioning preserves the belief values of focal elements containing the conditioning event, while setting all others to zero.

It can easily be shown that disjunctive conditioning is less committal than both Dempster and credal conditioning:

$$Bel_{\textcircled{\cup}}(A|B) \leq Bel_{\mathrm{Cr}}(A|B) \leq Pl_{\mathrm{Cr}}(A|B) \leq Pl_{\textcircled{\cup}}(A|B).$$

4.5.6 Conditional events as equivalence classes: Spies's definition

In an interesting paper [1749], Spies proposed a very original approach to belief function conditioning based on a link between conditional events and discrete random sets. Conditional events are defined there as *sets of equivalent events under the conditioning relation*. By applying a multivalued mapping (see Section 3.1.1) to them, we obtain a new, intriguing definition of a conditional belief function which lies within the random-set interpretation. An updating rule, which is equivalent to the law of total probability if all beliefs are probabilities, was also introduced.

In more detail, let $(\Omega, \mathcal{F}, P)$ and $\Gamma : \Omega \to 2^\Theta$ be the source probability space and the random set, respectively. The null sets for $P_A = P(.|A)$ are the collection of events with conditional probability 0: $\mathcal{N}(P(.|A)) = \{B : P(B|A) = 0\}$. Let $\triangle$ be the symmetric difference $A \triangle B = (A \cap \bar{B}) \cup (\bar{A} \cap B)$ of two sets A, Bel. We can prove the following Lemma.

Lemma 1. *If Z exists such that $B, C \in Z \triangle \mathcal{N}(P_A)$, then $P(B|A) = P(C|A)$.*

Roughly speaking, two events have the same conditional probability if they are both the symmetric difference between the same event and some null set. We can then define conditional events as the following equivalence classes.

Definition 53. *A conditional event $[B|A]$ with $A, B \subseteq \Theta$ is a set of events with the same conditional probability $P(B|A)$:*

$$[B|A] = B \triangle \mathcal{N}(P_A).$$

It can also be proven that $[B|A] = \{C : A \cap B \subseteq C \subseteq \bar{A} \cup B\}$.

We can now define a *conditional multivalued mapping* for $B \subseteq \Theta$ as $\Gamma_B(\omega) = [\Gamma(\omega)|B]$, where $\Gamma : \Omega \to 2^\Theta$. In other words, if $A = \Gamma(\omega)$, Γ_B maps ω to $[A|B]$. As a consequence, all elements of each conditioning event (an equivalence class) must be assigned equal belief/plausibility, and a conditional belief function is a 'second-order' belief function with values on *collections* of focal elements (the conditional events themselves).

Definition 54. *Given a belief function Bel, the conditional belief function given $B \subseteq \Theta$ is*

$$Bel([C|B]) = P(\{\omega : \Gamma_B(\omega) = [C|B]\}) = \frac{1}{K} \sum_{A \in [C|B]} m(A).$$

It is important to realise that, in this definition, a conditional belief function is *not* a BF on the subalgebra $\{Y = C \cap B, C \subseteq \Theta\}$. It can be proven that Spies's conditional belief functions are closed under Dempster's rule of combination and, therefore, once again, are coherent with the random-set interpretation of the theory.

A multivalued extension of conditional belief functions was also introduced by Slobodova in [1662]. In [1661], she described how conditional BFs (defined as in Spies's approach) fit into the framework of valuation-based systems.

4.5.7 Other work

Inevitably, the above review does not cover all the work that has been conducted on the issue.

Klopotek and Wierzchon [1004], for instance, provided in 1999 a frequency-based interpretation for conditional belief functions. Tang and Zheng [1800] discussed the issue of conditioning in a multidimensional space. Lehrer [1128] proposed a geometric approach to determining the conditional expectation of non-additive probabilities. Such a conditional expectation was then applied for updating, and to introduce a notion of independence.

4.5.8 Conditioning: A summary

Table 4.1 summarises the behaviour of the main conditioning operators in terms of the degrees of belief and plausibility of the resulting conditional belief function.

Table 4.1: Belief and plausibility values of the outcomes of various conditioning operators

Operator	**Belief value**	**Plausibility value**
Dempster's $\oplus$	$\dfrac{Pl(B) - Pl(B \setminus A)}{Pl(B)}$	$\dfrac{Pl(A \cap B)}{Pl(B)}$
Credal Cr	$\dfrac{Bel(A \cap B)}{Bel(A \cap B) + Pl(\bar{A} \cap B)}$	$\dfrac{Pl(A \cap B)}{Pl(A \cap B) + Bel(\bar{A} \cap B)}$
Geometric G	$\dfrac{Bel(A \cap B)}{Bel(B)}$	$\dfrac{Bel(B) - Bel(B \setminus A)}{Bel(B)}$
Conjunctive $\textcircled{\cap}$	$Bel(A \cup \bar{B}), A \cap B \neq \emptyset$	$Pl(A \cap B), A \not\supset B$
Disjunctive $\textcircled{\cup}$	$Bel(A), A \supset B$	$Pl(A), A \cap B = \emptyset$

It is clear from what we have seen that conditioning operators form a nested family, from the most committal to the least:

$$\begin{aligned} Bl_{\textcircled{\cup}}(\cdot|B) \leq Bl_{\text{Cr}}(\cdot|B) \leq Bl_{\oplus}(\cdot|B) \leq Bl_{\textcircled{\cap}}(\cdot|B) \\ \leq Pl_{\textcircled{\cap}}(\cdot|B) \leq Pl_{\oplus}(\cdot|B) \leq Pl_{\text{Cr}}(\cdot|B) \leq Pl_{\textcircled{\cup}}(\cdot|B). \end{aligned}$$

This supports the case for reasoning with intervals of belief functions, rather than individual measures, in particular the intervals associated with conjunctive and disjunctive combination.

An open question is whether geometric conditioning (Chapter 15) is induced by some combination rule dual to Dempster's, and how it fits within this picture. We will discuss this further in Chapter 17.

4.6 Manipulating (conditional) belief functions

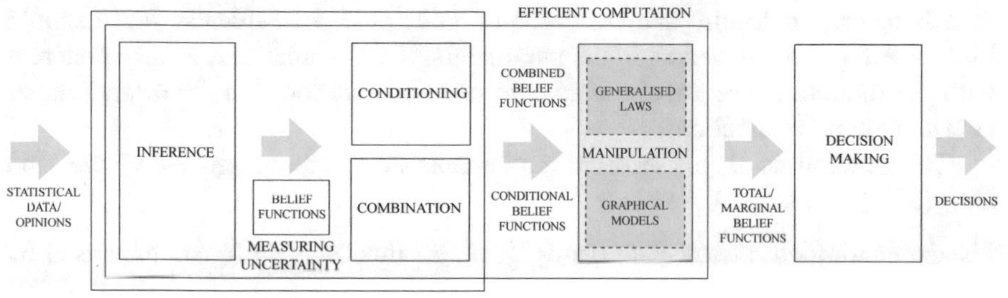

The manipulation of conditional belief functions is an essential feature of belief theory. An important contribution in this sense was made by Philippe Smets's generalised Bayes theorem [1718, 1725], which is discussed in Section 4.6.1. The author of this book has studied, among other things, the issue of combining conditional belief functions into a single total BF, via a generalisation of the classical total probability theorem (Section 4.6.2). Multivariate belief functions (Section 4.6.3) and their efficient treatment via suitable graphical models (Section 4.6.4) are also crucial elements of a reasoning chain within the framework of the theory of evidence.

4.6.1 The generalised Bayes theorem

Smets's generalised Bayes theorem is about generalising the framework of full Bayesian inference to the case of belief functions.

As we know, in classical Bayesian inference a likelihood function $p(.|\theta), \theta \in \Theta$, is known so that we use it to compute the likelihood of a new sample $p(x|\theta), x \in \mathbb{X}$. After observing x, the probability distribution on Θ is updated to the posterior via Bayes' theorem:

$$P(\theta|x) = \frac{P(x|\theta)P(\theta)}{P(x) = \sum_{\theta'} P(x|\theta')P(\theta')} \quad \forall \theta \in \Theta,$$

so that we obtain a posterior distribution on the parameter space.

In Smets's generalised setting, the input is a set of 'conditional' belief functions (rather than likelihoods $p(x|\theta)$, as in the classical case) on Θ,

$$Bel_{\mathbb{X}}(X|\theta) \quad X \subset \mathbb{X}, \theta \in \Theta, \tag{4.50}$$

each associated with a value $\theta \in \Theta$ of the parameter. Note that these are not the same conditional belief functions that we defined in Section 4.5, where a conditioning event $B \subset \Theta$ alters a prior belief function Bel_Θ, mapping it to $Bel_\Theta(.|B)$. Instead, they should be seen as a parameterised family of BFs on the data, which are somehow made available to us.

The desired output is another family of belief functions on Θ, parameterised by all *sets of measurements* X on $\mathbb{X}$,

$$Bel_\Theta(A|X) \quad \forall X \subset \mathbb{X},$$

as it is natural to require that each (conditional) piece of evidence (4.50) should have an effect on our beliefs in the parameters. This requirement is also coherent with the random-set setting, in which we need to condition on *set-valued*, rather than singleton, observations.

The generalised Bayes theorem implements an inference process of the kind $Bel_\mathbb{X}(X|\theta) \mapsto Bel_\Theta(A|X)$ by:

1. Computing an intermediate family of belief functions on $\mathbb{X}$ parameterised by *sets of parameter values* $A \subset \Theta$,

$$Bel_\mathbb{X}(X|A) = \circledcup_{\theta \in A} Bel_\mathbb{X}(X|\theta) = \prod_{\theta \in A} Bel_\mathbb{X}(X|\theta),$$

   via the disjunctive rule of combination $\circledcup$.
2. After this, assuming that $Pl_\Theta(A|X) = Pl_\mathbb{X}(X|A)\ \forall A \subset \Theta, X \subset \mathbb{X}$ immediately yields

$$Bel_\Theta(A|X) = \prod_{\theta \in \bar{A}} Bel_\mathbb{X}(\bar{X}|\theta).$$

The GBT generalises Bayes' rule (by replacing P with Pl) when priors are uniform, and follows from the following two requirements.

Axiom 1 *If we apply the generalised Bayes theorem to two distinct variables X and Y, we obtain the result that*

$$Bel_\Theta(A|X,Y) = Bel_\Theta(A|X) \circledcap Bel_\Theta(A|Y),$$

i.e., the conditional belief function on Θ is the conjunctive combination of the two.

Axiom 2 $Pl_\mathbb{X}(X|A)$ *is a function of* $\{Pl_\mathbb{X}(X|\theta), Pl_\mathbb{X}(\bar{X}|\theta) : \theta \in A\}$.

Under Axiom 1 of the GBT, the two variables are *conditional cognitive independent* (a notion which extends that of stochastic independence), namely

$$Pl_{\mathbb{X}\times\mathbb{Y}}(X \cap Y|\theta) = Pl_\mathbb{X}(X|\theta) \cdot Pl_\mathbb{Y}(Y|\theta) \quad \forall X \subset \mathbb{X}, Y \subset \mathbb{Y}, \theta \in \Theta.$$

Note that Shafer's proposal for statistical inference, in contrast,

$$Pl_\Theta(A|x) = \max_{\theta \in A} Pl_\Theta(\theta|x),$$

does not satisfy Axiom 1. Smets called (somewhat arguably) his Axiom 2 the 'generalised likelihood principle'.

4.6.2 Generalising total probability

A number of researchers have been working on the generalisation to belief functions of a fundamental result of probability theory: the *law of total probability*. This is sometimes called 'Jeffrey's rule' (see Chapter 1, Section 1.3.9), for it can be considered as a generalisation of Bayes' update rule to the case in which the new evidence comes in the form of a probability distribution.

Both laws concern the way non-conditional, joint models can be soundly built from conditional and unconditional information.

Law of total probability Namely, suppose P is defined on a σ-algebra $\mathbb{A}$, and that a new probability measure P' is defined on a subalgebra $\mathbb{B}$ of $\mathbb{A}$. We seek an updated probability P'' which:

- has the probability values specified by P' for events in the subalgebra $\mathbb{B}$;
- is such that $\forall\, B \in \mathbb{B}$, $X, Y \subset B$, $X, Y \in \mathbb{A}$,

$$\frac{P''(X)}{P''(Y)} = \begin{cases} P(X)/P(Y) & \text{if } P(Y) > 0, \\ 0 & \text{if } P(Y) = 0. \end{cases}$$

It can be proven that there is a unique solution to the above problem, given by Jeffrey's rule:

$$P''(A) = \sum_{B \in \mathbb{B}} P(A|B) P'(B). \tag{4.51}$$

The most compelling interpretation of such a scenario is that the initial probability measure *stands corrected by the second one* on a number of events (but not all). Therefore, the law of total probability generalises standard conditioning, which is just the special case in which $P'(B) = 1$ for some Bel and the subalgebra $\mathbb{B}$ reduces to a single event Bel.

Spies's solution As a matter of fact, Spies proved the existence of a solution to the generalisation of Jeffrey's rule to belief functions, within his conditioning framework (Section 4.5.6). The problem generalises as follows.

Let $\Pi = \{B_1, \ldots, B_n\}$ be a disjoint partition of Θ, and let:

- $m_1, \ldots, m_n$ be the mass assignments of a collection of conditional belief functions $Bel_1, \ldots, Bel_n$ on $B_1, \ldots, B_n$, respectively;
- m be the mass of an *unconditional* belief function Bel_B on the coarsening associated with the partition Π.

The following proposition is then true.

Proposition 27. *The belief function* $Bel_{tot} : 2^\Theta \to [0, 1]$ *such that*

$$Bel_{tot}(A) = \sum_{C \subseteq A} \left(m \oplus \bigoplus_{i}^{n} m_{B_i} \right)(C) \tag{4.52}$$

is a marginal belief function on Θ, such that $Bel_{tot}(.|B_i) = Bel_i$ $\forall i$ and the marginalisation of Bel_{tot} to the partition Π coincides with $Bel_{\mathbb{B}}$. Furthermore, if all the belief functions involved are probabilities, then Bel_{tot} reduces to the result of Jeffrey's rule of total probability.

In other words, (4.52) is an admissible solution to the generalised total probability problem. The bottom line of Proposition 27 is that by combining the unconditional, a priori belief function with all the conditionals we get an admissible marginal which generalises total probability.

Whether this is this the only admissible solution to the problem will be discussed in Section 17.1.3.

Smets's generalisations of Jeffrey's rule Philippe Smets also proposed generalisations of Jeffrey's rule based on geometric and Dempster conditioning.

Let $B(A)$ the smallest element of $\mathbb{B}$ containing A (the upper approximation of A in rough set theory: see Section 6.7), and let $\mathcal{B}(A)$ be the set of As which share the same $B(A)$. We seek an overall belief function Bel'' such that

- $Bel''(B) = Bel'(B)$ for all $B \in \mathbb{B}$;
- for all $B \in \mathbb{B}$, $X, Y \subset B$, $X, Y \in \mathbb{A}$,

$$\frac{Bel'(X)}{Bel''(Y)} = \begin{cases} \dfrac{Bel(X|B)}{Bel(Y|B)} & Bel(Y|B) > 0, \\ 0 & Bel(Y|B) = 0, \end{cases} \tag{4.53}$$

where $Bel(.|B)$ is defined by either geometric conditioning or Dempster conditioning.

In the former case, we obtain Smets's *Jeffrey–geometric* rule of conditioning. Recalling that, by (4.48),

$$\frac{Bel_{\mathrm{G}}(X|B)}{Bel_{\mathrm{G}}(Y|B)} = \frac{Bel(X \cap B)}{Bel(Y \cap B)} = \frac{Bel(X)}{Bel(Y)},$$

we have

$$m_{\mathrm{JG}}(A) = \frac{m(A)}{\sum_{X \in \mathcal{B}(A)} m(X)} m'(B(A)) \quad \forall A \in \mathbb{A} \text{ s.t. } \sum_{X \in \mathcal{B}(A)} m(X) \neq 0;$$

$m_{\mathrm{JG}}(A) = 0$ if A is such that $\sum_{X \in \mathcal{B}(A)} m(X) = 0$. Whenever $m'(B) = 1$ for a single Bel (Bel' is 'categorical' or 'logical'), this reduces to geometric conditioning.

When we plug Dempster conditioning into (4.53), we get the *Jeffrey–Dempster* rule of conditioning,

$$m_{\mathrm{JD}}(A) = \frac{m(A|B(A))}{\sum_{X \in \mathcal{B}(A)} m(X|B(A))} m'(B(A)) \; \forall A \in \mathbb{A} \text{ s.t. } \sum_{X \in \mathcal{B}(A)} m(X|B(A)) \neq 0,$$

and 0 otherwise. Again, whenever $m'(B) = 1$ for a single Bel, this reduces to Dempster conditioning.

Ruspini's work Ruspini [1514] also reported results on approximate deduction in the context of belief and interval probability theory, assuming approximate conditional knowledge about the truth of conditional propositions expressed as sets of possible values (actually numeric intervals) of conditional probabilities. Under different interpretations of this conditional knowledge, several formulae were produced to integrate unconditioned estimates (assumed to be given as sets of possible values of unconditioned probabilities) with conditional estimates.

4.6.3 Multivariate belief functions

In many applications, we need to express uncertain information about a number of distinct variables (e.g., X and Y) taking values in different domains (Θ_X and Θ_Y, respectively). The reasoning process then needs to take place in the Cartesian product of the frames of discernment associated with each individual variable (i.e., for the case of two variables, $\Theta_{XY} = \Theta_X \times \Theta_Y$). The resulting 'multivariate' analysis of belief functions is really a special case of the more general notion of belief functions defined on elements of a family of compatible frames (compare Chapter 2, Section 2.5.2). For instance, $\Theta_X = \{\text{red}, \text{blue}, \text{green}\}$ and $\Theta_Y = \{\text{small}$, medium, large$\}$ may be the domains of the two attributes X and Y describing, respectively, the colour and the size of an object of interest. In such a case, their minimal (common) refinement $\Theta_X \otimes \Theta_Y = \Theta_X \times \Theta_Y$ (compare Theorem 1) reduces to their Cartesian product.

Let Θ_X and Θ_Y then be two compatible frames associated with two distinct variables, and let m^{XY} be a mass function on $\Theta_X \times \Theta_Y$. The latter can be expressed in the coarser frame Θ_X, for instance, by transferring each mass $m^{XY}(A)$ to the projection of A on Θ_X (see Fig. 4.11(a)). We obtain a *marginal* mass function on Θ_X, denoted by $m^{XY\downarrow X}$:

$$m^{XY\downarrow X}(B) = \sum_{\{A \subseteq \Theta_{XY}, A\downarrow\Theta_X = B\}} m^{XY}(A) \quad \forall B \subseteq \Theta_X.$$

We can also interpret the vacuous extension (Definition 20) on Cartesian products of frames as the 'inverse' of marginalisation. Namely, a mass function m^X on Θ_X can be expressed in $\Theta_X \times \Theta_Y$ by transferring each mass $m_X(B)$ to the *cylindrical extension* of Bel (Fig. 4.11(b)). The vacuous extension of m_X onto $\Theta_X \times \Theta_Y$ will then be

$$m^{X\uparrow XY}(A) = \begin{cases} m^X(B) & \text{if } A = B \times \Omega_Y, \\ 0 & \text{otherwise.} \end{cases}$$

Multivariate belief function analysis is strongly linked to the notion of a *graphical model* (see Section 4.6.4), as these are structures designed to efficiently represent uncertainty measures defined over Cartesian products of sample spaces.

Belief function factorisation Multivariate analysis as a building block for a theory of graphical models for belief functions was mainly developed in the late 1980s and

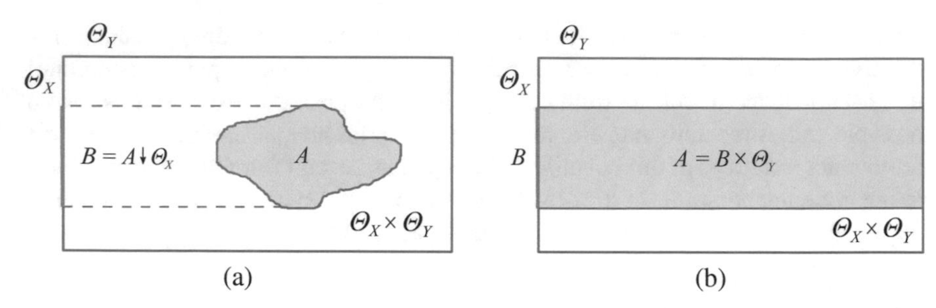

Fig. 4.11: (a) The process of marginalising a focal element of the product frame $\Theta_X \times \Theta_Y$. (b) The cylindrical extension of a focal element onto the product frame $\Theta_X \times \Theta_Y$.

early 1990s, starting from the seminal thesis by Augustine Kong [1033], centred on the notion of a *hypergraph*. Graphical models for belief functions were later studied by Shafer, Shenoy and Mellouli [1611, 1278], and Shenoy and Shafer [1630, 1609, 1631]. The issue is closely related to that of *local propagation*, which is elaborated upon in Section 4.7.4.

Following Kong's idea, Almond [28] proposed a simple procedure for transforming a model hypergraph into a tree of closures. In that author's words, this is a tree of 'chunks' of the original problem, in which each 'chunk' can be computed independently of all other chunks except its neighbours. Each node in the tree of closures passes messages (expressed as belief functions) to its neighbours, consisting of the local information fused with all the information that has propagated through the other branches of the tree, allowing one to efficiently compute marginal beliefs. A substantial portion of Almond's book [32] is devoted to a single example, of calculating the reliability of a complex system with the help of the graphical modelling of belief functions, incorporating practical notes from the author's own experience with the BELIEF software package. Graphical belief models are described there, together with the algorithms used in their manipulation.

Thoma [1813] proposed methods for storing a multivariate belief function efficiently, based on factorisations and decompositions of both the belief function and its individual focal elements. As a result, computations, such as combination and marginalisation, can be done locally, without first reconstructing the belief function from its storage components, on simplified hypergraphs. Thoma studied how such factorisation properties change as a belief function undergoes projection or is combined with other belief functions, and showed that commonality, one of the four combinatorial representations of a belief function, can be interpreted as the Fourier transform of the basic probability assignment. In analogy to the fast Fourier transform, then, a fast Möbius transform algorithm was developed (see Section 4.7.1).

Other related work was done by Xu [1987], who developed a theory of local computation on Markov trees and presented a method for computing the marginal for a subset which may not be contained in one vertex, unlike what is required in other existing techniques, but is a subset of the union of several vertices. Chen

[262], instead, discussed an application of multivariate belief functions to a spatial reasoning problem in which an expert wishes to identify anatomical structures in a set of correlated images acquired from X-ray CT and MRI. Different kinds of evidence provided by various knowledge sources form a hierarchy of frames, where multivariate belief functions are used to represent domain-specific knowledge.

4.6.4 Graphical models

Evolved from local propagation models, a number of graphical models for reasoning with conditional belief functions[39] were later developed. Among the most significant contributions, we should cite Cano et al.'s uncertainty propagation in directed acyclic networks [219], Xu and Smets's evidential networks with conditional belief functions [1995], Shenoy's valuation networks [1529, 1626], and Ben Yaghlane and Mellouli's directed evidential networks [116, 113].

Probabilistic graphical models Conditional independence relationships are the building block of probabilistic graphical models [1398, 1101, 1410], in which conditional probabilities are directly manipulated using Bayes' rule.

The support for a (probabilistic) graphical model is a *directed acyclic graph* (DAG) $G = (V, E)$, in which each node $v \in V$ is associated with a random variable X_v. The set of random variables $X = \{X_v, v \in V\}$ is a *Bayesian network* with respect to G if its joint probability density function is the product of the individual density functions, conditional on their parent variables, namely

$$p_X(x) = \prod_{v \in V} p\left(x_v \middle| x_{pa(v)}\right), \tag{4.54}$$

where $pa(v)$ is the set of parents of v. The condition (4.54) expresses the *conditional independence* of the variables from any of their non-descendants, given the values of their parent variables.

Belief propagation [1398], in particular, is a message-passing algorithm for performing inference on graphical models, which calculates the marginal distribution for each unobserved node, conditional on any observed nodes. The algorithm works by passing real-valued functions called *messages* $\mu_{v \to u}$, each message travelling from a node v to one of its neighbours $N(v)$, $u \in N(v)$, along an edge of the graph. Upon convergence, the estimated marginal distribution of each node is proportional to the product of all messages from adjoining factors (bar the normalisation constant):

$$p_{X_v}(x_v) \propto \prod_{u \in N(v)} \mu_{u \to v}(x_v).$$

[39] Graphical models for dealing with other types of uncertainty measures also exist, including for instance Cozman's *credal networks* [312] and Tessem's algorithm for interval probability propagation [1809]. A useful 2005 survey of graphical models for imprecise probabilities, also due to Cozman, can be found in [313].

Graphical models for belief functions Graphical models for reasoning with belief functions have also been developed. A significant difference is that, in networks using belief functions, relations among variables are usually represented by *joint* belief functions rather than conditional ones. Furthermore, such networks typically have *undirected graphs* as support, rather than DAGs (Fig. 4.12). Examples include hypertrees [1630], qualitative Markov trees [1611, 65], join trees [1628] and valuation networks [1626].

The origin of this line of research is a seminal result proven by Shenoy and Shafer, who showed that if combination and marginalisation satisfy three sensible axioms, then local computation becomes possible (see Section 4.7.4). Later on, Cano et al. [219] showed that adding three more axioms to the original ones allows us to use Shenoy and Shafer's axiomatic framework for propagation in DAGs, as in the case of classical Bayesian networks.

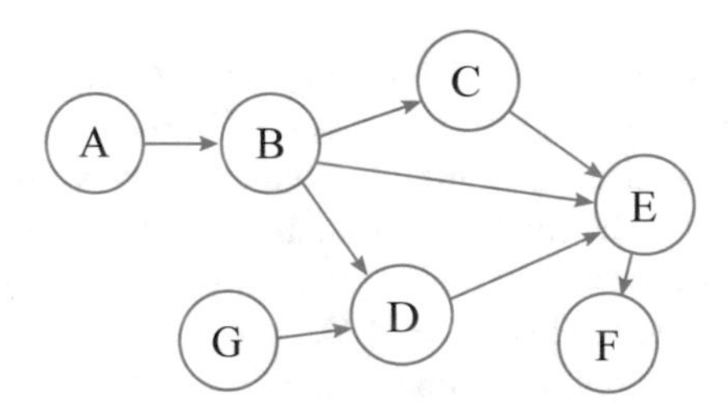

Fig. 4.12: An example of a directed acyclic graph.

Working with graphs of conditional belief function independence relations, however, is arguably more efficient [1626], for it allows an exponential number of conditional independence statements to be represented by a graph with a polynomial number of vertices. Owing to a lack of directed belief networks directly comparable to Bayesian networks, some recent publications have attempted to integrate belief function theory and Bayesian nets [1653]. Examples are Cobb and Shenoy's [291] work on plausibility transformation between models, and Simon and Weber's [1652] implementation of belief calculus in classical Bayesian networks.

Evidential networks with conditional belief functions An alternative line of research has evolved from Xu and Smets's notion of *evidential networks with conditional belief functions* (ENCs) [1995]. These networks use directed acyclic graphs in which, however, edges represent conditional relations (i.e., the fact that the values of a parent variable X_v generate conditional belief functions on the values of another variable X_u), rather than conditional independence relations (as in Bayesian networks). The generalised Bayesian theorem (see Section 4.6.1) is used for propagation on networks of conditional relations.

A strong limitation of propagation on ENCs is that they only apply to binary relations between nodes. In response, Ben Yaghlane and Mellouli proposed the class of *directed evidential networks* (DEVNs) [116, 113], which generalise ENCs to relations involving any number of nodes.

Directed evidential networks DEVNs are directed acyclic graphs in which directed arcs describe conditional dependence relations expressed by conditional belief functions for each node given its parents. New observations introduced into the network are represented by belief functions allocated to some nodes.

The problem is the following:[40] given a number of belief functions $\{Bel^0_v, v \in V\}$ presented to the nodes $v \in V$ of a DEVN, we seek the marginal on each node v of their joint belief function. If there is a conditional relation between two nodes v and u, the framework uses disjunctive combination (4.26) and the generalised Bayesian theorem to compute the posterior belief function $Bel_v(A|B)$ given the conditional $Bel_u(B|A)$, for all $A \subseteq \Theta_v$, $B \subseteq \Theta_u$, with Θ_v and Θ_u being the FoDs for variables X_v and X_u, respectively. The desired marginal is computed for each node v by combining by Dempster's rule $\oplus$ all the messages received from its neighbours u, each with prior belief Bel^u_0, and its own prior belief Bel^v_0:

$$Bel^v = Bel^v_0 \oplus Bel_{u \to v}, \quad Bel_{u \to v}(A) = \sum_{B \subseteq \Theta_u} m^u_0(B) Bel^v(A|B),$$

where $Bel^v(A|B)$ for all $A \subseteq \Theta_v$, $B \subseteq \Theta_u$ is given by the GBT.

We recall here the (simpler) propagation algorithm for polytrees [113]. Each variable of the network has a λ *value* and a π *value* associated with it (Algorithm 2). Whenever a new observation Bel^O is inputted into a node v, the state of the network is updated via Algorithm 3, recursively on the input nodes.

Ben Yaghlane et al. also proposed a simplified scheme for simply directed networks, and an extension to DEVNs by first transforming them to binary join trees.

Compositional models A recent research strand in multivariate belief function analysis is concerned with Radim Jirousek and co-authors' *compositional models* [911]. Compositional models were introduced for probability theory in [903] as an alternative to Bayesian networks for efficient representation of multidimensional measures. Jirusek showed in [910] that multidimensional basic assignments can be rather efficiently represented in the form of compositional models based on the iterative application of a composition operator.

Let $\Theta_1, \ldots, \Theta_n$ be the frames of discernment associated with the variables $X_1, \ldots, X_n$, and denote by Θ_S, $S \subseteq \{1, \ldots, n\}$, the Cartesian product of the frames associated with variables in the set S. Let the *join* $A \bowtie B$ of two sets $A \subseteq \Theta_K$ and $B \subseteq \Theta_L$ be the intersection of their cylindrical extension in $\Theta_{K \cup L}$:

$$A \bowtie B = \{\theta \in \Theta_{K \cup L} : \theta^{\downarrow K} \in A, \theta^{\downarrow L} \in B\}.$$

[40] This book's notation.

Algorithm 2 DEVN on polytrees – initialisation ([113], Section 5.1)

procedure DEVNINIT($Bel_v^0, v \in V$)

for each node v: $Bel_v \leftarrow Bel_v^0$, $\pi_v \leftarrow Bel_v$, $\lambda_v \leftarrow$ the vacuous BF;

for each root node, send a new message $\pi_{v \to u}$ to all children u of v as

$$\pi_{v \to u} = Bel_{v \to u}, \quad Bel_{v \to u}(A) = \sum_{B \subset \Theta_v} m_v^0(B) Bel_u(A|B) \; \forall A \subseteq \Theta_u;$$

node u waits for the messages from all its parents, then it:

- computes the new π_u value via

$$\pi_u = Bel_u \oplus \left(\bigoplus_{v \in pa(u)} \pi_{v \to u} \right);$$

- computes the new marginal belief $Bel_u \leftarrow \pi_u \oplus \lambda_u$;
- sends the new π_u message to all its children.

end procedure

Algorithm 3 DEVN on polytrees – updating ([113], Section 5.2)

procedure DEVNUPDATE

node v computes its new value $Bel_v = \pi_v \oplus \lambda_v$, where

$$\pi_v = Bel_v^0 \oplus \left(\oplus_{u \in pa(v)} \pi_{u \to v} \right), \quad \lambda_v = Bel_v^O \oplus \left(\oplus_{w \in ch(v)} \lambda_{w \to v} \right);$$

for every child node $w \in ch(v)$, we calculate and send the new message to all children not yet updated:

$$\pi_{v \to w} = Bel_{v \to w}, \quad Bel_{v \to w}(A) = \sum_{B \subset \Theta_v} m_v(B) Bel_w(A|B),$$

where $Bel_w(A|B)$ is the disjunctive combination of the initial conditionals;

for every parent node u, we compute a new message $\lambda_{v \to u}$ and send it to all parents not yet updated:

$$\lambda_{v \to u} = Bel_{v \to u}, \quad Bel_{v \to u}(A) = \sum_{B \subset \Theta_v} m_v(B) Bel_u(A|B),$$

where $A \subseteq \Theta_u$, and $Bel_u(A|B)$ is the posterior given by the GBT.

end procedure

Definition 55. *For two arbitrary basic assignments m_1 on Θ_K and m_2 on Θ_L, where K and L are two non-empty sets of variables, their* composition $m_1 \triangleright m_2$ *is defined for each $C \subseteq \Theta_{K\cup L}$ by one of the following expressions:*

1. *If $m_2^{\downarrow K\cap L}(C^{\downarrow K\cap L}) > 0$ and $C = C^{\downarrow K} \bowtie C^{\downarrow L}$, then*

$$m_1 \triangleright m_2(C) = \frac{m_1(C^{\downarrow K}) m_2(C^{\downarrow L})}{m_2^{\downarrow K\cap L}(C^{\downarrow K\cap L})}.$$

2. *If $m_2^{\downarrow K\cap L}(C^{\downarrow K\cap L}) = 0$ and $C = C^{\downarrow K} \times \Theta_{L\setminus K}$, then*

$$m_1 \triangleright m_2(C) = m_1(C^{\downarrow K}).$$

3. *In all other cases, $m_1 \triangleright m_2(C) = 0$.*

Case 1 copies the idea of probabilistic composition, while case 2 is the situation in which the two masses are in conflict, and the whole mass of m_1 is assigned to the respective least informative subset of $\Theta_{K\cup L}$, i.e, $C^{\downarrow K} \times \Theta_{L\setminus K}$. Note that, unlike combination operators, the operator of composition serves to compose pieces of local information, usually coming from one source, into a global model. The resulting mass $m_1 \triangleright m_2$ is a BPA on $\Theta_{K\cup L}$, whose projection onto Θ_K returns m_1. It is commutative iff $m_1^{\downarrow K\cap L} = m_2^{\downarrow K\cap L}$.

Factorisation is defined in the following way [905].

Definition 56. *A BPA m on $\Theta_{K\cup L}$* factorises with respect to the pair (K, L) *if there exist two non-negative set functions*

$$\phi : 2^{\Theta_K} \to [0, +\infty), \quad \psi : 2^{\Theta_L} \to [0, +\infty)$$

such that for all $A \subseteq \Theta_{K\cup L}$,

$$m(A) = \begin{cases} \phi(A^{\downarrow K})\psi(A^{\downarrow L}) & A^{\downarrow K} \bowtie A^{\downarrow L}, \\ 0 & \textit{otherwise}. \end{cases}$$

The above definition requires the mass value $m(A)$ to be a function of just two values, $\phi(A^{\downarrow K}), \psi(A^{\downarrow L})$, and generalises that of Dempster [414] and Walley and Fine [1878]:

$$m(A) = \begin{cases} m^{\downarrow 1}(A^{\downarrow 1}) m^{\downarrow 2}(A^{\downarrow 2}) & A = A^{\downarrow 1} \times A^{\downarrow 2}, \\ 0 & \text{otherwise}. \end{cases}$$

In [907], Jirousek and Shenoy compared the factorisation approaches based on Dempster's rule and on the composition operator, showing that both are equivalent in the restricted case of 'unconditional' factorisation.

The notion of factorisation is closely connected to that of conditional independence. An alternative definition of conditional independence can be expressed in terms of commonality functions: $Q^{\downarrow K\cup L}(A) = Q^{\downarrow K}(A^{\downarrow K}) Q^{\downarrow L}(A^{\downarrow L})$; however, this was proven by Studeny not to be consistent with marginalisation [1774]. In response, Jirousek and Vejnarov proposed the following definition [910].

Definition 57. *The groups of variables X_K and X_L are conditionally independent given X_M with respect to m, where m is a BPA on $\Theta_{\{1,\ldots,n\}}$ and $K, L, M \subset \{1, \ldots, n\}$ are disjoint, if for any $A \subseteq \Theta_{K \cup L \cup M}$ which is a join of the form $A = A^{\downarrow K \cup M} \bowtie A^{\downarrow L \cup M}$ we have*

$$m^{\downarrow K \cup L \cup M}(A) \cdot m^{\downarrow M}(A^{\downarrow M}) = m^{\downarrow K \cup M}(A^{\downarrow K \cup M}) \cdot m^{\downarrow L \cup M}(A^{\downarrow L \cup M}),$$

whereas $m^{\downarrow K \cup L \cup M}(A) = 0$ for all other $A \subseteq \Theta_{K \cup L \cup M}$.

Belief-theoretical graphical models possessing the same Markov properties as those that hold for probabilistic graphical models, under the assumption that one accepts a new definition of conditional independence, were developed in [904]. A more recent publication [906] has explored the possibility of transferring Lauritzen–Spiegelhalter local computation models from probability theory to the theory of evidence, by discussing the connection between model structure and systems of conditional independence relations, and showing that under additional conditions one can locally compute specific basic assignments which can be considered to be conditional.

4.7 Computing

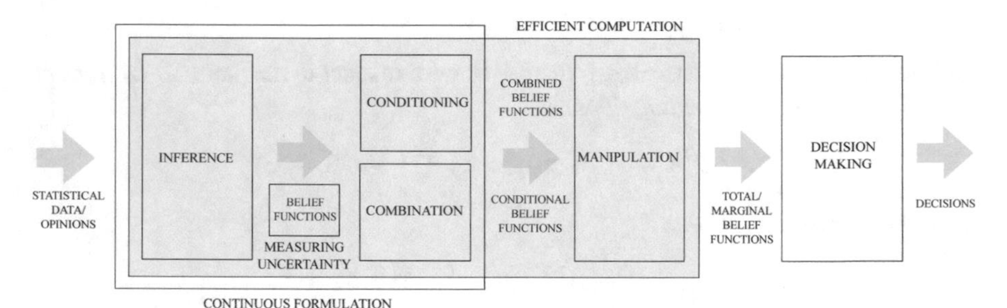

Belief functions are complex objects [1812]. Working with them in a naive way is computationally expensive, preventing their application to problems (such as some in engineering [72]) in which computational complexity is crucial. As pointed out by Orponen [1366], while belief, plausibility and commonality values $Bel(A)$, $Pl(A)$ and $Q(A)$ can be computed in polynomial time, the problems of computing the combinations $(Bel_1 \oplus \cdots \oplus Bel_n(A)$, $(Pl_1 \oplus \cdots \oplus Pl_n)(A)$ and $(Q_1 \oplus \cdots \oplus Q_n)(A)$ are $\#$P-complete. In an attempt to address this issue, Srivastava [1757] developed an alternative form of Dempster's rule of combination for binary variables, which provides a closed-form expression for efficient computation.

The issue has been recognised since the early days of the theory of evidence [88]. While some authors (e.g. Voorbraak) proposed to tackle it by transforming belief functions into suitable probability measures prior to their aggregation (compare Section 4.7.2), others focused on the efficient approximate implementation of the rule of combination [371, 1992, 1127]. As a result, Monte Carlo approaches (Section 4.7.3) were proposed [1318, 1950], together with local propagation schemes

(Section 4.7.4), which later evolved into message-passing frameworks having similarities to modern Bayesian networks.

Essentially, belief functions have a problem with computational complexity only if naively implemented. Just as Bayesian inference on graphical models is NP-hard, even in approximate form, but can nevertheless be approached by Monte Carlo methods, inference and updating with belief functions are implementable to arbitrary accuracy. We do not need to assign mass values to all subsets of a potentially infinite domain, but we do need to be allowed to assign mass to subsets whenever this is the correct thing to do. Indeed, elementary items of evidence induce simple belief functions, which can be combined very efficiently. Furthermore, the most plausible hypothesis $\arg\max_{x \in \Theta} Pl(\{x\})$ can be found without computing the whole combined belief function [89].

4.7.1 Efficient algorithms

All aspects of the decision/estimation process are affected, from the efficient encoding of focal elements [770] to the computation of (relative) plausibilities [89], the evidence combination stage and also decision making [1992].

A nice, albeit rather dated, survey of almost all computational aspects of belief theory can be found in [1946].

Representing focal elements Haenni and Lehmann discussed in [770] several implementation aspects of multivariate belief theory, with the aim of proposing an appropriate representation of mass functions and efficient data structures and algorithms for combination and marginalisation.

There, in particular, they proposed a representation of focal elements which allows us to efficiently compute the main operations $X_1 \cap X_2$, equality testing ($X_1 = X_2$), projection onto a coarsening and (cylindrical) extension. Using a binary encoding, the binary representation of an intersection is the logical 'and' of those associated with X_1 and X_2. Equality can be tested by a simple comparison of the corresponding bit strings (in MATLAB™, `https://uk.mathworks.com/products/matlab/`, this is done with a single instruction). Simple algorithms for projection and extension of focal elements represented as binary strings are described in [770] in terms of logical 'or' and bit shift operators.

For large sets of configurations, disjunctive (DNF) and conjunctive (CNF) normal forms become attractive in the special case in which only binary variables are considered. *Ordered binary decision diagrams* [196] are also appealing, since polytime algorithms exist for all the required operations (intersection, equality testing, projection and extension).

Recently (2014), Liu [1193] has proposed representing a focal element as an integer, in order to reduce set operations to efficient bitwise operations.

Reducing the list of focal elements Various truncated schemes in which only focal elements with a mass above a certain threshold and/or up to a maximum number are retained have also been proposed.

Algorithm 4 A truncated Dempster–Shafer algorithm [1838]

procedure TRUNCATEDDS
    All combined propositions with BPA > MAX_BPM are kept
    All combined propositions with BPA < MIN_BPM are discarded
    for increasing cardinality of focal elements i **do**
        if number of retained propositions is smaller than MAX_NUM **then**
            retain, by decreasing BPA, propositions of size i
        end if
    end for
    if number of retained propositions is still smaller than MAX_NUM **then**
        retain propositions by decreasing BPA regardless of length
    end if
end procedure

An example is the truncation scheme proposed by Valin and Boily in [1838] in a naval and airborne sensor fusion scenario. Their *truncated Dempster–Shafer* (TDS) scheme retains propositions according to the rules in Algorithm 4 [1838]. In the same spirit, Tessem [1810] incorporates only the highest-valued focal elements in his m_{klx} approximation. A similar approach inspired the *summarisation* technique of Lowrance et al. [1218]. The latter eliminates irrelevant details from a mass distribution by collecting together all of the extremely small amounts of mass attributed to propositions and attributing the sum to the disjunction of those propositions. More sophisticated schemes for approximating a belief structure by clustering its focal elements were proposed by Harmanec [797] and Petit-Renaud and Denoeux [1424].

Building on the idea of approximation by focal-element clustering, Denoeux's work on inner and outer clustering [449] is about describing the original belief function Bel in terms of an inner approximation Bel^- and an outer approximation Bel^+ such that $Pl^- \leq Pl \leq Pl^+$.

Definition 58. *Given a disjoint partition* $\Pi = \mathcal{F}_1, \ldots, \mathcal{F}_K$, $\mathcal{F}_i \cap \mathcal{F}_j = \emptyset$, *of the focal sets of* Bel, *the* strong outer approximation *[533] of* Bel *is the new belief function* Bel^+_Π *with focal elements* $G_k = \cup_{F \in \mathcal{F}_k} F$ *formed by the unions of focal elements in each partition,* $k = 1, \ldots, K$, *and mass*

$$m^+_\Pi(G_k) = \sum_{F \in \mathcal{F}_k} m(F) \quad k = 1, \ldots, K. \tag{4.55}$$

See Fig. 4.13 for an example. As noted by both Harmanec [797] and Petit-Renaud [1424], the resulting mass function m^+_Π *strongly includes* m ($m \subseteq m^+_\Pi$), i.e., there exists a mass redistribution from focal elements of m^+_Π to focal elements of m. Dually, a *strong inner approximation* $m^-_\Pi \subseteq m$ of m can be constructed by taking for each group of focal elements $\mathcal{F}_j$ in the partition Π the intersection $H_j = \cap_{F \in \mathcal{F}_j} F$ of the original focal elements there, and mass $m^-_\Pi(H_j) = \sum_{F \in \mathcal{F}_j} m(F)$.

The question of what partitions generate 'good' inner and outer approximations arises. Harmanec proposed to measure this loss of accuracy via either of

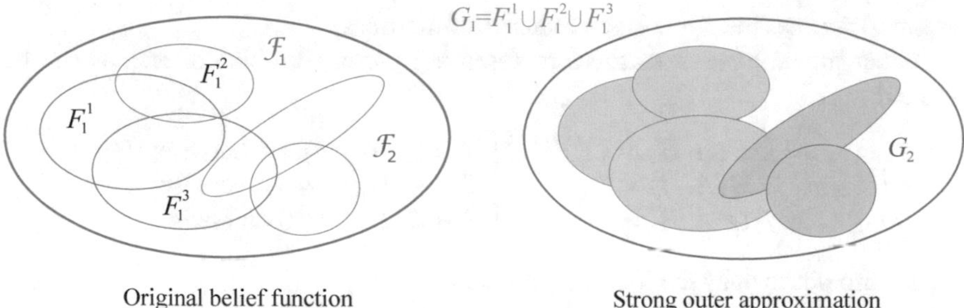

Fig. 4.13: An example of a strong outer approximation, in the case of $K = 2$ clusters $\mathcal{F}_1$, $\mathcal{F}_2$ of focal elements.

$$\sum_A Bel(A) - Bel_{\Pi}^{+/-}(A), \quad \sum_A Pl(A) - Pl_{\Pi}^{+/-}(A).$$

Alternatively, any measure of information content (compare Section 4.2) could be used. The question then reduces to a minimisation problem over the set of all possible partitions of focal elements of Bel, which is computationally intractable. A hierarchical clustering algorithm can, however, be employed to identify the optimal partition(s). At each step, two focal elements are merged, and the mass is transfered to their intersection or their union. The resulting approximations allow the calculation of lower and upper bounds on the belief and plausibility degrees induced by the conjunctive or disjunctive sum of any number of belief structures.

In a subsequent paper [110], Denoeux and Ben Yaghlane [466] reduced the problem of finding a suitable clustering of focal elements to that of classifying the columns of a binary data matrix, for which problem a hierarchical clustering procedure was proposed. The resulting inner and outer approximations can be combined efficiently in the coarsened frame using the fast Möbius transform (compare the next section).

Fast Möbius transform Given an ordering of the subsets of the frame of discernment Θ, mass, belief and plausibility functions can be represented as vectors, denoted by $\mathbf{m}$, $\mathbf{bel}$ and $\mathbf{pl}$. Various operations with belief functions can then be expressed via linear algebra operators acting on vectors and matrices [1722].

We can define the *negation* of a mass vector $\mathbf{m}$ as $\overline{m}(A) = m(\overline{A})$. Smets showed that $\overline{\mathbf{m}} = J\mathbf{m}$, where J is the matrix whose inverse diagonal is made of 1s. Given a mass vector $\mathbf{m}$, the vector $\mathbf{b}$ representing the *implicability* function $b(A) = \sum_{\emptyset\subseteq B\subseteq A} m(B)$ (recall that in general $b(A) \neq Bel(A) = \sum_{\emptyset\subsetneq B\subseteq A} m(B)$ for unnormalised belief functions) turns out to be given by $\mathbf{b} = BfrM\mathbf{m}$, where $BfrM$ is the transformation matrix such that $BfrM(A,B) = 1$ iff $B \subseteq A$ and 0 otherwise. Notably, such transformation matrices can be built recursively, as

$$BfrM_{i+1} = \begin{bmatrix} 1 & 0 \\ 1 & 1 \end{bmatrix} \otimes BfrM_i,$$

where $\otimes$ denotes the Kronecker product[41] of matrices.

Other transformation matrices representing various Möbius inversions can be defined:

$$
\begin{array}{lll}
MfrB(A,B) = (1)^{|A|-|B|} & \text{if } B \subseteq A, & = 0 \text{ otherwise;} \\
QfrM(A,B) = 1 & \text{if } A \subseteq B, & = 0 \text{ otherwise;} \\
MfrQ(A,B) = (1)^{|B|-|A|} & \text{if } A \subseteq B, & = 0 \text{ otherwise.}
\end{array}
$$

These turn out to obey the following relations:

$$
MfrB = BfrM^{-1}, \quad QfrM = JBfrMJ, \quad MfrQ = JBfrM^{-1}J.
$$

The vectors associated with classical, normalised belief functions and with plausibility functions can finally be computed as $\mathbf{Bel} = \mathbf{b} - b(\emptyset)\mathbf{1}$ and $\mathbf{pl} = \mathbf{1} - J\mathbf{b}$, respectively.

An interesting application of matrix calculus is the computation of the fast Möbius transform [954, 955], proposed by Kennes and Smets to efficiently compute the various Möbius transforms involved in belief calculus (e.g. from Bel to m). The fast Möbius transform consists of a series of recursive calculations, illustrated for the case of a frame $\Omega = \{a, b, c\}$ in Fig. 4.14.

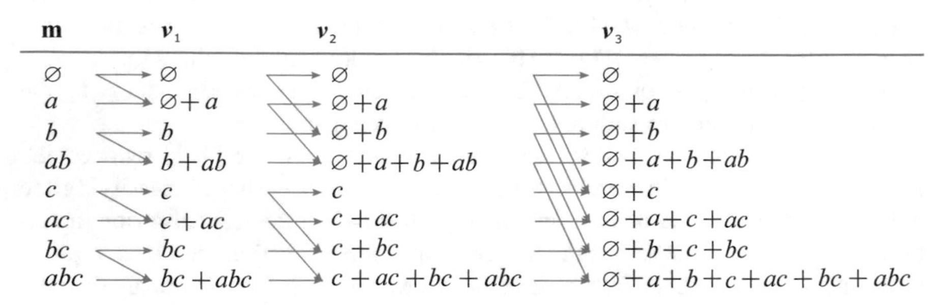

Fig. 4.14: Details of the fast Möbius transform when $\Omega = \{a, b, c\}$. The symbols a, ab etc. denote $m(a)$, $m(a, b)$ and so on.

The related series of computations can also be expressed in matrix form as $BfrM = M_3 \cdot M_2 \cdot M_1$, where the three matrices are as follows [1722]:

$$
M_1 = \begin{bmatrix}
1&0&0&0&0&0&0&0\\
1&1&0&0&0&0&0&0\\
0&0&1&0&0&0&0&0\\
0&0&1&1&0&0&0&0\\
0&0&0&0&1&0&0&0\\
0&0&0&0&1&1&0&0\\
0&0&0&0&0&0&1&0\\
0&0&0&0&0&0&1&1
\end{bmatrix}, \quad
M_2 = \begin{bmatrix}
1&0&0&0&0&0&0&0\\
0&1&0&0&0&0&0&0\\
1&0&1&0&0&0&0&0\\
0&1&0&1&0&0&0&0\\
0&0&0&0&1&0&0&0\\
0&0&0&0&0&1&0&0\\
0&0&0&0&1&0&1&0\\
0&0&0&0&0&1&0&1
\end{bmatrix}, \quad
M_3 = \begin{bmatrix}
1&0&0&0&0&0&0&0\\
0&1&0&0&0&0&0&0\\
0&0&1&0&0&0&0&0\\
0&0&0&1&0&0&0&0\\
1&0&0&0&1&0&0&0\\
0&1&0&0&0&1&0&0\\
0&0&1&0&0&0&1&0\\
0&0&0&1&0&0&0&1
\end{bmatrix}.
$$

[41] `https://en.wikipedia.org/wiki/Kronecker_product.`

Computing (relative) plausibilities In [89], a sufficient condition was developed for the equality of plausibility and commonality measures which, when met, allows for the efficient calculation of relative plausibility values.

Namely, let $\{m_1, \ldots, m_m\}$ be m mass functions to be combined. Then,

$$Pl(A) \approx Q(A) = K \cdot \prod_{1 \leq i \leq m} Q_i(A) \propto \prod_{1 \leq i \leq m} Q_i(A) \propto \prod_{1 \leq i \leq m} \sum_{B \in \mathcal{E}_i, B \supseteq A} m_i(B), \tag{4.56}$$

where $\mathcal{E}_i$ is the list of focal elements of m_i. In particular, the condition $Pl(A) = Q(A)$ is met whenever, for all $B \in \mathcal{E}_i$, $i = 1, \ldots, m$, either $A \subseteq B$ or $A \cap B = \emptyset$. A method based on (4.56) can be used to calculate the relative plausibility of atomic hypotheses ($A = \{x\}$) and, therefore, can be used to find the outcome that maximises this measure.

Computing mean and variance Kreinovich et al. [1072] argued that in real-life situations it is desirable to compute the ranges of possible values of the mean and variance for probability distributions in the credal set associated with a belief function. More precisely, they contemplated the case in which masses m_i are assigned to intervals $\mathbf{x}_i = [\underline{x}_i, \overline{x}_i]$, $i = 1, \ldots, n$, of the real line.

Efficient algorithms for computing such ranges were described by Langewisch and Choobineh in [1095], where they showed that they assume the form

$$\underline{E} = \sum_i m_i \underline{x}_i, \quad \overline{E} = \sum_i m_i \overline{x}_i,$$

and showed how to compute them in polynomial time. The upper bound $\overline{V}$ on the variance is equal to the maximum of the following concave optimisation problem:

$$\overline{V} = \max \left(\sum_i [\underline{m}_i (\underline{x}_i)^2 + \overline{m}_i (\overline{x}_i)^2] - \left(\sum_i (\underline{m}_i \underline{x}_i + \overline{m}_i \overline{x}_i) \right)^2 \right),$$

subject to $\underline{m}_i + \overline{m}_i = m_i$, $\underline{m}_i \geq 0$, $\overline{m}_i \geq 0$ $\forall i$. The lower bound $\underline{V}$, instead, is equal to the minimum of the following convex quadratic function:

$$\underline{V} = \min \left(\sum_i m_i (x_i)^2 - \left(\sum_i m_i x_i \right)^2 \right),$$

subject to $\underline{x}_i \leq x_i \leq \overline{x}_i \forall i$, which can be solved in time $O(n^2 \log(n))$. Kreinovich et al. [1072] showed that the above quadratic optimisation problem can actually be solved faster, in time $O(n \log(n))$.

4.7.2 Transformation approaches

One approach to efficient belief calculus that has been explored since the late 1980s consists in approximating belief functions by means of appropriate probability measures, prior to combining them for making decisions. This is known as the *probability transformation* problem [291].

A number of distinct transformations have been introduced: the most significant ones remain, however, Voorbraak's plausibility transform [1859], later supported by Shenoy [291], and Smets's pignistic transform [1678]. A dated survey by Bauer [100] reviewed in 1997 a number of approximation algorithms, empirically studying their appropriateness in decision-making situations. More recent proposals have been made by Daniel, Sudano and others.

Some approaches to probability transformation are explicitly aimed at reducing the complexity of belief calculus. Different approximations, however, appear to be aimed at different goals, besides that of reducing computational complexity. Here we review several of the most significant contributions in this area.

Probability transformation Given a frame of discernment Θ, let us denote by $\mathcal{B}$ the set of all belief functions on Θ, and by $\mathcal{P}$ the set of all probability measures there. According to [382], we define a *probability transform* as an operator

$$pt : \mathcal{B} \rightarrow \mathcal{P}, \quad Bel \mapsto pt[Bel] \tag{4.57}$$

mapping belief measures onto probability measures, such that

$$Bel(x) \leq pt[Bel](x) \leq Pl(x).$$

Note that such a definition requires the probability measure $pt[Bel]$ which results from the transform to be compatible with the upper and lower bounds that the original belief function Bel enforces *on the singletons only*, and not on all the focal sets as in (3.10). This is a minimal, sensible constraint which does not require probability transforms to adhere to the upper–lower probability semantics of belief functions. As a matter of fact, some important transforms are not compatible with such semantics.

A long list of papers have been published on the issue of probability transforms [1924, 1045, 100, 110, 449, 466, 769].

Pignistic transform In Smets's transferable belief model [1675, 1730], which we reviewed in Section 3.3.2, decisions are made by resorting to the *pignistic probability* (3.30), which we recall here:

$$BetP[Bel](x) = \sum_{A \supseteq \{x\}} \frac{m(A)}{|A|}, \tag{4.58}$$

which is the output of the *pignistic transform*, $BetP : \mathcal{B} \rightarrow \mathcal{P}$, $Bel \mapsto BetP[Bel]$.

Justified by a linearity axiom, the pignistic probability is the result of a redistribution process in which the mass of each focal element A is reassigned to all its elements $x \in A$ on an equal basis. It is therefore perfectly compatible with the upper–lower probability semantics of belief functions, as it is the centre of mass of the polytope (3.10) of consistent probabilities [245].

Plausibility transform Originally developed by Voorbraak [1859] as a probabilistic approximation intended to limit the computational cost of operating with belief functions, the *plausibility transform* [291] was later supported by Cobb and Shenoy by virtue of its commutativity properties with respect to the Dempster sum.

Although initially defined in terms of commonality values, the plausibility transform $\tilde{Pl} : \mathcal{B} \rightarrow \mathcal{P}, Bel \mapsto \tilde{Pl}[Bel]$ maps each belief function Bel to the probability distribution $\tilde{pl}[Bel]$ obtained by normalising the plausibility values $Pl(x)$[42] of the elements of the frame of discernment Θ:

$$\tilde{Pl}[Bel](x) = \frac{Pl(x)}{\sum_{y \in \Theta} Pl(y)}. \tag{4.59}$$

We call the output $\tilde{Pl}[Bel]$ (4.59) of the plausibility transform the *relative plausibility of singletons*. Voorbraak proved that his (in our terminology) relative plausibility of singletons $\tilde{Pl}[Bel]$ is a perfect representative of Bel when combined with other probabilities $P \in \mathcal{P}$ through Dempster's rule $\oplus$:

$$\tilde{Pl}[Bel] \oplus P = Bel \oplus P \quad \forall P \in \mathcal{P}. \tag{4.60}$$

Dually, a *relative belief transform* $\tilde{Bel} : \mathcal{B} \rightarrow \mathcal{P}, Bel \mapsto \tilde{Bel}[Bel]$ mapping each belief function Bel to the corresponding *relative belief of singletons* [337, 341, 765, 382],

$$\tilde{Bel}[Bel](x) = \frac{Bel(x)}{\sum_{y \in \Theta} Bel(y)}, \tag{4.61}$$

can be defined. The notion of a relative belief transform (under the name of 'normalised belief of singletons') was first proposed by Daniel in [382]. Some initial analyses of the relative belief transform and its close relationship with the (relative) plausibility transform were presented in [337, 341].

Sudano's transformations More recently, other proposals have been put forward by Dezert et al. [493], Burger [206] and Sudano [1778], based on redistribution processes similar to that of the pignistic transform.

The latter, in particular, has proposed the following four probability transforms[43]:

$$PrPl[Bel](x) \doteq \sum_{A \supseteq \{x\}} m(A) \frac{Pl(x)}{\sum_{y \in A} Pl(y)}, \tag{4.62}$$

$$PrBel[Bel](x) \doteq \sum_{A \supseteq \{x\}} m(A) \frac{Bel(x)}{\sum_{y \in A} Bel(y)} = \sum_{A \supseteq \{x\}} m(A) \frac{m(x)}{\sum_{y \in A} m(y)}, \tag{4.63}$$

[42] With a harmless abuse of notation, we will often denote the values of belief functions and plausibility functions on a singleton x by $m(x), Pl(x)$ rather than by $m(\{x\}), Pl(\{x\})$ in this book.

[43] Using this book's notation.

$$PrNPl[Bel](x) \doteq \frac{1}{\Delta} \sum_{A \cap \{x\} \neq \emptyset} m(A) = \tilde{Pl}[Bel](x), \tag{4.64}$$

$$PraPl[Bel](x) \doteq Bel(x) + \epsilon \cdot Pl(x), \quad \epsilon = \frac{1 - \sum_{y \in \Theta} Bel(y)}{\sum_{y \in \Theta} Pl(y)} = \frac{1 - k_{Bel}}{k_{Pl}}, \tag{4.65}$$

where

$$k_{Bel} \doteq \sum_{x \in \Theta} Bel(x), \quad k_{Pl} \doteq \sum_{x \in \Theta} Pl(x).$$

The first two transformations are clearly inspired by the pignistic function (3.30). While in the latter case the mass $m(A)$ of each focal element is redistributed homogeneously to all its elements $x \in A$, $PrPl[Bel]$ (4.62) redistributes $m(A)$ proportionally to the relative plausibility of a singleton x *inside* A. Similarly, $PrBel[Bel]$ (4.63) redistributes $m(A)$ proportionally to the relative *belief* of a singleton x within A.

The pignistic function $BetP[Bel]$, Sudano's $PrPl[Bel]$ and $PrBel[Bel]$, and the orthogonal projection $\pi[Bel]$ (see Chapter 11) arguably form a family of approximations inspired by the same notion, that of mass redistribution (or 'allocation': see (4.68)). As proven by (4.64), $PrNPl[Bel]$ is nothing but the relative plausibility of singletons (4.59), and suffers from the same limitations.

As far as the transformation $PraPl[Bel]$ (4.65) is concerned, since its definition only involves belief and plausibility values of *singletons*, it is more appropriate to think of it as of a probability transform of a probability interval system (Section 6.3), rather than a belief function approximation. Namely,

$$PraPl[(l, u)] \doteq l(x) + \frac{1 - \sum_y l(y)}{\sum_y u(y)} u(x),$$

where

$$\mathcal{P}(l, u) \doteq \Big\{ P : l(x) \leq P(x) \leq u(x), \forall x \in \Theta \Big\}$$

is a set of probability intervals.

The same can be said of the *intersection probability*, which we will study in detail in Chapter 11. By (11.12), the intersection probability can be expressed as an affine combination of the relative belief and plausibility of singletons as follows:

$$p[Bel](x) = (1 - \beta[Bel]) k_{Bel} \cdot \tilde{Bel}[Bel](x) + \beta[Bel] k_{Pl} \cdot \tilde{Pl}[Bel](x), \tag{4.66}$$

where $\beta[Bel] = (1 - k_{Bel})/(k_{Pl} - k_{Bel})$, so that

$$(1 - \beta[Bel]) k_{Bel} + \beta[Bel] k_{Pl} = \frac{k_{Pl} - 1}{k_{Pl} - k_{Bel}} k_{Bel} + \frac{1 - k_{Bel}}{k_{Pl} - k_{Bel}} k_{Pl} = 1.$$

The same holds for $PraPl[Bel]$, namely

$$\begin{aligned} PraPl[Bel](x) &= m(x) + \frac{1 - k_{Bel}}{k_{Pl}} Pl(x) \\ &= k_{Bel} \tilde{Bel}[Bel](x) + (1 - k_{Bel}) \tilde{Pl}[Bel](x). \end{aligned} \tag{4.67}$$

However, just as in the case of the relative belief and plausibility of singletons, (4.67) is not in general consistent with the original probability interval system (l, u). If there exists an element $x \in \Theta$ such that $l(x) = u(x)$ (the interval has zero width for that element), we have that

$$PraPl[(l,u)](x) = l(x) + \frac{1 - \sum_y l(y)}{\sum_y u(y)} u(x) = u(x) + \frac{1 - \sum_y l(y)}{\sum_y u(y)} u(x)$$

$$= u(x) \cdot \frac{\sum_y u(y) + 1 - \sum_y l(y)}{\sum_y u(y)} > u(x),$$

as $(\sum_y u(y) + 1 - \sum_y l(y))/\sum_y u(y) > 1$, and $PraPl[(l,u)]$ falls outside the interval.

Another fundamental objection against $PraPl[(l,u)]$ arises when we compare it with the intersection probability, thought of as a transform $p[(l,u)]$ of probability intervals (16.15). While the latter adds to the lower bound $l(x)$ an equal fraction of the uncertainty $u(x) - l(x)$ for all singletons (16.15), $PraPl[(l,u)]$ adds to the lower bound $l(x)$ an equal fraction *of the upper bound* $u(x)$, effectively counting twice the evidence represented by the lower bound $l(x)$ (4.67).

Entropy-related transforms A few authors have adopted entropy and uncertainty measures as a tool for addressing the transformation problem.

Meyerowitz et al. [1290] proposed two algorithms for selecting the maximum-entropy probability density consistent with a given belief function, a non-trivial task [1352]. The first algorithm converges to the optimum density, and gives the *allocation function* yielding the density. The allocation function

$$\alpha : \Theta \times 2^\Theta \to [0,1], \tag{4.68}$$

where $\alpha(x, A) > 0$ only if $x \in A$, represents the mass redistributed from A to x when generating a probability distribution consistent with Bel (compare Chapter 11). The second algorithm calculates the density function directly in a finite number of steps: an allocation, if required, can then be calculated by a linear program.

Harmanec and Klir [802] investigated uncertainty-invariant transformations between belief, possibility and probability measures, based on a measure of uncertainty in Dempster–Shafer theory defined as the maximum of the Shannon entropy for all probability distributions that conform to the given belief function. Their work was revisited in detail in Ronald Pryor's dissertation [1446]. As recently as 2010, however, Han et al. [789] questioned the approach to evaluating the quality of probability transformations based on entropy measures. As an alternative, the rationality of the uncertainty measure adopted was proposed as an evaluation criterion.

Other probability transforms In [379], Daniel proposed and analysed a number of alternative approaches to probability transforms, namely the cautious, plausibility, proportional and disjunctive probabilistic transformations. All were examined in terms of their consistency with the upper- and lower-probability semantics of belief functions, and with combination rules.

Pan and Yang [1384] introduced three new 'pignistic-style' probability transforms based on multiple belief functions, and compared them with other popular transform methods. The first two were

$$\begin{aligned} PrBP1(x) &= \sum_{A \supseteq \{x\}} \frac{Bel(x)Pl(x)}{\sum_{y \in A} Bel(y)Pl(y)} m(A), \\ PrBP2(x) &= \sum_{A \supseteq \{x\}} \frac{s(x)}{\sum_{y \in A} s(y)} m(A) \quad \text{whenever } m(x) \neq 0, \end{aligned}$$

where $s(x) = Bel(x)/(1 - Pl(x))$. A recursive definition similar to Sudano's [1776] was also provided:

$$PrBP3(x) = \sum_{A \supseteq \{x\}} \frac{s(x)}{\sum_{y \in A} s(y)} m(A), \quad s(x) = \frac{PrBP3(x)}{1 - PrBP3(x)},$$

which is initialised, once again, via $s(x) = Bel(x)/(1 - Pl(x))$.

Two new Bayesian approximations of belief functions were derived by the present author from purely geometric considerations [333], in the context of the geometric approach to uncertainty presented in this book [334]. They will be described in detail in Chapter 11. A similar approach based on minimising the Euclidean distance between measures was independently proposed by Weiler [1924]. The result was also compared with various alternative approaches, for example pignistic transformation for Dempster–Shafer belief and the Shapley value for fuzzy belief functions.

Possibility transformation Another way of reducing the computational complexity of reasoning with belief functions consists in mapping the latter to a consonant belief function, i.e., a BF whose focal elements are nested (see Section 2.8). As consonant belief functions have only N focal elements, where $N = |\Theta|$ is the cardinality of the frame Θ being considered, operating with them can be seen as an efficient way of reasoning with belief functions.

Outer consonant approximation As we know from Section 4.2.1, belief functions admit various order relations, including 'weak inclusion' (3.9): $Bel \leq Bel'$ if and only if $Bel(A) \leq Bel'(A)\ \forall A \subseteq \Theta$. One can then introduce the notion of *outer consonant approximations* [533] of a belief function Bel, i.e., those consonant BFs $Bel = Co$ such that $\forall A \subseteq \Theta\ Co(A) \leq Bel(A)$ (or, equivalently, $\forall A \subseteq \Theta\ Pl_{Co}(A) \geq Pl(A)$). These are the consonant belief functions which are less informative than Bel in the sense specified above – their geometry is studied in detail in Chapter 13.

Permutation families With the purpose of finding outer consonant approximations which are *maximal* with respect to the weak inclusion relation (3.9), Dubois and

Prade introduced two different families in [533]. The first family is obtained by considering all the permutations ρ of the elements $\{x_1, \ldots, x_n\}$ of the frame of discernment Θ, $\{x_{\rho(1)}, \ldots, x_{\rho(n)}\}$.

The following collection of nested sets can be then built:

$$\Big\{ S_1^\rho = \{x_{\rho(1)}\}, S_2^\rho = \{x_{\rho(1)}, x_{\rho(2)}\}, \ldots, S_n^\rho = \{x_{\rho(1)}, ..., x_{\rho(n)}\} \Big\},$$

so that given a belief function Bel with mass m and focal elements $\{E_1, \ldots, E_k\}$, a consonant BF Co^ρ can be defined with BPA

$$m^\rho(S_j^\rho) = \sum_{i:\min\{l:E_i \subseteq S_l^\rho\}=j} m(E_i). \tag{4.69}$$

Analogously, an iterative procedure can be defined in which all the permutations ρ of the focal elements of Bel are considered, $\{E_{\rho(1)}, \ldots, E_{\rho(k)}\}$, and the following family of sets is introduced,

$$\Big\{ S_1^\rho = E_{\rho(1)}, S_2^\rho = E_{\rho(1)} \cup E_{\rho(2)}, \ldots, S_k^\rho = E_{\rho(1)} \cup \cdots \cup E_{\rho(k)} \Big\},$$

so that a new belief function Co_ρ can be defined with BPA

$$m_\rho(S_j^\rho) = \sum_{i:\min\{l:E_i \subseteq S_l^\rho\}=j} m(E_i). \tag{4.70}$$

In general, approximations in the second family (4.70) are part of the first family (4.69) too [533, 91].

Isopignistic approximation A completely different approach was proposed by Aregui and Denoeux within Smets's transferable belief model [58].

Definition 59. *The* isopignistic approximation *of a belief function* $Bel : 2^\Omega \to [0,1]$ *is the unique consonant belief function* Co_{iso} *whose pignistic probability coincides with that of* Bel.

Its contour function (2.18) (or 'plausibility of singletons') is

$$pl_{\text{iso}}(x) = \sum_{x' \in \Theta} \min \Big\{ BetP[Bel](x), BetP[Bel](x') \Big\},$$

so that its mass assignment turns out to be

$$m_{\text{iso}}(A_i) = i \cdot (BetP[Bel](x_i) - BetP[Bel](x_{i+1})),$$

where $\{x_i\} = A_i \setminus A_{i-1}$. The isopignistic approximation was used by Daniel in his effort to decompose belief functions into a conflicting and a non-conflicting part.

More recently, Hernandez [823] presented a method for approximating a belief measure based on the concept of a fuzzy T-preorder, which ensures that the order defined by the degree of compatibility between the evidence and the singleton set is preserved. This approach allows us to define several equivalence criteria over the set of all basic probability assignment functions on a given domain.

Other approximation methods We conclude this part by listing a number of other approaches [561].

In his survey papers [101, 100] on the approximation problem, Bauer also introduced a *DI algorithm* designed to bring about minimal deviations in those values that are relevant to decision making. Grabisch gave in [719] a general necessary condition for a non-additive measure to be dominated by a k-additive measure. A k-*additive* belief function is one which assigns non-zero mass to focal elements of cardinality up to k only. The dominating measure can be seen as a linear transformation of the original measure – the case of belief functions was studied in [719] in detail. Jousselme et al. [935] analysed approximations based on computing the distance between the original belief function and the approximated one. This criterion was compared there with other error criteria, for example based on pignistic transformations. This work is much related to our analysis in Part III.

Haenni [769] proposed instead a belief function approximation method based on the concept of 'incomplete belief potentials'. The criterion used allows us to compute lower and upper bounds for belief and plausibility simultaneously and can be used for a resource-bounded propagation scheme in which the user determines in advance the maximum time available for the computation.

Widening the scope to the problem of approximating other uncertainty measures, Antonucci and Cuzzolin [48] posed the problem of approximating a general credal set with a lower probability, producing an outer approximation, with a bounded number of extreme points. This approximation was shown to be optimal in the sense that no other lower probabilities can specify smaller supersets of the original credal set. Notably, in order to be computed, the approximation does not need the extreme points of the credal set, but only its lower probabilities.

Baroni [93] proposed some general procedures for transforming imprecise probabilities into precise probabilities, possibilities or belief functions, which take account of the fact that information can be partial, i.e., may concern an arbitrary (finite) set of events.

Florea et al. [633] concerned themselves with the transformation of fuzzy membership functions into random sets (i.e., mapping possibility measures to belief measures). As such a transformation involves the creation of a large number of focal elements based on α-cuts of the fuzzy membership functions, in order to keep the process computationally tractable this large number of focal elements needs to be reduced by approximation techniques. Three approximation techniques were proposed and compared with classical approximation techniques used in evidence theory.

4.7.3 Monte Carlo approaches

Monte Carlo approaches also exist which allow us to address the issue arising with a naive implementation of Dempster's rule of combination [1940, 1070, 1317, 1950, 1060].

Wilson [1940], in particular, proposed a Monte Carlo algorithm which 'simulates' the random-set interpretation of belief functions, $Bel(A) = P(\Gamma(\omega) \subseteq$

Algorithm 5 A simple Monte Carlo algorithm for Dempster combination

1: **procedure** MONTECARLODEMPSTER($Bel_1, \ldots, Bel_m$)
2: **for** a large number of trials $n = 1 : N$ **do**
3: **for** $i = 1 : m$ **do**
4: randomly pick an element ω_i of Ω_i with probability $P_i(\omega_i)$
5: **end for**
6: let $\omega = (\omega_1, \ldots, \omega_m)$
7: **if** $\Gamma(\omega) = \emptyset$ **then**
8: restart trial
9: **end if**
10: **if** $\Gamma(\omega) \subseteq A$ **then**
11: trial succeeds, $T = 1$
12: **end if**
13: **end for**
14: **end procedure**

$A|\Gamma(\omega) \neq \emptyset)$. See also a related publication by Clarke and Wilson [288]. Namely, suppose that we seek $Bel = Bel_1 \oplus \ldots \oplus Bel_m$ on Θ, where the various pieces of evidence are induced by probability distributions P_i on Ω_i via multivalued mappings $\Gamma_i : \Omega_i \rightarrow 2^\Theta$. We can then formulate a simple Monte Carlo approach (Algorithm 5) in which 'codes' are randomly sampled from the source probability space, and the number of times their image implies $A \subseteq \Omega$ is counted to provide an estimator for the desired combination.

Computationally, the proportion of trials which succeed converges to $Bel(A)$: $E[\bar{T}] = Bel(A)$, Var$[\bar{T}] \leq 1/4N$. We say that the algorithm has accuracy k if $3\sigma[\bar{T}] \leq k$. Picking $\omega_i \in \Omega_i, i = 1, \ldots, m$, involves m random numbers, so it takes $K_1 \cdot m$, where K_1 is a constant. Testing whether $x_j \in \Gamma(\omega)$ (necessary to verify the tests on $\Gamma(\omega)$, lines 7 and 10 in Algorithm 5) takes less then $K_2 \cdot m$, with K_2 being another constant.

Therefore, the expected time of the algorithm is

$$\frac{N}{1-\kappa} m \cdot (K_1 + K_2|\Omega|),$$

where κ is Shafer's conflict measure (2.7). The expected time to achieve accuracy k turns out to be $((9/(4(1-\kappa)k^2)) \cdot m \cdot (K_1 + K_3|\Omega|)$ for constant K_3, better in the case of simple support functions. In conclusion, unless κ is close to 1 (highly conflicting evidence), Dempster combination is feasible for large values of m (the number of belief functions to be combined) and large cardinality of the hypothesis space Ω.

An improved version of this algorithm [1317] was proposed by Wilson and Moral for the case in which trials are not independent but form a Markov chain (Markov chain Monte Carlo, MCMC). This is based on a non-deterministic operation $OPERATION_i$, which changes at most the i-th coordinate $\omega'(i)$ of a code ω' to y, with chance $P_i(y)$:

$$Pr(OPERATION_i(\omega') = \omega) \propto P_i(\omega(i)) \text{ if } \omega(i) = \omega'(i), \text{ 0 otherwise.}$$

The MCMC algorithm illustrated in Algorithm 6 takes as input an m-dimensional code ω_0 and an event $A \subseteq \Theta$, and returns a value $BEL^N(\omega_0)$ which is the proportion of times over the N trials in which $\Gamma(\omega_c) \subseteq A$.

Algorithm 6 A Markov chain Monte Carlo algorithm for Dempster combination

```
procedure MCMCDEMPSTER(Bel_1, ..., Bel_m, A, ω_0)
    ω_c = ω_0
    S = 0
    for n = 1 : N do
        for i = 1 : m do
            ω_c = OPERATION_i(ω_c)
            if Γ(ω_c) ⊆ A then
                S = S + 1
            end if
        end for
    end forreturn S/Nm
end procedure
```

The following result can be proven.

Theorem 3. *If Ω is connected (i.e., any ω, ω' are linked by a chain of $OPERATION_i$), then, given ϵ, δ, there exist K', N' such that for all $K \geq K'$ and $N \geq N'$ and ω_0,*

$$Pr(|BEL_K^N(\omega_0)| < \epsilon) \geq 1 - \delta,$$

where $BEL_K^N(\omega_0)$ is the output of Algorithm 6.

A further step based on *importance sampling*, according to which samples $\omega^1, \ldots, \omega^N$ are picked according to an 'easy-to-handle' probability distribution P^*, was later proposed [1318]. In this approach, a weight $w_i = P(\omega)/P^*(\omega)$ is assigned to each sample ω. If $P(\omega) > 0$ implies $P^*(\omega) > 0$, then the average $(\sum_{\Gamma(\omega^i) \subseteq A} w_i)/N$ is an unbiased estimator of $Bel(A)$. Obviously, we want to try to use a P^* as close as possible to the real one.

A more sophisticated sampling-based computational strategy was presented by Helton et al. [818] as well. Results indicated that the strategy can be used to propagate belief functions in situations in which unsophisticated Monte Carlo procedures are impracticable owing to computational cost.

4.7.4 Local propagation

The complexity of Dempster's rule of computation is inherently exponential, owing to the necessity of considering all possible subsets of a frame. In fact, Orponen

[1366] proved that the problem of computing the orthogonal sum of a finite set of belief functions is NP-complete.

A short book by Shafer [1602], published in 1996[44], emphasised the basic computational principles that make probabilistic reasoning feasible in expert systems. The key to computation in these systems, he argued, is the modularity of the probabilistic model. Along this line of research, a number of local computation schemes have been proposed to tackle this issue [1758]. Here we review some of the most relevant such proposals, including Barnett's computational scheme, Gordon and Shortliffe's hierarchical evidence organised in diagnostic trees, Shafer and Logan's hierarchical evidence approach, and the Shafer–Shenoy architecture.

Barnett's computational scheme In Barnett's scheme [88, 89] computations are linear in the size of the frame of discernment Θ, if all the belief functions to be combined are simple support functions focused on singletons or their complements.

Now, assume that we have a belief function Bel_θ with only $\{\theta, \bar{\theta}, \Theta\}$ as focal elements for all singletons $\theta \in \Theta$, and we want to combine all $\{Bel_\theta, \theta \in \Theta\}$. The approximation scheme uses the fact that the plausibility of the combined belief function is a function of the commonalities $Q(A) = \sum_{B \supseteq A} m(B)$ of the input belief functions,

$$Pl(A) = \sum_{B \subseteq A, B \neq \emptyset} (-1)^{|B|+1} \prod_{\theta \in \Theta} Q_\theta(B).$$

After a few steps, we find that

$$Pl(A) = K \left(1 + \sum_{\theta \in A} \frac{Bel_\theta(\theta)}{1 - Bel_\theta(\theta)} - \prod_{\theta \in A} \frac{Bel_\theta(\bar{\theta})}{1 - Bel_\theta(\theta)} \right).$$

A similar result holds when the belief functions to be combined are dichotomous on elements of a coarsening of Θ.

The computation of a specific plausibility value $Pl(A)$ is therefore linear in the size of Θ (as only elements of A and not its subsets are involved). However, the number of events A themselves is still exponential – this was addressed by later authors.

Gordon and Shortliffe's diagnostic trees In [710], Gordon and Shortliffe posed the problem of computing degrees of belief only for events forming a hierarchy (a *diagnostic tree*: see Fig. 4.15), motivated by the fact that in some applications certain events are not relevant, for example some classes of diseases in medical diagnosis. Their scheme combines simple support functions focused on or against the nodes of the tree, and produces good approximations unless the evidence is highly conflicting (see the discussion on Monte Carlo methods in Section 4.7.3). However, intersection of complements may generate focal elements not associated with nodes in the tree.

The approximate algorithm can be summarised as follows:

[44] `http://www.glennshafer.com/books/pes.html`.

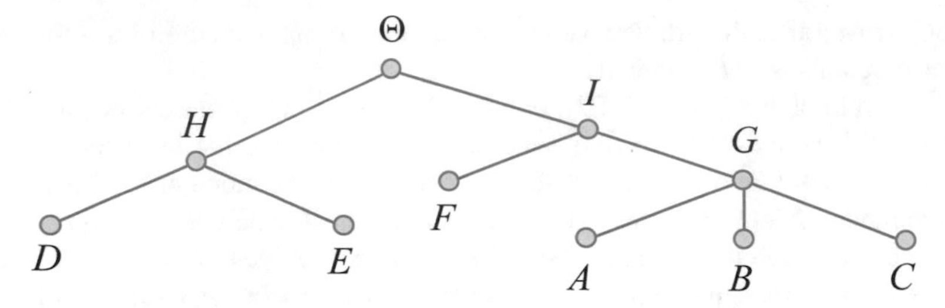

Fig. 4.15: An example of Gordon and Shortliffe's diagnostic tree, from [710], where, for example, $A \subset G$ and $H \subset \Theta$.

1. First, we combine all simple functions focused on the node events (by Dempster's rule).
2. Then, we successively (working down the tree) combine those focused on the complements of the nodes.
3. Whilst we are doing that, we replace each intersection of focal elements with the smallest node in the tree that contains it.

Obviously, the result depends on the order followed to implement the series of combinations in phase 2. There is no guarantee about the quality of the resulting approximation, and no degrees of belief are assigned to complements of node events. As a consequence, in particular, we cannot even compute the plausibilities of node events.

Shafer and Logan's hierarchical evidence In response, Shafer and Logan [1608] proposed an exact implementation with linear complexity for the combination of hierarchical evidence of a more general type. Indeed, although evidence in their scheme is still focused on nodes of a tree, it produces degrees of belief for a wider collection of hypotheses, including the plausibility values of the node events. The scheme operates on *local families* of hypotheses, formed by a node and its children.

Namely, suppose again that we have a dichotomous belief function Bel_A for every non-terminal node associated with the event $A \subseteq \Theta$, let ϑ be the set of non-terminal nodes, and let $Bel^{\downarrow}_A = \oplus\{Bel_B : B \subset A\}$ (or the vacuous belief function when A is a terminal node). Let $Bel^{\uparrow}_A = \oplus\{Bel_B : B \not\subset A, B \neq A\}$ (vacuous if $A = \Theta$ is the root node), and define

$$Bel^L_A = Bel_A \oplus Bel^{\downarrow}_A, \quad Bel^U_A = Bel_A \oplus Bel^{\uparrow}_A.$$

The goal of the scheme is to compute $Bel^T = \oplus\{Bel_A : A \in \vartheta\}$ (note that this is equal to $Bel_A \oplus Bel^{\downarrow}_A \oplus Bel^{\uparrow}_A$ for any node A).

Algorithm 7 starts from terminal nodes and works its way up the tree until we get

$$Bel^L_A(A),\ Bel^L_A(\bar{A}) \quad \forall A \in \vartheta.$$

Note that there is no need to apply Stage 1 to the root node Θ.

Algorithm 7 Hierarchical evidence – Stage 1 (up the tree)

procedure SHAFERLOGANPROPAGATIONUP
  let $\{A\} \cup S_A$ be a family composed of a node A and its children S_A for which $Bel_B^L(B)$, $Bel_B^L(\bar{B})$ are available for every $B \in S_A$
  compute $Bel_A^{\downarrow}(A) = \oplus\{Bel_B^L : B \in S_A\}(A)$
  same for $Bel_A^{\downarrow}(\bar{A})$
  compute $Bel_A^L(A) = (Bel_A \oplus Bel_A^{\downarrow})(A)$
  same for $Bel_A^L(\bar{A})$
end procedure

Now, let $\{A\} \cup S_A$ be a family for which $Bel_B^L(B)$, $Bel_B^L(\bar{B})$ for every $B \in S_A$ but also $Bel_A^U(A)$, $Bel_A^U(\bar{A})$ are available. In Stage 2 (Algorithm 8), we start from the family whose parent is Θ and work our way down the tree until we obtain

$$Bel_A^{\uparrow}(A),\ Bel_A^{\uparrow}(\bar{A}) \quad \forall A \in \vartheta.$$

There is no need to apply Stage 2 to terminal nodes. Finally, in Stage 3 (computing total beliefs), for each node A, we compute

$$Bel^T(A) = (Bel_A^{\uparrow} \oplus Bel_A^L)(A),$$

and similarly for $Bel^T(\bar{A})$. Throughout the algorithm, at each node A of the tree 12 belief values need to be stored.

Algorithm 8 Hierarchical evidence – Stage 2 (down the tree)

1: **procedure** SHAFERLOGANPROPAGATIONDOWN
2:   **for** each $B \in S_A$ **do**
3:     compute $Bel_B^{\downarrow}(B) = (Bel_A^U \oplus \{Bel_C^L : C \in S_A, C \neq B\})(B)$
4:     same for $Bel_B^{\downarrow}(\bar{B})$
5:     compute $Bel_B^U(B) = (Bel_B \oplus Bel_B^{\uparrow})(B)$
6:     same for $Bel_B^U(\bar{B})$
7:   **end for**
8: **end procedure**

Shafer–Shenoy architecture on qualitative Markov trees In their 1987 publication [1611], Shafer, Shenoy and Mellouli showed that an efficient implementation of Dempster's rule is possible if the questions of interest, represented as different partitions of an overall frame of discernment, are arranged in a *qualitative Markov tree* (QMT) [1280], a structure which generalises both diagnostic trees and Pearl's causal trees.

In this framework, relations between questions are encoded by *qualitative* conditional independence or dependence relations between the associated partitions. A propagation scheme which extends Pearl's belief propagation idea to belief functions on QMTs can then be devised.

Qualitative conditional independence

Definition 60. *A collection of partitions $\Psi_1, \ldots, \Psi_n$ of a frame Θ are* qualitatively conditionally independent *(QCI) given the partition Ψ if*

$$P \cap P_1 \cap \ldots \cap P_n \neq \emptyset$$

whenever $P \in \Psi$, $P_i \in \Psi_i$ and $P \cap P_i \neq \emptyset$ for all i.

For example, $\{\theta_1\} \times \{\theta_2\} \times \Theta_3$ and $\Theta_1 \times \{\theta_2\} \times \{\theta_3\}$ are QCI on $\Theta = \Theta_1 \times \Theta_2 \times \Theta_3$ given $\Theta_1 \times \{\theta_2\} \times \Theta_3$ for all $\theta_i \in \Theta_i$. This definition of independence does not involve probability, but only logical independence – nevertheless, stochastic conditional independence does imply QCI.

One can prove that if two belief functions Bel_1 and Bel_2 are 'carried' by partitions Ψ_1, Ψ_2 (i.e., they are vacuous extensions of belief functions on Ψ_1 and Ψ_2; see Chapter 1) which are QCI given Ψ, it follows that

$$(Bel_1 \oplus Bel_2)_\Psi = (Bel_1)_\Psi \oplus (Bel_2)_\Psi,$$

i.e., Dempster combination and restriction onto Ψ commute for them.

Qualitative Markov trees Given a tree, deleting a node and all incident edges yields a forest. Let us denote the collection of nodes of the j-th such subtree by $V_j(v)$.

Definition 61. *A* qualitative Markov tree *$QMT = (V, E)$ is a tree of partitions of a base frame of discernment Θ (each node $v \in V$ is associated with a partition Ψ_v of Θ), such that, for every node $v \in V$, the minimal refinements of partitions in $V_j(v)$ for $j = 1, \ldots, k$ are QCI given Ψ_v.*

A Bayesian causal tree becomes a qualitative Markov tree whenever we associate each node v with the partition Ψ_v induced by a random variable X_v. Furthermore, a QMT remains as such if we insert between a parent and a child a node associated with their common refinement.

Qualitative Markov trees can also be constructed from diagnostic trees – the same interpolation property holds in this case as well.

Shafer–Shenoy propagation Suppose a number of belief functions are inputted into a subset of nodes V'. We wish to compute $\oplus_{v \in V'} Bel_v$. However, rather than applying Dempster combination over the whole frame Θ, we can perform multiple Dempster combinations over partitions of Θ, under the restriction that each belief function to be combined is carried by a partition in the tree.

In a message-passing style, a 'processor' located at each node v combines belief functions using Ψ_v as a frame, and projects belief function onto its neighbours. The operations performed by each processor node are summarised in Algorithm 9 (see also Fig. 4.16).

Algorithm 9 Shafer–Shenoy propagation on qualitative Markov trees

procedure SHAFERSHENOYPROPAGATION($Bel_v, v \in V'$)
send Bel_v to its neighbours $N(v)$;
whenever a node v receives a new input, it computes:

$$(Bel^T)_{\Psi_v} \leftarrow (\oplus\{(Bel_u)_{\Psi_v} : u \in N(v)\} \oplus Bel_v)_{\Psi_v}$$

node v computes for each neighbour $w \in N(v)$:

$$Bel_{v \to w} \leftarrow (\oplus\{(Bel_u)_{\Psi_v} : u \in N(v) \setminus \{w\}\} \oplus Bel_v)_{\Psi_w}$$

and sends the result to w.
end procedure

Note that inputting new belief functions into the tree can take place asynchronously – the final result at each local processor v is the coarsening to that partition of the combination of all the inputted belief functions, $(\oplus_{u\in V'} Bel_u)_{\Psi_v}$. The total time to reach equilibrium is proportional to the tree's diameter. A general tool for propagating uncertainty based on the local computation technique of Shafer and Shenoy, called PULCINELLA, was developed by Saffiotti and Umkehrer in [1529].

Later on, Lauritzen [1102] studied a variant of some axioms originally developed by Shafer and Shenoy [1609], showing that extra assumptions are needed to perform similar local computations in a HUGIN-like architecture [896] or in the architecture of Lauritzen and Spiegelhalter [1103]. In particular, they showed that propagation of belief functions can be performed in those architectures.

Other architectures When the evidence is ordered in a *complete direct acyclic graph* it is possible to formulate algorithms with lower computational complexity, as shown by Bergsten et al. [130]. A very abstract description of propagation algorithms can be given in terms of the notion of a *valuation* [1623] (`http://ddl.escience.cn/f/MJ0u`). A number of architectures [1529, 1391] were proposed in the 1990s and 2000s which make use of valuations [1933], in particular for decision making [1991, 1988]. Haenni [764], for instance, put forward a generic approach to approximate inference based on the concept of valuation algebras. Convenient resource-bounded anytime algorithms were presented, in which the maximum computation time is determined by the user.

Bissig, Kohlas and Lehmann proposed instead a *fast-division architecture* [161] which has the advantage, when compared with the Shenoy–Shafer [1630] and Lauritzen–Spiegelhalter [1103] frameworks, of guaranteeing that the intermediate results will be belief functions. All these architectures have a Markov tree as the underlying computational structure. The main point is that belief functions, such as those living in domains which form the nodes of a Markov tree, often contain only a few focal sets, despite the size of their frame of discernment. The Lauritzen–

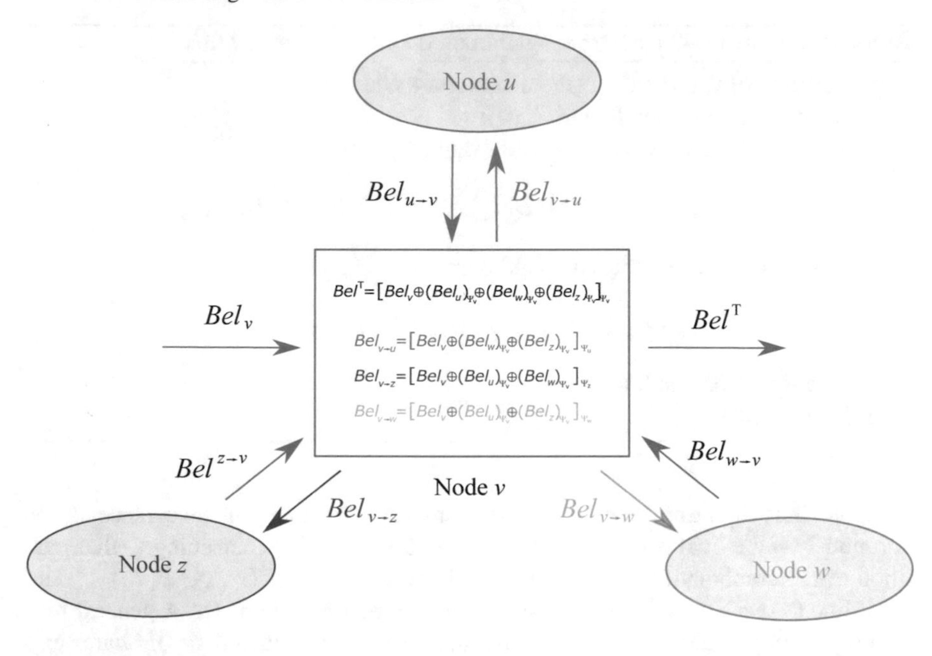

Fig. 4.16: Graphical representation of a local processor's operations in the Shenoy–Shafer architecture, for the case of three neighbours u, z and w of a node v (adapted from [1611]).

Spiegelhalter architecture does not profit from this observation, as it focuses more on the structure of the tree than on the information contained in its nodes.

4.8 Making decisions

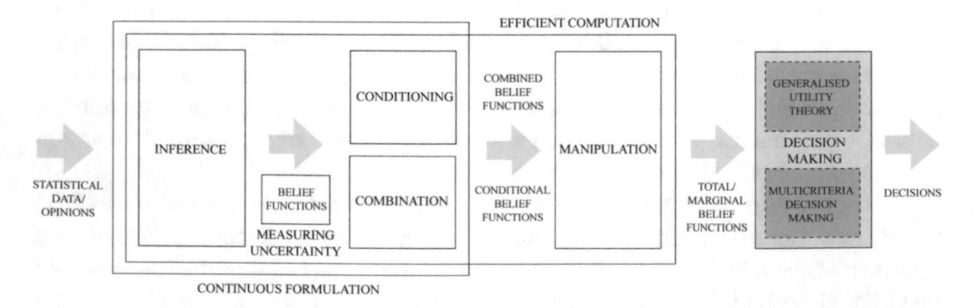

Decision making in the presence of partial evidence and subjective assessment is possibly the true, original rationale for the development of the theory of evidence. Consequently, decision making with belief functions has been consistently studied for the last three decades, originating a number of different approaches to the problem. A good, albeit dated, discussion of the meaning of belief functions in the context of decision making can be found, for instance, in [1711].

As we saw in the Introduction, belief functions are one mathematical tool for expressing what we have called 'Knightian' uncertainty. Therefore, decision making with belief functions fits within the wider topic of a theory of choice under uncertainty. A rather profound discussion of the matter, which gives rigorous expression to Frank Knight's distinction between risk and uncertainty, was presented by Bewley in [140]. Passing references to belief functions in the context of decision making can also be found in [324], among others.

A decision problem can be formalised by defining a set Ω of possible *states of the world*, a set $\mathcal{X}$ of *consequences* and a set $\mathcal{F}$ of *acts*, where an act is a function $f : \Omega \to \mathcal{X}$ mapping a world state to a consequence. Consequences are more or less 'desirable' to us, and making a decision amounts to selecting the act that produces the consequence which is most desirable.

4.8.1 Frameworks based on utilities

Expected utility theory Let $\succcurlyeq$ be a *preference relation* on $\mathcal{F}$, such that $f \succcurlyeq g$ means that f is at least as desirable as g. Savage [1537] showed that $\succcurlyeq$ satisfies a number of sensible rationality requirements iff there exists a probability measure P on Ω and a *utility function* $u : \mathcal{X} \to \mathbb{R}$ such that

$$\forall f, g \in \mathcal{F}, \quad f \succcurlyeq g \Leftrightarrow \mathbb{E}_P(u \circ f) \geq \mathbb{E}_P(u \circ g), \tag{4.71}$$

where $\mathbb{E}_P$ denotes the expectation with respect to P. Also, P and u are unique up to a positive affine transformation. In other words, the most desirable act is that which *maximises the expected utility* (MEU) of the outcome. Equation (4.71) is also called the *Laplace criterion*.

The MEU principle was first axiomatised by von Neumann and Morgenstern [1858].

Proposition 28. [1858] *The following statements are equivalent:*

1. *The preference relation $\succcurlyeq$ satisfies the axioms of preorder, continuity and independence.*
2. *There exists a utility function such that (4.71) holds.*

Given a probability distribution on Ω, an act $f : \Omega \to \mathcal{X}$ induces a probability measure on the set of consequences, called a *lottery*. Comparing the desirability of acts then reduces to comparing the desirability of lotteries.

Other 'rationality' criteria have also been proposed to derive a preference relation over acts [462], such as the *maximax* criterion,

$$f \succcurlyeq g \text{ iff } \max_{\omega \in \Omega} u \circ f(\omega) \geq \max_{\omega \in \Omega} u \circ g(\omega), \tag{4.72}$$

and the *maximin* (Wald) criterion [1869],

$$f \succcurlyeq g \text{ iff } \min_{\omega\in\Omega} u\circ f(\omega) \geq \min_{\omega\in\Omega} u\circ g(\omega). \tag{4.73}$$

A linear combination of the latter two is called the *Hurwicz criterion*: $f \succcurlyeq g$ iff

$$\alpha \min_{\omega\in\Omega} u\circ f(\omega) + (1-\alpha)\max_{\omega\in\Omega} u\circ f(\omega) \geq \alpha \min_{\omega\in\Omega} u\circ g(\omega) + (1-\alpha)\max_{\omega\in\Omega} u\circ g(\omega). \tag{4.74}$$

Finally, Savage's *minimax regret criterion* reads as follows:

$$f \succcurlyeq g \text{ iff } R[f] \leq R[g], \tag{4.75}$$

where

$$R[f] = \max_{\omega\in\Omega} r_f(\omega), \quad r_f(\omega) = \max_{g\in\mathcal{F}} u\circ g(\omega) - u\circ f(\omega) \tag{4.76}$$

is the maximal 'regret' for an act f, defined as the difference in utility from the best act, for a given state of nature.

Belief functions and decision making under uncertainty As pointed out in [462], belief functions become of a component of a decision problem if either of the following applies:

1. The decision maker's beliefs concerning the state of nature are described by a belief function Bel_Ω rather than by a probability measure on Ω.
2. The decision maker is unable to precisely describe the outcomes of (at least some) acts under each state of nature.

In the second case, an act may be formally represented by a multivalued mapping $f: \Omega \to 2^{\mathcal{X}}$, assigning a set of possible consequences $f(\omega) \subseteq \mathcal{X}$ to each state of nature ω, so that a probability distribution on Ω will induce a belief function on $\Theta = \mathcal{X}$. Denouex [462] calls a belief function on the set of consequences, however obtained, a *credibilistic* lottery. The expected utilities framework (Section 4.8.1) then needs to be extended to credibilistic (belief) lotteries.

Generalised Hurwicz criterion

Jaffray's work Jaffray [877, 875] was the first to apply von Neumann–Morgenstern linear utility theory (Proposition 28) to belief functions, leading to a generalised expected utility representation of preferences based on the Hurwicz criterion.

The Hurwicz criterion (4.74) can be generalised as follows:

$$\mathrm{e}_{m,\alpha} = \sum_{A\subset\mathcal{X}} m(A)\Big[\alpha \min_{x\in A} u(x) + (1-\alpha)\max_{x\in A} u(x)\Big] = \alpha\underline{\mathbb{E}}_m(u) + (1-\alpha)\overline{\mathbb{E}}_m(u), \tag{4.77}$$

where $\alpha \in [0,1]$ is a 'pessimism' index. This criterion was introduced and axiomatically justified by Jaffray [875], who argued that it was justified because of a property of mixtures of belief functions [879].

In [877], Jaffray showed that a preference relation among credibilistic lotteries is representable by a linear utility function if and only if it satisfies the von Neumann and Morgenstern axioms extended to credibilistic lotteries, i.e.:

1. *Transitivity and completeness*: $\succcurlyeq$ is a transitive and complete relation (i.e., is a weak order).
2. *Continuity*: for all m_1, m_2 and m_3 such that $m_1 \succ m_2 \succ m_3$, there exists $\alpha, \beta \in (0, 1)$ such that
$$\alpha m_1 + (1 - \alpha)m_3 \succ m_2 \succ \beta m_1 + (1 - \beta)m_3.$$
3. *Independence*: for all m_1, m_2 and m_3, and for all $\alpha \in (0, 1)$, $m_1 \succ m_2$ implies
$$\alpha m_1 + (1 - \alpha)m_3 \succ \alpha m_2 + (1 - \alpha)m_3.$$

In further studies [883], Jaffray and Wakker [1867, 1868] showed that uncertain information can be modelled through belief functions if and only if the non-probabilisable information is subject to the principles of complete ignorance. The representability of decisions by belief functions on outcomes was justified there by means of a 'neutrality' axiom. In more detail, these authors consider a subset $A \subseteq \Omega$ to be an 'ambiguous' event if there is a focal element of Bel which intersects both A and $\bar{A}$. A *weak sure-thing principle* is then satisfied if, for any two acts that have common outcomes outside an unambiguous event A, the preference does not depend on the level of those common outcomes. Its implications for decision making were identified in [883].

Hendon et al. [820] also worked to extend the von Neumann–Morgenstern expected utility theory to belief functions, and used this theory to characterise uncertainty neutrality and different degrees of uncertainty aversion.

Strat's decision apparatus Strat's decision apparatus [1768, 1770, 1771] is based on computing *intervals* of expected values for whole belief functions, and assumes that the decision frame Θ is itself a set of scalar values (e.g. dollar values, as in Fig. 4.17). In other words, his framework does not distinguish between utilities and elements of Θ (consequences), so that an interval of expected values can be computed, $E(\Theta) = [E_*(\Theta), E^*(\Theta)]$, where

$$E_*(\Theta) \doteq \sum_{A \subseteq \Theta} \inf(A) m(A), \quad E^*(\Theta) \doteq \sum_{A \subseteq \Theta} \sup(A) m(A),$$

and $\inf(A)$ and $\sup(A)$ are the minimum and maximum scalar values, respectively, within A. Strat argued, however, that this is not good enough to make a decision. For instance, should we pay 6$ for a ticket when the expected interval is $[\$5, \$8]$?

Consider a cloaked-carnival-wheel scenario, in which one of the sectors of the wheel is not visible to the player, so that we do not know the monetary outcome there (Fig. 4.17). This can be trivially modelled by a belief function over the frame of discernment of possible outcomes $\Theta = \{\$1, \$5, \$10, \$20\}$. Strat identified the probability ρ that the value assigned to the hidden sector was the one the player would choose (whereas $1 - \rho$ is the probability that the sector is chosen by the carnival hawker). The following proposition is then true [1770].

Proposition 29. *The expected value of the mass function of the wheel is*

$$E(\Theta) = E_*(\Theta) + \rho(E^*(\Theta) - E_*(\Theta)). \quad (4.78)$$

To decide whether to play the game, we only need to assess ρ. This can be reconduced to a specific probability transform (for instance, the pignistic one).

Note that (4.78) coincides exactly with Jaffray's generalised Hurwicz criterion (4.77). In other words, Strat interprets the coefficient α there as the subjective probability ρ that the ambiguity will be resolved unfavourably. In his doctoral dissertation, Lesh [1136] had also proposed a similar approach.

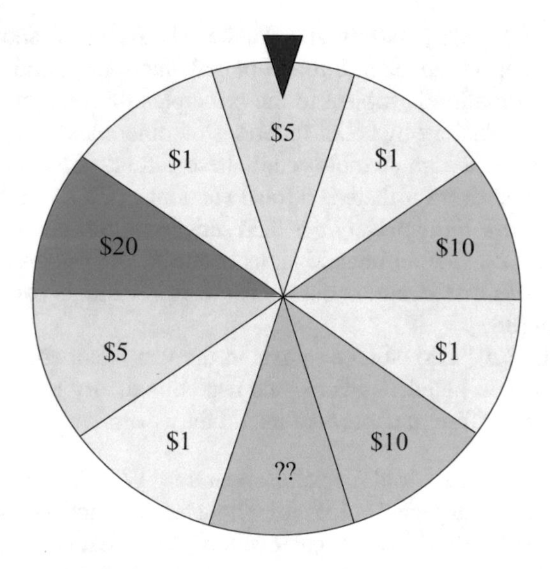

Fig. 4.17: Strat's cloaked carnival wheel.

Schubert [1548] subsequently showed that no assumption about the value of ρ in Strat's decision apparatus is necessary, as it is sufficient to assume a uniform probability distribution of the parameter to be able to discern the most preferable choice. Later on, Srivastava's proposal [1756] built on Strat's approach to explain Ellsberg's paradox and to model decision-making behaviour under ambiguity.

Decision making in the transferable belief model In the TBM, which we outlined in Chapter 3, Section 3.3.2, decision making is done by maximising the expected utility of actions based on the pignistic transform, rather than by generalising the expected utility framework to credibilistic lotteries.

The set $\mathcal{F}$ of possible actions and the set Θ of possible outcomes are distinct, and the utility function is defined on $\mathcal{F} \times \Theta$. In [1712, 1715], Smets proved the necessity of the pignistic transform by maximising the expected utility,

$$E_{BetP}[u] = \sum_{\theta \in \Theta} u(f, \theta) BetP(\theta),$$

by resorting to a linearity requirement, which (Smets argued) is unavoidable provided one accepts expected utility theory. In earlier work, [1678] he had advocated a so-called 'generalised insufficient reason' principle (see [1096] and Section 4.2.3) as a justification for extending the expected utility model to belief functions. The maximum pignistic expected utility criterion extends the Laplace (expected utility) criterion (4.71). A decision system based on the TBM was developed in [1991] and applied to a waste disposal problem by Xu et al.

In an interesting paper [1944], Wilson [1951] argued that Smets's decision-making method, however, is sensitive to the choice of the frame of discernment, which is often to a large extent arbitrary. Nevertheless, after considering all the refinements of a frame of discernment and their associated pignistic probability functions and decisions, he concluded that this 'robust' pignistic decision-making method is equivalent to the standard one. Alternative pignistic probability transforms for decision making have been proposed more recently by Pan et al. [1383].

Decision making in valuation-based systems A number of papers have been published on decision making with belief functions in the framework of *valuation-based* systems ([1991, 1988]; compare Section 4.7.4), in order to show that decision problems can be solved using local computations [2041, 1988].

The proposed calculus uses a weighting factor whose role is similar to Strat's probabilistic interpretation of an assumption that disambiguates decision problems represented with belief functions [1768]. The framework integrates a system for belief propagation and a system for Bayesian decision analysis, within the valuation-based-systems setting [1991, 1990, 1989]. The two levels are connected through the pignistic transformation. The system takes as inputs the user's beliefs and utilities, and suggests either the optimal decision or the optimal sequence of decisions.

In another series of publications, Elouedi, Smets et al. [570, 573] adapted the decision tree (see Section 5.2.6) technique to the case in which there is uncertainty about the class value, and where that uncertainty is represented by a belief function.

Choquet expected utilities Summarising what we have just discussed, does Savage's result (4.71) then imply that basing decisions on non-additive probabilities, such as belief functions, is irrational? The answer is clearly no, and indeed several authors have proposed decision-making frameworks under belief function uncertainty based on (generalisations of) utility theory.

The foundations for most of this work were laid axiomatically by Schmeidler [1539, 1540], Gilboa [689], and Gilboa and Schmeidler [690, 692]. A generalised expected utility model with non-additive probability measures based on Gilboa and Schmeidler's work was later proposed by Dow and Werlang [513].

In detail, Gilboa proposed a modification of Savage's axioms with, in particular, a weaker form of Axiom 2. As a consequence, a preference relation $\succcurlyeq$ meets these weaker requirements iff there exists a (*not necessarily additive*) measure μ and a *utility function* $u : \mathcal{X} \rightarrow \mathbb{R}$ such that

$$\forall f, g \in \mathcal{F}, \quad f \succcurlyeq g \Leftrightarrow C_\mu(u \circ f) \geq C_\mu(u \circ g),$$

where C_μ is the *Choquet integral* [283, 1242, 440, 1911, 1783], defined for $X : \Omega \to \mathbb{R}$ as

$$C_\mu(X) = \int_0^{+\infty} \mu(X(\omega) \geq t) \, \mathrm{d}t + \int_{-\infty}^0 \big[\mu(X(\omega) \geq t) - 1\big] \mathrm{d}t. \qquad (4.79)$$

Given a belief function Bel on Ω and a utility function u, this theorem supports making decisions based on the Choquet integral of u with respect to Bel or Pl.

Lower and upper expected utilities For finite Ω, it can be shown that

$$C_{Bel}(u \circ f) = \sum_{B \subseteq \Omega} m(B) \min_{\omega \in B} u(f(\omega)),$$

$$C_{Pl}(u \circ f) = \sum_{B \subseteq \Omega} m(B) \max_{\omega \in B} u(f(\omega)).$$

Let $\mathcal{P}(Bel)$, as usual, be the set of probability measures P compatible with Bel, i.e., those such that $Bel \leq P$. Then, it follows that

$$\begin{aligned} C_{Bel}(u \circ f) &= \min_{P \in \mathcal{P}(Bel)} \mathbb{E}_P(u \circ f) \doteq \underline{\mathbb{E}}(u \circ f), \\ C_{Pl}(u \circ f) &= \max_{P \in \mathcal{P}(Bel)} \mathbb{E}_P(u \circ f) \doteq \overline{\mathbb{E}}(u \circ f). \end{aligned}$$

Decision criteria For each act f we now have two expected utilities, $\underline{\mathbb{E}}(f)$ and $\overline{\mathbb{E}}(f)$. How do we make a decision? Various decision criteria can be formulated, based on interval dominance:

1. $f \succcurlyeq g$ iff $\underline{\mathbb{E}}(u \circ f) \geq \overline{\mathbb{E}}(u \circ g)$ (*conservative* strategy).
2. $f \succcurlyeq g$ iff $\underline{\mathbb{E}}(u \circ f) \geq \underline{\mathbb{E}}(u \circ g)$ (*pessimistic* strategy).
3. $f \succcurlyeq g$ iff $\overline{\mathbb{E}}(u \circ f) \geq \overline{\mathbb{E}}(u \circ g)$ (*optimistic* strategy).
4. $f \succcurlyeq g$ iff $\alpha\underline{\mathbb{E}}(u \circ f) + (1-\alpha)\overline{\mathbb{E}}(u \circ f) \geq \alpha\underline{\mathbb{E}}(u \circ g) + (1-\alpha)\overline{\mathbb{E}}(u \circ g)$ for some $\alpha \in [0, 1]$.

The conservative strategy 1 yields only a partial preorder: f and g are not comparable if $\underline{\mathbb{E}}(u \circ f) < \overline{\mathbb{E}}(u \circ g)$ and $\underline{\mathbb{E}}(u \circ g) < \overline{\mathbb{E}}(u \circ f)$. The pessimistic strategy corresponds to the maximin criterion (4.73), while the optimistic one corresponds to the maximax (4.72) criterion. Strategy 4, finally, generalises the Hurwicz criterion (4.74).

Going back to Ellsberg's paradox (see Chapter 1, Section 1.3.10), it is easy to see that the available evidence naturally translates into a belief function, with $m(\{R\}) = 1/3$ and $m(\{B, Y\}) = 2/3$. We can then compute lower and upper expected utilities for each action, as follows:

	R	Bel	Y	$\underline{\mathbb{E}}(u \circ f)$	$\overline{\mathbb{E}}(u \circ f)$
f_1	100	0	0	$u(100)/3$	$u(100)/3$
f_2	0	100	0	$u(0)$	$u(200)/3$
f_3	100	0	100	$u(100)/3$	$u(100)$
f_4	0	100	100	$u(200)/3$	$u(200)/3.$

The observed behaviour ($f_1 \succcurlyeq f_2$ and $f_4 \succcurlyeq f_3$) can then be explained by the pessimistic strategy 2.

Related work on the Choquet integral Various authors have analysed further or contributed to the framework of the generalised expected utility based on the Choquet integral.

Nguyen [1352], for instance, showed that (i) using expected utility with belief functions via the Choquet integral leads to the pessimistic strategy of minimising expected utilities, (ii) a more general framework can be established by considering expectations of functions of random sets, and (iii) the principle of insufficient reason can be properly generalised as a form of maximum entropy principle.

Ghirardato and Le Breton [682] provided a characterisation of the consequences of the assumption that a decision maker with a given utility function is Choquet rational, showing that this allows one to rationalise more choices than is possible when beliefs have to be additive. They found that, surprisingly, a considerable restriction on the types of belief allowed does not change the set of rational actions. They also discussed the relation between the Choquet expected utility model, the maxmin expected utility model and subjective expected utility when the risk attitude of the decision maker is not known.

Zhang [2105] provided an axiomatisation of the Choquet expected utility model where the capacity is an inner measure. The key axiom states that the decision maker uses unambiguous acts to approximate ambiguous ones. In addition, the notion of 'ambiguity' is subjective and derived from preferences.

More recently, Chu and Halpern proposed a new notion of generalised expected utility [285]. In [284], they showed that the latter provides a universal decision rule, in the sense that it essentially encompasses all other decision rules.

Closely related to this topic is an attractive axiomatic model of decision making under uncertainty in a credal set framework presented by Gajdos et al. [661]. The decision maker is told that the true probability law lies in a set $\mathcal{P}$, and is assumed to rank pairs of the form (P, f), where f is an act mapping states into outcomes. The key representation result delivers the maximin expected utility (MEU: see (4.73)), where the $\min$ operator ranges over a set of probability priors, just as in the work of Gilboa and Schmeidler [690]. However, unlike the MEU representation, the representation here also delivers a mapping ϕ which links the credal set $\mathcal{P}$, describing the available information, to the set of revealed priors, and represents the decision maker's attitude towards imprecise information. This allows both expected utility when the selected set is a singleton and extreme pessimism when the selected set is the same as $\mathcal{P}$.

E-capacities Eichberger and Kelsey [565], on their side, introduced *E-capacities* as a representation of beliefs which incorporates objective information about the probability of events. They showed that the Choquet integral of an E-capacity is Ellsberg's preference representation.

Namely, let $\mathcal{E} = \{E_1, \ldots, E_n\}$ be a partition of Ω with probabilities $P(E_i)$, and let $\beta_i(A) = 1$ if $E_i \subseteq A$, and 0 otherwise (i.e., the categorical belief function focused on E_i). The set of probabilities consistent with the partition $\mathcal{E}$ is defined as

$$\Pi(P) \doteq \left\{ p \in \mathcal{P}(\Omega) \middle| \sum_{\omega \in E_i} p(\omega) = P(E_i) \right\}.$$

Definition 62. *The* E-capacity $\nu(p, \rho)$ *based on an assessment* $p \in \Pi(P)$ *with degree of confidence* $\rho \in [0, 1]$ *is defined by*

$$\nu(A|p, \rho) \doteq \sum_{i=1}^{n} [\rho \cdot p(A \cap E_i) + (1 - \rho) \cdot P(E_i)\beta_i(A)] \tag{4.80}$$

for all $A \subseteq \Omega$.

An E-capacity can be viewed as a convex combination of the additive probability distribution p and the capacity μ, defined as

$$\mu(A) \doteq \sum_{i=1}^{n} P(E_i)\beta_i(A) \quad \forall A \subseteq \Omega.$$

Note that $\mu = Bel$ is really the belief function with focal elements equal to the elements of the given partition $\mathcal{E}$, and with mass assignment determined by P.

Lower and upper expected utilities *à la* Dempster In 1992, Caselton and Luo [234] analysed the application of belief theory to decision making, in the framework of Dempster's upper- and lower-probability formulation, and of the Borel extension of belief functions to intervals (see Section 4.9.2). Their decision-making framework was based upon work by Dempster and Kong [429].

Consider a simple decision problem where the uncertainties stem from the value of a parameter $\omega \in \Omega$ whose true value is unknown. Following conventional decision analysis, we need to specify the utility as a function $u(f, \omega)$ of an act f and of ω. A conventional Bayesian decision analysis would calculate the expected utility based on $u(f, \omega)$ and the posterior distribution of ω.

Whenever a mass assignment is available for intervals $[\underline{\omega}, \overline{\omega}]$ of values of ω, however, a pair of lower and upper expected utilities can be computed as

$$\begin{aligned} E^*[u(f)] &= \int_{[\underline{\omega},\overline{\omega}] \subset \Omega} m([\underline{\omega}, \overline{\omega}]) \sup_{\omega \in [\underline{\omega},\overline{\omega}]} u(f, \omega) \, \mathrm{d}[\underline{\omega}, \overline{\omega}], \\ E_*[u(f)] &= \int_{[\underline{\omega},\overline{\omega}] \subset \Omega} m([\underline{\omega}, \overline{\omega}]) \inf_{\omega \in [\underline{\omega},\overline{\omega}]} u(f, \omega) \, \mathrm{d}[\underline{\omega}, \overline{\omega}]. \end{aligned} \tag{4.81}$$

If the utility function is monotonically increasing, then (4.81) simplifies to

$$E^*[u(f)] = \int_{\omega \in \Omega} u(f, \omega) h(\omega) \, \mathrm{d}\omega, \quad E_*[u(f)] = \int_{\omega \in \Omega} u(f, \omega) g(\omega) \, \mathrm{d}\omega,$$

where

$$g(\underline{\omega}) = \int_{\underline{\omega}}^{\overline{\omega}_1} m([\underline{\omega}, \overline{\omega}]) \, \mathrm{d}\overline{\omega}, \quad h(\overline{\omega}) = \int_{\underline{\omega}_0}^{\overline{\omega}} m([\underline{\omega}, \overline{\omega}]) \, \mathrm{d}\underline{\omega}$$

are the marginal density functions defined for U and V, the random variables corresponding to the lower and upper bounds, respectively, of the interval.

Generalised minimax regret In [2022, 2028], Yager extended the minimax regret criterion (4.75) to belief functions, by defining the maximum regret for a *subset* of states of nature $A \subset \Theta$ as

$$r_f(A) = \max_{\omega \in A} r_f(\omega),$$

with $r_f(\omega)$ as in (4.76). The *expected maximum regret* for act f is therefore

$$\bar{R}[f] = \sum_{\emptyset \neq A \subseteq \Omega} m(A) r_f(A),$$

and an act f is preferred over an act g if $\bar{R}[f] \leq \bar{R}[g]$.

Aggregation operators The Laplace, maximax, maximin and Hurwicz criteria correspond to different ways of aggregating the utilities resulting from each act, using, respectively, the average, the maximum, the minimum, and a convex sum of the minimum and maximum operators. In fact, all these operators belong to a wider family called ordered weighted average (OWA) [2014, 2029] operators.

Definition 63. *An* OWA operator *of dimension* n *is a function* $F : \mathbb{R}^n \rightarrow \mathbb{R}$ *of the form*

$$F(x_1, \ldots, x_n) = \sum_{i=1}^{n} w_i x_{(i)}, \tag{4.82}$$

where $x_{(i)}$ *is the i-th largest element in the collection* $x_1, \ldots, x_n$*, and* $w_1, \ldots, w_n$ *are positive, normalised weights.*

A more general family of expected utility criteria can then be defined by aggregating the utility values within each focal set of a given belief function using OWA operators. Yager [2014, 2032], in particular, argued that OWA aggregation operators provide a unifying framework for decision making under ignorance. The OWA approach to decision making was supported by Engemann et al. [574]. In 2011, Tacnet and Dezert [1795] devised a new methodology called cautious ordered weighted averaging with evidential reasoning (COWA-ER), which cautiously merges Yager's approach with the efficient fusion of belief functions proposed in the Dezert–Smarandache theory.

Casanovas and Merigo's extensions Quite recently, Casanovas and Merigo [1288] revised further the concepts of induced ordered weighted averaging and geometric operators introduced by Yager. They focused on the aggregation step and examined some of its main properties, including the distinction between descending and ascending orders and different families of induced operators. The notion was extended to fuzzy OWA operators in [232, 233], and to linguistic aggregation operators in [1289], in which the authors of that publication suggested the use of a linguistic ordered weighted averaging belief structure called BS-LOWA and a linguistic hybrid averaging belief structure (called BS-LHA), together with a wide range of particular cases.

Other utility-based decision frameworks Several other authors have engaged with decision making with belief functions via generalised utility theory.

A framework for analysing decisions under risk was presented by Danielson [390], which allows the decision maker to be as deliberately imprecise as they feel is natural. In [252], a non-ad hoc decision rule based on the expected utility interval was proposed. Notably, Augustin [67] extended expected utility theory to interval probability (see Section 6.3). Since 2-monotone and totally monotone capacities are special cases of general interval probability, for which the Choquet integral and the interval-valued expectation are equivalent, this leads to an efficient way of dealing with the Choquet expected utility.

In a hybrid setting involving fuzzy set theory and Shafer's theory of evidence, Bordley [178] argued that the utility of a consequence of an action can be assessed as the membership function of the consequence in a fuzzy set 'satisfactory', while quantifying the degree of evidence in an event using belief functions.

Last but not least, Giang and Shenoy [686, 687] studied a utility theory decision framework for Walley's *partially consonant belief functions* (PCBs), i.e., belief functions whose set of foci is partitioned, and whose focal elements are nested within each partition. The framework is based on an axiomatic system analogous to von Neumann and Morgenstern's. A preference relation on partially consonant belief function lotteries was defined, and a representation theorem for this preference relation was proven, showing that any utility for a PCB lottery is a combination of a linear utility for a probabilistic lottery and a binary utility for a possibilistic lottery. Requirements for rational decision making under ignorance and sequential consistency were explored in [685] by the same authors, where they formalised the concept of sequential consistency of an evaluation model and proved results about the sequential consistency of the Jaffray–Wakker and Giang–Shenoy models under various conditions.

4.8.2 Frameworks not based on utilities

A number of interesting belief-theoretical decision-making approaches not based on some form of (generalised) utility theory have also been put forward.

Based on Boolean reasoning Skowron [1657], for instance, investigated the generation of optimal decision rules with some certainty coefficients based on belief and rough membership functions, and showed that the problem of rule generation can be solved by Boolean reasoning.

The decision rules considered there have the form $\tau \Rightarrow \tau'$, where τ is a Boolean combination of descriptors built from conditions, and τ' is an approximation to the expert's decision. These decision rules are generated with some 'certainty coefficients', expressed by basic probability assignments. Rule generation is based on the construction of appropriate Boolean functions from discernability matrices [1659]. Both locally optimal and globally optimal decision rules are considered.

Based on rough set theory As pointed out by Wu [1964], not all conditional attributes in an information system are necessary before decision rules are generated. The idea of knowledge reduction, in the sense of reducing attributes, is a contribution to data analysis from research on rough sets [1394].

In response, the author of [1964] introduced the concepts of *plausibility reduct* and *belief reduct* in incomplete information systems, and discussed their relationship with 'classical' concepts.

Definition 64. A complete information system *is a pair* (U, AT)*, where* U *is a finite universe of discourse and* $AT = \{a_1, \ldots, a_m\}$ *is a finite set of attributes* $a : U \to V_a$*, where* V_a *is the domain of attribute* a*.*

When we are not able to state with certainty what is the value taken by a given attribute a on a given object x, an information system is called an *incomplete* information system (IIS). Each non-empty subset of attributes $A \subseteq AT$ determines an 'indiscernibility' relation,

$$R_A = \{(x, y) \in U \times U | a(x) = a(y) \ \forall a \in A\}.$$

Definition 65. *Let* $S = (U, AT)$ *be an IIS. Then, an attribute subset* $A \subseteq AT$ *is referred to as:*

- *a* classical consistent set *of* S *if* $R_A = R_{AT}$*.* A classical reduct *of* S *is a minimal classical consistent set.*
- *a* belief consistent set *of* S *if* $Bel_A(X) = Bel_{AT}(X)$ *for all* $X \in U/R_{AT}$*, where* $Bel_A(X) \doteq |\underline{A}(X)|/|U|$ *and* $\underline{A}(X)$ *is the lower approximation of* X *with respect to the subset of attributes* A *(compare Section 6.7.1).* A belief reduct *of* S *is a minimal belief consistent set.*
- *a* plausibility consistent set *of* S *if* $Pl_A(X) = Pl_{AT}(X)$ *for all* $X \in U/R_{AT}$*, where* $Pl_A(X) \doteq |\overline{A}(X)|/|U|$ *and* $\overline{A}(X)$ *is the upper approximation of* X *with respect to the subset of attributes* A*.* A plausibility reduct *of* S *is a minimal plausibility consistent set.*

A consistent subset of attributes, in other words, has the same resolution power as the entire set of attributes AT.

The following proposition can be proven ([1964], Theorem 2).

Proposition 30. *A is a classical consistent set of an IIS S if and only if A is a belief consistent set of S. A is a classical reduct of S iff A is a belief reduct of S.*

As a result, attribute reduction in incomplete decision systems can be tackled within the theory of evidence.

Other approaches We finish by briefly recalling a few remaining related papers [1908, 409, 1926, 23].

Kleyle and de Corvin [979] proposed in 1990 a decision-making model based on an elimination process which uses sequentially acquired information. The procedure combines aspects of belief theory and a generalised information system proposed by Yovits et al. Belief theory was discussed by Sosnowski [1746] as a vehicle for constructing fuzzy decision rules. Chen [259] highlighted the necessity of applying rule-based inference with all possible consequent clauses when applying the procedure to more general decision-making problems. Sabbadin [1518] described a representation of decision problems under uncertainty based on *assumption-based truth maintenance systems* [405], in a logical setting. That author extended the ATMS framework so as to include, in addition to the usual assumption symbols, preference and decision symbols, allowing one to assign multiple (real or qualitative) values in order to model gradual uncertainty and preferences.

Finally, as recently as 2011, Peng [1152] proposed a new decision-making method using grey systems theory, with attribute values of corresponding alternatives in the form of intuitionistic fuzzy numbers.

4.8.3 Multicriteria decision making

A wealth of research has been conducted on the relation between *multicriteria decision making* (MCDM), also called *multi-attribute decision analysis* (MADA), and the theory of belief functions.

This discipline is concerned with the evaluation of multiple conflicting criteria in decision making. For instance, when purchasing a car, we may want to consider cost, comfort, safety and fuel economy as some of the main criteria. In their daily lives, people weigh multiple criteria implicitly. However, when deciding whether to build a new power plant, for instance, the problem needs to be more formally structured. MCDM generalises the decision-making setting described at the start of Section 4.8, in the sense that every act generates more than one consequence, with every consequence (attribute) living in general in a different space or possibly being of qualitative nature. On the other hand, the state of the world is supposed to be known with certainty: $\Omega = \{\omega\}$.

Mathematically, an MCDM problem can be written as

$$\max \mathbf{q}, \quad \text{subject to } \mathbf{q} \in Q, \tag{4.83}$$

where $\mathbf{q}$ is a vector of k criterion values, and Q is the feasible set of such vectors (alternatives). The issue lies in defining what the 'max' of a vector means. Multicriteria expected utility approaches exist, but estimating utilities for all alternatives (acts) and attributes (consequences) can easily become very expensive.

Typically, such problems do not admit a unique optimal solution, and it is necessary to use the decision maker's preferences to differentiate solutions. Solving can mean picking the 'best' alternative course of action, or selecting a small set of 'good' (non-dominated) options. A *non-dominated* solution $\mathbf{q}^*$ is such that it is not possible to move away from it without making sacrifice in at least one criterion: $\exists \mathbf{q} \neq \mathbf{q}^* : \mathbf{q} \geq \mathbf{q}^*$. Therefore, it makes sense for the decision maker to choose a solution from the non-dominated set.

The relationship between uncertainty and MCDM models becomes quite clear after reading [520], where Dubois, Prade, Grabisch and Modave pointed out the striking similarity between decision under uncertainty and multicriteria decision-making problems, and emphasised the remarkable formal equivalence between the postulates underlying the two approaches (e.g., between the 'sure-thing principle' and the mutual preferential independence of criteria). There, they also stressed the benefit of importing uncertainty-originated notions into multicriteria evaluation (in particular, for the weighing of the importance of (coalitions of) criteria).

As Bryson et al. noted [199], decision-making problems are complicated by at least two factors: (1) the qualitative/subjective nature of some criteria, which often results in uncertainty in the individual ratings, and (2) the fact that group decision making is required, so that some means of aggregating individual ratings is required.

The work in this area is mostly due to Yang and colleagues, Beynon and colleagues, and Fu and co-authors.

Yang's evidential reasoning Yang and co-authors, in particular, have developed over the years an *evidential reasoning*[45] (ER) framework [2041] for dealing with subjective assessments under uncertainty for MCDM, with both quantitative and qualitative attributes.

Subjective judgements may be used to differentiate one alternative from another on qualitative attributes. To evaluate the quality of the operation of a motorcycle, for example, typical judgements may be that 'the operation of a motor cycle is poor, good, or excellent to certain degrees'. In such judgements, poor, good and excellent denote distinctive evaluation grades. If a detailed concept is still too abstract to assess directly, it may be broken down further into more detailed ones. For instance, the concept y of brakes may be measured by stopping power (e_1), braking stability (e_2) and feel at the controls (e_3), which can be directly assessed and are therefore referred to as *basic attributes*.

ER algorithm Suppose there are L basic attributes $E = \{e_i, i = 1, \ldots, L\}$, associated with a general (higher-level) attribute y. Suppose the weights of the attributes are given as $\mathbf{w} = \{w_i, i = 1, \ldots, L\}$, $0 \leq w_i \leq 1$. As a matter of fact, weights may be estimated in rather sophisticated ways [2043]. Suppose also that N distinctive evaluation grades are defined, $H = \{H_1, \ldots, H_N\}$, from the least to the most preferable.

A given assessment for an alternative may be represented mathematically as

$$S(e_i) = \Big\{(H_n, \beta_{n,i}), n = 1, \ldots, N\Big\},$$

where $\beta_{n,i}$ denotes a degree of belief, $\sum_n \beta_{n,i} \leq 1$, $\beta_{n,i} \geq 0$. An assessment is complete if $\sum_n \beta_{n,i} = 1$, and incomplete otherwise. The problem becomes that of

[45] A rather unfortunate choice of words, as 'evidential reasoning' was used in earlier papers to denote the theory of evidence per se.

aggregating the assessments of the basic attributes e_i in order to generate the degree of belief β_n with which the general attribute y is assessed to have the grade H_n.

Suppose that the degree to which the i-th basic attribute e_i supports the hypothesis that the attribute y is assessed to have the n-th grade H_n is represented by a basic probability mass $m_{n,i} = w_i \beta_{n,i}$, where w_i is a normalised weight. Let $m_{H,i}$ be the mass left unassigned to any individual grade, as far as e_i is concerned.

Analogously, let $E_{I(i)} = \{e_1, \ldots, e_i\}$ be the subset of the first i basic attributes, $m_{n,I(i)}$ the degree to which all i attributes in $E_{I(i)}$ support the hypothesis (y, H_n), and $m_{H,I(i)}$ the mass left unassigned after all the attributes in $E_{I(i)}$ have been assessed.

Definition 66. *The* ER recursive algorithm *reads as follows:*

$$\begin{aligned} m_{n,I(i+1)} &= K_{I(i+1)}(m_{n,I(i)} m_{n,i+1} + m_{n,I(i)} m_{H,i+1} + m_{H,I(i)} m_{n,i+1}), \\ & \quad n = 1, \ldots, N, \\ m_{H,I(i+1)} &= K_{I(i+1)} m_{H,I(i)} m_{H,i+1}, \end{aligned} \tag{4.84}$$

where $K_{I(i+1)}$ is a normalisation factor, $m_{n,I(1)} = m_{n,1}$ for $n = 1, \ldots, N$ and $m_{H,I(1)} = m_{H,1}$.

Variants A number of variants of the ER algorithm exist. For instance, new schemes for weight normalisation and basic probability assignments were proposed in [2043]. In [2038] it was noted that, when assessing different qualitative attributes, different sets of evaluation grades may be needed. Some attributes may be quantitative and need to be assessed numerically. A generalised and extended decision matrix was then constructed and rule- and utility-based techniques developed for transforming various types of information within the matrix for the ER aggregation of attributes. In [2040], a generic rule-based inference methodology which uses evidential reasoning (called RIMER) was proposed. In this scheme, a rule base is designed with belief degrees embedded in all possible consequents of a rule, so that an input to an antecedent attribute is transformed into a belief distribution. Subsequently, inference in that rule base is implemented using ER.

Properties of the ER framework A number of aspects of the ER framework have been explored by the original authors and other researchers in a series of papers.

The nonlinear features of the information aggregation process, for instance, were studied in [2037, 2045], providing insights into the recursive nature of the ER approach. The model was extended further in [1902] to encompass new types of uncertainties, including interval belief degrees and interval data, which can occur in group decision-making situations. Two pairs of nonlinear optimisation models were constructed there to estimate the upper and lower bounds of the combined belief degrees and to compute the maximum and minimum expected utilities, respectively, of each alternative. Intelligent Decision System (IDS), a window-based software package that was also developed on the basis of the ER approach, was presented in [1975]. In [432], an approach to estimating the weights of the attributes in the ER framework, represented as intervals, was proposed by Deng et al., based on a preference given by the decision maker. Huynh et al. reanalysed the ER approach in [865], and proposed a new, general scheme of attribute aggregation in MADM under uncertainty.

More recently, Yao and Huang [2051] introduced a new type of approach to multiple-attribute aggregation called 'induced ordered weighted evidential reasoning' (IOWER), in which an ER belief decision matrix is combined with an induced ordered weighting vector for problem modelling, while attribute aggregation is done by Dempster's rule. The original ER algorithm turns out to be a special case of the IOWER algorithm.

Fuzzy and interval extensions The ER approach was developed further in [2042] to deal with MADA problems with both probabilistic and fuzzy uncertainties. There, precise data, ignorance and fuzziness were all modelled in the unified framework of a distributed fuzzy belief structure, leading to a fuzzy belief decision matrix. A new fuzzy ER algorithm was developed to aggregate multiple attributes using the information contained in the fuzzy belief matrix, resulting in an aggregated fuzzy distributed assessment for each alternative. In [754], the ER approach was enhanced further to deal with both interval uncertainty and fuzzy beliefs in assessing alternatives on an attribute (by a method called FIER). In [2122], both triangular fuzzy weights of criteria and fuzzy utilities assigned to evaluation grades were introduced into the ER approach. The Hadamard multiplicative combination of judgement matrices was extended there to the aggregation of triangular fuzzy judgement matrices, whose results were used as fuzzy weights in the fuzzy ER approach.

Group consensus with evidential reasoning Fu and Yang [646, 645] extended the ER approach to group consensus situations for multiple-attribute group decision analysis problems. A compatibility measure between two belief structures (see Section 4.10.1) was developed, so as to assess the compatibility between two experts' assessments using their utilities.

In [755], Guo et al. considered both interval weight assignments and interval belief degrees, as they appear in many decision situations such as group decision making. Similarly, in [647, 648] an ER-based consensus model (ERCM) was proposed to deal with situations in which experts' assessments are interval-valued (see Section 3.3.7) rather than precise. The case in which there is dependence among attributes was addressed in [650], which developed a dependence-based interval-valued ER (called DIER for short) approach based on Denoeux's cautious conjunctive rule. There, a pair of nonlinear optimisation problems considering the relative weights of attributes was constructed based on the cautious rule to aggregate dependent attributes.

Extensions of the analytic hierarchy process A different strand of work concerns the extension of the so-called *analytic hierarchy process* (AHP) to multicriteria decision making, originally developed to help decision makers rank information based on a number of criteria. Beynon et al. [143], in particular, extensively explored the potential of the theory of evidence as an alternative approach to multicriteria decision modelling within the AHP framework.

In their original paper [143], Beynon et al. discussed the incorporation of belief theory into the AHP, for generating a 'consensus' priority that represents a group's opinion with regard to the relative importance of criteria and alternatives. A mathematical analysis of the resulting Dempster–Shafer/AHP (DS/AHP) method was proposed in [141], where the functional form of the preference weightings given to groups of decision attributes was constructed. These functions allow an understanding of the appropriateness of the rating-scale values used in the DS/AHP method, through evaluating the range of uncertainty able to be expressed by the decision maker. A software expert system based on the DS/AHP method was outlined in [142]. The method allows a decision maker a considerably greater level of control (compared with conventional AHP methods) over the judgements made in identifying levels of favouritism towards groups of decision alternatives. For instance, DS/AHP allows one to assess the levels of uncertainty and conflict in the decisions made. The method was developed further for group decision making in [144], focusing on the aggregation of evidence from individual members of a decision-making group. Members were considered non-equivalent in their importance within the group, and a discount rate value was defined for each member of the group accordingly.

Interestingly, a DS-AHP method for MADM problems with incomplete information was separately developed by Hua et al. [854]. This approach first identifies all possible focal elements from an incomplete decision matrix, to later compute the basic probability assignment of each focal element and the belief interval of each decision alternative. Preference relations among all decision alternatives are determined by comparing their belief intervals. Bryson et al. [199] examined the group decision-making problem in the context where priorities are represented as numeric intervals. Dezert et al. [496] also presented an extension of multicriteria decision making based on the AHP. The combination of priority vectors corresponding to BPAs related to each (sub)criterion was performed there using the 'proportional conflict redistribution' rule of Dezert–Smarandache theory (recall Section 3.3.3).

Other work on MCDM Outside the boundaries of the ER and AHP-based frameworks, various other researchers have addressed multicriteria decision making in the framework of the theory of evidence.

Boujelben et al. [187] modelled imprecise evaluations of alternatives provided by one or several experts as basic belief assignments. Belief dominance was adopted to compare the different pairs of alternatives according to each criterion. Criteria weights were also expressed by means of a BPA.

Deng et al. [433] proposed an MCDM methodology which uses fuzzy and evidence theory, in which BPAs are determined by the distance to the ideal solution and the distance to the negative ideal solution. Dempster's combination rule is used to combine all the criterion data to get the final scores of the alternatives in the systems. Final decisions are made based on the pignistic probability transformation.

Huynh et al. [864] proposed several new attribute aggregation schemes for MADM under uncertainty. Sii et al. [1649] put forward a synthesis of the theory of evidence and the approach developed by Zadeh and Bellman in 1970, and later extended by Yager.

4.9 Continuous formulations

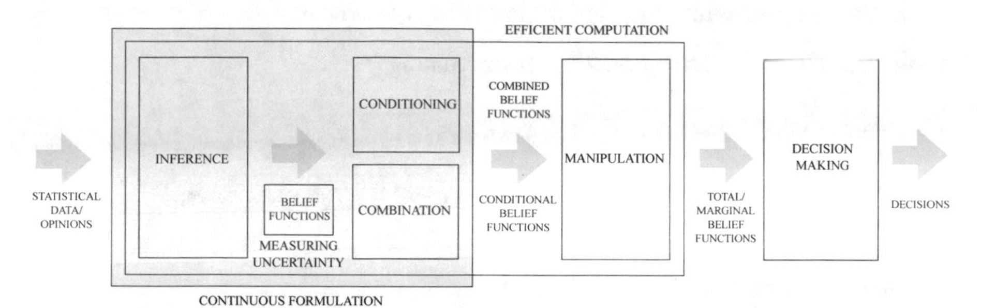

The need for a general formulation of the theory of evidence for continuous domains has long been recognised. The original formulation of the theory of evidence, summarised in Chapter 2, is inherently linked to finite frames of discernment. Numerous proposals have thus been made since, in order to extend the notion of a belief function to infinite hypothesis sets [1887, 1885, 611, 1843]. Among them have been Shafer's allocations of probability [1587], Nguyen's random sets [717, 1344, 1353, 826], Strat and Smets's random closed intervals [417, 1772, 1714], Liu's generalised evidence theory [1181] and Kroupa's belief functions on MV algebras [1077, 627, 626, 628].

4.9.1 Shafer's allocations of probabilities

The first attempt (1979) was due to Shafer himself, and goes under the name of *allocations of probabilities* [1587]. Shafer proved that every belief function can be represented as an allocation of probability, i.e., a $\cap$-homomorphism into a positive and completely additive probability algebra, deduced from the integral representation due to Choquet.

Basically, for every belief function Bel defined on a class of events $\mathcal{E} \subseteq 2^{\Theta}$ there exists a complete Boolean algebra $\mathcal{M}$, a positive measure μ and an allocation of probability ρ between $\mathcal{E}$ and $\mathcal{M}$ such that $Bel = \mu \circ \rho$. Canonical continuous extensions of belief functions defined on 'multiplicative subclasses' E to an arbitrary power set can then be introduced by allocation of probability. Such canonical extensions satisfy Shafer's notion of a belief function definable on infinitely many compatible frames, and show significant resemblance to the notions of the inner measure and extension of capacities.

Two regularity conditions for a belief function over an infinite domain were considered by Shafer: *continuity* and *condensability*.

Continuity and condensability

Definition 67. *A collection* $\mathcal{E} \subset 2^\Theta$ *is a* multiplicative subclass *of* 2^Θ *if* $A \cap B \in \mathcal{E}$ *for all* $A, B \in \mathcal{E}$. *A function* $Bel : \mathcal{E} \rightarrow [0,1]$ *such that* $Bel(\emptyset) = 0$, $Bel(\Theta) = 1$ *and* Bel *is monotone of order* ∞ *is a belief function.*

Equally, a plausibility function is alternating of order ∞.

Definition 68. *A capacity* μ *on* 2^Θ *is* n-alternating *if*

$$\begin{array}{l} \mu(A_1 \cap \ldots \cap A_n) \\ \leq \sum_i \mu(A_i) - \sum_{i<j} \mu(A_i \cup A_j) + ... + (-1)^{n+1} \mu(A_1 \cup ... \cup A_n) \end{array} \tag{4.85}$$

for all $A_1, \ldots, A_n \in 2^\Theta$.

Note the duality with the superadditivity property (3.6) of belief functions. When a capacity is n-alternating for every n, it is called an ∞-alternating capacity.

Definition 69. *A belief function on* 2^Θ *is* continuous *if*

$$Bel(\cap_i A_i) = \lim_{i \rightarrow \infty} Bel(A_i)$$

for every decreasing sequence of sets A_i. *A belief function on a multiplicative subclass* $\mathcal{E}$ *is continuous if it can be extended to a continuous one on* 2^Θ.

According to Shafer, continuity arises from partial beliefs on 'objective' probabilities.

Definition 70. *A belief function on* 2^Θ *is* condensable *if*

$$Bel(\cap \mathcal{A}) = \inf_{A \in \mathcal{A}} Bel(A)$$

for every downward net[46] $\mathcal{A}$ *in* 2^Θ. *A belief function on a multiplicative subclass* $\mathcal{E}$ *is condensable if it can be extended to a condensable one on* 2^Θ.

Condensability is related to Dempster's rule, as this property is required whenever we have an infinite number of belief functions to combine.

Choquet's integral representation Choquet's integral representation implies that every belief function can be represented by allocation of probability.

Definition 71. [283] *A function* $r : \mathcal{E} \rightarrow \mathcal{F}$ *is a* $\cap$-homomorphism *if it preserves* $\cap$.

The following result links belief functions to classical probability spaces via $\cap$-homomorphisms, as a direct consequence of Choquet's integral representation theorem [283].

[46] A *downward net* is such that, given two elements, there is always an element that is a subset of their intersection.

Proposition 31. *For every belief function Bel on a multiplicative subclass $\mathcal{E}$ of 2^Θ, there exist a set $\mathcal{X}$ and an algebra $\mathcal{F}$ of its subsets, a finitely additive probability measure μ on $\mathcal{F}$, and a $\cap$-homomorphism $r : \mathcal{E} \to \mathcal{F}$ such that $Bel = \mu \circ r$.*

Roughly speaking, a belief function can be thought of as generated by a probability measure, after some 'rearranging' of its focal elements in a way that preserves their set-theoretical intersection: $Bel(A) = \mu(r(A))$. An informal explanation can be derived from the multivalued-mapping representation of BFs: as $Bel(A) = \sum_{\omega \in \Omega : \Gamma(\omega) \subseteq A} \mu(\omega)$, if we define $r(A) \doteq \{\omega \in \Omega : \Gamma(\omega) \subseteq A\}$ then, obviously, $Bel(A) = \mu(r(A))$.

Shafer, instead, makes use of Choquet's representation theorem to prove that

$$Bel(A) = \mu(\{Bel_B : 2^\Theta \to [0,1] : Bel_B(A) = 1\}),$$

where Bel_B is the categorical belief function with focus on B, and one of the extreme points of the set of all belief functions on Θ. In this case $r(A)$ is the set of extreme (categorical) BFs (which are in 1–1 correspondence with filters in Θ) such that $B \subseteq A$[47].

If we replace the measure space $(\mathcal{X}, \mathcal{F}, \mu)$ with a probability algebra (a complete Boolean algebra $\mathcal{M}$ with a completely additive probability measure μ), we get Shafer's allocation of probability.

Proposition 32. ([1587], Theorem 3.1) *For every belief function Bel on a multiplicative subclass $\mathcal{E}$ of 2^Θ, there exists a $\cap$-homomorphism, termed an* allocation of probability, *$\rho : \mathcal{E} \to \mathcal{M}$ such that $Bel = \mu \circ \rho$.*

Non-zero elements of $\mathcal{M}$ can then be thought of as focal elements.

This approach was later reviewed by Jurg Kohlas [1018], who conducted an algebraic study of argumentation systems [1021, 1020] as methods for defining numerical degrees of support for hypotheses, by means of allocation of probability.

Canonical extension

Proposition 33. ([1587], Theorem 5.1) *A belief function on a multiplicative subclass $\mathcal{E}$ of 2^Θ can always be extended to a belief function on 2^Θ by* canonical extension*:*

$$\overline{Bel}(A) \doteq \sup_{\substack{n \geq 1, A_1, \ldots, A_n \in \mathcal{E} \\ A_i \subset A \forall i}} \left\{ \sum (-1)^{|I|+1} Bel\left(\bigcap_{i \in I} A_i \right) \middle| \emptyset \neq I \subset \{1, \ldots, n\} \right\}.$$

Indeed, there are many such extensions, of which $\overline{Bel}$ is the minimal one. The proof is based on the existence of an allocation of probability for the desired extension. Note the similarity with the superadditivity axiom – the notion is also related to that of an inner measure (Section 3.1.3), which provides approximate belief values for subsets outside the initial σ-algebra.

[47]As we will discuss later, Choquet's representation theorem could open the way to an extension of the geometric approach presented in Part II to continuous spaces.

What about evidence combination? The condensability property ensures that the Boolean algebra $\mathcal{M}$ represents intersection properly for arbitrary (not just finite) collections $\mathcal{B}$ of subsets:

$$\rho(\cap\mathcal{B}) = \bigwedge_{B\in\mathcal{B}} \rho(B) \quad \forall\mathcal{B}\subset 2^{\Theta},$$

where $\rho : 2^{\Theta} \rightarrow \mathcal{M}$ is the allocation and $\wedge$ is the meet operation in the Boolean algebra $\mathcal{M}$, allowing us to imagine Dempster combinations of infinitely many belief functions.

4.9.2 Belief functions on random Borel intervals

Almost at the same time, Strat [1772] and Smets [1714] had the idea of making continuous extensions of belief functions tractable via the standard methods of calculus, by allowing only focal elements which are closed intervals of the real line.

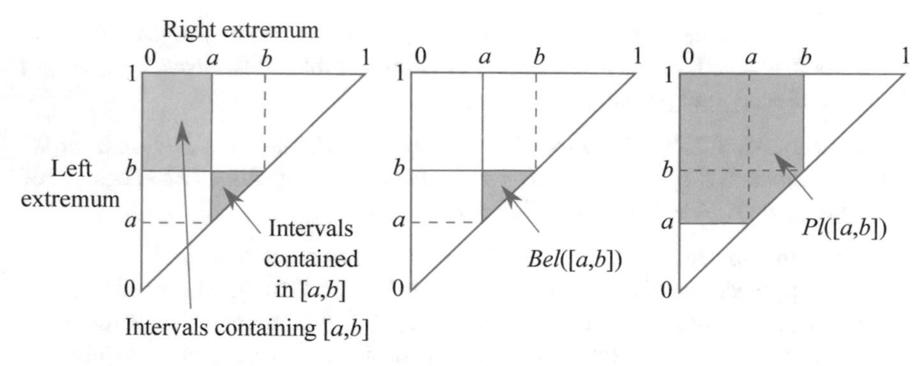

Fig. 4.18: Strat's representation of belief functions on intervals (from [1772]). Left: frame of discernment for subintervals $[a, b]$ of $[0, 1]$. Middle: the belief value for a subinterval $[a, b]$ is the integral of the mass distribution over the highlighted area. Right: support of the corresponding plausibility value.

Strat's initial idea was very simple. Take a real interval I (for instance $I = [0, 1]$) and split it into N pieces. Define as the frame of discernment the set of possible intervals with such extreme points: $[0, 1), [0, 2), [1, 4]$, etc. A belief function there has therefore $\sim N^2/2$ possible focal elements, so that its mass function lives on a discrete triangle, and one can compute belief and plausibility values simply by integration. This idea generalises to all arbitrary intervals of I as shown in Fig. 4.18. Belief and plausibility measures assume in this case the integral form

$$Bel([a, b]) = \int_a^b \int_x^b m(x, y)\, \mathrm{d}y\, \mathrm{d}x, \qquad Pl([a, b]) = \int_0^b \int_{\max(a,x)}^1 m(x, y)\, \mathrm{d}y\, \mathrm{d}x,$$

whereas Dempster's rule generalises as

$$Bel_1 \oplus Bel_2([a,b]) = \frac{1}{K} \int_0^a \int_b^1 \big[m_1(x,b)m_2(a,y) + m_2(x,b)m_1(a,y) + m_1(a,b)m_2(x,y) + m_2(a,b)m_1(x,y)\big] \, dy \, dx.$$

In this framework, Smets [1714] defined a *continuous pignistic PDF* as

$$Bet(a) \doteq \lim_{\epsilon \to 0} \int_0^a dx \int_{a+\epsilon}^1 \frac{m(x,y)}{y-x} \, dy. \tag{4.86}$$

This approach can easily be extended by considering belief functions defined on the *Borel σ-algebra*[48] of subsets of $\mathbb{R}$ generated by the collection $\mathcal{I}$ of closed intervals. The theory also provides a way of building a continuous belief function from a pignistic density, by applying the least commitment principle and assuming unimodal pignistic PDFs [1490]. We obtain

$$Bel(s) = -(s - \bar{s}) \frac{dBet(s)}{ds},$$

where $\bar{s}$ is such that $Bet(s) = Bet(\bar{s})$. For example, a normally distributed pignistic function $Bet(x) = \mathcal{N}(x, \mu, \sigma)$ generates a continuous belief function of the form $Bel(y) = \frac{2y}{\sqrt{2\pi}} e^{-y^2}$, where $y = (x - \mu)/\sigma$. In [510], the approach was further generalised and applied to consonant continuous belief functions in the case of mixtures of Gaussians.

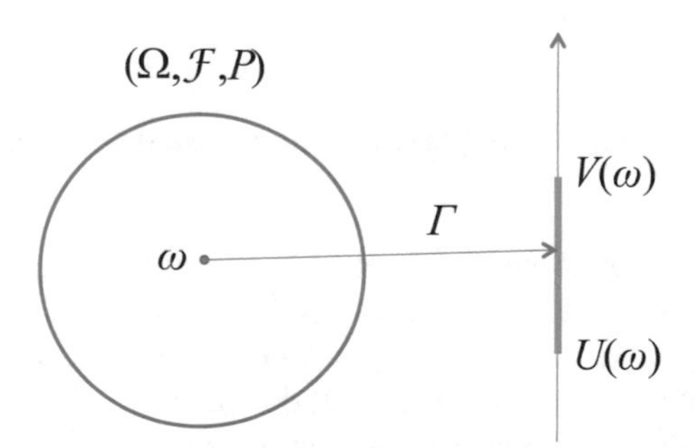

Fig. 4.19: Notion of a random closed interval.

Formally, let (U, V) be a two-dimensional random variable from $(\Omega, \mathcal{F}, P)$ to $(\mathbb{R}^2, \mathcal{B}(\mathbb{R}^2))$ such that $P(U \leq V) = 1$ and $\Gamma(\omega) = [U(\omega), V(\omega)] \subseteq \mathbb{R}$ (see

[48] A *Borel set* is any set in a topological space that can be formed from open sets (or, equivalently, from closed sets) through the operations of countable union, countable intersection and relative complement. For a topological space X, the collection of all Borel sets on X forms a σ-algebra, known as the Borel algebra or Borel σ-algebra.

Fig. 4.19). This setting defines a *random closed interval* (see Section 4.9.3), which induces a belief function on $(\mathbb{R}, \mathcal{B}(\mathbb{R}))$ defined by

$$Bel(A) = P([U, V] \subseteq A), \quad \forall A \in \mathcal{B}(\mathbb{R}).$$

Special cases of a random closed interval include, for instance:

- a fuzzy set $\pi(x)$ onto the real line, as this induces a mapping to a collection of nested intervals, parameterised by the level ω (Fig. 4.20(a));
- a *p-box*, i.e., a pair of upper and lower bounds F^*, F_* of a cumulative distribution function (see Section 6.8) also induces a family of intervals (Fig. 4.20(b)).

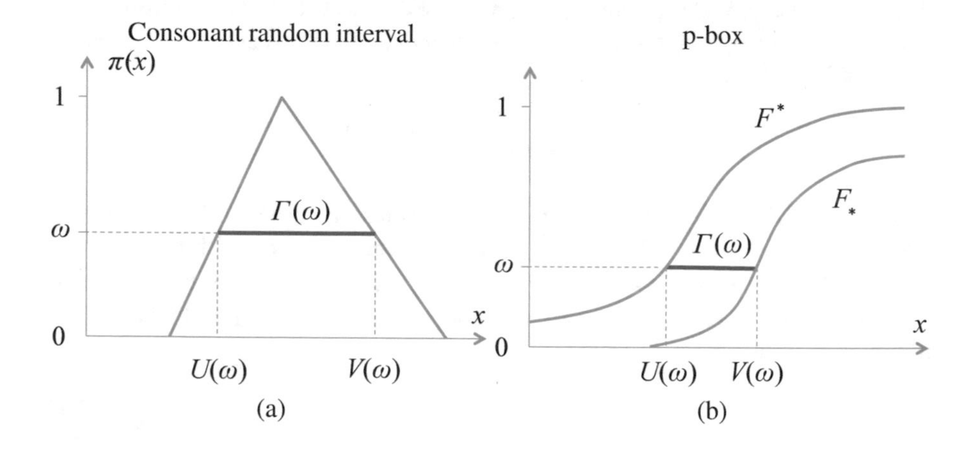

Fig. 4.20: Examples of random closed intervals.

The approach based on the Borel sets of the real line has proved more fertile so far than more general approaches, such as those based on random sets or allocations of probability. Generalisations of combination and conditioning rules follow quite naturally; inference with predictive belief functions on real numbers can be performed; and a pignistic probability for continuous BFs (4.86) can be straightforwardly defined [1490], allowing TBM-style decision making with continuous belief functions.

To cite a few contributions on continuous belief functions, a way to calculate pignistic probabilities with plausibility functions where the knowledge of the sources of information is represented by symmetric α-stable distributions was presented in [611]. Denoeux [457] extended the concept of stochastic ordering to belief functions on the real line defined by random closed intervals. Vannobel [1843] proposed a way to reconstruct the plausibility function curve in combination operations on continuous consonant basic belief densities, based on a graphical representation of the cross product of two focal sets originating from univariate Gaussian PDFs [1844].

4.9.3 Random sets

The most elegant formalism in which to formulate a continuous version of the theory of belief functions is perhaps the theory of random sets [1268, 1857, 826], i.e., probability measures over power sets, of which traditional belief functions are in fact a special case [1344].

In Chapter 3, Section 3.1.5, we have already recalled the fundamental definitions of upper and lower inverses and of a belief function induced by a random set, abstracted from the tutorial [1353]. Quinio [1449] argued that a unified topological approach to imprecision and uncertainty based on random closed sets solves many of the difficulties inherent in the combination of belief functions. The notion of condensability, which we reviewed in Definition 70 and is strictly related to this issue, can indeed be defined for upper probabilities generated by random sets [1349]. Joint belief functions can also be defined in terms of copulas [1350]. In an interesting piece of work, Kohlas [1019] argued that many properties of random sets, in particular those related to belief functions, can also be obtained by relaxing the algebraic structure of the domain. He then proceeded to study random variables with values in some kinds of graded semilattices, as a natural generalisation of evidence theory.

Applications From an application point of view, Vo et al. [1857] argued that random finite sets are natural representations of multitarget states and observations, allowing multisensor multitarget tracking to fit into the unifying random-set framework. The connection of Dempster–Shafer reasoning with random set theory in the context of multitarget tracking was highlighted in Matti Vihola's thesis [1856]. The random-set tracking model considered in [1856] included a model for sensors that produce at most one measurement per report. A sequential Monte Carlo implementation of the random-set tracking model was then derived. A random-set approach to data fusion was described by Mahler in [1234], based on a generalisation of the familiar Radon–Nikodym derivative from the theory of Lebesgue integrals.

Combination A serious stumbling block in the development of belief theory based on random sets, however, is the formulation of suitable aggregation operators. At the time they were formulated, Shafer's allocations of probability and Nguyen's random-set interpretations did not much mention combination rules – and 30 years have passed without many concrete steps towards such a goal. Nevertheless, for finite random sets (i.e., with a finite number of focal elements), under independence of variables, Dempster's rule can still be applied. For the case of dependent sources, Fetz and Oberguggenberger proposed an 'unknown interaction' model [606]. As for general, infinite random sets, Alvarez outlined in [36] an intriguing Monte Carlo sampling method, which we will review in Section 6.8.

Random sets and possibility theory Relationships exist between random set theory and possibility theory as well [1347]. In [1345], for instance, the formal connection between possibility distributions and the theory of random sets via Choquet's theorem was discussed, suggesting that plausible inferences and modelling of common sense can be derived from the statistics of random sets. Dubois and Prade stated in [537] a general expression for functions with random-set-valued arguments, which encompasses Zadeh's extension principle as well as functions of random variables and interval analysis. A monotonicity property was derived in [537] for algebraic operations performed between random-set-valued variables.

Molchanov's random set theory Random set theory has evolved autonomously by a large amount in recent years, in a topological rather than a statistical sense, mostly thanks to the work of Ilya Molchanov [1302, 1304].

Since the family of all sets is too large, we typically restrict ourselves to the case of random elements in the space of closed subsets of a certain topological space $\mathbb{E}$. In this section, the family of closed subsets of $\mathbb{E}$ is denoted by $\mathcal{C}$, while $\mathcal{K}$ denotes the family of all compact subsets of $\mathbb{E}$. It is also assumed that $\mathbb{E}$ is a *locally compact Hausdorff second countable* topological space.

Let $(\Omega, \mathcal{F}, P)$ be a probability space. The following definitions are taken from [1304].[49]

Definition 72. *A map* $X : \Omega \rightarrow \mathcal{C}$ *is called a* random closed set *if, for every compact set* K *in* $\mathbb{E}$,

$$\{\omega \in \Omega : X(\omega) \cap K \neq \emptyset\} \in \mathcal{F}.$$

Alternatively, we can define the Borel σ-algebra $\mathcal{B}(\mathcal{C})$ generated by $\{C \in \mathcal{C} : C \cap X \neq \emptyset\}$ for all compact sets K (an Effros σ-algebra).

Definition 73. *A map* $X : \Omega \rightarrow \mathcal{C}$ *is called a* random closed set *if* X *is measurable with respect to the Borel* σ*-algebra on* $\mathcal{C}$ *with respect to the Fell topology, namely*

$$X^{-1}(\mathcal{X}) = \{\omega : X(\omega) \in \mathcal{X}\} \in \mathcal{F}$$

for each $\mathcal{X} \in \mathcal{B}(\mathcal{C})$.

Example 14: random closed sets. (From [1304].) *If* ξ *is a random element in* $\mathbb{E}$ *(measurable with respect to the Borel* σ*-algebra on* $\mathbb{E}$*), then the singleton* $X = \{\xi\}$ *is a random closed set. If* ξ *is a random variable, then* $X = (-\infty, \xi]$ *is a random closed set on the line* $\mathbb{E} = \mathbb{R}^1$*. If* ξ_1, ξ_2 *and* ξ_3 *are three random vectors in* $\mathbb{R}^d$*, then the triangle with vertices* ξ_1, ξ_2 *and* ξ_3 *is a random closed set. If* ξ *is a random vector in* $\mathbb{R}^d$ *and* η *is a non-negative random variable, then the random ball* $B_\eta(\xi)$ *of radius* η *centred at* ξ *is a random closed set. Let* ζ_x*,* $x \in \mathbb{E}$ *be a real-valued stochastic process on* $\mathbb{E}$ *with continuous sample paths. Then its level set* $X = \{x : \zeta_x = t\}$ *is a random closed set for every* $t \in \mathbb{R}$*. Similarly,* $\{x : \zeta_x \leq t\}$ *and* $\{x : \zeta_x \geq t\}$ *are random closed sets (see Fig. 4.21).*

---

[49] We retain Molchanov's original notation X for random closed sets, in order to differentiate them from the random sets Γ in Dempster's formulation.

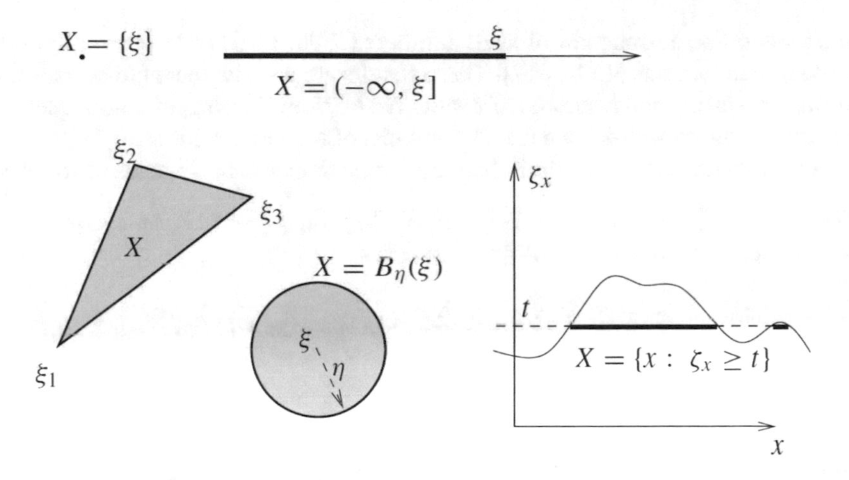

Fig. 4.21: Simple examples of random closed sets (from [1304]).

Capacity functionals The distribution of a random closed set X is determined by $P(\mathcal{X}) = P(\{\omega : X(\omega) \in \mathcal{X}\})$ for all $\mathcal{X} \in \mathcal{B}(\mathcal{C})$. In particular, we can consider the special case $\mathcal{X} = \mathcal{C}_K \doteq \{C \in \mathcal{C} : C \cap K \neq \emptyset\}$ and $P(\{X(\omega) \in \mathcal{C}_K\}) = P(\{X \cap K \neq \emptyset\})$, since the families $\mathcal{C}_K, K \in \mathcal{K}$, generate the Borel σ-algebra $\mathcal{B}(\mathcal{C})$.

Definition 74. *A functional* $T_X : \mathcal{K} \to [0,1]$ *given by*

$$T_X(K) = P(\{X \cap K \neq \emptyset\}), \quad K \in \mathcal{K}, \tag{4.87}$$

is termed the capacity functional *of* X.

We can write $T(K)$ instead of $T_X(K)$ where no ambiguity occurs. In particular, if $X = \{\xi\}$ is a random singleton, i.e., a classical random variable, then $T_X(K) = P(\{\xi \in K\})$, so that the capacity functional is the probability distribution of the random variable ξ.

The name 'capacity functional' follows from the fact that T_X is a functional on $\mathcal{K}$ which takes values in $[0,1]$, equals 0 on the empty set, is monotone and upper semicontinuous (i.e., T_X is a *capacity*: see Chapter 6, Definition 95) and is also completely alternating on $\mathcal{K}$ (Definition 68). It is easy to recognise in (4.87) the plausibility measure induced by the multivalued mapping X, restricted to compact subsets. The links betwen random closed sets and belief/plausibility functions, upper and lower probabilities, and contaminated models in statistics are very briefly hinted at in [1304], Chapter 1, Section 9.

Central limit theorem and Gaussian random sets A number of interesting topics were investigated by Molchanov, including the generalisation of the Radon–Nikodym theorem to capacities, convergence results, the notion of the expectation of a random set, a strong law of large numbers [1300, 1303] and a central limit theorem for random sets [1301, 817]. These results are closely related to probabilities in Banach spaces, and a number of results for Minkowski sums of random sets can be obtained using well-known results for sums of random functions.

For instance, the central limit theorem for random sets in $\mathbb{R}^d$ reads as follows.

Theorem 4. ([1304], Theorem 2.1) *Let $X, X_1, X_2, \ldots$ be i.i.d. square integrable random sets, i.e., such that $E\|X\|^2 < \infty$. Then*

$$\sqrt{n}\rho_H\left(\frac{X_1 + \cdots + X_n}{n}, EX\right) \to \sup_{u \in B_1} \|\zeta(u)\|,$$

where ρ_H is the Hausdorff metric and $\{\zeta(u), u \in B_1\}$ is a centred Gaussian random function in $C(B_1)$, the Banach space of continuous functions on the unit ball, with covariance

$$E[\zeta(u)\zeta(v)] = E[h(X,u)h(X,v)] - Eh(X,u)Eh(X,v).$$

As Theorem 4 shows, Gaussian random functions on the unit ball B_1 appear naturally in the limit theorem for Minkowski sums of random compact sets in $\mathbb{R}^d$. As Gaussian random elements in linear spaces can be defined through Gaussian distributions of all linear continuous functionals, a similar approach can be applied to the (nonlinear) space $\mathcal{K}$ of compact sets.

Let $Lip^+(\mathcal{K}, \mathbb{R})$ denote the family of functionals $g : \mathcal{K} \to \mathbb{R}$ which satisfy the following conditions:

1. g is *positively linear*, i.e., for all $a, b \geq 0$ and $K, L \in \mathcal{K}$, $g(aK + bL) = ag(K) + bg(L)$.
2. g is *Lipschitz* with respect to the Hausdorff metric.

Definition 75. *A random compact set X in $\mathbb{R}^d$ is said to be Gaussian if $g(X)$ is a Gaussian random variable for each $g \in Lip^+(\mathcal{K}, \mathbb{R})$.*

Further results on Minkowski sums can also be found in [1304]. Most relevantly to what we will discuss in Chapter 17, set-valued random processes and multivalued martingales are investigated there.

4.9.4 Kramosil's belief functions on infinite spaces

Within the compatibility relation interpretation of Section 3.1.1, Kramosil developed a measure-theoretic definition of a belief function which naturally copes with the case of infinite domains. The following summary is inspired by [1051], Chapter 9 – as usual, it has been translated into the notation of this book.

Measure-theoretic definition of belief function Let $(\Omega, \mathcal{F}, P)$ be a probability space, and let

$$\Gamma(\omega) = \{\theta \in \Theta : \rho(\omega, \theta) = 1\} \tag{4.88}$$

be defined for all $\omega \in \Omega$, where ρ is a compatibility relation $\rho : \Omega \times \Theta \to \{0, 1\}$ (compare Section 3.1.1). The mapping Γ is a *set-valued random variable*, which takes the probability space $(\Omega, \mathcal{F}, P)$ into a measurable space $(\mathcal{P}(\Theta), \mathcal{S})$ where $\mathcal{S} \subset \mathcal{P}(\mathcal{P}(\Theta))$ is a σ-field of *collections* of subsets of Θ, rather than a σ-field of *subsets* (as for conventional random variables).

For traditional random variables, a subset of Θ (typically the real line, $\Theta = \mathbb{R}$) is measurable whenever its anti-image through the mapping $X : \Omega \to \Theta$, $\omega \mapsto X(\omega) \in \Theta$ belongs to the σ-algebra $\mathcal{F}$. For set-valued random variables, however, measurability is a condition on any given *collection* of subsets of Θ. A collection $S \in \mathcal{S}$ of subsets of Θ belonging to $\mathcal{S}$ is called *measurable* whenever its anti-image

$$\Gamma^{-1}(S) \doteq \{\omega \in \Omega : \Gamma(\omega) \in S\}$$

belongs to the σ-algebra $\mathcal{F}$ of events $A \subset \Omega$ for which a probability value $P(A)$ exists, by virtue of the probability space defined there.

Existence of belief values and $\mathcal{S}$-regularity Given a set-valued random variable Γ, the (unnormalised) belief value

$$Bel(A) \doteq P(\{\omega \in \Omega : \emptyset \neq \Gamma(\omega) \subset A\}) \tag{4.89}$$

can be defined whenever the probability on the right-hand side exists. But this is equivalent to requiring that

$$\{\omega \in \Omega : \emptyset \neq \Gamma(\omega) \subset A\} = \Gamma^{-1}(\mathcal{P}(A)) \in \mathcal{F},$$

i.e., the anti-image of the collection of all non-empty subsets of $A \subset \Theta$ belongs to the σ-field $\mathcal{F}$. Such events A are called *$\mathcal{S}$-regular* ([1051], page 14).

Kohlas's simplified model A special case was presented by Kohlas in [1027], where the support Ω of the basic probability space is assumed to be finite and the σ-field $\mathcal{F}$ is identified with $\mathcal{P}(\Omega)$.

Under the second condition, in particular, any mapping (4.88) is measurable (since any anti-image of subsets of Θ must belong to the power set of Ω), and therefore a random variable. For each such mapping Γ, there exists a finite or countable system $\mathcal{B}_0$ of subsets of Θ such that

$$\mathcal{B}_0 = \Big\{B \subset \Theta : \{\omega \in \Omega : \Gamma(\omega) = B\} \neq \emptyset\Big\}.$$

Therefore,

$$P(\{\omega \in \Omega : \Gamma(\omega) = B\}) = \sum_{\omega : \Gamma(\omega) = B} P(\omega) = P(\Gamma^{-1}(\{B\})) \geq 0 \quad \text{if } B \in \mathcal{B}_0,$$

while $P(\Gamma^{-1}(\{B\})) = 0$ whenever $B \in \mathcal{P}(\Theta) \setminus \mathcal{B}_0$. Consequently, for each $A \subset \Theta$ the (unnormalised) belief value of A is

$$Bel(A) = P(\{\omega \in \Omega : \Gamma(\omega) \in \mathcal{P}(A) \setminus \emptyset\}) = \sum_{\emptyset \neq B \subset A, B \in \mathcal{B}_0} P(\Gamma^{-1}(\{B\})),$$

where $m(B) \doteq P(\Gamma^{-1}(\{B\}))$ plays the role of the basic probability assignment in the finite case.

Kohlas's model therefore outlines the domain where the combinatoric definitions of belief functions are immediately extendable to infinite sets, so that some algorithms or other implementation results can be directly applied to this wider class of situations.

DS-complete σ-field

Definition 76. *A σ-field $\mathcal{S} \subset \mathcal{P}(\mathcal{P}(\Theta))$ is called* Dempster–Shafer complete *(DS-complete) if every $A \subset \Theta$ is $\mathcal{S}$-regular.*

In other words, $\mathcal{P}(A) \in \mathcal{S}$ for all $A \subset \Theta$, and its belief value can be defined via (4.89). Kramosil proved the following proposition.

Proposition 34. ([1051], Theorem 9.1) *If $\mathcal{S} \subset \mathcal{P}(\mathcal{P}(\Theta))$ is DS-complete, any collection $S \subset \mathcal{P}(\Theta)$ of at most a countable number of subsets, each of at most countable cardinality, is measurable (i.e., $S \in \mathcal{S}$).*

Generalisations to non-$\mathcal{S}$-regular sets It is also possible to generalise belief functions to subsets of Θ that are not $\mathcal{S}$-regular, via the notions of the *inner measure* and *outer measure* (see Chapter 3, Section 3.1.3).

Let us suppose that the empty set of Θ is $\mathcal{S}$-regular, i.e., that $\mathcal{P}(\emptyset) = \{\emptyset\} \in \mathcal{S}$, and that $P(\{\omega \in \Omega : \Gamma(\omega) = \emptyset\}) < 1$. The following four generalisations of belief functions can be defined.

Definition 77. *Let $(\Omega, \mathcal{F}, P)$ be a probability space, let $(\mathcal{P}(\Theta), \mathcal{S})$ be a measurable space over a non-empty set $\mathcal{S}$, and let Γ be a set-valued random variable (4.88) defined on $(\Omega, \mathcal{F}, P)$ and taking its values in $(\mathcal{P}(\Theta), \mathcal{S})$.*

We can then define

$$\begin{aligned} Bel_+(A) &= \sup \Big\{ P(\{\omega \in \Omega : \Gamma(\omega) \in \mathcal{B}\}/\{\omega \in \Omega : \Gamma(\omega) \neq \emptyset\}) \Big| \\ &\qquad \mathcal{B} \in \mathcal{S}, \mathcal{B} \subset \mathcal{P}(A) \setminus \{\emptyset\} \Big\}, \\ Bel^+(A) &= \inf \Big\{ P(\{\omega \in \Omega : \Gamma(\omega) \in \mathcal{B}\}/\{\omega \in \Omega : \Gamma(\omega) \neq \emptyset\}) \Big| \\ &\qquad \mathcal{B} \in \mathcal{S}, \mathcal{B} \supset \mathcal{P}(A) \setminus \{\emptyset\} \Big\}, \\ Bel_{++}(A) &= \sup \{ Bel(B), B \subset A\}, \\ Bel^{++}(A) &= \inf \{ Bel(B), A \subset B \subset \Theta\}, \end{aligned} \tag{4.90}$$

where

$$Bel(A) \doteq P(\{\omega \in \Omega : \emptyset \neq \Gamma(\omega) \subset A\}/\{\omega \in \Omega : \Gamma(\omega) \neq \emptyset\}) \tag{4.91}$$

is the normalised belief value for $\mathcal{S}$-regular sets.

The interpretations of these objects are rather straightforward. $Bel_+(A)$ is the supremum of the belief values of a system of subsets that are measurable, and are included in the power set of the desired event A. $Bel^+(A)$ is the infimum of the belief values of a system of subsets that are measurable, and *contain* the power set of A. $Bel_{++}(A)$ and $Bel^{++}(A)$ are similarly computed, but over the subsets themselves, rather than the collections of subsets associated with their power sets.

The corresponding (generalised) plausibility functions can easily be calculated via $Pl(A) = 1 - Bel(A^c)$.

4.9.5 MV algebras

A recent interesting approach studies belief functions in a more general setting than that of a Boolean algebra of events, inspired by the generalisation of classical probability towards 'many-valued' events, such as those resulting from formulae in Lukasiewicz infinite-valued logic.

The work in this area is mainly due to Kroupa [1077, 1078] and Flaminio. Kroupa generalised belief functions to many-valued events, which are represented by elements of the Lindenbaum algebra of infinite-valued Lukasiewicz propositional logic. The following summary of MV algebras and their relationship with belief functions is abstracted from [1077].

Flaminio [627] built on Kroupa's work after noting that Kroupa's belief functions evaluate the degree of belief in the occurrence of fuzzy events by using classical sets as focal elements, and introduced a generalization of Kroupa's belief functions that allows us to deal with evidence on fuzzy subsets. In [626], the focus was on characterising normalised belief functions and their fusion by means of a generalised Dempster rule of combination. In [628], instead, a fuzzy modal logic to formalise reasoning with belief functions on MV algebras was introduced.

Definition of MV algebra An MV algebra is an algebra of many-valued events, upon which upper/lower probabilities and possibility measures can be defined.

Definition 78. *An MV algebra is an algebra* $\langle M, \oplus, \neg, 0 \rangle$ *with a binary operation* $\oplus$*, a unary operation* $\neg$ *and a constant 0 such that* $\langle M, \oplus, 0 \rangle$ *is an Abelian monoid and the following equations hold true for every* $f, g \in M$*:*

$$\neg\neg f = f, \quad f \oplus \neg 0 = \neg 0, \quad \neg(\neg f \oplus g) \oplus g = \neg(\neg g \oplus f) \oplus f.$$

Building on these base operators, one can also define

$$1 = \neg 0, \quad f \odot g = \neg(\neg f \oplus \neg g), \quad f \leq g \text{ if } \neg f \oplus g = 1.$$

If, in addition, we introduce two inf and sup operators as follows,

$$f \vee g = \neg(\neg f \oplus g) \oplus g, \quad f \wedge g = \neg(\neg f \vee \neg g),$$

we make $\langle M, \vee, \wedge, 0, 1 \rangle$ a *distributive lattice*.

Example 15. *For example, the so-called* standard *MV algebra is the real interval* $[0,1]$ *equipped with*

$$f \oplus g = \min(1, f+g), \quad \neg f = 1 - f, \quad f \odot g = \max(0, f+g-1).$$

In this case $\odot$ *and* $\oplus$ *are known as the* Lukasiewicz t-norm *and* t-conorm, *respectively.*

Boolean algebras are also a special case of MV algebras, with as $\oplus, \odot$ *and* $\neg$ *the union, intersection and complement Boolean operators. As we will see, a totally monotone function* $Bel : M \to [0,1]$ *can be defined on an MV algebra, by replacing* $\cup$ *with* $\vee$ *and* $\subset$ *with* $\leq$.

States as generalisations of finite probabilities *Semisimple* algebras are those MV algebras which are isomorphic to continuous functions onto $[0,1]$ on some compact Hausdorff space. These can be viewed as many-valued counterparts of algebras of sets, and are quite closely related to the imprecise-probabilistic notion of a *gamble* (see Section 6.1.2, Definition 90).

Definition 79. *A* state *is a mapping* $s : M \to [0,1]$ *such that* $s(1) = 1$ *and* $s(f + g) = s(f) + s(g)$ *whenever* $f \odot g = 0$.

Clearly, states are generalisations of finitely additive probability measures, once we replace events A, B with continuous functions f, g onto $[0,1]$ (elements of a semisimple MV algebra), and $\cap$ with $\odot$. In addition, states on semisimple MV algebras are integrals of a Borel probability measure on the Hausdorff space. Namely, $\forall f \in M$,

$$s(f) = \int f \, \mathrm{d}\mu.$$

Belief functions on MV algebras Now, consider the MV algebra $[0,1]^{\mathcal{P}(X)}$ of all functions $\mathcal{P}(X) \to [0,1]$, where X is finite. Define $\rho : [0,1]^X \to [0,1]^{\mathcal{P}(X)}$ as

$$\rho(f)(B) \doteq \begin{cases} \min\{f(x), x \in B\} & B \neq \emptyset, \\ 1 & \text{otherwise.} \end{cases}$$

If $f = 1_A$ (the indicator function of event A), then $\rho(1_A)(B) = 1$ iff $B \subseteq A$, and we can rewrite $Bel(A) = m(\rho(1_A))$, where m is defined on collections of events.

Definition 80. $Bel : [0,1]^X \to [0,1]$ *is a* belief function *on* $[0,1]^X$ *if there is a state on the MV algebra* $[0,1]^{\mathcal{P}(X)}$ *such that* $s(1_\emptyset) = 0$ *and* $Bel(f) = s(\rho(f))$*, for every* $f \in [0,1]^X$. *The state s is called a* state assignment.

Belief functions so defined have values on continuous functions of X (of which events are a special case). State assignments correspond to probability measures on Ω in the classical random-set interpretation (Fig. 4.22). There is an integral representation $\int$ by the Choquet integral of such belief functions – the whole approach is strongly linked to belief functions on fuzzy sets (Section 6.5.2).

All standard properties of classical belief functions are satisfied (e.g. superadditivity). In addition, the set of belief functions on $[0,1]^X$ is a simplex whose extreme points correspond to the generalisation of categorical BFs (see Chapter 7).

$$\begin{array}{ccc} \text{BF } Bel \text{ on } [0,1]^X & \longleftrightarrow & \text{BF } \beta \text{ on } \mathcal{P}(X) \\ \rho \updownarrow & & \updownarrow \rho \\ \text{State } s \text{ on } [0,1]^{\mathcal{P}(X)} & \stackrel{\int}{\longleftrightarrow} & \text{Probability } \mu \text{ on } \mathcal{P}(\mathcal{P}(X)) \end{array}$$

Fig. 4.22: Relationships between classical belief functions on $\mathcal{P}(X)$ and belief functions on $[0,1]^X$ (from [1077]).

4.9.6 Other approaches

Evidence theory generalised to Boolean algebras Guan and Bell [735] proposed a generalisation of evidence theory in which arbitrary Boolean algebras replace power sets of events, thus covering, for instance, infinite sets. A generalised orthogonal sum was theoretically justified there. An interpretation for the generalization of evidence theory by Guan and Bell was put forward by Liu in [1181] by generalising the concepts of upper and lower probabilities, and proving that the conditioning of belief functions defined by Guan and Bell is, in fact, a generalisation of Dempster's rule of conditioning.

Fuzzy belief functions in infinite spaces In an interesting paper of 2009 [1965], Wu et al. addressed the problem of generalising belief functions to an infinite space in a fuzzy context. They defined a general type of fuzzy belief structure determined by a fuzzy implication operator, and its induced dual pair of fuzzy belief and plausibility functions in infinite universes of discourse. In analogy to the classical case, the lower and upper fuzzy probabilities induced by the fuzzy belief space yield a dual pair of fuzzy belief and plausibility functions, which are, respectively, a fuzzy monotone Choquet capacity and a fuzzy alternating Choquet capacity of infinite order.

4.10 The mathematics of belief functions

Belief functions are rather complex mathematical objects: as a result, they possess links with a number of fields of (applied) mathematics, on one hand, and can lead to interesting generalisations of standard results of classical probability theory (e.g. Bayes' theorem, total probability), on the other.

To conclude this chapter, we briefly review past work on distances and the algebra of belief functions, and other notable mathematical properties.

4.10.1 Distances and dissimilarities

A number of norms for belief functions have been introduced [501, 902, 965, 1636], as a tool for assessing the level of conflict between different bodies of evidence, for approximating a BF with a different uncertainty measure and so on.

Families of dissimilarity measures Generalisations to belief functions of the classical Kullback–Leibler divergence $D_{\text{KL}}(P|Q) = \int_{-\infty}^{\infty} p(x) \log(p(x)/q(x)) dx$ of two probability distributions P, Q, for instance, have been proposed, together with measures based on information theory, such as fidelity, or entropy-based norms. Any exhaustive review would be impossible here: nevertheless, Jousselme and Maupin [938] managed in recent times to compile a very nice survey on the topic [630, 937], according to which tests on randomly generated belief functions reveal the existence of four families of dissimilarity measures, namely metric (i.e., proper distance functions), pseudo-metric (dissimilarities), non-structural (those which do not account for the structure of the focal elements) and non-metric ones.

Jousselme's distance The most popular and most cited measure of dissimilarity was proposed by Jousselme et al. [933] as a 'measure of performance' of algorithms (e.g., for object identification) in which successive evidence combination steps lead to convergence to the 'true' solution.

Jousselme's measure assumes that mass functions m are represented as vectors $\mathbf{m}$ (see Section 4.7.1), and reads as

$$d_{\text{J}}(m_1, m_2) \doteq \sqrt{\frac{1}{2}(\mathbf{m}_1 - \mathbf{m}_2)^T D(\mathbf{m}_1 - \mathbf{m}_2)},$$

where $D(A, B) = \frac{|A \cap B|}{|A \cup B|}$ for all $A, B \in 2^\Theta$. Jousselme's distance so defined (1) is positive definite (as proved by Bouchard et al. in [184]), and thus defines a metric distance; (2) takes into account the similarity among subsets (focal elements); and (3) is such that $D(A, B) < D(A, C)$ if C is 'closer' to A than B.

Other proposals Among others, it is worth mentioning the following proposals:

- the Dempster conflict κ and Ristic's closely related *additive global dissimilarity measure* [1491]: $-\log(1 - \kappa)$;
- the 'fidelity' or Bhattacharia coefficient [149] extended to belief functions, namely $\sqrt{\mathbf{m}_1}^T W \sqrt{\mathbf{m}_2}$, where W is positive definite and $\sqrt{\mathbf{m}}$ is the vector obtained by taking the square roots of each component of $\mathbf{m}$;
- Perry and Stephanou's distance [1421],

$$d_{\text{PS}}(m_1, m_2) = |\mathcal{E}_1 \cup \mathcal{E}_2| \left(1 - \frac{\mathcal{E}_1 \cap \mathcal{E}_2}{\mathcal{E}_1 \cup \mathcal{E}_2}\right) + (\mathbf{m}_{12} - \mathbf{m}_1)^T (\mathbf{m}_{12} - \mathbf{m}_2),$$

  where $\mathcal{E}$ is, as usual, the collection of focal elements of the belief function with mass assignment m and $\mathbf{m}_{12}$ is the mass vector of the Dempster combination of m_1 and m_2;
- Blackman and Popoli's *attribute distance* [169],

$$\begin{aligned} &d_{\text{BP}}(m_1, m_2) \\ &= -2 \log \left[\frac{1 - \kappa(m_1, m_2)}{1 - \max_i \{\kappa(m_i, m_i)\}} \right] + (\mathbf{m}_1 + \mathbf{m}_2)^T \mathbf{g}_A - \mathbf{m}_1^T G \mathbf{m}_2, \end{aligned}$$

where $\mathbf{g}_A$ is the vector with elements $\mathbf{g}_A(A) = \frac{|A|-1}{|\Theta|-1}$, $A \subset \Theta$, and

$$G(A,B) = \frac{(|A|-1)(|B|-1)}{(|\Theta|-1)^2}, \quad A, B \subset \Theta;$$

- the class of L_p (Minkowski) measures used by this author for geometric conditioning [356] and consonant/consistent approximation [360, 349, 352], as we will see in Parts III and IV,

$$d_{L_p}(m_1, m_2) = \left(\sum_{A \subseteq \Theta} |Bel_1(A) - Bel_2(A)|^p \right)^{1/p}$$

(note that the L_1 distance was used earlier by Klir [998] and Harmanec [797]);
- Fixen and Mahler's *Bayesian percent attribute miss* [623]:

$$\mathbf{m}_1' P \mathbf{m}_2,$$

where $P(A,B) = \frac{p(A \cap B)}{p(A)p(B)}$ and p is an a priori probability on Θ;
- Zouhal and Denoeux's inner product of pignistic functions [2135];
- the family of information-based distances also proposed by Denoeux [449], in which the distance between m_1 and m_2 is quantified by the difference between their information contents $U(m_1), U(m_2)$,

$$d_U(m_1, m_2) = |U(m_1) - U(m_2)|,$$

where U is any uncertainty measure for belief functions.

4.10.2 Algebra

Algebra of frames The algebra of frames of discernment forming a compatible family was studied extensively by Cuzzolin [363, 369, 329], but also by Kohlas and co-authors (see [1027], Chapter 7). In related work, Capotorti and Vantaggi [222] studied the consistency problem when dealing with pieces of evidence which provide partial evaluation on only a few events, by introducing computable consistency properties (as in de Finetti's coherence principle for probability assessments).

Lattice of frames

Definition 81. *A collection of compatible frames $\Theta_1, \ldots, \Theta_n$ are said to be* independent *[1583] ($\mathcal{IF}$) if:*

$$\rho_1(A_1) \cap \cdots \cap \rho_n(A_n) \neq \emptyset, \quad \forall \emptyset \neq A_i \subseteq \Theta_i,$$

where ρ_i is the refining from Θ_i to the minimal refinement (Theorem 1) of $\Theta_1, \ldots, \Theta_n$.

The notion comes from that of independence of Boolean subalgebras. One can prove that independence of sources is in fact *equivalent* to independence of frames [329]:

Proposition 35. *A set of compatible frames $\Theta_1, \ldots, \Theta_n$ are independent iff all the possible collections of belief functions $Bel_1, \ldots, Bel_n$ on $\Theta_1, \ldots, \Theta_n$ are Dempster-combinable on their minimal refinement $\Theta_1 \otimes \cdots \otimes \Theta_n$.*

In a family of frames, we can define the following order relation:

$$\Theta_1 \leq \Theta_2 \Leftrightarrow \exists \rho : \Theta_2 \to 2^{\Theta_1} \text{ refining}, \tag{4.92}$$

together with its dual $\leq^*$. Both $(\mathcal{F}, \leq)$ and $(\mathcal{F}, \leq^*)$ are lattices.

Definition 82. *A lattice L is* upper semimodular *if, for each pair x, y of elements of L, $x \succ x \wedge y$*[50] *implies $x \vee y \succ y$. A lattice L is* lower semimodular *if, for each pair x, y of elements of L, $x \vee y \succ y$ implies $x \succ x \wedge y$.*

Theorem 5. [329] *A family of compatible frames endowed with the order relation (4.92) $(\mathcal{F}, \leq)$ is an upper semimodular lattice. Dually, $(\mathcal{F}, \leq^*)$ is a lower semimodular lattice.*

Lattice-theoretic independence Now, abstract independence can be defined on collections $\{l_1, \ldots, l_n\}$ of non-zero elements of any semimodular lattice L with initial element $\mathbf{0} \doteq \bigwedge L$ as follows:

$$\mathcal{I}_1 : l_j \not\leq \bigvee_{i \neq j} l_i \; \forall j = 1, \ldots, n, \quad \mathcal{I}_2 : l_j \wedge \bigvee_{i<j} l_i = \mathbf{0} \; \forall j = 2, \ldots, n,$$

$$\mathcal{I}_3 : h\left(\bigvee_i l_i\right) = \sum_i h(l_i),$$

where the rank h of an element l is the length of the chain linking $\mathbf{0}$ and l. In particular, for the lower submodular lattice of compatible frames $\Theta_1, \ldots, \Theta_n$ with order relation $\leq^*$, these relations read as follows [351]:

$$\begin{array}{lcl} \{\Theta_1, \ldots, \Theta_n\} \in \mathcal{I}_1^* & \Leftrightarrow & \Theta_j \oplus \bigotimes_{i \neq j} \Theta_i \neq \Theta_j \; \forall j = 1, \ldots, n, \\ \{\Theta_1, \ldots, \Theta_n\} \in \mathcal{I}_2^* & \Leftrightarrow & \Theta_j \oplus \bigotimes_{i=1}^{j-1} \Theta_i = \mathbf{0}_{\mathcal{F}} \; \forall j = 2, \ldots, n, \\ \{\Theta_1, \ldots, \Theta_n\} \in \mathcal{I}_3^* & \Leftrightarrow & \left|\bigotimes_{i=1}^{n} \Theta_i\right| - 1 = \sum_{i=1}^{n} (|\Theta_i| - 1). \end{array}$$

Notably, the relation $\mathcal{I}_3^*$ is equivalent to saying that the dimension of the polytope of probability measures defined on the minimal refinement is the sum of the dimensions of the polytopes associated with the individual frames. Dual expressions hold for the upper semimodular lattice $(\mathcal{F}, \leq)$.

---

[50] x 'covers' y ($x \succ y$) if $x \geq y$ and there is no intermediate element in the chain linking the two elements.

The relationship between these lattice-theoretic forms of independence and independence of frames is summarised in Fig. 4.23 [351]. In the upper semimodular case (a), $\mathcal{IF}$ is mutually exclusive with all of the lattice-theoretic relations $\mathcal{I}_1, \mathcal{I}_2, \mathcal{I}_3$. In the lower semimodular case (b), $\mathcal{IF}$ is *a stronger condition* than both $\mathcal{I}_1^*$ and $\mathcal{I}_2^*$, whereas $\mathcal{IF}$ is mutually exclusive with the third independence relation (a form of matroidal [1374] independence).

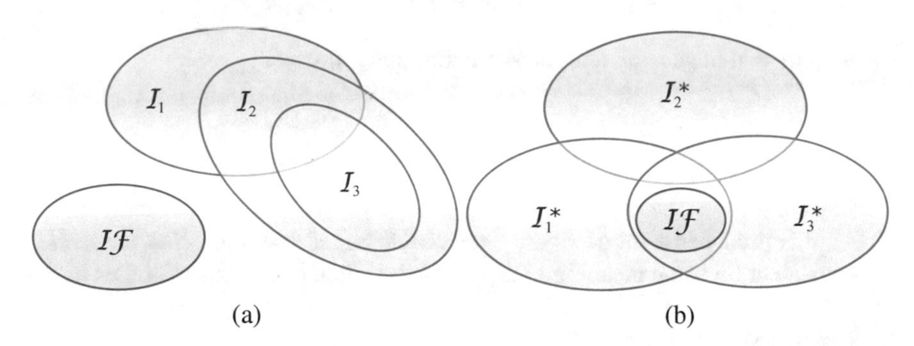

Fig. 4.23: Lattice-theoretical independence and independence of frames $\mathcal{IF}$ in the upper semimodular (a) and lower semimodular (b) cases.

An algebraic solution to conflict resolution This analysis hints at the possibility that independence of sources may be explained algebraically. Indeed, families of frames and projective geometries (sets of vector subspaces of a vector space) share the same kind of lattice structure, i.e., semimodularity. However, as we have seen, independence of sources is not a form of lattice-theoretic independence, nor it is a form of matroidal independence [351], but they are related in a rather complex way (Fig. 4.23).

Based on this analogy, a possible algebraic solution to the conflict problem (see Section 4.3.5) by means of a *generalised Gram–Schmidt* procedure (Algorithm 10) can be outlined.

Algebra of belief measures The algebraic structure of belief measures on the space of belief functions on a two-element frame, rather than the algebra of the frames on which BFs are defined, was studied by Daniel [389]. He introduced the *Dempster's semigroup* as the algebraic structure of binary belief functions endowed with Dempster's rule of combination.

Definition 83. *The (standard)* Dempster's semigroup $D = (\mathcal{B}_2^0, \oplus)$ *is the set* $\mathcal{B}_2^0$ *of all belief functions on a binary frame of discernment* $|\Theta| = 2$, *bar the Bayesian ones, endowed with the operation* $\oplus$ *(Dempster combination).*

An order relation was defined as follows.

Algorithm 10 Generalised Gram–Schmidt procedure

1: **procedure** GENERALISEDGRAMSCHMIDT($\{Bel_i, i = 1, \ldots, n\}$)
2: Consider a set of belief functions $Bel_i : 2^{\Theta_i} \rightarrow [0, 1]$ defined over $\Theta_1, \ldots, \Theta_n$, part of a family of compatible frames $\mathcal{F}$;
3: compute a new collection of *independent* frames of the same family $\mathcal{F}$:

$$\Theta_1, \ldots, \Theta_n \in \mathcal{F} \longrightarrow \Theta'_1, \ldots, \Theta'_m \in \mathcal{F},$$

with $m \neq n$ in general, and the same minimal refinement

$$\Theta_1 \otimes \cdots \otimes \Theta_n = \Theta'_1 \otimes \cdots \otimes \Theta'_m;$$

4: project the n original belief functions $Bel_1, \ldots, Bel_n$ onto the new set of frames;
5: this produces a set of *surely combinable* belief functions $Bel'_1, \ldots, Bel'_m$ equivalent (in some meaningful sense) to the initial collection of bodies of evidence.
6: **end procedure**

Definition 84. *$Bel \leq Bel'$ on $\Theta = \{x, y\}$ iff $h(Bel) \leq h(Bel')$ or $h(Bel) = h(Bel')$ and $Bel(x) \leq Bel'(x)$, where $h(Bel) \doteq (1 - Bel(y))/(2 - Bel(x) - Bel(y))$.*

The following proposition [389] could then be proved.

Proposition 36. *The Dempster semigroup endowed with the order relation $\leq$ of Definition 84 is an ordered commutative semigroup*[51] *with neutral element $\mathbf{0} = [0, 0]'$, and $\mathbf{0}' = [1/2, 1/2]$ as the only non-zero idempotent element.*

The analysis was later extended by Daniel to the disjunctive rule of combination [378]. An introduction to an algebra of belief functions on a three-element frame of discernment, in a quasi-Bayesian case, can be found in [387].

Brodzik and Enders [194] also used the language of semigroups to analyse the Dempster combination. They showed that the set of Bayesian belief functions and the set of categorical BFs, under either a mild restriction or a slight extension, are semigroups with respect to Dempster's rule. These two semigroups were shown to be related by a semigroup homomorphism, with elements of the Bayesian set acting as images of disjoint subsets of the set of belief functions. Subsequently, an inverse mapping from the Bayesian set onto the set of these subsets was identified and a procedure for computing certain elements of these subsets, acting as subset generators, was obtained.

[51] A commutative group is a structure $(X, \oplus, -, o)$ such that $(X, \oplus)$ is a commutative semigroup, o is a neutral element ($x \oplus o = x$ for all $x \in X$) and $-$ is a unary inverse operation, $x \oplus -x = o$. An ordered Abelian (semi)group consists of a commutative (semi)group X as above and a linear ordering $\leq$ of its elements satisfying monotonicity ($x \leq y$ implies $x \oplus z \leq y \oplus z$ for all $x, y, z \in X$).

4.10.3 Integration

Barres conducted interesting, albeit neglected, work on a theory of integration for belief measures [134].

Noting that elementwise multiplication is a binary operation on the set of belief functions, $Bel(A) = Bel_1(A) \cdot Bel_2(A)$, if belief and plausibility measures are defined on a locally compact and separable space X, a theorem of Choquet ensures that they can be represented by a probability measure on the space containing the closed subsets of X, the so-called basic probability assignment. Two new types of integrals with respect to a basic plausibility assignment m can thus be defined. One of them measures the degree of non-additivity of the belief/plausibility function, namely

$$\int f \otimes_\mu m \doteq \int_{\mathcal{F}} \left(\int_F f \, \mathrm{d}\mu \right) \mathrm{d}m(F),$$

where X is a locally compact and separable space, $\mathcal{F} = \mathcal{F}(X)$ is the class of closed subsets of X, $F \in \mathcal{F}$ is such a closed subset, m is a BPA on $\mathcal{F}$, μ is an inner regular measure and f is a μ-integrable function on X. One can prove that

$$\int f \otimes_\mu m = \int (f \cdot Pl)(x) \, \mathrm{d}\mu(x),$$

where Pl is the plausibility function induced by m.

The second integral is a generalisation of the Lebesgue integral,

$$\oint f \otimes_\mu m \doteq \int_{\mathcal{F}} \left(\int_F f \, \mathrm{d}\mu_F \right) \mathrm{d}m(F), \tag{4.93}$$

where, this time, $\mu_F(B) = \mu(B|F)$ is the conditional measure defined if $\mu(F) > 0$. The integral (4.93) reduces to the Lebesgue one for BPAs which represent probability measures ([134], Proposition 5.2). The latter was compared in [134] with Choquet's $\int_0^{\sup f} T(\{f > t\}) \, \mathrm{d}t$ and Sugeno's integrals for capacities T and non-additive set functions, showing that

$$\begin{aligned} \int_0^{\sup f} Bel(\{f > t\}) \, \mathrm{d}t &= \int \left(\inf_{x \in F} f(x) \right) \mathrm{d}m(F) \leq \oint f \otimes_\mu m \\ \leq \int_0^{\sup f} Pl(\{f > t\}) \, \mathrm{d}t &= \int \left(\sup_{x \in F} f(x) \right) \mathrm{d}m(F). \end{aligned}$$

4.10.4 Category theory

In an original piece of work, evidential reasoning was studied in a categorical perspective by Kennes [951], using a combination of the probability kinematics approach of Richard Jeffrey [893] (1965) and the maximum (cross-)entropy inference approach of E. T. Jaynes (1957). Categorical approaches to probability theory and to evidential reasoning had already been proposed by other authors, for example Giry

[697], Negoita [1341], Goodman and Nguyen [708], and Gardenfors [665]. In contrast to evidential reasoning, since the work of Goguen [699], the categorical study of fuzzy sets is and was then a well-established part of fuzzy set theory.

In particular, in [951] Kennes introduced the notion of *Dempster's category*, and proved the existence of and gave explicit formulae for conjunction and disjunction in the subcategory of separable belief functions.

Definition 85. *A* category *is a directed graph* (V, A)*, endowed with a* composition operator *$c : A \times A \to A$, $c(a_1, a_2) = a_1.a_2$ defined whenever the final node of arrow a_1 coincides with the initial node of arrow a_2, such that the following axioms are satisfied:*

1. *$(a_1.a_2).a_3 = a_1.(a_2.a_3)$ (associativity of composition).*
2. *$i_v.a = a = a.i_w$, where v is the initial node of a, w is its final node and i_v is the edge linking v with itself (identities are neutral for composition).*

Definition 86. Dempster's category *is defined on nodes which are mass distributions, and for which (1) the arrows $a : m_1 \to m_2$ are the BPAs such that $a \oplus m_1 = m_2$, (2) the composite of $a_1 : m_1 \to m_2$ and $a_2 : m_2 \to m_3$ is $a_1 \oplus a_2 : m_1 \to m_3$ and (3) the identity arrow at m is the vacuous belief function on the frame Θ: $1_m = m_\Theta$.*

Category theory can indeed provide natural definitions for logical connectives. In particular, disjunction and conjunction can be modelled by general categorical constructions known as products and coproducts.

Definition 87. *A* product *of two objects v, w of a category is an object, denoted $v \times w$, along with two arrows $p_v : v \times w \to v$, $p_w : v \times w \to w$ such that, for every object u of the category along with two arrows $a_v : u \to v$, $a_w : u \to w$, there exists a unique arrow $a : u \to v \times w$ such that $a.p_v = a_v$ and $a.p_w = a_w$.*

For two belief functions $v = Bel_1$, $w = Bel_2$, their product or 'disjunction' $Bel_1 \vee Bel_2$ is the most Dempster-updated BF such that Bel_1 and Bel_2 are both updated states of it. A 'conjunction' operator $\wedge$ can be dually defined. Categorical conjunction can be seen as the most cautious conjunction of beliefs, and thus no assumption about distinctness (of the sources) of beliefs is needed, as opposed to Dempster's rule of combination. In [951], explicit formulae for the categorical conjunction of separable belief functions were provided.

4.10.5 Other mathematical analyses

Dubois and Prade [527] adopted the view of bodies of evidence as generalised sets to study the algebraic structure of belief measures on a set, based on extended set union, intersection and complementation. Several notions of inclusion, used to compare a body of evidence with the product of its projections, were proposed and compared with each other. Lastly, approximations of a body of evidence in the form of fuzzy sets were derived, noting that focal elements can be viewed either as sets of actual values or as restrictions on the (unique) value of a variable.

The work of Roesmer [1494] deserves a note for its original connection between non-standard analysis and the theory of evidence.

Yager [2005] showed how variables whose values are represented by Dempster–Shafer structures (belief functions) can be combined under arithmetic operations such as addition. The procedure was generalised to allow for the combination of these types of variables under more general operations, noting that Dempster's rule is a special case of this situation under the intersection operation.

Wagner [1866] showed that a belief function consensus on a set with at least three members and a Bayesian belief function consensus on a set with at least four members must take the form of a weighted arithmetic mean. These results are unchanged when consensual uncertainty measures are allowed to take the form of Choquet capacities of low-order monotonicity.

Hernandez and Recasens [824] studied the notion of indistinguishability for belief functions, and provided effective definitions and procedures for computing the T-indistinguishability operator associated with a given body of evidence. They showed how these procedures may be adapted in order to provide a new method for tackling the problem of belief function approximation based on the concept of a T-preorder.

5

A toolbox for the working scientist

Machine learning is possibly the core of artificial intelligence, as it is concerned with the design of algorithms capable of allowing machines to learn from observations. Learning can be either *unsupervised* (when only observation values are available, without annotation) or *supervised* (when observations come with extra information such as class labels), although 'semi-supervised' situations in which annotation is only partially available are common. *Clustering* is a common example of unsupervised learning, while *classification* is the paramount application in supervised contexts. Ensemble classification, i.e., the combination of the outputs of several classifiers, is, for instance, a natural application of the theory of evidence. When the quantity to be learned is continuous, rather than categorical, the problem is termed *regression*.

An effective toolbox for the working scientist, however, cannot be limited to machine learning. As we argued in Chapter 1, most real-world applications are plagued by the presence of hard-to-quantify uncertainty. *Estimation*, in particular, plays an enormous role in many scientific disciplines, as the values of physical or economic quantities, for instance, need to be recovered from the available data. Estimation is central in control theory, for instance, where the state of a dynamical model (either deterministic or stochastic) needs to be evaluated. When the quantity of interest evolves in time, we are in the domain of *time series* analysis: for instance, the state of a hidden Markov model representing a person's gesture needs to be estimated instant by instant. When the quantity to be estimated lives in some future time, the problem is often called *prediction*.

Optimisation is another cross-disciplinary topic, as research questions are often formulated in terms of maximisation or minimisation problems, either constrained or unconstrained, and linear, quadratic or nonlinear. Machine learning itself is noth-

F. Cuzzolin, *The Geometry of Uncertainty*, Artificial Intelligence: Foundations, Theory, and Algorithms, https://doi.org/10.1007/978-3-030-63153-6_5

ing but an application of optimisation theory to the problem of designing machines which can learn from experience.

This chapter aims to provide a summary of the tools based on the theory of belief functions which have been developed to address machine learning problems such as unsupervised learning (e.g. clustering), supervised learning (e.g., classification and regression) and ranking aggregation, but also estimation, prediction, filtering and optimisation.

Chapter outline

The chapter is structured as follows.

We start with unsupervised learning (Section 5.1), focusing on evidential approaches to clustering, and hint at the most recent developments in the area. In Section 5.2, we briefly describe the main classification frameworks based on belief function theory, including the generalised Bayesian classifier (Section 5.2.1), evidential k-nearest neighbour classification (Section 5.2.2), classification in the transferable belief model (Section 5.2.3), multiclass SVM classification (Section 5.2.4), classification in the presence of partial data (Section 5.2.5), belief decision trees (Section 5.2.6), and frameworks fusing belief theory and neural networks (Section 5.2.7). We conclude the section by reviewing other classification approaches (Section 5.2.8), including the work by Isabelle Bloch, Mahler's modified Dempster–Shafer classification and credal bagging.

Section 5.3 is devoted to ensemble classification, i.e., how to combine the results of different classifiers using belief function theory. In particular, we recall the main contributions in the area by Xu et al., Rugova et al., and Ani and Deriche. In Section 5.4, we tackle ranking aggregation, whereas Section 5.5 is concerned with regression approaches, in particular Denoeux et al.'s fuzzy-belief non-parametric regression (Section 5.5.1) and belief modeling regression (Section 5.5.2).

The chapter concludes with a review of estimation methods (Section 5.6), including time series analysis (Section 5.6.2) and the generalisation of particle filtering (Section 5.6.3), and of optimisation strategies based on Dempster–Shafer theory (Section 5.7).

5.1 Clustering

In unsupervised learning, *clustering* provides a way of discovering the structure of data, by grouping data points into clusters according to their similarity, appropriately measured.

As pointed out by Denoeux,[52] the theory of belief functions provides a rich and flexible framework for representing uncertainty in clustering. The concept of a *credal partition* encompasses the main existing soft-clustering concepts (fuzzy, possibilistic, rough partitions). Efficient evidential clustering algorithms exist, which

[52] http://belief.utia.cz/slides/thierry.pdf.

can be applied to large datasets and large numbers of clusters, by carefully selecting appropriate focal sets. Finally, concepts from the theory of belief functions make it possible to compare and combine clustering structures generated by various soft-clustering algorithms.

Early work on the topic was conducted by Class et al. [289], who described an algorithm for soft-decision vector quantisation based on discrete-density hidden Markov models. This approach transformed a feature vector into a number of symbols associated with credibility values computed according to statistical models of distances and evidential reasoning.

5.1.1 Fuzzy, evidential and belief C-means

The well-known fuzzy C-means (FCM) algorithm for data clustering [145] was extended in [1262] by Masson and Denoeux to an *evidential C-means* (ECM) algorithm in order to work in the belief function framework with 'credal partitions'.

The problem is to generate from the observed attribute data $X = (\mathbf{x}_1, \ldots, \mathbf{x}_n)$, $\mathbf{x}_i \in \mathbb{R}^p$, a *credal partition* $M = (m_1, \ldots, m_n)$, i.e., a set of mass functions $m_i : 2^\Omega \to [0,1]$ on the set of clusters $\Omega = \{\omega_1, \ldots, \omega_c\}$, one for each data point. Each cluster is represented by a prototype. Each non-empty set of clusters A is also represented by a prototype, namely the centre of mass of the prototypes of the constituent clusters (see Fig. 5.1). A cyclic coordinate descent algorithm optimises

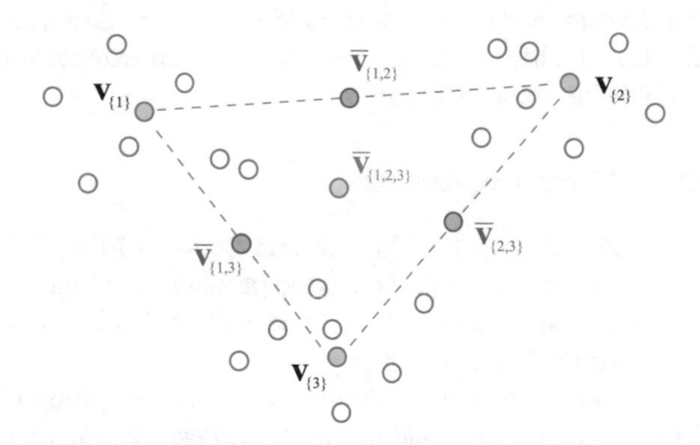

Fig. 5.1: Principle of the ECM clustering algorithm.

a cost function alternately with respect to the prototypes and to the credal partition. Once we have defined a set of focal elements (sets of clusters) $A_1, \ldots, A_f$, the cost function has the form

$$\sum_{i=1}^{n} \sum_{j=1}^{f} |A_j|^\alpha (m_i(A_j))^\beta d^2(\mathbf{x}_i, \mathbf{v}_j) + \delta^2 \sum_{i=1}^{n} (m_i(\emptyset))^\beta,$$

where $\mathbf{v}_j$ is the prototype for the set of clusters A_j and d is a distance function in $\mathbb{R}^p$. The parameter α controls the specificity of mass functions, β controls the 'hardness' of the credal partition and δ^2 controls the proportion of data considered as outliers.

Belief C-means (BCM) Liu et al. [657] noted that, however, in some cases the barycentres of some ECM clusters can become very close to each other. To circumvent this problem, they introduced the notion of an imprecise cluster. In their *belief C-means* (BCM) algorithm, objects lying in the middle of specific cluster barycentres must be committed with equal belief to each specific cluster, rather than belonging to an imprecise meta-cluster as in the ECM algorithm. Outlier objects far away from the centres of two or more specific clusters that are hard to distinguish are committed to an imprecise cluster, a disjunctive meta-cluster composed of these specific clusters. The mass of each specific cluster for each object is computed according to the distance between the object and the cluster centre. The distances between the object and the centres of the specific clusters and the distances between these centres are both taken into account in the determination of the mass of the meta-cluster, rather then using the barycentre of the meta-cluster as in ECM.

Other variants of ECM Several other variants of ECM have been proposed: a relational evidential C-means (RECM) algorithm for (metric) proximity data [1263]; ECM with adaptive metrics to obtain non-spherical clusters [47]; constrained evidential C-means (CECM) [46]; spatial evidential C-means (SECM) for image segmentation [1131]; credal C-means (CCM), which uses a different definition of the distance between a vector and a meta-cluster [659]; and median evidential C-means (MECM), which uses a different cost criterion, and is an extension of the median hard and fuzzy C-means algorithms [2120].

5.1.2 EVCLUS and later developments

In EVCLUS, originally developed by Masson and Denoeux [474, 473], given a matrix of dissimilarities between n objects, a basic probability assignment is assigned to each object in such a way that the degree of conflict between the masses assigned to pairs of objects reflects their dissimilarity.

Namely, let m_i and m_j be mass functions regarding the group membership of objects $\mathbf{x}_i$ and $\mathbf{x}_j$. The plausibility that this pair of objects belong to the same group is $1 - k_{ij} = \sum_{A \cap B \neq \emptyset} m_i(A) m_j(B)$. The approach is to minimise the discrepancy between the dissimilarities d_{ij} and the degrees of conflict k_{ij}, for instance by minimising

$$\sum_{i<j} (k_{ij} - \phi(d_{ij}))^2,$$

where ϕ is an increasing function from $[0, +\infty)$ to $[0, 1]$, for instance $\phi(d) = 1 - \exp(-\gamma d^2)$. The framework was later extended to interval-valued dissimilarities in [1261], so that the belief and plausibility that any two objects belong to the same cluster reflect, respectively, the observed lower and upper dissimilarity values.

Serir et al. [1578, 1577] recently introduced a new online clustering method called E2GK (evidential evolving Gustafson–Kessel), based on the concept of a credal partition, introduced in [474]. There, a credal partition was derived in an online fashion by applying an adapted version of the evolving Gustafson–Kessel (EGK) algorithm.

Ek-NNclus The Ek-NNclus approach starts from a different point of view [470].

Let us assume that there is a true unknown partition. The relevant frame of discernment should then be the set $\mathcal{R}$ of all equivalence relations (i.e., partitions) of the set of n observed objects. This set is huge: can we still reason on this frame of discernment using belief theory?

The input of the algorithm is an $n \times n$ matrix $D = (d_{ij})$ of dissimilarities between the n observed objects. Under the assumption that two objects have more chance to belonging to the same group if they are more similar, we can map d_{ij} to the mass function

$$m_{ij}(\{S\}) = \phi(d_{ij}), \quad m_{ij}(\{S, NS\}) = 1 - \phi(d_{ij})$$

over the frame $\{S, NS\}$ (objects i and j are similar; objects i and j are dissimilar). We want to combine these $n(n-1)/2$ mass functions to find the most plausible partition of the n objects.

Each mass function m_{ij} can be vacuously extended to the space $\mathcal{R}$ of equivalence relations, once $\{S\}$ has been mapped to the subset $\mathcal{R}_{ij}$ of partitions of the n objects such that objects o_i and o_j belong to the same group. Assuming that the mass functions m_{ij} are independent, the extended mass functions can then be combined on $\mathcal{R}$ by Dempster's rule. The resulting contour function (2.18) is

$$pl(R) \propto \prod_{i<j} (1 - \phi(d_{ij}))^{1-r_{ij}}, \quad R \in \mathcal{R},$$

where $r_{ij} = 1$ if o_i and o_j are in the same group, and $r_{ij} = 0$ otherwise.

After taking the logarithm of the contour function, finding the most plausible partition becomes a binary linear programming problem. The latter can be solved exactly only for small n. For large values of n, the problem can be solved approximately using a heuristic greedy search procedure: the *Ek-NNclus algorithm* (Algorithm 11).

Algorithm 11 Ek-NNclus algorithm

1: **procedure** EK-NNCLUS($D = (d_{ij})$)
2: starting from a random initial partition, classify each object in turn, using the Ek-NN rule;
3: the algorithm converges to a local maximum of the contour function $pl(R)$ if the number of neighbours k is equal to $n - 1$;
4: with $k < n-1$, the algorithm converges to a local maximum of an objective function that approximates $pl(R)$.
5: **end procedure**

Belief K-modes Ben Hariz et al. (2006) [108] developed a clustering method that they called *belief K-modes* (BKM, Algorithm 12), an extension of the K-modes algorithm in which object attribute values may be affected by uncertainty. This is encoded by a basic probability assignment defined on the set of possible attribute values.

Suppose each object $\mathbf{x}$ is characterised by s attribute values $x_1, \ldots, x_s$, where $x_i \in \Theta_i$. At learning time, given the number of clusters K, the following procedure is applied. The dissimilarity measure between any object $\mathbf{x}$ and a mode Q is defined

---

Algorithm 12 Belief K-modes algorithm

1: **procedure** BKM($m_{\Theta_j}[\mathbf{x}_i]$, $i = 1, \ldots, n$, $j = 1, \ldots, s$)
2: partition the objects in the training set $\mathcal{T}$ in K non-empty subsets $C_k = \{\mathbf{x}_1, \ldots, \mathbf{x}_{|C_k|}\}$;
3: compute seed points as the modes of the clusters of the current partition using the averaging rule of combination. Namely, the mode of C_k is a collection of belief functions, one for each attribute $j = 1, \ldots, s$,

$$Q_k = \{m_{\Theta_1}[Q_k], \ldots, m_{\Theta_s}[Q_k]\}$$

with mass assignment

$$m_{\Theta_j}[Q_k](A_j) = \frac{\sum_{i=1}^{|C_k|} m_{\Theta_j}[\mathbf{x}_i](A_j)}{|C_k|},$$

i.e., the average mass assigned to a subset A_j of attribute-j values by the objects in cluster C_k;
4: assign each object to the cluster whose mode is the nearest;
5: stop when no object has changed clusters as a result.
6: **end procedure**

---

as

$$D(\mathbf{x}, Q) = \sum_{j=1}^{s} d(m_{\Theta_j}[\mathbf{x}], m_{\Theta_j}[Q]),$$

where d is the Euclidean distance between two mass vectors.

Ensemble clustering The problem of ensemble clustering, i.e., the aggregation of multiple clusterings over the same data, was tackled in [1264] by Masson and Denoeux, by means of belief functions defined on the lattice of interval partitions of a set of objects. As clustering results can be represented as masses of evidence allocated to sets of partitions, a consensus belief function can be obtained using a suitable combination rule.

5.1.3 Clustering belief functions

An original line of research was pursued by Johan Schubert [1551, 1568, 1563], who studied deeply the problem of clustering $2^n - 1$ pieces of evidence into n clusters by minimising a *metaconflict* function [1567, 126].

The idea is that the available pieces of evidence may be associated with distinct events, so that we need to cluster them into disjoint subsets $\{\Pi_i\}$ according to the event they refer to. Schubert [1567, 126] uses the conflict associated with the Dempster sum as an indication of whether subsets of pieces of evidence belong together. This is formally described at a metalevel on a frame $\Theta = \{\text{AdP}, \neg \text{AdP}\}$, where AdP denotes 'adequate partition'. Each element of the partition (subset of pieces of evidence) then induces a belief function there, with mass $m_i(\neg \text{AdP}) \doteq \kappa(\oplus Bel_k, k \in \Pi_i)$, where the latter is the conflict level (2.7) of the Dempster combination of the belief functions in Π_i (see Section 2.3.2). We can then define a *metaconflict function* [1555] as

$$1 - (1 - c_0) \prod_i (1 - \kappa_i), \tag{5.1}$$

where c_0 is a conflict between the current hypothesis about the number of elements in the partition and our prior belief, and look for the partition which minimises (5.1). After simplifying the conflict of a sum of support functions as the sum of their pairwise conflicts, the minimisation problem can be described by a Potts model [1960].

A classification method [1550] based on comparison with prototypes representing clusters, instead of performing a full clustering of all the evidence, was later developed by Schubert based on his work on clustering of non-specific evidence. The resulting computational complexity was $\mathcal{O}(M \cdot N)$, where M is the maximum number of subsets and N the number of prototypes chosen for each subset.

A neural network structure was proposed for fast clustering in [1554], after noting that Schubert's previous proposal based on iterative optimisation was not feasible for large-scale problems. In [1568], Schubert extended his work in [1551] by letting the neural structure do the initial clustering in order to achieve high computational performance, to use its solution later to initialise the iterative optimisation procedure. A method for simultaneous clustering and determination of the number of clusters was developed in [1563]. The situation of weakly specified evidence, in the sense that it may not be certain to which event a proposition refers, was considered in [1546].

In more recent work [1556] (from 2008), the clustering of belief functions was performed by decomposing them into simple support and inverse simple support functions, which were then clustered based on their pairwise generalised weights of conflict, constrained by weights of attraction assigned to keep track of all decompositions.

5.2 Classification

In classification problems, the population is assumed to be partitioned into c groups or classes. Let $\Theta = \{\theta_1, \ldots, \theta_C\}$ denote the set of classes. Each instance of the problem is then described by a feature vector $\mathbf{x} \in \mathbb{R}^p$ and a class label $y \in \Theta$. Given a training set $\mathcal{T} = \{(\mathbf{x}_1, y_1), \ldots, (\mathbf{x}_n, y_n)\}$, the goal is to predict the class of a new instance described by $\mathbf{x}$.

Past publications on classification with belief functions have mainly dealt with two approaches:

1. *Ensemble classification*: the outputs from a number of standard classifiers are converted into belief functions and combined using Dempster's rule or some other alternative rule (e.g. [1241, 642, 1997]).
2. *Evidence-theoretic classifiers*: these have been developed to directly provide belief functions as outputs, in particular:
   - classifiers based on the *generalised Bayes theorem* (Section 4.6.1), which extends the classical Bayesian classifier when class densities and priors are ill-known [1718, 1725];
   - *distance-based approaches*, such as the evidential k-NN rule [2134] and the evidential neural network classifier [448].

Ensemble classification is discussed separately in Section 5.3. In the rest of this section, we review the main classification strategies developed within belief theory.

5.2.1 Generalised Bayesian classifier

A pioneering paper on multisource classification by Kim and Swain [968] assumed that the degrees of belief provided by each data source were represented as belief functions, inferred in partly consonant form using Shafer's likelihood-based proposal (recall Section 4.1.1).

The framework generalises Bayes' decision rule for the minimisation of expected loss, by replacing the latter with the lower and upper expected losses of making a decision $\theta_i \in \Theta$ after observing $\mathbf{x}$, namely

$$\underline{E}_i(\mathbf{x}) = \sum_{j=1}^{C} \lambda(\theta_i|\theta_j) Bel(\theta_j|\mathbf{x}), \quad \overline{E}_i(\mathbf{x}) = \sum_{j=1}^{C} \lambda(\theta_i|\theta_j) Pl(\theta_j|\mathbf{x}),$$

where $\lambda(\theta_i|\theta_j)$ is the loss incurred when the actual class is θ_i, after one has decided in favour of θ_j. Various decision rules which make use of the above interval of expected losses can be proposed. Eventually, Kim and Swain [968] tested rules based on the simple *minimum upper expected loss*

$$\hat{\theta}(\mathbf{x}) = \theta_i \quad \text{if} \quad \overline{E}_i(\mathbf{x}) \leq \overline{E}_j(\mathbf{x}) \quad \forall j \neq i$$

and *minimum lower expected loss*

$$\hat{\theta}(\mathbf{x}) = \theta_i \quad \text{if} \quad \underline{E}_i(\mathbf{x}) \leq \underline{E}_j(\mathbf{x}) \quad \forall j \neq i.$$

5.2.2 Evidential k-NN

Denoeux and Zouhal [443, 456] proposed a k-nearest-neighbour classifier based on Dempster–Shafer theory, where each neighbour of a sample is considered as an item of evidence supporting hypotheses about the class of the sample itself. The evidence of the k nearest neighbours is then pooled as usual by means of Dempster's rule. As this approach suffers from the problem of tuning a number of parameters, Zouhal and Denoeux [2135] later proposed to determine optimal or near-optimal parameter values from the data by minimising an appropriate error function.

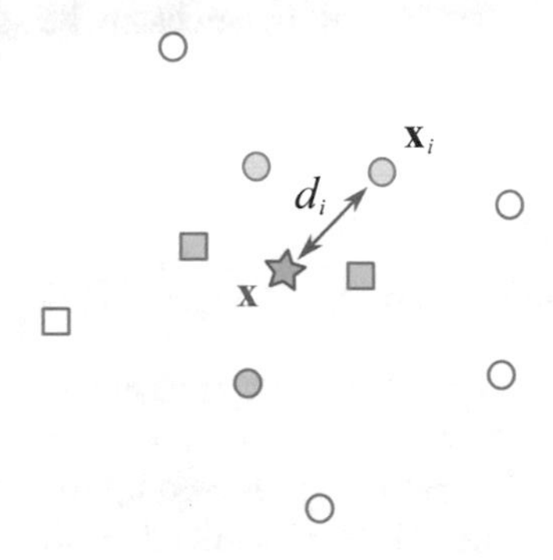

Fig. 5.2: Principle of the evidential k-nearest-neighbour (k-NN) classifier.

Namely, let $\mathcal{N}_k(\mathbf{x}) \subset \mathcal{T}$ denote the set of the k nearest neighbours of $\mathbf{x}$ in the training set $\mathcal{T}$, based on some appropriate distance measure d. Each $\mathbf{x}_i \in \mathcal{N}_k(\mathbf{x})$ can be considered as a piece of evidence regarding the class of $\mathbf{x}$ represented by a mass function m_i on Θ (see Fig. 5.2):

$$m_i(\{y_i\}) = \varphi\left(d_i\right), \quad m_i(\Theta) = 1 - \varphi\left(d_i\right).$$

The strength of this evidence decreases with the distance d_i between $\mathbf{x}$ and $\mathbf{x}_i$, since φ is assumed to be a decreasing function such that $\lim_{d \rightarrow +\infty} \varphi(d) = 0$. Evidence is then pooled as

$$m = \bigoplus_{\mathbf{x}_i \in \mathcal{N}_k(\mathbf{x})} m_i.$$

The function φ can be selected from among a family $\{\varphi_\lambda | \lambda \in \Lambda\}$ using, for instance, cross-validation. Finally, the class with the highest plausibility is selected.

Fuzzy extension A generalisation of evidential k-NN to the case in which the training class membership is described by a fuzzy set was proposed in [2136, 477]. Two options were put forward, based either on the transformation of each possibility distribution into a consonant belief function, or on the use of generalised belief structures with fuzzy focal elements. In each case, the expert's beliefs concerning the class membership of each new pattern was modelled by a belief function.

Evidential k-NN rule for partially supervised data In an even more general setting, the training set can be formalised as

$$\mathcal{T} = \{(\mathbf{x}_i, m_i), i = 1, \ldots, n\},$$

where $\mathbf{x}_i$ is the attribute vector for instance i, and m_i is a mass function representing uncertain expert knowledge about the class y_i of instance i. Special cases are:

- $m_i(\{\theta_{j_i}\}) = 1$ for all i and some j_i (*supervised* learning);
- $m_i(\Theta) = 1$ for all i (*unsupervised* learning).

The evidential k-NN framework [443, 456] can easily be adapted to handle such uncertain learning data [220]. Each mass function m_i is first 'discounted' (see Section 4.3.6) by a rate depending on the distance d_i:

$$m'_i(A) = \varphi(d_i)\, m_i(A), \; \forall A \subset \Theta, \quad m'_i(\Theta) = 1 - \textstyle\sum_{A \subset \Theta} m'_i(A).$$

k such discounted mass functions m'_i are then combined, as before.

Other k-NN approaches Other k-NN-inspired evidential approaches have been proposed by a number of authors.

Pal and Ghosh [1380], for instance, proposed up to five ways of integrating belief theory with rank nearest-neighbour classification rules, and compared the result with k-nearest neighbour, multivariate rank nearest-neighbour (m-MRNN) and Denoeux's evidential k-NN classifier [443, 456].

Zhu and Basir [2125] presented a fuzzy evidential reasoning algorithm in which the distance between the input pattern and each of its k nearest neighbours is used for mass determination, while the contextual information about the nearest neighbour in the training sample space is formulated by a fuzzy set to determine a fuzzy focal element. The evidence provided by the neighbours is pooled via fuzzy evidential reasoning, where feature selection is considered further through ranking and adaptive combination of neighbours. A fast implementation scheme for fuzzy evidential reasoning was also developed, and shown to outperform both evidence-theoretical k-nearest-neighbour classification [443, 456] and its fuzzy extension [2136, 477].

5.2.3 TBM model-based classifier

More recently, Denoeux and Smets [475] reviewed two main approaches to pattern classification developed within the transferable belief model framework: the *TBM model-based classifier*, developed by Appriou [49, 50], which relies on the generalised Bayes theorem, and the *TBM case-based classifier*, built on the similarity of the pattern to be classified to the training patterns (a.k.a., Denoeux's evidential k-NN classifier).

The GBT can be used to solve classification problems as follows. Let $m_{\mathbb{X}}[\theta_j] : 2^{\mathbb{X}} \rightarrow [0, 1]$, $j = 1, \ldots, C$, be the conditional mass of feature observations $\mathbf{x} \in \mathbb{X}$ given the assumption that the object's class is θ_j. We seek the agent's belief about the actual class c. This can be done by:

1. Extending the conditional BPAs $m_{\mathbb{X}}[\theta_j]$ to $\mathbb{X} \times \Theta$ by ballooning extension.
2. Combining the resulting masses conjunctively.
3. Conditioning the result on $\Theta \times \{\mathbf{x}\}$ by Dempster's rule of conditioning.
4. Marginalising the result on Θ.

In [1689], the result was shown to be equal to

$$m_\Theta[\mathbf{x}](A) = \prod_{\theta_j \in A} Pl_{\mathbb{X}}[\theta_j](\mathbf{x}) \prod_{\theta_j \in A^c} (1 - Pl_{\mathbb{X}}[\theta_j](\mathbf{x})), \quad A \subset \Theta.$$

Denoeux and Smets showed that the evidential k-NN classifier also proceeds from the same underlying principle, i.e., the GBT. The two approaches differ essentially in the nature of the information assumed to be available. Both methods collapse to a kernel rule in the case of precise and categorical learning data under certain initial assumptions.

5.2.4 SVM classification

Support vector machines constituted the dominant paradigm in supervised classification for at least two decades [1790], before being supplanted by deep learning approaches (see `www.deeplearning.net`, [1109]). Nevertheless, during all this time limited effort was directed towards adding uncertainty-theoretical elements to the SVM classification framework. A significant exception is a paper from 2005 by Liu and Zheng [1207], which we describe here.

One issue with SVM classification is that it is inherently related to a binary setting in which only two classes are discriminated. Extension to multiclass classification is classically addressed by decomposing the problem into a series of two-class problems, to which a one-against-all SVM strategy is applied.

Suppose we have a test sample $\mathbf{x}$ and C one-against-all classifiers SVM$_j$, each with its own decision function f_j. The hypothesis set is thus $\Theta = \{\theta_j, j = 1, \ldots, C\} = \{\text{sample } \mathbf{x} \text{ belongs to class } j, j = 1, \ldots, C\}$. When each SVM$_j$ is applied to $\mathbf{x}$, the result produces a piece of evidence supporting certain hypotheses. A mass function associated with a positive j-th classification result ($y_{f_j} = 1$) can then be defined as [1207]

$$m_j(A) = \begin{cases} 1 - \exp(-|f_j(\mathbf{x})|) & A = \{\theta_j\}, \\ \exp(-|f_j(\mathbf{x})|) & A = \Theta, \\ 0 & \text{otherwise.} \end{cases}$$

If $y_{f_j} = 0$, instead, we have

$$m_j(A) = \begin{cases} 1 - \exp(-|f_j(\mathbf{x})|) & A = \{\theta_j\}^c, \\ \exp(-|f_j(\mathbf{x})|) & A = \Theta, \\ 0 & \text{otherwise.} \end{cases}$$

By combining these pieces of evidence via Dempster's rule, we obtain a belief function Bel on Θ. Whatever the responses of the individual SVMs, optimising the belief value of the hypotheses yields the classical one-against-all decision rule,

$$\arg \max_{j=1,\dots,C} Bel(\{\theta_j\}) = \arg \max_{i=1,\dots,C} f_j(\mathbf{x}).$$

The bottom line is that, since it is not a wise strategy to completely trust the decisions made by the SVMs, reliability measures should be introduced and added to the above mass assignments. Liu and Zheng [1207] proposed to use either a static (SRM) or a dynamic (DRM) reliability measure. SRM works on a collective basis and yields a constant value regardless of the location of the test sample. DRM, on the other hand, accounts for the spatial variation of the classifier's performance.

5.2.5 Classification with partial training data

A few papers have been published on classification under partially specified training data. This may mean a restricted number of examples, training sets which do not include all possible classes or training sets whose sample classes are not known exactly, but only in an 'imprecise' fashion.

Back in 1997, Denoeux [444] analysed situations in which the training set is 'small', or when the latter does not contain samples from all classes. He examined various strategies, whose decision rules were assessed under different assumptions concerning the completeness of the training set. As we have seen in Section 5.2.2, evidential k-NN can be easily adapted to the case of soft labelling ([220]; see Fig. 5.2). More recently (2009), the issue was addressed by Côme et al. [303, 304], who assumed that training samples are assigned a 'soft' label, defined as a basic belief assignment over the set of classes. The training vectors are assumed to be generated by a mixture model. Using Smets's generalised Bayes theorem (see Section 4.6.1), these authors derived a criterion which generalises the likelihood function. Finally, a variant of the expectation maximisation algorithm was proposed for the optimisation of this criterion, allowing us to estimate model parameters.

Vannoorenberghe and Smets [1848] also proposed a credal EM (CrEM) approach for partially supervised learning, where the uncertainty is represented by belief functions as understood in the transferable belief model. They showed how the EM algorithm can be applied to the classification of objects when the learning set is imprecise, i.e., each object is only known to belong to a subset of classes, and/or uncertain (the knowledge about the actual class is represented by a probability function or a belief function).

More recently, Tabassian et al. [1793, 1794] proposed an approach in which the initial labels of the training data are ignored, and each training pattern is reassigned to one class or a subset of the main classes based on the level of ambiguity concerning its class label using the classes' prototypes. A multilayer perceptron is employed to learn the characteristics of the data with new labels, and for a given test pattern its outputs are considered as the basic belief assignment.

5.2.6 Decision trees

Interesting work has been aimed at generalising the framework of *decision trees* to situations in which uncertainty is represented by belief functions, mainly by Elouedi and co-authors. [53]

Classical definition A *decision tree* is a classifier composed of a number of *decision nodes*, in which an observation is tested in terms of one of its attributes. Decision nodes are arranged into a tree, in which an edge between a node and its children corresponds to one of the possible values of the test attribute. *Leaf nodes* are associated with a class label.

During learning, a decision tree is built based on the available training data following a divide-and-conquer strategy. The latter proceeds by recursively subdividing the training data, where each subdivision corresponds to a question asked about a specific attribute. Namely, at each step:

1. An attribute is selected to partition the training set in an optimal manner.
2. The current training set is split into training subsets according to the values of the selected attribute.

A typical attribute selection criterion is based on the *information gain*, $\text{Info}(\mathcal{T}) - \text{Info}_A(\mathcal{T})$. Information is measured via the classical Shannon entropy,

$$\text{Info}(\mathcal{T}) \doteq -\sum_{c \in \Theta} p_c \log_2 p_c, \tag{5.2}$$

where $\Theta = \{1, \ldots, C\}$ is the set of classes, $\mathcal{T}$ is the section of the training set examined at the current node, p_c is the proportion of objects in $\mathcal{T}$ with class label c and

$$\text{Info}_A(\mathcal{T}) = \sum_{a \in \text{range}(A)} \frac{|\mathcal{T}_{A=a}|}{|\mathcal{T}|} \text{Info}(\mathcal{T}_{A=a}), \tag{5.3}$$

where $\mathcal{T}_{A=a}$ is the subset of the current training set for which attribute A, with range of possible values $\text{range}(A)$, has value a.

Belief decision trees A *belief decision tree* [573, 570], originally defined by Elouedi et al., is composed of the same elements as a traditional decision tree but, at each step, class information about the items of the dataset is expressed by a basic probability assignment over the set Θ of possible classes for each object. As a consequence, each leaf node is associated with a BPA over the set of class values, expressing the uncertain class of objects belonging to that leaf node.

In [571], two different approaches to building a belief decision tree were proposed: an 'averaging' method and a so-called TBM conjunctive method. According to the former, the Shannon entropy is applied to the average pignistic probability within each split (see Algorithm 13). In the conjunctive approach, instead, the intra-group distance between attribute values is used for selection.

[53] Note that related work on the application of uncertainty measures for credal sets to the construction of classification trees was conducted by Abellán and Moral [4].

Algorithm 13 Belief decision trees: training by averaging

1: **procedure** BDTTRAINING
2: at each step, the pignistic probability $BetP[\mathbf{x}] : \Theta \to [0,1]$ is computed for each instance $\mathbf{x} \in \mathcal{T}$ in the current split of the training set (this is a probability distribution over the set of possible classes for that instance);
3: the average pignistic probability is computed for all instances in the split;
4: the entropy of the average $\text{Info}(\mathcal{T})$ is computed via the classical entropy (5.2);
5: the entropy $\text{Info}_A(\mathcal{T})$ of the average pignistic probability for each of the splits $A = a$ is computed via (5.3);
6: the attribute with the highest information gain is selected.
7: **end procedure**

Aggregation of belief decision trees Whereas [571] applied the classical entropy to pignistic functions in a TBM framework, Denoeux and Bjanger [467] replaced the Shannon entropy by an evidence-theoretic uncertainty measure taking into account not only the class proportions, but also the number of objects in each node, allowing the processing of training data whose class membership is only partially specified (in the form of a belief function). More precisely, Klir's composite uncertainty measure ((4.14), compare Section 4.2.2) is applied to the belief function

$$m_\Theta(\theta_c) = \frac{n_c(t)}{n(t)+1}, \quad m_\Theta(\Theta) = \frac{1}{n(t)+1},$$

where $n_c(t)$ is the number of training samples in the split with class c, and $n(t) = |\mathcal{T}|$.

Slightly later (2002), Vannoorenberghe and Denoeux [1847] worked towards extending [467], which was designed for building decision trees for two-class problems only, to multiclass problems, by combining trees obtained from various two-class coarsenings of the initial frame of object classes. The evidence provided by individual binary trees was projected onto the set of all classes by vacuous extension, and combined there via the averaging operator ([1328]: see Section 4.3.2), as such pieces of evidence cannot be considered as independent.

In [1845], Vannoorenberghe investigated further the aggregation of belief decision trees using machine learning techniques such as bagging and randomisation.

Recent work As recently as 2016 [1822], Trabelsi et al. developed a further version of belief-theoretical decision trees to handle the case where uncertainty is present only at the level of the attribute values. The same authors [1825] had previously developed pruning methods for belief decision trees induced by averaging and conjunctive approaches, with the aim of mitigating the problem of overfitting.

Separately, Sutton-Charani et al. [1789] proposed three possible extensions of the belief decision tree framework to multiclass problems, based on either combining multiple two-class trees together or directly extending the estimation of belief functions within the tree to the multiclass setting.

5.2.7 Neural networks

Neural network classification is once again an emerging trend in artificial intelligence, thanks to the development of 'deep' neural structures [1109]. Some (limited) effort was in fact directed in the past towards the application of the theory of evidence to neural network classifiers [465, 507]. Since they are both suitable for solving classification problems, neural networks and belief functions are sometimes integrated to yield more robust systems [1882, 1881].

The main work in the area is due to Denoeux, who in 2000 [448] proposed an adaptive version of his evidential k-NN classifier in which reference patterns are used as items of evidence pointing to the class membership of the pattern to be classified, as usual represented as BPAs and combined via Dempster's rule, implemented as a multilayer neural network. The idea is to use these 'prototypes' instead of nearest neighbours to mitigate the computational complexity of the classifier. After noting that the resulting classification framework is similar to that of radial basis functions [1436], Denoeux put forward a 'connectionist' view of his evidence-theoretic classifier, seen as a neural network with an input layer, two hidden layers and an output layer (see Fig. 5.3). The prototype vectors correspond in this interpretation to the weights of the connections between neurons in the input layer and those in the first hidden layer.

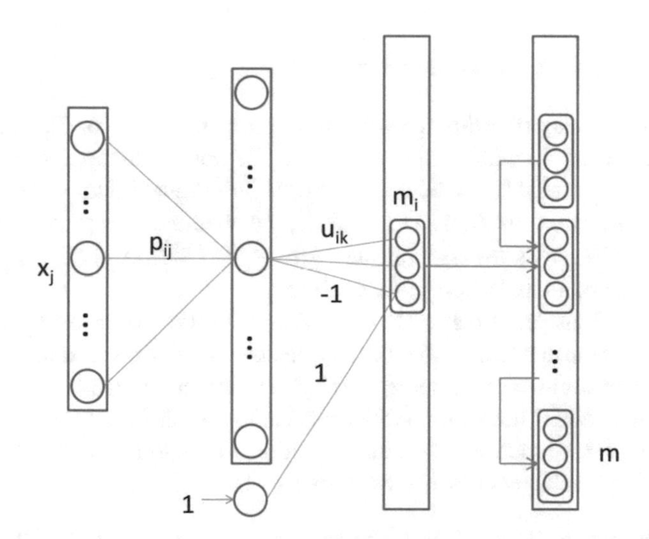

Fig. 5.3: Neural network architecture of the adaptive classifier proposed in [448].

The network can then be trained to optimise classification performance. In particular, this allows one to recover the best prototype vectors for a given training set ([448], Section III-C).

Separately, Binaghi et al. [155] proposed a neural model for fuzzy Dempster–Shafer classifiers, in which the learning task can be formulated as a search for the most adequate 'ingredients' of fuzzy and belief theories, such as (i) fuzzy aggregation operators for fusing data from different sources, and (ii) basic probability assignments describing the contributions of the evidence to the inference scheme.

The empirical performance of belief- and neural-network-based approaches to pattern classification were also compared in some papers. Loonis et al. [1212], for instance, compared a multiclassifier neural network fusion scheme with the straightforward application of Dempster's rule. Giacinto et al. compared neural networks and belief-based approaches to pattern recognition in the context of earthquake risk evaluation [683].

More recently, Fay et al. [593] proposed to combine hierarchical neural networks with the ability of belief theory to represent evidence at different levels of abstraction. Basically, this falls within the framework of ensemble classification in the special case of hierarchical neural networks. These authors consider that while fuzzy k-nearest neighbour classifiers produce outputs which meet the constraints for basic probability assignments, others (such as radial basis functions) do not. A ramp function is thus applied to each classifier output to normalise the result. Discounting is also applied to take into account the fact that, within the hierarchy, several classifiers have to provide results for samples belonging to classes they have not been trained with. Discounting thus weakens insular strong responses.

5.2.8 Other classification approaches

Bloch's work on classification in multisource sensing In 1996, Bloch [171] compared a number of decision rules based on a Dempster–Shafer representation of evidence for data fusion in medical imaging, including maximum belief, maximum plausibility (supported by Appriou [49, 50]), the absolute rule proposed by Suh et al. [1785] and Wesley's average rule $\arg\max \frac{Bel(A)+Pl(A)}{2}$ [1927]. In particular, the role of compound hypotheses was analysed.

Le Hégarat-Mascle, Bloch and Vidal-Madjar [1106] followed up in 1997 by showing that Dempster–Shafer theory may be successfully applied to unsupervised classification in multisource remote sensing, via the comparison of single-source classification results. These authors focused mainly on different sensible strategies for generating mass functions from the conditional probabilities of pixels being assigned to different clusters by different sensors.

Modified Dempster–Shafer classification Fixsen and Mahler [623, 624] described a 'modified Dempster–Shafer' (MDS) approach to object identification which incorporates Bayesian prior distributions into a modified rule of combination which reduces, under strong independence assumptions, to a special case of Bayes' rule (recall Chapter 3, Section 3.3.8). The MDS approach is applied there to a practical classification algorithm which uses an information-theoretic technique to limit the combinatorial explosion of evidence.

Fuzzy Dempster–Shafer rule-based classifiers Binaghi and Madella [156] proposed a fuzzy Dempster–Shafer model for multisource classification purposes. The salient aspect of this work is the definition of an empirical learning strategy for the automatic generation of fuzzy Dempster–Shafer classification rules from a set of exemplified training data.

ICM algorithm In an interesting paper, Foucher et al. [638] proposed a modification of the multiscale *iterated conditional mode* (ICM) algorithm using a local relaxation of Bayesian decision making based on belief theory, in an image-processing context.

Each pixel in an image is to be assigned a class. Whereas annealing methods, such as the Gibbs sampler, ensure convergence to a global MAP solution but are computationally expensive, deterministic methods such as ICM are much faster but remain suboptimal. The ICM method estimates a local MAP solution for the label by maximising $P(c_i|X, c_j \forall j \neq i)$, where X is the feature evidence and c_i is the class of the i-th pixel.

Credal bagging François et al. [641] showed that the quality and reliability of the outputs of a classifier may be improved using a variant of *bagging*.

'Bagging' [193] is an acronym for 'bootstrap aggregating'. From an original decision rule, a bagged estimator is produced by aggregating, using a majority vote, several replicates of the rule, trained on bootstrap resamples of the learning set. A bootstrap sample is created by drawing with replacement N examples from the learning set (so that, although it has the same size as the original training set, it may contain replicates). In [641], given B bootstrap training sets $\mathcal{T}_b$, each of the latter produces, via an evidential k-NN classifier, a belief measure m_b on Θ for all $b \in B$. These are finally aggregated into the average measure,

$$m_B(A) = \frac{1}{B}\sum_{b=1}^{B} m_b(A).$$

Aggregation takes place at the credal level via averaging, instead of via voting by majority rule as in classical bagging.

Other related work A number of other contributions are worth mentioning [947, 1271]. Zhu and Basir [2127], for instance, proposed in 2003 a *proportional difference evidence structure constructing scheme* (PDESCS) for representing probabilistic and fuzzy evidence as belief functions in a pattern classification context. Given a set of non-decreasing positive scores $s_1, \ldots, s_n$ associated with classes $c_1, \ldots, c_n$, PDESCS encodes them as the belief function with mass

$$m(\{c_1, \ldots, c_j\}) = \frac{s_j - s_{j+1}}{s_1}. \tag{5.4}$$

If multiple sources of information are available about the object of interest's class, the belief functions (5.4) are combined by Dempster's rule. When applied to posterior probability distributions within a maximum-commonality decision-making scheme, this scheme turns out to be equivalent to the Bayesian MAP approach.

Erkmen and Stephanou [577] proposed a model of belief functions based on fractal theory, and applied it to the design of an evidential classifier. Denoeux [469] presented a pattern recognition approach to system diagnosis, based on supervised and unsupervised procedures, both relying on the theory of evidence. Lefevre and Colot [1124, 1117] proposed a classification method based on Dempster–Shafer theory and information criteria, in which an attenuation factor based on the dissimilarity between probability distributions was introduced. Vannoorenberghe [1846] presented and compared several evidential classifiers, including Appriou's separable method [51] and the evidential k-NN classifier. He noted that these models can be derived from two decision rules, based on the minimisation of the lower and the pignistic expected loss, respectively.

Perry and Stephanou [1421] put forward an original technique for classifying observations expressed as aggregates of disparate belief functions. The classification process consists of applying a divergence (uncertainty) measure to the evidential aggregate of the belief functions and a set of prototype aggregate belief functions in a knowledge base. The divergence measures the difference between the information present in the combination of the two belief functions and each of the constituents. The knowledge base consists of inference classes, each formed by a set of aggregate prototypes typifying the class to a degree expressed by a fuzzy epitome coefficient.

More recently, Dempster and Chiu [427] considered a classification framework in which the inputs are uncertain observations, and the outputs consist of uncertain inferences concerning degrees of agreement between a new object and previously identified object classes, represented by belief functions. This leads to marginal belief functions concerning how well the new object matches objects in memory, so that beliefs and plausibilities that the new object lies in defined subsets of classes can be computed. Trabelsi et al. [1823, 1824] presented two classification approaches based on rough sets (compare Chapter 6, Section 6.7) that are able to learn decision rules from uncertain data. They assumed that uncertainty exists only in the decision attribute values, and is represented by the belief functions. Fiche et al. [610, 609] focused on the problem of fitting feature values to an α-stable distribution to classify imperfect data, in a supervised pattern recognition framework which rests on the theory of continuous belief functions. The distributions of the features were supposed there to be unimodal, and were estimated by a single Gaussian and an α-stable model. Mass functions were calculated from plausibility functions via the generalised Bayes theorem.

5.3 Ensemble classification

The work on multiclass SVMs [1207] that we reviewed in Section 5.2.4 can be seen as a particular instance of *ensemble classification*, i.e., the issue of combining the results of different classifiers on the same problem (see Fig. 5.4) [971]. Indeed, quite a lot of work on the application of belief functions to machine learning, and classification in particular, has focused on this question [642]. The reason is that combination rules (and Dempster's in particular, as we have seen in Section 5.2.4) easily lend themselves to the situation.

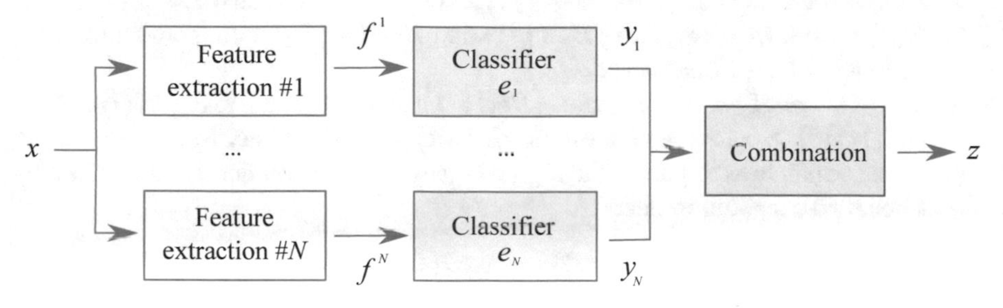

Fig. 5.4: Principle of ensemble classification.

5.3.1 Distance-based classification fusion

Xu's approach In a very highly cited publication [1997], Xu, Krzyzak and Suen investigated various possible methods of combining different classifiers, characterised by the level at which the fusion takes place. The entire Section VI of the paper was devoted to a method based on Dempster–Shafer theory, briefly illustrated below.

Given a set of classes Θ, the task of a classifier e is to assign to each input vector $\mathbf{x}$ a label $j = 1, \ldots, C, C+1$, where C is the number of classes and $C+1$ denotes the outcome in which e cannot support any specific class. Consider, then, an ensemble of K classifiers $e_1, \ldots, e_K$, each providing a piece of evidence of the form $e_k(\mathbf{x}) = j_k$, $j_k \in \{1, \ldots, C+1\}$. We can define a BPA m_k on the set of labels $\Theta = \{1, \ldots, C\}$ for each piece of evidence $e_k(\mathbf{x}) = j_k$ in the following way:

- when $j_k = C+1$, $m_k(\Theta) = 1$;
- when $j_k \in \Theta$,

$$m_k(A) = \begin{cases} \epsilon_r^k & A = \{j_k\}, \\ \epsilon_s^k & A = \{j_k\}^c, \\ 1 - \epsilon_r^k - \epsilon_s^k & A = \Theta, \end{cases} \tag{5.5}$$

where ϵ_r^k and ϵ_s^k are the *recognition rate* and the *substitution rate*, respectively, of the k-th classifier [1997], i.e., the rates at which the classifiers produce a correct/incorrect outcome; $\epsilon_r^k + \epsilon_s^k < 1$, as classifiers are allowed to be 'undecided'.

Just as in [1207], we wish then to compute the Dempster combination $m = m_1 \oplus \cdots \oplus m_K$ of the pieces of evidence provided by each individual classifier, and compute the belief and plausibility values of each class $i \in \Theta$. As vacuous belief functions do not alter the result of the combination, the evidence provided by 'indecisive' classifiers can be discarded. Also, assuming that no classifier is such that $\epsilon_r^k = 1$ or $\epsilon_s^k = 1$, the individual pieces of evidence are not in conflict and can be combined. Formulae for the resulting belief and plausibility values were provided in [1997]. A number of different decision rules making use of such values can be proposed:

1. $E(\mathbf{x}) = m$ if $m = \arg\max_j Bel(\{j\})$, $E(\mathbf{x}) = C + 1$ otherwise.
2. $E(\mathbf{x}) = m$ if $m = \arg\max_j Bel(\{j\})$ and $Bel(\{m\}) \geq \alpha$ for some threshold α, $E(\mathbf{x}) = C + 1$ otherwise.
3. $E(\mathbf{x}) = m$ if $m = \arg\max_j [Bel(\{j\}) - Bel(\{j\}^c)]$ and $Bel(\{m\}) - Bel(\{m\}^c) \geq \alpha$ for some threshold α, $E(\mathbf{x}) = C + 1$ otherwise.
4. A rule which tries to pursue the highest recognition rate under the constraint of a bounded substitution rate:

$$E(\mathbf{x}) = \begin{cases} m & Bel(\{m\}) = \max_j \{Bel(\{j\}) : Bel(\{j\}^c) \leq \alpha \ \forall j\}, \\ C+1 & \text{otherwise}. \end{cases}$$

The drawback of this method is the way evidence is measured. Firstly, several classifiers produce probability-like outputs rather than binary ones. Secondly, classifiers do not normally perform in the same way across different classes, a fact which was ignored in [1997] but was later addressed by Ani and Deriche ([22], see below).

Zhang and Srihari [2102] later argued that using the classifiers' classwise performances as BPAs outperforms the approach of Xu et al. [1997], which employs the classifiers' global performance. Just as in [1997], these authors' method considers as evidence only the top choice of each classifier. However, the BPA encoding the evidence for classifier k is obtained from its classwise performance parameters $\epsilon_r^k(\{j\})$ and $\epsilon_s^k(\{j\})$, $j = 1, \ldots, C$, which in (5.5) replace the overall performance parameters ϵ_r^k and ϵ_s^k of each classifier.

Finally, Mercier et al. [1281] considered the fusion of classifiers organised in a hierarchy, in which each classifier has the possibility to choose a class, a set of classes or a 'reject' option. They presented a method to combine these decisions based on the transferable belief model (Section 3.3.2), which extends the classical proposal of Xu et al. [1997]. A rational decision model allowing different levels of decision was also presented.

Rogova's approach Rogova [1495] later presented a combination of neural network classifiers based on Dempster–Shafer theory which uses statistical information about the relative classification strengths of several classifiers. Rogova's work was based on an earlier original idea proposed in 1988 by Mandler and Schurmann for combining nearest-neighbour classifiers with different distance measures [1241]. There, a statistical analysis of distances between learning data and a number of reference points in the input space was carried out to estimate distributions of intraclass and interclass distances. These distributions were used to calculate class-conditional probabilities that were transformed into pieces of evidence and then combined.

As pointed out in [1495], this raises the issues of the choice of appropriate reference vectors and of distance functions. Moreover, approximations associated with the estimation of statistical model parameters can lead to inaccurate measures of evidence.

Let X_j be the set of training points associated with class j, let $\mathbf{y}_k = f_k(\mathbf{x})$ be the output of the k-th classifier on input $\mathbf{x}$ and let $\bar{f}_k^j$ be the mean of $\{f_k(\mathbf{x}), \mathbf{x} \in X_j\}$. Let $d_k^j = \phi(\bar{f}_k^j, \mathbf{y}_k)$ be a proximity measure for the k-th classifier.

Given the frame of discernment $\Theta = \{1, \ldots, C\}$ of all classes, the k-th classifier induces for each class $j \in \Theta$ the following pair of BPAs there. The first one is in support of the hypothesis that j is the correct class, namely

$$m_k^j(\{j\}) = d_k^j, \quad m_k^j(\Theta) = 1 - d_k^j.$$

The second mass is in support of j not being the correct class:

$$m_k^{\neg j}(\{j\}^c) = 1 - \prod_{m \neq j} d_k^m, \quad m_k^{\neg j}(\Theta) = \prod_{m \neq j} d_k^m.$$

The sum $m_k^j \oplus m_k^{\neg j}$ combines the knowledge about hypothesis j provided by classifier k. The overall combination $\oplus_k (m_k^j \oplus m_k^{\neg j})$ provides a measure of the confidence in each class j given the input vector $\mathbf{x}$. After normalisation, the resulting belief function is Bayesian, with masses assigned to singleton hypotheses only. The optimal class can then be found as that which maximises

$$E(\mathbf{x}) = m = \arg\max_{j \in \Theta} \prod_k \frac{d_k^j \prod_{i \neq j}(1 - d_k^i)}{1 - d_k^j[1 - \prod_{i \neq j}(1 - d_k^i)]}.$$

In [1495], several alternative expressions for the dissimilarity function ϕ were empirically tested. As pointed out in [22], the main drawback of Rogova's method is the way reference vectors are calculated as means of output vectors. Also, just trying different proximity measures to pick the one that gives the highest classification accuracy is rather questionable.

Ani and Deriche Ani and Deriche [20, 22] considered the combination of multiple classifiers trained on distinct sets of features $\mathbf{f}^1(\mathbf{x}), \ldots, \mathbf{f}^N(\mathbf{x})$ extracted from the same observation vector $\mathbf{x}$, the output of each classifier e_k, $k = 1, \ldots, K$ being a real vector $\mathbf{y}_k$ of scores . They proposed a technique in which the evidence supporting each individual classifier is assessed. The mass assignment $m_k(\theta_j)$ encoding the belief in class label j produced by classifier e_k is

$$m_k(\theta_j) = \frac{d_k(\theta_j)}{\sum_{j=1}^C d_k(\theta_j) + g_k}, \quad m_k(\Theta) = \frac{g_k}{\sum_{j=1}^C d_k(\theta_j) + g_k}, \tag{5.6}$$

where

$$d_k(\theta_j) = \exp(-\|\mathbf{w}_k^j - \mathbf{y}_k\|^2)$$

and $\mathbf{w}_k^j$ is a reference vector of scores for class j and classifier k. The evidence produced by all classifiers is combined by Dempster's rule, obtaining a vector of masses of individual classes,

$$z(\theta_j) \doteq m_1(\theta_j) \oplus \cdots \oplus m_K(\theta_j).$$

The implementation adapts to the training data so that the overall mean square error is minimised. Namely, $\mathbf{w}_k^j$ and g_k are initialised at random, and then iteratively adjusted in order to minimise the mean square error $\sum_{\mathbf{x} \in \mathcal{T}} \|\mathbf{z}(\mathbf{x}) - \mathbf{t}(\mathbf{x})\|^2$, where $\mathbf{t}(\mathbf{x})$ is the target class vector for the training sample $\mathbf{x}$.

The computational cost is bounded by the fact that training needs to be performed only once, in an offline fashion.

5.3.2 Empirical comparison of fusion schemes

The empirical comparison of various fusion schemes, including some derived from Dempster–Shafer theory, has been the focus of some research [1880].

An interesting and very highly cited experimental comparison of various classifier fusion techniques, among them Dempster–Shafer fusion (implemented as in [1495] by Rogova et al.), was run by Kuncheva et al. [1085], where they were compared with these authors' *decision template* [1085] (DT) rule for adapting the class combiner to the application. Dempster–Shafer fusion was assessed to be the closest to the DT rule. Later (2000), Bahler and Navarro [73] also focused on the theoretical and experimental comparison of five classifier fusion methods, including one based on Dempster–Shafer theory (that of Xu et al., Section 5.3.1), but also Bayesian ensembles, behaviour-knowledge-space methods and logistic regression ensembles. Interestingly, the experiments showed that when the members of the ensemble had rejection rates greater than zero the performance of Dempster–Shafer combination improved, while that of behaviour-knowledge-space methods decreased and that of Bayesian ensembles was not affected.

5.3.3 Other classifier fusion schemes

Many other classifier fusion schemes have been proposed over the years [1221, 1104, 1811].

In an early paper by Shlien [1644], the dependence of binary decision trees on very large training sets for their discriminative power was addressed by using a Dempster combination of decision trees to achieve higher classification accuracies with a given training set. Bell et al. [103] looked at ways of combining classifier mass functions for text categorisation. The classifiers investigated included support vector machines, k-nearest neighbour classifiers and the k-NN-model-based approach (kNNM). Basically, though, their combination stage uses Dempster's rule, *sic et simpliciter*: these authors merely explicitly derived formulae for a number of relevant special cases involving belief functions that are discounted probabilities ('two-points', 'three-points', etc. in the authors' terminology).

Altnay [34] argued that, in classifier combination, the relative values of beliefs assigned to different hypotheses are more important than the accurate estimation of the combined belief function representing the joint observation space. They thus investigated whether there was a set of *dependent* classifiers which provides a better combined accuracy than independent classifiers when Dempster's rule of combination is used. Their analysis, carried out for three different representations of statistical evidence, supported this conjecture. The same authors [33] had earlier proposed a dynamic integration of boosting-based ensembles to take into account the heterogeneity of the input sets. An evidence-theoretic framework was developed to take into account the weights and distances of the neighbouring training samples in both training and testing boosting-based ensembles.

Data equalisation is a technique which is applied to the output nodes of individual classifiers in a multiclassifier system, such that the average difference of the output activation values is minimised, helping with the overall competitiveness of the output nodes of individual classifiers. In [676], Ng argued that this improves the accuracy rate of the resulting combined classifier. In [15, 14], Ahmadzadeh and Petrou analysed the combination of the results of a Bayesian network classifier and a fuzzy-logic-based classifier in the Dempster–Shafer context, in the case in which the two classifiers use two different sets of classes. Within belief theory, the problem of classifiers using different output classes can be addressed by simply introducing a common refinement of the two class sets. The relative reliability of the classifiers can also be taken into consideration. Bi [150, 151] explored the role of 'diversity measures' for classifiers within an ensemble, and proposed a formulation of classifier outputs as triplets of mass functions. In [152], Bi et al. designed a 'class-indifferent' method for combining classifier decisions represented by evidential structures called triplets and quartets, using Dempster's rule of combination.

More recently, Burger et al. [207] noted that in multiclass SVM classification implemented by fusing several binary classifiers, two types of mistakes tend to happen: (1) the response of each classifier does not use the entire information provided by the SVM, and (2) the decision method does not use the entire information associated with the classifier's response. They thus presented a method which partially prevents these two losses of information by applying belief theory to SVM fusion. In a closely related study, Martin and Quidu [1255] proposed an original approach to combining binary classifiers arising from different kinds of strategies, such as one-versus-one or one-versus-rest, of the kind usually employed in SVM classification. The decision functions coming from these binary classifiers were interpreted in terms of belief functions prior to being combined.

Sannen et al. [1532] proposed a trainable classifier fusion evidential method, which takes into account both the accuracy of the classifiers and information about their 'most typical' outputs, which are handled via the orthogonal sum and discounting. Aregui and Denoeux [57] proposed a method for converting a 'novelty' measure, such as that produced by one-class SVMs or kernel principal component analysis, into a belief function on a well-defined frame of discernment. This makes it possible to combine one-class classifiers with other information expressed in the same framework, such as expert opinions or multiclass classifiers. Reformat and Yager [1475] combined the outputs of individual classifiers based on an aggregation process which can be seen as a fusion of Dempster's rule with a generalised form of OWA operator (see Section 4.8.1), providing an added degree of flexibility in the aggregation process. Finally, Quost et al. [1452, 1451, 1453] proposed an alternative strategy for classifier combination which, rather than adopting Dempster's rule, explores the use of the cautious rule and of general t-norm-based rules. Two strategies were investigated for learning an optimal combination scheme, based on a parameterised family of t-norms.

5.4 Ranking aggregation

Consider a set of alternatives $O = \{o_1, o_2, \ldots, o_n\}$ and an unknown linear order (a transitive, antisymmetric and complete relation) on O. Typically, this linear order corresponds to preferences held by an agent or a group of agents – thus, $o_i \succ o_j$ is interpreted as 'alternative o_i is preferred to alternative o_j' (compare Section 4.1.2).

Suppose also that a source of information (elicitation procedure, classifier) provides us with $n(n-1)/2$ pairwise comparisons, affected by uncertainty. The problem is to derive the most plausible linear order from this uncertain (and possibly conflicting) information.

Example 16: Tritchler and Lockwood, 1991. (From [1265].) *Consider four scenarios $O = \{A, B, C, D\}$ describing ethical dilemmas in health care. Suppose two experts have given their preferences for all six possible scenario pairs with confidence degrees as in Fig. 5.5. Assuming the existence of a unique consensus linear*

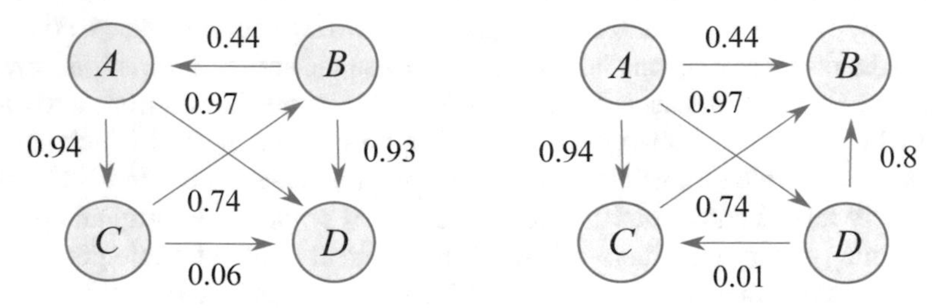

Fig. 5.5: Pairwise preferences in the example from Tritchler and Lockwood [1826].

ordering L^* and seeing the expert assessments as sources of information, what can we say about L^*?

In this problem, the frame of discernment is the set $\mathcal{L}$ of linear orders over O. Each pairwise comparison (o_i, o_j) yields a *pairwise mass function* $m^{\Theta_{ij}}$ on a coarsening $\Theta_{ij} = \{o_i \succ o_j, o_j \succ o_i\}$ with [1265]

$$m^{\Theta_{ij}}(o_i \succ o_j) = \alpha_{ij}, \quad m^{\Theta_{ij}}(o_j \succ o_i) = \beta_{ij}, \quad m^{\Theta_{ij}}(\Theta_{ij}) = 1 - \alpha_{ij} - \beta_{ij}.$$

The mass assignment $m^{\Theta_{ij}}$ may come from a single expert (e.g., an evidential classifier) or from the combination of the evaluations of several experts.

Let $\mathcal{L}_{ij} = \{L \in \mathcal{L} | (o_i, o_j) \in L\}$. Vacuously extending $m^{\Theta_{ij}}$ in $\mathcal{L}$ yields

$$m^{\Theta_{ij}\uparrow\mathcal{L}}(\mathcal{L}_{ij}) = \alpha_{ij}, \quad m^{\Theta_{ij}\uparrow\mathcal{L}}(\overline{\mathcal{L}_{ij}}) = \beta_{ij}, \quad m^{\Theta_{ij}\uparrow\mathcal{L}}(\mathcal{L}) = 1 - \alpha_{ij} - \beta_{ij}.$$

Subsequently combining the pairwise mass functions using Dempster's rule produces

$$m^{\mathcal{L}} = \bigoplus_{i<j} m^{\Theta_{ij}\uparrow\mathcal{L}}.$$

The plausibility of the combination $m^{\mathcal{L}}$ is

$$Pl(L) = \frac{1}{1-\kappa} \prod_{i<j} (1 - \beta_{ij})^{\ell_{ij}} (1 - \alpha_{ij})^{1-\ell_{ij}},$$

where $\ell_{ij} = 1$ if $(o_i, o_j) \in L$, and 0 otherwise (an algorithm for computing the degree of conflict κ was given in [1826]). Its logarithm can be maximised by solving the following binary integer programming problem:

$$\max_{\ell_{ij} \in \{0,1\}} \sum_{i<j} \ell_{ij} \ln \left(\frac{1 - \beta_{ij}}{1 - \alpha_{ij}} \right),$$

subject to

$$\begin{cases} \ell_{ij} + \ell_{jk} - 1 \leq \ell_{ik}, & \forall i < j < k \quad (1), \\ \ell_{ik} \leq \ell_{ij} + \ell_{jk}, & \forall i < j < k \quad (2). \end{cases}$$

Constraint (1) ensures that if $\ell_{ij} = 1$ and $\ell_{jk} = 1$ then $\ell_{ik} = 1$, while (2) ensures that $\ell_{ij} = 0$ and $\ell_{jk} = 0 \Rightarrow \ell_{ik} = 0$.

Argentini and Blanzieri proposed a different solution to the ranking aggregation problem, which they called the belief ranking estimator (BRE). This approach takes into account two aspects still unexplored in ranking combination: the approximation quality of the experts, and the level of uncertainty related to the position of each item in the ranking. BRE estimates the true ranking in an unsupervised way, given a set of rankings that are diverse quality estimations of an unknown, 'true' ranking. The uncertainty in the item positions in each ranking is modelled within the belief function framework, allowing us to encode different sources of a priori knowledge about the correctness of the ranking positions and to weigh, at the same time, the reliability of the experts involved in the combination.

5.5 Regression

Regression, as opposed to classification, is about learning a mapping from a set of input observations $\mathbb{X}$ to a domain which is generally assumed to be continuous (i.e., some subset Y of $\mathbb{R}^n$, for some n).

Regression was considered by belief theorists only relatively recently. The first related paper is due to Denoeux [464], who proposed in 1997 a functional regression approach based on the transferable belief model. This method uses reference vectors for computing a belief structure that quantifies the uncertainty attached to the prediction of the target data, given the input data. The procedure may be implemented in a neural network with a specific architecture and adaptive weights, and allows us to compute an imprecise assessment of the target data in the form of lower and upper expectations. The width of the resulting interval reflects the partial indeterminacy of the prediction resulting from the relative scarcity of the training observations.

Substantial work was later conducted by Petit-Renaud and Denoeux (nonparametric regression analysis; see Section 5.5.1) and by Cuzzolin and co-authors (belief-modelling regression; see Section 5.5.2). These two approaches are described in more detail here. More recently, Laanaya et al. (2010) [1089] proposed a regression approach based on the statistical learning theory of Vapnik, in which the regression of membership functions and of belief functions yields a fuzzy SVM and a belief SVM classifier, respectively. From the training data, the membership and belief functions are generated using two classical approaches, given respectively by fuzzy and belief k-nearest-neighbour methods.

5.5.1 Fuzzy-belief non-parametric regression

In the approach of Petit-Renaud and Denoeux [1423], the training data is supposed to come in the form $\mathcal{T} = \{(\mathbf{x}_i, m_i), i = 1, \ldots, N\}$, where m_i is a fuzzy belief assignment (FBA) (a collection of fuzzy focal elements (sets of values) with associated belief masses [447]) over a continuous domain Y, which quantifies our partial knowledge of the response to input $\mathbf{x}_i$. A prediction of the output value $y \in Y$ given an input vector $\mathbf{x} \in \mathbb{X}$ is also given in the form of an FBA, computed using a non-parametric, instance-based approach.

Given a test input $\mathbf{x}$, the FBA m_i associated with each training sample $\mathbf{x}_i$ is discounted as a function of the distance between $\mathbf{x}$ and $\mathbf{x}_i$. The resulting pieces of evidence are pooled via conjunctive combination to generate an FBA m_Y over Y. Using the (fuzzy) pignistic function to represent the result, summary statistics such as the median or the expectation can be computed. Upper and lower expectations can also be computed as follows:

$$\hat{y}^*(\mathbf{x}) = \sum_{A \in \mathcal{E}} m_Y(A) \sup_{y \in A} y, \quad \hat{y}_*(\mathbf{x}) = \sum_{A \in \mathcal{E}} m_Y(A) \inf_{y \in A} y, \tag{5.7}$$

where m_Y is the overall FBA over Y, and $\mathcal{E}$ its list of focal elements.

The method can cope with heterogeneous training data, including numbers, intervals and fuzzy numbers, and was compared in [1423] with standard regression techniques on standard datasets.

5.5.2 Belief-modelling regression

Albeit initially formulated in the context of object pose estimation in computer vision, the *belief-modelling regression* approach of Cuzzolin and co-authors [365,

366, 370, 704] is in fact very general, and shares several similarities with the work of Petit-Renaud et al. [1423].

In example-based object pose estimation, the available evidence comes in the form of a training set of images containing sample poses of an unspecified object, described as D-dimensional vectors $q \in \mathcal{Q} \subset \mathbb{R}^D$. As described in [704], at training time, an 'oracle' provides for each training image I_k the configuration q_k of the object portrayed there, and its location within the image in the form of a bounding box. The object explores its range of possible configurations, while both a set of sample poses $\tilde{\mathcal{Q}}$ and N series of feature values are collected:

$$\tilde{\mathcal{Q}} \doteq \Big\{ q_k, k = 1, \ldots, T \Big\}, \quad \tilde{\mathcal{Y}} \doteq \Big\{ y_i(k), k = 1, \ldots, T \Big\}. \tag{5.8}$$

During testing, a supervised localisation algorithm is employed to locate the object within the test image. The features extracted from the detected bounding box are then exploited to produce an estimate of the object's configuration.

Belief-modelling regression (BMR) is based on learning from the training data an approximation of the unknown mapping between each feature space $\mathcal{Y}_i$ and the pose space $\mathcal{Q}$. EM clustering [1365] is applied to the N training sequences of feature values, to obtain a mixture-of-Gaussians representation of the feature space

$$\Big\{ \Gamma_i^j, j = 1, \ldots, n_i \Big\}, \quad \Gamma_i^j \sim \mathcal{N}(\mu_i^j, \Sigma_i^j),$$

which generates an approximate feature space

$$\Theta_i \doteq \Big\{ \mathcal{Y}_i^1, \ldots, \mathcal{Y}_i^{n_i} \Big\}, \tag{5.9}$$

where $\mathcal{Y}_i^j$ is the region of the i-th feature space dominated by the j-th Gaussian component Γ_i^j of the mixture, and an approximate feature–pose map of the form (see Fig. 5.6)

$$\rho_i : \mathcal{Y}_i^j \mapsto \tilde{\mathcal{Q}}_i^j \doteq \Big\{ q_k \in \tilde{\mathcal{Q}} : y_i(k) \in \mathcal{Y}_i^j \Big\}. \tag{5.10}$$

When a new image is acquired at test time, new visual features $y_1, \ldots, y_N$ are extracted from the detection box and mapped to a collection of belief functions $Bel_1, \ldots, Bel_N$ on the set of sample poses $\tilde{\mathcal{Q}}$. They can then be combined by conjunctive combination, to yield a belief estimate $Bel = \textcircled{\cap} Bel_i$ of the pose on the set of sample poses $\tilde{\mathcal{Q}}$. At that stage we can either compute the expected pose associated with each vertex of the credal set $\mathcal{P}[Bel]$ on $\tilde{\mathcal{Q}}$ associated with the belief estimate Bel,

$$\hat{q} = \sum_{k=1}^{T} p(q_k) q_k, \quad p \text{ vertex of } \mathcal{P}[Bel],$$

or approximate Bel with an appropriate probability measure on $\tilde{\mathcal{Q}}$ (e.g. the pignistic function).

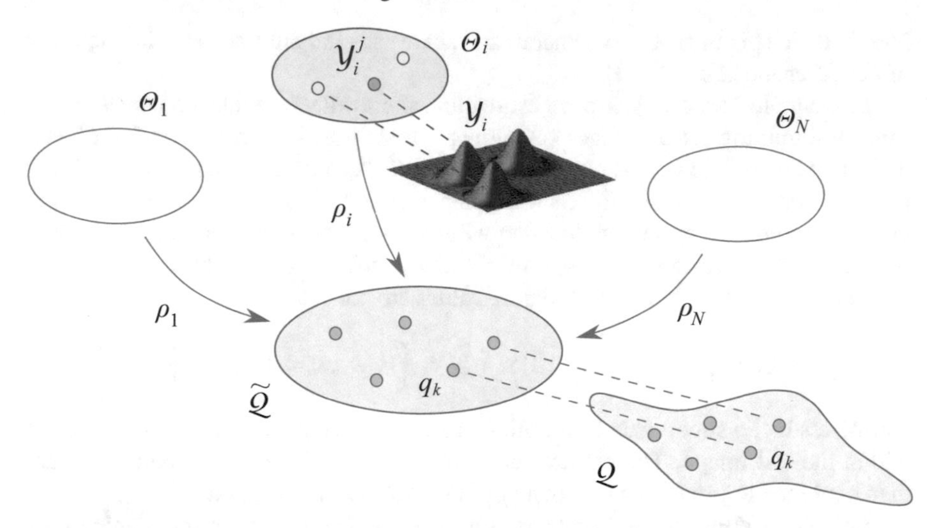

Fig. 5.6: An *evidential model* of an object is a collection of approximate feature spaces (5.9) $\Theta_1, \ldots, \Theta_N$, together with their refining maps (5.10) to their common refinement (see Definition 14, Section 2.5.2), the collection $\tilde{Q}$ (5.8) of training poses, seen as an approximation of the true, unknown pose space Q.

In [704], an extensive empirical comparison was conducted on video sequences captured by DV[54] cameras from two different views. The accuracy of BMR was favourably compared with that of common traditional regression frameworks such as relevant vector machine (RVM) [1818] and Gaussian process regression (GPR) [1464].

5.6 Estimation, prediction and identification

Significant work has been conducted on state estimation, time series prediction and system identification via belief function theory.

5.6.1 State estimation

An early paper by Reece [1472] (from 1997), for instance, studied model-based parameter estimation for noisy processes when the process models are incomplete or imprecise. Reece adopted the qualitative representation advocated by Kuipers, in terms of operations on intervals of real values. A weighted opinion pool formalism for multisensor data fusion was investigated, a definition for unbiased estimation on quantity spaces developed and a consistent mass assignment function for mean estimators for two-state systems derived. This was also extended to representations involving more than two states by utilising the notion of frame refining (see

[54] `https://en.m.wikipedia.org/wiki/DV`.

Chapter 2, Section 2.5). Dempster–Shafer theory was generalised to a finite set of theories, showing how an extreme theory can be used to develop mean minimum-mean-square-error estimators applicable to situations with correlated noise.

Rombaut et al. [1497] proposed a method based on evidence theory for determining the state of a dynamical system modelled by a Petri net, using observations of its inputs. Whereas in [887] a belief Petri net model using the formalism of evidence theory was defined, and the resolution of the system was done heuristically by adapting the classical evolution equations of Petri nets, in [1497] a more principled approach based on the transferable belief model was adopted, leading to simpler computations. In a related publication [1338], Nassreddine et al. outlined a state estimation method for multiple model systems using belief function theory, which can handle the case of systems with an unknown and variant motion model. First, a set of candidate models is selected and an associated mass function is computed based on the measurement likelihood of possible motion models. The estimated state of the system is then derived by computing the expectation with respect to the pignistic probability.

5.6.2 Time series analysis

Time series analysis is another important research area which is affected by uncertainty. Until now, however, the application of belief functions to state estimation has been rather limited [1339].

A number of papers have focused on evidential versions of the Kalman filter [944]. In 2001, Dempster [423] showed that computations in the Kalman filter can be interpreted in a graphical model in terms of the propagation of Gaussian belief functions (a special kind of belief function characterised by a normal density over linear subspaces). Smets and Ristic [1733] provided another interpretation of the Kalman filter in a belief-function paradigm, in the context of an approach to joint tracking and classification based on continuous belief functions. Mahler [1238] also discussed various extensions of the Kalman filter that make it possible to handle uncertainties expressed by belief functions.

In the wider context of time series analysis, Nassreddine et al. [1339] presented an approach to nonlinear state estimation based on belief function theory and interval analysis, which uses belief structures composed of a finite number of axis-aligned boxes with associated masses. Such belief structures can represent partial information about model and measurement uncertainties more accurately than can the bounded-error approach alone. Focal sets are propagated in system equations using interval arithmetics and constraint satisfaction techniques.

Ramasso et al. [1457] separately proposed a tool called a *temporal credal filter with conflict-based model change* (TCF-CMC) to 'smooth' belief functions online in the transferable-belief-model framework. The TCF-CMC takes the temporal aspects of belief functions into account, and relies on the explicit modelling of conflict information. The same lead author [1456] also presented 'credal forward', 'credal backward' and 'credal Viterbi' algorithms for hidden-Markov-model analysis which allow one to filter temporal belief functions and to assess state sequences in the TBM framework.

Quite recently, Niu and Yang [1359] developed a Dempster–Shafer regression technique for time series prediction, which includes a strategy for iterated, multistep-ahead prediction to keep track of the rapid variation of time series signals during a data-monitoring process in an industrial plant.

5.6.3 Particle filtering

Particle filtering [511] is a well-established Bayesian Monte Carlo technique for estimating the current state of a hidden Markov process using a fixed number of samples, called 'particles'. Extensions of the particle filtering technique to belief theory were proposed by Muñoz-Salinas et al. [1360] and Reineking [1478].

Rafael Muñoz-Salinas et al.'s proposal [1360] solves the multitarget tracking problem by combining pieces of evidence from multiple heterogeneous and unreliable sensors. For each particle, the evidence for finding the person being tracked at the particle location is calculated by each sensor, each sensor coming with a degree of reliability computed based on the visible portion of the targets and their occlusions. All the pieces of evidence collected by a set of cameras are fused, taking reliability into consideration, to identify the best hypothesis.

Thomas Reineking [1478] also derived a particle filtering algorithm within the Dempster–Shafer framework. The aim was to maintain a belief distribution over the state space of a hidden Markov process by deriving the corresponding recursive update equations, which turns out to be a strict generalisation of Bayesian filtering. The solution of these equations can be efficiently approximated via particle filtering based on importance sampling, which makes the Dempster–Shafer approach tractable even for large state spaces. In [1478], the performance of the resulting algorithm was compared with exact evidential as well as Bayesian inference.

5.6.4 System identification

The identification of the parameters of a dynamical model, either deterministic (e.g. auto-regressive moving average models, ARMA) or stochastic (such as hidden Markov models), is also typically affected by imperfect knowledge.

Expectation maximisation [430], in particular, is a very commonly used tool for estimating the parameters of a probability distribution, such as that generated by a stochastic model. In a classification context, Jraidi and Elouedi [940] proposed to generalise the EM algorithm to cope with imperfect knowledge at two levels: the attribute values of the observations, and their class labels. Knowledge was represented by belief functions as understood in the transferable belief model. In [188], instead, Boukharouba et al. put forward an approach to the identification of piecewise autoregressive exogenous (ARX) systems from input-output data. The method was able to solve simultaneously data assignment, parameter estimation and the number of submodels thanks to an evidential procedure. As the data were locally linear, the strategy aimed at simultaneously minimising the error between the measured output and each submodel's output and the distance between data belonging to the same submodel. A soft multicategory support vector classifier was used to find a complete polyhedral partition of the regressor set by discriminating all the classes simultaneously.

5.7 Optimisation

Belief theory can also be useful when dealing with optimisation problems in the presence of uncertainty.

The most significant effort in the area is due to Limbourg [1165], in the context of *multi-objective evolutionary algorithms* (MOEAs), a powerful tool for the global optimisation of deterministic functions. In many real-world problems, uncertainty about the correctness of the system model and environmental factors do not allow us to determine clear objective values. As stochastic sampling represents only aleatory, as opposed to epistemic, uncertainty, Limbourg proposed in [1165] some extensions of MOEAs to handle epistemic uncertainty in the form of belief functions.

A few other authors have contributed to this topic with, however, limited impact. As early as 1993, Cucka and Rosenfeld [320] proposed the use of Dempster–Shafer theory for pooling intrinsic support in the relaxation labelling problem, whereas a heuristic search algorithm based on belief functions was outlined by Perneel in 1994 [1419]. A modified Dempster–Shafer methodology was instead proposed by Chen and Rao [258] for solving iterative multicriteria design optimisation problems. Based on the information generated at each iteration, this method handles multiple design criteria, which are often conflicting and non-commensurable, by constructing belief structures that can quantitatively evaluate the effectiveness of each design in the range 0 to 1. An overall satisfaction function is then defined for converting the original multicriteria design problem into a single-criterion problem so that standard single-objective programming techniques can be employed.

Monte Carlo methods can, as we know, be used to obtain the minimum of a function or energy. Resconi et al. [1486] presented in 1998 a speed-up of classical Monte Carlo methods based on the observation that the Boltzmann transition probability and the concept of local thermodynamic equilibrium give rise to an initial state of maximum entropy, which is subsequently modified by using information about the internal structure of the system. As the classical thermodynamic model does not take into account any internal structure within the system, these authors proposed instead a physical model of belief measures. Harmanec and Klir's algorithm for computing probability distributions consistent with a belief measure was exploited to utilise the Boltzmann distribution under an arbitrary probability distribution to guide the Monte Carlo iterative method in order to obtain the global minimum value of the energy.

Mueller and Piché [1323] applied Dempster–Shafer theory to surrogate model selection and combination in the framework of the 'response surface' approach to global optimisation, studying various conflict redistribution rules and the implications of the chosen surrogate model type (i.e., combined, single or hybrid).

Finally, *belief linear programming* is an uncertain linear programming method where uncertainty is expressed by belief functions, developed by Reformat et al. [1474].

6

The bigger picture

As we have seen in the Introduction, several different mathematical theories of uncertainty are competing to be adopted by practitioners in all fields of applied science [1874, 2057, 1615, 784, 1229, 665]. The emerging consensus is that there is no such thing as *the* best mathematical description of uncertainty (compare [931, 583, 1635, 984, 986, 445], to cite a few influential papers) – the choice of the most suitable methodology should depend on the actual problem at hand. Whenever a probability measure can be estimated, most authors suggest the use of a classical Bayesian approach. If probability values cannot be reliably estimated, upper and lower probabilities may instead be preferred. The thesis of this book is that, indeed, most situations do require one to assess the uncertainty associated with set-valued observations, and therefore random sets (and their belief function incarnation) are a fundamental tool when dealing with such problems.

Scholars have discussed and compared rather extensively the various approaches to uncertainty theory. Notably, Klir [987] surveyed in 2004 various theories of imprecise probabilities, with the aim of proposing suitable unifying principles. Smets [1696, 1706], for his part, contributed to the analysis of the difference between imprecision and uncertainty [962], and compared the applicability of various models of uncertainty. More recently, Destercke et al. [486] explored unifying principles for uncertainty representations in the context of their 'generalised p-boxes' proposal. A number of papers, in particular, ran theoretical and empirical [814] comparisons between belief function theory and other mathematical models of uncertainty [1111, 2026, 1238, 819, 1476, 1029, 545], especially in the sensor fusion context [172, 239, 192, 830, 1815, 204]. Finally, a number of attempts were made at unifying most of the approaches to uncertainty theory into a single coherent framework [987, 1874], including Walley's theory of imprecise probability [1874, 1877], Klir's

F. Cuzzolin, *The Geometry of Uncertainty*, Artificial Intelligence: Foundations, Theory, and Algorithms, https://doi.org/10.1007/978-3-030-63153-6_6

generalised information theory [987] and Zadeh's generalised theory of uncertainty [2093]. A fascinating effort towards the identification of such a unifying framework has been made by Resconi et al. [1484], using as a unifying language the syntactic and semantic structures of modal logic.

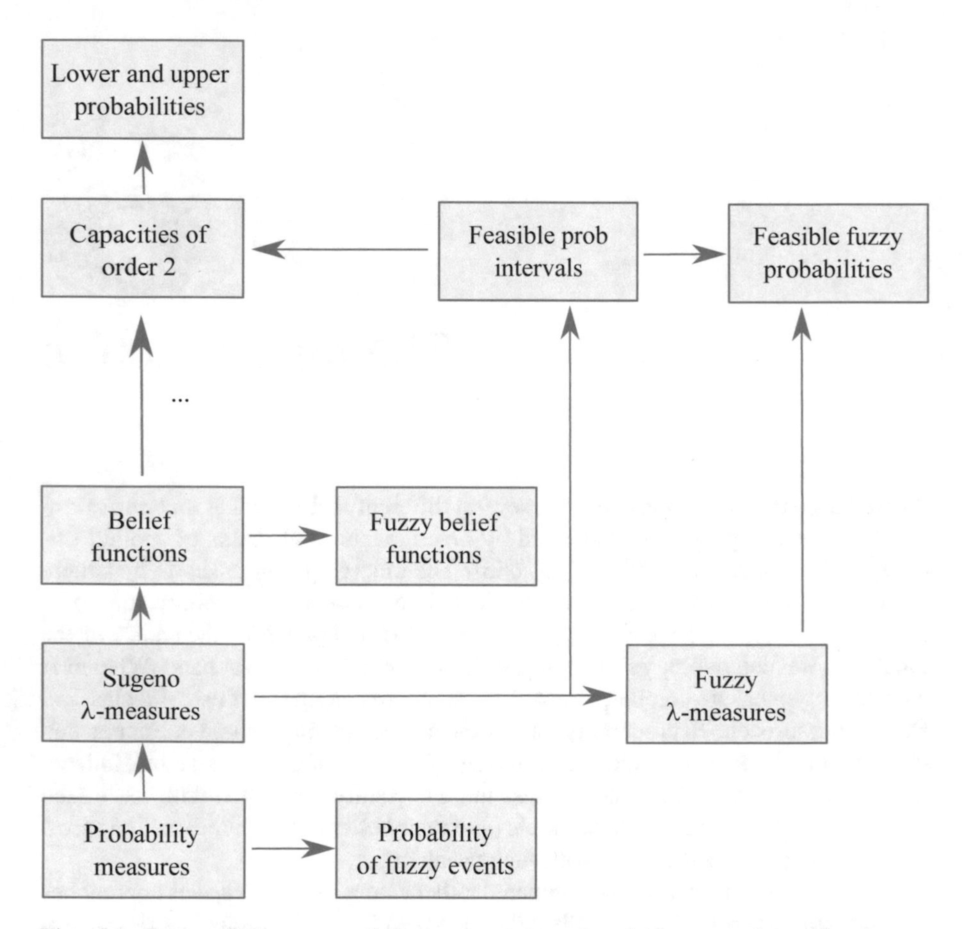

Fig. 6.1: Some of the approaches to uncertainty theories surveyed in this chapter are arranged here into a hierarchy, with less general frameworks at the bottom and more general ones at the top. A link between them indicates that the top formalism comprises the bottom one as a special case. Adapted from [987].

Our understanding of what belief functions are would be significantly limited if we did not expand our field of view to encompass the rich tapestry of methodologies which go under the name of 'theories of uncertainty'. Figure 6.1 arranges them into a hierarchy, according to their generality. Similar diagrams, for a smaller subset of methodologies, appear in [486] and [987], among others.

In this chapter, therefore, we will briefly survey the other major approaches to uncertainty theory in a rather exhaustive way, focusing in particular on their relationship (if any) with the theory of evidence.

Chapter outline

We start by summarising the basic notions of Peter Walley's theory of imprecise probabilities, arguably the broadest attempt yet to produce a general theory of uncertainty, and its behavioural rationale (Section 6.1).

The other most general framework for the description of uncertainty is probably the theory of capacities, also called (for historical reasons) 'fuzzy measures' – these are introduced in Section 6.2. A special case of capacities, but a very general one, is 2-monotone capacities or 'probability intervals' (Section 6.3), which include belief functions (as ∞-monotone capacities) as a subgroup. In Section 6.4, higher-order probabilities (and second-order ones, in particular) are introduced and their relation with belief functions is discussed. Fuzzy theory (Section 6.5) includes, in a wide sense, both possibility theory (Section 6.5.1) and the various extensions of belief functions with values on fuzzy sets that have been proposed. Logic (Section 6.6) is much intertwined with belief theory, a fact we have already hinted at in the previous chapter. In particular, modal logic interpretations (Section 6.6.7), which subsume Pearl's probability of probability semantics (Section 6.6.8), exhibit very strong links with Dempster–Shafer theory. Other major uncertainty-related concepts with significant links to belief theory are Pawlak's rough sets (Section 6.7) and the notion of a probability box, or 'p-box' (Section 6.8). A theory of epistemic beliefs quite resembling Shafer's theory of belief functions is that of Spohn, outlined in Section 6.9. Section 6.10 is devoted to Zadeh's 'generalised uncertainty theory' framework, in which various uncertainty theories are unified in terms of generalised constraints acting on 'granules'. Baoding Liu's 'uncertainty theory' formalism is briefly reviewed in Section 6.11. Yakov Ben-Haim's info-gap theory, an interesting framework for assessing the robustness of dynamical models, is recalled in Section 6.12. Finally, Vovk and Shafer's game-theoretical interpretation of probability is summarised in Section 6.13.

A comprehensive survey of other mathematical formalisms for the description of uncertainty, including granular computing, Laskey's assumptions, Grosof's evidential confirmation theory, Groen and Mosleh's extension of Bayesian theory, and neighbourhood systems, concludes the chapter.

6.1 Imprecise probability

Imprecise probability is a term that, in a broader sense, refers to the approaches which make use of *collections* of probability measures, rather than single distributions, to model a problem in a robust statistical sense. In a narrower sense, it specifically denotes the framework put forward by Walley [1877, 1874] to unify all these approaches in a coherent setting. The latter is perhaps, to date, the most extensive attempt to formulate a general theory of imprecise probabilities, whose generality is comparable to that of the theory based on arbitrary closed and convex probability distributions, and can be formalised in terms of *lower and upper previsions* [311].

A recent survey of the theory of coherent lower previsions has been given by Miranda [1292].

6.1.1 Lower and upper probabilities

A *lower probability* [617] $\underline{P}$ is a function from 2^Θ, the power set of Θ, to the unit interval $[0,1]$. With any lower probability $\underline{P}$ is associated a dual upper probability function $\overline{P}$, defined for any $A \subseteq \Theta$ as $\overline{P}(A) = 1 - \underline{P}(A^c)$, where A^c is the complement of A.

With any lower probability $\underline{P}$ we can also associate a closed convex set

$$\mathcal{P}(\underline{P}) = \Big\{ P : P(A) \geq \underline{P}(A), \forall A \subseteq \Theta \Big\} \tag{6.1}$$

of probability measures P dominating $\underline{P}$. Such a polytope or convex set of probability distributions is usually called a *credal set* [1141]. When the lower probability is a belief measure, $\underline{P} = Bel$, we simply get the credal set of probabilities consistent with Bel (3.10).

As pointed out by Walley [1874], not all convex sets of probabilities can be described by merely focusing on events.

Definition 88. *A lower probability $\underline{P}$ on Θ is called 'consistent' ('avoids sure loss' in Walley's terminology [1874]) if $\mathcal{P}(\underline{P}) \neq \emptyset$ or, equivalently,*

$$\sup_{\theta \in \Theta} \sum_{i=1}^n 1_{E_i}(\theta) \geq \sum_{i=1}^n \underline{P}(E_i), \tag{6.2}$$

whenever n is a non-negative integer, $E_1, \ldots, E_n \in \mathcal{F}$ are events of the σ-algebra $\mathcal{F}$ on which $\underline{P}$ is defined and 1_{E_i} is the characteristic (indicator) function of the event E_i on Θ.

Definition 89. *A lower probability $\underline{P}$ is called 'tight' ('coherent' in Walley's terminology) if*

$$\inf_{P \in \mathcal{P}(\underline{P})} P(A) = \underline{P}(A)$$

or, equivalently,

$$\sup_{\theta \in \Theta} \left[\sum_{i=1}^n 1_{E_i}(\theta) - m \cdot 1_{E_0}(\theta) \right] \geq \sum_{i=1}^n \underline{P}(E_i) - m \cdot \underline{P}(E_0), \tag{6.3}$$

whenever n, m are non-negative integers, $E_0, E_1, \ldots, E_n \in \mathcal{F}$ and 1_{E_i} is again the characteristic function of the event E_i.

Consistency means that the lower-bound constraints $\underline{P}(A)$ can indeed be satisfied by some probability measure, while tightness indicates that $\underline{P}$ is the lower envelope on subsets of $\mathcal{P}(\underline{P})$. Any coherent lower probability is monotone and superadditive.

6.1.2 Gambles and behavioural interpretation

The concepts of avoiding sure loss and coherence are also applicable to any functional defined on a class of bounded functions on Θ, called *gambles*. According to this point of view, a lower probability is a functional defined on the class of all characteristic (indicator) functions of sets.

The behavioural rationale for general imprecise probability theory derives from equating 'belief' to 'inclination to act'. An agent believes in a certain outcome to the extent it is willing to accept a gamble on that outcome. The pioneer of this interpretation of probability and belief is Bruno de Finetti [403, 404], as we recalled in the Introduction. A 'gamble' is a decision which generates different utilities in different states (outcomes) of the world. The following outline is abstracted from [394].

Definition 90. *Let Ω be the set of possible outcomes ω. A* gamble *is a bounded real-valued function on Ω, $X : \Omega \to \mathbb{R}$, $\omega \mapsto X(\omega)$.*

Clearly, the notion of a gamble is very close to that of utility (see Section 4.8.1). Note that gambles are not constrained to be normalised or non-negative. Whether one is willing to accept a gamble depends on one's belief in the outcome.

Let us denote an agent's *set of desirable gambles* by $\mathcal{D} \subseteq \mathcal{L}(\Omega)$, where $\mathcal{L}(\Omega)$ is the set of all bounded real-valued functions on Ω. Since whether a gamble is desirable depends on the agent's belief in the outcome, $\mathcal{D}$ can be used as a model of the agent's uncertainty about the problem.

Definition 91. *A set $\mathcal{D}$ of desirable gambles is* coherent *iff:*

1. *0 (the constant gamble $X(\omega) = 0$ for all $\omega \in \Omega$) $\notin \mathcal{D}$.*
2. *If $X > 0$ (i.e., $X(\omega) > 0$ for all $\omega \in \Omega$), then $X \in \mathcal{D}$.*
3. *If $X, Y \in \mathcal{D}$, then $X + Y \in \mathcal{D}$.*
4. *If $X \in \mathcal{D}$ and $\lambda > 0$, then $\lambda X \in \mathcal{D}$.*

As a consequence, if $X \in \mathcal{D}$ and $Y > X$, then $Y \in \mathcal{D}$. In other words, a coherent set of desirable gambles is a *convex cone* (i.e., it is closed under convex combination): see Fig. 6.2.

6.1.3 Lower and upper previsions

Now, suppose the agent buys a gamble X for a price μ. This yields a new gamble $X - \mu$.

Definition 92. *The lower prevision $\underline{P}(X)$ of a gamble X,*

$$\underline{P}(X) \doteq \sup\Big\{\mu : X - \mu \in \mathcal{D}\Big\},$$

is the supremum acceptable price for buying X.

In the same way, selling a gamble X for a price μ yields a new gamble $\mu - X$.

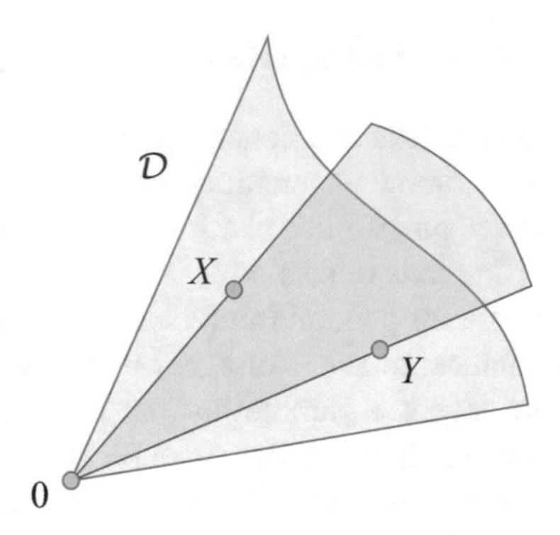

Fig. 6.2: Coherent sets of desirable gambles form convex cones.

Definition 93. *The upper prevision* $\overline{P}(X)$ *of a gamble* X,

$$\overline{P}(X) \doteq \inf\left\{\mu : \mu - X \in \mathcal{D}\right\},$$

is the supremum acceptable price for selling X.

By definition, $\overline{P}(X) = -\underline{P}(-X)$. When the lower and upper previsions coincide, $\overline{P}(X) = \underline{P}(X) = P(X)$ is called the (precise) *prevision* of X, or a 'fair price' in de Finetti's sense [403].

A graphical interpretation of lower, upper and precise previsions is given in Fig. 6.3. Specifying a precise prevision for X amounts to being able, for any real price p, to decide whether we want to buy or sell gamble X. When only lower and

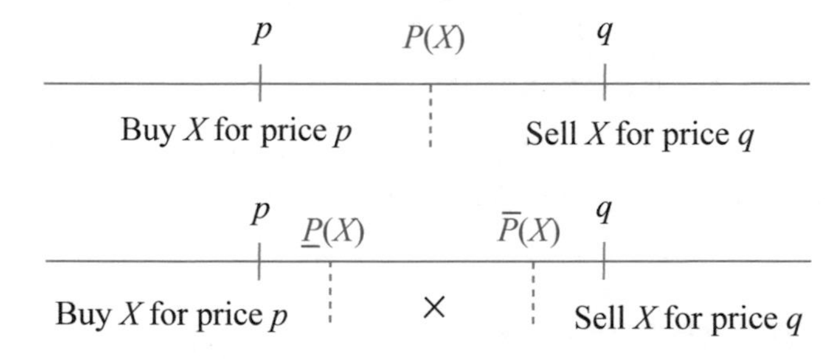

Fig. 6.3: Interpretation of precise previsions (top) versus lower and upper previsions (bottom) in term of acceptability of gambles (adapted from [394]).

upper previsions can be defined, we remain undecided for any price in the interval $[\underline{P}(X), \overline{P}(X)]$.

6.1.4 Events as indicator gambles

As events $A \subset \Omega$ are nothing but special indicator gambles of the form

$$1_A(\omega) = \begin{cases} 1 & \omega \in A, \\ 0 & \omega \notin A, \end{cases}$$

the lower/upper probability of an event can be defined trivially as the lower/upper prevision of the corresponding indicator gamble: $\underline{P}(A) = \underline{P}(1_A), \overline{P}(A) = \overline{P}(1_A)$.

The indicator gamble expresses the fact that we (the agent) are rewarded whatever the outcome in A (in an equal way), and not rewarded if the outcome is outside A. The corresponding lower and upper previsions (probabilities) then measure the evidence for and against the event A.

6.1.5 Rules of rational behaviour

Lower and upper previsions represent commitments to act in certain ways under certain circumstances. Rational rules of behaviour require that:

1. The agent does not specify betting rates such that it loses utility whatever the outcome (a principle we have called 'avoiding sure loss') – when the gamble X_k is an indicator function 1_{E_i}, this is expressed by Definition 88.
2. The agent is fully aware of the consequences of its betting rates ('coherence') – when the gamble X_k is an indicator function 1_{E_i}, this is expressed by Definition 89.

If the first condition is not met, there exists a positive combination of gambles which the agent finds individually desirable but which is not desirable to it. One consequence of avoiding sure loss is that $\underline{P}(A) \leq \overline{P}(A)$. A consequence of coherence is that lower previsions are subadditive: $\underline{P}(A) + \underline{P}(A) \leq \underline{P}(A \cup B)$ for $A \cap B \neq 0$. A precise prevision P is coherent iff: (i) $P(\lambda X + \mu Y) = \lambda P(X) + \mu P(Y)$; (ii) if $X > 0$, then $P(X) \geq 0$; and (iii) $P(\Omega) = 1$, and coincides with de Finetti's notion of a *coherent prevision*.

Special cases of coherent lower/upper previsions include probability measures, de Finetti previsions, 2-monotone capacities, Choquet capacities, possibility and necessity measures, belief/plausibility measures and random sets, but also probability boxes, (lower and upper envelopes of) credal sets, and robust Bayesian models.

6.1.6 Natural and marginal extension

The *natural extension* operator addresses the problem of extending a coherent lower prevision defined on a collection of gambles to a lower prevision on all gambles, under the constraints of the extension being coherent and conservative (least committal). Given a set $\mathcal{D}$ of gambles the agent has judged desirable, the natural extension $\mathcal{E}$ of $\mathcal{D}$ is the smallest coherent set of desirable gambles that includes $\mathcal{D}$, i.e. the smallest extension of $\mathcal{D}$ to a convex cone of gambles that contains all positive gambles but not the zero gamble.

For the special case of lower probabilities, it is defined as follows.

Definition 94. *Let $\underline{P}$ be a lower probability on Ω that avoids sure loss, and let $\mathcal{L}$ be the set of all bounded functions on Ω. The functional $\underline{E}$ defined on $\mathcal{L}$ as*

$$\underline{E}(f) = \sup\left\{ \sum_{i=1}^{n} \lambda_i \underline{P}(E_i) + c \,\middle|\, f \geq \sum_{i=1}^{n} \lambda_i 1_{E_i} + c, \;\; n \geq 0, E_i \subseteq \Omega, \lambda_i \geq 0, \right.$$
$$\left. c \in \{-\infty, +\infty\} \right\} \quad \forall f \in \mathcal{L} \tag{6.4}$$

is called the natural extension *of $\underline{P}$.*

When $\underline{P}$ is a classical ('precise') probability, the natural extension agrees with the expectation. Also, $\underline{E}(1_A) = \underline{P}(A)$ for all A iff $\underline{P}$ is coherent.

A *marginal extension* operator [1294] can be introduced to cope with the aggregation of conditional lower previsions.

6.1.7 Belief functions and imprecise probabilities

Belief functions as coherent lower probabilities Wang and Klir [1912] showed that on a frame of just two elements, coherence and complete monotonicity reduce to superadditivity: hence, any coherent lower probability is a belief measure. Even on frames of three elements, however, there exist 2-monotone (coherent) lower probabilities which are not 3-monotone (belief measures). Belief functions are in fact a special type of coherent lower probabilities which, as we just saw, can be seen as a special class of lower previsions (see [1874], Section 5.13). Walley proved that coherent lower probabilities are closed under convex combination: the relationship of belief functions to convexity will be discussed in Part II.

The class of lower previsions induced by multivalued mappings was considered in [1295].

Natural extension and Choquet integrals of belief measures Both the Choquet integral (4.79) with respect to monotone set functions (such as belief functions) and the natural extension of lower probabilities are generalisations of the Lebesgue integral with respect to σ-additive measures. Wang and Klir [1911] investigated the relations between Choquet integrals, natural extensions and belief measures, showing that the Choquet integral with respect to a belief measure is always greater than or equal to the corresponding natural extension.

More precisely, the Choquet integral $\int f \, \mathrm{d}\underline{P}$ for all $f \in \mathcal{L}$ is a nonlinear functional on $\mathcal{L}$, and $(X, \mathcal{L}, \int f \, \mathrm{d}\underline{P})$ is a lower prevision [1874]. It can be proven that the latter is coherent when $\underline{P}$ is a belief measure, and that the following proposition is true.

Proposition 37. $\underline{E}(f) \leq \int f \, \mathrm{d}\underline{P}$ *for any $f \in \mathcal{L}$ whenever $\underline{P}$ is a belief measure.*

Conceptual autonomy of belief functions In terms of semantics, a study by Baroni and Vicig [94] argued against the existence of appreciable relationships between the mass assignment interpretation of belief functions, as in the transferable-belief-model framework, and coherent lower probabilities, which formally encompass them as a special case, confirming the conceptual autonomy of belief functions with respect to imprecise probability.

These authors asked whether, in particular, belief functions, with their non-negative masses, can be characterised as a distance subclass of coherent lower probabilities. Their conclusions, although based on specific examples, were that from a TBM point of view, any search for a common conceptual basis with any form of probability theory is doomed to fail, since the two theories have different roots. For instance, one can state the following proposition.

Proposition 38. *If $\underline{P}$ is a coherent lower probability on Θ with $|\Theta| = 3$, then the Möbius inverse m of $\underline{P}$ is such that $m(\Theta) \geq -\frac{1}{2}$.*

An intriguing behavioural interpretation of belief functions which is directly linked to Walley's was recently (2017) proposed by Kerkvliet in his PhD dissertation [957] – we briefly recalled it in Chapter 3, Section 3.1.6.

6.2 Capacities (a.k.a. fuzzy measures)

The *theory of capacities* or *fuzzy measure theory* [1910, 1782, 724] is a generalisation of classical measure theory upon which classical mathematical probability is constructed, rather than a generalisation of probability theory itself. It considers generalised measures in which the additivity property (see Definition 2) is replaced by the weaker property of monotonicity.

The central concept of a *fuzzy measure* or *capacity* [283] was introduced by Choquet in 1953 and independently defined by Sugeno in 1974 [1781] in the context of fuzzy integrals. A number of uncertainty measures can be seen as special cases of fuzzy measures, including belief functions, possibilities and probability measures [1091].

Definition 95. *Given a domain Θ and a non-empty family $\mathcal{F}$ of subsets of Θ, a* monotone measure *(often called a* monotone capacity *or* fuzzy measure*) μ on $(\Theta, \mathcal{F})$ is a function $\mu : \mathcal{F} \rightarrow [0,1]$ which satisfies the following conditions:*

1. *$\mu(\emptyset) = 0$.*
2. *If $A \subseteq B$, then $\mu(A) \leq \mu(B)$ for every $A, B \in \mathcal{F}$ ('monotonicity').*
3. *For any increasing sequence $A_1 \subseteq A_2 \subseteq \cdots$ of subsets in $\mathcal{F}$,*

$$\textit{if } \bigcup_{i=1}^{\infty} A_i \in \mathcal{F}, \textit{ then } \lim_{i \rightarrow \infty} \mu(A_i) = \mu\Big(\bigcup_{i=1}^{\infty} A_i\Big)$$

('continuity from below').

4. *For any decreasing sequence* $A_1 \supseteq A_2 \supseteq \cdots$ *of subsets in* $\mathcal{F}$,

$$\text{if } \bigcap_{i=1}^{\infty} A_i \in \mathcal{F} \text{ and } \mu(A_1) < \infty, \text{ then } \lim_{i \to \infty} \mu(A_i) = \mu\Big(\bigcap_{i=1}^{\infty} A_i\Big)$$

('continuity from above').

When Θ is finite, the last two requirements are trivially satisfied and can be disregarded. Monotone *decreasing* measures can be obtained by replacing $\leq$ with $\geq$ in condition 2.

6.2.1 Special types of capacities

The relation between different classes of fuzzy measures, including Sugeno's λ-measures (see below), Shafer's belief functions and Zadeh's possibility measures, was studied by Banon in the finite-domain setting in [85]. A later review by Lamata [1091], although based on that of Banon, considered additional kinds of fuzzy measures, and applied classification to pairs of dual fuzzy measures.

Capacities and belief functions

Definition 96. *A capacity* μ *is said to be of* order k *if it satisfies the inequalities*

$$\mu\left(\bigcup_{j=1}^{k} A_j\right) \geq \sum_{\emptyset \neq K \subseteq [1,\ldots,k]} (-1)^{|K|+1} \mu\left(\bigcap_{j \in K} A_j\right) \tag{6.5}$$

for all collections of k *subsets* A_j, $j \in K$, *of* Θ.

Clearly, if $k' > k$, the resulting theory is less general than a theory of capacities of order k (i.e., it contemplates fewer measures). The less general theory of that kind is that of infinitely monotone capacities [1670].

Proposition 39. *Belief functions are infinitely monotone capacities.*

We just need to compare (6.5) with the superadditivity property of belief functions (3.6). The Möbius transforms of a capacity μ can be computed as

$$m(A) = \sum_{B \subset A} (-1)^{|A-B|} \mu(B), \tag{6.6}$$

just as for belief functions (2.3). For infinitely monotone capacities, as we know, the Möbius inverse (the basic probability assignment) is non-negative.

Klir et al. published an excellent discussion [997] of the relations between belief and possibility theory [402, 1113], and examined different methods for constructing fuzzy measures in the context of expert systems. The product of capacities representing belief functions was instead studied in [821]. The result ([821], Equation (12)) is nothing but the unnormalised Dempster combination (or, equivalently, a disjunctive combination in which mass zero is assigned to the empty set), and was proved to satisfy a linearity property (commutativity with convex combination). In [2018], Yager analysed a class of fuzzy measures generated by a belief measure, seen as providing partial information about an underlying fuzzy measure. An entire class of such fuzzy measures exists – the notion of the entropy of a fuzzy measure was used in [2018] to select significant representatives from this class.

Sugeno's λ-measures The λ-measures g_λ, introduced by Sugeno [1782], are special regular monotone measures that satisfy the requirement

$$g_\lambda(A \cup B) = g_\lambda(A) + g_\lambda(B) + \lambda g_\lambda(A) g_\lambda(B) \tag{6.7}$$

for any given pair of disjoint sets $A, B \in 2^\Theta$, where $\lambda \in (-1, \infty)$ is a parameter by which individual measures in this class are distinguished. It is well known [1910] that each λ-measure is uniquely determined by values $g_\lambda(\theta)$, $\theta \in \Theta$, subject to the condition that at least two of these values are non-zero. The parameter λ can then be uniquely recovered from them as follows:

$$1 + \lambda = \prod_{\theta \in \Theta} [1 + \lambda g_\lambda(\theta)]. \tag{6.8}$$

Given $g_\lambda(\theta)$ for all $\theta \in \Theta$ and λ, the values $g_\lambda(A)$ of the λ-measure for all subsets $A \in 2^\Theta$ are then determined by (6.7).

The following three cases must be distinguished:

1. If $\sum_\theta g_\lambda(\theta) < 1$, then g_λ is a lower probability and, thus, a superadditive measure; λ is determined by the root of (6.8) in the interval $(0, \infty)$, which is unique.
2. If $\sum_\theta g_\lambda(\theta) = 1$, then g_λ is a probability measure; $\lambda = 0$, and is the only root of (6.8).
3. If $\sum_\theta g_\lambda(\theta) > 1$, then g_λ is an upper probability and, hence, a subadditive measure; λ is determined by the root of (6.8) in the interval $(-1, 0)$, which is unique.

Finally [1910], lower and upper probabilities based on λ-measures are special belief and plausibility measures [135], respectively.

Interval-valued probability distributions Systems of probability intervals (see Section 6.3) are also a special case of monotone capacities.

Proposition 40. *[393] The lower and upper probability measures associated with a feasible ('reachable') set of probability intervals are Choquet capacities of order 2, namely*

$$\begin{aligned} l(A \cup B) + l(A \cap B) &\geq l(A) + l(B) \quad \forall A, B \subseteq \Theta, \\ u(A \cup B) + u(A \cap B) &\leq u(A) + u(B) \quad \forall A, B \subseteq \Theta. \end{aligned} \tag{6.9}$$

6.3 Probability intervals (2-monotone capacities)

Dealing with general lower probabilities defined on 2^Θ can be difficult when Θ is large: it may then be interesting for practical applications to focus on simpler models.

A *set of probability intervals* or *interval probability system* [939, 1808, 393] is a system of constraints on the probability values of a probability distribution[55] $p : \Theta \to [0,1]$ on a finite domain Θ of the form

$$\mathcal{P}(l,u) \doteq \Big\{ p : l(x) \leq p(x) \leq u(x), \forall x \in \Theta \Big\}. \tag{6.10}$$

Probability intervals [321, 1955, 1797] were introduced as a tool for uncertain reasoning in [393, 1315], where combination and marginalisation of intervals were studied in detail. The authors of [393, 1315] also studied the specific constraints such intervals ought to satisfy in order to be consistent and tight. As pointed out, for instance, in [863], probability intervals typically arise through measurement errors. As a matter of fact, measurements can be inherently of interval nature (owing to the finite resolution of the instruments). In that case the *probability* interval of interest is the class of probability measures consistent with the *measured* interval.

A set of constraints of the form (6.10) also determines a credal set: credal sets generated by probability intervals are a subclass of all credal sets generated by lower and upper probabilities [1673]. Their vertices can be computed as in [393], p. 174.

A set of probability intervals may be such that some combinations of values taken from the intervals do not correspond to any probability distribution function, indicating that the intervals are unnecessarily broad.

Definition 97. *A set of probability intervals is called* feasible *if and only if, for each $x \in \Theta$ and every value $v(x) \in [l(u), u(x)]$, there exists a probability distribution function $p : \Theta \to [0,1]$ for which $p(x) = v(x)$.*

If $\mathcal{P}(l,u)$ is not feasible, it can be converted to a set of feasible intervals via

$$l'(x) = \max \Big\{ l(x), 1 - \sum_{y \neq x} u(y) \Big\}, \quad u'(x) = \min \Big\{ u(x), 1 - \sum_{y \neq x} l(y) \Big\}.$$

In a similar way, given a set of bounds $\mathcal{P}(l,u)$, we can obtain lower and upper probability values $\underline{P}(A)$ on any subset $A \subseteq \Theta$ by using the following simple formulae:

$$\begin{aligned} \underline{P}(A) &= \max \Big\{ \sum_{x \in A} l(x), 1 - \sum_{x \notin A} u(x) \Big\}, \\ \underline{P}(A) &= \min \Big\{ \sum_{x \in A} u(x), 1 - \sum_{x \notin A} l(x) \Big\}. \end{aligned} \tag{6.11}$$

[55] In this book, we generally denote a probability measure by P (upper case), and the associated probability distribution by p (lower case).

Belief functions are also associated with a set of lower- and upper-probability constraints of the form (6.10): they correspond therefore to a special class of interval probability systems, associated with credal sets of a specific form.

Combination, marginalisation and conditioning operators for probability intervals can be defined: for the details, we refer the reader to [393]. A generalised Bayesian inference framework based on interval probabilities was separately proposed by Pan in [1385].

6.3.1 Probability intervals and belief measures

Besides describing at length the way two compatible sets of probability intervals can be combined via disjunction or conjunction, and marginalisation and conditioning operators for them, [393] delved into the relationship between probability intervals and belief/plausibility measures, and considered the problem of approximating a belief function with a probability interval.

Given a pair (Bel, Pl) of belief and plausibility functions on Θ, we wish to find the set of probability intervals $\mathcal{P}^*(l^*, u^*)$ such that:

- $(Bel, Pl) \subset \mathcal{P}^*(l^*, u^*)$ (as credal sets);
- for every $\mathcal{P}(l, u)$ such that $(Bel, Pl) \subset \mathcal{P}(l, u)$, it is also the case that $\mathcal{P}^*(l^*, u^*) \subset \mathcal{P}(l, u)$,

i.e., $\mathcal{P}^*(l^*, u^*)$ is the smallest set of probability intervals containing (Bel, Pl).

Proposition 41. ([1134]; [393], Proposition 13) *For all* $x \in \Theta$

$$l^*(x) = Bel(x), \quad u^*(x) = Pl(x), \tag{6.12}$$

i.e., the minimal probability interval containing a pair of belief and plausibility functions is that whose lower bound is the belief of singletons, and whose upper bound is the plausibility of singletons.

The opposite problem of finding, given an arbitrary set of probability intervals, a belief function such that (6.12) is satisfied can only be solved whenever ([393], Proposition 14)

$$\sum_x l(x) \leq 1, \quad \sum_{y \neq x} l(y) + u(x) \leq 1 \;\forall x \in \Theta, \quad \sum_x l(x) + \sum_x u(x) \geq 2.$$

In that case several pairs (Bel, Pl) exist which satisfy (6.12): Lemmer and Kyburg proposed an algorithm for selecting one in [1134]. When proper and reachable sets are considered, the first two conditions are trivially satisfied.

The problem of approximating an arbitrary probability interval with a pair of belief and plausibility functions was also considered in [393] – it turns out that such approximations only have focal elements of size less than or equal to 2 ([393], Proposition 16).

6.4 Higher-order probabilities

We have seen that belief functions, if the updating process is not taken into account, are in 1–1 correspondence with convex sets of probabilities. As noted by Ferson et al. [605], 'Bounding probability is different from the approach of second-order or two-dimensional probability [e.g. [831, 322]] in which uncertainty about probabilities is itself modeled with probability.' Nevertheless, as we have also pointed out in Chapter 3, Section 3.1.7, the two notions are related, as credal sets correspond to the support of indicator functions on the probability simplex.

The question of the meaningfulness of higher-order probabilities has in fact been discussed by a number of authors, including Savage [580]. Gaifman [660] and Domotor [505], for instance, provided both rigorous axiomatisations and model-theoretic semantics for their systems. Jeffrey, for whom probabilities are essentially derivable from preferences, considered instead higher-order preferences [891], from which one might imagine getting higher-order probabilities. Skyrms [1660] argued that higher-order probabilities are essential for a correct representation of belief.

6.4.1 Second-order probabilities and belief functions

As pointed out by Baron [90], a *second-order probability* $Q(P)$ may be understood as the probability that the true probability of something has the value P, where 'true' may be interpreted as the value that would be assigned if certain information were available. Baron proceeded to derive a rule for combining evidence from two independent sources whenever each source i generates a second-order probability $Q_i(P)$, and showed that Dempster's rule is a special case of the rule derived in connection with second-order probabilities. Belief functions, then, represent a restriction of a full Bayesian analysis.

Josang [921] showed that a bijective mapping exists between Dirichlet distributions and (Dirichlet) belief functions, namely those BFs whose focal elements are of size either 1 or $|\Theta|$. Such a link can be exploited to apply belief-based reasoning to statistical data, or to apply statistical and probabilistic analysis to belief functions.

6.4.2 Gaifman's higher-order probability spaces

An agenda for a formal theory of higher-order probabilities was proposed by Gaifman in 1988 [660]. There, a simple HOP (*higher-order probability space*) consists of a probability space and an operation PR, such that, for every event A and every real closed interval Δ, $PR(A, \Delta)$ is the event that A's 'true' probability (in the sense described above) lies in Δ. In a general HOP, the operation PR includes an additional argument ranging over an ordered set of time points, so that $PR(A, t, \Delta)$ is the event that A's probability at time t lies in Δ. The analysis follows from a number of intuitively justified axioms – various connections with modal logic are pointed out.

6.4.3 Kyburg's analysis

A rather profound discussion of the notion of higher-order probability was published by Kyburg in [1087]. There, it was claimed that higher-order probabilities can always be replaced by the marginal distributions of a joint probability distribution defined on $I \times \Omega$, where Ω is the sample space of lower-order probabilities, and I is a finite set indexing all possible probability distributions P_i (i.e., the sample space for the second-order probability Q). This follows from the principle that the value of the first-order probability $P(\omega)$ must be equal to the expectation of the second-order probability applied to first-order probabilities [1660]:

$$P(\omega) = \sum_{i \in I} Q(P_i) P_i(\omega) = E[P_i(\omega)],$$

combined with the use of expected utilities for decision making purposes.

Both the case in which higher-order probabilities are of the same kind as lower-order ones and that in which lower-order probabilities are construed as frequencies while higher-order probabilities are construed as subjective degrees of belief were considered in [1087]. In neither case, Kyburg claimed, do higher-order probabilities appear to offer any conceptual or computational advantage.

6.4.4 Fung and Chong's metaprobability

A work by Fung and Chong [654] discussed second-order probability (or, in these authors' language, *metaprobability theory*) as a way to provide soft or hard constraints on beliefs in much the same manner as the Dempster–Shafer theory provides constraints on probability masses on subsets of the state space. As Fung and Chong point out, second-order probabilities lack practical motivations for their use, while 'methodological issues are concerned mainly with controlling the combinatorics of metaprobability state spaces.'

In these authors' work, the metaprobabilistic updating of beliefs is still based on Bayes' rule, namely

$$p^2(p|D, Pr) \propto p^2(D|p, Pr) \cdot p^2(p|Pr),$$

where D is the data (evidence) and Pr a prior on space of (first-order) probability distributions p.

6.5 Fuzzy theory

The concept of a *fuzzy set* was introduced by Lotfi Zadeh [2083] and Dieter Klaua [973] in 1965 as an extension of the classical notion of a set.

While in classical set theory an element either belongs or does not belong to the set, fuzzy set theory allows a more gradual assessment of the membership of elements in a set. The degree of membership is described by a 'membership function', a function from the domain of the set to the real unit interval [0, 1]. Zadeh later introduced *possibility theory* as an extension of fuzzy set theory, in order to provide a graded semantics for natural language statements.[56] The theory was later developed further thanks to the contributions of Didier Dubois [531] and Henri Prade. Indeed, possibility measures are also the basis of a mathematical theory of partial belief.

6.5.1 Possibility theory

Definition 98. *A* possibility measure *on a domain Θ is a function $Pos : 2^\Theta \rightarrow [0,1]$ such that $Pos(\emptyset) = 0$, $Pos(\Theta) = 1$ and*

$$Pos\left(\bigcup_i A_i\right) = \sup_i Pos(A_i)$$

for every family of subsets $\{A_i | A_i \in 2^\Theta, i \in I\}$, where I is an arbitrary set index.

Each possibility measure is uniquely characterised by a *membership function* or *possibility distribution* $\pi : \Theta \rightarrow [0,1]$ such that $\pi(x) \doteq Pos(\{x\})$ via the formula

$$Pos(A) = \sup_{x \in A} \pi(x).$$

The dual quantity $Nec(A) = 1 - Pos(A^c)$ is called the *necessity measure*.

Many studies have pointed out that necessity measures coincide in the theory of evidence with the class of consonant belief functions. By condition 4 of Proposition 16, we obtain the following result.

Proposition 42. *The plausibility function Pl associated with a belief function Bel on a domain Θ is a possibility measure iff Bel is consonant, in which case its membership function coincides with Bel's contour function: $\pi = pl$. Equivalently, a belief function Bel is a necessity measure iff Bel is consonant.*

Possibility theory (in the finite case) is then embedded in the ToE. The points of contact between the evidential formalism in the implementation of the transferable belief model and possibility theory was (briefly) investigated in [1679].

Related work Authors such as Heilpern [816], Yager [2018, 2007], Palacharla [1381], Romer [1500], Kreinovich [1071] and many others [1424, 707, 598, 1167, 632, 1580] have studied the connection between fuzzy and Dempster–Shafer theory.

An early publication by Ferier [402], for instance, focused on the interpretation of membership functions of fuzzy sets in terms of plausibility and belief. Around the same time, a highly technical paper by Goodman [707] explored the mathematical relationship between fuzzy and random sets, showing that the membership function of any fuzzy subset of a space is the common 'one-point coverage function' of an equivalence class of (in general, infinitely many) random subsets of that space.

---

[56] `http://www.scholarpedia.org/article/Possibility_theory`.

Fuzzy sets can therefore be seen as equivalence classes of random sets. In 1985, Wang [1909] gave a necessary and sufficient condition for a mapping from an arbitrary non-empty class of subsets of Θ into the unit interval $[0,1]$ to be extended to a possibility measure on its power set 2^Θ and, analogously, a necessary and sufficient condition for extending it to a consonant belief function.

Shafer himself studied the connection between the theories of belief functions and possibility measures in [1596], showing how both use compatibility relations linking different frames of discernment, with his 'canonical examples' as the background. Yager used in [2031] the Dempster–Shafer framework to provide a machinery for including randomness in the fuzzy-systems modelling process in fuzzy logic control. In [1580] the theory of evidence, in its probabilistic interpretation, was adopted to justify and construct methods for interval and fuzzy number comparison.

In 1990, Smets [1679] highlighted the difference between degrees of possibility (or necessity) and degrees of belief (or plausibility), the idea being that degrees of possibility and of necessity are extensions of the modal concepts of possibility and necessity. Ten years later [1710], he provided a semantics for possibility measures based, once again, on the transferable belief model. The conjunctive combination of two possibility measures was shown to correspond to the 'hyper-cautious' conjunctive combination of the belief functions induced by the possibility measures.

Fuzzy set connectives were expressed by Dubois et al. as combinations of belief structures in [550]. The same authors [521] later compared belief function and possibility theories on the problem of assessing the value of a candidate, viewed as a multiple combination problem.

As recently as 2012, Feng et al. [598] studied the reduction of a fuzzy covering and the fusion of multiple fuzzy covering systems based on evidence theory and rough set theory. A pair of belief and plausibility functions was defined, and the reduction of a fuzzy covering based on these functions was presented.

6.5.2 Belief functions on fuzzy sets

Possibility measures are equivalent, then, to consonant belief functions. In addition, belief functions can be generalised to assume values on fuzzy sets, rather than on traditional 'crisp' ones [1670, 2063, 153, 447]. Following Zadeh's work, Ishizuka et al. [872], Ogawa and Fu [1364], Yager [2000], Yen [2062, 2064] and recently Biacino [153] extended Dempster–Shafer theory to fuzzy sets by defining an appropriate measure of inclusion between fuzzy sets. Zhang [749] also separately introduced the concepts of possibility measures and consonant belief functions on fuzzy sets, obtaining results analogous to those of Wang [1909].

Implication operators As noted by Romer and Candell [1498], in the generalisation of belief theory to fuzzy sets a central notion is played by the notion of a fuzzy *implication operator*.

A belief measure can be defined on fuzzy sets as follows:

$$Bel(A) = \sum_{B \in \mathcal{M}} \mathcal{I}(B \subseteq A) m(B), \tag{6.13}$$

where $\mathcal{M}$ is the collection of all fuzzy subsets of Θ, A and B are two such fuzzy subsets, m is a mass function, defined this time on the collection of fuzzy subsets on Θ (rather than the collection of crisp subsets, or power set), and $\mathcal{I}(B \subseteq A)$ is a measure of how much the fuzzy set B is included in the fuzzy set A.

In fact, defining the notion of inclusion for fuzzy sets is not trivial – various measures of inclusion can be and have been proposed. Just as a fuzzy set is completely determined by its membership function, different measures of inclusion between fuzzy sets are associated with a function $I : [0,1] \times [0,1] \to [0,1]$ (a fuzzy implication operator), from which one can get [1965]

$$\mathcal{I}(A \subseteq B) = \bigwedge_{x \in \Theta} I(A(x), B(x)),$$

where $A(x)$ and $B(x)$ are the degrees of membership of element x in the fuzzy sets A and B, respectively. Among the most popular implication operators, we can cite that of Lukasiewicz, $I(x,y) = \min\{1, 1 - x - y\}$, proposed by Ishizuka [872], and that of Kleene-Dienes, $I(x,y) = \max\{1 - x, y\}$, the latter supported by Yager [2000].

Yen's linear optimisation problem In one of the most significant contributions to the issue, Yen [2064] showed that computing the degree of belief in a hypothesis in Dempster–Shafer theory can be formulated as an optimisation problem, opening the way to extending the notion of belief function by generalising the objective function and the constraints of such an optimisation problem.

Dempster's rule was extended in [2064] by (1) combining generalised compatibility relations based on possibility theory, and (2) normalising the combination results to account for partially conflicting evidence.

Definition 99. ([2064], Definition 1) *A generalised (fuzzy) compatibility relation between Ω and Θ is a fuzzy relation $C : 2^{\Omega \times \Theta} \to [0,1]$ that represents the joint possibility distribution of the two domains, i.e., $C(\omega, \theta) = \Pi_{X,Y}(\omega, \theta)$, where X and Y are variables whose ranges are Ω and Θ, respectively.*

Since, as we know, belief functions are lower envelopes of the sets of probability values induced by redistributing the mass of each focal element to its singleton elements, a belief value $Bel(A)$ can be viewed as the optimal solution to the following linear programming problem:

$$\min \sum_{x \in A} \sum_{B} m(x : B) \tag{6.14}$$

subject to

$$m(x : B) \geq 0 \;\forall x, B, \quad m(x : B) = 0 \;\forall x \notin B, \quad \sum_{x} m(x : B) = m(B) \;\forall B,$$

where $m(x : B)$ denotes the mass assigned to $x \in B$ by an admissible redistribution process. A similar maximisation problem will yield the corresponding plausibility value $Pl(A)$.

Yen modified the objective function (6.14) to account for the membership function μ_A of each (fuzzy) subset A, thus generalising the notion of belief values to fuzzy sets, namely

$$\min \sum_{x \in A} \sum_{B} m(x : B) \cdot \mu_A(x),$$

whose optimal solution is obtained by assigning all the mass of B to one of its elements with the lowest degree of membership in A.

Recent work Although these extensions of belief theory all arrive at frameworks within which both probabilistic and vague information can be handled, they are all restricted to finite frames of discernment. Moreover, it is unclear whether or not the belief and plausibility functions so obtained satisfy subadditivity (or, respectively, superadditivity) (3.6) in the fuzzy environment. Biacino [153] thus studied fuzzy belief functions induced by an infinitely monotone inclusion and proved that they are indeed lower probabilities.

Separately, Wu et al. [1965] recently developed a theory of fuzzy belief functions on infinite spaces. A fuzzy implication operator is said to be *semicontinuous* if

$$I(\vee_j a_j, \wedge_k b_k) = \bigwedge_{j,k} I(a_j, b_k).$$

Their main result then reads as follows ([1965], Theorem 10).

Proposition 43. *Let Θ be a non-empty set that may be infinite and I a semicontinuous implicator on $[0,1]$. If $Bel, Pl : \mathcal{F}(\Theta) \rightarrow [0,1]$ are the fuzzy belief and plausibility functions induced by a fuzzy belief structure $(\mathcal{M}, m)$, then Bel is a fuzzy monotone Choquet capacity of infinite order on Θ and Pl is a fuzzy alternating Choquet capacity of infinite order there.*

6.5.3 Vague sets

In vague set theory [669], a *vague set* A is characterised by a truth-membership function $t_A(x)$ and a false-membership function $f_A(x)$ for all elements $x \in U$ of a universe of discourse U, further generalising classical fuzzy set theory, as $t_A(x)$ is a lower bound on the grade of membership of x generated by the evidence in favour of x, and $f_A(x)$ is a lower bound on the negation of x derived from the evidence against x. In general,

$$t_A(x) + f_A(x) \leq 1,$$

where the gap $1 - (t_A(x) + f_A(x))$ represents our ignorance about the element x, and is zero for traditional fuzzy sets. Basically, a vague set is a pair of lower and upper bounds on the membership function of a fuzzy set.

The authors of [1151] showed that belief theory is a special case of vague set theory [669], as the belief value of a proposition and the grade membership of an element in a vague set share a similar form. When the elements in a vague set are subsets of a total set (frame of discernment), when $x \in U \leftrightarrow A \subset \Theta$ and when the grade membership of the subsets is redefined according to Dempster–Shafer theory, $t_A(x) = Bel(A)$ and $f_A(x) = Pl(A)$, vague set theory and belief theory coincide. In the present author's view, however, the analysis conducted in [1151] is simplistic and possibly flawed.

Intuitionistic fuzzy sets Atanassov's *intuitionistic fuzzy sets* [62], an approach mathematically equivalent to vague sets (see above), but formulated earlier, can also be interpreted in the framework of belief theory, so that all mathematical operations on intuitionistic fuzzy values can be represented as operations on belief intervals, allowing us to use Dempster's rule of combination to aggregate intuitionistic fuzzy values for decision making [560].

6.5.4 Other fuzzy extensions of the theory of evidence

Numerous other fuzzy extensions of belief theory have been proposed [2047, 871]. Constraints on belief functions imposed by fuzzy random variables, for instance, were studied in [1499, 1042]. Fuzzy evidence theory was used for decision making in [2052].

We briefly list the other major proposals below.

Zadeh's extension Zadeh [2090] was maybe the first to propose the extension of evidence theory to fuzzy sets. Zadeh's model is a generalisation of Dempster's idea of belief functions induced by multivalued mappings.

A body of evidence is represented by the probability distribution P_X of a random variable X with values in $\Omega = \{\omega_1, \ldots, \omega_n\}$, and a possibility distribution $\Pi_{Y|X}$ of Y (with values in Θ) given X. For a given value $\omega_i \in \Omega$, the latter defines a fuzzy set F_i of Θ such that $\Pi_{Y|X=\omega_i} = F_i$. The probability distribution of X induces on Θ a belief function m whose focal elements are the fuzzy sets F_i, $i = 1, \ldots, n$ (called a *fuzzy belief structure* by Yager [2000]), and $m(F_i) = P_X(\omega_i)$ for all $\omega_i \in \Omega$.

Subsequently [2031], Yager [2006, 2018] and Filev proposed a combined fuzzy-evidential framework for fuzzy modelling. In another paper [2017], Yager investigated the issue of normalisation (i.e., the assignment of non-zero values to empty sets as a consequence of evidence combination) in the fuzzy Dempster–Shafer theory of evidence, proposing in response a technique which he called 'smooth normalisation'.

Denoeux's fuzzy belief structures Denoeux introduced in [447] the concepts of *interval-valued* (already recalled in Section 3.3.7) and *fuzzy-valued* belief structures, defined, respectively, as crisp and fuzzy sets of belief structures satisfying hard or elastic constraints.

Definition 100. A fuzzy-valued belief structure *is a normal fuzzy subset* $\tilde{\mathrm{m}}$ *of* $\mathcal{M}_\Theta$ *(the set of all mass functions on* Θ*) such that there exist* n *fuzzy subsets* $F_1, \ldots, F_n$ *of* Θ*, and* n *non-null fuzzy numbers* $\tilde{m}_i$*,* $i = 1, \ldots, n$*, with supports from* $[0, 1]$*, such that, for every* $m \in \mathcal{M}_\Theta$*,*

$$\mu_{\tilde{\mathbf{m}}}(m) \doteq \begin{cases} \min_{i=1,\ldots,n} \mu_{\tilde{m}_i}(m(F_i)) & \sum_i m(F_i) = 1, \\ 0 & \textit{otherwise}, \end{cases}$$

where $\mu_{\tilde{m}_i}$ *denotes the membership function of the fuzzy number* $\tilde{m}_i$.

Various concepts of Dempster–Shafer theory, including belief and plausibility functions, combination rules, and normalisation procedures were generalised [447]. Most calculations implied by the manipulation of these concepts are based on simple forms of linear programming problems for which analytical solutions exist, making the whole scheme computationally tractable.

Fuzzy conditioned Dempster–Shafer (FCDS) theory Mahler [1235] formulated a *fuzzy conditioned Dempster–Shafer* theory as a probability-based calculus for dealing with imprecise and vague evidence. The theory uses a finite-level Zadeh fuzzy logic and a Dempster-like combination operator, which is 'conditioned' to reflect the influence of any a priori knowledge which can be modelled by a belief measure on finite-level fuzzy sets. This author showed that the FCDS theory is grounded in the theory of random fuzzy sets, and is a generalisation of Bayesian theory to the case in which both evidence and a priori knowledge are imprecise and vague.

Lucas and Araabi's fuzzy-valued measure In [1223], Lucas and Araabi proposed their own generalisation of Dempster–Shafer theory [2062, 2064] to a fuzzy-valued measure, obtaining functions which are all equivalent to one another and to the original fuzzy body of evidence.

Fuzzy interval-valued belief degrees Finally, Aminravan et al. [37] introduced an extended fuzzy Dempster–Shafer scheme based on the simultaneous use of fuzzy interval-grade and interval-valued belief degrees (referred to as IGIB). The latter facilitates the modelling of uncertainties in terms of local ignorance associated with expert knowledge, whereas the former allows us to handle the lack of information about belief degree assignments. Generalised fuzzy sets can also be readily transformed into the proposed fuzzy IGIB structure.

6.6 Logic

Many generalisations of classical logic exist in which propositions are assigned probability values[57] [1358], rather than truth values (0 or 1). As belief functions naturally generalise probability measures, it is quite natural to define non-classical logic frameworks in which propositions are assigned belief, rather than probability, values. The theoretical basis for using probability theory as a general logic of plausible inference is given by Cox's theorem [1842]. The latter states that any system for plausible reasoning that satisfies certain qualitative requirements intended to ensure

[57] https://en.wikipedia.org/wiki/Probabilistic_logic.

consistency with classical deductive logic and correspondence with common-sense reasoning is isomorphic to probability theory. Nevertheless, as pointed out by Van Horn [1842], the requirements used to obtain this result are debatable.

Important work on the logical foundations of belief theory was presented, in particular, by Ruspini [1510, 1511], Saffiotti [1526], Josang [918] and Haenni [767], among others. A great many other logic-based frameworks have been proposed [1503, 77, 1510, 2116, 1513, 124, 1442, 408, 730, 1388, 1842, 786, 775, 776, 136, 2087, 43, 862, 787, 762, 1337, 1326, 2128, 1525, 1274, 1198, 121, 777, 264, 1935, 279, 282, 280], so many that it would be impossible to review them all here.

6.6.1 Saffiotti's belief function logic

In propositional logic, propositions or formulae are either true or false, i.e., their truth value is either 0 or 1 [1267]. Formally, an *interpretation* or *model* of a formula ϕ is a valuation function mapping ϕ to the truth value 'true' (i.e., 1). Each formula can therefore be associated with the set of interpretations or models (or 'hyper-interpretations' [1526]) under which its truth value is 1. If we define a frame of discernment formed by all possible interpretations, each formula ϕ is associated with the subset $A(\phi)$ of this frame which collects together all its interpretations. If the available evidence allows one to define a belief function (or 'bf-interpretation' [1526]) on this frame of possible interpretations, each formula $A(\phi) \subseteq \Theta$ is then naturally assigned a degree of belief $Bel(A(\phi))$ between 0 and 1 [1526, 767], measuring the total amount of evidence supporting the proposition 'ϕ is true'.

Alessandro Saffiotti, in particular, built in 1992 a hybrid logic attaching belief values to classical first-order logic, which he called *belief function logic* (BFL) [1526], giving new perspectives on the role of Dempster's rule. Many formal properties of first-order logic directly generalise to BFL. Formally, BFL works with formulae of the form $F : [a, b]$, where F is a sentence of a first-order language, and $0 \leq a \leq b \leq 1$. Roughly speaking, a is the degree of belief that F is true, and $(1 - b)$ the degree of belief that F is false.

Definition 101. *A belief function Bel is a* bf-model *of a belief function (bf-)formula* $F : [a, b]$ *iff* $Bel(F) \geq a$ *and* $Bel(F) \leq b$.

Definition 102. *A bf-formula* Φ bf-entails *another bf-formula* Ψ *iff every bf-model of* Φ *is also a bf-model of* Ψ.

A number of properties of bf-entailment were proved in [1526]. Given a set of bf-formulae, a *D-model* for the set can be obtained by Dempster-combining the models of the individual formulae. Saffiotti proved that if the set of formulae is coherent (in a first-order-logic sense), then it is D-consistent, i.e., the combined model assigns zero mass to the empty set (in which case it is the least informative bf-model for the set of formulae).

In related work, Andersen and Hooker [43] proved probabilistic logic and Dempster–Shafer theory to be instances of a certain type of linear programming model, with exponentially many variables.

6.6.2 Josang's subjective logic

In a number of publications [918, 914, 922], Audun Josang laid the foundations of a *subjective logic* [918] for uncertain probabilities, much related to the notion of belief function, and compatible with binary logic and probability calculus. Some elements of this framework were presented when we discussed the associated 'consensus' operator (Chapter 4, Section 4.3.2).

There, the term *opinion* was used to denote the representation of a subjective belief, and a set of operations for combining opinions were proposed [914], namely propositional conjunction ([918], Theorem 3), and propositional disjunction ([918], Theorem 4), which were beliefs about the conjunction and disjunction, respectively, of distinct propositions, both commutative and associative. Propositional conjunction and disjunction are equivalent to the AND and OR operators of Baldwin's support logic [77], except for the relative atomicity parameter, which is absent in Baldwin's logic. The negation of an opinion can also be defined. Multiplication and comultiplication of beliefs as generalisations of multiplication and comultiplication of probabilities, as well as of binary-logic AND and OR, were investigated in [922].

In summary, subjective logic is an extension of standard logic that uses continuous uncertainty and belief parameters instead of discrete truth values. It can also be seen as an extension of classical probability calculus by using a second-order probability representation instead of the standard first-order representation. In addition to the standard logical operations, subjective logic contains some operations specific to belief theory, such as consensus and recommendation.

6.6.3 Fagin and Halpern's axiomatisation

Fagin and Halpern [586] designed a language for reasoning about probability which allows us to make statements such as 'the probability of E_1 is less than 1/3', or 'the probability of E_1 is at least twice the probability of E_2', where E_1 and E_2 are arbitrary events.

They considered both the case in which all events are measurable, and the more general case in which they may not be measurable. While the measurable case is essentially a formalisation of Nilsson's probabilistic logic [1358], they argued that the general (non-measurable) case corresponds precisely to replacing probability measures with belief functions. They provided a complete axiomatisation for both cases, and showed that the problem of deciding satisfiability is NP-complete, no worse than that of propositional logic.

In [786], Halpern introduced a logic for reasoning about evidence that essentially views evidence as a function from prior beliefs (before making an observation) to posterior beliefs (after making the observation). In [787], he presented a propositional logic for reasoning about expectation, where the semantics depends on the underlying representation of uncertainty. Complete axiomatisations for this logic were given in the case of belief functions, among others.

6.6.4 Probabilistic argumentation systems

A series of publications by Haenni and Lehmann [44, 771, 1125, 767] introduced the concept of *probabilistic argumentation systems* (PASs), in which uncertainty is expressed using so-called *assumptions* [1125].

The main goal of assumption-based reasoning is to determine the set of all supporting arguments for a given hypothesis. The assignment of probabilities to assumptions leads to the framework of probabilistic argumentation systems and allows an additional quantitative judgement about a given hypothesis. Rather than computing first the set of supporting arguments and then deriving the desired degree of support for the hypothesis, Lehmann [1125] avoided symbolic computations right away thanks to the fact that the sought degree of support corresponds to the notion of normalised belief in Dempster–Shafer theory. A probabilistic argumentation system can then be transformed into a set of independent mass functions. In Lehmann's thesis [1125], a formal language called ABEL [44] was also introduced, which allows one to express probabilistic argumentation systems in a convenient way. As shown in [771], PASs can be seen as a powerful modelling language to be put on top of belief theory, while Dempster–Shafer theory can be used as an efficient computational tool for numerical computations in probabilistic argumentation systems.

A new perspective on the relationship between logic and probability was introduced in [767], which sees logical and probabilistic reasoning as merely two opposite, extreme cases of one and the same universal theory of reasoning: probabilistic argumentation.

6.6.5 Default logic

In a couple of interesting papers [1939, 1945], Nic Wilson developed a new version of Reiter's *default logic* [1481] which is computationally much simpler, and can be shown to be a limiting case of a Dempster–Shafer framework.

Default logic is a logic for reasoning with default rules, i.e., rules which hold *in general* (by default) but not always, as they may have exceptions. A *default rule* is a rule of the form 'if we know a, then deduce c, as long as b is consistent with our current beliefs', or '$a : b/c$' for short. Wilson's results demonstrate a strong connection between two apparently very different approaches to reasoning with uncertainty, and open up the possibility of mixing default and numerical rules within the same framework. Indeed, in Wilson's work, Dempster–Shafer theory is a formally justified as well as computationally efficient way of representing if–then rules [1939]. For example, simple implications of the form 'if a then c, with reliability α' can be realised in belief theory by allocating mass α to the implication $a \rightarrow c$, and $1 - \alpha$ to the tautology, an interpretation that was, however, criticised by Pearl. Nevertheless, Wilson argued, since an uncertain rule can be thought of as approximating a certain rule whose antecendent is unknown, the former can be interpreted as the statement 'we know that $a \rightarrow c$ is true in a proportion α of worlds'.

Separately, Benferhat, Saffiotti and Smets [125, 124] proposed an approach to dealing with default information in the framework of belief theory based on *epsilon-*

belief assignments, structures inspired by Adams's epsilon semantics [10, 1361] in which mass values are either (infinitesimally) close to 0 or close to 1. Let $\mathbb{E} = \mathbb{E}_0 \cup \mathbb{E}_1 \cup \{0\} \cup \{1\}$ be the set containing the values 0, 1 and the quantities infinitesimally close to the two. An *infinitesimal* $\epsilon \in \mathbb{E}_0$ can be formally defined as a real continuous function from $(0, 1)$ to $(0, 1)$ such that $\lim_{\eta \to 0} \epsilon(\eta) = 0$ and $\lim_{\eta \to 0} \epsilon(\eta)/\eta^k \in \mathbb{R} \setminus \{0\}$ for some non-negative integer k.

Definition 103. *Let Θ be a finite non-empty set. An* ϵ-mass assignment *on Θ is a function $m_{\mathcal{E}} : 2^\Theta \to \mathbb{E}$ such that, for all $\eta \in (0, 1)$,*

- $m_{\mathcal{E}}(\emptyset)(\eta) = 0$;
- $\sum_{A \subseteq \Theta} m_{\mathcal{E}}(A)(\eta) = 1$.

Benferhat et al. showed that these structures can be used to give a uniform semantics to several popular non-monotonic systems, including Kraus, Lehmann and Magidor's system **P**, Goldszmidt and Pearl's $\mathbf{Z}^+$ [701], Brewka's preferred subtheories, Geffner's conditional entailment, Pinkas's penalty logic [1434], possibilistic logic, and the lexicographic approach.

6.6.6 Ruspini's logical foundations

A pair of foundational papers by Ruspini [1510, 1513] developed, in 1986–1987, a generalisation of Carnap's approach [228] to the development of logical bases for probability theory [2116], using formal structures that are based on *epistemic logic*,[58] a form of modal logic (see Section 6.6.7) introduced to deal with the knowledge state that rational agents hold about the real world.

The method rests on an enhanced notion of a 'possible world', which incorporates descriptions of knowledge states. An interpretation W for a sentence space S is a mapping from S to $\{T, F\}$. An interpretation is called a 'possible world' if it satisfies a number of axioms ([1510], II-2-2), equivalent to the modal logic system S5. We define K as a unary operator which represents the *knowledge* an agent has about the truth value of the sentence S.

Definition 104. *Two possible worlds W_1, W_2 for the sentence space S are said to be* epistemically equivalent, *$W_1 \sim W_2$, if, for any sentence $\mathcal{E}$, the sentence $K\mathcal{E}$ is true in W_1 iff it is also true in W_2.*

In other words, the two worlds are equivalent if they carry the same knowledge about the truth values of the sentences.

Definition 105. *The quotient space $\mathcal{U}(S)$ of all possible worlds for S under epistemic equivalence $\sim$ is called the* epistemic space *of S, and its members (equivalent classes) are called* epistemic states.

Within such 'epistemic' universes, classes of sets called *support sets* are identified as those that map a given sentence $K\mathcal{E}$ to the truth value T. The *epistemic*

[58] https://en.wikipedia.org/wiki/Epistemic_modal_logic.

algebra of $\mathcal{U}(S)$ is the smallest subset algebra of $\mathcal{U}(S)$ which contains the class of all support sets, and has the structure of a σ-algebra.

Probabilities defined over epistemic algebras were shown by Ruffini to be equivalent to belief and basic probability functions. It was also shown that any extensions of a probability function defined on an epistemic algebra (representing different states of knowledge) to a truth algebra (representing true states of the real world) must satisfy the interval probability bounds of Dempster–Shafer theory.

In [1510, 1513], the problem of combining the knowledge of several agents was also addressed, resulting in an 'additive combination' formula

$$m(p) = \kappa \sum_{p_1 \wedge p_2 = p} P(e_1(p_1) \cap e_2(p_2)), \tag{6.15}$$

where $e_1(p), e_2(p)$ denote the epistemic sets for the proposition (sentence) p associated with two epistemic operators K_1, K_2, and P is a probability defined over the epistemic algebra. Under appropriate independence assumptions, (6.15) obviously generalises Dempster's rule of combination.

In a closely related paper, Wilkins and Lavington [1935] also described a formal framework for uncertain reasoning in the possible-worlds paradigm, showing that the model theory of belief functions is simpler than the model theory of lower-order monotone capacities. They looked for generalisations of belief functions which admit a conjunction operator, and identified collections of belief functions that have a full propositional logic. The issue of finding a conjunction operator was, however, not completely solved.

6.6.7 Modal logic interpretation

In close relation to the work of Ruspini [1510, 1511], which is based on a form of epistemic logic, Resconi, Harmanec et al. [1484, 800, 799, 1485, 804] proposed the semantics of *propositional modal logic* as a unifying framework for various uncertainty theories, such as fuzzy set, possibility and evidential theory, establishing in this way an interpretation of belief measures on infinite sets. In opposition to Ruspini, Harmanec et al. used a more general system of modal logic, addressing the issue of the completeness of the interpretation.

Modal logic is a type of formal logic, developed primarily in the 1960s, that extends classical propositional logic to include operators expressing modality.[59] Modalities are formalised via *modal operators*. In particular, the modalities of truth include *possibility* ('It is possible that p', $\diamond p$) and *necessity* ('It is necessary that p', $\Box p$). These notions are often expressed using the idea of possible worlds: necessary propositions are those which are true in all possible worlds, whereas possible propositions are those which are true in at least one possible world.

Formally, the language of modal logic consists of a set of atomic propositions, logical connectives $\neg, \vee, \wedge, \rightarrow, \leftrightarrow$, and modal operators of possibility $\diamond$ and necessity $\Box$. Sentences or propositions of the language are of the following form:

[59] https://en.wikipedia.org/wiki/Modal_logic.

1. Atomic propositions.
2. If p and q are propositions, so are $\neg p$, $p \wedge q$, $p \vee q$, $p \rightarrow q$, $p \leftrightarrow q$, $\Box p$ and $\diamond p$.

A *standard model* of modal logic is a triplet $M = \langle W, R, V \rangle$, where W denotes a set of possible worlds, R is a binary relation on W called the *accessibility relation* (e.g. world v is accessible from world w when wRv) and V is the value assignment function $V(w,p) \in \{T, F\}$, whose output is the truth value of proposition p in world w. The accessibility relation expresses the fact that some things may be possible in one world and impossible from the standpoint of another. Different restrictions on the accessibility relation yield different classes of standard models. A standard model M is called a *T-model* if R is reflexive. The notation $\|p\|^M$ denotes the *truth set* of a proposition p (what we call above the 'hyper-interpretation'), i.e., the set of all worlds in which p is true:

$$\|p\|^M = \{w | w \in W, V(w,p) = T\}. \tag{6.16}$$

In [800, 804], a modal logic interpretation of Dempster–Shafer theory on finite universes or decision spaces Θ was proposed in terms of finite models, i.e., models with a finite set of worlds. Consider propositions e_A of the form 'a given incompletely characterised element θ is classified in set A', where $\theta \in \Theta$, $A \in 2^\Theta$.

Proposition 44. [1483] *A finite T-model $M = \langle W, R, V \rangle$ that satisfies the* singleton valuation assumption *(SVA) (one and only one proposition $e_{\{\theta\}}$ is true in each world) induces a plausibility measure Pl_M and a belief measure Bel_M on 2^Θ, defined by*

$$Bel_M(A) = \frac{|\|\Box e_A\|^M|}{|W|}, \quad Pl_M(A) = \frac{|\| \diamond e_A\|^M|}{|W|}. \tag{6.17}$$

Proposition 44 states that the belief value of A is the proportion of worlds in which the proposition 'Θ belongs to A' is considered necessary, while its plausibility is the proportion of worlds in which this is considered possible. The SVA amounts to saying that each world in model M gives its own unique answer to the classification question. Furthermore, we can state the following propositions.

Proposition 45. [800] *A finite T-model $M = \langle W, R, V \rangle$ that satisfies the SVA induces a basic probability assignment m_M on 2^Θ, defined by*

$$m_M(A) = \frac{|\|E_A\|^M|}{|W|},$$

where

$$E_A = \Box e_A \wedge \left(\bigwedge_{B \subset A} (\neg(\Box e_B)) \right).$$

Proposition 46. [800] *The modal logic interpretation of basic probability assignments introduced in Proposition 45 is complete, i.e., for every rational-valued basic probability assignment m on 2^Θ, there exists a finite T-model M satisfying the SVA such that $m_M = m$.*

In order to develop a modal logic interpretation on arbitrary universes, one needs to add to a model M a probability measure on the set of possible worlds [804].

In a series of papers by Tsiporkova et al. [1830, 1829, 1831], the modal logic interpretation of Harmanec et al. was developed further by building multivalued mappings inducing belief measures. A modal logic interpretation of Dempster's rule was also proposed there.

At about the same time as Harmanec et al., Murai et al. [1326] also presented a basis for unifying modal logic and belief theory, by proving soundness and completeness theorems for several systems of modal logic with respect to classes of newly defined belief function models. Their conclusion was that, indeed, modal-logical structure is intrinsic to Dempster-Shafer theory.

Other publications on the modal logic interpretation include [698, 84, 1485, 804].

6.6.8 Probability of provability

The Dempster–Shafer theory of evidence can be conceived as a theory of probability of provability, as it can be developed on the basis of assumption-based reasoning [1405, 1682, 137]. This interpretation was first put forward by Pearl [1405], although Ruspini [1510] had proposed a similar 'probability of knowing' concept. Conceptually, as Smets noted, the probability-of-provability approach is not different from the original framework of Dempster [415], but can better explain the origin of the rule of conditioning.

Within this approach, Besnard and Kohlas [137] modelled reasoning by consequence relations in the sense of Tarski, showing that it is possible to construct evidence theory on top of the very general logics defined by these consequence relations. Support functions can be derived, which are, as usual, set functions, monotone of infinite order. Plausibility functions can also be defined: however, as negation need not be defined in these general logics, the usual duality relations between the support and plausibility functions of Dempster–Shafer theory do not hold in general.

Hajek [777] extended the work of Resconi and Harmanec [803] to show that each belief function on a finite frame (including unnormalised ones) can be understood as a probability of provability, the latter understood as a 'modality' in the sense of Gödel–Löb modal provability logic. His work was later picked up by Esteva and Godo [579].

6.6.9 Other logical frameworks

Incidence calculus *Incidence calculus* [205] is a probabilistic logic for dealing with uncertainty in intelligent systems. Incidences are assigned to formulae: they are the logic conditions under which the formula is true. Probabilities are assigned to incidences, and the probability of a formula is computed from the sets of incidences assigned to it.

Definition 106. *An* incidence map *is a function from the set of axioms $\mathcal{A}$ of a propositional language to the power set of $\mathcal{W}$, the set of possible worlds, such that*

$$i(\phi) = \{w \in \mathcal{W} | w \models \phi\},$$

where $\phi \in \mathcal{A}$ is an axiom and $i(\phi)$ is called the incidence set *of ϕ.*

In [1205], in particular, Liu, Bundy et al. proposed a method for discovering incidences that can be used to calculate mass functions for belief functions, based on the notion of a *basic incidence assignment* ([1205], Definition 3.3).

Baldwin's evidential support logic programming In 1987, Baldwin [77] described a *support logic programming* system which generalises logic programming by incorporating various forms of uncertainty. In this system a conclusion does not logically follow from some axioms, but is supported to a certain degree by means of evidence (as in other logical frameworks discussed earlier). The negation of the conclusion is also supported to a certain degree, and the two supports do not necessarily add up to one.

A calculus for such a support logic programming system was described in [77].

Provan's analysis In [1442], Dempster–Shafer theory was formulated by Provan (1990) in terms of propositional logic using the notion of provability, via the tuple (Σ, ρ), where Σ is a set of propositional clauses and ρ is an assignment of mass to each clause $\Sigma_i \in \Sigma$. It was shown that the disjunction of minimal support clauses for a clause Σ_i with respect to a set Σ of propositional clauses, when represented in terms of symbols for the $\rho(\Sigma_i)$s, corresponds to a symbolic representation of the belief function for Σ_i. Belief function combination by Dempster's rule then corresponds to a combination of the corresponding support clauses.

In [1442], the disjointness of the Boolean formulae representing belief functions was shown to be necessary. Because of intractability even for moderately sized problem instances, efficient approximation methods for Dempster–Shafer computations were proposed.

Penalty logic *Penalty logic* [1434], introduced by Pinkas, associates each formula of a knowledge base with a price to be paid if that formula is violated. Penalties may be used as a criterion for selecting preferred consistent subsets in an inconsistent knowledge base, thus inducing a non-monotonic inference relation.

A precise formalisation and the main properties of penalty logic and its associated non-monotonic inference relation were given by Dupin, Lang and Schiex in [408], where they also showed that penalty logic and Dempster–Shafer theory are related, especially in the infinitesimal case.

Other contributions A number of other logical frameworks have been proposed for working with belief functions, so many that it would be impossible to assign to each to them the space they deserve. We shall just briefly recall them here.

Grosof [730] investigated a family of probabilistic logics, generalising the work of Nilsson [1358]. He developed a variety of logical notions for probabilistic reasoning, including soundness, completeness and convergence, and showed that Dempster–Shafer theory is formally identical to a special case of his generalised probabilistic logic. A similar analysis was conducted by Chen [264]. In his note on the Dutch book method [1388], Paris considered generalising the classical Dutch book argument for identifying degree of belief with probability to yield corresponding analogues of probability functions for modal, intuitionistic and paraconsistent logics. Hjek [775, 776] surveyed the topic of the logical analysis of reasoning under uncertainty and vagueness, using many-valued and modal logics and their generalisations, with an emphasis on the distinction between degrees of belief and degrees of truth. Bertschy and Monney [136] considered the following problem from reliability theory. Given a disjunctive normal form $\phi = \phi_1 \wedge \cdots \wedge \phi_r$, we want to find a representation of ϕ in disjoint formulae, simple enough that the computation of their probabilities is a simple task. The problem also appears in the calculation of degrees of support in the theory of evidence [1027]. A new method to solve this problem was then presented in [136]. Zadeh [2087] investigated a generalisation of syllogistic reasoning in which numerical quantifiers assume the role of probabilities, showing that six basic syllogisms are sufficient to provide a systematic framework for the computation of certainty factors. Interestingly, Andersen and Hooker [43] showed that several logics of uncertainty, including belief function theory, are instances of a certain type of linear programming model, typically with exponentially many variables. Hunter [862] compared evidence combination in Dempster–Shafer theory with the combination of evidence in probabilistic logic. A minimal set of sufficient conditions for these two methods to agree were stated.

Laurence Cholvy wrote a series of papers on the logical interpretation of the theory of evidence. In 2000 [279], she tried to provide people used to manipulating logical formalisms with a key for understanding the theory of evidence and its use in multisensor data fusion, by giving a logical interpretation of this formalism when the numbers are rational. In [280], she presented another logical interpretation of masses in which the degree of belief for a proposition A is interpreted as the proportion, in a given set, of proofs of A (as opposed to $Bel(A)$ being viewed as the probability that an agent knows (or can prove) A). As this new interpretation, however, cannot correctly explain Dempster's rule, a new rule of combination which fits this interpretation was proposed.

Benavoli et al. [118] explored uncertain implication rules, and presented a transformation for converting uncertain implication rules into an evidence theory framework, showing that it satisfies the main properties of logical implication, namely reflexivity, transitivity and contrapositivity.

Narens's book [1337] examined the logical and qualitative foundations of probability theories, including belief theory. Rather than using event spaces formed by Boolean algebras, as in classical propositional logic, the analysis was conducted in a different event algebra closely linked to intuitionistic logic. Qing [2128] proposed a general framework for uncertainty reasoning based on Dempster–Shafer theory in the context of logic calculus. Saffiotti coupled belief theory with *knowledge representation* in [1525], centred around the notion of 'Dempster–Shafer belief bases', abstract data types representing uncertain knowledge that use belief function theory for representing degrees of belief about our knowledge, and the linguistic structures of an arbitrary knowledge representation system for representing the knowledge itself.

Nilsson's entailment scheme for probabilistic logic [1358], which can predict the probability of an event when the probabilities of certain other connected events are known, and uses a maximum entropy method proposed by Cheeseman [257], was extended by McLeish in [1274]. This new entailment scheme for belief functions overcomes the problem with Nilsson's scheme, that only vectors associated with 'consistent' worlds can be used. Bharadwaj et al. [147] addressed the problem of reasoning under time constraints with incomplete and uncertain information, based on the idea of *variable precision logic*, introduced by Michalski and Winston. The latter suggested the *censored production rule* (CPR) as an underlying representational and computational mechanism to enable logic-based systems to exhibit variable precision. As an extension of CPR, a *hierarchical censored production rules* system of knowledge representation was put forward which exhibits both variable certainty and variable specificity [147].

Finally, Kroupa [1076] generalised belief functions to formulae expressed in Lukasiewicz logic.

6.7 Rough sets

First described by the Polish computer scientist Zdzisaw I. Pawlak, rough sets [1393] are a very popular mathematical description of uncertainty, which is strongly linked to the idea of a partition of the universe of hypotheses. They provide a formal approximation of a crisp set (i.e., a traditional set) in terms of a pair of sets which give a lower and an upper approximation of the original one.

The concept of a rough set [1396] overlaps, to some extent, with that of a belief function [1658]. According to Pawlak et al. [1397], the main difference is that whereas Dempster–Shafer theory uses belief functions as its main tool, rough set theory makes use of lower and upper approximations of sets, without requiring any additional information (e.g. in the form of a membership function or a basic probability assignment).

6.7.1 Pawlak's algebras of rough sets

Let Θ be a finite universe, and let $\mathcal{R} \subseteq \Theta \times \Theta$ be an equivalence relation which partitions it into a family of disjoint subsets $\Theta/\mathcal{R}$, called *elementary sets*. We can then introduce *definable* or *measurable* sets $\sigma(\Theta/\mathcal{R})$ as the unions of one or more elementary sets, plus the empty set $\emptyset$.

The *lower approximation* $\underline{apr}(A)$ of $A \subset \Theta$ makes use of the definable elements whose equivalence classes are contained in A. Dually, its *upper approximation* $\overline{apr}(A)$ is composed of the definable elements whose equivalence classes have a non-empty intersection with A. Lower and upper approximations are pictorially described in Fig. 6.4.

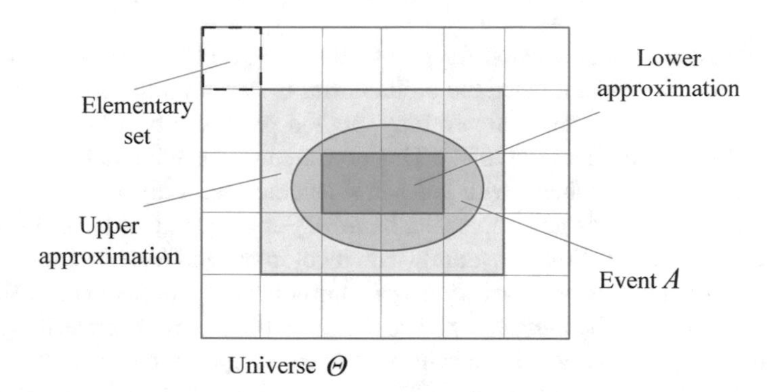

Fig. 6.4: Lower and upper approximations by elementary sets in rough set theory.

6.7.2 Belief functions and rough sets

The relationship between belief functions and rough set algebras was studied by Yao and Lingras [2057]. Several highly cited papers focused on this topic [1166, 1968, 1966, 1395, 1182, 2107, 2056, 1005, 1642]. An early contribution in the area (1998) was made by Kopotek [1005], who presented a view of belief functions as measures of diversity in relational databases, in which Dempster's rule corresponds to the join operator of relational database theory.

In a Pawlak rough set algebra, the *qualities* of the lower and upper approximations $\underline{apr}(A)$ and $\overline{apr}(A)$ of a subset $A \subseteq \Theta$ are defined as

$$\underline{q}(A) \doteq \frac{|\underline{apr}(A)|}{|\Theta|}, \quad \overline{q}(A) \doteq \frac{|\overline{apr}(A)|}{|\Theta|},$$

and measure the fraction of definable elements, over all those definable on Θ, involved in the two approximations. This recalls Harmanec's modal logic interpretation of belief functions (6.17). Indeed, it can be proven that Pawlak's rough set algebra corresponds to the modal logic S5 [2053], in which the lower and upper approximation operators correspond to the necessity and possibility operators. Furthermore, we have the following result.

Proposition 47. *The quality of the lower approximation $\underline{q}$ is a belief function, with basic probability assignment $m(E) = |E|/|\Theta|$ for all $E \in \Theta/\mathcal{R}$, and 0 otherwise.*

One issue with this interpretation is that belief and plausibility values obviously need to be rational numbers. Therefore, given an arbitrary belief function Bel, it may not be possible to build a rough set algebra such that $\underline{q}(A) = Bel(A)$. The following result establishes a sufficient condition under which this is possible.

Proposition 48. *Suppose Bel is a belief function on Θ with mass m such that:*

1. *The set of focal elements of Bel is a partition of Θ.*
2. *$m(A) = |A|/|\Theta|$ for every focal element of Bel.*

Then there exists a rough set algebra such that $\underline{q}(A) = Bel(A)$.

A more general condition can be established via inner and outer measures (Chapter 4, Section 3.1.3). Given a σ-algebra $\mathcal{F}$ of subsets of Θ, one can construct a rough set algebra such that $\mathcal{F} = \sigma(\Theta/\mathcal{R})$. Suppose P is a probability on $\mathcal{F}$ – then it can be extended to 2^Θ using inner and outer measures as follows:

$$P_*(A) = \sup\left\{P(X)\middle|X \in \sigma(\Theta/\mathcal{R}), X \subseteq A\right\} = P(\underline{apr}(A)),$$
$$P^*(A) = \inf\left\{P(X)\middle|X \in \sigma(\Theta/\mathcal{R}), X \supseteq A\right\} = P(\overline{apr}(A)).$$

Pawlak calls these 'rough probabilities' of A but, in fact, these are a pair of belief and plausibility functions.

Note that the set of focal elements Θ/R is a partition of the universe (frame) Θ. Therefore, Pawlak rough set algebras can only interpret belief functions whose focal elements form a partition of the frame of discernment. Nevertheless, further generalisations via serial rough algebras and interval algebras can be achieved [2057].

Further related work Subsequently, Yao et al. [2056] proposed a model of granular computing based on reformulating, reinterpreting and combining results from rough sets, quotient space theory and belief functions. In the model two operations, called 'zooming-in' and 'zooming-out', are used to study connections between the elements of a universe and the elements of a granulated universe, as well as connections between computations in the two universes.

As already discussed in Section 4.8.2 in the context of decision theory, Wu et al. introduced in [1968] the concepts of plausibility reduct and belief reduct in incomplete information systems, and discussed their relationship with 'classical' concepts [1964], one of the main issues in the study of rough set theory. The connections between rough set theory and the theory of evidence were studied further by the same authors in [1966]. The basic concept and properties of knowledge reduction based on inclusion degree and evidence-reasoning theory were also touched on by Zhang et al. [2107]. More recently, Liu [1182] also studied the connection between rough set and belief theory, by extending some results to arbitrary binary relations on two universal sets.

6.8 Probability boxes

Probability boxes [605, 2027] arise from the need in reliability analysis to assess the probability of failure of a system, expressed as

$$P_X(F) = \int_F f(x)\, dx,$$

where $f(x)$ is a probability density function of the variables x representing the materials and structure, and F is the failure region of values in which the structure is unsafe. Unfortunately, the available information is usually insufficient to define accurately the sought joint PDF f. Random sets (Section 3.1.5) and imprecise probability theories can then be useful for modelling this uncertainty.

Recall that, if P is a probability measure on the real line $\mathbb{R}$, its cumulative distribution function is a non-decreasing mapping from $\mathbb{R}$ to $[0,1]$, denoted by F_P, such that for any $r \in \mathbb{R}$, $F_P(r) = P((-\infty, r])$. It is thus quite natural to describe the uncertainty in f as a pair of lower and upper bounds on the associated CDF, representing the epistemic uncertainty about the random variable.

Definition 107. *A* probability box, *or* p-box *[604, 605], $\langle \underline{F}, \overline{F} \rangle$ is a class of cumulative distribution functions*

$$\langle \underline{F}, \overline{F} \rangle = \left\{ F \text{ CDF} \,\middle|\, \underline{F} \leq F \leq \overline{F} \right\},$$

delimited by lower and upper bounds $\underline{F}$ and $\overline{F}$.

6.8.1 Probability boxes and belief functions

P-boxes and random sets/belief functions are very closely related. Indeed, every pair of belief/plausibility functions Bel, Pl defined on the real line $\mathbb{R}$ (a random set) generates a unique p-box whose CDFs are all those consistent with the evidence generating the belief function,

$$\underline{F}(x) = Bel((-\infty, x]), \quad \overline{F}(x) = Pl((-\infty, x]). \tag{6.18}$$

Conversely, every probability box generates an entire equivalence class of random intervals consistent with it [928]. A p-box can be discretised to obtain from it a random set which approximates it, but this discretisation is not unique [604, 781]. For instance [928], given a probability box $\langle \underline{F}, \overline{F} \rangle$, a random set on the real line can be obtained with the following infinite collection of intervals of $\mathbb{R}$ as focal elements:

$$\mathcal{F} = \left\{ \gamma = [\overline{F}^{-1}(\alpha), \underline{F}^{-1}(\alpha)] \ \forall \alpha \in [0,1] \right\}, \tag{6.19}$$

where

$$\overline{F}^{-1}(\alpha) \doteq \inf\{\overline{F}(x) \geq \alpha\}, \quad \underline{F}^{-1}(\alpha) \doteq \inf\{\underline{F}(x) \geq \alpha\}$$

are the 'quasi-inverses' of the upper and lower CDFs $\overline{F}$ and $\underline{F}$, respectively.

6.8.2 Approximate computations for random sets

In the case of an infinite random set, belief and plausibility values are computed via the following integrals:

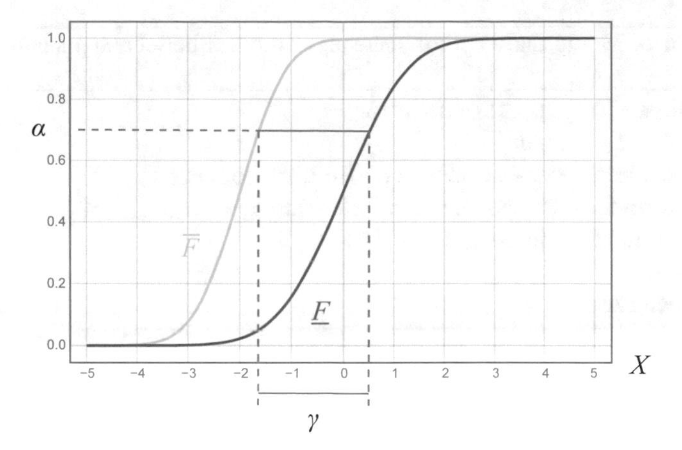

Fig. 6.5: A p-box amounts to a multivalued mapping Γ associating values $\alpha \in [0,1]$ with closed intervals $\gamma = \Gamma(\alpha)$ of $\mathbb{R}$, i.e., focal elements of the underlying random set [36].

$$Bel(A) = \int_{\omega \in \Omega} I[\Gamma(\omega) \subset A] \, \mathrm{d}P(\omega) \quad Pl(A) = \int_{\omega \in \Omega} I[\Gamma(\omega) \cap A \neq \emptyset] \, \mathrm{d}P(\omega), \tag{6.20}$$

where $\Gamma : \Omega \to 2^\Theta$ is the multivalued mapping generating the random set (see Chapter 3, Section 3.1.5), and $I[\cdot]$ is the indicator function. This is not trivial at all – however, we can use the p-box representation of infinite random sets (6.18), with the set of focal elements (6.19), to compute approximations of similar integrals [36]. The idea is to index each of its focal elements by a number $\alpha \in [0,1]$.

Consider then the unique p-box (6.18) associated with the random set Bel. If there exists a cumulative distribution function F_α for α over $[0,1]$ we can draw values of α at random from it, obtaining sample focal elements of the underlying random set (Fig. 6.5). We can then compute the belief and plausibility integrals (6.20) by adding the masses of the sample intervals.

Using this sampling representation, we can also approximately compute the Dempster combination of d input random sets. Each selection of one focal element from each random set is denoted by the vector α of corresponding indices α_i: $\alpha = [\alpha_1, \ldots, \alpha_d] \in (0,1]^d$. Suppose a *copula* C (i.e., a probability distribution whose marginals are uniform) is defined on the unit hypercube where α lives. We can then use it to compute the desired integrals as follows:

$$P_\Gamma(G) = \int_{\alpha \in G} \mathrm{d}C(\alpha), \quad G \subset (0,1]^d.$$

A joint focal element can be represented either by the hypercube $\gamma = \times_{i=1}^d \gamma_i \subseteq X$ or by the point $\alpha = [\alpha_1, \ldots, \alpha_d] \in (0,1]^d$.

If all input random sets are independent, these integrals decompose into a series of d nested integrals (see [36], Equation (36)). Alvarez [36] proposed a Monte

Algorithm 14 Monte Carlo approximate algorithm for belief and plausibility calculation

```
procedure MONTECARLOALVAREZ
    for j = 1, ..., n do
        randomly extract a sample α_j from the copula C;
        form the corresponding focal element A_j = ×_{i=1,...,d} γ_i^d;
        assign to it mass m(A_j) = 1/n.
    end for
end procedure
```

Carlo approach to their calculation (Algorithm 14). It can be proven that such an approximation converges as $n \to +\infty$ almost surely to the actual random set.

6.8.3 Generalised probability boxes

According to Destercke et al. [486], probability boxes are not adequate for computin the probability that some output remains close to a reference value ρ, which corresponds to computing upper and lower estimates of the probability of events of the form $|\tilde{x} - \rho| < \epsilon$. In response, they developed generalisations of p-boxes to arbitrary (finite) spaces, which can address such types of query.

Note that any two cumulative distribution functions F, F' modelling a p-box are comonotonic, i.e., for any $x, y \in X$, we have that $F(x) < F(y)$ implies $F'(x) < F'(y)$.

Definition 108. *A generalised p-box $\langle \underline{F}, \overline{F} \rangle$ on X is a pair of comonotonic mappings $\underline{F} : X \to [0,1]$, $\overline{F} : X \to [0,1]$ such that $\underline{F}(x) \leq \overline{F}(x)$ for all $x \in X$, and there exists at least one element $x \in X$ such that $\underline{F}(x) = \overline{F}(x) = 1$.*

Basically, there exists a permutation of the elements of X such that the bounds of a generalised p-box become CDFs defining a 'traditional' p-box. A generalised p-box is associated with a collection of nested sets[60] $A_y = \{x \in X : \underline{F}(x) \leq \underline{F}(y), \overline{F}(x) \leq \overline{F}(y)\}$, which are naturally associated with a possibility distribution.

While generalising p-boxes, these objects are a special case of random sets (∞-monotone capacities) and thus a special case of probability intervals (Fig. 6.6).

6.9 Spohn's theory of epistemic beliefs

In Spohn's theory of epistemic beliefs, in an *epistemic state* for a variable X, some propositions are believed to be true (or 'believed'), while some others are believed to be false (or 'disbelieved'), and the remainder are neither believed nor disbelieved.

[60]The present author's notation.

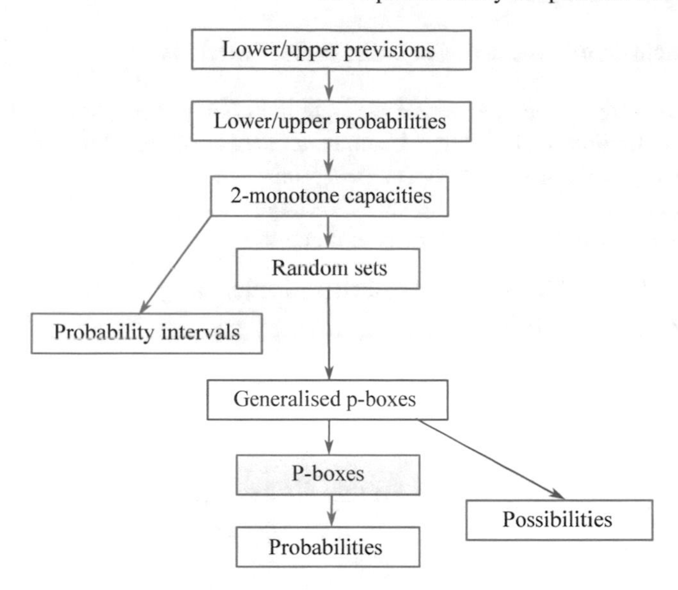

Fig. 6.6: Generalised p-boxes in the (partial) hierarchy of uncertainty measures [486].

6.9.1 Epistemic states

A number of conditions are required to guarantee logical consistency. Let Θ_X be the set of possible values of X.

Definition 109. *An epistemic state is said to be* consistent *if the following five axioms are satisfied:*

1. *For any proposition A, exactly one of the following conditions holds: (i) A is believed; (ii) A is disbelieved; (iii) A is neither believed nor disbelieved.*
2. *Θ_X is (always) believed.*
3. *A is believed if and only if A^c is disbelieved.*
4. *If A is believed and $B \supseteq A$, then B is believed.*
5. *If A and B are believed, then $A \cap B$ is believed.*

Let $\mathcal{B}$ denote the set of all subsets of Θ_X that are believed in a given epistemic state. We can then state the following proposition.

Proposition 49. *The epistemic state is consistent if and only if there exists a unique non-empty subset C of Θ_X such that $\mathcal{B} = \{A \subseteq \Theta_X : A \supseteq C\}$.*

We can note the similarity with the definition of *consistent belief functions* (Section 10.4), whose focal elements are also constrained to contain a common intersection subset.

6.9.2 Disbelief functions and Spohnian belief functions

The basic representation of an epistemic state in Spohn's theory is called an *ordinal conditional function* in [1752], p. 115, and a *natural* conditional function in [1753], p. 316. Shenoy calls this function a *disbelief* function.

Formally, let g denote a finite set of variables, and Θ_g the joint frame (set of possible values) for the variables in the collection g.

Definition 110. ([1753], p. 316) *A* disbelief function *for g is a function $\delta : 2^{\Theta_g} \rightarrow \mathbb{N}^+$ such that*

1. *$\delta(\theta) \in \mathbb{N}$ for all $\theta \in \Theta_g$.*
2. *There exists $\theta \in \Theta_g$ such that $\delta(\theta) = 0$.*
3. *For any $A \subset \Theta_g$, $A \neq \emptyset$,*

$$\delta(A) = \min\left\{\delta(\theta), \theta \in A\right\}.$$

4. *$\delta(\emptyset) = +\infty$.*

Note that, just like a possibility measure, a disbelief function is completely determined by its values on the singletons of the frame, Θ_g.

A proposition A is believed in the epistemic state represented by δ iff $A \supseteq C$, where $C = \{\theta : \delta(\theta) = 0\}$ (or, equivalently, iff $\delta(A^c) > 0$). A is disbelieved whenever $\delta(A) > 0$, and the proposition is neither believed nor disbelieved iff $\delta(A) = \delta(A^c) = 0$. The quantity $\delta(A^c)$ can thus be interpreted as the degree of belief for A. As a consequence, a disbelief function models degrees of disbelief for disbelieved propositions directly, whereas it models degrees of belief for believed propositions only indirectly. *Spohnian* belief functions can model both beliefs and disbeliefs directly.

Definition 111. *A* Spohnian belief function *for g is a function $\beta : 2^{\Theta_g} \rightarrow \mathbb{Z}^+$ such that*

$$\beta(A) = \begin{cases} -\delta(A) & \delta(A) > 0, \\ \delta(A^c) & \delta(A) = 0 \end{cases}$$

for all $A \subseteq \Theta_g$, where δ is some disbelief function for g.

A disbelief function can be uniquely recovered from a given (Spohnian) belief function – the latter has a number of desirable properties [1625].

6.9.3 α-conditionalisation

Spohn proposed the following rule for modifying a disbelief function in the light of new information.

Definition 112. ([1752], p. 117) *Suppose δ is a disbelief function for g representing our initial epistemic state. Suppose we learn something about a contingent proposition A (or A^c) that consequently leads us to believe A to a degree α (or, equivalently,* *disbelieve A^c to a degree α), where $\alpha \in \mathbb{N}$. The resulting epistemic state, called the A, α-conditionalisation of δ and denoted by the disbelief function $\delta_{A,\alpha}$, is defined as*

$$\delta_{A,\alpha}(\theta) = \begin{cases} \delta(\theta) - \delta(A) & \theta \in A, \\ \delta(\theta) + \alpha - \delta(A^c) & \theta \notin A, \end{cases}$$

for all $\theta \in \Theta_g$.

As stated by Shenoy [1625], Spohn's theory of epistemic beliefs shares the essential abstract features of both probability theory and Dempster–Shafer theory, in particular (1) a functional representation of knowledge (or beliefs), (2) a rule of marginalisation and (3) a rule of combination. Shenoy [1625] went on to show that disbelief functions can also be propagated via local computations, as shown in Chapter 4, Section 4.7.4 for belief functions.

6.10 Zadeh's generalised theory of uncertainty

In Zadeh's *generalised* (or perception-based [2091]) *theory of uncertainty* (GTU) [2093], the assumption that information is statistical in nature is replaced by a much more general thesis that information is a generalised constraint (as opposed to standard constraints of the form $X \in C$), with statistical uncertainty being a special, albeit important case.

A *generalised constraint* has the form $GC(X)$: X is$r$ R, where $r \in \{$ blank, probabilistic, veristic, random set, fuzzy graph, etc. $\}$ is a label which determines the type of constraint, and R is a constraining relation of that type (it could be a probability distribution, a random set, etc.). This allows us to operate on perception-based information, for instance 'Usually Robert returns from work at about 6 p.m.' or 'It is very unlikely that there will be a significant increase in the price of oil in the near future.' Generalised constraints serve to define imprecise probabilities, utilities and other constructs, and generalised constraint propagation is employed as a mechanism for reasoning with imprecise probabilities as well as for computation with perception-based information.

Given the way it is defined, a generalised constraint is obviously a generalisation of the notion of a belief function as well, albeit a rather nomenclative one. Zadeh claims that, in this interpretation, the theory of evidence is a theory of a mixture of probabilistic and possibilistic constraints [2093], whereas his GTU embraces all possible mixtures and therefore accommodates most theories of uncertainty. Furthermore, bivalence is abandoned throughout the GTU, and its foundation is shifted from bivalent logic to fuzzy logic. As a consequence, in the GTU everything is or is allowed to be a matter of degree or, equivalently, fuzzy. All variables are or are allowed to be 'granular', with a granule being 'a clump of values ... which are drawn together by indistinguishability, equivalence, similarity, proximity or functionality' (see Fig. 6.7).

One of the main objectives of the GTU is the capability to operate on information described in natural language. As a result, a *generalised constraint language* (GCL)

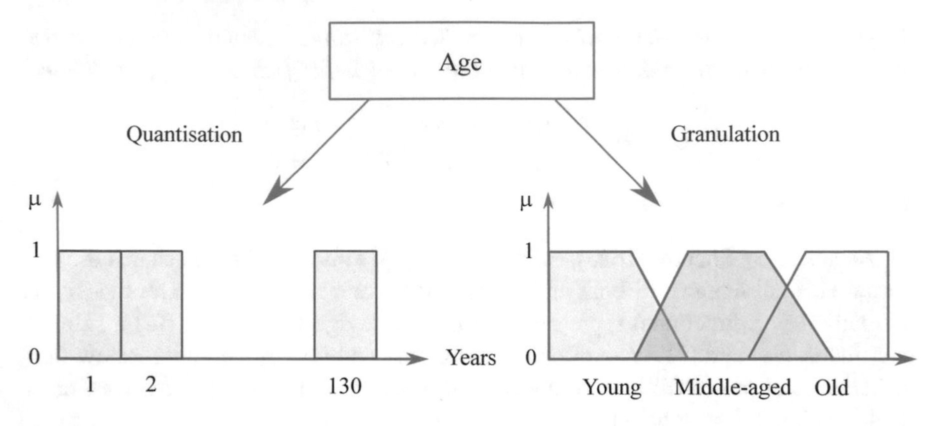

Fig. 6.7: Quantisation (left) versus granulation (right) of a variable 'Age' (from [2092]).

is defined as the set of all generalised constraints together with the rules governing syntax, semantics and generation. Examples of GCL elements are (X is small) is likely, and (X,Y) isp $A) \wedge (X$ is $B)$, where 'isp' denotes a probabilistic constraint, 'is' denotes a possibilistic constraint and $\wedge$ denotes conjunction.

Finally, in the GTU, computation/deduction is treated as an instance of question-answering. Given a system of propositions described in a natural language p, and a query q, likewise expressed in a natural language, the GTU performs generalised constraint propagation governed by deduction rules that, in Zadeh's words, are 'drawn from the Computation/Deduction module. The Computation/Deduction module comprises a collection of agent-controlled modules and submodules, each of which contains protoformal deduction rules drawn from various fields and various modalities of generalised constraints'.

In this author's view, generality is achieved by the GTU in a rather nomenclative way, which explains the complexity and lack of naturalness of the formalism.

6.11 Baoding Liu's uncertainty theory

Liu's *uncertainty theory* [1177, 1178] is based on the notion of an *uncertain measure*, defined as a function $\mathcal{M}$ on the σ-algebra $\mathcal{F}$ of events over a non-empty set Θ[61] which obeys the following axioms:

1. $\mathcal{M}(\Theta) = 1$ (normality).
2. $\mathcal{M}(A_1) \leq \mathcal{M}(A_2)$ whenever $A_1 \subset A_2$ (monotonicity).
3. $\mathcal{M}(A) + \mathcal{M}(A^c) = 1$ ('self-duality').
4. For every countable sequence of events $\{A_i\}$, we have (countable subadditivity)

[61] The original notation has been changed to fit that adopted in this book.

$$\mathcal{M}\left(\bigcup_{i=1}^{\infty} A_i\right) \leq \sum_{i=1}^{\infty} \mathcal{M}(A_i).$$

Liu's uncertain measures are supposed to formalise subjective degrees of belief, rather than empirical frequencies, and his theory is therefore an approach to subjective probability. Self-duality (axiom 3) is justified by the author as consistent with the law of the excluded middle.

Clearly, uncertain measures are monotone capacities (see Definition 95). Just as clearly, probability measures satisfy these axioms. However, Liu claims, probability theory is not a special case of his formalism, since probabilities do not satisfy the *product axiom*,

$$\mathcal{M}\left(\prod_{k=1}^{\infty} A_k\right) = \bigwedge_{k=1}^{\infty} \mathcal{M}_k(A_k),$$

for all Cartesian products of events from individual uncertain spaces $(\Theta_k, \mathcal{F}_k, \mathcal{M}_k)$. However, the product axiom was introduced by Liu only in 2009 [1178] (long after his introduction of uncertainty theory in 2002 [1179]). Also, the extension of uncertain measures to any subset of a product algebra is rather cumbersome and unjustified (see Equation (1.10) in [1177], or Fig. 6.8 in this book, extracted from Fig. 1.1 in [1177]). More generally, the justification that Liu provides for his choice of axioms is somewhat lacking.

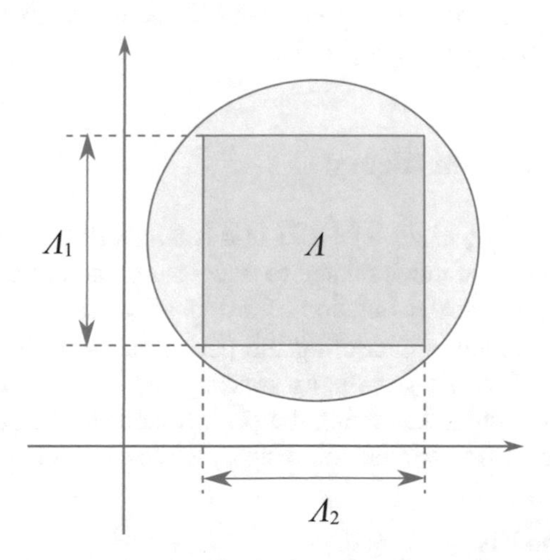

Fig. 6.8: Extension of rectangle to product algebras in Liu's uncertain theory (from [1177]). The uncertain measure of Λ (the disc) is the size of the inscribed rectangle $\Lambda_1 \times \Lambda_2$ if the latter is greater than 0.5. Otherwise, if the size of the inscribed rectangle of Λ^c is greater than 0.5, then $\mathcal{M}(\Lambda^c)$ is just its inscribed rectangle and $\mathcal{M}(\Lambda) = 1 - \mathcal{M}(\Lambda^c)$. If no inscribed rectangle exists for either Λ or Λ^c which is greater than 0.5, then we set $\mathcal{M}(\Lambda) = 0.5$.

Based on such measures, a straightforward generalisation of random variables can then be defined ('uncertain variables'), as measurable (in the usual sense) functions from an uncertainty space $(\Theta, \mathcal{F}, \mathcal{M})$ to the set of real numbers. A set $\xi_1, \ldots, \xi_n$ of uncertain variables are said to be independent if

$$\mathcal{M}\left(\bigcap_{i=1}^{m}\{\xi_i \in B_i\}\right) = \min_i \mathcal{M}(\{\xi_i \in B_i\}).$$

Just as a random variable may be characterised by a probability density function, and a fuzzy variable is described by a membership function, uncertain variables are characterised by *identification functions* (λ, p), which do not exist, however, for every uncertain variable and are subject to rather complex and seemingly arbitrary axioms (see [1178], Definition 4), for example

$$\sup_B \lambda(x) + \int_B \rho(x)\,\mathrm{d}x \geq 0.5 \quad \text{and/or} \quad \sup_{B^c} \lambda(x) + \int_{B^c} \rho(x)\,\mathrm{d}x \geq 0.5, \quad (6.21)$$

where λ and ρ and non-negative functions of the real line, and B is any Borel set of real numbers. Equation (6.21) mirrors the extension method of Fig. 6.8. While an uncertain entropy and an uncertain calculus were built by Liu on this basis, the general lack of rigour and of a convincing justification for a number of elements of the theory leaves the author of this book quite unimpressed with this work. No mention of belief functions or other well-established alternative representations of subjective probabilities is made in [1177].

6.12 Info-gap decision theory

Yakov Ben-Haim's *info-gap theory* [107] is a framework for assessing the robustness of dynamical models, much related to uncertainty quantification.

Let $F(\omega, \delta)$ be the transfer function of a dynamical system, with parameters δ, and let $\hat{f}(F, \delta)$ be a function measuring the performance of the system. The basic robustness question that info-gap theory investigates is the following: how wrong can our model be without jeopardising the performance of the system? The bottom line of the approach is to maximise robustness, while sacrificing performance.

6.12.1 Info-gap models

More precisely, an info-gap model describes, non-probabilistically, the uncertain difference between the best-estimated model $\tilde{F}(\omega, \delta)$ and other possible models $F(\omega, \delta)$ for the system at hand.

Definition 113. *An* info-gap model *for the uncertainty in the dynamic behaviour of a structure is an unbounded family of nested sets of dynamic models.*

Info-gap models come in many variants. The *uniform-bound info-gap model* is defined as

$$\mathcal{F}(\alpha, \tilde{F}) = \left\{ F(\omega, \delta) : |F(\omega, \delta) - \tilde{F}(\omega, \delta)| \leq \alpha \right\}, \quad \alpha \geq 0. \tag{6.22}$$

More realistic info-gap models which incorporate information about constrained rates of variation read as follows:

$$\mathcal{F}(\alpha, \tilde{F}) = \left\{ F(\omega, c, \delta) : \left| \frac{c_i - \tilde{c}_i}{\tilde{c}_i} \right| \leq \alpha, i = 1, \ldots, J \right\}, \quad \alpha \geq 0.$$

From (6.22), it is apparent that an info-gap model entails two separate levels of uncertainty: one described by the set of possible dynamic models within $\mathcal{F}(\alpha, \tilde{F})$, and the other by the uncertainty in the value of α (called the *horizon of uncertainty*). In info-gap theory uncertainty is unbounded, rather than, say, convex.

6.12.2 Robustness of design

In this formalism, the performance of a system is expressed by a vector $\hat{f}(F, \delta)$ of R components, subject to performance requirements of the general form

$$\hat{f}_i(F, \delta) \leq f_{c,i}, \quad i = 1, \ldots, R.$$

Definition 114. *The* robustness of design δ, *to uncertainty in the dynamics, is the greatest horizon of uncertainty* α *up to which the response to all dynamic models* $F(\omega, \delta)$ *is acceptable according to each of the performance requirements:*

$$\hat{\alpha}(\delta, f_c) = \max \left\{ \alpha \middle| \max_{F \in \mathcal{F}(\alpha, \tilde{F})} \hat{f}_i(F, \delta) \leq f_{c,i}, i = 1, \ldots, R \right\}.$$

An important application of info-gap theory is climate change mitigation [782].

6.13 Vovk and Shafer's game-theoretical framework

Game-theoretic probability [1615, 1863, 1862] is an approach to epistemic probability based on a game-theoretical scenario in which, in a repetitive structure, one can make make good probability forecasts relative to whatever state of knowledge one has.

Within this theory, probabilistic predictions are proven by constructing a betting strategy which allows one to multiply capital indefinitely if the prediction fails. Probabilistic theories are tested by betting against its predictions. To make probability forecasts, one needs to construct a forecasting strategy that defeats all reasonable betting strategies. The theory thus provides an original, novel understanding of epistemic probability; one can use defensive forecasting [1861, 1864] to get additive probabilities. At the same time, as Vovk and Shafer suggest, just as in Kolmogorov's measure-theoretical probability, this framework is open to diverse interpretations [1615].

Mathematically, the approach builds on the earlier theory of *prequential probability* [392]. Dawid's prequential principle states that the forecasting success of a probability distribution for a sequence of events should be evaluated using only the actual outcomes and the sequence of forecasts (conditional probabilities) to which these outcomes give rise, without reference to other aspects of the probability distribution.

According to the present author's understanding, a further extension of this reasoning to non-repetitive situations, such as those contemplated in belief theory, among others, is still being sought. Nevertheless, a betting interpretation of probabilities and Dempster–Shafer degrees of belief was attempted by Shafer in [1605]. Interesting relations with Walley's imprecise probability, on the other hand, were pointed out by several authors [397, 1292].

6.13.1 Game-theoretic probability

Game-theoretic probability goes back in fact to the work of Blaise Pascal ('Probability is about betting'), Antoine Cournot ('Events of small probability do not happen'), and Jean Ville [1604]. The latter, in particular, showed in 1939 that the laws of probability can be derived from this principle: *You will not multiply the capital you risk by a large factor* (the so-called *Cournot's principle*).

Algorithm 15 Game-theoretic probability protocol

```
procedure GAMETHEORETICALPROTOCOL
    K_0 = 1
    for n = 1, ..., N do
        Forecaster announces prices for various payoffs.
        Skeptic decides which payoffs to buy.
        Reality determines the payoffs.
        Skeptic's capital changes as K_n = K_{n-1} + net gain or loss.
    end for
end procedure
```

Suppose you, the agent, possess a certain amount of capital $\mathcal{K}$, and that you gamble on the outcomes of a random variable without risking more than your initial capital. Your resulting wealth is a non-negative random variable X with expected value $E(X)$ equal to your initial capital $\mathcal{K}_0$. Markov's inequality,

$$P\left(X \geq \frac{E(X)}{\epsilon}\right) \leq \epsilon,$$

states that you have probability ϵ or less of multiplying your initial capital by $1/\epsilon$ or more. Ville proved what is now called, in standard probability theory, *Doob's inequality*, which generalises Markov's inequality to a sequence of bets:

$$P\left(\sup_{0\leq t\leq T} X_t \geq C\right) \leq \frac{E[X_T^p]}{C^p}, \tag{6.23}$$

for all $C > 0$, $p \geq 1$, where X is a *submartingale* taking non-negative real values, i.e., a sequence $X_1, X_2, X_3, \ldots$ of integrable random variables satisfying

$$E[X_{n+1}|X_1, \ldots, X_n] \geq X_n.$$

The Ville/Vovk perfect-information protocol for probability can be stated in terms of a game involving three players, Forecaster, Skeptic and Reality, as in Algorithm 15. Using Shafer and Vovk's weather-forecasting example, Forecaster may be a very complex computer program that escapes precise mathematical definition as it is constantly under development, which uses information external to the model to announce every evening a probability for rain on the following day. Skeptic decides whether to bet for or against rain and how much, and Reality decides whether it in fact rains. As shown by Ville, the fundamental hypothesis that Skeptic cannot get rich can be tested by any strategy for betting at Forecaster's odds. The bottom line of Shafer and Vovk's theory is that Forecaster can beat Skeptic [1864].

6.13.2 Ville/Vovk game-theoretic testing

In Ville's theory (Algortithm 16), Forecaster is a known probability distribution for Reality's move. It always gives conditional probabilities for Reality's next move given her past moves.

Consider a probability distribution P for $y_1, y_2, \ldots$

Proposition 50. *(Ville) Skeptic has a winning strategy.*

In Vovk's generalisation, Forecaster does not necessarily use a known probability distribution, but only provides a series of real numbers $p_n \in [0, 1]$, with no additional assumptions. Skeptic still has a winning strategy, under Vovk's weak law of large numbers[62] [1604]. Suppose N rounds of Vovk's game (Algorithm 17) are conducted. If winning is defined as $\mathcal{K}_n$ being never negative, and either $\mathcal{K}_N \geq C$ or $|\frac{1}{N}\sum_{n=1}^{N}(y_n - p_n)| < \epsilon$, then we can state the following proposition.

Proposition 51. *(Vovk) Skeptic has a winning strategy if* $N \geq C/4\epsilon^2$.

In measure-theoretic probability there is arguably a shift in emphasis from sums of independent random variables, once a central object of study, to stochastic processes in which the probabilities of events depend on preceding outcomes in complex ways, and in particular *martingales*.

---

[62] In related work, game-theoretic versions of the strong law of large numbers for unbounded variables were studied by Kumon et al. [1084, 1604].

Algorithm 16 Ville's setting

procedure VILLE
$\mathcal{K}_0 = 1$
for $n = 1, 2, \ldots$ **do**
Skeptic announces $s_n \in \mathbb{R}$
Reality announces $y_n \in \{0, 1\}$
$\mathcal{K}_n = \mathcal{K}_{n-1} + s_n\big(y_n - P(y_n = 1|y_1, \ldots, y_{n-1})\big)$
end for
Skeptic wins if $\mathcal{K}_n$ is never negative, and either

$$\lim_{n\to\infty} \frac{1}{n}\sum_{i=1}^{n} \big(y_i - P(y_i = 1|y_1, \ldots, y_{n-1})\big) = 0 \tag{6.24}$$

or $\lim_{n\to\infty} \mathcal{K}_n = \infty$
end procedure

Algorithm 17 Vovk's setting

procedure VOVK
$\mathcal{K}_0 = 1$
for $n = 1, 2, \ldots$ **do**
Forecaster announces $p_n \in [0, 1]$
Skeptic announces $s_n \in \mathbb{R}$
Reality announces $y_n \in \{0, 1\}$
$\mathcal{K}_n = \mathcal{K}_{n-1} + s_n(y_n - p_n)$
end for
Skeptic wins if $\mathcal{K}_n$ is never negative, and either

$$\lim_{n\to\infty} \frac{1}{n}\sum_{i=1}^{n}(y_i - p_i) = 0$$

or $\lim_{n\to\infty} \mathcal{K}_n = \infty$
end procedure

Definition 115. *A discrete-time martingale is a discrete-time stochastic process (i.e., a sequence of random variables) $X_1, X_2, X_3, \ldots$ that satisfies for any time n*

$$E(|X_n|) < \infty, \quad E(X_{n+1}|X_1, \ldots, X_n) = X_n$$

or, equivalently,

$$E(X_{n+1} \mid X_1, \ldots, X_n) - X_n = 0. \tag{6.25}$$

Shafer and Vovk's game-theoretical framework puts martingales first, as comparing (6.24) with (6.25) shows that they are the process $\{\mathcal{K}_n\}$ describing the changes in Skeptics capital.

6.13.3 Upper price and upper probability

Assuming the initial capital is $\mathcal{K}_0 = \alpha$, the *upper price* and *upper probability* can be defined as follows.

Definition 116. *For any real-valued function* X *on* $([0,1] \times \{0,1\})^N$, $\overline{E}X = \inf\{\alpha |$ *Skeptic has a strategy guaranteeing* $\mathcal{K}_n \geq X(p_1, y_1, \ldots, p_N, y_N)\}$.

Definition 117. *For any subset* $A \subseteq ([0,1] \times \{0,1\})^N$, $\overline{P}A = \inf\{\alpha |$ *Skeptic has a strategy guaranteeing* $\mathcal{K}_N \geq 1$ *if* A *happens, and* $\mathcal{K}_N \geq 0$ *otherwise*$\}$.

Proposition 51 can then be put in terms of the upper probability, as follows.

Proposition 52.

$$\overline{P}\left(\left| \frac{1}{N} \sum_{n=1}^{N} (y_n - p_n) \right| \geq \epsilon \right) \leq \frac{1}{4N\epsilon^2}.$$

Defensive forecasting is not covered in this brief summary. We refer the reader to [1864].

6.14 Other formalisms

6.14.1 Endorsements

In 1983, Cohen and Grinberg proposed a theory for reasoning about uncertainty [297, 296], resting on a representation of states of certainty called *endorsements*. These authors claimed that 'numerical representations of certainty hide the reasoning that produces them and thus limit one's reasoning about uncertainty', and that, while easy to propagate, numbers have unclear meanings. Numerical approaches to reasoning under uncertainty, in their view, are restricted because the set of numbers is not a sufficiently rich representation to support considerable heuristic knowledge about uncertainty and evidence. Cohen and Grinberg's main claim is that there is more about evidence than by 'how much' it is believed, and that other aspects need to be taken into account. Namely, *justifications* need to be provided in support of evidence. They argued that different kinds of evidence should be distinguished by an explicit record of what makes them different, not by numbers between 0 and 1.

6.14.2 Fril-fuzzy theory

Baldwin et al. [83, 79] presented a theory of uncertainty, consistent with and combining the theories of probability and fuzzy sets. This theory extends the logic programming form of knowledge representation to include uncertainties such as probabilistic knowledge and fuzzy incompleteness. Applications to knowledge engineering, including expert and decision support systems, evidential and case-based reasoning, fuzzy control, and databases were illustrated.

6.14.3 Granular computing

Yao [2055] surveyed *granular computing*, intended as a set of theories and techniques which make use of granules, i.e., groups or clusters of concepts.

In [2055], Yao discussed the basic issues of granular computing, focussing in particular on the construction of and computation with granules. A set-theoretic model of granular computing was proposed, based on the notion of power algebras. The main idea is that any binary operation $\circ$ on a universe U can be lifted to a binary operation $\circ^+$ (the associated *power operation*) on subsets of U as follows:

$$X \circ^+ Y = \{x \circ y | x \in X, y \in Y\}$$

for any $X, Y \subset U$. For instance, one can perform arithmetic operations on interval numbers by lifting arithmetic operations on real numbers, obtaining an interval algebra which may serve as a basis for interval reasoning with numeric truth values, such as interval fuzzy reasoning [2058], interval probabilistic reasoning [1450] and reasoning with granular probabilities [985].

6.14.4 Laskey's assumptions

In [1100], Laskey demonstrated a formal equivalence between belief theory and *assumption-based truth maintenance systems* [372], so that any Dempster–Shafer inference network can be represented as a set of ATMS justifications with probabilities attached to assumptions.

A proposition's belief is equal to the probability of its label conditioned on label consistency. In [1100], an algorithm was given for computing these beliefs. When ATMSs are used to manage beliefs, non-independencies between nodes are automatically and correctly accounted for.

6.14.5 Harper's Popperian approach to rational belief change

Around the same time as Shafer's PhD thesis (1975), Harper [806] proposed in a very comprehensive and interesting paper to use Popper's treatment of probability (*propensity*; see Section 1.2.4) and an epistemic constraint on probability assignments to conditionals to extend the Bayesian representation of rational belief so that revision of previously accepted evidence is allowed for.

Definition 118. *A function $P : \mathcal{F} \times \mathcal{F} \rightarrow [0,1]$ mapping two elements of a minimal algebra, endowed with a binary operation AB and a unary one $\bar{A}$, to the reals is a* Popper probability function *if, for all $A, B, C \in \mathcal{F}$:*

1. $0 \leq P(B|A) \leq P(A|A) = 1$.
2. *If* $P(A|B) = 1 = P(B|A)$ *then* $P(C|A) = P(C|B)$.
3. *If* $P(C|A) \neq 1$, *then* $P(\bar{B}|A) = 1 - P(B|A)$.
4. $P(AB|C) = P(A|C) \cdot P(B|AC)$.
5. $P(AB|C) \leq P(B|C)$.

In Popperian probability, conditional probability is primitive. Absolute probability can be represented as

$$P(A) \doteq P(A|T), \quad T = \overline{A\bar{A}}.$$

Note that on traditional σ-algebras, $T = (A \cap A^c)^c = \Omega$. When $P(A)$, as defined above, is greater than 0 the Popper conditional probability is the ratio of absolute probabilities $P(B|A) = P(AB)/P(A)$, i.e, it obeys the classical definition of conditional probability. The most salient difference is that, in Popper's theory, $P(B|A)$ exists even when $P(A) = 0$! As a result, previously accepted evidence can be revised, in order to condition on events which currently have 0 probability.

The results of Harper's extension included an epistemic semantics for Lewis's theory of *counterfactual conditionals* [1144, 1145].

6.14.6 Shastri's evidential reasoning in semantic networks

In his PhD thesis, Shastri [1620] argued that the best way to cope with partial and incomplete information is to adopt an evidential form of reasoning, wherein inference does not involve establishing the truth of a proposition but rather finding the most likely hypothesis from among a set of alternatives. In order for inference to take place in real time, one must provide a computational account of how this may be performed in an acceptable time frame. *Inheritance* and *categorisation* within a conceptual hierarchy were identified by Shastri as two operations that humans perform very fast, and which lie at the core of intelligent behaviour and are precursors to more complex reasoning. These considerations led to an evidential framework for representing conceptual knowledge, wherein the principle of maximum entropy is applied to deal with uncertainty and incompleteness. In [1620] it was demonstrated that such a framework offers a uniform treatment of inheritance and categorisation, and can be encoded as an interpreter-free, connectionist network.

In [1621], Shastri and Feldman proposed an evidence combination rule which is incremental, commutative and associative and hence shares most of the attractive features of Dempster's rule, while being 'demonstrably better' (in these authors' words) than Dempster's rule in the context considered there.

6.14.7 Evidential confirmation theory

Grosof [732] considered the issue of aggregating measures of confirmatory and disconfirmatory evidence for a common set of propositions.

He showed that a revised MYCIN certainty factor [812] (an ad hoc method for managing uncertainty then widely used in rule-based expert systems) and the PROSPECTOR [554] methods are special cases of Dempster–Shafer theory. His paper also showed that by using a nonlinear but invertible transformation, we can interpret a special case of Dempster's rule in terms of conditional independence. This unified approach, Grosof argued, would resolve the 'take-them-or-leave-them' problem with priors: MYCIN had to leave them out, while PROSPECTOR had to have them in.

6.14.8 Groen's extension of Bayesian theory

Groen and Mosleh [727] proposed an extension of Bayesian theory based on a view of inference according to which observations are used to rule out possible valuations of the variables. This extension is different from probabilistic approaches such as Jeffrey's rule (see Section 4.6.2), in which certainty in a single proposition A is replaced by a probability on a disjoint partition of the universe, and Cheeseman's rule of distributed meaning [256], while non-probabilistic analogues are found in evidence and possibility theory.

An interpretation I of an observation O is defined as the union of those values of a variable of interest X (e.g., the luminosity of an object) that are consistent with, i.e., not contradicted by, an uncertain observation O (e.g., the object is 'very dark'). Evidence uncertainty is defined as the uncertainty regarding the interpretation of an observation, and can be expressed either as a PDF on the space $\mathcal{I}$ of possible interpretations, $\pi(I|H)$, $I \in \mathcal{I}$, or by introducing an *interpretation function*

$$\rho(x) \doteq Pr(x \in I|H), \quad 0 \leq \rho(x) \leq 1,$$

where $x \in I$ denotes the event that value x is not contradicted by the observation, $\rho(x)$ is the probability, or degree of belief, that this is the case, and H denotes our prior knowledge. The two definitions can be mapped onto each other by writing

$$\rho(x) = Pr(x \in I|H) = \sum_{I \in \mathcal{I}} Pr(x \in I|I, H)\pi(I|H) = \sum_{I \in \mathcal{I}: x \in I} \pi(I|H).$$

Prior to making an observation, our initial state of uncertainty about the pair 'observable variable' $x \in X$, variable of interest $\theta \in \Theta$ can be represented by the probability distribution

$$\pi(x, \theta|H) = \pi(x|\theta, H)\pi(\theta|H)$$

defined over $X \times \Theta$. Under the constraint that interpretations are certain, i.e., representations are either contradicted by observations O or not, the interpretation function assumes the form

$$\rho(x) = \begin{cases} 0 & x \text{ is contradicted by } O, \\ 1 & x \text{ is not contradicted by } O. \end{cases}$$

Full Bayesian inference (1.2) can then be modified by introducing the above interpretation function into Bayes' rule. The posterior joint distribution of the observations x and the parameter θ will thus follow the law

$$\pi(x, \theta|H, O) = \frac{\rho(x)\pi(x|\theta, H)\pi(\theta|H)}{\int_{x,\theta} \rho(x)\pi(x|\theta, H)\pi(\theta|H) \, \mathrm{d}x \, \mathrm{d}\theta}.$$

The rationale is that the observation provides no basis for a preference among representations that are not contradicted. Consequently, the relative likelihoods of the remaining representations should not be affected by the observation.

6.14.9 Padovitz's unifying model

In 2006, Padovitz et al. [1375] proposed a novel approach to representing and reasoning about context in the presence of uncertainty, based on multi-attribute utility theory as a means to integrate heuristics about the relative importance, inaccuracy and characteristics of sensory information.

The authors qualitatively and quantitatively compared their reasoning approach with Dempster-Shafer sensor data fusion.

6.14.10 Similarity-based reasoning

Similarity-based reasoning (SBR) is based on the principle that 'similar causes bring about similar effects'. The author of [857] proposed a probabilistic framework for SBR, based on a 'similarity profile' which provides a probabilistic characterisation of the similarity relation between observed cases (instances).

Definition 119. *A* similarity-based inference *(SBI) set-up is a 6-tuple*

$$\Sigma = \langle (\mathcal{S}, \mu_{\mathcal{S}}), \mathcal{R}, \phi, \sigma_{\mathcal{S}}, \sigma_{\mathcal{R}}, \mathcal{M} \rangle,$$

where $\mathcal{S}$ is a finite set of situations endowed with a probability measure $\mu_{\mathcal{S}}$ (defined on $2^{\mathcal{S}}$), $\mathcal{R}$ is a set of results, and the function $\phi : \mathcal{S} \to \mathcal{R}$ assigns (unique) results to situations. The functions $\sigma_{\mathcal{S}} : \mathcal{S} \times \mathcal{S} \to [0,1]$ and $\sigma_{\mathcal{R}} : \mathcal{R} \times \mathcal{R} \to [0,1]$ define (reflexive, symmetric and normalised) similarity measures over the set of situations and the set of results, respectively, and $\mathcal{M}$ is a finite memory of cases $c = (s, \phi(s))$,

$$\mathcal{M} = \{(s_1, r_1), \ldots, (s_n, r_n)\}.$$

The cases in the memory are assumed to be samples repeatedly and independently taken from $\mu_{\mathcal{S}}$, a common assumption in machine learning. Given an SBI setup, the problem is to find, given a new situation $s_0 \in \mathcal{S}$, to predict the associated result $r_0 = \phi(s_0)$. Clearly, the set-up can be assimilated into regression in machine learning.

The author of [857] further developed an inference scheme in which instance-based evidence is represented in the form of belief functions, casting the combination of evidence derived from individual cases as an information fusion problem. Namely, she defined an *imperfect specification* of the outcome $\phi(s_0)$ induced by a case $(s, r) \in \mathcal{M}$ as a multivalued mapping from a set of contexts C to the set of possible responses $\mathcal{R}$, i.e., a random set (belief function) $\Gamma : C \to 2^{\mathcal{R}}$ with

$$\begin{aligned} C &= \{\sigma_{\mathcal{R}}(\phi(s), \phi(s')) | s, s' \in \mathcal{S}\} \subset [0,1], \\ \Gamma(c) &= \sigma_{\mathcal{R}}^{(-1)}(r, c) \doteq \{r' \in \mathcal{R} | \sigma_{\mathcal{R}}(r, r') = c\} \subset \mathcal{R} \end{aligned}$$

and $P(c) = \mu(c)$, where μ is a probability measure over C derived from the case (s, r), mapping a similarity value to a set of responses.

6.14.11 Neighbourhoods systems

A *neighbourhood system* mathematically formalises the notion of a 'negligible quantity', and is formally defined as the collection

$$\mathcal{NS} = \left\{ NS(p) \middle| p \in U \right\},$$

where $NS(p)$ is a maximal family of neighbourhoods of $p \in U$, and U is the universe of discourse. A minimal neighbourhood system can be constructed by selecting all the minimal neighbourhoods of all the points of U. A *basic* neighbourhood system is a minimal system in which every point has a single minimal neighbourhood, and can be seen as a mapping assigning a subset $A(p) \subset U$ to each $p \in U$.

Neighbourhood systems span topology (topological neighbourhood systems), rough sets and binary relations (basic neighbourhood systems). In [1170], the authors studied real-valued functions based on neighbourhood systems, showing how they covered many important uncertainty quantities such as belief functions, measures and probability distributions.

If we assign a basic probability to each basic neighbourhood (and zero to all other sets), we immediately get a belief function on U. More generally, if we assign a basic probability to all minimal neighbourhoods (and zero to all other sets), we again get a belief function on U. The converse is also true [1170].

Proposition 53. *Let U be a finite space. Then U has a belief function iff there is a neighbourhood system on U.*

6.14.12 Comparative belief structures

Comparative belief is a generalisation of Fine's *comparative probability* [615], which asserts that, given any two propositions A and B, it is necessary to be able to say whether $Bel(A)$ is greater than, less than or equal to $Bel(B)$, but it is not necessary to express numerically how much more one believes in one proposition than in another.

The authors of [1957] provided an axiomatic system for belief relations, i.e., binary relations $\succ$ on the subsets of a frame of discernment Θ, and showed that within this system there are belief functions which 'almost agree' with comparative belief relations. Namely,

$$A \succ B \Rightarrow Bel(A) > Bel(B) \quad \forall A, B \in 2^{\Theta}.$$

Part II

The geometry of uncertainty

7

The geometry of belief functions

Belief measures are complex objects, even when classically defined on finite sample spaces. Because of this greater complexity, manipulating them, as we saw in Chapter 4, opens up a wide spectrum of options, whether we consider the issues of conditioning and combination, we seek to map belief measures onto different types of measures, or we are concerned with decision making, and so on.

In particular, in estimation problems it is often required to compute a pointwise estimate of the quantity of interest: object tracking [365, 366], for instance, is a typical example. The problem consists in estimating at each time instant the current configuration or 'pose' of a moving object from a sequence of images. Image features can be represented as belief functions and combined to produce a belief estimate of the object's pose (compare [364], Chapter 8). Deriving a pointwise estimate from such a belief estimate allows us to provide an expected pose estimate. This can be done, for example, by finding the most appropriate probabilistic approximation of the current belief measure and computing the corresponding expected pose. As we saw in Chapter 4, Section 4.7.2, the link between belief and probability measures is the foundation of a popular approach to the theory of evidence, the transferable belief model [1675], as well as a central issue in belief theory. In fact, the probability transformation problem can be seen from a different perspective by investigating the space belief functions live in, and what class of functions is the most suitable to measure distances between belief measures, or between belief functions and other classes of uncertainty measures.

Probability distributions on finite domains admit a straightforward representation in terms of points of a *simplex* with n vertices, where n is the size of the sample space. As we will see in this chapter and the following ones, belief functions admit a similar geometric representation describable in the language of convex geometry. As

F. Cuzzolin, *The Geometry of Uncertainty*, Artificial Intelligence: Foundations, Theory, and Algorithms, https://doi.org/10.1007/978-3-030-63153-6_7

each belief function $Bel : 2^\Theta \rightarrow [0, 1]$ on a given frame of discernment Θ is completely specified by its $2^{|\Theta|} - 2$ belief values $\{Bel(A), \forall A \subset \Theta, \emptyset \subsetneq A \subsetneq \Theta\}$, Bel can be thought of as a vector $\mathbf{bel} = [Bel(A), \emptyset \subsetneq A \subsetneq \Theta]'$ of $\mathbb{R}^N$, $N = 2^{|\Theta|} - 2$ (compare Section 4.7.1). The collection $\mathcal{B}$ of all vectors of $\mathbb{R}^N$ which do correspond to belief functions is also a simplex, which we call a *belief space*. A whole geometric approach to belief theory, which can be naturally extended to general uncertainty theory, can then be developed in which belief measures are manipulated as geometric entities, and reasoning with belief functions is studied in this geometric language.

Albeit originally motivated by the transformation problem [382], this geometric approach is really just a manifestation of the close links which exist between combinatorics, convex geometry and subjective probability. These links have never been systematically explored, although some work has been done in this direction, especially by Grabisch [721, 720, 835] and Yao [2057, 2054]. A few other authors have approached, in particular, the study of the interplay between belief functions [1883] and probabilities in a geometric set-up [757, 165, 379]. In robust Bayesian statistics, more generally, an extensive literature exists on the study of convex sets of probability distributions [240, 310, 129, 825, 1575].

The closest reference is, possibly, the work of P. Black, who devoted his doctoral thesis [164] to the study of the geometry of belief functions and other monotone capacities. An abstract of his results can be found in [165], where he used the shapes of geometric loci to give a direct visualisation of the distinct classes of monotone capacities. In particular, a number of results on the lengths of the edges of the credal sets representing monotone capacities were given there, together with their 'size', interpreted as the sum of those lengths.

Whereas Black's work focused on a geometric analysis of belief functions as special types of credal sets *in the probability simplex*, in the geometric approach to belief and uncertainty theory introduced here, belief measures and the corresponding basic probability assignments are represented by points in a (convex) *belief space*, embedded in a Cartesian space.

The notion of representing uncertainty measures such as belief functions as points of a geometric space is also appealing because of its potential ability to provide a comprehensive framework capable of encompassing (most) mathematical descriptions of uncertainty (which we thoroughly reviewed in Chapter 6) in a single framework. We will return to this idea in Chapter 17.

Outline of Part II

From this perspective, Part II of this book is devoted to introducing the basic notions of the geometric approach to uncertainty, starting from the case of belief and possibility measures.

As the first pillar of the theory of evidence is the notion of a basic probability assignment, i.e., the concept of assigning masses directly to subsets rather than to elements of a sample space (frame), we first need to express basic probability assignments in a convex-geometric language. We do this in this chapter, which is centred on the notion of a *belief space* $\mathcal{B}$, i.e., the polytope of all belief functions definable on a given frame. This chapter also proposes a complementary, differential-geometric view of a belief space in terms of a recursive bundle space decomposition, associated with assigning mass recursively to subsets of increasing cardinality.

Belief functions, however, are useful only when combined in an evidence revision process. The original, and still by far the most popular, mechanism shaping this process is, in the theory of evidence, Dempster's rule. In Chapter 8, we proceed to study the behaviour of the rule of combination in our geometric framework. Both the global effect of Dempster's rule and its pointwise behaviour, when applied to single belief functions, is studied. While the former allows us to define the notion of a 'conditional subspace' as the locus of all possible futures of the current belief function, the latter results in an elegant geometric construction for Dempster combination which makes use of the 'foci' of these conditional subspaces.

Chapter 9 applies this geometric analysis to the other measures which carry the same information as a belief function, i.e., plausibility and commonality. We show there that sets of plausibility and commonality measures share a similar simplicial geometry with belief functions, by deriving explicit expressions for their Möbius transforms. The associated simplices can be shown to be congruent to the belief space associated with the same frame of discernment.

Chapter 9 widens the geometric approach to possibility theory. The geometry of consonant belief functions, as the counterparts of necessity measures in the theory of evidence, is described in terms of simplicial complexes. Consistent belief functions, the analogues of consistent knowledge bases in classical logic, are shown to have a similar geometric structure. The decomposition of belief functions into natural consistent components is outlined.

A full study of the transformation problem within the geometric approach is conducted in Part III.

Chapter outline

A central role is played by the notion of a belief space, as the space of all the belief functions one can define on a given frame of discernment. In Section 7.1, convexity and symmetries of the belief space are first studied by means of the Möbius inversion lemma.

In Section 7.2, instead, we show that every belief function can be uniquely decomposed as a convex combination of categorical belief functions (see Section 2.4.4), i.e., BFs assigning mass 1 to a single event, giving the belief space the form of a *simplex*, i.e., the convex closure of a set of $m+1$ (affinely independent) points embedded in $\mathbb{R}^m$. Various faces of $\mathcal{B}$ turn out to be associated with different classes of belief functions.

A differential-geometric view of the belief space is presented in Section 7.3, starting from the case of a ternary frame. With the aid of well-known combinatorial results, a recursive bundle structure of $\mathcal{B}$ is proved (Section 7.4), and an interpretation of its components (bases and fibres) in term of important classes of belief functions is provided.

As in all the following chapters, a list of open research questions (Section 7.5), whose solutions have the potential to lead to further significant developments, concludes Chapter 7.

7.1 The space of belief functions

Consider a frame of discernment Θ, and introduce in the Euclidean space $\mathbb{R}^{|2^\Theta|}$ an orthonormal reference frame $\{\mathbf{x}_A\}_{A\in 2^\Theta}$. Each vector $\mathbf{v} \in \mathbb{R}^{|2^\Theta|}$ can then be expressed in terms of this basis of vectors as

$$\mathbf{v} = \sum_{A\subseteq\Theta} v_A \mathbf{x}_A = [v_A, A \subseteq \Theta]'.$$

For instance, if the frame of discernment has cardinality 3, so that $\Theta = \{x, y, z\}$, each such vector has the form

$$\mathbf{v} = \Big[v_{\{x\}}, v_{\{y\}}, v_{\{z\}}, v_{\{x,y\}}, v_{\{x,z\}}, v_{\{y,z\}}, v_\Theta\Big]'.$$

As each belief function $Bel : 2^\Theta \to [0, 1]$ is completely specified by its belief values $Bel(A)$ on all the subsets of Θ, any such vector $\mathbf{v}$ is potentially a belief function, its component v_A measuring the belief value of A: $v_A = Bel(A)\ \forall A \subseteq \Theta$.

Definition 120. *The* belief space *associated with Θ is the set $\mathcal{B}_\Theta$ of vectors $\mathbf{v}$ of $\mathbb{R}^{|2^\Theta|}$ such that there exists a belief function $Bel : 2^\Theta \to [0, 1]$ whose belief values correspond to the components of $\mathbf{v}$, for an appropriate ordering of the subsets of Θ.*

We denote the vector of $\mathbb{R}^{|2^\Theta|}$ which corresponds to a belief function Bel by $\mathbf{bel}$. In the following, whenever no ambiguity exists about the frame of discernment being considered, we will drop the dependency on the underlying frame Θ, and denote the belief space by $\mathcal{B}$.

7.1.1 The simplex of dominating probabilities

To have a first idea of the shape of the belief space, it can be useful to start by understanding the geometric properties of Bayesian belief functions.

Lemma 2. *Whenever $Bel = P : 2^\Theta \to [0, 1]$ is a Bayesian belief function defined on a frame Θ, and Bel is an arbitrary subset of Θ, we have that*

$$\sum_{A\subseteq B} P(A) = 2^{|B|-1} P(B).$$

Proof. The sum can be rewritten as $\sum_{\theta\in B} k_\theta P(\{\theta\})$, where k_θ is the number of subsets A of Bel containing θ. But $k_\theta = 2^{|B|-1}$ for each singleton, so that

$$\sum_{A\subseteq B} P(A) = 2^{|B|-1} \sum_{\theta\in B} P(\{\theta\}) = 2^{|B|-1} P(B).$$

□

As a consequence, all Bayesian belief functions are constrained to belong to a well-determined region of the belief space.

Corollary 1. *The set $\mathcal{P}$ of all the Bayesian belief functions which can be defined on a frame of discernment Θ is a subset of the $(|\Theta| - 1)$-dimensional region*

$$\mathcal{L} = \left\{ Bel : 2^\Theta \to [0,1] \in \mathcal{B} \text{ s.t. } \sum_{A \subseteq \Theta} Bel(A) = 2^{|\Theta|-1} \right\} \tag{7.1}$$

of the belief space $\mathcal{B}$, which we call the limit simplex.[63]

Theorem 6. *Given a frame of discernment Θ, the corresponding belief space $\mathcal{B}$ is a subset of the region of $\mathbb{R}^{|2^\Theta|}$ 'dominated' by the limit simplex $\mathcal{L}$:*

$$\sum_{A \subseteq \Theta} Bel(A) \leq 2^{|\Theta|-1},$$

where the equality holds iff Bel is Bayesian.

Proof. The sum $\sum_{A \subseteq \Theta} Bel(A)$ can be written as

$$\sum_{A \subseteq \Theta} Bel(A) = \sum_{i=1}^{k} a_i \cdot m(A_i),$$

where k is the number of focal elements of Bel and a_i is the number of subsets of Θ which include the i-th focal element A_i, namely $a_i = |\{B \subset \Theta \text{ s.t. } B \supseteq A_i\}|$. Obviously, $a_i = 2^{|\Theta \setminus A_i|} \leq 2^{|\Theta|-1}$ and the equality holds iff $|A_i| = 1$. Therefore

$$\sum_{A \subseteq \Theta} Bel(A) = \sum_{i=1}^{k} m(A_i) 2^{|\Theta \setminus A_i|} \leq 2^{|\Theta|-1} \sum_{i=1}^{k} m(A_i) = 2^{|\Theta|-1} \cdot 1 = 2^{|\Theta|-1},$$

where the equality holds iff $|A_i| = 1$ for every focal element of Bel, i.e., Bel is Bayesian.

□

It is important to point out that $\mathcal{P}$ does not, in general, exhaust the limit simplex $\mathcal{L}$. Similarly, the belief space does not necessarily coincide with the entire region bounded by $\mathcal{L}$.

[63] It can be proved that $\mathcal{L}$ is indeed a simplex: http://www.cis.upenn.edu/~cis610/geombchap2.pdf.

7.1.2 Dominating probabilities and L_1 norm

Another hint about the structure of $\mathcal{B}$ comes from a particular property of Bayesian belief functions with respect to the classical L_1 distance in the Cartesian space $\mathbb{R}^{|\Theta|}$. Let $\mathcal{C}_{Bel}$ denote the core (the union (2.1) of its focal elements) of a belief function Bel, and let us introduce the following order relation:

$$Bel \geq Bel' \quad \Leftrightarrow \quad Bel(A) \geq Bel'(A) \; \forall A \subseteq \Theta. \tag{7.2}$$

Lemma 3. *If $Bel \geq Bel'$, then $\mathcal{C}_{Bel} \subseteq \mathcal{C}_{Bel'}$.*

Proof. Trivially, since $Bel(A) \geq Bel'(A)$ for every $A \subseteq \Theta$, this holds for $\mathcal{C}_{Bel'}$ too, so that $Bel(\mathcal{C}_{Bel'}) = 1$. But, then, $\mathcal{C}_{Bel} \subseteq \mathcal{C}_{Bel'}$.

□

Theorem 7. *If $Bel : 2^\Theta \rightarrow [0,1]$ is a belief function defined on a frame Θ, then*

$$\|Bel - P\|_{L_1} = \sum_{A \subseteq \Theta} |Bel(A) - P(A)| = const$$

for every Bayesian belief function $P : 2^\Theta \rightarrow [0,1]$ dominating Bel according to the order relation (7.2).

Proof. Lemma 3 guarantees that $\mathcal{C}_P \subseteq \mathcal{C}_{Bel}$, so that $P(A) - Bel(A) = 1 - 1 = 0$ for every $A \supseteq \mathcal{C}_{Bel}$. On the other hand, if $A \cap \mathcal{C}_{Bel} = \emptyset$, then $P(A) - Bel(A) = 0 - 0 = 0$. We are left with sets which amount to the union of a non-empty proper subset of $\mathcal{C}_{Bel}$ and an arbitrary subset of $\Theta \setminus \mathcal{C}_{Bel}$. Given $A \subseteq \mathcal{C}_{Bel}$, there exist $2^{|\Theta \setminus \mathcal{C}_{Bel}|}$ subsets of the above type which contain A. Therefore,

$$\sum_{A \subseteq \Theta} |Bel(A) - P(A)| = 2^{|\Theta \setminus \mathcal{C}_{Bel}|} \left[\sum_{A \subseteq \mathcal{C}_{Bel}} P(A) - \sum_{A \subseteq \mathcal{C}_{Bel}} Bel(A) \right].$$

Finally, by Lemma 2 the latter is equal to

$$f(Bel) \doteq 2^{|\Theta \setminus \mathcal{C}_{Bel}|} \left[2^{|\mathcal{C}_{Bel}|-1} - 1 - \sum_{A \subseteq \mathcal{C}_{Bel}} Bel(A) \right]. \tag{7.3}$$

□

The L_1 distance (7.3) between a belief function Bel and any Bayesian BF P dominating it is not a function of P, but depends only on Bel. A probability distribution satisfying the hypothesis of Theorem 7 is said to be *consistent* with Bel [1086]. Ha et al. [758] proved that the set $\mathcal{P}[Bel]$ of probability measures consistent with a given belief function Bel can be expressed (in the probability simplex $\mathcal{P}$) as the sum of the probability simplices associated with its focal elements $A_i, \; i = 1, \ldots, k$, weighted by the corresponding masses:

$$\mathcal{P}[Bel] = \sum_{i=1}^{k} m(A_i) \, \text{conv}(A_i),$$

where $\text{conv}(A_i)$ is the convex closure of the probability measures $\{P_\theta : \theta \in A_i\}$ assigning mass 1 to a single element θ of A_i. The analytical form of the set $\mathcal{P}[Bel]$ of consistent probabilities was studied further in [357].

7.1.3 Exploiting the Möbius inversion lemma

These preliminary results suggest that the belief space may indeed have the form of a simplex. To continue our analysis, we need to resort to the axioms of basic probability assignments (Definition 4).

Given a belief function Bel, the corresponding basic probability assignment can be found by applying the Möbius inversion lemma (2.3), which we recall here:

$$m(A) = \sum_{B \subseteq A} (-1)^{|A \setminus B|} Bel(B). \tag{7.4}$$

To determine whether a vector $\mathbf{v} \in \mathbb{R}^{|2^\Theta|}$ corresponds to a belief function Bel, we can simply compute its Möbius transform,

$$m_{\mathbf{v}}(A) = \sum_{B \subseteq A} (-1)^{|A \setminus B|} v_B,$$

and check whether the resulting $m_{\mathbf{v}}$ satisfies the axioms that basic probability assignments must obey.

In detail, the normalisation constraint $\sum_{A \subseteq \Theta} m_{\mathbf{v}}(A) = 1$ translates into $\mathcal{B} \subseteq \{\mathbf{v} : v_\Theta = 1\}$. The *positivity* condition is more interesting, for it implies an inequality which echoes the third axiom of belief functions (see Definition 27 or [1583]):

$$v_A - \cdots + (-1)^{|A \setminus B|} \sum_{|B|=k} v_B + \cdots + (-1)^{|A|-1} \sum_{\theta \in \Theta} v_{\{\theta\}} \geq 0 \quad \forall A \subseteq \Theta. \tag{7.5}$$

Example 17: ternary frame. *Let us see how these constraints act on the belief space in the case of a ternary frame* $\Theta = \{\theta_1, \theta_2, \theta_3\}$. *After introducing the notation*

$$x = v_{\{\theta_1\}}, \; y = v_{\{\theta_2\}}, \; z = v_{\{\theta_3\}}, \; u = v_{\{\theta_1,\theta_2\}}, \; v = v_{\{\theta_1,\theta_3\}}, \; w = v_{\{\theta_2,\theta_3\}},$$

the positivity constraint (7.5) can be rewritten as

$$\mathcal{B} : \begin{cases} x \geq 0, \quad u \geq (x+y), \\ y \geq 0, \quad v \geq (x+z), \\ z \geq 0, \quad w \geq (y+z), \\ 1 - (u+v+w) + (x+y+z) \geq 0. \end{cases} \tag{7.6}$$

Note that v_Θ is not needed as a coordinate, for it can be recovered by normalisation. It follows that the belief space $\mathcal{B}$ is the set of vectors $\mathbf{bel} = \mathbf{v} = [x, y, z, u, v, w]'$ of $\mathbb{R}^6$ which satisfy (7.6).

7.1.4 Convexity of the belief space

Now, all the positivity constraints of (7.5) (which determine the shape of the belief space $\mathcal{B}$) are of the form

$$\sum_{A \in \mathcal{A}} v_A \geq \sum_{B \in \mathcal{B}} v_B, \tag{7.7}$$

where $\mathcal{A}$ and $\mathcal{B}$ are two disjoint collections of subsets of Θ, as the above example and (7.6) confirm. We immediately obtain the following result.

Theorem 8. *The belief space $\mathcal{B}$ is convex.*

Proof. Let us consider two points $\mathbf{bel}^0, \mathbf{bel}^1 \in \mathcal{B}$ of the belief space, corresponding to two belief functions, and prove that all the points $\mathbf{bel}^\alpha$ of the segment $\mathbf{bel}^0 + \alpha(\mathbf{bel}^1 - \mathbf{bel}^0)$, $0 \leq \alpha \leq 1$, belong to $\mathcal{B}$.

Since $\mathbf{bel}^0$ and $\mathbf{bel}^1$ belong to $\mathcal{B}$, they both obey each of the constraints (7.7):

$$\sum_{A \in \mathcal{A}} v_A^0 \geq \sum_{B \in \mathcal{B}} v_B^0, \quad \sum_{A \in \mathcal{A}} v_A^1 \geq \sum_{B \in \mathcal{B}} v_B^1.$$

Hence, for every point $\mathbf{bel}^\alpha$ with coordinates v_A^α, $A \subset \Theta$, we have that

$$\begin{aligned} \sum_{A \in \mathcal{A}} v_A^\alpha &= \sum_{A \in \mathcal{A}} [v_A^0 + \alpha(v_A^1 - v_A^0)] = \sum_{A \in \mathcal{A}} v_A^0 + \alpha \sum_{A \in \mathcal{A}} (v_A^1 - v_A^0) \\ &= (1-\alpha) \sum_{A \in \mathcal{A}} v_A^0 + \alpha \sum_{A \in \mathcal{A}} v_A^1 \geq (1-\alpha) \sum_{B \in \mathcal{B}} v_B^0 + \alpha \sum_{B \in \mathcal{B}} v_B^1 \\ &= \sum_{B \in \mathcal{B}} [v_B^0 + \alpha(v_B^1 - v_B^0)] = \sum_{B \in \mathcal{B}} v_B^\alpha. \end{aligned}$$

Hence, $\mathbf{bel}^\alpha$ also satisfies the constraints whenever $\alpha \in (0, 1)$. Therefore, $\mathcal{B}$ is convex.

□

It is well known that belief functions are a special type of coherent lower probabilities (see Chapter 6, Section 6.1.1), which in turn can be seen as a subclass of lower previsions (see [1874], Section 5.13). Walley proved that coherent lower probabilities are closed under convex combination – this implies that convex combinations of belief functions (completely monotone lower probabilities) are still coherent. Theorem 8 is a stronger result, stating that they are also completely monotone.

7.2 Simplicial form of the belief space

In fact, for any given frame of discernment, the belief space $\mathcal{B}$ is a *simplex*, i.e., the convex closure of a collection of affinely independent[64] points (vertices). Its vertices are the special belief functions which assign unitary mass to a single subset of the frame of discernment.

We define the *categorical* belief function focused on $A \subseteq \Theta$, which we denote by Bel_A, as the unique belief function with BPA $m_{Bel_A}(A) = 1$, $m_{Bel_A}(B) = 0$ for all $B \neq A$. As usual, we denote by $\mathbf{bel}$ the vector of $\mathbb{R}^{N-2}$, $N = |2^\Theta|$, in 1–1 correspondence with the belief function Bel on Θ.

Theorem 9. *Every vector representation* $\mathbf{bel}$ *of a belief function* $Bel \in \mathcal{B}$ *can be uniquely expressed as a convex combination of the vectors* $\mathbf{bel}_A$, $\emptyset \subsetneq A \subsetneq \Theta$, *representing the categorical belief functions on* Θ*:*

$$\mathbf{bel} = \sum_{\emptyset \subsetneq A \subsetneq \Theta} m(A)\, \mathbf{bel}_A, \tag{7.8}$$

with coefficients given by the basic probability assignment m *of Bel.*

Proof. Every belief function Bel in $\mathcal{B}$ is represented by the vector of $\mathbb{R}^{N-2}$

$$\mathbf{bel} = \left[\sum_{B \subseteq A} m(B),\ \emptyset \subsetneq A \subsetneq \Theta \right]' = \sum_{\emptyset \subsetneq A \subsetneq \Theta} m(A) \big[\delta(B),\ \emptyset \subsetneq B \subsetneq \Theta\big]',$$

where $\delta(B) = 1$ iff $B \supseteq A$. As the vector $\big[\delta(B),\ \emptyset \subsetneq B \subsetneq \Theta\big]' = \mathbf{bel}_A$ is, by definition, the vector of belief values associated with the categorical belief function Bel_A, we have the thesis.

□

This 'convex decomposition' property is easily generalised in the following way.

Theorem 10. *The set of all the belief functions with focal elements in a given collection* $\mathcal{X} \subset 2^{2^\Theta}$ *is closed and convex in* $\mathcal{B}$*, namely*

$$\{\mathbf{bel} : \mathcal{E}_{Bel} \subset \mathcal{X}\} = Cl(\{\mathbf{bel}_A : A \in \mathcal{X}\}),$$

where $\mathcal{E}_{Bel}$ *denotes the list of focal elements of Bel, and* Cl *is the convex closure of a set of points of a Cartesian space* $\mathbb{R}^m$*:*

$$Cl(\mathbf{v}_1, \ldots, \mathbf{v}_k) \doteq \left\{ \mathbf{v} \in \mathbb{R}^m : \mathbf{v} = \alpha_1 \mathbf{v}_1 + \cdots + \alpha_k \mathbf{v}_k,\ \sum_i \alpha_i = 1,\ \alpha_i \geq 0\ \forall i \right\}. \tag{7.9}$$

[64] A collection of vectors $\mathbf{v}_0, \mathbf{v}_1, \ldots, \mathbf{v}_k$ are said to be *affinely independent* iff the difference vectors $\mathbf{v}_1 - \mathbf{v}_0, \ldots, \mathbf{v}_k - \mathbf{v}_0$ are linearly independent.

Proof. By definition,

$$\{\mathbf{bel} : \mathcal{E}_{Bel} \subset \mathcal{X}\} = \left\{ \mathbf{bel} = \left[\sum_{B \subseteq A, B \in \mathcal{E}_{Bel}} m(B), \emptyset \subsetneq A \subsetneq \Theta \right]', \mathcal{E}_{Bel} \subset X \right\}.$$

But

$$\begin{aligned} \mathbf{bel} = \left[\sum_{B \subseteq A, B \in \mathcal{E}_{Bel}} m(B), \emptyset \subsetneq A \subsetneq \Theta \right]' &= \sum_{B \in \mathcal{E}_{Bel}} m(B)\, \mathbf{bel}_B \\ &= \sum_{B \in \mathcal{X}} m(B)\, \mathbf{bel}_B, \end{aligned}$$

after trivially extending m to the elements $B \in \mathcal{X} \setminus \mathcal{E}_{Bel}$, by enforcing the condition $m(B) = 0$ for those elements. Since m is a basic probability assignment, $\sum_{B \in \mathcal{X}} m(B) = \sum_{B \in \mathcal{E}_{Bel}} m(B) + \sum_{B \in \mathcal{X} \setminus \mathcal{E}_{Bel}} m(B) = 1 + 0 = 1$ and the thesis follows.

□

As a direct consequence, we have the following corollary.

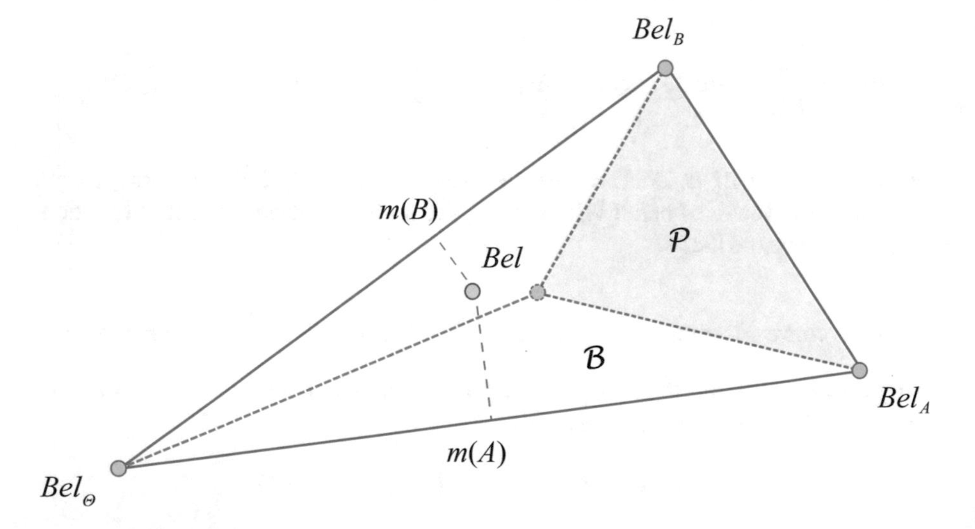

Fig. 7.1: Illustration of the concept of a *belief space*, the space of all belief measures on a given frame of discernment.

Corollary 2. *The belief space $\mathcal{B}$ is the convex closure (7.9) of all the (vectors representing) categorical belief functions, namely*

$$\mathcal{B} = Cl(\mathbf{bel}_A, \emptyset \subsetneq A \subseteq \Theta). \tag{7.10}$$

As it is easy to see that the vectors $\{\mathbf{bel}_A, \emptyset \subsetneq A \subsetneq \Theta\}$ associated with all categorical belief functions (except the vacuous one) are linearly independent, the vectors $\{\mathbf{bel}_A - \mathbf{bel}_\Theta = \mathbf{bel}_A, \emptyset \subsetneq A \subsetneq \Theta\}$ (since $\mathbf{bel}_\Theta = \mathbf{0}$ is the origin of $\mathbb{R}^{N-2}$) are also linearly independent, i.e., the vertices $\{\mathbf{bel}_A, \emptyset \subsetneq A \subseteq \Theta\}$ of the belief space (7.10) are affinely independent. Hence we have the following result.

Corollary 3. *$\mathcal{B}$ is a simplex.*

The simplicial form of the belief space is illustrated in Fig. 7.1.

Example 18: simplicial structure on a binary frame. *As an example, let us consider a frame of discernment containing only two elements, $\Theta_2 = \{x, y\}$. In this very simple case each belief function $Bel : 2^{\Theta_2} \rightarrow [0,1]$ is completely determined by its belief values*[65] *$Bel(x), Bel(y)$, as $Bel(\Theta) = 1$ and $Bel(\emptyset) = 0 \; \forall Bel \in \mathcal{B}$.*

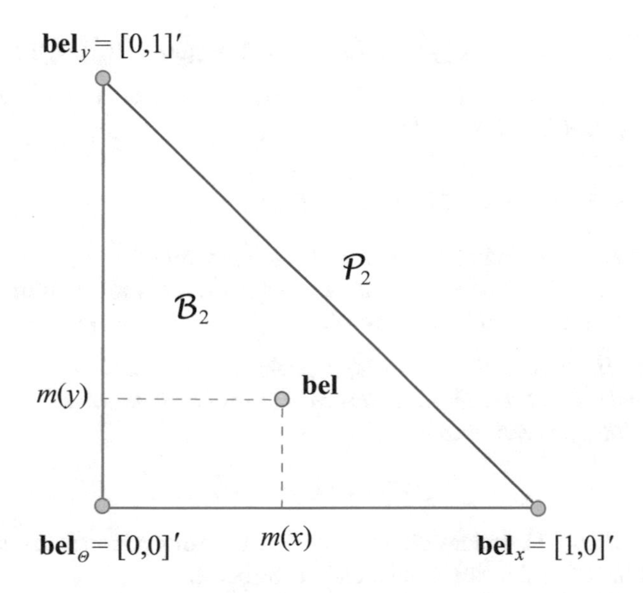

Fig. 7.2: The belief space $\mathcal{B}_2$ for a binary frame is a triangle in $\mathbb{R}^2$ whose vertices are the vectors $\mathbf{bel}_x, \mathbf{bel}_y, \mathbf{bel}_\Theta$ associated with the categorical belief functions Bel_x, Bel_y, Bel_Θ focused on $\{x\}, \{y\}, \Theta$, respectively.

We can therefore collect them in a vector of $\mathbb{R}^{N-2} = \mathbb{R}^2$ (since $N = 2^2 = 4$):

$$\mathbf{bel} = [Bel(x) = m(x), Bel(y) = m(y)]' \in \mathbb{R}^2. \tag{7.11}$$

[65] With a harmless abuse of notation we denote the value of a belief function on a singleton x by $Bel(x)$ rather than $Bel(\{x\})$. Accordingly, we will write $m(x)$ instead of $m(\{x\})$, and denote the vector of the belief space associated with the categorical BF focused on x by $\mathbf{bel}_x$.

Since $m(x) \geq 0$, $m(y) \geq 0$ and $m(x) + m(y) \leq 1$, we can easily infer that the set $\mathcal{B}_2$ of all the possible belief functions on Θ_2 can be depicted as the triangle in the Cartesian plane shown in Fig. 7.2, whose vertices are the points

$$\mathbf{bel}_\Theta = [0,0]', \quad \mathbf{bel}_x = [1,0]', \quad \mathbf{bel}_y = [0,1]'$$

(compare (7.10)). These correspond (through (7.11)) to the 'vacuous' belief function Bel_Θ ($m_\Theta(\Theta) = 1$), the categorical Bayesian BF Bel_x with $m_x(x) = 1$ and the categorical Bayesian belief function Bel_y with $m_y(y) = 1$, respectively.

Bayesian belief functions on Θ_2 obey the constraint $m(x) + m(y) = 1$, and are therefore located on the segment $\mathcal{P}_2$ joining $\mathbf{bel}_x = [1,0]'$ and $\mathbf{bel}_y = [0,1]'$. Clearly, the L_1 distance between Bel and any Bayesian BF dominating it is constant and equal to $1 - m(x) - m(y)$ (see Theorem 7).

The limit simplex (2.17) is the locus of all the set functions $\varsigma : 2^\Theta \rightarrow \mathbb{R}$ such that

$$\varsigma(\emptyset) + \varsigma(x) + \varsigma(y) + \varsigma(\{x,y\}) = 1 + \varsigma(x) + \varsigma(y) = 2,$$

i.e. $\varsigma(x) + \varsigma(y) = 1$. Clearly, as $P(x) + P(y) = 1$, $\mathcal{P}_2$ is a proper[66] *subset of the limit simplex (recall Section 7.1.1).*

7.2.1 Faces of $\mathcal{B}$ as classes of belief functions

Obviously, a Bayesian belief function (a finite probability) is a BF with focal elements in the collection of singletons $\mathcal{X} = \{\{x_1\}, \ldots, \{x_n\}\}$. Immediately, by Theorem 10, we have the following corollary.

Corollary 4. *The region of the belief space which corresponds to probability measures is the part of its boundary which is the convex combination of all the categorical probabilities, i.e., the simplex*

$$\mathcal{P} = Cl(\mathbf{bel}_x, x \in \Theta).$$

$\mathcal{P}$ is then an $(n-1)$-dimensional face of $\mathcal{B}$ (whose dimension is instead $N-2 = 2^n - 2$, as it has $2^n - 1$ affinely independent vertices).

Some one-dimensional faces of the belief space also have an intuitive meaning in terms of belief values. Consider the segments $Cl(\mathbf{bel}_\Theta, \mathbf{bel}_A)$ joining the vacuous belief function $\mathbf{bel}_\Theta$ ($m_\Theta(\Theta) = 1$, $m_\Theta(B) = 0 \; \forall B \neq \Theta$) with a categorical BF $\mathbf{bel}_A$. Points $\mathbf{bel}$ of these segments can be written as a convex combination as $\mathbf{bel} = \alpha \; \mathbf{bel}_A + (1-\alpha) \; \mathbf{bel}_\Theta$. The corresponding belief function Bel has BPA $m_{Bel}(A) = \alpha$, $m_{Bel}(\Theta) = 1 - \alpha$, i.e., Bel is a simple support function focused on A (Chapter 2).

The union of these segments for all events A,

$$\mathcal{S} = \bigcup_{A \subset \Theta} Cl(\mathbf{bel}_\Theta, \mathbf{bel}_A),$$

[66] The limit simplex is in fact the region of normalised sum functions (Section 7.3.3) ς that satisfy the constraint $\sum_{x \in \Theta} m_\varsigma(x) = 1$.

is the region $\mathcal{S}$ of all simple support belief functions on Θ. In the binary case (Fig. 7.2) simple support functions focused on $\{x\}$ lie on the horizontal segment $Cl(\mathbf{bel}_\Theta, \mathbf{bel}_x)$, whereas simple support belief functions focused on $\{y\}$ form the vertical segment $Cl(\mathbf{bel}_\Theta, \mathbf{bel}_y)$.

7.3 The differential geometry of belief functions

We have learned that belief functions can be identified with vectors of a sufficiently large Cartesian space ($\mathbb{R}^N$, where $N = 2^{|\Theta|} - 2$ and $|\Theta|$ is the cardinality of the frame on which the belief functions are defined). More precisely, the set of vectors $\mathcal{B}$ of $\mathbb{R}^N$ which correspond to belief functions, or the 'belief space', is a simplex whose vertices are the categorical belief functions assigning mass 1 to a single event.

As we will show in this second part of the chapter, we can also think of the mass $m(A)$ given to each event A as *recursively* assigned to subsets of increasing size. Geometrically, this translates as a recursive decomposition of the space of belief functions, which can be formally described through the differential-geometric notion of a *fibre bundle* [551]. A fibre bundle is a generalisation of the familiar idea of the Cartesian product, in which each point of the (total) space under consideration can be smoothly projected onto a *base space*, defining a number of *fibres* formed by points which project onto the same element of the base. In our case, as we will see in the following, $\mathcal{B}$ can be decomposed $n = |\Theta|$ times into bases and fibres, which are themselves simplices and possess natural interpretations in terms of degrees of belief. Each level $i = 1, \ldots, n$ of this decomposition reflects nothing but the assignment of basic probabilities to size-i events.

After giving an informal presentation of the way the BPA mechanism induces a recursive decomposition of $\mathcal{B}$, we will analyse a simple case study of a ternary frame (Section 7.3.1) to get an intuition about how to prove our conjecture about the bundle structure of the belief space in the general case, and give a formal definition of a smooth fibre bundle (Section 7.3.2). After noticing that points of $\mathbb{R}^{N-2}$ outside the belief space can be seen as (normalised) sum functions (Section 7.3.3), we will proceed to prove the recursive bundle structure of the space of all sum functions (Section 7.4). As $\mathcal{B}$ is immersed in this Cartesian space, it inherits a 'pseudo'-bundle structure (Section 7.4.2) in which bases and fibres are no longer vector spaces but are simplices in their own right (Section 7.4.3), and possess meanings in terms of i-additive belief functions.

7.3.1 A case study: The ternary case

Let us first consider the structure of the belief space for a frame of cardinality $n = 3$, $\Theta = \{x, y, z\}$, according to the principle of assigning mass recursively to subsets of increasing size. In this case each belief function Bel is represented by the vector

$$\mathbf{bel} = [Bel(x), Bel(y), Bel(z), Bel(\{x,y\}), Bel(\{x,z\}), Bel(\{y,z\})]' \in \mathbb{R}^6.$$

If the mass not assigned to singletons, $1 - m(x) - m(y) - m(z)$, is attributed to $A = \Theta$, we have that $Bel(\{x, y\}) = m(\{x, y\}) + m(x) + m(y) = m(x) + m(y)$ and so on for all size-2 events, so that Bel belongs to the three-dimensional region

$$\Big\{\mathbf{bel} : 0 \leq Bel(x) + Bel(y) + Bel(z) \leq 1, Bel(\{x, y\}) = Bel(x) + Bel(y), \\ Bel(\{x, z\}) = Bel(x) + Bel(z), Bel(\{y, z\}) = Bel(y) + Bel(z) \Big\}, \tag{7.12}$$

which we can denote by $\mathcal{D}$. It is easy to realise that any $\mathbf{bel} \in \mathcal{B}_3$ on Θ_3 can be mapped onto a point $\pi[\mathbf{bel}]$ of $\mathcal{D}$, equal to

$$\Big[Bel(x), Bel(y), Bel(z), Bel(x) + Bel(y), Bel(x) + Bel(z), Bel(y) + Bel(z)\Big]', \tag{7.13}$$

through a projection map $\pi : \mathcal{B} \to \mathcal{D}$. Let us call $\mathcal{D}$ the *base* of $\mathcal{B}_3$. Such a base admits as a coordinate chart the basic probabilities of the singletons, as each point $\mathbf{d} \in \mathcal{D}$ can be written as $\mathbf{d} = [m(x), m(y), m(z)]'$.

Given a point $\mathbf{d} \in \mathcal{D}$ of the base, it is interesting to study the set of belief functions $Bel \in \mathcal{B}_3$ which have $\mathbf{d}$ as a projection, $\mathcal{F}(\mathbf{d}) \doteq \{\mathbf{bel} : \pi[\mathbf{bel}] = \mathbf{d}\}$. By virtue of the constraint acting on BPAs, such belief functions have to satisfy the following constraints:

$$\begin{cases} m(\{x, y\}) \geq 0 \equiv \; Bel(\{x, y\}) \geq m(x) + m(y), \\ m(\{x, z\}) \geq 0 \equiv \; Bel(\{x, z\}) \geq m(x) + m(z), \\ m(\{y, z\}) \geq 0 \equiv \; Bel(\{y, z\}) \geq m(y) + m(z), \\ m(\Theta) \geq 0 \\ \quad \equiv Bel(\{x, y\}) + Bel(\{x, z\}) + Bel(\{y, z\}) \leq 1 + m(x) + m(y) + m(z), \end{cases}$$

where $\equiv$ denotes equivalence. Each BF $\mathbf{d} \in \mathcal{D}$ on the base is associated with a whole 'fibre' $\mathcal{F}(\mathbf{d})$ of belief functions projecting onto $\mathbf{d}$ (as they have the same BPA on singletons),

$$\mathcal{F}(\mathbf{d}) = \Big\{ \mathbf{bel} \in \mathcal{B}_3 : Bel(x) = m(x), Bel(y) = m(y), Bel(z) = m(z), \\ Bel(\{x, y\}) \geq m(x) + m(y), \\ Bel(\{x, z\}) \geq m(x) + m(z), Bel(\{y, z\}) \geq m(y) + m(z) \Big\}. \tag{7.14}$$

Belief functions on $\mathcal{F}(\mathbf{d})$ can be parameterised by the three coordinates $m(\{x, y\})$, $m(\{x, z\})$ and $m(\{y, z\})$, the basic probabilities of events of size greater than 1 (see Fig. 7.3, right).

Given the nature of BPAs as simplicial coordinates (7.8), we can infer that the base (7.12) of the belief space $\mathcal{B}_3$ is in fact the three-dimensional simplex $\mathcal{D} = Cl(\mathbf{bel}_x, \mathbf{bel}_y, \mathbf{bel}_z, \mathbf{bel}_\Theta)$, as all the mass is distributed among $\{x\}, \{y\}, \{z\}$ and Θ in all possible ways (see Fig. 7.3, left). The base $\mathcal{D}$ is the simplex of all *quasi-Bayesian* or *discounted* belief functions on Θ_3, i.e., the belief functions for which $m(A) \neq 0$ iff $|A| = 1, n$. For each point $\mathbf{d}$ of the base, the corresponding fibre (7.14) is also a simplex of dimension 3 (Fig. 7.3, right).

Summarising, we have learned that (at least in the ternary case):

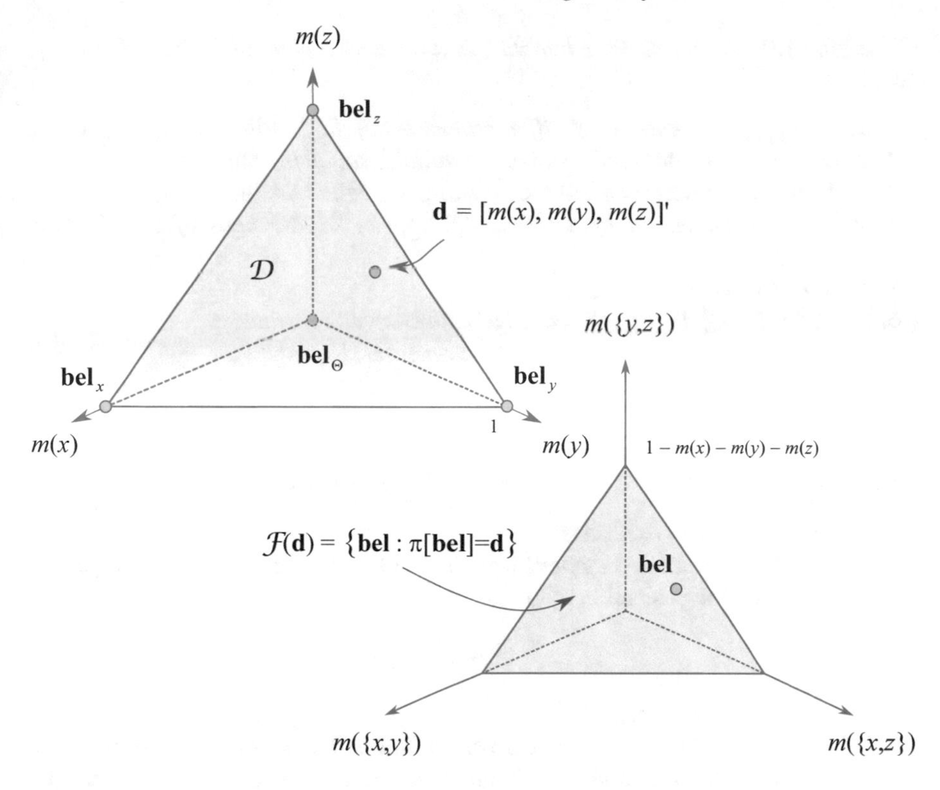

Fig. 7.3: Bundle structure of the belief space in the case of ternary frames $\Theta_3 = \{x, y, z\}$.

- the belief space can be decomposed into a base, i.e., the set of belief functions assigning mass zero to events A of size $1 < |A| < n$,

$$\mathcal{D} \doteq \Big\{ \mathbf{bel} \in \mathcal{B} : m(A) = 0, \forall \ A : 1 < |A| < n \Big\}, \tag{7.15}$$

  and a number of fibres $\mathcal{F}(\mathbf{d})$, each passing through a point $\mathbf{d}$ of the base;
- points of the base are parameterised by the masses assigned to singletons, $\mathbf{d} = [m(A), |A| = 1]'$, whereas points on the fibres have as coordinates the mass values assigned to higher-cardinality events, $[m(A), 1 < |A| < n]'$;
- both the base and the fibres are simplices.

As we will see in the following, the same type of decomposition applies recursively to general belief spaces, for an increasing size of the events we assign mass to. We first need to introduce the relevant mathematical tool.

7.3.2 Definition of smooth fibre bundles

Fibre bundles [551] generalise of the notion of the Cartesian product.

Definition 121. *A smooth fibre bundle ξ is a composed object $\{E, B, \pi, F, G, \mathcal{U}\}$, where:*

1. *E is an $(s+r)$-dimensional differentiable manifold,*[67] *called the* total space.
2. *B is an r-dimensional differentiable manifold, called the* base space.
3. *F is an s-dimensional differentiable manifold, called the* fibre.
4. *$\pi : E \to B$ is a smooth application of full rank r in each point of B, called the* projection.
5. *G is the* structure group.
6. *The atlas $\mathcal{U} = \{(U_\alpha, \phi_\alpha)\}$ defines a* bundle structure, *namely:*
   - *the base B admits a covering with open sets U_α such that*
   - *$E_\alpha \doteq \pi^{-1}(U_\alpha)$ is equipped with* smooth direct product coordinates

$$\begin{array}{rl} \phi_\alpha : \pi^{-1}(U_\alpha) \to & U_\alpha \times F \\ e & \mapsto (\phi'_\alpha(e), \phi''_\alpha(e)), \end{array} \tag{7.16}$$

satisfying two conditions:

- *the coordinate component with values in the base space is* compatible with the projection map

$$\pi \circ \phi_\alpha^{-1}(x, f) = x \tag{7.17}$$

or, equivalently, $\phi'_\alpha(e) = \pi(e)$;

- *the coordinate component with values on the fibre can be transformed, jumping from one coordinate chart into another, by means of elements of the structure group. Formally, the applications*

$$\begin{array}{rl} \lambda_{\alpha\beta} \doteq \phi_\beta \phi_\alpha^{-1} : U_{\alpha\beta} \times F \to & U_{\alpha\beta} \times F, \\ (x, f) & \mapsto (x, T^{\alpha\beta}(x) f), \end{array}$$

called gluing functions, *are implemented by means of transformations $T^{\alpha\beta}(x) : F \to F$ defined by applications from a domain $U_{\alpha\beta}$ to the structure group*

$$T^{\alpha\beta} : U_{\alpha\beta} \to G,$$

satisfying the following conditions:

$$T^{\alpha\beta} = (T^{\beta\alpha})^{-1}, \quad T^{\alpha\beta} T^{\beta\gamma} T^{\gamma\alpha} = 1. \tag{7.18}$$

Intuitively, the base space is covered by a number of open neighbourhoods $\{U_\alpha\}$, which induce a similar covering $\{E_\alpha = \pi^{-1}(U_\alpha)\}$ on the total space E. Points e of each neighbourhood E_α of the total space admit coordinates separable into two parts: the first one, $\phi'(e) = \pi(e)$, is the projection of e onto the base B, while the second part is its coordinate on the fibre F. Fibre coordinates are such that

[67] A *differentiable manifold* is a topological manifold with a globally defined differential structure. A *topological manifold* is a locally Euclidean Hausdorff space, i.e., such that each point of the manifold has a neighbourhood which is isomorphic to $\mathbb{R}^n$.

in the intersection of two different charts $E_\alpha \cap E_\beta$, they can be transformed into each other by means of the action of a group G.

In the following, all the manifolds involved will be linear spaces, so that each of them can be covered by a single chart. This makes the bundle structure trivial, i.e., the identity transformation. The reader can then safely ignore the gluing conditions on ϕ''_α.

7.3.3 Normalised sum functions

As the belief space does not exhaust the whole of $\mathbb{R}^{N-2}$ it is natural to wonder whether arbitrary points of $\mathbb{R}^{N-2}$, possibly 'outside' $\mathcal{B}$, have any meaningful interpretation in this framework [327]. In fact, each vector $\mathbf{v} = [v_A, \emptyset \subsetneq A \subseteq \Theta]' \in \mathbb{R}^{N-1}$ can be thought of as a set function $\varsigma : 2^\Theta \setminus \emptyset \to \mathbb{R}$ such that $\varsigma(A) = v_A$. By applying the Möbius transformation (2.3) to such functions ς, we obtain another set function $m_\varsigma : 2^\Theta \setminus \emptyset \to \mathbb{R}$ such that $\varsigma(A) = \sum_{B \subseteq A} m_\varsigma(B)$. In other words, each vector $\mathbf{v} \in \mathbb{R}^{N-1}$ can be thought of as a sum function. However, in contrast to basic probability assignments, the Möbius inverses m_ς of generic sum functions $\varsigma \in \mathbb{R}^{N-1}$ are not guaranteed to satisfy the non-negativity constraint $m_\varsigma(A) \not\geq 0$ $\forall A \subseteq \Theta$.

Now, the section $\{\mathbf{v} \in \mathbb{R}^{N-1} : v_\Theta = 1\}$ of $\mathbb{R}^{N-1}$ corresponds to the constraint $\varsigma(\Theta) = 1$. Therefore, all the points of this section are sum functions satisfying the normalisation axiom $\sum_{A \subset \Theta} m_\varsigma(A) = 1$, or *normalised sum functions* (NSFs). NSFs are the natural extensions of belief functions in our geometric framework.

7.4 Recursive bundle structure

7.4.1 Recursive bundle structure of the space of sum functions

We can now reinterpret our analysis of the ternary case by means of the formal definition of a smooth fibre bundle. The belief space $\mathcal{B}_3$ can in fact be equipped with a base (7.12), and a projection (7.13) from the total space $\mathcal{R}^6$ to the base, which generates fibres of the form (7.14). However, the original definition of a fibre bundle requires the spaces involved to be *manifolds*, whereas our analysis of the ternary case suggests we have to deal with *simplices* here.

It can be noticed, however, that the idea of recursively assigning mass to subsets of increasing size does not necessarily require the mass itself to be positive. In other words, this procedure can be applied to normalised sum functions, yielding a 'classical' fibre bundle structure for the space $\mathcal{S} = \mathbb{R}^{N-2}$ of all NSFs on Θ, in which all the bases and fibres involved are linear spaces. We will see in the following what happens when we consider proper belief functions.

Theorem 11. *The space $\mathcal{S} = \mathbb{R}^{N-2}$ of all the sum functions ς with their domain on a finite frame Θ of cardinality $|\Theta| = n$ has a recursive fibre bundle structure, i.e., there exists a sequence of smooth fibre bundles*

$$\xi_i = \left\{ \mathcal{F}_{\mathcal{S}}^{(i-1)}, \mathcal{D}_{\mathcal{S}}^{(i)}, \mathcal{F}_{\mathcal{S}}^{(i)}, \pi_i \right\}, \quad i = 1, \ldots, n-1,$$

where $\mathcal{F}_{\mathcal{S}}^{(0)} = \mathcal{S} = \mathbb{R}^{N-2}$, *and the* total space $\mathcal{F}_{\mathcal{S}}^{(i-1)}$, *the* base space $\mathcal{D}_{\mathcal{S}}^{(i)}$ *and the* fibre $\mathcal{F}_{\mathcal{S}}^{(i)}$ *of the i-th bundle level are linear subspaces of* $\mathbb{R}^{N-2}$ *of dimensions* $\sum_{k=i}^{n-1} \binom{n}{k}$, $\binom{n}{i}$ *and* $\sum_{k=i+1}^{n-1} \binom{n}{k}$, *respectively.*

Both $\mathcal{F}_{\mathcal{S}}^{(i-1)}$ *and* $\mathcal{D}_{\mathcal{S}}^{(i)}$ *admit a* global *coordinate chart. As*

$$\dim \mathcal{F}_{\mathcal{S}}^{(i-1)} = \sum_{k=i,\ldots,n-1} \binom{n}{k} = \left| \left\{ A \subset \Theta : i \leq |A| < n \right\} \right|,$$

each point ς^{i-1} *of* $\mathcal{F}_{\mathcal{S}}^{(i-1)}$ *can be written as*

$$\varsigma^{i-1} = \left[\varsigma^{i-1}(A), A \subset \Theta, i \leq |A| < n \right]'$$

and the smooth direct product coordinates *(7.16) at the i-th bundle level are*

$$\phi'(\varsigma^{i-1}) = \left\{ \varsigma^{i-1}(A), \ |A| = i \right\}, \quad \phi''(\varsigma^{i-1}) = \left\{ \varsigma^{i-1}(A), \ i < |A| < n \right\}.$$

The projection map π_i *of the i-th bundle level is a full-rank differentiable application*

$$\begin{array}{rcl} \pi_i : \mathcal{F}_{\mathcal{S}}^{(i-1)} & \to & \mathcal{D}_{\mathcal{S}}^{(i)}, \\ \varsigma^{i-1} & \mapsto & \pi_i[\varsigma^{i-1}], \end{array}$$

whose expression in this coordinate chart is

$$\pi_i[\varsigma^{i-1}] = [\varsigma^{i-1}(A), |A| = i]'. \tag{7.19}$$

Bases and fibres are simply geometric counterparts of the mass assignment mechanism. Once we have assigned a certain amount of mass to subsets of size smaller than i, the fraction of mass attributed to size-i subsets determines a point on a linear space $\mathcal{D}_{\mathcal{S}}^{(i)}$. For each point of $\mathcal{D}_{\mathcal{S}}^{(i)}$, the remaining mass can 'float' among the higher-size subsets, describing again a vector space $\mathcal{F}_{\mathcal{S}}^{(i)}$.

7.4.2 Recursive bundle structure of the belief space

As we have seen in the ternary example in Section 7.3.1, as the belief space is a simplex immersed in $\mathcal{S} = \mathbb{R}^{N-2}$, the fibres of $\mathbb{R}^{N-2}$ intersect the space of belief functions too. $\mathcal{B}$ then inherits some sort of bundle structure from the Cartesian space in which it is immersed. The belief space can also be recursively decomposed into fibres associated with events A of the same size. As one can easily conjecture, the intersections of the fibres of $\mathbb{R}^{N-2}$ with the simplex $\mathcal{B}$ are themselves simplices: the bases and fibres in the case of the belief space are therefore polytopes instead of linear spaces. Owing to the decomposition of $\mathbb{R}^{N-2}$ into bases and fibres, we can apply the non-negativity and normalisation constraints which distinguish belief functions from NSFs *separately at each level*, eliminating at each step the fibres passing through points of the base that do not meet these conditions.

We first need a simple combinatorial result.

Lemma 4. *The following inequality holds:*

$$\sum_{|A|=i} Bel(A) \leq 1 + \sum_{m=1}^{i-1} (-1)^{i-(m+1)} \binom{n-(m+1)}{i-m} \cdot \sum_{|B|=m} Bel(B),$$

and the upper bound is reached whenever

$$\sum_{|A|=i} m(A) = 1 - \sum_{|A|<i} m(A).$$

The bottom line of Lemma 4 is that, given a mass assignment for events of size $1, \ldots, i-1$, the upper bound for $\sum_{|A|=i} Bel(A)$ is obtained by assigning all the remaining mass to the collection of size i subsets.

Theorem 12. *The belief space $\mathcal{B} \subset \mathcal{S} = \mathbb{R}^{N-2}$ inherits by intersection with the recursive bundle structure of $\mathcal{S}$ a 'convex' bundle decomposition. Each i-th-level 'fibre' can be expressed as*

$$\mathcal{F}_{\mathcal{B}}^{(i-1)}(\mathbf{d}^1, \ldots, \mathbf{d}^{i-1}) = \Big\{ \mathbf{bel} \in \mathcal{B} \Big| V_i \wedge \cdots \wedge V_{n-1}(\mathbf{d}^1, \ldots, \mathbf{d}^{i-1}) \Big\}, \tag{7.20}$$

where $V_i(\mathbf{d}^1, \ldots, \mathbf{d}^{i-1})$ denotes the system of constraints

$$V_i(\mathbf{d}^1, \ldots, \mathbf{d}^{i-1}) : \begin{cases} m(A) \geq 0 & \forall A \subseteq \Theta : |A| = i, \\ \sum_{|A|=i} m(A) \leq 1 - \sum_{|A|<i} m(A), & \end{cases} \tag{7.21}$$

and depends on the mass assigned to lower-size subsets $\mathbf{d}^m = [m(A), |A| = m]'$, $m = 1, \ldots, i-1$.

The corresponding 'base' $\mathcal{D}_{\mathcal{B}}^{(i)}(\mathbf{d}^1, \ldots, \mathbf{d}^{i-1})$ is expressed in terms of basic probability assignments as the collection of $\mathbf{bel} \in \mathcal{F}^{(i-1)}(\mathbf{d}^1, \ldots, \mathbf{d}^{i-1})$ such that

$$\begin{cases} m(A) = 0 & \forall A : i < |A| < n, \\ m(A) \geq 0 & \forall A : |A| = i, \\ \sum_{|A|=i} m(A) \leq 1 - \sum_{|A|<i} m(A). & \end{cases} \tag{7.22}$$

7.4.3 Bases and fibres as simplices

Simplicial and bundle structures coexist in the space of belief functions, both of them consequences of the interpretation of belief functions as sum functions, and of the basic probability assignment machinery. It is then natural to conjecture that the bases and fibres of the recursive bundle decomposition of $\mathcal{B}$ must also be simplices of some kind, as suggested by the ternary example (Section 7.3.1).

Let us work recursively, and suppose we have already assigned a mass $k < 1$ to the subsets of size smaller than i:

$$m(A) = \text{const} = m_A, \; |A| < i. \tag{7.23}$$

All the admissible belief functions constrained by this mass assignment are then forced to live in the following $(i-1)$-th-level fibre of $\mathcal{B}$:

$$\mathcal{F}_{\mathcal{B}}^{(i-1)}(\mathbf{d}^1, \ldots, \mathbf{d}^{i-1}), \quad \mathbf{d}^j = [m_A, |A| = j]',$$

which, as the proof of Theorem 12 suggests, is a function of the mass assigned to lower-size events.

We have seen that such a fibre $\mathcal{F}_{\mathcal{B}}^{(i-1)}(\mathbf{d}^1, \ldots, \mathbf{d}^{i-1})$ admits a pseudo-bundle structure whose pseudo-base space is $\mathcal{D}_{\mathcal{B}}^{(i)}(\mathbf{d}^1, \ldots, \mathbf{d}^{i-1})$, given by (7.22). Let us denote by $k = \sum_{|A|<i} m_A$ the total mass already assigned to lower-size events, and call

$$\mathcal{P}^{(i)}(\mathbf{d}^1, \ldots, \mathbf{d}^{i-1}) \doteq \left\{ \mathbf{bel} \in \mathcal{F}_{\mathcal{B}}^{(i-1)}(\mathbf{d}^1, \ldots, \mathbf{d}^{i-1}) : \sum_{|A|=i} m(A) = 1 - k \right\}$$

$$\mathcal{O}^{(i)}(\mathbf{d}^1, \ldots, \mathbf{d}^{i-1}) \doteq \left\{ \mathbf{bel} \in \mathcal{F}_{\mathcal{B}}^{(i-1)}(\mathbf{d}^1, \ldots, \mathbf{d}^{i-1}) : m(\Theta) = 1 - k \right\}$$

the collections of belief functions on the fibre $\mathcal{F}_{\mathcal{B}}^{(i-1)}(\mathbf{d}^1, \ldots, \mathbf{d}^{i-1})$ assigning all the remaining basic probability $1-k$ to subsets of size i or to Θ, respectively.

As the simplicial coordinates of a BF in $\mathcal{B}$ are given by its basic probability assignment (7.8), each $\mathbf{bel} \in \mathcal{F}_{\mathcal{B}}^{(i-1)}(\mathbf{d}^1, \ldots, \mathbf{d}^{i-1})$ on such a fibre can be written as

$$\begin{aligned}
\mathbf{bel} &= \sum_{A \subseteq \Theta} m(A) \, \mathbf{bel}_A = \sum_{|A|<i} m_A \, \mathbf{bel}_A + \sum_{|A| \geq i} m(A) \, \mathbf{bel}_A \\
&= \frac{k}{k} \sum_{|A|<i} m_A \, \mathbf{bel}_A + \frac{1-k}{1-k} \sum_{|A| \geq i} m(A) \, \mathbf{bel}_A \\
&= \frac{k}{\sum_{|A|<i} m_A} \sum_{|A|<i} m_A \, \mathbf{bel}_A + \frac{1-k}{\sum_{|A| \geq i} m(A)} \sum_{|A| \geq i} m(A) \, \mathbf{bel}_A.
\end{aligned}$$

We can therefore introduce two new belief functions Bel_0, Bel' associated with any $\mathbf{bel} \in \mathcal{F}_{\mathcal{B}}^{(i-1)}(\mathbf{d}^1, \ldots, \mathbf{d}^{i-1})$, with basic probability assignments

$$\begin{aligned}
m_0(A) &\doteq \frac{m_A}{\sum_{|B|<i} m_B} \;\; |A| < i, \quad m_0(A) = 0 \; |A| \geq i, \\
m'(A) &\doteq \frac{m(A)}{\sum_{|B| \geq i} m(B)} \;\; |A| \geq i, \; m'(A) = 0 \; |A| < i,
\end{aligned}$$

respectively, and decompose $\mathbf{bel}$ as follows:

$$\mathbf{bel} = k \sum_{|A|<i} m_0(A) \, \mathbf{bel}_A + (1-k) \sum_{|A| \geq i} m'(A) \, \mathbf{bel}_A = k \, \mathbf{bel}_0 + (1-k) \, \mathbf{bel}',$$

where $\mathbf{bel}_0 \in Cl(\mathbf{bel}_A : |A| < i)$, $\mathbf{bel}' \in Cl(\mathbf{bel}_A : |A| \geq i)$. As

$$\sum_{A\subseteq\Theta} m_0(A) = \sum_{|A|<i} m_0(A) = \frac{\sum_{|A|<i} m_A}{\sum_{|A|<i} m_A} = 1, \quad \sum_{A\subseteq\Theta} m'(A) = 1,$$

both Bel_0 and Bel' are indeed admissible belief functions, Bel_0 assigning non-zero mass to subsets of size smaller than i only, Bel' assigning mass to subsets of size i or higher.

However, Bel_0 is the same for all belief functions on $\mathcal{F}_{\mathcal{B}}^{(i-1)}(\mathbf{d}^1,\ldots,\mathbf{d}^{i-1})$, as it is determined by the mass assignment (7.23). The other component Bel' (or, better, its vector representation $\mathbf{bel}'$) is instead free to vary in $Cl(\mathbf{bel}_A : |A| \geq i)$. Hence, we get the following convex expressions for $\mathcal{F}_{\mathcal{B}}^{(i-1)}$, $\mathcal{P}^{(i)}$ and $\mathcal{O}^{(i)}$ (neglecting for the sake of simplicity the dependence on $\mathbf{d}^1,\ldots,\mathbf{d}^{i-1}$ or, equivalently, on Bel_0):

$$\begin{aligned}
\mathcal{F}_{\mathcal{B}}^{(i-1)} &= \left\{ \mathbf{bel} = k\ \mathbf{bel}_0 + (1-k)\ \mathbf{bel}', \mathbf{bel}' \in Cl(\mathbf{bel}_A, |A| \geq i) \right\} \\
&= k\ \mathbf{bel}_0 + (1-k) Cl(\mathbf{bel}_A, |A| \geq i), \\
\mathcal{P}^{(i)} &= k\ \mathbf{bel}_0 + (1-k) Cl(\mathbf{bel}_A : |A| = i), \\
\mathcal{O}^{(i)} &= k\ \mathbf{bel}_0 + (1-k)\ \mathbf{bel}_\Theta.
\end{aligned} \tag{7.24}$$

By definition, the i-th base $\mathcal{D}_{\mathcal{B}}^{(i)}$ is the collection of belief functions such that

$$m(A) = 0 \quad i < |A| < n, \qquad m(A) = \text{const} = m_A \quad |A| < i,$$

so that points on $\mathcal{D}_{\mathcal{B}}^{(i)}$ are free to distribute the remaining mass to Θ or to size-i events only. Therefore, we obtain the following convex expression for the i-th-level base space $\mathcal{D}_{\mathcal{B}}^{(i)}$ of $\mathcal{B}$:

$$\begin{aligned}
\mathcal{D}_{\mathcal{B}}^{(i)} &= k\ \mathbf{bel}_0 + (1-k) Cl(\mathbf{bel}_A : |A| = i \text{ or } A = \Theta) \\
&= Cl(k\ \mathbf{bel}_0 + (1-k)\ \mathbf{bel}_A : |A| = i \text{ or } A = \Theta) \\
&= Cl(k\ \mathbf{bel}_0 + (1-k)\ \mathbf{bel}_\Theta, k\ \mathbf{bel}_0 + (1-k)\ \mathbf{bel}_A : |A| = i) \\
&= Cl(\mathcal{O}^{(i)}, \mathcal{P}^{(i)}).
\end{aligned}$$

In the ternary case in Section 7.3.1 we find, in particular, that

$$\mathcal{O}^{(1)} = \mathbf{bel}_\Theta,\ \mathcal{P}^{(1)} = \mathcal{P} = Cl(\mathbf{bel}_x, \mathbf{bel}_y, \mathbf{bel}_z),\ \mathcal{D}^{(1)} = Cl(\mathcal{O}^{(1)}, \mathcal{P}^{(1)}).$$

The elements of this bundle decomposition possess a natural meaning in terms of belief values. In particular, $\mathcal{P}^{(1)} = \mathcal{P}$ is the set of all the Bayesian belief functions, while $\mathcal{D}^{(1)}$ is the collection of all the *discounted* probabilities [1583], i.e., belief functions of the form $(1-\epsilon)P + \epsilon\ \mathbf{bel}_\Theta$, with $0 \leq \epsilon \leq 1$ and $P \in \mathcal{P}$.

On the other hand, in the literature, belief functions assigning mass to events of cardinality smaller than a certain size i are called *i-additive belief functions* [1296]. It is clear that the set $\mathcal{P}^{(i)}$ (7.24) is nothing but the collection of all i-additive BFs. The i-th-level base of $\mathcal{B}$ can then be interpreted as the region of all 'discounted' i-additive belief functions.

7.5 Research questions

This geometric analysis of belief functions on finite domains, although rigorous and quite exhaustive, opens up a number of intriguing research questions.

Question 1. First of all, is it possible to extend the results of Chapter 7 to the case of infinite sample spaces?

In Section 4.9, we singled out random closed intervals, MV algebras and random sets as the most promising continuous formulations of belief measures. Whatever the case, functional analysis seems to be a requirement for any geometric description of continuous belief functions. In the Borel interval case, for instance, a belief (mass) function is in 1–1 correspondence with an integrable function from a triangle onto $[0, 1]$.

We have seen that probability measures, as a special case of belief functions, are naturally contemplated in the framework of the belief space. In Chapter 10, we will see how this holds for possibility/necessity measures as well. However, in Chapter 6 we appreciated how most other uncertainty measures are either unrelated to or, actually, more general than belief structures.

Question 2. How can we extend the geometric language to other classes of uncertainty measures?

Two suitable candidates appear to be monotone capacities (Section 6.2), in particular systems of probability intervals or 2-monotone capacities (Section 6.3), and gambles/lower previsions. Any geometric representation of previsions, in particular, is likely to resemble that of MV algebras, on one hand, and that of belief functions on fuzzy sets, on the other, as they are all instances of functionals assigning values to (continuous) functions. In Chapter 17, Section 17.2.3, we will outline the geometry of 2-monotone capacities and draw conclusions towards a general geometric theory of capacities.

Appendix: Proofs

Proof of Theorem 11

The bottom line of the proof is that the mass associated with a sum function can be recursively assigned to subsets of increasing size. We prove Theorem 11 by induction.

First level of the bundle structure. As mentioned, each normalised sum function $\varsigma \in \mathbb{R}^{N-2}$ is uniquely associated with a mass function m_ς through the Möbius inversion lemma. To define a base space for the first level, we set to zero the mass of all events of size $1 < |A| < n$. This determines a linear space $\mathcal{D}_\mathcal{S} \subset \mathcal{S} = \mathbb{R}^{N-2}$ defined by the system of linear equations

$$\mathcal{D}_{\mathcal{S}} \doteq \left\{ \varsigma \in \mathbb{R}^{N-2} : m_\varsigma(A) = \sum_{B \subset A} (-1)^{|A-B|} \varsigma(B) = 0,\ 1 < |A| < n \right\}$$

of dimension $\dim \mathcal{D}_{\mathcal{S}} = n = |\Theta|$ (as there are n unconstrained variables corresponding to the singletons). As $\mathcal{D}_{\mathcal{S}}$ is linear, it admits a global coordinate chart. Each point $\mathbf{d} \in \mathcal{D}$ is parameterised by the mass values that the corresponding sum function ς assigns to singletons:

$$\mathbf{d} = [m_\varsigma(A) = \varsigma(A), |A| = 1]'.$$

The second step is to identify a projection map between the total space $\mathcal{S} = \mathbb{R}^{N-2}$ and the base $\mathcal{D}_{\mathcal{S}}$. The Möbius inversion lemma (2.3) indeed induces a projection map from $\mathcal{S}$ to $\mathcal{D}_{\mathcal{S}}$,

$$\begin{array}{rcl} \pi : \mathcal{S} = \mathbb{R}^{N-2} & \to & \mathcal{D}_{\mathcal{S}} \subset \mathbb{R}^{N-2}, \\ \varsigma & \mapsto & \pi[\varsigma], \end{array}$$

mapping each NSF $\varsigma \in \mathbb{R}^{N-2}$ to a point $\pi[\varsigma]$ of the base space $\mathcal{D}$:

$$\pi[\varsigma](A) = [\varsigma(A), |A| = 1]'. \tag{7.25}$$

Finally, to define a bundle structure, we need to describe the fibres of the total space $\mathcal{S} = \mathbb{R}^{N-2}$, i.e., the vector subspaces of $\mathbb{R}^{N-2}$ which project onto a given point $\mathbf{d} \in \mathcal{D}$ of the base. Each $\mathbf{d} \in \mathcal{D}$ is of course associated with the linear space of all the NSFs $\varsigma \in \mathbb{R}^{N-2}$ whose projection $\pi[\varsigma]$ on $\mathcal{D}$ is $\mathbf{d}$,

$$\mathcal{F}_{\mathcal{S}}(\mathbf{d}) \doteq \left\{ \varsigma \in \mathcal{S} : \pi[\varsigma] = \mathbf{d} \in \mathcal{D} \right\}.$$

It is easy to see that as $\mathbf{d}$ varies on the base space $\mathcal{D}$, the linear spaces we obtain are all diffeomorphic to $\mathcal{F} \doteq \mathbb{R}^{N-2-n}$.

According to Definition 121, this defines a bundle structure, since:

- $E \doteq \mathcal{S} = \mathbb{R}^{N-2}$ is a smooth manifold, in particular a linear space;
- $B \doteq \mathcal{D}_{\mathcal{S}}$, the base space, is a smooth (linear) manifold;
- $F = \mathcal{F}_{\mathcal{S}}$, the fibre, is a smooth manifold, again a linear space.

The projection $\pi : \mathcal{S} = \mathbb{R}^{N-2} \to \mathcal{D}_{\mathcal{S}}$ is differentiable (as it is a linear function of the coordinates $\varsigma(A)$ of ς) and has full rank n in every point $\varsigma \in \mathbb{R}^{N-2}$. This is easy to see when we represent π as a matrix (for as ς is a vector, a linear function of ς can always be thought of as a matrix),

$$\pi[\varsigma] = \Pi \varsigma,$$

where

$$\Pi = \begin{bmatrix} 1 & 0 & \cdots 0 & 0 \cdots & 0 \\ 0 & 1 & 0 & 0\ 0 \cdots & 0 \\ & & \cdots & \cdots & \\ 0 \cdots & & 0 & 1\ 0 \cdots & 0 \end{bmatrix} = [I_n | 0_{n \times (N-2-n)}]$$

according to (7.25), and the rows of Π are obviously linearly independent.

As mentioned above, the bundle structure (Definition 121, item 6) is trivial, since $\mathcal{D}_\mathcal{S}$ is linear and can be covered by a single coordinate system (7.16). The direct product coordinates are

$$\begin{array}{rcl} \phi : \mathcal{S} = \mathbb{R}^{N-2} & \to & \mathcal{D}_\mathcal{S} \times \mathcal{F}_\mathcal{S}, \\ \varsigma & \mapsto & (\pi[\varsigma], f[\varsigma]), \end{array}$$

where the coordinates of ς on the fibre $\mathcal{F}_\mathcal{S}$ are the mass values it assigns to higher-size events,

$$f[\varsigma] = [m_\varsigma(A), 1 < |A| < n]'.$$

Bundle structure of level i. By induction, let us suppose that $\mathcal{S}$ admits a recursive bundle structure for all cardinalities from 1 to $i-1$, characterised according to the hypotheses, and prove that $\mathcal{F}_\mathcal{S}^{(i-1)}$ can in turn be decomposed in the same way into a linear base space and a collection of diffeomorphic fibres. By the inductive hypothesis, $\mathcal{F}_\mathcal{S}^{(i-1)}$ has dimension $N-2-\sum_{k=1}^{i-1}\binom{n}{k}$ and each point $\varsigma^{i-1} \in \mathcal{F}_\mathcal{S}^{(i-1)}$ has coordinates[68]

$$\varsigma^{i-1} = [\varsigma^{i-1}(A), i \leq |A| < n]'.$$

We can then apply the constraints $\varsigma^{i-1}(A) = 0$, $i < |A| < n$, which identify the linear variety

$$\mathcal{D}_\mathcal{S}^{(i)} \doteq \left\{ \varsigma^{i-1} \in \mathcal{F}_\mathcal{S}^{(i-1)} : \varsigma^{i-1}(A) = 0,\ i < |A| < n \right\} \tag{7.26}$$

embedded in $\mathcal{F}_\mathcal{S}^{(i-1)}$, of dimension $\binom{n}{i}$ (the number of size-i subsets of Θ).

The projection map (7.19) induces in $\mathcal{F}_\mathcal{S}^{(i-1)}$ fibres of the form

$$\mathcal{F}_\mathcal{S}^{(i)} \doteq \left\{ \varsigma^{i-1} \in \mathcal{F}_\mathcal{S}^{(i-1)} : \pi_i[\varsigma^{i-1}] = \text{const} \right\},$$

which are also linear manifolds. These induce, in turn, a trivial bundle structure in $\mathcal{F}_\mathcal{S}^{(i-1)}$:

$$\begin{array}{rcl} \phi : \mathcal{F}_\mathcal{S}^{(i-1)} & \to & \mathcal{D}_\mathcal{S}^{(i)} \times \mathcal{F}_\mathcal{S}^{(i)}, \\ \varsigma^{i-1} & \mapsto & (\phi'(\varsigma^{i-1}), \phi''(\varsigma^{i-1})), \end{array}$$

with $\phi'(\varsigma^{i-1}) = \pi_i[\varsigma^{i-1}] = [\varsigma^{i-1}(A), |A| = i]'$. Again, the map (7.19) is differentiable and has full rank, for its $\binom{n}{i}$ rows are independent.

The decomposition ends when $\dim \mathcal{F}_\mathcal{S}^{(n)} = 0$, and all fibres reduce to points of $\mathcal{S}$.

[68] The quantity $\varsigma^{i-1}(A)$ is in fact the mass $m_\varsigma(A)$ that the original NSF ς attaches to A, but this is irrelevant for the purpose of the decomposition.

Proof of Lemma 4

Since $\binom{n-m}{i-m}$ is the number of subsets of size i containing a fixed set B, where $|B| = m$ in a frame with n elements, we can write

$$\begin{aligned}\sum_{|A|=i} Bel(A) &= \sum_{|A|=i} \sum_{B\subseteq A} m(B) = \sum_{m=1}^{i} \sum_{|B|=m} \binom{n-m}{i-m} m(B) \\ &= \sum_{|B|=i} m(B) + \sum_{m=1}^{i-1} \sum_{|B|=m} \binom{n-m}{i-m} m(B) \\ &\leq 1 - \sum_{|B|<i} m(B) + \sum_{m=1}^{i-1} \sum_{|B|=m} \binom{n-m}{i-m} m(B),\end{aligned} \tag{7.27}$$

as $\sum_{|B|=i} m(B) = 1 - \sum_{|B|<i} m(B)$ by normalisation. By Möbius inversion (2.3),

$$\begin{aligned}\sum_{|A|<i} m(A) &= \sum_{|A|<i} \sum_{B\subseteq A} (-1)^{|A-B|} Bel(B) \\ &= \sum_{|A|=m=1}^{i-1} \sum_{|B|=l=1}^{m} (-1)^{m-l} \binom{n-l}{m-l} \sum_{|B|=l} Bel(B)\end{aligned} \tag{7.28}$$

for, again, $\binom{n-l}{m-l}$ is the number of subsets of size m containing a fixed set B, where $|B| = l$ in a frame with n elements.

The roles of the indices m and l can be exchanged, so that we obtain

$$\sum_{|B|=l=1}^{i-1} m(B) = \sum_{|B|=l=1}^{i-1} \left[\sum_{|B|=l} Bel(B) \cdot \sum_{m=l}^{i-1} (-1)^{m-l} \binom{n-l}{m-l} \right]. \tag{7.29}$$

Now, a well-known combinatorial identity ([715], Volume 3, Equation (1.9)) states that, for $i - (l+1) \geq 1$,

$$\sum_{m=l}^{i-1} (-1)^{m-l} \binom{n-l}{m-l} = (-1)^{i-(l+1)} \binom{n-(l+1)}{i-(l+1)}. \tag{7.30}$$

By applying (7.30) to the last equality, (7.28) becomes

$$\sum_{|B|=l=1}^{i-1} \left[\sum_{|B|=l} Bel(B) \cdot (-1)^{i-(l+1)} \binom{n-(l+1)}{i-(l+1)} \right]. \tag{7.31}$$

Similarly, by (7.29) we have

$$\begin{aligned}\sum_{m=1}^{i-1} \sum_{|B|=m} \binom{n-m}{i-m} m(B) &= \sum_{l=1}^{i-1} \sum_{|B|=l} Bel(B) \cdot \sum_{m=l}^{i-1} (-1)^{m-l} \binom{n-l}{m-l} \binom{n-m}{i-m} \\ &= \sum_{l=1}^{i-1} \sum_{|B|=l} Bel(B) \cdot \sum_{m=l}^{i-1} (-1)^{m-l} \binom{i-l}{m-l} \binom{n-l}{i-l},\end{aligned}$$

as it is easy to verify that

$$\binom{n-l}{m-l}\binom{n-m}{i-m} = \binom{i-l}{m-l}\binom{n-l}{i-l}.$$

By applying (7.30) again to the last equality, we get

$$\sum_{m=1}^{i-1}\sum_{|B|=m}\binom{n-m}{i-m}m(B) = \sum_{l=1}^{i-1}\sum_{|B|=l}(-1)^{i-(l+1)}\binom{n-l}{i-l}. \tag{7.32}$$

By replacing (7.29) and (7.32) in (7.27), we get the thesis.

Proof of Theorem 12

To understand the effect on $\mathcal{B} \subset \mathcal{S}$ of the bundle decomposition of the space of normalised sum functions $\mathcal{S} = \mathbb{R}^{N-2}$ in which it is immersed, we need to consider the effect of the non-negativity and normalisation conditions $m_\varsigma \geq 0$ and $\sum_A m_\varsigma(A) = 1$, for they constrain the admissible values of the coordinates of points of $\mathcal{S}$.

First of all, we need to notice that these constraints are separable into groups that apply to subsets of the same size. The set of conditions

$$\begin{cases} m(A) \geq 0, & \forall A \subseteq \Theta, \\ \sum_{A\subseteq\Theta} m(A) = 1 \end{cases}$$

can in fact be decomposed as $V_1 \wedge \cdots \wedge V_{n-1}$, where the system of constraints V_i is given by (7.21). The bottom inequality in (7.21) implies that, given a mass assignment for events of size $1, \ldots, i-1$ ($\sum_{|A|<i} m(A)$), the upper bound for $\sum_{|A|=i} Bel(A)$ is obtained by assigning all the remaining mass to the collection of size-i subsets: $\sum_{|A|=i} m(A) = 1 - \sum_{|A|<i} m(A)$ (Lemma 4).

Let us see the effect of these constraints on the bundle structure of $\mathcal{S}$.

Level 1. By definition, $\mathcal{B} = \{\varsigma \in \mathcal{S} : V_1 \wedge \cdots \wedge V_{n-1}\}$. As the coordinates of the points of $\mathcal{S}$ are decomposed into coordinates on the base $[m_\varsigma(A), |A| = 1]'$ and coordinates on the fibre $[m_\varsigma(A), 1 < |A| < n]'$, it is easy to see that V_1,

$$\begin{cases} m(A) \geq 0, & |A| = 1, \\ \sum_{|A|=1} m(A) \leq 1, \end{cases} \tag{7.33}$$

acts only on the base $\mathcal{D}^{(1)}_{\mathcal{S}}$, yielding a new set

$$\mathcal{D}^{(1)}_{\mathcal{B}} = \left\{ \mathbf{bel} \in \mathcal{B} : m(A) = 0 \ 1 < |A| < n, m(A) \geq 0 \ |A| = 1, \sum_{|A|=1} m(A) \leq 1 \right\}$$

of the form of (7.22) for $i = 1$.

This, in turn, selects the fibres of $\mathcal{S}$ passing through $\mathcal{D}^{(1)}_{\mathcal{B}}$ and discards the others. As a matter of fact, there cannot be admissible belief functions within fibres passing through points outside this region, since all points of a fibre $\mathcal{F}^{(1)}_{\mathcal{S}}$ share the same level-1 coordinates $\mathbf{d}^1$: when the basis point $\mathbf{d}^1$ does not satisfy the inequalities (7.33), none of them can.

Therefore, the remaining constraints $V_2 \wedge \cdots \wedge V_{n-1}$ act on the fibres $\mathcal{F}^{(1)}_{\mathcal{S}}$ of $\mathcal{S}$ passing through $\mathcal{D}^{(1)}_{\mathcal{B}}$. However, (7.21) shows that those higher-size constraints $V_2, \ldots, V_{n-1}$ in fact depend on the point $\mathbf{d}^1 = [m(A), |A| = 1]' \in \mathcal{D}^{(1)}_{\mathcal{B}}$ on the base space. Each admissible fibre $\mathcal{F}^{(1)}_{\mathcal{S}} \sim \mathbb{R}^{N-2-n}$ of $\mathcal{S}$ is then subject to a different system of constraints $V_2 \wedge \cdots \wedge V_{n-1}(\mathbf{d}^1)$, yielding the corresponding first-level fibre of $\mathcal{B}$ (see (7.20)),

$$\mathcal{F}^{(1)}_{\mathcal{B}}(\mathbf{d}^1) = \Big\{ \mathbf{bel} \in \mathcal{F}^{(1)}_{\mathcal{S}}(\mathbf{d}^1) : V_2 \wedge \cdots \wedge V_{n-1}(\mathbf{d}^1) \Big\}.$$

Level i. Let us now suppose, by induction, that we have a family of constraints $V_i \wedge \cdots \wedge V_{n-1}(\mathbf{d}^1, \ldots, \mathbf{d}^{i-1})$ of the form of (7.21), acting on $\mathcal{F}^{(i-1)}_{\mathcal{S}}(\mathbf{d}^1, \ldots, \mathbf{d}^{i-1})$. The points of $\mathcal{F}^{(i-1)}_{\mathcal{S}}(\mathbf{d}^1, \ldots, \mathbf{d}^{i-1})$ have coordinates which can be decomposed into coordinates $\mathbf{d}^i = [m_\varsigma(A), |A| = i]'$ on the base $\mathcal{D}^{(i)}_{\mathcal{S}}$ and coordinates on the fibre $\mathcal{F}^{(i)}_{\mathcal{S}}$. Again, the set of constraints V_i acts on coordinates associated with size-i events only, i.e., it acts on $\mathcal{D}^{(i)}_{\mathcal{S}}$ and not on $\mathcal{F}^{(i)}_{\mathcal{S}}$.

Furthermore, constraints of the type (7.21) for $k > i$ become trivial when acting on $\mathcal{D}^{(i)}_{\mathcal{S}}$. In fact, inequalities of the form $m(A) \geq 0$, $|A| > i$, are satisfied by $\mathcal{D}^{(i)}_{\mathcal{S}}$ by definition, since it imposes the condition $m_\varsigma(A) = 0$, $|A| > i$. On the other hand, all inequalities corresponding to the second row of (7.21) for $k > i$ reduce to the corresponding inequality for size-i subsets. Instead of displaying a long combinatorial proof, we can just recall the meaning of Lemma 4: the upper bound on $\sum_{|A|=i} Bel(A)$ is obtained by assigning maximum mass to the collection of size-i subsets. But, then, points in $\mathcal{D}^{(i)}_{\mathcal{S}}$ correspond to a zero-assignment for higher-size events $m_\varsigma(A) = 0, i < |A| < n$, and all those upper bounds are automatically satisfied.

We then get the i-th-level base for $\mathcal{B}$: $\mathcal{D}^{(i)}_{\mathcal{B}}(\mathbf{d}^1, \ldots, \mathbf{d}^{i-1})$ is the set of belief functions $\mathbf{bel} \in \mathcal{F}^{(i-1)}_{\mathcal{S}}(\mathbf{d}^1, \ldots, \mathbf{d}^{i-1})$ such that the conditions (7.22) are satisfied. The remaining constraints $V_{i+1} \wedge \ldots \wedge V_{n-1}(\mathbf{d}^1, \ldots, \mathbf{d}^i)$ act on the fibres $\mathcal{F}^{(i)}_{\mathcal{S}}$ of $\mathcal{S}$ passing through points $\mathbf{d}^i$ of $\mathcal{D}^{(i)}_{\mathcal{B}}(\mathbf{d}^1, \ldots, \mathbf{d}^{i-1})$, yielding a collection of level-i fibres for $\mathcal{B}$,

$$\mathcal{F}^{(i)}_{\mathcal{B}}(\mathbf{d}^1, \ldots, \mathbf{d}^i) = \Big\{ \mathbf{bel} \in \mathcal{F}^{(i)}_{\mathcal{S}} : V_{i+1} \wedge \ldots \wedge V_{n-1}(\mathbf{d}^1, \ldots, \mathbf{d}^i) \Big\}.$$

8

Geometry of Dempster's rule

As we have seen in Chapter 7, belief functions can be seen as points of a simplex $\mathcal{B}$ called the 'belief space'. It is therefore natural to wonder whether the orthogonal sum operator (2.6), a mapping from a pair of prior belief functions to a posterior belief function on the same frame, can also be interpreted as a geometric operator in $\mathcal{B}$. The answer is positive, and in this chapter we will indeed understand this property of Dempster's rule in this geometric setting. As we point out when we conclude our discussion of the geometry of the orthogonal sum in the belief space, such an analysis can obviously be extended to the many other combination rules proposed in the past (see Chapter 4, Section 4.3).

The key observation which allows us to conduct our analysis is that the objects we obtain by relaxing the constraint of mass assignment being non-negative, which we called in Section 7.3.3 *normalised sum functions* or 'pseudo-belief functions' (the points of the embedding Cartesian space which lie outside the belief space $\mathcal{B}$), admit a straightforward extension of Dempster's rule, originally defined for proper belief functions (Section 8.1). This leads to an analysis of the behaviour of the orthogonal sum when applied to whole affine subspaces (Section 8.2) and convex combinations of (pseudo-)belief functions. In particular, this allows us to derive a 'convex decomposition' of Dempster's rule of combination in terms of Bayes' rule of conditioning (Section 8.3), and prove that under specific conditions the orthogonal sum and affine closure *commute* (Section 8.4).

In Section 8.5, we exploit the commutativity property to introduce the notion of a *conditional subspace* $\langle Bel \rangle$ generated by an arbitrary belief function Bel, i.e., the set of all combinations of Bel with any other combinable BF. Conditional subspaces describe the 'global' behaviour of the rule of combination, and can be interpreted as

F. Cuzzolin, *The Geometry of Uncertainty*, Artificial Intelligence: Foundations, Theory, and Algorithms, https://doi.org/10.1007/978-3-030-63153-6_8

the set of possible 'futures' of a given belief function (seen as an uncertain knowledge state). Geometrically, they once again have the form of convex sets.

The second part of the chapter, instead, is dedicated to an analysis of the 'pointwise' behaviour of Dempster's rule. We first discuss (Section 8.6) a toy problem, the geometry of $\oplus$ in the binary belief space $\mathcal{B}_2$, to gather useful intuition about the general case. We observe that Dempster's rule exhibits a rather elegant behaviour when applied to collections of belief functions assigning the same mass k to a fixed subset A (*constant-mass loci*), which turn out to be affine subspaces of normalised sum functions. As a consequence, their images under the mapping $Bel \oplus (.)$ can be derived by applying the commutativity results of Section 8.4.

Perhaps the most striking result of our geometric analysis of Dempster's rule states that, for each subset A, the resulting mapped affine spaces have a common intersection for all $k \in [0,1]$, a geometric entity which is therefore characteristic of the belief function Bel being combined. We call the latter the A-th *focus* of the conditional subspace $\langle Bel \rangle$. In Section 8.7, we formally prove the existence and study the geometry of such foci. This eventually leads us to an interesting algorithm for the geometric construction of the orthogonal sum of two belief functions. A list of open research questions (Section 8.8) concludes the chapter.

The material presented here is a re-elaboration of results first published in [327]. As usual, all proofs have been collected together in an appendix at the end of the chapter. In this chapter and the following ones, we will use Bel to denote both a belief function and the corresponding vector in the belief space, whenever no confusion is possible.

8.1 Dempster combination of pseudo-belief functions

As mentioned above, we start by observing that Dempster's rule can be easily extended to normalised sum functions. This is necessary since, as we will see later on, the geometry of the orthogonal sum can only be appropriately described in terms of whole affine spaces which do not fit within the confines of the belief space, a simplex.

Theorem 13. *The application of Dempster's rule as defined as in (2.6) to a pair of normalised sum functions $\varsigma_1, \varsigma_2 : 2^\Theta \to \mathbb{R}$ yields another normalised sum function, which we denote by $\varsigma_1 \oplus \varsigma_2$.*

Just like we do for proper belief functions, we say that two NSFs ς_1, ς_2 are *not combinable* if the denominator of (2.6) is zero:

$$\Delta(\varsigma_1, \varsigma_2) \doteq \sum_{A \subseteq \Theta, B \subseteq \Theta : A \cap B \neq \emptyset} m_{\varsigma_1}(A) m_{\varsigma_2}(B) = 0, \tag{8.1}$$

where m_{ς_1} and m_{ς_1} denote the Möbius transforms of the two normalised sum functions ς_1 and ς_2, respectively.

Note that in the case of NSFs *the normalisation factor* $\Delta(\varsigma_1, \varsigma_2)$ *can be zero even in the presence of non-empty intersections between focal elements of* ς_1, ς_2. This becomes clear as soon as we rewrite it in the form

$$\Delta(\varsigma_1, \varsigma_2) = \sum_{C \neq \emptyset} \sum_{A,B \subseteq \Theta : A \cap B = C} m_{\varsigma_1}(A) m_{\varsigma_2}(B),$$

since there can exist non-zero products $m_{\varsigma_1}(A) m_{\varsigma_2}(B)$ whose overall sum is zero (since $m_{\varsigma_1}(A), m_{\varsigma_2}(B)$ are arbitrary real numbers).

Example 19: zero normalisation factor. *A simple example may be useful for helping us to grasp this point more easily. Consider a sum function* ς_1 *with focal elements* A_1, A_2, A_3 *and masses* $m_1(A_1) = 1, m_1(A_2) = -1, m_1(A_3) = 1$ *such that* $A_2 \subsetneq A_1$*, as in Fig. 8.1. If we combine* ς_1 *with a new NSF* ς_2 *with a single focal element* B*,* $m_2(B) = 1$ *(which, incidentally, is a belief function), we can see that, although* $A_1 \cap B \doteq D \neq \emptyset$ *and* $A_2 \cap B = D \neq \emptyset$*, the denominator of (2.6) is equal to* $1 \cdot (-1) + 1 \cdot 1 = 0$ *and the two functions turn out not to be combinable.*

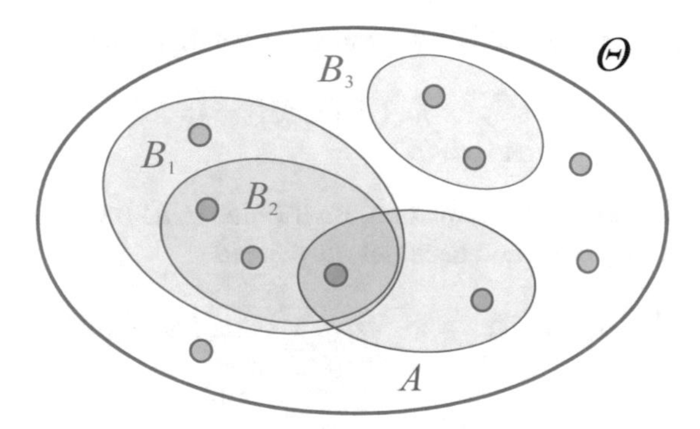

Fig. 8.1: Example of a pair of non-combinable normalised sum functions whose focal elements nevertheless have non-empty intersections.

8.2 Dempster sum of affine combinations

The extension of Dempster's rule to normalised sum functions (Theorem 13) allows us to work with the entire Cartesian space the belief space is embedded in, rather than with $\mathcal{B}$. In Chapter 7, we have seen that any belief function can be seen as a convex closure of categorical BFs. A similar relation exists between NSFs and affine closures. In fact, any affine combination $\varsigma = \sum_i \alpha_i \varsigma_i$ with $\sum_i \alpha_i = 1$ of a collection of normalised sum functions $\{\varsigma_1, \ldots, \varsigma_n\}$ is still an NSF, since

$$\sum_{A\neq\emptyset} m_\varsigma(A) = \sum_{A\neq\emptyset} \sum_{B\subset A} (-1)^{|A-B|} \sum_i \alpha_i \varsigma_i(B)$$
$$= \sum_i \alpha_i \sum_{A\neq\emptyset} \sum_{B\subset A} (-1)^{|A-B|} \varsigma_i(B)$$
$$= \sum_i \alpha_i \sum_{A\neq\emptyset} m_{\varsigma_i}(A) = \sum_i \alpha_i = 1.$$

We can then proceed to show how Dempster's rule applies to affine combinations of pseudo-belief functions, and to convex closures of (proper) belief functions in particular. We first consider the issue of combinability.

Lemma 5. *Consider a collection of normalised sum functions* $\{\varsigma_1, \ldots, \varsigma_n\}$, *and their affine combination* $\varsigma = \sum_i \alpha_i \varsigma_i$, $\sum_i \alpha_i = 1$. *A normalised sum function* ς *is combinable with* $\sum_i \alpha_i \varsigma_i$ *iff*

$$\sum_i \alpha_i \Delta_i \neq 0,$$

where $\Delta_i = \sum_{A\cap B\neq\emptyset} m_\varsigma(A) m_{\varsigma_i}(B)$.

Proof. By definition (8.1), two NSFs ς and τ are combinable iff

$$\sum_{A\cap B\neq\emptyset} m_\varsigma(A) m_\tau(B) \neq 0.$$

If $\tau = \sum_i \alpha_i \varsigma_i$ is an affine combination, its Möbius transform is $m_\tau = \sum_i \alpha_i m_{\varsigma_i}$ and the combinability condition becomes, as desired

$$\sum_{A\cap B\neq\emptyset} m_\varsigma(A) \Big(\sum_i \alpha_i m_{\varsigma_i}(B) \Big) = \sum_i \alpha_i \sum_{A\cap B\neq\emptyset} m_\varsigma(A) m_{\varsigma_i}(B) = \sum_i \alpha_i \Delta_i \neq 0.$$

□

A couple of remarks. If $\Delta_i = 0$ for all i, then $\sum_i \alpha_i \Delta_i = 0$, so that if ς is not combinable with any ς_i then the combination $\varsigma \oplus \sum_i \alpha_i \varsigma_i$ does not exist, in accordance with our intuition. On the other hand, even if *all* the NSFs ς_i are combinable with ς, there is always a choice of the coefficients α_i of the affine combination such that $\sum_i \alpha_i \Delta_i = 0$, so that ς is still not combinable with the affine combination. This remains true when affine combinations of belief functions (for which $\Delta_i > 0 \ \forall i$) are considered.

Lemma 6. *Consider a collection of normalised sum functions* $\{\varsigma, \varsigma_1, \ldots, \varsigma_n\}$. *For any set of real numbers* $\alpha_1, \ldots, \alpha_n$ *such that* $\sum_i \alpha_i = 1$, $\sum_i \alpha_i \Delta_i \neq 0$, ς *is combinable with the affine combination* $\sum_i \alpha_i \varsigma_i$, *and the mass assignment (Möbius transform) of their orthogonal sum is given by*

$$m_{\varsigma \oplus \sum_i \alpha_i \varsigma_i}(C) = \sum_i \frac{\alpha_i N_i(C)}{\sum_j \alpha_j \Delta_j},$$

where

$$N_i(C) \doteq \sum_{B \cap A = C} m_{\varsigma_i}(B) m_\varsigma(A)$$

is the numerator of $m_{\varsigma \oplus \varsigma_i}(C)$.

Under specific conditions, the orthogonal sum of an affine combination of (pseudo-)belief functions can be expressed as an affine combination of the partial combinations. Such conditions are specified in the following theorem.

Theorem 14. *Consider a collection* $\{\varsigma, \varsigma_1, \ldots, \varsigma_n\}$ *of normalised sum functions such that* $\sum_i \alpha_i = 1$, $\sum_i \alpha_i \Delta_i \neq 0$, *i.e., the NSF* ς *is combinable with the affine combination* $\sum_i \alpha_i \varsigma_i$. *If* ς_i *is combinable with* ς *for each* $i = 1, \ldots, n$ *(and in particular, when all the normalised sum functions involved,* $\{\varsigma, \varsigma_1, \ldots, \varsigma_n\} = \{Bel, Bel_1, \ldots, Bel_n\}$*, are belief functions), then* $\varsigma \oplus \sum_i \alpha_i \varsigma_i$ *is still an affine combination of the partial sums* $\varsigma \oplus \varsigma_i$*:*

$$\varsigma \oplus \sum_i \alpha_i \varsigma_i = \sum_i \beta_i (\varsigma \oplus \varsigma_i), \tag{8.2}$$

with coefficients given by

$$\beta_i = \frac{\alpha_i \Delta_i}{\sum_{j=1}^n \alpha_j \Delta_j}. \tag{8.3}$$

Proof. By definition, $m_{\varsigma \oplus \varsigma_i}(C) = N_i(C)/\Delta_i$. If ς is combinable with ς_i $\forall i$, then $\Delta_i \neq 0$ $\forall i$ so that $N_i(C) = 0$ iff $\Delta_i \cdot m_{\varsigma \oplus \varsigma_i}(C) = 0$, and we can write, by Lemma 6,

$$m_{\varsigma \oplus \sum_i \alpha_i \varsigma_i}(C) = \frac{\sum_i \alpha_i N_i(C)}{\sum_j \alpha_j \Delta_j} = \frac{\sum_i \alpha_i \Delta_i m_{\varsigma \oplus \varsigma_i}(C)}{\sum_j \alpha_j \Delta_j} = \sum_i \beta_i m_{\varsigma \oplus \varsigma_i}(C). \tag{8.4}$$

The Möbius transform (2.3) immediately yields (8.2).

If Bel and Bel_i are both belief functions, $\Delta_i = 0$ implies $N_i(C) = 0$ for all $C \subset \Theta$. We can then still write $N_i(C) = \Delta_i \cdot m_{\varsigma \oplus \varsigma_i}(C)$, so that (8.4) holds.

□

When we consider convex combinations of proper belief functions only, the combinability condition of Lemma 5 simplifies as follows.

Lemma 7. *If* $\sum_i \alpha_i = 1$ *and* $\alpha_i > 0$ *for all* i*, then*

$$\exists \, Bel \oplus \sum_i \alpha_i \, Bel_i \Leftrightarrow \exists i : \exists Bel \oplus Bel_i,$$

i.e., Bel is combinable with the affine combination if and only if it is combinable with at least one *of the belief functions* Bel_i*,*

This is due to the fact that if $\alpha_i \Delta_i > 0$, then $\sum_i \alpha_i \Delta_i > 0$.

Theorem 14 then specialises in the following way.

Corollary 5. *Consider a collection of belief functions* $\{Bel, Bel_1, \ldots, Bel_n\}$ *such that there exists at least one BF* Bel_j *combinable with* Bel. *If* $\sum_i \alpha_i = 1$, $\alpha_i > 0$ *for all* $i = 1, \ldots, n$, *then*

$$Bel \oplus \sum_i \alpha_i Bel_i = \sum_i \beta_i Bel \oplus Bel_i,$$

where β_i *is again defined by (8.3).*

8.3 Convex formulation of Dempster's rule

An immediate consequence of the properties of Dempster's rule with respect to affine (and therefore convex) combination is an interesting convex decomposition of the orthogonal sum itself.

Theorem 15. *The orthogonal sum* $Bel \oplus Bel'$ *of two belief functions can be expressed as a convex combination of the results* $Bel \oplus Bel_A$ *of Bayes conditioning of* Bel *with respect to all the focal elements of* Bel', *namely*

$$Bel \oplus Bel' = \sum_{A \in \mathcal{E}_{Bel'}} \frac{m'(A)Pl(A)}{\sum_{B \in \mathcal{E}_{Bel'}} m'(B)Pl(B)} Bel \oplus Bel_A, \tag{8.5}$$

where $\mathcal{E}_{Bel}$ *denotes, as usual, the collection of focal elements of a belief function* Bel.

Proof. We know from Chapter 7 that any belief function $Bel' \in \mathcal{B}$ can be written as a convex sum of the categorical BFs Bel_A (7.8). We can therefore apply Corollary 5 to (7.8), obtaining

$$Bel \oplus Bel' = Bel \oplus \sum_{A \in \mathcal{E}_{Bel'}} m'(A) Bel_A = \sum_{A \in \mathcal{E}_{Bel'}} \mu(A) Bel \oplus Bel_A,$$

where

$$\mu(A) \doteq \frac{m'(A)\Delta_A}{\sum_{B \in \mathcal{E}_{Bel'}} m'(B)\Delta_B}. \tag{8.6}$$

Here Δ_A is the normalisation factor for $Bel \oplus Bel_A$, namely

$$\Delta_A = \sum_{B: B \cap A \neq \emptyset} m(B) = 1 - \sum_{B \subset A^c} m(B) = 1 - Bel(A^c) = Pl(B).$$

By plugging the result into (8.6), we have (8.5), as desired.

□

We can simplify (8.5) after realising that some of the partial combinations $Bel \oplus Bel_A$ may in fact coincide. Since $Bel \oplus Bel_A = Bel \oplus Bel_B$ iff $A \cap \mathcal{C}_{Bel} = B \cap \mathcal{C}_{Bel}$, we can write

$$Bel \oplus Bel' = \sum_{\substack{A=A'\cap \mathcal{C}_{Bel} \\ A' \in \mathcal{E}_{Bel'}}} Bel \oplus Bel_A \frac{\sum_{\substack{B\cap \mathcal{C}_{Bel}=A \\ B\in \mathcal{E}_{Bel'}}} m'(B)Pl(B)}{\sum_{B\in\mathcal{E}_{Bel'}} m'(B)Pl(B)}. \tag{8.7}$$

It is well known that Dempster sums involving categorical belief functions, $Bel \oplus Bel_A$, can be thought of as applications of (a generalised) Bayes' rule to the original belief function Bel, when conditioning with respect to an event A: $Bel \oplus Bel_A \doteq Bel|A$ (recall Section 4.5.1). More precisely, if $Bel = P$ is a probability measure, $Bel \oplus Bel_A = P(.|A)$.

Theorem 15, therefore, highlights a convex decomposition of Dempster's rule of combination in terms of Dempster's rule of conditioning (Section 4.5.1):

$$Bel \oplus Bel' = \sum_A \frac{m'(A)Pl(A)}{\sum_B m'(B)Pl(B)} Bel|_{\oplus} A.$$

8.4 Commutativity

In Chapter 7, we have seen that the basic probability assignment mechanism is represented in the belief space framework by the convex closure operator. Theorem 14 in fact treats in full generality *affine* combinations of points, for these prove to be more relevant to a geometric description of the rule of combination. The next natural step, therefore, is to analyse Dempster combinations of affine closures, i.e., sets of affine combinations of points.

Let us denote by $v(\varsigma_1, \ldots, \varsigma_n)$ the affine subspace generated by a collection of normalised sum functions $\{\varsigma_1, \ldots, \varsigma_n\}$:

$$v(\varsigma_1, \ldots, \varsigma_n) \doteq \left\{ \varsigma : \varsigma = \sum_{i=1}^n \alpha_i \varsigma_i, \ \sum_i \alpha_i = 1 \right\}.$$

Theorem 16. *Consider a collection of pseudo-belief functions $\{\varsigma, \varsigma_1, \ldots, \varsigma_n\}$ defined on the same frame of discernment. If ς_i is combinable with ς ($\Delta_i \neq 0$) for all i, then*

$$v(\varsigma \oplus v(\varsigma_1, \ldots, \varsigma_n)) = v(\varsigma \oplus \varsigma_1, \ldots, \varsigma \oplus \varsigma_n).$$

More precisely,

$$v(\varsigma \oplus \varsigma_1, \ldots, \varsigma \oplus \varsigma_n) = \varsigma \oplus v(\varsigma_1, \ldots, \varsigma_n) \cup \mathcal{M}(\varsigma, \varsigma_1, \ldots, \varsigma_n),$$

where $\mathcal{M}(\varsigma, \varsigma_1, \ldots, \varsigma_n)$ is the following affine subspace:

$$v\left(\frac{\Delta_j}{\Delta_j - \Delta_n} \varsigma \oplus \varsigma_j - \frac{\Delta_n}{\Delta_j - \Delta_n} \varsigma \oplus \varsigma_i \middle| \forall j : \Delta_j \neq \Delta_n, \forall i : \Delta_i = \Delta_n \right). \tag{8.8}$$

If $\{\varsigma, \varsigma_1, \ldots, \varsigma_n\} = \{Bel, Bel_1, \ldots, Bel_n\}$ are all belief functions, then

$$v(Bel \oplus v(Bel_1, \ldots, Bel_n)) = v(Bel \oplus Bel_{i_1}, \ldots, Bel \oplus Bel_{i_m}),$$

where $\{Bel_{i_1}, \ldots, Bel_{i_m}\}$, $m \leq n$, *are all the belief functions combinable with* Bel *in the collection* $\{Bel_1, \ldots, Bel_n\}$.

8.4.1 Affine region of missing points

Theorem 16 states that Dempster's rule maps affine spaces to affine spaces, *except* for a lower-dimensional subspace. From its proof (see the appendix to this chapter), the affine coordinates $\{\alpha_i\}$ of a point $\tau \in v(\varsigma_1, \ldots, \varsigma_n)$ correspond to the affine coordinates $\{\beta_i\}$ of the sum $\varsigma \oplus \tau \in v(\varsigma \oplus \varsigma_1, \ldots, \varsigma \oplus \varsigma_n)$ through the following equation:

$$\alpha_i = \frac{\beta_i}{\Delta_i} \frac{1}{\sum_j \beta_j / \Delta_j}. \tag{8.9}$$

Hence, the values of the affine coordinates β_i of $v(\varsigma \oplus \varsigma_1, \ldots, \varsigma \oplus \varsigma_n)$ which are not associated with affine coordinates of $v(\varsigma_1, \ldots, \varsigma_n)$ are given by (8.28):

$$\sum_i \frac{\beta_i}{\Delta_i} = 0. \tag{8.10}$$

If the map $\varsigma \oplus (.)$ is injective, then the points of the subspace $\mathcal{M}(\varsigma, \varsigma_1, \ldots, \varsigma_n)$ associated with the affine coordinates β_i satisfying (8.10) are not images through $\varsigma \oplus (.)$ of any points of $v(\varsigma_1, \ldots, \varsigma_n)$: we call them *missing points*.

However, if the map in *not* injective, points in the original affine space with admissible coordinates can be mapped onto $\mathcal{M}(\varsigma, \varsigma_1, \ldots, \varsigma_n)$. In other words, missing coordinates do not necessarily determine missing points.

If we restrict our attention to convex combinations ($\alpha_i \geq 0 \; \forall i$) only of belief functions ($\Delta_i \geq 0$), Theorem 16 implies the following corollary.

Corollary 6. *Cl and* $\oplus$ *commute, i.e., if Bel is combinable with* Bel_i $\forall i = 1, \ldots, n$, *then*

$$Bel \oplus Cl(Bel_1, \ldots, Bel_n) = Cl(Bel \oplus Bel_1, \ldots, Bel \oplus Bel_n).$$

8.4.2 Non-combinable points and missing points: A duality

Even when all the pseudo-belief functions ς_i of Theorem 16 are combinable with ς, the affine space $v(\varsigma_1, \ldots, \varsigma_n)$ generated by them includes an affine subspace of non-combinable functions, namely those satisfying the following constraint:

$$\sum_i \alpha_i \Delta_i = 0, \tag{8.11}$$

where Δ_i is the degree of conflict between ς and ς_i.

There exists a sort of duality between these non-combinable points and the missing points in the image subspace $v(\varsigma \oplus \varsigma_1, \ldots, \varsigma \oplus \varsigma_n)$. In fact, by (8.9),

$\mathcal{M}(\varsigma, \varsigma_1, \ldots, \varsigma_n)$ turns out to be the image of the infinite point of $v(\varsigma_1, \ldots, \varsigma_n)$ via $\varsigma \oplus (.)$, since $\sum_i \beta_i / \Delta_i = 0$ implies $\alpha_i \to \infty$. On the other hand, the non-combinable points in $v(\varsigma_1, \ldots, \varsigma_n)$ satisfy (8.11), so that (8.10) yields $\beta_j \to \infty$. Non-combinable points are hence mapped to the infinite point of $v(\varsigma \oplus \varsigma_1, \ldots, \varsigma \oplus \varsigma_n)$.

This geometric duality is graphically represented in Fig. 8.2.

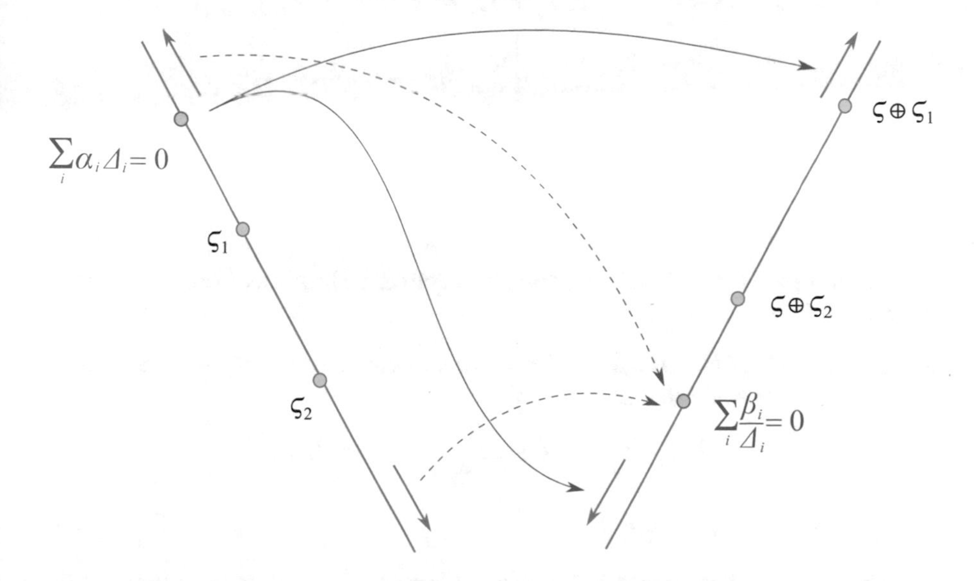

Fig. 8.2: The dual role of non-combinable and missing points in Theorem 16, and their relation with the infinite points of the associated affine spaces.

8.4.3 The case of unnormalised belief functions

The results in this section greatly simplify when we consider *unnormalised* belief functions, i.e., belief functions assigning non-zero mass to the empty set too (recall Section 3.3.2). The meaning of the basic probability value of $\emptyset$ has been studied by Smets [1685], as a measure of the internal conflict present in a BPA m. It is easy to see that, for the Dempster sum of two belief functions Bel_1 and Bel_2, we get $m_{Bel_1 \oplus Bel_2}(\emptyset) = 1 - \Delta(Bel_1, Bel_2)$ with $\Delta(Bel_1, Bel_2)$ as above.

Clearly, Dempster's rule can be naturally modified to cope with such functions. Equation (2.6) simplifies in the following way: if m_1, m_2 are the BPAs of two unnormalised belief functions Bel_1, Bel_2, their Dempster combination becomes

$$m_{Bel_1 \circledcap Bel_2}(C) = \sum_{A \cap B = C} m_1(A) m_2(B). \tag{8.12}$$

This is the *conjunctive combination* operator (4.25) introduced by Smets in his transferable belief model [1677] (see Section 3.3.2 as well).

Obviously enough, UBFs are *always* combinable through (8.12). If we denote by $\textcircled{\cap}$ the conjunctive combination operator, given a collection of UBFs $\tilde{Bel}, \tilde{Bel}_1, \ldots, \tilde{Bel}_n$, we have that

$$\begin{aligned} m_{\tilde{Bel} \textcircled{\cap} \sum_i \alpha_i \tilde{Bel}_i}(C) &= \sum_{B \cap A = C} m_{\sum_i \alpha_i \tilde{Bel}_i}(B) \tilde{m}(A) \\ &= \sum_{B \cap A = C} \left[\sum_i \alpha_i \tilde{m}_i(B) \right] \tilde{m}(A) \\ &= \sum_i \alpha_i \sum_{B \cap A = C} \tilde{m}_i(B) m_{\tilde{Bel}}(A) \\ &= \sum_i \alpha_i m_{\tilde{Bel} \textcircled{\cap} \tilde{Bel}_i}(C) = m_{\sum_i \alpha_i \tilde{Bel} \textcircled{\cap} \tilde{Bel}_i}(C), \end{aligned}$$

where $\tilde{m}, \tilde{m}_i$ are the BPAs of $\tilde{Bel}$, $\tilde{Bel}_i$, respectively. Therefore, Corollary 5 transforms as follows.

Proposition 54. *If $\tilde{Bel}, \tilde{Bel}_1, \ldots, \tilde{Bel}_n$ are unnormalised belief functions defined on the same frame of discernment, then*

$$\tilde{Bel} \textcircled{\cap} \sum_i \alpha_i \tilde{Bel}_i = \sum_i \alpha_i \tilde{Bel} \textcircled{\cap} \tilde{Bel}_i,$$

whenever $\sum_i \alpha_i = 1$, $\alpha_i \geq 0 \; \forall i$.

The commutativity results of Section 8.4 remain valid too. Indeed, Proposition 54 implies that

$$\begin{aligned} &\tilde{Bel} \textcircled{\cap} Cl(\tilde{Bel}_1, \ldots, \tilde{Bel}_n) = \tilde{Bel} \textcircled{\cap} \left\{ \sum_i \alpha_i \tilde{Bel}_i : \sum_i \alpha_i = 1, \alpha_i \geq 0 \right\} \\ &= \left\{ \tilde{Bel} \textcircled{\cap} \sum_i \alpha_i \tilde{Bel}_i : \sum_i \alpha_i = 1, \alpha_i \geq 0 \right\} \\ &= \left\{ \sum_i \alpha_i \tilde{Bel} \textcircled{\cap} \tilde{Bel}_i : \sum_i \alpha_i = 1, \alpha_i \geq 0 \right\} = Cl(\tilde{Bel} \textcircled{\cap} \tilde{Bel}_1, \ldots, \tilde{Bel} \textcircled{\cap} \tilde{Bel}_n) \end{aligned}$$

for any collection of UBFs $\{\tilde{Bel}, \tilde{Bel}_1, \ldots, \tilde{Bel}_n\}$, since their combinability under conjunctive combination is always guaranteed.

8.5 Conditional subspaces

8.5.1 Definition

The commutativity results that we proved in Section 8.2 are rather powerful, as they specify how the rule of combination works when applied to entire regions of the

Cartesian space, and in particular to affine closures of normalised sum functions. Since the belief space itself is a convex region of $\mathbb{R}^N$, it is easy to realise that these results can help us draw a picture of the 'global' behaviour of $\oplus$ within our geometric approach to the theory of evidence.

Definition 122. *Given a belief function $Bel \in \mathcal{B}$, we define the* conditional subspace $\langle Bel \rangle$ *as the set of all Dempster combinations of Bel with* any *other combinable belief function on the same frame, namely*

$$\langle Bel \rangle \doteq \Big\{ Bel \oplus Bel', \ Bel' \in \mathcal{B} \ s.t. \ \exists \ Bel \oplus Bel' \}. \tag{8.13}$$

Roughly speaking, $\langle Bel \rangle$ is the set of possible 'futures' of Bel under the assumption that new evidence is combined with Bel via Dempster's rule.

Since not all belief functions are combinable with a given Bel, we need to understand the geometric structure of such combinable BFs. We define the *compatible subspace* $C(Bel)$ associated with a belief function Bel as the collection of all the belief functions with focal elements included in the core of Bel:

$$C(Bel) \doteq \Big\{ Bel' : \mathcal{C}_{Bel'} \subset \mathcal{C}_{Bel} \Big\}.$$

The conditional subspace associated with Bel is nothing but the combination of Bel with its compatible subspace.

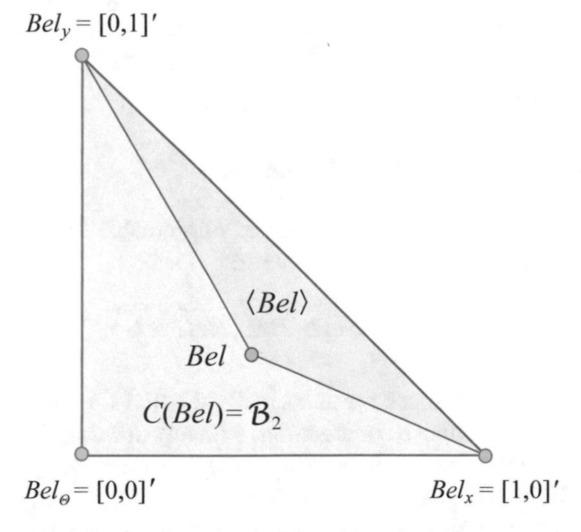

Fig. 8.3: Conditional and compatible subspaces for a belief function Bel in the binary belief space $\mathcal{B}_2$. The coordinate axes measure the belief values of $\{x\}$ and $\{y\}$, respectively. The vertices of $\langle Bel \rangle$ are Bel, Bel_x and Bel_y, since $Bel \oplus Bel_x = Bel_x \ \forall Bel \neq Bel_y$, and $Bel \oplus Bel_y = Bel_y \ \forall Bel \neq Bel_x$.

Theorem 17. $\langle Bel \rangle = Bel \oplus C(Bel) = Cl\{Bel \oplus Bel_A,\ A \subseteq \mathcal{C}_{Bel}\}$.

Proof. Let $\mathcal{E}_{Bel'} = \{A_i\}$ and $\mathcal{E}_{Bel} = \{B_j\}$ be the collections of focal elements of two belief functions Bel and Bel, respectively, defined on the same frame of discernment, and such that Bel' is combinable with Bel.

Obviously, $B_j \cap A_i = (B_j \cap \mathcal{C}_{Bel}) \cap A_i = B_j \cap (A_i \cap \mathcal{C}_{Bel})$. Therefore, once we have defined a new belief function Bel'' with focal elements $\{F_j, j = 1, \ldots, m\} \doteq \{A_i \cap \mathcal{C}_{Bel},\ i = 1, \ldots, n\}$ and basic probability assignment

$$m''(F_j) = \sum_{i: A_i \cap \mathcal{C}_{Bel} = F_j} m'(A_i),$$

we have that $Bel \oplus Bel' = Bel \oplus Bel''$. In other words, any point of $\langle Bel \rangle$ is a point of $Bel \oplus C(Bel)$. The reverse implication is trivial.

Finally, Theorem 10 ensures that $C(Bel) = Cl(Bel_A, A \subseteq \mathcal{C}_{Bel})$, so that Corollary 6 eventually yields the desired expression for $\langle Bel \rangle$ (since Bel_A is combinable with Bel for all $A \subset \mathcal{C}_{Bel}$).

□

Figure 8.3 illustrates the form of the conditional subspaces in the belief space related to the simplest, binary frame. The original belief function Bel is always a vertex of its own conditional subspace $\langle Bel \rangle$, as the result of the combination of itself with the BF focused on its core, $Bel \oplus Bel_{\mathcal{C}_{Bel}} = Bel$. Moreover, the conditional subspace is a subset of the compatible subspace, i.e., $\langle Bel \rangle \subseteq C(Bel)$, since if $Bel'' = Bel \oplus Bel'$ for some $Bel' \in C(Bel)$ then $\mathcal{C}_{Bel''} \subseteq \mathcal{C}_{Bel}$, i.e., Bel'' is combinable with Bel as well.

8.5.2 The case of unnormalised belief functions

The notion of a conditional subspace is directly applicable to unnormalised belief functions as well. More precisely, we can write

$$\langle \tilde{Bel} \rangle = \left\{ \tilde{Bel} \oplus \tilde{Bel}', \tilde{Bel}' \in \tilde{\mathcal{B}} \right\},$$

where $\tilde{\mathcal{B}}$ denotes the belief space for unnormalised belief functions, since all UBFs are combinable (via conjunctive combination) with any arbitrary UBF $\tilde{Bel}$. The idea of a compatible subspace retains its validity, though, as the empty set is a subset of the core of any UBF. Note that in this case, however, if $\mathcal{C}_{\tilde{Bel}} \cap \mathcal{C}_{\tilde{Bel}'} = \emptyset$, then the combination $\tilde{Bel} \oplus \tilde{Bel}'$ reduces to $Bel_\emptyset$, the UBF assigning mass 1 to $\emptyset$.

The proof of Theorem 17 still works for UBFs too, so that we can write

$$\langle \tilde{Bel} \rangle = \tilde{Bel} \oplus C(\tilde{Bel}),$$

where $C(\tilde{Bel}) = Cl(Bel_A : \emptyset \subseteq A \subseteq \mathcal{C}_{\tilde{Bel}})$ ($A = \emptyset$ this time included).

8.5.3 Vertices of conditional subspaces

The vertices of a conditional subspace possess an interesting structure. Indeed, (2.6) implies

$$Bel \oplus Bel_A(B) = \frac{\sum_{E \in \mathcal{E}_{Bel}: E \cap A \subset B} m(E)}{1 - \sum_{E \in \mathcal{E}_{Bel}: E \cap A = \emptyset} m(E)}$$

(since $E \cap A \subset B$ implies $E \cap (A \cap B) \neq \emptyset$ but $E \cap (A \setminus B) = \emptyset$; see Fig. 8.4)

$$= \frac{Bel((A \setminus B)^c) - Bel(A^c)}{Pl(A)} = \frac{Pl(A) - Pl(A \setminus B)}{Pl(A)}.$$

Therefore,

$$Bel \oplus Bel_A = \frac{1}{Pl(A)} \sum_{B \subset \Theta} \mathbf{v}_B (Pl(A) - Pl(A \setminus B)), \tag{8.14}$$

where, as usual, we have denoted by $\mathbf{v}_B$ the B-th axis of the orthonormal reference frame in $\mathbb{R}^{2^{|\Theta|}-2}$ with respect to which we measure belief coordinates. Note that $Pl(A) \neq 0$ for every $A \subseteq \mathcal{C}_{Bel}$.

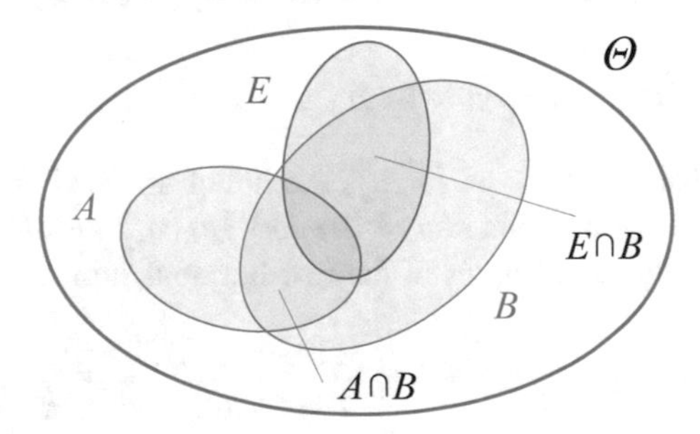

Fig. 8.4: One of the sets E involved in the computation of the vertices of $\langle Bel \rangle$.

Equation (8.14) suggests an interesting result.

Theorem 18.

$$Bel = \sum_{\emptyset \subsetneq A \subseteq \mathcal{C}_{Bel}} (-1)^{|\mathcal{C}_{Bel} \setminus A|+1} Pl(A) Bel \oplus Bel_A + (-1)^{|\mathcal{C}_{Bel}|-1} m(\mathcal{C}_{Bel}) \mathbf{1}, \tag{8.15}$$

where $\mathbf{1}$ *denotes the vector of* $\mathbb{R}^{2^{|\Theta|}-2}$ *whose components are all equal to 1.*

Any belief function Bel can be decomposed into an affine combination of its own Dempster combinations with all the categorical belief functions that agree with it (up to a constant $m(\mathcal{C}_{Bel})$ that measures the uncertainty of the model). The coefficients of this decomposition are nothing but the plausibilities of the events $A \subseteq \mathcal{C}_{Bel}$ given the evidence represented by Bel.

8.6 Constant-mass loci

As pointed out before, Theorem 17 depicts, in a sense, the *global* action of Dempster's rule in the belief space. Theorem 15, instead, expresses its *pointwise* behaviour in the language of affine geometry. It would be interesting to give a geometric interpretation to (8.5), too. The commutativity results we proved in Section 8.4 could clearly be very useful, but we still need an intuition to lead us in this effort.

In the remainder of the chapter, we will find this source of intuition in the discussion of the pointwise geometry of Dempster's rule in the simplest, binary frame and make inferences about the general case. We will realise that Dempster's rule exhibits a very elegant behaviour when applied to sets of belief functions that assign the same mass k to a fixed subset A, or *constant-mass loci*. Such loci turn out to be convex sets, while they assume the form of affine subspaces when we extend our analysis to normalised sum functions. This allows us to apply the commutativity results of Theorem 16 to compute their images through the map $Bel \oplus (.)$.

Strikingly, for any subset A, the resulting affine spaces *have a common intersection for all* $k \in [0,1]$, which is therefore characteristic of Bel. We call this the A-th *focus* of the conditional subspace. In the following sections, we will prove the existence and study the geometry of these foci. This will in turn lead us to an interesting, general geometric construction for the orthogonal sum of two belief functions.

8.6.1 Geometry of Dempster's rule in $\mathcal{B}_2$

We already know that when $\Theta_2 = \{x, y\}$, the belief space $\mathcal{B}_2$ is two-dimensional. Hence, given two belief functions $Bel = [m_b(x), m_b(y)]'$ and $Bel' = [k, l]'$ on Θ_2, it is simple to derive the coordinates of their orthogonal sum as

$$\begin{aligned} m_{Bel \oplus Bel'}(x) &= 1 - \frac{(1 - m(x))(1 - k)}{1 - m(x)l - m(y)k}, \\ m_{Bel \oplus Bel'}(y) &= 1 - \frac{(1 - m(y))(1 - l)}{1 - m(x)l - m(y)k}. \end{aligned} \tag{8.16}$$

At first glance, this expression does not suggest any particular geometrical intuition. Let us then keep the first operand Bel fixed, and analyse the behaviour of $Bel \oplus Bel'$ as a function of the two variables k, l.

We need to distinguish two cases, $m(\Theta_2) \neq 0$ and $m(\Theta_2) = 0$, since in the first case,

$$\langle Bel \rangle = Cl(Bel, Bel \oplus Bel_x, Bel \oplus Bel_y) = Cl(Bel, Bel_x, Bel_y),$$

whereas if $Bel = P$ is Bayesian,

$$\langle P \rangle = Cl(P, P \oplus Bel_x, P \oplus Bel_y) = Cl(P, Bel_x, Bel_y) = \mathcal{P}_2.$$

Foci in the case $m(\Theta_2) \neq 0$ If $m(\Theta_2) \neq 0$, when k is kept constant in (8.16), the resulting combination describes a line segment in the belief space. If we instead allow Bel' to be a normalised sum function and apply the extended Dempster's rule, the locus of all combinations becomes the line containing the same segment, *except* for a single point with coordinates

$$F_x(Bel) = \left(1, -\frac{m(\Theta_2)}{m(x)}\right), \tag{8.17}$$

which incidentally coincides with the limit of $Bel \oplus Bel'$ for $l \to \pm\infty$ (we omit the details). As shown in Fig. 8.5, this is true for every $k \in [0, 1]$. Simple manipulations

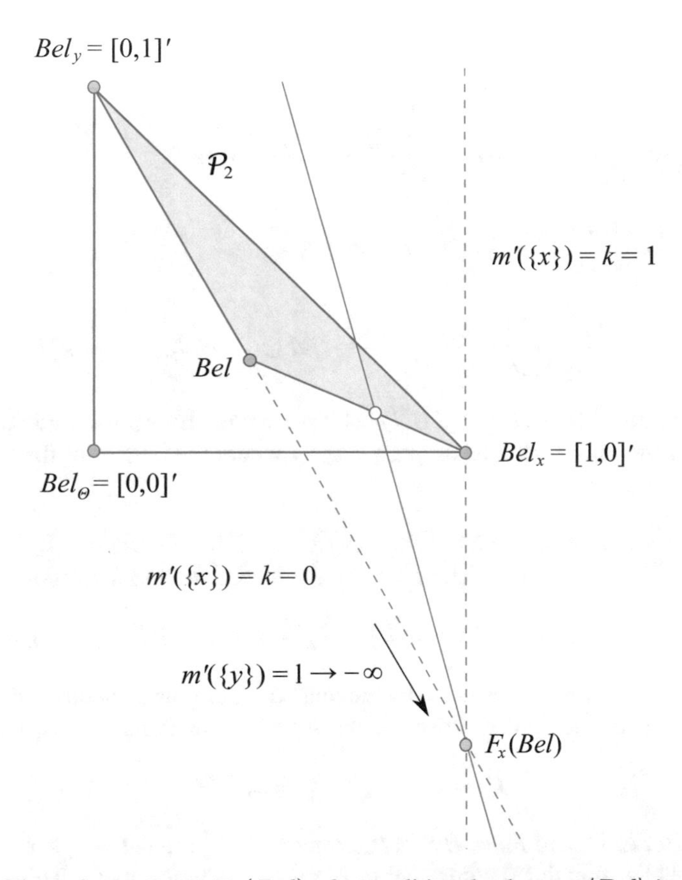

Fig. 8.5: The x-focus (8.17) $F_x(Bel)$ of a conditional subspace $\langle Bel \rangle$ in the binary belief space for $m(\Theta_2) \neq 0$. The circle placed on $F_x(Bel)$ indicates that the latter is a missing point for each of the lines representing images of constant-mass loci according to Bel'.

of (8.16) will help us realise that all the collections of Dempster sums $Bel \oplus Bel'$

(where Bel' is an NSF) with $Bel'(x) = k = \text{const}$ have a common intersection at the point (8.17), located outside the belief space. In the same way, this holds for the sets $\{Bel \oplus Bel' : Bel'(y) = l = \text{const}\}$, which each form a distinct line passing through a twin point:

$$F_y(Bel) = \left(-\frac{m(\Theta)}{m(y)}, 1 \right).$$

We call $F_x(Bel), F_y(Bel)$ the *foci* of the conditional subspace $\langle Bel \rangle$.

Foci as intersections of lines Note that $F_x(Bel)$ can be located by intersecting the two lines for $k = 0$ and $k = 1$. It is also worth noticing that the shape of these loci is in agreement with the prediction of Theorem 16 for $\varsigma_1 = Bel_1 = k\ Bel_x$ and $\varsigma_2 = Bel_2 = k\ Bel_x + (1-k)Bel_y$.

Indeed, according to (8.8), the missing points have coordinates

$$\begin{aligned} & v \left(\frac{\Delta_1}{\Delta_1 - \Delta_2} Bel \oplus Bel_1 - \frac{\Delta_2}{\Delta_1 - \Delta_2} Bel \oplus Bel_2 \right) \\ & = v \left(\frac{1 - k + k(1 - m(y))}{m(x)(1-k)} \cdot Bel \oplus k\ Bel_x \right. \\ & \left. + \frac{(m(x) - 1)(1-k) - k(1 - m(y))}{m(x)(1-k)} Bel \oplus \Big[k\ Bel_x + (1-k) Bel_y \Big] \right), \end{aligned}$$

which coincide with those of $F_x(Bel)$ (as it is easy to check). It is quite interesting to note that the intersection takes place exactly where the images of the lines $\{k = \text{const}\}$ do not exist.

Geometric construction in $\mathcal{B}_2$ When $m(\Theta_2) \neq 0$, for all $Bel' = [k, l]' \in \mathcal{B}_2$ the sum $Bel \oplus Bel'$ is uniquely determined by the intersection of the following lines:

$$l_x \doteq Bel \oplus \{Bel'' : m''(x) = k\}, \quad l_y \doteq Bel \oplus \{Bel'' : m''(y) = l\}.$$

These lines are, in turn, determined by the related focus plus an additional point. We can, for instance, choose their intersections with the probabilistic subspace $\mathcal{P}$,

$$P_x \doteq l_x \cap \mathcal{P}, \quad P_y \doteq l_y \cap \mathcal{P},$$

for $P_x = Bel \oplus P'_x$ and $P_y = Bel \oplus P'_y$, where P'_x, P'_y are the unique probabilities (Bayesian belief functions) with $m(x) = k$, $m(y) = l$, respectively. This suggests a geometric construction for the orthogonal sum of a pair of belief functions Bel, Bel' in $\mathcal{B}_2$ (Algorithm 18), illustrated in Fig. 8.6.

Foci for higher-size events The notion of a focus $F_A(Bel)$ can also be defined for $A = \Theta_2$. For instance, in the case $m(\Theta_2) \neq 0$, a few steps yield the following coordinates for $F_{\Theta_2}(Bel)$:

Algorithm 18 Dempster's rule: geometric construction in $\mathcal{B}_2$

1: **procedure** GEODEMPSTER2(Bel, Bel')
2: compute the foci $F_x(Bel), F_y(Bel)$ of the conditional subspace $\langle Bel \rangle$;
3: project Bel' onto $\mathcal{P}$ along the orthogonal directions, obtaining P'_x and P'_y;
4: combine Bel with P'_x and P'_y to get P_x and P_y;
5: draw the lines $\overline{P_x F_x(Bel)}$ and $\overline{P_y F_y(Bel)}$: their intersection is the desired orthogonal sum $Bel \oplus Bel'$.
6: **end procedure**

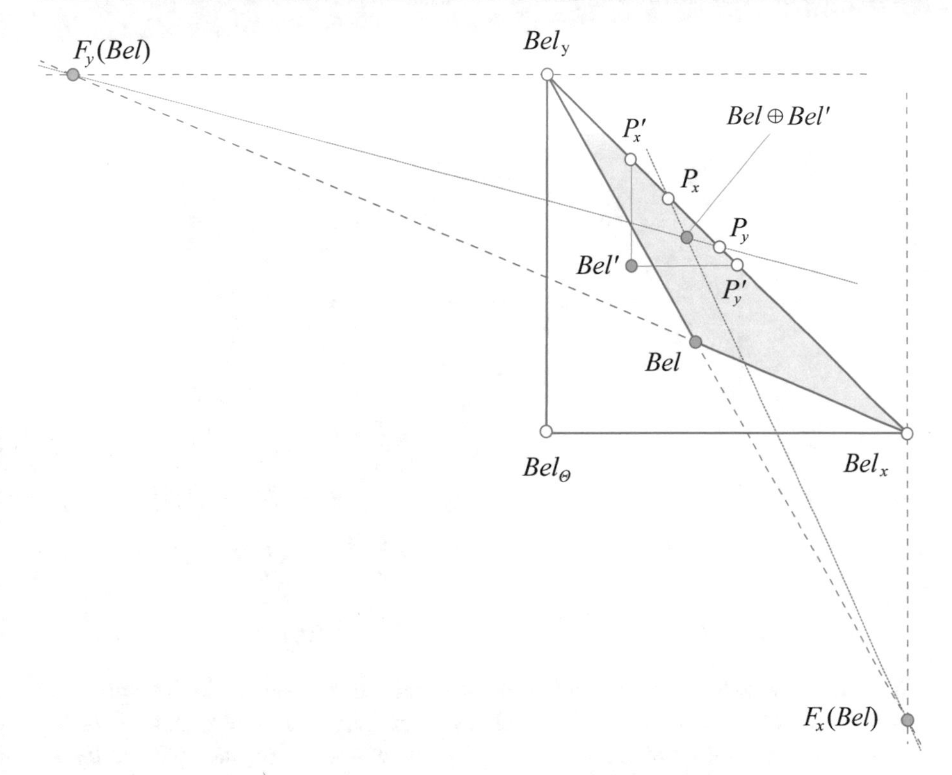

Fig. 8.6: Graphical construction of Dempster's orthogonal sum in the binary belief space $\mathcal{B}_2$ (details in Section 8.7.2).

$$F_{\Theta_2}(Bel) = \left(\frac{1 - m(y)}{m(x) - m(y)}, \frac{1 - m(x)}{m(y) - m(x)} \right), \tag{8.18}$$

which turn out to belong to $v(\mathcal{P})$. At any rate, the point $F_\Theta(Bel)$ plays no role in the geometric construction of Dempster's rule.

Case $m(\Theta_2) = 0$ If $m(\Theta_2) = 0$, the situation is slightly different. The combination locus turns out to be $v(Bel_x, Bel_y) \setminus \{Bel_x\}$ for every $k \in [0, 1)$. Note that in this case,

$$F_x(Bel) = (1, -m(\Theta)/m(x)) = [1,0]' = Bel_x.$$

For $k = 1$, instead, (8.16) yields

$$Bel \oplus [1 \cdot Bel_x + l \cdot Bel_y] = \begin{cases} Bel_x & l \neq 1, \\ \emptyset & l = 1. \end{cases}$$

Incidentally, in this case the missing coordinate $l = 1$ (see Section 8.4) does not correspond to an actual missing point. The situation is represented in Fig. 8.7.

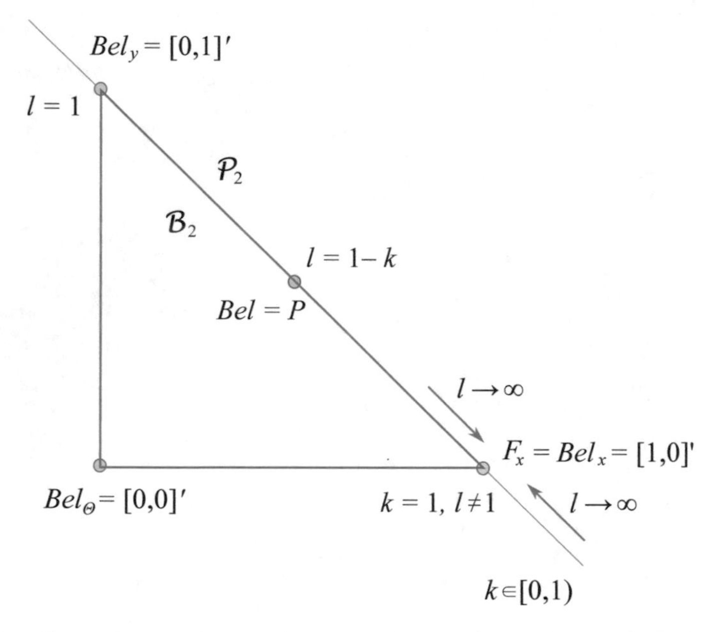

Fig. 8.7: The x-focus of a conditional subspace in the binary belief space for $m(\Theta_2) = 0$ ($Bel = P \in \mathcal{P}$). For each value of k in $[0,1)$, the image of the locus k = const through the map $Bel \oplus (.)$ coincides with the line spanned by $\mathcal{P}$, with missing point Bel_x. The value of the parameter l of this line is shown for some relevant points. For $k = 1$, the locus reduces to the point Bel_x for all values of l.

Algorithm 18 *does not work* when $m(\Theta_2) = 0$. Indeed, in this case the intersection of the lines

$$\overline{P_x F_x(Bel)} = \overline{P_x Bel_x} = v(\mathcal{P}), \quad \overline{P_y F_y(Bel)} = \overline{P_y Bel_y} = v(\mathcal{P})$$

is clearly not unique.

8.6.2 Affine form of constant-mass loci

Our study of the binary belief space suggests how to proceed in the general case. We first need to specify the shape of constant-mass loci, and the way Dempster's rule acts on them. After proving the existence of the intersections of their images and understanding their geometry, we will finally be able to formulate a geometric construction for the orthogonal sum in a generic belief space.

Let us introduce the following notation:

$$H_A^k \doteq \{Bel : m(A) = k\}, k \in [0,1], \quad \mathcal{H}_A^k \doteq \{\varsigma : m_\varsigma(A) = k\}, k \in \mathbb{R},$$

for constant-mass loci related to belief functions and normalised sum functions, respectively. Their dimension is of course $\dim(\mathcal{B}) - 1$, which for $\mathcal{B}_2$ becomes $\dim(\mathcal{H}_A^k) = 4 - 2 - 1 = 1$, so that any constant-mass locus is a line, as we have seen in the previous case study.

Theorem 19.

$$\mathcal{H}_A^k = v(k\ Bel_A + \gamma_B\ Bel_B : B \subseteq \Theta, B \neq A), \quad k \in \mathbb{R}, \tag{8.19}$$

where $\gamma_B \in \mathbb{R}$ is an arbitrary real number for any $B \subseteq \Theta$, $B \neq A$, whereas

$$H_A^k = Cl(k\ Bel_A + (1-k)Bel_B : B \subseteq \Theta, B \neq A), \quad k \in [0,1]. \tag{8.20}$$

Proof. For (8.20), by (7.8), $m(A) = k$ iff

$$Bel = k\ Bel_A + \sum_{B \neq A} \alpha_B\ Bel_B, \quad \sum_{B \neq A} \alpha_B + k = 1, \tag{8.21}$$

with $\alpha_B \geq 0\ \forall B \neq A$. Trivially,

$$Bel = k\ Bel_A + (1-k) \sum_{B \neq A} \alpha'_B\ Bel_B, \quad \sum_{B \neq A} \alpha'_B = 1, \quad \alpha'_B \geq 0\ \forall B \neq A,$$

so that

$$Bel \in k\ Bel_A + (1-k)Cl(Bel_B, B \neq A) = Cl(k\ Bel_A + (1-k)Bel_B, B \neq A)$$

and (8.20) is proved.

A similar argument, but with $\alpha_B \in \mathbb{R}$, shows that if $\varsigma(A) = k$, (8.19) holds. To show that the opposite implication holds, we note that any $\varsigma \in v(k\ Bel_A + \gamma_B\ Bel_B : \gamma_B \in \mathbb{R}\ \forall B \subset \Theta, B \neq A)$ can be written as

$$\begin{aligned}
\varsigma &= \sum_{B \neq A} \beta_B (k\ Bel_A + \gamma_B\ Bel_B) = \sum_{B \neq A} \beta_B k\ Bel_A + \sum_{B \neq A} \beta_B \gamma_B\ Bel_B \\
&= k\ Bel_A + \sum_{B \neq A} \beta_B \gamma_B\ Bel_B \\
&= k\ Bel_A + \sum_{B \neq A} \beta_B \gamma_B\ Bel_B + \left(1 - k - \sum_{B \neq A} \beta_B \gamma_B\right) Bel_\Theta,
\end{aligned}$$

since $Bel_\Theta = \mathbf{0}$ and $\sum_{B \neq A} \beta_B = 1$. The last expression is equivalent to

$$\varsigma = k\ Bel_A + \sum_{B \neq A, \Theta} \beta_B \gamma_B\ Bel_B + \left(1 - k - \sum_{B \neq A, \Theta} \beta_B \gamma_B\right) Bel_\Theta,$$

i.e., ς is an affine combination of Bel_F, $\emptyset \subsetneq F \subseteq \Theta$, with coefficients $\alpha_A = k$, $\alpha_B = \beta_B \gamma_B$ for $B \neq A, \Theta$, and $\alpha_\Theta = 1 - k - \sum_{B \neq A, \Theta} \beta_B \gamma_B$, and thus $\varsigma(A) = k$. □

8.6.3 Action of Dempster's rule on constant-mass loci

Once we have expressed constant-mass loci as affine closures, we can exploit the commutativity property to compute their images through Dempster's rule.

Since $\langle Bel \rangle = Bel \oplus C(Bel)$, our intuition suggests that we should only consider constant-mass loci related to subsets of $\mathcal{C}_{Bel}$. In fact, given a belief function Bel' with basic probability assignment m', it is clear that, by definition,

$$Bel' = \bigcap_{A \subseteq \Theta} \mathcal{H}_A^{m'(A)},$$

for there cannot be several distinct normalised sum functions with the same mass assignment. From Theorem 17, we know that $Bel \oplus Bel' = Bel \oplus Bel''$, where $Bel'' = Bel' \oplus Bel_{\mathcal{C}_{Bel}}$ is a new belief function with probability assignment $m''(A) = \sum_{B: B \cap \mathcal{C}_{Bel} = A} m'(B)$. After introducing the notation

$$H_A^k(Bel) \doteq \left\{ Bel' \in \mathcal{B} : Bel' \in C(Bel), m_{Bel'}(A) = k \right\} = H_A^k \cap C(Bel),$$

$$\mathcal{H}_A^k(Bel) \doteq \left\{ \varsigma \in \mathbb{R}^N : \varsigma \in v(C(Bel)), m_\varsigma(A) = k \right\} = \mathcal{H}_A^k \cap v(C(Bel)),$$

with $A \subseteq \mathcal{C}_{Bel}$, we can write

$$Bel'' = \bigcap_{A \subseteq \mathcal{C}_{Bel}} \mathcal{H}_A^{m''(A)}(Bel),$$

since $Bel'' \in \mathcal{H}_A^{m''(A)}$ for all $A \subseteq \mathcal{C}_{Bel}$ by definition, and the intersection is unique in $v(C(Bel))$. We are then only interested in computing loci of the form

$$Bel \oplus H_A^k(Bel) = Bel \oplus Cl(k\ Bel_A + (1-k) Bel_B : B \subseteq \mathcal{C}_{Bel}, B \neq A),$$

as a natural consequence of Theorem 19.

To get a more comprehensive view of the problem, let us consider the expression

$$Bel \oplus H_A^k = Bel \oplus Cl(k\ Bel_A + (1-k) Bel_B : B \subseteq \Theta, B \neq A),$$

and assume $\mathcal{C}_{Bel} \neq \Theta$. Theorem 14 allows us to write, whenever $\Delta_B^k \doteq k\ Pl(A) + (1-k) Pl(B) \neq 0$,

$$\begin{aligned}
&Bel \oplus \Big[k\ Bel_A + (1-k)Bel_B\Big] \\
&= \frac{k\ Pl(A)}{\Delta^k_B} Bel \oplus Bel_A + \frac{(1-k)Pl(B)}{\Delta^k_B} Bel \oplus Bel_B \qquad (8.22) \\
&= Bel \oplus \Big[k\ Bel_{A\cap \mathcal{C}_{Bel}} + (1-k)Bel_{B\cap \mathcal{C}_{Bel}}\Big],
\end{aligned}$$

for $Bel \oplus Bel_X = Bel \oplus Bel_{X\cap \mathcal{C}_{Bel}}$ and $Pl(X) = Pl(X \cap \mathcal{C}_{Bel})$.

When $A\cap \mathcal{C}_{Bel} = \emptyset$, Theorem 16 yields the following for (8.22) (since $k\ Bel_A + (1-k)Bel_B$ is combinable with Bel iff $B \cap \mathcal{C}_{Bel} \neq \emptyset$):

$$\begin{aligned}
Bel \oplus H^k_A &= Cl\Big(Bel \oplus [k\ Bel_A + (1-k)Bel_B] : B \cap \mathcal{C}_{Bel} \neq \emptyset, B \neq A\Big) \\
&= Cl\Big(Bel \oplus [k\ Bel_A + (1-k)Bel_B] : B \subseteq \mathcal{C}_{Bel}\Big) \\
&= Cl(Bel \oplus Bel_B : B \subseteq \mathcal{C}_{Bel}) = \langle Bel \rangle,
\end{aligned}$$

since $B \subseteq \mathcal{C}_{Bel}$ implies $B \neq A$. The set trivially coincides with the whole conditional subspace.

If, on the other hand, $A \cap \mathcal{C}_{Bel} \neq \emptyset$, the image of H^k_A becomes

$$Bel \oplus H^k_A = Cl\Big(Bel \oplus Bel_A, Bel \oplus [k\ Bel_A + (1-k)Bel_B] : B \subseteq \mathcal{C}_{Bel}, B \neq A\Big)$$

regardless of whether $A \subset \mathcal{C}_{Bel}$ or $A \not\subset \mathcal{C}_{Bel}$. Indeed, if $A \subseteq \mathcal{C}_{Bel}$, then

$$Bel \oplus H^k_A = Cl\Big(Bel \oplus [k\ Bel_A + (1-k)Bel_B] : B \subseteq \mathcal{C}_{Bel}\Big),$$

since $B = A \cup \mathcal{C}^c_{Bel} \neq A$ is associated with the point

$$Bel \oplus [k\ Bel_A + (1-k)Bel_{A\cup \mathcal{C}^c_{Bel}}] = Bel \oplus [k\ Bel_A + (1-k)Bel_A] = Bel \oplus Bel_A$$

by (8.22). If, instead, $A \not\subset \mathcal{C}_{Bel}$, the set $B = A\cap \mathcal{C}_{Bel} \neq A$ corresponds to the point

$$\begin{aligned}
&Bel \oplus [k\ Bel_A + (1-k)Bel_{A\cap \mathcal{C}_{Bel}}] \\
&= Bel \oplus [k\ Bel_{A\cap \mathcal{C}_{Bel}} + (1-k)Bel_{A\cap \mathcal{C}_{Bel}}] = Bel \oplus Bel_{A\cap \mathcal{C}_{Bel}} = Bel \oplus Bel_A.
\end{aligned}$$

In conclusion, $Bel \oplus H^k_A$ has the undesirable property of assuming different shapes according to whether $\mathcal{C}_{Bel} = \Theta$ or $\mathcal{C}_{Bel} \neq \Theta$. Moreover, in the latter case the notion of a focus vanishes and a geometric construction of Dempster's rule makes no sense (just recall the 2D example in Section 8.6.1).

In the following, we will then operate on images of *restrictions* of constant-mass loci to the compatible subspace $C(Bel)$. Again, since $H^k_A(Bel) = Cl(k\ Bel_A + (1-k)Bel_B : B \subset \mathcal{C}_{Bel}, B \neq A)$, Corollary 6 yields

$$Bel \oplus H^k_A(Bel) = Cl\Big(Bel \oplus [k\ Bel_A + (1-k)Bel_B] : B \subseteq \mathcal{C}_{Bel}, B \neq A\Big),$$

where $Bel \oplus [k\ Bel_A + (1-k)Bel_B]$ is again given by (8.22). When $k = 0$, in particular,

$$Bel \oplus H_A^0(Bel) = Cl(Bel, Bel \oplus Bel_B : B \neq A),$$

i.e., we get the *antipodal face* of $\langle Bel \rangle$ with respect to the event A (see [326]).

8.7 Geometric orthogonal sum

8.7.1 Foci of conditional subspaces

We have conjectured that the two-dimensional example presented in Section 8.6.1 is representative of the general case, and that it is always true that all the affine subspaces $v(Bel \oplus \mathcal{H}_A^k(Bel))$ (the images of the constant-mass loci for normalised sum functions) related to the same subset $A \subseteq \mathcal{C}_{Bel}$ have a common intersection, which is then characteristic of the conditional subspace $\langle Bel \rangle$ itself.

Accordingly, we have the following definition.

Definition 123. *We define the A-th* focus *of the conditional subspace $\langle Bel \rangle$, $A \subseteq \mathcal{C}_{Bel}$, as the linear variety*

$$\mathcal{F}_A(Bel) \doteq \bigcap_{k \in [0,1]} v(Bel \oplus \mathcal{H}_A^k(Bel)).$$

As mentioned before, we have no interest in the focus $\mathcal{F}_{\mathcal{C}_{Bel}}(Bel)$. Hence, in the following discussion we will assume $A \neq \mathcal{C}_{Bel}$.

The toy problem of Section 8.6.1 also suggests that an analysis based exclusively on belief functions could lead to wrong conclusions. In such a case, in fact, since $v(Bel \oplus H_A^1(Bel))$ reduces to the single point $Bel \oplus Bel_A$, we would only able to compute the intersection

$$\bigcap_{k \in [0,1)} v(Bel \oplus H_A^k(Bel)),$$

which is in general different from the actual focus $\mathcal{F}_A(Bel)$, as defined above. Recall, for instance, the case $m(\Theta_2) = 0$ in the previous analysis of the binary belief space.

Our conjecture is indeed supported by a formal analysis based on the affine methods we introduced in Section 8.2 and the results of Section 8.6.3. We note first that when $k \in [0,1)$, Theorem 19 yields (since we can choose $\gamma_B = 1-k$ for all $B \neq A$)

$$Bel \oplus \mathcal{H}_A^k(Bel) = Bel \oplus v(k\ Bel_A + (1-k)Bel_B,\ B \subset \mathcal{C}_{Bel}\ B \neq A).$$

As the generators of $\mathcal{H}_A^k(Bel)$ are all combinable with Bel, we can apply Theorem 16 to obtain

$$v(Bel \oplus \mathcal{H}_A^k(Bel)) = v(Bel \oplus [k\ Bel_A + (1-k)Bel_B] : B \subseteq \mathcal{C}_{Bel}, B \neq A)$$
$$= v(Bel \oplus H_A^k(Bel)). \tag{8.23}$$

Let us then take a first step towards a proof of existence of $\mathcal{F}_A(Bel)$.

Theorem 20. *For all $A \subseteq \mathcal{C}_{Bel}$, the family of affine spaces $\{v(Bel \oplus \mathcal{H}_A^k(Bel)) : 0 \leq k < 1\}$ has a non-empty common intersection*

$$\mathcal{F}'_A(Bel) \doteq \bigcap_{k \in [0,1)} v(Bel \oplus \mathcal{H}_A^k(Bel)),$$

and

$$\mathcal{F}'_A(Bel) \supset v(\varsigma_B | B \subseteq \mathcal{C}_{Bel}, B \neq A),$$

where

$$\varsigma_B = \frac{1}{1 - Pl(B)} Bel + \frac{Pl(B)}{Pl(B) - 1} Bel \oplus Bel_B. \tag{8.24}$$

The proof of Theorem 20 can be easily modified to cope with the case $A = \mathcal{C}_{Bel}$. The system (8.33) is still valid, so we just need to modify the last part of the proof by replacing $A = \mathcal{C}_{Bel}$ with another arbitrary subset $C \subsetneq \mathcal{C}_{Bel}$. This yields a family of generators for $\mathcal{F}'_{\mathcal{C}_{Bel}}(Bel)$, whose shape,

$$\frac{Pl(C)}{Pl(C) - Pl(B)} Bel \oplus Bel_C + \frac{Pl(B)}{Pl(B) - Pl(C)} Bel \oplus Bel_B,$$

turns out to be slightly different from that of (8.24). Clearly, when applied to the binary belief space, this formula yields (8.18).

Now, the binary case suggests that the focus should be uniquely determined by the intersection of just two subspaces, one of which is associated with some $k \in [0,1)$, the other being $v(Bel \oplus \mathcal{H}_A^1(Bel))$. We can simplify the maths by choosing $k = 0$: the result is particularly attractive.

Theorem 21. $v(Bel \oplus \mathcal{H}_A^0(Bel)) \cap v(Bel \oplus \mathcal{H}_A^1(Bel)) = v(\varsigma_B | B \subseteq \mathcal{C}_{Bel}, B \neq A)$.

Again, the proof can be modified to include the case $A = \mathcal{C}_{Bel}$. An immediate consequence of Theorems 21 and 20 is that

$$v(Bel \oplus \mathcal{H}_A^0(Bel)) \cap v(Bel \oplus \mathcal{H}_A^1(Bel)) \subset \bigcap_{k \in [0,1)} v(Bel \oplus \mathcal{H}_A^k(Bel)),$$

so that, in turn,

$$\begin{aligned} &\mathcal{F}_A(Bel) \\ &= \bigcap_{k \in [0,1]} v(Bel \oplus \mathcal{H}_A^k(Bel)) \\ &= v(Bel \oplus \mathcal{H}_A^1(Bel)) \cap \bigcap_{k \in [0,1)} v(Bel \oplus \mathcal{H}_A^k(Bel)) \\ &= v(Bel \oplus \mathcal{H}_A^0(Bel)) \cap v(Bel \oplus \mathcal{H}_A^1(Bel)) \cap \bigcap_{k \in [0,1)} v(Bel \oplus \mathcal{H}_A^k(Bel)) \\ &= v(Bel \oplus \mathcal{H}_A^0(Bel)) \cap v(Bel \oplus \mathcal{H}_A^1(Bel)). \end{aligned}$$

In other words, we have the following corollary.

Corollary 7. *Given a belief function Bel, the A-th focus of its conditional subspace $\langle Bel \rangle$ is the affine subspace*

$$\mathcal{F}_A(Bel) = v(\varsigma_B | B \subseteq \mathcal{C}_{Bel}, B \neq A) \tag{8.25}$$

generated by the collection of points (8.24). It is natural to call them focal points *of the conditional subspace $\langle Bel \rangle$.*

Note that the coefficient of Bel in (8.24) is non-negative, while the coefficient of $Bel \oplus Bel_B$ is non-positive. Hence, focal points cannot belong to the belief space, i.e., they are not admissible belief functions. Nevertheless, they possess a very intuitive meaning in terms of mass assignment, namely

$$\varsigma_B = \lim_{k \to +\infty} Bel \oplus (1-k) Bel_B.$$

Indeed, $\lim_{k \to +\infty} Bel \oplus (1-k) Bel_B$ is equal to

$$= \lim_{k \to +\infty} [k \; Bel_\Theta + (1-k) Bel_B]$$

$$= \lim_{k \to +\infty} \left(\frac{k \; Bel}{k + (1-k) Pl(B)} + \frac{(1-k) Pl(B) Bel \oplus Bel_B}{k + (1-k) Pl(B)} \right)$$

$$= \frac{1}{1 - Pl(B)} Bel - \frac{Pl(B)}{1 - Pl(B)} Bel \oplus Bel_B = \varsigma_B.$$

The B-th focal point can then be obtained as the limit of the combination of Bel with the simple belief function having B as its only non-trivial focal element, when the mass of B tends towards $-\infty$.

An even more interesting relationship connects the focal points of $\langle Bel \rangle$ to the missing points of $v(Bel \oplus \mathcal{H}_A^k(Bel))$.

Theorem 22. *$\mathcal{F}_A(Bel)$ coincides with the missing-point subspace for each locus $v(Bel \oplus \mathcal{H}_A^k(Bel))$, $k \in [0,1]$.*

Proof. The collection of missing points of $v(Bel \oplus \mathcal{H}_A^k(Bel))$ is determined by the limits $\lim_{\gamma_B \to \infty} Bel \oplus [k \; Bel_A + \gamma_B \; Bel_B]$ for every $B \subset \mathcal{C}_{Bel}$, $B \neq A$. But, then,

$$\lim_{\gamma_B \to \infty} Bel \oplus [k \; Bel_A + \gamma_B \; Bel_B]$$

$$= \lim_{\gamma_B \to \infty} \left[\frac{k \; Pl(A) Bel \oplus Bel_A}{k \; Pl(A) + \gamma_B \; Pl(B) + 1 - k - \gamma_B} \right.$$

$$\left. + \frac{\gamma_B \; Pl(B) Bel \oplus Bel_B}{k \; Pl(A) + \gamma_B \; Pl(B) + 1 - k - \gamma_B} + \frac{(1 - k - \gamma_B) Bel}{k \; Pl(A) + \gamma_B \; Pl(B) + 1 - k - \gamma_B} \right]$$

$$= \varsigma_B$$

for every $B \neq A$.

□

This again confirms what we have observed in the binary case, where the focus $\mathcal{F}_x(Bel)$ turned out to be located in correspondence with the missing points of the images $v(Bel \oplus [k\ Bel_x + Bel_y], Bel \oplus [k\ Bel_x + Bel_\Theta])$ (see Figs. 8.5 and 8.7).

8.7.2 Algorithm

We are finally ready to formulate a geometric algorithm for the Dempster combination of belief functions in the general belief space. Let us then consider the orthogonal sum $Bel \oplus Bel' = Bel \oplus Bel''$ of a pair Bel, Bel' of belief functions, where Bel'' is the projection of Bel' onto $C(Bel)$. As we have proved in Section 8.6.3, the second belief function is uniquely identified as the intersection

$$Bel'' = \bigcap_{A \subset \mathcal{C}_{Bel}} \mathcal{H}_A^{m''(A)}(Bel).$$

Now,

$$Bel \oplus Bel'' \in \bigcap_{A \subset \mathcal{C}_{Bel}} Bel \oplus \mathcal{H}_A^{m''(A)}(Bel),$$

which is, in turn, equal to

$$\left\{ Bel''' \in C(Bel) \middle| \ \forall A \subset \mathcal{C}_{Bel}\ \exists Bel'_A : m'_A(A) = m''(A) \right. \\ \left. \text{s.t. } Bel''' = Bel \oplus Bel'_A \right\}.$$

If, in addition, the map $Bel \oplus (.)$ is injective (i.e., if $\dim(\langle Bel \rangle) = \dim(C(Bel))$), this intersection is unique, as there can be only one such $Bel'' = Bel'_A$ for all A. In other words, $\oplus$ and $\cap$ *commute* and we can write, by (8.23),

$$\begin{aligned} & Bel \oplus Bel'' \\ &= \bigcap_{A \subset \mathcal{C}_{Bel}} Bel \oplus \mathcal{H}_A^{m''(A)}(Bel) = \bigcap_{A \subset \mathcal{C}_{Bel}} v(Bel \oplus \mathcal{H}_A^{m''(A)}(Bel)) \\ &= \bigcap_{A \subset \mathcal{C}_{Bel}} v\Big(Bel \oplus [m''(A) Bel_A + (1 - m''(A)) Bel_B] : B \subset \mathcal{C}_{Bel}, B \neq A\Big). \end{aligned}$$

At this point, the geometric algorithm for the orthogonal sum $Bel \oplus Bel'$ is easily outlined (Algorithm 19). We just need one last result.

Theorem 23.

$$v(Bel \oplus \mathcal{H}_A^k(Bel)) = v(\mathcal{F}_A(Bel), Bel \oplus k\ Bel_A).$$

The proof (see the appendix) is valid for $k < 1$ since, when $k = 1$, the combination is trivial, but it can be easily modified to cope with unitary masses.

Algorithm 19 Dempster's rule: geometric construction

1: **procedure** GEODEMPSTER(Bel, Bel')
2: First, all the foci $\{\mathcal{F}_A(Bel), A \subseteq \mathcal{C}_{Bel}\}$ of the subspace $\langle Bel \rangle$ conditioned by the first belief function Bel are computed by calculating the corresponding focal points (8.24);
3: then, an additional point $Bel \oplus m''(A)Bel_A$ for each $A \subseteq \mathcal{C}_{Bel}$ is detected, thus identifying the subspace

$$
\begin{aligned}
& v(Bel \oplus \mathcal{H}_A^{m''(A)}(Bel)) \\
& = v\Big(Bel \oplus [m''(A)Bel_A + (1 - m''(A))Bel_B] : B \subset \mathcal{C}_{Bel}, B \neq A\Big);
\end{aligned}
$$

4: the intersection of all these subspaces is computed, eventually yielding the desired combination $Bel \oplus Bel' = Bel \oplus Bel''$.
5: **end procedure**

It is interesting to note that the focal points ς_B have to be computed just once, as trivial functions of the upper probabilities $Pl(B), \forall B \subseteq \mathcal{C}_{Bel}$. In fact, each focus is nothing more than a particular selection of $2^{|\mathcal{C}_{Bel}|} - 3$ focal points out of a collection of $2^{|\mathcal{C}_{Bel}|} - 2$. Different groups of points are selected for each focus with no need for further calculations. Furthermore, the computation of each focal point ς_B involves just Dempster conditioning (as in $Bel|B = Bel \oplus Bel_B$) rather than the more computationally expensive Dempster combination.

8.8 Research questions

The geometric treatment of Dempster's rule of combination stimulates a number of open questions, as the natural prosecution of this line of research.

Question 3. Can the Dempster inversion problem, or 'canonical decomposition' (see Chapter 2, Proposition 4), be solved by geometric means?

A hint of how to address Question 3 comes from the following analysis of the binary case, abstracted from [364].

We seek the canonical decomposition of a generic belief function $Bel : 2^{\Theta_2} \rightarrow [0, 1]$ defined on a binary frame of discernment $\Theta_2 = \{x, y\}$, represented by the vector $\mathbf{bel} \in \mathcal{B}_2$. Note that any belief function in $\mathcal{B}_2$ is separable, except for Bel_x and Bel_y. Let us denote by $\mathcal{CO}_x$ and $\mathcal{CO}_y$ the sets of simple support functions focusing on $\{x\}$ and $\{y\}$, respectively.[69]

Proposition 55. [364] *For all* $\mathbf{bel} \in \mathcal{B}_2$*, there exist two uniquely determined simple support functions* $\mathbf{co}_x \in \mathcal{CO}_x$ *and* $\mathbf{co}_y \in \mathcal{CO}_y$ *such that*

[69] This notation comes from the fact that simple support functions coincide with consonant belief functions on a binary frame: see Chapter 2, Section 2.8.

$$\mathbf{bel} = \mathbf{co}_x \oplus \mathbf{co}_y.$$

The two simple support functions are geometrically defined as the intersections

$$\mathbf{co}_x = \overline{\mathbf{bel}_y\mathbf{bel}} \cap \mathcal{CO}_x, \quad \mathbf{co}_y = \overline{\mathbf{bel}_x\mathbf{bel}} \cap \mathcal{CO}_y, \tag{8.26}$$

where $\overline{\mathbf{bel}_x\mathbf{bel}}$ *denotes the line passing through the vectors* $\mathbf{bel}_x$ *and* $\mathbf{bel}$ *(see Fig. 8.8).*

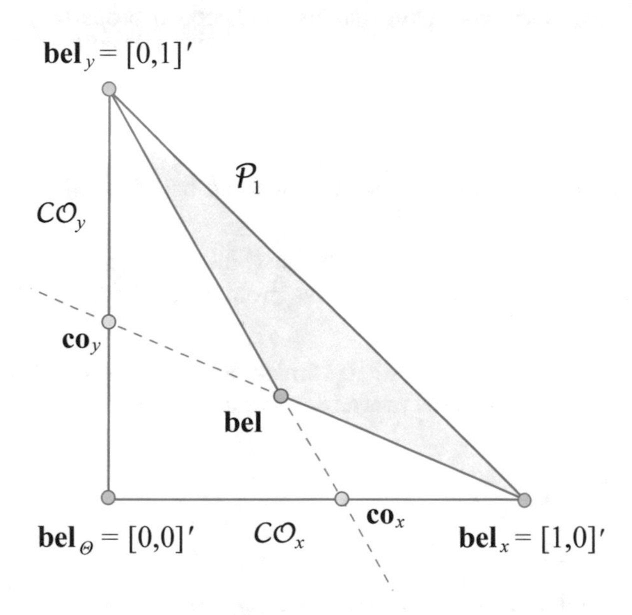

Fig. 8.8: Geometric canonical decomposition of a belief function in $\mathcal{B}_2$.

Proposition 4 suggests the possibility of exploiting our knowledge of the geometry of conditional subspaces to generalise Proposition 55 to arbitrary belief spaces. Indeed, (8.26) can be expressed in the following way:

$$\begin{aligned}\mathbf{co}_{x/y} &= Cl(\mathbf{bel}, \mathbf{bel} \oplus \mathbf{bel}_{x/y}) \cap Cl(\mathbf{bel}_\Theta, \mathbf{bel}_{x/y}) \\ &= Cl(\mathbf{bel}, \mathbf{bel}_{x/y}) \cap Cl(\mathbf{bel}_\Theta, \mathbf{bel}_{x/y}).\end{aligned}$$

This shows that the geometric language introduced here, based on the two operators of the convex closure and the orthogonal sum, is potentially powerful enough to provide a geometric solution to the canonical decomposition problem, as an alternative to both Smets's [1698] and Kramosil's [1064] solutions (see Section 4.3.2).

Note that the inversion problem for the disjunctive rule and other rules of combination may also be posed in geometric terms, once an analysis of their geometry has been conducted.

Question 4. What is the geometric behaviour of the other major combination rules?

Based on our survey of combination rules in Chapter 4, the most interesting candidates are Smets's conjunctive and disjunctive combination rules, together with Denoeux's bold and cautious rules. A geometric description of the former two, in particular, would provide as well a characterisation of belief function 'tubes' in the belief space. We have already taken some steps in this direction when discussing the case of unnormalised belief functions. In the final chapter of this book, we will analyse the geometric behaviour of the conjunctive combination rule and of Yager's combination rule, make conjectures about their general properties, and outline a geometric construction for them, at least in the simple case of binary frames.

The remark that this chapter has mostly focused on classical, normalised belief functions stimulates the following further question.

Question 5. Does our geometric construction of Dempster's rule apply, with suitable amendments, to the case of the unnormalised belief functions contemplated under the open-world assumption?

We will conduct a preliminary analysis of this issue in Chapter 17, Section 17.2.1. A more challenging and speculative problem can be reserved for the more distant future.

Question 6. How does the geometric treatment of rules of combination, including Dempster's rule, extend to belief measures on infinite domains?

Appendix: Proofs

Proof of Theorem 13

Let $m_{\varsigma_1}, m_{\varsigma_2}$ be the Möbius transforms of ς_1, ς_2, respectively. The application of (2.6) to ς_1, ς_2 yields a mass that satisfies the normalisation constraint.

In detail,

$$\sum_{C \neq \emptyset} m_{\varsigma_1 \oplus \varsigma_2}(C) = \sum_{C \neq \emptyset} \frac{\displaystyle\sum_{A \cap B = C} m_{\varsigma_1}(A) m_{\varsigma_2}(B)}{1 - \displaystyle\sum_{A \cap B = \emptyset} m_{\varsigma_1}(A) m_{\varsigma_2}(B)}$$

$$= \frac{\displaystyle\sum_{C \neq \emptyset} \sum_{A \cap B = C} m_{\varsigma_1}(A) m_{\varsigma_2}(B)}{1 - \displaystyle\sum_{A \cap B = \emptyset} m_{\varsigma_1}(A) m_{\varsigma_2}(B)} = \frac{\displaystyle\sum_{A \cap B \neq \emptyset} m_{\varsigma_1}(A) m_{\varsigma_2}(B)}{1 - \displaystyle\sum_{A \cap B = \emptyset} m_{\varsigma_1}(A) m_{\varsigma_2}(B)}.$$

Since $\sum_{A \subseteq \Theta} m_{\varsigma_1}(A) = 1$ and $\sum_{A \subseteq \Theta} m_{\varsigma_2}(A) = 1$ (for ς_1, ς_2 are NSFs), we have that $\sum_{A \subseteq \Theta, B \subseteq \Theta} m_{\varsigma_1}(A) m_{\varsigma_2}(B) = 1$. Therefore

$$\sum_{A \cap B \neq \emptyset} m_{\varsigma_1}(A) m_{\varsigma_2}(B) = 1 - \sum_{A \cap B = \emptyset} m_{\varsigma_1}(A) m_{\varsigma_2}(B).$$

Proof of Lemma 6

By Lemma 5, $\sum_i \alpha_i \varsigma_i$ is combinable with ς. Hence, remembering that

$$m_{\sum_i \alpha_i \varsigma_i}(B) = \sum_{X \subset B} (-1)^{|B \setminus X|} \sum_i \alpha_i \varsigma_i(X)$$

$$= \sum_i \alpha_i \sum_{X \subset B} (-1)^{|B \setminus X|} \varsigma_i(X) = \sum_i \alpha_i m_{\varsigma_i}(B),$$

applying Dempster's rule yields

$$m_{\varsigma \oplus \sum_i \alpha_i \varsigma_i}(C)$$

$$= \frac{\sum_{B \cap A = C} m_{\sum_i \alpha_i \varsigma_i}(B) m_\varsigma(A)}{1 - \sum_{B \cap A = \emptyset} m_{\sum_i \alpha_i \varsigma_i}(B) m_\varsigma(A)} = \frac{\sum_{B \cap A = C} m_\varsigma(A) \sum_i \alpha_i m_{\varsigma_i}(B)}{1 - \sum_{B \cap A = \emptyset} m_\varsigma(A) \sum_i \alpha_i m_{\varsigma_i}(B)}$$

$$= \frac{\sum_i \alpha_i \sum_{B \cap A = C} m_{\varsigma_i}(B) m_\varsigma(A)}{\sum_i \alpha_i - \sum_i \alpha_i \sum_{B \cap A = \emptyset} m_{\varsigma_i}(B) m_\varsigma(A)} = \frac{\sum_i \alpha_i \sum_{B \cap A = C} m_{\varsigma_i}(B) m_\varsigma(A)}{\sum_i \alpha_i \left[1 - \sum_{B \cap A = \emptyset} m_{\varsigma_i}(B) m_\varsigma(A) \right]}$$

$$= \sum_i \frac{\alpha_i N_i(C)}{\sum_j \alpha_j \Delta_j},$$

since $\sum_i \alpha_i = 1$.

Proof of Theorem 16

By definition,

$$\varsigma \oplus v(\varsigma_1, ..., \varsigma_n)$$

$$= \varsigma \oplus \left\{ \sum_i \alpha_i \varsigma_i, \sum_i \alpha_i = 1 \right\} = \left\{ \varsigma \oplus \sum_i \alpha_i \varsigma_i : \sum_i \alpha_i = 1, \exists \varsigma \oplus \sum_i \alpha_i \varsigma_i \right\}$$

$$= \left\{ \varsigma \oplus \sum_i \alpha_i \varsigma_i : \sum_i \alpha_i = 1, \sum_i \alpha_i \Delta_i \neq 0 \right\},$$

by Lemma 5. If $\{\varsigma_1, \ldots, \varsigma_n\}$ are all combinable NSFs or belief functions, Theorem 14 applies and we can write

$$\varsigma \oplus v(\varsigma_1, \ldots, \varsigma_n) = \left\{ \sum_i \beta_i \varsigma \oplus \varsigma_i, \ \beta_i = \frac{\alpha_i \Delta_i}{\sum_j \alpha_j \Delta_j} : \sum_i \alpha_i = 1, \sum_i \alpha_i \Delta_i \neq 0 \right\}.$$

Since $\varDelta_i = 0$ implies $\beta_i = 0$, we have that

$$\varsigma \oplus v(\varsigma_1, \ldots, \varsigma_n) = \left\{ \sum_{i:\varDelta_i \neq 0} \frac{\alpha_i \varDelta_i}{\sum_{j:\varDelta_j \neq 0} \alpha_j \varDelta_j} \varsigma \oplus \varsigma_i : \sum_i \alpha_i = 1, \sum_{i:\varDelta_i \neq 0} \alpha_i \varDelta_i \neq 0 \right\},$$

and as $\sum_i \beta_i = 1$ all these points belong to $v(\varsigma \oplus \varsigma_i, i : \varDelta_i \neq 0)$.

For the converse to be true, we have to prove that, given a set of non-zero normalisation factors $\{\varDelta_{i_1}, \ldots, \varDelta_{i_m}\}$, for any collection of real numbers $\{\beta_1, \ldots, \beta_m\}$ such that $\sum_{j=1}^m \beta_j = 1$ there exists another collection $\{\alpha_1, \ldots, \alpha_n\}$ with $n \geq m$ and $\sum_{i=1}^n \alpha_i = 1$, $\sum_{j=1}^m \alpha_{i_j} \varDelta_{i_j} \neq 0$ such that

$$\beta_j = \frac{\alpha_{i_j} \varDelta_{i_j}}{\sum_{k=1}^m \alpha_{i_k} \varDelta_{i_k}} \quad \forall j = 1, \ldots, m. \tag{8.27}$$

This implies

$$\alpha_{i_j} = \frac{\beta_j}{\varDelta_{i_j}} \sum_{k=1}^m \alpha_{i_k} \varDelta_{i_k} \propto \frac{\beta_j}{\varDelta_{i_j}} \quad \forall j = 1, \ldots, m.$$

Hence, if we take $\alpha_{i_j} \doteq \beta_j / \varDelta_{i_j}$ for $j = 1, \ldots, m$ the system (8.27) is satisfied, since $\beta_j = \beta_j / \sum_k \beta_k = \beta_j$, and

$$\sum_{j=1}^m \alpha_{i_j} \varDelta_{i_j} = \sum_{j=1}^m \beta_j = 1 \neq 0,$$

i.e., the combinability condition is also satisfied.

To satisfy the normalisation constraint, we just need to choose the other $n - m$ coefficients in such a way that

$$\sum_{i:\varDelta_i = 0} \alpha_i = 1 - \sum_{i:\varDelta_i \neq 0} \alpha_i = 1 - \sum_{j=1}^m \alpha_{i_j},$$

which is always possible when $n > m$ since these coefficients do not play any role in the other two constraints. If, instead, $n = m$ (i.e., when we consider only *combinable* functions), we have no choice but to normalise the coefficients $\alpha_{i_j} = \alpha_j$, obtaining in conclusion

$$\alpha'_j = \frac{\alpha_j}{\sum_{i=1}^n \alpha_i} = \frac{\beta_j}{\varDelta_j \sum_{i=1}^n \beta_i / \varDelta_i}, \quad j = 1, ..., n.$$

However, this is clearly impossible iff $\sum_{i=1}^n \beta_i / \varDelta_i = 0$, which is equivalent to

$$\sum_{i=1}^{n-1} \beta_i \frac{\varDelta_i - \varDelta_n}{\varDelta_i} = 1, \quad \sum_{i=1}^n \beta_i = 1, \tag{8.28}$$

which reduces further to

$$\sum_{j:\Delta_j \neq \Delta_n} \beta_j \frac{\Delta_j - \Delta_n}{\Delta_j} = 1, \quad \sum_{i=1}^n \beta_i = 1.$$

A set of basis solutions of the system (8.28) is therefore given by

$$\beta_j = \frac{\Delta_j}{\Delta_j - \Delta_n}, \; \beta_i = 0 \; \forall i \neq j : \Delta_i \neq \Delta_n, \; \sum_{i:\Delta_i = \Delta_n} \beta_i = 1 - \frac{\Delta_j}{\Delta_j - \Delta_n}$$

for every $j : \Delta_j \neq \Delta_n$. Each basis solution corresponds to an affine subspace of $v(\varsigma \oplus \varsigma_1, \ldots, \varsigma \oplus \varsigma_n)$, namely

$$\left\{ \frac{\Delta_j}{\Delta_j - \Delta_n} \varsigma \oplus \varsigma_j + \left(1 - \frac{\Delta_j}{\Delta_j - \Delta_n}\right) \sum_{i:\Delta_i = \Delta_n} \hat{\beta}_i \varsigma \oplus \varsigma_i \middle| \sum_{i:\Delta_i = \Delta_n} \hat{\beta}_i = 1 \right\}$$

for every $j : \Delta_j \neq \Delta_n$. The above subspace can be also expressed as

$$\begin{aligned} & \frac{\Delta_j}{\Delta_j - \Delta_n} \varsigma \oplus \varsigma_j - \frac{\Delta_n}{\Delta_j - \Delta_n} v(\varsigma \oplus \varsigma_i | i : \Delta_i = \Delta_n) \\ = v & \left(\frac{\Delta_j}{\Delta_j - \Delta_n} \varsigma \oplus \varsigma_j - \frac{\Delta_n}{\Delta_j - \Delta_n} \varsigma \oplus \varsigma_i \middle| i : \Delta_i = \Delta_n \right). \end{aligned}$$

Since the general solution of the system (8.28) is an arbitrary affine combination of the basis ones, the region of the points in $v(\varsigma \oplus \varsigma_1, \ldots, \varsigma \oplus \varsigma_n)$ that correspond to inadmissible coordinates $\{\beta_i\}$ such that $\sum_i \beta_i / \Delta_i = 0$ is, finally,

$$v \left(\frac{\Delta_j}{\Delta_j - \Delta_n} \varsigma \oplus \varsigma_j - \frac{\Delta_n}{\Delta_j - \Delta_n} \varsigma \oplus \varsigma_i \middle| \forall j : \Delta_j \neq \Delta_n, \forall i : \Delta_i = \Delta_n \right).$$

Proof of Theorem 18

Let us consider the following expression:

$$\sum_{\substack{A \subset \mathcal{C}_{Bel} \\ A \neq \emptyset}} (-1)^{|A|} Pl(A) Bel \oplus Bel_A = \sum_{\substack{A \subset \mathcal{C}_{Bel} \\ A \neq \emptyset}} (-1)^{|A|} \sum_{B \subset \Theta} \mathbf{v}_B [Pl(A) - Pl(A \setminus B)]. \tag{8.29}$$

Since

$$\sum_{B \subset \Theta} \mathbf{v}_B [Pl(\emptyset) - Pl(\emptyset \setminus B)] = \sum_{B \subset \Theta} \mathbf{v}_B [Pl(\emptyset) - Pl(\emptyset)] = 0,$$

(8.29) becomes

$$\sum_{A\subset\mathcal{C}_{Bel}}(-1)^{|A|}Pl(A)\sum_{B\subset\Theta}\mathbf{v}_B-\sum_{A\subset\mathcal{C}_{Bel}}(-1)^{|A|}\sum_{B\subset\Theta}\mathbf{v}_B Pl(A\setminus B)$$
$$=\mathbf{1}\sum_{A\subset\mathcal{C}_{Bel}}(-1)^{|A|}Pl(A)-\sum_{B\subset\Theta}\mathbf{v}_B\sum_{A\subset\mathcal{C}_{Bel}}(-1)^{|A|}Pl(A\setminus B), \tag{8.30}$$

where $\mathbf{1}$ is the (2^n-2)-dimensional vector whose entries are all equal to 1. Also,

$$\sum_{A\subset\mathcal{C}_{Bel}}(-1)^{|A|}Pl(A)=\sum_{A\subset\mathcal{C}_{Bel}}(-1)^{|A|}(1-Bel(A^c))$$
$$=\sum_{A\subset\mathcal{C}_{Bel}}(-1)^{|A|}-\sum_{A\subset\mathcal{C}_{Bel}}(-1)^{|A|}Bel(A^c), \tag{8.31}$$

where

$$\sum_{A\subset\mathcal{C}_{Bel}}(-1)^{|A|}=\sum_{|A|=k=0}^{|\mathcal{C}_{Bel}|}(-1)^k 1^{|\mathcal{C}_{Bel}|-k}\binom{\mathcal{C}_{Bel}}{k}=0,$$

using the binomial expression for the power $(-1+1)^{|\mathcal{C}_{Bel}|}$.

As for the second addend of (8.31), since

$$Bel(A^c)=\sum_{B\subset A^c,B\subset\Theta}m(B)=\sum_{B\subset A^c,B\subset\mathcal{C}_{Bel}}m(B)=\sum_{B\subset\mathcal{C}_{Bel}\setminus A}m(B)$$
$$=Bel(\mathcal{C}_{Bel}\setminus A)$$

(as $m(B)=0$ for $B\not\subset\mathcal{C}_{Bel}$), we have that

$$-\sum_{A\subset\mathcal{C}_{Bel}}(-1)^{|A|}Bel(A^c)=-\sum_{A\subset\mathcal{C}_{Bel}}(-1)^{|A|}Bel(\mathcal{C}_{Bel}\setminus A)$$
$$=-\sum_{B\subset\mathcal{C}_{Bel}}(-1)^{|\mathcal{C}_{Bel}\setminus B|}Bel(B)=-m(\mathcal{C}_{Bel})$$

by the Möbius inversion formula (2.3), having defined $B\doteq\mathcal{C}_{Bel}\setminus A$.

As for the second part of (8.30), we can note that for every $X\subset B$ any set of the form $A=C+X$, with $C\subset\mathcal{C}_{Bel}\setminus B$, yields the same difference

$$A\setminus B=C+X\setminus B=C.$$

Hence, if we fix $C\doteq A\setminus B$ and let X vary, we get, for all $B\subset\Theta$,

$$\sum_{A\subset\mathcal{C}_{Bel}}(-1)^{|A|}Pl(A\setminus B)=\sum_{C\subset\mathcal{C}_{Bel}\setminus B}Pl(C)\sum_{X\subset B\cap\mathcal{C}_{Bel}}(-1)^{|C+X|}$$
$$=\sum_{C\subset\mathcal{C}_{Bel}\setminus B}Pl(C)\sum_{|X|=k=0}^{|B\cap\mathcal{C}_b|}(-1)^{|C|+|X|}\binom{|B\cap\mathcal{C}_b|}{|X|}$$
$$=\sum_{C\subset\mathcal{C}_{Bel}\setminus B}(-1)^{|C|}Pl(C)\sum_{k=0}^{|B\cap\mathcal{C}_b|}(-1)^k\binom{|B\cap\mathcal{C}_b|}{k},$$

which is equal to 0 by Newton's binomial theorem again.

In conclusion, (8.29) becomes

$$\sum_{A \subset \mathcal{C}_{Bel}, A \neq \emptyset} (-1)^{|A|} Pl(A) Bel \oplus Bel_A = -\mathbf{1} m(\mathcal{C}_{Bel}),$$

whose immediate consequence is (8.15).

Proof of Theorem 20

By (8.23), if $k \neq 1$,

$$v(Bel \oplus \mathcal{H}_A^k(Bel)) = \sum_{B \subset \mathcal{C}_{Bel}, B \neq A} \alpha_B Bel \oplus [k\ Bel_A + (1-k) Bel_B], \tag{8.32}$$

with

$$\sum_{B \subset \mathcal{C}_{Bel}, B \neq A} \alpha_B = 1.$$

Corollary 5 thus yields, after defining $\Delta_B^k \doteq k\ Pl(A) + (1-k) Pl(B)$,

$$\begin{aligned} & v(Bel \oplus \mathcal{H}_A^k(Bel)) \\ &= \sum_{B \subset \mathcal{C}_{Bel}, B \neq A} \alpha_B \left[\frac{k\ Pl(A)}{\Delta_B^k} Bel \oplus Bel_A + \frac{(1-k) Pl(B)}{\Delta_B^k} Bel \oplus Bel_B \right] \\ &= Bel \oplus Bel_A \sum_{\substack{B \subset \mathcal{C}_{Bel} \\ B \neq A}} \frac{\alpha_B k\ Pl(A)}{\Delta_B^k} + \sum_{\substack{B \subset \mathcal{C}_{Bel} \\ B \neq A}} Bel \oplus Bel_B \frac{\alpha_B\ Pl(B)(1-k)}{\Delta_B^k}. \end{aligned}$$

As the vectors $\{Bel \oplus Bel_B, B \subset \mathcal{C}_{Bel}\}$ are generators of $\langle Bel \rangle$, we just need to find values of the scalars α_B such that the coefficients of the generators do not depend on k. After introducing the notation $\alpha_B = \alpha'_B \Delta_B^k$, we can write the following system of conditions:

$$\begin{cases} k\ Pl(A) \displaystyle\sum_{B \subset \mathcal{C}_{Bel}, B \neq A} \alpha'_B \in \mathbb{R}, \\ (1-k) Pl(B) \alpha'_B \in \mathbb{R} \qquad\qquad \forall\ B \subset \mathcal{C}_{Bel}, B \neq A, \\ \displaystyle\sum_{B \subset \mathcal{C}_{Bel}, B \neq A} \alpha'_B \Delta_B^k = 1. \end{cases}$$

From the second and third equations, we have that $\alpha'_B = \frac{x_B}{1-k}\ \forall B \subset \mathcal{C}_{Bel}, B \neq A$, for some $x_B \in \mathbb{R}$. By replacing this expression in the normalisation constraint, we obtain

$$\sum_{B \subset \mathcal{C}_{Bel}, B \neq A} x_B [k \, Pl(A) + (1-k) Pl(B)]$$
$$= k \, Pl(A) \sum_{B \subset \mathcal{C}_{Bel}, B \neq A} x_B + (1-k) \sum_{B \subset \mathcal{C}_{Bel}, B \neq A} Pl(B) = 1-k,$$

so that the real vector $[x_B, B \subset \mathcal{C}_{Bel}, B \neq A]'$ has to be a solution of the system

$$\begin{cases} \displaystyle\sum_{B \subset \mathcal{C}_{Bel}, B \neq A} x_B Pl(B) = 1, \\ \displaystyle\sum_{B \subset \mathcal{C}_{Bel}, B \neq A} x_B = 0, \end{cases}$$

where the last equality ensures that

$$k \, Pl(A) \sum_B \alpha'_B = \frac{k \, Pl(A)}{1-k} \sum_B x_B = 0 \in \mathbb{R}.$$

After subtracting the above equations, we get the system

$$\begin{cases} \displaystyle\sum_{B \subseteq \mathcal{C}_{Bel}, B \neq A} x_B (Pl(B) - 1) = 1, \\ \displaystyle\sum_{B \subset \mathcal{C}_{Bel}, B \neq A} x_B = 0, \end{cases} \tag{8.33}$$

since $Pl(\mathcal{C}_{Bel}) - 1 = 0$. Whenever $A \neq \mathcal{C}_{Bel}$, $x_{\mathcal{C}_{Bel}}$ appears in the second equation only. The admissible solutions of the system are then all the affine combinations of the following basis solutions:

$$x_{\bar{B}} = \frac{1}{Pl(\bar{B}) - 1}, \quad x_{\mathcal{C}_{Bel}} = \frac{1}{1 - Pl(\bar{B})}, \quad x_B = 0 \; \forall B \subseteq \mathcal{C}_{Bel}, B \neq A, \bar{B}.$$

Each basis solution generates the following values of the coefficients $\{\alpha_B\}$:

$$\alpha_{\bar{B}} = \frac{x_B \Delta^k_{\bar{B}}}{1-k}, \quad \alpha_{\mathcal{C}_{Bel}} = \frac{x_{\mathcal{C}_{Bel}} \Delta^k_{\mathcal{C}_{Bel}}}{1-k}, \quad \alpha_B = 0 \; \forall B \subseteq \mathcal{C}_{Bel}, B \neq A, \bar{B},$$

associated in turn via (8.32) with the point

$$k \, Pl(A) \left(\frac{\alpha_{\bar{B}}}{\Delta^k_{\bar{B}}} + \frac{\alpha_{\mathcal{C}_{Bel}}}{\Delta^k_{\mathcal{C}_{Bel}}} \right) Bel \oplus Bel_A + (1-k) \frac{\alpha_{\bar{B}}}{\Delta^k_{\bar{B}}} Bel \oplus Bel_{\bar{B}}$$
$$+ (1-k) \frac{\alpha_{\mathcal{C}_{Bel}}}{\Delta^k_{\mathcal{C}_{Bel}}} Bel$$
$$= \frac{k \, Pl(A)}{1-k} (x_{\bar{B}} + x_{\mathcal{C}_{Bel}}) Bel \oplus Bel_A + x_{\bar{B}} \, Pl(\bar{B}) Bel \oplus Bel_{\bar{B}} + x_{\mathcal{C}_{Bel}} \, Bel$$
$$= \frac{Pl(\bar{B})}{Pl(\bar{B}) - 1} Bel \oplus Bel_{\bar{B}} + \frac{1}{1 - Pl(\bar{B})} Bel.$$

The affine subspace generated by all these points then belongs to $\bigcap_{k\in[0,1)} v(Bel \oplus \mathcal{H}^k_A(Bel))$, even if it is not guaranteed to exhaust the whole of $\mathcal{F}'_A(Bel)$.

Proof of Theorem 21

We first need to compute the explicit form of $v(Bel \oplus \mathcal{H}^1_A(Bel))$. After recalling that

$$\mathcal{H}^1_A(Bel) = v(Bel_A + \gamma_B\, Bel_B : B \subset \mathcal{C}_{Bel}, B \neq A),$$

we notice that for any B there exists a value of γ_B such that the point $Bel_A + \gamma_B\, Bel_B$ is combinable with Bel, i.e., $\Delta_B = Pl(A) + \gamma_B\, Pl(B) - \gamma_B \neq 0$. Since Bel_A, Bel_B and Bel_Θ are all belief functions, Theorem 14 applies and we get

$$\begin{aligned} &Bel \oplus [Bel_A + \gamma_B\, Bel_B - \gamma_B\, Bel_\Theta] \\ &= \frac{Pl(A) Bel \oplus Bel_A + \gamma_B\, Pl(B) Bel \oplus Bel_B - \gamma_B\, Bel}{1 \cdot Pl(A) + \gamma_B\, Pl(B) - \gamma_B}, \end{aligned}$$

so that it suffices to ensure that $\gamma_B \neq \frac{Pl(A)}{1-Pl(B)}$.

Let us then choose a suitable value to simplify this expression, for instance $\gamma_B = -\frac{Pl(A)}{1-Pl(B)}$ (note that $Pl(A) \neq 0$ for $A \subseteq \mathcal{C}_{Bel}$) . We get

$$\frac{1}{2}\left[Bel \oplus Bel_A + Bel \oplus Bel_B \frac{Pl(B)}{Pl(B)-1} + Bel \frac{1}{1-Pl(B)}\right].$$

For $B = \mathcal{C}_{Bel}$, instead, $Bel \oplus [Bel_A + \gamma_B\, Bel_B - \gamma_B\, Bel_\Theta] = Bel \oplus Bel_A$. Thus we can write

$$\begin{aligned} &v(Bel \oplus \mathcal{H}^1_A(Bel)) \\ &= v\Bigg(Bel \oplus Bel_A, \frac{1}{2}\left[Bel \oplus Bel_A + Bel \oplus Bel_B \frac{Pl(B)}{Pl(B)-1} + Bel \frac{1}{1-Pl(B)}\right] : \\ &\quad : B \subseteq \mathcal{C}_{Bel}, B \neq A\Bigg). \end{aligned}$$

On the other hand,

$$v(Bel \oplus \mathcal{H}^0_A(Bel)) = v(Bel \oplus H^0_A(Bel)) = v(Bel \oplus Bel_B : B \subset \mathcal{C}_{Bel}, B \neq A).$$

A sum function ς belonging to both subspaces must then satisfy the following pair of constraints:

$$\varsigma = \sum_{B \subset \mathcal{C}_{Bel}, B \neq A} \alpha_B\, Bel \oplus Bel_B, \quad \sum_B \alpha_B = 1,$$

($\varsigma \in v(Bel \oplus \mathcal{H}^0_A(Bel))$), and

$$\varsigma = \beta_{\mathcal{C}_{Bel}} Bel \oplus Bel_A$$

$$+\frac{1}{2} \sum_{B \subseteq \mathcal{C}_{Bel}, B \neq A} \beta_B \left[Bel \oplus Bel_A + Bel \oplus Bel_B \frac{Pl(B)}{Pl(B) - 1} + \frac{Bel}{1 - Pl(B)} \right]$$

$$= Bel \oplus Bel_A \left(\beta_{\mathcal{C}_{Bel}} + \sum_{B \subseteq \mathcal{C}_{Bel}, B \neq A} \frac{\beta_B}{2} \right)$$

$$+\frac{Bel}{2} \sum_{B \subseteq \mathcal{C}_{Bel}, B \neq A} \frac{\beta_B}{1 - Pl(B)} + \sum_{B \subseteq \mathcal{C}_{Bel}, B \neq A} Bel \oplus Bel_B \frac{\beta_B Pl(B)}{2(Pl(B) - 1)}$$

($\varsigma \in Bel \oplus \mathcal{H}_A^1(Bel)$), where $\sum_{B \subset \mathcal{C}_{Bel}, B \neq A} \beta_B = 1$. By comparison, we get

$$\begin{cases} \beta_{\mathcal{C}_{Bel}} + \frac{1}{2} \sum_{B \subseteq \mathcal{C}_{Bel}, B \neq A} \beta_B = 0, \\ \alpha_B = \frac{1}{2} \beta_B \frac{Pl(B)}{Pl(B) - 1} & B \subseteq \mathcal{C}_{Bel}, B \neq A, \\ \alpha_{\mathcal{C}_{Bel}} = \frac{1}{2} \sum_{B \subseteq \mathcal{C}_{Bel}, B \neq A} \frac{\beta_B}{1 - Pl(B)}. \end{cases} \tag{8.34}$$

Therefore, as $\beta_B = \frac{2\alpha_B (Pl(B) - 1)}{Pl(B)}$ for $B \neq A, \mathcal{C}_{Bel}$, the last equation becomes

$$\alpha_{\mathcal{C}_{Bel}} + \sum_{B \subseteq \mathcal{C}_{Bel}, B \neq A} \alpha_B \frac{1}{Pl(B)} = 0 \equiv \sum_{B \subset \mathcal{C}_{Bel}, B \neq A} \frac{\alpha_B}{Pl(B)} = 0.$$

In conclusion, the points of the intersection $v(Bel \oplus \mathcal{H}_A^0(Bel)) \cap v(Bel \oplus \mathcal{H}_A^1(Bel))$ are associated with affine coordinates $\{\alpha_B\}$ of $v(Bel \oplus \mathcal{H}_A^0(Bel))$ satisfying the constraints

$$\sum_{B \subset \mathcal{C}_{Bel}, B \neq A} \alpha_B = 1, \quad \sum_{B \subset \mathcal{C}_{Bel}, B \neq A} \alpha_B \frac{1}{Pl(B)} = 0.$$

To recover the actual shape of this subspace, we just need to take their difference, which yields

$$\sum_{B \subseteq \mathcal{C}_{Bel}, B \neq A} \alpha_B \frac{Pl(B) - 1}{Pl(B)} = 1, \quad \alpha_{\mathcal{C}_{Bel}} = 1 - \sum_{B \subseteq \mathcal{C}_{Bel}, B \neq A} \alpha_B. \tag{8.35}$$

Note that the first constraint of (8.34) can be written as

$$\sum_{B \subseteq \mathcal{C}_{Bel}, B \neq A} \beta_B = 2$$

and is automatically satisfied by the last system's solutions. If we choose for any $\bar{B} \subseteq \mathcal{C}_{Bel}, \bar{B} \neq A$ the following basis solution of the system (8.35),

$$\begin{cases} \alpha_{\bar{B}} = \dfrac{Pl(\bar{B})}{Pl(\bar{B}) - 1}, & \\ \alpha_B = 0 & B \subseteq \mathcal{C}_{Bel}, B \neq A, \bar{B}, \\ \alpha_{\mathcal{C}_{Bel}} = \dfrac{1}{1 - Pl(\bar{B})}, & \end{cases}$$

we get a set of generators ς_B for $v(Bel \oplus \mathcal{H}^0_A(Bel)) \cap v(Bel \oplus \mathcal{H}^1_A(Bel))$, with ς_B given by (8.24).

Proof of Theorem 23

It suffices to show that each point

$$\begin{aligned} & Bel \oplus [k\ Bel_A + (1-k)Bel_{\bar{B}}] \\ & = Bel \oplus Bel_A \frac{k\ Pl(A)}{k\ Pl(A) + (1-k)Pl(\bar{B})} \\ & \quad + Bel \oplus Bel_{\bar{B}} \frac{(1-k)Pl(\bar{B})}{k\ Pl(A) + (1-k)Pl(\bar{B})} \end{aligned} \tag{8.36}$$

for any $\bar{B} \subset \mathcal{C}_{Bel}, \bar{B} \neq A$, can be generated by the collection $\{Bel \oplus kBel_A, \varsigma_B : B \subseteq \mathcal{C}_{Bel}, B \neq A\}$. We have that

$$\begin{aligned} & \beta\ Bel \oplus k\ Bel_A + \sum_{B \subseteq \mathcal{C}_{Bel}, B \neq A} \beta_B \varsigma_B \\ & = \beta \left(\frac{k\ Pl(A)}{k\ Pl(A) + (1-k)} Bel \oplus Bel_A + \frac{(1-k)}{k\ Pl(A) + (1-k)} Bel \right) \\ & \quad + \sum_{B \subseteq \mathcal{C}_{Bel}, B \neq A} \beta_B \left(\frac{1}{1 - Pl(B)} Bel + \frac{Pl(B)}{Pl(B) - 1} Bel \oplus Bel_B \right) \\ & = \beta \frac{k\ Pl(A)}{k\ Pl(A) + (1-k)} Bel \oplus Bel_A \\ & \quad + \left(\beta \frac{(1-k)}{k\ Pl(A) + (1-k)} + \sum_{B \subseteq \mathcal{C}_{Bel}, B \neq A} \beta_B \frac{1}{1 - Pl(B)} \right) Bel \\ & \quad + \sum_{B \subseteq \mathcal{C}_{Bel}, B \neq A} \beta_B \frac{Pl(B)}{Pl(B) - 1} Bel \oplus Bel_B. \end{aligned}$$

If we choose $\beta_B = 0$ for $B \neq \bar{B}$, we get

$$\beta\; Bel \oplus k\; Bel_A + \sum_{B \subseteq \mathcal{C}_{Bel}, B \neq A} \beta_B \varsigma_B$$

$$= \frac{\beta k\; Pl(A)}{k\; Pl(A) + (1-k)} Bel \oplus Bel_A + \left(\frac{(1-k)\beta}{k\; Pl(A) + (1-k)} + \frac{\beta_{\bar{B}}}{1 - Pl(\bar{B})} \right) Bel$$

$$+ \frac{\beta_{\bar{B}}\; Pl(\bar{B})}{Pl(\bar{B}) - 1} Bel \oplus Bel_{\bar{B}}.$$

Thus, whenever

$$\beta = \frac{k\; Pl(A) + (1-k)}{k\; Pl(A) + (1-k) Pl(B)}, \quad \beta_{\bar{B}} = \frac{(1-k)(Pl(\bar{B}) - 1)}{k\; Pl(A) + (1-k) Pl(\bar{B})}$$

$(\beta + \beta_{\bar{B}} = 1)$, the coefficient of Bel vanishes and we get the point (8.36).

9

Three equivalent models

As we have seen in Chapter 2, plausibility functions

$$Pl : 2^\Theta \rightarrow [0,1], \quad Pl(A) = 1 - Bel(A^c) = \sum_{B \cap A \neq \emptyset} m(B)$$

and commonality functions

$$Q : 2^\Theta \rightarrow [0,1], \quad Q(A) = \sum_{B \supseteq A} m(B)$$

are both equivalent representations of the evidence carried by a belief measure. It is therefore natural to wonder whether they share with belief functions the combinatorial form of a 'sum function' on the power set 2^Θ.

In this chapter, we show that we can indeed represent the same evidence in terms of a *basic plausibility* (or *commonality*) *assignment* on the power set, and compute the related plausibility (or commonality) set function by integrating the basic assignment over similar intervals [347]. Proving that both plausibility and commonality functions share with belief measures the structure of a sum function amounts, in a sense, to introducing two alternative combinatorial formulations of the theory of evidence. The notions of basic plausibility and commonality assignments turn out also to be useful in situations involving the combination of plausibility or commonality functions. This is the case for the probability transformation problem [1218, 1675, 1859, 1810, 100, 291, 333] (recall Chapter 4, Section 4.7.2), or when one is computing the canonical decomposition of support functions [1698, 1052]. We will appreciate this in more detail in Part III.

F. Cuzzolin, *The Geometry of Uncertainty*, Artificial Intelligence: Foundations, Theory, and Algorithms, https://doi.org/10.1007/978-3-030-63153-6_9

In the second part of the chapter, we will also see that the geometric approach introduced in Chapter 7 [334] can naturally be extended to these alternative combinatorial models. Just as belief functions can be seen as points of a simplex whose simplicial coordinates are provided by their Möbius inverse (the basic probability assignment), plausibility and commonality functions possess a similar simplicial geometry, with their own Möbius transforms playing, once again, the role of simplicial coordinates.

The equivalence of the associated formulations of belief theory is geometrically mirrored by the congruence of their simplices. In particular, the relation between upper and lower probabilities (so important in subjective probability) can be geometrically expressed as a simple rigid transformation.

Chapter outline

We first introduce the notions of basic plausibility (Section 9.1) and commonality (9.2) assignments, as the Möbius transforms of the plausibility and commonality functions, respectively.

Later, we show that the geometric approach to uncertainty can be extended to plausibility (Section 9.3) and commonality (Section 9.4) functions, in such a way that the simplicial structure of the related spaces can be recovered as a function of their Möbius transforms. We then show (Section 9.5) that the equivalence of the proposed alternative formulations is reflected by the congruence of the corresponding simplices in our geometric framework. The pointwise geometry of the triplet (Bel, Pl, Q) in terms of the rigid transformation mapping them onto each other, as a geometric nexus between the proposed models, is discussed in Section 9.6.

9.1 Basic plausibility assignment

Let us define the Möbius inverse $\mu : 2^\Theta \to \mathbb{R}$ of a plausibility function Pl as

$$\mu(A) \doteq \sum_{B \subseteq A} (-1)^{|A \setminus B|} Pl(B), \tag{9.1}$$

so that

$$Pl(A) = \sum_{B \subseteq A} \mu(B). \tag{9.2}$$

It is natural to call the function $\mu : 2^\Theta \to \mathbb{R}$ defined by (9.1) a *basic plausibility assignment* (BPlA). Plausibility functions are then sum functions on 2^Θ of the form (9.2), whose Möbius inverse is the BPlA (9.1).

Basic probabilities and plausibilities are obviously related.

Theorem 24. *Given a belief function Bel with basic probability assignment m, the corresponding basic plausibility assignment can be expressed in terms of m as follows:*

$$\mu(A) = \begin{cases} (-1)^{|A|+1} \sum_{C \supseteq A} m(C) & A \neq \emptyset, \\ 0 & A = \emptyset. \end{cases} \tag{9.3}$$

Like BPAs, basic plausibility assignments satisfy the normalisation constraint. In other words, plausibility functions are *normalised sum functions* (see Section 7.3.3) [18]

$$\sum_{A \subseteq \Theta} \mu(A) = -\sum_{\emptyset \subsetneq A \subseteq \Theta} (-1)^{|A|} \sum_{C \supseteq A} m(C) = -\sum_{C \subseteq \Theta} m(C) \sum_{\emptyset \subsetneq A \subseteq C} (-1)^{|A|} = 1,$$

since

$$-\sum_{\emptyset \subsetneq A \subseteq C} (-1)^{|A|} = -(0 - (-1)^0) = 1$$

by Newton's binomial theorem,

$$\sum_{k=0}^{n} \binom{n}{k} p^k q^{n-k} = (p+q)^n. \tag{9.4}$$

However, unlike its counterpart m, μ is not guaranteed to be non-negative.

Example 20: basic plausibility assignment. *Let us consider, as an example, a belief function Bel on a binary frame $\Theta_2 = \{x, y\}$ with basic probability assignment*

$$m_b(x) = \frac{1}{3}, \quad m(\Theta) = \frac{2}{3}.$$

Using (9.1), we can compute its basic plausibility assignment as follows:

$$\begin{aligned}
\mu(x) &= (-1)^{|x|+1} \sum_{C \supseteq \{x\}} m(C) = (-1)^2 (m(x) + m(\Theta)) = 1, \\
\mu(y) &= (-1)^{|y|+1} \sum_{C \supseteq \{y\}} m(C) = (-1)^2 m(\Theta) = \frac{2}{3}, \\
\mu(\Theta) &= (-1)^{|\Theta|+1} \sum_{C \supseteq \Theta} m(C) = (-1) m(\Theta) = -\frac{2}{3} < 0,
\end{aligned}$$

confirming that basic plausibility assignments satisfy the normalisation constraint but not the non-negativity one.

9.1.1 Relation between basic probability and plausibility assignments

Basic probability and plausibility assignments are linked by a rather elegant relation.

Theorem 25. *Given a belief function $Bel : 2^\Theta \to [0,1]$, for each element $x \in \Theta$ of the frame of discernment, the sum of the basic plausibility values of all the events containing x equals its basic probability assignment:*

$$\sum_{A \supseteq \{x\}} \mu(A) = m(x). \tag{9.5}$$

Proof.

$$\sum_{A \supseteq \{x\}} \mu(A) = \sum_{A \supseteq \{x\}} (-1)^{|A|+1} \left(\sum_{B \supseteq A} m(B) \right) = \sum_{B \supseteq \{x\}} m(B) \sum_{\{x\} \subseteq A \subseteq B} (-1)^{|A|},$$

where, by Newton's binomial theorem, $\sum_{k=0}^{n} 1^{n-k}(-1)^k = 0$, and so

$$\sum_{\{x\} \subseteq A \subseteq B} (-1)^{|A|} = \begin{cases} 0 & B \neq \{x\}, \\ -1 & B = \{x\}. \end{cases}$$

□

9.2 Basic commonality assignment

It is straightforward to prove that commonality functions are also sum functions and show some interesting similarities with plausibility functions.

Let us define the Möbius inverse $q : 2^\Theta \to \mathbb{R}$, $B \mapsto q(B)$ of a commonality function Q as

$$q(B) = \sum_{\emptyset \subseteq A \subseteq B} (-1)^{|B \setminus A|} Q(A). \tag{9.6}$$

It is natural to call the quantity (9.6) the *basic commonality assignment* associated with a belief function Bel. To arrive at its explicit form, we just need to substitute the definition of $Q(A)$ into (9.6). We obtain

$$\begin{aligned} q(B) &= \sum_{\emptyset \subseteq A \subseteq B} (-1)^{|B \setminus A|} \left(\sum_{C \supseteq A} m(C) \right) \\ &= \sum_{\emptyset \subsetneq A \subseteq B} (-1)^{|B \setminus A|} \left(\sum_{C \supseteq A} m(C) \right) + (-1)^{|B|-|\emptyset|} \sum_{C \supseteq \emptyset} m(C) \\ &= \sum_{B \cap C \neq \emptyset} m(C) \left(\sum_{\emptyset \subsetneq A \subseteq B \cap C} (-1)^{|B \setminus A|} \right) + (-1)^{|B|}. \end{aligned}$$

But now, since $B \setminus A = B \setminus C + B \cap C \setminus A$, we have that

$$\sum_{\substack{A\subseteq(B\cap C)\\A\neq\emptyset}} (-1)^{|B\setminus A|} = (-1)^{|B\setminus C|} \sum_{\emptyset\subsetneq A\subseteq(B\cap C)} (-1)^{|B\cap C|-|A|}$$
$$= (-1)^{|B\setminus C|}\Big[(1-1)^{|B\cap C|} - (-1)^{|B\cap C|-|\emptyset|}\Big] = (-1)^{|B|+1}.$$

Therefore, a basic commonality value $q(B)$ can be expressed as

$$\begin{aligned} q(B) &= (-1)^{|B|+1} \sum_{B\cap C\neq\emptyset} m(C) + (-1)^{|B|} = (-1)^{|B|} \left(1 - \sum_{B\cap C\neq\emptyset} m(C)\right) \\ &= (-1)^{|B|}(1 - Pl(B)) = (-1)^{|B|} Bel(B^c) \end{aligned} \tag{9.7}$$

(note that $q(\emptyset) = (-1)^{|\emptyset|} Bel(\emptyset) = 1$).

9.2.1 Properties of basic commonality assignments

Basic commonality assignments *do not* satisfy the normalisation axiom, as

$$\sum_{\emptyset\subseteq B\subseteq\Theta} q(B) = Q(\Theta) = m(\Theta).$$

In other words, whereas belief functions are normalised sum functions with a non-negative Möbius inverse, and plausibility functions are normalised sum functions, commonality functions are combinatorially *unnormalised* sum functions.

Going back to Example 20, the basic commonality assignment associated with $m(x) = 1/3$, $m(\Theta) = 2/3$ is, by (9.7),

$$\begin{aligned} q(\emptyset) &= (-1)^{|\emptyset|} Bel(\Theta) = 1, & q(x) &= (-1)^{|x|} Bel(y) = -m(y) = 0, \\ q(\Theta) &= (-1)^{|\Theta|} Bel(\emptyset) = 0, & q(y) &= (-1)^{|y|} Bel(x) = -m(x) = -\frac{1}{3}, \end{aligned}$$

so that

$$\sum_{\emptyset\subseteq B\subseteq\Theta} q(B) = 1 - \frac{1}{3} = \frac{2}{3} = m(\Theta) = Q(\Theta).$$

9.3 The geometry of plausibility functions

The theory of evidence can thus be given alternative combinatorial formulations in terms of plausibility and commonality assignments. As a consequence, plausibility and commonality functions (just like belief functions [334, 333, 327]) also possess simple but elegant geometric descriptions in terms of simplices.

It is useful, first, to extend our geometric representation to unnormalised belief functions. UBFs are naturally associated with vectors with $N = 2^{|\Theta|}$ coordinates, as $Bel(\emptyset)$ cannot be neglected any more. We can then extend the set of categorical belief functions as follows:

$$\{Bel_A \in \mathbb{R}^N, \emptyset \subseteq A \subseteq \Theta\},$$

this time including a new vector $Bel_\emptyset \doteq [1\ 0\ \cdots 0]'$. Note also that in this case $Bel_\Theta = [0 \cdots 0\ 1]' \neq \mathbf{0}$. The space of UBFs is again a simplex, this time in $\mathbb{R}^N$, namely

$$\mathcal{B}^U = Cl(Bel_A, \emptyset \subseteq A \subseteq \Theta).$$

As for plausibility functions, these are completely specified by their $N-2$ plausibility values $\{Pl(A), \emptyset \subsetneq A \subsetneq \Theta\}$. Just as for belief functions, then, they can be represented as vectors of $\mathbb{R}^{N-2}$. We can therefore associate a pair of belief and plausibility functions Bel and Pl with the following vectors, which (as in Chapter 8) we still denote by Bel and Pl:

$$Bel = \sum_{\emptyset \subsetneq A \subseteq \Theta} Bel(A)\mathbf{v}_A, \quad Pl = \sum_{\emptyset \subsetneq A \subsetneq \Theta} Pl(A)\mathbf{v}_A, \tag{9.8}$$

where $\{\mathbf{v}_A : \emptyset \subsetneq A \subsetneq \Theta\}$ is, as usual, a reference frame in the Cartesian space $\mathbb{R}^{N-2}$ (see Chapter 7, Section 7.1).

9.3.1 Plausibility assignment and simplicial coordinates

Since the categorical belief functions $\{Bel_A : \emptyset \subsetneq A \subsetneq \Theta\}$ also form a set of independent vectors in $\mathbb{R}^{N-2}$, the collections $\{\mathbf{v}_A\}$ and $\{\mathbf{bel}_A\} = \{Bel_A\}$ represent two distinct coordinate frames in the same Cartesian space. To understand where a plausibility vector is located in the categorical reference frame $\{Bel_A, \emptyset \subsetneq A \subsetneq \Theta\}$, we need to compute the coordinate change between these frames.

Lemma 8. *The coordinate change between the two coordinate frames* $\{\mathbf{v}_A : \emptyset \subsetneq A \subsetneq \Theta\}$ *and* $\{Bel_A : \emptyset \subsetneq A \subsetneq \Theta\}$ *is given by*

$$\mathbf{v}_A = \sum_{B \supseteq A} (-1)^{|B \setminus A|} Bel_B. \tag{9.9}$$

We can indeed use Lemma 8 to find the coordinates of a plausibility function in the categorical reference frame, by putting the corresponding vector Pl (9.8) in the form of (7.8). By replacing the expression (9.9) for $\mathbf{v}_A$ in (9.8), we get

$$\begin{aligned} Pl = \sum_{\emptyset \subsetneq A \subsetneq \Theta} Pl(A)\mathbf{v}_A &= \sum_{\emptyset \subsetneq A \subsetneq \Theta} Pl(A) \left(\sum_{B \supseteq A} Bel_B (-1)^{|B \setminus A|} \right) \\ &= \sum_{\emptyset \subsetneq B \subsetneq \Theta} Bel_B \left(\sum_{A \subseteq B} (-1)^{|B-A|} Pl(A) \right) = \sum_{\emptyset \subsetneq A \subsetneq \Theta} \mu(A) Bel_A, \end{aligned} \tag{9.10}$$

where we have used the definition (9.1) of a basic plausibility assignment and inverted the roles of A and B for sake of homogeneity of the notation.

Incidentally, as $Bel_\Theta = [0, \ldots, 0]' = \mathbf{0}$ is the origin of $\mathbb{R}^{N-2}$, we can also write

$$Pl = \sum_{\emptyset \subsetneq A \subseteq \Theta} \mu(A) Bel_A$$

(this time including Θ). Analogously to what happens in (7.8), the coordinates of Pl in the categorical reference frame are given by the values of its Möbius inverse, the basic plausibility assignment μ.

9.3.2 Plausibility space

Let us define the *plausibility space* as the region $\mathcal{PL}$ of $\mathbb{R}^{N-2}$ whose points correspond to admissible plausibility functions.

Theorem 26. *The plausibility space $\mathcal{PL}$ is a simplex $\mathcal{PL} = Cl(Pl_A, \emptyset \subsetneq A \subseteq \Theta)$ whose vertices can be expressed in terms of the categorical belief functions (the vertices of the belief space) as*

$$Pl_A = - \sum_{\emptyset \subsetneq B \subseteq A} (-1)^{|B|} Bel_B. \tag{9.11}$$

Note that

$$Pl_x = -(-1)^{|x|} Bel_x = Bel_x \quad \forall x \in \Theta,$$

so that $\mathcal{B} \cap \mathcal{PL} \supset \mathcal{P}$.

The vertices of the plausibility space have a natural interpretation.

Theorem 27. *The vertex Pl_A of the plausibility space is the plausibility vector associated with the categorical belief function Bel_A: $Pl_A = Pl_{Bel_A}$.*

When considering the case of unnormalised belief functions, whose role is so important in the transferable belief model, it is easy to see that Theorems 24 and 27 fully retain their validity. In the case of Theorem 26, however, as in general $m(\emptyset) \neq 0$, we need to modify (9.18) by adding a term related to the empty set. This yields

$$Pl = \sum_{\emptyset \subsetneq C \subseteq \Theta} m(C) Pl_C + m(\emptyset) Pl_\emptyset,$$

where Pl_C, $C \neq \emptyset$, is still given by (9.11), and $Pl_\emptyset = \mathbf{0}$ is the origin of $\mathbb{R}^N$. Note that even in the case of unnormalised belief functions (9.11), the empty set need not be considered, for $\mu(\emptyset) = 0$.

Example 21: belief and plausibility spaces in the binary case. *Figure 9.1 shows the geometry of the belief and plausibility spaces in the familiar case study of a binary frame $\Theta_2 = \{x, y\}$, where the belief and plausibility vectors are points of a plane $\mathbb{R}^2$ with coordinates*

$$\begin{array}{ll} Bel = [Bel(x) = m(x), Bel(y) = m(y)]', \\ Pl \ = [Pl(x) = 1 - m(y), Pl(y) = 1 - m(x)]', \end{array}$$

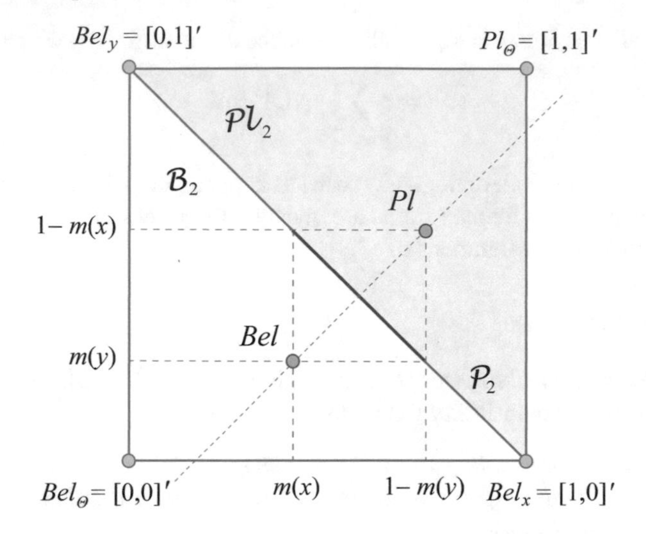

Fig. 9.1: In the binary case, the belief and plausibility spaces $\mathcal{B}_2$ and $\mathcal{PL}_2$ are congruent and lie in symmetric locations with respect to the axis of symmetry formed by the probability simplex $\mathcal{P}_2$.

respectively. They form two simplices (in this special case, two triangles)

$$\begin{array}{ll}\mathcal{B}_2 & = Cl(Bel_\Theta = [0,0]' = \mathbf{0}, Bel_x, Bel_y), \\ \mathcal{PL}_2 & = Cl(Pl_\Theta = [1,1]' = \mathbf{1}, Pl_x = Bel_x, Pl_y = Bel_y),\end{array}$$

which are symmetric with respect to the probability simplex $\mathcal{P}_2$ (in this case a line segment) and congruent, so that they can be moved onto each other by means of a rigid transformation. In this simple case, such a transformation is just a reflection through the Bayesian segment $\mathcal{P}_2$.

From Fig. 9.1, it is clear that each pair of belief/plausibility functions (Bel, Pl) determines a line $a(Bel, Pl)$ which is orthogonal to $\mathcal{P}$, on which they lie in symmetric positions on the two sides of the Bayesian segment.

9.4 The geometry of commonality functions

In the case of commonality functions, as

$$Q(\emptyset) = \sum_{A \supseteq \emptyset} m(A) = \sum_{A \subseteq \Theta} m(A) = 1, \quad Q(\Theta) = \sum_{A \supseteq \Theta} m(A) = m(\Theta),$$

any commonality function Q can be represented using $2^{|\Theta|} = N$ coordinates. The geometric counterpart of a commonality function is therefore the following vector of $\mathbb{R}^N$:

$$Q = \sum_{\emptyset \subseteq A \subseteq \Theta} Q(A) \mathbf{v}_A,$$

where $\{\mathbf{v}_A : \emptyset \subseteq A \subseteq \Theta\}$ is the extended reference frame introduced in the case of unnormalised belief functions ($A = \Theta, \emptyset$ this time included).

Just as before, we can use Lemma 8 to transform the reference frame and obtain the coordinates of Q with respect to the base $\{Bel_A, \emptyset \subseteq A \subseteq \Theta\}$ formed by all the categorical UBFs. We get

$$\begin{aligned} Q &= \sum_{\emptyset \subseteq A \subseteq \Theta} Q(A) \left(\sum_{B \supseteq A} (-1)^{|B \setminus A|} Bel_B \right) \\ &= \sum_{\emptyset \subseteq B \subseteq \Theta} Bel_B \left(\sum_{A \subseteq B} (-1)^{|B \setminus A|} Q(A) \right) = \sum_{\emptyset \subseteq B \subseteq \Theta} q(B) Bel_B, \end{aligned}$$

where q is the basic commonality assignment (9.6).

Once again, we can use the explicit form (9.7) of a basic commonality assignment to recover the shape of the space $\mathcal{Q} \subset \mathbb{R}^N$ of all the commonality functions. We obtain

$$\begin{aligned} Q &= \sum_{\emptyset \subseteq B \subseteq \Theta} (-1)^{|B|} Bel_B \left(\sum_{\emptyset \subseteq A \subseteq B^c} m(A) \right) \\ &= \sum_{\emptyset \subseteq A \subseteq \Theta} m(A) \left(\sum_{\emptyset \subseteq B \subseteq A^c} (-1)^{|B|} Bel_B \right) = \sum_{\emptyset \subseteq A \subseteq \Theta} m(A) Q_A, \end{aligned}$$

where

$$Q_A \doteq \sum_{\emptyset \subseteq B \subseteq A^c} (-1)^{|B|} Bel_B \tag{9.12}$$

is the A-th vertex of the *commonality space*. The latter is hence given by

$$\mathcal{Q} = Cl(Q_A, \emptyset \subseteq A \subseteq \Theta).$$

Again, Q_A is the commonality function associated with the categorical belief function Bel_A, i.e.,

$$Q_{Bel_A} = \sum_{\emptyset \subseteq B \subseteq \Theta} q_{Bel_A}(B) Bel_B.$$

In fact, $q_{Bel_A}(B) = (-1)^{|B|}$ if $B^c \supseteq A$ (i.e., $B \subseteq A^c$), while $q_{Bel_A}(B) = 0$ otherwise, so that the two quantities coincide:

$$Q_{Bel_A} = \sum_{\emptyset \subseteq B \subseteq A^c} (-1)^{|B|} Bel_B = Q_A.$$

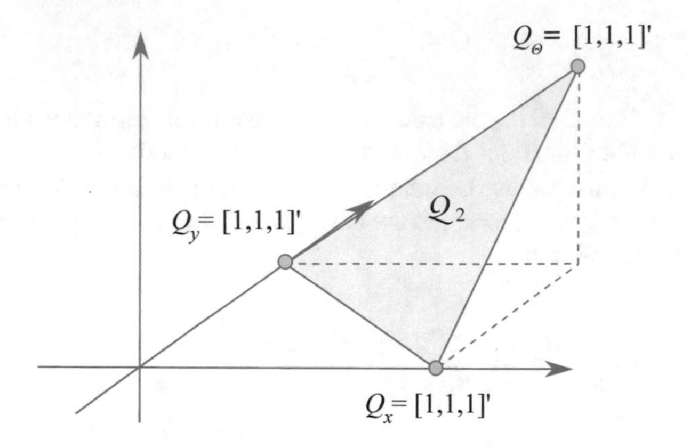

Fig. 9.2: Commonality space in the binary case.

Example 22: commonality space in the binary case. *In the binary case, the commonality space $\mathcal{Q}_2$ needs $N = 2^2 = 4$ coordinates to be represented. Each commonality vector $Q = [Q(\emptyset), Q(x), Q(y), Q(\Theta)]'$ is such that*

$$Q(\emptyset) = 1, \qquad Q(x) = \sum_{A \supseteq \{x\}} m(A) = Pl(x),$$

$$Q(\Theta) = m(\Theta), \quad Q(y) = \sum_{A \supseteq \{y\}} m(A) = Pl(y).$$

The commonality space $\mathcal{Q}_2$ can then be drawn (if we neglect the coordinate $Q(\emptyset)$, which is constant for all belief functions Bel) as in Fig. 9.2.

The vertices of $\mathcal{Q}_2$ are, according to (9.12),

$$Q_\emptyset = \sum_{\emptyset \subseteq B \subseteq \Theta} (-1)^{|B|} Bel_B = Bel_\emptyset + Bel_\Theta - Bel_x - Bel_y$$

$$= [1,1,1,1]' + [0,0,0,1]' - [0,1,0,1]' - [0,0,1,1]' = [1,0,0,0]' = Q_{Bel_\emptyset},$$

$$Q_x = \sum_{\emptyset \subseteq B \subseteq \{y\}} (-1)^{|B|} Bel_B = Bel_\emptyset - Bel_y = [1,1,1,1]' - [0,0,1,1]'$$

$$= [1,1,0,0]' = Q_{Bel_x},$$

$$Q_y = \sum_{\emptyset \subseteq B \subseteq \{x\}} (-1)^{|B|} Bel_B = Bel_\emptyset - Bel_x = [1,1,1,1]' - [0,1,0,1]'$$

$$= [1,0,1,0]' = Q_{Bel_y}.$$

9.5 Equivalence and congruence

Summarising, both plausibility and commonality functions can then be thought of as sum functions on the partially ordered set 2^Θ (although, whereas belief and plausibility functions are normalised sum functions, commonality functions are not). This in turn allows us to describe them as points of appropriate simplices $\mathcal{B}, \mathcal{PL}$ and $\mathcal{Q}$ in a Cartesian space.

In fact, it turns out that the equivalence of such alternative models of the theory of evidence is geometrically mirrored by the congruence of the associated simplices.

9.5.1 Congruence of belief and plausibility spaces

We have seen that in the case of a binary frame of discernment, $\mathcal{B}$ and $\mathcal{PL}$ are congruent, i.e., they can be superposed by means of a rigid transformation (see Example 21). Indeed, the congruence of belief, plausibility and commonality spaces is a general property.

Theorem 28. *The one-dimensional faces $Cl(Bel_A, Bel_B)$ and $Cl(Pl_A, Pl_B)$ of the belief and plausibility spaces, respectively, are congruent. Namely*

$$\|Pl_B - Pl_A\|_p = \|Bel_A - Bel_B\|_p,$$

where $\| \|_p$ denotes the classical norm $\|\mathbf{v}\|_p \doteq \sqrt[p]{\sum_{i=1}^N |v_i|^p}$, $p = 1, 2, \ldots, +\infty$.

Proof. This a direct consequence of the definition of the plausibility function. Let us denote by C, D two generic subsets of Θ. As $Pl_A(C) = 1 - Bel_A(C^c)$, we have that $Bel_A(C^c) = 1 - Pl_A(C)$, which in turn implies

$$Bel_A(C^c) - Bel_B(C^c) = 1 - Pl_A(C) - 1 + Pl_B(C) = Pl_B(C) - Pl_A(C).$$

Therefore, $\forall p$

$$\sum_{C \subseteq \Theta} |Pl_B(C) - Pl_A(C)|^p = \sum_{C \subseteq \Theta} |Bel_A(C^c) - Bel_B(C^c)|^p$$

$$= \sum_{D \subseteq \Theta} |Bel_A(D) - Bel_B(D)|^p.$$

□

Notice that the proof of Theorem 28 holds no matter whether the pair $(\emptyset, \emptyset^c = \Theta)$ is considered or not (i.e., it is immaterial whether classical or unnormalised belief functions are considered).

A straightforward consequence is the following.

Corollary 8. *$\mathcal{B}$ and $\mathcal{PL}$ are congruent; $\mathcal{B}^U$ and $\mathcal{PL}^U$ are congruent.*

This is true because their corresponding one-dimensional faces have the same length. This is due to a generalisation of a well-known theorem of Euclid, which states that triangles whose sides are of the same length are congruent. It is worth noticing that, although this holds for *simplices* (generalised triangles), the same is not true for *polytopes* in general, i.e., convex closures of a number of vertices greater than $n+1$, where n is the dimension of the Cartesian space in which they are defined (think, for instance, of a square and a rhombus both with sides of length 1).

Example 23: congruence in the binary case. *In the case of* unnormalised *belief functions, the belief, plausibility and commonality spaces all have* $N = 2^{|\Theta|}$ *vertices and dimension* $N - 1$*:*

$$\mathcal{B}^U = Cl(Bel_A, \emptyset \subseteq A \subseteq \Theta), \qquad \mathcal{PL}^U = Cl(Pl_A, \emptyset \subseteq A \subseteq \Theta),$$

$$\mathcal{Q}^U = Cl(Q_A, \emptyset \subseteq A \subseteq \Theta).$$

For a frame $\Theta_2 = \{x, y\}$ *of cardinality 2, they form the following three-dimensional simplices embedded in a four-dimensional Cartesian space:*

$$\begin{aligned} \mathcal{B}_2^U &= Cl\Big(Bel_\emptyset = [1,1,1,1]',\ Bel_x = [0,1,0,1]',\ Bel_y = [0,0,1,1]', \\ &\qquad Bel_\Theta = [0,0,0,1]'\Big), \\ \mathcal{PL}_2^U &= Cl\Big(Pl_\emptyset = [0,0,0,0]',\quad Pl_x = [0,1,0,1]',\quad Pl_y = [0,0,1,1]', \\ &\qquad Pl_\Theta = [0,1,1,1]'\Big), \\ \mathcal{Q}_2^U &= Cl\Big(Q_\emptyset = [1,0,0,0]',\quad Q_x = [1,1,0,0]',\quad Q_y = [1,0,1,0]', \\ &\qquad Q_\Theta = [1,1,1,1]'\Big). \end{aligned} \tag{9.13}$$

We know from Example 21 that $\mathcal{PL}_2$ *and* $\mathcal{B}_2$ *are congruent, at least in the normalised case. By (9.13), it follows that*

$$\begin{aligned} \|Bel_\emptyset - Bel_x\|_2 &= \|[1,0,1,0]'\|_2 = \sqrt{2} = \|[0,1,0,1]'\|_2 = \|Pl_x - Pl_\emptyset\|_2, \\ \|Bel_y - Bel_\Theta\|_2 &= \|[0,0,1,0]'\|_2 = 1 \quad = \|[0,1,0,0]'\|_2 = \|Pl_\Theta - Pl_y\|_2 \end{aligned}$$

etc., and as $\mathcal{B}_2^U$ *and* $\mathcal{PL}_2^U$ *are simplices they are also congruent in the unnormalised case as well.*

9.5.2 Congruence of plausibility and commonality spaces

A similar result holds for plausibility and commonality spaces.

We first need to point out the relationship between the vertices of the plausibility and commonality spaces in the unnormalised case, as

$$Pl_A = - \sum_{\emptyset \subsetneq B \subseteq A} (-1)^{|B|} Bel_B,$$

while

$$Q_A = \sum_{\emptyset \subseteq B \subseteq A^c} (-1)^{|B|} Bel_B = \sum_{\emptyset \subsetneq B \subseteq A^c} (-1)^{|B|} Bel_B + Bel_\emptyset = -Pl_{A^c} + Bel_\emptyset. \tag{9.14}$$

Theorem 29. *The one-dimensional faces $Cl(Q_B, Q_A)$ and $Cl(Pl_{B^c}, Pl_{A^c})$ of the commonality and the plausibility space, respectively, are congruent. Namely,*

$$\|Q_B - Q_A\|_p = \|Pl_{B^c} - Pl_{A^c}\|_p.$$

Proof. Since $Q_A = Bel_\emptyset - Pl_{A^c}$, then

$$Q_A - Q_B = Bel_\emptyset - Pl_{A^c} - Bel_\emptyset + Pl_{B^c} = Pl_{B^c} - Pl_{A^c},$$

and the two faces are trivially congruent.

□

Therefore, the following map between vertices of $\mathcal{PL}^U$ and $\mathcal{Q}^U$,

$$Q_A \mapsto Pl_{A^c}, \tag{9.15}$$

maps one-dimensional faces of the commonality space to congruent faces of the plausibility space, $Cl(Q_A, Q_B) \mapsto Cl(Pl_{A^c}, Pl_{B^c})$. Therefore, the two simplices are congruent.

However, (9.15) clearly acts as a 1–1 application of *unnormalised* categorical commonality and plausibility functions (as the complement of $\emptyset$ is Θ, so that $Q_\Theta \mapsto Pl_\emptyset$). Therefore we can also claim the following.

Corollary 9. *$\mathcal{Q}^U$ and $\mathcal{PL}^U$ are congruent.*

Of course, by virtue of Corollary 1, we also have the following result.

Corollary 10. *$\mathcal{Q}^U$ and $\mathcal{B}^U$ are congruent.*

Example 24: congruence of commonality and plausibility spaces. *Let us return to the binary example, $\Theta_2 = \{x, y\}$. It is easy to see from Figs. 9.1 and 9.2 that $\mathcal{PL}_2$ and $\mathcal{Q}_2$ are* not *congruent in the case of* normalised *belief functions, as $\mathcal{Q}_2$ is an equilateral triangle with sides of length $\sqrt{2}$, while $\mathcal{PL}_2$ has two sides of length 1. In the unnormalised case, instead, recalling (9.13), we have that*

$$\begin{array}{ll} Q_\Theta - Q_\emptyset = [0,1,1,1]', & Pl_\Theta - Pl_\emptyset = [0,1,1,1]', \\ Q_x - Q_y = [0,1,-1,0]', & Pl_x - Pl_y = [0,1,-1,0]', \\ Q_x - Q_\Theta = [0,0,-1,-1]', & Pl_\emptyset - Pl_y = [0,0,-1,-1]' \end{array} \tag{9.16}$$

and so on, confirming that $\mathcal{Q}_2^U$ and $\mathcal{PL}_2^U$ are indeed congruent.

9.6 Pointwise rigid transformation

Belief, plausibility and commonality functions form simplices which can be moved onto each other by means of a rigid transformation, as a reflection of the equivalence of the associated models.

Let us also analyse the geometric behaviour of individual functions of the three types, i.e., of each triplet of related non-additive measures (Bel, Pl, Q). As we have observed, in the binary case the pointwise geometry of a plausibility vector can be described in terms of a reflection with respect to the probability simplex $\mathcal{P}$. In the general case, as the simplices $\mathcal{B}^U$, $\mathcal{PL}^U$ and $\mathcal{Q}^U$ are all congruent, logic dictates that there exists a Euclidean transformation $\tau \in E(N)$, where $E(N)$ denotes the Euclidean group in $\mathbb{R}^N$, mapping each simplex onto one of the others.

9.6.1 Belief and plausibility spaces

In the case of the belief and plausibility spaces (in the standard, normalised case), the rigid transformation is obviously encoded by (3.16): $Pl(A) = 1 - Bel(A^c)$. Since $Pl = \sum_{\emptyset \subsetneq A \subseteq \Theta} Pl(A)\mathbf{v}_A$, (3.16) implies that

$$Pl = \mathbf{1} - Bel^c,$$

where Bel^c is the unique belief function whose belief values are the same as Bel's on the complement of each event A: $Bel^c(A) \doteq Bel(A^c)$.

As in the normalised case $\mathbf{1} = Pl_\Theta$ and $\mathbf{0} = Bel_\Theta$, the above relation reads as

$$Pl = \mathbf{1} - Bel^c = \mathbf{0} + \mathbf{1} - Bel^c = Bel_\Theta + Pl_\Theta - Bel^c.$$

As a consequence, the segments $Cl(Bel_\Theta, Pl_\Theta)$ and $Cl(Bel^c, Pl)$ have the same centre of mass, for

$$\frac{Bel^c + Pl}{2} = \frac{Bel_\Theta + Pl_\Theta}{2}.$$

In other words, we have the following theorem.

Theorem 30. *The plausibility vector Pl associated with a belief function Bel is the reflection in $\mathbb{R}^{N-2}$ through the segment $Cl(Bel_\Theta, Pl_\Theta) = Cl(\mathbf{0}, \mathbf{1})$ of the 'complement' belief function Bel^c.*

Geometrically, Bel^c is obtained from Bel by means of another reflection (by swapping the coordinates associated with the reference axes $\mathbf{v}_A$ and $\mathbf{v}_{A^c}$), so that the desired rigid transformation is completely determined. Figure 9.3 illustrates the nature of the transformation, and its instantiation in the binary case for normalised belief functions.

In the case of unnormalised belief functions ($Bel_\emptyset = \mathbf{1}$, $Pl_\emptyset = \mathbf{0}$), we have

$$Pl = Pl_\emptyset + Bel_\emptyset - Bel^c,$$

i.e., Pl is the reflection of Bel^c through the segment $Cl(Bel_\emptyset, Pl_\emptyset) = Cl(\mathbf{0}, \mathbf{1})$.

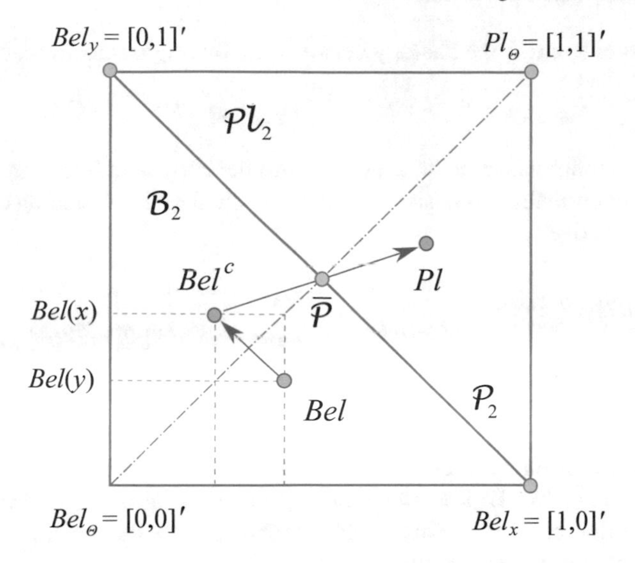

Fig. 9.3: The pointwise rigid transformation mapping Bel onto Pl in the normalised case. In the binary case, the middle point of the segment $Cl(\mathbf{0}, \mathbf{1})$ is the mean probability $\overline{\mathcal{P}}$.

9.6.2 Commonality and plausibility spaces

The form of the desired pointwise transformation is also quite simple in the case of the pair $(\mathcal{PL}^U, \mathcal{Q}^U)$. We can use (9.14), getting

$$
\begin{aligned}
Q &= \sum_{\emptyset \subseteq A \subseteq \Theta} m(A) Q_A = \sum_{\emptyset \subseteq A \subseteq \Theta} m(A)(Bel_\emptyset - Pl_{A^c}) \\
&= Bel_\emptyset - \sum_{\emptyset \subseteq A \subseteq \Theta} m(A) Pl_{A^c} = Bel_\emptyset - Pl^{m^c},
\end{aligned}
$$

where Bel^{m^c} is the unique belief function whose BPA is $m^c(A) \doteq m(A^c)$ (as opposed to Bel^c, for which $Bel^c(A) = Bel(A^c)$). But then, since $Pl_\emptyset = \mathbf{0} = [0, \cdots, 0]'$ for unnormalised belief functions (remember the binary example), we can rewrite the above equation as

$$
Q = Pl_\emptyset + Bel_\emptyset - Pl^{m^c}.
$$

In conclusion, we have the following result.

Theorem 31. *The commonality vector associated with a belief function Bel is the reflection in $\mathbb{R}^N$ through the segment $Cl(Pl_\emptyset, Bel_\emptyset) = Cl(\mathbf{0}, \mathbf{1})$ of the plausibility vector Pl^{m^c} associated with the belief function Bel^{m^c}.*

In this case, however, Bel^{m^c} is obtained from Bel by swapping the coordinates with respect to the base $\{Bel_A, \emptyset \subseteq A \subseteq \Theta\}$. A pictorial representation for the binary case (similar to Fig. 9.3) is more difficult in this case, as $\mathbb{R}^4$ is involved.

It is natural to stress the analogy between the two rigid transformations

$$\tau_{\mathcal{B}^U\mathcal{PL}^U} : \mathcal{B}^U \to \mathcal{PL}^U, \quad \tau_{\mathcal{PL}^U\mathcal{Q}^U} : \mathcal{PL}^U \to \mathcal{Q}^U,$$

mapping an unnormalised belief function onto the corresponding plausibility function, and an unnormalised plausibility function onto the corresponding commonality function, respectively:

$$\begin{array}{ll} \tau_{\mathcal{B}^U\mathcal{PL}^U} : Bel & \stackrel{Bel(A)\mapsto Bel(A^c)}{\longrightarrow} Bel^c \stackrel{\text{reflection through } Cl(\mathbf{0},\mathbf{1})}{\longrightarrow} Pl, \\ \tau_{\mathcal{PL}^U\mathcal{Q}^U} : Pl & \stackrel{m(A)\mapsto m(A^c)}{\longrightarrow} Pl^{m^c} \stackrel{\text{reflection through } Cl(\mathbf{0},\mathbf{1})}{\longrightarrow} Q. \end{array}$$

They are both the composition of two reflections: a swap of the axes of the coordinate frame $\{\mathbf{v}_A, A \subset \Theta\}$ ($\{Bel_A, A \subset \Theta\}$) induced by the set-theoretic complement, plus a reflection with respect to the centre of the segment $Cl(\mathbf{0}, \mathbf{1})$.

We will see in Part III how the alternative models introduced here, and in particular the notion of a basic plausibility assignment, can be put to good use in the probability transformation problem.

Appendix: Proofs

Proof of Theorem 24

The definition (3.16) of the plausibility function yields

$$\mu(A) = \sum_{B\subseteq A}(-1)^{|A\setminus B|} Pl(B) = \sum_{B\subseteq A}(-1)^{|A\setminus B|}(1 - Bel(B^c))$$

$$= \sum_{B\subseteq A}(-1)^{|A\setminus B|} - \sum_{B\subseteq A}(-1)^{|A\setminus B|} Bel(B^c)$$

$$= 0 - \sum_{B\subseteq A}(-1)^{|A\setminus B|} Bel(B^c) = -\sum_{B\subseteq A}(-1)^{|A\setminus B|} Bel(B^c),$$

since by Newton's binomial theorem $\sum_{B\subseteq A}(-1)^{|A\setminus B|} = 0$ if $A \neq \emptyset$ and $(-1)^{|A|}$ otherwise.

If $B \subseteq A$, then $B^c \supseteq A^c$, so that the above expression becomes

$$\begin{aligned}\mu(A) &= -\sum_{\emptyset\subsetneq B\subseteq A}(-1)^{|A\setminus B|}\left(\sum_{C\subseteq B^c} m(C)\right)\\ &= -\sum_{C\subseteq\Theta} m(C)\left(\sum_{B:B\subseteq A, B^c\supseteq C}(-1)^{|A\setminus B|}\right)\\ &= -\sum_{C\subseteq\Theta} m(C)\left(\sum_{B\subseteq A\cap C^c}(-1)^{|A\setminus B|}\right),\end{aligned} \tag{9.17}$$

for $B^c \supseteq C, B \subseteq A$ is equivalent to $B \subseteq C^c, B \subseteq A \equiv B \subseteq (A\cap C^c)$.

Let us now analyse the following function of C:

$$f(C) \doteq \sum_{B\subseteq A\cap C^c}(-1)^{|A\setminus B|}.$$

If $A\cap C^c = \emptyset$, then $B = \emptyset$ and the sum is equal to $f(C) = (-1)^{|A|}$. If $A\cap C^c \neq \emptyset$, instead, then we can write $D \doteq A\cap C^c$ to obtain

$$f(C) = \sum_{B\subseteq D}(-1)^{|A\setminus B|} = \sum_{B\subseteq D}(-1)^{|A\setminus D|+|D\setminus B|},$$

since $B \subseteq D \subseteq A$ and $|A| - |B| = |A| - |D| + |D| - |B|$. But then

$$f(C) = (-1)^{|A|-|D|}\sum_{B\subseteq D}(-1)^{|D|-|B|} = 0,$$

given that $\sum_{B\subseteq D}(-1)^{|D|-|B|} = 0$ by Newton's binomial formula again.

In conclusion, $f(C) = 0$ if $C^c\cap A \neq \emptyset$, and $f(C) = (-1)^{|A|}$ if $C^c \cap A = \emptyset$. We can then rewrite (9.17) as

$$
\begin{aligned}
-\sum_{C\subseteq\Theta} m(C)f(C) &= -\sum_{C^c\cap A\neq\emptyset} m(C)\cdot 0 - \sum_{C^c\cap A=\emptyset} m(C)\cdot(-1)^{|A|} \\
&= (-1)^{|A|+1}\sum_{C^c\cap A=\emptyset} m(C) = (-1)^{|A|+1}\sum_{C\supseteq A} m(C).
\end{aligned}
$$

Proof of Lemma 8

We first need to recall that (the vector associated with) a categorical belief function can be expressed as

$$
Bel_A = \mathbf{bel}_A = \sum_{C\supseteq A} \mathbf{v}_C.
$$

The entry of the vector Bel_A associated with the event $\emptyset \subsetneq B \subsetneq \Theta$ is, by definition,

$$
Bel_A(B) = \begin{cases} 1 & B\supseteq A, \\ 0 & B\not\supseteq A. \end{cases}
$$

As $\mathbf{v}_C(B) = 1$ iff $B = C$, and 0 otherwise, the corresponding entry of the vector $\sum_{C\supseteq A}\mathbf{v}_C$ is also

$$
\sum_{C\supseteq A} \mathbf{v}_C(B) = \begin{cases} 1 & B\supseteq A, \\ 0 & B\not\supseteq A. \end{cases}
$$

Therefore, if (9.9) is true we have that

$$
\begin{aligned}
Bel_A = \sum_{C\supseteq A} \mathbf{v}_C &= \sum_{C\supseteq A}\sum_{B\supseteq C} Bel_B(-1)^{|B\setminus C|} \\
&= \sum_{B\supseteq A} Bel_B \left(\sum_{A\subseteq C\subseteq B} (-1)^{|B\setminus C|} \right).
\end{aligned}
$$

Let us then consider the factor $\sum_{A\subseteq C\subseteq B}(-1)^{|B\setminus C|}$. When $A = B$, $C = A = B$ and the coefficient becomes 1. On the other hand, when $B \neq A$ we have that

$$
\sum_{A\subseteq C\subseteq B} (-1)^{|B\setminus C|} = \sum_{D\subseteq B\setminus A} (-1)^{D} = 0
$$

by Newton's binomial theorem ($\sum_{k=0}^{n} 1^{n-k}(-1)^k = [1+(-1)]^n = 0$). Hence $Bel_A = Bel_A$, and we have the thesis.

Proof of Theorem 26

We just need to rewrite (9.10) as a convex combination of points.

By (9.3), we get

$$\begin{aligned}Pl &= \sum_{\emptyset\subsetneq A\subseteq\Theta} \mu(A)Bel_A = \sum_{\emptyset\subsetneq A\subseteq\Theta} (-1)^{|A|+1}\left(\sum_{C\supseteq A} m(C)\right)Bel_A \\ &= \sum_{\emptyset\subsetneq A\subseteq\Theta} (-1)^{|A|+1}Bel_A\left(\sum_{C\supseteq A} m(C)\right) \\ &= \sum_{\emptyset\subsetneq C\subseteq\Theta} m(C)\left(\sum_{\emptyset\subsetneq A\subseteq C} (-1)^{|A|+1}Bel_A\right) = \sum_{\emptyset\subsetneq C\subseteq\Theta} m(C)Pl_C.\end{aligned} \tag{9.18}$$

The latter is indeed a convex combination, since basic probability assignments are non-negative (but $m(\emptyset = 0)$) and have a unitary sum. It follows that

$$\begin{aligned}\mathcal{PL} &= \{Pl_{Bel}, Bel \in \mathcal{B}\} \\ &= \left\{\sum_{\emptyset\subsetneq C\subseteq\Theta} m(C)Pl_C,\ \sum_C m(C) = 1,\ \ m(C) \geq 0\ \forall C \subseteq \Theta\right\} \\ &= Cl(Pl_A, \emptyset \subsetneq A \subseteq \Theta),\end{aligned}$$

after swapping C and A to keep the notation consistent.

Proof of Theorem 27

Equation (9.11) is equivalent to

$$Pl_A(C) = -\sum_{\emptyset\subsetneq B\subseteq A} (-1)^{|B|}Bel_B(C) \quad \forall C \subseteq \Theta.$$

But since $Bel_B(C) = 1$ if $C \supseteq B$, and 0 otherwise, we have that

$$Pl_A(C) = -\sum_{B\subseteq A, B\subseteq C, B\neq\emptyset} (-1)^{|B|} = -\sum_{\emptyset\subsetneq B\subseteq A\cap C} (-1)^{|B|}.$$

Now, if $A \cap C = \emptyset$ then there are no addends in the above sum, which is then zero. Otherwise, by Newton's binomial formula (9.4), we have

$$Pl_A(C) = -\left\{[1 + (-1)]^{|A\cap C|} - (-1)^0\right\} = 1.$$

On the other hand, by the definition of the plausibility function,

$$Pl_{Bel_A}(C) = \sum_{B\cap C\neq\emptyset} m_{Bel_A}(B) = \begin{cases} 1 & A \cap C \neq \emptyset, \\ 0 & A \cap C = \emptyset, \end{cases}$$

and the two quantities coincide.

10

The geometry of possibility

Possibility measures [531], just like probability measures, are a special case of belief functions when defined on a finite frame of discernment. More precisely, 'necessity' measures, i.e., measures of the form $Nec(A) = 1 - Pos(A^c)$, $A \subseteq \Theta$, where Pos is a possibility measure, have as counterparts in the theory of evidence consonant belief functions [533, 930, 546, 91], i.e., belief functions whose focal elements are nested (see Proposition 16, and Section 6.5.1).

Studying the geometry of consonant belief functions amounts therefore to investigating the geometry of possibility theory. In this chapter, we then move forward to analyse their convex geometry, as a step towards a unified geometric picture of a wider class of uncertainty measures. In particular, in the first part we show that consonant BFs are in correspondence with chains of subsets of their domain, and hence are located in a collection of convex regions of the belief space which has the form of a *simplicial complex*, i.e., a structured collection of simplices meeting a few intuitive requirements. In the second half of the chapter, consistent belief functions are introduced, as the set of BFs whose focal elements have non-empty intersection. Consistent belief functions are a subclass of consonant ones, and possess an intriguing interpretation as the counterparts of consistent knowledge bases in classical logic (under a basic belief logic interpretation). Consistent belief functions are shown also to belong to a simplicial complex, highlighting a decomposition of arbitrary belief functions into n consistent ones, where n is the cardinality of the frame of discernment they are defined on.

Based on this convex analysis of possibility theory, Chapters 13 and 14 will investigate the problem of transforming a belief function into a consonant and a consistent one, respectively.

F. Cuzzolin, *The Geometry of Uncertainty*, Artificial Intelligence: Foundations, Theory, and Algorithms, https://doi.org/10.1007/978-3-030-63153-6_10

Chapter outline

We first recall in Section 10.1 the relationship between consonant belief functions and necessity measures. We then move on to study the geometry of the space of consonant belief functions, or the *consonant subspace* $\mathcal{CO}$ (Section 10.2). After observing the correspondence between consonant belief functions and maximal chains of events, we look for useful insights by studying the case of ternary frames, which leads us to prove that the consonant subspace has the form of a *simplicial complex* [551], a structured collection of simplices. In Section 10.3 we investigate the convex geometry of the components of $\mathcal{CO}$ in more detail, proving that they are all congruent to each other, and can be decomposed into faces which are right triangles.

In the second half of the chapter, we introduce the notion of a consistent belief function as the natural generalisation in the context of belief theory of consistent knowledge bases in classical logic (Section 10.4). In Section 10.5, following the intuition provided by the simple case of binary frames, we prove that the set of consistent BFs (just like consonant belief functions do) form a simplicial complex in the space of all belief functions, and that the maximal simplices of such a complex are all congruent to each other. Finally, in Section 10.5.3 we show that each belief function can be decomposed into a number of consistent components living in the consistent complex, closely related to the pignistic transformation [1944, 1730].

To improve readability, several proofs have been collected together in an appendix. This chapter is a re-elaboration of material first published in [328, 360, 345].

10.1 Consonant belief functions as necessity measures

Recall that a belief function is said to be *consonant* if its focal elements are nested.

The following conditions are equivalent [1583]:

1. Bel is consonant.
2. $Bel(A \cap B) = \min(Bel(A), Bel(B))$ for every $A, B \subset \Theta$.
3. $Pl(A \cup B) = \max(Pl(A), Pl(B))$ for every $A, B \subset \Theta$.
4. $Pl(A) = \max_{x \in A} Pl(\{x\})$ for all non-empty $A \subset \Theta$.
5. There exists a positive integer n and simple support functions $Bel_1, \ldots, Bel_n$ such that $Bel = Bel_1 \oplus \cdots \oplus Bel_n$ and the focus of Bel_i is contained in the focus of Bel_j whenever $i < j$.

Consonant belief functions represent bodies of evidence all pointing in the same direction. However, their constituent pieces of evidence do not need to be completely nested for the belief function resulting from their aggregation to be consonant, as the next proposition states.

Proposition 56. *Suppose $Bel_1, \ldots, Bel_n$ are non-vacuous simple support functions with foci $\mathcal{C}_1, \ldots, \mathcal{C}_n$, respectively, and $Bel = Bel_1 \oplus \cdots \oplus Bel_n$ is consonant. If $\mathcal{C}$ denotes the core of Bel, then the sets $\mathcal{C}_i \cap \mathcal{C}$ are nested.*

By condition 2 above, it follows that

$$0 = Bel(\emptyset) = Bel(A \cap A^c) = \min(Bel(A), Bel(A^c)),$$

i.e., either $Bel(A) = 0$ or $Bel(A^c) = 0$ for every $A \subseteq \Theta$. This result and Proposition 13 explain why we said in Chapter 2 that consonant and quasi-support functions represent 'opposite' classes of belief functions.

As we recalled in Chapter 6, Section 6.5.1, we have the following definition.

Definition 124. A possibility measure *on a domain Θ is a function $Pos : 2^\Theta \to [0,1]$ such that $Pos(\emptyset) = 0$, $Pos(\Theta) = 1$ and*

$$Pos\left(\bigcup_i A_i\right) = \sup_i Pos(A_i)$$

for any family $\{A_i | A_i \in 2^\Theta, i \in I\}$, where I is an arbitrary set index.

Each possibility measure is uniquely characterised by a membership function $\pi : \Theta \to [0,1]$ s.t. $\pi(x) \doteq Pos(\{x\})$ via the formula $Pos(A) = \sup_{x \in A} \pi(x)$.

From condition 4, it follows immediately that the plausibility function Pl associated with a belief function Bel on a domain Θ is a possibility measure iff Bel is consonant, with Bel's contour function pl (recall Section 2.7.3, (2.18)) playing the role of the membership function: $\pi = pl$. Studying the geometry of consonant belief functions amounts, therefore, to studying the geometry of possibility.

10.2 The consonant subspace

To gather useful intuition, we start as usual from the familiar running example of belief functions defined on a binary frame $\Theta_2 = \{x, y\}$, continuing from the example in Chapter 7, Example 18 (see Fig. 10.1).

We know that the region $\mathcal{P}_2$ of all Bayesian belief functions on Θ_2 is, in this case, the diagonal line segment $Cl(Bel_x, Bel_y)$. On the other hand, simple support functions focused on $\{x\}$ lie on the horizontal segment $Cl(Bel_\Theta, Bel_x)$, while simple support functions focused on $\{y\}$ form the vertical segment $Cl(Bel_\Theta, Bel_y)$. On $\Theta = \Theta_2 = \{x, y\}$, consonant belief functions can have as their chain of focal elements either $\{\{x\}, \Theta\}$ or $\{\{y\}, \Theta\}$. As a consequence, all consonant belief functions on Θ_2 are simple support functions, and their region $\mathcal{CO}_2$ is the union of two segments:

$$\mathcal{CO}_2 = \mathcal{S}_2 = \mathcal{CO}^{\{x,\Theta\}} \cup \mathcal{CO}^{\{y,\Theta\}} = Cl(Bel_\Theta, Bel_x) \cup Cl(Bel_\Theta, Bel_y).$$

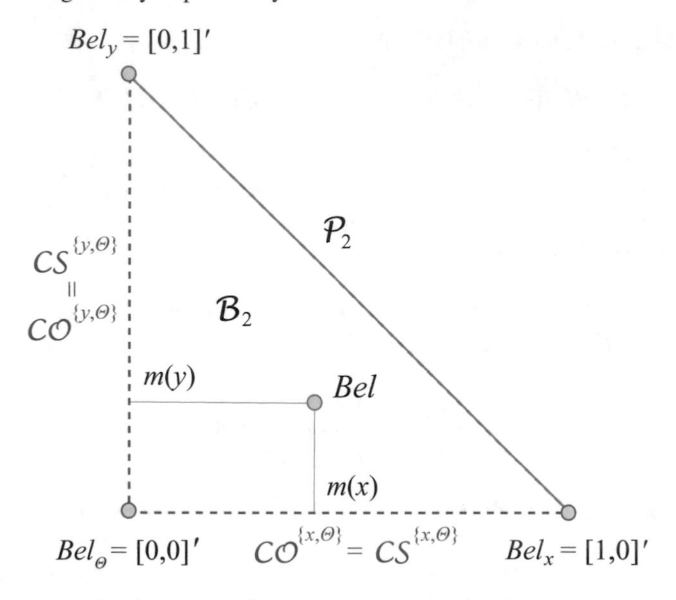

Fig. 10.1: The belief space $\mathcal{B}$ for a binary frame is a triangle in $\mathbb{R}^2$ whose vertices are the categorical belief functions focused on $\{x\}, \{y\}$ and Θ (Bel_x, Bel_y, Bel_Θ), respectively. The probability region is the segment $Cl(Bel_x, Bel_y)$, whereas consonant (and consistent; see Section 10.4) belief functions are constrained to belong to the union of the two segments $\mathcal{CS}^{\{x,\Theta\}} = \mathcal{CO}^{\{x,\Theta\}} = Cl(Bel_\Theta, Bel_x)$ and $\mathcal{CS}^{\{y,\Theta\}} = \mathcal{CO}^{\{y,\Theta\}} = Cl(Bel_\Theta, Bel_y)$.

10.2.1 Chains of subsets as consonant belief functions

In general terms, while arbitrary belief functions do not admit restrictions on their list of focal elements, consonant BFs are characterised by the fact that their focal elements can be rearranged into a totally ordered set by set inclusion.

The power set 2^Θ of a frame of discernment is a partially ordered set (poset) with respect to set-theoretic inclusion: $\subseteq$ has the three properties of reflexivity (whenever $A \subseteq \Theta$, $A \subseteq A$), antisymmetry ($A \subseteq B$ and $B \subseteq A$ implies $A = B$) and transitivity ($A \subseteq B$ and $B \subseteq C$ implies $A \subseteq C$). A *chain* of a poset is a collection of pairwise comparable elements (a *totally ordered set*). All the possible lists of focal elements associated with consonant belief functions correspond therefore to all the possible chains of subsets $A_1 \subset \cdots \subset A_m$ in the partially ordered set $(2^\Theta, \subseteq)$.

Now, Theorem 10 implies that the BFs whose focal elements belong to a chain $\mathcal{C} = \{A_1, \ldots, A_m\}$ form in the belief space the simplex $Cl(Bel_{A_1}, \ldots, Bel_{A_m})$ (remember that the vectors $\{Bel_A, A \subset \Theta\}$ representing categorical belief functions are affinely independent in the embedding Cartesian space). Let us denote by $n \doteq |\Theta|$ the cardinality of the frame of discernment Θ. Since each chain in $(2^\Theta, \subseteq)$ is a subset of a maximal one (a chain including subsets of any size from 1 to n), the region of consonant belief functions turns out to be the union of a collection of simplices, each of them associated with a maximal chain $\mathcal{C}$:

$$\mathcal{CO} = \bigcup_{\mathcal{C}=\{A_1\subset\cdots\subset A_n\}} \mathcal{CO}^{\mathcal{C}} \doteq \bigcup_{\mathcal{C}=\{A_1\subset\cdots\subset A_n\}} Cl(Bel_{A_1},\ldots,Bel_{A_n}). \qquad (10.1)$$

The number of such maximal simplices in $\mathcal{CO}$ is equal to the number of maximal chains in $(2^\Theta, \subseteq)$, i.e.,

$$\prod_{k=1}^{n} \binom{k}{1} = n!,$$

since, given a size-k set we can build a new set containing it by just choosing one of the remaining elements. Since the length of a maximal chain is $|\Theta| = n$ and the vectors $\{Bel_A, A \subset \Theta\}$ are affinely independent, the dimension of the vector spaces generated by these convex components is the same and equal to $\dim Cl(Bel_{A_1},\ldots,Bel_{A_n}) = n-1$.

Each categorical belief function Bel_A obviously belongs to several distinct components. In particular, if $|A| = k$ the total number of maximal chains containing A is $(n-k)!k!$ – indeed, in the power set of A, the number of maximal chains is $k!$, while to form a chain from A to Θ we just need to add an element of $A^c = \Theta \setminus A$ (whose size is $n-k$) at each step. The integer $(n-k)!k!$ is then also the number of maximal simplices of $\mathcal{CO}$ containing Bel_A.

In particular, each vertex Bel_x of the probability simplex $\mathcal{P}$ (for which $|\{x\}| = k = 1$) belongs to a sheaf of $(n-1)!$ convex components of the consonant subspace. An obvious remark is that $\mathcal{CO}$ is *connected*: each maximal convex component $\mathcal{CO}^{\mathcal{C}}$ is obviously connected, and each pair of such components has at least Bel_Θ as common intersection.

Example 25: ternary case. *Let us consider, as a more significant illustrative example, the case of a frame of size 3, $\Theta = \{x, y, z\}$.*

Belief functions $Bel \in \mathcal{B}_3$ can be written as six-dimensional vectors

$$Bel = [Bel(x), Bel(y), Bel(z), Bel(\{x,y\}), Bel(\{x,z\}), Bel(\{y,z\})]'.$$

All the maximal chains $\mathcal{C}$ of 2^Θ are listed below:

$$\begin{array}{lll} \{x\} \subset \{x,z\} \subset \Theta, & \{y\} \subset \{x,y\} \subset \Theta, & \{z\} \subset \{y,z\} \subset \Theta, \\ \{x\} \subset \{x,y\} \subset \Theta, & \{y\} \subset \{y,z\} \subset \Theta, & \{z\} \subset \{x,z\} \subset \Theta. \end{array}$$

Each element of Θ is then associated with two chains, and the total number of maximal convex components, whose dimension is $|\Theta| - 1 = 2$, is $3! = 6$:

$$\begin{array}{ll} Cl(Bel_x, Bel_{\{x,z\}}, Bel_\Theta), & Cl(Bel_y, Bel_{\{x,y\}}, Bel_\Theta), \\ Cl(Bel_x, Bel_{\{x,y\}}, Bel_\Theta), & Cl(Bel_y, Bel_{\{y,z\}}, Bel_\Theta), \end{array}$$
$$Cl(Bel_z, Bel_{\{y,z\}}, Bel_\Theta),$$
$$Cl(Bel_z, Bel_{\{x,z\}}, Bel_\Theta).$$

Each two-dimensional maximal simplex (for instance $Cl(Bel_x, Bel_{\{x,z\}}, Bel_\Theta)$) has an intersection of dimension $|\Theta| - 2 = 1$ (in the example, $Cl(Bel_{\{x,z\}}, Bel_\Theta)$) with a single other maximal component ($Cl(Bel_z, Bel_{\{x,z\}}, Bel_\Theta)$), associated with a different element of Θ.

In conclusion, the geometry of the ternary frame can be represented as in Fig. 10.2, where the belief space

$$\mathcal{B}_3 = Cl(Bel_x, Bel_y, Bel_z, Bel_{\{x,y\}}, Bel_{\{x,z\}}, Bel_{\{y,z\}}, Bel_\Theta)$$

is six-dimensional, its probabilistic face is a simplex $\mathcal{P}_3 = Cl(Bel_x, Bel_y, Bel_z)$ of dimension 2, and the consonant subspace $\mathcal{CO}_3$, also part of the boundary of $\mathcal{B}_3$, is given by the union of the maximal simplices listed above.

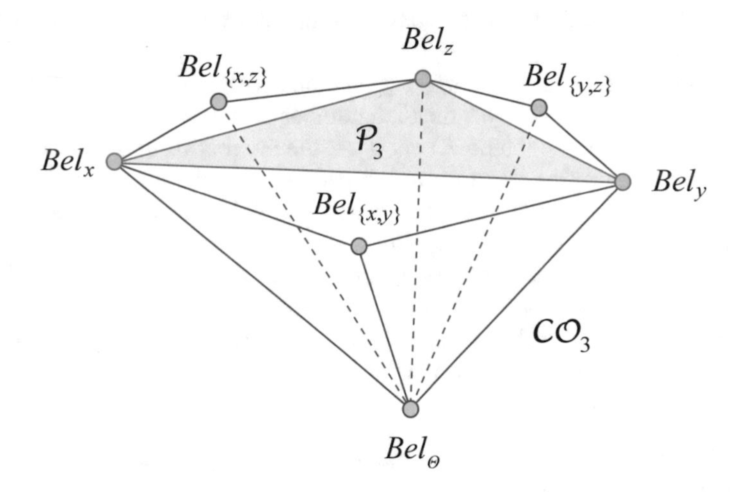

Fig. 10.2: The simplicial complex $\mathcal{CO}_3$ of all the consonant belief functions for a ternary frame Θ_3. The complex is composed of $n! = 3! = 6$ maximal simplicial components of dimension $n - 1 = 2$, each vertex of $\mathcal{P}_3$ being shared by $(n-1)! = 2! = 2$ of them. The region is connected, and is part of the boundary $\partial\mathcal{B}_3$ of the belief space $\mathcal{B}_3$.

10.2.2 The consonant subspace as a simplicial complex

These properties of $\mathcal{CO}$ can be summarised via a useful concept of convex geometry, which generalises that of a simplex [551].

Definition 125. *A* simplicial complex *is a collection Σ of simplices of arbitrary dimensions possessing the following properties:*

1. *If a simplex belongs to Σ, then all its faces of any dimension belong to Σ.*
2. *The intersection of any two simplices in the complex is a face of both.*

Let us consider, for instance, the case of two triangles (two-dimensional simplices) in $\mathbb{R}^2$. Roughly speaking, condition 2 asks for the intersection of the two triangles not to contain points of their interiors (Fig. 10.3, left). The intersection cannot just be any subset of their borders either (middle), but has to be a face (right, in this case a single vertex). Note that if two simplices intersect in a face τ, they

Fig. 10.3: Intersection of simplices in a complex. Only the right-hand pair of triangles satisfies condition 2 of the definition of a simplicial complex (Definition 125).

obviously intersect in every face of τ.

Theorem 32. *The consonant subspace $\mathcal{CO}$ is a simplicial complex embedded in the belief space $\mathcal{B}$.*

Proof. Property 1 of Definition 125 is trivially satisfied. As a matter of fact, if a simplex $Cl(Bel_{A_1}, \ldots, Bel_{A_n})$ corresponds to a chain $A_1 \subset \cdots \subset A_n$ in the poset $(2^\Theta, \subseteq)$, each face of this simplex is in correspondence with a subchain in 2^Θ, and therefore with a simplex of consonant belief functions.

For property 2, let us consider the intersection of two arbitrary simplices in the complex,

$$Cl(Bel_{A_1}, \ldots, Bel_{A_{n_1}}) \cap Cl(Bel_{B_1}, \ldots, Bel_{B_{n_2}}),$$

associated with chains $\mathcal{A} = \{A_1, \ldots, A_{n_1}\}$ and $\mathcal{B} = \{B_1, \ldots, B_{n_2}\}$, respectively. As the vectors $\{Bel_A, \emptyset \subsetneq A \subsetneq \Theta\}$ are linearly independent in $\mathbb{R}^{N-2}$, no linear combination of the vectors Bel_{B_i} can yield an element of $v(Bel_{A_1}, \ldots, Bel_{A_{n_1}})$ (the linear space generated by $Bel_{A_1}, \ldots, Bel_{A_{n_1}}$), unless some of those vectors coincide. The desired intersection is therefore

$$Cl(Bel_{C_{i_1}}, \ldots, Bel_{C_{i_k}}), \tag{10.2}$$

where

$$\left\{C_{i_j}, j = 1, ..., k\right\} = \mathcal{C} = \mathcal{A} \cap \mathcal{B},$$

with $k < n_1, n_2$. But then $\mathcal{C}$ is a subchain of both $\mathcal{A}$ and $\mathcal{B}$, so that (10.2) is a face of both $Cl(Bel_{A_1}, \ldots, Bel_{A_{n_1}})$ and $Cl(Bel_{B_1}, \ldots, Bel_{B_{n_2}})$.

□

As Figure 10.2 shows, the probability simplex $\mathcal{P}$ and the maximal simplices of $\mathcal{CO}$ have the same dimension, and are both part of the boundary $\partial\mathcal{B}$ of the belief space.

10.3 Properties of the consonant subspace

More can be said about the geometry of the consonant subspace, and in particular about the features of its constituent maximal simplices.

10.3.1 Congruence of the convex components of $\mathcal{CO}$

Indeed, as the binary case study suggests, all maximal simplices of the consonant complex are *congruent*, i.e., they can be mapped onto each other by means of rigid transformations. In the binary case, for instance, the two components $\mathcal{CO}^{\{x,\Theta\}} = Cl(Bel_\Theta, Bel_x)$ and $\mathcal{CO}^{\{y,\Theta\}} = Cl(Bel_\Theta, Bel_y)$ are segments of the same (Euclidean) length, namely

$$\|\mathcal{CO}^{\{x,\Theta\}}\| = \|Bel_x - Bel_\Theta\| = \|[1,0]'\| = 1 = \|Bel_y - Bel_\Theta\| = \|\mathcal{CO}^{\{y,\Theta\}}\|$$

(see Fig. 10.1 again).

We can get an intuition about how to prove that this is true in the general case by studying the more significant ternary case.

Example 26: congruence in the ternary case. *Continuing from Example 25,*

$$\mathcal{CO}^{\{x,\{x,y\},\Theta\}} = Cl(Bel_x, Bel_{\{x,y\}}, Bel_\Theta),$$

$$\begin{array}{rl} Cl(Bel_x, Bel_\Theta) & \|Bel_x - Bel_\Theta\| = \|[1\,0\,0\,1\,1\,0]'\| = \sqrt{3}, \\ Cl(Bel_{\{x,y\}}, Bel_\Theta) \leftrightarrow & \|Bel_{\{x,y\}} - Bel_\Theta\| = \|[0\,0\,0\,1\,0\,0]'\| = 1, \\ Cl(Bel_x, Bel_{\{x,y\}}) & \|Bel_x - Bel_{\{x,y\}}\| = \|[1\,0\,0\,0\,1\,0]'\| = \sqrt{2}, \end{array}$$

$$\mathcal{CO}^{\{x,\{x,z\},\Theta\}} = Cl(Bel_x, Bel_{\{x,z\}}, Bel_\Theta),$$

$$\begin{array}{rl} Cl(Bel_x, Bel_\Theta) & \|Bel_x - Bel_\Theta\| = \|[1\,0\,0\,1\,1\,0]'\| = \sqrt{3}, \\ Cl(Bel_{\{x,z\}}, Bel_\Theta) \leftrightarrow & \|Bel_{\{x,z\}} - Bel_\Theta\| = \|[0\,0\,0\,0\,1\,0]'\| = 1, \\ Cl(Bel_x, Bel_{\{x,z\}}) & \|Bel_x - Bel_{\{x,z\}}\| = \|[1\,0\,0\,1\,0\,0]'\| = \sqrt{2}, \end{array}$$

$$\mathcal{CO}^{\{y,\{x,y\},\Theta\}} = Cl(Bel_y, Bel_{\{x,y\}}, Bel_\Theta),$$

$$\begin{array}{rl} Cl(Bel_y, Bel_\Theta) & \|Bel_y - Bel_\Theta\| = \|[0\,1\,0\,1\,0\,1]'\| = \sqrt{3}, \\ Cl(Bel_{\{x,y\}}, Bel_\Theta) \leftrightarrow & \|Bel_{\{x,y\}} - Bel_\Theta\| = \|[0\,0\,0\,1\,0\,0]'\| = 1, \\ Cl(Bel_y, Bel_{\{x,y\}}) & \|Bel_y - Bel_{\{x,y\}}\| = \|[0\,1\,0\,0\,0\,1]'\| = \sqrt{2}, \end{array}$$

$$\mathcal{CO}^{\{z,\{x,z\},\Theta\}} = Cl(Bel_z, Bel_{\{x,z\}}, Bel_\Theta),$$

$$\begin{array}{lcl} Cl(Bel_z, Bel_\Theta) & & \|Bel_z - Bel_\Theta\| = \|[0\,0\,1\,0\,1\,1]'\| = \sqrt{3}, \\ Cl(Bel_{\{x,z\}}, Bel_\Theta) & \leftrightarrow & \|Bel_{\{x,z\}} - Bel_\Theta\| = \|[0\,0\,0\,0\,1\,0]'\| = \ 1, \\ Cl(Bel_z, Bel_{\{x,z\}}) & & \|Bel_z - Bel_{\{x,z\}}\| = \|[0\,0\,1\,0\,0\,1]'\| = \sqrt{2}. \end{array}$$

it is clear that the one-dimensional faces of each pair of maximal simplices can be put into 1–1 correspondence, based on their having the same norm. For instance, for the pair of triangles

$$Cl(Bel_x, Bel_{\{x,y\}}, Bel_\Theta), \quad Cl(Bel_z, Bel_{\{x,z\}}, Bel_\Theta),$$

the desired correspondences are as follows: $Cl(Bel_x, Bel_\Theta) \leftrightarrow Cl(Bel_z, Bel_\Theta)$, $Cl(Bel_{\{x,y\}}, Bel_\Theta) \leftrightarrow Cl(Bel_{\{x,z\}}, Bel_\Theta)$ *and* $Cl(Bel_x, Bel_{\{x,y\}}) \leftrightarrow Cl(Bel_z, Bel_{\{x,z\}})$*, for such pairs of segments have the same norm.*

This can be proven in the general case as well.

Theorem 33. *All the maximal simplices of the consonant subspace are congruent.*

Proof. To get a proof for the general case, we need to find a 1–1 map between one-dimensional sides of any two maximal simplices. Let $\mathcal{A} = \{A_1 \subset \cdots \subset A_i \subset \cdots \subset A_n = \Theta\}$, $\mathcal{B} = \{B_1 \subset \cdots \subset B_i \subset \cdots \subset B_n = \Theta\}$ be the associated maximal chains.

The trick consists in associating pairs of events with the same cardinality:

$$Cl(Bel_{A_i}, Bel_{A_j}) \leftrightarrow Cl(Bel_{B_i}, Bel_{B_j}), \quad |A_i| = |B_i| = i,\ |A_j| = |B_j| = j > i.$$

The categorical belief function Bel_{A_i} is such that $Bel_{A_i}(B) = 1$ when $B \supseteq A_i$, and $Bel_{A_i}(B) = 0$ otherwise. On the other hand, $Bel_{A_j}(B) = 1$ when $B \supseteq A_j \supset A_i$, and $Bel_{A_j}(B) = 0$ otherwise, since $A_j \supset A_i$ by hypothesis. Hence,

$$|Bel_{A_i} - Bel_{A_j}(B)| = 1 \Leftrightarrow B \supseteq A_i, B \not\supset A_j,$$

so that

$$\|Bel_{A_i} - Bel_{A_j}\|_2 = \sqrt{|\{B \subseteq \Theta : B \supseteq A_i, B \not\supset A_j\}|} = \sqrt{|A_j \setminus A_i|}.$$

But this is true for each similar pair in any other maximal chain, so that

$$\|Bel_{A_i} - Bel_{A_j}\|_2 = \|Bel_{B_i} - Bel_{B_j}\|_2 \quad \forall i, j \in [1, \ldots, n]$$

for each pair of maximal simplices of $\mathcal{CO}$. By a generalisation of a well-known theorem of Euclid, this implies that the two simplices are congruent:

$$Cl(Bel_{A_1}, \ldots, Bel_\Theta) \sim Cl(Bel_{B_1}, \ldots, Bel_\Theta).$$

□

It is easy to see that the components of $\mathcal{CO}$ are *not* congruent with $\mathcal{P}$, even though they both have dimension $n-1$. In the binary case, for instance,

$$\mathcal{P} = Cl(Bel_x, Bel_y), \quad \|\mathcal{P}\| = \|Bel_y - Bel_x\| = \sqrt{2},$$

whereas $\|\mathcal{CO}^{\{x,\Theta\}}\| = \|\mathcal{CO}^{\{y,\Theta\}}\| = 1$.

10.3.2 Decomposition of maximal simplices into right triangles

An analysis of the norm of the difference of two categorical belief functions can give us additional information about the nature and structure of the maximal simplices of the consonant subspace.

We know from [345] that in $\mathbb{R}^{N-2}$, each triangle

$$Cl(Bel_\Theta, Bel_B, Bel_A)$$

with $\Theta \supsetneq B \supsetneq A$ is a right triangle with $\widehat{Bel_\Theta Bel_B Bel_A}$ as the right angle. Indeed, we can prove a much more general result here.

Theorem 34. *If $A_i \supsetneq A_j \supsetneq A_k$, then $\widehat{Bel_{A_i} Bel_{A_j} Bel_{A_k}} = \pi/2$.*

Proof. As $A_i \supsetneq A_j \supsetneq A_k$, we can write

$$Bel_{A_j} - Bel_{A_i}(B) = \begin{cases} 1 & B \supseteq A_j, B \not\supset A_i, \\ 0 & \text{otherwise}, \end{cases}$$

$$Bel_{A_k} - Bel_{A_i}(B) = \begin{cases} 1 & B \supseteq A_k, B \not\supset A_i, \\ 0 & \text{otherwise}, \end{cases}$$

$$Bel_{A_k} - Bel_{A_j}(B) = \begin{cases} 1 & B \supseteq A_k, B \not\supset A_j, \\ 0 & \text{otherwise}. \end{cases}$$

This implies

$$Bel_{A_i} - Bel_{A_j}(B) = 1 \Rightarrow B \not\supset A_i, B \supseteq A_j \Rightarrow Bel_{A_j} - Bel_{A_k}(B) = 0$$

and viceversa, so that the inner product $\langle Bel_{A_i} - Bel_{A_j}, Bel_{A_j} - Bel_{A_k} \rangle = 0$ is zero, and therefore $\widehat{Bel_{A_i} Bel_{A_j} Bel_{A_k}}$ is $\pi/2$.

□

All triangles $Cl(Bel_{A_i}, Bel_{A_j}, Bel_{A_k})$ in $\mathcal{CO}$ such that $A_i \supsetneq A_j \supsetneq A_k$ are right triangles. But, as each maximal simplicial component $\mathcal{CO}^{\mathcal{C}}$ of the consonant complex has vertices associated with the elements $A_1 \subsetneq \cdots \subsetneq A_n$ of a maximal chain, any three of them will also form a chain. Hence all two-dimensional faces of any maximal component of $\mathcal{CO}$ are right triangles. All its three-dimensional faces (tetrahedrons) have right triangles as faces (Fig. 10.4), and so on.

10.4 Consistent belief functions

Consonant belief functions are not the most general class of belief functions associated with collections of consistent, non-contradictory pieces of evidence. Here, we introduce here this new class of belief functions starting from an analogy with the notion of consistency in classical logic.

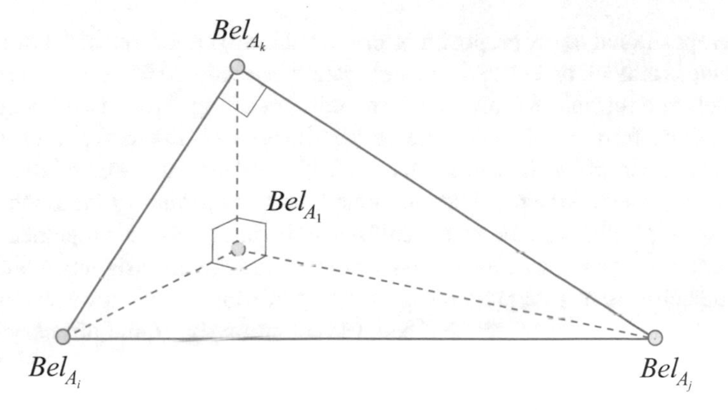

Fig. 10.4: Each tetrahedron $Cl(Bel_{A_i}, Bel_{A_j}, Bel_{A_k}, Bel_{A_l})$ formed by vertices of a maximal simplex of the consonant subspace, $A_i \subsetneq A_j \subsetneq A_k \subsetneq A_l$, has faces that are all right triangles.

10.4.1 Consistent knowledge bases in classical logic

In classical logic, a set Φ of formulae, or *knowledge base*, is said to be *consistent* if and only if there does not exist another formula ϕ such that the knowledge base implies both such a formula and its negation: $\Phi \vdash \phi$, $\Phi \vdash \neg\phi$. In other words, it is impossible to derive incompatible conclusions from the set of propositions that form a consistent knowledge base. The application of inference rules to inconsistent collections of formulae may lead to incompatible conclusions, depending on the subset of assumptions one starts one's reasoning from [1389].

A variety of approaches have been proposed in the context of classical logic to address the issue of inconsistent knowledge bases, such as fragmenting the latter into maximally consistent subsets, limiting the power of the formalism or adopting non-classical semantics [1440, 97]. Even when a knowledge base is formally inconsistent, however, it may still contain potentially useful information. Paris [1389], for instance, has tackled the problem not by assuming each proposition in the knowledge base as a fact, but by attributing to it a certain degree of belief in a probabilistic logic approach. This leads to something very similar to a belief function.

10.4.2 Belief functions as uncertain knowledge bases

This parallelism with classical logic reminds us of the fact that belief functions are also collections of disparate pieces of evidence, incorporated over time as they become available. As a result, each belief function is likely to contain self-contradictory information, which is in turn associated with a degree of 'internal' conflict. As we have seen in Chapter 4, conflict and combinability play a central role in the theory of evidence [2009, 1670, 1460], and have recently been subject to novel analyses [1196, 860, 1209].

In propositional logic, propositions or formulae are either true or false, i.e., their truth value is either 0 or 1 [1267]. As we recalled in Section 6.6.1, an 'interpretation' or 'model' of a formula ϕ is a valuation function mapping ϕ to the truth value 'true' (i.e., 1). Each formula can therefore be associated with the set of interpretations or models under which its truth value is 1. If we define a frame of discernment collecting together all the possible interpretations, each formula ϕ is associated with the subset $A(\phi)$ of this frame which collects all its interpretations together.

A straightforward extension of classical logic consists in assigning a probability value to such sets of interpretations, i.e., to each formula. If, however, the available evidence allows us to define a belief function on the frame of possible interpretations, each formula $A(\phi) \subseteq \Theta$ is then naturally assigned a degree of belief $Bel(A(\phi))$ between 0 and 1 [1526, 767], measuring the total amount of evidence supporting the proposition 'ϕ is true'.

A belief function can therefore be seen in this context as a generalisation of a knowledge base [1526, 767], i.e., a set of propositions together with their non-zero belief values, $Bel = \{A \subseteq \Theta : Bel(A) \neq 0\}$.

10.4.3 Consistency in belief logic

To determine what consistency amounts to in a belief function logic framework, however, we need to make precise the notion of 'proposition implied by a belief function', just as a (classical) knowledge base Φ implies a classical proposition ϕ. A number of options can be explored.

One way to define implication is to decide that $Bel \vdash B \subseteq \Theta$ if B is implied by all its focal elements, namely

$$Bel \vdash B \Leftrightarrow A \subseteq B \; \forall A : m(A) \neq 0.$$

Clearly, this is equivalent to saying that the propositions implied by Bel are those whose belief value is 1:

$$Bel \vdash B \Leftrightarrow Bel(B) = 1. \tag{10.3}$$

Note that if we wished to impose the condition that

$$Bel \vdash B \Leftrightarrow A \subseteq B \; \forall A : Bel(A) \neq 0$$

instead, since $Bel(\Theta) = 1 \neq 0$ we would have $\Theta \subset B$, i.e., only $B = \Theta$ would be implied by any arbitrary belief function Bel.

A 'dual' implication principle can be formulated in terms of plausibility as

$$Bel \vdash B \Leftrightarrow Pl(B) = 1. \tag{10.4}$$

Finally, an alternative definition requires the proposition to receive non-zero support by the belief function Bel:

$$Bel \vdash B \Leftrightarrow Bel(B) \neq 0. \tag{10.5}$$

Whatever our preferred choice of implication, we can define the class of consistent belief functions as the set of BFs which cannot imply contradictory propositions.

Definition 126. *A belief function Bel on Θ is termed* consistent *if there exists no proposition $B \subset \Theta$ such that both B and its negation B^c are implied by Bel.*

Depending on the various implication relations (10.3), (10.4), (10.5) listed above, we obtain different classes of 'consistent' belief functions.

Let us call the intersection of all focal elements of Bel the *conjunctive core* $\mathcal{C}^\cap$ of the belief function Bel:

$$\mathcal{C}^\cap \doteq \bigcap_{m(A)\neq 0} A. \tag{10.6}$$

In opposition, we can term the union of its focal elements (traditionally called just the 'core') the *disjunctive core* of the belief function Bel: $\mathcal{C}^\cup = \cup_{m(A)\neq 0} A$.

When we use the implication (10.3), all (normalised) belief functions are 'consistent'. Since $Bel(B) = 1$ is equivalent to $\mathcal{C}^\cup \subset B$, if both B and B^c were implied by Bel we would have $\mathcal{C}^\cup \subset B$, $\mathcal{C}^\cup \subset B^c$, and thus necessarily $\mathcal{C}^\cup = \emptyset$. Therefore, the only non-consistent belief function is $Bel_\emptyset$ (an unnormalised belief function).

The implication (10.5) carries much more significant consequences.

Theorem 35. *A belief function is* consistent *under the implication relation (10.5), namely there do not exist two complementary propositions $B, B^c \subseteq \Theta$ which both have non-zero support from Bel, $Bel(B) \neq 0$, $Bel(B^c) \neq 0$, if and only if its conjunctive core is non-empty:*

$$\mathcal{C}^\cap = \bigcap_{A:m(A)\neq 0} A \neq \emptyset. \tag{10.7}$$

Proof. $\Leftarrow$: If there existed B and B^c both implied by Bel, we would have

$$\begin{aligned}\bigcap\{A : Bel(A) \neq 0\} &= \bigcap\{A : Bel(A) \neq 0, A \neq B, B^c\} \cap B \cap B^c \\ &= \bigcap\{A : Bel(A) \neq 0, A \neq B, B^c\} \cap \emptyset = \emptyset,\end{aligned}$$

which goes against the hypothesis that $\mathcal{C}^\cap \neq \emptyset$, since

$$\bigcap\{A : Bel(A) \neq 0\} = \bigcap\{A : \exists C \subseteq A, m(C) \neq 0\} = \bigcap_{m(C)\neq 0} C = \mathcal{C}^\cap.$$

$\Rightarrow$: If $\mathcal{C}^\cap = \emptyset$, we can pick as B any focal element of Bel, which by definition is such that $Bel(B) \geq m(B) \neq 0$. Then B^c will contain at least one other focal element, otherwise the conjunctive core would be non-empty, so that $Bel(B^c) \neq 0$. But then Bel is inconsistent, against the hypothesis.

□

In the case of the implication (10.4), instead, the condition (10.7) does not guarantee consistency, unless $|\mathcal{C}^\cap| = 1$. Indeed, if the conjunctive core is not a singleton, we can split it into two disjoint subsets A and B, both with plausibility 1, so that $B^c = A \cup (\mathcal{C}^\cap)^c$ also has plausibility 1. In other words, under the implication relation (10.4), to be consistent, belief functions do not just need to have intersecting focal elements, but the intersection needs to be a singleton.

Based on the above argument, it makes sense to use the term *consistent belief functions* for those which are consistent with respect to the implication (10.4), i.e., those with a non-empty conjunctive core (by Theorem 35). If we define the amount of *internal conflict* of a belief function as

$$c(Bel) \doteq \sum_{A,B \subseteq \Theta : A \cap B = \emptyset} m(A)m(B), \qquad (10.8)$$

the next theorem immediately follows.

Theorem 36. *A belief function $Bel : 2^\Theta \rightarrow [0,1]$ is consistent if and only if its internal conflict is zero.*

As consistent belief functions only support (to some degree) either a proposition or its complement, they can be seen as the counterpart of consistent knowledge bases in belief function logic. As such, working with consistent belief functions is desirable to avoid inconsistent inferences (flagged by a non-zero internal conflict (10.8)). In Chapter 14, we will see how our geometric approach can be exploited to transform an arbitrary belief function into a consistent one. In order to achieve this result, however, we need to study the geometry of consistent BFs.

10.5 The geometry of consistent belief functions

As a matter of fact, just like consonant BFs, consistent belief functions live in a simplicial complex whose faces possess certain geometrical features. A natural decomposition of any arbitrary belief function into $|\Theta|$ 'consistent components', each of them living in a maximal simplex of the consistent complex, can then be proposed. Such components can thus be seen as natural consistent transformations, with different conjunctive cores, of an arbitrary belief function.

10.5.1 The region of consistent belief functions

Example 27: consistent belief functions in the binary frame. *In our running example of a frame of discernment of cardinality 2, consistent belief functions may obviously have as a collection of focal elements either $\mathcal{E} = \{\{x\}, \Theta\}$ or $\mathcal{E} = \{\{y\}, \Theta\}$. Thus, by Theorem 10, all consistent BFs on $\Theta_2 = \{x, y\}$ live in the union of two convex components,*

$$\mathcal{CS}_2 = \mathcal{CS}^{\{x,\Theta\}} \cup \mathcal{CS}^{\{y,\Theta\}} = Cl(Bel_\Theta, Bel_x) \cup Cl(Bel_\Theta, Bel_y),$$

which coincide with the region of consonant belief functions, as depicted in Fig. 10.1.

In the general case, consistent belief functions are characterised, as we know, by the fact that their focal elements have non-empty intersection. All possible lists of focal elements associated with consistent BFs then correspond to all possible collections of intersecting events,

$$\left\{ A_1, \ldots, A_m \subseteq \Theta : \bigcap_{i=1}^{m} A_i \neq \emptyset \right\}.$$

Just as in the consonant case, Theorem 10 implies that all the belief functions whose focal elements belong to such a collection, no matter what the actual values of their basic plausibility assignment may be, form a simplex $Cl(Bel_{A_1}, \ldots, Bel_{A_m})$. Such a collection is 'maximal' when it is not possible to add a further focal element A_{m+1} such that $\cap_{i=1}^{m+1} A_i \neq \emptyset$.

It is easy to see that collections of events with non-empty intersection are maximal iff they have the form $\{A \subseteq \Theta : A \ni x\}$, where $x \in \Theta$ is a singleton; in other words, they are *principal filters* in the poset $(2^{\Theta}, \subset)$ associated with singleton subsets $\{x\}$. Consequently, the region of consistent belief functions is the union of the collection of maximal simplices associated with principal filters,

$$\mathcal{CS} = \bigcup_{x \in \Theta} Cl(Bel_A, A \ni x). \tag{10.9}$$

There are obviously $n \doteq |\Theta|$ such maximal simplices in $\mathcal{CS}$. Each of them has

$$|\{A : A \ni x\}| = |\{A \subseteq \Theta : A = \{x\} \cup B, B \subset \{x\}^c\}| = 2^{|\{x\}^c|} = 2^{n-1}$$

vertices, so that their dimension as simplices in the belief space is $2^{n-1} - 1 = \dim \mathcal{B}/2$ (as the dimension of the whole belief space is $\dim \mathcal{B} = 2^n - 2$).

Clearly, $\mathcal{CS}$ is connected, as as each maximal simplex is by definition connected, and Bel_{Θ} belongs to all maximal simplices.

10.5.2 The consistent complex

Just as in the consonant case, the region (10.9) of consistent belief functions is an instance of a simplicial complex (see Definition 125) [551].

Theorem 37. *$\mathcal{CS}$ is a simplicial complex in the belief space $\mathcal{B}$.*

As with consonant belief functions, more can be said about the geometry of the maximal faces of $\mathcal{CS}$. In $\Theta = \{x, y, z\}$, for instance, the consistent complex $\mathcal{CS}$ is composed of three maximal simplices of dimension $|\{A \ni x\}| - 1 = 3$ (see Section 10.3.2):

$$\begin{array}{l} Cl(Bel_A : A \ni x) = Cl(Bel_x, Bel_{\{x,y\}}, Bel_{\{x,z\}}, Bel_{\Theta}), \\ Cl(Bel_A : A \ni y) = Cl(Bel_y, Bel_{\{x,y\}}, Bel_{\{y,z\}}, Bel_{\Theta}), \\ Cl(Bel_A : A \ni z) = Cl(Bel_z, Bel_{\{x,z\}}, Bel_{\{y,z\}}, Bel_{\Theta}). \end{array} \tag{10.10}$$

Once again, the vertices of each pair of such maximal simplices can be put into 1–1 correspondence.

Consider, for instance, both $\mathcal{CS}^{\{A\ni x\}} \doteq Cl(Bel_x, Bel_{\{x,y\}}, Bel_{\{x,z\}}, Bel_\Theta)$ and $\mathcal{CS}^{\{A\ni z\}} \doteq Cl(Bel_z, Bel_{\{x,z\}}, Bel_{\{y,z\}}, Bel_\Theta)$. The desired mapping is

$$x \leftrightarrow z,\ \{x,z\} \leftrightarrow \{x,z\},\ \{x,y\} \leftrightarrow \{y,z\},\ \Theta \leftrightarrow \Theta,$$

for corresponding segments in the two simplices have the same length. For example, $Cl(Bel_x, Bel_\Theta)$ is congruent with $Cl(Bel_z, Bel_\Theta)$, as

$$\|Bel_x - Bel_\Theta\| = \|[1\ 0\ 0\ 1\ 1\ 0]'\| = \sqrt{3} = \|[0\ 0\ 1\ 0\ 1\ 1]'\| = \|Bel_z - Bel_\Theta\|.$$

In the same way, $Cl(Bel_{\{x,z\}}, Bel_{\{x,y\}})$ is congruent with $Cl(Bel_{\{x,z\}}, Bel_{\{y,z\}})$:

$$\begin{aligned}\|Bel_{\{x,z\}} - Bel_{\{x,y\}}\| &= \|[0\ 0\ 0\ -1\ 1\ 0]'\| = \sqrt{2} \\ &= \|[0\ 0\ 0\ 0\ 1\ -1]'\| = \|Bel_{\{x,z\}} - Bel_{\{y,z\}}\|.\end{aligned}$$

This is true in the general case.

Theorem 38. *All maximal simplices of the consistent complex are congruent.*

10.5.3 Natural consistent components

We wish to conclude this chapter devoted to possibility theory by showing that each belief function Bel can be decomposed into $n = |\Theta|$ consistent components with distinct conjunctive cores, which can be interpreted as natural projections of Bel onto the maximal components of the consistent simplicial complex. Interestingly, this decomposition turns out to be closely related to the pignistic transformation [1730],

$$BetP[Bel](x) = \sum_{A \supseteq \{x\}} \frac{m(A)}{|A|}.$$

Indeed, we can write

$$\begin{aligned}Bel &= \sum_{A\subseteq\Theta} m(A) Bel_A = \sum_{x\in\Theta}\sum_{A\ni x} \frac{m(A)}{|A|} Bel_A \\ &= \sum_{x\in\Theta} BetP[Bel](x) \frac{\sum_{A\ni x} \frac{m(A)}{|A|} Bel_A}{BetP[Bel](x)} = \sum_{x\in\Theta} BetP[Bel](x) Bel^x,\end{aligned} \tag{10.11}$$

once we define

$$Bel^x \doteq \frac{1}{BetP[Bel](x)} \sum_{A\ni x} \frac{m(A)}{|A|} Bel_A, \quad x \in \Theta.$$

These $n = |\Theta|$ consistent belief functions can be considered as 'consistent components' of Bel on the maximal components $\mathcal{CS}^{\{A\ni x\}}$, $x \in \Theta$ of the consistent complex.

As a result, each arbitrary belief function Bel lives in the $(n-1)$-dimensional simplex $\mathcal{P}^{Bel} \doteq Cl(Bel^x, x \in \Theta)$ with these n consistent components as vertices (see Fig. 10.5). Strikingly, its convex coordinates in this simplex coincide with the coordinates of the pignistic probability in the probability simplex $\mathcal{P}$, namely

$$BetP[Bel] = \sum_{x \in \Theta} BetP[Bel](x) Bel_x \quad \longleftrightarrow \quad Bel = \sum_{x \in \Theta} BetP[Bel](x) Bel^x.$$

Obviously, if $Bel = P \in \mathcal{P}$ then $Bel^x = Bel_x \; \forall x \in \Theta$.

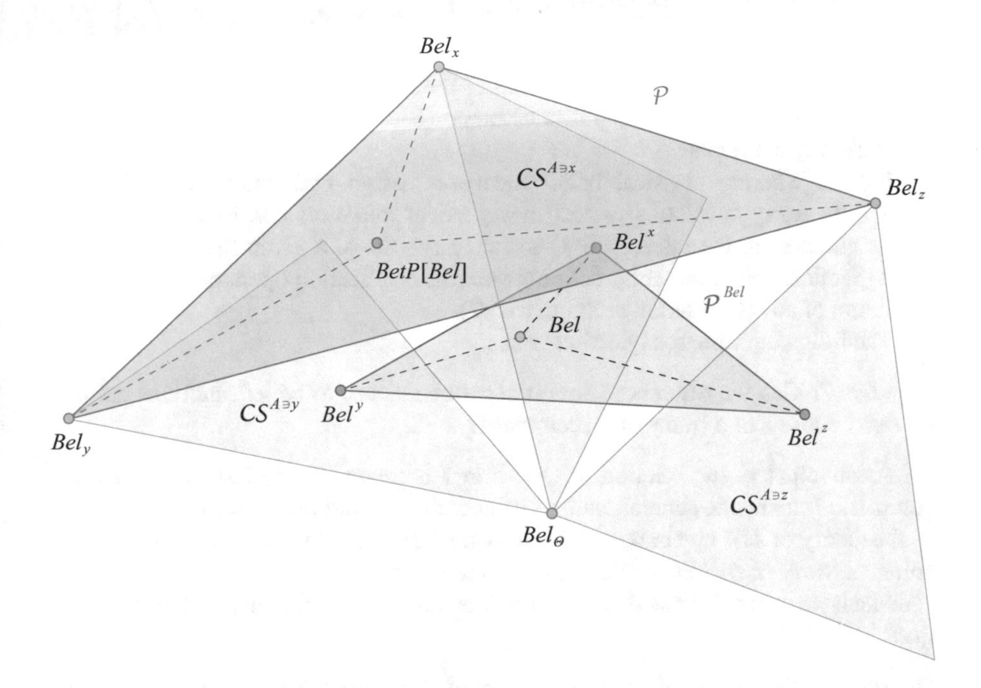

Fig. 10.5: Pictorial representation of the role of the pignistic values $\{BetP[Bel](x), x \in \Theta\}$ for a belief function and the related pignistic function. Both Bel and $BetP[Bel]$ live in a simplex ($\mathcal{P} = Cl(Bel_x, x \in \Theta)$ and $\mathcal{P}^{Bel} = Cl(Bel^x, x \in \Theta)$, respectively), on which they possess the same convex coordinates $\{BetP[Bel](x), x \in \Theta\}$. The vertices Bel^x, $x \in \Theta$ of the simplex $\mathcal{P}^{Bel}$ can be interpreted as consistent components of the belief function Bel on the simplicial complex of consistent belief functions $\mathcal{CS}$.

Equation (10.11) provides a bridge between the notions of belief, probability and possibility, by associating each belief function with its 'natural' probabilistic components (the pignistic function) and possibilistic components (the quantities Bel^x).

It is natural to wonder whether the consistent components Bel^x of Bel can be interpreted as actual consistent approximations of Bel, i.e., whether they minimise some sort of distance between Bel and the consistent complex. It is immediate to see that

$$BetP[b] = \arg\min_{P \in \mathcal{P}} d(P, Bel),$$

whenever d is any function of the convex coordinates of Bel in $\mathcal{P}^{Bel}$ (as they coincide with the pignistic values for both Bel and $BetP[b]$).

Consistent approximation will be analysed in more detail in Chapter 13.

10.6 Research questions

As we recalled in Chapter 6, Section 6.5.1, the links between belief and possibility/fuzzy theory go further than the equivalence between consonant belief functions and necessity measures.

Indeed, whereas classical belief measures assign real numbers to classical ('crisp', in fuzzy terminology) sets, fuzzy belief measures can be defined which assign numbers in the interval $[0, 1]$ to fuzzy subsets of a given frame of discernment (Section 6.5.2), either by defining suitable implication operators (6.5.2) or as a solution of an optimisation problem (6.5.2).

The following research question arises.

Question 7. Can we extend the notion of a belief space to belief functions defined on fuzzy subsets of a frame of discernment?

As anticipated, the solution to Question 7 is related to the issue of creating a geometric framework general enough to encompass continuous formulations such as the theory of MV algebras, as well as competing uncertainty approaches such as coherent lower previsions on gambles (recall Section 6.1.3).

Much work is still to be done to complete our analysis of consistent belief functions.

Question 8. The notion of consistency is central in imprecise probability (see Section 6.1): how does the concept of consistency for belief functions relate to the former?

Question 9. If consistency is desirable, for the reasons explained in Section 10.4, should we seek a combination rule able to preserve the consistency of a belief structure?

Last but not least, the justification of consistent belief functions in terms of knowledge bases requires additional work, in the light of the extensive efforts conducted in the area of logic interpretation of the theory of evidence, which we reviewed in Section 6.6.

Appendix: Proofs

Proof of Theorem 37

Property 1 of Definition 125 is trivially satisfied. As a matter of fact, if a simplex $Cl(Bel_{A_1}, \ldots, Bel_{A_n})$ corresponds to focal elements with non-empty intersection, points of any face of this simplex (obtained by selecting a subset of vertices) will clearly be belief functions with a non-empty conjunctive core, and will therefore correspond to consistent BFs.

As for property 2, consider the intersection of two maximal simplices of $\mathcal{CS}$ associated with two distinct conjunctive cores $\mathcal{C}_1, \mathcal{C}_2 \subseteq \Theta$,

$$Cl(Bel_A : A \supseteq \mathcal{C}_1) \cap Cl(Bel_A : A \supseteq \mathcal{C}_2).$$

Now, each convex closure of points $Bel_1, \ldots, Bel_m$ in a Cartesian space is trivially included in the *affine* space they generate:

$$\begin{aligned} Cl(Bel_1, \ldots, Bel_m) &\subsetneq a(Bel_1, \ldots, Bel_m) \\ &\doteq \left\{ Bel : Bel = \alpha_1 \; Bel_1 + \cdots + \alpha_m \; Bel_m, \sum_i \alpha_i = 1 \right\} \end{aligned}$$

(we just need to relax the positivity constraint on the coefficients α_i). But the categorical belief functions $\{Bel_A : \emptyset \subsetneq A \subsetneq \Theta\}$ are linearly independent (as it is straightforward to check), so that $a(Bel_A, A \in L_1) \cap a(Bel_A, A \in L_2) \neq \emptyset$ (where L_1, L_2 are lists of subsets of Θ) if and only if $L_1 \cap L_2 \neq \emptyset$. Here $L_1 = \{A \subseteq \Theta : A \supseteq \mathcal{C}_1\}$, $L_2 = \{A \subseteq \Theta : A \supseteq \mathcal{C}_2\}$, so that the condition reads as

$$\{A \subseteq \Theta : A \supseteq \mathcal{C}_1\} \cap \{A \subseteq \Theta : A \supseteq \mathcal{C}_2\} = \{A \subseteq \Theta : A \supseteq \mathcal{C}_1 \cup \mathcal{C}_2\} \neq \emptyset.$$

As $\mathcal{C}_1 \cup \mathcal{C}_2 \supseteq \mathcal{C}_1, \mathcal{C}_2$, we have that $Cl(Bel_A, A \supseteq \mathcal{C}_1 \cup \mathcal{C}_2)$ is a face of both simplices.

Proof of Theorem 38

We need to find a 1–1 map between the vertices of any two maximal simplices $Cl(Bel_A, A \ni x)$, $Cl(Bel_A : A \supseteq y)$ of $\mathcal{CS}$ such that corresponding sides are congruent.

To do this, we need to rewrite the related collections of events as

$$\begin{aligned} \{A \subseteq \Theta : A \ni x\} &= \{A \subseteq \Theta : A = B \cup \{x\}, B \subseteq \{x\}^c\}, \\ \{A \subseteq \Theta : A \ni y\} &= \{A \subseteq \Theta : A = B \cup \{y\}, B \subseteq \{y\}^c\}. \end{aligned} \tag{10.12}$$

But, in turn, $\{B \subseteq \{x\}^c\} = \{B \subseteq \{x,y\}^c\} \cup \{B \not\ni x, B \ni y\}$ and $\{B \subseteq \{y\}^c\} = \{B \subseteq \{x,y\}^c\} \cup \{B \not\ni y, B \ni x\}$. Therefore,

$$\begin{aligned} \{B \subseteq \{x\}^c\} &= \{B \subseteq \{x,y\}^c\} \cup \{B = C \cup \{y\}, C \subseteq \{x,y\}^c\}, \\ \{B \subseteq \{y\}^c\} &= \{B \subseteq \{x,y\}^c\} \cup \{B = C \cup \{x\}, C \subseteq \{x,y\}^c\}. \end{aligned}$$

Let us then define the following map between events of the two collections (10.12):

$$\begin{array}{rcl} \{A \subseteq \Theta, A \ni x\} & \to & \{A \subseteq \Theta, A \ni y\}, \\ A = B \cup \{x\} & \mapsto & A' = B' \cup \{y\}, \end{array} \qquad (10.13)$$

where

$$\begin{cases} B \mapsto B' = B & B \subseteq \{x,y\}^c, \\ B = C \cup \{y\} \mapsto B' = C \cup \{x\} & B \not\subseteq \{x,y\}^c. \end{cases} \qquad (10.14)$$

We can prove that (10.13) preserves the length of the segments in the corresponding maximal simplices $Cl(Bel_A, A \ni x)$, $Cl(Bel_A, A \ni y)$. We first need to find an explicit expression for $\|Bel_A - Bel_{A'}\|$, $A, A' \subseteq \Theta$.

Again, each categorical belief function Bel_A is such that

$$Bel_A(B) = 1 \; B \supseteq A, \quad Bel_A(B) = 0 \text{ otherwise.}$$

If $A' \supseteq A$, then $Bel_{A'}(B) = 1$ if $B \supseteq A' \supseteq A$, and $Bel_{A'}(B) = 0$ otherwise. Hence

$$Bel_A - Bel_{A'}(B) \neq 0 \Leftrightarrow Bel_A - Bel_{A'}(B) = 1 \Leftrightarrow B \supseteq A, B \not\supseteq A'$$

and

$$\|Bel_A - Bel_{A'}\| = \sqrt{|\{B \supseteq A, B \not\supseteq A'\}|} = \sqrt{|A' \setminus A|}.$$

For each pair of vertices $A_1 = B_1 \cup \{x\}$, $A_2 = B_2 \cup \{x\}$ in the first component, we can distinguish four cases:

1. $B_1 \subseteq \{x,y\}^c$, $B_2 \subseteq \{x,y\}^c$, in which case $B_1' = B_1$, $B_2' = B_2$ and

$$\begin{aligned} |A_2' \setminus A_1'| &= |(B_2' \cup \{y\}) \setminus (B_1' \cup \{y\})| = |B_2' \setminus B_1'| = |B_2 \setminus B_1| \\ &= |B_2' \cup \{x\} \setminus B_1' \cup \{x\}| = |A_2 \setminus A_1|, \end{aligned}$$

   so that

$$\|Bel_{A_2'} - Bel_{A_1'}\| = \|Bel_{A_2} - Bel_{A_1}\|. \qquad (10.15)$$

2. $B_1 \subseteq \{x,y\}^c$ but $B_2 \not\subseteq \{x,y\}^c$, $B_2 = C_2 \cup \{y\}$, in which case $B_1' = B_1$, $B_2' = C_2 \cup \{x\}$, which implies

$$\begin{aligned} A_2' \setminus A_1' &= B_2' \setminus B_1' = (C_2 \cup \{x\}) \setminus B_1 = (C_2 \setminus B_1) \cup \{x\}, \\ A_2 \setminus A_1 &= B_2 \setminus B_1 = (C_2 \cup \{y\}) \setminus B_1 = (C_2 \setminus B_1) \cup \{y\}. \end{aligned}$$

   But then $|A_2' \setminus A_1'| = |A_2 \setminus A_1|$, so that once again (10.15) holds.
3. $B_1 \not\subseteq \{x,y\}^c$, $B_1 = C_1 \cup \{y\}$ but $B_2 \subseteq \{x,y\}^c$, which by the symmetry of the $\setminus$ operator yields (10.15) again.
4. $B_1 \not\subseteq \{x,y\}^c$, $B_1 = C_1 \cup \{y\}$, $B_2 \not\subseteq \{x,y\}^c$, $B_2 = C_2 \cup \{y\}$, in which case $B_1' = C_1 \cup \{x\}$, $B_2' = C_2 \cup \{x\}$, so that

$$\begin{aligned} B_2' \setminus B_1' &= (C_2 \cup \{x\}) \setminus (C_1 \cup \{x\}) = C_2 \setminus C_1 \\ &= (C_2 \cup \{y\}) \setminus (C_1 \cup \{y\}) = B_2 \setminus B_1. \end{aligned}$$

Summarising, in all cases $\|Bel_{A_2'} - Bel_{A_1'}\| = \|Bel_{A_2} - Bel_{A_1}\|$, for pairs of segments $Cl(A_1, A_2)$, $Cl(A_1', A_2')$ in the two maximal components associated through the mapping (10.14) introduced above. By the usual generalisation of Euclid's theorem, this implies that the two simplices are congruent.

Part III

Geometric interplays

Probability transforms: The affine family

As we have seen in Chapter 4, the relation between belief and probability in the theory of evidence has been and continues to be an important subject of study. The reason is that a probability transform mapping belief functions to probability measures is instrumental in addressing a number of issues: mitigating the inherently exponential complexity of belief calculus (Section 4.7), making decisions via the probability distributions obtained in a utility theory framework (Section 4.8) and obtaining pointwise estimates of quantities of interest from belief functions (e.g. the pose of an articulated object in computer vision: see [364], Chapter 8, or [366]).

Since, as we learned in Part II, both belief and probability measures can be assimilated into points of a Cartesian space (Chapter 7), the problem can be posed in a geometric setting. In Part III, we apply the geometric approach introduced in Part II to the study of the problem of transforming an uncertainty measure (e.g. a belief function) into a different type of measure (namely, a probability or a possibility). Without loss of generality, we can define a probability transform as a mapping from the belief space to the probability simplex (see Section 4.7.2)

$$\begin{array}{rl} \mathcal{PT}: \mathcal{B} & \rightarrow \mathcal{P}, \\ Bel \in \mathcal{B} & \mapsto \mathcal{PT}[Bel] \in \mathcal{P}, \end{array}$$

such that an appropriate distance function or similarity measure d from Bel is minimised [382]:

$$\mathcal{PT}[Bel] = \arg\min_{P \in \mathcal{P}} d(Bel, P) \tag{11.1}$$

(compare our review of dissimilarity measures in Section 4.10.1).

F. Cuzzolin, *The Geometry of Uncertainty*, Artificial Intelligence: Foundations, Theory, and Algorithms, https://doi.org/10.1007/978-3-030-63153-6_11

A minimal, sensible requirement is for the probability which results from the transform to be compatible with the upper and lower bounds that the original belief function Bel enforces on the singletons only, rather than on all the focal sets as in (3.10). Thus, this does not require probability transforms to adhere to the much-debated upper–lower probability semantics of belief functions (see Chapter 3). As a matter of fact, some important transforms of this kind are not compatible with such semantics, as we will see here.

Approximation and conditioning approaches explicitly based on traditional Minkowski norms are proposed and developed in Chapters 13, 14 and 15. Here, however, we wish to pursue a wider understanding of the geometry of a number of probability transforms, and their classification into families composed of transforms which exhibit common properties.

In fact, only a few authors have in the past posed the study of the connections between belief functions and probabilities in a geometric setting. In particular, Ha and Haddawy [758] proposed an 'affine operator', which can be considered a generalisation of both belief functions and interval probabilities, and can be used as a tool for constructing convex sets of probability distributions. In their work, uncertainty is modelled as sets of probabilities represented as 'affine trees', while actions (modifications of the uncertain state) are defined as tree manipulators. In a later publication [757], the same authors presented an interval generalisation of the probability cross-product operator, called the 'convex-closure' (cc) operator, analysed the properties of the cc operator relative to manipulations of sets of probabilities and presented interval versions of Bayesian propagation algorithms based on it. Probability intervals were represented there in a computationally efficient fashion by means of a data structure called a 'pcc-tree', in which branches are annotated with intervals, and nodes with convex sets of probabilities.

The topic of this chapter is somewhat related to Ha's cc operator, as we deal here with probability transforms which commute (at least under certain conditions) with affine combination. We call this group of transforms the *affine family*, of which Smets's pignistic transform (Section 4.7.2) is the foremost representative. We introduce two new probability transformations of belief functions, both of them derived from purely geometric considerations, that can be grouped together with the pignistic function in the affine family.

Chapter outline

As usual, we first look for insight by considering the simplest case of a binary frame (Section 11.1). Each belief function Bel is associated there with three different geometric entities, namely the simplex of consistent probabilities $\mathcal{P}[Bel] = \{P \in \mathcal{P} : P(A) \geq Bel(A) \; \forall A \subset \Theta\}$ (see Section 3.1.4, Chapter 4 and [245]), the line (Bel, Pl) joining Bel to the related plausibility function Pl, and the orthogonal complement $\mathcal{P}^{\perp}$ of the probabilistic subspace $\mathcal{P}$. These in turn determine three different probability measures associated with Bel, i.e., the barycentre of $\mathcal{P}[Bel]$ or *pignistic function* $BetP[Bel]$, the *intersection probability* $p[Bel]$ and the *orthogo-*

nal projection $\pi[Bel]$ of Bel onto $\mathcal{P}$. In the binary case, all these Bayesian belief functions coincide.

In Section 11.2 we prove that, even though the ('dual') line (Bel, Pl) is always orthogonal to $\mathcal{P}$, it does not in general intersect the Bayesian simplex. However, it does intersect the region of Bayesian normalised sum functions (compare Chapter 7, Section 7.3.3), i.e., the generalisations of belief functions obtained by relaxing the positivity constraint for masses. This intersection yields a Bayesian NSF $\varsigma[Bel]$.

We see later, in Section 11.3, that $\varsigma[Bel]$ is in turn associated with a proper Bayesian *belief* function $p[Bel]$, which we call the *intersection probability*. We provide two different interpretations of the way this probability distributes the masses of the focal elements of Bel to the elements of Θ, both functions of the difference between the plausibility and the belief of singletons, and compare the combinatorial and geometric behaviour of $p[Bel]$ with that of the pignistic function and the relative plausibility of singletons.

Section 11.4 concerns the study of the orthogonal projection of Bel onto the probability simplex $\mathcal{P}$, i.e., the transform (11.1) associated with the classical L_2 distance. We show that $\pi[Bel]$ always exists and is indeed a probability function. After deriving the condition under which a belief function Bel is orthogonal to $\mathcal{P}$, we give two equivalent expressions for the orthogonal projection. We see that $\pi[Bel]$ can be reduced to another probability signalling the distance of Bel from orthogonality, and that this 'orthogonality flag' can in turn be interpreted as the result of a mass redistribution process (often called 'specialisation' in the literature: see Definition 45) analogous to that associated with the pignistic transform. We prove that, just as $BetP[Bel]$ does, $\pi[Bel]$ commutes with the affine combination operator, and can therefore be expressed as a convex combination of basis pignistic functions, making the orthogonal projection and pignistic function fellow members of a common *affine family* of probability transformations.

For the sake of completeness, the case of unnormalised belief functions (see Section 11.5) is also discussed. We argue that, while the intersection probability $p[Bel]$ is not defined for a generic UBF Bel, the orthogonal projection $\pi[Bel]$ does exist and retains its properties.

Finally, in Section 11.6, more general conditions under which the three affine transformations coincide are analysed.

Most of the material of this chapter was originally published in [327]. As usual, proofs have been moved to an appendix for clarity of exposition.

11.1 Affine transforms in the binary case

In Chapter 9, Example 21 (Fig. 11.1), we extensively illustrated the geometry of belief functions living on a binary frame $\Theta_2 = \{x, y\}$. There, both the belief space $\mathcal{B}$ and the plausibility space $\mathcal{PL}$ are simplices with vertices $\{Bel_\Theta = [0,0]', Bel_x = [1,0]', Bel_y = [0,1]'\}$ and $\{Pl_\Theta = [1,1]', Pl_x = Bel_x, Pl_y = Bel_y\}$, respectively.

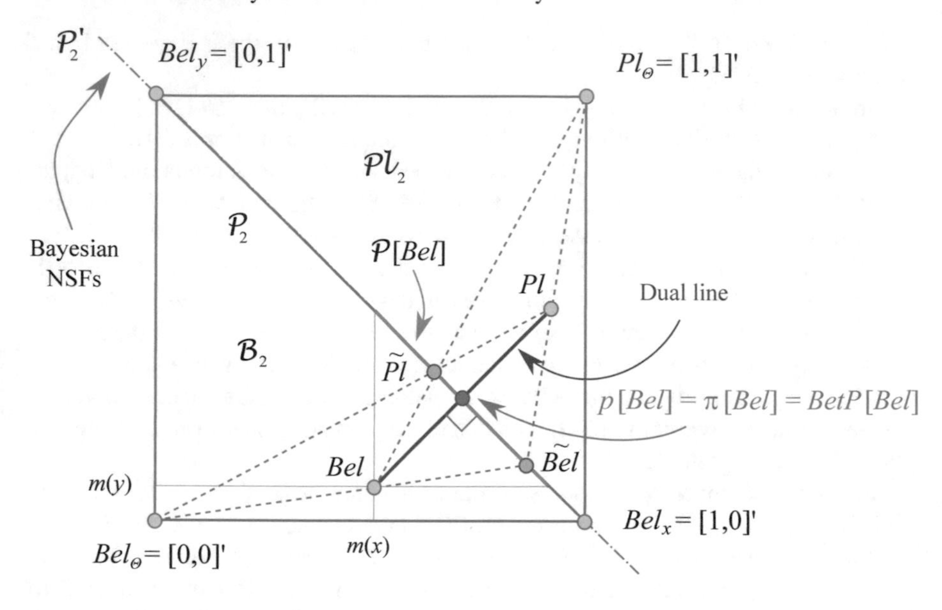

Fig. 11.1: In a binary frame $\Theta_2 = \{x, y\}$, a belief function Bel and the corresponding plausibility function Pl are always located in symmetric positions with respect to the segment $\mathcal{P}$ of all the probabilities defined on Θ_2. The associated relative plausibility $\tilde{Pl}$ and belief $\tilde{Bel}$ of singletons are just the intersections of the probability simplex $\mathcal{P}$ with the line passing through Pl and $Bel_\Theta = [0, 0]'$ and that joining Bel and Bel_Θ, respectively. The pignistic function, orthogonal projection and intersection probability all coincide with the centre of the segment of probabilities $\mathcal{P}[Bel]$ which dominate Bel (shown in red).

Let us first recall the definition (9.1) of a basic plausibility assignment (the Möbius transform of the plausibility function: see Chapter 9, Section 9.1), and its expression (9.3) in terms of the associated basic probability assignment,

$$\mu(A) = (-1)^{|A|+1} \sum_{B \supseteq A} m(B),$$

whenever A is non-empty. We can then compute the BPlA of an arbitrary belief function Bel on Θ_2 as

$$\mu(x) = (-1)^2 \sum_{B \ni x} m(B) = m(x) + m(\Theta) = Pl(x),$$

$$\mu(y) = (-1)^2 \sum_{B \ni y} m(B) = m(y) + m(\Theta) = Pl(y).$$

Thus, the point of $\mathbb{R}^2$ which represents its (dual) plausibility function is simply (see Fig. 11.1 again)

$$Pl = Pl(x)Bel_x + Pl(y)Bel_y.$$

As we first noticed in Chapter 9, the belief and plausibility spaces lie in symmetric locations with respect to the Bayesian simplex $\mathcal{P} = Cl(Bel_x, Bel_y)$. Furthermore, each pair of measures (Bel, Pl) determines a line orthogonal to $\mathcal{P}$, where Bel and Pl lie in symmetric positions on the two sides of $\mathcal{P}$ itself.

Clearly, in the binary case the set $\mathcal{P}[Bel] = \{P \in \mathcal{P} : P(A) \geq Bel(A)\ \forall A \subseteq \Theta\}$ of all the probabilities dominating Bel is a segment whose centre of mass $\overline{\mathcal{P}}[Bel]$ is known [245, 547, 357] to be Smets's *pignistic function* [1678, 1730],

$$\begin{aligned} BetP[Bel] &= \sum_{x \in \Theta} Bel_x \sum_{A \supset x} \frac{m(A)}{|A|} \\ &= Bel_x \left(m(x) + \frac{m(\Theta)}{2} \right) + Bel_y \left(m(y) + \frac{m(\Theta)}{2} \right), \end{aligned} \tag{11.2}$$

and coincides with both the orthogonal projection $\pi[Bel]$ of Bel onto $\mathcal{P}$ and the intersection $p[Bel]$ of the line $a(Bel, Pl)$ with the Bayesian simplex $\mathcal{P}$,

$$p[Bel] = \pi[Bel] = BetP[Bel] = \overline{\mathcal{P}}[Bel].$$

Inherently epistemic notions such as 'consistency' and 'linearity' (one of the rationality principles behind the pignistic transform [1717]) seem to be related to geometric properties such as orthogonality. It is natural to wonder whether this is true in general, or is just an artefact of the binary frame.

Incidentally, neither the relative plausibility (4.59) nor the relative belief (4.61) of singletons (see Section 4.7.2 or the original paper [1859]) follows the same pattern. We will consider their behaviour separately in Chapter 12.

11.2 Geometry of the dual line

In the binary case, the plane $\mathbb{R}^2$ in which both $\mathcal{B}$ and $\mathcal{PL}$ are embedded is the space of all the normalised sum functions on Θ_2 (compare Section 7.3.3). The region $\mathcal{P}'$ of all the Bayesian NSFs one can define on $\Theta = \{x, y\}$ is the line

$$\mathcal{P}' = \left\{ \varsigma \in \mathbb{R}^2 : m_\varsigma(x) + m_\varsigma(y) = 1 \right\} = a(\mathcal{P}),$$

i.e., the affine space $a(\mathcal{P}) = a(Bel_x, x \in \Theta)$ generated by $\mathcal{P}$.[70]

---

[70] Here $a(v_1 \ldots, v_k)$ denotes the affine subspace of a Cartesian space $\mathbb{R}^m$ generated by a collection of points $v_1, \ldots, v_k \in \mathbb{R}^m$, i.e. the set

$$a(v_1, \ldots, v_k) \doteq \left\{ v \in \mathbb{R}^m : v = \alpha_1 v_1 + \cdots + \alpha_k v_k, \sum_i \alpha_i = 1 \right\}.$$

11.2.1 Orthogonality of the dual line

We first note that $\mathcal{P}'$ can be written as a translated version of a vector space as follows:

$$a(\mathcal{P}) = Bel_x + \text{span}(Bel_y - Bel_x, \forall y \in \Theta, y \neq x),$$

where span$(Bel_y - Bel_x)$ denotes the vector space generated by the $n-1$ difference vectors $Bel_y - Bel_x$ ($n = |\Theta|$), and Bel_x is the categorical belief function focused on x (Section 7.2). Since a categorical BF Bel_B focused on a specific subset B is such that

$$Bel_B(A) = \begin{cases} 1 & A \supseteq B, \\ 0 & \text{otherwise}, \end{cases} \tag{11.3}$$

these vectors show a rather peculiar symmetry, namely

$$Bel_y - Bel_x(A) = \begin{cases} 1 & A \supset \{y\}, A \not\supset \{x\}, \\ 0 & A \supset \{x\}, \{y\} \text{ or } A \not\supset \{x\}, \{y\}, \\ -1 & A \not\supset \{y\}, A \supset \{x\}. \end{cases} \tag{11.4}$$

The latter can be exploited to prove the following lemma.

Lemma 9. $[Bel_y - Bel_x](A^c) = -[Bel_y - Bel_x](A)\ \forall A \subseteq \Theta$.

Proof. By (11.3), $[Bel_y - Bel_x](A) = 1$ implies

$$A \supset \{y\}, A \not\supset \{x\} \Rightarrow A^c \supset \{x\}, A^c \not\supset \{y\} \Rightarrow [Bel_y - Bel_x](A^c) = -1,$$

and viceversa. On the other hand, $[Bel_y - Bel_x](A) = 0$ implies $A \supset \{y\}, A \supset \{x\}$ or $A \not\supset \{y\}, A \not\supset \{x\}$. In the first case $A^c \not\supset \{x\}, \{y\}$, and in the second case $A^c \supset \{x\}, \{y\}$. In both cases, $[Bel_y - Bel_x](A^c) = 0$.

□

Lemma 9 allows us to prove the following theorem, just as in the binary case (see the appendix).

Theorem 39. *The line connecting Pl and Bel is orthogonal to the affine space generated by the probabilistic simplex. Namely,*

$$a(Bel - Pl) \perp a(\mathcal{P}).$$

11.2.2 Intersection with the region of Bayesian normalised sum functions

One might be tempted to conclude that, since $a(Bel, Pl)$ and $\mathcal{P}$ are always orthogonal, their intersection *is* the orthogonal projection of Bel onto $\mathcal{P}$ as in the binary case. Unfortunately, this is not the case, for in the case of arbitrary frames of discernment they *do not intersect* each other.

As a matter of fact, Bel and Pl belong to an $N - 2 = (2^n - 2)$-dimensional Euclidean space (recall that we neglect the trivially constant components associated with the empty set $\emptyset$ and the entire frame Θ, see Chapter 7), while the simplex $\mathcal{P}$

generates a vector space whose dimension is only $n-1$. If $n=2$, then $n-1=1$ and $2^n-2=2$, so that $a(\mathcal{P})$ divides the plane into two half-planes with Bel on one side and Pl on the other side (see Fig. 11.1 again).

Formally, for a point on the line $a(Bel, Pl)$ to be a probability measure, we need to find a value of α such that $Bel + \alpha(Pl - Bel) \in \mathcal{P}$. Its components are obviously $Bel(A) + \alpha[Pl(A) - Bel(A)]$ for any subset $A \subset \Theta$, $A \neq \Theta, \emptyset$. In particular, when $A = \{x\}$ is a singleton,

$$Bel(x) + \alpha[Pl(x) - Bel(x)] = Bel(x) + \alpha[1 - Bel(\{x\}^c) - Bel(x)]. \quad (11.5)$$

In order for this point to belong to $\mathcal{P}$, it needs to satisfy the normalisation constraint for singletons, namely

$$\sum_{x\in\Theta} Bel(x) + \alpha \sum_{x\in\Theta} [1 - Bel(\{x\}^c) - Bel(x)] = 1.$$

The latter yields a single candidate value $\beta[Bel]$ for the line coordinate of the desired intersection; more precisely,

$$\alpha = \frac{1 - \sum_{x\in\Theta} Bel(x)}{\sum_{x\in\Theta}[1 - Bel(\{x\}^c) - Bel(x)]} \doteq \beta[Bel]. \quad (11.6)$$

Using the terminology of Section 7.3.3, the candidate projection

$$\varsigma[Bel] \doteq Bel + \beta[Bel](Pl - Bel) = a(Bel, Pl) \cap \mathcal{P}' \quad (11.7)$$

(where $\mathcal{P}'$ denotes once again the set of all Bayesian normalised sum functions in $\mathbb{R}^{N-2}$) is a Bayesian NSF, but is not guaranteed to be a Bayesian *belief function.*

For normalised sum functions, the normalisation condition $\sum_{x\in\Theta} m_\varsigma(x) = 1$ implies $\sum_{|A|>1} m_\varsigma(A) = 0$, so that $\mathcal{P}'$ can be written as

$$\mathcal{P}' = \left\{ \varsigma = \sum_{A\subset\Theta} m_\varsigma(A) Bel_A \in \mathbb{R}^{N-2} : \sum_{|A|=1} m_\varsigma(A) = 1, \sum_{|A|>1} m_\varsigma(A) = 0 \right\}. \quad (11.8)$$

Theorem 40. *The coordinates of $\varsigma[Bel]$ in the reference frame of the categorical Bayesian belief functions $\{Bel_x, x \in \Theta\}$ can be expressed in terms of the basic probability assignment m of Bel as follows:*

$$m_{\varsigma[Bel]}(x) = m(x) + \beta[Bel] \sum_{A\supsetneq x} m(A), \quad (11.9)$$

where

$$\beta[Bel] = \frac{1 - \sum_{x\in\Theta} m(x)}{\sum_{x\in\Theta} (Pl(x) - m(x))} = \frac{\sum_{|B|>1} m(B)}{\sum_{|B|>1} m(B)|B|}. \quad (11.10)$$

Equation (11.9) ensures that $m_{\varsigma[Bel]}(x)$ is positive for each $x \in \Theta$. A more symmetrical-looking version of (11.9) can be obtained after realising that

$$\frac{\sum_{|B|=1} m(B)}{\sum_{|B|=1} m(B)|B|} = 1,$$

so that we can write

$$m_{\varsigma[Bel]}(x) = Bel(x)\frac{\sum_{|B|=1} m(B)}{\sum_{|B|=1} m_B(B)|B|} + [Pl - Bel](x)\frac{\sum_{|B|>1} m(B)}{\sum_{|B|>1} m(B)|B|}. \tag{11.11}$$

It is easy to prove that the line $a(Bel, Pl)$ intersects the actual probability simplex *only for 2-additive belief functions* (see the appendix of this chapter as usual).

Theorem 41. *The Bayesian normalised sum function* $\varsigma[Bel]$ *is a probability measure,* $\varsigma[Bel] \in \mathcal{P}$*, if and only if Bel is 2-additive, i.e.,* $m(A) = 0$ $|A| > 2$*. In the latter case Pl is the reflection of Bel through* $\mathcal{P}$*.*

For 2-additive belief functions, $\varsigma[Bel]$ is nothing but the *mean probability* function $(Bel + Pl)/2$. In the general case, however, the reflection of Bel through $\mathcal{P}$ not only does not coincide with Pl, but is not even a plausibility function [358].

11.3 The intersection probability

Summarising, although the dual line $a(Bel, Pl)$ is always orthogonal to $\mathcal{P}$, it does not intersect the probabilistic subspace in general, while it does intersect the region of Bayesian normalised sum functions in a point which we have denoted by $\varsigma[Bel]$ (11.7).

But, of course, since $\sum_x m_{\varsigma[Bel]}(x) = 1$, $\varsigma[Bel]$ is naturally associated with a Bayesian *belief* function, assigning an equal amount of mass to each singleton and 0 to each $A : |A| > 1$. Namely, we can define the probability measure

$$p[Bel] \doteq \sum_{x \in \Theta} m_{\varsigma[Bel]}(x) Bel_x, \tag{11.12}$$

where $m_{\varsigma[Bel]}(x)$ is given by (11.9). Trivially, $p[Bel]$ is a probability measure, since by definition $m_{p[Bel]}(A) = 0$ for $|A| > 1$, $m_{p[Bel]}(x) = m_{\varsigma[Bel]}(x) \geq 0$ $\forall x \in \Theta$, and, by construction,

$$\sum_{x \in \Theta} m_{p[Bel]}(x) = \sum_{x \in \Theta} m_{\varsigma[Bel]}(x) = 1.$$

We call $p[Bel]$ the *intersection probability* associated with Bel.

The relative geometry of $\varsigma[Bel]$ and $p[Bel]$ with respect to the regions of Bayesian belief and normalised sum functions, respectively, is outlined in Fig. 11.2.

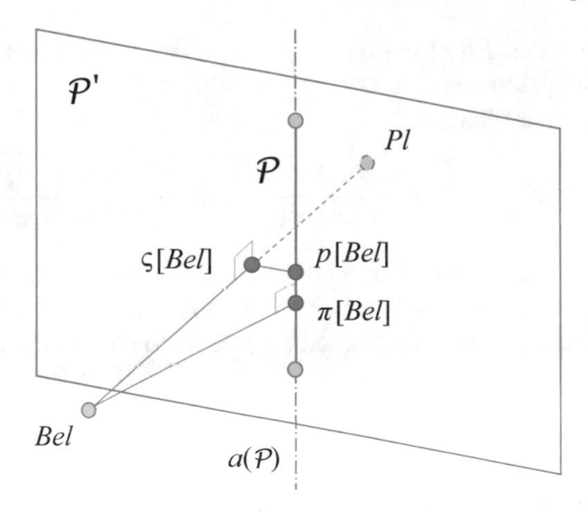

Fig. 11.2: The geometry of the line $a(Bel, Pl)$ and the relative locations of $p[Bel]$, $\varsigma[Bel]$ and $\pi[Bel]$ for a frame of discernment of arbitrary size. Each belief function Bel and the related plausibility function Pl lie on opposite sides of the hyperplane $\mathcal{P}'$ of all Bayesian normalised sum functions, which divides the space $\mathbb{R}^{N-2}$ of all NSFs into two halves. The line $a(Bel, Pl)$ connecting them always intersects $\mathcal{P}'$, but not necessarily $a(\mathcal{P})$ (vertical line). This intersection $\varsigma[Bel]$ is naturally associated with a probability $p[Bel]$ (in general distinct from the orthogonal projection $\pi[Bel]$ of Bel onto $\mathcal{P}$), having the same components in the base $\{Bel_x, x \in \Theta\}$ of $a(\mathcal{P})$. $\mathcal{P}$ is a simplex (a segment in the figure) in $a(\mathcal{P})$: $\pi[Bel]$ and $p[Bel]$ are both 'true' probabilities.

11.3.1 Interpretations of the intersection probability

Non-Bayesianity flag A first interpretation of this probability transform follows from noticing that

$$\beta[Bel] = \frac{1 - \sum_{x\in\Theta} m(x)}{\sum_{x\in\Theta} Pl(x) - \sum_{x\in\Theta} m(x)} = \frac{1 - k_{Bel}}{k_{Pl} - k_{Bel}},$$

where

$$k_{Bel} \doteq \sum_{x\in\Theta} m(x), \quad k_{Pl} \doteq \sum_{x\in\Theta} Pl(x) = \sum_{A\subset\Theta} m(A)|A| \tag{11.13}$$

are the total mass (belief) of singletons and the total plausibility of singletons, respectively (or equivalently, the normalisation factors for the relative belief $\tilde{Bel}$ and the relative plausibility $\tilde{Pl}$ of singletons, respectively). Consequently, the intersection probability $p[Bel]$ can be rewritten as

$$p[Bel](x) = m(x) + (1 - k_{Bel})\frac{Pl(x) - m(x)}{k_{Pl} - k_{Bel}}. \tag{11.14}$$

When Bel is Bayesian, $Pl(x) - m(x) = 0 \ \forall x \in \Theta$. If Bel is not Bayesian, there exists at least a singleton x such that $Pl(x) - m(x) > 0$.

The Bayesian belief function

$$R[Bel](x) \doteq \frac{\sum_{A \supsetneq x} m(A)}{\sum_{|A|>1} m(A)|A|} = \frac{Pl(x) - m(x)}{\sum_{y \in \Theta} \left(Pl(y) - m(y) \right)}$$

thus measures the relative contribution of each singleton x to the non-Bayesianity of Bel. Equation (11.14) shows, in fact, that the non-Bayesian mass $1 - k_{Bel}$ of the original belief function Bel is reassigned by $p[Bel]$ to each singleton according to its relative contribution $R[Bel](x)$ to the non-Bayesianity of Bel.

Intersection probability and epistemic transforms Clearly, from (11.14) the flag probability $R[Bel]$ also relates the intersection probability $p[Bel]$ to other two classical Bayesian approximations, the relative plausibility $\tilde{Pl}$ and belief $\tilde{Bel}$ of singletons (see Section 4.7.2). We just need to rewrite (11.14) as

$$p[Bel] = k_{Bel} \, \tilde{Bel} + (1 - k_{Bel}) R[Bel]. \tag{11.15}$$

Since $k_{Bel} = \sum_{x \in \Theta} m(x) \leq 1$, (11.15) implies that the intersection probability $p[Bel]$ belongs to the segment linking the flag probability $R[Bel]$ to the relative belief of singletons $\tilde{Bel}$. Its convex coordinate on this segment is the total mass of singletons k_{Bel}.

The relative plausibility function $\tilde{Pl}$ can also be written in terms of $\tilde{Bel}$ and $R[Bel]$ as, by definition (4.59),

$$\begin{aligned} R[Bel](x) &= \frac{Pl(x) - m(x)}{k_{Pl} - k_{Bel}} = \frac{Pl(x)}{k_{Pl} - k_{Bel}} - \frac{m(x)}{k_{Pl} - k_{Bel}} \\ &= \tilde{Pl}(x) \frac{k_{Pl}}{k_{Pl} - k_{Bel}} - \tilde{Bel}(x) \frac{k_{Bel}}{k_{Pl} - k_{Bel}}, \end{aligned}$$

since $\tilde{Pl}(x) = Pl(x)/k_{Pl}$ and $\tilde{Bel}(x) = m(x)/k_{Bel}$. Therefore,

$$\tilde{Pl} = \left(\frac{k_{Bel}}{k_{Pl}} \right) \tilde{Bel} + \left(1 - \frac{k_{Bel}}{k_{Pl}} \right) R[Bel]. \tag{11.16}$$

In conclusion, both the relative plausibility of singletons $\tilde{Pl}$ and the intersection probability $p[Bel]$ belong to the segment $Cl(R[Bel], \tilde{Bel})$ joining the relative belief $\tilde{Bel}$ and the probability flag $R[Bel]$ (see Fig. 11.3). The convex coordinate of $\tilde{Pl}$ in $Cl(R[Bel], \tilde{Bel})$ (11.16) measures the ratio between the total mass and the total plausibility of the singletons, while that of $\tilde{Bel}$ measures the total mass of the singletons k_{Bel}. However, since $k_{Pl} = \sum_{A \subset \Theta} m(A)|A| \geq 1$, we have that $k_{Bel}/k_{Pl} \leq k_{Bel}$: hence the relative plausibility function of singletons $\tilde{Pl}$ is closer to $R[Bel]$ than $p[Bel]$ is (Figure 11.3 again).

Obviously, when $k_{Bel} = 0$ (the relative belief of singletons $\tilde{Bel}$ does not exist, for Bel assigns no mass to singletons), the remaining probability approximations coincide: $p[Bel] = \tilde{Pl} = R[Bel]$ by (11.14).

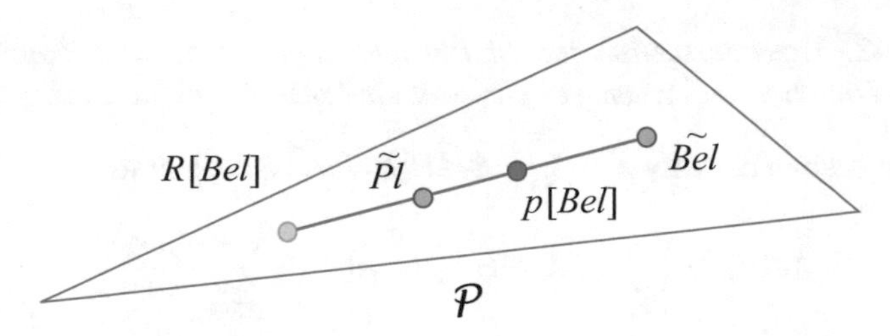

Fig. 11.3: Location in the probability simplex $\mathcal{P}$ of the intersection probability $p[Bel]$ and the relative plausibility of singletons $\tilde{Pl}$ with respect to the non-Bayesianity flag $R[Bel]$. They both lie on the segment joining $R[Bel]$ and the relative belief of singletons $\tilde{Bel}$, but $\tilde{Pl}$ is closer to $R[Bel]$ than $p[Bel]$ is.

Meaning of the ratio $\beta[Bel]$ and the pignistic function To shed more light on $p[Bel]$, and to get an alternative interpretation of the intersection probability, it is useful to compare $p[Bel]$ as expressed in (11.14) with the pignistic function,

$$BetP[Bel](x) \doteq \sum_{A \supseteq x} \frac{m(A)}{|A|} = m(x) + \sum_{A \supset x, A \neq x} \frac{m(A)}{|A|}.$$

We can notice that in $BetP[Bel]$ the mass of each event A, $|A| > 1$, is considered *separately*, and its mass $m(A)$ is shared *equally* among the elements of A. In $p[Bel]$, instead, it is the *total* mass $\sum_{|A|>1} m(A) = 1 - k_{Bel}$ of non-singletons which is considered, and this total mass is distributed *proportionally* to their non-Bayesian contribution to each element of Θ.

How should $\beta[Bel]$ be interpreted, then? If we write $p[Bel](x)$ as

$$p[Bel](x) = m(x) + \beta[Bel](Pl(x) - m(x)), \tag{11.17}$$

we can observe that a fraction, measured by $\beta[Bel]$, of its non-Bayesian contribution $Pl(x) - m(x)$ is *uniformly* assigned to each singleton. This leads to another parallelism between $p[Bel]$ and $BetP[Bel]$. It suffices to note that, if $|A| > 1$,

$$\beta[Bel_A] = \frac{\sum_{|B|>1} m(B)}{\sum_{|B|>1} m(B)|B|} = \frac{1}{|A|},$$

so that both $p[Bel](x)$ and $BetP[Bel](x)$ assume the form

$$m(x) + \sum_{A \supset x, A \neq x} m(A) \beta_A,$$

where $\beta_A = \text{const} = \beta[Bel]$ for $p[Bel]$, while $\beta_A = \beta[Bel_A]$ in the case of the pignistic function.

Under what conditions do the intersection probability and pignistic function coincide? A sufficient condition can be easily given for a special class of belief functions.

Theorem 42. *The intersection probability and pignistic function coincide for a given belief function Bel whenever the focal elements of Bel have size 1 or k only.*

Proof. The desired equality $p[Bel] = BetP[Bel]$ is equivalent to

$$m(x) + \sum_{A \supsetneq x} m(A)\beta[Bel] = m(x) + \sum_{A \supsetneq x} \frac{m(A)}{|A|},$$

which in turn reduces to

$$\sum_{A \supsetneq x} m(A)\beta[Bel] = \sum_{A \supsetneq x} \frac{m(A)}{|A|}.$$

If $\exists\, k : m(A) = 0$ for $|A| \neq k$, $|A| > 1$, then $\beta[Bel] = 1/k$ and the equality is satisfied.

□

In particular, this is true when Bel is 2-additive.

Example 28: intersection probability. *Let us briefly discuss these two interpretations of* $p[Bel]$ *in a simple example. Consider a ternary frame* $\Theta = \{x, y, z\}$, *and a belief function Bel with BPA*

$$\begin{array}{c} m(x) = 0.1,\, m(y) = 0,\, m(z) = 0.2, \\ m(\{x,y\}) = 0.3,\, m(\{x,z\}) = 0.1,\, m(\{y,z\}) = 0,\, m(\Theta) = 0.3. \end{array} \tag{11.18}$$

The related basic plausibility assignment is, according to (9.1),

$$\begin{aligned} \mu(x) &= (-1)^{|x|+1} \sum_{B \supseteq \{x\}} m(B) \\ &= m(x) + m(\{x,y\}) + m(\{x,z\}) + m(\Theta) = 0.8, \end{aligned}$$

$$\mu(y) = 0.6, \qquad \mu(z) = 0.6, \qquad \mu(\{x,y\}) = -0.6,$$

$$\mu(\{x,z\}) = -0.4, \;\; \mu(\{y,z\}) = -0.3, \;\; \mu(\Theta) = 0.3.$$

Figure 11.4 depicts the subsets of Θ *with non-zero BPA (left) and BPlA (middle) induced by the belief function (11.18): dashed ellipses indicate a negative mass. The total mass that (11.18) accords to singletons is* $k_{Bel} = 0.1 + 0 + 0.2 = 0.3$. *Thus, the line coordinate* $\beta[Bel]$ *of the intersection* $\varsigma[Bel]$ *of the line* $a(Bel, Pl)$ *with* $\mathcal{P}'$ *is*

$$\beta[Bel] = \frac{1 - k_{Bel}}{m(\{x,y\})|\{x,y\}| + m(\{x,z\})|\{x,z\}| + m(\Theta)|\Theta|} = \frac{0.7}{1.7}.$$

By (11.9), the mass assignment of $\varsigma[Bel]$ *is therefore*

$$m_{\varsigma[Bel]}(x) = m(x) + \beta[Bel](\mu(x) - m(x)) = 0.1 + 0.7 \cdot \frac{0.7}{1.7} = 0.388,$$

$$m_{\varsigma[Bel]}(y) = 0 + 0.6 \cdot \frac{0.7}{1.7} = 0.247, \qquad m_{\varsigma[Bel]}(z) = 0.2 + 0.4 \cdot \frac{0.7}{1.7} = 0.365,$$

$$m_{\varsigma[Bel]}(\{x,y\}) = 0.3 - 0.9 \cdot \frac{0.7}{1.7} = -0.071,$$

$$m_{\varsigma[Bel]}(\{x,z\}) = 0.1 - 0.5 \cdot \frac{0.7}{1.7} = -0.106,$$

$$m_{\varsigma[Bel]}(\{y,z\}) = 0 - 0.3 \cdot \frac{0.7}{1.7} = -0.123, \quad m_{\varsigma[Bel]}(\Theta) = 0.3 + 0 \cdot \frac{0.7}{1.7} = 0.3.$$

We can verify that all singleton masses are indeed non-negative and add up to one, while the masses of the non-singleton events add up to zero,

$$-0.071 - 0.106 - 0.123 + 0.3 = 0,$$

confirming that $\varsigma[Bel]$ *is a Bayesian normalised sum function. Its mass assignment has signs which are still described by Fig. 11.4 (middle) although, as* $m_{\varsigma[Bel]}$ *is a weighted average of* m *and* μ*, its mass values are closer to zero.*

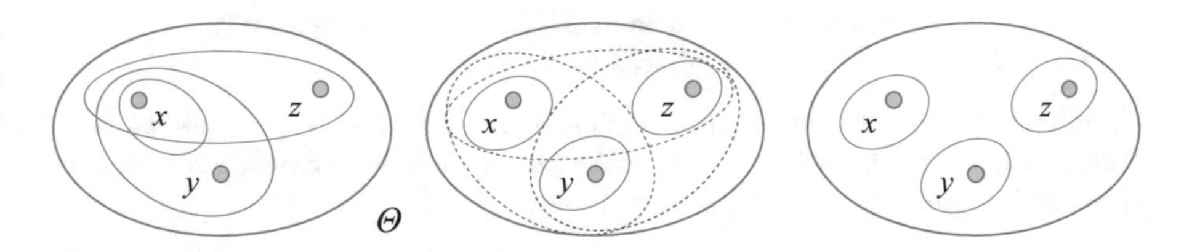

Fig. 11.4: Signs of non-zero masses assigned to events by the functions discussed in Example 28. Left: BPA of the belief function (11.18), with five focal elements. Middle: the associated BPlA assigns positive masses (solid ellipses) to all events of size 1 and 3, and negative ones (dashed ellipses) to all events of size 2. This is the case for the mass assignment associated with ς (11.9) too. Right: the intersection probability $p[Bel]$ (11.12) retains, among the latter, only the masses assigned to singletons.

In order to compare $\varsigma[Bel]$ *with the intersection probability, we need to recall (11.14): the non-Bayesian contributions of* x, y, z *are, respectively,*

$$\begin{aligned} Pl(x) - m(x) &= m(\Theta) + m(\{x,y\}) + m(\{x,z\}) = 0.7, \\ Pl(y) - m(y) &= m(\{x,y\}) + m(\Theta) = 0.6, \\ Pl(z) - m(z) &= m(\{x,z\}) + m(\Theta) = 0.4, \end{aligned}$$

so that the non-Bayesianity flag is $R(x) = 0.7/1.7$*,* $R(y) = 0.6/1.7$*,* $R(z) = 0.4/1.7$*. For each singleton* θ*, the value of the intersection probability results from* *adding to the original BPA $m(\theta)$ a share of the mass of the non-singleton events $1 - k_{Bel} = 0.7$ proportional to the value of $R(\theta)$ (see Fig. 11.4, right):*

$$\begin{array}{l} p[Bel](x) = m(x) + (1 - k_{Bel})R(x) = 0.1 + 0.7 * 0.7/1.7 = 0.388, \\ p[Bel](y) = m(y) + (1 - k_{Bel})R(y) = 0 + 0.7 * 0.6/1.7 \quad = 0.247, \\ p[Bel](z) = m(z) + (1 - k_{Bel})R(z) = 0.2 + 0.7 * 0.4/1.7 = 0.365. \end{array}$$

We can see that $p[Bel]$ coincides with the restriction of $\varsigma[Bel]$ to singletons.

Equivalently, $\beta[Bel]$ measures the share of $Pl(x) - m(x)$ assigned to each element of the frame of discernment:

$$\begin{array}{l} p[Bel](x) = m(x) + \beta[Bel](Pl(x) - m(x)) = 0.1 + 0.7/1.7 * 0.7, \\ p[Bel](y) = m(y) + \beta[Bel](Pl(y) - m(y)) = 0 + 0.7/1.7 * 0.6, \\ p[Bel](z) = m(z) + \beta[Bel](Pl(z) - m(z)) = 0.2 + 0.7/1.7 * 0.4. \end{array}$$

11.3.2 Intersection probability and affine combination

We have seen that $p[Bel]$ and $BetP[Bel]$ are closely related probability transforms, linked by the role of the quantity $\beta[Bel]$. It is natural to wonder whether $p[Bel]$ exhibits a similar behaviour with respect to the convex closure operator (see Chapter 4 and (3.30)). Indeed, although the situation is a little more complex in this second case, $p[Bel]$ turns also out to be related to $Cl(.)$ in a rather elegant way.

Let us introduce the notation $\beta[Bel_i] = N_i/D_i$.

Theorem 43. *Given two arbitrary belief functions Bel_1, Bel_2 defined on the same frame of discernment, the intersection probability of their affine combination $\alpha_1\ Bel_1 + \alpha_2\ Bel_2$ is, for any $\alpha_1 \in [0,1]$, $\alpha_2 = 1 - \alpha_1$,*

$$\begin{aligned} p[\alpha_1\ Bel_1 + \alpha_2\ Bel_2] = \widehat{\alpha_1 D_1}\big(\alpha_1 p[Bel_1] + \alpha_2 T[Bel_1, Bel_2]\big) \\ + \widehat{\alpha_2 D_2}\big(\alpha_1 T[Bel_1, Bel_2]) + \alpha_2 p[Bel_2]\big), \end{aligned} \qquad (11.19)$$

where $\widehat{\alpha_i D_i} = \frac{\alpha_i D_i}{\alpha_1 D_1 + \alpha_2 D_2}$, $T[Bel_1, Bel_2]$ is the probability with values

$$T[Bel_1, Bel_2](x) \doteq \hat{D}_1 p[Bel_2, Bel_1] + \hat{D}_2 p[Bel_1, Bel_2], \qquad (11.20)$$

with $\hat{D}_i \doteq \frac{D_i}{D_1 + D_2}$, and

$$\begin{aligned} p[Bel_2, Bel_1](x) &\doteq m_2(x) + \beta[Bel_1]\big(Pl_2(x) - m_2(x)\big), \\ p[Bel_1, Bel_2](x) &\doteq m_1(x) + \beta[Bel_2]\big(Pl_1(x) - m_1(x)\big). \end{aligned} \qquad (11.21)$$

Geometrically, $p[\alpha_1\ Bel_1 + \alpha_2\ Bel_2]$ can be constructed as in Fig. 11.5 as a point of the simplex $Cl(T[Bel_1, Bel_2], p[Bel_1], p[Bel_2])$. The point

$$\alpha_1 T[Bel_1, Bel_2] + \alpha_2 p[Bel_2]$$

is the intersection of the segment $Cl(T, p[Bel_2])$ with the line l_2 passing through $\alpha_1 p[Bel_1] + \alpha_2 p[Bel_2]$ and parallel to $Cl(T, p[Bel_1])$. Dually, the point

$$\alpha_2 T[Bel_1, Bel_2] + \alpha_1 p[Bel_1]$$

is the intersection of the segment $Cl(T, p[Bel_1])$ with the line l_1 passing through $\alpha_1 p[Bel_1] + \alpha_2 p[Bel_2]$ and parallel to $Cl(T, p[Bel_2])$. Finally, $p[\alpha_1 Bel_1 + \alpha_2 Bel_2]$ is the point of the segment

$$Cl(\alpha_1 T + \alpha_2 p[Bel_2], \alpha_2 T + \alpha_1 p[Bel_1])$$

with convex coordinate $\widehat{\alpha_1 D_1}$ (or, equivalently, $\widehat{\alpha_2 D_2}$).

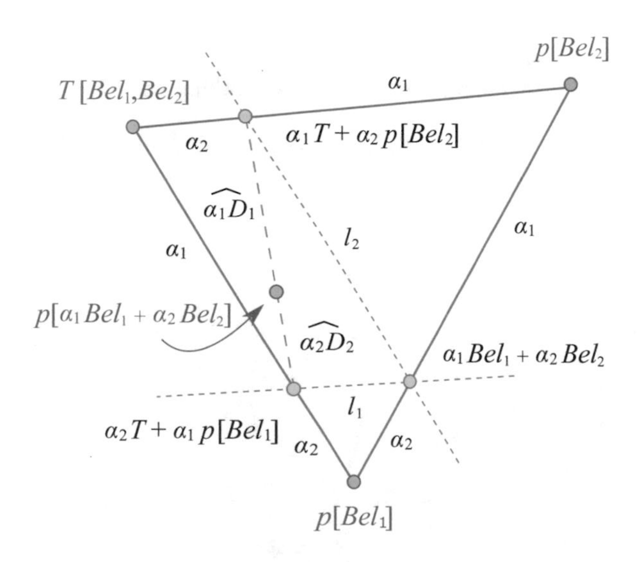

Fig. 11.5: Behaviour of the intersection probability $p[Bel]$ under affine combination. $\alpha_2 T + \alpha_1 p[Bel_1]$ and $\alpha_1 T + \alpha_2 p[Bel_2]$ lie in inverted locations on the segments joining $T[Bel_1, Bel_2]$ and $p[Bel_1]$, $p[Bel_2]$, respectively: $\alpha_i p[Bel_i] + \alpha_j T$ is the intersection of the line $Cl(T, p[Bel_i])$ with the line parallel to $Cl(T, p[Bel_j])$ passing through $\alpha_1 p[Bel_1] + \alpha_2 p[Bel_2]$. The quantity $p[\alpha_1 Bel_1 + \alpha_2 Bel_2]$ is, finally, the point of the segment joining them with convex coordinate $\widehat{\alpha_i D_i}$.

Location of $T[Bel_1, Bel_2]$ in the binary case As an example, let us consider the location of $T[Bel_1, Bel_2]$ in the binary belief space $\mathcal{B}_2$ (Fig. 11.6), where

$$\beta[Bel_1] = \beta[Bel_2] = \frac{m_i(\Theta)}{2m_i(\Theta)} = \frac{1}{2} \quad \forall Bel_1, Bel_2 \in \mathcal{B}_2,$$

and $p[Bel]$ always commutes with the convex closure operator. Accordingly,

$$
\begin{aligned}
T[Bel_1, Bel_2](x) &= \frac{m_1(\Theta)}{m_1(\Theta) + m_2(\Theta)} \left[m_2(x) + \frac{m_2(\Theta)}{2} \right] \\
&\quad + \frac{m_2(\Theta)}{m_1(\Theta) + m_2(\Theta)} \left[m_1(x) + \frac{m_1(\Theta)}{2} \right] \\
&= \frac{m_1(\Theta)}{m_1(\Theta) + m_2(\Theta)} p[Bel_2] + \frac{m_2(\Theta)}{m_1(\Theta) + m_2(\Theta)} p[Bel_1].
\end{aligned}
$$

Looking at Fig. 11.6, simple trigonometric considerations show that the segment $Cl(p[Bel_i], T[Bel_1, Bel_2])$ has length $m_i(\Theta)/(\sqrt{2} \tan \phi)$, where ϕ is the angle between the segments $Cl(Bel_i, T)$ and $Cl(p[Bel_i], T)$. $T[Bel_1, Bel_2]$ is then the unique point of $\mathcal{P}$ such that the angles $\widehat{Bel_1 T p[Bel_1]}$ and $\widehat{Bel_2 T p[Bel_2]}$ coincide, i.e., T is the intersection of $\mathcal{P}$ with the line passing through Bel_i and the reflection of Bel_j through $\mathcal{P}$. As this reflection (in $\mathcal{B}_2$) is nothing but Pl_j,

$$
T[Bel_1, Bel_2] = Cl(Bel_1, Pl_2) \cap \mathcal{P} = Cl(Bel_2, Pl_1) \cap \mathcal{P}.
$$

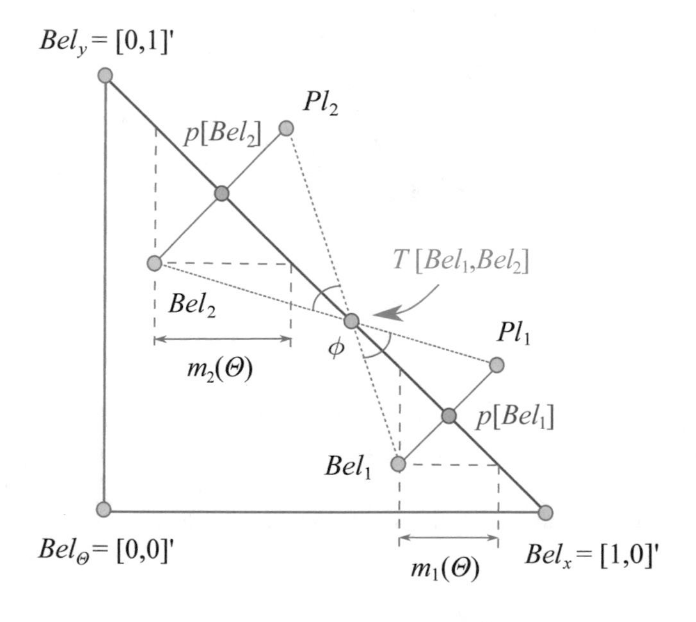

Fig. 11.6: Location of the probability function $T[Bel_1, Bel_2]$ in the binary belief space $\mathcal{B}_2$.

11.3.3 Intersection probability and convex closure

Although the intersection probability does not commute with affine combination (Theorem 43), $p[Bel]$ can still be assimilated into the orthogonal projection and the pignistic function. Theorem 44 states the conditions under which $p[Bel]$ and convex closure (Cl) commute.

Theorem 44. *The intersection probability and convex closure commute iff*

$$T[Bel_1, Bel_2] = \hat{D}_1 p[Bel_2] + \hat{D}_2 p[Bel_1]$$

or, equivalently, either $\beta[Bel_1] = \beta[Bel_2]$ *or* $R[Bel_1] = R[Bel_2]$.

Geometrically, only when the two lines l_1, l_2 in Fig. 11.5 are parallel to the affine space $a(p[Bel_1], p[Bel_2])$ (i.e., $T[Bel_1, Bel_2] \in Cl(p[Bel_1], p[Bel_2])$; compare the above) does the desired quantity $p[\alpha_1 Bel_1 + \alpha_2 Bel_2]$ belong to the line segment $Cl(p[Bel_1], p[Bel_2])$ (i.e., it is also a convex combination of $p[Bel_1]$ and $p[Bel_2]$).

Theorem 44 reflects the two complementary interpretations of $p[Bel]$ we gave in terms of $\beta[Bel]$ and $R[Bel]$ ((11.14) and (11.17)):

$$p[Bel] = m(x) + (1 - k_{Bel}) R[Bel](x), \;\; p[Bel] = m(x) + \beta[Bel](Pl(x) - m(x)).$$

If $\beta[Bel_1] = \beta[Bel_2]$, both belief functions assign to each singleton the same share of their non-Bayesian contribution. If $R[Bel_1] = R[Bel_2]$, the non-Bayesian mass $1 - k_{Bel}$ is distributed in the same way to the elements of Θ.

A sufficient condition for the commutativity of $p[.]$ and $Cl(.)$ can be obtained via the following decomposition of $\beta[Bel]$:

$$\beta[Bel] = \frac{\sum_{|B|>1} m(B)}{\sum_{|B|>1} m(B)|B|} = \frac{\sum_{k=2}^{n} \sum_{|B|=k} m(B)}{\sum_{k=2}^{n} k \cdot \sum_{|B|=k} m(B)} = \frac{\sigma_2 + \cdots + \sigma_n}{2\sigma_2 + \cdots + n\sigma_n}, \tag{11.22}$$

where $\sigma_k \doteq \sum_{|B|=k} m(B)$.

Theorem 45. *If the ratio between the total masses of focal elements of different cardinality is the same for all the belief functions involved, namely*

$$\frac{\sigma_1^l}{\sigma_1^m} = \frac{\sigma_2^l}{\sigma_2^m} \quad \forall l, m \geq 2 \text{ s.t. } \sigma_1^m, \sigma_2^m \neq 0, \tag{11.23}$$

then the intersection probability (considered as an operator mapping belief functions to probabilities) and convex combination commute.

11.4 Orthogonal projection

Although the intersection of the line $a(Bel, Pl)$ with the region $\mathcal{P}'$ of Bayesian NSFs is not always in $\mathcal{P}$, an orthogonal projection $\pi[Bel]$ of Bel onto $a(\mathcal{P})$ is obviously guaranteed to exist, as $a(\mathcal{P})$ is nothing but a linear subspace in the space of normalised sum functions (such as Bel). An explicit calculation of $\pi[Bel]$, however, requires a description of the orthogonal complement of $a(\mathcal{P})$ in $\mathbb{R}^{N-2}$.

11.4.1 Orthogonality condition

We seek a necessary and sufficient condition[71] for an arbitrary vector of $\mathbb{R}^N$, $\mathbf{v} = \sum_{A\subset\Theta} v_A \mathbf{x}_A$ (where $\{\mathbf{x}_A, A \in 2^\Theta\}$ is the usual orthonormal reference frame there), to be orthogonal to the probabilistic subspace $a(\mathcal{P})$.

The scalar product between $\mathbf{v}$ and the generators $Bel_y - Bel_x$ of $a(P), x, y \in \Theta$, is

$$\langle \mathbf{v}, Bel_y - Bel_x \rangle = \left\langle \sum_{A\subset\Theta} v_A \mathbf{x}_A, Bel_y - Bel_x \right\rangle = \sum_{A\subset\Theta} v_A [Bel_y - Bel_x](A),$$

which, recalling (11.4), becomes

$$\langle \mathbf{v}, Bel_y - Bel_x \rangle = \sum_{A\ni y, A\not\ni x} v_A - \sum_{A\ni x, A\not\ni y} v_A.$$

The orthogonal complement $a(\mathcal{P})^\perp$ of $a(\mathcal{P})$ can then be expressed as

$$v(\mathcal{P})^\perp = \left\{ \mathbf{v} \middle| \sum_{A\ni y, A\not\ni x} v_A = \sum_{A\ni x, A\not\ni y} v_A \; \forall y \neq x, \; x, y \in \Theta \right\}. \tag{11.24}$$

Lemma 10. *A belief function Bel defined on a frame Θ belongs to the orthogonal complement (11.24) of $a(\mathcal{P})$ if and only if*

$$\sum_{B\ni y, B\not\ni x} m(B) 2^{1-|B|} = \sum_{B\ni x, B\not\ni y} m(B) 2^{1-|B|} \quad \forall y \neq x, \; x, y \in \Theta. \tag{11.25}$$

By Lemma 10, we can prove the following theorem.

Theorem 46. *The orthogonal projection $\pi[Bel]$ of Bel onto $a(\mathcal{P})$ can be expressed in terms of the basic probability assignment m of Bel in two equivalent forms:*

$$\pi[Bel](x) = \sum_{A\ni x} m(A) 2^{1-|A|} + \sum_{A\subset\Theta} m(A) \left(\frac{1 - |A| 2^{1-|A|}}{n} \right), \tag{11.26}$$

$$\pi[Bel](x) = \sum_{A\ni x} m(A) \left(\frac{1 + |A^c| 2^{1-|A|}}{n} \right) + \sum_{A\not\ni x} m(A) \left(\frac{1 - |A| 2^{1-|A|}}{n} \right). \tag{11.27}$$

Equation (11.27) shows that $\pi[Bel]$ *is indeed a probability*, since $1+|A^c|2^{1-|A|} \geq 0$ and $1 - |A|2^{1-|A|} \geq 0 \; \forall |A| = 1, \ldots, n$. This is not at all trivial, as $\pi[Bel]$ is the projection of Bel onto the *affine* space $a(\mathcal{P})$, and could have in principle assigned negative masses to one or more singletons.

This makes the orthogonal projection a valid probability transform.

[71] The proof is valid for $A = \Theta, \emptyset$ too; see Section 11.5.

11.4.2 Orthogonality flag

Theorem 46 does not provide any clear intuition about the meaning of $\pi[Bel]$ in terms of degrees of belief. In fact, if we process (11.27), we can reduce it to a new Bayesian belief function strictly related to the pignistic function.

Theorem 47. *The orthogonal projection of Bel onto $\mathcal{P}$ can be decomposed as:*

$$\pi[Bel] = \overline{\mathcal{P}}(1 - k_O[Bel]) + k_O[Bel]O[Bel],$$

where $\overline{\mathcal{P}}$ is the uniform probability and

$$O[Bel](x) = \frac{\bar{O}[Bel](x)}{k_O[Bel]} = \frac{\sum_{A \ni x} m(A)2^{1-|A|}}{\sum_{A \subset \Theta} m(A)|A|2^{1-|A|}} = \frac{\displaystyle\sum_{A \ni x} \frac{m(A)}{2^{|A|}}}{\displaystyle\sum_{A \subset \Theta} \frac{m(A)|A|}{2^{|A|}}} \tag{11.28}$$

is a Bayesian belief function.

As $0 \leq |A|2^{1-|A|} \leq 1$ for all $A \subset \Theta$, $k_O[Bel]$ assumes values in the interval $[0, 1]$. By Theorem 47, then the orthogonal projection is always located on the line segment $Cl(\overline{\mathcal{P}}, O[Bel])$ joining the uniform, non-informative probability on Θ to the Bayesian belief function $O[Bel]$.

The interpretation of $O[Bel]$ becomes clear when we notice that the condition (11.25) (under which a belief function Bel is orthogonal to $a(\mathcal{P})$) can be rewritten as

$$\sum_{B \ni y} m(B)2^{1-|B|} = \sum_{B \ni x} m(B)2^{1-|B|},$$

which is in turn equivalent to $O[Bel](x) = \text{const} = \overline{\mathcal{P}}$ for all singletons $x \in \Theta$. Therefore $\pi[Bel] = \overline{\mathcal{P}}$ if and only if $Bel \perp a(\mathcal{P})$, and $O - \overline{\mathcal{P}}$ measures the non-orthogonality of Bel with respect to $\mathcal{P}$.

The Bayesian belief function $O[Bel]$ then deserves the name of *orthogonality flag*.

11.4.3 Two mass redistribution processes

A compelling link can be drawn between the orthogonal projection and the pignistic function via the orthogonality flag $O[Bel]$.

Let us introduce the following two belief functions associated with Bel:

$$Bel_{||} \doteq \frac{1}{k_{||}} \sum_{A \subset \Theta} \frac{m(A)}{|A|} Bel_A, \quad Bel_{2||} \doteq \frac{1}{k_{2||}} \sum_{A \subset \Theta} \frac{m(A)}{2^{|A|}} Bel_A,$$

where $k_{||}$ and $k_{2||}$ are the normalisation factors needed to make them admissible.

Theorem 48. *$O[Bel]$ is the relative plausibility of singletons of $Bel_{2||}$; $BetP[Bel]$ is the relative plausibility of singletons of $Bel_{||}$.*

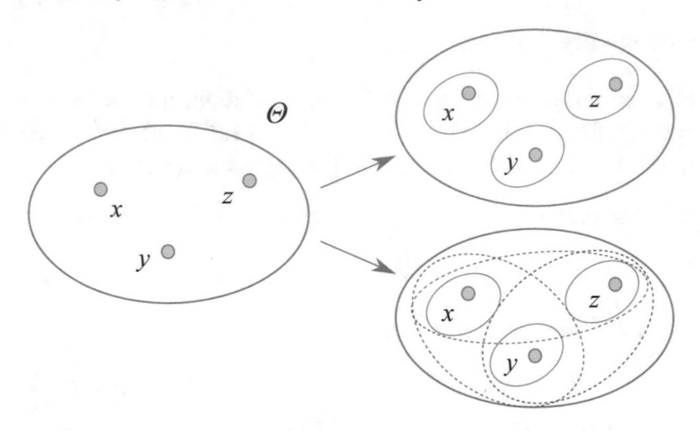

Fig. 11.7: Redistribution processes associated with the pignistic transformation and orthogonal projection. In the pignistic transformation (top), the mass of each focal element A is distributed among its elements. In the orthogonal projection (bottom), instead (through the orthogonality flag), the mass of each focal element A is divided among all its subsets $B \subseteq A$. In both cases, the related relative plausibility of singletons yields a Bayesian belief function.

The two functions $Bel_{||}$ and $Bel_{2||}$ represent two different processes acting on Bel (see Fig. 11.7). The first one redistributes equally the mass of each focal element among its *singletons* (yielding directly the Bayesian belief function $BetP[Bel]$). The second one redistributes equally the mass of each focal element A to its *subsets* $B \subset A$ ($\emptyset$, A included). In this second case we get an unnormalised [1685] belief function Bel^U with mass assignment

$$m^U(A) = \sum_{B \supset A} \frac{m(B)}{2^{|B|}},$$

whose relative belief of singletons $\tilde{Bel}^U$ is the orthogonality flag $O[Bel]$.

Example 29: orthogonality flag and orthogonal projection. *Let us consider again, as an example, the belief function Bel on the ternary frame* $\Theta = \{x, y, z\}$ *considered in Example 28:*

$$m(x) = 0.1,\ m(y) = 0,\ m(z) = 0.2,$$
$$m(\{x,y\}) = 0.3,\ m(\{x,z\}) = 0.1,\ m(\{y,z\}) = 0,\ m(\Theta) = 0.3.$$

To compute the orthogonality flag $O[Bel]$*, we need to apply the redistribution process of Fig. 11.7 (bottom) to each focal element of Bel. In this case their masses are divided among their subsets as follows:*

$$\begin{array}{ll} m(x) = 0.1 & \mapsto m'(x) = m'(\emptyset) = 0.1/2 = 0.05, \\ m(z) = 0.2 & \mapsto m'(z) = m'(\emptyset) = 0.2/2 = 0.1, \\ m(\{x,y\}) = 0.3 & \mapsto m'(\{x,y\}) = m'(x) = m'(y) = m'(\emptyset) = 0.3/4 = 0.075, \\ m(\{x,z\}) = 0.1 & \mapsto m'(\{x,z\}) = m'(x) = m'(z) = m'(\emptyset) = 0.1/4 = 0.025, \\ m(\Theta) = 0.3 & \mapsto m'(\Theta) = m'(\{x,y\}) = m'(\{x,z\}) = m'(\{y,z\}) \\ & \quad = m'(x) = m'(y) = m'(z) = m'(\emptyset) = 0.3/8 = 0.0375. \end{array}$$

By summing all the contributions related to singletons, we get

$$\begin{aligned} m^U(x) &= 0.05 + 0.075 + 0.025 + 0.0375 = 0.1875, \\ m^U(y) &= 0.075 + 0.0375 = 0.1125, \\ m^U(z) &= 0.1 + 0.025 + 0.0375 = 0.1625, \end{aligned}$$

whose sum is the normalisation factor $k_O[Bel] = m^U(x) + m^U(y) + m^U(z) = 0.4625$. *After normalisation, we get* $O[Bel] = [0.405, 0.243, 0.351]'$.

The orthogonal projection $\pi[Bel]$ *is, finally, the convex combination of* $O[Bel]$ *and* $\overline{\mathcal{P}} = [1/3, 1/3, 1/3]'$ *with coefficient* $k_O[Bel]$*:*

$$\begin{aligned} \pi[Bel] &= \overline{\mathcal{P}}(1 - k_O[Bel]) + k_O[Bel]O[Bel] \\ &= [1/3, 1/3, 1/3]' \, (1 - 0.4625) + 0.4625 \, [0.405, 0.243, 0.351]' \\ &= [0.366, 0.291, 0.342]'. \end{aligned}$$

11.4.4 Orthogonal projection and affine combination

As strong additional evidence of their close relationship, the orthogonal projection and the pignistic function both commute with affine combination.

Theorem 49. *The orthogonal projection and affine combination commute. Namely, if* $\alpha_1 + \alpha_2 = 1$*, then*

$$\pi[\alpha_1 \, Bel_1 + \alpha_2 \, Bel_2] = \alpha_1 \pi[Bel_1] + \alpha_2 \pi[Bel_2].$$

This property can be used to find an alternative expression for the orthogonal projection as a convex combination of the pignistic functions associated with all the categorical belief functions.

Lemma 11. *The orthogonal projection of a categorical belief function* Bel_A *is*

$$\pi[Bel_A] = (1 - |A|2^{1-|A|})\overline{\mathcal{P}} + |A|2^{1-|A|}\overline{\mathcal{P}}_A,$$

where $\overline{\mathcal{P}}_A = (1/|A|) \sum_{x \in A} Bel_x$ *is the centre of mass of all the probability measures with support in A.*

Proof. By (11.28), $k_O[Bel_A] = |A|2^{1-|A|}$, so that

$$\bar{O}[Bel_A](x) = \begin{cases} 2^{1-|A|} & x \in A \\ 0 & x \notin A \end{cases} \Rightarrow O[Bel_A](x) = \begin{cases} (1/|A|) & x \in A, \\ 0 & x \notin A, \end{cases}$$

i.e., $O[Bel_A] = (1/|A|) \sum_{x \in A} Bel_x = \overline{\mathcal{P}}_A$.

□

Theorem 50. *The orthogonal projection can be expressed as a convex combination of all non-informative probabilities with support on a single event A as follows:*

$$\pi[Bel] = \overline{\mathcal{P}}\left(1 - \sum_{A \neq \Theta} \alpha_A\right) + \sum_{A \neq \Theta} \alpha_A \overline{\mathcal{P}}_A, \quad \alpha_A \doteq m(A)|A|2^{1-|A|}. \quad (11.29)$$

11.4.5 Orthogonal projection and pignistic function

As $\overline{\mathcal{P}}_A = BetP[Bel_A]$, we can appreciate that

$$\begin{aligned} BetP[Bel] &= \sum_{A \subset \Theta} m(A) BetP[Bel_A], \\ \pi[Bel] &= \sum_{A \neq \Theta} \alpha_A \, BetP[Bel_A] + \left(1 - \sum_{A \neq \Theta} \alpha_A\right) BetP[Bel_\Theta], \end{aligned} \quad (11.30)$$

where $\alpha_A = m(A)k_O[Bel_A]$. Both the orthogonal projection and the pignistic function are thus convex combinations of the collection of categorical pignistic functions. However, as $k_O[Bel_A] = |A|2^{1-|A|} < 1$ for $|A| > 2$, the orthogonal projection turns out to be closer (when compared with the pignistic probability) to the vertices associated with events of lower cardinality (see Fig. 11.8).

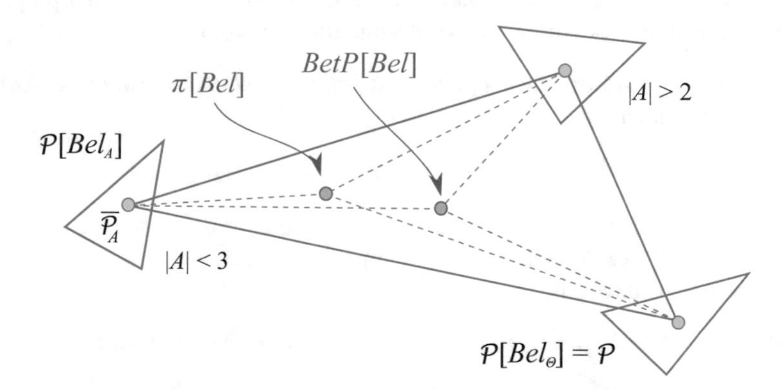

Fig. 11.8: The orthogonal projection $\pi[Bel]$ and the pignistic function $BetP[Bel]$ both lie in the simplex whose vertices are the categorical pignistic functions, i.e., the uniform probabilities with support on a single event A. However, as the convex coordinates of $\pi[Bel]$ are weighted by a factor $k_O[Bel_A] = |A|2^{1-|A|}$, the orthogonal projection is closer to vertices related to lower-size events.

Let us consider as an example the usual ternary frame $\Theta_3 = \{x, y, x\}$, and a belief function defined there with focal elements

$$m(x) = \tfrac{1}{3},\ m(\{x,z\}) = \tfrac{1}{3},\ m(\Theta_3) = \tfrac{1}{3}. \tag{11.31}$$

According to (11.29), we have

$$\begin{aligned}
\pi[Bel] &= \frac{1}{3}\overline{\mathcal{P}}_{\{x\}} + \frac{1}{3}\overline{\mathcal{P}}_{\{x,z\}} + \left(1 - \frac{1}{3} - \frac{1}{3}\right)\overline{\mathcal{P}} \\
&= \frac{1}{3}Bel_x + \frac{1}{3}\frac{Bel_x + Bel_z}{2} + \frac{1}{3}\frac{Bel_x + Bel_y + Bel_z}{3} \\
&= \frac{11}{18}Bel_x + \frac{1}{9}Bel_y + \frac{5}{18}Bel_z,
\end{aligned}$$

and $\pi[Bel]$ is the barycentre of the simplex $Cl(\overline{\mathcal{P}}_{\{x\}}, \overline{\mathcal{P}}_{\{x,z\}}, \overline{\mathcal{P}})$ (see Fig. 11.9). On the other hand,

$$BetP[Bel](x) = \frac{m(x)}{1} + \frac{m(x,z)}{2} + \frac{m(\Theta_3)}{3} = \frac{11}{18},$$

$$BetP[Bel](y) = \frac{1}{9}, \quad BetP[Bel](z) = \frac{1}{6} + \frac{1}{9} = \frac{5}{18},$$

i.e., the pignistic function and orthogonal projection coincide: $BetP[Bel] = \pi[Bel]$.

In fact, this is true for every belief function $Bel \in \mathcal{B}_3$ defined on a ternary frame, since by (11.30), when $|\Theta| = 3$, $\alpha_A = m(A)$ for $|A| \leq 2$, and $1 - \sum_A \alpha_A = 1 - \sum_{A \neq \Theta} m(A) = m(\Theta)$.

11.5 The case of unnormalised belief functions

The above results have been obtained for 'classical' belief functions, where the mass assigned to the empty set is 0: $Bel(\emptyset) = m(\emptyset) = 0$. However, as discussed in Chapter 4, there are situations ('open-world' scenarios) in which it makes sense to work with unnormalised belief functions [1685], namely BFs admitting non-zero support $m(\emptyset) \neq 0$ for the empty set [1722]. The latter is an indicator of the amount of internal conflict carried by a belief function Bel, but can also be interpreted as the chance that the existing frame of discernment does not exhaust all the possible outcomes of the problem.

UBFs are naturally associated with vectors with $N = 2^{|\Theta|}$ coordinates. A coordinate frame of basis UBFs can be defined as follows:

$$\{Bel_A \in \mathbb{R}^N, \emptyset \subseteq A \subseteq \Theta\},$$

this time including a vector $Bel_\emptyset \doteq [1\ 0\ \cdots 0]'$. Note also that in this case $Bel_\Theta = [0 \cdots 0\ 1]'$ is not the null vector.

It is natural to wonder whether the above definitions and properties of $p[Bel]$ and $\pi[Bel]$ retain their validity. Let us consider again the binary case.

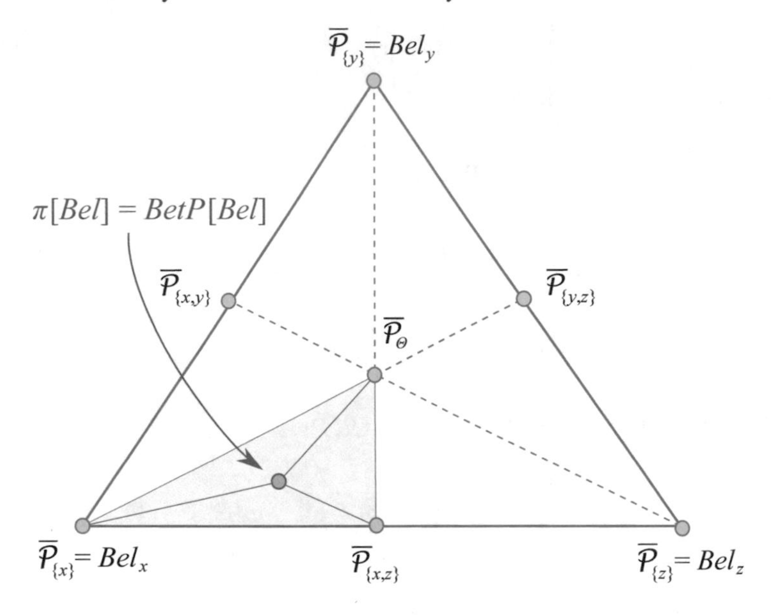

Fig. 11.9: Orthogonal projection and pignistic function for the belief function (11.31) on the ternary frame $\Theta_3 = \{x, y, z\}$.

Example 30: unnormalised belief functions, binary case. *We now have to use four coordinates, associated with all the subsets of* Θ*:* $\emptyset$, $\{x\}$, $\{y\}$ *and* Θ *itself. Remember that, for unnormalised belief functions,*

$$Bel(A) = \sum_{\emptyset \subsetneq B \subseteq A} m(B),$$

i.e., the contribution of the empty set is not considered when computing the belief value of an event $A \subset \Theta$*. One can, however, define the associated* implicability function *as*

$$b(A) = \sum_{\emptyset \subseteq B \subseteq A} m(B) = Bel(A) + m(\emptyset), \tag{11.32}$$

this time including $B = \emptyset$ *in the sum. The four-dimensional vectors corresponding to the categorical belief and plausibility functions are therefore*

$$\begin{array}{ll} Bel_\emptyset = [0,0,0,0]', & Pl_\emptyset = [0,0,0,0]', \\ Bel_x = [0,1,0,1]', & Pl_x = [0,1,0,1]' = Bel_x, \\ Bel_y = [0,0,1,1]', & Pl_y = [0,0,1,1]' = Bel_y, \\ Bel_\Theta = [0,0,0,1]', & Pl_\Theta = [0,1,1,1]'. \end{array}$$

A striking difference from the 'classical' case is that $Bel(\Theta) = 1 - m(\emptyset) = Pl(\Theta)$, which implies that both the belief and the plausibility spaces are *not* in general subsets of the section $\{\mathbf{v} \in \mathbb{R}^N : v_\Theta = 1\}$ of $\mathbb{R}^N$. In other words, unnormalised belief functions and unnormalised plausibility functions are *not* normalised sum functions.

As a consequence, the line $a(Bel, Pl)$ is not guaranteed to intersect the affine space $\mathcal{P}'$ of the Bayesian NSFs. For instance, the line connecting $Bel_\emptyset$ and $Pl_\emptyset$ in the binary case reduces to $Bel_\emptyset = Pl_\emptyset = \mathbf{0}$, which clearly does not belong to $\mathcal{P}' = \{[a, b, (1-b), -a]', a, b \in \mathbb{R}\}$. Simple calculations show that in fact $a(Bel, Pl) \cap \mathcal{P}' \neq \emptyset$ iff $m(\emptyset) = 0$ (i.e., Bel is 'classical') or (trivially) $Bel \in \mathcal{P}$. This is true in the general case.

Proposition 57. *The intersection probability is well defined for classical belief functions only.*

It is interesting to note, however, that the orthogonality results of Section 11.2.1 are still valid, since Lemma 9 does not involve the empty set, while the proof of Theorem 39 is valid for the components $A = \emptyset, \Theta$ too (as $Bel_y - Bel_x(A) = 0$ for $A = \emptyset, \Theta$).

Therefore we have the following proposition.

Proposition 58. *The dual line $a(Bel, Pl)$ is orthogonal to $\mathcal{P}$ for each unnormalised belief function Bel, although $\varsigma[Bel] = a(Bel, Pl) \cap \mathcal{P}'$ exists if and only if Bel is a classical belief function.*

Analogously, the orthogonality condition (11.25) is not affected by the mass of the empty set. The orthogonal projection $\pi[Bel]$ of a UBF Bel is then well defined (check the proof of Theorem 46), and it is still given by (11.26) and (11.27), with the caveat that the summations on the right-hand side include $\emptyset$ as well:

$$\pi[Bel](x) = \sum_{A \ni x} m(A) 2^{1-|A|} + \sum_{\emptyset \subseteq A \subset \Theta} m(A) \left(\frac{1 - |A| 2^{1-|A|}}{n} \right),$$

$$\pi[Bel](x) = \sum_{A \ni x} m(A) \left(\frac{1 + |A^c| 2^{1-|A|}}{n} \right) + \sum_{\emptyset \subseteq A \not\ni x} m(A) \left(\frac{1 - |A| 2^{1-|A|}}{n} \right).$$

11.6 Comparisons within the affine family

By virtue of its neat relation to affine combination, the intersection probability can be considered a member of a family of Bayesian transforms which also includes the pignistic function and the orthogonal projection: we call it the *affine family* of probability transforms.

To complete our analysis, we seek to identify conditions under which the intersection probability, the orthogonal projection and the pignistic function coincide. As a matter of fact, preliminary sufficient conditions have been already devised [333].

Proposition 59. *The intersection probability and orthogonal projection coincide if Bel is 2-additive, i.e., whenever $m(A) = 0$ for all $A : |A| > 2$.*

A similar sufficient condition for the pair $p[Bel]$, $BetP[Bel]$ can be found by using the following decomposition of $\beta[Bel]$:

$$\beta[Bel] = \frac{\sum_{|B|>1} m(B)}{\sum_{|B|>1} m(B)|B|} = \frac{\sum_{k=2}^{n} \sum_{|B|=k} m(B)}{\sum_{k=2}^{n} (k \sum_{|B|=k} m(B))} = \frac{\sigma^2 + \cdots + \sigma^n}{2\,\sigma^2 + \cdots + n\,\sigma^n}, \tag{11.33}$$

where, as usual, $\sigma^k \doteq \sum_{|B|=k} m(B)$.

Proposition 60. *The intersection probability and pignistic function coincide if* $\exists$ $k \in [2, \ldots, n]$ *such that* $\sigma^i = 0$ *for all* $i \neq k$, *i.e., the focal elements of* Bel *have size 1 or* k *only.*

This is the case for binary frames, in which all belief functions satisfy the conditions of both Proposition 59 and Proposition 60. As a result, $p[Bel] = BetP[Bel] = \pi[Bel]$ for all the BFs defined on $\Theta = \{x, y\}$ (see Fig. 9.1 again).

More stringent conditions can, however, be formulated in terms of equal distribution of masses among focal elements.

Theorem 51. *If a belief function* Bel *is such that its mass is equally distributed among focal elements of the same size, namely for all* $k = 2, \ldots, n$,

$$m(A) = const \quad \forall A : |A| = k, \tag{11.34}$$

then its pignistic and intersection probabilities coincide: $BetP[Bel] = p[Bel]$.

The dondition (11.34) is sufficient to guarantee the equality of the intersection probability and orthogonal projection too.

Theorem 52. *If a belief function* Bel *satisfies the condition (11.34) (i.e., its mass is equally distributed among focal elements of the same size), then the related orthogonal projection and intersection probability coincide.*

Example 31: ternary frame. *In the special case of a ternary frame* $\pi[Bel] = BetP[Bel]$ *[333], so that checking whether* $p[Bel] = \pi[Bel]$ *is equivalent to checking whether* $p[Bel] = BetP[Bel]$, *one can prove the following proposition [333].*

Proposition 61. *For belief functions* Bel *defined on a ternary frame, the* L_p *distance* $\|p[Bel] - BetP[Bel]\|_p$ *between the intersection probability and the pignistic function in the probability simplex has three maxima, corresponding to the three belief functions with basic probability assignments*

$$\begin{aligned} m_1 &= [0, 0, 0, 3 - \sqrt{6}, 0, 0, \sqrt{6} - 2]', \\ m_2 &= [0, 0, 0, 0, 3 - \sqrt{6}, 0, \sqrt{6} - 2]', \\ m_3 &= [0, 0, 0, 0, 0, 3 - \sqrt{6}, \sqrt{6} - 2]', \end{aligned}$$

regardless of the norm ($p = 1, 2, \infty$) *chosen.*

Proposition 61 opens the way to a more complete quantitative analysis of the differences between the intersection probability and the other Bayesian transforms in the same family.

Appendix: Proofs

Proof of Theorem 39

Having denoted the A-th axis of the orthonormal reference frame $\{\mathbf{x}_A : \emptyset \subsetneq A \subsetneq \Theta\}$ in $\mathbb{R}^{N-2}$ by $\mathbf{x}_A$ as usual, (see Chapter 7, Section 7.1), we can write the difference $Bel - Pl$ as

$$Pl - Bel = \sum_{\emptyset \subsetneq A \subsetneq \Theta} [Pl(A) - Bel(A)]\mathbf{x}_A,$$

where

$$\begin{aligned}[Pl - Bel](A^c) &= Pl(A^c) - Bel(A^c) = 1 - Bel(A) - Bel(A^c) \\ &= 1 - Bel(A^c) - Bel(A) = Pl(A) - Bel(A) \\ &= [Pl - Bel](A).\end{aligned} \tag{11.35}$$

The scalar product $\langle \cdot, \cdot \rangle$ between the vector $Pl - Bel$ and any arbitrary basis vector $Bel_y - Bel_x$ of $a(\mathcal{P})$ is therefore

$$\langle Pl - Bel, Bel_y - Bel_x \rangle = \sum_{\emptyset \subsetneq A \subsetneq \Theta} [Pl - Bel](A) \cdot [Bel_y - Bel_x](A),$$

which, by (11.35), becomes

$$\sum_{|A| \leq \lfloor |\Theta|/2 \rfloor, A \neq \emptyset} [Pl - Bel](A) \Big\{ [Bel_y - Bel_x](A) + [Bel_y - Bel_x](A^c) \Big\}.$$

By Lemma, 9 all the addends in the above expression are zero.

Proof of Theorem 40

The numerator of (11.6) is, trivially, $\sum_{|B|>1} m(B)$. On the other hand,

$$1 - Bel(\{x\}^c) - Bel(x) = \sum_{B \subset \Theta} m(B) - \sum_{B \subset \{x\}^c} m(B) - m(x) = \sum_{B \supsetneq \{x\}} m(B),$$

so that the denominator of $\beta[Bel]$ becomes

$$\begin{aligned}\sum_{y \in \Theta} [Pl(y) - Bel(y)] &= \sum_{y \in \Theta} (1 - Bel(\{y\}^c) - Bel(y)) \\ &= \sum_{y \in \Theta} \sum_{B \supsetneq \{y\}} m(B) = \sum_{|B|>1} m(B)|B|,\end{aligned}$$

yielding (11.10). Equation (11.9) comes directly from (11.5) when we recall that $Bel(x) = m(x)$, $\varsigma(x) = m_\varsigma(x)$ $\forall x \in \Theta$.

Proof of Theorem 41

By definition (11.7), $\varsigma[Bel]$ reads as follows in terms of the reference frame $\{Bel_A, A \subset \Theta\}$:

$$\begin{aligned}\varsigma[Bel] &= \sum_{A\subset\Theta} m(A)Bel_A + \beta[Bel]\left(\sum_{A\subset\Theta}\mu(A)Bel_A - \sum_{A\subset\Theta} m(A)Bel_A\right)\\ &= \sum_{A\subset\Theta} Bel_A\Big[m(A) + \beta[Bel](\mu(A) - m(A))\Big],\end{aligned}$$

since $\mu(.)$ is the Möbius inverse of $Pl(.)$. For $\varsigma[Bel]$ to be a Bayesian *belief* function, accordingly, all the components related to non-singleton subsets need to be zero:

$$m(A) + \beta[Bel](\mu(A) - m(A)) = 0 \quad \forall A : |A| > 1.$$

This condition reduces to (after recalling the expression (11.10) for $\beta[Bel]$)

$$\mu(A)\sum_{|B|>1} m(B) + m(A)\sum_{|B|>1} m(B)(|B|-1) = 0 \quad \forall A : |A| > 1. \tag{11.36}$$

But, as we can write

$$\sum_{|B|>1} m(B)(|B|-1) = \sum_{|B|>1} m(B) + \sum_{|B|>2} m(B)(|B|-2),$$

(11.36) reads as

$$[\mu(A) + m(A)]\sum_{|B|>1} m(B) + m(A)\sum_{|B|>2} m(B)(|B|-2) = 0$$

or, equivalently,

$$[m(A) + \mu(A)]M_1[Bel] + m(A)M_2[Bel] = 0 \quad \forall A : |A| > 1, \tag{11.37}$$

after defining

$$M_1[Bel] \doteq \sum_{|B|>1} m(B), \quad M_2[Bel] \doteq \sum_{|B|>2} m(B)(|B|-2).$$

Clearly,

$$\begin{aligned}M_1[Bel] = 0 &\Leftrightarrow m(B) = 0\ \forall B : |B| > 1 \Leftrightarrow Bel \in \mathcal{P},\\ M_2[Bel] = 0 &\Leftrightarrow m(B) = 0\ \forall B : |B| > 2,\end{aligned}$$

as all the terms inside the summations are non-negative by the definition of the basic probability assignment.

We can distinguish three cases: (1) $M_1 = 0 = M_2$ ($Bel \in \mathcal{P}$), (2) $M_1 \neq 0$ but $M_2 = 0$ and (3) $M_1 \neq 0 \neq M_2$. If (1) holds, then Bel is, trivially, a probability measure. If (3) holds, (11.37) implies $m(A) = \mu(A) = 0$, $|A| > 1$, i.e., $Bel \in \mathcal{P}$, which is a contradiction.

The only non-trivial case is then (2), $M_2 = 0$. There, the condition (11.37) becomes

$$M_1[Bel]\,[m(A) + \mu(A)] = 0, \quad \forall A : |A| > 1.$$

For all $|A| > 2$, we have that $m(A) = \mu(A) = 0$ (since $M_2 = 0$) and the constraint is satisfied. If $|A| = 2$, instead,

$$\mu(A) = (-1)^{|A|+1} \sum_{B \supset A} m(B) = (-1)^{2+1} m(A) = -m(A)$$

(since $m(B) = 0\ \forall B \supset A$, $|B| > 2$), so that $\mu(A) + m(A) = 0$ and the constraint is again satisfied.

Finally, as the coordinate $\beta[Bel]$ of $\varsigma[Bel]$ on the line $a(Bel, Pl)$ can be rewritten as a function of $M_1[Bel]$ and $M_2[Bel]$ as follows,

$$\beta[Bel] = \frac{M_1[Bel]}{M_2[Bel] + 2M_1[Bel]}, \tag{11.38}$$

if $M_2 = 0$, then $\beta[Bel] = 1/2$ and $\varsigma[Bel] = (b + Pl)/2$.

Proof of Theorem 43

By definition, the quantity $p[\alpha_1 Bel_1 + \alpha_2 Bel_2](x)$ can be written as

$$m_{\alpha_1 Bel_1 + \alpha_2 Bel_2}(x) + \beta[\alpha_1 Bel_1 + \alpha_2 Bel_2] \sum_{A \supsetneq \{x\}} m_{\alpha_1 Bel_1 + \alpha_2 Bel_2}(A), \tag{11.39}$$

where

$$\begin{aligned} & \beta[\alpha_1 Bel_1 + \alpha_2 Bel_2] \\ & = \frac{\sum_{|A|>1} m_{\alpha_1 Bel_1 + \alpha_2 Bel_2}(A)}{\sum_{|A|>1} m_{\alpha_1 Bel_1 + \alpha_2 Bel_2}(A)|A|} = \frac{\alpha_1 \sum_{|A|>1} m_1(A) + \alpha_2 \sum_{|A|>1} m_2(A)}{\alpha_1 \sum_{|A|>1} m_1(A)|A| + \alpha_2 \sum_{|A|>1} m_2(A)|A|} \\ & = \frac{\alpha_1 N_1 + \alpha_2 N_2}{\alpha_1 D_1 + \alpha_2 D_2} = \frac{\alpha_1 D_1 \beta[Bel_1] + \alpha_2 D_2 \beta[Bel_2]}{\alpha_1 D_1 + \alpha_2 D_2} \\ & = \widehat{\alpha_1 D_1} \beta[Bel_1] + \widehat{\alpha_2 D_2} \beta[Bel_2], \end{aligned}$$

once we introduce the notation $\beta[Bel_i] = N_i / D_i$.

Substituting this decomposition for $\beta[\alpha_1 Bel_1 + \alpha_2 Bel_2]$ into (11.39) yields

$$
\begin{aligned}
&\alpha_1 m_1(x) + \alpha_2 m_2(x) \\
&+\left(\widehat{\alpha_1 D_1}\beta[Bel_1] + \widehat{\alpha_2 D_2}\beta[Bel_2]\right)\left(\alpha_1 \sum_{A \supsetneq \{x\}} m_1(A) + \alpha_2 \sum_{A \supsetneq \{x\}} m_2(A)\right) \\
= &\frac{(\alpha_1 D_1 + \alpha_2 D_2)(\alpha_1 m_1(x) + \alpha_2 m_2(x))}{\alpha_1 D_1 + \alpha_2 D_2} \\
&+\frac{\alpha_1 D_1 \beta[Bel_1] + \alpha_2 D_2 \beta[Bel_2]}{\alpha_1 D_1 + \alpha_2 D_2}\left(\alpha_1 \sum_{A \supsetneq \{x\}} m_1(A) + \alpha_2 \sum_{A \supsetneq \{x\}} m_2(A)\right),
\end{aligned} \tag{11.40}
$$

which can be reduced further to

$$
\begin{aligned}
&\frac{\alpha_1 D_1}{\alpha_1 D_1 + \alpha_2 D_2}\Big(\alpha_1 m_1(x) + \alpha_2 m_2(x)\Big) \\
&+\frac{\alpha_1 D_1}{\alpha_1 D_1 + \alpha_2 D_2}\beta[Bel_1]\left(\alpha_1 \sum_{A \supsetneq \{x\}} m_1(A) + \alpha_2 \sum_{A \supsetneq \{x\}} m_2(A)\right) \\
&+\frac{\alpha_2 D_2}{\alpha_1 D_1 + \alpha_2 D_2}\Big(\alpha_1 m_1(x) + \alpha_2 m_2(x)\Big) \\
&+\frac{\alpha_2 D_2}{\alpha_1 D_1 + \alpha_2 D_2}\beta[Bel_2]\left(\alpha_1 \sum_{A \supsetneq \{x\}} m_1(A) + \alpha_2 \sum_{A \supsetneq \{x\}} m_2(A)\right) \\
= &\frac{\alpha_1 D_1}{\alpha_1 D_1 + \alpha_2 D_2}\alpha_1\left(m_1(x) + \beta[Bel_1]\sum_{A \supsetneq \{x\}} m_1(A)\right) \\
&+\frac{\alpha_1 D_1}{\alpha_1 D_1 + \alpha_2 D_2}\alpha_2\left(m_2(x) + \beta[Bel_1]\sum_{A \supsetneq \{x\}} m_2(A)\right) \\
&+\frac{\alpha_2 D_2}{\alpha_1 D_1 + \alpha_2 D_2}\alpha_1\left(m_1(x) + \beta[Bel_2]\sum_{A \supsetneq \{x\}} m_1(A)\right) \\
&+\frac{\alpha_2 D_2}{\alpha_1 D_1 + \alpha_2 D_2}\alpha_2\left(m_2(x) + \beta[Bel_2]\sum_{A \supsetneq \{x\}} m_2(A)\right) \\
= &\frac{\alpha_1^2 D_1}{\alpha_1 D_1 + \alpha_2 D_2}p[Bel_1] + \frac{\alpha_2^2 D_2}{\alpha_1 D_1 + \alpha_2 D_2}p[Bel_2] \\
&+\frac{\alpha_1 \alpha_2}{\alpha_1 D_1 + \alpha_2 D_2}\Big(D_1 p[Bel_2, Bel_1](x) + D_2 p[Bel_1, Bel_2](x)\Big),
\end{aligned} \tag{11.41}
$$

after recalling (11.21).

We can notice further that the function

$$F(x) \doteq D_1 p[Bel_2, Bel_1](x) + D_2 p[Bel_1, Bel_2](x)$$

is such that

$$\begin{aligned}
&\sum_{x\in\Theta} F(x) \\
&= \sum_{x\in\Theta} \big[D_1 m_2(x) + N_1(Pl_2 - m_2(x)) + D_2 m_1(x) + N_2(Pl_1 - m_1(x))\big] \\
&= D_1(1 - N_2) + N_1 D_2 + D_2(1 - N_1) + N_2 D_1 = D_1 + D_2
\end{aligned}$$

(making use of (11.10)). Thus, $T[Bel_1, Bel_2](x) = F(x)/(D_1 + D_2)$ is a probability (as $T[Bel_1, Bel_2](x)$ is always non-negative), expressed by (11.20). By (11.40), the quantity $p[\alpha_1 Bel_1 + \alpha_2 Bel_2](x)$ can be expressed as

$$\frac{\alpha_1^2 D_1 p[Bel_1](x) + \alpha_2^2 D_2 p[Bel_2](x) + \alpha_1\alpha_2(D_1 + D_2)T[Bel_1, Bel_2](x)}{\alpha_1 D_1 + \alpha_2 D_2},$$

i.e., (11.19).

Proof of Theorem 44

By (11.19), we have that

$$\begin{aligned}
&p[\alpha_1 Bel_1 + \alpha_2 Bel_2] - \alpha_1 p[Bel_1] - \alpha_2 p[Bel_2] \\
&= \widehat{\alpha_1 D_1}\alpha_1 p[Bel_1] + \widehat{\alpha_1 D_1}\alpha_2 T + \widehat{\alpha_2 D_2}\alpha_1 T + \widehat{\alpha_2 D_2}\alpha_2 p[Bel_2] \\
&\quad -\alpha_1 p[Bel_1] - \alpha_2 p[Bel_2] \\
&= \alpha_1 p[Bel_1](\widehat{\alpha_1 D_1} - 1) + \widehat{\alpha_1 D_1}\alpha_2 T + \widehat{\alpha_2 D_2}\alpha_1 T + \alpha_2 p[Bel_2](\widehat{\alpha_2 D_2} - 1) \\
&= -\alpha_1 p[Bel_1]\widehat{\alpha_2 D_2} + \widehat{\alpha_1 D_1}\alpha_2 T + \widehat{\alpha_2 D_2}\alpha_1 T - \alpha_2 p[Bel_2]\widehat{\alpha_1 D_1} \\
&= \widehat{\alpha_1 D_1}\Big(\alpha_2 T - \alpha_2 p[Bel_2]\Big) + \widehat{\alpha_2 D_2}\Big(\alpha_1 T - \alpha_1 p[Bel_1]\Big) \\
&= \frac{\alpha_1\alpha_2}{\alpha_1 D_1 + \alpha_2 D_2}\Big[D_1(T - p[Bel_2]) + D_2(T - p[Bel_1])\Big].
\end{aligned}$$

This is zero iff

$$T[Bel_1, Bel_2](D_1 + D_2) = p[Bel_1]D_2 + p[Bel_2]D_1,$$

which is equivalent to

$$T[Bel_1, Bel_2] = \hat{D}_1 p[Bel_2] + \hat{D}_2 p[Bel_1],$$

as $(\alpha_1\alpha_2)/(\alpha_1 D_1 + \alpha_2 D_2)$ is always non-zero in non-trivial cases. This is equivalent to (after replacing the expressions for $p[Bel]$ (11.17) and $T[Bel_1, Bel_2]$ (11.20))

$$D_1(Pl_2 - m_2(x))(\beta[Bel_2] - \beta[Bel_1]) + D_2(Pl_1 - m_1(x))(\beta[Bel_1] - \beta[Bel_2]) = 0,$$

which is in turn equivalent to

$$(\beta[Bel_2] - \beta[Bel_1])\Big[D_1(Pl_2(x) - m_2(x)) - D_2(Pl_1(x) - m_1(x))\Big] = 0.$$

Obviously, this is true iff $\beta[Bel_1] = \beta[Bel_2]$ *or* the second factor is zero, i.e.,

$$D_1 D_2 \frac{Pl_2(x) - m_2(x)}{D_2} - D_1 D_2 \frac{Pl_1(x) - m_1(x)}{D_1}$$

$$= D_1 D_2 (R[Bel_2](x) - R[Bel_1](x)) = 0$$

for all $x \in \Theta$, i.e., $R[Bel_1] = R[Bel_2]$.

Proof of Theorem 45

By (11.22), the equality $\beta[Bel_1] = \beta[Bel_2]$ is equivalent to

$$(2\sigma_2^2 + \cdots + n\sigma_2^n)(\sigma_1^2 + \cdots + \sigma_1^n) = (2\sigma_1^2 + \cdots + n\sigma_1^n)(\sigma_2^2 + \cdots + \sigma_2^n).$$

Let us assume that there exists a cardinality k such that $\sigma_1^k \neq 0 \neq \sigma_2^k$. We can then divide the two sides by σ_1^k and σ_2^k, obtaining

$$\left(2\frac{\sigma_2^2}{\sigma_2^k} + \cdots + k + \cdots + n\frac{\sigma_2^n}{\sigma_2^k}\right)\left(\frac{\sigma_1^2}{\sigma_1^k} + \cdots + 1 + \cdots + \frac{\sigma_1^n}{\sigma_1^k}\right)$$

$$= \left(2\frac{\sigma_1^2}{\sigma_1^k} + \cdots + k + \cdots + n\frac{\sigma_1^n}{\sigma_1^{k_1}}\right)\left(\frac{\sigma_2^2}{\sigma_2^k} + \cdots + 1 + \cdots + \frac{\sigma_2^n}{\sigma_2^k}\right).$$

Therefore, if $\sigma_1^j/\sigma_1^k = \sigma_2^j/\sigma_2^k \; \forall j \neq k$, the condition $\beta[Bel_1] = \beta[Bel_2]$ is satisfied. But this is equivalent to (11.23).

Proof of Lemma 10

When the vector $\mathbf{v}$ in (11.24) is a belief function ($v_A = Bel(A)$), we have that

$$\sum_{A \ni y, A \not\ni x} Bel(A) = \sum_{A \ni y, A \not\ni x} \sum_{B \subset A} m(B) = \sum_{B \subset \{x\}^c} m(B) 2^{n-1-|B \cup \{y\}|},$$

since $2^{n-1-|B\cup\{y\}|}$ is the number of subsets A of $\{x\}^c$ containing both B and $\{y\}$. The orthogonality condition then becomes

$$\sum_{B\subset\{x\}^c} m(B)2^{n-1-|B\cup\{y\}|} = \sum_{B\subset\{y\}^c} m(B)2^{n-1-|B\cup\{x\}|} \quad \forall y \neq x.$$

Now, sets B such that $B \subset \{x,y\}^c$ appear in both summations, with the same coefficient (since $|B\cup\{x\}| = |B\cup\{y\}| = |B|+1$).

After simplifying the common factor 2^{n-2}, we get (11.25).

Proof of Theorem 46

Finding the orthogonal projection $\pi[Bel]$ of Bel onto $a(\mathcal{P})$ is equivalent to imposing the condition $\langle \pi[Bel] - Bel, Bel_y - Bel_x\rangle = 0 \ \forall y \neq x$. Replacing the masses of $\pi - Bel$, namely

$$\begin{cases} \pi(x) - m(x), \ x \in \Theta, \\ -m(A), \qquad |A| > 1 \end{cases}$$

in (11.25) yields, after extracting the singletons x from the summation, the following system of equations:

$$\begin{cases} \pi(y) = \pi(x) + \displaystyle\sum_{A\ni y, A\not\ni x, |A|>1} m(A)2^{1-|A|} + m(y) \\ \qquad -m(x) - \displaystyle\sum_{A\ni x, A\not\ni y, |A|>1} m(A)2^{1-|A|} \qquad \forall y \neq x, \\ \displaystyle\sum_{z\in\Theta} \pi(z) = 1. \end{cases} \tag{11.42}$$

By replacing the first $n-1$ equations of (11.42) in the normalisation constraint (bottom equation), we get

$$\pi(x) + \sum_{y\neq x} \left[\pi(x) + m(y) - m(x) + \sum_{\substack{A\ni y, A\not\ni x \\ |A|>1}} m(A)2^{1-|A|} - \sum_{\substack{A\ni x, A\not\ni y \\ |A|>1}} m(A)2^{1-|A|} \right] = 1,$$

which is equivalent to

$$n\pi(x) = 1 + (n-1)m(x) - \sum_{y\neq x} m(y) + \sum_{y\neq x} \sum_{A\ni x, A\not\ni y, |A|>1} m(A)2^{1-|A|} - \sum_{y\neq x} \sum_{A\ni y, A\not\ni x, |A|>1} m(A)2^{1-|A|}.$$

Considering the last addend, we have that

$$\sum_{y\neq x}\ \sum_{A\ni y, A\not\ni x, |A|>1} m(A)2^{1-|A|} = \sum_{A\not\ni x, |A|>1} m(A)2^{1-|A|}|A|,$$

as all the events A not containing x do contain some $y \neq x$, and they are counted $|A|$ times (i.e., once for each element they contain). For the second last addend, instead,

$$\sum_{y\neq x}\ \sum_{A\ni x, A\not\ni y, |A|>1} m(A)2^{1-|A|} = \sum_{y\neq x}\ \sum_{A\supsetneq\{x\}, A\not\ni y} m(A)2^{1-|A|}$$

$$= \sum_{A\supset\{x\}, 1<|A|<n} m(A)2^{1-|A|}(n-|A|) = \sum_{A\supsetneq\{x\}} m(A)2^{1-|A|}(n-|A|),$$

for $n - |A| = 0$ when $A = \Theta$. Hence,

$$\pi(x) = \frac{1}{n}\Bigg[n\cdot m(x) + 1 - \sum_{y\in\Theta} m(y) + n\sum_{A\supsetneq\{x\}} m(A)2^{1-|A|}$$
$$- \sum_{A\supsetneq\{x\}} m(A)2^{1-|A|}|A| - \sum_{A\not\supset\{x\}, |A|>1} m(A)2^{1-|A|}|A|\Bigg].$$

We then just need to note that

$$-\sum_{y\in\Theta} m(y) = -\sum_{|A|=1} m(A)|A|2^{1-|A|},$$

so that the orthogonal projection can be finally expressed as

$$\pi(x) = \frac{1}{n}\left[n\cdot m(x) + n\sum_{A\supsetneq\{x\}} m(A)2^{1-|A|} + 1 - \sum_{A\subset\Theta} m(A)|A|2^{1-|A|}\right]$$
$$= m(x) + \sum_{A\supsetneq\{x\}} m(A)2^{1-|A|} + \sum_{A\subset\Theta} m(A)\left(\frac{1-|A|2^{1-|A|}}{n}\right),$$

namely (11.26). Since

$$2^{1-|A|} + \frac{1}{n} - \frac{|A|}{n}2^{1-|A|} = \frac{1+2^{1-|A|}(n-|A|)}{n} = \frac{1+2^{1-|A|}|A^c|}{n},$$

the second form (11.27) follows.

Proof of Theorem 47

By (11.27), we can write

$$\pi[Bel](x) = \bar{O}[Bel](x) + \frac{1}{n}\left(\sum_{A\subset\Theta} m(A) - \sum_{A\subset\Theta} m(A)|A|2^{1-|A|}\right)$$
$$= \bar{O}[Bel](x) + \frac{1}{n}\Big(1 - k_O[Bel]\Big).$$

But, since

$$\sum_{x\in\Theta} \bar{O}[Bel](x) = \sum_{x\in\Theta}\sum_{A\ni x} m(A)2^{1-|A|} = \sum_{A\subset\Theta} m(A)|A|2^{1-|A|} = k_O[Bel], \tag{11.43}$$

i.e., $k_O[Bel]$ is the normalisation factor for $\bar{O}[Bel]$, the function (11.28) is a Bayesian belief function, and we can write, as desired (since $\overline{\mathcal{P}}(x) = 1/n$),

$$\pi[Bel] = (1 - k_O[Bel])\overline{\mathcal{P}} + k_O[Bel]O[Bel].$$

Proof of Theorem 48

By the definition of a plausibility function and (11.43), it follows that

$$Pl_{2||}(x) = \sum_{A\ni x} m_{2||}(A) = \frac{1}{k_{2||}}\sum_{A\ni x}\frac{m(A)}{2^{|A|}} = \frac{\bar{O}[Bel]}{2k_{2||}},$$

$$\sum_{x\in\Theta} Pl_{2||}(x) = \frac{1}{k_{2||}}\sum_{x\in\Theta}\sum_{A\ni x}\frac{m(A)}{2^{|A|}} = \frac{k_O[Bel]}{2k_{2||}}.$$

Hence $\tilde{Pl}_{2||}(x) = \bar{O}[Bel]/k_O[Bel] = O[Bel]$. Similarly,

$$Pl_{||}(x) = \sum_{A\ni x} m_{||}(A) = \frac{1}{k_{||}}\sum_{A\ni x}\frac{m(A)}{|A|} = \frac{1}{k_{||}}BetP[Bel](x)$$

and, since $\sum_x BetP[Bel](x) = 1$, $\tilde{Pl}_{||}(x) = BetP[Bel](x)$.

Proof of Theorem 49

By Theorem 47, $\pi[Bel] = (1 - k_O[Bel])\overline{\mathcal{P}} + \bar{O}[Bel]$, where

$$k_O[Bel] = \sum_{A\subset\Theta} m(A)|A|2^{1-|A|}$$

and $\bar{O}[Bel](x) = \sum_{A\ni x} m(A)2^{1-|A|}$. Hence,

$$\begin{aligned} k_O[\alpha_1 Bel_1 + \alpha_2 Bel_2] &= \sum_{A\subset\Theta} (\alpha_1 m_1(A) + \alpha_2 m_2(A))|A|2^{1-|A|} \\ &= \alpha_1 k_O[Bel_1] + \alpha_2 k_O[Bel_2], \end{aligned}$$

$$\begin{aligned} \bar{O}[\alpha_1 Bel_1 + \alpha_2 Bel_2](x) &= \sum_{A\ni x} (\alpha_1 m_1(A) + \alpha_2 m_2(A))2^{1-|A|} \\ &= \alpha_1\bar{O}[Bel_1] + \alpha_2\bar{O}[Bel_2], \end{aligned}$$

which in turn implies (since $\alpha_1 + \alpha_2 = 1$)

$$\begin{aligned}
&\pi[\alpha_1 Bel_1 + \alpha_2 Bel_2] \\
&= \Big(1 - \alpha_1 k_O[Bel_1] - \alpha_2 k_O[Bel_2]\Big)\overline{\mathcal{P}} + \alpha_1 \bar{O}[Bel_1] + \alpha_2 \bar{O}[Bel_2] \\
&= \alpha_1 \Big[(1 - k_O[Bel_1])\overline{\mathcal{P}} + \bar{O}[Bel_1]\Big] + \alpha_2 \Big[(1 - k_O[Bel_2])\overline{\mathcal{P}} + \bar{O}[Bel_2]\Big] \\
&= \alpha_1 \pi[Bel_1] + \alpha_2 \pi[Bel_2].
\end{aligned}$$

Proof of Theorem 50

By Theorem 49,

$$\pi[Bel] = \pi\left[\sum_{A\subset\Theta} m(A) Bel_A\right] = \sum_{A\subset\Theta} m(A)\pi[Bel_A],$$

which, by Lemma 11, becomes

$$\begin{aligned}
\pi[Bel] &= \sum_{A\subset\Theta} m(A)\left[(1 - |A|2^{1-|A|})\overline{\mathcal{P}} + |A|2^{1-|A|}\overline{\mathcal{P}}_A\right] \\
&= \left(1 - \sum_{A\subset\Theta} m(A)|A|2^{1-|A|}\right)\overline{\mathcal{P}} + \sum_{A\subset\Theta} m(A)|A|2^{1-|A|}\overline{\mathcal{P}}_A \\
&= \sum_{A\neq\Theta} m(A)|A|2^{1-|A|}\overline{\mathcal{P}}_A + \left(1 - \sum_{A\subset\Theta} m(A)|A|2^{1-|A|}\right)\overline{\mathcal{P}} \\
&\quad + m(\Theta)|\Theta|2^{1-|\Theta|}\overline{\mathcal{P}},
\end{aligned}$$

i.e., (11.29).

Proof of Theorem 51

If Bel satisfies (11.34), then the values of the intersection probability are, $\forall x \in \Theta$,

$$p[Bel](x) = m(x) + \beta[Bel] \sum_{A\supsetneq\{x\}} m(A) = m(x) + \beta[Bel] \sum_{k=2}^{n} \sigma^k \frac{\binom{n-1}{k-1}}{\binom{n}{k}}$$

(as there are $\binom{n-1}{k-1}$ events of size k containing x, and $\binom{n}{k}$ events of size k in total)

$$\begin{aligned}
&= m(x) + \beta[Bel] \sum_{k=2}^{n} \sigma^k \frac{k}{n} = m(x) + \frac{1}{n} \frac{\sigma^2 + \cdots + \sigma^n}{2\sigma^2 + \cdots + n\sigma^n}(2\sigma^2 + \cdots + n\sigma^n) \\
&= m(x) + \frac{1}{n}(\sigma^2 + \cdots + \sigma^n),
\end{aligned}$$

after recalling the decomposition (11.33) of $\beta[Bel]$.

On the other hand, under the hypothesis, the pignistic function reads as

$$
\begin{aligned}
BetP[Bel](x) &= m(x) + \sum_{k=2}^{n} \sum_{A \supsetneq \{x\}, |A|=k} \frac{m(A)}{k} = m(x) + \sum_{k=2}^{n} \frac{\sigma^k}{k} \frac{\binom{n-1}{k-1}}{\binom{n}{k}} \\
&= m(x) + \sum_{k=2}^{n} \frac{\sigma^k}{k} \frac{k}{n} = m(x) + \sum_{k=2}^{n} \frac{\sigma^k}{n},
\end{aligned}
\tag{11.44}
$$

and the two functions coincide.

Proof of Theorem 52

The orthogonal projection of a belief function Bel on the probability simplex $\mathcal{P}$ has the following expression [333] (11.27):

$$
\pi[Bel](x) = \sum_{A \supseteq \{x\}} m(A) \left(\frac{1 + |A^c| 2^{1-|A|}}{n} \right) + \sum_{A \not\supset \{x\}} m(A) \left(\frac{1 - |A| 2^{1-|A|}}{n} \right).
$$

Under the condition (11.34), it becomes

$$
\begin{aligned}
\pi[Bel](x) = m(x) &+ \sum_{k=2}^{n} \left(\frac{1 + (n-k)2^{1-k}}{n} \right) \sum_{A \supseteq \{x\}, |A|=k} m(A) \\
&+ \sum_{k=2}^{n} \left(\frac{1 - (n-k)2^{1-k}}{n} \right) \sum_{A \not\supset \{x\}, |A|=k} m(A),
\end{aligned}
\tag{11.45}
$$

where, again,

$$
\sum_{A \supseteq \{x\}, |A|=k} m(A) = \sigma^k \frac{k}{n},
$$

whereas

$$
\sum_{A \not\supset \{x\}, |A|=k} m(A) = \sigma^k \frac{\binom{n-1}{k}}{\binom{n}{k}} = \sigma^k \frac{(n-1)!}{k!(n-k-1)!} \frac{k!(n-k)!}{n!} = \sigma^k \frac{n-k}{n}.
$$

Replacing these expressions in (11.45) yields

$$
\begin{aligned}
m(x) &+ \sum_{k=2}^{n} \left(\frac{1 + (n-k)2^{1-k}}{n} \right) \sigma^k \frac{k}{n} + \sum_{k=2}^{n} \left(\frac{1 - (n-k)2^{1-k}}{n} \right) \sigma^k \frac{n-k}{n} \\
&= m(x) + \sum_{k=2}^{n} \left(\sigma^k \frac{k}{n^2} + \sigma^k \frac{n-k}{n^2} \right) = m(x) + \frac{1}{n} \sum_{k=2}^{n} \sigma^k,
\end{aligned}
$$

i.e., the value (11.44) of the intersection probability under the same assumptions.

12

Probability transforms: The epistemic family

Chapter 11 has taught us that the pignistic transform is just a member of what we have called the 'affine' group of probability transforms, characterised by their commutativity with affine combination (a property called 'linearity' in the TBM). This family also includes the intersection probability and the orthogonal projection, the first one closely associated with probability intervals and the second generated by minimising the classical Minkowski L_2 distance.

Whereas linearity and affine combination are important, especially in our geometric framework, it is natural to wonder what probability transforms exhibit natural relationships with evidence combination, in particular the original Dempster's rule of combination. This chapter thus studies the probability transforms which commute with Dempster's rule, and are therefore considered by some scholars as more consistent with the original Dempster–Shafer framework. In particular, the relative plausibility and relative belief transforms belong to this group, which we call the *epistemic family* of probability transforms.

Chapter outline

As belief functions have different, rather conflicting interpretations (see Chapter 3), in Section 12.1 we first discuss the semantics of relative belief and plausibility in both the probability-bound and Shafer's versions of the theory. Within the probability-bound interpretation (Section 12.1.1), neither of the transforms is consistent with the original belief functions, as they cannot be associated with a valid redistribution of the mass of the focal elements to the singletons (or 'specialisation',

F. Cuzzolin, *The Geometry of Uncertainty*, Artificial Intelligence: Foundations, Theory, and Algorithms, https://doi.org/10.1007/978-3-030-63153-6_12

in Smets's terminology). In Shafer's formulation of the theory of evidence as an evidence combination process, instead, the arguments proposed for the plausibility transform can be extended to the case of the relative belief transform (Section 12.1.2).

Indeed, we argue here that the relative plausibility [358] and belief transforms are closely related probability transformations (Section 12.2). Not only can the relative belief of singletons be seen as the relative plausibility of singletons of the associated plausibility function (Section 12.2.2), but both transforms have a number of dual properties with respect to Dempster's rule of combination (Section 12.2.3). In particular, while $\tilde{Pl}$ commutes with the Dempster sum *of belief functions*, $\tilde{Bel}$ commutes with orthogonal sums of *plausibility functions* (compare Chapter 9, for the extension of Dempster's rule to normalised sum functions and plausibility functions in particular). Similarly, whereas $\tilde{Pl}$ perfectly represents a belief function Bel when combined with any probability distribution (4.60), $\tilde{Bel}$ perfectly represents the associated plausibility function Pl when combined with a probability through the natural extension of the Dempster sum (Section 12.2.4). The resulting duality is summarised in the following table:

$$
\begin{array}{ccc}
Bel & \longleftrightarrow & Pl \\
\tilde{Pl} & \longleftrightarrow & \tilde{Bel} \\
Bel \oplus P = \tilde{Pl} \oplus P \;\; \forall P \in \mathcal{P} & \longleftrightarrow & Pl \oplus P = \tilde{Bel} \oplus P \;\; \forall P \in \mathcal{P} \\
\tilde{Pl}[Bel_1 \oplus Bel_2] & & \tilde{Bel}[Pl_1 \oplus Pl_2] \\
\| & \longleftrightarrow & \| \\
\tilde{Pl}[Bel_1] \oplus \tilde{Pl}[Bel_2] & & \tilde{Bel}[Pl_1] \oplus \tilde{Bel}[Pl_2].
\end{array}
$$

The symmetry/duality between (relative) plausibility and belief is broken, however, as the existence of the relative belief of singletons is subject to a strong condition,

$$\sum_{x \in \Theta} m(x) \neq 0, \tag{12.1}$$

stressing the issue of its applicability (Section 12.4). Even though this situation is 'singular' (in the sense that it excludes most belief and probability measures: Section 12.4.1), in practice the situation in which the mass of all singletons is nil is not so uncommon. However, in Section 12.4.2 we point out that the relative belief is only a member of a class of *relative mass* transformations, which can be interpreted as low-cost proxies for both the plausibility and the pignistic transforms (Section 12.4.3). We discuss their applicability as approximate transformations in two significant scenarios.

The second part of the chapter is devoted to the study of the geometry of epistemic transforms, in both the space of all pseudo-belief functions (Section 12.5), in which the belief space is embedded, and the probability simplex (Section 12.6). In fact, the geometry of relative belief and plausibility can be reduced to that of two specific pseudo-belief functions, called the 'plausibility of singletons' (12.15) and the 'belief of singletons' (12.17), which are introduced in Sections 12.5.1 and

12.5.2, respectively. Their geometry can be described in terms of three planes (Section 12.5.3) and angles (Section 12.5.4) in the belief space. Such angles are, in turn, related to a probability distribution which measures the relative uncertainty in the probabilities of singletons determined by Bel, and can be considered as the third member of the epistemic family of transformations. As $\tilde{Bel}$ does not exist when $\sum_x m(x) = 0$, this singular case needs to be discussed separately (Section 12.5.5). Several examples illustrate the relation between the geometry of the functions involved and their properties in terms of degrees of belief.

As probability transforms map belief functions onto probability distributions, it makes sense to study their behaviour in the simplex of all probabilities as well. We will get some insight into this in Section 12.6, at least in the case study of a frame of size 3.

Finally, as a step towards a complete understanding of the probability transformation problem, we discuss (Section 12.7) what we have learned about the relationship between the affine and epistemic families of probability transformations. Inspired by a binary case study, we provide sufficient conditions under which all transforms coincide, in terms of equal distribution of masses and equal contribution to the plausibility of singletons.

12.1 Rationale for epistemic transforms

As we know well by now, the original semantics of belief functions derives from Dempster's analysis of the effect of multivalued mappings $\Gamma : \Omega \to 2^\Theta, \omega \in \Omega \mapsto \Gamma(\omega) \subseteq \Theta$ on evidence available in the form of a probability measure on a 'source' domain Ω on a set of alternative hypotheses Θ (Section 3.1.1). As such, belief values are probabilities of events implying other events.

In some of his papers [425], however, Dempster himself claimed that the mass $m(A)$ associated with a non-singleton event $A \subseteq \Theta$ may be understood as a 'floating probability mass' which cannot be attached to any particular singleton event $x \in A$ because of the lack of precision of the (multivalued) operator which quantifies our knowledge via the mass function. This has originated a popular but controversial interpretation of belief functions as coherent sets of probabilities determined by sets of lower and upper bounds on their probability values (Section 3.1.4).

As Shafer has admitted in the past, there is a sense in which a single belief function can indeed be interpreted as a consistent system of probability bounds. However, the issue with the probability-bound interpretation of belief functions becomes evident when considering two or more belief functions addressing the same question but representing conflicting items of evidence, i.e., when Dempster's rule is applied to aggregate evidence. In [1583, 1607], Shafer disavowed any probability-bound interpretation, a position later seconded by Dempster [426].

We will come back to this point in Section 12.1.2, in which we will link the relative belief transform to Cobb and Shenoy's arguments [291] in favour of the plausibility transform as a link between Shafer's theory of evidence (endowed with Dempster's rule) and Bayesian reasoning. To corroborate this argument, in Section

12.1.1 we show that both the plausibility and the relative belief transforms (unlike Smets's pignistic transform) are not consistent with a probability-bound interpretation of belief functions.[72]

12.1.1 Semantics within the probability-bound interpretation

In their static, probability-bound interpretation, belief functions $Bel : 2^\Theta \rightarrow [0,1]$ each determine a convex set $\mathcal{P}[Bel]$ of 'consistent' probability distributions (3.10). These are the result of a redistribution process, in which the mass of each focal element is shared between its elements in an arbitrary proportion [339]. One such probability is the pignistic one (3.30). The pignistic transform was originally based on the principle of insufficient reason proposed by Bernoulli, Laplace [1096] and Keynes (see Section 4.2.3), which states that 'if there is no known reason for predicating of our subject one rather than another of several alternatives, then relatively to such knowledge the assertions of each of these alternatives have an equal probability'. A direct consequence of the principle of insufficient reason[73] in the probability-bound interpretation of belief functions is that, when considering how to redistribute the mass of an event A, it is wise to assume equiprobability amongst its singletons – yielding the pignistic probability.

It is easy to prove that the relative belief and plausibility of singletons are not the result of such a redistribution process, and therefore are not consistent with the original belief function in the sense defined above. Indeed, the relative plausibility of singletons (4.59) is the result of a process in which:

- for each singleton $x \in \Theta$, a mass reassignment strategy (there could be more than one) is selected in which the mass of all the events containing it is reassigned to x, yielding $\{Pl(x), x \in \Theta\}$;
- however, as different reassignment strategies are supposed to hold for different singletons (many of which belong to the same higher-size focal elements), this scenario is not compatible with the existence of a single coherent redistribution of mass from focal elements to singletons, as the basic probability of the same higher-cardinality event is assigned to different singletons;
- the plausibility values $Pl(x)$ obtained are nevertheless normalised to yield a formally admissible probability distribution.

Similarly, for the relative belief of singletons (4.61):

- for each singleton $x \in \Theta$, a mass reassignment strategy is selected in which only the mass of $\{x\}$ itself is reassigned to x, yielding $\{Bel(x) = m(x), x \in \Theta\}$;

[72]Even in this interpretation, however, a rationale for such transformations can be given via a utility-theoretical argument as in the case of the pignistic probability. We will discuss this in Chapter 16.

[73]Later on, however, Smets [1717] argued that the principle of insufficient reason could not by itself justify the uniqueness of the pignistic transform, and proposed a justification based on a number of axioms.

- once again, this scenario does not correspond to a valid redistribution process, as the mass of all higher-size focal elements is not assigned to any singletons;
- the values $Bel(x)$ obtained are nevertheless normalised to produce a valid probability distribution.

The fact that both of these probability transforms come from jointly assuming a number of incompatible redistribution processes is reflected by the fact that the resulting probability distributions are not guaranteed to belong to the set of probabilities (3.10) consistent with Bel.

Theorem 53. *The relative belief of singletons of a belief function Bel is not always consistent with Bel.*

Theorem 54. *The relative plausibility of singletons of a belief function Bel is not always consistent with Bel.*

Example 32: lack of consistency. *As an example, consider a belief function on $\Theta = \{x_1, x_2, \ldots, x_n\}$ with two focal elements:*

$$m(x_1) = 0.01, \quad m(\{x_2, \ldots, x_n\}) = 0.99. \tag{12.2}$$

This can be interpreted as the following real-world situation. A number of people $x_2, \ldots, x_n$ have no money of their own but are all candidates to inherit the wealth of a very rich relative. Person x_1 is not, but has some small amount of money of their own. Note that it is not correct to interpret $x_2, \ldots, x_n$ as assured joint owners of a certain amount of wealth (say, shares in the same company), as (12.2) is in fact consistent (in the probability-bound interpretation) with a distribution which assigns probability 0.99 to a single person in the group $x_2, \ldots, x_n$.

The relative belief of singletons associated with (12.2) is

$$\tilde{Bel}(x_1) = 1, \quad \tilde{Bel}(x_i) = 0 \;\; \forall i = 2, \ldots, n. \tag{12.3}$$

Clearly, this is not a good representative of the set of probabilities consistent with the above belief function, as it does not contemplate at all the chance that the potential heirs $x_2, \ldots, x_n$ have of gaining a substantial amount of money. Indeed, according to Theorem 53, (12.3) is not at all consistent with (12.2).

12.1.2 Semantics within Shafer's interpretation

Shafer has strongly argued against a probability-bound interpretation of belief functions. When these are not taken in isolation but as pieces of evidence to be combined, such an interpretation forces us to consider only groups of belief functions whose degrees of belief, when interpreted as probability bounds, can be satisfied simultaneously (in other words, when their sets of consistent probabilities have non-empty intersection). In Shafer's (and Shenoy's) view, though, when belief functions are combined via Dempster's rule this is irrelevant, although consistent probabilities that simultaneously bound all the belief functions being combined as well as the resulting BF do exist when no renormalisation is required in their Dempster combination. Consequently, citing Shafer, authors who support a probability-bound interpretation of belief functions are generally uncomfortable with renormalisation [2089].

In this context, Cobb and Shenoy [291] argued in favour of the plausibility transform as a link between Shafer's theory of evidence (endowed with Dempster's rule) and Bayesian reasoning. Besides some general arguments supporting probability transformations of belief functions in general, their points more specifically about the plausibility transform can be summarised as follows:

- a probability transformation consistent with Dempster's rule can improve our understanding of the theory of evidence by providing probabilistic semantics for belief functions, i.e., 'meanings' of basic probability assignments in the context of betting for hypotheses in the frame Θ;
- in opposition to some literature on belief functions suggesting that the theory of evidence is more expressive than probability theory since the probability model obtained by using the pignistic transformation leads to non-intuitive results [176], Cobb and Shenoy showed that by using the plausibility transformation method the original belief function model and the corresponding probability model yield the same qualitative results;
- a probability transformation consistent with Dempster's rule allows us to build probabilistic models by converting/transforming belief function models obtained by using the belief function semantics of distinct evidence [1629].

Mathematically, Cobb and Shenoy proved [295] that the plausibility transform commutes with Dempster's rule, and satisfies a number of additional properties which they claim 'allow an integration of Bayesian and D-S reasoning that takes advantage of the efficiency in computation and decision-making provided by Bayesian calculus while retaining the flexibility in modeling evidence that underlies D-S reasoning'.

In this chapter, we prove that a similar set of (dual) properties hold for the relative belief transform, associating the relative belief and relative plausibility transforms in a family of probability transformations strongly related to Shafer's interpretation of the theory of evidence via Dempster's rule.

12.2 Dual properties of epistemic transforms

The relative belief and plausibility of singletons are, as we show here, linked by a form of duality, as $\tilde{Bel}$ can be interpreted as the relative plausibility of singletons *of the plausibility function* Pl associated with Bel. Furthermore, $\tilde{Bel}$ and $\tilde{Pl}$ share a close relationship with Dempster's combination rule $\oplus$, as they possess a set of dual properties with respect to $\oplus$. This suggests a classification of all the probability transformations of belief functions in terms of the operator they relate to.

12.2.1 Relative plausibility, Dempster's rule and pseudo-belief functions

Cobb and Shenoy [295] proved that the relative plausibility function $\tilde{Pl}$ (4.59) commutes with Dempster's rule, and possesses a number of additional properties.[74]

Proposition 62. *The following statements hold:*

1. *If $Bel = Bel_1 \oplus \cdots \oplus Bel_m$, then $\tilde{Pl} = \tilde{Pl}_1 \oplus \cdots \oplus \tilde{Pl}_m$: the Dempster sum and the relative plausibility commute.*
2. *If m is idempotent with respect to Dempster's rule, i.e., if $m \oplus m = m$, then $\tilde{Pl}$ is idempotent with respect to Bayes' rule.*
3. *Let us define the limit of a belief function Bel as*

$$Bel^\infty \doteq \lim_{n\to\infty} Bel^n \doteq \lim_{n\to\infty} Bel \oplus \cdots \oplus Bel \quad (n \text{ times}). \tag{12.4}$$

If $\exists x \in \Theta$ such that $Pl(x) > Pl(y)$ for all $y \neq x, y \in \Theta$, then

$$\tilde{Pl}^\infty(x) = 1, \quad \tilde{Pl}^\infty(y) = 0 \quad \forall y \neq x.$$

4. *If $\exists A \subseteq \Theta$, $|A| = k$, such that*

$$Pl(x) = Pl(y)\ \forall x, y \in A, \quad Pl(x) > Pl(z)\ \forall x \in A, z \in A^c,$$

then

$$\tilde{Pl}^\infty(x) = \tilde{Pl}^\infty(y) = \frac{1}{k}\ \forall x, y \in A, \quad \tilde{Pl}^\infty(z) = 0\ \forall z \in A^c.$$

On his side, Voorbraak [1859] showed that the following proposition holds.

Proposition 63. *The relative plausibility of singletons $\tilde{Pl}$ is a perfect representative of Bel in the probability space when combined through Dempster's rule:*

$$Bel \oplus P = \tilde{Pl} \oplus P, \quad \forall P \in \mathcal{P}.$$

The relative belief of singletons possesses similar dual properties. Their study, however, requires one to extend the analysis to normalised sum functions (also called 'pseudo-belief functions': see Section 7.3.3).

12.2.2 A (broken) symmetry

A direct consequence of the duality between belief and plausibility measures is the existence of a striking symmetry between the (relative) plausibility and the (relative) belief transform. A formal proof of this symmetry is based on Theorem 25 [347]:

$$\sum_{A \supseteq \{x\}} \mu(A) = m(x), \tag{12.5}$$

where μ is the basic plausibility assignment associated with a belief function Bel (see Section 9.1.1).

[74] The original statements in [293] have been reformulated according to the notation used in this book.

Theorem 55. *Given a pair of belief/plausibility functions $Bel, Pl : 2^\Theta \to [0,1]$, the relative belief transform of the belief function Bel coincides with the plausibility transform of the associated plausibility function Pl (interpreted as a pseudo-belief function):*

$$\tilde{Bel}[Bel] = \tilde{Pl}[Pl].$$

The symmetry between the relative plausibility and the relative belief of singletons is broken by the fact that the latter is not defined for belief functions with no singleton focal sets. By Theorem 55, $\tilde{Bel}$ is itself an instance of a relative plausibility (in particular, of a plausibility function Pl). The fact that $\tilde{Pl}$ always exists seems to contradict the result of Theorem 55.

This seeming paradox can be explained by the combinatorial nature of belief, plausibility and commonality functions. As we proved in Chapter 9 [347], whereas belief measures are sum functions of the form $Bel(A) = \sum_{B \subset A} m(B)$ whose Möbius transform m is both normalised and non-negative, plausibility measures are sum functions whose Möbius transform μ is not necessarily non-negative (commonality functions are not even normalised). As a consequence, the quantity

$$\sum_x Pl_{Pl}(x) = \sum_x \sum_{A \supseteq \{x\}} \mu(A) = \sum_{A \supseteq \Theta} \mu(A)|A|$$

can be equal to zero, in which case $\tilde{Pl}_{Pl} = \tilde{Bel}$ does not exist.

12.2.3 Dual properties of the relative belief operator

The duality between $\tilde{Bel}$ and $\tilde{Pl}$ (albeit to some extent imperfect) extends the behaviour of the pair of transformations with respect to Dempster's rule (2.6).

We have seen in Chapter 8, Section 8.1 that the orthogonal sum can be naturally extended to a pair ς_1, ς_2 of normalised sum functions (pseudo-belief functions) [327], by simply applying (2.6) to their Möbius inverses $m_{\varsigma_1}, m_{\varsigma_2}$.

Proposition 64. *Dempster's rule defined as in (2.6), when applied to a pair of pseudo-belief functions ς_1, ς_2, yields again a pseudo-belief function.*

We still denote the orthogonal sum of two NSFs ς_1, ς_2 by $\varsigma_1 \oplus \varsigma_2$.

As plausibility functions are indeed NSFs, Dempster's rule can then be formally applied to them as well. It is convenient to introduce a dual form of the relative belief operator, mapping a plausibility function to the corresponding relative belief of singletons, $\tilde{Bel} : \mathcal{PL} \to \mathcal{P}, Pl \mapsto \tilde{Bel}[Pl]$, where

$$\tilde{Bel}[Pl](x) \doteq \frac{m(x)}{\sum_{y \in \Theta} m(y)} \quad \forall x \in \Theta \tag{12.6}$$

is defined as usual for belief functions Bel such that $\sum_y m(y) \neq 0$.

Indeed, as Bel and Pl are in 1–1 correspondence, we can think indifferently of the same operator as mapping a belief function Bel to its relative belief $\tilde{Bel}$, or as mapping the unique plausibility function Pl associated with Bel to $\tilde{Bel}$.

The following commutativity theorem follows, as the dual of point 1 in Proposition 62.

Theorem 56. *The relative belief operator commutes with respect to the Dempster combination* of plausibility functions*:*

$$\tilde{Bel}[Pl_1 \oplus Pl_2] = \tilde{Bel}[Pl_1] \oplus \tilde{Bel}[Pl_2].$$

Theorem 56 implies that

$$\tilde{Bel}[(Pl)^n] = (\tilde{Bel}[Pl])^n. \tag{12.7}$$

As an immediate consequence, an idempotence property which is the dual of property 2 of Proposition 62 holds for the relative belief of singletons.

Corollary 11. *If Pl is idempotent with respect to Dempster's rule, i.e., if $Pl \oplus Pl = Pl$, then $\tilde{Bel}[Pl]$ is itself idempotent: $\tilde{Bel}[Pl] \oplus \tilde{Bel}[Pl] = \tilde{Bel}[Pl]$.*

Proof. By Theorem 56, $\tilde{Bel}[Pl] \oplus \tilde{Bel}[Pl] = \tilde{Bel}[Pl \oplus Pl]$, and if $Pl \oplus Pl = Pl$ the thesis immediately follows.

□

The dual results of the remaining two statements of Proposition 62 can be proven in a similar fashion.

Theorem 57. *If $\exists x \in \Theta$ such that $Bel(x) > Bel(y)\ \forall y \neq x, y \in \Theta$, then*

$$\tilde{Bel}[Pl^\infty](x) = 1, \quad \tilde{Bel}[Pl^\infty](y) = 0 \quad \forall y \neq x.$$

A similar proof can be provided for the following generalisation of Theorem 57.

Corollary 12. *If $\exists A \subseteq \Theta$, $|A| = k$, such that $Bel(x) = Bel(y)\ \forall x, y \in A$, $Bel(x) > Bel(z)\ \forall x \in A, z \in A^c$, then*

$$\tilde{Bel}[Pl^\infty](x) = \tilde{Bel}[Pl^\infty](y) = \frac{1}{k}\ \forall x, y \in A, \quad \tilde{Bel}[Pl^\infty](z) = 0\ \forall z \in A^c.$$

Example 33: a numerical example. *It is crucial to point out that commutativity (Theorem 56) and idempotence (Corollary 11) hold for combinations of* plausibility functions, *not of belief functions.*

Let us consider as an example the belief function Bel on the frame of size 4, $\Theta = \{x, y, z, w\}$, determined by the following basic probability assignment:

$$m(\{x, y\}) = 0.4, \quad m(\{y, z\}) = 0.4, \quad m(w) = 0.2. \tag{12.8}$$

Its BPlA is, according to (9.1), given by

$$\begin{array}{lll} \mu(x) = 0.4, & \mu(y) = 0.8, & \mu(z) = 0.4, \\ \mu(w) = 0.2, & \mu(\{x, y\}) = -0.4, & \mu(\{y, z\}) = -0.4 \end{array} \tag{12.9}$$

(only non-zero values are reported).

To check the validity of Theorems 56 and 57, let us analyse the two series of probability measures $(\tilde{Bel}[Pl])^n$ *and* $\tilde{Bel}[(Pl)^n]$. *By applying Dempster's rule to the BPlA (12.9)* ($Pl^2 = Pl \oplus Pl$), *we get a new BPlA* μ^2 *with values*

$$\mu^2(x) = \frac{4}{7}, \quad \mu^2(y) = \frac{8}{7}, \quad \mu^2(z) = \frac{4}{7}, \quad \mu^2(w) = -\frac{1}{7},$$
$$\mu^2(\{x,y\}) = -\frac{4}{7}, \quad \mu^2(\{y,z\}) = -\frac{4}{7}$$

(see Fig. 12.1). To compute the corresponding relative belief of singletons $\tilde{Bel}[Pl^2]$,

{y,z}		{y}	{z}		{y}	{y,z}
{x,y}	{x}	{y}			{x,y}	{y}
{w}				{w}		
{z}			{z}			{z}
{y}		{y}			{y}	{y}
{x}	{x}				{x}	
	{x}	{y}	{z}	{w}	{x,y}	{y,z}

Fig. 12.1: Intersection of focal elements in the Dempster combination of the BPlA (12.9) with itself. Non-zero mass events for each addend $\mu_1 = \mu_2 = \mu$ correspond to rows and columns of the table, each entry of the table hosting the related intersection.

we first need to compute the plausibility values,

$$Pl^2(\{x,y,z\}) = \mu^2(x) + \mu^2(y) + \mu^2(z) + \mu^2(\{x,y\}) + \mu^2(\{y,z\}) = \frac{8}{7},$$
$$Pl^2(\{x,y,w\}) = 1, \quad Pl^2(\{x,z,w\}) = 1, \quad Pl^2(\{y,z,w\}) = 1.$$

The latter imply (as, by definition, $Pl(A) \doteq 1 - Bel(A^c)$*)*

$$Bel^2(w) = -\frac{1}{7}, \quad Bel^2(z) = Bel^2(y) = Bel^2(x) = 0.$$

Therefore $\tilde{Bel}[Pl^2] = [0,0,0,1]'$ *(when we represent probability distributions as vectors of the form* $[p(x), p(y), p(z), p(w)]'$*).*

Theorem 56 is confirmed as, by (12.8) (since $\{w\}$ *is the only singleton with non-zero mass),* $\tilde{Bel} = [0,0,0,1]'$ *so that* $\tilde{Bel} \oplus \tilde{Bel} = [0,0,0,1]'$ *and* $\tilde{Bel}[.]$ *commutes with* $Pl\oplus$.

By Dempster-combining Pl^2 *with* Pl *one more time, we get the BPlA*

$$\mu^3(x) = \frac{16}{31}, \quad \mu^3(y) = \frac{32}{31}, \quad \mu^3(z) = \frac{16}{31}, \quad \mu^3(w) = -\frac{1}{31},$$

$$\mu^3(\{x,y\}) = -\frac{16}{31}, \quad \mu^3(\{y,z\}) = -\frac{16}{31},$$

which corresponds to

$$Pl^3(\{x,y,z\}) = \frac{32}{31}, \; Pl^3(\{x,y,w\}) = Pl^3(\{x,z,w\}) = Pl^3(\{y,z,w\}) = 1.$$

Therefore,

$$Bel^3(w) = -\frac{1}{31}, \quad Bel^3(z) = Bel^3(y) = Bel^3(x) = 0,$$

and $\tilde{Bel}[Pl^3] = [0,0,0,1]'$*, which is again equal to* $\tilde{Bel} \oplus \tilde{Bel} \oplus \tilde{Bel}$ *as Theorem 56 guarantees.*

The series of basic plausibility assignments $(\mu)^n$ *clearly converges to*

$$\mu^n(x) \to \frac{1}{2}^+, \quad \mu^3(y) \to 1^+, \quad \mu^3(z) \to \frac{1}{2}^+, \quad \mu^3(w) \to 0^-,$$
$$\mu^3(\{x,y\}) \to -\frac{1}{2}^-, \quad \mu^3(\{y,z\}) \to -\frac{1}{2}^-,$$

associated with the following plausibility values: $\lim_{n\to\infty} Pl^n(\{x,y,z\}) = 1^+$, $Pl^n(\{x,y,w\}) = Pl^n(\{x,z,w\}) = Pl^n(\{y,z,w\}) = 1 \; \forall n \geq 1$*. These correspond to the following values of belief of singletons:* $\lim_{n\to\infty} Bel^n(w) = 0^-$, $Bel^n(z) = Bel^n(y) = Bel^n(x) = 0 \; \forall n \geq 1$*, so that*

$$\lim_{n\to\infty} \tilde{Bel}[Pl^\infty](w) = \lim_{n\to\infty} \frac{Bel^n(w)}{Bel^n(w)} = 1,$$

$$\begin{aligned}\lim_{n\to\infty} \tilde{Bel}[Pl^\infty](x) &= \lim_{n\to\infty} \tilde{Bel}[Pl^\infty](y) = \lim_{n\to\infty} \tilde{Bel}[Pl^\infty](z) \\ &= \lim_{n\to\infty} \frac{0}{Bel^n(w)} = \lim_{n\to\infty} 0 = 0,\end{aligned}$$

in perfect agreement with Theorem 57.

12.2.4 Representation theorem for relative beliefs

A dual of the representation theorem (Proposition 63) for the relative belief transform can also be proven, once we recall the following result for the Dempster sum of affine combinations [327] (see Chapter 8, Theorem (14)).

Proposition 65. *The orthogonal sum* $Bel \oplus \sum_i \alpha_i Bel_i$, $\sum_i \alpha_i = 1$ *of a belief function* Bel *with any*[75] *affine combination of belief functions is itself an affine combination of the partial sums* $Bel \oplus Bel_i$*, namely*

[75] In fact, the collection $\{Bel_i, i\}$ is required to include *at least* a belief function which is combinable with Bel [327].

$$Bel \oplus \sum_i \alpha_i Bel_i = \sum_i \gamma_i (Bel \oplus Bel_i), \tag{12.10}$$

where $\gamma_i = \frac{\alpha_i k(Bel, Bel_i)}{\sum_j \alpha_j k(Bel, Bel_j)}$ *and* $k(Bel, Bel_i)$ *is the normalisation factor of the partial Dempster sum* $Bel \oplus Bel_i$*.*

Again, the duality between $\tilde{Bel}$ and $\tilde{Pl}$ suggests that the relative belief of singletons represents the associated *plausibility* function Pl, rather than the corresponding belief function Bel: $\tilde{Bel} \oplus P \neq Bel \oplus P$, $P \in \mathcal{P}$.

Theorem 58. *The relative belief of singletons* $\tilde{Bel}$ *perfectly represents the corresponding plausibility function* Pl *when combined with any probability through (the extended) Dempster's rule:*

$$\tilde{Bel} \oplus P = Pl \oplus P$$

for all Bayesian belief functions $P \in \mathcal{P}$*.*

Theorem 58 can be obtained from Proposition 63 by replacing Bel with Pl and $\tilde{Pl}$ with $\tilde{Bel}$, by virtue of their duality.

Example 34: the previous example continued. *Once again, the representation theorem in Theorem 58 is about combinations of plausibility functions (as pseudo-belief functions),* not *combinations of belief functions. Going back to the previous example, the combination* $Bel^2 \doteq Bel \oplus Bel$ *of* Bel *with itself has basic probability assignment*

$$m^2(\{x,y\}) = \frac{m(\{x,y\}) \cdot m(\{x,y\})}{k(Bel, Bel)} = \frac{0.16}{0.68} = 0.235,$$

$$m^2(\{y,z\}) = \frac{m(\{y,z\}) \cdot m(\{y,z\})}{k(Bel, Bel)} = \frac{0.16}{0.68} = 0.235,$$

$$m^2(w) = \frac{m(w) \cdot m(w)}{k(Bel, Bel)} = \frac{0.04}{0.68} = 0.058,$$

$$m^2(y) = \frac{m(\{x,y\}) \cdot m(\{y,z\}) + m(\{y,z\}) \cdot m(\{x,y\})}{k(Bel, Bel)} = 0.47,$$

which obviously yields

$$\widetilde{Bel \oplus Bel} = \left[0, \frac{0.47}{0.528}, 0, \frac{0.058}{0.528}\right]' \neq \tilde{Bel} \oplus \tilde{Bel} = [0, 0, 0, 1]'.$$

The main reason for this is that the plausibility function of a sum of two belief functions is not *the sum of the associated plausibilities:*

$$[Pl_1 \oplus Pl_2] \neq Pl_{Bel_1 \oplus Bel_2}.$$

12.2.5 Two families of Bayesian approximations

The following table summarises the duality results we have just presented:

$$\begin{array}{ccc}
Bel & \longleftrightarrow & Pl \\
\tilde{Pl} & \longleftrightarrow & \tilde{Bel} \\
Bel \oplus P = \tilde{Pl} \oplus P \ \forall P \in \mathcal{P} & \longleftrightarrow & Pl \oplus P = \tilde{Bel} \oplus P \ \forall P \in \mathcal{P} \\
\tilde{Pl}[Bel_1 \oplus Bel_2] & & \tilde{Bel}[Pl_1 \oplus Pl_2] \\
\shortparallel & \longleftrightarrow & \shortparallel \\
\tilde{Pl}[Bel_1] \oplus \tilde{Pl}[Bel_2] & & \tilde{Bel}[Pl_1] \oplus \tilde{Bel}[Pl_2] \\
Bel \oplus Bel = Bel & & Pl \oplus Pl = Pl \\
\top & \longleftrightarrow & \top \\
\tilde{Pl}[Bel] \oplus \tilde{Pl}[Bel] = \tilde{Pl}[Bel] & & \tilde{Bel}[Pl] \oplus \tilde{Bel}[Pl] = \tilde{Bel}[Pl].
\end{array}$$

Note that, just as Voorbraak's and Cobb and Shenoy's results are not valid for all pseudo-belief functions but only for proper BFs., the above dual results do not hold for all pseudo-belief functions either, but only for those which are plausibility functions.

These results lead to a classification of all probability transformations into two families, related to the Dempster sum and affine combination, respectively. In fact, the notion that there exist two distinct families of probability transformations, each determined by the operator they commute with, was already implicitly present in the literature. Smets's linearity axiom [1730], which lies at the foundation of the pignistic transform, obviously corresponds (even though expressed in a somewhat different language) to commutativity with the affine combination of belief functions. To address the criticism that this axiom was subject to, Smets introduced later a formal justification based on an expected utility argument in the presence of conditional evidence [1717]. On the other hand, Cobb and Shenoy argued in favour of commutativity with respect to Dempster's rule, on the basis that the Dempster–Shafer theory of evidence is a coherent framework of which Dempster's rule is an integral part, and that a Dempster-compatible transformation can provide a useful probabilistic semantics for belief functions.

Incidentally, there seems to be a flaw in Smets's argument that the pignistic transform is uniquely determined as the probability transformation which commutes with affine combination: in [333] and Chapter 11, we indeed proved that the orthogonal transform (Section 11.4) also possesses the same property. Analogously, we showed here that the plausibility transform is not unique as a probability transformation which commutes with $\oplus$ (although, in this latter case, the transformation is applied to different objects).

12.3 Plausibility transform and convex closure

We add a further element to this ongoing debate by proving that the plausibility transform, although it does not obviously commute with *affine combination*, does commute with the *convex closure* (7.9) of belief functions in the belief space $\mathcal{B}$:

$$Cl(Bel_1, \ldots, Bel_k) = \left\{ Bel \in \mathcal{B} : Bel = \sum_{i=1}^{k} \alpha_i Bel_i, \sum_i \alpha_i = 1, \ \alpha_i \geq 0 \ \forall i \right\}.$$

Let us first study its behaviour with respect to affine combination.

Lemma 12. *For all $\alpha \in \mathbb{R}$, we have that*

$$\tilde{Pl}[\alpha Bel_1 + (1-\alpha) Bel_2] = \beta_1 \tilde{Pl}[Bel_1] + \beta_2 \tilde{Pl}[Bel_2],$$

where

$$\beta_1 = \frac{\alpha k_{Pl_1}}{\alpha k_{Pl_1} + (1-\alpha) k_{Pl_2}}, \quad \beta_2 = \frac{\alpha k_{Pl_2}}{\alpha k_{Pl_1} + (1-\alpha) k_{Pl_2}},$$

and k_{Pl_i} denotes the total plausibility of singletons (11.13) for the belief function Bel_i, $i = 1, 2$.

This leads to the following theorem.

Theorem 59. *The relative plausibility operator commutes with convex closure in the belief space: $\tilde{Pl}[Cl(Bel_1, \ldots, Bel_k)] = Cl(\tilde{Pl}[Bel_1], \ldots, \tilde{Pl}[Bel_k])$.*

The behaviour of the plausibility transform, in this respect, is similar to that of Dempster's rule (Theorem 6; [327]), supporting the argument that the plausibility transform is indeed naturally associated with the original Dempster–Shafer framework.

A dual result holds for the relative belief transform.

12.4 Generalisations of the relative belief transform

A serious issue with the relative belief transform is its applicability, as $\tilde{Bel}[\cdot]$ does not exist for a large class of belief functions (those which assign no mass to singletons). Although this singular case involves only a small fraction of all belief measures (as we show in Section 12.4.1), this issue arises in many practical cases, for instance when fuzzy membership functions are employed to model the available evidence.

12.4.1 Zero mass assigned to singletons as a singular case

Let us first consider the set of belief functions for which a relative belief of singletons does not exist. In the binary case $\Theta = \{x, y\}$, the existence constraint (12.1) implies that the only belief function which does not admit relative belief of singletons is the vacuous one, Bel_Θ, for which $m_\Theta(\Theta) = 1$. Symmetrically, the pseudo-belief function $\varsigma = Pl_\Theta$ (for which $Pl_\Theta(x) = Pl_\Theta(y) = 1$) is such that $Pl_{Pl_\Theta} = Bel_\Theta$, so that $\tilde{Pl}_{Pl_\Theta}$ does not exist either. Figure 12.2 (left) illustrates the geometry of the relative belief operator in the binary case – the dual singular points $Bel_\Theta, \varsigma = Pl_\Theta$ are highlighted.

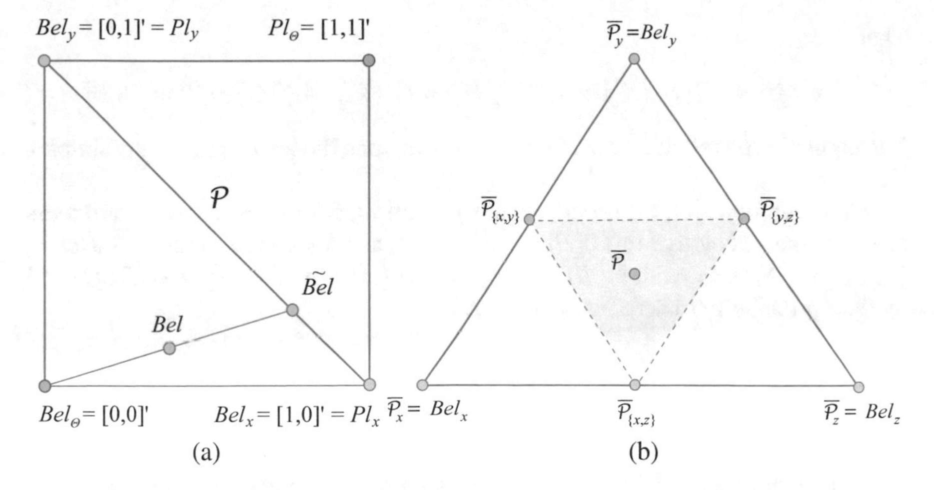

Fig. 12.2: (a) The location of the relative belief of singletons $\tilde{Bel} = [\frac{m(x)}{m(x)+m(y)}, \frac{m(y)}{m(x)+m(y)}]'$ associated with an arbitrary belief function Bel on $\{x,y\}$ is shown. The singular points $Bel_\Theta = [0,0]'$ and $Pl_\Theta = [1,1]'$ are marked by purple circles. (b) The images under the pignistic and plausibility transforms of the subset of belief functions $\{Bel : m(x) + m(y) + m(z) = 0\}$ span only a proper subset of the probability simplex. This region is shown here in the ternary case $\Theta = \{x,y,z\}$ (the triangle delimited by dashed lines).

The analysis of the binary case shows that the set of belief functions for which $\tilde{Bel}$ does not exist is a lower-dimensional subset of the belief space $\mathcal{B}$. To support this point, we determine here the region spanned by the most common probability transformations: the plausibility and pignistic transforms.

Theorem 59 proves that the plausibility transform commutes with convex closure. As (by Proposition 65; [333]) the pignistic transform (3.30) commutes with affine combination, we have that $BetP$ also commutes with Cl:

$$BetP[Cl(Bel_1,\ldots,Bel_k)] = Cl(BetP[Bel_i], i = 1,\ldots,k).$$

To determine the image under both probability transforms of any convex set of belief functions, $Cl(Bel_1,\ldots,Bel_k)$, it then suffices to compute the images of its vertices.

The space of all belief functions $\mathcal{B} \doteq \{Bel : 2^\Theta \to [0,1]\}$, in particular, is the convex closure of all the categorical BFs Bel_A: $\mathcal{B} = Cl(Bel_A, A \subseteq \Theta)$ [334] (see Theorem 10). The image of a categorical belief function Bel_A (a vertex of $\mathcal{B}$) under either the plausibility or the pignistic transform is

$$\tilde{Pl}_A(x) = \frac{\sum_{B\supseteq\{x\}} m_A(B)}{\sum_{B\supseteq\{x\}} m_A(B)|B|} = \begin{cases} \frac{1}{|A|} & x \in A \\ 0 & \text{otherwise} \end{cases} \doteq \overline{\mathcal{P}}_A = \sum_{B\supseteq\{x\}} \frac{m_A(B)}{|B|}$$

$$= BetP[Bel_A](x).$$

Therefore,

$$BetP[\mathcal{B}] = Cl(BetP[Bel_A], A \subseteq \Theta) = Cl(\overline{\mathcal{P}}_A, A \subseteq \Theta) = \mathcal{P} = \tilde{Pl}[\mathcal{B}].$$

The pignistic and relative plausibility transforms span the *whole* probability simplex $\mathcal{P}$.

Consider, however, the set of (singular) belief functions which assign zero mass to singletons. They live in $Cl(Bel_A, |A| > 1)$, as they can be written as $Bel = \sum_{|A|>1} m(A) Bel_A$, with $m(A) \geq 0$, $\sum_{|A|>1} m(A) = 1$. The region of $\mathcal{P}$ spanned by their probability transforms is therefore

$$\begin{array}{l} \tilde{Pl}[Cl(Bel_A, |A| > 1)] \\ = Cl(\tilde{Pl}_A, |A| > 1) = Cl(\overline{\mathcal{P}}_A, |A| > 1) \\ = Cl(BetP[Bel_A], |A| > 1) = BetP[Cl(Bel_A, |A| > 1)]. \end{array}$$

If (12.1) is not satisfied, both probability transforms span only a limited region of the probability simplex. In the case of a ternary frame, this yields the triangle

$$Cl(\overline{\mathcal{P}}_{\{x,y\}}, \overline{\mathcal{P}}_{\{x,z\}}, \overline{\mathcal{P}}_{\{y,z\}}, \overline{\mathcal{P}}_{\Theta}) = Cl(\overline{\mathcal{P}}_{\{x,y\}}, \overline{\mathcal{P}}_{\{x,z\}}, \overline{\mathcal{P}}_{\{y,z\}}),$$

delimited by dashed lines in Fig. 12.2 (right).

12.4.2 The family of relative mass probability transformations

One may argue that although the 'singular' case concerns only a small fraction of all belief and probability measures, in many practical application there is a bias towards some particular models which are most exposed to the problem.

For example, uncertainty is often represented using a fuzzy membership function [999]. If the membership function has only a finite number of values, then it is equivalent to a belief function whose focal sets are linearly ordered under set inclusion, $A_1 \subseteq \cdots \subseteq A_n = \Theta$, $|A_i| = i$, or a 'consonant' belief function (see Chapter 2 and [1583, 531]). In consonant belief functions, at most one focal element A_1 is a singleton; hence most information is stored in the non-singleton focal elements.

This train of thought leads to the realisation that the relative belief transform is merely one representative of an entire family of probability transformations. Indeed, it can be thought of as the probability transformation which, given a belief functions Bel, does the following:

1. It retains the focal elements of size 1 only, yielding an unnormalised belief function.
2. It computes (indifferently) the latter's relative plausibility/pignistic transform,

$$\tilde{Bel}(x) = \frac{\sum_{A \supseteq x, |A|=1} m(A)}{\sum_y \sum_{A \supseteq x, |A|=1} m(A)} = \frac{m(x)}{k_{Bel}} = \frac{\sum_{A \supseteq x, |A|=1} \frac{m(A)}{|A|}}{\sum_y \sum_{A \supseteq x, |A|=1} \frac{m(A)}{|A|}}.$$

A family of natural generalisations of the relative belief transform is thus obtained in the following way, given an arbitrary belief function Bel:

1. Retaining the focal elements of size s only.
2. Computing either the resulting relative plausibility ...
3. ... or the associated pignistic transformation.

Intriguingly, both option 2 and option 3 yield the same probability distribution. Indeed, the application of the relative plausibility transform yields

$$P(x) = \frac{\sum_{A\supseteq\{x\}:|A|=s} m(A)}{\sum_{y\in\Theta}\sum_{A\supseteq\{y\}:|A|=s} m(A)} = \frac{\sum_{A\supseteq\{x\}:|A|=s} m(A)}{\sum_{A\subseteq\Theta:|A|=s} m(A)|A|}$$
$$= \frac{\sum_{A\supseteq\{x\}:|A|=s} m(A)}{s\sum_{A\subseteq\Theta:|A|=s} m(A)},$$

whereas applying the pignistic transform produces:

$$P(x) = \frac{\sum_{A\supseteq\{x\}:|A|=s} \frac{m(A)}{|A|}}{\sum_{y\in\Theta}\sum_{A\supseteq\{y\}:|A|=s} \frac{m(A)}{|A|}} = \frac{s\sum_{A\supseteq\{x\}:|A|=s} m(A)}{s\sum_{y\in\Theta}\sum_{A\supseteq\{y\}:|A|=s} m(A)}, \quad (12.11)$$

i.e., the very same result. The following natural extension of the relative belief operator is thus well defined.

Definition 127. *Given any belief function $Bel : 2^\Theta \rightarrow [0,1]$ with basic probability assignment m, we define the* relative mass transformation *of level s as the unique transform which maps Bel to the probability distribution (12.11).*

We denote by $\tilde{m}_s$ the output of the relative mass transform of level s.

12.4.3 Approximating the pignistic probability and relative plausibility

Classical transforms as convex combinations of relative mass transformations It is easy to see that both the relative plausibility of singletons and the pignistic probability are convex combinations of all the (n) relative mass probabilities $\{\tilde{m}_s, s = 1,\ldots,n\}$.

Namely, let us denote by $k^s_{Bel} = k^s = \sum_{A\subseteq\Theta:|A|=s} m(A)$ the total mass of focal elements of size s, and by $Pl(x;k) = \sum_{A\supseteq\{x\}:|A|=s} m(A)$ the contribution to the plausibility of x from the same size-s focal elements. Immediately,

$$\sum_y Pl(y) = \sum_y \sum_{A\supseteq\{y\}} m(A) = \sum_{A\subseteq\Theta} m(A)|A|$$
$$= \sum_{r=1}^{|\Theta|} r \left(\sum_{A\subseteq\Theta,|A|=r} m(A) \right) = \sum_{r=1}^{|\Theta|} rk^r.$$

This yields the following convex decomposition of the relative plausibility of singletons into relative mass probabilities $\tilde{m}_s$:

$$\tilde{Pl}(x) = \frac{Pl(x)}{\sum_y Pl(y)} = \frac{\sum_s Pl(x;s)}{\sum_r r k^r} = \sum_s \frac{Pl(x;s)}{\sum_r r k^r} = \sum_s \frac{Pl(x;s)}{s k^s} \frac{s k^s}{\sum_r r k^r}$$
$$= \sum_s \alpha_s \tilde{m}_s(x), \tag{12.12}$$

as $\tilde{m}_s(x) = \frac{Pl(x;s)}{sk^s}$. The coefficients

$$\alpha_s = \frac{sk^s}{\sum_r r k^r} \propto sk^s = \sum_y Pl(y;s)$$

of the convex combination measure, for each level s, the total plausibility contribution of the focal elements of size s.

In the case of the pignistic probability, we get

$$BetP[Bel](x) = \sum_{A \supseteq \{x\}} \frac{m(A)}{|A|} = \sum_s \frac{1}{s} \sum_{A \supseteq \{x\}, |A|=s} m(A)$$
$$= \sum_s \frac{1}{s} Pl(x;s) = \sum_s k^s \frac{Pl(x;s)}{sk^s} = \sum_s k^s \tilde{m}_s(x), \tag{12.13}$$

with coefficients $\beta_s = k^s$ measuring, for each level s, the mass contribution of the focal elements of size s.

Relative mass transforms as low-cost proxies: Approximation criteria Accordingly, the relative mass probabilities can be seen as basic components of both the pignistic and the plausibility transform, associated with the evidence carried by focal elements of a specific size. As such transforms can be computed just by considering size-s focal elements, they can also be thought of as low-cost proxies for both the relative plausibility and the pignistic probability, since only the $\binom{n}{s}$ size-s focal elements (instead of the initial 2^n) have to be stored, while all the others can be dropped without further processing.

We can think of two natural criteria for such an approximation of $\tilde{Pl}$, $BetP$ via the relative mass transforms:

- (C1) we retain the component s whose coefficient α_s/β_s is the largest in the convex decomposition (12.12)/(12.13);
- (C2) we retain the component associated with the minimal-size focal elements.

Clearly, the second criterion delivers the classical relative belief transform whenever $\sum_x m(x) \neq 0$. When the mass of singletons is zero, instead, C2 amounts to a natural extension of the relative belief operator,

$$\tilde{Bel}^{\text{ext}}(x) \doteq \frac{\sum_{A \supseteq \{x\}: |A| = \min} m(A)}{|A|_{\min} \sum_{A \subseteq \Theta: |A| = \min} m(A)}. \tag{12.14}$$

In fact, the two approximation criteria favour different aspects of the original belief function. C1 focuses on the strength of the evidence carried by focal elements of equal size. Note that the optimal C1 approximations of the plausibility and pignistic transforms are, in principle, distinct:

$$\hat{s}[\tilde{Pl}] = \arg\max_s s k^s, \quad \hat{s}[BetP] = \arg\max_s k^s.$$

The optimal approximation of the pignistic probability will not necessarily be the best approximation of the relative plausibility of singletons as well.

Criterion C2 favours instead the *precision* of the pieces of evidence involved. Let us compare these two approaches in two simple scenarios.

Two opposite scenarios While C1 appears to be a sensible, rational principle (the selected proxy must be the greatest contributor to the actual classical probability transformation), C2 seems harder to justify. Why should one retain only the smallest focal elements, regardless of their mass? The attractive feature of the relative belief of singletons, among all possible C2 approximations, is its simplicity: the original mass is directly redistributed onto the singletons. What about the 'extended' operator (12.14)?

Example 35: scenario 1. *Suppose we wish to approximate the plausibility/pignistic transform of a belief function $Bel : 2^\Theta \to [0,1]$, with BPA $m(A = \{x,y\}) = m(B = \{y,z\}) = \epsilon$, $|A| = |B| = 2$, and $m(\Theta) = 1 - 2\epsilon \gg m(A)$ (Fig. 12.3, left). Its relative plausibility of singletons is given by*

$$\begin{array}{ll} \tilde{Pl}(x) \propto m(A) + m(\Theta), & \tilde{Pl}(y) \propto m(A) + m(B) + m(\Theta), \\ \tilde{Pl}(z) \propto m(B) + m(\Theta), & \tilde{Pl}(w) \propto m(\Theta) \ \forall w \neq x, y, z. \end{array}$$

Its pignistic probability reads as

$$BetP(x) = \frac{m(A)}{2} + \frac{m(\Theta)}{n}, \quad BetP(y) = \frac{m(A) + m(B)}{2} + \frac{m(\Theta)}{n},$$

$$BetP(z) = \frac{m(B)}{2} + \frac{m(\Theta)}{n}, \quad BetP(w) = \frac{m(\Theta)}{n} \ \forall w \neq x, y, z.$$

Both transformations have a profile similar to that of Fig. 12.3 (right) (where we have assumed $m(A) > m(B)$).

Now, according to criterion C1, the best approximation (among all relative mass transforms) of both $\tilde{Pl}$ and $BetP[Bel]$ is given by selecting the focal element of size $n = |\Theta|$, i.e., Θ, as the greatest contributor to both of the convex sums (12.12) and (12.13). However, it is easy to see that this yields as an approximation the uniform probability $p(w) = 1/n$, which is the least informative probability distribution. In particular, the fact that the available evidence supports to a limited extent the singletons x, y and z is completely discarded, and no decision is possible.

If, on the other hand, we operate according to criterion C2, we end up selecting the size-2 focal elements A and B. The resulting approximation is

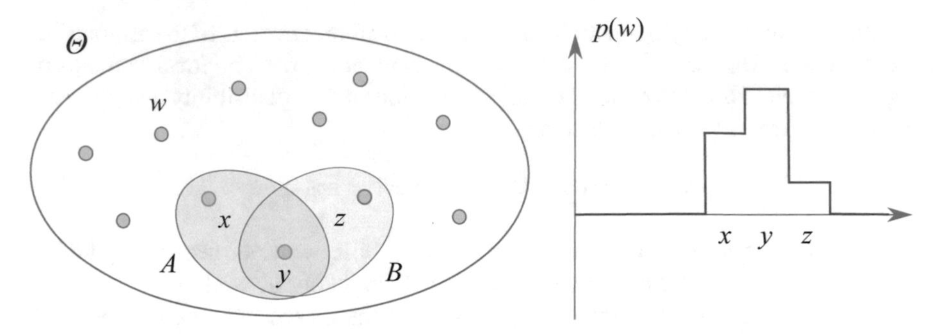

Fig. 12.3: Left: the original belief function in scenario 1. Right: corresponding profile of both the relative plausibility of singletons and the pignistic probability.

$$\tilde{m}_2(x) \propto m(A), \quad \tilde{m}_2(y) \propto m(A) + m(B), \quad \tilde{m}_2(z) \propto m(B),$$

$\tilde{m}_2(w) = 0 \; \forall w \neq x, y, z$. *This mass assignment has the same profile as that of* $\tilde{Pl}$ *or* $BetP[Bel]$ *(Fig. 12.3, right): any decision made according to the latter will correspond to one made on the basis of* $\tilde{Pl}$ *or* $BetP[Bel]$.

In a decision-making sense, therefore, $\tilde{m}_2 = \tilde{Bel}^{ext}$ *is the most correct approximation of both the plausibility and the pignistic transforms. We end up making the same decisions, at a much lower (in general) computational cost.*

Example 36: scenario 2. *Consider now a second scenario, involving a belief function with only two focal elements* A *and* B*, with* $|A| > |B|$ *and* $m(A) \gg m(B)$ *(Fig. 12.4, left). Both the relative plausibility and the pignistic probability have the*

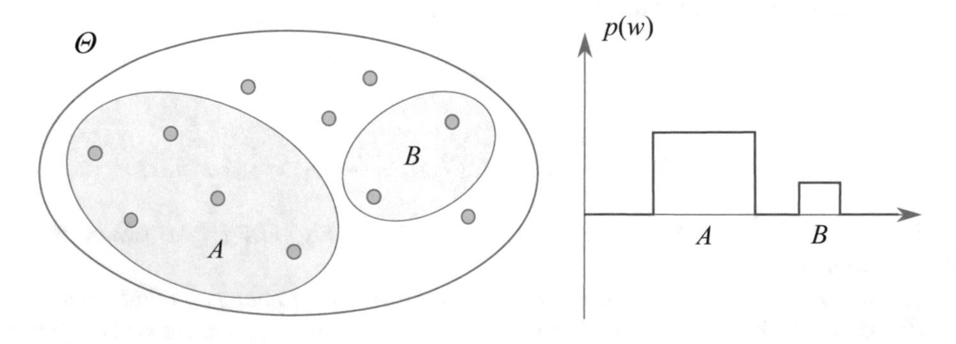

Fig. 12.4: Left: the belief function in the second scenario. Right: corresponding profile of both the relative plausibility of singletons and the pignistic probability.

following values,

$$\tilde{Pl}(w) = BetP(w) \propto m(A) \; w \in A, \quad \tilde{Pl}(w) = BetP(w) \propto m(B) \; w \in B,$$

and correspond to the profile shown in Fig. 12.4 (right).

In this second case, C1 and C2 generate the uniform probability with support in A (as $m(A) \gg m(B)$) and the uniform probability with support in B (as $|B| < |A|$), respectively. Therefore, it is C1 that yields the best approximation of both the plausibility and the pignistic transforms in a decision-making perspective.

A critical look In this discussion, the second scenario corresponds to a situation in which the evidence is highly conflicting. In such a case we are given two opposite decision alternatives, and it is quite difficult to say which one makes more sense. Should we privilege precision or support by evidence?

Some insight into this issue comes from recalling that higher-size focal elements are an expression of 'epistemic' uncertainty (in Smets's terminology), as they come from missing data/lack of information about the problem at hand. Besides, by their nature, they allow a lower resolution for decision-making purposes (in the second scenario above, if we trust C1, we are left uncertain about whether to pick one of $|A|$ outcomes, while if we adopt C2, the uncertainty is restricted to $|B|$ outcomes).

In conclusion, it is not irrational, in the case of conflicting evidence, to judge larger-size focal elements 'less reliable' (as carriers of greater ignorance) than more focused focal elements. It follows that there should be a preference for the approximation criterion C2, which ultimately supports the case for the relative belief operator and its natural extension (12.14).

12.5 Geometry in the space of pseudo-belief functions

After studying the dual properties of the pair of epistemic transforms, and proposing a generalisation of the relative belief operator in singular cases, it is time to complete our understanding of the geometry of probability transformations by considering the geometric behaviour of epistemic mappings.

In Section 11.1, we had a quick glance at their geometry in the binary case. Using the terminology we acquired in the last two chapters, we learned that transformations in the affine family coincide on a binary frame. On the other hand, we saw that the members of the epistemic family, the relative belief and the relative plausibility of singletons, do not follow the same pattern. To understand the geometry of transformations in the epistemic family, we need to introduce a pair of pseudo-belief functions related to them.

12.5.1 Plausibility of singletons and relative plausibility

Let us define the *plausibility of singletons* as the pseudo-belief function $\overline{Pl} : 2^\Theta \to [0,1]$ with Möbius inverse $\overline{m}_{\overline{Pl}} : 2^\Theta \to \mathbb{R}$ given by

$$\overline{m}_{\overline{Pl}}(x) = Pl(x) \; \forall x \in \Theta, \qquad \overline{m}_{\overline{Pl}}(\Theta) = 1 - \sum_x Pl(x) = 1 - k_{Pl},$$
$$\overline{m}_{\overline{Pl}}(A) = 0 \quad \forall A \subseteq \Theta : |A| \neq 1, n.$$

$\overline{m}_{\overline{Pl}}$ satisfies the normalisation constraint:

$$\sum_{A \subseteq \Theta} \overline{m}_{\overline{Pl}}(A) = \sum_{x \in \Theta} Pl(x) + \left(1 - \sum_{x \in \Theta} Pl(x)\right) = 1.$$

Then, as $1 - k_{Pl} \leq 0$, $\overline{Pl}$ is a pseudo-belief function (Section 7.1). Note that $\overline{Pl}$ is *not*, however, a plausibility function.

In the belief space, $\overline{Pl}$ is represented by the vector

$$\overline{Pl} = \sum_{x \in \Theta} Pl(x)\ Bel_x + (1 - k_{Pl})\ Bel_\Theta = \sum_{x \in \Theta} Pl(x)\ Bel_x, \tag{12.15}$$

as $Bel_\Theta = \mathbf{0}$ is the origin of the reference frame in $\mathbb{R}^{N-2}$.

Theorem 60. *$\widetilde{Pl}$ is the intersection of the line joining the vacuous belief function Bel_Θ and the plausibility of singletons $\overline{Pl}$ with the probability simplex.*[76]

Proof. By (4.59) and (12.15), we have that

$$\widetilde{Pl} = \sum_{x \in \Theta} \widetilde{Pl}(x) Bel_x = \frac{\overline{Pl}}{k_{Pl}}.$$

Since $Bel_\Theta = \mathbf{0}$ is the origin of the reference frame, $\widetilde{Pl}$ lies on the segment $Cl(\overline{Pl}, Bel_\Theta)$. This, in turn, implies $\widetilde{Pl} = Cl(\overline{Pl}, Bel_\Theta) \cap \mathcal{P}$.

□

The geometry of $\widetilde{Pl}$ depends on that of $\overline{Pl}$ through Theorem 60. In the binary case $\overline{Pl} = Pl$, and we go back to the situation of Fig. 11.1.

12.5.2 Belief of singletons and relative belief

By the definition of the intersection probability $p[Bel]$ (11.12), given in Section 12.2.5, it follows that

$$\begin{aligned} p[Bel] &= \sum_{x \in \Theta} m(x) Bel_x + \beta[Bel] \sum_{x \in \Theta} (Pl(x) - m(x)) Bel_x \\ &= (1 - \beta[Bel]) \sum_{x \in \Theta} m(x) Bel_x + \beta[Bel] \sum_{x \in \Theta} Pl(x) Bel_x. \end{aligned} \tag{12.16}$$

Analogously to what was done for the plausibility of singletons, we can define the belief function (*belief of singletons*) $\overline{Bel} : 2^\Theta \to [0,1]$ with as basic probability assignment

$$m_{\overline{Bel}}(x) = m(x), \quad m_{\overline{Bel}}(\Theta) = 1 - k_{Bel}, \quad m_{\overline{Bel}}(A) = 0\ \forall A \subseteq \Theta : |A| \neq 1, n,$$

[76]This result, at least in the binary case, appeared also in [382].

where the scalar quantity $k_{Bel} \doteq \sum_{x\in\Theta} m(x)$ (11.13) measures, as usual, the total belief (and mass) of singletons. The belief of singletons assigns to Θ all the mass that Bel gives to non-singletons. In the belief space, $\overline{Bel}$ is represented by the vector

$$\overline{Bel} = \sum_{x\in\Theta} m(x) Bel_x + (1 - k_{Bel}) Bel_\Theta = \sum_{x\in\Theta} m(x) Bel_x, \tag{12.17}$$

(as, again, $Bel_\Theta = \mathbf{0}$). Equation (12.16) can then be written as

$$p[Bel] = (1 - \beta[Bel])\, \overline{Bel} + \beta[Bel]\, \overline{Pl}. \tag{12.18}$$

Namely, the intersection probability is the convex combination of the belief and plausibility of singletons with coefficient $\beta[Bel]$.

In the binary case, $\overline{Bel} = Bel$ and $\overline{Pl} = Pl$, so that the plausibility of singletons is a plausibility function (Fig. 11.1).

12.5.3 A three-plane geometry

The geometry of the relative plausibility and belief of singletons can therefore be reduced to that of $\overline{Pl}, \overline{Bel}$.

As we know, a belief function Bel and the corresponding plausibility function Pl have the same coordinates with respect to the vertices Bel_A, Pl_A of the belief and the plausibility space, respectively:

$$Bel = \sum_{\emptyset \neq A \subseteq \Theta} m(A) Bel_A \quad \longleftrightarrow \quad Pl = \sum_{\emptyset \neq A \subseteq \Theta} m(A) Pl_A.$$

Just as the latter form a pair of 'dual' vectors in the respective spaces, the plausibility $\overline{Pl}$ and belief $\overline{Bel}$ of singletons have duals (which we can denote by $\widehat{Pl}$ and $\widehat{Bel}$) characterised by having the same coordinates in the plausibility space: $\overline{Bel} \leftrightarrow \widehat{Bel}$, $\overline{Pl} \leftrightarrow \widehat{Pl}$. They can be written as

$$\begin{aligned} \widehat{Bel} &= \sum_{x\in\Theta} m(x) Pl_x + (1 - k_{Bel}) Pl_\Theta = \overline{Bel} + (1 - k_{Bel}) Pl_\Theta, \\ \widehat{Pl} &= \sum_{x\in\Theta} Pl(x) Pl_x + (1 - k_{Pl}) Pl_\Theta = \overline{Pl} + (1 - k_{Pl}) Pl_\Theta \end{aligned} \tag{12.19}$$

(as $Pl_x = Bel_x$ for all $x \in \Theta$), where, again, $Pl_\Theta = \mathbf{1}$.

We can prove the following theorem (see the appendix to this chapter).

Theorem 61. *The line passing through the duals (12.19) of the plausibility of singletons (12.15) and the belief of singletons (12.17) crosses* $p[Bel]$ *too, and we have that*

$$\beta[Bel]\, (\widehat{Pl} - \widehat{Bel}) + \widehat{Bel} = p[Bel] = \beta[Bel]\, (\overline{Pl} - \overline{Bel}) + \overline{Bel}. \tag{12.20}$$

If $k_{Bel} \neq 0$, the geometry of the relative plausibility and belief of singletons can therefore be described in terms of the three planes

$$a(\overline{Pl}, p[Bel], \widehat{Pl}), \quad a(Bel_\Theta, \widetilde{Pl}, Pl_\Theta), \quad a(Bel_\Theta, \widetilde{Bel}, Pl_\Theta)$$

(see Fig. 12.5), where $\widetilde{Bel} = \overline{Bel}/k_{Bel}$ is the relative belief of singletons. More

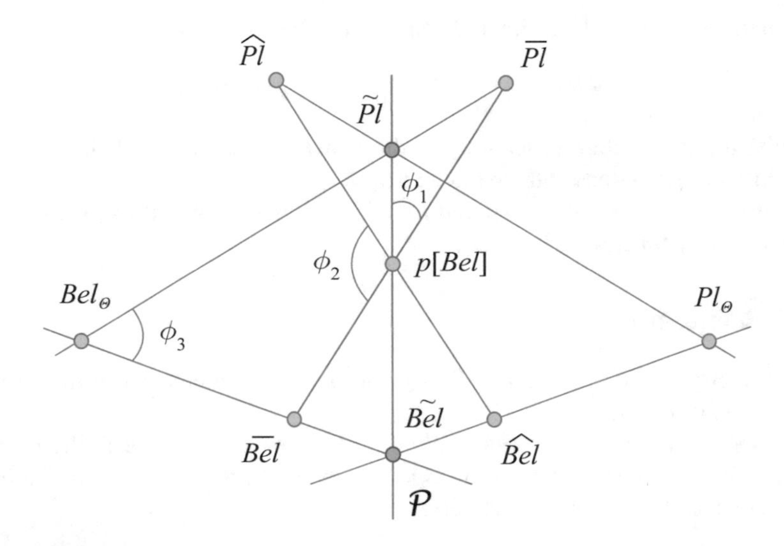

Fig. 12.5: Planes and angles describing the geometry of the relative plausibility and belief of singletons, in terms of the plausibility of singletons $\overline{Pl}$ and the belief of singletons $\overline{Bel}$. Geometrically, two lines or three points are sufficient to uniquely determine a plane passing through them. The two lines $a(\overline{Bel}, \overline{Pl})$ and $a(\widehat{Bel}, \widehat{Pl})$ uniquely determine a plane $a(\overline{Bel}, p[Bel], \widehat{Bel})$. Two other planes are uniquely determined by the origins of the belief and plausibility spaces, Bel_Θ and Pl_Θ, together with either the relative plausibility of singletons $\widetilde{Pl}$ or the relative belief of singletons $\widetilde{Bel}$: $a(Bel_\Theta, \widetilde{Pl}, Pl_\Theta)$ (top of the diagram) and $a(Bel_\Theta, \widetilde{Bel}, Pl_\Theta)$ (bottom), respectively. The angles $\phi_1[Bel], \phi_2[Bel], \phi_3[Bel]$ are all independent, as the value of each of them reflects a different property of the original belief function Bel. The original belief function Bel and plausibility function Pl do not appear here, for the sake of simplicity. They play a role only through the related plausibility of singletons (12.15) and belief of singletons (12.17).

particularly:

1. $p[Bel]$ is the intersection of $a(\overline{Bel}, \overline{Pl})$ and $a(\widehat{Bel}, \widehat{Pl})$, and has the same affine coordinate on the two lines (Section 12.5.1). Those two lines then span a plane, which we can denote by

$$a(\overline{Bel}, p[Bel], \widehat{Bel}) = a(\overline{Pl}, p[Bel], \widehat{Pl}).$$

2. Furthermore, by definition,

$$\widetilde{Pl} - Bel_\Theta = \frac{\overline{Pl} - Bel_\Theta}{k_{Pl}}, \tag{12.21}$$

whereas (12.19) implies $\widetilde{Pl} = \overline{Pl}/k_{Pl} = [\widehat{Pl} - (1-k_{Pl})Pl_\Theta]/k_{Pl}$, so that

$$\widetilde{Pl} - Pl_\Theta = \frac{\widehat{Pl} - Pl_\Theta}{k_{Pl}}. \tag{12.22}$$

By comparing (12.21) and (12.22), we realise that $\widetilde{Pl}$ has the same affine coordinate on the two lines $a(Bel_\Theta, \overline{Pl})$ and $a(Pl_\Theta, \widehat{Pl})$, which intersect exactly in $\widetilde{Pl}$. The functions $Bel_\Theta, Pl_\Theta, \widetilde{Pl}, \overline{Pl}$ and $\widehat{Pl}$ therefore determine another plane, which we can denote by

$$a(Bel_\Theta, \widetilde{Pl}, Pl_\Theta).$$

3. Analogously, by definition, $\widetilde{Bel} - Bel_\Theta = (\overline{Bel} - Bel_\Theta)/k_{Bel}$, while (12.19) yields $\widetilde{Bel} - Pl_\Theta = (\widehat{Bel} - Pl_\Theta)/k_{Bel}$. The relative belief of singletons then has the same affine coordinate on the two lines $a(Bel_\Theta, \overline{Bel})$ and $a(Pl_\Theta, \widehat{Bel})$. The latter intersect exactly in $\widetilde{Bel}$. The quantities $Bel_\Theta, Pl_\Theta, \widetilde{Bel}, \overline{Bel}$ and $\widehat{Bel}$ thus determine a single plane, denoted by

$$a(Bel_\Theta, \widetilde{Bel}, Pl_\Theta).$$

12.5.4 A geometry of three angles

In the binary case, $Bel = \overline{Bel} = \widehat{Pl} = [m(x), m(y)]'$, $Pl = \overline{Pl} = \widehat{Bel} = [1 - m(y), 1 - m(x)]'$ and all these quantities are coplanar. This suggests a description of the geometry of $\widetilde{Pl}, \widetilde{Bel}$ in terms of the three angles

$$\phi_1[Bel] = \widehat{\widetilde{Pl}\ p[Bel]\ \overline{Pl}}, \ \phi_2[Bel] = \widehat{\overline{Bel}\ p[Bel]\ \widehat{Pl}}, \ \phi_3[Bel] = \widehat{\widetilde{Bel}\ Bel_\Theta\ \widetilde{Pl}} \tag{12.23}$$

(see Fig. 12.5 again). These angles are all independent, and each of them has a distinct interpretation in terms of degrees of belief, as different values of them reflect different properties of the belief function Bel and the associated probability transformations.

Orthogonality condition for $\phi_1[Bel]$ We know that the dual line $a(Bel, Pl)$ is always orthogonal to $\mathcal{P}$ (Section 11.2). The line $a(\overline{Bel}, \overline{Pl})$, though, is *not* in general orthogonal to the probabilistic subspace.

Formally, the simplex $\mathcal{P} = Cl(Bel_x, x \in \Theta)$ determines an affine (or vector) space $a(\mathcal{P}) = a(Bel_x, x \in \Theta)$. A set of generators for $a(\mathcal{P})$ is formed by the $n-1$ vectors $Bel_y - Bel_x$, $\forall y \in \Theta$, $y \neq x$, after we have chosen an arbitrary element $x \in \Theta$ as a reference. The non-orthogonality of $a(\overline{Bel}, \overline{Pl})$ and $a(\mathcal{P})$ can therefore be expressed by saying that, for at least one such basis vector, the scalar product $\langle \cdot \rangle$ with the difference vector $\overline{Pl} - \overline{Bel}$ (which generates the line $a(\overline{Bel}, \overline{Pl})$) is non-zero:

$$\exists y \neq x \in \Theta \text{ s.t. } \langle \overline{Pl} - \overline{Bel}, Bel_y - Bel_x \rangle \neq 0. \tag{12.24}$$

Recall that $\phi_1[Bel]$, as defined in (12.23), is the angle between $a(\overline{Bel}, \overline{Pl})$ and the line $a(\widetilde{Bel}, \widetilde{Pl})$, which lies in the probabilistic subspace.

The condition under which orthogonality holds has a significant interpretation in terms of the uncertainty expressed by the belief function Bel about the probability value of each singleton.

Theorem 62. *The line $a(\overline{Bel}, \overline{Pl})$ is orthogonal to the vector space generated by $\mathcal{P}$ (and therefore $\phi_1[Bel] = \pi/2$) if and only if*

$$\sum_{A \supsetneq \{x\}} m(A) = Pl(x) - m(x) = const \quad \forall x \in \Theta.$$

Relative uncertainty of singletons If Bel is Bayesian, $Pl(x) - m(x) = 0 \; \forall x \in \Theta$. If Bel is *not* Bayesian, there exists at least a singleton x such that $Pl(x) - m(x) > 0$.

In this case we can define the probability function

$$R[Bel] \doteq \sum_{x \in \Theta} \frac{Pl(x) - m(x)}{k_{Pl} - k_{Bel}} Bel_x = \frac{\overline{Pl} - \overline{Bel}}{k_{Pl} - k_{Bel}}. \tag{12.25}$$

The value $R[Bel](x)$ indicates how much the uncertainty $Pl(x) - m(x)$ in the probability value on x 'weighs' in the total uncertainty in the probabilities of singletons. It is then natural to call it the *relative uncertainty in the probabilities of singletons*. When Bel is Bayesian, $R[Bel]$ does not exist.

Corollary 13. *The line $a(\overline{Bel}, \overline{Pl})$ is orthogonal to $\mathcal{P}$ iff the relative uncertainty in the probabilities of singletons is the uniform probability: $R[Bel](x) = 1/|\Theta|$ for all $x \in \Theta$.*

If this holds, the evidence carried by Bel yields the same uncertainty in the probability value of all singletons. By the definition of $p[Bel]$ (11.12), we have that

$$\begin{aligned} p[Bel](x) &= m(x) + \beta[Bel](Pl(x) - m(x)) \\ &= m(x) + \frac{1 - k_{Bel}}{\sum_{y \in \Theta}(Pl(y) - m(y))}(Pl(x) - m(x)) \\ &= m(x) + (1 - k_{Bel})R[Bel](x) = m(x) + \frac{1 - k_{Bel}}{n}, \end{aligned}$$

namely, the intersection probability reassigns the mass originally given by Bel to non-singletons to each singleton on an equal basis.

Dependence of ϕ_2 on the relative uncertainty The value of $\phi_2[Bel]$ also depends on the relative uncertainty in the probabilities of singletons.

Theorem 63. *Let us denote by $\mathbf{1} = Pl_\Theta$ the vector $[1, \ldots, 1]'$. Then*

$$\cos(\pi - \phi_2[Bel]) = 1 - \frac{\langle \mathbf{1}, R[Bel] \rangle}{\|R[Bel]]\|^2}, \tag{12.26}$$

where $\langle \mathbf{1}, R[Bel] \rangle$ denotes the scalar product between the unit vector $\mathbf{1} = [1, \ldots, 1]'$ and the vector $R[Bel] \in \mathbb{R}^{N-2}$.

We can observe that:

1. $\phi_2[Bel] = \pi$ ($\cos = 1$) iff $\langle \mathbf{1}, R[Bel] \rangle = 0$. But this never actually happens, as $\langle \mathbf{1}, P \rangle = 2^{n-1} - 1 \; \forall P \in \mathcal{P}$ (see the proof of Theorem 63).
2. $\phi_2[Bel] = 0$ ($\cos = -1$) iff $\|R[Bel]\|^2 = \langle \mathbf{1}, R[Bel] \rangle / 2$. This situation also never materialises for belief functions defined on non-trivial frames of discernment.

Theorem 64. *$\phi_2[Bel] \neq 0$ and the lines $a(\overline{Bel}, \overline{Pl})$, $a(\widehat{Bel}, \widehat{Pl})$ never coincide for any $Bel \in \mathcal{B}$ whenever $|\Theta| > 2$; on the contrary, $\phi_2[Bel] = 0 \; \forall Bel \in \mathcal{B}$ whenever $|\Theta| \leq 2$.*

Example 37: binary versus ternary frame. *Let us confirm this by comparing the situations for the twp-element and three-element frames. If $\Theta = \{x, y\}$, we have that $Pl(x) - m(x) = m(\Theta) = Pl(y) - m(y)$, and the relative uncertainty function is*

$$R[Bel] = \frac{1}{2} Bel_x + \frac{1}{2} Bel_y = \overline{\mathcal{P}} \quad \forall Bel \in \mathcal{B}_2$$

(where $\overline{\mathcal{P}}$ denotes the uniform probability on Θ; see Fig. 12.6), and $R[Bel] = \frac{1}{2}\mathbf{1} = \frac{1}{2} Pl_\Theta$. In the binary case, the angle $\phi_2[Bel]$ is zero for all belief functions. As we have learned, $Bel = \overline{Bel} = \widehat{Pl}$, $Pl = \overline{Pl} = \widehat{Bel}$ and the geometry of the epistemic family is planar.

On the other hand, if $\Theta = \{x, y, z\}$, not even the vacuous belief function Bel_Θ satisfies condition 2. In that case, $R[Bel_\Theta] = \overline{\mathcal{P}} = \frac{1}{3} Bel_x + \frac{1}{3} Bel_y + \frac{1}{3} Bel_z$ and R is still the uniform probability. But $\langle R[Bel_\Theta], \mathbf{1} \rangle = 3$, while

$$\langle R[Bel_\Theta], R[Bel_\Theta] \rangle = \langle \overline{\mathcal{P}}, \overline{\mathcal{P}} \rangle = \left\langle \left[\frac{1}{3} \frac{1}{3} \frac{1}{3} \frac{2}{3} \frac{2}{3} \frac{2}{3} \right]', \left[\frac{1}{3} \frac{1}{3} \frac{1}{3} \frac{2}{3} \frac{2}{3} \frac{2}{3} \right]' \right\rangle = \frac{15}{9}.$$

Unifying condition for the epistemic family The angle $\phi_3[Bel]$ is related to the condition under which the relative plausibility of singletons and relative belief of singletons coincide. In fact, this angle is zero iff $\widetilde{Bel} = \widetilde{Pl}$, which is equivalent to

$$\frac{m(x)}{k_{Bel}} = \frac{Pl(x)}{k_{Pl}} \quad \forall x \in \Theta.$$

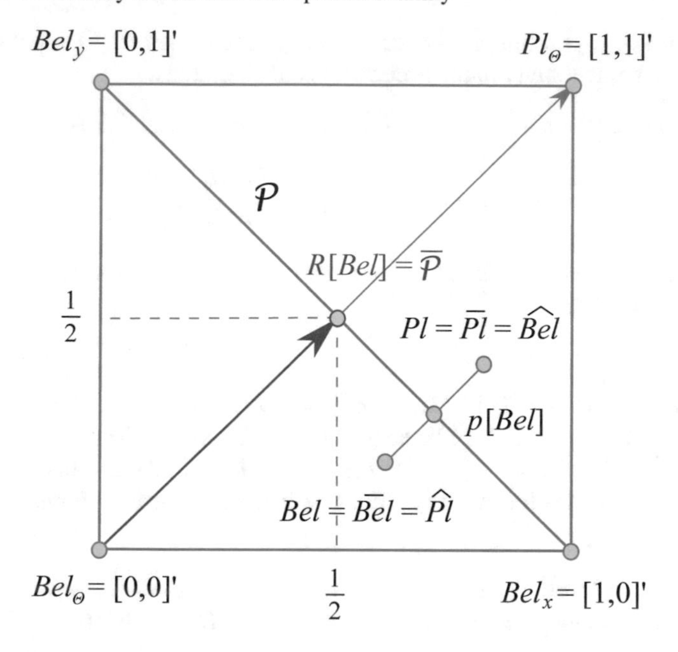

Fig. 12.6: The angle $\phi_2[Bel]$ is zero for all belief functions in the size-2 frame $\Theta = \{x, y\}$, as $R[Bel] = [\frac{1}{2}, \frac{1}{2}]'$ is parallel to $Pl_\Theta = \mathbf{1}$ for all Bel.

Again, this necessary and sufficient condition for $\phi_3[Bel] = 0$ can be expressed in terms of the relative uncertainty in the probabilities of singletons, as

$$
\begin{aligned}
R[Bel](x) &= \frac{Pl(x) - m(x)}{k_{Pl} - k_{Bel}} \\
&= \frac{1}{k_{Pl} - k_{Bel}} \left(\frac{k_{Pl}}{k_{Bel}} m(x) - m(x) \right) = \frac{m(x)}{k_{Bel}} \quad \forall x \in \Theta,
\end{aligned}
\tag{12.27}
$$

i.e., $R[Bel] = \widetilde{Bel}$, with $R[Bel]$ 'squashing' $\widetilde{Pl}$ onto $\widetilde{Bel}$ from the outside. In this case the quantities $\overline{Pl}, \widehat{Pl}, \widetilde{Pl}, p[Bel], \overline{Bel}, \widehat{Bel}, \widetilde{Bel}$ all lie in the same plane.

12.5.5 Singular case

We need to pay some attention to the singular case (from a geometric point of view) in which the relative belief of singletons does not exist, $k_{Bel} = \sum_x m(x) = 0$.

As a matter of fact, the belief of singletons $\overline{Bel}$ still exists even in this case, and by (12.17), $\overline{Bel} = Bel_\Theta$ whereas, by duality, $\widehat{Bel} = Pl_\Theta$. Recall the description in terms of planes that we gave in Section 12.5.3. In this case the first two planes,

$$
\begin{aligned}
a(Bel, p[Bel], \widehat{Bel}) &= a(a(\widehat{Bel}, \widehat{Pl}), a(\overline{Bel}, \overline{Pl})) \\
&= a(a(Bel_\Theta, \widehat{Pl}), a(Pl_\Theta, \overline{Pl})) = a(Bel_\Theta, \widetilde{Pl}, Pl_\Theta)
\end{aligned}
$$

coincide, while the third one, $a(Bel_\Theta, \widetilde{Bel}, Pl_\Theta)$, simply does not exist. The geom-

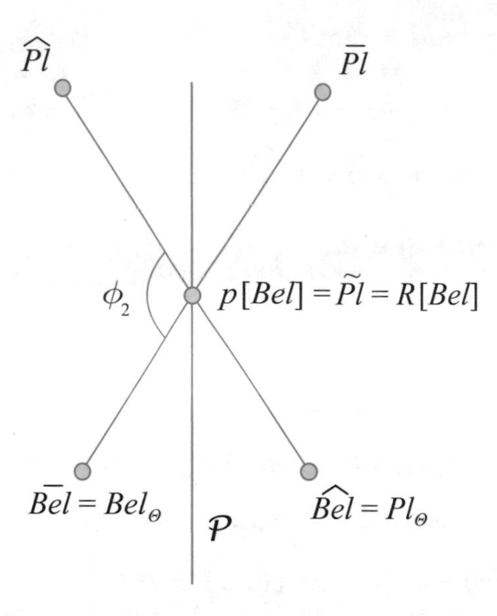

Fig. 12.7: Geometry of the relative plausibility of singletons and relative uncertainty in the probabilities of singletons in the singular case when $k_{Bel} = \sum_x m(x) = 0$.

etry of the epistemic family reduces to a planar one (see Fig. 12.7), which depends only on the angle $\phi_2[Bel]$. It is remarkable that, in this case,

$$p[Bel](x) = m(x) + \frac{1 - k_{Bel}}{k_{Pl} - k_{Bel}}[Pl(x) - m(x)] = \frac{1}{k_{Pl}} Pl(x) = \widetilde{Pl}(x).$$

Theorem 65. *If a belief function Bel does not admit a relative belief of singletons (as Bel assigns zero mass to all singletons), then its relative plausibility of singletons and intersection probability coincide.*

Also, in this case the relative uncertainty in the probabilities of singletons coincides with the relative plausibility of singletons too: $R[Bel] = \widetilde{Pl} = p[Bel]$ (see (12.25)).

12.6 Geometry in the probability simplex

The geometry of the relative belief and plausibility of singletons in the space of all (pseudo-)belief functions is a function of three angles and planes. It is also interesting, however, to see how they behave as distributions in the probability simplex.

We can observe for instance that, as

$$\begin{aligned}R[Bel](k_{Pl} - k_{Bel}) &= \overline{Pl} - \overline{Bel} = \widetilde{Pl} \cdot k_{Pl} - \widetilde{Bel} \cdot k_{Bel} \\ &= \widetilde{Pl} \cdot k_{Pl} - \widetilde{Bel} \cdot k_{Bel} + k_{Pl} \cdot \widetilde{Bel} - k_{Pl} \cdot \widetilde{Bel} \\ &= k_{Pl}(\widetilde{Pl} - \widetilde{Bel}) + \widetilde{Bel}(k_{Pl} - k_{Bel}),\end{aligned}$$

$R[Bel]$ lies on the line joining $\widetilde{Bel}$ and $\widetilde{Pl}$:

$$R[Bel] = \widetilde{Bel} + \frac{k_{Pl}}{k_{Pl} - k_{Bel}}(\widetilde{Pl} - \widetilde{Bel}). \tag{12.28}$$

Let us study the situation in a simple example.

Example 38: geometry in the three-element frame. *Consider a belief function Bel_1 with basic probability assignment*

$$m_1(x) = 0.5, \quad m_1(y) = 0.1, \quad m_1(\{x,y\}) = 0.3, \quad m_1(\{y,z\}) = 0.1$$

on $\Theta = \{x, y, z\}$. The probability intervals of the singletons have widths

$$\begin{aligned}Pl_1(x) - m_1(x) &= m_1(\{x,y\}) = 0.3, \\ Pl_1(y) - m_1(y) &= m_1(\{x,y\}) + m_1(\{y,z\}) = 0.4, \\ Pl_1(z) - m_1(z) &= m_1(\{y,z\}) = 0.1.\end{aligned}$$

Their relative uncertainty is therefore

$$R[Bel_1](x) = \frac{3}{8}, \quad R[Bel_1](y) = \frac{1}{2}, \quad R[Bel_1](z) = \frac{1}{8}.$$

$R[Bel_1]$ is plotted as a point of the probability simplex $\mathcal{P} = Cl(Bel_x, Bel_y, Bel_z)$ in Fig. 12.8. Its (Euclidean) distance from the uniform probability $\overline{\mathcal{P}} = [\frac{1}{3}, \frac{1}{3}, \frac{1}{3}]'$ in $\mathcal{P}$ is

$$\begin{aligned}\|\overline{\mathcal{P}} - R[Bel_1]\| &= \left[\sum_x \left(\frac{1}{3} - R[Bel_1](x)\right)^2\right]^{1/2} \\ &= \left[\left(\frac{1}{3} - \frac{3}{8}\right)^2 + \left(\frac{1}{3} - \frac{1}{2}\right)^2 + \left(\frac{1}{3} - \frac{1}{8}\right)^2\right]^{1/2} = 0.073.\end{aligned}$$

The related intersection probability (as $k_{Bel_1} = 0.6$, $k_{Pl_1} = 0.8 + 0.5 + 0.1 = 1.4$ and $\beta[Bel_1] = (1 - 0.6)/(1.4 - 0.6) = \frac{1}{2}$),

$$p[Bel_1](x) = 0.5 + \frac{1}{2}0.3 = 0.65, \quad p[Bel_1](y) = 0.1 + \frac{1}{2}0.4 = 0.3,$$
$$p[Bel_1](z) = 0 + \frac{1}{2}0.1 = 0.05,$$

is plotted as a square (second from the left) on the edge of the dotted triangle in Fig. 12.8.

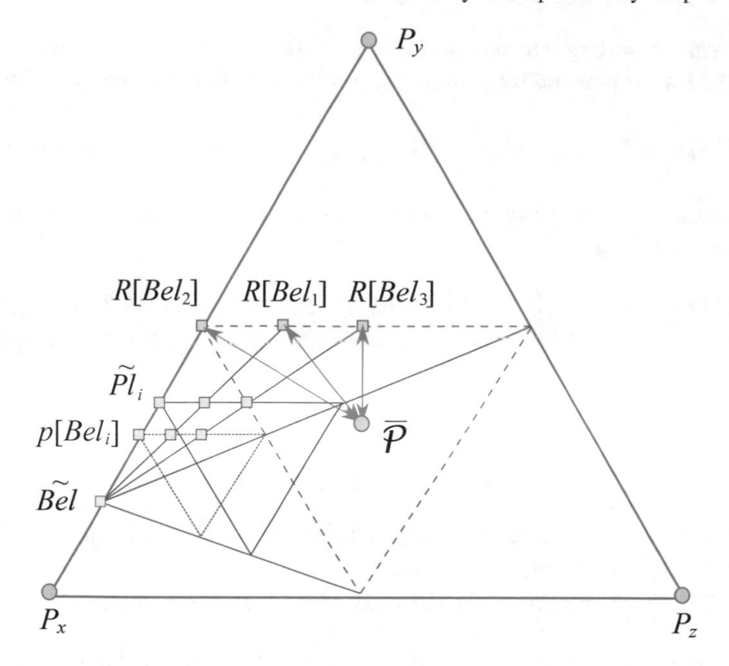

Fig. 12.8: Locations of the members of the epistemic family in the probability simplex $\mathcal{P} = Cl(Bel_x, Bel_y, Bel_z)$ for a three-element frame $\Theta = \{x, y, z\}$. The relative uncertainty in the probability of singletons $R[Bel]$, the relative plausibility of singletons $\widetilde{Pl}$ and the intersection probability $p[Bel]$ for the family of belief functions on the three-element frame defined by the mass assignment (12.29) lie in the dashed, solid and dotted triangles, respectively. The locations of $R[Bel_1]$, $R[Bel_2]$ and $R[Bel_3]$ for the three belief functions Bel_1, Bel_2 and Bel_3 discussed in the example are shown. The relative plausibility of singletons and the intersection probability for the same BFs appear in the corresponding triangles in the same order. The relative belief of singletons $\widetilde{Bel}$ lies on the bottom-left square for all the belief functions of the family (12.29) considered.

A larger uncertainty in the probability of singletons is associated with Bel_2,

$$m_2(x) = 0.5, \quad m_2(y) = 0.1, \quad m_2(z) = 0, \quad m_2(\{x, y\}) = 0.4,$$

in which all the higher-size mass is assigned to a single focal element $\{x, y\}$. In that case $Pl_2(x) - m_2(x) = 0.4$, $Pl_2(y) - m_2(y) = 0.4$, $Pl_2(z) - m_2(z) = 0$, so that the relative uncertainty in the probabilities of singletons is $R[Bel_2](x) = \frac{1}{2}$, $R[Bel_2](y) = \frac{1}{2}$, $R[Bel_2](z) = 0$, with a Euclidean distance from $\overline{\mathcal{P}}$ equal to

$$d_2 = \left[\left(\frac{1}{6}\right)^2 + \left(\frac{1}{6}\right)^2 + \left(\frac{1}{3}\right)^2 \right]^{1/2} = 0.408.$$

The corresponding intersection probability (as $\beta[Bel_2] = (1 - 0.6)/0.8$ is still equal to $\frac{1}{2}$) is the first square from the left on the edge of the above dotted triangle:

$$p[Bel_2](x) = 0.5+\frac{1}{2}0.4 = 0.7, \ \ p[Bel_2](y) = 0.1+\frac{1}{2}0.4 = 0.3, \ \ p[Bel_2](z) = 0.$$

If we spread the mass of non-singletons onto two focal elements to get a third belief function Bel_3,

$$m_3(x) = 0.5, \quad m_3(y) = 0.1, \quad m_3(\{x,y\}) = 0.2, \quad m_3(\{y,z\}) = 0.2,$$

we get the following uncertainty intervals,

$$pl_3(x) - m_3(x) = 0.2, \quad Pl_3(y) - m_3(y) = 0.4,$$

$$Pl_3(z) - m_3(z) = 0.2,$$

which correspond to $R[Bel_3](x) = \frac{1}{4}$, $R[Bel_3](y) = \frac{1}{2}$, $R[Bel_3](z) = \frac{1}{4}$ and a distance from $\overline{\mathcal{P}}$ of 0.2041.

Finally, the intersection probability assumes the following values:

$$p[Bel_3](x) = 0.5 + \frac{1}{2}0.2 = 0.6, \quad p[Bel_3](y) = 0.1 + \frac{1}{2}0.4 = 0.3,$$
$$p[Bel_3](z) = 0 + \frac{1}{2}0.2 = 0.1.$$

Assigning a certain mass to the singletons determines a set of belief functions compatible with such a probability assignment. In our example, Bel_1, Bel_2 and Bel_3 all belong to the following such set:

$$\left\{ Bel : m(x) = 0.5, m(y) = 0.1, m(z) = 0, \sum_{|A|>1} m(A) = 0.4 \right\}. \qquad (12.29)$$

The corresponding relative uncertainty in the probability of singletons is constrained to live in the simplex delimited by the dashed lines in Fig. 12.8. Of the three belief functions considered here, Bel_2 corresponds to the maximal imbalance between the masses of size-2 focal elements, as it assigns the whole mass to $\{x,y\}$. As a result, $R[Bel_2]$ has the maximum distance from the uniform probability $\overline{\mathcal{P}}$. The belief function Bel_3, instead, spreads the mass equally between $\{x,y\}$ and $\{y,z\}$. As a result, $R[Bel_3]$ has the minimum distance from $\overline{\mathcal{P}}$. Similarly, the intersection probability (12.16) is constrained to live in the simplex delimited by the dotted lines.

All these belief functions have, by definition (4.61), the same relative belief $\widetilde{Bel}$. The lines determined by $R[Bel]$ and $p[Bel]$ for each admissible belief function Bel in the set (12.29) intersect in

$$\widetilde{Bel}(x) = \frac{5}{6}, \quad \widetilde{Bel}(y) = \frac{1}{6}, \quad Bel(z) = 0$$

(square at the bottom left). This is due to the fact that

$$p[Bel] = \sum_x \Big[m(x) + (1 - k_{Bel}) R[Bel](x)\Big] Bel_x$$
$$= \sum_x m(x) Bel_x + (1 - k_{Bel}) R[Bel] = k_{Bel} \widetilde{Bel} + (1 - k_{Bel}) R[Bel],$$

so that $\widetilde{Bel}$ is collinear with $R[Bel], p[Bel]$.

Finally, the associated relative plausibilities of singletons also live in a simplex (delimited by the solid lines in Fig. 12.8). The probabilities $\widetilde{Pl}_1$, $\widetilde{Pl}_2$ and $\widetilde{Pl}_3$ are identified in the figure by squares, located in the same order as above. According to (12.28), $\widetilde{Pl}$, $\widetilde{Bel}$ and $R[Bel]$ are also collinear for all belief functions Bel.

Example 39: singular case in the three-element frame. *Let us pay attention to the singular case. For each belief function Bel such that $m(x) = m(y) = m(z) = 0$, the plausibilities of the singletons of a size-3 frame are*

$$\begin{aligned} Pl(x) &= m(\{x,y\}) + m(\{x,z\}) + m(\Theta) = 1 - m(\{y,z\}), \\ Pl(y) &= m(\{x,y\}) + m(\{y,z\}) + m(\Theta) = 1 - m(\{x,z\}), \\ Pl(z) &= m(\{x,z\}) + m(\{y,z\}) + m(\Theta) = 1 - m(\{x,y\}). \end{aligned}$$

Furthermore, by hypothesis, $Pl(w) - m(w) = Pl(w)$ for all $w \in \Theta$, so that

$$\sum_w \Big(Pl(w) - m(w)\Big) = Pl(x) + Pl(y) + Pl(z)$$

$$= 2\Big(m(\{x,y\}) + m(\{x,z\}) + m(\{y,z\})\Big) + 3m(\Theta) = 2 + m(\Theta),$$

and we get

$$\beta[Bel] = \frac{1 - \sum_w m(w)}{\sum_w (Pl(w) - m(w))} = \frac{1}{\sum_w Pl(w)} = \frac{1}{2 + m(\Theta)}.$$

Therefore

$$R[Bel](x) = \frac{Pl(x) - m(x)}{\sum_w (Pl(w) - m(w))} = \frac{1 - m(\{y,z\})}{2 + m(\Theta)},$$

$$R[Bel](y) = \frac{1 - m(\{x,z\})}{2 + m(\Theta)}, \quad R[Bel](z) = \frac{1 - m(\{x,y\})}{2 + m(\Theta)},$$

$$p[Bel](x) = m(x) + \beta[Bel](Pl(x) - m(x)) = \beta[Bel]Pl(x) = \frac{1 - m(\{y,z\})}{2 + m(\Theta)},$$

$$p[Bel](y) = \frac{1 - m(\{x,z\})}{2 + m(\Theta)}, \quad p[Bel](z) = \frac{1 - m(\{x,y\})}{2 + m(\Theta)},$$

$$\widetilde{Pl}(x) = \frac{Pl(x)}{\sum_w Pl(w)} = \frac{1 - m(\{y,z\})}{2 + m(\Theta)},$$

$$\widetilde{Pl}(y) = \frac{1 - m(\{x,z\})}{2 + m(\Theta)}, \quad \widetilde{Pl}(z) = \frac{1 - m(\{x,y\})}{2 + m(\Theta)}$$

and $R[Bel] = \widetilde{Pl} = p[Bel]$ *as stated by Theorem 65. While in the non-singular case all these quantities live in different simplices that 'converge' to* $\widetilde{Bel}$ *(Fig. 12.8), when* $\widetilde{Bel}$ *does not exist all such simplices coincide.*

For a given value of $m(\Theta)$*, this is the triangle with vertices*

$$\left[\frac{1}{2+m(\Theta)}, \frac{1}{2+m(\Theta)}, \frac{m(\Theta)}{2+m(\Theta)}\right]', \left[\frac{1}{2+m(\Theta)}, \frac{m(\Theta)}{2+m(\Theta)}, \frac{1}{2+m(\Theta)}\right]',$$
$$\left[\frac{m(\Theta)}{2+m(\Theta)}, \frac{1}{2+m(\Theta)}, \frac{1}{2+m(\Theta)}\right]'. \tag{12.30}$$

As a reference, for $m(\Theta) = 0$ *the latter is the triangle delimited by the points* p_1, p_2, p_3 *in Fig. 12.9 (solid lines). For* $m(\Theta) = 1$*, we get a single point,* $\overline{\mathcal{P}}$ *(the central black square in the figure). For* $m(\Theta) = \frac{1}{2}$*, instead, (12.30) yields*

$$Cl(p_1', p_2', p_3') = Cl\left(\left[\frac{2}{5}, \frac{2}{5}, \frac{1}{5}\right]', \left[\frac{2}{5}, \frac{1}{5}, \frac{2}{5}\right]', \left[\frac{1}{5}, \frac{2}{5}, \frac{2}{5}\right]'\right)$$

(the dotted triangle in Fig. 12.9). For comparison, let us compute the values of Smets's pignistic probability (which in the three-element case coincides with the orthogonal projection [333]; see Section 11.4.5). We get

$$BetP[Bel](x) = \frac{m(\{x,y\}) + m(\{x,z\})}{2} + \frac{m(\Theta)}{3},$$

$$BetP[Bel](y) = \frac{m(\{x,y\}) + m(\{y,z\})}{2} + \frac{m(\Theta)}{3},$$

$$BetP[Bel](z) = \frac{m(\{x,z\}) + m(\{y,z\})}{2} + \frac{m(\Theta)}{3}.$$

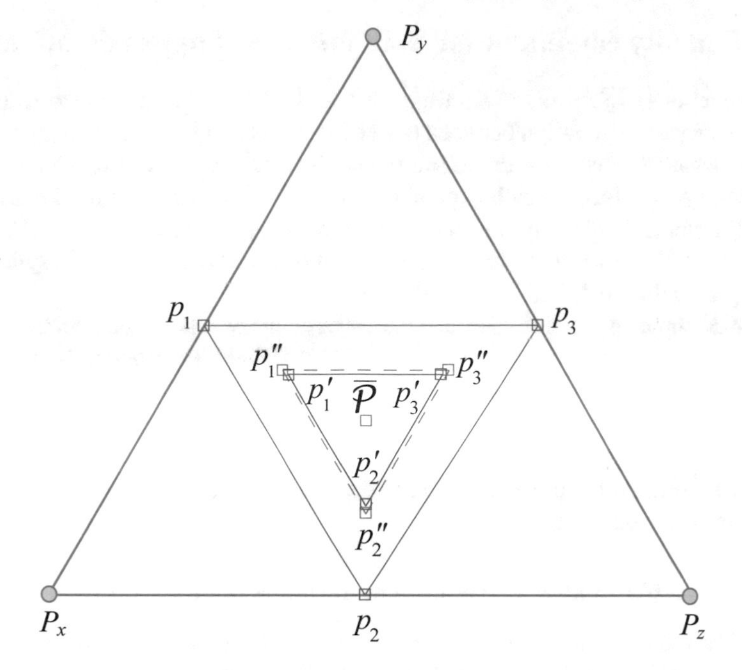

Fig. 12.9: Simplices spanned by $R[Bel] = p[Bel] = \widetilde{Pl}$ and $BetP[Bel] = \pi[Bel]$ in the probability simplex for the cardinality-3 frame in the singular case $m(x) = m(y) = m(z) = 0$, for different values of $m(\Theta)$. The triangle spanned by $R[Bel] = p[Bel] = \widetilde{Pl}$ (solid lines) coincides with that spanned by $BetP[Bel] = \pi[Bel]$ for all Bel such that $m(\Theta) = 0$. For $m(\Theta) = \frac{1}{2}$, $R[Bel] = p[Bel] = \widetilde{Pl}$ spans the triangle $Cl(p_1', p_2', p_3')$ (dotted lines), while $BetP[Bel] = \pi[Bel]$ spans the triangle $Cl(p_1'', p_2'', p_3'')$ (dashed lines). For $m(\Theta) = 1$, both groups of transformations reduce to a single point $\overline{\mathcal{P}}$.

Thus, the simplices spanned by the pignistic function for the same sample values of $m(\Theta)$ are (Fig. 12.9 again) $m(\Theta) = 1 \rightarrow \overline{\mathcal{P}}$, $m(\Theta) = 0 \rightarrow Cl(p_1, p_2, p_3)$ and $m(\Theta) = \frac{1}{2} \rightarrow Cl(p_1'', p_2'', p_3'')$, where

$$p_1'' = \left[\frac{5}{12}, \frac{5}{12}, \frac{1}{6}\right]', p_2'' = \left[\frac{5}{12}, \frac{1}{6}, \frac{5}{12}\right]', p_3'' = \left[\frac{1}{6}, \frac{5}{12}, \frac{5}{12}\right]'$$

(the vertices of the dashed triangle in the figure).

The behaviour of the two families of probability transformations is rather similar, at least in the singular case. In both cases approximations are allowed to span only a proper subset of the probability simplex $\mathcal{P}$, stressing the pathological situation of the singular case itself.

12.7 Equality conditions for both families of approximations

The rich tapestry of results in Sections 12.5 and 12.6 completes our knowledge of the geometry of the relation between belief functions and their probability transformations which started with the affine family in Chapter 11. The epistemic family is formed by transforms which depend on the balance between the total plausibility k_{Pl} of the elements of the frame and the total mass k_{Bel} assigned to them. This measure of the relative uncertainty in the probabilities of the singletons is symbolised by the probability distribution $R[Bel]$.

The examples in Section 12.6 shed some light on the relative behaviour of all the major probability transformations, at least in the probability simplex. It is now time to understand under what conditions the probabilities generated by transformations of different families reduce to the same distribution. Theorem 65 is a first step in this direction: when Bel does not admit relative belief, its relative plausibility $\widetilde{Pl}$ and intersection probability $p[Bel]$ coincide. Once again, we will start by gathering insight from the binary case.

12.7.1 Equal plausibility distribution in the affine family

Let us first focus on functions of the affine family. In particular, let us consider the orthogonal projection (11.27) of Bel onto $\mathcal{P}$ [333], recalled below as

$$\pi[Bel](x) = \sum_{A \supseteq \{x\}} m(A) \left(\frac{1 + |A^c| 2^{1-|A|}}{n} \right) + \sum_{A \not\supseteq \{x\}} m(A) \left(\frac{1 - |A| 2^{1-|A|}}{n} \right),$$

and the pignistic transform (3.30). We can prove the following lemma.

Lemma 13. *The difference $\pi[Bel](x) - BetP[Bel](x)$ between the probability values of the orthogonal projection and the pignistic function is equal to*

$$\sum_{A \subseteq \Theta} m(A) \left(\frac{1 - |A| 2^{1-|A|}}{n} \right) - \sum_{A \supseteq \{x\}} m(A) \left(\frac{1 - |A| 2^{1-|A|}}{|A|} \right). \tag{12.31}$$

The following theorem is an immediate consequence of Lemma 13.

Theorem 66. *The orthogonal projection and the pignistic function coincide iff*

$$\sum_{A \supseteq \{x\}} m(A) \left(1 - |A| 2^{1-|A|}\right) \frac{|A^c|}{|A|} = \sum_{A \not\supseteq \{x\}} m(A) \left(1 - |A| 2^{1-|A|}\right) \quad \forall x \in \Theta. \tag{12.32}$$

Theorem 66 gives an exhaustive but rather arid description of the relation between $\pi[Bel]$ and $BetP[Bel]$. More significant sufficient conditions can be given in terms of belief values. Let us denote by

$$Pl(x; k) \doteq \sum_{A \supset \{x\}, |A| = k} m(A)$$

the support that focal elements of size k provide to each singleton x.

Corollary 14. *Each of the following is a sufficient condition for the equality of the pignistic and orthogonal transforms of a belief function Bel ($BetP[Bel] = \pi[Bel]$):*

1. *$m(A) = 0$ for all $A \subseteq \Theta$ such that $|A| \neq 1, 2, n$.*
2. *The mass of Bel is equally distributed among all the focal elements $A \subseteq \Theta$ of the same size $|A| = k$, for all sizes $k = 3, \ldots, n-1$:*

$$m(A) = \frac{\sum_{|B|=k} m(B)}{\binom{n}{k}} \quad \forall A : |A| = k, \forall k = 3, \ldots, n-1.$$

3. *For all singletons $x \in \Theta$, and for all $k = 3, \ldots, n-1$,*

$$Pl(x; k) = const = Pl(\cdot; k). \tag{12.33}$$

Whenever mass is equally distributed among higher-size events, the orthogonal projection *is* the pignistic function (condition 2). The probability closest to Bel (in the Euclidean sense) is also the barycentre of the simplex $\mathcal{P}[Bel]$ of consistent probabilities. This is also the case when events of the same size contribute by the same amount to the plausibility of each singleton (condition 3).

It is easy to see that condition 1 implies (is stronger than) condition 2, which in turn implies condition 3. All of them are satisfied by belief functions on size-2 frames. In particular, Corollary 14 implies the following result.

Corollary 15. *$BetP[Bel] = \pi[Bel]$ for $|\Theta| \leq 3$.*

12.7.2 Equal plausibility distribution as a general condition

As a matter of fact, the condition (12.33) on the equal distribution of plausibility provides an equality condition for probability transformations of Bel of both families.

Consider again the binary case shown in Fig. 11.1. We can appreciate that belief functions such that $m(x) = m(y)$ lie on the bisector of the first quadrant, which is orthogonal to $\mathcal{P}$. Their relative plausibility is then equal to their orthogonal projection $\pi[Bel]$. Theorem 62 can indeed be interpreted in terms of an equal distribution of plausibility among singletons. If (12.33) is satisfied for all $k = 2, \ldots, n-1$ (this is trivially true for $k = n$), then the uncertainty $Pl(x) - m(x) = \sum_{A \supsetneq \{x\}} m(A)$ in the probability value of each singleton $x \in \Theta$ becomes

$$\sum_{A \supsetneq \{x\}} m(A) = \sum_{k=2}^{n} \sum_{|A|=k, A \supset \{x\}} m(A) = m(\Theta) + \sum_{k=2}^{n-1} Pl(\cdot; k), \tag{12.34}$$

which is constant for all singletons $x \in \Theta$. The following is then a consequence of Theorem 62 and (12.34).

Corollary 16. *If $Pl(x; k) = \text{const}$ for all $x \in \Theta$ and for all $k = 2, \ldots, n-1$, then the line $a(\overline{Bel}, \overline{Pl})$ is orthogonal to $\mathcal{P}$, and the relative uncertainty in the probabilities of the singletons is the uniform probability $R[Bel] = \overline{\mathcal{P}}$.*

The quantity $Pl(x;k)$ seems then to be connected to geometric orthogonality in the belief space. We say that a belief function $Bel \in \mathcal{B}$ is orthogonal to $\mathcal{P}$ when the vector $\overrightarrow{\mathbf{0}\ Bel}$ joining the origin $\mathbf{0}$ of $\mathbb{R}^{N-2}$ with Bel is orthogonal to it. We showed in Chapter 11 that this is the case if and only if (11.25)

$$\sum_{A \supset \{y\}, A \not\supset \{x\}} m(A) 2^{1-|A|} = \sum_{A \supset \{x\}, A \not\supset \{y\}} m(A) 2^{1-|A|}$$

for each pair of distinct singletons $x, y \in \Theta$, $x \neq y$. For instance, the uniform Bayesian belief function $\overline{\mathcal{P}}$ is orthogonal to $\mathcal{P}$.

Again, a sufficient condition for (11.25) can be given in terms of an equal distribution of plausibility. Confirming the intuition given by the binary case, in this case all probability transformations of Bel converge to the same probability.

Theorem 67. *If $Pl(x;k) = \text{const} = Pl(\cdot;k)$ for all $k = 1, \ldots, n-1$, then Bel is orthogonal to $\mathcal{P}$, and*

$$\widetilde{Pl} = R[Bel] = \pi[Bel] = BetP[Bel] = \overline{\mathcal{P}}. \tag{12.35}$$

We can summarise our findings by stating that, if focal elements of the same size contribute equally to the plausibility of each singleton ($Pl(x;k) = \text{const}$), the following consequences for the relation between all probability transformations and their geometry hold, as a function of the range of values of $|A| = k$ for which the hypothesis is true:

$$\begin{array}{l} \forall k = 3, \ldots, n : BetP[Bel] = \pi[Bel], \\ \forall k = 2, \ldots, n : a(\overline{Bel}, \overline{Pl}) \perp \mathcal{P}, \\ \forall k = 1, \ldots, n : Bel \perp \mathcal{P}, \widetilde{Pl} = \widetilde{Bel} = R[Bel] = \overline{\mathcal{P}} \\ \qquad\qquad\qquad\; = BetP[Bel] = p[Bel] = \pi[Bel]. \end{array}$$

Less binding conditions may be harder to formulate – we plan on studying them in the near future.

Appendix: Proofs

Proof of Theorem 53

We just need a simple counter-example. Consider a belief function $Bel : 2^\Theta \rightarrow [0,1]$ on $\Theta = \{x_1, x_2, \ldots, x_n\}$, where $k_{Bel} \doteq \sum_{x \in \Theta} m(x)$ is the total mass that it assigns to singletons, with BPA $m(x_i) = k_{Bel}/n$ for all i, $m(\{x_1, x_2\}) = 1 - k_{Bel}$. Then

$$\begin{array}{ll} Bel(\{x_1, x_2\}) & = 2 \cdot \dfrac{k_{Bel}}{n} + 1 - k_{Bel} = 1 - k_{Bel}\left(\dfrac{n-2}{n}\right), \\ \tilde{Bel}(x_1) = \tilde{Bel}(x_2) & = \dfrac{1}{n} \Rightarrow \tilde{Bel}(\{x_1, x_2\}) = \dfrac{2}{n}. \end{array}$$

For $\tilde{Bel}$ to be consistent with Bel (3.10), it is necessary that $\tilde{Bel}(\{x_1, x_2\}) \geq Bel(\{x_1, x_2\})$, namely

$$\frac{2}{n} \geq 1 - k_{Bel}\frac{n-2}{n} \equiv k_{Bel} \geq 1,$$

which in turn reduces to $k_{Bel} = 1$. If $k_{Bel} < 1$ (Bel is not a probability), its relative belief of singletons is not consistent.

Proof of Theorem 54

Let us pick for simplicity a frame of discernment with just three elements, $\Theta = \{x_1, x_2, x_3\}$, and the following BPA:

$$m(\{x_i\}^c) = \frac{k}{3} \; \forall i = 1, 2, 3, \quad m(\{x_1, x_2\}^c) = m(\{x_3\}) = 1 - k.$$

In this case, the plausibility of $\{x_1, x_2\}$ is obviously $Pl(\{x_1, x_2\}) = 1 - (1 - k) = k$, while the plausibilities of the singletons are $Pl(x_1) = Pl(x_2) = \frac{2}{3}k$, $Pl(x_3) = 1 - \frac{1}{3}k$. Therefore $\sum_{x \in \Theta} Pl(x) = 1 + k$, and the relative plausibility values are

$$\tilde{Pl}(x_1) = \tilde{Pl}(x_2) = \frac{\frac{2}{3}k}{1+k}, \quad \tilde{Pl}(x_3) = \frac{1 - \frac{1}{3}k}{1+k}.$$

For $\tilde{Pl}$ to be consistent with Bel, we would need

$$\tilde{Pl}(\{x_1, x_2\}) = \tilde{Pl}(x_1) + \tilde{Pl}(x_2) = \frac{4}{3}k\frac{1}{1+k} \leq Pl(\{x_1, x_2\}) = k,$$

which happens if and only if $k \geq \frac{1}{3}$. Therefore, for $k < \frac{1}{3}$, $\tilde{Pl} \notin \mathcal{P}[Bel]$.

Proof of Theorem 55

Each pseudo-belief function admits a (pseudo-)plausibility function, as in the case of standard BFs, which can be computed as $Pl_\varsigma(A) = \sum_{B \cap A \neq \emptyset} m_\varsigma(B)$. For the class of pseudo-belief functions ς which correspond to the plausibility of some belief function Bel ($\varsigma = Pl$ for some $Bel \in \mathcal{B}$), their pseudo-plausibility function is $Pl_{Pl}(A) = \sum_{B \cap A \neq \emptyset} \mu(B)$, as μ (9.1) is the Möbius inverse of Pl. When applied to the elements $x \in \Theta$ of the frame on which both Bel and Pl are defined, this yields $Pl_{Pl}(x) = \sum_{B \ni x} \mu(B) = m(x)$ by (12.5), which implies

$$\tilde{Pl}[Pl](x) = \frac{Pl_{Pl}(x)}{\sum_{y \in \Theta} Pl_{Pl}(y)} = \frac{m(x)}{\sum_{y \in \Theta} m(y)} = \tilde{Bel}[Bel].$$

Proof of Theorem 56

The basic plausibility assignment of $Pl_1 \oplus Pl_2$ is, according to (2.6),

$$\mu_{Pl_1 \oplus Pl_2}(A) = \frac{1}{k(Pl_1, Pl_2)} \sum_{X \cap Y = A} \mu_1(X)\mu_2(Y).$$

Therefore, according to (12.5), the corresponding relative belief of singletons $\tilde{Bel}[Pl_1 \oplus Pl_2](x)$ (12.6) is proportional to

$$\begin{aligned} m_{Pl_1 \oplus Pl_2}(x) &= \sum_{A \supseteq \{x\}} \mu_{Pl_1 \oplus Pl_2}(A) \\ &= \frac{\sum_{A \supseteq \{x\}} \sum_{X \cap Y = A} \mu_1(X)\mu_2(Y)}{k(Pl_1, Pl_2)} = \frac{\sum_{X \cap Y \supseteq \{x\}} \mu_1(X)\mu_2(Y)}{k(Pl_1, Pl_2)}, \end{aligned} \tag{12.36}$$

where $m_{Pl_1 \oplus Pl_2}(x)$ denotes the BPA of the (pseudo-)belief function which corresponds to the plausibility function $Pl_1 \oplus Pl_2$.

On the other hand, as $\sum_{X \supseteq \{x\}} \mu(X) = m(x)$,

$$\tilde{Bel}[Pl_1](x) \propto m_1(x) = \sum_{X \supseteq \{x\}} \mu_1(X), \quad \tilde{Bel}[Pl_2](x) \propto m_2(x) = \sum_{X \supseteq \{x\}} \mu_2(X).$$

Their Dempster combination is therefore

$$\begin{aligned} (\tilde{Bel}[Pl_1] \oplus \tilde{Bel}[Pl_2])(x) &\propto \left(\sum_{X \supseteq \{x\}} \mu_1(X) \right) \left(\sum_{Y \supseteq \{x\}} \mu_2(Y) \right) \\ &= \sum_{X \cap Y \supseteq \{x\}} \mu_1(X)\mu_2(Y), \end{aligned}$$

and, by normalising, we get (12.36).

Proof of Theorem 57

By taking the limit on both sides of (12.7), we get

$$\tilde{Bel}[Pl^\infty] = (\tilde{Bel}[Pl])^\infty. \tag{12.37}$$

Let us consider the quantity $(\tilde{Bel}[Pl])^\infty = \lim_{n \to \infty} (\tilde{Bel}[Pl])^n$ on the right-hand side. Since $(\tilde{Bel}[Pl])^n(x) = K(Bel(x))^n$ (where K is a constant independent of x), and x is the unique most believed state, it follows that

$$(\tilde{Bel}[Pl])^\infty(x) = 1, \quad (\tilde{Bel}[Pl])^\infty(y) = 0 \;\; \forall y \neq x. \tag{12.38}$$

Hence, by (12.37), $\tilde{Bel}[Pl^\infty](x) = 1$ and $\tilde{Bel}[Pl^\infty](y) = 0$ for all $y \neq x$.

Proof of Theorem 58

By virtue of (9.10), we can express a plausibility function as an affine combination of all the categorical belief functions Bel_A.

We can then apply the commutativity property (12.10), obtaining

$$Pl \oplus P = \sum_{A \subseteq \Theta} \nu(A) P \oplus Bel_A, \tag{12.39}$$

where

$$\nu(A) = \frac{\mu(A) k(P, Bel_A)}{\sum_{B \subseteq \Theta} \mu(B) k(P, Bel_B)}, \quad P \oplus Bel_A = \frac{\sum_{x \in A} P(x) Bel_x}{k(P, Bel_A)},$$

with $k(P, Bel_A) = \sum_{x \in A} P(x)$.
By substituting these expressions into (12.39), we get

$$\begin{aligned} Pl \oplus P &= \frac{\sum_{A \subseteq \Theta} \mu(A) \left(\sum_{x \in A} P(x) Bel_x \right)}{\sum_{B \subseteq \Theta} \mu(B) \left(\sum_{y \in B} P(y) \right)} \\ &= \frac{\sum_{x \in \Theta} P(x) \left(\sum_{A \supseteq \{x\}} \mu(A) \right) Bel_x}{\sum_{y \in \Theta} P(y) \left(\sum_{B \supseteq \{y\}} \mu(B) \right)} = \frac{\sum_{x \in \Theta} P(x) m(x) Bel_x}{\sum_{y \in \Theta} P(y) m(y)}, \end{aligned}$$

once again by (12.5). But this is exactly $\tilde{Bel} \oplus P$, as a direct application of Dempster's rule (2.6) shows.

Proof of Lemma 12

We first need to analyse the behaviour of the plausibility transform with respect to affine combination of belief functions.

By definition, the plausibility values of the affine combination $\alpha Bel_1 + (1 - \alpha) Bel_2$ are

$$\begin{aligned} & pl[\alpha Bel_1 + (1-\alpha) Bel_2](x) \\ &= \sum_{A \supseteq \{x\}} m_{\alpha Bel_1 + (1-\alpha) Bel_2}(A) = \sum_{A \supseteq \{x\}} [\alpha m_1(A) + (1-\alpha) m_2(A)] \\ &= \alpha \sum_{A \supseteq \{x\}} m_1(A) + (1-\alpha) \sum_{A \supseteq \{x\}} m_2(A) = \alpha Pl_1(x) + (1-\alpha) Pl_2(x). \end{aligned}$$

Hence, after denoting by $k_{Pl_i} = \sum_{y \in \Theta} Pl_i(y)$ the total plausibility of the singletons with respect to Bel_i, the values of the relative plausibility of singletons can be computed as

$$
\begin{aligned}
&\tilde{Pl}[\alpha Bel_1 + (1-\alpha)Bel_2](x) \\
&= \frac{\alpha Pl_1(x) + (1-\alpha)Pl_2(x)}{\sum_{y\in\Theta}[\alpha Pl_1(y) + (1-\alpha)Pl_2(y)]} = \frac{\alpha Pl_1(x) + (1-\alpha)Pl_2(x)}{\alpha k_{Pl_1} + (1-\alpha)k_{Pl_2}} \\
&= \frac{\alpha Pl_1(x)}{\alpha k_{Pl_1} + (1-\alpha)k_{Pl_2}} + \frac{(1-\alpha)Pl_2(x)}{\alpha k_{Pl_1} + (1-\alpha)k_{Pl_2}} \\
&= \frac{\alpha k_{Pl_1}}{\alpha k_{Pl_1} + (1-\alpha)k_{Pl_2}}\tilde{Pl}_1(x) + \frac{(1-\alpha)k_{Pl_2}}{\alpha k_{Pl_1} + (1-\alpha)k_{Pl_2}}\tilde{Pl}_2(x) \\
&= \beta_1\tilde{Pl}_1(x) + \beta_2\tilde{Pl}_2(x).
\end{aligned}
$$

Proof of Theorem 59

The proof follows the structure of that of Theorem 3 and Corollary 3 in [327], on the commutativity of Dempster's rule and convex closure.

Formally, we need to prove that:

1. Whenever $Bel = \sum_k \alpha_k Bel_k$, $\alpha_k \geq 0$, $\sum_k \alpha_k = 1$, we have that $\tilde{Pl}[Bel] = \sum_k \beta_k \tilde{Pl}[Bel_k]$ for some convex coefficients β_k.
2. Whenever $P \in Cl(\tilde{Pl}[Bel_k], k)$ (i.e., $P = \sum_k \beta_k \tilde{Pl}[Bel_k]$ with $\beta_k \geq 0$, $\sum_k \beta_k = 1$), there exists a set of convex coefficients $\alpha_k \geq 0$, $\sum_k \alpha_k = 1$ such that $P = \tilde{Pl}[\sum_k \alpha_k Bel_k]$.

Now, condition 1 follows directly from Lemma 12. Condition 2, instead, amounts to proving that there exist $\alpha_k \geq 0$, $\sum_k \alpha_k = 1$ such that

$$
\beta_k = \frac{\alpha_k k_{Pl_k}}{\sum_j \alpha_j k_{Pl_j}} \forall k, \tag{12.40}
$$

which, in turn, is equivalent to

$$
\alpha_k = \frac{\beta_k}{k_{Pl_k}} \cdot \sum_j \alpha_j k_{Pl_j} \propto \frac{\beta_k}{k_{Pl_k}} \forall k
$$

as $\sum_j \alpha_j k_{Pl_j}$ does not depend on k. If we pick $\alpha_k = \beta_k / k_{Pl_k}$, the system (12.40) is satisfied: by further normalisation, we obtain the desired result.

Proof of Theorem 61

By (12.19), $\widehat{Bel} - \overline{Bel} = (1 - k_{Bel})Pl_\Theta$ and $\widehat{Pl} - \overline{Pl} = (1 - k_{Pl})Pl_\Theta$. Hence,

$$\begin{aligned}
&\beta[Bel](\widehat{Pl} - \widehat{Bel}) + \widehat{Bel} \\
&= \beta[Bel]\big[\overline{Pl} + (1 - k_{Pl})Pl_\Theta - \overline{Bel} - (1 - k_{Bel})Pl_\Theta\big] + \overline{Bel} + (1 - k_{Bel})Pl_\Theta \\
&= \beta[Bel]\big[\overline{Pl} - \overline{Bel} + (k_{Bel} - k_{Pl})Pl_\Theta\big] + \overline{Bel} + (1 - k_{Bel})Pl_\Theta \\
&= \overline{Bel} + \beta[Bel](\overline{Pl} - \overline{Bel}) + Pl_\Theta\big[\beta[Bel](k_{Bel} - k_{Pl}) + 1 - k_{Bel}\big].
\end{aligned}$$

But, by the definition of $\beta[Bel]$ (11.10),

$$\beta[Bel](k_{Bel} - k_{Pl}) + 1 - k_{Bel} = \frac{1 - k_{Bel}}{k_{Pl} - k_{Bel}}(k_{Bel} - k_{Pl}) + 1 - k_{Bel} = 0,$$

and (12.20) is satisfied.

Proof of Theorem 62

By the definition of Bel_A ($Bel_A(C) = 1$ if $C \supseteq A$, and 0 otherwise), we have that

$$\langle Bel_A, Bel_B \rangle = \sum_{C \subseteq \Theta} Bel_A(C) Bel_B(C) = \sum_{C \supseteq A,B} 1 \cdot 1 = \|Bel_{A \cup B}\|^2.$$

The scalar products of interest can then be written as

$$\begin{aligned}
\langle \overline{Pl} - \overline{Bel}, Bel_y - Bel_x \rangle &= \left\langle \sum_{z \in \Theta} (Pl(z) - m(z))\, Bel_z, Bel_y - Bel_x \right\rangle \\
&= \sum_{z \in \Theta} (Pl(z) - m(z)) \Big[\langle Bel_z, Bel_y \rangle - \langle Bel_z, Bel_x \rangle \Big] \\
&= \sum_{z \in \Theta} (Pl(z) - m(z)) \Big[\|Bel_{z \cup y}\|^2 - \|Bel_{z \cup x}\|^2 \Big]
\end{aligned}$$

for all $y \neq x$. We can distinguish three cases:

- if $z \neq x, y$, then $|z \cup x| = |z \cup y| = 2$ and the difference $\|Bel_{z \cup x}\|^2 - \|Bel_{z \cup y}\|^2$ goes to zero;
- if $z = x$, then $\|Bel_{z \cup x}\|^2 - \|Bel_{z \cup y}\|^2 = \|Bel_x\|^2 - \|Bel_{x \cup y}\|^2 = (2^{n-2} - 1) - (2^{n-1} - 1) = -2^{n-2}$, where $n = |\Theta|$;
- if, instead, $z = y$, then $\|Bel_{z \cup x}\|^2 - \|Bel_{z \cup y}\|^2 = \|Bel_{x \cup y}\|^2 - \|Bel_y\|^2 = 2^{n-2}$.

Hence, for all $y \neq x$,

$$\langle \overline{Pl} - \overline{Bel}, Bel_y - Bel_x \rangle = 2^{n-2}(Pl(y) - m(y)) - 2^{n-2}(Pl(x) - m(x)),$$

and, as $\sum_{A \supsetneq x} m(A) = Pl(x) - m(x)$, the thesis follows.

Proof of Corollary 13

As a matter of fact,

$$\sum_{x\in\Theta}\sum_{A\supsetneq\{x\}} m(A) = \sum_{x\in\Theta}(Pl(x) - m(x)) = k_{Pl} - k_{Bel},$$

so that the condition of Theorem 62 can be written as

$$Pl(x) - m(x) = \sum_{A\supsetneq\{x\}} m(A) = \frac{k_{Pl} - k_{Bel}}{n} \ \forall x.$$

Replacing this in (12.25) yields

$$R[Bel] = \sum_{x\in\Theta} \frac{1}{n} Bel_x.$$

Proof of Theorem 63

By (12.20), $p[Bel] = \overline{Bel} + \beta[Bel](\overline{Pl} - \overline{Bel})$. After recalling that $\beta[Bel] = (1 - k_{Bel})/(k_{Pl} - k_{Bel})$, we can write

$$\begin{aligned}\overline{Pl} - p[Bel] &= \overline{Pl} - \left[\overline{Bel} + \beta[Bel](\overline{Pl} - \overline{Bel})\right] = (1 - \beta[Bel])(\overline{Pl} - \overline{Bel}) \\ &= \frac{k_{Pl} - 1}{k_{Pl} - k_{Bel}}(\overline{Pl} - \overline{Bel}) = (k_{Pl} - 1)\ R[Bel],\end{aligned} \tag{12.41}$$

by the definition (12.25) of $R[Bel]$. Moreover, as $\widehat{Pl} = \overline{Pl} + (1 - k_{Pl})Pl_\Theta$ by (12.19), we get

$$\widehat{Pl} - p[Bel] = (k_{Pl} - 1)(R[Bel] - Pl_\Theta). \tag{12.42}$$

Combining (12.42) and (12.41) then yields

$$\begin{aligned}\langle \widehat{Pl} - p[Bel], \overline{Pl} - p[Bel]\rangle &= \Big\langle (k_{Pl} - 1)(R[Bel] - Pl_\Theta), (k_{Pl} - 1)R[Bel]\Big\rangle \\ &= (k_{Pl} - 1)^2 \langle R[Bel] - Pl_\Theta, R[Bel]\rangle \\ &= (k_{Pl} - 1)^2 \Big(\langle R[Bel], R[Bel]\rangle - \langle Pl_\Theta, R[Bel]\rangle\Big) \\ &= (k_{Pl} - 1)^2 \Big(\langle R[Bel], R[Bel]\rangle - \langle \mathbf{1}, R[Bel]\rangle\Big).\end{aligned}$$

But now

$$\cos(\pi - \phi_2) = \frac{\langle \widehat{Pl} - p[Bel], \overline{Pl} - p[Bel]\rangle}{\|\widehat{Pl} - p[Bel]\| \|\overline{Pl} - p[Bel]\|},$$

where the leftmost factor in the denominator reads as

$$\begin{aligned}\|\widehat{Pl} - p[Bel]\| &= \Big[\langle \widehat{Pl} - p[Bel], \widehat{Pl} - p[Bel]\rangle\Big]^{1/2} \\ &= (k_{Pl} - 1)\Big[\langle R[Bel] - Pl_\Theta, R[Bel] - Pl_\Theta\rangle\Big]^{1/2} \\ &= (k_{Pl} - 1)\Big[\langle R[Bel], R[Bel]\rangle + \langle Pl_\Theta, Pl_\Theta\rangle - 2\langle R[Bel], Pl_\Theta\rangle\Big]^{1/2}\end{aligned}$$

and $\|\overline{Pl} - p[Bel]\| = (k_{Pl} - 1)\|R[Bel]\|$ by (12.41). Hence,

$$\cos(\pi - \phi_2[Bel]) = \frac{\|R[Bel]\|^2 - \langle \mathbf{1}, R[Bel]\rangle}{\|R[Bel]]\|\sqrt{\|R[Bel]\|^2 + \langle \mathbf{1}, \mathbf{1}\rangle - 2\langle R[Bel], \mathbf{1}\rangle}}. \quad (12.43)$$

We can simplify this expression further by noticing that for each probability $P \in \mathcal{P}$ we have $\langle \mathbf{1}, P\rangle = 2^{|\{x\}^c|} - 1 = 2^{n-1} - 1$, while $\langle \mathbf{1}, \mathbf{1}\rangle = 2^n - 2$, so that $\langle \mathbf{1}, \mathbf{1}\rangle - 2\langle P, \mathbf{1}\rangle = 0$. As $R[Bel]$ is a probability, we get (12.26).

Proof of Theorem 64

For this proof, we make use of (12.43). As $\phi_2[Bel] = 0$ iff $\cos(\pi - \phi_2[Bel]) = -1$, the desired condition is

$$-1 = \frac{\|R[Bel]\|^2 - \langle \mathbf{1}, R[Bel]\rangle}{\|R[Bel]]\|\sqrt{\|R[Bel]\|^2 + \langle \mathbf{1}, \mathbf{1}\rangle - 2\langle R[Bel], \mathbf{1}\rangle}}$$

i.e., after squaring both the numerator and the denominator,

$$\begin{aligned}&\|R[Bel]\|^2(\|R[Bel]\|^2 + \langle \mathbf{1}, \mathbf{1}\rangle - 2\langle R[Bel], \mathbf{1}\rangle) \\ &\quad = \|R[Bel]\|^4 + \langle \mathbf{1}, R[Bel]\rangle^2 - 2\langle \mathbf{1}, R[Bel]\rangle\|R[Bel]\|^2.\end{aligned}$$

After erasing the common terms, we find that $\phi_2[Bel]$ is zero if and only if

$$\langle \mathbf{1}, R[Bel]\rangle^2 = \|R[Bel]\|^2\langle \mathbf{1}, \mathbf{1}\rangle. \quad (12.44)$$

The condition (12.44) has the form

$$\langle A, B\rangle^2 = \|A\|^2\|B\|^2\cos^2(\widehat{AB}) = \|A\|^2\|B\|^2,$$

i.e., $\cos^2(\widehat{AB}) = 1$, with $A = Pl_\Theta$, $B = R[Bel]$. This yields $\cos(\widehat{R[Bel]Pl_\Theta}) = 1$ or $\cos(\widehat{R[Bel]Pl_\Theta}) = -1$, i.e., $\phi_2[Bel] = 0$ if and only if $R[Bel]$ is (anti-)parallel to $Pl_\Theta = \mathbf{1}$. But this means $R[Bel] = \alpha Pl_\Theta$ for some scalar value $\alpha \in \mathbb{R}$, namely

$$R[Bel] = -\alpha \sum_{A \subseteq \Theta} (-1)^{|A|} Bel_A$$

(since $Pl_\Theta = -\sum_{A \subseteq \Theta}(-1)^{|A|} Bel_A$ by (9.11)). But $R[Bel]$ is a probability (i.e., a linear combination of categorical probabilities Bel_x only) and, since the vectors $\{Bel_A, A \subsetneq \Theta\}$ which represent all categorical belief functions are linearly independent, the two conditions are never jointly met, unless $|\Theta| = 2$.

Proof of Lemma 13

Using the form (11.27) of the orthogonal projection, we get

$$\pi[Bel] - BetP[Bel](x)$$

$$= \sum_{A \supseteq \{x\}} m(A) \left(\frac{1 + |A^c| 2^{1-|A|}}{n} - \frac{1}{|A|} \right) + \sum_{A \not\supset \{x\}} m(A) \left(\frac{1 - |A| 2^{1-|A|}}{n} \right).$$

After we note that

$$\frac{1 + |A^c| 2^{1-|A|}}{n} - \frac{1}{|A|} = \frac{|A| + |A|(n - |A|) 2^{1-|A|} - n}{n|A|}$$

$$= \frac{(|A| - n)(1 - |A| 2^{1-|A|})}{n|A|} = \left(\frac{1}{n} - \frac{1}{|A|} \right) \left(1 - |A| 2^{1-|A|} \right),$$

we can write

$$\pi[Bel](x) - BetP[Bel](x)$$

$$= \sum_{A \supseteq \{x\}} m(A) \left(\frac{1 - |A| 2^{1-|A|}}{n} \right) \left(1 - \frac{n}{|A|} \right) + \sum_{A \not\supset \{x\}} m(A) \left(\frac{1 - |A| 2^{1-|A|}}{n} \right) \tag{12.45}$$

or, equivalently, (12.31).

Proof of Theorem 66

By (12.31), the condition $\pi[Bel](x) - BetP[Bel](x) = 0$ for all $x \in \Theta$ reads as

$$\sum_{A \subseteq \Theta} m(A) \left(\frac{1 - |A| 2^{1-|A|}}{n} \right) = \sum_{A \supseteq \{x\}} m(A) \left(\frac{1 - |A| 2^{1-|A|}}{|A|} \right) \quad \forall x \in \Theta,$$

i.e.,

$$\sum_{A \not\supset \{x\}} m(A) \left(\frac{1 - |A| 2^{1-|A|}}{n} \right) = \sum_{A \supseteq \{x\}} m(A) (1 - |A| 2^{1-|A|}) \left(\frac{1}{|A|} - \frac{1}{n} \right),$$

for all singletons $x \in \Theta$, i.e., (12.32).

Proof of Corollary 14

Let us consider all claims. Equation (12.32) can be expanded as follows:

$$\sum_{k=3}^{n-1} (1 - k2^{1-k}) \frac{n-k}{k} \sum_{\substack{A \supset \{x\} \\ |A|=k}} m(A) = \sum_{k=3}^{n-1} (1 - k2^{1-k}) \sum_{\substack{A \not\supset \{x\} \\ |A|=k}} m(A),$$

which is equivalent to

$$\sum_{k=3}^{n-1} (1 - k2^{1-k}) \left[\frac{n-k}{k} \sum_{A \supset \{x\}, |A|=k} m(A) - \sum_{A \not\supset \{x\}, |A|=k} m(A) \right] = 0,$$

or

$$\sum_{k=3}^{n-1} \left(\frac{1 - k2^{1-k}}{k} \right) \left[n \sum_{A \supset \{x\}, |A|=k} m(A) - k \sum_{|A|=k} m(A) \right] = 0,$$

after noticing that $1 - k \cdot 2^{1-k} = 0$ for $k = 1, 2$ and the coefficient of $m(\Theta)$ in (12.32) is zero, since $|\Theta^c| = |\emptyset| = 0$. The condition of Theorem 66 can then be rewritten as

$$n \sum_{k=3}^{n-1} \left(\frac{1 - k2^{1-k}}{k} \right) \sum_{\substack{A \supset \{x\} \\ |A|=k}} m(A) = \sum_{k=3}^{n-1} (1 - k2^{1-k}) \sum_{|A|=k} m(A) \qquad (12.46)$$

for all $x \in \Theta$. Condition 1 follows immediately from (12.46).

As for condition 2, the equation becomes

$$\sum_{k=3}^{n-1} \frac{n}{k} (1 - k2^{1-k}) \binom{n-1}{k-1} \frac{\sum_{|A|=k} m(A)}{\binom{n}{k}} = \sum_{k=3}^{n-1} (1 - k2^{1-k}) \sum_{|A|=k} m(A),$$

which is satisfied since

$$\frac{n}{k} \binom{n-1}{k-1} = \binom{n}{k}.$$

Finally, let us consider condition 3. Under (12.33), the system of equations (12.46) reduces to a single equation,

$$\sum_{k=3}^{n-1} (1 - k2^{1-k}) \frac{n}{k} Pl(.; k) = \sum_{k=3}^{n-1} (1 - k2^{1-k}) \sum_{|A|=k} m(A).$$

The latter is satisfied if

$$\sum_{|A|=k} m(A) = \frac{n}{k} Pl(.; k) \text{ for all } k = 3, \ldots, n-1,$$

which is in turn equivalent to $n \; Pl(.; k) = k \sum_{|A|=k} m(A)$ for all $k = 3, \ldots, n-1$. Under the hypothesis of the theorem, we find that

$$n \; Pl(.; k) = k \sum_{|A|=k} m(A) = \sum_{|A|=k} m(A)|A| = \sum_{x \in \Theta} \sum_{A \supset \{x\}, |A|=k} m(A).$$

Proof of Theorem 67

The condition (11.25) is equivalent to

$$\sum_{A\supseteq\{y\}} m(A)2^{1-|A|} = \sum_{A\supseteq\{x\}} m(A)2^{1-|A|}$$

$$\equiv \sum_{k=1}^{n-1} \frac{1}{2^k} \sum_{|A|=k, A\supset\{y\}} m(A) = \sum_{k=1}^{n-1} \frac{1}{2^k} \sum_{|A|=k, A\supset\{x\}} m(A)$$

$$\equiv \sum_{k=1}^{n-1} \frac{1}{2^k} Pl(y;k) = \sum_{k=1}^{n-1} \frac{1}{2^k} Pl(x;k)$$

for all $y \neq x$. If $Pl(x;k) = Pl(y;k)\ \forall y \neq x$, the equality is satisfied.

To prove (12.35), let us rewrite the values of the pignistic function $BetP[Bel](x)$ in terms of $Pl(x;k)$ as

$$BetP[Bel](x) = \sum_{A\supseteq\{x\}} \frac{m(A)}{|A|} = \sum_{k=1}^{n} \sum_{A\supset\{x\},|A|=k} \frac{m(A)}{k} = \sum_{k=1}^{n} \frac{Pl(x;k)}{k}.$$

The latter is constant under the hypothesis, yielding $BetP[Bel] = \overline{\mathcal{P}}$. Also, as

$$Pl(x) = \sum_{A\supseteq\{x\}} m(A) = \sum_{k=1}^{n} \sum_{A\supset\{x\},|A|=k} m(A) = \sum_{k=1}^{n} Pl(x;k),$$

we get

$$\widetilde{Pl}(x) = \frac{Pl(x)}{k_{Pl}} = \frac{\sum_{k=1}^{n} Pl(x;k)}{\sum_{x\in\Theta}\sum_{k=1}^{n} Pl(x;k)},$$

which is equal to $1/n$ if $Pl(x;k) = Pl(\cdot;k)\ \forall k, x$.

Finally, under the same condition,

$$\begin{aligned} p[Bel](x) &= m(x) + \beta[Bel]\Big(Pl(x) - m(x)\Big) \\ &= Pl(\cdot;1) + \beta[Bel]\sum_{k=2}^{n} Pl(\cdot;k) = \frac{1}{n}, \end{aligned}$$

$$\widetilde{Bel}(x) = \frac{m(x)}{k_{Bel}} = \frac{Pl(x;1)}{\sum_{y\in\Theta} Pl(y;1)} = \frac{Pl(\cdot;1)}{nPl(\cdot;1)} = \frac{1}{n}.$$

13

Consonant approximation

As we have learned in the last two chapters, probability transforms are a very well-studied topic in belief calculus, as they are useful as a means to reduce the computational complexity of the framework (Section 4.7), they allow us to reduce decision making with belief functions to the classical utility theory approach (Section 4.8.1) and they are theoretically interesting for understanding the relationship between Bayesian reasoning and belief theory [291].

Less extensively studied is the problem of mapping a belief function to possibility measures, which (as we learned in Chapter 10) have (the plausibility functions associated with) consonant belief functions as counterparts in the theory of evidence. Approximating a belief function by a necessity measure is then equivalent to mapping it to a consonant BF [533, 930, 927, 91]. As possibilities are completely determined by their values on the singletons $Pos(x)$, $x \in \Theta$, they are less computationally expensive than belief functions (indeed, their complexity is linear in the size of the frame of discernment, just like standard probabilities), making the approximation process interesting for many applications.

Just as in the case of Bayesian belief functions, the study of possibility transforms can shed light on the relation between belief and possibility theory. Dubois and Prade [533], in particular, conducted extensive work on the consonant approximation problem [930, 927], introducing, in particular, the notion of an 'outer consonant approximation' (recall Section 4.7.2). Dubois and Prade's work was later extended by Baroni [91] to capacities.

In [345], the author of this book provided a comprehensive picture of the geometry of the set of outer consonant approximations. For each possible maximal chain $\{A_1 \subset \cdots \subset A_n = \Theta\}$ of focal elements, i.e., a collection of nested subsets of all possible cardinalities from 1 to $|\Theta|$, a maximal outer consonant approximation with

F. Cuzzolin, *The Geometry of Uncertainty*, Artificial Intelligence: Foundations, Theory, and Algorithms, https://doi.org/10.1007/978-3-030-63153-6_13

mass assignment

$$m'(A_i) = Bel(A_i) - Bel(A_{i-1}) \tag{13.1}$$

can be singled out. The latter mirrors the behaviour of the vertices of the credal set of probabilities dominating either a belief function or a 2-alternating capacity [245, 1293].

Another interesting approximation was studied in the context of Smets's transferable belief model [1730], where the pignistic transform assumes a central role in decision making. One can then define an 'isopignistic' approximation (see Definition 59) as the unique consonant belief function whose pignistic probability is identical to that of the original belief function [524, 546]. An expression for the isopignistic consonant BF associated with a unimodal probability density, in particular, was derived in [1714], whereas in [58] consonant belief functions were constructed from sample data using confidence sets of pignistic probabilities.

The geometric approach to approximation

In more recent times, the opportunities for seeking probability or consonant approximations/transformations of belief functions by minimising appropriate distance functions has been explored by this author [333, 360, 349]. As to what distances are the most appropriate, as we have seen in Section 4.10.1, Jousselme et al. [937] conducted in 2010 an interesting survey of the distance or similarity measures so far introduced for belief functions, and proposed a number of generalisations of known measures. Other similarity measures between belief functions were proposed by Shi et al. [1636], Jiang et al. [902] and others [965, 501]. In principle, many of these measures could be employed to define conditional belief functions, or to approximate belief functions by necessity or probability measures.

As we have proved in Chapter 10 [345], consonant belief functions (or their vector counterparts in the belief space) live in a structured collection of simplices or 'simplicial complex'. Each maximal simplex of the consonant complex is associated with a maximal chain of nested focal elements $\mathcal{C} = \{A_1 \subset A_2 \subset \cdots \subset A_n = \Theta\}$. Computing the consonant belief function(s) at a minimal distance from a given BF therefore involves (1) computing a partial solution for each possible maximal chain and (2) identifying the global approximation among all the partial ones.

Note, however, that geometric approximations can be sought in different Cartesian spaces. A belief function can be represented either by the vector of its belief values or by the vector of its mass values. We call the set of vectors of the first kind the *belief space* $\mathcal{B}$ [334, 368] (Chapter 7), and the collection of vectors of the second kind the *mass space* $\mathcal{M}$ [356]. In both cases, the region of consonant belief functions is a simplicial complex. In the mass space representation, however, we further also to consider the fact that, because of normalisation, only $N-2$ mass values (where $N = 2^{|\Theta|}$) are sufficient to determine a belief function. Approximations can then be computed in vector spaces of dimension $N-1$ or $N-2$, leading to different but correlated results.

Chapter content

The theme of this chapter is to conduct an exhaustive analytical study of all the consonant approximations of belief functions. The first part is devoted to understanding the geometry of classical outer consonant approximations, while in the second part geometric consonant approximations which minimise appropriate distances from the original belief function are characterised in detail.

We focus in particular on approximations induced by minimising L_1, L_2 or L_∞ (Minkowski) distances, in both the belief and the mass space, and in both representations of the latter. Our aim is to initiate a theoretical study of the nature of consonant approximations induced by geometric distance minimisation, starting with L_p norms, as a stepping stone to a more extensive line of research. We characterise their semantics in terms of degrees of belief and their mutual relationships, and analytically compare them with existing approximations. A landscape emerges in which belief-, mass- and pignistic-based approximations form distinct families with different semantics.

In some cases, improper partial solutions (in the sense that they potentially include negative mass assignments) can be generated by the L_p minimisation process. The resulting set of approximations, in other words, may fall partly outside the simplex of proper consonant belief functions for a given desired chain of focal elements. This situation is not entirely new, as outer approximations themselves include infinitely many improper solutions. Nevertheless, only the subset of acceptable solutions is retained. As we see here, the set of all (admissible or not) solutions is typically much simpler to describe geometrically, in terms of simplices or polytopes. Computing the set of proper approximations would in all cases require significant further effort, which for reasons of clarity and length we reserve for the near future.

Additionally, in this chapter only normalised belief functions (i.e., BFs whose mass of the empty set is zero) are considered. As we have seen in Section 11.5, however, unnormalised belief functions play an important role in the TBM [1685], as the mass of the empty set is an indicator of conflicting evidence. The analysis of the unnormalised case is also left for future work for lack of sufficient space here.

Summary of main results

We show that outer consonant approximations form a convex subset of the consonant complex, for every choice of the desired maximal chain $\mathcal{C} = \{A_1 \subset \cdots \subset A_n\}$ of focal elements A_i. In particular, the set of outer consonant approximations of a belief function Bel with chain $\mathcal{C}$, which we denote by $\mathcal{O}^{\mathcal{C}}[Bel]$, is a polytope whose vertices are indexed by all the functions that reassign the mass of each focal element to elements of the chain containing it (which we term *assignment functions*). In particular, the maximal outer approximation is the vertex of this polytope associated with the permutation of singletons which produces the desired maximal chain $\mathcal{C}$. Two distinct sets of results are reported for geometric approximations.

Partial approximations in the mass space $\mathcal{M}$ amount to redistributing in various ways the mass of focal elements outside the desired maximal chain to elements of the chain itself (compare [356]). In the $(N-1)$-dimensional representation, the L_1 (partial) consonant approximations are such that their mass values are greater than those of the original belief function on the desired maximal chain. They form a simplex which is entirely admissible, and whose vertices are obtained by reassigning all the mass originally outside the desired maximal chain $\mathcal{C}$ to a single focal element of the chain itself. The barycentre of this simplex is the L_2 partial approximation, which redistributes the mass outside the chain to all the elements of $\mathcal{C}$ on an equal basis. The simplex of $L_1, \mathcal{M}$ approximations, in addition, exhibits interesting relations with outer consonant approximations.
When the partial L_∞ approximation is unique, it coincides with the L_2 approximation and the barycenter of the set of L_1 approximations, and it is obviously admissible. When it is not unique, it is a simplex whose vertices assign to each element of the chain (except one) the maximum mass outside the chain: this set is, in general, not entirely admissible. The L_1 and L_2 partial approximations calculated when one adopts an $(N-2)$ section of $\mathcal{M}$ coincide. For each possible neglected component $\bar{A} \in \mathcal{C}$, they describe all the vertices of the simplex of $L_1, \mathcal{M}$ partial approximations in the $(N-1)$-dimensional representation. In each such section, the L_∞ partial approximations form instead a (partly admissible) region whose size is determined by the largest mass outside the desired maximal chain.

Finally, the global approximations in the L_1, L_2 and L_∞ cases span the simplicial components of $\mathcal{CO}$ whose chains minimise the sum of the mass, the sum of squared masses and the maximum mass outside the desired maximal chain, respectively.

In the belief space $\mathcal{B}$, all L_p (Minkowski) approximations amount to picking different representatives from the following n lists of belief values:

$$\mathcal{L}^i = \Big\{ Bel(A), A \supseteq A_i, A \not\supset A_{i+1} \Big\} \quad \forall i = 1, \ldots, n.$$

Belief functions are defined on a partially ordered set, the power set $\{A \subseteq \Theta\}$, of which a maximal chain is a maximal totally ordered subset. Therefore, given two elements of the chain $A_i \subset A_{i+1}$, there are a number of 'intermediate' focal elements A which contain the latter but not the former. This list is uniquely determined by the desired chain.

Indeed, all partial L_p approximations in the belief space have mass $m'(A_i) = f(\mathcal{L}^i) - f(\mathcal{L}^{i-1})$, where f is a simple function of the belief values in the list, such as the maximum, average or median. Classical maximal outer and 'contour-based' approximations (Definition 131) can also be expressed in the same way. As they would all reduce to the maximal outer approximation (13.1) if the power set was totally ordered, all these consonant approximations can be considered as generalisations of the latter. Sufficient conditions for their admissibility can be given in terms of the (partial) plausibility values of the singletons.

Table 13.1: Properties of the geometric consonant approximations studied in the second part of this chapter, in terms of multiplicity and admissibility of partial solutions, and the related global solutions

Approximation	Multiplicity of partial solutions	Admissibility of partial solutions	Global solution(s)
$L_1, \mathcal{M}$	Simplex	Entirely	$\arg\min_C \sum_{B \not\subseteq C} m(B)$
$L_2, \mathcal{M}$	Point, barycentre of $L_1, \mathcal{M}$	Yes	$\arg\min_C \sum_{B \not\subseteq C} (m(B))^2$
$L_\infty, \mathcal{M}$	Point/simplex	Yes/partial	$\arg\min_C \sum_{B \not\subseteq C} m(B)$ / $\arg\min_C \max_{B \not\subseteq C} m(B)$
$L_1, \mathcal{M} \setminus \bar{A}$	Point, vertex of $L_1, \mathcal{M}$	Yes	As in $L_1, \mathcal{M}$
$L_2, \mathcal{M} \setminus \bar{A}$	Point, as in $L_1, \mathcal{M} \setminus \bar{A}$	Yes	As in $L_2, \mathcal{M}$
$L_\infty, \mathcal{M} \setminus \bar{A}$	Polytope	Not entirely	$\arg\min_C \max_{B \not\subseteq C} m(B)$
$L_1, \mathcal{B}$	Polytope	Depends on contour function	Not easy to interpret
$L_2, \mathcal{B}$	Point	Depends on contour function	Not known
$L_\infty, \mathcal{B}$	Polytope	Depends on contour function	$\arg\max_C Pl(A_1)$

As for global approximations, in the L_∞ case they fall on the component(s) associated with the maximum-plausibility singleton(s). In the other two cases they are, for now, more difficult to interpret.

Table 13.1 illustrates the behaviour of the different geometric consonant approximations explored here, in terms of multiplicity, admissibility and global solutions.

Chapter outline

First (Section 13.1), we study the geometry of the polytope of outer consonant approximations and its vertices. We then provide the necessary background on the geometric representation of belief and mass and the geometric approach to the approximation problem (Section 13.2).

In Section 13.3, we approach the problem in the mass space by analytically computing the approximations induced by the L_1, L_2 and L_∞ norms (Section 13.3.1), discussing their interpretation in terms of mass reassignment and the relationship between the results in the mass space and those on its sections (Section 13.3.2), analysing the computability and admissibility of the global approximations (Section 13.3.3), studying the relation of the approximations obtained to classical outer consonant approximations (Section 13.3.4), and, finally, illustrating the results in the significant ternary case.

In the last main part of the chapter (Section 13.4), we analyse the L_p approximation problem in the belief space. Again, we compute the approximations induced by the L_1 (Section 13.4.1), L_2 (Section 13.4.2) and L_∞ (Section 13.4.3) norms, we propose a comprehensive view of all approximations in the belief space via lists of belief values determined by the desired maximal chain (Section 13.4.4), and we draw some conclusions about the behaviour of geometric approximations in the belief and mass spaces (Section 13.5). An earlier, more condensed version of this material was first published in [352].

To improve readability, as usual all proofs have been collected together in an appendix to be found at the end of the chapter.

13.1 Geometry of outer consonant approximations in the consonant simplex

We first seek a geometric interpretation of the outer consonant approximation (see Chapter 4, Section 4.7.2).

13.1.1 Geometry in the binary case

For the binary belief space $\mathcal{B}_2$, the set $O[Bel]$ of all the outer consonant approximations of Bel is depicted in Fig. 13.1 (solid lines). This set is the intersection of the region (shaded in the diagram) of the points Bel' such that, $\forall A \subseteq \Theta$, $Bel'(A) \leq Bel(A)$, and the complex $\mathcal{CO} = \mathcal{CO}^{\{x,\Theta\}} \cup \mathcal{CO}^{\{y,\Theta\}}$ of consonant belief functions (see Fig. 10.1). Among them, the consonant belief functions generated by the $6 = 3!$ possible permutations of the three focal elements $\{x\}, \{y\}, \{x,y\}$ of Bel (4.70) correspond to the orthogonal projections of Bel onto $\mathcal{CO}^{\{x,\Theta\}}$ and $\mathcal{CO}^{\{y,\Theta\}}$, plus the vacuous belief function $Bel_\Theta = \mathbf{0}$.

Let us denote by $O^{\mathcal{C}}[Bel]$ the intersection of the set $O[Bel]$ of all outer consonant approximations of Bel with the component $\mathcal{CO}^{\mathcal{C}}$ of the consonant complex, with $\mathcal{C}$ a maximal chain of 2^Θ. We can notice that, for each maximal chain $\mathcal{C}$:

1. $O^{\mathcal{C}}[Bel]$ is convex (in this case $\mathcal{C} = \{x,\Theta\}$ or $\{y,\Theta\}$).
2. $O^{\mathcal{C}}[Bel]$ is in fact a polytope, i.e., the convex closure of a number of vertices: in particular, a segment in the binary case ($O^{\{x,\Theta\}}[Bel]$ or $O^{\{y,\Theta\}}[Bel]$).
3. The maximal (with respect to weak inclusion (3.9)) outer approximation of Bel is one of the vertices of this polytope $O^{\mathcal{C}}[Bel]$ (Co^ρ, (4.69)), that associated with the permutation ρ of singletons which generates the chain.

In the binary case there are just two such permutations, $\rho_1 = \{x,y\}$ and $\rho_2 = \{y,x\}$, which generate the chains $\{x,\Theta\}$ and $\{y,\Theta\}$, respectively.

All these properties hold in the general case as well.

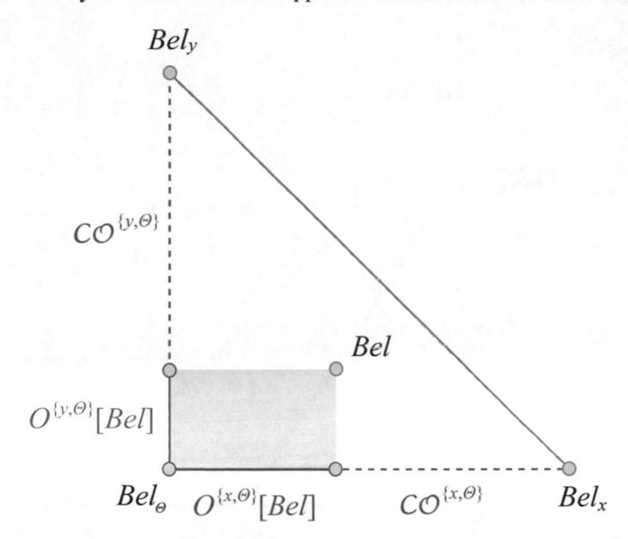

Fig. 13.1: Geometry of the outer consonant approximations of a belief function $Bel \in \mathcal{B}_2$.

13.1.2 Convexity

Theorem 68. *Let Bel be a belief function on Θ. For each maximal chain $\mathcal{C}$ of 2^Θ, the set of outer consonant approximations $O^{\mathcal{C}}[Bel]$ of Bel which belong to the simplicial component $\mathcal{CO}^{\mathcal{C}}$ of the consonant space $\mathcal{CO}$ is convex.*

Proof. Consider two belief functions Bel_1, Bel_2 weakly included in Bel. Then

$$\alpha_1 Bel_1(A)+\alpha_2 Bel_2(A) \leq \alpha_1 Bel(A)+\alpha_2 Bel(A) = (\alpha_1+\alpha_2)Bel(A) = Bel(A)$$

whenever $\alpha_1 + \alpha_2 = 1$, $\alpha_i \geq 0$. If α_1, α_2 are not guaranteed to be non-negative, the sum $\alpha_1 Bel_1(A)+\alpha_2 Bel_2(A)$ can be greater than $Bel(A)$ (see Fig. 13.2). Now, this holds in particular if the two belief functions are consonant – their convex combination, though, is obviously not guaranteed to be consonant. If they both belong to the same maximal simplex of the consonant complex, however, their convex combination still lives in the simplex and $\alpha_1 Bel_1 + \alpha_2 Bel_2$ is both consonant and weakly included in Bel.

□

13.1.3 Weak inclusion and mass reassignment

A more cogent statement about the shape of $O[Bel]$ can be proven by means of the following result on the basic probability assignment of consonant belief functions weakly included in Bel.

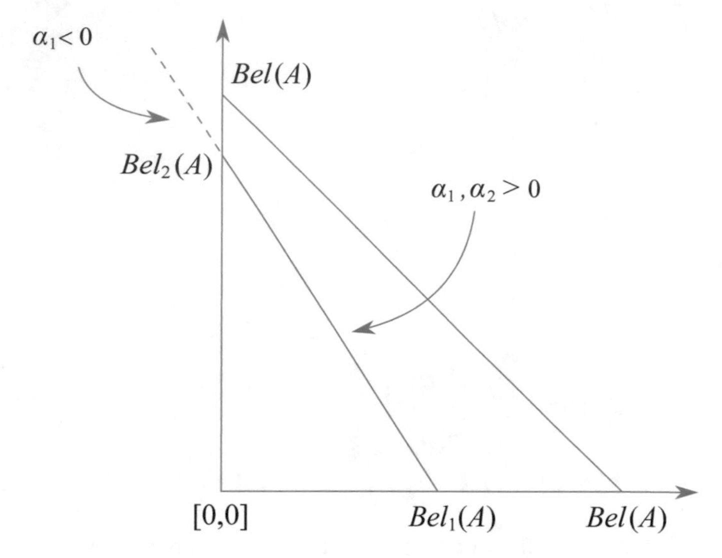

Fig. 13.2: The convex combination of two belief functions Bel_1, Bel_2 weakly included in Bel is still weakly included in Bel: this does not hold for affine combinations (dashed line).

Lemma 14. *Consider a belief function Bel with basic probability assignment m. A consonant belief function Co is weakly included in Bel, for all $A \subseteq \Theta$ $Co(A) \leq Bel(A)$, if and only if there is a choice of coefficients $\{\alpha_A^B, B \subseteq \Theta, A \supseteq B\}$ satisfying*

$$\forall B \subseteq \Theta, \forall A \supseteq B, \;\; 0 \leq \alpha_A^B \leq 1; \qquad \forall B \subseteq \Theta, \; \sum_{A \supseteq B} \alpha_A^B = 1, \tag{13.2}$$

such that Co has basic probability assignment:

$$m_{Co}(A) = \sum_{B \subseteq A} \alpha_A^B m(B). \tag{13.3}$$

Lemma 14 states that the BPA of any outer consonant approximation of Bel is obtained by reassigning the mass of each focal element A of Bel to some $B \supseteq A$.

We will use this result extensively in what follows.

13.1.4 The polytopes of outer approximations

Let us call $\mathcal{C} = \{B_1, \ldots, B_n\}$ ($|B_i| = i$) the chain of focal elements of a consonant belief function weakly included in Bel.

It is natural to conjecture that, for each maximal simplex $\mathcal{CO}^{\mathcal{C}}$ of $\mathcal{CO}$ associated with a maximal chain $\mathcal{C}$, $O^{\mathcal{C}}[Bel]$ is the convex closure of the consonant belief functions $Co^{\mathbf{B}}[Bel]$ with basic probability assignment

$$m^{\mathbf{B}}[Bel](B_i) = \sum_{A \subseteq \Theta : \mathbf{B}(A) = B_i} m(A). \tag{13.4}$$

Each of these vertex consonant belief functions is associated with an *assignment function*

$$\begin{array}{rl} \mathbf{B} : 2^{\Theta} & \rightarrow \mathcal{C}, \\ A & \mapsto \mathbf{B}(A) \supseteq A, \end{array} \tag{13.5}$$

which maps each subset A to one of the focal elements of the chain $\mathcal{C} = \{B_1 \subset \ldots \subset B_n\}$ which contains it.

Theorem 69. *For each simplicial component $\mathcal{CO}^{\mathcal{C}}$ of the consonant space associated with any maximal chain of focal elements $\mathcal{C} = \{B_1, \ldots, B_n\}$, the set of outer consonant approximations of an arbitrary belief function Bel is the convex closure*

$$O^{\mathcal{C}}[Bel] = Cl(Co^{\mathbf{B}}[Bel], \forall \mathbf{B})$$

of the consonant BFs (13.4), indexed by all admissible assignment functions (13.5).

In other words, $O^{\mathcal{C}}[Bel]$ is a polytope, the convex closure of a number of belief functions in 1–1 correspondence with the assignment functions (13.5). Each $\mathbf{B}$ is characterised by assigning each event A to an element $B_i \supseteq A$ of the chain $\mathcal{C}$.

As we will see in Example 40, the points (13.4) are not guaranteed to be proper vertices of the polytope $O^{\mathcal{C}}[Bel]$, as some of them can be written as convex combinations of the others.

13.1.5 Maximal outer approximations

We can prove instead that the outer approximation (4.69), obtained by permuting the singletons of Θ as described in Section 4.7.2, is actually a vertex of $O^{\mathcal{C}}[Bel]$.

More precisely, all possible permutations of the elements of Θ generate exactly $n!$ different outer approximations of Bel, each of which lies on a single simplicial component $\mathcal{CO}^{\mathcal{C}}$ of the consonant complex. Each such permutation ρ generates a maximal chain $\mathcal{C}_\rho = \{S_1^\rho, \ldots, S_n^\rho\}$ of focal elements, so that the corresponding belief function will lie on $\mathcal{CO}^{\mathcal{C}_\rho}$.

Theorem 70. *The outer consonant approximation Co^ρ (4.69) generated by a permutation ρ of the singleton elements of Θ is a vertex of $O^{\mathcal{C}_\rho}[Bel]$.*

Furthermore, we can prove the following corollary.

Corollary 17. *The maximal outer consonant approximation with maximal chain $\mathcal{C}$ of a belief function Bel is the vertex (4.69) of $O^{\mathcal{C}_\rho}[Bel]$ associated with the permutation ρ of the singletons which generates $\mathcal{C} = \mathcal{C}_\rho$.*

By definition (4.69), Co^ρ assigns the mass $m(A)$ of each focal element A to the smallest element of the chain containing A. By Lemma 14, each outer consonant approximation of Bel with chain $\mathcal{C}$, $Co \in O^{\mathcal{C}_\rho}[Bel]$, is the result of redistributing the mass of each focal element A to all its supersets in the chain $\{B_i \supseteq A, B_i \in \mathcal{C}\}$. But then each such Co is weakly included in Co^ρ, for its BPA can be obtained by redistributing the mass of the minimal superset B_j, where $j = \min\{i : B_i \subseteq A\}$, to all supersets of A. Hence, Co^ρ is the maximal outer approximation with chain $\mathcal{C}_\rho$.

13.1.6 Maximal outer approximations as lower chain measures

A different perspective on maximal outer consonant approximations is provided by the notion of a *chain measure* [195].

Let $\mathcal{S}$ be a family of subsets of a non-empty set Θ containing $\emptyset$ and Θ itself. The *inner extension* of a monotone set function $\mu : \mathcal{S} \to [0,1]$ (such that $A \subseteq B$ implies $\mu(A) \leq \mu(B)$) is

$$\mu_*(A) = \sup_{B \in \mathcal{S}, B \subset A} \mu(B). \tag{13.6}$$

An *outer extension* can be dually defined.

Definition 128. *A monotone set function $\nu : \mathcal{S} \to [0,1]$ is called a* lower chain measure *if there exists a chain with respect to set inclusion $\mathcal{C} \subset \mathcal{S}$, which includes $\emptyset$ and Θ, such that*

$$\nu = (\nu|\mathcal{C})_*|\mathcal{S},$$

i.e., ν is the inner extension of its own restriction to the elements of the chain.

We can prove that, for a lower chain measure ν on $\mathcal{S}$, the following holds:

$$\nu\left(\bigcap_{A \in \mathcal{A}}\right) = \inf_{A \in \mathcal{A}} \nu(A)$$

for all finite set systems $\mathcal{A}$ such that $\cap_{A \in \mathcal{A}} A \in \mathcal{S}$. If this property holds for *arbitrary* $\mathcal{A}$, and $\mathcal{S}$ is closed under arbitrary intersection, then ν is called a *necessity measure*. Any necessity measure is a lower chain measure, but the converse does not hold. However, the class of necessity measures coincides with the class of lower chain measures if Θ is finite. As consonant belief functions are necessity measures on finite domains, they are trivially also lower chain measures, and viceversa.

Now, let Bel be a belief function and $\mathcal{C}$ a maximal chain in 2^Θ. Then we can build a chain measure (a consonant belief function) associated with Bel as follows:

$$Bel_{\mathcal{C}}(A) = \max_{B \in \mathcal{C}, B \subseteq A} Bel(B). \tag{13.7}$$

We can prove the following.

Theorem 71. *The chain measure (13.7) associated with the maximal chain $\mathcal{C}$ coincides with the vertex Co_ρ (4.69) of the polytope of outer consonant approximations $O^{\mathcal{C}_\rho}[Bel]$ of Bel associated with the permutation ρ of the elements of Θ which generates $\mathcal{C} = \mathcal{C}_\rho$.*

The chain measure associated with a belief function Bel and a maximal chain $\mathcal{C}$ *is* the maximal outer consonant approximation of Bel.

Example 40: outer approximations on the ternary frame. *Let us consider as an example a belief function Bel on a ternary frame $\Theta = \{x, y, z\}$ and study the polytope of outer consonant approximations with focal elements*

$$\mathcal{C} = \Big\{ \{x\}, \{x,y\}, \{x,y,z\} \Big\}.$$

According to Theorem 69, the polytope is the convex closure of the consonant approximations associated with all the assignment functions $\mathbf{B} : 2^\Theta \to \mathcal{C}$. *On a ternary frame, there are* $\prod_{k=1}^{3} k^{2^{3-k}} = 1^4 \cdot 2^2 \cdot 3^1 = 12$ *such functions. We list them here as vectors of the form*

$$\mathbf{B} = \Big[\mathbf{B}(\{x\}), \mathbf{B}(\{y\}), \mathbf{B}(\{z\}), \mathbf{B}(\{x,y\}), \mathbf{B}(\{x,z\}), \mathbf{B}(\{y,z\}), \mathbf{B}(\{x,y,z\}) \Big]',$$

namely

$$\begin{array}{ll} \mathbf{B}_1 = & \big[\{x\}, \quad \{x,y\}, \Theta, \{x,y\}, \Theta, \Theta, \Theta\big]', \\ \mathbf{B}_2 = & \big[\{x\}, \quad \{x,y\}, \Theta, \Theta, \quad \Theta, \Theta, \Theta\big]', \\ \mathbf{B}_3 = & \big[\{x\}, \quad \Theta, \quad \Theta, \{x,y\}, \Theta, \Theta, \Theta\big]', \\ \mathbf{B}_4 = & \big[\{x\}, \quad \Theta, \quad \Theta, \Theta, \quad \Theta, \Theta, \Theta\big]', \\ \mathbf{B}_5 = & \big[\{x,y\}, \{x,y\}, \Theta, \{x,y\}, \Theta, \Theta, \Theta\big]', \\ \mathbf{B}_6 = & \big[\{x,y\}, \{x,y\}, \Theta, \Theta, \quad \Theta, \Theta, \Theta\big]', \\ \mathbf{B}_7 = & \big[\{x,y\}, \Theta, \quad \Theta, \{x,y\}, \Theta, \Theta, \Theta\big]', \\ \mathbf{B}_8 = & \big[\{x,y\}, \Theta, \quad \Theta, \Theta, \quad \Theta, \Theta, \Theta\big]', \\ \mathbf{B}_9 = & \big[\Theta, \quad \{x,y\}, \Theta, \{x,y\}, \Theta, \Theta, \Theta\big]', \\ \mathbf{B}_{10} = & \big[\Theta, \quad \{x,y\}, \Theta, \Theta, \quad \Theta, \Theta, \Theta\big]', \\ \mathbf{B}_{11} = & \big[\Theta, \quad \Theta, \quad \Theta, \{x,y\}, \Theta, \Theta, \Theta\big]', \\ \mathbf{B}_{12} = & \big[\Theta, \quad \Theta, \quad \Theta, \Theta, \quad \Theta, \Theta, \Theta\big]'. \end{array}$$

They correspond to the following consonant belief functions $Co^{\mathbf{B}_i}$ *with BPA* $m^{\mathbf{B}_i} = [m^{\mathbf{B}_i}(\{x\}), m^{\mathbf{B}_i}(\{x,y\}), m^{\mathbf{B}_i}(\Theta)]'$:

$$\begin{array}{lllll} m^{\mathbf{B}_1} & = [m(x), & m(y)+m(x,y), & 1-Bel(x,y) &]', \\ m^{\mathbf{B}_2} & = [m(x), & m(y), & 1-m(x)-m(y) &]', \\ m^{\mathbf{B}_3} & = [m(x), & m(x,y), & 1-m(x)-m(x,y) &]', \\ m^{\mathbf{B}_4} & = [m(x), & 0, & 1-m(x) &]', \\ m^{\mathbf{B}_5} & = [0, & Bel(x,y), & 1-Bel(x,y) &]', \\ m^{\mathbf{B}_6} & = [0, & m(x)+m(y), & 1-m(x)-m(y) &]', \\ m^{\mathbf{B}_7} & = [0, & m(x)+m(x,y), & 1-m(x)-m(x,y) &]', \\ m^{\mathbf{B}_8} & = [0, & m(x), & 1-m(x) &]', \\ m^{\mathbf{B}_9} & = [0, & m(y)+m(x,y), & 1-m(y)-m(x,y) &]', \\ m^{\mathbf{B}_{10}} & = [0, & m(y), & 1-m(y) &]', \\ m^{\mathbf{B}_{11}} & = [0, & m(x,y), & 1-m(x,y) &]', \\ m^{\mathbf{B}_{12}} & = [0, & 0, & 1 &]'. \end{array} \tag{13.8}$$

Figure 13.3 shows the resulting polytope $O^{\mathcal{C}}[Bel]$ *for a belief function*

$$m(x) = 0.3, \quad m(y) = 0.5, \quad m(\{x,y\}) = 0.1, \quad m(\Theta) = 0.1, \tag{13.9}$$

in the component $\mathcal{CO}^{\mathcal{C}} = Cl(Bel_x, Bel_{\{x,y\}}, Bel_\Theta)$ *of the consonant complex (the triangle in the figure). The polytope* $O^{\mathcal{C}}[Bel]$ *is plotted in magenta, together with all of the 12 points (13.8) (magenta squares). Many of them lie on a side of the polytope. However, the point obtained by permutation of singletons (4.69) is a vertex (green square): it is the first item,* $m^{\mathbf{B}_1}$, *in the list (13.8).*

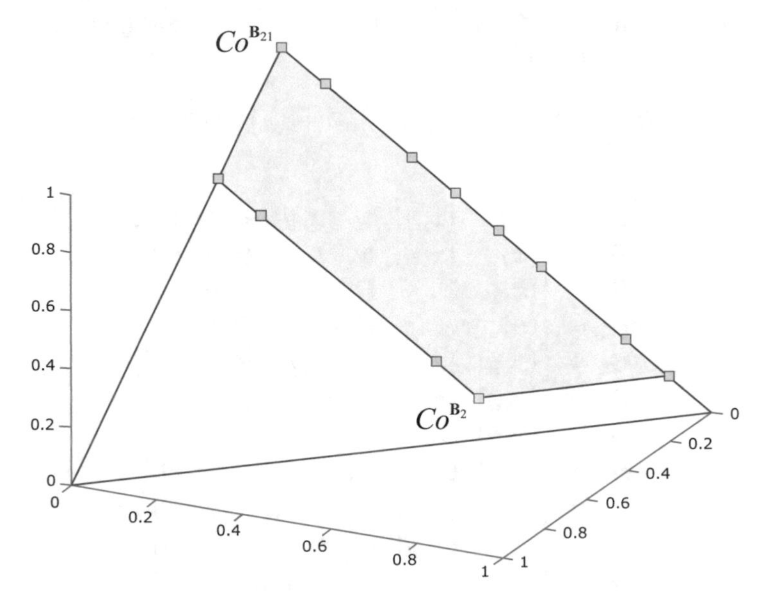

Fig. 13.3: Not all of the points (13.4) associated with assignment functions are actually vertices of $O^{\mathcal{C}}[Bel]$. Here, the polytope $O^{\mathcal{C}}[Bel]$ of outer consonant approximations with $\mathcal{C} = \{\{x\}, \{x, y\}, \Theta\}$ for the belief function (13.9) on $\Theta = \{x, y, z\}$ is plotted in magenta, together with all 12 points (13.8) (magenta squares). The minimal and maximal outer approximations with respect to weak inclusion are $Co^{\mathbf{B}_{12}}$ and $Co^{\mathbf{B}_1}$, respectively.

It is interesting to point out that the points (13.8) are ordered with respect to weak inclusion (we just need to apply its definition, or the redistribution property of Lemma 14). The result is summarised in the graph in Fig. 13.4. We can appreciate that the vertex $Co^{\mathbf{B}_1}$ *generated by singleton permutation is indeed the maximal outer approximation of* Bel, *as stated in Corollary 17.*

13.2 Geometric consonant approximation

While in the first part of the chapter we have studied the geometry of the most popular approach to consonant approximation (outer approximation), in the second part we pursue a different approach based on minimising appropriate distances between the given belief function and the consonant complex (see Chapter 10).

We first briefly outline an alternative geometric representation of belief functions, in terms of mass (rather than belief) vectors.

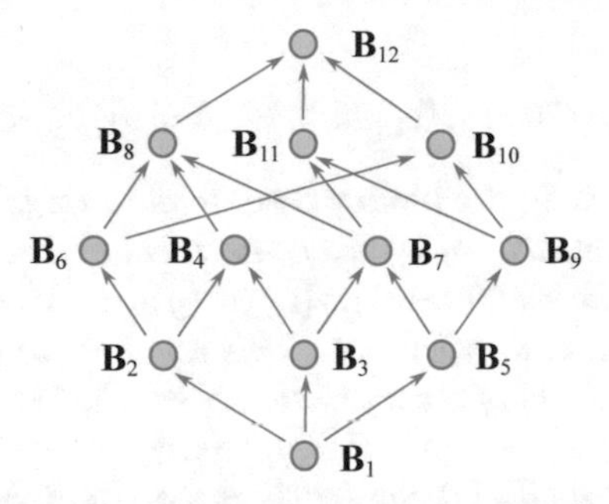

Fig. 13.4: Partial order of the points (13.8) with respect to weak inclusion. For the sake of simplicity, the consonant belief functions $o^{\mathbf{B}_i}$ associated with the assignment function $\mathbf{B}_i$ is denoted here by B_i. An arrow from $\mathbf{B}_i$ to $\mathbf{B}_j$ stands for $Co^{\mathbf{B}_j} \leq Co^{\mathbf{B}_i}$.

13.2.1 Mass space representation

Just as belief functions can be represented as vectors of a sufficiently large Cartesian space (see Chapter 7), they can be associated with the related vector of mass values

$$\mathbf{m} = \sum_{\emptyset \subsetneq A \subseteq \Theta} m(A) \mathbf{m}_A \in \mathbb{R}^{N-1}, \tag{13.10}$$

(compare (7.8) and Theorem 9), where $\mathbf{m}_A$ is the vector of mass values of the categorical belief function Bel_A: $m_A(A) = 1$, $m_A(B) = 0 \; \forall B \neq A$. Note that in $\mathbb{R}^{N-1}$ the vacuous belief function is associated with the vector $\mathbf{m}_\Theta = [0, \ldots, 0, 1]'$, rather than the origin, and cannot be neglected in (13.10).

Since the mass of any focal element $\bar{A}$ is uniquely determined by all the other masses by virtue of the normalisation constraint, we can also choose to represent BPAs as vectors of $\mathbb{R}^{N-2}$ of the form

$$\mathbf{m} = \sum_{\emptyset \subsetneq A \subset \Theta, A \neq \bar{A}} m(A) \mathbf{m}_A, \tag{13.11}$$

in which one component, $\bar{A}$, is neglected. This leads to two possible approaches to consonant approximations in the mass space. We will consider both in the following. As we have done for the belief space representation, to improve readability in what follows we will use the notation m to denote indifferently a basic probability assignment and the associated mass vector $\mathbf{m}$, whenever this causes no confusion.

Whatever the representation chosen, it is not difficult to prove that the collection $\mathcal{M}$ of vectors of the Cartesian space which represent valid basic probability assignments is also a simplex, which we can call the *mass space*. Depending on whether we choose either the first or the second lower-dimensional representation, $\mathcal{M}$ is either the convex closure

$$\mathcal{M} = Cl(m_A, \emptyset \subsetneq A \subset \Theta) \subset \mathbb{R}^{N-1}$$

or

$$\mathcal{M} = Cl(m_A, \emptyset \subsetneq A \subset \Theta, A \neq \bar{A}) \subset \mathbb{R}^{N-2}.$$

Example 41: mass space in the binary case. *In the case of a frame of discernment containing only two elements, $\Theta_2 = \{x, y\}$, each belief function $Bel : 2^{\Theta_2} \to [0,1]$ can be represented by the vector $m = [m(x), m(y)]' \in \mathbb{R}^2$, as $m(\Theta) = 1 - m(x) - m(y)$ by normalisation, and $m(\emptyset) = 0$. This amounts to adopting the $(N-2)$-dimensional version (13.11) of the mass space, which, in this case, coincides with the belief space, as $Bel(x) = m(x)$, $Bel(y) = m(y)$.*

As we know from Example 18, the set $\mathcal{B}_2 = \mathcal{M}_2$ of all basic probability assignments on Θ_2 is the triangle depicted in Fig. 13.5, with vertices

$$Bel_\Theta = m_\Theta = [0,0]', \quad Bel_x = m_x = [1,0]', \quad Bel_y = m_y = [0,1]'.$$

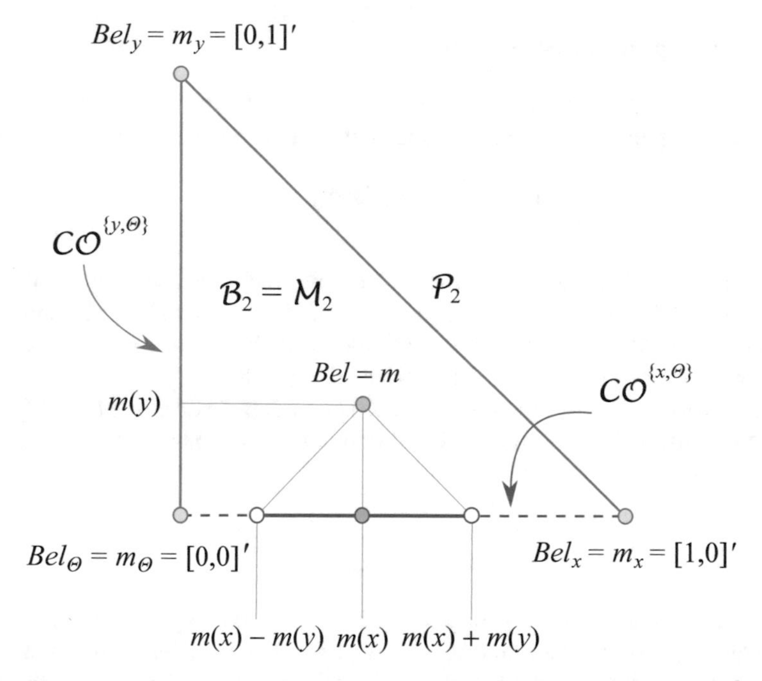

Fig. 13.5: The mass space $\mathcal{M}_2$ for a binary frame is a triangle in $\mathbb{R}^2$ whose vertices are the mass vectors associated with the categorical belief functions focused on $\{x\}, \{y\}$ and Θ: m_x, m_y, m_Θ. The belief space $\mathcal{B}_2$ coincides with $\mathcal{M}_2$ when $\Theta = \Theta_2 = \{x, y\}$. Consonant BFs live in the union of the segments $\mathcal{CO}^{\{x,\Theta\}} = Cl(m_x, m_\Theta)$ (in dashed purple) and $\mathcal{CO}^{\{y,\Theta\}} = Cl(m_y, m_\Theta)$. The unique $L_1 = L_2$ consonant approximation (purple circle) and the set of L_∞ consonant approximations (solid purple segment) on $\mathcal{CO}^{\{x,\Theta\}}$ are shown.

The region $\mathcal{P}_2$ of all Bayesian belief functions on Θ_2 is the diagonal line segment $Cl(m_x, m_y) = Cl(Bel_x, Bel_y)$. On $\Theta_2 = \{x, y\}$, consonant belief functions can have as their chain of focal elements either $\{\{x\}, \Theta_2\}$ or $\{\{y\}, \Theta_2\}$. Therefore, they live in the union of two segments (see Fig. 13.5)

$$\mathcal{CO}_2 = \mathcal{CO}^{\{x,\Theta\}} \cup \mathcal{CO}^{\{y,\Theta\}} = Cl(m_x, m_\Theta) \cup Cl(m_y, m_\Theta).$$

13.2.2 Approximation in the consonant complex

We have seen in Chapter 10 [345] (10.1) that the region $\mathcal{CO} = \mathcal{CO}_\mathcal{B}$ of consonant belief functions in the belief space is a simplicial complex, the union of a collection of (maximal) simplices, each of them associated with a maximal chain $\mathcal{C} = \{A_1 \subset \cdots \subset A_n\}$, $|A_i| = i$, of subsets of the frame Θ:

$$\mathcal{CO}_\mathcal{B} = \bigcup_{\mathcal{C}=\{A_1\subset\cdots\subset A_n\}} Cl(Bel_{A_1}, \ldots, Bel_{A_n}).$$

Analogously, the region $\mathcal{CO}_\mathcal{M}$ of consonant belief functions in the mass space $\mathcal{M}$ is the simplicial complex

$$\mathcal{CO}_\mathcal{M} = \bigcup_{\mathcal{C}=\{A_1\subset\cdots\subset A_n\}} Cl(m_{A_1}, \ldots, m_{A_n}).$$

Definition 129. *Given a belief function Bel, we define the* consonant approximation of Bel induced by a distance function d *in $\mathcal{M}$ ($\mathcal{B}$) as the belief function or functions $Co_{\mathcal{M}/\mathcal{B},d}[m/Bel]$ which minimise the distance $d(m, \mathcal{CO}_\mathcal{M})$ ($d(Bel, \mathcal{CO}_\mathcal{B})$) between m (Bel) and the consonant simplicial complex in $\mathcal{M}$ ($\mathcal{B}$),*

$$Co_{\mathcal{M},d}[m] = \arg \min_{m_{Co}\in\mathcal{CO}_\mathcal{M}} d(m, m_{Co}) \Big/ Co_{\mathcal{B},d}[Bel] = \arg \min_{Co\in\mathcal{CO}_\mathcal{B}} d(Bel, Co) \tag{13.12}$$

(see Fig. 13.6).

Choice of norm Consonant belief functions are the counterparts of necessity measures in the theory of evidence, so that their plausibility functions are possibility measures, which in turn are inherently related to L_∞, as $Pos(A) = \max_{x\in A} Pos(x)$ (see Section 10.1). It therefore makes sense to conjecture that a consonant transformation obtained by picking as the distance function d in (13.12) one of the classical L_p norms would be meaningful.

For vectors $m, m' \in \mathcal{M}$ representing the basic probability assignments of two belief functions Bel, Bel', these norms read as

$$\begin{aligned}
\|m - m'\|_{L_1} &\doteq \sum_{\emptyset\subsetneq B\subseteq\Theta} |m(B) - m'(B)|, \\
\|m - m'\|_{L_2} &\doteq \sqrt{\sum_{\emptyset\subsetneq B\subseteq\Theta} (m(B) - m'(B))^2}, \\
\|m - m'\|_{L_\infty} &\doteq \max_{\emptyset\subsetneq B\subseteq\Theta} |m(B) - m'(B)|,
\end{aligned} \tag{13.13}$$

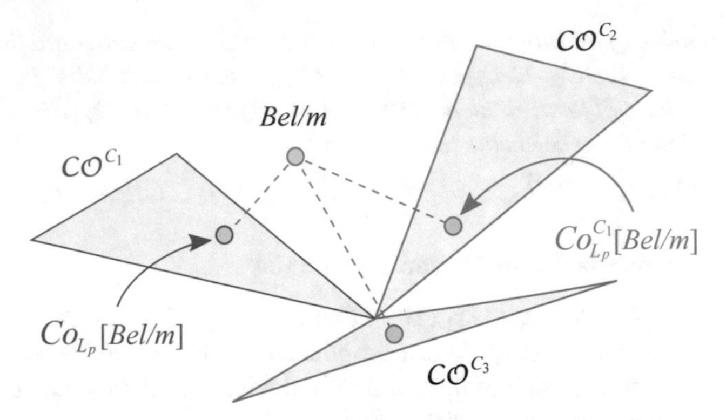

Fig. 13.6: In order to minimise the distance of a belief/mass vector from the consonant simplicial complex, we need to find all the partial solutions (13.15) (blue circles) on all the maximal simplices which form the complex, and compare these partial solutions to select a global one (purple circle).

while the same norms in the belief space read as

$$
\begin{aligned}
\|Bel - Bel'\|_{L_1} &\doteq \sum_{\emptyset \subsetneq B \subseteq \Theta} |Bel(B) - Bel'(B)|, \\
\|Bel - Bel'\|_{L_2} &\doteq \sqrt{\sum_{\emptyset \subsetneq B \subseteq \Theta} (Bel(B) - Bel'(B))^2}, \\
\|Bel - Bel'\|_{L_\infty} &\doteq \max_{\emptyset \subsetneq B \subseteq \Theta} |Bel(B) - Bel'(B)|.
\end{aligned}
\tag{13.14}
$$

L_p norms have been recently successfully employed in the probability transformation problem [333] and for conditioning [327, 356]. Recall that L_2 distance minimisation induces the orthogonal projection of Bel onto $\mathcal{P}$ (Chapter 11).

Clearly, however, a number of other norms can be used to define consonant (or Bayesian) approximations. For instance, generalisations to belief functions of the classical Kullback–Leibler divergence of two probability distributions P, Q,

$$
D_{KL}(P|Q) = \int_{-\infty}^{\infty} p(x) \log \left(\frac{p(x)}{q(x)} \right) \mathrm{d}x,
$$

or other measures based on information theory such as fidelity and entropy-based norms [938] can be studied. Many other similarity measures have indeed been proposed [1636, 902, 965, 501]. The application of similarity measures more specific to belief functions or inspired by classical probability to the approximation problem is left for future work.

Distance of a point from a simplicial complex As the consonant complex $\mathcal{CO}$ is a collection of simplices which generate distinct linear spaces (in both the belief and the mass space), solving the consonant approximation problem involves first finding a number of partial solutions

$$
\begin{aligned}
Co^{\mathcal{C}}_{\mathcal{B},L_p}[Bel] &= \arg\min_{Co\in\mathcal{CO}^{\mathcal{C}}_{\mathcal{B}}} \|Bel - Co\|_{L_p}, \\
Co^{\mathcal{C}}_{\mathcal{M},L_p}[m] &= \arg\min_{m_{Co}\in\mathcal{CO}^{\mathcal{C}}_{\mathcal{M}}} \|m - m_{Co}\|_{L_p},
\end{aligned} \tag{13.15}
$$

(see Fig. 13.6), one for each maximal chain $\mathcal{C}$ of subsets of Θ. Then, the distance of Bel from all such partial solutions needs to be assessed in order to select a global optimal approximation.

In the following we start, as usual, from the simple but interesting binary case (Fig. 13.5). Some of its features are retained in the general case; others are not. Note also that, in the binary case, the consonant and consistent [360] (Chapter 14) approximations coincide, and there is no difference between the belief and the mass space [356] representations.

13.2.3 Möbius inversion and preservation of norms, induced orderings

Given a belief function Bel, the corresponding basic probability assignment m can be obtained via Möbius inversion as $m(A) = \sum_{B\subseteq A}(-1)^{|A\setminus B|}Bel(B)$. More generally, a Möbius inverse exists for any sum function $f(A) = \sum_{B\leq A} g(B)$ defined on a partially ordered set with ordering $\leq$, and is the combinatorial analogue of the derivative operator in calculus [18]. If a norm d existed for belief functions that was preserved by Möbius inversion, $d(Bel, Bel') = d(m, m')$, then the approximation problems (13.15) in $\mathcal{B}$ and $\mathcal{M}$ would obviously yield the same result(s). The same would be true if Möbius inversion preserved the *ordering* induced by d:

$$
d(Bel, Bel_1) \leq d(Bel, Bel_2) \Leftrightarrow d(m, m_1) \leq d(m, m_2) \;\; \forall Bel, Bel_1, Bel_2 \in \mathcal{B}.
$$

Unfortunately, this is not the case, at least for any of the above L_p norms. Let us consider again the binary example, and measure the distance between the categorical belief function $Bel = Bel_y$ and the segment $Cl(Bel_x, Bel_\Theta)$ of consonant BFs with chain of focal elements $\{x\} \subset \Theta$. When using the L_2 distance in the belief space, we obtain $\|Bel_y - Bel_\Theta\|_{L_2} = \|[0,1,1]' - [0,0,1]'\| = 1 < \|Bel_y - Bel_x\|_{L_2} = \|[0,1,1]' - [1,0,1]'\| = \sqrt{2}$, and Bel_Θ is closer to Bel_y than Bel_x is. In the mass space embedded in $\mathbb{R}^3$, instead, we have $\|m_y - m_\Theta\|_{L_2} = \|[0,1,0]' - [0,0,1]'\| = \sqrt{2} = \|m_y - m_x\|_{L_2} = \|[0,1,0]' - [1,0,0]'\| = \sqrt{2}$, whereas $\|m_y - (m_x + m_\Theta)/2\|_{L_2} = \sqrt{3/2} < \sqrt{2}$. The L_2 partial consonant approximation in the first case is Bel_Θ, and in the second case $(m_x + m_\Theta)/2$. Similar results can be shown for L_1 and L_∞.

In conclusion, separate approximation problems (13.15) have to be set up in the belief and the mass space. Indeed, an interesting question is whether there actually is a norm whose induced ordering is preserved by Möbius inversion. This is an extremely challenging open problem, which, to the best of our knowledge, has not been studied so far and cannot be quickly addressed here, but we intend to tackle it, among others, in the near future.

13.3 Consonant approximation in the mass space

Let us thus compute the analytical form of all L_p consonant approximations in the mass space, in both its $\mathbb{R}^{N-1}$ and its $\mathbb{R}^{N-2}$ forms (see (13.10) and (13.11)). We start by analysing the difference vector $m - m_{Co}$ between the original mass vector and its approximation.

In the complete, $(N-1)$-dimensional version $\mathcal{M}$ of the mass space (see (13.10)), the mass vector associated with an arbitrary consonant belief function Co with maximal chain of focal elements $\mathcal{C}$ reads as $m_{Co} = \sum_{A\in\mathcal{C}} m_{Co}(A) m_A$. The sought difference vector is therefore

$$m - m_{Co} = \sum_{A\in\mathcal{C}} (m(A) - m_{Co}(A)) m_A + \sum_{B\not\in\mathcal{C}} m(B) m_B. \tag{13.16}$$

When we choose an $(N-2)$-dimensional section of the mass space (see (13.11)), instead, we need to distinguish between the cases where the missing focal element $\bar{A}$ is an element of the desired maximal chain $\mathcal{C}$ and where it is not. In the former case, the mass vector associated with the same, arbitrary consonant belief function Co with maximal chain $\mathcal{C}$ is $m_{Co} = \sum_{A\in\mathcal{C}, A\neq\bar{A}} m_{Co}(A) m_A$, and the difference vector is

$$m - m_{Co} = \sum_{A\in\mathcal{C}, A\neq\bar{A}} (m(A) - m_{Co}(A)) m_A + \sum_{B\not\in\mathcal{C}} m(B) m_B. \tag{13.17}$$

If, instead, the missing component is *not* an element of $\mathcal{C}$, the arbitrary consonant BF is $m_{Co} = \sum_{A\in\mathcal{C}} m_{Co}(A) m_A$, while the difference vector becomes

$$m - m_{Co} = \sum_{A\in\mathcal{C}} (m(A) - m_{Co}(A)) m_A + \sum_{B\not\in\mathcal{C}, B\neq\bar{A}} m(B) m_B. \tag{13.18}$$

One can observe that, since (13.18) coincides with (13.16) (factoring out the missing component $\bar{A}$), minimising the L_p norm of the difference vector in an $(N-2)$-dimensional section of the mass space which leaves out a focal element outside the desired maximal chain yields the same results as in the complete mass space.[77] Therefore, in what follows we consider only consonant approximations in $(N-2)$-dimensional sections obtained by excluding a component associated with an element $\bar{A} \in \mathcal{C}$ of the desired maximal chain.

[77] The absence of the missing component $m(\bar{A})$ in (13.18) implies, in fact, a small difference when it comes to the L_∞ approximation: Theorem 74 and (13.23), concerning the vertices of the polytope of L_∞ approximations, remain valid as long as we replace $\max_{B\not\in\mathcal{C}} m(B)$ with $\max_{B\not\in\mathcal{C}, B\neq\bar{A}} m(B)$.

In the following, we denote by $\mathcal{CO}^{\mathcal{C}}_{\mathcal{M}\setminus\bar{A},L_p}[m]$ (upper case) the set of partial L_p approximations of Bel with maximal chain $\mathcal{C}$ in the section of the mass space which excludes $\bar{A} \in \mathcal{C}$. We drop the superscript $\mathcal{C}$ for global solutions, drop $\setminus\bar{A}$ for solutions in the complete mass space and use $Co^{\mathcal{C}}_{\mathcal{M}\setminus\bar{A},L_p}[m]$ for pointwise solutions and the barycentres of sets of solutions.

13.3.1 Results of Minkowski consonant approximation in the mass space

L_1 approximation Minimising the L_1 norm of the difference vectors (13.16), (13.17) yields the following result.

Theorem 72. *Given a belief function $Bel : 2^\Theta \to [0,1]$ with basic probability assignment m, the partial L_1 consonant approximations of Bel with a maximal chain of focal elements $\mathcal{C}$ in the complete mass space $\mathcal{M}$ are the set of consonant belief functions Co with chain $\mathcal{C}$ such that $m_{Co}(A) \geq m(A)\ \forall A \in \mathcal{C}$. They form a simplex*

$$\mathcal{CO}^{\mathcal{C}}_{\mathcal{M},L_1}[m] = Cl\big(m^{\bar{A}}_{L_1}[m], \bar{A} \in \mathcal{C}\big), \tag{13.19}$$

whose vertices have BPA

$$m^{\bar{A}}_{L_1}[m](A) = \begin{cases} m(A) + \sum_{B \notin \mathcal{C}} m(B) & A = \bar{A}, \\ m(A) & A \in \mathcal{C}, A \neq \bar{A}, \end{cases} \tag{13.20}$$

and whose barycentre has the mass assignment

$$Co^{\mathcal{C}}_{\mathcal{M},L_1}[m](A) = m(A) + \frac{1}{n} \sum_{B \notin \mathcal{C}} m(B) \qquad \forall A \in \mathcal{C}. \tag{13.21}$$

The set of global L_1 approximations of Bel is the union of the simplices (13.19) associated with the maximal chains which maximise their total original mass:

$$\mathcal{CO}_{\mathcal{M},L_1}[m] = \bigcup_{\mathcal{C} \in \arg\max_{\mathcal{C}} \sum_{A \in \mathcal{C}} m(A)} \mathcal{CO}^{\mathcal{C}}_{\mathcal{M},L_1}[m].$$

The partial L_1 consonant approximation $Co^{\mathcal{C}}_{\mathcal{M}\setminus\bar{A},L_1}[m]$ of Bel in the section of the mass space $\mathcal{M}$ with missing component $\bar{A} \in \mathcal{C}$ is unique and has the BPA (13.20). The global approximations of this kind are also associated with the maximal chains $\arg\max_{\mathcal{C}} \sum_{A \in \mathcal{C}} m(A)$.

L_2 approximation In order to find the L_2 consonant approximation(s) in $\mathcal{M}$, instead, it is convenient to recall that the minimal L_2 distance between a point and a vector space is attained by the point of the vector space V such that the difference vector is orthogonal to all the generators $\mathbf{g}_i$ of V:

$$\arg\min_{\mathbf{q} \in V} \|P - \mathbf{q}\|_{L_2} = \hat{q} \in V : \langle P - \hat{q}, \mathbf{g}_i \rangle = 0 \ \ \forall i,$$

whenever $P \in \mathbb{R}^m$, $V = \mathrm{span}(\{\mathbf{g}_i, i\})$. Instead of minimising the L_2 norm of the difference vector $\|m - m_{Co}\|_{L_2}$, we can thus impose a condition of orthogonality between the difference vector itself, $m - m_{Co}$, and each component $\mathcal{CO}^{\mathcal{C}}_{\mathcal{M}}$ of the consonant complex in the mass space.

Theorem 73. *Given a belief function $Bel : 2^\Theta \rightarrow [0,1]$ with basic probability assignment m, the partial L_2 consonant approximation of Bel with a maximal chain of focal elements $\mathcal{C}$ in the complete mass space $\mathcal{M}$ has mass assignment (13.21)*

$$Co^{\mathcal{C}}_{\mathcal{M},L_2}[m] = Co^{\mathcal{C}}_{\mathcal{M},L_1}[m].$$

The set of all global L_2 approximations is

$$\mathcal{CO}_{\mathcal{M},L_2}[m] = \bigcup_{\mathcal{C} \in \arg\min_{\mathcal{C}} \sum_{B \not\in \mathcal{C}} (m(B))^2} Co^{\mathcal{C}}_{\mathcal{M},L_2}[m],$$

i.e., the union of the partial solutions associated with maximal chains of focal elements which minimise the sum of squared masses outside the chain.

The partial L_2 consonant approximation of Bel in the section of the mass space with missing component $\bar{A} \in \mathcal{C}$ is unique, and coincides with the L_1 partial consonant approximation in the same section (13.20):

$$Co^{\mathcal{C}}_{\mathcal{M}\setminus\bar{A},L_2}[m] = Co^{\mathcal{C}}_{\mathcal{M}\setminus\bar{A},L_1}[m].$$

L_2 global approximations in the section form the union of the related partial approximations associated with the chains $\arg\min_{\mathcal{C}} \sum_{B \not\in \mathcal{C}} (m(B))^2$. Note that the global solutions in the L_1 and L_2 cases fall, in general, on different simplicial components of $\mathcal{CO}$.

L_∞ approximation

Theorem 74. *Given a belief function $Bel : 2^\Theta \rightarrow [0,1]$ with basic probability assignment m, the partial L_∞ consonant approximations of Bel with a maximal chain of focal elements $\mathcal{C}$ in the complete mass space $\mathcal{M}$ form a simplex*

$$\mathcal{CO}^{\mathcal{C}}_{\mathcal{M},L_\infty}[m] = Cl\big(m^{\bar{A}}_{L_\infty}[m], \bar{A} \in \mathcal{C}\big), \tag{13.22}$$

whose vertices have BPA

$$m^{\bar{A}}_{L_\infty}[m](A)$$
$$= \begin{cases} m(A) + \max\limits_{B \not\in \mathcal{C}} m(B) & A \in \mathcal{C}, A \neq \bar{A}, \\ m(\bar{A}) + \max\limits_{B \not\in \mathcal{C}} m(B) + \left(\sum\limits_{B \not\in \mathcal{C}} m(B) - n \max\limits_{B \not\in \mathcal{C}} m(B) \right) & A = \bar{A}, \end{cases} \tag{13.23}$$

whenever the belief function to be approximated is such that

$$\max_{B \not\in \mathcal{C}} m(B) \geq \frac{1}{n} \sum_{B \not\in \mathcal{C}} m(B). \tag{13.24}$$

When the opposite is true, the sought partial L_∞ consonant approximation reduces to a single consonant belief function, the barycentre of the above simplex, located on the partial L_2 approximation (and the barycentre of the L_1 partial approximations) (13.21).

When (13.24) holds, the global L_∞ consonant approximations are associated with the maximal chain(s) of focal elements

$$\arg\min_{\mathcal{C}} \max_{B \not\in \mathcal{C}} m(B); \tag{13.25}$$

otherwise, they correspond to the maximal chains

$$\arg\min_{\mathcal{C}} \sum_{B \not\in \mathcal{C}} m(B).$$

The partial L_∞ consonant approximations of Bel in the section of the mass space $\mathcal{M}$ with missing component $\bar{A} \in \mathcal{C}$ form a set $\mathcal{CO}^{\mathcal{C}}_{\mathcal{M} \setminus \bar{A}, L_\infty}[m]$ whose elements have a BPA m_{Co} such that

$$m(A) - \max_{B \not\in \mathcal{C}} m(B) \leq m_{Co}(A) \leq m(A) + \max_{B \not\in \mathcal{C}} m(B) \quad \forall A \in \mathcal{C}, A \neq \bar{A}. \tag{13.26}$$

Its barycentre reassigns all the mass originally outside the desired maximal chain $\mathcal{C}$ to $\bar{A}$, leaving the masses of the other elements of the chain untouched (13.20):

$$Co^{\mathcal{C}}_{\mathcal{M} \setminus \bar{A}, L_\infty}[m] = Co^{\mathcal{C}}_{\mathcal{M} \setminus \bar{A}, L_2}[m] = m^{\bar{A}}_{L_1}[m].$$

The related global approximations of Bel are associated with the optimal chain(s) (13.25).

13.3.2 Semantics of partial consonant approximations in the mass space

$N-1$ representation Summarising, the partial L_p approximations of an arbitrary mass function m in the complete mass space $\mathcal{M}$ are

$$\begin{aligned}
\mathcal{CO}^{\mathcal{C}}_{\mathcal{M}, L_1}[m] &= Cl(m^{\bar{A}}_{L_1}[m], \bar{A} \in \mathcal{C}) \\
&= \left\{ Co \in \mathcal{CO}^{\mathcal{C}}_{\mathcal{M}} : m_{Co}(A) \geq m(A) \; \forall A \in \mathcal{C} \right\}, \\
Co^{\mathcal{C}}_{\mathcal{M}, L_2}[m] &= Co^{\mathcal{C}}_{\mathcal{M}, L_1}[m] : m_{Co}(A) = m(A) + \frac{1}{n} \sum_{B \not\in \mathcal{C}} m(B), \\
\mathcal{CO}^{\mathcal{C}}_{\mathcal{M}, L_\infty}[m] &= \begin{cases} Cl(m^{L_\infty}_{\bar{A}}, \bar{A} \in \mathcal{C}) & \text{if (13.24) holds,} \\ Co^{\mathcal{C}}_{\mathcal{M}, L_2}[m] & \text{otherwise.} \end{cases}
\end{aligned} \tag{13.27}$$

We can observe that, for each desired maximal chain of focal elements $\mathcal{C}$:

1. The L_1 partial approximations of Bel are those consonant belief functions whose mass assignment dominates that of Bel over all the elements of the chain.
2. This set is a fully admissible simplex, whose vertices are obtained by re-assigning all the mass outside the desired chain to a single focal element of the chain itself (see (13.20)).
3. Its barycentre coincides with the L_2 partial approximation with the same chain, which redistributes the original mass of focal elements outside the chain to all the elements of the chain on an equal basis (13.21).
4. When the partial L_∞ approximation is unique, it coincides with the L_2 approximation and the barycentre of the L_1 approximations.
5. When it is not unique, it is a simplex whose vertices assign to each element of the chain (except one) the maximum mass outside the chain, and whose barycentre is again the L_2 approximation.

Note that the simplex of L_∞ partial solutions (point 5) may fall outside the simplex of consonant belief functions with the same chain – therefore, some of those approximations will not be admissible.

$N-2$ representation When we adopt an $(N-2)$-dimensional section $\mathcal{M} \setminus \bar{A}$ of the mass space instead, the partial L_p approximations are

$$\begin{aligned}
Co^{\mathcal{C}}_{\mathcal{M}\setminus\bar{A},L_1}[m] &= Co^{\mathcal{C}}_{\mathcal{M}\setminus\bar{A},L_2}[m] \\
&= Co^{\mathcal{C}}_{\mathcal{M}\setminus\bar{A},L_\infty}[m] \quad : \begin{cases} m_{Co}(A) = m(A), \; A \in \mathcal{C}, A \neq \bar{A} \\ m_{Co}(\bar{A}) = m(\bar{A}) + \sum_{B \not\in \mathcal{C}} m(B), \end{cases} \\
\mathcal{CO}^{\mathcal{C}}_{\mathcal{M}\setminus\bar{A},L_\infty}[m] &= \Big\{ Co \in \mathcal{CO}^{\mathcal{C}}_{\mathcal{M}} \quad : \big| m_{Co}(A) - m(A) \big| \leq \max_{B \not\in \mathcal{C}} m(B) \\
&\qquad \forall A \in \mathcal{C}, A \neq \bar{A} \Big\}.
\end{aligned} \tag{13.28}$$

Therefore, for each desired maximal chain $\mathcal{C}$:

- The L_1 and L_2 partial approximations are uniquely determined, and coincide with the barycentre of the set of L_∞ partial approximations.
- Their interpretation is straightforward: all the mass outside the chain is reassigned to a single focal element of the chain $\bar{A} \in \mathcal{C}$.
- The set of L_∞ (partial) approximations falls entirely inside the simplex of admissible consonant belief functions only if each focal element in the desired chain has a mass greater than that of all focal elements outside the chain:

$$\min_{A \in \mathcal{C}} m(A) \geq \max_{B \not\in \mathcal{C}} m(B).$$

- The latter forms a generalised rectangle in the mass space $\mathcal{M}$, whose size is determined by the largest mass outside the desired maximal chain.

Comparison and general overview As a general trait, approximations in the mass space amount to some redistribution of the original mass to focal elements of the desired maximal chain. The relationships between the different L_p consonant approximations in the full mass space and those in any arbitrary $(N-2)$-dimensional section of $\mathcal{M}$ are summarised in the diagram in Fig. 13.7: whilst both are acceptable geometric representations of mass vectors, the two approaches generate different but related results.

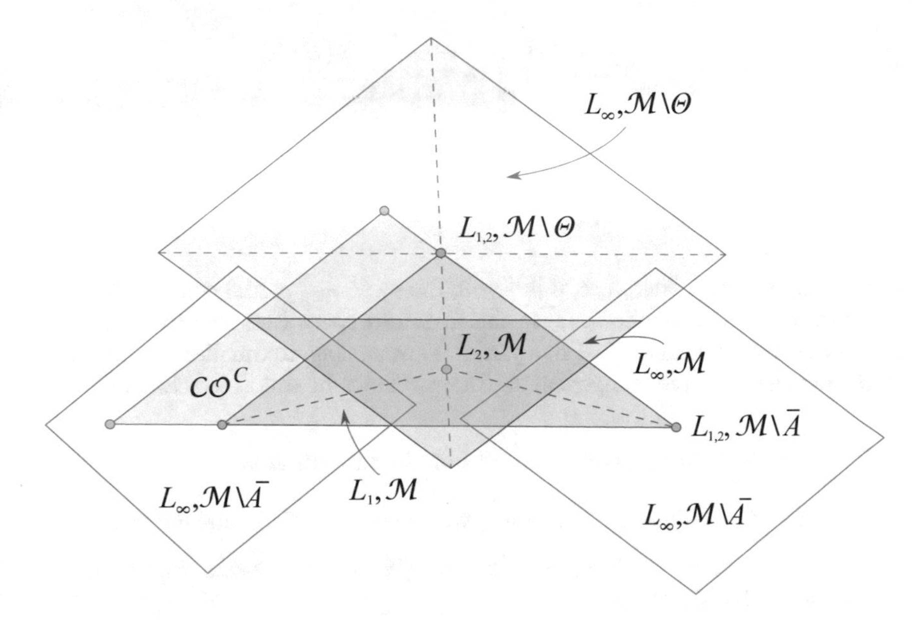

Fig. 13.7: Graphical representation of the relationships between the different (partial) L_p consonant approximations with desired maximal chain $\mathcal{C}$, in the related simplex $\mathcal{CO}^{\mathcal{C}}_{\mathcal{M}}$ of the consonant complex $\mathcal{CO}$. Approximations in the full mass space $\mathcal{M}$ and approximations computed in an $(N-2)$-dimensional section with missing component $\bar{A} \in \mathcal{C}$ are compared. In the latter case, the special case $\bar{A} = \Theta$ is highlighted.

By (13.28), the L_1/L_2 approximations in all the $(N-2)$-dimensional sections of the mass space (and the barycentres of the related sets of L_∞ approximations) track all the vertices of the L_1 simplex in $\mathcal{M}$. As the section of the component $\bar{A}$ to be neglected is quite arbitrary, the latter simplex (and its barycentre) seem to play a privileged role.

L_∞ approximations in any such sections are not entirely admissible, and do not show any particular relation to the simplex of $L_1, \mathcal{M}$ solutions. The relation between the L_∞ partial solutions in the full mass space and those computed in its sections $\mathcal{M} \setminus \bar{A}$ remains to be determined.

Theorem 75. *Given a belief function $Bel : 2^\Theta \to [0,1]$ with BPA m and a maximal chain of focal elements $\mathcal{C}$, the partial L_∞ consonant approximations of Bel in the complete mass space $\mathcal{CO}^{\mathcal{C}}_{\mathcal{M},L_\infty}[m]$ are not necessarily partial L_∞ approximations $\mathcal{CO}^{\mathcal{C}}_{\mathcal{M}\setminus\bar{A},L_\infty}[m]$ in the section excluding $\bar{A}$. However, for all $\bar{A} \in \mathcal{C}$, the two sets of approximations share the vertex (13.23).*

Notice that, in the ternary case ($n = 3$), the condition (13.72),

$$\max_{B \notin \mathcal{C}} m(B) > \frac{1}{n-2} \sum_{B \notin \mathcal{C}} m(B)$$

(see the proof of Theorem 75 in the appendix), becomes

$$\max_{B \notin \mathcal{C}} m(B) > \sum_{B \notin \mathcal{C}} m(B),$$

which is impossible. Therefore, if $|\Theta| = 3$, the set of L_∞ partial consonant approximations of a belief function Bel in the full mass space (the blue triangle in Fig. 13.7) is a subset of the set of its L_∞ partial consonant approximations in the section of $\mathcal{B}$ which neglects the component $\bar{A}$, for any choice of $\bar{A} \in \mathcal{C}$ (see Example 42).

13.3.3 Computability and admissibility of global solutions

As far as global solutions are concerned, we can observe the following facts:

- in the L_1 case, in both the $(N-1)$- and the $(N-2)$-dimensional representations, the optimal chain(s) are

$$\arg\min_{\mathcal{C}} \sum_{B \notin \mathcal{C}} m(B) = \arg\max_{\mathcal{C}} \sum_{A \in \mathcal{C}} m(A);$$

- in the L_2 case, again in both representations, these are

$$\arg\min_{\mathcal{C}} \sum_{B \notin \mathcal{C}} (m(B))^2;$$

- in the L_∞ case, the optimal chain(s) are

$$\arg\min_{\mathcal{C}} \max_{B \notin \mathcal{C}} m(B),$$

unless the approximation is unique in the full mass space, in which case the optimal chains behave as in the L_1 case.

Admissibility of partial and global solutions Concerning their admissibility, we know that all L_1/L_2 partial solutions are always admissible, in both representations of mass vectors. For the L_∞ case, in the full mass space not even global solutions are guaranteed to have vertices that are all admissible (13.23). Indeed,

$$\sum_{B \not\in C} m(B) - n \cdot \max_{B \not\in C} m(B) \leq 0,$$

as the condition (13.24) holds; therefore $m_{L_\infty}^{\bar{A}}[m](\bar{A})$ can be negative.

In $\mathcal{M} \setminus \bar{A}$, by (13.26), the set of L_∞ approximations is entirely admissible iff

$$\min_{A \in C, A \neq \bar{A}} m(A) \geq \max_{B \not\in C} m(B). \tag{13.29}$$

A counter-example shows that minimising $\max_{B \not\in C} m(B)$ (i.e., considering global L_∞ solutions) does not necessarily imply (13.29): think of a belief function Bel with no focal elements of cardinality 1, but several focal elements of cardinality 2.

The computation of the admissible part of this set of solutions is not trivial, and is left for future work.

Computational complexity of global solutions In terms of computability, finding the global L_1/L_2 approximations involves therefore finding the maximum-mass (or squared mass) chain(s). This is expensive, as we have to examine all $n!$ of them. The most favourable case (in terms of complexity) is the L_∞ one, as all the chains which do not contain the maximum-mass element(s) are optimal. Looking for the maximum-mass focal elements requires a single pass of the list of focal elements, with complexity $O(2^n)$ rather than $O(n!)$. On the other hand, in this case the global consonant approximations are spread over a potentially large number of simplicial components of $\mathcal{CO}$, and are therefore less informative.

13.3.4 Relation to other approximations

This behaviour compares rather unfavourably with that of two other natural consonant approximations.

Isopignistic approximation

Definition 130. *Given a belief function $Bel : 2^\Theta \rightarrow [0,1]$, its* isopignistic *consonant approximation [546] is defined as the unique consonant belief function $Co_{\text{iso}}[Bel]$ such that $BetP[Co_{\text{iso}}[Bel]] = BetP[Bel]$. Its contour function (2.18) is*

$$pl_{Co_{\text{iso}}[Bel]}(x) = \sum_{x' \in \Theta} \min \Big\{ BetP[Bel](x), BetP[Bel](x') \Big\}. \tag{13.30}$$

It is well known that, given the contour function $pl(x) \doteq Pl(\{x\})$ of a consistent belief function $Bel : 2^\Theta \rightarrow [0,1]$ (such that $\max_x pl(x) = 1$), we can obtain the unique consonant BF which has pl as its contour function via the following formulae:

$$m_{Co}(A_i) = \begin{cases} pl(x_i) - pl(x_{i+1}) & i = 1, \ldots, n-1, \\ pl(x_n) & i = n, \end{cases} \tag{13.31}$$

where $x_1, \ldots, x_n$ are the singletons of Θ sorted by plausibility value, and $A_i = \{x_1, \ldots, x_i\}$ for all $i = 1, \ldots, n$. Such a unique transformation is not in general feasible for arbitrary belief functions.

Contour-based approximation The isopignistic transform builds a contour function (possibility distribution) from the pignistic values of the singletons, in the following way. Given the list of singletons $x_1, \ldots, x_n$ ordered by pignistic value, (13.30) reads as

$$\begin{aligned} pl_{Co_{\text{iso}}[Bel]}(x_i) &= 1 - \sum_{j=1}^{i-1} \Big(BetP[Bel](x_j) - BetP[Bel](x_i) \Big) \\ &= \sum_{j=i}^{n} BetP[Bel](x_j) + (i-1) BetP[Bel](x_i). \end{aligned}$$

By applying (13.31), we obtain the following mass values:

$$m_{Co_{\text{iso}}[Bel]}(A_i) = i \cdot \Big(BetP[Bel](x_i) - BetP[Bel](x_{i+1}) \Big), \quad i = 1, \ldots, n-1, \tag{13.32}$$

with $m_{Co_{\text{iso}}[Bel]}(A_n) = n \cdot BetP[Bel](x_n)$.

Definition 131. *Given a belief function $Bel : 2^\Theta \to [0,1]$, its* contour-based *consonant approximation with maximal chain of focal elements $\mathcal{C} = \{A_1 \subset \cdots \subset A_n\}$ has mass assignment*

$$m_{Co_{\text{con}}[Bel]}(A_i) = \begin{cases} 1 - pl(x_2) & i = 1, \\ pl(x_i) - pl(x_{i+1}) & i = 2, \ldots, n-1, \\ pl(x_n) & i = n, \end{cases} \tag{13.33}$$

where $\{x_1\} = A_1$, $\{x_i\} \doteq A_i \setminus A_{i-1}$ for all $i = 2, \ldots, n$.

Such an approximation uses the (unnormalised) contour function of an arbitrary belief function Bel *as if* it was a possibility distribution, by replacing the plausibility of the maximal element with 1, and applying the mapping (13.31).

In order to guarantee their admissibility, both the isopignistic and the contour-based approximations require sorting the pignistic and the plausibility values, respectively, of the singletons (an operation whose complexity is $O(n \log n)$). On top of that, though, one must add the complexity of actually computing the value of $BetP[Bel](x)$ ($pl(x)$) from a mass vector, which requires n scans (one for each singleton x) with an overall complexity of $n \cdot 2^n$.

Outer consonant approximation An interesting relationship between the outer consonant and L_1 consonant approximations in the mass space $\mathcal{M}$ also emerges.

Theorem 76. *Given a belief function $Bel : 2^\Theta \rightarrow [0,1]$, the set of partial L_1 consonant approximations $\mathcal{CO}^{\mathcal{C}}_{\mathcal{M},L_1}[m]$ with maximal chain of focal elements $\mathcal{C}$ in the complete mass space and the set $\mathcal{O}^{\mathcal{C}}[Bel]$ of its partial outer consonant approximations with the same chain have a non-empty intersection. This intersection contains at least the convex closure of the candidate vertices of $\mathcal{O}^{\mathcal{C}}[Bel]$ whose assignment functions are such that $\mathbf{B}(A_i) = A_i$ for all $i = 1, \ldots, n$.*

Proof. Clearly, if $\mathbf{B}(A_i) = A_i$ for all $i = 1, \ldots, n$, then the mass $m(A_i)$ is reassigned to A_i itself for each element A_i of the chain. Hence $m_{Co}(A_i) \geq m(A_i)$, and the resulting consonant belief function belongs to $\mathcal{CO}^{\mathcal{C}}_{\mathcal{M},L_1}[m]$ (see (13.27)). □

In particular, both $Co^{\mathcal{C}}_{\max}[Bel]$ (4.69) and

$$Co^{\mathcal{C}}_{\mathcal{M}\setminus\Theta,L_{1/2}}[m] : \begin{cases} m_{Co}(A) = m(A) & A \in \mathcal{C}, A \neq \Theta, \\ m_{Co}(\Theta) = m(\Theta) + \sum_{B \notin \mathcal{C}} m(B) & A = \Theta, \end{cases} \tag{13.34}$$

belong to both the (partial) outer and the $L_1, \mathcal{M}$ consonant approximations. The quantity (13.34) is generated by the trivial assignment function assigning all the mass $m(B)$, $B \subseteq \Theta$, $B \notin \mathcal{C}$ to $A_n = \Theta$: $\mathbf{B}(B) = \Theta$ for all $B \notin \mathcal{C}$.

A negative result can, on the other hand, be proven for L_∞ approximations.

Theorem 77. *Given a belief function $Bel : 2^\Theta \rightarrow [0,1]$, the set of its (partial) outer consonant approximations $\mathcal{O}^{\mathcal{C}}[Bel]$ with maximal chain $\mathcal{C}$ and the set of its partial L_∞ approximations (in the complete mass space) with the same chain may have an empty intersection.*

In particular, $Co^{\mathcal{C}}_{\max}[Bel]$ is not necessarily an $L_\infty, \mathcal{M}$ approximation of Bel.

Example 42: ternary example, mass space. *It can be useful to compare the different approximations in the toy case of a ternary frame, $\Theta = \{x, y, z\}$.*

Let the desired consonant approximation have maximal chain $\mathcal{C} = \{\{x\} \subset \{x, y\} \subset \Theta\}$. Figure 13.8 illustrates the different partial L_p consonant approximations in $\mathcal{M}$ in the simplex of consonant belief functions with chain $\mathcal{C}$, for a belief function Bel with masses

$$m(x) = 0.2, \quad m(y) = 0.3, \quad m(x,z) = 0.5. \tag{13.35}$$

Notice that only L_p approximations in the section with $\bar{A} = \Theta$ are shown, for the sake of simplicity.

According to the formulae on page 8 of [342] (see also Example 40), the set of outer consonant approximations of (13.35) with chain $\{\{x\}, \{x,y\}, \Theta\}$ is the convex closure of the points

$$\begin{aligned} m^{\mathbf{B}_1/\mathbf{B}_2} &= [m(x), m(y), 1 - m(x) - m(y)]', \\ m^{\mathbf{B}_3/\mathbf{B}_4} &= [m(x), 0, 1 - m(x)]', \\ m^{\mathbf{B}_5/\mathbf{B}_6} &= [0, m(x) + m(y), 1 - m(x) - m(y)]', \\ m^{\mathbf{B}_7/\mathbf{B}_8} &= [0, m(x), 1 - m(x)]', \\ m^{\mathbf{B}_9/\mathbf{B}_{10}} &= [0, m(y), 1 - m(y)]', \\ m^{\mathbf{B}_{11}/\mathbf{B}_{12}} &= [0, 0, 1]'. \end{aligned} \tag{13.36}$$

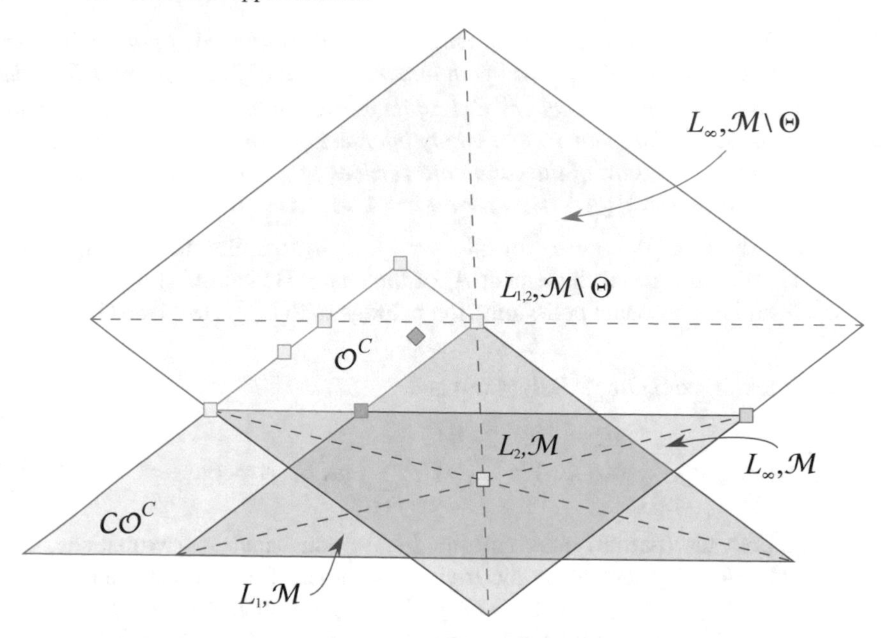

Fig. 13.8: The simplex $\mathcal{CO}^{\mathcal{C}}$ in the mass space of consonant belief functions with maximal chain $\mathcal{C} = \{\{x\} \subset \{x, y\} \subset \Theta\}$ defined on $\Theta = \{x, y, z\}$, and the L_p partial consonant approximations in $\mathcal{M}$ of the belief function Bel with basic probabilities (13.35). The $L_2, \mathcal{M}$ approximation is plotted as a light blue square, as the barycentre of the sets of both the $L_1, \mathcal{M}$ (purple triangle) and the $L_\infty, \mathcal{M}$ (blue triangle) approximations. The contour-based approximation is a vertex of the triangle $L_\infty, \mathcal{M}$ and is denoted by a green square. The various L_p approximations are also depicted for the section $\mathcal{M} \setminus \Theta$ of the mass space: the unique L_1/L_2 approximation is a vertex of $L_1, \mathcal{M}$, while the polytope of L_∞ approximations in the section is depicted in light green. The related set $\mathcal{O}^{\mathcal{C}}[Bel]$ of partial outer consonant approximations (13.36) is also shown for comparison (light yellow), together with the points (13.36), shown as yellow squares. The maximal outer approximation is denoted by an ochre square. The isopignistic function is represented by an orange diamond.

These points are plotted in Fig. 13.8 as yellow squares. We can observe that, as can be proven by Theorem 76, both $Co^{\mathcal{C}}_{\max}[Bel]$ (4.69) and $Co^{\mathcal{C}}_{\mathcal{M}\setminus\Theta, L_{1/2}}[m]$ (13.34) belong to the intersection of the (partial) outer and $L_1, \mathcal{M}$ consonant approximations.

This example also suggests that (partial) outer consonant approximations are included in L_∞ consonant approximations, calculated by neglecting the component $\bar{A} = \Theta$. However, this is not so, as attested by the binary case $\Theta = \{x, y\}$, for which the $L_\infty, \mathcal{M} \setminus \Theta$ solutions satisfy, for the maximal chain $\mathcal{C} = \{\{x\} \subset \Theta\}$, $m(x) - m(y) \leq m_{Co}(x) \leq m(x) + m(y)$, while the outer approximations are such that $0 \leq m_{Co}(x) \leq m(x)$.

As for the isopignistic and contour-based approximations, they coincide in this case with the vectors

$$\begin{aligned} m_{\text{iso}} &= [0.15, 0.1, 0.75]', \\ m_{\text{con}} &= [1 - pl(y), pl(y) - pl(z), pl(z)]' = [0.7, -0.2, 0.5]'. \end{aligned}$$

The pignistic values of the elements in this example are $BetP[Bel](x) = 0.45$, $BetP[Bel](y) = 0.3$, $BetP[Bel](z) = 0.25$, *so that the chain associated with the isopignistic approximation is indeed* $\{\{x\}, \{x, y\}, \Theta\}$. *Notice, though, that 'pseudo-' isopignistic approximations can be computed for all chains via (13.32), none of which are admissible. The contour-based approximation is not admissible in this case, as singletons have a different plausibility ordering.*

While no relationship whatsoever seems to link the isopignistic and L_p *consonant approximations (as expected), the former appears to be an outer approximation as well. The contour-based approximation coincides in this example with a vertex of the set of* $L_\infty, \mathcal{M}$ *approximations. However, this is not generally true: just compare (13.23) and (13.33).*

13.4 Consonant approximation in the belief space

Consonant approximations in the mass space have a quite natural semantics in terms of mass redistributions. As we will see in this section, instead, (partial) L_p approximations in the belief space are closely associated with lists of belief values determined by the desired maximal chain, and through them with the maximal outer approximation (4.69), as we will understand in Section 13.4.4.

We first need to make explicit the analytical form of the difference vector $Bel - Co$ between the original belief function Bel and the desired approximation Co.

Lemma 15. *Given a belief function* $Bel : 2^\Theta \to [0, 1]$ *and an arbitrary consonant BF* Co *defined on the same frame with maximal chain of focal elements* $\mathcal{C} = \{A_1 \subset \cdots \subset A_n\}$, *the difference between the corresponding vectors in the belief space is*

$$Bel - Co = \sum_{A \not\supset A_1} Bel(A) \mathbf{x}_A + \sum_{i=1}^{n-1} \sum_{\substack{A \supset A_i \\ A \not\supset A_{i+1}}} \mathbf{x}_A \left[\gamma(A_i) + Bel(A) - \sum_{j=1}^{i} m(A_j) \right], \tag{13.37}$$

where

$$\gamma(A) \doteq \sum_{B \subseteq A, B \in \mathcal{C}} (m(B) - m_{Co}(B)),$$

and $\{\mathbf{x}_A, \emptyset \neq A \subsetneq \Theta\}$ *is the usual orthonormal reference frame in the belief space* $\mathcal{B}$ *(Section 7.1).*

13.4.1 L_1 approximation

Theorem 78. *Given a belief function $Bel : 2^\Theta \rightarrow [0,1]$ and a maximal chain of focal elements $\mathcal{C} = \{A_1 \subset \cdots \subset A_n\}$ in Θ, the partial L_1 consonant approximations $\mathcal{CO}^{\mathcal{C}}_{\mathcal{B},L_1}[Bel]$ in the belief space with maximal chain $\mathcal{C}$ have mass vectors forming the following convex closure:*

$$
\begin{array}{l}
Cl\Big(\big[Bel^1, Bel^2 - Bel^1, \ldots, Bel^i - Bel^{i-1}, \ldots, 1 - Bel^{n-1}\big]', \\
\qquad Bel^i \in \big\{\gamma^i_{\text{int1}}, \gamma^i_{\text{int2}}\big\} \; \forall i = 1, \ldots, n-1 \Big),
\end{array}
\tag{13.38}
$$

where $\gamma^i_{\text{int1}}, \gamma^i_{\text{int2}}$ are the innermost (median) elements of the list of belief values

$$
\mathcal{L}_i = \Big\{ Bel(A), A \supseteq A_i, A \not\supset A_{i+1} \Big\}. \tag{13.39}
$$

In particular, $Bel^{n-1} = \gamma^{n-1}_{\text{int1}} = \gamma^{n-1}_{\text{int2}} = Bel(A_{n-1})$. Note that, although L_p approximations in this section are computed in $\mathcal{B}$, we present the results in terms of mass assignments as they are simpler and easier to interpret.

Owing to the nature of the partially ordered set 2^Θ, the innermost values of the above lists (13.39) cannot be analytically identified in full generality (although they can easily be computed numerically). Nevertheless, the partial L_1 approximations in $\mathcal{B}$ can be derived analytically in some cases. By (13.38), the barycentre of the set of partial L_1 consonant approximations in $\mathcal{B}$ has as mass vector

$$
\begin{array}{l}
m_{Co^{\mathcal{C}}_{\mathcal{B},L_1}[Bel]} \\
= \left[\dfrac{\gamma^1_{\text{int1}} + \gamma^1_{\text{int2}}}{2}, \dfrac{\gamma^2_{\text{int1}} + \gamma^2_{\text{int2}}}{2} - \dfrac{\gamma^1_{\text{int1}} + \gamma^1_{\text{int2}}}{2}, \ldots, 1 - Bel(A_{n-1}) \right]'.
\end{array}
\tag{13.40}
$$

Bel's global L_1 approximation(s) can easily be derived from the expression for the norm of the difference vector (see proof of Theorem 78, (13.75)).

Theorem 79. *Given a belief function $Bel : 2^\Theta \rightarrow [0,1]$, its global L_1 consonant approximations $\mathcal{CO}_{\mathcal{B},L_1}[Bel]$ in $\mathcal{B}$ live in the collection of partial approximations of that kind associated with maximal chain(s) which maximise the cumulative lower halves of the lists of belief values $\mathcal{L}^i$ (13.39),*

$$
\arg\max_{\mathcal{C}} \sum_i \sum_{Bel(A) \in \mathcal{L}_i, Bel(A) \leq \gamma_{\text{int1}}} Bel(A). \tag{13.41}
$$

13.4.2 (Partial) L_2 approximation

To find the partial consonant approximation(s) at the minimum L_2 distance from Bel in $\mathcal{B}$, we need to impose the orthogonality of the difference vector $Bel - Co$ with respect to any given simplicial component $\mathcal{CO}^{\mathcal{C}}_{\mathcal{B}}$ of the complex $\mathcal{CO}_{\mathcal{B}}$:

$$
\langle Bel - Co, Bel_{A_i} - Bel_\Theta \rangle = \langle Bel - Co, Bel_{A_i} \rangle = 0 \quad \forall A_i \in \mathcal{C}, i = 1, \ldots, n-1, \tag{13.42}
$$

as $Bel_\Theta = \mathbf{0}$ is the origin of the Cartesian space in $\mathcal{B}$, and $Bel_{A_i} - Bel_\Theta$ for $i = 1, \ldots, n-1$ are the generators of the component $\mathcal{CO}^{\mathcal{C}}_{\mathcal{B}}$.

Using once again (13.37), the orthogonality conditions (13.42) translate into the following linear system of equations:

$$\left\{ \sum_{A \not\in \mathcal{C}} m(A) \langle Bel_A, Bel_{A_i} \rangle + \sum_{A \in \mathcal{C}, A \neq \Theta} (m(A) - m_{Co}(A)) \langle Bel_A, Bel_{A_i} \rangle = 0 \right. \tag{13.43}$$

for all $i = 1, \ldots, n-1$. This is a linear system in $n-1$ unknowns $m_{Co}(A_i)$, $i = 1, \ldots, n-1$, and $n-1$ equations. The resulting L_2 partial approximation of Bel is also a function of the lists of belief values (13.39).

Theorem 80. *Given a belief function $Bel : 2^\Theta \to [0,1]$, its partial*[78] *L_2 consonant approximation $Co^{\mathcal{C}}_{\mathcal{B},L_2}[Bel]$ in $\mathcal{B}$ with maximal chain $\mathcal{C} = \{A_1 \subset \cdots \subset A_n\}$ is unique, and has basic probability assignment*

$$m_{Co^{\mathcal{C}}_{\mathcal{B},L_2}[Bel]}(A_i) = \mathrm{ave}(\mathcal{L}_i) - \mathrm{ave}(\mathcal{L}_{i-1}) \quad \forall i = 1, ..., n, \tag{13.44}$$

where $\mathcal{L}_0 \doteq \{0\}$, and $\mathrm{ave}(\mathcal{L}_i)$ is the average of the list of belief values $\mathcal{L}_i$ (13.39):

$$\mathrm{ave}(\mathcal{L}_i) = \frac{1}{2^{|A^c_{i+1}|}} \sum_{A \supseteq A_i, A \not\supset A_{i+1}} Bel(A). \tag{13.45}$$

13.4.3 L_∞ approximation

Partial approximations The partial L_∞ approximations in $\mathcal{B}$ also form a polytope, this time with 2^{n-1} vertices.

Theorem 81. *Given a belief function $Bel : 2^\Theta \to [0,1]$, its partial L_∞ consonant approximations in the belief space $\mathcal{CO}^{\mathcal{C}}_{\mathcal{B},L_\infty}[Bel]$ with maximal chain of focal elements $\mathcal{C} = \{A_1 \subset \cdots \subset A_n\}$, $A_i = \{x_1, \ldots, x_i\}$ for all $i = 1, \ldots, n$, have mass vectors which live in the following convex closure of 2^{n-1} vertices:*

$$Cl\Bigg([Bel^1, Bel^2 - Bel^1, \ldots, Bel^i - Bel^{i-1}, \ldots, 1 - Bel^{n-1}]' \Bigg| \\ \forall i = 1, \ldots, n-1,\ Bel^i \in \left\{ \frac{Bel(A_i) + Bel(\{x_{i+1}\}^c)}{2} \pm Bel(A_1^c) \right\} \Bigg). \tag{13.46}$$

The barycentre $Co^{\mathcal{C}}_{\mathcal{B},L_\infty}[Bel]$ of this set has mass assignment

[78] The computation of the global L_2 approximation(s) is rather involved. We plan to address this issue in the near future.

$$m_{co^{\mathcal{C}}_{\mathcal{B},L_\infty}[Bel]}(A_i)$$

$$= \begin{cases} \dfrac{Bel(A_1) + Bel(\{x_2\}^c)}{2} & i = 1, \\ \dfrac{Bel(A_i) - Bel(A_{i-1})}{2} + \dfrac{Pl(\{x_i\}) - Pl(\{x_{i+1}\})}{2} & i = 2, \ldots, n-1, \\ 1 - Bel(A_{n-1}) & i = n. \end{cases} \tag{13.47}$$

Note that, since $Bel(A_1^c) = 1 - Pl(A_1) = 1 - Pl(\{x_1\})$, the size of the polytope (13.46) of partial L_1 approximations of Bel is a function of the plausibility of the innermost desired focal element only. As expected, it reduces to zero only when Bel is a consistent belief function (Section 10.4) and $A_1 = \{x_1\}$ has plausibility 1.

A straightforward interpretation of the barycentre of the partial L_∞ approximations in $\mathcal{B}$ in terms of degrees of belief is possible when we notice that, for all $i = 1, \ldots, n$,

$$m_{Co}(A_i) = \frac{m_{Co^{\mathcal{C}}_{\max}[Bel]}(A_i) + m_{Co_{\text{con}}[Bel]}(A_i)}{2}$$

(recall (4.69) and (13.33)), i.e., (13.47) is the average of the maximal outer consonant approximation and what we have called the 'contour-based' consonant approximation (Definition 131).

Global approximations To compute the global L_∞ approximation(s) of the original belief function Bel in $\mathcal{B}$, we need to locate (as usual) the partial solution(s) whose L_∞ distance from Bel is the smallest.

Given the expression (13.73) for the L_∞ norm of the difference vector (see the proof of Theorem 81), this partial distance is (for each maximal chain $\mathcal{C} = \{A_1 \subset \cdots \subset A_n = \Theta\}$) equal to $Bel(A_1^c)$. Therefore the global L_∞ consonant approximations of Bel in the belief space are associated with the chains of focal elements

$$\arg\min_{\mathcal{C}} Bel(A_1^c) = \arg\min_{\mathcal{C}} (1 - Pl(A_1)) = \arg\max_{\mathcal{C}} Pl(A_1).$$

Theorem 82. *Given a belief function $Bel : 2^\Theta \to [0, 1]$, the set of global L_∞ consonant approximations of Bel in the belief space is the collection of partial approximations associated with maximal chains whose smallest focal element is the maximum-plausibility singleton:*

$$\mathcal{CO}_{\mathcal{B},L_\infty}[Bel] = \bigcup_{\mathcal{C}: A_1 = \{\arg\max_{x\in\Theta} Pl(\{x\})\}} \mathcal{CO}^{\mathcal{C}}_{\mathcal{B},L_\infty}[Bel].$$

13.4.4 Approximations in the belief space as generalised maximal outer approximations

From Theorems 78, 80 and 81, a comprehensive view of our results on L_p consonant approximation in the belief space can be given in terms of the lists of belief values (13.39)

$$\mathcal{L}_0 \doteq \{Bel(\emptyset) = 0\}, \quad \mathcal{L}_i = \Big\{Bel(A), A \supseteq A_i, A \not\supset A_{i+1}\Big\} \, \forall i = 1, \ldots, n,$$

and $\mathcal{L}_n \doteq \{Bel(\Theta) = 1\}$.

The basic probability assignments of all the partial approximations in the belief space are differences of simple functions of belief values taken from these lists, which are uniquely determined by the desired chain of non-empty focal elements $A_1 \subset \cdots \subset A_n$. Namely,

$$\begin{aligned}
m_{Co^{\mathcal{C}}_{\max}[Bel]}(A_i) &= \min(\mathcal{L}_i) - \min(\mathcal{L}_{i-1}), \\
m_{Co^{\mathcal{C}}_{\text{con}}[Bel]}(A_i) &= \max(\mathcal{L}_i) - \max(\mathcal{L}_{i-1}), \\
m_{Co^{\mathcal{C}}_{\mathcal{B},L_1}[Bel]}(A_i) &= \frac{\text{int}_1(\mathcal{L}_i) + \text{int}_2(\mathcal{L}_i)}{2} - \frac{\text{int}_1(\mathcal{L}_{i-1}) + \text{int}_2(\mathcal{L}_{i-1})}{2}, \\
m_{Co^{\mathcal{C}}_{\mathcal{B},L_2}[Bel]}(A_i) &= \text{ave}(\mathcal{L}_i) - \text{ave}(\mathcal{L}_{i-1}), \\
m_{Co^{\mathcal{C}}_{\mathcal{B},L_\infty}[Bel]}(A_i) &= \frac{\max(\mathcal{L}_i) + \min(\mathcal{L}_i)}{2} - \frac{\max(\mathcal{L}_{i-1}) + \min(\mathcal{L}_{i-1})}{2},
\end{aligned} \tag{13.48}$$

for all $i = 1, \ldots, n$, where the expression for $Co^{\mathcal{C}}_{\mathcal{B},L_\infty}[Bel]$ comes directly from (13.47). For each vertex of the L_1 polytope, one of the innermost elements of the i-th list, $\text{int}_1(\mathcal{L}_i)$, $\text{int}_2(\mathcal{L}_i)$, is picked from the list $\mathcal{L}_i$, for each component of the mass vector. This yields

$$m_{Co}(A_i) = \text{int}_1(\mathcal{L}_i)/\text{int}_2(\mathcal{L}_i) - \text{int}_1(\mathcal{L}_{i-1})/\text{int}_2(\mathcal{L}_{i-1}),$$

(where / denotes the choice of one alternative). For each vertex of the L_∞ polytope, either $\max(\mathcal{L}_i)$ or $\min(\mathcal{L}_i)$ is selected, yielding

$$m_{Co}(A_i) = \max(\mathcal{L}_i)/\min(\mathcal{L}_i) - \max(\mathcal{L}_{i-1})/\min(\mathcal{L}_{i-1}).$$

The different approximations in $\mathcal{B}$ (13.48) correspond, therefore, to different choices of a representative for the each of the lists $\mathcal{L}_i$.

The maximal outer approximation $Co^{\mathcal{C}}_{\max}[Bel]$ is obtained by picking as the representative $\min(\mathcal{L}_i)$, $Co^{\mathcal{C}}_{\text{con}}[Bel]$ amounts to picking $\max(\mathcal{L}_i)$, the barycentre of the L_1 approximations amounts to choosing the average innermost (median) value, the barycentre of the L_∞ approximations amounts to picking the average outermost value, and L_2 amounts to picking the overall average value of the list. Each vertex of the L_1 solutions amounts to selecting, for each component, one of the innermost values; each vertex of the L_∞ polytope results from selecting one of the outermost values.

Interpretation of the list $\mathcal{L}_i$ Belief functions are defined on a partially ordered set, the power set $2^\Theta = \{A \subseteq \Theta\}$, of which a maximal chain is a maximal totally ordered subset. Therefore, given two elements $A_i \subset A_{i+1}$ of the chain, there are a number of 'intermediate' focal elements A which contain the latter element but not the former. If 2^Θ was a totally ordered set, the list $\mathcal{L}_i$ would contain a single element $Bel(A_i)$ and all the L_p approximations (13.48) would reduce to the maximal outer consonant approximation $Co^{\mathcal{C}}_{\max}[Bel]$, with BPA $m_{Co^{\mathcal{C}}_{\max}[Bel]}(A_i) = Bel(A_i) - Bel(A_{i-1})$. The variety of L_p approximations in $\mathcal{B}$ is therefore a consequence of belief functions being defined on partially ordered sets: together with the contour-based approximation (13.33), such quantities can all be seen as distinct generalisations of the maximal outer consonant approximation.

Relations between approximations in the belief space The list $\mathcal{L}_i$ is composed of $2^{|A^c_{i+1}|} = 2^{n-(i+1)}$ elements, for all $i = 1, \ldots, n$. Obviously, $|\mathcal{L}_0| = 1$ by definition.

We can therefore infer the following relationships between the various L_p approximations in $\mathcal{B}$:

- the barycentre of the set of L_∞ approximations is always the average of the maximal outer and contour-based approximations;
- $\text{ave}(\mathcal{L}_i) = (\max(\mathcal{L}_i)+\min(\mathcal{L}_i))/2 = (\text{int}_1(\mathcal{L}_i)+\text{int}_2(\mathcal{L}_i))/2$ whenever $|\mathcal{L}_i| \leq 2$, i.e., for $i \geq n-2$ or $i = 0$; thus, the last two components of the L_1 barycentre, L_2 and L_∞ barycentre approximations coincide, namely
$$m_{Co^{\mathcal{C}}_{\mathcal{B},L_1}[Bel]}(A_i) = m_{Co^{\mathcal{C}}_{\mathcal{B},L_2}[Bel]}(A_i) = m_{Co^{\mathcal{C}}_{\mathcal{B},L_\infty}[Bel]}(A_i)$$
for $i = n-1, n$;
- in particular, $Co^{\mathcal{C}}_{\mathcal{B},L_1}[Bel] = Co^{\mathcal{C}}_{\mathcal{B},L_2}[Bel] = Co^{\mathcal{C}}_{\mathcal{B},L_\infty}[Bel]$ whenever $|\Theta| = n \leq 3$;
- the last component of all the pointwise approximations in (13.48) is the same:
$$\begin{aligned} m_{Co^{\mathcal{C}}_{\max}[Bel]}(A_n) &= m_{Co^{\mathcal{C}}_{\text{con}}[Bel]}(A_n) = m_{Co^{\mathcal{C}}_{\mathcal{B},L_1}[Bel]}(A_n) \\ &= m_{Co^{\mathcal{C}}_{\mathcal{B},L_2}[Bel]}(A_n) = m_{Co^{\mathcal{C}}_{\mathcal{B},L_\infty}[Bel]}(A_n) \\ &= 1 - Bel(A_{n-1}). \end{aligned}$$

Admissibility As is clear from the table of results in (13.48), all the L_p approximations in the belief space are differences of vectors of values that are all positive; indeed, differences of shifted versions of the same positive vector. As such vectors $[(\text{int}_1(\mathcal{L}_i) + \text{int}_2(\mathcal{L}_i))/2, i = 1, \ldots, n]'$, $[(\max(\mathcal{L}_i) + \min(\mathcal{L}_i))/2, i = 1, \ldots, n]'$, $[\text{ave}(\mathcal{L}_i), i = 1, \ldots, n]'$ are not guaranteed to be monotonically increasing for any arbitrary maximal chain $\mathcal{C}$, none of the related partial approximations are guaranteed to be entirely admissible. However, sufficient conditions under which they are admissible can be worked out by studying the structure of the list of belief values (13.39).

Let us first consider $Co^{\mathcal{C}}_{\max}$ and $Co^{\mathcal{C}}_{\text{con}}$. As

$$\min(\mathcal{L}_{i-1}) = Bel(A_{i-1}) \leq Bel(A_i) = \min(\mathcal{L}_i) \quad \forall i = 2, \ldots, n,$$

the maximal partial outer approximation is admissible for all maximal chains $\mathcal{C}$. For the contour-based approximation,

$$\max(\mathcal{L}_i) = Bel(A_i + A_{i+1}^c) = Bel(x_{i+1}^c) = 1 - Pl(x_{i+1}),$$

whereas $\max(\mathcal{L}_{i-1}) = 1 - Pl(x_i)$, so that

$$\max(\mathcal{L}_i) - \max(\mathcal{L}_{i-1}) = Pl(x_i) - Pl(x_{i+1}),$$

which is guaranteed to be non-negative if the chain $\mathcal{C}$ is generated by singletons sorted by their plausibility values. Thus, as

$$m_{Co^{\mathcal{C}}_{\mathcal{B},L_\infty}[Bel]}(A_i) = \frac{\max(\mathcal{L}_i) - \max(\mathcal{L}_{i-1})}{2} + \frac{\min(\mathcal{L}_i) - \min(\mathcal{L}_{i-1})}{2},$$

the barycentre of the set of $L_\infty, \mathcal{B}$ approximations is also admissible on the same chain(s).

Example 43: ternary example, belief space. *As we have done in the mass space case, it can be helpful to visualise the outcomes of the L_p consonant approximation process in the belief space in the case in which $\Theta = \{x, y, z\}$, and compare them with the approximations in the mass space considered in Example 42 (Fig. 13.9). To*

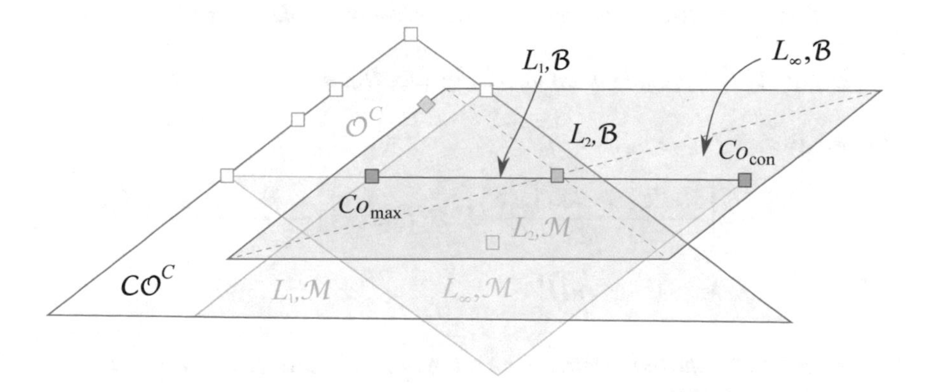

Fig. 13.9: Comparison between L_p partial consonant approximations in the mass and belief spaces $\mathcal{M}$ and $\mathcal{B}$ for the belief function with basic probabilities (13.35) on $\Theta = \{x, y, z\}$. The $L_2, \mathcal{B}$ approximation is plotted as a blue square, as the barycentre of the sets of both the $L_1, \mathcal{B}$ (purple segment) and the $L_\infty, \mathcal{B}$ (light blue quadrangle) approximations. The contour-based and maximal outer approximations are, in this example, the extremal points of the segment $L_1, \mathcal{B}$ (purple squares). The polytope $\mathcal{O}^{\mathcal{C}}$ of partial outer consonant approximations, the isopignistic approximation and the various L_p partial approximations in $\mathcal{M}$ (retained from Fig. 13.8) are also visible in less saturated colours.

ensure a fair comparison, we have plotted both sets of approximations in the belief and mass spaces as vectors of mass values.

When $\Theta = \{x, y, z\}$ and $A_1 = \{x\}$, $A_2 = \{x, y\}$, $A_3 = \{x, y, z\}$, the relevant lists of belief values are $\mathcal{L}_1 = \{Bel(x), Bel(\{x, z\})\}$ and $\mathcal{L}_2 = \{Bel(\{x, y\})\}$, so that

$$
\begin{aligned}
\min(\mathcal{L}_1) &= \text{int}_1(\mathcal{L}_1) = Bel(x), \\
\max(\mathcal{L}_1) &= \text{int}_2(\mathcal{L}_1) = Bel(\{x, z\}), \\
\text{ave}(\mathcal{L}_1) &= \frac{Bel(x) + Bel(\{x, z\})}{2}, \\
\min(\mathcal{L}_2) &= \text{int}_1(\mathcal{L}_2) = \max(\mathcal{L}_2) = \text{int}_2(\mathcal{L}_2) = \text{ave}(\mathcal{L}_2) = Bel(\{x, y\}).
\end{aligned}
$$

Therefore, the set of L_1 partial consonant approximations is, by (13.38), a segment with vertices (the purple squares in Fig. 13.9)

$$
\begin{array}{l}
\big[Bel(x), Bel(\{x, y\}) - Bel(x), 1 - Bel(\{x, y\})\big]', \\
\big[Bel(\{x, z\}), Bel(\{x, y\}) - Bel(\{x, z\}), 1 - Bel(\{x, y\})\big]',
\end{array} \tag{13.49}
$$

which coincide with the maximal outer approximation and the contour-based approximation, respectively. This set is not entirely admissible, not even in the special ternary case.

The partial L_2 approximation in $\mathcal{B}$ is, by (13.48), unique, with mass vector

$$
\begin{aligned}
m_{Co_{\mathcal{B}}, L_2}[Bel] &= m_{Co_{\mathcal{B}}, L_\infty}[Bel] \\
&= \left[\frac{Bel(x) + Bel(\{x, z\})}{2}, Bel(\{x, y\}) - \frac{Bel(x) + Bel(\{x, z\})}{2}, \right. \\
&\qquad \left. 1 - Bel(\{x, y\}) \right]',
\end{aligned} \tag{13.50}
$$

and coincides with the barycentre of the set of partial L_∞ approximations (note that this is not so in the general case).

The full set of partial L_∞ approximations has vertices (13.46)

$$\Big[\tfrac{Bel(x)+Bel(\{x,z\})-2Bel(\{y,z\})}{2},\ Bel(\{x,y\}) - \frac{Bel(x)+Bel(\{x,z\})}{2},$$
$$1 - Bel(\{x,y\}) + Bel(\{y,z\})\Big]',$$

$$\Big[\tfrac{Bel(x)+Bel(\{x,z\})-2Bel(\{y,z\})}{2},\ \tfrac{2Bel(\{x,y\})-Bel(x)+Bel(\{x,z\})+4Bel(\{y,z\})}{2},$$
$$1 - Bel(\{x,y\}) - Bel(\{y,z\})\Big]',$$

$$\Big[\tfrac{Bel(x)+Bel(\{x,z\})+2Bel(\{y,z\})}{2},\ \tfrac{2Bel(\{x,y\})-Bel(x)-Bel(\{x,z\})-4Bel(\{y,z\})}{2},$$
$$1 - Bel(\{x,y\}) + Bel(\{y,z\})\Big]',$$

$$\Big[\tfrac{Bel(x)+Bel(\{x,z\})+2Bel(\{y,z\})}{2},\ Bel(\{x,y\}) - \frac{Bel(x)+Bel(\{x,z\})}{2},$$
$$1 - Bel(\{x,y\}) - Bel(\{y,z\})\Big]',$$

which, as expected, are not all admissible (see the light blue quadrangle in Fig. 13.9 again).

The example hints at the possibility that the contour-based approximation and/or the L_2, L_∞ barycentre approximations in the belief space may be related to the set of L_1 approximations in the full mass space: this deserves further analysis. On the other hand, we know that the maximal partial outer approximation (4.69) is not in general a vertex of the polygon of L_1 partial approximations in $\mathcal{B}$, unlike what the ternary example (for which $\mathrm{int}_1(\mathcal{L}_1) = Bel(x)$) suggests.

13.5 Belief versus mass space approximations

We can draw some conclusions by comparing the results of Sections 13.3 and 13.4:

- L_p consonant approximations in the mass space are, basically, associated with different but related mass redistribution processes: the mass outside the desired chain of focal elements is reassigned in some way to the elements of the chain;
- their relationships with classical outer approximations (on one hand) and approximations based on the pignistic transform (on the other) are rather weak;
- the different L_p approximations in $\mathcal{M}$ are characterised by natural geometric relations;
- consonant approximation in the belief space is inherently linked to the lists of belief values of focal elements 'intermediate' between each pair of elements of the desired chain;
- the contour-based approximation, together with the L_p approximations in the belief space, can be seen as distinct generalisations of the maximal outer approximation, induced by the nature of the power set as a partially ordered set;

- in the mass space, some partial approximations are always entirely admissible and should be preferred (this is the case for the L_1 and L_2 approximations in $\mathcal{M}$), but some others are not;
- for the belief case, although all partial L_p approximations are differences between shifted versions of the same positive vector, admissibility is not guaranteed for all maximal chains; however, sufficient conditions exist.

Table 13.1 summarises the behaviour of the different geometric consonant approximations studied in this chapter, in terms of multiplicity, admissibility and global solutions.

13.5.1 On the links between approximations in $\mathcal{M}$ and $\mathcal{B}$

Approximations in $\mathcal{B}$ and approximations in $\mathcal{M}$ do not coincide. This is a direct consequence of the fact that Möbius inversion does not preserve either L_p norms or the ordering induced by them, as was clear from the counter-examples discussed in Section 13.2.3.

That said, links of some sort between L_p approximations in the two spaces may still exist. Let us consider, in particular, partial approximations. The ternary example in Fig. 13.9 suggests the following conjectures:

1. The L_2 partial approximation in $\mathcal{B}$ is one of the L_1 partial approximations in $\mathcal{M}$.
2. The L_2 partial approximation in $\mathcal{B}$ is one of the L_∞ partial approximations in $\mathcal{M}$, and possibly belongs to the boundary of the simplex (13.22).
3. The L_1 partial approximation in $\mathcal{B}$ is also an element of (13.22).

Unfortunately, counter-examples can be provided in all these cases. Let us express $m^{\mathcal{C}}_{\mathcal{B},L_2}[Bel](A_i)$[79] (13.44) as a function of the masses of the original belief function Bel. For all $i = 1, \ldots, n-1$,

$$
\begin{aligned}
\text{ave}(\mathcal{L}_i) &= \frac{1}{2^{|A^c_{i+1}|}} \sum_{A \supseteq A_i, A \not\supseteq A_{i+1}} Bel(A) = \frac{1}{2^{|A^c_{i+1}|}} \sum_{A \supseteq A_i, A \not\supseteq A_{i+1}} \sum_{\emptyset \neq B \subseteq A} m(B) \\
&= \frac{1}{2^{|A^c_{i+1}|}} \sum_{B \not\supseteq \{x_{i+1}\}} m(B) \cdot \Big| \{A = A_i \cup C : \emptyset \subseteq C \subseteq A^c_{i+1}, A \supseteq B\} \Big| \\
&= \frac{1}{2^{|A^c_{i+1}|}} \sum_{B \not\supseteq \{x_{i+1}\}} m(B) \cdot 2^{|A^c_{i+1} \setminus (B \cap A^c_{i+1})|} = \sum_{B \not\supseteq \{x_{i+1}\}} \frac{m(B)}{2^{|B \cap A^c_{i+1}|}},
\end{aligned}
$$

so that, for all $i = 2, \ldots, n-1$,

$$
\begin{aligned}
m^{\mathcal{C}}_{\mathcal{B},L_2}[Bel](A_i) &= \text{ave}(\mathcal{L}_i) - \text{ave}(\mathcal{L}_{i-1}) = \sum_{B \not\ni x_{i+1}} \frac{m(B)}{2^{|B \cap A^c_{i+1}|}} - \sum_{B \not\ni x_i} \frac{m(B)}{2^{|B \cap A^c_i|}} \\
&= \sum_{B \ni x_i, B \not\ni x_{i+1}} \frac{m(B)}{2^{|B \cap A^c_{i+1}|}} - \sum_{B \ni x_{i+1}, B \not\ni x_i} \frac{m(B)}{2^{|B \cap A^c_i|}}.
\end{aligned}
\tag{13.51}
$$

[79] Simplified notation.

Note that $m(A_i)$ is one of the terms of the first summation. Now, conjecture 1 requires the above mass to be greater than or equal to $m(A_i)$ for all i (Theorem 72): clearly, though, by (13.51), the difference $m^{\mathcal{C}}_{\mathcal{B},L_2}[Bel](A_i) - m(A_i)$ is not guaranteed to be positive.

As for conjecture 2, the set of $L_\infty, \mathcal{M}$ partial approximations is determined by the constraints (13.74) (see the proof of Theorem 77):

$$\begin{cases} m_{Co}(A) - m(A) \leq \max_{B \not\in \mathcal{C}} m(B) & A \in \mathcal{C}, A \neq \bar{A}, \\ \sum_{A \in \mathcal{C}, A \neq \bar{A}} (m_{Co}(A) - m(A)) \geq \sum_{B \not\in \mathcal{C}} m(B) - \max_{B \not\in \mathcal{C}} m(B). \end{cases} \tag{13.52}$$

Now, suppose that Bel is such that $m(B) = 0$ for all $B \ni x_{i+1}$, $B \not\ni x_i$. Then,

$$m^{\mathcal{C}}_{\mathcal{B},L_2}[Bel](A_i) = \sum_{B \ni x_i, B \not\ni x_{i+1}} \frac{m(B)}{2^{|B \cap A^c_{i+1}|}},$$

which contains, among other addends, $\sum_{B \subseteq A_i, B \ni x_i} m(B)$ (for, if $B \subseteq A_i$, we have $2^{|B \cap A^c_{i+1}|} = 2^{|\emptyset|} = 1$). Clearly, if $\arg\max_{B \not\in \mathcal{C}} m(B)$ is a proper subset of A_i containing $\{x_i\}$, and other subsets $B \ni x_i, B \not\ni x_{i+1}$ distinct from the latter and A_i have non-zero mass, the first constraint of (13.52) is not satisfied.

Finally, consider conjecture 3. By Theorem 78, all the vertices of the set of $L_1, \mathcal{B}$ partial approximations have as their mass $m_{Co}(A_1)$ either one of the median elements of the list $\mathcal{L}_1 = \{Bel(A) : A \supseteq A_1, \not\supseteq A_2\}$. These median elements are of the form $Bel(A_1 \cup C)$, for some $C \subseteq A^c_2$, namely

$$Bel(A_1 \cup C) = m(A_1) + \sum_{B \subseteq A_1 \cup C, C \subseteq A^c_2, B \neq A_1} m(B),$$

so that

$$m_{Co}(A_1) - m(A_1) = \sum_{B \subseteq A_1 \cup C, C \subseteq A^c_2, B \neq A_1} m(B).$$

Once again, if Bel is such that $\arg\max_{B \not\in \mathcal{C}} m(A)$ is one of these subsets $B \subseteq A_1 \cup C$, $C \subseteq A^c_2$, and it is not the only one with non-zero mass, the first constraint in (13.52) is not satisfied for $A = A_1$ by any of the vertices of the set of $L_1, \mathcal{B}$ approximations. Hence, the latter has an empty intersection with the set of $L_\infty, \mathcal{M}$ partial approximations.

In conclusion, not only are approximations in $\mathcal{M}$ and $\mathcal{B}$ distinct, owing to the properties of Möbius inversion, but they are also not related in any apparent way.

13.5.2 Three families of consonant approximations

Indeed, approximations in the mass and the belief space turn out to be inherently related to completely different philosophies of the consonant approximation problem: mass redistribution versus generalised maximal outer approximation. Whereas mass space proxies correspond to different mass redistribution processes, L_p consonant approximation in the belief space amounts to generalising in different but related ways the classical approach embodied by the maximal outer approximation (4.69). The latter, together with the contour-based approximation (13.33), form therefore a different, coherent family of consonant approximations. As for the isopignistic approximation, this seems to be completely unrelated to approximations in either the mass or the belief space, as it is derived in the context of the transferable belief model.

The isopignistic, mass space and belief space consonant approximations form three distinct families of approximations, with fundamentally different rationales: which approach to be used will therefore vary according to the framework chosen, and the problem at hand.

Appendix: Proofs

Proof of Lemma 14

(1) Sufficiency. If (13.3) holds for all focal elements $A \subseteq \Theta$, then

$$Co(A) = \sum_{X \subseteq A} m_{Co}(X) = \sum_{X \subseteq A} \sum_{B \subseteq X} \alpha_X^B m(B) = \sum_{B \subseteq A} m(B) \sum_{B \subseteq X \subseteq A} \alpha_X^B$$

where, by the condition (13.2),

$$\sum_{B \subseteq X \subseteq A} \alpha_X^B \leq \sum_{X \supseteq B} \alpha_X^B = 1.$$

Therefore,

$$Co(A) \leq \sum_{B \subseteq A} m(B) = Bel(A),$$

i.e., Co is weakly included in Bel.

(2) Necessity. Let us denote by $\mathcal{C} = \{B_1, \ldots, B_n\}$, $n = |\Theta|$ the chain of focal elements of Co, and consider first the subsets $A \subseteq \Theta$ such that $A \not\supset B_1$ ($A \notin \mathcal{C}$). In this case $Co(A) = 0 \leq m(A)$ whatever the mass assignment of Bel: we then just need to focus on the elements $A = B_i \in \mathcal{C}$ of the chain.

We need to prove that, for all $B_i \in \mathcal{C}$,

$$m_{Co}(B_i) = \sum_{B \subseteq B_i} \alpha_{B_i}^B m(B) \quad \forall i = 1, \ldots, n. \tag{13.53}$$

Let us introduce the notation $\alpha_i^B \doteq \alpha_{B_i}^B$ for the sake of simplicity. For each i, we can sum the first i equations of the system (13.53) and obtain the equivalent system of equations

$$Co(B_i) = \sum_{B \subseteq B_i} \beta_i^B m(B) \quad \forall i = 1, \ldots, n \tag{13.54}$$

as $Co(B_i) = \sum_{j=1}^{i} m_{Co}(B_j)$, for Co is consonant. For all $B \subseteq \Theta$, the coefficients $\beta_i^B \doteq \sum_{j=1}^{i} \alpha_j^B$ need to satisfy

$$0 \leq \beta_{i_{\min}}^B \leq \cdots \leq \beta_n^B = 1, \tag{13.55}$$

where $i_{\min} = \min\{j : B_j \supseteq B\}$.

We can prove by induction on i that if Co is weakly included in Bel, i.e., $Co(B_i) \leq Bel(B_i)$ for all $i = 1, \ldots, n$, then there exists a solution $\{\beta_i^B, B \subseteq B_i, i = 1, \ldots, n\}$ to the system (13.54) which satisfies the constraint (13.55).

Let us look for solutions of the form

$$\left\{ \beta_i^B, B \subseteq B_{i-1}; \beta_i^B = \frac{Co(B_i) - \sum_{X \subseteq B_{i-1}} \beta_i^X m(X)}{\sum_{X \subseteq B_i, X \not\subset B_{i-1}} m(X)}, B \subseteq B_i, B \not\subset B_{i-1} \right\} \tag{13.56}$$

in which the coefficients (variables) β_i^B associated with subsets of the previous focal element B_{i-1} are left unconstrained, while all the coefficients associated with subsets that are in B_i but not in B_{i-1} are set to a common value, which depends on the free variables β_i^X, $X \subseteq B_{i-1}$.

Step $i = 1$. We get

$$\beta_1^{B_1} = \frac{Co(B_1)}{m(B_1)},$$

which is such that $0 \leq \beta_1^{B_1} \leq 1$ as $Co(B_1) \leq m(B_1)$, and trivially satisfies the first equation of the system (13.54): $Co(B_1) = \beta_1^{B_1} m(B_1)$.

Step i. Suppose there exists a solution (13.56) for $\{B \subseteq B_j, j = 1, \ldots, i-1\}$. We first have to show that all solutions of the form (13.56) for i solve the i-th equation of the system (13.54).

When we substitute (13.56) into the i-th equation of (13.54), we get (as the variables β_i^B in (13.56) do not depend on B for all $B \subseteq B_i, B \not\subset B_{i-1}$)

$$Co(B_i) = \sum_{B \subseteq B_{i-1}} \beta_i^B m(B) + \left(\frac{Co(B_i) - \sum_{X \subseteq B_{i-1}} \beta_i^X m(X)}{\sum_{X \subseteq B_i, X \not\subset B_{i-1}} m(X)} \right) \sum_{\substack{B \subseteq B_i \\ B \not\subset B_{i-1}}} m(B),$$

i.e., $Co(B_i) = Co(B_i)$ and the equation is satisfied.

We also need to show, though, that there exist solutions of the above form (13.56) that satisfy the ordering constraint (13.55), i.e.,

$$0 \leq \beta_i^B \leq 1, \ B \subseteq B_i, B \not\subset B_{i-1}; \quad \beta_{i-1}^B \leq \beta_i^B \leq 1, \ B \subseteq B_{i-1}. \tag{13.57}$$

The inequalities (13.57) generate, in turn, constraints on the free variables in (13.56), i.e., $\{\beta_i^B, B \subseteq B_{i-1}\}$. Given the form of (13.56), those conditions (in the same order as in (13.57)) assume the form

$$\begin{cases} \sum_{B\subseteq B_{i-1}} \beta_i^B m(B) \leq Co(B_i), \\ \sum_{B\subseteq B_{i-1}} \beta_i^B m(B) \geq Co(B_i) - \sum_{B\subseteq B_i, B\not\subseteq B_{i-1}} m(B), \\ \beta_i^B \geq \beta_{i-1}^B & \forall B\subseteq B_{i-1}, \\ \beta_i^B \leq 1 & \forall B\subseteq B_{i-1}. \end{cases} \quad (13.58)$$

Let us denote the above constraints on the free variables $\{\beta_i^B, B\subseteq B_{i-1}\}$ by 1, 2, 3 and 4.

- 1 and 2 are trivially compatible;
- 1 is compatible with 3, as substituting $\beta_i^B = \beta_{i-1}^B$ into 1 yields (owing to the $(i-1)$-th equation of the system)

$$\sum_{B\subseteq B_{i-1}} \beta_i^B m(B) = \sum_{B\subseteq B_{i-1}} \beta_{i-1}^B m(B) = Co(B_{i-1}) \leq Co(B_i);$$

- 4 is compatible with 2, as substituting $\beta_i^B = 1$ into 2 yields

$$\sum_{B\subseteq B_{i-1}} \beta_i^B m(B) = \sum_{B\subseteq B_{i-1}} m(B) \geq Co(B_i) - \sum_{B\subseteq B_i, B\not\subseteq B_{i-1}} m(B),$$

  which is equivalent to

$$\sum_{B\subseteq B_i} m(B) = Bel(B_i) \leq Co(B_i),$$

  which is in turn true by hypothesis;
- 4 and 1 are clearly compatible, as we just need to choose β_i^B small enough;
- 2 and 3 are compatible, as we just need to choose β_i^B large enough.

In conclusion, all the constraints in (13.58) are mutually compatible. Hence there exists an admissible solution to the i-th equation of the system (13.54), which proves the induction step.

Proof of Theorem 69

We need to prove that:

1. Each consonant belief function $Co \in \mathcal{CO}^{\mathcal{C}}$ such that $Co(A) \leq Bel(A)$ for all $A\subseteq \Theta$ can be written as a convex combination of the points (13.4)

$$Co = \sum_{\mathbf{B}} \alpha^{\mathbf{B}} Co^{\mathbf{B}}[Bel], \quad \sum_{\mathbf{B}} \alpha^{\mathbf{B}} = 1, \alpha^{\mathbf{B}} \geq 0 \;\forall \mathbf{B}.$$

2. Conversely, each convex combination of the $Co^{\mathbf{B}}[Bel]$ satisfies the condition $\sum_{\mathbf{B}} \alpha^{\mathbf{B}} Co^{\mathbf{B}}[Bel](A) \leq Bel(A)$ for all $A \subseteq \Theta$.

Let us consider item 2 first. By the definition of a belief function,

$$Co^{\mathbf{B}}[Bel](A) = \sum_{B \subseteq A, B \in \mathcal{C}} m^{\mathbf{B}}[Bel](B),$$

where

$$m^{\mathbf{B}}[Bel](B) = \sum_{X \subseteq B : \mathbf{B}(X) = B} m(X).$$

Therefore,

$$Co^{\mathbf{B}}[Bel](A) = \sum_{B \subseteq A, B \in \mathcal{C}} \; \sum_{X \subseteq B : \mathbf{B}(X) = B} m(X) = \sum_{X \subseteq B_i : \mathbf{B}(X) = B_j, j \leq i} m(X), \tag{13.59}$$

where B_i is the largest element of the chain $\mathcal{C}$ included in A. Since $B_i \subseteq A$, the quantity (13.59) is obviously no greater than $\sum_{B \subseteq A} m(B) = Bel(A)$. Hence,

$$\sum_{\mathbf{B}} \alpha^{\mathbf{B}} Co^{\mathbf{B}}[Bel](A) \leq \sum_{\mathbf{B}} \alpha^{\mathbf{B}} Bel(A) = Bel(A) \sum_{\mathbf{B}} \alpha^{\mathbf{B}} = Bel(A) \quad \forall A \subseteq \Theta.$$

Let us now prove item 1. According to Lemma 14, if $\forall A \subseteq \Theta \; Co(A) \leq Bel(A)$, then the mass $m_{Co}(B_i)$ of each event B_i of the chain is

$$m_{Co}(B_i) = \sum_{A \subseteq B_i} m(A) \alpha^A_{B_i}. \tag{13.60}$$

We then need to write (13.60) as a convex combination of the $m^{\mathbf{B}}[Bel](B_i)$, i.e.,

$$\sum_{\mathbf{B}} \alpha^{\mathbf{B}} Co^{\mathbf{B}}[Bel](B_i) = \sum_{\mathbf{B}} \alpha^{\mathbf{B}} \sum_{\substack{X \subseteq B_i \\ \mathbf{B}(X) = B_i}} m(X) = \sum_{X \subseteq B_i} m(X) \sum_{\mathbf{B}(X) = B_i} \alpha^{\mathbf{B}}.$$

In other words, we need to show that the following system of equations,

$$\left\{ \alpha^A_{B_i} = \sum_{\mathbf{B}(A) = B_i} \alpha^{\mathbf{B}} \quad \forall i = 1, \ldots, n; \; \forall A \subseteq B_i, \right. \tag{13.61}$$

has at least one solution $\{\alpha^{\mathbf{B}}, \mathbf{B}\}$ such that $\sum_{\mathbf{B}} \alpha^{\mathbf{B}} = 1$ and $\forall \mathbf{B} \; \alpha^{\mathbf{B}} \geq 0$. The normalisation constraint is in fact trivially satisfied, as from (13.61) it follows that

$$\sum_{B_i \supseteq A} \alpha^A_{B_i} = 1 = \sum_{B_i \supseteq A} \sum_{\mathbf{B}(A) = B_i} \alpha^{\mathbf{B}} = \sum_{\mathbf{B}} \alpha^{\mathbf{B}}.$$

Using the normalisation constraint, the system of equations (13.61) reduces to

$$\begin{cases} \alpha^A_{B_i} = \sum_{\mathbf{B}(A)=B_i} \alpha^{\mathbf{B}} \quad \forall i=1,\ldots,n-1; \ \ \forall A \subseteq B_i. \end{cases} \tag{13.62}$$

We can show that each equation in the reduced system (13.62) involves at least one variable $\alpha^{\mathbf{B}}$ which is not present in any other equation.

Formally, the set of assignment functions which satisfy the constraint of the equation for A, B_i but not all others is not empty:

$$\left\{ \mathbf{B} \,\middle|\, \mathbf{B}(A) = B_i \bigwedge_{\substack{\forall j=1,\ldots,n-1 \\ j \neq i}} \mathbf{B}(A) \neq B_j \bigwedge_{\substack{\forall A' \neq A \\ \forall j=1,\ldots,n-1}} \mathbf{B}(A') \neq B_j \right\} \neq \emptyset. \tag{13.63}$$

But the assignment functions $\mathbf{B}$ such that $\mathbf{B}(A) = B_i$ and $\forall A' \neq A \; \mathbf{B}(A') = \Theta$ all satisfy the condition (13.63). Indeed, they obviously satisfy $\mathbf{B}(A) \neq B_j$ for all $j \neq i$, while, clearly, for all $A' \subseteq \Theta$, $\mathbf{B}(A') = \Theta \neq B_j$, as $j < n$ so that $B_j \neq \Theta$.

A non-negative solution of (13.62) (and hence of (13.61)) can be obtained by setting for each equation one of the variables $\alpha^{\mathbf{B}}$ to be equal to the left-hand side $\alpha^A_{B_i}$, and all the others to zero.

Proof of Theorem 70

The proof is divided into two parts.

1. We first need to find an assignment $\mathbf{B} : 2^\Theta \to \mathcal{C}_\rho$ which generates Co^ρ.

Each singleton x_i is mapped by ρ to the position j: $i = \rho(j)$. Then, given any event $A = \{x_{i_1}, \ldots, x_{i_m}\}$, its elements are mapped to the new positions $x_{j_{i_1}}, \ldots, x_{j_{i_m}}$, where $i_1 = \rho(j_{i_1}), \ldots, i_m = \rho(j_{i_m})$. But then the map

$$\mathbf{B}_\rho(A) = \mathbf{B}_\rho(\{x_{i_1}, \ldots, x_{i_m}\}) = S^\rho_j \doteq \{x_{\rho(1)}, \ldots, x_{\rho(j)}\},$$

where

$$j \doteq \max\{j_{i_1}, \ldots, j_{i_m}\},$$

maps each event A to the smallest S^ρ_i in the chain which contains A: $j = \min\{i : A \subseteq S^\rho_i\}$. Therefore it generates a consonant belief function with BPA (4.69), i.e., Co^ρ.

2. In order for Co^ρ to be an actual vertex, we need to ensure that it cannot be written as a convex combination of the other (pseudo-)vertices $Co^{\mathbf{B}}[Bel]$:

$$Co^\rho = \sum_{\mathbf{B} \neq \mathbf{B}_\rho} \alpha^{\mathbf{B}} Co^{\mathbf{B}}[Bel], \quad \sum_{\mathbf{B} \neq \mathbf{B}_\rho} \alpha^{\mathbf{B}} = 1, \ \forall \mathbf{B} \neq \mathbf{B}_\rho \ \alpha^{\mathbf{B}} \geq 0.$$

As $m_{Co^{\mathbf{B}}}(B_i) = \sum_{A:\mathbf{B}(A)=B_i} m(A)$, the above condition reads as

$$\left\{ \sum_{A \subseteq B_i} m(A) \left(\sum_{\mathbf{B}:\mathbf{B}(A)=B_i} \alpha^{\mathbf{B}} \right) = \sum_{A \subseteq B_i : \mathbf{B}_\rho(A) = B_i} m(A) \quad \forall B_i \in \mathcal{C}. \right.$$

Remembering that $\mathbf{B}_\rho(A) = B_i$ iff $A \subseteq B_i, A \not\subset B_{i-1}$, we find that

$$\left\{ \sum_{A \subseteq B_i} m(A) \left(\sum_{\mathbf{B}:\mathbf{B}(A)=B_i} \alpha^{\mathbf{B}} \right) = \sum_{A \subseteq B_i, A \not\subset B_{i-1}} m(A) \quad \forall B_i \in \mathcal{C}. \right.$$

For $i = 1$, the condition is $m(B_1)\left(\sum_{\mathbf{B}:\mathbf{B}(B_1)=B_1} \alpha^{\mathbf{B}}\right) = m(B_1)$, namely

$$\sum_{\mathbf{B}:\mathbf{B}(B_1)=B_1} \alpha^{\mathbf{B}} = 1, \qquad \sum_{\mathbf{B}:\mathbf{B}(B_1)\neq B_1} \alpha^{\mathbf{B}} = 0.$$

Substituting the above equalities into the second constraint, $i = 2$, yields

$$m(B_2 \setminus B_1) \left(\sum_{\substack{\mathbf{B}:\mathbf{B}(B_1)=B_1 \\ \mathbf{B}(B_2 \setminus B_1) \neq B_2}} \alpha^{\mathbf{B}} \right) + m(B_2) \left(\sum_{\substack{\mathbf{B}:\mathbf{B}(B_1)=B_1 \\ \mathbf{B}(B_2) \neq B_2}} \alpha^{\mathbf{B}} \right) = 0,$$

which implies $\alpha^{\mathbf{B}} = 0$ for all the assignment functions $\mathbf{B}$ such that $\mathbf{B}(B_2 \setminus B_1) \neq B_2$ or $\mathbf{B}(B_2) \neq B_2$. The only non-zero coefficients can then be the $\alpha^{\mathbf{B}}$ such that $\mathbf{B}(B_1) = B_1$, $\mathbf{B}(B_2 \setminus B_1) = B_2$, $\mathbf{B}(B_2) = B_2$.

By induction, we find that $\alpha^{\mathbf{B}} = 0$ for all $\mathbf{B} \neq \mathbf{B}_\rho$.

Proof of Theorem 71

Let us denote by $\{B_1, \ldots, B_n\}$ the elements of the maximal chain $\mathcal{C}$, as usual. By definition, the masses that Co_ρ assigns to the elements of the chain are

$$m_{Co_\rho}(B_i) = \sum_{B \subseteq B_i, B \not\subset B_{i-1}} m(B).$$

Hence the belief value of Co_ρ on an arbitrary event $A \subseteq \Theta$ can be written as

$$\begin{aligned} Co_\rho(A) &= \sum_{B_i \subseteq A, B_i \in \mathcal{C}} m_{Co_\rho}(B_i) = \sum_{B_i \subseteq A} \sum_{B \subseteq B_i, B \not\subset B_{i-1}} m(B) \\ &= \sum_{B \subseteq B_{i_A}} m(B) = Bel(B_{i_A}), \end{aligned}$$

where B_{i_A} is the largest element of the chain included in A. But then, as the elements $B_1 \subset \cdots \subset B_n$ of the chain are nested and any belief function Bel is monotone,

$$Co_\rho(A) = Bel(B_{i_A}) = \max_{B_i \in \mathcal{C}, B_i \subseteq A} Bel(B_i),$$

i.e., Co_ρ is indeed given by (13.7).

Proof of Theorem 72

$\mathbb{R}^{N-1}$ representation The L_1 norm of the difference vector (13.16) is

$$\|m - m_{Co}\|_{L_1} = \sum_{A\in\mathcal{C}} |m(A) - m_{Co}(A)| + \sum_{B\notin\mathcal{C}} m(B)$$

$$= \sum_{A\in\mathcal{C}} |\beta(A)| + \sum_{B\notin\mathcal{C}} m(B),$$

as a function of the variables $\{\beta(A) \doteq m(A) - m_{Co}(A), A \in \mathcal{C}, A \neq \Theta\}$. Since

$$\sum_{A\in\mathcal{C}} \beta(A) = \sum_{A\in\mathcal{C}} (m(A) - m_{Co}(A)) = \sum_{A\in\mathcal{C}} m(A) - 1 = -\sum_{B\notin\mathcal{C}} m(B),$$

we have that

$$\beta(\Theta) = -\sum_{B\notin\mathcal{C}} m(B) - \sum_{A\in\mathcal{C},A\neq\Theta} \beta(A).$$

Therefore, the above norm reads as

$$\|m - m_{Co}\|_{L_1}$$

$$= \left| -\sum_{B\notin\mathcal{C}} m(B) - \sum_{A\in\mathcal{C},A\neq\Theta} \beta(A) \right| + \sum_{A\in\mathcal{C},A\neq\Theta} |\beta(A)| + \sum_{B\notin\mathcal{C}} m(B). \quad (13.64)$$

The norm (13.64) is a function of the form

$$\sum_i |x_i| + \left| -\sum_i x_i - k \right|, \quad k \geq 0, \quad (13.65)$$

which has an entire simplex of minima, namely $x_i \leq 0 \; \forall i$, $\sum_i x_i \geq -k$. The minima of the L_1 norm (13.64) are therefore the solutions to the following system of constraints:

$$\begin{cases} \beta(A) \leq 0 & \forall A \in \mathcal{C}, A \neq \Theta, \\ \displaystyle\sum_{A\in\mathcal{C},A\neq\Theta} \beta(A) \geq -\sum_{B\notin\mathcal{C}} m(B). & \end{cases} \quad (13.66)$$

This reads, in terms of the mass assignment m_{Co} of the desired consonant approximation, as

$$\begin{cases} m_{Co}(A) \geq m(A) & \forall A \in \mathcal{C}, A \neq \Theta, \\ \displaystyle\sum_{A\in\mathcal{C},A\neq\Theta} (m(A) - m_{Co}(A)) \geq -\sum_{B\notin\mathcal{C}} m(B). & \end{cases} \quad (13.67)$$

Note that the last constraint reduces to

$$\sum_{A\in\mathcal{C},A\neq\Theta} m(A) - 1 + m_{Co}(\Theta) \geq \sum_{A\in\mathcal{C}} m(A) - 1,$$

i.e., $m_{Co}(\Theta) \geq m(\Theta)$. Therefore the partial L_1 approximations in $\mathcal{M}$ are those consonant belief functions Co such that $m_{Co}(A) \geq m(A)$ $\forall A \in \mathcal{C}$. The vertices of the set of partial approximations (13.66) are given by the vectors of variables $\{\beta_{\bar{A}}, \bar{A} \in \mathcal{C}\}$ such that $\beta_{\bar{A}}(\bar{A}) = m(B)$, for $\beta_{\bar{A}}(A) = 0$ for $A \neq \bar{A}$ whenever $A \neq \Theta$, while $\beta_\Theta = \mathbf{0}$. Immediately, in terms of masses, the vertices of the set of partial L_1 approximations have BPA (13.20) and barycentre (13.21).

To find the global L_1 consonant approximation(s) over the whole consonant complex, we need to locate the component $\mathcal{CO}^{\mathcal{C}}_{\mathcal{M}}$ at the minimum L_1 distance from m. All the partial approximations (13.67) in $\mathcal{CO}^{\mathcal{C}}_{\mathcal{M}}$ have an L_1 distance from m equal to $2\sum_{B\not\in\mathcal{C}} m(B)$. Therefore, the minimum-distance component(s) of the complex are those whose maximal chains originally have maximum mass with respect to m.

$\mathbb{R}^{N-2}$ representation Consider now the difference vector (13.17). Its L_1 norm is

$$\|m - m_{Co}\|_{L_1} = \sum_{A\in\mathcal{C},A\neq\bar{A}} |m(A) - m_{Co}(A)| + \sum_{B\not\in\mathcal{C}} m(B),$$

which is obviously minimised by $m(A) = m_{Co}(A)$ $\forall A \in \mathcal{C}, A \neq \bar{A}$, i.e., (13.20). Such a (unique) partial approximation onto $\mathcal{CO}^{\mathcal{C}}_{\mathcal{M}}$ has an L_1 distance from m given by $\sum_{B\not\in\mathcal{C}} m(B)$. Therefore, the minimum-distance component(s) of the consonant complex are, once again, those associated with the maximal chains

$$\arg\min_{\mathcal{C}} \sum_{B\not\in\mathcal{C}} m(B) = \arg\max_{\mathcal{C}} \sum_{A\in\mathcal{C}} m(A).$$

Proof of Theorem 73

As the generators of $\mathcal{CO}^{\mathcal{C}}_{\mathcal{M}}$ are the vectors in $\mathcal{M}$ $\{m_A - m_\Theta, A \in \mathcal{C}, A \neq \Theta\}$, we need to impose the condition

$$\langle m - m_{Co}, m_A - m_\Theta \rangle = 0$$

for all $A \in \mathcal{C}$, $A \neq \Theta$.

$\mathbb{R}^{N-1}$ representation In the complete mass space, the vector $m_A - m_\Theta$ is such that $m_A - m_\Theta(B) = 1$ if $B = A$, $= -1$ if $B = \Theta$ and $= 0$ if $B \neq A, \Theta$. Hence, the orthogonality condition becomes $\beta(A) - \beta(\Theta) = 0$ for all $A \in \mathcal{C}, A \neq \Theta$, where again $\beta(A) = m(A) - m_{Co}(A)$. Since

$$\beta(\Theta) = -\sum_{B\not\in\mathcal{C}} m(B) - \sum_{A\in\mathcal{C},A\neq\Theta} \beta(A)$$

(see the proof of Theorem 72), the orthogonality condition becomes

$$2\beta(A) + \sum_{B \notin \mathcal{C}} m(B) + \sum_{B \in \mathcal{C}, B \neq A, \Theta} \beta(B) = 0 \quad \forall A \in \mathcal{C}, A \neq \Theta.$$

Its solution is clearly

$$\beta(A) = \frac{-\sum_{B \notin \mathcal{C}} m(B)}{n} \quad \forall A \in \mathcal{C},\ A \neq \Theta,$$

as, by the substitution

$$-\frac{2}{n} \sum_{B \notin \mathcal{C}} m(B) + \sum_{B \notin \mathcal{C}} m(B) - \frac{n-2}{n} \sum_{B \notin \mathcal{C}} m(B) = 0,$$

we obtain (13.21).

To find the global L_2 approximation(s), we need to compute the L_2 distance of m from the closest such partial solution. We have

$$\begin{aligned} \|m - m_{Co}\|_{L_2}^2 &= \sum_{A \subseteq \Theta} (m(A) - m_{Co}(A))^2 \\ &= \sum_{A \in \mathcal{C}} \left(\frac{\sum_{B \notin \mathcal{C}} m(B)}{n} \right)^2 + \sum_{B \notin \mathcal{C}} (m(B))^2 \\ &= \frac{\left(\sum_{B \notin \mathcal{C}} m(B) \right)^2}{n} + \sum_{B \notin \mathcal{C}} (m(B))^2, \end{aligned}$$

which is minimised by the component $\mathcal{CO}_{\mathcal{M}}^{\mathcal{C}}$ that minimises $\sum_{B \notin \mathcal{C}} (m(B))^2$.

$\mathbb{R}^{N-2}$ representation In the case of a section of the mass space with a missing component $\bar{A} \in \mathcal{C}$, as $m_{\bar{A}} = \mathbf{0}$ there, the orthogonality condition reads as

$$\langle m - m_{Co}, m_A \rangle = \beta(A) = 0 \quad \forall A \in \mathcal{C}, A \neq \Theta,$$

i.e., $\beta(A) = 0\ \forall A \in \mathcal{C},\ A \neq \bar{A}$, and we get (13.20) once again.

The optimal distance is, in this case,

$$\begin{aligned} \|m - m_{Co}\|_{L_2}^2 &= \sum_{A \subseteq \Theta} (\beta(A))^2 = \sum_{B \notin \mathcal{C}} (m(B))^2 + \beta(\bar{A}) \\ &= \sum_{B \notin \mathcal{C}} (m(B))^2 + \left(m(\bar{A}) - m(\bar{A}) - \sum_{B \notin \mathcal{C}} m(B) \right)^2 \\ &= \sum_{B \notin \mathcal{C}} (m(B))^2 + \left(\sum_{B \notin \mathcal{C}} m(B) \right)^2, \end{aligned}$$

which is once again minimised by the maximal chain(s) $\arg\min_{\mathcal{C}} \sum_{B \notin \mathcal{C}} (m(B))^2$.

Proof of Theorem 74

$\mathbb{R}^{N-1}$ representation In the complete mass space, the L_∞ norm of the difference vector is

$$\|m - m_{Co}\|_{L_\infty} = \max\left\{ \max_{A \in \mathcal{C}} |\beta(A)|, \max_{B \not\in \mathcal{C}} m(B) \right\}.$$

As $\beta(\Theta) = \sum_{B \in \mathcal{C}} m(B) - 1 - \sum_{B \in \mathcal{C}, B \neq \Theta} \beta(B)$, we have that

$$|\beta(\Theta)| = \left| \sum_{B \not\in \mathcal{C}} m(B) + \sum_{B \in \mathcal{C}, B \neq \Theta} \beta(B) \right|$$

and the norm to be minimised becomes

$$\begin{array}{l} \|m - m_{Co}\|_{L_\infty} \\ = \max\left\{ \max_{A \in \mathcal{C}, A \neq \Theta} |\beta(A)|, \left| \sum_{B \not\in \mathcal{C}} m(B) + \sum_{B \in \mathcal{C}, B \neq \Theta} \beta(B) \right|, \max_{B \not\in \mathcal{C}} m(B) \right\}. \end{array} \tag{13.68}$$

This is a function of the form

$$\max\left\{ |x_1|, |x_2|, |x_1 + x_2 + k_1|, k_2 \right\}, \quad 0 \leq k_2 \leq k_1 \leq 1. \tag{13.69}$$

Such a function has two possible behaviours in terms of its minimal points in the plane x_1, x_2.

Case 1. If $k_1 \leq 3k_2$, its contour function has the form shown in Fig. 13.10 (left). The set of minimal points is given by $x_i \geq -k_2$, $x_1 + x_2 \leq k_2 - k_1$. In the general case of an arbitrary number $m - 1$ of variables $x_1, \ldots, x_{m-1}$ such that $x_i \geq -k_2$, $\sum_i x_i \leq k_2 - k_1$, the set of minimal points is a simplex with m vertices: each vertex v^i is such that

$$v^i(j) = -k_2 \;\; \forall j \neq i; \quad v^i(i) = -k_1 + (m-1)k_2$$

(obviously, $v^m = [-k_2, \ldots, -k_2]$).

Concerning (13.68), in the first case ($\max_{B \not\in \mathcal{C}} m(B) \geq \frac{1}{n} \sum_{B \not\in \mathcal{C}} m(B)$) the set of partial L_∞ approximations is given by the following system of inequalities:

$$\begin{cases} \beta(A) \geq -\max_{B \not\in \mathcal{C}} m(B) & A \in \mathcal{C}, A \neq \Theta, \\ \sum_{B \in \mathcal{C}, B \neq \Theta} \beta(B) \leq \max_{B \not\in \mathcal{C}} m(B) - \sum_{B \not\in \mathcal{C}} m(B). & \end{cases}$$

This determines a simplex of solutions $Cl(m^{\bar{A}}_{L_\infty}[m], \bar{A} \in \mathcal{C})$ with vertices

$$m^{\bar{A}}_{L_\infty}[m] : \begin{cases} \beta_{\bar{A}}(A) = -\max_{B \not\in \mathcal{C}} m(B) & A \in \mathcal{C}, A \neq \bar{A}, \\ \beta_{\bar{A}}(\bar{A}) = -\sum_{B \not\in \mathcal{C}} m(B) + (n-1) \max_{B \not\in \mathcal{C}} m(B), & \end{cases}$$

or, in terms of their BPAs, (13.23). Its barycentre has mass assignment, $\forall A \in \mathcal{C}$,

$$\frac{\sum_{\bar{A}\in\mathcal{C}} m^{\bar{A}}_{L_\infty}[m](A)}{n} = \frac{n\cdot m(A) + \sum_{B\not\in\mathcal{C}} m(B)}{n} = m(A) + \frac{\sum_{B\not\in\mathcal{C}} m(B)}{n}$$

i.e., the L_2 partial approximation (13.21). The corresponding minimal L_∞ norm of the difference vector is, according to (13.68), equal to $\max_{B\not\in\mathcal{C}} m(B)$.

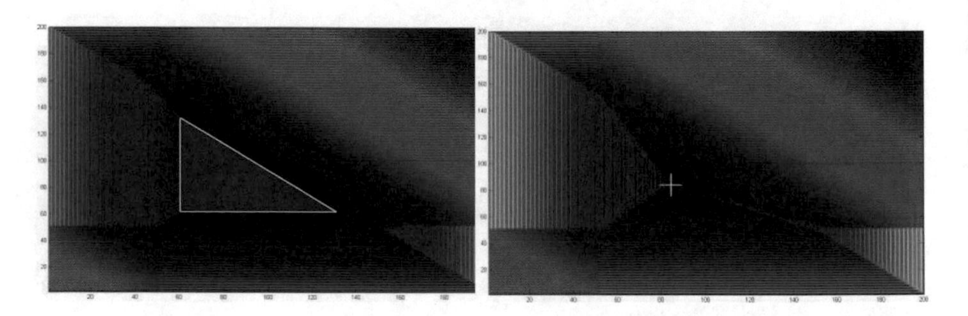

Fig. 13.10: Left: contour function (level sets) and minimal points (white triangle) of a function of the form (13.69), when $k_1 \leq 3k_2$. In this example, $k_2 = 0.4$ and $k_1 = 0.5$. Right: contour function and minimal point of a function of the form (13.69), when $k_1 > 3k_2$. In this example, $k_2 = 0.1$ and $k_1 = 0.5$.

Case 2. In the second case, $k_1 > 3k_2$, i.e., for the norm (13.68),

$$\max_{B\not\in\mathcal{C}} m(B) < \frac{1}{n}\sum_{B\not\in\mathcal{C}} m(B),$$

the contour function of (13.69) is as in Fig. 13.10 (right). There is a single minimal point, located at $[-1/3k_1, -1/3k_1]$.

For an arbitrary number $m-1$ of variables, the minimal point is

$$\Big[(-1/m)k_1, \ldots, (-1/m)k_1\Big]',$$

i.e., for the system (13.68),

$$\beta(A) = -\frac{1}{n}\sum_{B\not\in\mathcal{C}} m(B) \quad \forall A \in \mathcal{C}, A \neq \Theta.$$

In terms of basic probability assignments, this yields (13.21) (the mass of Θ is obtained by normalisation). The corresponding minimal L_∞ norm of the difference vector is $\frac{1}{n}\sum_{B\not\in\mathcal{C}} m(B)$.

$\mathbb{R}^{N-2}$ representation In the section of the mass space with missing component $\bar{A} \in \mathcal{C}$, the L_∞ norm of the difference vector (13.17) is

$$\begin{aligned} &\|m - m_{Co}\|_{L_\infty} \\ &= \max_{\emptyset \subsetneq A \subsetneq \Theta} |m(A) - m_{Co}(A)| = \max \left\{ \max_{A \in \mathcal{C}, A \neq \bar{A}} |\beta(A)|, \max_{B \not\in \mathcal{C}} m(B) \right\}, \end{aligned} \tag{13.70}$$

which is minimised by

$$|\beta(A)| \leq \max_{B \not\in \mathcal{C}} m(B) \quad \forall A \in \mathcal{C}, A \neq \bar{A}, \tag{13.71}$$

i.e., in the mass coordinates m_{Co}, (13.26). According to (13.70), the corresponding minimal L_∞ norm is $\max_{B \not\in \mathcal{C}} m(B)$. Clearly, the vertices of the set (13.71) are all the vectors of β variables such that $\beta(A) = +/- \max_{B \not\in \mathcal{C}} m(B)$ for all $A \in \mathcal{C}$, $A \neq \bar{A}$. Its barycentre is given by $\beta(A) = 0$ for all $A \in \mathcal{C}$, $A \neq \bar{A}$, i.e., (13.20).

Proof of Theorem 75

By (13.23), the vertex $m^{\bar{A}}_{L_\infty}[m]$ of $\mathcal{CO}^{\mathcal{C}}_{\mathcal{M},L_\infty}[m]$ satisfies the constraints (13.26) for $\mathcal{CO}^{\mathcal{C}}_{\mathcal{M}\setminus\bar{A},L_\infty}[m]$.

For the other vertices of $\mathcal{CO}^{\mathcal{C}}_{\mathcal{M},L_\infty}[m]$ (13.23), let us check under what conditions on the quantity

$$\Delta \doteq \sum_{B \not\in \mathcal{C}} m(B) - n \max_{B \not\in \mathcal{C}} m(B)$$

$m^{\bar{A}}_{L_\infty}[m]$ satisfies (13.26). If Δ is positive,

$$n \max_{B \not\in \mathcal{C}} m(B) < \sum_{B \not\in \mathcal{C}} m(B) \equiv \max_{B \not\in \mathcal{C}} m(B) < \frac{1}{n},$$

which cannot happen, by the constraint (13.24). Therefore, Δ is non-positive. In order for the vertex not to belong to (13.26), we need $m(\bar{A}) + \max_{B \not\in \mathcal{C}} m(B) + \Delta < m(\bar{A}) - \max_{B \not\in \mathcal{C}} m(B)$, i.e.,

$$\max_{B \not\in \mathcal{C}} m(B) > \frac{1}{n-2} \sum_{B \not\in \mathcal{C}} m(B), \tag{13.72}$$

which cannot be ruled out under the condition (13.24).

Proof of Lemma 15

In the belief space, the original belief function Bel and the desired consonant approximation Co are written as

$$Bel = \sum_{\emptyset \subsetneq A \subsetneq \Theta} Bel(A) \mathbf{x}_A, \quad Co = \sum_{A \supseteq A_1} \left(\sum_{B \subseteq A, B \in \mathcal{C}} m_{Co}(B) \right) \mathbf{x}_A.$$

Their difference vector is therefore

$$\begin{aligned}
& Bel - Co \\
&= \sum_{A \not\supset A_1} Bel(A)\mathbf{x}_A + \sum_{A \supseteq A_1} \mathbf{x}_A \left[Bel(A) - \sum_{B \subseteq A, B \in \mathcal{C}} m_{Co}(B) \right] \\
&= \sum_{A \not\supset A_1} Bel(A)\mathbf{x}_A + \sum_{A \supseteq A_1} \mathbf{v}_A \left[\sum_{\emptyset \subsetneq B \subseteq A} m(B) - \sum_{B \subseteq A, B \in \mathcal{C}} m_{Co}(B) \right] \\
&= \sum_{A \not\supset A_1} Bel(A)\mathbf{x}_A + \sum_{A \supseteq A_1} \mathbf{x}_A \left[\sum_{B \subseteq A, B \in \mathcal{C}} (m(B) - m_{Co}(B)) + \sum_{B \subseteq A, B \notin \mathcal{C}} m(B) \right] \\
&= \sum_{A \not\supset A_1} Bel(A)\mathbf{x}_A + \sum_{A \supseteq A_1} \mathbf{x}_A \left[\gamma(A) + \sum_{B \subseteq A, B \notin \mathcal{C}} m(B) \right] \\
&= \sum_{A \not\supset A_1} Bel(A)\mathbf{x}_A + \sum_{A \supseteq A_1} \mathbf{x}_A \left[\gamma(A) + Bel(A) - \sum_{j=1}^{i} m(A_j) \right],
\end{aligned} \tag{13.73}$$

after introducing the auxiliary variables

$$\gamma(A) \doteq \sum_{B \subseteq A, B \in \mathcal{C}} (m(B) - m_{Co}(B)).$$

All the terms in (13.73) associated with subsets $A \supseteq A_i, A \not\supset A_{i+1}$ depend on the same auxiliary variable $\gamma(A_i)$, while the difference in the component $\mathbf{x}_\Theta$ is, trivially, $1 - 1 = 0$. Therefore, we obtain (13.37).

Proof of Theorem 77

To understand the relationship between the sets $\mathcal{CO}^{\mathcal{C}}_{\mathcal{M},L_\infty}[m]$ and $\mathcal{O}^{\mathcal{C}}[Bel]$, let us rewrite the system of constraints for L_∞ approximations in $\mathcal{M}$ under the condition (13.24) as

$$\begin{cases}
m_{Co}(A) - m(A) \leq \max_{B \notin \mathcal{C}} m(B) & A \in \mathcal{C}, A \neq \bar{A}, \\
\sum_{A \in \mathcal{C}, A \neq \bar{A}} (m_{Co}(A) - m(A)) \geq \left(\sum_{B \notin \mathcal{C}} m(B) - \max_{B \notin \mathcal{C}} m(B) \right).
\end{cases} \tag{13.74}$$

In fact, when (13.24) does not hold, $Co^{\mathcal{C}}_{\mathcal{M},L_\infty}[m] = Co^{\mathcal{C}}_{\mathcal{M},L_2}[m]$, which is in general outside $\mathcal{O}^{\mathcal{C}}[Bel]$.

To be a pseudo-vertex of the set of partial outer approximations, a consonant belief function Co must be the result of reassigning the mass of each focal element to an element of the chain which contains it. Imagine that all the focal elements not in the desired chain $\mathcal{C}$ have the same mass: $m(B) = \text{const}$ for all $B \notin \mathcal{C}$. Then, only up to $n-1$ of them can be reassigned to elements of the chain different from $\bar{A}$. Indeed, if one were to reassign n outside focal elements to such elements of the chain (in the absence of mass redistribution internal to the chain), some element $A \in \mathcal{C}$ of the chain would certainly violate the first constraint in (13.74), as it would receive mass from at least two outside focal elements, yielding

$$m_{Co}(A) - m(A) \geq 2 \max_{B \notin \mathcal{C}} m(B) > \max_{B \notin \mathcal{C}} m(B).$$

In fact, this is true even if mass redistribution does take place within the chain. Suppose that some mass $m(A)$, $A \in \mathcal{C}$, is reassigned to some other $A' \in \mathcal{C}$. By the first constraint in (13.74), this is allowed only if $m(A) \leq \max_{B \notin \mathcal{C}} m(B)$. Therefore the mass of just one outside focal element can still be reassigned to A, while now none can be reassigned to A'. In both cases, since the number of elements outside the chain, $m = 2^n - 1 - n$, is greater than n (unless $n \leq 2$), the second equation of (13.74) implies

$$(n-1) \max_{B \notin \mathcal{C}} m(B) \geq (m-1) \max_{B \notin \mathcal{C}} m(B),$$

which cannot hold under (13.24).

Proof of Theorem 78

After we recall the expression (13.37) for the difference vector $Bel - Co$ in the belief space, the latter's L_1 norm reads as

$$\begin{aligned} \|Bel - Co\|_{L_1} = & \sum_{i=1}^{n-1} \sum_{A \supseteq A_i, A \not\supseteq A_{i+1}} \left| \gamma(A_i) + Bel(A) - \sum_{j=1}^{i} m(A_j) \right| \\ & + \sum_{A \not\supseteq A_1} |Bel(A)|. \end{aligned} \tag{13.75}$$

The norm (13.75) can be decomposed into a number of summations which depend on a single auxiliary variable $\gamma(A_i)$. Such components are of the form $|x + x_1| + \ldots + |x + x_n|$, with an even number of 'nodes' $-x_i$.

Let us consider the simple function shown in Fig. 13.11 (left): it is easy to see that similar functions are minimised by the interval of values between their two innermost nodes, i.e., in the case of the norm (13.75),

$$\sum_{j=1}^{i} m(A_j) - \gamma^i_{\text{int}_1} \le \gamma(A_i) \le \sum_{j=1}^{i} m(A_j) - \gamma^i_{\text{int}_2} \quad \forall i = 1, \ldots, n-1. \quad (13.76)$$

This is equivalent to

$$\gamma^i_{\text{int}_1} \le \sum_{j=1}^{i} m_{Co}(A_j) \le \gamma^i_{\text{int}_2} \quad \forall i = 1, \ldots, n-2, \quad (13.77)$$

while $m_{Co}(A_{n-1}) = Bel(A_{n-1})$, as by definition (13.39) $\gamma^{n-1}_{\text{int}_1} = \gamma^{n-1}_{\text{int}_2} = Bel(A_{n-1})$.

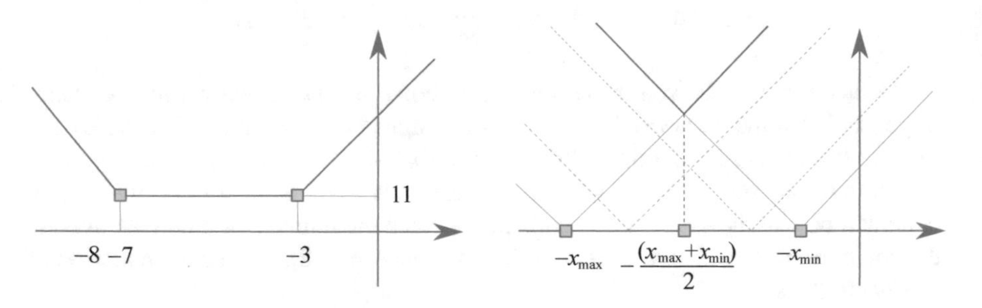

Fig. 13.11: Left: minimising the L_1 distance from the consonant subspace involves functions such as the one depicted above, $|x+1|+|x+3|+|x+7|+|x+8|$, which is minimised by $3 \le x \le 7$. Right: minimising the L_∞ distance from the consonant subspace involves functions of the form $\max\{|x+x_1|, ..., |x+x_n|\}$ (solid lines).

This is a set of constraints of the form $l_1 \le x \le u_1$, $l_2 \le x+y \le u_2$, $l_3 \le x+y+z \le u_3, \ldots$, which can also be expressed as $l_1 \le x \le u_1, l_2 - x \le y \le u_2 - x$, $l_3 - (x+y) \le z \le u_3 - (x+y), \ldots$. This is a polytope whose 2^{n-2} vertices are obtained by assigning to x, $x+y$, $x+y+z$ and so on either their lower or their upper bound. For the specific set (13.77), this yields exactly (13.38).

Proof of Theorem 79

The minimum value of a function of the form $|x+x_1| + \ldots + |x+x_n|$ is

$$\sum_{i \ge \text{int}_2} x_i - \sum_{i \le \text{int}_1} x_i.$$

In the case of the L_1 norm (13.75), this minimum attained value is

$$\sum_{A: A \supseteq A_i, A \not\supseteq A_{i+1}, Bel(A) \ge \gamma_{\text{int}_2}} Bel(A) - \sum_{A: A \supseteq A_i, A \not\supseteq A_{i+1}, Bel(A) \le \gamma_{\text{int}_1}} Bel(A),$$

since the addends $\sum_{j=1}^{i} m(A_j)$ disappear in the difference.

Overall, the minimal L_1 norm is

$$\sum_{i=1}^{n-2} \left(\sum_{\substack{A:A\supseteq A_i, A\not\supset A_{i+1} \\ Bel(A)\geq\gamma_{\text{int}_2}}} Bel(A) - \sum_{\substack{A:A\supseteq A_i, A\not\supset A_{i+1} \\ Bel(A)\leq\gamma_{\text{int}_1}}} Bel(A) \right) + \sum_{A\not\supset A_1} Bel(A)$$

$$= \sum_{\emptyset\subsetneq A\subsetneq\Theta, A\neq A_{n-1}} Bel(A) - 2\sum_{i=1}^{n-2} \sum_{\substack{A:A\supseteq A_i, A\not\supset A_{i+1} \\ Bel(A)\leq\gamma_{\text{int}_1}}} Bel(A),$$

which is minimised by the chains (13.41).

Proof of Theorem 80

By substituting the hypothesised solution (13.44) for the L_2 approximation in $\mathcal{B}$ into the system of constraints (13.43) we get, for all $j = 1, \ldots, n-1$,

$$\begin{cases} \sum_{A\subsetneq\Theta} m(A)\langle Bel_A, Bel_{A_j}\rangle - \text{ave}(\mathcal{L}_{n-1})\langle Bel_{A_{n-1}}, Bel_{A_{n-1}}\rangle \\ \qquad - \sum_{i=1}^{n-2} \text{ave}(\mathcal{L}_i)\Big(\langle Bel_{A_i}, Bel_{A_j}\rangle - \langle Bel_{A_{i+1}}, Bel_{A_j}\rangle\Big) = 0, \end{cases}$$

where $\langle Bel_{A_{n-1}}, Bel_{A_{n-1}}\rangle = 1$ for all j. Furthermore, since $\langle Bel_A, Bel_B\rangle = |\{C \subsetneq \Theta : C \supseteq A, B\}| = 2^{|(A\cup B)^c|} - 1$, we have that

$$\langle Bel_{A_i}, Bel_{A_j}\rangle - \langle Bel_{A_{i+1}}, Bel_{A_j}\rangle = \langle Bel_{A_j}, Bel_{A_j}\rangle - \langle Bel_{A_j}, Bel_{A_j}\rangle = 0$$

whenever $i < j$, and

$$\begin{aligned} &\langle Bel_{A_i}, Bel_{A_j}\rangle - \langle Bel_{A_{i+1}}, Bel_{A_j}\rangle \\ &= \big(|\{A \supseteq A_i, A_j\}| - 1\big) - \big(|\{A \supseteq A_{i+1}, A_j\}| - 1\big) \\ &= |\{A \supseteq A_i\}| - |\{A \supseteq A_{i+1},\}| = 2^{|A_{i+1}^c|} \end{aligned}$$

whenever $i \geq j$. The system of constraints becomes (as $2^{A_n^c} = 2^{|\emptyset|} = 1$)

$$\begin{cases} \sum_{A\subsetneq\Theta} m(A)\langle Bel_A, Bel_{A_j}\rangle - \sum_{i=j}^{n-1} \text{ave}(\mathcal{L}_i) 2^{|A_{i+1}^c|} = 0 \quad \forall j = 1, \ldots, n-1, \end{cases}$$

which, given the expression (13.45) for $\text{ave}(\mathcal{L}_i)$, reads as

$$\begin{cases} \sum_{A\subsetneq\Theta} m(A)\langle Bel_A, Bel_{A_j}\rangle - \sum_{i=j}^{n-1} \sum_{\substack{A\supseteq A_i \\ A\not\supset A_{i+1}}} Bel(A) = 0 \quad \forall j = 1, \ldots, n-1. \end{cases} \tag{13.78}$$

Let us study the second addend of each equation above. We get

$$\sum_{i=j}^{n-1} \sum_{A \supseteq A_i, A \not\supset \{x_{i+1}\}} Bel(A) = \sum_{A_j \subseteq A \subsetneq \Theta} Bel(A),$$

as any $A \supseteq A_j$, $A \neq \Theta$ is such that $A \supseteq A_i$ and $A \not\supset A_{i+1}$ for some A_i in the desired maximal chain which contains A_j. Let us define x_{i+1} as the lowest-index element (according to the ordering associated with the desired focal chain $A_1 \subset \cdots \subset A_n$, i.e., $x_j \doteq A_j \setminus A_{j-1}$) among those singletons in A^c. By construction, $A \supseteq A_i$ and $A \not\supset \{x_{i+1}\}$.

Finally,

$$\begin{aligned}\sum_{A_j \subseteq A \subsetneq \Theta} Bel(A) &= \sum_{A_j \subseteq A \subsetneq \Theta} \sum_{C \subseteq A} m(C) \\ &= \sum_{C \subsetneq \Theta} m(C) \Big| \{A : C \subseteq A \subsetneq \Theta, A \supseteq A_j\} \Big|,\end{aligned}$$

where

$$\begin{aligned}\big|\{A : C \subseteq A \subsetneq \Theta, A \supseteq A_j\}\big| &= \big|\{A : A \supseteq (C \cup A_j), A \neq \Theta\}\big| \\ &= 2^{|(C \cup A_j)^c|} - 1 = \langle Bel_C, Bel_{A_j} \rangle.\end{aligned}$$

Therefore, summarising,

$$\sum_{i=j}^{n-1} \sum_{A \supseteq A_i, A \not\supset \{x_{i+1}\}} Bel(A) = \sum_{C \subsetneq \Theta} m(C) \langle Bel_C, Bel_{A_j} \rangle.$$

By substituting the latter into (13.78), we obtain the trivial identity $0 = 0$.

Proof of Theorem 81

Given the expression (13.37) for the difference vector of interest in the belief space, we can compute the explicit form of its L_∞ norm as

$$\begin{aligned}&\|Bel - Co\|_\infty \\ &= \max\left\{ \max_i \max_{\substack{A \supseteq A_i \\ A \not\supset A_{i+1}}} \left| \gamma(A_i) + Bel(A) - \sum_{j=1}^{i} m(A_j) \right|, \max_{A \not\supset A_1} \left| \sum_{B \subseteq A} m(B) \right| \right\} \\ &= \max\left\{ \max_i \max_{A \supseteq A_i, A \not\supset A_{i+1}} \left| \gamma(A_i) + Bel(A) - \sum_{j=1}^{i} m(A_j) \right|, Bel(A_1^c) \right\},\end{aligned} \tag{13.79}$$

as $\max_{A \not\supset A_1} \left| \sum_{B \subseteq A} m(B) \right| = Bel(A_1^c)$. Now, (13.79) can be minimised separately for each $i = 1, \ldots, n-1$. Clearly, the minimum is attained when the variable elements in (13.79) are not greater than the constant element $Bel(A_1^c)$:

$$\max_{A \supseteq A_i, A \not\supset A_{i+1}} \left| \gamma(A_i) + Bel(A) - \sum_{j=1}^{i} m(A_j) \right| \leq Bel(A_1^c). \tag{13.80}$$

The left-hand side of (13.80) is a function of the form $\max \left\{ |x + x_1|, \ldots, |x + x_n| \right\}$ (see Fig. 13.11, right). Such functions are minimised by $x = -\frac{x_{min} + x_{max}}{2}$. In the case of (13.80), the minimum and maximum offset values are, respectively,

$$\gamma^i_{\mathrm{min}} = Bel(A_i) - \sum_{j=1}^{i} m(A_j),$$

$$\gamma^i_{\mathrm{max}} = Bel(\{x_{i+1}\}^c) - \sum_{j=1}^{i} m(A_j) = Bel(A_i + A_{i+1}^c) - \sum_{j=1}^{i} m(A_j),$$

once $\{x_{i+1}\} = A_{i+1} \setminus A_i$ is introduced. As, for each value of γ, $|\gamma(A_i) + \gamma|$ is dominated by either $|\gamma(A_i) + \gamma^i_{\mathrm{min}}|$ or $|\gamma(A_i) + \gamma^i_{\mathrm{max}}|$, the norm of the difference vector is minimised by the values of $\gamma(A_i)$ such that

$$\max \left\{ |\gamma(A_i) + \gamma^i_{\mathrm{min}}|, |\gamma(A_i) + \gamma^i_{\mathrm{max}}| \right\} \leq Bel(A_1^c) \quad \forall i = 1, \ldots, n-1,$$

i.e.,

$$-\frac{\gamma^i_{\mathrm{min}} + \gamma^i_{\mathrm{max}}}{2} - Bel(A_1^c) \leq \gamma(A_i) \leq -\frac{\gamma^i_{\mathrm{min}} + \gamma^i_{\mathrm{max}}}{2} + Bel(A_1^c) \; \forall i = 1, \ldots, n-1.$$

In terms of mass assignments, this is equivalent to

$$\begin{aligned} -Bel(A_1^c) + \frac{Bel(A_i) + Bel(\{x_{i+1}\}^c)}{2} &\leq \sum_{j=1}^{i} m_{Co}(A_i) \\ &\leq Bel(A_1^c) + \frac{Bel(A_i) + Bel(\{x_{i+1}\}^c)}{2}. \end{aligned} \tag{13.81}$$

Once again, this is a set of constraints of the form $l_1 \leq x \leq u_1$, $l_2 \leq x + y \leq u_2$, $l_3 \leq x+y+z \leq u_3, \ldots$, which can also be expressed as $l_1 \leq x \leq u_1$, $l_2 - x \leq y \leq u_2 - x$, $l_3 - (x+y) \leq z \leq u_3 - (x+y), \ldots$. This is a polytope with vertices obtained by assigning to x, $x+y$, $x+y+z$ etc. either their lower or their upper bound. This generates 2^{n-1} possible combinations, which for the specific set (13.81) yields (see the proof of Theorem 78) (13.46).

For the barycentre of (13.46), we have that

$$
\begin{aligned}
m_{Co}(A_1) &= \frac{Bel(A_1) + Bel(\{x_2\}^c)}{2} \\
m_{Co}(A_i) &= \frac{Bel(A_i) + Bel(\{x_{i+1}\}^c)}{2} - \frac{Bel(A_{i-1}) + Bel(\{x_i\}^c)}{2} \\
&= \frac{Bel(A_i) - Bel(A_{i-1})}{2} + \frac{Pl(\{x_i\}) - Pl(\{x_{i+1}\})}{2},
\end{aligned}
$$

$$
\begin{aligned}
m_{Co}(A_n) = 1 - \sum_{i=2}^{n-1} \left[\frac{Bel(A_i) + Bel(\{x_{i+1}\}^c)}{2} - \frac{Bel(A_{i-1}) + Bel(\{x_i\}^c)}{2} \right] \\
- \frac{Bel(A_1) + Bel(\{x_2\}^c)}{2} \\
= 1 - Bel(A_{n-1}).
\end{aligned}
$$

14

Consistent approximation

As we know, belief functions are complex objects, in which different and sometimes contradictory bodies of evidence may coexist, as they describe mathematically the fusion of possibly conflicting expert opinions and/or imprecise/corrupted measurements, among other things. As a consequence, making decisions based on such objects can be misleading. As we discussed in the second part of Chapter 10, this is a well-known problem in classical logic, where the application of inference rules to inconsistent knowledge bases (sets of propositions) may lead to incompatible conclusions [1389]. We have also seen that belief functions can be interpreted as generalisations of knowledge bases in which a belief value, rather than a truth value, is attributed to each formula (interpreted as the set of worlds for which that formula is true). Finally, we have identified consistent belief functions, belief functions whose focal elements have non-empty intersection, as the natural counterparts of consistent knowledge bases in belief theory.

Analogously to consistent knowledge bases, consistent belief functions are characterised by null internal conflict. It may therefore be desirable to transform a generic belief function to a consistent one prior to making a decision or choosing a course of action. This is all the more valuable as several important operators used to update or elicit evidence represented as belief measures, such as the Dempster sum [416] and disjunctive combination [1689] (see Section 4.3), do not preserve consistency. We have seen in this book how the transformation problem is spelled out in the probabilistic [382, 333] (Chapters 11 and 12) and possibilistic [533] (Chapter 13) cases. As we have argued for probability transforms, consistent transformations can be defined as the solutions to a minimisation problem of the form

$$Cs[Bel] = \arg\min_{Cs \in \mathcal{CS}} dist(Bel, Cs), \tag{14.1}$$

F. Cuzzolin, *The Geometry of Uncertainty*, Artificial Intelligence: Foundations, Theory, and Algorithms, https://doi.org/10.1007/978-3-030-63153-6_14

where Bel is the original belief function, $dist$ is an appropriate distance measure in the belief space, and $\mathcal{CS}$ denotes the collection of all consistent belief functions. We call (14.1) the *consistent transformation problem*. Once again, by plugging different distance functions into (14.1), we get distinct consistent transformations. We refer to Section 4.10.1 for a review of dissimilarity measures for belief functions.

As we did for consonant belief functions in Chapter 13, in this chapter we focus on what happens when we apply classical L_p (Minkowski) norms to the consistent approximation problem. The L_∞ norm, in particular, appears to be closely related to consistent belief functions, as the region of consistent BFs can be expressed as

$$\mathcal{CS} = \left\{ Bel : \max_{x \in \Theta} pl(x) = 1 \right\},$$

i.e., the set of belief functions for which the L_∞ norm of the contour function $pl(x)$ is equal to 1. In addition, consistent belief functions relate to possibility distributions, and possibility measures Pos are inherently associated with L_∞, as $Pos(A) = \max_{x \in A} Pos(x)$.

Chapter content

In this chapter, therefore, we address the L_p consistent transformation problem in full generality, and discuss the semantics of the results. Since (Chapter 10) consistent belief functions live in a simplicial complex $\mathcal{CS}$ [551, 338], a partial solution has to be found separately for each maximal simplex $\mathcal{CS}^{\{A \ni x\}}$ of the consistent complex, each associated with an ultrafilter $\{A \ni x\}$, $x \in \Theta$ of focal elements. Global solutions are identified as the partial solutions at minimum distance from the original belief function. We conduct our analysis in both the mass space $\mathcal{M}$ of basic probability vectors and the belief space $\mathcal{B}$ of vectors of belief values.

In the mass space representation, the partial L_1 consistent approximation focused on a certain element x of the frame is simply obtained by reassigning all the mass outside the filter $\{A \supseteq \{x\}\}$ to Θ. Global L_1 approximations are associated, as expected, with conjunctive cores (10.6) containing the maximum-plausibility element(s) of Θ. L_∞ approximation generates a 'rectangle' of partial approximations, with their barycentre in the L_1 partial approximation. The corresponding global approximations span the components focused on the element(s) x such that $\max_{B \not\ni x} m(B)$ is minimum. The L_2 partial approximation coincides with the L_1 one if the mass vectors include $m(\Theta)$ as a component. Otherwise, the L_2 partial approximation reassigns the mass $Bel(\{x\}^c)$ outside the desired filter to each element of the filter focused on x on an equal basis.

In the belief space representation, the partial approximations determined by both the L_1 and the L_2 norms are unique and coincide, besides having a rather elegant interpretation in terms of classical inner approximations [533, 91]. The L_1/L_2 consistent approximation process generates, for each component $\mathcal{CS}^{\{A \ni x\}}$ of $\mathcal{CS}$, the *consistent transformation focused on* x, i.e., a new belief function whose focal elements have the form $A' = A \cup \{x\}$, whenever A is a focal element of the original BF

Bel. The associated global L_1/L_2 solutions do not lie, in general, on the component of the consistent complex related to the maximum-plausibility element.

Finally, the L_∞ norm determines an entire polytope of solutions whose barycentre lies in the L_1/L_2 approximation, and which is natural to associate with the polytope of inner Bayesian approximations. Global optimal L_∞ approximations focus on the maximum-plausibility element, and their centre of mass is the consistent transformation focused on it.

Chapter outline

We briefly recall in Section 14.1 how to solve the transformation problem separately for each maximal simplex of the consistent complex. We then proceed to solve the L_1, L_2 and L_∞ consistent approximation problems in full generality, in both the mass (Section 14.2) and the belief (Section 14.3) space representations. In Section 14.4, we compare and interpret the outcomes of L_p approximations in the two frameworks, with the help of the ternary example.

14.1 The Minkowski consistent approximation problem

As we have seen in Chapter 10, the region $\mathcal{CS}$ of consistent belief functions in the belief space is the union

$$\mathcal{CS}_\mathcal{B} = \bigcup_{x\in\Theta} \mathcal{CS}_\mathcal{B}^{\{A\ni x\}} = \bigcup_{x\in\Theta} Cl(Bel_A, A \ni x)$$

of a number of (maximal) simplices, each associated with a maximal ultrafilter $\{A \supseteq \{x\}\}$, $x \in \Theta$ of subsets of Θ (those containing a given element x). It is not difficult to see that the same holds in the mass space, where the consistent complex is the union

$$\mathcal{CS}_\mathcal{M} = \bigcup_{x\in\Theta} \mathcal{CS}_\mathcal{M}^{\{A\ni x\}} = \bigcup_{x\in\Theta} Cl(m_A, A \ni x)$$

of maximal simplices $Cl(m_A, A \ni x)$ formed by the mass vectors associated with all the belief functions with a conjunctive core containing a particular element x of Θ.

As in the consonant case (see Section 13.2.2), solving (14.1), in particular in the case of Minkowski (L_p) norms, involves finding a number of partial solutions

$$\begin{aligned} Cs^x_{L_p}[Bel] &= \arg\min_{Cs\in\mathcal{CS}_\mathcal{B}^{\{A\ni x\}}} \|Bel - Cs\|_{L_p}, \\ Cs^x_{L_p}[m] &= \arg\min_{m_{Cs}\in\mathcal{CS}_\mathcal{M}^{\{A\ni x\}}} \|m - m_{Cs}\|_{L_p} \end{aligned} \tag{14.2}$$

in the belief and mass spaces, respectively. Then, the distance of Bel from all such partial solutions needs to be assessed to select the global, optimal approximation(s).

14.2 Consistent approximation in the mass space

Let us therefore compute the analytical form of all L_p consistent approximations in the mass space. We start by describing the difference vector $m - m_{Cs}$ between the original mass vector and its approximation. Using the notation

$$m_{Cs} = \sum_{B \supseteq \{x\}, B \neq \Theta} m_{Cs}(B) m_B, \quad m = \sum_{B \subsetneq \Theta} m(B) m_B$$

(as in $\mathbb{R}^{N-2}$ $m(\Theta)$ is not included, by normalisation), the difference vector can be expressed as

$$m - m_{Cs} = \sum_{B \supseteq \{x\}, B \neq \Theta} (m(B) - m_{Cs}(B)) m_B + \sum_{B \not\supseteq \{x\}} m(B) m_B. \quad (14.3)$$

Its L_p norms are then given by the following expressions:

$$\|m - m_{Cs}\|_{L_1}^{\mathcal{M}} = \sum_{B \supseteq \{x\}, B \neq \Theta} |m(B) - m_{Cs}(B)| + \sum_{B \not\supseteq \{x\}} |m(B)|,$$

$$\|m - m_{Cs}\|_{L_2}^{\mathcal{M}} = \sqrt{\sum_{B \supseteq \{x\}, B \neq \Theta} |m(B) - m_{Cs}(B)|^2 + \sum_{B \not\supseteq \{x\}} |m(B)|^2},$$

$$\|m - m_{Cs}\|_{L_\infty}^{\mathcal{M}} = \max \left\{ \max_{B \supseteq \{x\}, B \neq \Theta} |m(B) - m_{Cs}(B)|, \max_{B \not\supseteq \{x\}} |m(B)| \right\}. \quad (14.4)$$

14.2.1 L_1 approximation

Let us first tackle the L_1 case. After introducing the auxiliary variables $\beta(B) \doteq m(B) - m_{Cs}(B)$, we can write the L_1 norm of the difference vector as

$$\|m - m_{Cs}\|_{L_1}^{\mathcal{M}} = \sum_{B \supseteq \{x\}, B \neq \Theta} |\beta(B)| + \sum_{B \not\supseteq \{x\}} |m(B)|, \quad (14.5)$$

which is obviously minimised by $\beta(B) = 0$, for all $B \supseteq \{x\}$, $B \neq \Theta$. We thus have the following theorem.

Theorem 83. *Given an arbitrary belief function $Bel : 2^\Theta \to [0,1]$ and an element $x \in \Theta$ of its frame of discernment, its unique L_1 consonant approximation $Cs^x_{L_1,\mathcal{M}}[m]$ in $\mathcal{M}$ with a conjunctive core containing x is the consonant belief function whose mass distribution coincides with that of Bel on all the subsets containing x:*

$$m_{Cs^x_{L_1,\mathcal{M}}[m]}(B) = \begin{cases} m(B) & \forall B \supseteq \{x\}, B \neq \Theta, \\ m(\Theta) + Bel(\{x\}^c) & B = \Theta. \end{cases} \quad (14.6)$$

The mass of all the subsets not in the desired principal ultrafilter $\{B \supseteq \{x\}\}$ is simply reassigned to Θ.

Global approximation The global L_1 consistent approximation in $\mathcal{M}$ coincides with the partial approximation (14.6) at the minimum distance from the original mass vector m. By (14.5), the partial approximation focused on x has distance $Bel(\{x\}^c) = \sum_{B \not\supset \{x\}} m(B)$ from m. The global L_1 approximation $m_{Cs_{L_1,\mathcal{M}}[m]}$ is therefore the (union of the) partial approximation(s) associated with the maximum-plausibility singleton(s),

$$\hat{x} = \arg\min_x Bel(\{x\}^c) = \arg\max_x Pl(x).$$

14.2.2 L_∞ approximation

In the L_∞ case the desired norm of the difference vector is

$$\|m - m_{Cs}\|_{L_\infty}^{\mathcal{M}} = \max \left\{ \max_{B \supseteq \{x\}, B \neq \Theta} |\beta(B)|, \max_{B \not\supset \{x\}} m(B) \right\}.$$

The above norm is trivially minimised by $\{\beta(B)\}$ such that

$$|\beta(B)| \leq \max_{B \not\supset \{x\}} m(B) \quad \forall B \supseteq \{x\}, B \neq \Theta,$$

namely

$$-\max_{B \not\supset \{x\}} m(B) \leq m(B) - m_{Cs}(B) \leq \max_{B \not\supset \{x\}} m(B) \quad \forall B \supseteq \{x\}, B \neq \Theta.$$

Theorem 84. *Given an arbitrary belief function $Bel : 2^\Theta \to [0,1]$ and an element $x \in \Theta$ of its frame of discernment, its L_∞ consistent approximations in $\mathcal{M}$ with a conjunctive core containing x, $Cs^x_{L_\infty,\mathcal{M}}[m]$, are those whose mass values on all the subsets containing x differ from the original ones by the maximum mass of the subsets not in the ultrafilter. Namely, for all $B \supset \{x\}$, $B \neq \Theta$,*

$$m(B) - \max_{C \not\supset \{x\}} m(C) \leq m_{Cs^x_{L_\infty,\mathcal{M}}[m]}(B) \leq m(B) + \max_{C \not\supset \{x\}} m(C). \tag{14.7}$$

Clearly this set of solutions can also include pseudo-belief functions.

Global approximation Once again, the global L_∞ consistent approximation in $\mathcal{M}$ coincides with the partial approximation (14.7) at minimum distance from the original BPA m. The partial approximation focused on x has distance $\max_{B \not\supset \{x\}} m(B)$ from m. The global L_∞ approximation $m_{Cs_{L_\infty,\mathcal{M}}[m]}$ is therefore the (union of the) partial approximation(s) associated with the singleton(s) such that

$$\hat{x} = \arg\min_x \max_{B \not\supset \{x\}} m(B).$$

14.2.3 L_2 approximation

To find the L_2 partial consistent approximation(s) in $\mathcal{M}$, we resort as usual to enforcing the condition that the difference vector is orthogonal to all the generators of the linear space involved. In the consistent case, we need to impose the orthogonality of the difference vector $m - m_{Cs}$ with respect to the subspace $\mathcal{CS}^{\{A \ni x\}}$ associated with consistent mass functions focused on $\{x\}$.

The generators of (the linear space spanned by) $\mathcal{CS}^{\{A \ni x\}}$ are the vectors $m_B - m_x$, for all $B \supsetneq \{x\}$. The desired orthogonality condition therefore reads as

$$\langle m - m_{Cs}, m_B - m_x \rangle = 0,$$

where $m - m_{Cs}$ is given by (14.3), while $m_B - m_x(C) = 1$ if $C = B$, $= -1$ if $C = \{x\}$, and 0 elsewhere. Therefore, using once again the variables $\{\beta(B)\}$, the condition simplifies as follows:

$$\langle m - m_{Cs}, m_B - m_x \rangle = \begin{cases} \beta(B) - \beta(\{x\}) = 0 \; \forall B \supsetneq \{x\}, B \neq \Theta, \\ -\beta(x) = 0 \qquad\qquad\qquad B = \Theta. \end{cases} \tag{14.8}$$

Notice that, when we use vectors m of $\mathbb{R}^{N-1}$ (including $B = \Theta$; compare (13.10))

$$m = \sum_{\emptyset \subsetneq B \subseteq \Theta} m(B) m_B \tag{14.9}$$

to represent belief functions, the orthogonality condition reads instead as

$$\langle m - m_{Cs}, m_B - m_x \rangle = \beta(B) - \beta(\{x\}) = 0 \quad \forall\, B \supsetneq \{x\}. \tag{14.10}$$

Theorem 85. *Given an arbitrary belief function $Bel : 2^\Theta \to [0, 1]$ and an element $x \in \Theta$ of its frame of discernment, its unique L_2 partial consistent approximation in $\mathcal{M}$ with a conjunctive core containing x, $Cs^x_{L_2,\mathcal{M}}[m]$, coincides with its partial L_1 approximation $Cs^x_{L_1,\mathcal{M}}[m]$.*

However, when the mass representation (14.9) in $\mathbb{R}^{N-1}$ is used, the partial L_2 approximation is obtained by redistributing equally to each element of the ultrafilter $\{B \supseteq \{x\}\}$ an equal fraction of the mass of focal elements not in it:

$$m_{Cs^x_{L_2,\mathcal{M}}[m]}(B) = m(B) + \frac{Bel(\{x\}^c)}{2^{|\Theta|-1}} \quad \forall B \supseteq \{x\}. \tag{14.11}$$

The partial L_2 approximation in $\mathbb{R}^{N-1}$ redistributes the mass equally to all the elements of the ultrafilter.

Global approximation The global L_2 consistent approximation in $\mathcal{M}$ is again given by the partial approximation (14.11) at the minimum L_2 distance from m. In the $N - 2$ representation, by the definition of the L_2 norm in $\mathcal{M}$ (14.4), the partial approximation focused on x has distance from m

$$(Bel(\{x\}^c))^2 + \sum_{B \not\supset \{x\}} (m(B))^2 = \left(\sum_{B \not\supset \{x\}} m(B) \right)^2 + \sum_{B \not\supset \{x\}} (m(B))^2,$$

which is minimised by the element(s) $\hat{x} \in \Theta$ such that

$$\hat{x} = \arg\min_x \sum_{B \not\supset \{x\}} (m(B))^2.$$

In the $(N-1)$-dimensional representation, instead, the partial approximation focused on x has distance from m

$$\sum_{B \supseteq \{x\}, B \neq \Theta} \left[m(B) - \left(m(B) + \frac{Bel(\{x\}^c)}{2^{|\Theta|-1}} \right) \right]^2 + \sum_{B \not\supset \{x\}} \Big(m(B) \Big)^2$$

$$= \frac{\left(\sum_{B \not\supset \{x\}} m(B) \right)^2}{2^{|\Theta|-1}} + \sum_{B \not\supset \{x\}} (m(B))^2,$$

which is minimised by the same singleton(s). Note that, although (in the $N-2$ representation) the partial L_1 and L_2 approximations coincide, the global approximations in general may fall on different components of the consonant complex.

14.3 Consistent approximation in the belief space

14.3.1 L_1/L_2 approximations

We have seen that in the mass space (at least in its $N-2$ representation, Theorem 85), the L_1 and L_2 approximations coincide. This is true in the belief space in the general case as well. We will gather some intuition about the general solution by considering first the slightly more complex case of a ternary frame, $\Theta = \{x, y, z\}$.

We will use the notation

$$Cs = \sum_{B \supseteq \{x\}} m_{Cs}(B) Bel_B, \quad Bel = \sum_{B \subsetneq \Theta} m(B) Bel_B.$$

Linear system for the L_2 case A consistent belief function $Cs \in \mathcal{CS}^{\{A \ni x\}}$ is a solution of the L_2 approximation problem if $Bel - Cs$ is orthogonal to all the generators $\{Bel_B - Bel_\Theta = Bel_B, \{x\} \subseteq B \subsetneq \Theta\}$ of $\mathcal{CS}^{\{A \ni x\}}$:

$$\langle Bel - Cs, Bel_B \rangle = 0 \quad \forall B : \{x\} \subseteq B \subsetneq \Theta.$$

As

$$Bel - Cs = \sum_{A \subsetneq \Theta} \big(m(A) - m_{Cs}(A) \big) Bel_A = \sum_{A \subsetneq \Theta} \beta(A) Bel_A,$$

the condition becomes, $\forall B : \{x\} \subseteq B \subsetneq \Theta$,

$$\left\{ \sum_{A \supseteq \{x\}} \beta(A) \langle Bel_A, Bel_B \rangle + \sum_{A \not\supset \{x\}} m(A) \langle Bel_A, Bel_B \rangle = 0. \right. \qquad (14.12)$$

Linear system for the L_1 case In the L_1 case, the minimisation problem to be solved is

$$\arg\min_{\alpha} \left\{ \sum_{A \supseteq \{x\}} \left| \sum_{B \subseteq A} m(B) - \sum_{B \subseteq A, B \supseteq \{x\}} m_{Cs}(B) \right| \right\}$$

$$= \arg\min_{\beta} \left\{ \sum_{A \supseteq \{x\}} \left| \sum_{B \subseteq A, B \supseteq \{x\}} \beta(B) + \sum_{B \subseteq A, B \not\supseteq \{x\}} m(B) \right| \right\},$$

which is clearly solved by setting all addends to zero. This yields the following linear system:

$$\left\{ \sum_{B \subseteq A, B \supseteq \{x\}} \beta(B) + \sum_{B \subseteq A, B \not\supseteq \{x\}} m(B) = 0 \quad \forall A : \{x\} \subseteq A \subsetneq \Theta. \right. \tag{14.13}$$

Example 44: linear transformation in the ternary case. *An interesting fact emerges when we compare the linear systems for L_1 and L_2 in the ternary case $\Theta = \{x, y, x\}$:*

$$\begin{cases} 3\beta(x) + \beta(x,y) + \beta(x,z) \\ \qquad +m(y) + m(z) = 0, \\ \beta(x) + \beta(x,y) + m(y) = 0, \\ \beta(x) + \beta(x,z) + m(z) = 0, \end{cases} \qquad \begin{cases} \beta(x) = 0, \\ \beta(x) + \beta(x,y) + m(y) = 0, \\ \beta(x) + \beta(x,z) + m(z) = 0. \end{cases} \tag{14.14}$$

The solution is the same for both, as the second linear system can be obtained from the first by a simple linear transformation of rows (we just need to replace the first equation, e_1, of the first system with the difference $e_1 - e_2 - e_3$).

Linear transformation in the general case This holds in the general case as well.

Lemma 16.

$$\sum_{B \supseteq A} \langle Bel_B, Bel_C \rangle (-1)^{|B \setminus A|} = \begin{cases} 1 & C \subseteq A, \\ 0 & \textit{otherwise}. \end{cases}$$

Corollary 18. *The linear system (14.12) can be reduced to the system (14.13) through the following linear transformation of rows:*

$$\text{row}_A \mapsto \sum_{B \supseteq A} \text{row}_B (-1)^{|B \setminus A|}. \tag{14.15}$$

To obtain both the L_2 and the L_1 consistent approximations of Bel, it thus suffices to solve the system (14.13) associated with the L_1 norm.

Theorem 86. *The unique solution of the linear system (14.13) is*

$$\beta(A) = -m(A \setminus \{x\}) \quad \forall A : \{x\} \subseteq A \subsetneq \Theta.$$

The theorem is proved by simple substitution.

Therefore, the partial consistent approximations of Bel on the maximal simplicial component $\mathcal{CS}^{\{A \ni x\}}$ of the consistent complex have basic probability assignment

$$m_{Cs^x_{L_1}[Bel]}(A) = m_{Cs^x_{L_2}[Bel]}(A) = m(A) - \beta(A) = m(A) + m(A \setminus \{x\})$$

for all events A such that $\{x\} \subseteq A \subsetneq \Theta$. The value of $m_{Cs}(\Theta)$ can be obtained by normalisation, as follows:

$$\begin{aligned} m_{Cs}(\Theta) &= 1 - \sum_{\{x\} \subseteq A \subsetneq \Theta} m_{Cs}(A) = 1 - \sum_{\{x\} \subseteq A \subsetneq \Theta} (m(A) + m(A \setminus \{x\})) \\ &= 1 - \sum_{\{x\} \subseteq A \subsetneq \Theta} m(A) - \sum_{\{x\} \subseteq A \subsetneq \Theta} m(A \setminus \{x\}) \\ &= 1 - \sum_{A \neq \Theta, \{x\}^c} m(A) = m(\{x\}^c) + m(\Theta), \end{aligned}$$

as all events $B \not\supset \{x\}$ can be written as $B = A \setminus \{x\}$ for $A = B \cup \{x\}$.

Corollary 19.

$$m_{Cs^x_{L_1}[Bel]}(A) = m_{Cs^x_{L_2}[Bel]}(A) = m(A) + m(A \setminus \{x\})$$

$\forall x \in \Theta$, and for all A such that $\{x\} \subseteq A \subseteq \Theta$.

Interpretation as focused consistent transformations The expression for the basic probability assignment of the L_1/L_2 consistent approximations of Bel (Corollary 19) is simple and elegant. It also has a straightforward interpretation: to get a consistent belief function focused on a singleton x, the mass contribution of all events B such that $B \cup \{x\} = A$ is assigned to A. But there are just two such events: A itself, and $A \setminus \{x\}$.

Example 45: focused consistent transformations. *The partial consistent approximation with conjunctive core $\{x\}$ of an example belief function defined on a frame $\Theta = \{x, y, z, w\}$ is illustrated in Fig. 14.1. The BF with focal elements $\{y\}$, $\{y, z\}$ and $\{x, z, w\}$ is transformed by the mapping*

$$\begin{array}{ll} \{y\} & \mapsto \{x\} \cup \{y\} = \{x, y\}, \\ \{y, z\} & \mapsto \{x\} \cup \{y, z\} = \{x, y, z\}, \\ \{x, z, w\} & \mapsto \{x\} \cup \{x, z, w\} = \{x, z, w\} \end{array}$$

into a consistent belief function with focal elements $\{x, y\}$, $\{x, y, z\}$ and $\{x, z, w\}$ and the same basic probability assignment.

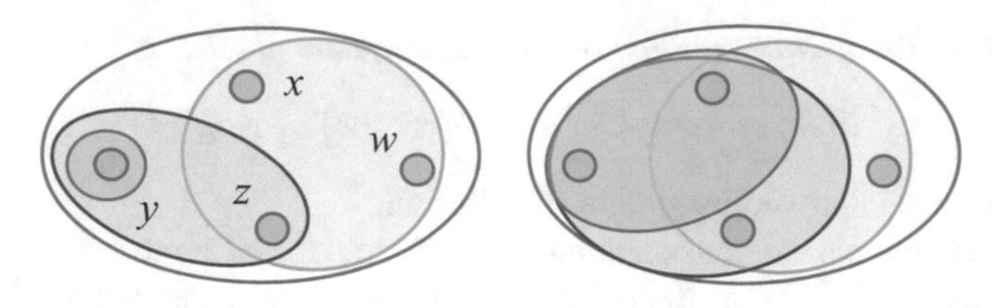

Fig. 14.1: A belief function on $\Theta = \{x, y, z, w\}$ (left) and its partial L_1/L_2 consistent approximation in $\mathcal{B}$ with core $\{x\}$ (right). Focal elements before and after the mapping share the same colour.

The partial solutions of the L_1/L_2 consistent approximation problem turn out to be related to the classical inner consonant approximations of a belief function Bel, i.e., the set of consonant belief functions Co such that $Co(A) \geq Bel(A)$ for all $A \subseteq \Theta$. Dubois and Prade [533] proved that such approximations exist iff Bel is consistent. When Bel is *not* consistent, a *focused consistent transformation* can be applied to get a new belief function Bel' such that

$$m'(A \cup \{x_i\}) = m(A) \quad \forall A \subseteq \Theta,$$

and x_i is the element of Θ with the highest plausibility.

Corollary 19 then states that the L_1/L_2 consistent approximation on each component $\mathcal{CS}^{\{A \ni x\}}$ of the consistent simplex $\mathcal{CS}$ is simply the consistent transformation focused on x in the sense specified above.

Global optimal solution for the L_1 case To find the global consistent approximation of Bel, we need to seek the partial approximation(s) $Cs^x_{L_{1/2}}[Bel]$ at minimum distance from Bel.

In the L_1 case, we obtain the following result.

Theorem 87. *The global L_1 consistent approximation of a belief function Bel is*

$$Cs_{L_1}[Bel] \doteq \arg \min_{Cs \in \mathcal{CS}} \|Bel - Cs^x_{L_1}[Bel]\| = cs^{\hat{x}}_{L_1}[Bel],$$

i.e., the partial approximation associated with the singleton element

$$\hat{x} = \arg\min_{x \in \Theta} \sum_{A \subseteq \{x\}^c} Bel(A). \tag{14.16}$$

In the binary case $\Theta = \{x, y\}$, the condition of Theorem 87 reduces to

$$\hat{x} = \arg\min_{x} \sum_{A \subseteq \{x\}^c} Bel(A) = \arg\min_{x} m(\{x\}^c) = \arg\max_{x} pl(x),$$

and the global approximation falls on the component of the consistent complex associated with the element of maximum plausibility. Unfortunately, this does not necessarily hold in the case of an arbitrary frame Θ:

$$\arg\min_{x\in\Theta} \sum_{A\subseteq\{x\}^c} Bel(A) \neq \arg\max_{x\in\Theta} pl(x),$$

as a simple counter-example can prove.

Global optimal solution for the L_2 case

Theorem 88. *The global L_2 consistent approximation of a belief function $Bel : 2^\Theta \rightarrow [0,1]$ is*

$$Cs_{L_2}[Bel] \doteq \arg\min_{Cs\in\mathcal{CS}} \|Bel - Cs^x_{L_2}[Bel]\| = Cs^{\hat{x}}_{L_2}[Bel],$$

i.e., the partial approximation associated with the singleton element

$$\hat{x} = \arg\min_{x\in\Theta} \sum_{A\subseteq\{x\}^c} \big(Bel(A)\big)^2.$$

Once again, in the binary case the condition of Theorem 88 reads as

$$\hat{x} = \arg\min_{x} \sum_{A\subseteq\{x\}^c} (Bel(A))^2 = \arg\min_{x}(m(\{x\}^c))^2 = \arg\max_{x} pl(x),$$

and the global approximation for L_2 also falls on the component of the consistent complex associated with the element of maximum plausibility. However, this is not generally true for an arbitrary frame.

14.3.2 L_∞ consistent approximation

As observed in the binary case, for each component $\mathcal{CS}^{\{A\ni x\}}$ of the consistent complex the set of partial L_∞ approximations form a polytope whose centre of mass is equal to the partial L_1/L_2 approximation.

Theorem 89. *Given an arbitrary belief function $Bel : 2^\Theta \rightarrow [0,1]$ and an element $x \in \Theta$ of its frame of discernment, its L_∞ partial consistent approximation with a conjunctive core containing x in the belief space, $CS^x_{L_\infty,\mathcal{B}}[m]$, is determined by the following system of constraints:*

$$-Bel(\{x\}^c) - \sum_{B\subseteq A, B\not\supseteq\{x\}} m(B) \leq \gamma(A) \leq Bel(\{x\}^c) - \sum_{B\subseteq A, B\not\supseteq\{x\}} m(B), \tag{14.17}$$

where

$$\gamma(A) \doteq \sum_{B\subseteq A, B\supseteq\{x\}} \beta(B) = \sum_{B\subseteq A, B\supseteq\{x\}} (m(B) - m_{Cs}(B)). \tag{14.18}$$

This defines a high-dimensional 'rectangle' in the space of solutions $\{\gamma(A), \{x\} \subseteq A \subsetneq \Theta\}$.

Corollary 20. *Given a belief function $Bel : 2^\Theta \rightarrow [0,1]$, its partial L_1/L_2 approximation on any given component $\mathcal{CS}_{\mathcal{B}}^{\{A \ni x\}}$ of the consistent complex $\mathcal{CS}_{\mathcal{B}}$ in the belief space is the geometric barycentre of the set of its L_∞ consistent approximations on the same component.*

As the L_∞ distance between Bel and $\mathcal{CS}^{\{A \ni x\}}$ is minimum for the singleton element(s) x which minimise $\|Bel - Cs^x_{L_\infty}\|_\infty = Bel(\{x\}^c)$, we also have the following result.

Corollary 21. *The global L_∞ consistent approximations of a belief function $Bel : 2^\Theta \rightarrow [0,1]$ in $\mathcal{B}$ form the union of the partial L_∞ approximations of Bel on the component(s) of the consistent complex associated with the maximum-plausibility element(s):*

$$CS_{L_\infty,\mathcal{B}}[m] = \bigcup_{x = \arg\max pl(x)} CS^x_{L_\infty,\mathcal{B}}[m].$$

14.4 Approximations in the belief versus the mass space

Summarising, in the mass space $\mathcal{M}$:

- the partial L_1 consistent approximation focused on a certain element x of the frame is obtained by reassigning all the mass $Bel(\{x\}^c)$ outside the filter to Θ;
- the global approximation is associated, as expected, with conjunctive cores containing the maximum-plausibility element(s) of Θ;
- the L_∞ approximation generates a polytope of partial approximations, with its barycentre in the L_1 partial approximation;
- the corresponding global approximations span the component(s) focused on the element(s) x such that $\max_{B \not\ni x} m(B)$ is minimal;
- the L_2 partial approximation coincides with the L_1 one in the $(N-2)$-dimensional mass representation;
- in the $(N-1)$-dimensional representation, the L_2 partial approximation reassigns the mass outside the desired filter ($Bel(\{x\}^c)$) to each element of the filter focused on x on an equal basis;
- global approximations in the L_2 case are more difficult to interpret.

In the belief space:

- the partial L_1 and L_2 approximations coincide on each component of the consistent complex, and are unique;
- for each $x \in \Theta$, they coincide with the consistent transformation [533] focused on x: for all events A such that $\{x\} \subseteq A \subseteq \Theta$,

$$m_{Cs^x_{L_1}[Bel]}(A) = m_{Cs^x_{L_2}[Bel]}(A) = m(A) + m(A \setminus \{x\});$$

- the L_1 global consistent approximation is associated with the singleton(s) $x \in \Theta$ such that

$$\hat{x} = \arg\min_x \sum_{A \subseteq \{x\}^c} Bel(A),$$

while the L_2 global approximation is associated with

$$\hat{x} = \arg\min_x \sum_{A \subseteq \{x\}^c} \left(Bel(A)\right)^2;$$

neither appear to have simple epistemic interpretations;
- the set of partial L_∞ solutions form a polytope on each component of the consistent complex, whose centre of mass lies on the partial L_1/L_2 approximation;
- the global L_∞ solutions fall on the component(s) associated with the maximum-plausibility element(s), and their centre of mass, when such an element is unique, is the consistent transformation focused on the maximum-plausibility singleton [533].

The approximations in both the mass and the belief space reassign the total mass $Bel(\{x\}^c)$ outside the filter focused on x, albeit in different ways. However, mass space consistent approximations do so either on an equal basis or by favouring no particular focal element in the filter (i.e., by reassigning the entire mass to Θ). They do not distinguish focal elements by virtue of their set-theoretic relationships with subsets $B \not\ni x$ outside the filter. In contrast, approximations in the belief space do so according to the focused consistent transformation principle.

Example 46: comparison in the ternary case. *It can be useful to illustrate the different approximations in the toy case of a ternary frame, $\Theta = \{x, y, z\}$, for the sake of completeness. Assuming we want the consistent approximation to focus on x, Fig. 14.2 illustrates the different partial consistent approximations in the simplex $\mathcal{CS}^{\{A \ni x\}} = Cl(m_x, m_{\{x,y\}}, m_{\{x,z\}}, m_\Theta)$ of consistent belief functions focused on x in a ternary frame, for the belief function with masses*

$$\begin{array}{ccc} m(x) = 0.2, & m(y) = 0.1, & m(z) = 0, \\ m(\{x, y\}) = 0.4, & m(\{x, z\}) = 0, & m(\{y, z\}) = 0.3. \end{array} \quad (14.19)$$

This is a tetrahedron with four vertices, shown in blue.

The set of mass space partial L_∞ approximations is represented in the figure as a purple cube. As expected, it does not entirely fall inside the tetrahedron of admissible consistent belief functions. Its barycentre (purple square) coincides with the (mass space) L_1 partial consistent approximation. The L_2, $\mathbb{R}^{N-2}$ approximation also coincides, as expected, with the L_1 approximation there. There seems to exist a strong case for the latter, as it possesses a natural interpretation in terms of mass assignment: all the mass outside the filter is reassigned to Θ, increasing the overall uncertainty of the belief state. The mass space L_2 partial approximation in the $\mathbb{R}^{N-1}$ representation (green square) is distinct from the latter, but still falls inside the polytope of L_∞ partial approximations (purple cube) and is admissible, as it falls in the interior of the simplicial component $\mathcal{CS}^{\{A \ni x\}}$. It possesses quite a strong interpretation, as it splits the mass not in the filter focused on x equally among all the subsets in the filter.

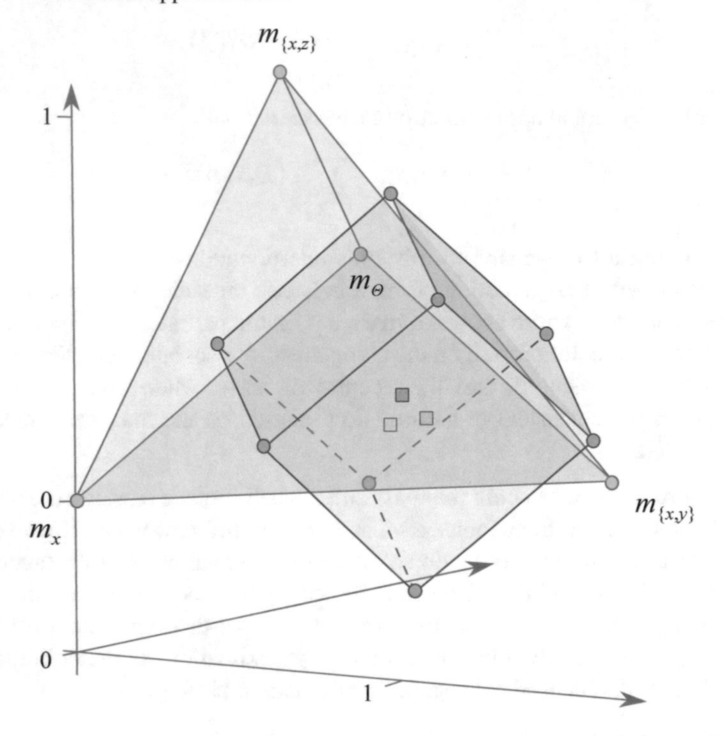

Fig. 14.2: The simplex (blue tetrahedron) $Cl(m_x, m_{x,y}, m_{x,z}, m_\Theta)$ of consistent belief functions focused on x on the ternary frame $\Theta = \{x, y, z\}$, and the associated L_p partial consistent approximations for the example belief function (14.19).

The unique L_1/L_2 partial approximation in the belief space $\mathcal{B}$ is shown as an ochre square. It has something in common with the $L_1/L_2, \mathbb{R}^{N-2}$ mass space approximation (purple square), as they both live in the same face of the simplex of consistent belief functions focused on x (highlighted in a darker shade of blue). They both assign zero mass to $\{x, z\}$, which is not supported by any focal element of the original belief function.

Appendix: Proofs

Proof of Theorem 85

In the $N-2$ representation, by (14.8) we have that $\beta(B) = 0$, i.e., $m_{Cs}(B) = m(B)$ $\forall B \supseteq \{x\}$, $B \neq \Theta$. By normalisation, we get $m_{Cs}(\Theta) = m(\Theta) + m(\{x\}^c)$: but this is exactly the L_1 approximation (14.6).

In the $N-1$ representation, by (14.10) we have that $m_{Cs}(B) = m_{Cs}(x) + m(B) - m(x)$ for all $B \supsetneq \{x\}$. By normalising, we get

$$\sum_{\{x\}\subseteq B\subseteq \Theta} m_{Cs}(B) = m_{Cs}(x) + \sum_{\{x\}\subsetneq B\subseteq \Theta} m_{Cs}(B)$$

$$= 2^{|\Theta|-1} m_{Cs}(x) + pl(x) - 2^{|\Theta|-1} m(x) = 1,$$

i.e., $m_{Cs}(x) = m(x) + (1 - pl(x))/2^{|\Theta|-1}$, as there are $2^{|\Theta|-1}$ subsets in the ultrafilter containing x. By substituting the value of $m_{Cs}(x)$ into the first equation, we get (14.11).

Proof of Corollary 18

If we apply the linear transformation (14.15) to the system (14.12), we get

$$\sum_{B\supseteq A} \left[\sum_{C\supseteq \{x\}} \beta(C)\langle Bel_B, Bel_C\rangle + \sum_{C\not\supset \{x\}} m(C)\langle Bel_B, Bel_C\rangle \right] (-1)^{|B\setminus A|}$$

$$= \sum_{C\supseteq\{x\}} \beta(C) \sum_{B\supseteq A} \langle Bel_B, Bel_C\rangle (-1)^{|B\setminus A|}$$

$$+ \sum_{C\not\supset\{x\}} m(C) \sum_{B\supseteq A} \langle Bel_B, Bel_C\rangle (-1)^{|B\setminus A|}$$

$\forall A : \{x\} \subseteq A \subsetneq \Theta$. Therefore, by Lemma 16, we obtain

$$\left\{ \sum_{C\supseteq\{x\}, C\subseteq A} \beta_C + \sum_{C\not\supset\{x\}, C\subseteq A} m(C) = 0 \quad \forall A : \{x\} \subseteq A \subsetneq \Theta, \right.$$

i.e., the system of equations (14.13).

Proof of Lemma 16

We first note that, by the definition of a categorical belief function Bel_A,

$$\langle Bel_B, Bel_C\rangle = \sum_{D\supseteq B,C; D\neq\Theta} 1 = \sum_{E\subsetneq (B\cup C)^c} 1 = 2^{|(B\cup C)^c|} - 1.$$

Therefore,

$$\begin{aligned} \sum_{B\subseteq A} \langle Bel_B, Bel_C\rangle (-1)^{|B\setminus A|} &= \sum_{B\subseteq A} (2^{|(B\cup C)^c|} - 1)(-1)^{|B\setminus A|} \\ &= \sum_{B\subseteq A} 2^{|(B\cup C)^c|}(-1)^{|B\setminus A|} - \sum_{B\subseteq A} (-1)^{|B\setminus A|} \\ &= \sum_{B\subseteq A} 2^{|(B\cup C)^c|}(-1)^{|B\setminus A|}, \end{aligned} \quad (14.20)$$

as

$$\sum_{B\subseteq A} (-1)^{|B\setminus A|} = \sum_{k=0}^{|B\setminus A|} 1^{|A^c|-k}(-1)^k = 0$$

by Newton's binomial theorem (9.4).

As both $B \supseteq A$ and $C \supseteq A$, the set B can be decomposed into the disjoint sum $B = A + B' + B''$, where $\emptyset \subseteq B' \subseteq C \setminus A$, $\emptyset \subseteq B'' \subseteq (C \cup A)^c$. The quantity (14.20) can then be written as

$$\begin{aligned} &\sum_{\emptyset\subseteq B'\subseteq C\setminus A} \sum_{\emptyset\subseteq B''\subseteq (C\cup A)^c} 2^{|(A\cup C)^c|-|B''|}(-1)^{|B'|+|B''|} \\ &= \sum_{\emptyset\subseteq B'\subseteq C\setminus A} (-1)^{|B'|} \sum_{\emptyset\subseteq B''\subseteq (C\cup A)^c} (-1)^{|B''|} 2^{|(A\cup C)^c|-|B''|}, \end{aligned}$$

where

$$\sum_{\emptyset\subseteq B''\subseteq (C\cup A)^c} (-1)^{|B''|} 2^{|(A\cup C)^c|-|B''|} = [2+(-1)]^{|(A\cup C)^c|} = 1^{|(A\cup C)^c|} = 1,$$

again by Newton's binomial theorem. We therefore obtain

$$\sum_{\emptyset\subseteq B'\subseteq C\setminus A} (-1)^{|B'|},$$

which is zero when $C \setminus A \neq \emptyset$ and equal to 1 when $C \setminus A = \emptyset$, i.e., for $C \subseteq A$.

Proof of Theorem 87

The L_1 distance between the partial approximation and Bel can be easily computed:

$$\|Bel - Cs^x_{L_1}[Bel]\|_{L_1} = \sum_{A \subseteq \Theta} |Bel(A) - Cs^x_{L_1}[Bel](A)|$$

$$= \sum_{A \not\supseteq \{x\}} |Bel(A) - 0| + \sum_{A \supseteq \{x\}} \left| Bel(A) - \sum_{B \subseteq A, B \supseteq \{x\}} m_{Cs}(B) \right|$$

$$= \sum_{A \not\supseteq \{x\}} Bel(A) + \sum_{A \supseteq \{x\}} \left| \sum_{B \subseteq A} m(B) - \sum_{B \subseteq A, B \supseteq \{x\}} \Big(m(B) + m(B \setminus \{x\}) \Big) \right|$$

$$= \sum_{A \not\supseteq \{x\}} Bel(A) + \sum_{A \supseteq \{x\}} \left| \sum_{B \subseteq A, B \not\supseteq \{x\}} m(B) - \sum_{B \subseteq A, B \supseteq \{x\}} m(B \setminus \{x\}) \right|$$

$$= \sum_{A \not\supseteq \{x\}} Bel(A) + \sum_{A \supseteq \{x\}} \left| \sum_{C \subseteq A \setminus \{x\}} m(C) - \sum_{C \subseteq A \setminus \{x\}} m(C) \right|$$

$$= \sum_{A \not\supseteq \{x\}} Bel(A) = \sum_{A \subseteq \{x\}^c} Bel(A).$$

Proof of Theorem 88

The L_2 distance between the partial approximation and Bel is

$$\|Bel - Cs^x_{L_2}[Bel]\|^2 = \sum_{A \subseteq \Theta} \Big(Bel(A) - cs^x_{L_2}[Bel](A) \Big)^2$$

$$= \sum_{A \subseteq \Theta} \left(\sum_{B \subseteq A} m(B) - \sum_{B \subseteq A, B \supseteq \{x\}} m_{Cs}(B) \right)^2$$

$$= \sum_{A \subseteq \Theta} \left(\sum_{B \subseteq A} m(B) - \sum_{B \subseteq A, B \supseteq \{x\}} m(B) - \sum_{B \subseteq A, B \supseteq \{x\}} m(B \setminus \{x\}) \right)^2$$

$$= \sum_{A \not\supseteq \{x\}} \big(Bel(A)\big)^2 + \sum_{A \supseteq \{x\}} \left(\sum_{B \subseteq A, B \not\supseteq \{x\}} m(B) - \sum_{B \subseteq A, B \supseteq \{x\}} m(B \setminus \{x\}) \right)^2$$

$$= \sum_{A \not\supseteq \{x\}} \big(Bel(A)\big)^2 + \sum_{A \supseteq \{x\}} \left(\sum_{C \subseteq A \setminus \{x\}} m(C) - \sum_{C \subseteq A \setminus \{x\}} m(C) \right)^2,$$

so that

$$\|Bel - Cs^x_{L_2}[Bel]\|^2 = \sum_{A \not\supseteq \{x\}} \big(Bel(A)\big)^2 = \sum_{A \subseteq \{x\}^c} \big(Bel(A)\big)^2.$$

Proof of Theorem 89

The desired set of approximations is, by definition,

$$CS^x_{L_\infty}[Bel] = \arg\min_{m_{Cs}(.)} \max_{A \subsetneq \Theta} \left\{ \left| \sum_{B \subseteq A} m(B) - \sum_{B \subseteq A, B \supseteq \{x\}} m_{Cs}(B) \right| \right\}.$$

The quantity $\max_{A \subsetneq \Theta}$ has as its lower bound the value associated with the largest norm which does not depend on $m_{Cs}(.)$, i.e.,

$$\max_{A \subsetneq \Theta} \left\{ \left| \sum_{B \subseteq A} m(B) - \sum_{B \subseteq A, B \supseteq \{x\}} m_{Cs}(B) \right| \right\} \geq Bel(\{x\}^c).$$

Equivalently, we can write

$$\max_{A \subsetneq \Theta} \left\{ \left| \sum_{B \subseteq A, B \supseteq \{x\}} \beta(B) + \sum_{B \subseteq A, B \not\supset \{x\}} m(B) \right| \right\} \geq Bel(\{x\}^c).$$

In the above constraint, only the expressions associated with $A \supseteq \{x\}$ contain variable terms $\{\beta(B), B\}$. Therefore, the desired optimal values are such that

$$\left\{ \left| \sum_{B \subseteq A, B \supseteq \{x\}} \beta(B) + \sum_{B \subseteq A, B \not\supset \{x\}} m(B) \right| \leq Bel(\{x\}^c), \quad \{x\} \subseteq A \subsetneq \Theta. \right. \tag{14.21}$$

After introducing the change of variables (14.18), the system (14.21) reduces to

$$\left\{ \left| \gamma(A) + \sum_{B \subseteq A, B \not\supset \{x\}} m(B) \right| \leq Bel(\{x\}^c), \quad \{x\} \subseteq A \subsetneq \Theta, \right.$$

whose solution is (14.17).

Proof of Corollary 20

The centre of mass of the set (14.17) of solutions of the L_∞ consistent approximation problem is given by

$$\gamma(A) = - \sum_{B \subseteq A, B \not\supset \{x\}} m(B), \quad \{x\} \subseteq A \subsetneq \Theta,$$

which, in the space of the variables $\{\beta(A), \{x\} \subseteq A \subsetneq \Theta\}$, reads as

$$\left\{ \sum_{B \subseteq A, B \supseteq \{x\}} \beta(B) = - \sum_{B \subseteq A, B \not\supset \{x\}} m(B), \quad \{x\} \subseteq A \subsetneq \Theta. \right.$$

But this is exactly the linear system (14.13) which determines the L_1/L_2 consistent approximation $Cs^x_{L_{1/2}}[Bel]$ of Bel on $\mathcal{CS}^{\{A \ni x\}}$.

Part IV

Geometric reasoning

15

Geometric conditioning

As we have seen in Chapter 4, Section 4.5, various distinct definitions of conditional belief functions have been proposed in the past. As Dempster's original approach [414] in his multivalued-mapping framework was almost immediately and strongly criticised, a string of subsequent proposals [1086, 245, 584, 880, 691, 439, 2080, 873, 1800] in different mathematical set-ups were made. A good overview of such approaches was given in Section 4.5.

Quite recently, the idea of formulating the problem geometrically [165, 391, 1229] has emerged. Lehrer [1128], in particular, advanced a geometric approach to determine the conditional expectation of non-additive probabilities (such as belief functions). As we appreciated extensively in Chapters 13 and 14, the use of norm minimisation for posing and solving problems such as probability and possibility transforms has been gaining traction. In a similar way, conditional belief functions can be defined by minimising a suitable distance function between the original BF Bel and the *conditioning simplex* $\mathcal{B}_A$ associated with the conditioning event A, i.e., the set of belief functions whose BPA assigns mass to subsets of A only:

$$Bel_d(.|A) \doteq \arg \min_{Bel' \in \mathcal{B}_A} d(Bel, Bel'). \tag{15.1}$$

Such a geometrical approach to conditioning provides an additional option in terms of approaches to the problem and, as we will see in this chapter, is a promising candidate for the role of a general framework for conditioning.

As in the consonant and consistent approximation problems, any dissimilarity measure [501, 902, 937, 1636, 965] could in principle be plugged into the above minimisation problem (15.1) to define conditional belief functions. In [356], the author of this book computed some conditional belief functions generated by min-

F. Cuzzolin, *The Geometry of Uncertainty*, Artificial Intelligence: Foundations, Theory, and Algorithms, https://doi.org/10.1007/978-3-030-63153-6_15

imising L_p norms in the mass space (see Chapter 13, Section 13.2.1), where belief functions are represented by the vectors of their basic probabilities.

Chapter content

In this chapter, we explore the geometric conditioning approach in both the mass space $\mathcal{M}$ and the belief space $\mathcal{B}$. We adopt, once again, distance measures d of the classical L_p family, as a first step towards a complete analysis of the geometric approach to conditioning. We show that geometric conditional belief functions in $\mathcal{B}$ are more complex than in the mass space, less naive objects whose interpretation in terms of degrees of belief is, however, less natural.

Conditioning in the mass space In summary, the L_1 conditional belief functions in $\mathcal{M}$ with conditioning event A form a polytope in which each vertex is the BF obtained by reassigning the entire mass not contained in A to a single subset of A, $B \subseteq A$. In turn, the unique L_2 conditional belief function is the barycentre of this polytope, i.e., the belief function obtained by reassigning the mass $\sum_{B \not\subset A} m(B)$ to each focal element $B \subseteq A$ on an equal basis. Such results can be interpreted as a generalisation of Lewis's imaging approach to belief revision, originally formulated in the context of probabilities [1143].

Conditioning in the belief space Conditional belief functions in the belief space seem to have rather less straightforward interpretations. The barycentre of the set of L_∞ conditional belief functions can be interpreted as follows: the mass of all the subsets whose intersection with A is $C \subsetneq A$ is reassigned by the conditioning process *half to* C, and *half to* A *itself*. While in the $\mathcal{M}$ case the barycentre of the L_1 conditional BFs is obtained by reassigning the mass of all $B \not\subset A$ to each $B \subsetneq A$ on an equal basis, for the barycentre of the L_∞ conditional belief functions in $\mathcal{B}$ normalisation is achieved by adding or subtracting their masses according to the cardinality of C (even or odd). As a result, the mass function obtained is not necessarily non-negative. Furthermore, while being quite similar to it, the L_2 conditional belief function in $\mathcal{B}$ is distinct from the barycentre of the L_∞ conditional BFs. In the L_1 case, not only are the resulting conditional pseudo-belief functions not guaranteed to be proper belief functions, but it also appears difficult to find simple interpretations of these results in terms of degrees of belief.

A number of interesting cross-relations between conditional belief functions from the two representation domains appear to exist from an empirical comparison, and require further investigation.

Chapter outline

We commence by defining in Section 15.1 the notion of a geometric conditional belief function. In Section 15.2, we pick the 'mass' representation of BFs, and prove the analytical forms of the L_1, L_2 and L_∞ conditional belief functions in $\mathcal{M}$. We discuss their interpretation in terms of degrees of belief, and hint at an interesting link with Lewis's imaging [1143], when generalised to belief functions. Section 15.3 is devoted to the derivation of geometric conditional belief functions in the belief space. In Section 15.3.1, we prove the analytical form of L_2 conditional BFs in $\mathcal{B}$, and propose a preliminary interpretation for them. We do the same in Sections 15.3.2 and 15.3.3 for L_1 and L_∞ conditional belief functions, respectively. In Section 15.4, a critical comparison of conditioning in the mass and belief spaces is done, with the help of the usual ternary case study.

Finally, in Section 15.5 a number of possible future developments of the geometric approach to conditioning are discussed, outlining a programme of research for the future.

15.1 Geometric conditional belief functions

Given an arbitrary conditioning event $A \subseteq \Theta$, the vector m_a associated with any belief function Bel_a whose mass supports only focal elements $\{\emptyset \subsetneq B \subseteq A\}$ included in a given event A can be decomposed as

$$m_a = \sum_{\emptyset \subsetneq B \subseteq A} m_a(B) m_B. \tag{15.2}$$

The set of all such vectors is the simplex

$$\mathcal{M}_A \doteq Cl(m_B, \ \emptyset \subsetneq B \subseteq A).$$

The same is true in the belief space, where (the vector associated with) each belief function Bel_a assigning mass to focal elements included in A only is decomposable as

$$Bel_a = \sum_{\emptyset \subsetneq B \subseteq A} Bel_a(B) Bel_B.$$

These vectors live in a simplex $\mathcal{B}_A \doteq Cl(Bel_B, \ \emptyset \subsetneq B \subseteq A)$. We call $\mathcal{M}_A$ and $\mathcal{B}_A$ the *conditioning simplices* in the mass and in the belief space, respectively.

Definition 132. *Given a belief function $Bel : 2^\Theta \rightarrow [0, 1]$, we define the* geometric conditional belief function induced by a distance function *d in $\mathcal{M}$ (or $\mathcal{B}$) as the belief function(s) $Bel_{d,\mathcal{M}}(.|A)$ (or $Bel_{d,\mathcal{B}}(.|A)$) on Θ which minimise(s) the distance $d(m, \mathcal{M}_A)$ (or $d(Bel, \mathcal{B}_A)$) between the mass (or belief) vector representing Bel and the conditioning simplex associated with A in $\mathcal{M}$ (or $\mathcal{B}$).*

As recalled above, a large number of proper distance functions or mere dissimilarity measures for belief functions have been proposed in the past, and many others can be conjectured or designed [937]. As we did in Chapters 13 and 14, we consider here as distance functions the three major L_p norms $d = L_1$, $d = L_2$ and $d = L_\infty$ in both the mass (13.13) and the belief (13.14) spaces.

15.2 Geometric conditional belief functions in $\mathcal{M}$

15.2.1 Conditioning by L_1 norm

Given a belief function Bel with a basic probability assignment m collected in a vector $m \in \mathcal{M}$, its L_1 conditional version(s) $Bel_{L_1,\mathcal{M}}(.|A)$ have a basic probability assignment $m_{L_1,\mathcal{M}}(.|A)$ such that

$$m_{L_1,\mathcal{M}}(.|A) \doteq \arg \min_{m_a \in \mathcal{M}_A} \|m - m_a\|_{L_1}. \tag{15.3}$$

Using the expression (13.13) for the L_1 norm in the mass space $\mathcal{M}$, (15.3) becomes

$$\arg \min_{m_a \in \mathcal{M}_A} \|m - m_a\|_{L_1} = \arg \min_{m_a \in \mathcal{M}_A} \sum_{\emptyset \subsetneq B \subseteq \Theta} |m(B) - m_a(B)|.$$

By exploiting the fact that the candidate solution m_a is an element of $\mathcal{M}_A$ (15.2), we can greatly simplify this expression.

Lemma 17. *The difference vector $m - m_a$ in $\mathcal{M}$ has the form*

$$\begin{array}{l} m - m_a = \displaystyle\sum_{\emptyset \subsetneq B \subsetneq A} \beta(B) m_B + \left(Bel(A) - 1 - \sum_{\emptyset \subsetneq B \subsetneq A} \beta(B) \right) m_A \\ \qquad + \displaystyle\sum_{B \not\subset A} m(B) m_B, \end{array} \tag{15.4}$$

where $\beta(B) \doteq m(B) - m_a(B)$.

Theorem 90. *Given a belief function $Bel : 2^\Theta \to [0,1]$ and an arbitrary non-empty focal element $\emptyset \subsetneq A \subseteq \Theta$, the set of L_1 conditional belief functions $Bel_{L_1,\mathcal{M}}(.|A)$ with respect to A in $\mathcal{M}$ is the set of BFs with a (disjunctive) core in A such that their mass dominates that of Bel over all the proper subsets of A:*

$$Bel_{L_1,\mathcal{M}}(.|A) = \left\{ Bel_a : 2^\Theta \to [0,1] : \mathcal{C}_a \subseteq A,\ m_a(B) \geq m(B)\ \forall \emptyset \subsetneq B \subseteq A \right\}. \tag{15.5}$$

Geometrically, the set of L_1 conditional belief functions in $\mathcal{M}$ has the form of a simplex.

Theorem 91. *Given a belief function $Bel : 2^\Theta \to [0,1]$ and an arbitrary non-empty focal element $\emptyset \subsetneq A \subseteq \Theta$, the set of L_1 conditional belief functions $Bel_{L_1,\mathcal{M}}(.|A)$ with respect to A in $\mathcal{M}$ is the simplex*

$$\mathcal{M}_{L_1,A}[Bel] = Cl(m[Bel]|_{L_1}^B A, \emptyset \subsetneq B \subseteq A), \tag{15.6}$$

whose vertex $m[Bel]|_{L_1}^B A$, $\emptyset \subsetneq B \subseteq A$, has coordinates $\{m_a(X), \emptyset \subsetneq X \subsetneq A\}$ such that

$$\begin{cases} m_a(B) = m(B) + 1 - Bel(A) = m(B) + Pl(A^c), \\ m_a(X) = m(X) & \forall \emptyset \subsetneq X \subsetneq A, X \neq B. \end{cases} \quad (15.7)$$

It is important to notice that all the vertices of the simplex of L_1 conditional belief functions fall inside $\mathcal{M}_A$ proper (as the mass assignment (15.7) is non-negative for all subsets X). A priori, some of them could have belonged to the linear space generated by $\mathcal{M}_A$ whilst falling outside the simplex $\mathcal{M}_A$ (i.e., some of the solutions $m_a(B)$ might have been negative). This is indeed the case for geometrical belief functions induced by other norms, as we will see in the following.

15.2.2 Conditioning by L_2 norm

Let us now compute the analytical form of the L_2 conditional belief function(s) in the mass space.

We make use of the form (15.4) of the difference vector $m - m_a$, where again m_a is an arbitrary vector of the conditioning simplex $\mathcal{M}_A$. As usual, rather than minimising the norm of the difference, we seek the point of the conditioning simplex such that the difference vector is orthogonal to all the generators of $a(\mathcal{M}_A)$ (the affine space spanned by $\mathcal{M}_A$).

Theorem 92. *Given a belief function $Bel : 2^\Theta \to [0,1]$ and an arbitrary non-empty focal element $\emptyset \subsetneq A \subseteq \Theta$, the unique L_2 conditional belief function $Bel_{L_2,\mathcal{M}}(.|A)$ with respect to A in $\mathcal{M}$ is the BF whose basic probability assignment redistributes the mass $1 - Bel(A)$ to each focal element $B \subseteq A$ in an equal way. Namely,*

$$m_{L_2,\mathcal{M}}(B|A) = m(B) + \frac{\sum_{B \not\subset A} m(B)}{2^{|A|} - 1} = m(B) + \frac{Pl(A^c)}{2^{|A|} - 1} \quad \forall \emptyset \subsetneq B \subseteq A. \quad (15.8)$$

According to (15.8), the L_2 conditional belief function is unique, and corresponds to the mass function which redistributes the mass that the original belief function assigns to focal elements not included in A to each and all the subsets of A in an equal, even way.

The L_2 and L_1 conditional belief functions in $\mathcal{M}$ exhibit a strong relationship.

Theorem 93. *Given a belief function $Bel : 2^\Theta \to [0,1]$ and an arbitrary non-empty focal element $\emptyset \subsetneq A \subseteq \Theta$, the L_2 conditional belief function $Bel_{L_2,\mathcal{M}}(.|A)$ with respect to A in $\mathcal{M}$ is the centre of mass of the simplex $\mathcal{M}_{L_1,A}[Bel]$ (15.6) of L_1 conditional belief functions with respect to A in $\mathcal{M}$.*

Proof. By definition, the centre of mass of $\mathcal{M}_{L_1,A}[Bel]$, whose vertices are given by (15.7), is the vector

$$\frac{1}{2^{|A|} - 1} \sum_{\emptyset \subsetneq B \subseteq A} m[Bel]|_{L_1}^{B} A,$$

whose entry B is given by

$$\frac{1}{2^{|A|}-1}\Big[m(B)(2^{|A|}-1)+(1-Bel(A))\Big],$$

i.e., (15.8).

□

15.2.3 Conditioning by L_∞ norm

Similarly, we can use (15.4) to minimise the L_∞ distance between the original mass vector m and the conditioning subspace $\mathcal{M}_A$.

Theorem 94. *Given a belief function $Bel : 2^\Theta \to [0,1]$ with BPA m, and an arbitrary non-empty focal element $\emptyset \subsetneq A \subseteq \Theta$, the L_∞ conditional belief functions $m_{L_\infty,\mathcal{M}}(.|A)$ with respect to A in $\mathcal{M}$ form the simplex*

$$\mathcal{M}_{L_\infty,A}[Bel] = Cl(m[Bel]|^{\bar{B}}_{L_\infty} A,\ \bar{B} \subseteq A),$$

with vertices

$$\begin{cases} m[Bel]|^{\bar{B}}_{L_\infty}(B|A) = m(B) + \max\limits_{C \not\subseteq A} m(C) & \forall B \subseteq A, B \neq \bar{B} \\ m[Bel]|^{\bar{B}}_{L_\infty}(\bar{B}|A) = m(\bar{B}) + \sum\limits_{C \not\subseteq A} m(C) - (2^{|A|}-2)\max\limits_{C \not\subseteq A} m(C), \end{cases} \tag{15.9}$$

whenever

$$\max_{C \not\subseteq A} m(C) \geq \frac{1}{2^{|A|}-1} \sum_{C \not\subseteq A} m(C). \tag{15.10}$$

This reduces to the single belief function

$$m_{L_\infty,\mathcal{M}}(B|A) = m(B) + \frac{1}{2^{|A|}-1} \sum_{C \not\subseteq A} m(C) \quad \forall B \subseteq A$$

whenever

$$\max_{C \not\subseteq A} m(C) < \frac{1}{2^{|A|}-1} \sum_{C \not\subseteq A} m(C). \tag{15.11}$$

The latter is the barycentre of the simplex of L_∞ conditional belief functions in the former case, and coincides with the L_2 conditional belief function (15.8).

Note that, as (15.9) is not guaranteed to be non-negative, the simplex of L_∞ conditional belief functions in $\mathcal{M}$ does not necessarily fall entirely inside the conditioning simplex $\mathcal{M}_A$, i.e., it may include pseudo-belief functions. Its vertices are obtained by assigning the maximum mass not in the conditioning event to all its subsets indifferently. Normalisation is then achieved, not by dividing the result by $1-\kappa$ (as in Dempster's rule), but by *subtracting* the total mass in excess of 1 in the specific component $\bar{B}$. This pattern of behaviour is exhibited by other geometric conditional belief functions, as shown in the following.

Example 47: the ternary frame. *If* $|A| = 2$, $A = \{x, y\}$, *the conditioning simplex is two-dimensional, with three vertices* m_x, m_y *and* $m_{\{x,y\}}$. *For a belief function* Bel *on* $\Theta = \{x, y, z\}$, *Theorem 90 states that the vertices of the simplex* $\mathcal{M}_{L_1,A}$ *of* L_1 *conditional belief functions in* $\mathcal{M}$ *are*

$$\begin{array}{ll} m[Bel]|_{L_1}^{\{x\}}\{x,y\} & = [m(x) + Pl(z), m(y), \quad m(\{x,y\}) \quad]', \\ m[Bel]|_{L_1}^{\{y\}}\{x,y\} & = [m(x), \quad m(y) + Pl(z), m(\{x,y\}) \quad]', \\ m[Bel]|_{L_1}^{\{x,y\}}\{x,y\} & = [m(x), \quad m(y), \quad m(\{x,y\}) + Pl(z)\,]'. \end{array}$$

Figure 15.1 shows such a simplex in the case of a belief function Bel *on the ternary frame* $\Theta = \{x, y, z\}$ *with a basic probability assignment*

$$m = [0.2,\ 0.3,\ 0,\ 0,\ 0.5,\ 0]', \tag{15.12}$$

i.e., $m(x) = 0.2$, $m(y) = 0.3$, $m(\{x, z\}) = 0.5$.

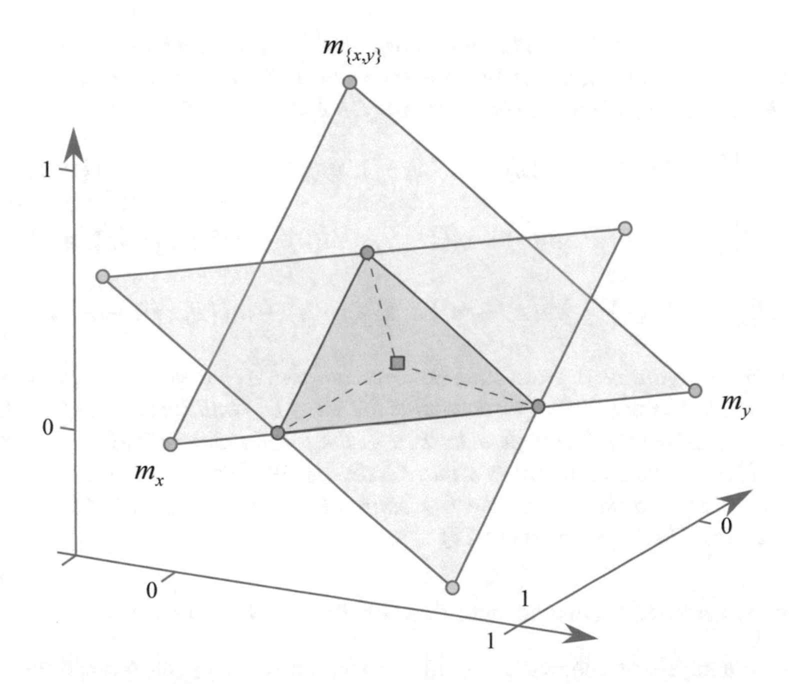

Fig. 15.1: The simplex (purple triangle) of L_1 conditional belief functions in $\mathcal{M}$ associated with the belief function with mass assignment (15.12) in $\Theta = \{x, y, z\}$. The related unique L_2 conditional belief function in $\mathcal{M}$ is also plotted as a purple square, and coincides with the centre of mass of the L_1 set. The set of L_∞ conditional (pseudo-)belief functions is also depicted (green triangle).

In the case of the belief function (15.12), by (15.8), its L_2 conditional belief function in $\mathcal{M}$ has BPA

$$\begin{aligned} m(x) &= m(x) + \frac{1 - Bel(\{x,y\})}{3} = m(x) + \frac{Pl(z)}{3}, \\ m(y) &= m(y) + \frac{Pl(z)}{3}, \\ m(\{x,y\}) &= m(\{x,y\}) + \frac{Pl(z)}{3}. \end{aligned} \tag{15.13}$$

Figure 15.1 visually confirms that this L_2 conditional belief function (purple square) lies in the barycentre of the simplex of the related L_1 conditional BFs (purple triangle).

Concerning L_∞ conditional belief functions, the BF (15.12) is such that

$$\begin{aligned} \max_{C \not\subset A} m(C) &= \max\Big\{m(z), m(\{x,z\}), m(\{y,z\}), m(\Theta)\Big\} = m(\{x,z\}) \\ &= 0.5 \geq \frac{1}{2^{|A|}-1} \sum_{C \not\subset A} m(C) = \frac{1}{3} m(\{x,z\}) = \frac{0.5}{3}. \end{aligned}$$

We hence fall under the condition (15.10), and there is a whole simplex of L_∞ conditional belief functions (in $\mathcal{M}$). According to (15.9), this simplex has $2^{|A|} - 1 = 3$ vertices, namely (taking into account the zero masses in (15.12)),

$$\begin{aligned} m[Bel]|^{\{x\}}_{L_\infty,\mathcal{M}}\{x,y\} &= [m(x) - m(\{x,z\}), m(y) + m(\{x,z\}), m(\{x,z\}) \;\;]', \\ m[Bel]|^{\{y\}}_{L_\infty,\mathcal{M}}\{x,y\} &= [m(x) + m(\{x,z\}), m(y) - m(\{x,z\}), m(\{x,z\}) \;\;]', \\ m[Bel]|^{\{x,y\}}_{L_\infty,\mathcal{M}}\{x,y\} &= [m(x) + m(\{x,z\}), m(y) + m(\{x,z\}), -m(\{x,z\})\,]', \end{aligned} \tag{15.14}$$

drawn in the figure as a green triangle. We can notice that this set is not entirely admissible, but its admissible part contains the set of L_1 conditional belief functions. Indeed, the latter is the triangle inscribed in the former, determined by its median points. This amounts therefore to a more 'conservative' approach to conditioning. Note also that both the L_1 and the L_∞ simplices have the same barycentre in the L_2 conditional belief function (15.13).

15.2.4 Features of geometric conditional belief functions in $\mathcal{M}$

From the analysis of geometric conditioning in the space of mass functions $\mathcal{M}$, a number of facts arise:

- L_p conditional belief functions, albeit obtained by minimising purely geometric distances, possess very simple and elegant interpretations in terms of degrees of belief;
- while some of them correspond to pointwise conditioning, some others form entire polytopes of solutions whose vertices also have simple interpretations;

- the conditional belief functions associated with the major L_1, L_2 and L_∞ norms are closely related to each other;
- in particular, while distinct, both the L_1 and the L_∞ simplices have their barycentre in (or coincide with) the L_2 conditional belief function;
- they are all characterised by the fact that, in the way they reassign mass from focal elements $B \not\subset A$ not in A to focal elements in A, they do not distinguish between subsets which have non-empty intersection with A from those which have not.

The last point is quite interesting: mass-based geometric conditional belief functions do not seem to care about the contribution focal elements make to the *plausibility* of the conditioning event A, but only about whether or not they contribute to the *degree of belief* of A. The reason is, roughly speaking, that in mass vectors m the mass of a given focal element appears only in the corresponding entry of m. In opposition, belief vectors Bel are such that each entry $Bel(B) = \sum_{X \subseteq B} m(X)$ of them contains information about the mass of all the subsets of B. As a result, it is to be expected that geometric conditioning in the belief space $\mathcal{B}$ will see the mass redistribution process function in a manner linked to the contribution of each focal element to the plausibility of the conditioning event A.

We will see this in detail in Section 15.3.

15.2.5 Interpretation as general imaging for belief functions

In addition, an interesting interpretation of geometric conditional belief functions in the mass space can be provided in the framework of the *imaging* approach [1417].

Example 48: the urn. *Suppose we briefly glimpse a transparent urn filled with black or white balls, and are asked to assign a probability value to the possible configurations of the urn. Suppose also that we are given three options: 30 black balls and 30 white balls (state a), 30 black balls and 20 white balls (state b) and 20 black balls and 20 white balls (state c). Hence, $\Theta = \{a, b, c\}$. Since the observation only gave us the vague impression of having seen approximately the same numbers of black and white balls, we would probably deem the states a and c to be equally likely, but at the same time we would tend to deem the event 'a or c' twice as likely as the state b. Hence, we assign probability $\frac{1}{3}$ to each of the states. Now, we are told that state c is false. How do we revise the probabilities of the two remaining states a and Bel?*

Lewis [1143] argued that, upon observing that a certain state $x \in \Theta$ is impossible, we should transfer the probability originally allocated to x to the remaining state deemed the 'most similar' to x. In this case, a is the state most similar to c, as they both consider an equal number of black and white balls. We obtain $(\frac{2}{3}, \frac{1}{3})$ as the probability values of a and b, respectively. Gärdenfors further extended Lewis's idea (*general imaging*, [665]) by allowing one to transfer a part λ of the probability $\frac{1}{3}$ initially assigned to c to state a, and the remaining part $1 - \lambda$ to state b. These fractions should be independent of the initial probabilistic state of belief.

What happens when our state of belief is described by a belief function, and we are told that A is true? In the general imaging framework, we need to reassign the mass $m(C)$ of each focal element not included in A to all the focal elements $B \subseteq A$, according to some weights $\{\lambda(B), B \subseteq A\}$.

Suppose there is no reason to attribute larger weights to any focal element in A. One option is to represent our complete ignorance about the similarities between C and each $B \subseteq A$ as a vacuous belief function on the set of weights. If applied to all the focal elements C not included in A, this results in an entire polytope of revised belief functions, each associated with an arbitrary normalised weighting. It is not difficult to see that this coincides with the set of L_1 conditional belief functions $Bel_{L_1,\mathcal{M}}(.|A)$ of Theorem 90. On the other hand, we can represent the same ignorance as a uniform probability distribution on the set of weights $\{\lambda(B), B \subseteq A\}$, for all $C \not\subseteq A$. Again, it is easy to see that general imaging produces in this case a single revised belief function, the L_2 conditional belief functions $Bel_{L_2,\mathcal{M}}(.|A)$ of Theorem 92.

As a final remark, the 'information order independence' axiom of belief revision [1417] states that the revised belief should not depend on the order in which the information is made available. In our case, the revised (conditional) belief function obtained by observing first an event A and later another event A' should be the same as the ones obtained by revising first with respect to A' and then A. Both the L_1 and the L_2 geometric conditioning operators presented here satisfy this axiom, supporting the case for their rationality.

15.3 Geometric conditioning in the belief space

To analyse the problem of geometric conditioning by projecting a belief function Bel represented by the corresponding vector Bel of belief values onto an appropriate conditioning simplex $\mathcal{B}_A = Cl(Bel_B, \emptyset \subsetneq B \subseteq A)$, let us write the difference vector between Bel and an arbitrary point Bel_a of $\mathcal{B}_A$ as

$$
\begin{aligned}
Bel - Bel_a &= \sum_{\emptyset \subsetneq B \subseteq \Theta} m(B) Bel_B - \sum_{\emptyset \subsetneq B \subseteq A} m_a(B) Bel_B \\
&= \sum_{\emptyset \subsetneq B \subseteq A} \big(m(B) - m_a(B)\big) Bel_B + \sum_{B \not\subseteq A} m(B) Bel_B \\
&= \sum_{\emptyset \subsetneq B \subseteq A} \beta(B) Bel_B + \sum_{B \not\subseteq A} m(B) Bel_B,
\end{aligned}
\tag{15.15}
$$

where, once again, $\beta(B) = m(B) - m_a(B)$.

15.3.1 L_2 conditioning in $\mathcal{B}$

We start with the L_2 norm, as this seems to have a more straightforward interpretation in the belief space.

Theorem 95. *Given a belief function $Bel : 2^\Theta \rightarrow [0,1]$ with basic probability assignment m, and an arbitrary non-empty focal element $\emptyset \subsetneq A \subseteq \Theta$, the L_2 conditional belief function $Bel_{L_2,\mathcal{B}}(.|A)$ with respect to A in the belief space $\mathcal{B}$ is unique, and has basic probability assignment*

$$m_{L_2,\mathcal{B}}(C|A) = m(C) + \sum_{B \subseteq A^c} m(B \cup C) 2^{-|B|} + (-1)^{|C|+1} \sum_{B \subseteq A^c} m(B) 2^{-|B|} \tag{15.16}$$

for each proper subset $\emptyset \subsetneq C \subsetneq A$ of the event A.

Example 49: the ternary frame. *In the ternary case, the unique L_2 mass space conditional belief function has a BPA m_a such that*

$$\begin{aligned} m_a(x) \quad &= m(x) + \frac{m(z) + m(\{x,z\})}{2}, \\ m_a(y) \quad &= m(y) + \frac{m(z) + m(\{y,z\})}{2}, \\ m_a(\{x,y\}) &= m(\{x,y\}) + m(\Theta) + \frac{m(\{x,z\}) + m(\{y,z\})}{2}. \end{aligned} \tag{15.17}$$

At first glance, each focal element $B \subseteq A$ seems to be assigned to a fraction of the original mass $m(X)$ of all focal elements X of Bel such that $X \subseteq B \cup A^c$. This contribution seems to be proportional to the size of $X \cap A^c$, i.e., to how much the focal element of Bel falls outside the conditioning event A.

Notice that Dempster conditioning $Bel_\oplus(.|A) = Bel \oplus Bel_A$ yields in this case

$$m_\oplus(x|A) = \frac{m(x) + m(\{x,z\})}{1 - m(z)}, \quad m_\oplus(\{x,y\}|A) = \frac{m(\{x,y\}) + m(\Theta)}{1 - m(z)}.$$

L_2 conditioning in the belief space differs from its 'sister' operation in the mass space (Theorem 92) in that it makes use of the set-theoretic relations between focal elements, just as Dempster's rule does. However, in contrast to Dempster conditioning, no normalisation is enforced, as even subsets of A^c ($\{z\}$ in this case) contribute as addends to the mass of the resulting conditional belief function.

As for the general case (15.16), we can notice that the (unique) L_2 conditional belief function in the belief space is not guaranteed to be a proper belief function, as some masses can be negative, owing to the addend

$$(-1)^{|C|+1} \sum_{B \subseteq A^c} m(B) 2^{-|B|}.$$

The quantity shows, however, an interesting connection with the redistribution process associated with the orthogonal projection $\pi[Bel]$ of a belief function onto the probability simplex ([333]; Section 11.4), in which the mass of each subset A is redistributed among all its subsets $B \subseteq A$ on an equal basis.

Here (15.16), the mass of each focal element not included in A, is also broken into $2^{|B|}$ parts, equal to the number of its subsets. Only one such part is reattributed to $C = B \cap A$, while the rest is redistributed to A itself.

15.3.2 L_1 conditioning in $\mathcal{B}$

To discuss L_1 conditioning in the belief space, we need to write explicitly the difference vector $Bel - Bel_a$.

Lemma 18. *The L_1 norm of the difference vector $Bel - Bel_a$ can be written as*

$$\|Bel - Bel_a\|_{L_1} = \sum_{\emptyset \subsetneq B \cap A \subsetneq A} \Big| \gamma(B \cap A) + Bel(B) - Bel(B \cap A) \Big|,$$

so that the L_1 conditional belief functions in $\mathcal{B}$ are the solutions of the following minimisation problem:

$$\arg\min_{\gamma} \sum_{\emptyset \subsetneq B \cap A \subsetneq A} \Big| \gamma(B \cap A) + Bel(B) - Bel(B \cap A) \Big|,$$

where $\beta(B) = m(B) - m_a(B)$ and $\gamma(B) = \sum_{C \subseteq B} \beta(B)$.

As we also noticed for the L_1 minimisation problem in the mass space, each group of addends which depend on the same variable $\gamma(X)$, $\emptyset \subsetneq X \subsetneq A$, can be minimised separately. Therefore, the set of L_1 conditional belief functions in the belief space $\mathcal{B}$ is determined by the following minimisation problems:

$$\arg\min_{\gamma(X)} \sum_{B : B \cap A = X} \Big| \gamma(X) + Bel(B) - Bel(X) \Big| \quad \forall \emptyset \subsetneq X \subsetneq A. \tag{15.18}$$

The functions appearing in (15.18) are of the form $|x+k_1|+\ldots+|x+k_m|$, where m is even. Such functions are minimised by the interval determined by the two central 'nodes' $-k_{\text{int}_1} \leq -k_{\text{int}_2}$ (compare Fig. 13.11 in the proof of Theorem 78, Chapter 13).

In the case of the system (15.18), this yields

$$Bel(X) - Bel(B^X_{\text{int}_1}) \leq \gamma(X) \leq Bel(X) - Bel(B^X_{\text{int}_2}), \tag{15.19}$$

where $B^X_{\text{int}_1}$ and $B^X_{\text{int}_2}$ are the events corresponding to the central, median values of the collection $\{Bel(B), B \cap A = X\}$. Unfortunately, it is not possible, in general, to determine the median values of such a collection of belief values, as belief functions are defined on a partially (rather than totally) ordered set (the power set 2^Θ).

Special case This is possible, however, in the special case in which $|A^c| = 1$ (i.e., the conditioning event is of cardinality $n-1$). In this case,

$$B^X_{\text{int}_1} = X + A^c, \quad B^X_{\text{int}_2} = X,$$

so that the solution in the variables $\{\gamma(X)\}$ is

$$Bel(X) - Bel(X + A^c) \leq \gamma(X) \leq 0, \quad \emptyset \subsetneq X \subsetneq A.$$

It is not difficult to see that, in the variables $\{\beta(X)\}$, the solution reads as

$$Bel(X) - Bel(X + A^c) \leq \beta(X) \leq - \sum_{\emptyset \subsetneq B \subsetneq X} \Big(Bel(B) - Bel(B + A^c)\Big),$$

$\emptyset \subsetneq X \subsetneq A$, i.e., in the mass of the desired L_1 conditional belief function,

$$m(X) + \sum_{\emptyset \subsetneq B \subsetneq X} \Big(Bel(B) - Bel(B + A^c)\Big) \\ \leq m_a(X) \leq m(X) + \Big(Bel(X + A^c) - Bel(X)\Big).$$

Not only are the resulting conditional (pseudo-)belief functions not guaranteed to be proper belief functions, but it is also difficult to find straightforward interpretations of these results in terms of degrees of belief. On these grounds, we might be tempted to conclude that the L_1 norm is not suitable for inducing conditioning in belief calculus.

However, the analysis of the ternary case seems to hint otherwise. The L_1 conditional belief functions $Bel_{L_1,\mathcal{B}_3}(.|\{x,y\})$ with respect to $A = \{x,y\}$ in $\mathcal{B}_3$ are all those BFs with a core included in A such that the conditional mass of $B \subsetneq A$ falls between $Bel(B)$ and $Bel(B \cup A^c)$:

$$Bel(B) \leq m_{L_1,\mathcal{B}}(B|A) \leq Bel(B \cup A^c).$$

The barycentre of the L_1 solutions is

$$\beta(x) = -\frac{m(z) + m(\{x,z\})}{2}, \quad \beta(y) = -\frac{m(z) + m(\{y,z\})}{2},$$

i.e., the L_2 conditional belief function (15.17), just as in the case of geometric conditioning in the mass space $\mathcal{M}$. The same can be easily proved for all $A \subseteq \{x,y,z\}$.

Theorem 96. *For every belief function $Bel : 2^{\{x,y,z\}} \to [0,1]$, the unique L_2 conditional belief function $Bel_{L_2,\mathcal{B}_3}(.|\{x,y\})$ with respect to $A \subseteq \{x,y,z\}$ in $\mathcal{B}_3$ is the barycentre of the polytope of L_1 conditional belief functions with respect to A in $\mathcal{B}_3$.*

15.3.3 L_∞ conditioning in $\mathcal{B}$

Finally, let us approach the problem of finding L_∞ conditional belief functions given an event A, starting with the ternary case study.

Example 50: the ternary case. *In the ternary case $\Theta = \{x,y,z\}$, we have, for the conditioning event $A = \{x,y\}$,*

$$\|Bel - Bel_a\|_{L_\infty} = \max_{\emptyset \subsetneq B \subsetneq \Theta} |Bel(B) - Bel_a(B)|$$

$$
\begin{aligned}
&= \max\Big\{ |Bel(x) - Bel_a(x)|, |Bel(y) - Bel_a(y)|, |Bel(z)|, \\
&\quad |Bel(\{x,y\}) - Bel_a(\{x,y\})|, |Bel(\{x,z\}) - Bel_a(\{x,z\})|, \\
&\quad |Bel(\{y,z\}) - Bel_a(\{y,z\})| \Big\} \\
&= \max\Big\{ |m(x) - m_a(x)|, |m(y) - m_a(y)|, |m(z)|, \\
&\quad |m(x) + m(y) + m(\{x,y\}) - m_a(x) - m_a(y) - m_a(\{x,y\})|, \\
&\quad |m(x) + m(z) + m(\{x,z\}) - m_a(x)|, \\
&\quad |m(y) + m(z) + m(\{y,z\}) - m_a(y)| \Big\} \\
&= \max\Big\{ |\beta(x)|, |\beta(y)|, m(z), 1 - Bel(\{x,y\}), |\beta(x) + m(z) + m(\{x,z\})|, \\
&\quad |\beta(y) + m(z) + m(\{y,z\})| \Big\},
\end{aligned}
$$

which is minimised by (as $1 - Bel(x,y) \geq m(z)$)

$$
\begin{aligned}
\beta(x) &: \max\Big\{ |\beta(x)|, |\beta(x) + m(z) + m(x,z)| \Big\} \leq 1 - Bel(x,y), \\
\beta(y) &: \max\Big\{ |\beta(y)|, |\beta(y) + m(z) + m(y,z)| \Big\} \leq 1 - Bel(x,y).
\end{aligned}
$$

On the left-hand side, we have functions of the form $\max\{|x|, |x+k|\}$. *The interval of values in which such a function is below a certain threshold $k' \geq k$ is $[-k', k'-k]$. This yields*

$$
\begin{aligned}
Bel(\{x,y\}) - 1 \leq \beta(x) \leq 1 - Bel(\{x,y\}) - (m(z) + m(\{x,z\})), \\
Bel(\{x,y\}) - 1 \leq \beta(y) \leq 1 - Bel(\{x,y\}) - (m(z) + m(\{y,z\})).
\end{aligned}
\tag{15.20}
$$

The solution in the masses of the sought L_∞ conditional belief function reads as

$$
\begin{aligned}
m(x) - m(\{y,z\}) - m(\Theta) \leq m_a(x) \leq 1 - (m(y) + m(\{x,y\})), \\
m(y) - m(\{x,z\}) - m(\Theta) \leq m_a(y) \leq 1 - (m(x) + m(\{x,y\})).
\end{aligned}
\tag{15.21}
$$

Its barycentre is clearly given by

$$
\begin{aligned}
m_a(x) &= m(x) + \frac{m(z) + m(\{x,z\})}{2} \\
m_a(y) &= m(y) + \frac{m(z) + m(\{y,z\})}{2} \\
m_a(\{x,y\}) &= 1 - m_a(x) - m_a(y) \\
&= m(\{x,y\}) + m(\Theta) + \frac{m(\{x,z\}) + m(\{y,z\})}{2}
\end{aligned}
\tag{15.22}
$$

i.e., the L_2 conditional belief function (15.17) as computed in the ternary case.

The general case From the expression (15.15) for the difference $Bel - Bel_a$ we get, after introducing the variables $\gamma(C) = \sum_{X \subseteq C} \beta(X)$, $\emptyset \subsetneq C \subseteq A$,

$$\max_{\emptyset \subsetneq B \subsetneq \Theta} |Bel(B) - Bel_a(B)|$$

$$= \max_{\emptyset \subsetneq B \subsetneq \Theta} \left| \sum_{C \subseteq A \cap B} \beta(C) + \sum_{\substack{C \subseteq B \\ C \not\subset A}} m(C) \right| = \max_{\emptyset \subsetneq B \subsetneq \Theta} \left| \gamma(A \cap B) + \sum_{\substack{C \subseteq B \\ C \not\subset A}} m(C) \right|$$

$$= \max \left\{ \max_{B: B \cap A = \emptyset} \left| \sum_{\substack{C \subseteq B \\ C \not\subset A}} m(C) \right|, \max_{B: B \cap A \neq \emptyset, A} \left| \gamma(A \cap B) + \sum_{\substack{C \subseteq B \\ C \not\subset A}} m(C) \right|, \right.$$

$$\left. \max_{B: B \cap A = A, B \neq \Theta} \left| \gamma(A) + \sum_{C \subseteq B, C \not\subset A} m(C) \right| \right\}, \tag{15.23}$$

where, once again, $\gamma(A) = Bel(A) - 1$.

Lemma 19. *The values $\gamma^*(X)$ which minimise (15.23) are, $\forall \ \emptyset \subsetneq X \subsetneq A$,*

$$-\big(1 - Bel(A)\big) \leq \gamma^*(X) \leq \big(1 - Bel(A)\big) - \sum_{\substack{C \cap A^c \neq \emptyset \\ C \cap A \subseteq X}} m(C). \tag{15.24}$$

Lemma 19 can be used to prove the following form of the set of L_∞ conditional belief functions in $\mathcal{B}$.

Theorem 97. *Given a belief function $Bel : 2^\Theta \rightarrow [0,1]$ and an arbitrary non-empty focal element $\emptyset \subsetneq A \subseteq \Theta$, the set of L_∞ conditional belief functions $Bel_{L_\infty, \mathcal{B}}(.|A)$ with respect to A in $\mathcal{B}$ is the set of BFs with focal elements in $\{X \subseteq A\}$ which satisfy the following constraints for all $\emptyset \subsetneq X \subseteq A$:*

$$m(X) + \sum_{\substack{C \cap A^c \neq \emptyset \\ \emptyset \subseteq C \cap A \subseteq X}} m(C) + (2^{|X|} - 1)(1 - Bel(A)) \leq m_a(X)$$

$$\leq m(X) + (2^{|X|} - 1)(1 - Bel(A)) - \sum_{\substack{C \cap A^c \neq \emptyset \\ \emptyset \subseteq C \cap A \subsetneq X}} m(C) - (-1)^{|X|} \sum_{B \subseteq A^c} m(B). \tag{15.25}$$

This result appears rather difficult to interpretat in terms of mass allocation. Nevertheless, Example 51 suggests that this set, or at least its admissible part, has some nice properties worth exploring.

For instance, its barycentre has a much simpler form.

Barycentre of the L_∞ solution The barycentre of (15.24) is

$$\gamma(X) = -\frac{1}{2} \sum_{C\cap A^c \neq \emptyset, C\cap A \subseteq X} m(C),$$

a solution which corresponds, in the set of variables $\{\beta(X)\}$, to the system

$$\left\{ \sum_{\emptyset \subsetneq C \subseteq X} \beta(C) + \frac{1}{2} \sum_{C\cap A^c \neq \emptyset, C\cap A \subseteq X} m(C) = 0, \quad \forall \emptyset \subsetneq X \subsetneq A. \right. \quad (15.26)$$

The following result proceeds from the latter expression.

Theorem 98. *The centre of mass of the set of L_∞ conditional belief functions $Bel_{L_\infty,\mathcal{B}}(.|A)$ with respect to A in the belief space $\mathcal{B}$ is the unique solution of the system of equations (15.26), and has basic probability assignment*

$$\begin{aligned} m_{L_\infty,\mathcal{B}}(C|A) &= m(C) + \frac{1}{2} \sum_{\emptyset \subsetneq B \subseteq A^c} \Big[m(B\cup C) + (-1)^{|C|+1} m(B) \Big] \\ &= m(C) + \frac{1}{2} \sum_{\emptyset \subsetneq B \subseteq A^c} m(B+C) + \frac{1}{2}(-1)^{|C|+1} Bel(A^c). \end{aligned} \quad (15.27)$$

15.4 Mass space versus belief space conditioning

To conclude this overview of geometric conditioning via L_p norms, it is worth comparing the outcomes of mass space versus belief space L_p conditioning.

15.4.1 Geometric conditioning: A summary

Given a belief function $Bel : 2^\Theta \rightarrow [0,1]$ and an arbitrary non-empty conditioning focal element $\emptyset \subsetneq A \subseteq \Theta$:

1. The set of L_1 conditional belief functions $Bel_{L_1,\mathcal{M}}(.|A)$ with respect to A in $\mathcal{M}$ is the set of BFs with a core in A such that their mass dominates that of Bel over all the subsets of A:

   $$Bel_{L_1,\mathcal{M}}(.|A) = \Big\{ Bel_a : \mathcal{C}_a \subseteq A, m_a(B) \geq m(B) \;\; \forall \emptyset \subsetneq B \subseteq A \Big\}.$$

   This set is a simplex $\mathcal{M}_{L_1,A}[Bel] = Cl(m[Bel]|^B_{L_1} A, \emptyset \subsetneq B \subseteq A)$, whose vertices $m_a = m[Bel]|^B_{L_1} A$ have BPA

   $$\begin{cases} m_a(B) = m(B) + 1 - Bel(A) = m(B) + Pl(A^c), \\ m_a(X) = m(X) \qquad\qquad\qquad\qquad\qquad\qquad \forall \emptyset \subsetneq X \subsetneq A, X \neq B. \end{cases}$$

2. The unique L_2 conditional belief function $Bel_{L_2,\mathcal{M}}(.|A)$ with respect to A in $\mathcal{M}$ is the BF whose BPA redistributes the mass $1 - Bel(A) = Pl(A^c)$ to each focal element $B \subseteq A$ in an equal way,

$$m_{L_2,\mathcal{M}}(B|A) = m(B) + \frac{Pl(A^c)}{2^{|A|} - 1}, \qquad (15.28)$$

$\forall \emptyset \subsetneq B \subseteq A$, and corresponds to the centre of mass of the simplex $\mathcal{M}_{L_1,A}[Bel]$ of L_1 conditional belief functions.
3. The L_∞ conditional belief functions either coincides with the L_2 one, or forms a simplex obtained by assigning the maximum mass outside A (rather than the sum of such masses $Pl(A^c)$) to all subsets of A (but one) indifferently.

L_1 and L_2 conditioning are closely related in the mass space, and have a compelling interpretation in terms of general imaging [1417, 665].

The L_2 and $\overline{L_\infty}$ conditional belief functions computed in the belief space are, instead,

$$m_{L_2,\mathcal{B}}(B|A) = m(B) + \sum_{C \subseteq A^c} m(B+C)2^{-|C|} + (-1)^{|B|+1} \sum_{C \subseteq A^c} m(C)2^{-|C|},$$

$$m_{\overline{L_\infty},\mathcal{B}}(B|A) = m(B) + \frac{1}{2} \sum_{\emptyset \subsetneq C \subseteq A^c} m(B+C) + \frac{1}{2}(-1)^{|B|+1} Bel(A^c).$$

For the L_2 case, the result makes a lot of sense in the ternary case, but it is difficult to interpret in its general form (above). It seems to be related to the process of mass redistribution among all subsets, as happens with the (L_2-induced) orthogonal projection of a belief function onto the probability simplex. In both expressions above, we can note that normalisation is achieved by alternately subtracting and summing a quantity, rather than via a ratio or, as in (15.28), by reassigning the mass of all $B \not\subset A$ to each $B \subsetneq A$ on an equal basis.

We can interpret the barycentre of the set of L_∞ conditional belief functions as follows: the mass of all the subsets whose intersection with A is $C \subsetneq A$ is reassigned by the conditioning process half to C, and half to A itself. In the case of $C = A$ itself, by normalisation, all the subsets $D \supseteq A$, including A, have their whole mass reassigned to A, consistently with the above interpretation. The mass $Bel(A^c)$ of the subsets which have no relation to the conditioning event A is used to guarantee the normalisation of the resulting mass distribution. As a result, the mass function obtained is not necessarily non-negative: again, such a version of geometrical conditioning may generate pseudo-belief functions.

The L_1 case is also intriguing, as in that case it appears impossible to obtain a general analytic expression, whereas in the special cases in which this is possible the result has potentially interesting interpretations, as confirmed by the empirical comparison in Example 51.

Generally speaking, though, L_p conditional belief functions in the belief space seem to have rather less straightforward interpretations than the corresponding quantities in the mass space.

Example 51: comparison for the ternary example. *We conclude by comparing the different approximations in the case study of the ternary frame, $\Theta = \{x, y, z\}$, already introduced in Example 47.*

Assuming again that the conditioning event is $A = \{x, y\}$, the unique L_2 conditional belief function in $\mathcal{B}$ is given by (15.17), while the L_∞ conditional belief functions form the set determined by (15.21), with barycentre in (15.22). By Theorem 90, the vertices of $\mathcal{M}_{L_1,\{x,y\}}[Bel]$ are instead

$$
\begin{aligned}
m[Bel]|_{L_1}^{\{x\}}\{x,y\} &= [m(x) + Pl(z), m(y), m(\{x,y\})]', \\
m[Bel]|_{L_1}^{\{y\}}\{x,y\} &= [m(x), m(y) + Pl(z), m(\{x,y\})]', \\
m[Bel]|_{L_1}^{\{x,y\}}\{x,y\} &= [m(x), m(y), m(\{x,y\}) + Pl(z)]'.
\end{aligned}
$$

By Theorem 92, the L_2 conditional belief function given $\{x, y\}$ in $\mathcal{M}$ has BPA

$$
\begin{array}{ll}
m'(x) = m(x) + \dfrac{1 - Bel(\{x,y\})}{3} = m(x) + \dfrac{Pl(z)}{3}, & \\
m'(y) = m(y) + \dfrac{Pl(z)}{3}, & m'(\{x,y\}) = m(\{x,y\}) + \dfrac{Pl(z)}{3}.
\end{array}
$$

Figure 15.2 illustrates the different geometric conditional belief functions given $A = \{x, y\}$ for the belief function with masses as in (15.12), i.e., $m(x) = 0.2$, $m(y) = 0.3$, $m(\{x, z\}) = 0.5$, drawn in the mass space. As in Example 47, the conditioning simplex is two-dimensional, with three vertices m_x, m_y and $m_{\{x,y\}}$. We already know that $m_{L_2,\mathcal{M}}(.|A)$ lies in the barycentre of the simplex of the $L_1, \mathcal{M}$ conditional belief functions. The same is true (at least in the ternary case) for $m_{L_2,\mathcal{B}}(.|A)$ (the pink square), which is the barycentre of the (orange) polytope of $m_{L_\infty,\mathcal{B}}(.|A)$ conditional belief functions.

The set of L_1 conditional belief functions in $\mathcal{B}$ is, instead, a line segment whose barycentre is $m_{L_2,\mathcal{B}}(.|A)$. Such a set is:

- *entirely included in the set of L_∞ approximations in both $\mathcal{B}$ and $\mathcal{M}$, thus representing a more conservative approach to conditioning;*
- *entirely admissible.*

It seems that, hard as it is to compute, L_1 conditioning in the belief space delivers interesting results. A number of interesting cross-relations between conditional belief functions in the two representation domains appear to exist:

1. *$m_{L_\infty,\mathcal{B}}(.|A)$ seems to contain $m_{L_1,\mathcal{M}}(.|A)$.*
2. *The two L_2 conditional belief functions $m_{L_2,\mathcal{M}}(.|A)$ and $m_{L_2,\mathcal{B}}(.|A)$ appear to lie on the same line joining opposite vertices of $m_{L_\infty,\mathcal{B}}(.|A)$.*
3. *$m_{L_\infty,\mathcal{B}}(.|A)$ and $m_{L_\infty,\mathcal{M}}(.|A)$ have several vertices in common.*

There is probably more to these conditioning approaches than has been shown by the simple comparison done here. Finding the admissible parts of $m_{L_\infty,\mathcal{B}}(.|A)$ and $m_{L_\infty,\mathcal{M}}(.|A)$, for instance, remains an open problem.

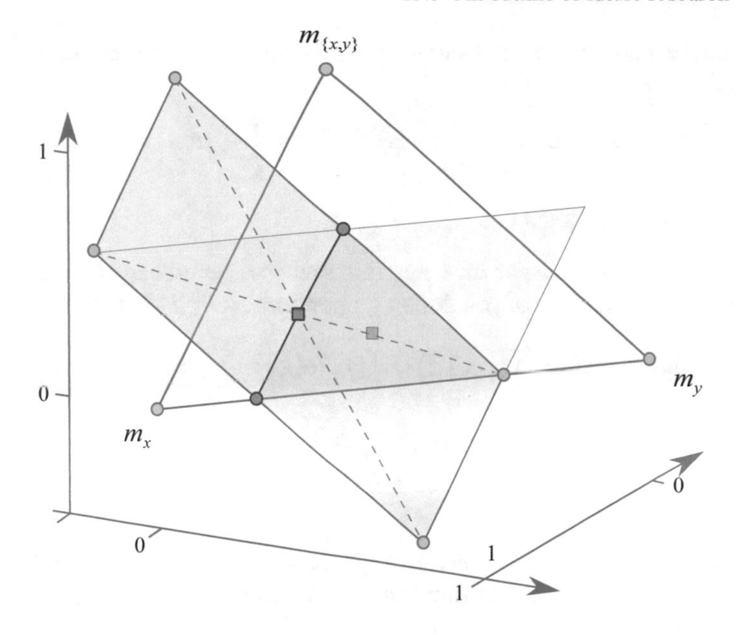

Fig. 15.2: The set of $L_\infty, \mathcal{B}$ conditional belief functions is drawn here as an orange rectangle for the belief function with mass assignment (15.12) in $\Theta = \{x, y, z\}$, with conditioning event $A = \{x, y\}$, considered in Example 47. The region falls partly outside the conditioning simplex (blue triangle). The set of L_1 conditional belief functions in $\mathcal{B}$ is a line segment (in pink) with its barycentre in the L_2 conditional BF (pink square). In the ternary case, $L_2, \mathcal{B}$ is the barycentre of the $L_\infty, \mathcal{B}$ rectangle. The L_p conditional belief functions in the mass space, already considered in Fig. 15.1, are still visible in the background. Interesting cross-relations between conditional functions in $\mathcal{M}$ and $\mathcal{B}$ seem to emerge, which are not clearly reflected by their analytical expressions computed here.

15.5 An outline of future research

This chapter's sketch of conditioning by geometric means opens up a number of interesting questions.

Question 10. What classes of conditioning rules can be generated by a distance minimisation process, such as that introduced here? Do they span all known definitions of conditioning (Section 4.5), once one applies a sufficiently general class of dissimilarity measures?

In particular, is Dempster conditioning itself a special case of geometric conditioning?

A related question links geometric conditioning with combination rules. Indeed, we have seen in Chapter 8 that Dempster's combination rule can be decomposed into a convex combination of Dempster conditioning with respect to all possible events A:

$$Bel \oplus Bel' = Bel \oplus \sum_{A \subseteq \Theta} m'(A) Bel_A = \sum_{A \subseteq \Theta} \mu(A) Bel \oplus Bel_A,$$

where $\mu(A) \propto m'(A) Pl(A)$.

Question 11. Can we imagine reversing this link, and generating combination rules $\uplus$ as convex combinations of conditioning operators $Bel|_A^{\uplus}$? That is,

$$Bel \uplus Bel' = \sum_{A \subseteq \Theta} m'(A) Bel \uplus Bel_A = \sum_{A \subseteq \Theta} m'(A) Bel|_A^{\uplus}.$$

Additional constraints may have to be imposed in order to obtain a unique result, for instance commutativity with affine combination (or *linearity*, in Smets's terminology [1697]).

In the near future, we thus plan to explore the world of combination rules induced by conditioning rules, starting from the different geometric conditioning processes introduced here.

Appendix: Proofs

Proof of Lemma 17

By definition,

$$m - m_a = \sum_{\emptyset \subsetneq B \subseteq \Theta} m(B) m_B - \sum_{\emptyset \subsetneq B \subseteq A} m_a(B) m_B.$$

A change of variables $\beta(B) \doteq m(B) - m_a(B)$ further yields

$$m - m_a = \sum_{\emptyset \subsetneq B \subseteq A} \beta(B) m_B + \sum_{B \not\subseteq A} m(B) m_B. \tag{15.29}$$

Observe that the variables $\{\beta(B), \emptyset \subsetneq B \subseteq A\}$ are not all independent. Indeed,

$$\sum_{\emptyset \subsetneq B \subseteq A} \beta(B) = \sum_{\emptyset \subsetneq B \subseteq A} m(B) - \sum_{\emptyset \subsetneq B \subseteq A} m_a(B) = Bel(A) - 1,$$

as $\sum_{\emptyset \subsetneq B \subseteq A} m_a(B) = 1$ by definition (since $m_a \in \mathcal{M}_A$). As a consequence, in the optimisation problem (15.3) only $2^{|A|} - 2$ variables are independent (as $\emptyset$ is not included), while

$$\beta(A) = Bel(A) - 1 - \sum_{\emptyset \subsetneq B \subsetneq A} \beta(B).$$

By substituting the above equality into (15.29), we get (15.4).

Proof of Theorem 90

The minima of the L_1 norm of the difference vector are given by the set of constraints

$$\begin{cases} \beta(B) \leq 0 & \forall \emptyset \subsetneq B \subsetneq A, \\ \sum_{\emptyset \subsetneq B \subsetneq A} \beta(B) \geq Bel(A) - 1. \end{cases} \tag{15.30}$$

In the original simplicial coordinates $\{m_a(B), \emptyset \subsetneq B \subseteq A\}$ of the candidate solution m_a in $\mathcal{M}_A$, this system reads as

$$\begin{cases} m(B) - m_a(B) \leq 0 & \forall \emptyset \subsetneq B \subsetneq A, \\ \sum_{\emptyset \subsetneq B \subsetneq A} \big[m(B) - m_a(B)\big] \geq Bel(A) - 1, \end{cases}$$

i.e., $m_a(B) \geq m(B)\ \forall \emptyset \subsetneq B \subseteq A$.

Proof of Theorem 91

It is easy to see that, by (15.30), the $2^{|A|}-2$ vertices of the simplex of L_1 conditional belief functions in $\mathcal{M}$ (denoted by $m[Bel]|_{L_1}^B A$, where $\emptyset \subsetneq B \subseteq A$) are determined by the following solutions:

$$m[Bel]|_{L_1}^A A : \{\, \beta(X) = 0 \quad \forall \emptyset \subsetneq X \subsetneq A,$$

$$m[Bel]|_{L_1}^B A : \begin{cases} \beta(B) = Bel(A) - 1, & \\ \beta(X) = 0 & \forall \emptyset \subsetneq X \subsetneq A, X \neq B \end{cases} \forall \emptyset \subsetneq B \subsetneq A.$$

In the $\{m_a(B)\}$ coordinates, the vertex $m[Bel]|_{L_1}^B A$ is the vector $m_a \in \mathcal{M}_A$ defined by (15.7).

Proof of Theorem 92

The generators of $\mathcal{M}_A$ are all the vectors $m_B - m_A$, $\forall \emptyset \subsetneq B \subsetneq A$, and have the following structure:

$$[0, \ldots, 0, 1, 0, \ldots, 0, -1, 0, \ldots, 0]'$$

with all entries zero except for entry B (equal to 1) and entry A (equal to -1). Making use of (15.29), the condition $\langle m - m_a, m_B - m_A \rangle = 0$ thus assumes the form

$$\beta(B) - Bel(A) + 1 + \sum_{\emptyset \subsetneq X \subsetneq A, X \neq B} \beta(X) = 0$$

for all possible generators of $\mathcal{M}_A$, i.e.,

$$2\beta(B) + \sum_{\emptyset \subsetneq X \subsetneq A, X \neq B} \beta(X) = Bel(A) - 1 \quad \forall \emptyset \subsetneq B \subsetneq A. \tag{15.31}$$

The system (15.31) is a linear system of $2^{|A|} - 2$ equations in $2^{|A|} - 2$ variables (the $\beta(X)$), which can be written as $\mathcal{A}\beta = (Bel(A) - 1)\mathbf{1}$, where $\mathbf{1}$ is the vector of the appropriate size with all entries equal to 1. Its unique solution is, trivially, $\beta = (Bel(A) - 1) \cdot \mathcal{A}^{-1}\mathbf{1}$. The matrix $\mathcal{A}$ and its inverse are

$$\mathcal{A} = \begin{bmatrix} 2 \; 1 \cdots 1 \\ 1 \; 2 \cdots 1 \\ \cdots \\ 1 \; 1 \cdots 2 \end{bmatrix}, \quad \mathcal{A}^{-1} = \frac{1}{d+1} \begin{bmatrix} d & -1 & \cdots & -1 \\ -1 & d & \cdots & -1 \\ & & \cdots & \\ -1 & -1 & \cdots & d \end{bmatrix},$$

where d is the number of rows (or columns) of $\mathcal{A}$. It is easy to see that $\mathcal{A}^{-1}\mathbf{1} = \frac{1}{d+1}\mathbf{1}$, where, in our case, $d = 2^{|A|} - 2$.

The solution to (15.31) is then, in matrix form,

$$\beta = \mathcal{A}^{-1}\mathbf{1} \cdot \big(Bel(A) - 1\big) = \frac{1}{2^{|A|} - 1}\mathbf{1}\big(Bel(A) - 1\big)$$

or, more explicitly,

$$\beta(B) = \frac{Bel(A) - 1}{2^{|A|} - 1} \quad \forall \emptyset \subsetneq B \subsetneq A.$$

Thus, in the $\{m_a(B)\}$ coordinates the L_2 conditional belief function reads as

$$m_a(B) = m(B) + \frac{1 - Bel(A)}{2^{|A|} - 1} = m(B) + \frac{Pl(A^c)}{2^{|A|} - 1} \quad \forall \emptyset \subsetneq B \subseteq A,$$

A included.

Proof of Theorem 94

The L_∞ norm of the difference vector (15.4) reads as

$$\begin{aligned} &\|m - m_a\|_{L_\infty} \\ &= \max \left\{ |\beta(B)|, \emptyset \subsetneq B \subsetneq A; \; |m(B)|, B \not\subset A; \; \left| Bel(A) - 1 - \sum_{\emptyset \subsetneq B \subsetneq A} \beta(B) \right| \right\}. \end{aligned}$$

As

$$\left| Bel(A) - 1 - \sum_{\emptyset \subsetneq B \subsetneq A} \beta(B) \right| = \left| \sum_{B \not\subset A} m(B) + \sum_{\emptyset \subsetneq B \subsetneq A} \beta(B) \right|,$$

the above norm simplifies to

$$
\max \left\{ |\beta(B)|, \emptyset \subsetneq B \subsetneq A; \max_{B \not\subset A}\{m(B)\}; \left| \sum_{B \not\subset A} m(B) + \sum_{\emptyset \subsetneq B \subsetneq A} \beta(B) \right| \right\}. \tag{15.32}
$$

This is a function of the form

$$
f(x_1, \ldots, x_{K-1}) = \max \left\{ |x_i| \; \forall i, \left| \sum_i x_i + k_1 \right|, k_2 \right\}, \tag{15.33}
$$

with $0 \leq k_2 \leq k_1 \leq 1$. These functions were studied in the proof of Theorem 74 (Chapter 13), and illustrated in Fig. 13.10. For the norm (15.32), the condition $k_2 \geq k_1/K$ for functions of the form (15.33) reads as

$$
\max_{C \not\subset A} m(C) \geq \frac{1}{2^{|A|} - 1} \sum_{C \not\subset A} m(C). \tag{15.34}
$$

In such a case (see the proof of Theorem 74 again), the set of L_∞ conditional belief functions is given by the constraints $x_i \geq -k_2$, $\sum_i x_i \leq k_2 - k_1$, namely

$$
\begin{cases}
\beta(B) \geq -\max_{C \not\subset A} m(C) & \forall B \subsetneq A, \\
\sum_{B \subsetneq A} \beta(B) \leq \max_{C \not\subset A} m(C) - \sum_{C \not\subset A} m(C).
\end{cases}
$$

This is a simplex $Cl(m[Bel]|_{\bar{B}}^{L_\infty} A, \bar{B} \subseteq A)$, where each vertex $m[Bel]|_{\bar{B}}^{L_\infty} A$ is characterised by the following values $\beta_{\bar{B}}$ of the auxiliary variables:

$$
\begin{cases}
\beta_{\bar{B}}(B) = -\max_{C \not\subset A} m(C) & \forall B \subseteq A, B \neq \bar{B}, \\
\beta_{\bar{B}}(\bar{B}) = -\sum_{C \not\subset A} m(C) + (2^{|A|} - 2) \max_{C \not\subset A} m(C),
\end{cases}
$$

or, in terms of their basic probability assignments, (15.9).

The barycentre of this simplex can be computed as follows:

$$
\begin{aligned}
m_{\overline{L_\infty}, \mathcal{M}}(B|A) &\doteq \frac{\sum_{\bar{B} \subseteq A} m[Bel]|_{\bar{B}}^{L_\infty}(B|A)}{2^{|A|} - 1} \\
&= \frac{(2^{|A|} - 1) m(B) + \sum_{C \not\subset A} m(C)}{2^{|A|} - 1} = m(B) + \frac{\sum_{C \not\subset A} m(C)}{2^{|A|} - 1},
\end{aligned}
$$

i.e., the L_2 conditional belief function (15.8). The corresponding minimal L_∞ norm of the difference vector is, according to (15.32), equal to $\max_{C \not\subset A} m(C)$.

When (15.34) does not hold,

$$
\max_{C \not\subset A} m(C) < \frac{1}{2^{|A|} - 1} \sum_{C \not\subset A} m(C), \tag{15.35}
$$

the system (15.32) has a unique solution, namely

$$\beta(B) = -\frac{1}{2^{|A|}-1} \sum_{C \not\subset A} m(C) \quad \forall B \subsetneq A.$$

In terms of basic probability assignments, this reads as

$$m_{L_\infty,\mathcal{M}}(B|A) = m(B) + \frac{1}{2^{|A|}-1} \sum_{C \not\subset A} m(C) \quad \forall B \subseteq A.$$

The corresponding minimal L_∞ norm of the difference vector is

$$\frac{1}{2^{|A|}-1} \sum_{C \not\subset A} m(C).$$

Proof of Theorem 95

The orthogonality of the difference vector with respect to the generators $Bel_C - Bel_A$, $\emptyset \subsetneq C \subsetneq A$, of the conditioning simplex

$$\langle Bel - Bel_a, Bel_C - Bel_A \rangle = 0 \quad \forall \emptyset \subsetneq C \subsetneq A,$$

where $Bel - Bel_a$ is given by (15.15), reads as

$$\begin{cases} \sum_{B \not\subset A} m(B) \Big[\langle Bel_B, Bel_C \rangle - \langle Bel_B, Bel_A \rangle \Big] \\ \qquad + \sum_{B \subseteq A} \beta(B) \Big[\langle Bel_B, Bel_C \rangle - \langle Bel_B, Bel_A \rangle \Big] = 0 \end{cases} \forall \emptyset \subsetneq C \subsetneq A.$$

Now, categorical belief functions are such that

$$\langle Bel_B, Bel_C \rangle = |\{Y \supseteq B \cup C, Y \neq \Theta\}| = 2^{|(B \cup C)^c|} - 1 \tag{15.36}$$

and $\langle Bel_B, Bel_A \rangle = 2^{|(B \cup A)^c|} - 1$. As $(B \cup A)^c = A^c$ when $B \subseteq A$, the system of orthogonality conditions is equivalent to, $\forall \emptyset \subsetneq C \subsetneq A$,

$$\Big\{ \sum_{B \not\subset A} m(B) \Big[2^{|(B \cup C)^c|} - 2^{|(B \cup A)^c|} \Big] + \sum_{B \subsetneq A} \beta(B) \Big[2^{|(B \cup C)^c|} - 2^{|A^c|} \Big] = 0. \tag{15.37}$$

This is a system of $2^{|A|} - 2$ equations in the $2^{|A|} - 2$ variables $\{\beta(B), B \subsetneq A\}$.
In the $\{\beta(B)\}$ variables, (15.16) reads as

$$\beta(C) = -\sum_{B \subseteq A^c} m(B \cup C) 2^{-|B|} + (-1)^{|C|} \sum_{B \subseteq A^c} m(B) 2^{-|B|}.$$

To prove Theorem 95, we just need to substitute the above expression into the system of constraints (15.37). We obtain, for all $\emptyset \subsetneq C \subsetneq A$,

$$\sum_{B\not\subset A} m(B)\left[2^{|(B\cup C)^c|} - 2^{|(B\cup A)^c|}\right]$$

$$+\sum_{B\subsetneq A}\left[-\sum_{X\subseteq A^c} m(X\cup B)2^{-|X|} + (-1)^{|B|}\sum_{X\subseteq A^c} m(X)2^{-|X|}\right]$$

$$\times\left[2^{|(B\cup C)^c|} - 2^{|A^c|}\right] = 0.$$

Now, whenever $B \not\subset A$, it can bc decomposed as $B = X + Y$, with $\emptyset \subsetneq X \subseteq A^c$, $\emptyset \subseteq Y \subseteq A$. Therefore $B \cup C = (Y \cup C) + X$, $B \cup A = A + X$ and, since

$$2^{-|X|}(2^{|(Y\cup C)^c|} - 2^{|A^c|}) = 2^{|[(Y\cup C)+X]^c|} - 2^{|(A+X)^c|},$$

we can write the above system of constraints as

$$\sum_{\substack{\emptyset\subsetneq X\subseteq A^c\\ \emptyset\subseteq Y\subseteq A}} m(X+Y)\left(2^{|[(Y\cup C)+X]^c|} - 2^{|(A+X)^c|}\right)$$

$$+\sum_{\substack{\emptyset\subsetneq X\subseteq A^c\\ \emptyset\subsetneq Y\subsetneq A}} \Big((-1)^{|Y|}m(X) - m(X\cup Y)\Big)\,2^{-|X|}\left(2^{|(Y\cup C)^c|} - 2^{|A^c|}\right) = 0.$$

As $2^{-|X|}(2^{|(Y\cup C)^c|}-2^{|A^c|}) = 2^{n-|Y\cup C|-|X|}-2^{n-|A|-|X|}2^{|[(Y\cup C)+X]^c|}-2^{|(A+X)^c|}$, the system simplifies further to

$$\sum_{\substack{\emptyset\subsetneq X\subseteq A^c\\ \emptyset\subseteq Y\subseteq A}} m(X+Y)\left(2^{|[(Y\cup C)+X]^c|} - 2^{|(A+X)^c|}\right)$$

$$+\sum_{\substack{\emptyset\subsetneq X\subseteq A^c\\ \emptyset\subsetneq Y\subsetneq A}} \Big((-1)^{|Y|}m(X) - m(X+Y)\Big)\left(2^{|[(Y\cup C)+X]^c|} - 2^{|(A+X)^c|}\right) = 0.$$

After separating the contributions of $Y = \emptyset$ and $Y = A$ in the first sum, noting that $A \cup C = A$ as $C \subset A$, and splitting the second sum into a part which depends on $m(X)$ and one which depends on $m(X + Y)$, the system of constraints becomes, again for all $\emptyset \subsetneq C \subsetneq A$,

$$\sum_{\emptyset\subsetneq X\subseteq A^c}\sum_{\emptyset\subsetneq Y\subsetneq A} m(X+Y)\Big(2^{|[(Y\cup C)+X]^c|}-2^{|(A+X)^c|}\Big)$$

$$+\sum_{\emptyset\subsetneq X\subseteq A^c} m(X)\cdot\Big(2^{|(X+C)^c|}-2^{|(X+A)^c|}\Big)$$

$$+\sum_{\emptyset\subsetneq X\subseteq A^c} m(X+A)\Big(2^{|(X+A)^c|}-2^{|(X+A)^c|}\Big)$$

$$+\sum_{\emptyset\subsetneq X\subseteq A^c}\sum_{\emptyset\subsetneq Y\subsetneq A}(-1)^{|Y|}m(X)\Big(2^{|[(Y\cup C)+X]^c|}-2^{|(A+X)^c|}\Big)$$

$$-\sum_{\emptyset\subsetneq X\subseteq A^c}\sum_{\emptyset\subsetneq Y\subsetneq A} m(X+Y)\Big(2^{|[(Y\cup C)+X]^c|}-2^{|(A+X)^c|}\Big)=0.$$

By further simplification, we obtain

$$\begin{aligned}&\sum_{\emptyset\subsetneq X\subseteq A^c} m(X)\left(2^{|(X+C)^c|}-2^{|(X+A)^c|}\right)\\&+\sum_{\emptyset\subsetneq X\subseteq A^c}\sum_{\emptyset\subsetneq Y\subsetneq A}(-1)^{|Y|}m(X)\left(2^{|[(Y\cup C)+X]^c|}-2^{|(A+X)^c|}\right)=0.\end{aligned}\tag{15.38}$$

The first addend is easily reduced to

$$\sum_{\emptyset\subsetneq X\subseteq A^c} m(X)\Big(2^{|(X+C)^c|}-2^{|(X+A)^c|}\Big)=\sum_{\emptyset\subsetneq X\subseteq A^c} m(X)2^{-|X|}\Big(2^{|C^c|}-2^{|A^c|}\Big).$$

For the second addend, we have

$$\begin{aligned}&\sum_{\emptyset\subsetneq X\subseteq A^c}\sum_{\emptyset\subsetneq Y\subsetneq A}(-1)^{|Y|}m(X)\Big(2^{|[(Y\cup C)+X]^c|}-2^{|(A+X)^c|}\Big)\\
&=\sum_{\emptyset\subsetneq X\subseteq A^c} m(X)2^{-|X|}\sum_{\emptyset\subsetneq Y\subsetneq A}(-1)^{|Y|}\Big(2^{|(Y\cup C)^c|}-2^{|A^c|}\Big)\\
&=\sum_{\emptyset\subsetneq X\subseteq A^c} m(X)2^{-|X|}\Bigg[\sum_{\emptyset\subseteq Y\subseteq A}(-1)^{|Y|}\Big(2^{|(Y\cup C)^c|}-2^{|A^c|}\Big)\\
&\qquad-\Big(2^{|(\emptyset\cup C)^c|}-2^{|A^c|}\Big)-(-1)^{|A|}\Big(2^{|(A\cup C)^c|}-2^{|A^c|}\Big)\Bigg]\\
&=\sum_{\emptyset\subsetneq X\subseteq A^c} m(X)2^{-|X|}\left[\sum_{\emptyset\subseteq Y\subseteq A}(-1)^{|Y|}\Big(2^{|(Y\cup C)^c|}-2^{|A^c|}\Big)-\Big(2^{|C^c|}-2^{|A^c|}\Big)\right].\end{aligned}\tag{15.39}$$

At this point we can notice that

$$\sum_{\emptyset \subseteq Y \subseteq A} (-1)^{|Y|} \left(2^{|(Y \cup C)^c|} - 2^{|A^c|} \right)$$

$$= \sum_{\emptyset \subseteq Y \subseteq A} (-1)^{|Y|} 2^{|(Y \cup C)^c|} - 2^{|A^c|} \sum_{\emptyset \subseteq Y \subseteq A} (-1)^{|Y|} = \sum_{\emptyset \subseteq Y \subseteq A} (-1)^{|Y|} 2^{|(Y \cup C)^c|},$$

since $\sum_{\emptyset \subseteq Y \subseteq A} (-1)^{|Y|} = 0$ by Newton's binomial theorem.

For the remaining term in (15.39), using a standard technique, we can decompose Y into the disjoint sum $Y = (Y \cap C) + (Y \setminus C)$ and rewrite the term as

$$\sum_{\emptyset \subseteq Y \subseteq A} (-1)^{|Y|} 2^{|(Y \cup C)^c|}$$

$$= \sum_{|Y \cap C|=0}^{|C|} \binom{|C|}{|Y \cap C|} \sum_{|Y \setminus C|}^{|A \setminus C|} \binom{|A \setminus C|}{|Y \setminus C|} (-1)^{|Y \cap C| + |Y \setminus C|} 2^{n - |Y \setminus C| - |C|}$$

$$= 2^{n-|C|} \sum_{|Y \cap C|=0}^{|C|} \binom{|C|}{|Y \cap C|} (-1)^{|Y \cap C|} \sum_{|Y \setminus C|}^{|A \setminus C|} \binom{|A \setminus C|}{|Y \setminus C|} (-1)^{|Y \setminus C|} 2^{-|Y \setminus C|},$$

where

$$\sum_{|Y \setminus C|}^{|A \setminus C|} \binom{|A \setminus C|}{|Y \setminus C|} (-1)^{|Y \setminus C|} 2^{-|Y \setminus C|} = \left(-1 + \frac{1}{2} \right)^{|A \setminus C|} = -2^{-|A \setminus C|}$$

by Newton's binomial theorem. Hence we obtain

$$\sum_{\emptyset \subseteq Y \subseteq A} (-1)^{|Y|} 2^{|(Y \cup C)^c|} = -2^{n-|C|-|A \setminus C|} \sum_{|Y \cap C|=0}^{|C|} \binom{|C|}{|Y \cap C|} (-1)^{|Y \cap C|} = 0,$$

again by Newton's binomial theorem. By substituting this result in cascade into (15.39) and (15.38), we have that the system of constraints is always satisfied, as it reduces to the equality $0 = 0$.

Proof of Lemma 18

After introducing the auxiliary variables $\beta(B) = m(B) - m_a(B)$ and $\gamma(B) = \sum_{C \subseteq B} \beta(B)$, the desired norm becomes

$$
\begin{aligned}
&\|Bel - Bel_a\|_{L_1} \\
&= \sum_{\emptyset \subsetneq B \subsetneq \Theta} |Bel(B) - Bel_a(B)| = \sum_{\emptyset \subsetneq B \subsetneq \Theta} \left| \sum_{\emptyset \subsetneq C \subseteq A \cap B} \beta(C) + \sum_{C \subseteq B, C \not\subset A} m(C) \right| \\
&= \sum_{\emptyset \subsetneq B \subsetneq \Theta} \left| \gamma(A \cap B) + \sum_{C \subseteq B, C \not\subset A} m(C) \right| \\
&= \sum_{B : B \cap A = \emptyset} \left| \sum_{C \subseteq B, C \not\subset A} m(C) \right| + \sum_{B : B \cap A \neq \emptyset, \Theta} \left| \gamma(A \cap B) + \sum_{C \subseteq B, C \not\subset A} m(C) \right| \\
&\quad + \sum_{B : B \cap A = A, B \neq \Theta} \left| \gamma(A) + \sum_{C \subseteq B, C \not\subset A} m(C) \right|,
\end{aligned}
$$

where

$$
\gamma(A) = \sum_{C \subseteq A} \beta(C) = \sum_{C \subseteq A} m(C) - \sum_{C \subseteq A} m_a(C) = Bel(A) - 1.
$$

Thus, the first and the third addend above are constant, and since

$$
\sum_{C \subseteq B, C \not\subset A} m(C) = Bel(B) - Bel(B \cap A),
$$

we obtain, as desired,

$$
\arg\min_{\gamma} \|Bel - Bel_a\|_{L_1} = \arg\min_{\gamma} \sum_{\emptyset \subsetneq B \cap A \subsetneq A} \left| \gamma(B \cap A) + \sum_{C \subseteq B, C \not\subset A} m(C) \right|.
$$

Proof of Lemma 19

The first term in (15.23) is such that

$$
\max_{B : B \cap A = \emptyset} \left| \sum_{C \subseteq B, C \not\subset A} m(C) \right| = \max_{B \subseteq A^c} \left| \sum_{C \subseteq B} m(C) \right| = \max_{B \subseteq A^c} Bel(B) = Bel(A^c).
$$

For the third one, we have instead

$$
\sum_{C \subseteq B, C \not\subset A} m(C) \leq \sum_{C \cap A^c} m(C) = Pl(A^c) = 1 - Bel(A),
$$

which is maximised when $B = A$, in which case it is equal to

$$\left| Bel(A) - 1 + \sum_{C \subseteq A, C \not\subseteq A} m(C) \right| = |Bel(A) - 1 + 0| = 1 - Bel(A).$$

Therefore, the L_∞ norm (15.23) of the difference $Bel - Bel_a$ reduces to

$$\max_{\emptyset \subsetneq B \subsetneq \Theta} |Bel - Bel_a(B)|$$

$$= \max \left\{ \max_{B: B \cap A \neq \emptyset, A} \left| \gamma(A \cap B) + \sum_{C \subseteq B, C \not\subseteq A} m(C) \right|, 1 - Bel(A) \right\}, \tag{15.40}$$

which is obviously minimised by all the values of $\gamma^*(X)$ such that

$$\max_{B: B \cap A \neq \emptyset, A} \left| \gamma^*(A \cap B) + \sum_{C \subseteq B, C \not\subseteq A} m(C) \right| \leq 1 - Bel(A).$$

The variable term in (15.40) can be decomposed into collections of terms which depend on the same individual variable $\gamma(X)$:

$$\max_{B: B \cap A \neq \emptyset, A} \left| \gamma(A \cap B) + \sum_{C \subseteq B, C \not\subseteq A} m(C) \right|$$

$$= \max_{\emptyset \subsetneq X \subsetneq A} \max_{\emptyset \subseteq Y \subseteq A^c} \left| \gamma(X) + \sum_{\emptyset \subsetneq Z \subseteq Y} \sum_{\emptyset \subseteq W \subseteq X} m(Z + W) \right|,$$

where $B = X + Y$, with $X = A \cap B$ and $Y = B \cap A^c$. Note that $Z \neq \emptyset$, as $C = Z + W \not\subseteq A$. Therefore, the global optimal solution decomposes into a collection of solutions $\{\gamma^*(X), \emptyset \subsetneq X \subsetneq A\}$ for each individual problem, where

$$\gamma^*(X) : \max_{\emptyset \subseteq Y \subseteq A^c} \left| \gamma^*(X) + \sum_{\emptyset \subsetneq Z \subseteq Y} \sum_{\emptyset \subseteq W \subseteq X} m(Z + W) \right| \leq 1 - Bel(A). \tag{15.41}$$

We need to distinguish three cases:

1. If $\gamma^*(X) \geq 0$, we have that

$$\gamma^*(X) : \max_{\emptyset \subseteq Y \subseteq A^c} \left\{ \gamma^*(X) + \sum_{\emptyset \subsetneq Z \subseteq Y} \sum_{\emptyset \subseteq W \subseteq X} m(Z + W) \right\}$$

$$= \gamma^*(X) + \sum_{\emptyset \subsetneq Z \subseteq A^c} \sum_{\emptyset \subseteq W \subseteq X} m(Z + W)$$

$$= \gamma^*(X) + \sum_{C \cap A^c \neq \emptyset, C \cap A \subseteq X} m(C) \leq 1 - Bel(A),$$

since when $\gamma^*(X) \geq 0$ the argument to be maximised is non-negative, and its maximum is trivially achieved by $Y = A^c$. Hence, all the

$$\gamma^*(X) : \gamma^*(X) \leq 1 - Bel(A) - \sum_{C \cap A^c \neq \emptyset, C \cap A \subseteq X} m(C) \tag{15.42}$$

are optimal.

2. If $\gamma^*(X) < 0$, the maximum in (15.41) can be achieved by either $Y = A^c$ or $Y = \emptyset$, and we are left with the two corresponding terms in the max:

$$\max_{\emptyset \subseteq Y \subseteq A^c} \left\{ \left| \gamma^*(X) + \sum_{C \cap A^c \neq \emptyset, C \cap A \subseteq X} m(C) \right|, -\gamma^*(X) \right\} \leq 1 - Bel(A). \tag{15.43}$$

Now, either

$$\left| \gamma^*(X) + \sum_{C \cap A^c \neq \emptyset, C \cap A \subseteq X} m(C) \right| \geq -\gamma^*(X)$$

or vice versa. In the first case, since the argument of the absolute value has to be non-negative,

$$\gamma^*(X) \geq -\frac{1}{2} \sum_{C \cap A^c \neq \emptyset, C \cap A \subseteq X} m(C).$$

Furthermore, the optimality condition is satisfied when

$$\gamma^*(X) + \sum_{C \cap A^c \neq \emptyset, C \cap A \subseteq X} m(C) \leq 1 - Bel(A),$$

which is equivalent to

$$\gamma^*(X) \leq 1 - Bel(A) - \sum_{C \cap A^c \neq \emptyset, C \cap A \subseteq X} m(C) = \sum_{C \cap A^c \neq \emptyset, C \cap (A \setminus X) \neq \emptyset} m(C),$$

which in turn is trivially true for $\gamma^*(X) < 0$ and $m(C) \geq 0$ for all C. Therefore, all

$$0 \geq \gamma^*(X) \geq -\frac{1}{2} \sum_{C \cap A^c \neq \emptyset, C \cap A \subseteq X} m(C) \tag{15.44}$$

are optimal as well.

3. In the last case,

$$\left| \gamma^*(X) + \sum_{C \cap A^c \neq \emptyset, C \cap A \subseteq X} m(C) \right| \leq -\gamma^*(X),$$

i.e.,

$$\gamma^*(X) \leq -\frac{1}{2} \sum_{C \cap A^c \neq \emptyset, C \cap A \subseteq X} m(C).$$

The optimality condition is satisfied for

$$-\gamma^*(X) \leq 1 - Bel(A) \equiv \gamma^*(X) \geq Bel(A) - 1,$$

which is satisfied for all

$$Bel(A) - 1 \leq \gamma^*(X) \leq -\frac{1}{2} \sum_{C \cap A^c \neq \emptyset, C \cap A \subseteq X} m(C). \tag{15.45}$$

Putting (15.42), (15.44) and (15.45) together, we have the thesis.

Proof of Theorem 97

Following Lemma 19, it is not difficult to see by induction that, in the original auxiliary variables $\{\beta(X)\}$, the set of L_∞ conditional belief functions in $\mathcal{B}$ is determined by the following constraints:

$$-K(X) + (-1)^{|X|} \sum_{\substack{C \cap A^c \neq \emptyset \\ C \cap A \subseteq X}} m(C) \leq \beta(X) \leq K(X) - \sum_{\substack{C \cap A^c \neq \emptyset \\ C \cap A \subseteq X}} m(C), \tag{15.46}$$

where we have defined

$$K(X) \doteq (2^{|X|} - 1)(1 - Bel(A)) - \sum_{C \cap A^c \neq \emptyset, \emptyset \subseteq C \cap A \subsetneq X} m(C).$$

In the masses of the sought L_∞ conditional belief functions, (15.46) becomes

$$\begin{cases} m(X) - K(X) + \sum_{\emptyset \subsetneq B \subseteq A^c} m(X + B) \leq m_a(X), \\ m_a(X) \leq m(X) + K(X) - (-1)^{|X|} \sum_{B \subseteq A^c} m(B), \end{cases}$$

which reads, after replacing the expression for $K(X)$, as

$$m(X) + (2^{|X|} - 1)(1 - Bel(A)) + \sum_{\substack{C \cap A^c \neq \emptyset \\ \emptyset \subseteq C \cap A \subsetneq X}} m(C) + \sum_{\emptyset \subsetneq B \subseteq A^c} m(X + B)$$

$$\leq m_a(X) \leq$$

$$m(X) + (2^{|X|} - 1)(1 - Bel(A)) - \sum_{\substack{C \cap A^c \neq \emptyset \\ \emptyset \subseteq C \cap A \subsetneq X}} m(C) - (-1)^{|X|} \sum_{B \subseteq A^c} m(B).$$

By further trivial simplification, we have the result as desired.

Proof of Theorem 98

The proof is by substitution. In the $\{\beta(B)\}$ variables, the thesis reads as

$$\beta(C) = \frac{1}{2} \sum_{\emptyset \subsetneq B \subseteq A^c} \Big[(-1)^{|C|} m(B) - m(B \cup C) \Big]. \tag{15.47}$$

By substituting (15.47) into (15.26), we get, since

$$\sum_{\emptyset \subsetneq C \subseteq X} (-1)^{|C|} = 0 - (-1)^0 = -1$$

by Newton's binomial theorem,

$$\begin{aligned}
&\frac{1}{2} \sum_{\substack{\emptyset \subsetneq C \subseteq X \\ \emptyset \subsetneq B \subseteq A^c}} \Big[(-1)^{|C|} m(B) - m(B \cup C) \Big] + \frac{1}{2} \sum_{C \cap A^c \neq \emptyset, C \cap A \subseteq X} m(C) \\
&= \frac{1}{2} \sum_{\emptyset \subsetneq B \subseteq A^c} m(B) \sum_{\emptyset \subsetneq C \subseteq X} (-1)^{|C|} - \frac{1}{2} \sum_{\substack{\emptyset \subsetneq B \subseteq A^c \\ \emptyset \subsetneq C \subseteq X}} m(B \cup C) + \frac{1}{2} \sum_{\substack{C \cap A^c \neq \emptyset \\ C \cap A \subseteq X}} m(C) \\
&= -\frac{1}{2} \sum_{\emptyset \subsetneq B \subseteq A^c} m(B) - \frac{1}{2} \sum_{\substack{\emptyset \subsetneq B \subseteq A^c \\ \emptyset \subsetneq C \subseteq X}} m(B \cup C) + \frac{1}{2} \sum_{\substack{C \cap A^c \neq \emptyset \\ C \cap A \subseteq X}} m(C) \\
&= -\frac{1}{2} \sum_{\emptyset \subsetneq B \subseteq A^c} \sum_{\emptyset \subseteq C \subseteq X} m(B \cup C) + \frac{1}{2} \sum_{C \cap A^c \neq \emptyset, C \cap A \subseteq X} m(C) \\
&= -\frac{1}{2} \sum_{C \cap A^c \neq \emptyset, C \cap A \subseteq X} m(C) + \frac{1}{2} \sum_{C \cap A^c \neq \emptyset, C \cap A \subseteq X} m(C) = 0
\end{aligned}$$

for all $\emptyset \subsetneq X \subsetneq A$, and the system (15.26) is satisfied.

16

Decision making with epistemic transforms

As we learned in Chapter 4, decision making with belief functions has been extensively studied. Approaches based on (upper/lower) expected utility (e.g. Strat's approach) and multicriteria decision making, in particular, have attracted much attention.

In the transferable belief model [1730, 1717], in particular, decision making is done by maximising the expected utility of actions based on the pignistic transform,

$$E[u] = \sum_{\omega \in \Omega} u(f, \omega) BetP(\omega),$$

where Ω is the collection of all the possible outcomes ω, $\mathcal{F}$ is the set of possible actions f and the utility function u is defined on $\mathcal{F} \times \Omega$. As we know, besides satisfying a number of sensible rationality principles, this probability transform has a nice geometric interpretation in the probability simplex as the barycentre of the credal set of probability measures consistent with the original belief function Bel,

$$\mathcal{P}[Bel] \doteq \Big\{ P \in \mathcal{P} : P(A) \geq Bel(A) \; \forall A \subseteq \Theta \Big\}.$$

Betting and credal semantics seem to be connected, in the case of the pignistic transform. Unfortunately, while their geometry in the belief space is well understood, a credal semantics is still lacking for most of the transforms that we studied in Part III.

We address this issue here in the framework of *probability intervals* [1808, 393] (Chapter 6, Section 6.3), which we briefly recall here.

F. Cuzzolin, *The Geometry of Uncertainty*, Artificial Intelligence: Foundations, Theory, and Algorithms, https://doi.org/10.1007/978-3-030-63153-6_16

A *set of probability intervals* or *interval probability system* is a system of constraints on the probability values of a probability distribution $p : \Theta \rightarrow [0, 1]$ on a finite domain Θ of the form

$$\mathcal{P}[(l, u)] \doteq \Big\{p : l(x) \leq p(x) \leq u(x), \forall x \in \Theta\Big\}. \tag{16.1}$$

Probability intervals were introduced as a tool for uncertain reasoning in [393], where combination and marginalisation of intervals were studied in detail. Specific constraints for such intervals to be consistent and tight were also given. A typical way in which probability intervals arise is through measurement errors, for measurements can be inherently of interval nature (owing to the finite resolution of the instruments) [863]. In such a case, the probability interval of interest is the class of probability measures consistent with the measured interval.

A set of constraints of the form (6.10) determines a convex set of probabilities, or 'credal set' [1141]. Lower and upper probabilities (see Section 6.1) determined by $\mathcal{P}[(l, u)]$ on any event $A \subseteq \Theta$ can be easily obtained from the lower and upper bounds (l, u) as follows:

$$\begin{aligned} \underline{P}(A) &= \max\left\{\sum_{x \in A} l(x), 1 - \sum_{x \notin A} u(x)\right\}, \\ \underline{P}(A) &= \min\left\{\sum_{x \in A} u(x), 1 - \sum_{x \notin A} l(x)\right\}. \end{aligned} \tag{16.2}$$

Making decisions based on credal sets is not trivial, for the natural extensions of the classical expected utility rule amount to multiple potentially optimal decisions [1827]. As an alternative, similarly to what is done for belief functions, a single probability measure may be sought to represent the credal set associated with a set of probability intervals.

Chapter content

As we show here, the credal set associated with a probability interval possesses an interesting structure, as it can be decomposed into a pair of simplices [344].

Indeed, the probabilities consistent with a certain interval system (6.10) lie in the intersection of two simplices: a *lower simplex* $T[l]$ determined by the lower bound $l(x) \leq p(x)$, and an *upper simplex* $T[u]$ determined by the upper constraint $p(x) \leq u(x)$:

$$T[l] \doteq \Big\{p : p(x) \geq l(x) \; \forall x \in \Theta\Big\}, \quad T[u] \doteq \Big\{p : p(x) \leq u(x) \;\; \forall x \in \Theta\Big\}. \tag{16.3}$$

For interval probability systems associated with random sets (belief functions), the two simplices read as

$$T^1[Bel] \doteq \Big\{p : p(x) \geq Bel(x) \; \forall x \in \Theta\Big\},$$

$$T^{n-1}[Bel] \doteq \Big\{p : p(x) \leq Pl(x) \; \forall x \in \Theta\Big\}.$$

Note that, in both cases, these regions are simplices, rather than just polytopes, only when 'pseudo'-probabilities not satisfying the non-negativity axiom are considered (see Section 16.1.1).

This geometric characterisation allows us to provide probability transforms of the epistemic family (Chapter 12) with a credal semantics similar to that of the pignistic function. Each such transform can be described as the (special) *focus* f of a pair S, T of simplices, i.e., the unique probability measure falling on the intersection of the lines joining corresponding vertices of S and T (see Fig. 16.1), for an appropriate permutation of these vertices, and has the same simplicial coordinates in the two simplices.

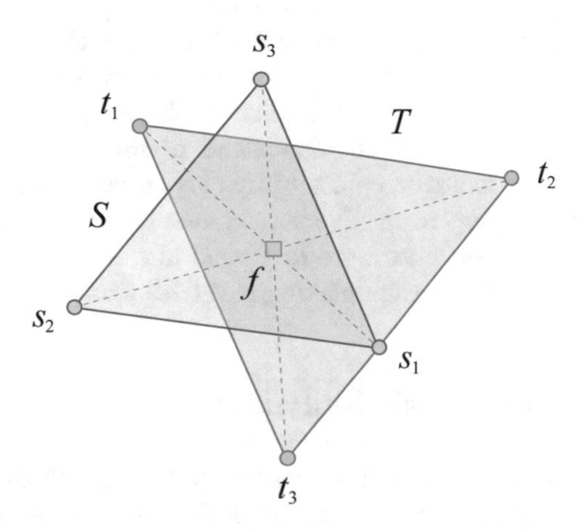

Fig. 16.1: When it exists, the focus of a pair of simplices S, T is, for an appropriate permutation of their vertices, the unique intersection of the lines joining their corresponding vertices s_i, t_i.

In particular we prove in this chapter that, while the relative belief of singletons is the focus of $\{\mathcal{P}, T^1[Bel]\}$, the relative plausibility of singletons is the focus of $\{\mathcal{P}, T^{n-1}[Bel]\}$, and the intersection probability is that of $\{T^1[Bel], T^{n-1}[Bel]\}$. Their focal coordinates encode major features of the underlying belief function: the total mass it assigns to singletons, their total plausibility and the fraction of the related probability interval which determines the intersection probability. As the centre of mass is a special case of a focus, this credal interpretation of epistemic transforms potentially paves the way for TBM-like frameworks based on those transformations.

Chapter outline

We start by proving that the credal set associated with a system of probability intervals can be decomposed in terms of a pair of upper and lower simplices (Section 16.1). We point out that the intersection probability, although originally defined for belief functions (Chapter 11), is closely linked to the notion of an interval probability system and can be seen as the natural representative of the associated credal set (Section 16.2).

Drawing inspiration from an analysis of the ternary case (Section 16.3), we prove in Section 16.4 that all the probability transformations considered (relative belief and plausibility of singletons, and intersection probability) are geometrically the (special) foci of different pairs of simplices, and discuss the meaning of the mapping associated with a focus in terms of mass assignment. We prove that the upper and lower simplices can themselves be interpreted as the sets of probabilities consistent with the belief and plausibility of singletons.

We conclude by discussing decision making with epistemic transforms (Section 16.5). More particularly, in Section 16.5.1 decision-making frameworks based on the credal interpretations of upper and lower probability constraints introduced and the associated probability transformations which generalise the transferable belief model are outlined. In Section 16.5.2, instead, preliminary results are discussed which show that relative belief and plausibility play an interesting role in determining the safest betting strategy in an adversarial game scenario in which the decision maker has to minimise their maximum loss/maximise their minimum return, in a modified Wald approach to decision making (see Chapter 4, Section 4.8.1).

16.1 The credal set of probability intervals

Just as belief functions do, probability interval systems admit a credal representation, which for intervals associated with belief functions is also strictly related to the credal set $\mathcal{P}[Bel]$ of all consistent probabilities.

By the definition (3.10) of $\mathcal{P}[Bel]$, it follows that the polytope of consistent probabilities can be decomposed into a number of component polytopes, namely

$$\mathcal{P}[Bel] = \bigcap_{i=1}^{n-1} \mathcal{P}^i[Bel], \tag{16.4}$$

where $\mathcal{P}^i[Bel]$ is the set of probabilities that satisfy the lower probability constraint for size-i events,

$$\mathcal{P}^i[Bel] \doteq \Big\{ P \in \mathcal{P} : P(A) \geq Bel(A), \forall A : |A| = i \Big\}. \tag{16.5}$$

Note that for $i = n$ the constraint is trivially satisfied by all probability measures P: $\mathcal{P}^n[Bel] = \mathcal{P}$.

16.1.1 Lower and upper simplices

A simple and elegant geometric description of interval probability systems can be provided if, instead of considering the polytopes (16.5), we focus on the credal sets

$$T^i[Bel] \doteq \Big\{P' \in \mathcal{P}' : P'(A) \geq Bel(A), \forall A : |A| = i\Big\}.$$

Here $\mathcal{P}'$ denotes the set of all *pseudo-probability measures* P' on Θ, whose distribution $p' : \Theta \to \mathbb{R}$ satisfy the normalisation constraint $\sum_{x\in\Theta} p'(x) = 1$ but not necessarily the non-negativity one – there may exist an element x such that $p'(x) < 0$. In particular, we focus here on the set of pseudo-probability measures which satisfy the lower constraint on singletons,

$$T^1[Bel] \doteq \Big\{p' \in \mathcal{P}' : p'(x) \geq Bel(x) \;\; \forall x \in \Theta\Big\}, \tag{16.6}$$

and the set $T^{n-1}[Bel]$ of pseudo-probability measures which satisfy the analogous constraint on events of size $n-1$,

$$\begin{aligned} T^{n-1}[Bel] &\doteq \Big\{P' \in \mathcal{P}' : P'(A) \geq Bel(A) \;\; \forall A : |A| = n-1\Big\} \\ &= \Big\{P' \in \mathcal{P}' : P'(\{x\}^c) \geq Bel(\{x\}^c) \;\; \forall x \in \Theta\Big\} \\ &= \Big\{p' \in \mathcal{P}' : p'(x) \leq Pl(x) \;\; \forall x \in \Theta\Big\}, \end{aligned} \tag{16.7}$$

i.e., the set of pseudo-probabilities which satisfy the upper constraint on singletons.

16.1.2 Simplicial form

The extension to pseudo-probabilities allows us to prove that the credal sets (16.6) and (16.7) have the form of simplices.

Theorem 99. *The credal set $T^1[Bel]$, or* lower simplex, *can be written as*

$$T^1[Bel] = Cl(t^1_x[Bel], x \in \Theta), \tag{16.8}$$

namely as the convex closure of the vertices

$$t^1_x[Bel] = \sum_{y\neq x} m(y) Bel_y + \left(1 - \sum_{y \neq x} m(y)\right) Bel_x. \tag{16.9}$$

Dually, the upper simplex $T^{n-1}[Bel]$ *reads as the convex closure*

$$T^{n-1}[Bel] = Cl(t^{n-1}_x[Bel], x \in \Theta) \tag{16.10}$$

of the vertices

$$t^{n-1}_x[Bel] = \sum_{y\neq x} Pl(y) Bel_y + \left(1 - \sum_{y \neq x} Pl(y)\right) Bel_x. \tag{16.11}$$

By (16.9), each vertex $t_x^1[Bel]$ of the lower simplex is a pseudo-probability that adds the total mass $1 - k_{Bel}$ of non-singletons to that of the element x, leaving all the others unchanged:

$$m_{t_x^1[Bel]}(x) = m(x) + 1 - k_{Bel}, \quad m_{t_x^1[Bel]}(y) = m(y) \; \forall y \neq x.$$

In fact, as $m_{t_x^1[Bel]}(z) \geq 0$ for all $z \in \Theta$ and for all $x \in \Theta$ (all $t_x^1[Bel]$ are actual probabilities), we have that

$$T^1[Bel] = \mathcal{P}^1[Bel], \tag{16.12}$$

and $T^1[Bel]$ is completely included in the probability simplex.

On the other hand, the vertices (16.11) of the upper simplex are not guaranteed to be valid probabilities.
Each vertex $t_x^{n-1}[Bel]$ assigns to each element of Θ different from x its plausibility $Pl(y) = pl(y)$, while it subtracts from $Pl(x)$ the 'excess' plausibility $k_{Pl} - 1$:

$$\begin{aligned} m_{t_x^{n-1}[Bel]}(x) &= Pl(x) + (1 - k_{Pl}), \\ m_{t_x^{n-1}[Bel]}(y) &= Pl(y) \qquad\qquad\qquad \forall y \neq x. \end{aligned}$$

Now, as $1 - k_{Pl}$ can be a negative quantity, $m_{t_x^{n-1}[Bel]}(x)$ can also be negative and $t_x^{n-1}[Bel]$ is not guaranteed to be a 'true' probability.

We will have confirmation of this fact in the example in Section 16.3.

16.1.3 Lower and upper simplices and probability intervals

By comparing (6.10), (16.6) and (16.7), it is clear that the credal set $\mathcal{P}[(l, u)]$ associated with a set of probability intervals (l, u) is nothing but the intersection

$$\mathcal{P}[(l, u)] = T[l] \cap T[u]$$

of the lower and upper simplices (16.3) associated with its lower- and upper-bound constraints, respectively. In particular, when these lower and upper bounds are those enforced by a pair of belief and plausibility measures on the singleton elements of a frame of discernment, $l(x) = Bel(x)$ and $u(x) = Pl(x)$, we get

$$\mathcal{P}[(Bel, Pl)] = T^1[Bel] \cap T^{n-1}[Bel].$$

16.2 Intersection probability and probability intervals

There are clearly many ways of selecting a single measure to represent a collection of probability intervals (6.10). Note, however, that each of the intervals $[l(x), u(x)]$, $x \in \Theta$, carries the same weight within the system of constraints (6.10), as there is no reason for the different elements x of the domain to be treated differently. It is then sensible to require that the desired representative probability should behave homogeneously in each element x of the frame Θ.

Mathematically, this translates into seeking a probability distribution $p : \Theta \to [0, 1]$ such that

$$p(x) = l(x) + \alpha(u(x) - l(x))$$

for all the elements x of Θ, and some constant value $\alpha \in [0, 1]$ (see Fig. 16.2). This value needs to be between 0 and 1 in order for the sought probability distribution p to belong to the interval. It is easy to see that there is indeed a *unique* solution to

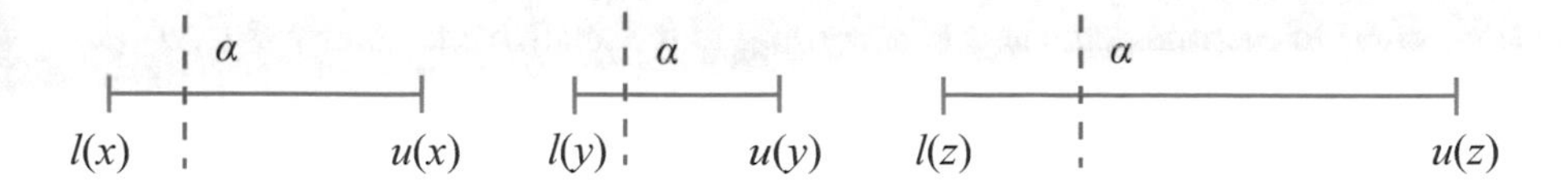

Fig. 16.2: An illustration of the notion of the intersection probability for an interval probability system (l, u) on $\Theta = \{x, y, z\}$ (6.10).

this problem. It suffices to enforce the normalisation constraint

$$\sum_x p(x) = \sum_x \Big[l(x) + \alpha(u(x) - l(x)) \Big] = 1$$

to understand that the unique value of α is given by

$$\alpha = \beta[(l, u)] \doteq \frac{1 - \sum_{x \in \Theta} l(x)}{\sum_{x \in \Theta} \big(u(x) - l(x)\big)}. \tag{16.13}$$

Definition 133. *The* intersection probability $p[(l, u)] : \Theta \to [0, 1]$ *associated with the interval probability system (6.10) is the probability distribution*

$$p[(l, u)](x) = \beta[(l, u)]u(x) + (1 - \beta[(l, u)])l(x), \tag{16.14}$$

with $\beta[(l, u)]$ *given by (16.13).*

The ratio $\beta[(l, u)]$ (16.13) measures the fraction of each interval $[l(x), u(x)]$ which we need to add to the lower bound $l(x)$ to obtain a valid probability function (adding up to one).

It is easy to see that when (l, u) are a pair of belief/plausibility measures (Bel, Pl), we obtain the intersection probability that we defined for belief functions (Section 11.3). Although originally defined by geometric means, the intersection probability is thus in fact 'the' rational probability transform for general interval probability systems.

As was the case for $p[Bel]$, $p[(l, u)]$ can also be written as

$$p[(l,u)](x) = l(x) + \left(1 - \sum_x l(x)\right) R[(l,u)](x), \tag{16.15}$$

where

$$R[(l,u)](x) \doteq \frac{u(x) - l(x)}{\sum_{y\in\Theta}(u(y) - l(y))} = \frac{\Delta(x)}{\sum_{y\in\Theta}\Delta(y)}. \tag{16.16}$$

Here $\Delta(x)$ measures the width of the probability interval for x, whereas $R[(l,u)] : \Theta \to [0,1]$ measures how much the uncertainty in the probability value of each singleton 'weighs' on the total width of the interval system (6.10). We thus term it the *relative uncertainty* of singletons. Therefore, $p[(l,u)]$ distributes the mass $(1 - \sum_x l(x))$ to each singleton $x \in \Theta$ according to the relative uncertainty $R[(l,u)](x)$ it carries for the given interval.

16.3 Credal interpretation of Bayesian transforms: Ternary case

Let us first consider the case of a frame of cardinality 3, $\Theta = \{x,y,z\}$, and a belief function Bel with mass assignment

$$\begin{array}{ccc} m(x) = 0.2, & m(y) = 0.1, & m(z) = 0.3, \\ m(\{x,y\}) = 0.1, & m(\{y,z\}) = 0.2, & m(\Theta) = 0.1. \end{array} \tag{16.17}$$

Figure 16.3 illustrates the geometry of the related credal set $\mathcal{P}[Bel]$ in the simplex $Cl(P_x, P_y, P_z)$ of all the probability measures on Θ.

It is well known (see Section 3.1.4) that the credal set associated with a belief function is a polytope whose vertices are associated with all possible permutations of singletons.

Proposition 66. *Given a belief function $Bel : 2^\Theta \to [0,1]$, the simplex $\mathcal{P}[Bel]$ of the probability measures consistent with Bel is the polytope*

$$\mathcal{P}[Bel] = Cl(P^\rho[Bel]\ \forall \rho),$$

where ρ is any permutation $\{x_{\rho(1)}, \ldots, x_{\rho(n)}\}$ of the singletons of Θ, and the vertex $P^\rho[Bel]$ is the Bayesian belief function such that

$$P^\rho[Bel](x_{\rho(i)}) = \sum_{A \ni x_\rho(i), A \not\ni x_\rho(j)\ \forall j<i} m(A). \tag{16.18}$$

By Proposition 66, for the example belief function (16.17), $\mathcal{P}[Bel]$ has as vertices the probabilities $P^{\rho^1}, P^{\rho^2}, P^{\rho^3}, P^{\rho^4}, P^{\rho^5}[Bel]$ identified by purple squares in Fig. 16.3, namely

$$\begin{array}{ll} \rho^1 = (x,y,z): & P^{\rho^1}[Bel](x) = .4,\ P^{\rho^1}[Bel](y) = .3,\ P^{\rho^1}[Bel](z) = .3, \\ \rho^2 = (x,z,y): & P^{\rho^2}[Bel](x) = .4,\ P^{\rho^2}[Bel](y) = .1,\ P^{\rho^2}[Bel](z) = .5, \\ \rho^3 = (y,x,z): & P^{\rho^3}[Bel](x) = .2,\ P^{\rho^3}[Bel](y) = .5,\ P^{\rho^3}[Bel](z) = .3, \\ \rho^4 = (z,x,y): & P^{\rho^4}[Bel](x) = .3,\ P^{\rho^4}[Bel](y) = .1,\ P^{\rho^4}[Bel](z) = .6, \\ \rho^5 = (z,y,x): & P^{\rho^5}[Bel](x) = .2,\ P^{\rho^5}[Bel](y) = .2,\ P^{\rho^5}[Bel](z) = .6 \end{array} \tag{16.19}$$

(as the permutations (y, x, z) and (y, z, x) yield the same probability distribution).

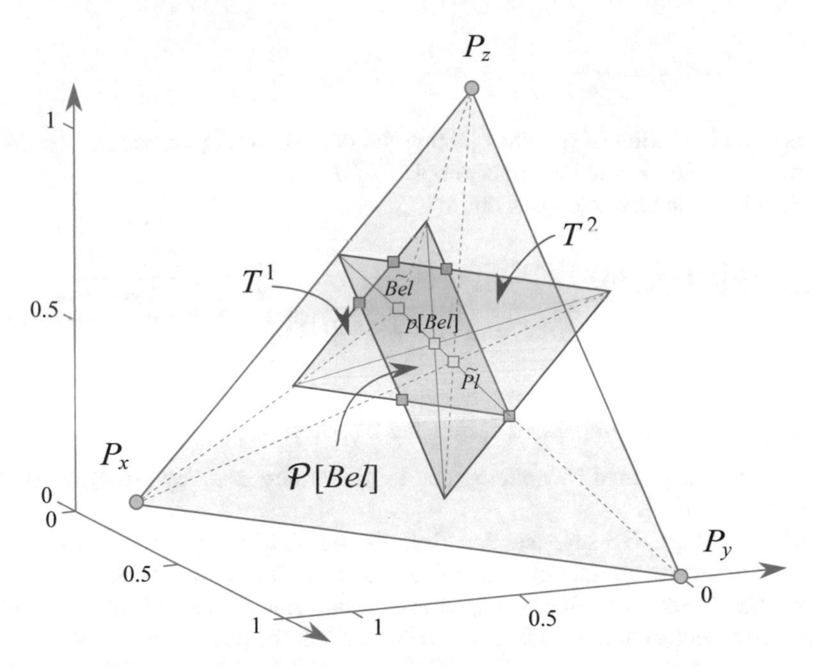

Fig. 16.3: The polytope $\mathcal{P}[Bel]$ of probabilities consistent with the belief function Bel (16.17) defined on $\{x, y, z\}$ is shown in a darker shade of purple, in the probability simplex $\mathcal{P} = Cl(P_x, P_y, P_z)$ (shown in light blue). Its vertices (purple squares) are given in (16.19). The intersection probability, relative belief and relative plausibility of singletons (green squares) are the foci of the pairs of simplices $\{T^1[Bel], T^2[Bel]\}$, $\{T^1[Bel], \mathcal{P}\}$ and $\{\mathcal{P}, T^2[Bel]\}$, respectively. In the ternary case, $T^1[Bel]$ and $T^2[Bel]$ (light purple) are regular triangles. Geometrically, their focus is the intersection of the lines joining their corresponding vertices (dashed lines for $\{T^1[Bel], \mathcal{P}\}$,$\{\mathcal{P}, T^2[Bel]\}$; solid lines for $\{T^1[Bel], T^2[Bel]\}$).

We can notice a number of interesting facts:

1. $\mathcal{P}[Bel]$ (the polygon delimited by the purple squares) is the intersection of the two triangles (two-dimensional simplices) $T^1[Bel]$ and $T^2[Bel]$.
2. The relative belief of singletons,

$$\tilde{Bel}(x) = \frac{.2}{.6} = \frac{1}{3}, \quad \tilde{Bel}(y) = \frac{.1}{.6} = \frac{1}{6}, \quad \tilde{Bel}(z) = \frac{.3}{.6} = \frac{1}{2},$$

   is the intersection of the lines joining the corresponding vertices of the probability simplex $\mathcal{P}$ and the lower simplex $T^1[Bel]$.
3. The relative plausibility of singletons,

$$\tilde{Pl}(x) = \frac{m(x) + m(\{x,y\}) + m(\Theta)}{k_{Pl} - k_{Bel}} = \frac{.4}{.4 + .5 + .6} = \frac{4}{15},$$

$$\tilde{Pl}(y) = \frac{.5}{.4 + .5 + .6} = \frac{1}{3}, \quad \tilde{Pl}(z) = \frac{2}{5},$$

is the intersection of the lines joining the corresponding vertices of the probability simplex $\mathcal{P}$ and the upper simplex $T^2[Bel]$.

4. Finally, the intersection probability,

$$p[Bel](x) = m(x) + \beta[Bel]\big(m(\{x,y\}) + m(\Theta)\big) = .2 + \frac{.4 * 0.2}{1.5 - 0.4} = .27,$$
$$p[Bel](y) = .1 + \frac{.4}{1.1} 0.4 = .245, \quad p[Bel](z) = .485,$$

is the unique intersection of the lines joining the corresponding vertices of the upper and lower simplices $T^2[Bel]$ and $T^1[Bel]$.

Fact 1 can be explained by noticing that in the ternary case, by (16.4), $\mathcal{P}[Bel] = T^1[Bel] \cap T^2[Bel]$.

Although Fig. 16.3 suggests that $\tilde{Bel}$, $\tilde{Pl}$ and $p[Bel]$ might be consistent with Bel, this is a mere artefact of this ternary example, for we proved in Theorem 53 that neither the relative belief of singletons nor the relative plausibility of singletons necessarily belongs to the credal set $\mathcal{P}[Bel]$. Indeed, the point of this chapter is that the epistemic transforms $\tilde{Bel}$, $\tilde{Pl}$, $p[Bel]$ are instead consistent with the interval probability system $\mathcal{P}[(Bel, Pl)]$ associated with the original belief function Bel:

$$\tilde{Bel}, \tilde{Pl}, p[Bel] \in \mathcal{P}[(Bel, Pl)] = T^1[Bel] \cap T^{n-1}[Bel].$$

Their geometric behaviour as described by facts 2, 3 and 4 still holds in the general case, as we will see in Section 16.4.

16.4 Credal geometry of probability transformations

16.4.1 Focus of a pair of simplices

Definition 134. *Consider two simplices in* $\mathbb{R}^{n-1}$, $S = Cl(s_1, \ldots, s_n)$ *and* $T = Cl(t_1, \ldots, t_n)$, *with the same number of vertices. If there exists a permutation* ρ *of* $\{1, \ldots, n\}$ *such that the intersection*

$$\bigcap_{i=1}^{n} a(s_i, t_{\rho(i)}) \tag{16.20}$$

of the lines joining corresponding vertices of the two simplices exists and is unique, then $p = f(S, T) \doteq \bigcap_{i=1}^{n} a(s_i, t_{\rho(i)})$ *is termed the* focus *of the two simplices* S *and* T.

Not all pairs of simplices admit a focus. For instance, the pair of simplices (triangles) $S = Cl(s_1 = [2,2]', s_2 = [5,2]', s_3 = [3,5]')$ and $T = Cl(t_1 = [3,1]', t_2 = [5,6]', t_3 = [2,6]')$ in $\mathbb{R}^2$ does not admit a focus, as no matter what permutation of the order of the vertices we consider, the lines joining corresponding vertices do not intersect. Geometrically, all pairs of simplices admitting a focus can be constructed by considering all possible stars of lines, and all possible pairs of points on each line of each star as pairs of corresponding vertices of the two simplices.

We call a focus *special* if the affine coordinates of $f(S,T)$ on the lines $a(s_i, t_{\rho(i)})$ all coincide, namely $\exists \alpha \in \mathbb{R}$ such that

$$f(S,T) = \alpha s_i + (1-\alpha) t_{\rho(i)} \quad \forall i = 1, \ldots, n. \tag{16.21}$$

Not all foci are special. As an example, the simplices $S = Cl(s_1 = [-2,-2]', s_2 = [0,3]', s_3 = [1,0]')$ and $T = Cl(t_1 = [-1,0]', t_2 = [0,-1]', t_3 = [2,2]')$ in $\mathbb{R}^2$ admit a focus, in particular for the permutation $\rho(1) = 3, \rho(2) = 2, \rho(3) = 1$ (i.e., the lines $a(s_1, t_3)$, $a(s_2, t_2)$ and $a(s_3, t_1)$ intersect in $\mathbf{0} = [0,0]'$). However, the focus $f(S,T) = [0,0]'$ is not special, as its simplicial coordinates in the three lines are $\alpha = \frac{1}{2}$, $\alpha = \frac{1}{4}$ and $\alpha = \frac{1}{2}$, respectively.

On the other hand, given any pair of simplices in $\mathbb{R}^{n-1}$ $S = Cl(s_1, \ldots, s_n)$ and $T = Cl(t_1, \ldots, t_n)$, for any permutation ρ of the indices there always exists a linear variety of points which have the same affine coordinates in both simplices,

$$\left\{ p \in \mathbb{R}^{n-1} \middle| p = \sum_{i=1}^{n} \alpha_i s_i = \sum_{j=1}^{n} \alpha_j t_{\rho(j)}, \sum_{i=1}^{n} \alpha_i = 1. \right\} \tag{16.22}$$

Indeed, the conditions on the right-hand side of (16.22) amount to a linear system of n equations in n unknowns (α_i, $i = 1, \ldots, n$). Thus, there exists a linear variety of solutions to such a system, whose dimension depends on the rank of the matrix of constraints in (16.22).

It is rather easy to prove the following theorem.

Theorem 100. *Any special focus $f(S,T)$ of a pair of simplices S, T has the same affine coordinates in both simplices, i.e.,*

$$f(S,T) = \sum_{i=1}^{n} \alpha_i s_i = \sum_{j=1}^{n} \alpha_j t_{\rho(j)}, \quad \sum_{i=1}^{n} \alpha_i = 1,$$

where ρ is the permutation of indices for which the intersection of the lines $a(s_i, t_{\rho(i)})$, $i = 1, \ldots, n$ exists.

Note that the affine coordinates $\{\alpha_i, i\}$ associated with a focus can be negative, i.e., the focus may be located outside one or both simplices.

Notice also that the barycentre itself of a simplex is a special case of a focus. In fact, the centre of mass b of a d-dimensional simplex S is the intersection of the medians of S, i.e., the lines joining each vertex with the barycentre of the opposite ($(d-1)$-dimensional) face (see Fig. 16.4). But those barycentres, for all $(d-1)$-dimensional faces, themselves constitute the vertices of a simplex T.

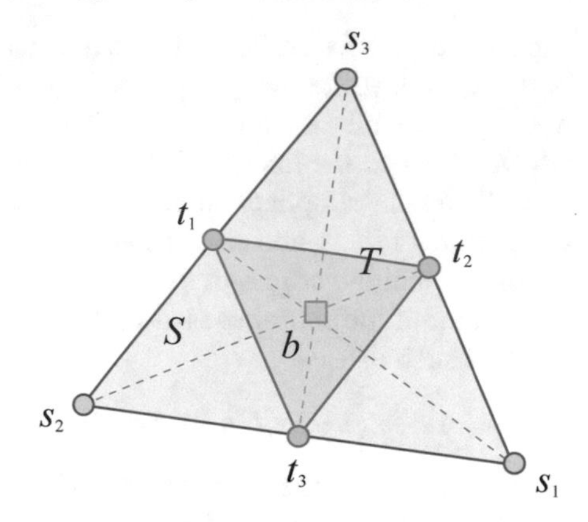

Fig. 16.4: The barycentre of a simplex is itself a special case of a focus, as shown here for a two-dimensional example.

16.4.2 Probability transformations as foci

The notion of a focus of a pair of simplices provides a unified geometric interpretation of the coherent family of Bayesian approximations formed by the relative belief and plausibility of singletons, and the intersection probability.

Theorem 101. *The relative belief of singletons has the same affine coordinates in* $\mathcal{P}$ *and* $T^1[Bel]$.

Dually, we have the following theorem.

Theorem 102. *The relative plausibility of singletons has the same affine coordinates in* $\mathcal{P}$ *and* $T^{n-1}[Bel]$.

Proof. We just need to replace $m(x)$ with $Pl(x)$ in the proof of Theorem 101. □

In fact, we have the following results.

Theorem 103. *The relative belief of singletons* $\tilde{Bel}$ *is the special focus of the pair of simplices* $\{\mathcal{P}, T^1[Bel]\}$*, with its affine coordinate on the corresponding intersecting lines equal to the reciprocal* $1/k_{Bel}$ *of the total mass of singletons.*

Theorem 104. *The relative plausibility of singletons* $\tilde{Pl}$ *is the special focus of the pair of simplices* $\{\mathcal{P}, T^{n-1}[Bel]\}$*, with its affine coordinate on the corresponding intersecting lines equal to the reciprocal* $1/k_{Pl}$ *of the total plausibility of singletons.*

Similar results hold for the intersection probability.

Theorem 105. *For each belief function* Bel*, the intersection probability* $p[Bel]$ *has the same affine coordinates in the lower and upper simplices* $T^1[Bel]$ *and* $T^{n-1}[Bel]$*, respectively.*

Theorem 106. *The intersection probability is the special focus of the pair of lower and upper simplices $\{T^1[Bel], T^{n-1}[Bel]\}$, with its affine coordinate on the corresponding intersecting lines equal to $\beta[Bel]$ (11.10).*

The fraction $\alpha = \beta[Bel]$ of the width of the probability interval that generates the intersection probability can be read in the probability simplex as its coordinate on any of the lines determining the special focus of $\{T^1[Bel], T^{n-1}[Bel]\}$.

16.4.3 Semantics of foci and a rationality principle

The pignistic function adheres to sensible rationality principles and, as a consequence, it has a clear geometrical interpretation as the centre of mass of the credal set associated with a belief function Bel. Similarly, the intersection probability has an elegant geometric behaviour with respect to the credal set associated with an interval probability system, being the (special) focus of the related upper and lower simplices.

Now, selecting the special focus of two simplices representing two different constraints (i.e., the point with the same convex coordinates in the two simplices) means adopting the single probability distribution which satisfies both constraints in exactly the same way. If we assume homogeneous behaviour in the two sets of constraints $\{P(x) \geq Bel(x) \ \forall x\}$, $\{P(x) \leq Pl(x) \ \forall x\}$ as a rationality principle for the probability transformation of an interval probability system, then the intersection probability necessarily follows as the unique solution to the problem.

16.4.4 Mapping associated with a probability transformation

Interestingly, each pair of simplices $S = Cl(s_1, \ldots, s_n)$, $T = Cl(t_1, \ldots, t_n)$ in $\mathbb{R}^{n-1}$ is naturally associated with a mapping, from each point p of $\mathbb{R}^{n-1}$ with simplicial coordinates α_i in S to the point of $\mathbb{R}^{n-1}$ with the same simplicial coordinates α_i in T. For any permutation of indices $\rho(i)$, a different unique mapping exists, namely

$$\begin{aligned} F^\rho_{S,T} : \mathbb{R}^{n-1} & \rightarrow \mathbb{R}^{n-1}, \\ p = \sum_{i=1}^n \alpha_i s_i & \mapsto F^\rho_{S,T}(p) = \sum_{i=1}^n \alpha_i t_{\rho(i)}. \end{aligned} \tag{16.23}$$

Whenever there is such a permutation ρ such that a special focus exists, the latter is a fixed point of this transformation: $F^\rho_{S,T}(f(S,T)) = f(S,T)$.

Each Bayesian transformation in 1–1 correspondence with a pair of simplices (relative plausibility, relative belief and intersection probability) is therefore associated with a mapping of probabilities to probabilities.

Assuming a trivial permutation of the vertices, the mapping (16.23) induced by the relative belief of singletons is actually quite interesting. Any probability distribution $p = \sum_x p(x) Bel_x \in \mathcal{P}$ is mapped by $F_{\mathcal{P},T^1[Bel]}$ to the probability distribution

$$
\begin{aligned}
F_{\mathcal{P},T^1[Bel]}(p) &= \sum_x p(x) t^1_x[Bel] \\
&= \sum_{x\in\Theta} p(x) \left[\sum_{y\neq x} m(y) Bel_y + \left(1 - \sum_{y\neq x} m(y)\right) Bel_x \right] \\
&= \sum_{x\in\Theta} Bel_x \left[\left(1 - \sum_{y\neq x} m(y)\right) p(x) + m(x)(1-p(x)) \right] \\
&= \sum_{x\in\Theta} Bel_x \left[p(x) - p(x) \sum_{y\in\Theta} m(y) + m(x) \right] \\
&= \sum_{x\in\Theta} Bel_x \Big[m(x) + p(x)(1-k_{Bel}) \Big],
\end{aligned}
\tag{16.24}
$$

namely the probability obtained by adding to the belief value of each singleton x a fraction $p(x)$ of the mass $(1 - k_{Bel})$ of non-singletons. In particular, (16.24) maps the relative uncertainty of singletons $R[Bel]$ to the intersection probability $p[Bel]$:

$$
\begin{aligned}
F_{\mathcal{P},T^1[Bel]}(R[Bel]) &= \sum_{x\in\Theta} Bel_x \big[m(x) + R[Bel](x)(1-k_{Bel}) \big] \\
&= \sum_{x\in\Theta} Bel_x \, p[Bel](x) = p[Bel].
\end{aligned}
$$

In a similar fashion, the relative plausibility of singletons is associated with the mapping

$$
\begin{aligned}
F_{\mathcal{P},T^{n-1}[Bel]}(p) &= \sum_{x\in\Theta} p(x) t^{n-1}_x[Bel] \\
&= \sum_{x\in\Theta} p(x) \left[\sum_{y\neq x} Pl(y) Bel_y + \left(1 - \sum_{y\neq x} Pl(y)\right) Bel_x \right] \\
&= \sum_{x\in\Theta} Bel_x \left[\left(1 - \sum_{y\neq x} Pl(y)\right) p(x) + Pl(x)(1-p(x)) \right] \\
&= \sum_{x\in\Theta} Bel_x \left[p(x) - p(x) \sum_{y\in\Theta} Pl(y) + Pl(x) \right] \\
&= \sum_{x\in\Theta} Bel_x \Big[Pl(x) + p(x)(1-k_{Pl}) \Big],
\end{aligned}
\tag{16.25}
$$

which generates a probability by subtracting from the plausibility of each singleton x a fraction $p(x)$ of the plausibility $k_{Pl} - 1$ in 'excess'.

It is curious to note that the map associated with $\tilde{Pl}$ also maps $R[Bel]$ to $p[Bel]$. In fact,

$$
\begin{aligned}
F_{\mathcal{P},T^{n-1}[Bel]}(R[Bel]) &= \sum_{x \in \Theta} Bel_x \left[Pl(x) + R[Bel](x)(1 - k_{Pl}) \right] \\
&= \sum_{x \in \Theta} \left[Pl(x) + \frac{1 - k_{Pl}}{k_{Pl} - k_{Bel}} (Pl(x) - m(x)) \right] \\
&= \sum_{x \in \Theta} Bel_x \left[Pl(x) + (\beta[Bel] - 1)(Pl(x) - m(x)) \right] \\
&= \sum_{x \in \Theta} \left[\beta[Bel] Pl(x) + (1 - \beta[Bel]) m(x) \right] \\
&= \sum_{x \in \Theta} Bel_x p[Bel](x) = p[Bel].
\end{aligned}
$$

A similar mapping exists for the intersection probability too.

16.4.5 Upper and lower simplices as consistent probabilities

The relative belief and plausibility are then the (special) foci associated with the lower and upper simplices $T^1[Bel]$ and $T^{n-1}[Bel]$, the geometric incarnations of the lower and upper constraints on singletons. We can close the circle opened by the analogy with the pignistic transformation by showing that those two simplices can in fact also be interpreted as the sets of probabilities consistent with the plausibility (12.15) and belief (12.17) of singletons, respectively (see Chapter 12).

In fact, the set of pseudo-probabilities consistent with a pseudo-belief function ς can be defined as

$$
\mathcal{P}[\varsigma] \doteq \left\{ P' \in \mathcal{P}' : P'(A) \geq \varsigma(A) \, \forall A \subseteq \Theta \right\},
$$

just as we did for 'standard' belief functions. We can then prove the following result.

Theorem 107. *The simplex $T^1[Bel] = \mathcal{P}^1[Bel]$ associated with the lower probability constraint for singletons (16.6) is the set of probabilities consistent with the belief of singletons $\overline{Bel}$:*

$$
T^1[Bel] = \mathcal{P}[\overline{Bel}].
$$

The simplex $T^{n-1}[Bel]$ associated with the upper probability constraint for singletons (16.7) is the set of pseudo-probabilities consistent with the plausibility of singletons $\overline{Pl}$:

$$
T^{n-1}[Bel] = \mathcal{P}[\overline{Pl}].
$$

A straightforward consequence is the following.

Corollary 22. *The barycentre $t^1[Bel]$ of the lower simplex $T^1[Bel]$ is the output of the pignistic transform of $\overline{Bel}$:*

$$t^1[Bel] = BetP[\overline{Bel}].$$

The barycentre $t^{n-1}[Bel]$ of the upper simplex $T^{n-1}[Bel]$ is the output of the pignistic transform of $\overline{Pl}$:

$$t^{n-1}[Bel] = BetP[\overline{Pl}].$$

Proof. As the pignistic function is the centre of mass of the simplex of consistent probabilities, and the upper and lower simplices are the sets of probabilities consistent with $\overline{Bel}$ and $\overline{Pl}$, respectively (by Theorem 107), the thesis immediately follows.

□

Another corollary stems from the fact that the pignistic function and affine combination commute:

$$BetP[\alpha_1 Bel_1 + \alpha_2 Bel_2] = \alpha_1 BetP[Bel_1] + \alpha_2 BetP[Bel_2]$$

whenever $\alpha_1 + \alpha_2 = 1$.

Corollary 23. *The intersection probability is the convex combination of the barycentres of the lower and upper simplices, with coefficient (11.10)*

$$p[Bel] = \beta[Bel] t^{n-1}[Bel] + (1 - \beta[Bel]) t^1[Bel].$$

16.5 Decision making with epistemic transforms

16.5.1 Generalisations of the TBM

In summary, all the Bayesian transformations of a belief function considered here (pignistic function, relative plausibility, relative belief and intersection probability) possess a simple credal interpretation in the probability simplex. Indeed, they can all be linked to different (credal) sets of probabilities, in this way extending the classical interpretation of the pignistic transformation as the barycentre of the polygon of consistent probabilities.

As $\mathcal{P}[Bel]$ is the credal set associated with a belief function Bel, the upper and lower simplices geometrically embody the probability interval associated with Bel:

$$\mathcal{P}[(Bel, Pl)] = \Big\{ p \in \mathcal{P} : Bel(x) \leq p(x) \leq Pl(x), \forall x \in \Theta \Big\}.$$

By applying the notion of a focus to all possible pairs of simplices in the triad $\{\mathcal{P}, T^1[Bel], T^{n-1}[Bel]\}$, we obtain, in turn, all the different Bayesian transformations considered here:

$$
\begin{array}{lll}
\left\{\mathcal{P}, T^1[Bel]\right\} & : f(\mathcal{P}, T^1[Bel]) & = \tilde{Bel}, \\
\left\{\mathcal{P}, T^{n-1}[Bel]\right\} & : f(\mathcal{P}, T^{n-1}[Bel]) & = \tilde{Pl}, \\
\left\{T^1[Bel], T^{n-1}[Bel]\right\} & : f(T^1[Bel], T^{n-1}[Bel]) & = p[Bel].
\end{array} \tag{16.26}
$$

Their coordinates as foci encode major features of the underlying belief function: the total mass it assigns to singletons, their total plausibility and the fraction β of the related probability interval which yields the intersection probability.

The credal interpretation of upper, lower and interval probability constraints on singletons gives us a perspective on the foundations of the formulation of TBM-like frameworks for such systems. We can think of the TBM as a pair $\{\mathcal{P}[Bel], BetP[Bel]\}$ formed by a credal set linked to each belief function Bel (in this case the polytope of consistent probabilities) and a probability transformation (the pignistic function). As the barycentre of a simplex is a special case of a focus, the pignistic transformation is just another probability transformation induced by the focus of two simplices.

The results in this chapter therefore suggest similar frameworks,

$$
\left\{\{\mathcal{P}, T^1[Bel]\}, \tilde{Bel}\right\}, \quad \left\{\{\mathcal{P}, T^{n-1}[Bel]\}, \tilde{Pl}\right\},
$$

$$
\left\{\{T^1[Bel], T^{n-1}[Bel]\}, p[Bel]\right\},
$$

in which lower, upper and interval constraints on probability distributions on $\mathcal{P}$ are represented by similar pairs, formed by the associated credal set (in the form of a pair of simplices) and by the probability transformation determined by their focus. Decisions are then made based on the appropriate focus probability: relative belief, plausibility or interval probability, respectively.

In the TBM [1689], disjunctive/conjunctive combination rules are applied to belief functions to update or revise our state of belief according to new evidence. The formulation of similar alternative frameworks for lower, upper and interval probability systems would then require us to design specific evidence elicitation/revision operators for such credal sets or, in other words, for lower, upper and interval probability systems.

Question 12. What are the most appropriate elicitation operators for lower, upper and interval probability systems? How are they related to combination rules for belief functions in the case of probability intervals induced by belief functions?

16.5.2 A game/utility theory interpretation

From this perspective, an interesting interpretation of the relative belief and plausibility of singletons can be provided in a game/utility theory context [1858, 1769, 883].

In expected utility theory [1858] (see Section 4.8.1), a decision maker can choose between a number of 'lotteries' (probability distributions) L_i in order to maximise their expected return or utility, calculated as

$$E(L_i) = \sum_{x \in \Theta} u(x) \cdot p_i(x),$$

where u is a utility function $u : \Theta \to \mathbb{R}^+$ which measures the relative satisfaction (for the decision maker) of the different outcomes $x \in \Theta$ of the lottery, and $p_i(x)$ is the probability of x under lottery L_i.

The cloaked-carnival-wheel scenario Consider instead the following game theory scenario, inspired by Strat's expected utility approach to decision making with belief functions [1768, 1548], which we reviewed in Section 4.8.1.

At a country fair, people are asked to bet on one of the possible outcomes of a spinning carnival wheel. Suppose the outcomes are $\{\clubsuit, \diamondsuit, \heartsuit, \spadesuit\}$, and that they each have the same utility (return) to the player. This is equivalent to a lottery (probability distribution), in which each outcome has a probability proportional to the area of the corresponding sectors on the wheel. However, the fair manager decides to make the game more interesting by covering part of the wheel. Players are still asked to bet on a single outcome, knowing that the manager is allowed to rearrange the hidden sector of the wheel as he/she pleases (see Fig. 16.5, a variant of Fig. 4.17).

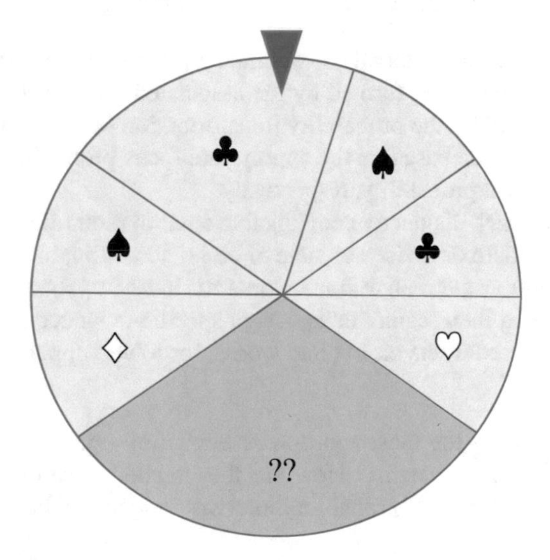

Fig. 16.5: The modified carnival wheel, in which part of the spinning wheel is cloaked.

Clearly, this situation can be described as a belief function, in particular one in which the fraction of the area associated with the hidden sector is assigned as mass to the whole decision space $\{\clubsuit, \diamondsuit, \heartsuit, \spadesuit\}$. If additional (partial) information is provided, for instance that $\diamondsuit$ cannot appear in the hidden sector, different belief functions must be chosen instead.

Regardless of the particular belief function Bel (set of probabilities) at hand, the rule allowing the manager to pick an arbitrary distribution of outcomes in the hidden sector translates mathematically into allowing him/her to choose *any* probability distribution $p \in \mathcal{P}[Bel]$ consistent with Bel in order to damage the player. Supposing the aim of the player is to maximise their minimum chance of winning the bet, which outcome (singleton) should they pick?

A minimax/maximin decision strategy In the probability-bound interpretation (discussed in Section 3.1.4), the belief value of each singleton $x \in \Theta$ measures the minimum support that x can receive from a distribution of the family associated with the belief function Bel:

$$Bel(x) = \min_{P \in \mathcal{P}[Bel]} P(x).$$

Hence $x_{\text{maximin}} \doteq \arg\max_{x \in \Theta} Bel(x)$ is the outcome which maximises this minimum support. In the example in Fig. 16.5, as $\clubsuit$ is the outcome which occupies the largest share of the visible part of the wheel, the safest bet (the one which guarantees the maximum chance in the worst case) is $\clubsuit$. In more formal language, $\clubsuit$ is the singleton with the largest belief value. Now, if we normalise to compute the related relative belief of singletons, this outcome is obviously preserved:

$$x_{\text{maximin}} = \arg\max_{x \in \Theta} \tilde{Bel}(x) = \arg\max_{x \in \Theta} \min_{P \in \mathcal{P}[Bel]} P(x).$$

In conclusion, if the utility function is constant (i.e., no element of Θ can be preferred over the others), x_{maximin} (the peak(s) of the relative belief of singletons) represents the best possible defensive strategy aimed at maximising the minimum utility of the possible outcomes.

Dually, $Pl(x)$ measures the maximum possible support for x by a distribution consistent with Bel, so that

$$x_{\text{minimax}} = \arg\min_{x \in \Theta} \tilde{Pl}(x) = \arg\min_{x \in \Theta} \max_{P \in \mathcal{P}[Bel]} P(x)$$

is the outcome which minimises the maximum possible support.

Suppose for the sake of simplicity that the loss function $l : \Theta \rightarrow \mathbb{R}^+$ which measures the relative dissatisfaction for the outcomes is constant, and that in the same game theory set-up our opponent is (again) free to pick a consistent probability distribution $P \in \mathcal{P}[Bel]$. Then the element with the minimum relative plausibility is the best possible defensive strategy aimed at minimising the maximum possible loss.

Note that when the utility function is not constant, the above minimax and maximin problems naturally generalise as

$$x_{\text{maximin}} = \arg\max_{x \in \Theta} \tilde{Bel}(x) u(x), \quad x_{\text{minimax}} = \arg\min_{x \in \Theta} \tilde{Pl}(x) l(x).$$

Whereas in classical utility theory the decision maker has to select the best 'lottery' (probability distribution) in order to maximise their expected utility, here the 'lottery' is chosen by their opponent (given the available partial evidence), and the decision maker is left with betting on the safest strategy (element of Θ).

In conclusion, the relative belief and plausibility of singletons play an important role in determining the safest betting strategy in an adversarial scenario in which the decision maker has to minimise their maximum loss or maximise their minimum return.

Appendix: Proofs

Proof of Theorem 99

Lemma 20. *The points $\{t^1_x[Bel], x \in \Theta\}$ are affinely independent.*

Proof. Let us suppose, contrary to the thesis, that there exists an affine decomposition of one of the points, say $t_x[Bel]$, in terms of the others:

$$t^1_x[Bel] = \sum_{z \neq x} \alpha_z t^1_z[Bel], \quad \alpha_z \geq 0 \; \forall z \neq x, \; \sum_{z \neq x} \alpha_z = 1.$$

But then we would have, by the definition of $t^1_z[Bel]$,

$$\begin{aligned} t^1_x[Bel] &= \sum_{z \neq x} \alpha_z t^1_z[Bel] \\ &= \sum_{z \neq x} \alpha_z \left(\sum_{y \neq z} m(y) Bel_y \right) + \sum_{z \neq x} \alpha_z (m(z) + 1 - k_{Bel}) Bel_z \\ &= m(x) Bel_x \sum_{z \neq x} \alpha_z + \sum_{z \neq x} Bel_z m(z)(1 - \alpha_z) \\ &\quad + \sum_{z \neq x} \alpha_z m(z) Bel_z + (1 - k_{Bel}) \sum_{z \neq x} \alpha_z Bel_z \\ &= \sum_{z \neq x} m(z) Bel_z + m(x) Bel_x + (1 - k_{Bel}) \sum_{z \neq x} \alpha_z Bel_z . \end{aligned}$$

The latter is equal to (16.9)

$$t^1_x[Bel] = \sum_{z \neq x} m(z) Bel_z + (m(x) + 1 - k_{Bel}) Bel_x$$

if and only if $\sum_{z \neq x} \alpha_z Bel_z = Bel_x$. But this is impossible, as the categorical probabilities Bel_x are trivially affinely independent.

□

Let us give details of the proof for $T^1[Bel]$. We need to show that:

1. All the points $p' \in \mathcal{P}'$ which belong to $Cl(t^1_x[Bel], x \in \Theta)$ satisfy $p'(x) \geq m(x)$ too.
2. All the points which do not belong to the above polytope do not satisfy the constraint either.

Concerning item 1, as

$$t^1_x[Bel](y) = \begin{cases} m(y) & x \neq y, \\ 1 - \sum_{z \neq y} m(z) = m(y) + 1 - k_{Bel} & x = y, \end{cases}$$

the condition $p' \in Cl(t^1_x[Bel], x \in \Theta)$ is equivalent to

$$p'(y) = \sum_{x \in \Theta} \alpha_x t^1_x[Bel](y) = m(y) \sum_{x \neq y} \alpha_x + (1 - k_{Bel})\alpha_y + m(y)\alpha_y \quad \forall y \in \Theta,$$

where $\sum_x \alpha_x = 1$ and $\alpha_x \geq 0 \; \forall x \in \Theta$. Therefore,

$$p'(y) = m(y)(1-\alpha_y) + (1-k_{Bel})\alpha_y + m(y)\alpha_y = m(y) + (1-k_{Bel})\alpha_y \geq m(y),$$

as $1 - k_{Bel}$ and α_y are both non-negative quantities.

Concerning item 2, if $p' \notin Cl(t^1_x[Bel], x \in \Theta)$, then $p' = \sum_{x \in \Theta} \alpha_x t^1_x[Bel]$, where $\exists z \in \Theta$ such that $\alpha_z < 0$. But, then,

$$p'(z) = m(z) + (1 - k_{Bel})\alpha_z < m(z),$$

as $(1-k_{Bel})\alpha_z < 0$, unless $k_{Bel} = 1$, in which case Bel is already a probability. By Lemma 20, the points $\{t^1_x[Bel], x \in \Theta\}$ are affinely independent; hence $T^1[Bel]$ is a simplex.

Dual proofs for Lemma 20 and Theorem 99 can be provided for $T^{n-1}[Bel]$ by simply replacing the belief values of singletons with their plausibility values.

Proof of Theorem 100

Suppose that p is a special focus of S and T, namely $\exists \alpha \in \mathbb{R}$ such that

$$p = \alpha s_i + (1 - \alpha) t_{\rho(i)} \quad \forall i = 1, \ldots, n,$$

for some permutation ρ of $\{1, \ldots, n\}$. Then, necessarily,

$$t_{\rho(i)} = \frac{1}{1 - \alpha}[p - \alpha s_i] \quad \forall i = 1, \ldots, n.$$

If p has coordinates $\{\alpha_i, i = 1, \ldots, n\}$ in T, $p = \sum_{i=1}^n \alpha_i t_{\rho(i)}$, it follows that

$$p=\sum_{i=1}^{n}\alpha_i t_{\rho(i)}=\frac{1}{1-\alpha}\sum_i \alpha_i\left[p-\alpha s_i\right]$$

$$=\frac{1}{1-\alpha}\left[p\sum_i \alpha_i-\alpha\sum_i \alpha_i s_i\right]=\frac{1}{1-\alpha}\left[p-\alpha\sum_i \alpha_i s_i\right].$$

The latter implies that $p=\sum_i \alpha_i s_i$, i.e., p has the same simplicial coordinates in S and T.

Proof of Theorem 101

By definition (4.61), $\tilde{Bel}$ can be expressed in terms of the vertices of the probability simplex $\mathcal{P}$ as

$$\tilde{Bel}=\sum_{x\in\Theta}\frac{m(x)}{k_{Bel}}Bel_x.$$

We can prove that $\tilde{Bel}$ can be written as the same affine combination

$$\tilde{Bel}=\sum_{x\in\Theta}\frac{m(x)}{k_{Bel}}t^1_x[Bel]$$

in terms of the vertices $t^1_x[Bel]$ of $T^1[Bel]$, and thus under the (trivial) correspondence $Bel_x\leftrightarrow t^1_x[Bel]$ between vertices of the two simplices.

Substituting (16.9) in the above equation yields

$$\sum_{x\in\Theta}\frac{m(x)}{k_{Bel}}t^1_x[Bel]=\sum_{x\in\Theta}\frac{m(x)}{k_{Bel}}\left[\sum_{y\neq x}m(y)Bel_y+\left(1-\sum_{y\neq x}m(y)\right)Bel_x\right]$$

$$=\sum_{x\in\Theta}Bel_x\left(\frac{m(x)}{k_{Bel}}\sum_{y\neq x}m(y)\right)+\sum_{x\in\Theta}\frac{m(x)}{k_{Bel}}Bel_x$$

$$-\sum_{x\in\Theta}Bel_x\left(\frac{m(x)}{k_{Bel}}\sum_{y\neq x}m(y)\right)=\sum_{x\in\Theta}\frac{m(x)}{k_{Bel}}Bel_x=\tilde{Bel}.$$

Proof of Theorem 103

We need to verify the condition (16.21). The latter assumes the following form (since $s_i=Bel_x$, $t_{\rho(i)}=t^1_x[Bel]$), for all $x\in\Theta$:

$$\sum_{x\in\Theta} \frac{m(x)}{k_{Bel}} Bel_x$$

$$= t^1_x[Bel] + \alpha(Bel_x - t^1_x[Bel]) = (1-\alpha)t^1_x[Bel] + \alpha Bel_x$$

$$= (1-\alpha)\left[\sum_{y\neq x} m(y)Bel_y + \big(1 - k_{Bel} + m(x)\big)Bel_x\right] + \alpha Bel_x$$

$$= Bel_x\Big[(1-\alpha)\big(1 - k_{Bel} + m(x)\big) + \alpha\Big] + \sum_{y\neq x} m(y)(1-\alpha)Bel_y.$$

Clearly, for $1-\alpha = \frac{1}{k_{Bel}}$, $\alpha = \frac{k_{Bel}-1}{k_{Bel}}$ (a quantity which does not depend on x), the condition is satisfied for all singletons $x \in \Theta$.

Proof of Theorem 104

Similarly, we need to verify (16.21) for $s_i = Bel_x$, $t_{\rho(i)} = t^{n-1}_x[Bel]$. Namely, for all $x \in \Theta$,

$$\sum_{x\in\Theta} \frac{Pl(x)}{k_{Pl}} Bel_x$$

$$= t^{n-1}_x[Bel] + \alpha(Bel_x - t^{n-1}_x[Bel]) = (1-\alpha)t^{n-1}_x[Bel] + \alpha Bel_x$$

$$= (1-\alpha)\left[\sum_{y\neq x} Pl(y)Bel_y + \big(1 - k_{Pl} + Pl(x)\big)Bel_x\right] + \alpha Bel_x$$

$$= Bel_x\Big[(1-\alpha)\big(1 - k_{Pl} + Pl(x)\big) + \alpha\Big] + \sum_{y\neq x} Pl(y)(1-\alpha)Bel_y.$$

For $1-\alpha = \frac{1}{k_{Pl}}$, $\alpha = \frac{k_{Pl}-1}{k_{Pl}}$ the condition is satisfied for all singletons.

Proof of Theorem 105

The common simplicial coordinates of $p[Bel]$ in $T^1[Bel]$ and $T^{n-1}[Bel]$ turn out to be the values of the relative uncertainty function (16.16) for Bel,

$$R[Bel](x) = \frac{Pl(x) - m(x)}{k_{Pl} - k_{Bel}}. \tag{16.27}$$

Recalling the expression (16.9) for the vertices of $T^1[Bel]$, the point of the simplex $T^1[Bel]$ with coordinates (16.27) is

$$\begin{aligned}
&\sum_x R[Bel](x) t^1_x[Bel] \\
&= \sum_x R[Bel](x) \left[\sum_{y \neq x} m(y) Bel_y + \left(1 - \sum_{y \neq x} m(y) \right) Bel_x \right] \\
&= \sum_x R[Bel](x) \left[\sum_{y \in \Theta} m(y) Bel_y + (1 - k_{Bel}) Bel_x \right] \\
&= \sum_x Bel_x \left[(1 - k_{Bel}) R[Bel](x) + m(x) \sum_y R[Bel](y) \right] \\
&= \sum_x Bel_x \Big[(1 - k_{Bel}) R[Bel](x) + m(x) \Big],
\end{aligned}$$

as $R[Bel]$ is a probability ($\sum_y R[Bel](y) = 1$). By Equation (16.15), the above quantity coincides with $p[Bel]$.

The point of $T^{n-1}[Bel]$ with the same coordinates, $\{R[Bel](x), x \in \Theta\}$, is

$$\begin{aligned}
&\sum_x R[Bel](x) t^{n-1}_x[Bel] \\
&= \sum_x R[Bel](x) \left[\sum_{y \neq x} Pl(y) Bel_y + \left(1 - \sum_{y \neq x} Pl(y) \right) Bel_x \right] \\
&= \sum_x R[Bel](x) \left[\sum_{y \in \Theta} Pl(y) Bel_y + (1 - k_{Pl}) Bel_x \right] \\
&= \sum_x Bel_x \left[(1 - k_{Pl}) R[Bel](x) + Pl(x) \sum_y R[Bel](y) \right] \\
&= \sum_x Bel_x \Big[(1 - k_{Pl}) R[Bel](x) + Pl(x) \Big] \\
&= \sum_x Bel_x \left[Pl(x) \frac{1 - k_{Bel}}{k_{Pl} - k_{Bel}} - m(x) \frac{1 - k_{Bel}}{k_{Pl} - k_{Bel}} \right],
\end{aligned}$$

which is equal to $p[Bel]$ by (16.27).

Proof of Theorem 106

Again, we need to impose the condition (16.21) on the pair $\{T^1[Bel], T^{n-1}[Bel]\}$, namely

$$p[Bel] = t_x^1[Bel] + \alpha(t_x^{n-1}[Bel] - t_x^1[Bel]) = (1-\alpha)t_x^1[Bel] + \alpha t_x^{n-1}[Bel]$$

for all the elements $x \in \Theta$ of the frame, α being some constant real number. This is equivalent to (after substituting the expressions (16.9), (16.11) for $t_x^1[Bel]$ and $t_x^{n-1}[Bel]$)

$$\begin{aligned}
&\sum_{x\in\Theta} Bel_x \Big[m(x) + \beta[Bel](Pl(x) - m(x)) \Big] \\
&= (1-\alpha) \left[\sum_{y\in\Theta} m(y) Bel_y + (1 - k_{Bel}) Bel_x \right] \\
&\quad + \alpha \left[\sum_{y\in\Theta} Pl(y) Bel_y + (1 - k_{Pl}) Bel_x \right] \\
&= Bel_x \Big[(1-\alpha)(1-k_{Bel}) + (1-\alpha)m(x) + \alpha Pl(x) + \alpha(1-k_{Pl}) \Big] \\
&\quad + \sum_{y\neq x} Bel_y \Big[(1-\alpha)m(y) + \alpha Pl(y) \Big] \\
&= Bel_x \Big\{ (1-k_{Bel}) + m(x) + \alpha \big[Pl(x) + (1-k_{Pl}) - m(x) - (1-k_{Bel}) \big] \Big\} \\
&\quad + \sum_{y\neq x} Bel_y \Big[m(y) + \alpha \big(Pl(y) - m(y) \big) \Big].
\end{aligned}$$

If we set $\alpha = \beta[Bel] = (1-k_{Bel})/(k_{Pl} - k_{Bel})$, we get the following for the coefficient of Bel_x in the above expression (i.e., the probability value of x):

$$\begin{aligned}
&\frac{1-k_{Bel}}{k_{Pl}-k_{Bel}} \big[Pl(x) + (1-k_{Pl}) - m(x) - (1-k_{Bel}) \big] + (1-k_{Bel}) + m(x) \\
&= \beta[Bel][Pl(x) - m(x)] + (1-k_{Bel}) + m(x) - (1-k_{Bel}) = p[Bel](x).
\end{aligned}$$

On the other hand,

$$m(y) + \alpha(Pl(y) - m(y)) = m(y) + \beta[Bel](Pl(y) - m(y)) = p[Bel](y)$$

for all $y \neq x$, no matter what the choice of x.

Proof of Theorem 107

For each belief function Bel, the vertices of the consistent polytope $\mathcal{P}[Bel]$ are generated by a permutation ρ of the elements of Θ (3.12). This is true for the belief function $\overline{Bel}$ too, i.e., the vertices of $\mathcal{P}[\overline{Bel}]$ are also generated by permutations of singletons.

In this case, however:

- given such a permutation $\rho = (x_{\rho(1)}, \ldots, x_{\rho(n)})$, the mass of Θ (the only non-singleton focal element of $\overline{Bel}$) is assigned according to the mechanism of Proposition 66 to $x_{\rho(1)}$, while all the other elements receive only their original mass $m(x_{\rho(j)})$, $j > 1$;
- therefore, all the permutations ρ putting the same element in the first place yield the same vertex of $\mathcal{P}[\overline{Bel}]$;
- hence there are just n such vertices, one for each choice of the first element $x_{\rho(1)} = x$;
- but this vertex, a probability distribution, has mass values (simplicial coordinates in $\mathcal{P}$)

$$m(x) = m(x) + (1 - k_{Bel}), \quad m(y) = m(y) \; \forall y \neq x,$$

as $(1 - k_{Bel})$ is the mass that $\overline{Bel}$ assigns to Θ;
- the latter clearly corresponds to $t^1_x[Bel]$ (16.9).

A similar proof holds for $\overline{Pl}$, as Proposition 1 remains valid for pseudo-belief functions too.

Proof of Corollary 23

By (12.18), the intersection probability $p[Bel]$ lies on the line joining $\overline{Pl}$ and $\overline{Bel}$, with coordinate $\beta[Bel]$:

$$p[Bel] = \beta[Bel]\overline{Pl} + (1 - \beta[Bel])\overline{Bel}.$$

If we apply the pignistic transformation, we get directly

$$\begin{aligned} BetP[p[Bel]] = p[Bel] &= BetP[\beta[Bel]\overline{Bel} + (1 - \beta[Bel])\overline{Pl}] \\ &= \beta[Bel]BetP[\overline{Bel}] + (1 - \beta[Bel])BetP[\overline{Pl}] \\ &= \beta[Bel]t^{n-1}[Bel] + (1 - \beta[Bel])t^1[Bel] \end{aligned}$$

by Corollary 22.

Part V

The future of uncertainty

17 An agenda for the future

As we have seen in this book, the theory of belief functions is a modelling language for representing and combining elementary items of evidence, which do not necessarily come in the form of sharp statements, with the goal of maintaining a mathematical representation of an agent's beliefs about those aspects of the world which the agent is unable to predict with reasonable certainty. While arguably a more appropriate mathematical description of uncertainty than classical probability theory, for the reasons we have thoroughly explored in this book, the theory of evidence is relatively simple to understand and implement, and does not require one to abandon the notion of an event, as is the case, for instance, for Walley's imprecise probability theory. It is grounded in the beautiful mathematics of random sets, which constitute the natural continuous extensions of belief functions, and exhibits strong relationships with many other theories of uncertainty. As mathematical objects, belief functions have fascinating properties in terms of their geometry, algebra and combinatorics. This book is dedicated in particular to the geometric approach to belief and other uncertainty measures proposed by the author.

Despite initial concerns about the computational complexity of a naive implementation of the theory of evidence, evidential reasoning can actually be implemented on large sample spaces and in situations involving the combination of numerous pieces of evidence. Elementary items of evidence often induce simple belief functions, which can be combined very efficiently with complexity $O(n+1)$. We do not need to assign mass to all subsets, but we do need to be allowed to do so when necessary (e.g., in the case of missing data). Most relevantly, the most plausible hypotheses can be found without actually computing the overall belief function. At any rate, Monte Carlo approximations can easily be implemented when an explicit result for the combination is required. Last but not least, local propagation schemes

F. Cuzzolin, *The Geometry of Uncertainty*, Artificial Intelligence: Foundations, Theory, and Algorithms, https://doi.org/10.1007/978-3-030-63153-6_17

allow for the parallelisation of belief function reasoning, similarly to what happens with Bayesian networks.

As we saw in Chapter 4, statistical evidence can be represented in belief theory in several ways, for example the following (this list is not exhaustive):

- by likelihood-based belief functions, in a way that generalises both likelihood-based and Bayesian inference;
- via Dempster's inference approach, which makes use of auxiliary variables in a fiducial setting, and its natural evolution, weak belief;
- in the framework of the beneralised Bayesian theorem proposed by Smets.

Decision-making strategies based on intervals of expected utilities can be formulated which allow more cautious decisions than those based on traditional approaches, and are able to explain the empirical aversion to second-order uncertainty highlighted in Ellsberg's paradox. A straightforward extension of the theory, originally formulated for finite sample spaces, to continuous domains can be achieved via the Borel interval representation initially put forward by Strat and Smets, when the analysis is restricted to intervals of real values. In the more general case of arbitrary subsets of the real domain, the theory of random sets is, in our view, the natural mathematical framework to adopt. Finally, an extensive array of estimation, classification and regression tools based on the theory of belief functions has already been developed, and further contributions can be envisaged.

Open issues

In their foreword to the volume Classic Works of the Dempster–Shafer Theory of Belief Functions [2033], Dempster and Shafer considered that, although flourishing by some measures, belief function theory had still not addressed questions such as deciding whether bodies of evidence were independent, or what to do if they were dependent. They lamented the ongoing confusion and disagreement about how to interpret the theory, and its limited acceptance in the field of mathematical statistics, where it first began. In response, they proposed an agenda to move the theory forward, centred on the following three elements:

- a richer understanding of the uses of probability, for they believed the theory was best regarded as a way of using probability [1607, 422, 1613];
- a deeper understanding of statistical modelling itself [468], which would go beyond a traditional statistical analysis which begins by specifying probabilities that are supposed to be known except for certain parameters;
- in-depth examples of sensible Dempster–Shafer analyses of a variety of problems of real scientific and technological importance.

As for the last point, interesting examples in which the use of belief functions provides sound and elegant solutions to real-life problems, essentially characterised by 'missing' information, had already been given by Smets in [1708]. These include classification problems in which the training set is such that the classes are only partially known, an information retrieval system handling interdocument relationships, the combination of data from sensors providing data about partially overlapping frames, and the determination of the number of sources in a multisensor environment by studying sensor conflict. We reviewed a number of these problems in Chapter 5.

The correct epistemic interpretation of belief function theory certainly needs to be clarified once and for all. In this book we have argued that belief measures should be seen as *random variables for set-valued observations* (recall the random die example in Chapter 1). In an interesting, although rather overlooked, recent publication [553], instead, Dubucs and Shafer highlighted two ways of interpreting numerical degrees of belief in terms of betting, in a manner that echoes Shafer and Vovk's game-theoretical interpretation of probability (Section 6.13): (i) you can offer to bet at the odds defined by the degrees of belief, or (ii) you can make a judgement that a strategy for taking advantage of such betting offers will not multiply the capital it risks by a large factor. Both interpretations, the authors argue, can be applied to ordinary probabilities and used to justify updating by conditioning, whereas only the second can be applied to belief functions and used to justify Dempster's rule of combination.

The most appropriate mechanism for evidence combination is also still being debated. The reason is that the choice seems to depend on meta-information about the reliability and independence of the sources involved, which is hardly accessible. As we argue here (and we have hinted at in Chapter 4), *working with intervals of belief functions* may be the way forward, as this acknowledges the meta-uncertainty about the nature of the sources generating the evidence. The same holds for conditioning, as opposed to combining, belief functions.

Finally, although the theory of belief functions on Borel intervals of the real line is rather elegant, it cannot achieve full generality. Here we support the view that the way forward is to ground the theory in the mathematics of random sets (see Section 3.1.5).

A research programme

We thus deem it appropriate to conclude this book by outlining what, in our view, is the research agenda for the future development of random set and belief function theory. For obvious reasons, we will only be able to touch on a few of the most interesting avenues of investigation, and even those will not be discussed beyond a certain level of detail. We hope, however, that this brief survey will stimulate the reader to pursue some of these research directions and to contribute to the further maturation of the theory in the near future.

Although random set theory as a mathematical formalism is quite well developed, thanks in particular to the work of Ilya Molchanov [1302, 1304], a theory of statistical inference with random sets is not yet in sight. In Section 17.1 of this final chapter we briefly consider, in particular, the following questions:

- the introduction of the notions of lower and upper likelihoods (Section 17.1.1), in order to move beyond belief function inference approaches which take the classical likelihood function at face value;
- the formulation of a framework for (generalised) logistic regression with belief functions, making use of such generalised lower and upper likelihoods (Section 17.1.2) and their decomposition properties;
- the generalisation of the law of total probability to random sets (Section 17.1.3), starting with belief functions [364];
- the generalisation of classical limit theorems (central limit theorem, law of large numbers) to the case of random sets (Section 17.1.4): this would allow, for instance, a rigorous definition of Gaussian random sets and belief functions alternative to that proposed by Molchanov (see Section 4.9.3, Definition 75);
- the introduction (somewhat despite Dempster's aversion to them) of parametric models based on random sets (Section 17.1.5), which would potentially allow us to perform robust hypothesis testing (Section 17.1.5), thus laying the foundations for a theory of frequentist inference with random sets (Section 17.1.5);
- the development of a theory of random variables and processes in which the underlying probability space is replaced by a random-set space (Section 17.1.6): in particular, this requires the generalisation of the notion of the Radon–Nikodym derivative to belief measures.

The geometric approach to uncertainty proposed in this book is also open to a number of further developments (Section 17.2), including:

- the geometry of combination rules other than Dempster's (Section 17.2.1), and the associated conditioning operators (17.2.2);
- an exploration of the geometry of continuous extensions of belief functions, starting from the geometry of belief functions on Borel intervals to later tackle the general random-set representation;
- a geometric analysis of uncertainty measures beyond those presented in Chapters 10 and 13, where we provided an initial extension to possibility theory, including major ones such as capacities and gambles (Section 17.2.3);
- newer geometric representations (Section 17.2.4), based on (among others) isoperimeters of convex bodies or exterior algebras.

More speculatively, the possibility exists of conducting inference in a purely geometric fashion, by finding a common representation for both belief measures and the data that drive the inference.

Further theoretical advances are necessary, in our view. In particular:

- a set of prescriptions for reasoning with intervals of belief functions: as we saw in Chapter 4, this seems to be a natural way to avoid the tangled issue of choosing a combination rule and handling metadata;
- the further development of machine learning tools based on belief theory (see Chapter 5), able to connect random set theory with current trends in the field, such as transfer learning and deep learning.

Last but not least, future work will need to be directed towards tackling high-impact applications by means of random set/belief function theory (Section 17.3). Here we will briefly consider, in particular:

- the possible creation of a framework for climate change predictions based on random sets (Section 17.3.1), able to overcome the limitations of existing (albeit neglected) Bayesian approaches;
- the generalisation of max-entropy classifiers and log-linear models to the case of belief measures (Section 17.3.2), as an example of the fusion of machine learning with uncertainty theory;
- new robust foundations for machine learning itself (Section 17.3.3), obtained by generalising Vapnik's *probably approximately correct* (PAC) analysis to the case in which the training and test distributions, rather than coinciding, come from the same random set.

17.1 A statistical random set theory

17.1.1 Lower and upper likelihoods

The traditional likelihood function (Chapter 1, Section 1.2.3) is a conditional probability of the data given a parameter $\theta \in \Theta$, i.e., a family of PDFs over the measurement space $\mathbb{X}$ parameterised by θ. Most of the work on belief function inference just takes the notion of a likelihood as a given, and constructs belief functions from an input likelihood function (see Chapter 4, Section 4.1.1).

However, there is no reason why we should not formally define a 'belief likelihood function' mapping a sample observation $x \in \mathbb{X}$ to a real number, rather than use the conventional likelihood to construct belief measures. It is natural to define such a belief likelihood function as a family of belief functions on $\mathbb{X}$, $Bel_{\mathbb{X}}(.|\theta)$, parameterised by $\theta \in \Theta$. An expert in belief theory will note that such a parameterised family is the input to Smets's generalised Bayesian theorem [1689], a collection of 'conditional' belief functions. Such a belief likelihood takes values on *sets* of outcomes, $A \subset \Theta$, of which individual outcomes are just a special case.

This seems to provide a natural setting for computing likelihoods of set-valued observations, in accordance with the random-set philosophy that underpins this book.

Belief likelihood function of repeated trials Let $Bel_{\mathbb{X}_i}(A|\theta)$, for $i = 1, 2, \ldots, n$ be a parameterised family of belief functions on $\mathbb{X}_i$, the space of quantities that can be observed at time i, depending on a parameter $\theta \in \Theta$. A series of repeated trials then assumes values in $\mathbb{X}_1 \times \cdots \times \mathbb{X}_n$, whose elements are tuples of the form $\mathbf{x} = (x_1, \ldots, x_n) \in \mathbb{X}_1 \times \cdots \times \mathbb{X}_n$. We call such tuples 'sharp' samples, as opposed to arbitrary subsets $A \subset \mathbb{X}_1 \times \cdots \times \mathbb{X}_n$ of the space of trials. Note that we are not assuming the trials to be equally distributed at this stage, nor do we assume that they come from the same sample space.

Definition 135. *The belief likelihood function* $Bel_{\mathbb{X}_1 \times \cdots \times \mathbb{X}_n} : 2^{\mathbb{X}_1 \times \cdots \times \mathbb{X}_n} \rightarrow [0,1]$ *of a series of repeated trials is defined as*

$$Bel_{\mathbb{X}_1 \times \cdots \times \mathbb{X}_n}(A|\theta) \doteq Bel_{\mathbb{X}_1}^{\uparrow \times_i \mathbb{X}_i} \odot \cdots \odot Bel_{\mathbb{X}_n}^{\uparrow \times_i \mathbb{X}_i}(A|\theta), \tag{17.1}$$

where $Bel_{\mathbb{X}_j}^{\uparrow \times_i \mathbb{X}_i}$ *is the vacuous extension (Definition 20) of* $Bel_{\mathbb{X}_j}$ *to the Cartesian product* $\mathbb{X}_1 \times \cdots \times \mathbb{X}_n$ *where the observed tuples live,* $A \subset \mathbb{X}_1 \times \cdots \times \mathbb{X}_n$ *is an arbitrary subset of series of trials and* $\odot$ *is an arbitrary combination rule.*

In particular, when the subset A reduces to a sharp sample, $A = \{\mathbf{x}\}$, we can define the following generalisations of the notion of a likelihood.

Definition 136. *We call the quantities*

$$\begin{aligned} \underline{L}(\mathbf{x}) &\doteq Bel_{\mathbb{X}_1 \times \cdots \times \mathbb{X}_n}(\{(x_1, \ldots, x_n)\}|\theta), \\ \overline{L}(\mathbf{x}) &\doteq Pl_{\mathbb{X}_1 \times \cdots \times \mathbb{X}_n}(\{(x_1, \ldots, x_n)\}|\theta) \end{aligned} \tag{17.2}$$

the lower likelihood *and* upper likelihood, *respectively, of the sharp sample* $A = \{\mathbf{x}\} = \{(x_1, \ldots, x_n)\}$.

Binary trials: The conjunctive case Belief likelihoods factorise into simple products whenever conjunctive combination is employed (as a generalisation of classical stochastic independence) in Definition 135, and the case of trials with binary outcomes is considered.

Focal elements of the belief likelihood Let us first analyse the case $n = 2$. We seek the Dempster sum $Bel_{\mathbb{X}_1} \oplus Bel_{\mathbb{X}_2}$, where $\mathbb{X}_1 = \mathbb{X}_2 = \{T, F\}$.

Figure 17.1(a) is a diagram of all the intersections of focal elements of the two input belief functions on $\mathbb{X}_1 \times \mathbb{X}_2$. There are $9 = 3^2$ distinct non-empty intersections, which correspond to the focal elements of $Bel_{\mathbb{X}_1} \oplus Bel_{\mathbb{X}_2}$. According to (2.6), the mass of the focal element $A_1 \times A_2$, $A_1 \subseteq \mathbb{X}_1$, $A_2 \subseteq \mathbb{X}_2$, is then

$$m_{Bel_{\mathbb{X}_1} \oplus Bel_{\mathbb{X}_2}}(A_1 \times A_2) = m_{\mathbb{X}_1}(A_1) \cdot m_{\mathbb{X}_2}(A_2). \tag{17.3}$$

Note that this result holds when the conjunctive rule $\textcircled{\cap}$ is used as well (4.25), for none of the intersections are empty, and hence no normalisation is required. Nothing is assumed about the mass assignment of the two belief functions $Bel_{\mathbb{X}_1}$ and $Bel_{\mathbb{X}_2}$.

We can now prove the following lemma.

Lemma 21. *For any* $n \in \mathbb{Z}$, *the belief function* $Bel_{\mathbb{X}_1} \oplus \cdots \oplus Bel_{\mathbb{X}_n}$, *where* $\mathbb{X}_i = \mathbb{X} = \{T, F\}$, *has* 3^n *focal elements, namely all possible Cartesian products* $A_1 \times \cdots \times A_n$ *of* n *non-empty subsets* A_i *of* $\mathbb{X}$, *with BPA*

$$m_{Bel_{\mathbb{X}_1} \oplus \cdots \oplus Bel_{\mathbb{X}_n}}(A_1 \times \cdots \times A_n) = \prod_{i=1}^{n} m_{\mathbb{X}_i}(A_i).$$

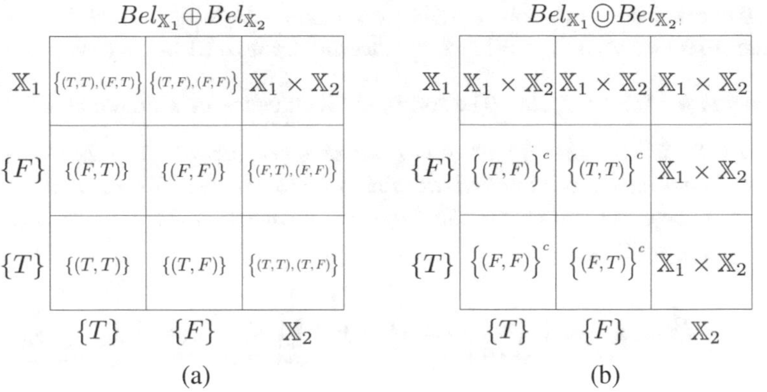

$Bel_{\mathbb{X}_1} \oplus Bel_{\mathbb{X}_2}$

$\mathbb{X}_1$	$\{(T,T),(F,T)\}$	$\{(T,F),(F,F)\}$	$\mathbb{X}_1 \times \mathbb{X}_2$
$\{F\}$	$\{(F,T)\}$	$\{(F,F)\}$	$\{(F,T),(F,F)\}$
$\{T\}$	$\{(T,T)\}$	$\{(T,F)\}$	$\{(T,T),(T,F)\}$
	$\{T\}$	$\{F\}$	$\mathbb{X}_2$

(a)

$Bel_{\mathbb{X}_1} \circledcirc Bel_{\mathbb{X}_2}$

$\mathbb{X}_1$	$\mathbb{X}_1 \times \mathbb{X}_2$	$\mathbb{X}_1 \times \mathbb{X}_2$	$\mathbb{X}_1 \times \mathbb{X}_2$
$\{F\}$	$\{(T,F)\}^c$	$\{(T,T)\}^c$	$\mathbb{X}_1 \times \mathbb{X}_2$
$\{T\}$	$\{(F,F)\}^c$	$\{(F,T)\}^c$	$\mathbb{X}_1 \times \mathbb{X}_2$
	$\{T\}$	$\{F\}$	$\mathbb{X}_2$

(b)

Fig. 17.1: (a) Graphical representation of the Dempster combination $Bel_{\mathbb{X}_1} \oplus Bel_{\mathbb{X}_2}$ on $\mathbb{X}_1 \times \mathbb{X}_2$, in the binary case in which $\mathbb{X}_1 = \mathbb{X}_2 = \{T, F\}$. (b) Graphical representation of the disjunctive combination $Bel_{\mathbb{X}_1} \circledcirc Bel_{\mathbb{X}_2}$ there.

Proof. The proof is by induction. The thesis was shown to be true for $n = 2$ in (17.3). In the induction step, we assume that the thesis is true for n, and prove it for $n + 1$.

If $Bel_{\mathbb{X}_1} \oplus \cdots \oplus Bel_{\mathbb{X}_n}$, defined on $\mathbb{X}_1 \times \cdots \times \mathbb{X}_n$, has as focal elements the n-products $A_1 \times \cdots \times A_n$ with $A_i \in \big\{\{T\}, \{F\}, \mathbb{X}\big\}$ for all i, its vacuous extension to $\mathbb{X}_1 \times \cdots \times \mathbb{X}_n \times \mathbb{X}_{n+1}$ will have as focal elements the $n+1$-products of the form $A_1 \times \cdots \times A_n \times \mathbb{X}_{n+1}$, with $A_i \in \big\{\{T\}, \{F\}, \mathbb{X}\big\}$ for all i.

The belief function $Bel_{\mathbb{X}_{n+1}}$ is defined on $\mathbb{X}_{n+1} = \mathbb{X}$, with three focal elements $\{T\}$, $\{F\}$ and $\mathbb{X} = \{T, F\}$. Its vacuous extension to $\mathbb{X}_1 \times \cdots \times \mathbb{X}_n \times \mathbb{X}_{n+1}$ thus has the following three focal elements: $\mathbb{X}_1 \times \cdots \times \mathbb{X}_n \times \{T\}$, $\mathbb{X}_1 \times \cdots \times \mathbb{X}_n \times \{F\}$ and $\mathbb{X}_1 \times \cdots \times \mathbb{X}_n \times \mathbb{X}_{n+1}$. When computing $(Bel_{\mathbb{X}_1} \oplus \cdots \oplus Bel_{\mathbb{X}_n}) \oplus Bel_{\mathbb{X}_{n+1}}$ on the common refinement $\mathbb{X}_1 \times \cdots \times \mathbb{X}_n \times \mathbb{X}_{n+1}$, we need to compute the intersection of their focal elements, namely

$$\big(A_1 \times \cdots \times A_n \times \mathbb{X}_{n+1}\big) \cap \big(\mathbb{X}_1 \times \cdots \times \mathbb{X}_n \times A_{n+1}\big) = A_1 \times \cdots \times A_n \times A_{n+1}$$

for all $A_i \subseteq \mathbb{X}_i$, $i = 1, \ldots, n + 1$. All such intersections are distinct for distinct focal elements of the two belief functions to be combined, and there are no empty intersections. By Dempster's rule (2.6), their mass is equal to the product of the original masses:

$$\begin{aligned} & m_{Bel_{\mathbb{X}_1} \oplus \cdots \oplus Bel_{\mathbb{X}_{n+1}}}(A_1 \times \cdots \times A_n \times A_{n+1}) \\ &= m_{Bel_{\mathbb{X}_1} \oplus \cdots \oplus Bel_{\mathbb{X}_n}}(A_1 \times \cdots \times A_n) \cdot m_{Bel_{\mathbb{X}_{n+1}}}(A_{n+1}). \end{aligned}$$

Since we have assumed that the factorisation holds for n, the thesis easily follows.

□

As no normalisation is involved in the combination $Bel_{\mathbb{X}_1} \oplus \cdots \oplus Bel_{\mathbb{X}_n}$, Dempster's rule coincides with the conjunctive rule, and Lemma 21 holds for $\textcircled{\cap}$ as well.

Factorisation for 'sharp' tuples The following then becomes a simple corollary.

Theorem 108. *When either* $\textcircled{\cap}$ *or* $\oplus$ *is used as a combination rule in the definition of the belief likelihood function, the following decomposition holds for tuples* $(x_1, \ldots, x_n)$, $x_i \in \mathbb{X}_i$, *which are the singleton elements of* $\mathbb{X}_1 \times \cdots \times \mathbb{X}_n$, *with* $\mathbb{X}_1 = \cdots = \mathbb{X}_n = \{T, F\}$:

$$Bel_{\mathbb{X}_1 \times \cdots \times \mathbb{X}_n}(\{(x_1, \ldots, x_n)\}|\theta) = \prod_{i=1}^{n} Bel_{\mathbb{X}_i}(\{x_i\}|\theta). \tag{17.4}$$

Proof. For the singleton elements of $\mathbb{X}_1 \times \cdots \times \mathbb{X}_n$, since $\{(x_1, \ldots, x_n)\} = \{x_1\} \times \cdots \times \{x_n\}$, (17.4) becomes

$$\begin{aligned} & Bel_{\mathbb{X}_1 \times \cdots \times \mathbb{X}_n}(\{(x_1, \ldots, x_n)\}) \\ &= m_{Bel_{\mathbb{X}_1} \oplus \cdots \oplus Bel_{\mathbb{X}_n}}(\{(x_1, \ldots, x_n)\}) = \prod_{i=1}^{n} m_{\mathbb{X}_i}(\{x_i\}) = \prod_{i=1}^{n} Bel_{\mathbb{X}_i}(\{x_i\}), \end{aligned}$$

where the mass factorisation follows from Lemma 21, as on singletons mass and belief values coincide.

□

There is evidence to support the following conjecture as well.

Conjecture 1. When either $\textcircled{\cap}$ or $\oplus$ is used as a combination rule in the definition of the belief likelihood function, the following decomposition holds for the associated plausibility values on tuples $(x_1, \ldots, x_n)$, $x_i \in \mathbb{X}_i$, which are the singleton elements of $\mathbb{X}_1 \times \cdots \times \mathbb{X}_n$, with $\mathbb{X}_1 = \ldots = \mathbb{X}_n = \{T, F\}$:

$$Pl_{\mathbb{X}_1 \times \cdots \times \mathbb{X}_n}(\{(x_1, \ldots, x_n)\}|\theta) = \prod_{i=1}^{n} Pl_{\mathbb{X}_i}(\{x_i\}|\theta). \tag{17.5}$$

In fact, we can write

$$\begin{aligned} Pl_{\mathbb{X}_1 \times \cdots \times \mathbb{X}_n}(\{(x_1, \ldots, x_n)\}) &= 1 - Bel_{\mathbb{X}_1 \times \cdots \times \mathbb{X}_n}(\{(x_1, \ldots, x_n)\}^c) \\ &= 1 - \sum_{B \subseteq \{(x_1, \ldots, x_n)\}^c} m_{Bel_{\mathbb{X}_1} \oplus \cdots \oplus Bel_{\mathbb{X}_n}}(B). \end{aligned} \tag{17.6}$$

By Lemma 21, all the subsets B with non-zero mass are Cartesian products of the form $A_1 \times \cdots \times A_n$, $\emptyset \neq A_i \subseteq \mathbb{X}_i$. We then need to understand the nature of the focal elements of $Bel_{\mathbb{X}_1 \times \cdots \times \mathbb{X}_n}$ which are subsets of an arbitrary singleton complement $\{(x_1, \ldots, x_n)\}^c$.

For binary spaces $\mathbb{X}_i = \mathbb{X} = \{T, F\}$, by the definition of the Cartesian product, each such $A_1 \times \cdots \times A_n \subseteq \{(x_1, \ldots, x_n)\}^c$ is obtained by replacing a number

$1 \leq k \leq n$ of components of the tuple $(x_1, \ldots, x_n) = \{x_1\} \times \cdots \times \{x_n\}$ with a different subset of $\mathbb{X}_i$ (either $\{x_i\}^c = \mathbb{X}_i \setminus \{x_i\}$ or $\mathbb{X}_i$). There are $\binom{n}{k}$ such sets of k components in a list of n. Of these k components, in general $1 \leq m \leq k$ will be replaced by $\{x_i\}^c$, while the other $1 \leq k - m < k$ will be replaced by $\mathbb{X}_i$. Note that not all k components can be replaced by $\mathbb{X}_i$, since the resulting focal element would contain the tuple $\{(x_1, \ldots, x_n)\} \in \mathbb{X}_1 \times \cdots \times \mathbb{X}_n$.

The following argument can be proved for $(x_1, ..., x_n) = (T, ..., T)$, under the additional assumption that $Bel_{\mathbb{X}_1} \cdots Bel_{\mathbb{X}_n}$ are equally distributed with $p \doteq Bel_{\mathbb{X}_i}(\{T\})$, $q \doteq Bel_{\mathbb{X}_i}(\{F\})$ and $r \doteq Bel_{\mathbb{X}_i}(\mathbb{X}_i)$.
If this is the case, for fixed values of m and k all the resulting focal elements have the same mass value, namely $p^{n-k} q^m r^{k-m}$. As there are exactly $\binom{k}{m}$ such focal elements, (17.6) can be written as

$$1 - \sum_{k=1}^{n} \binom{n}{k} \sum_{m=1}^{k} \binom{k}{m} p^{n-k} q^m r^{k-m},$$

which can be rewritten as

$$1 - \sum_{m=1}^{n} q^m \sum_{k=m}^{n} \binom{n}{k} \binom{k}{m} p^{n-k} r^{k-m}.$$

A change of variable $l = n - k$, where $l = 0$ when $k = n$, and $l = n - m$ when $k = m$, allows us to write this as

$$1 - \sum_{m=1}^{n} q^m \sum_{l=0}^{n-m} \binom{n}{n-l} \binom{n-l}{m} p^l r^{(n-m)-l},$$

since $k - m = n - l - m$, $k = n - l$. Now, as

$$\binom{n}{n-l} \binom{n-l}{m} = \binom{n}{m} \binom{n-m}{l},$$

we obtain

$$1 - \sum_{m=1}^{n} q^m \binom{n}{m} \sum_{l=0}^{n-m} \binom{n-m}{l} p^l r^{(n-m)-l}.$$

By Newton's binomial theorem, the latter is equal to

$$1 - \sum_{m=1}^{n} q^m \binom{n}{m} (p+r)^{n-m} = 1 - \sum_{m=1}^{n} q^m \binom{n}{m} (1-q)^{n-m},$$

since $p + r = 1 - q$. As $\sum_{m=0}^{n} \binom{n}{m} q^m (1-q)^{n-m} = 1$, again by Newton's binomial theorem, we get

$$Pl_{\mathbb{X}_1 \times \cdots \times \mathbb{X}_n}(\{(T, \ldots, T)\}) = 1 - [1 - (1-q)^n] = (1-q)^n = \prod_{i=1}^{n} Pl_{\mathbb{X}_i}(\{T\}).$$

Question 13. Does the decomposition (17.6) hold for any sharp sample and arbitrary belief functions $Bel_{\mathbb{X}_1}, \ldots, Bel_{\mathbb{X}_n}$ defined on arbitrary frames of discernment $\mathbb{X}_1, \ldots, \mathbb{X}_n$?

Factorisation for Cartesian products The decomposition (17.4) is equivalent to what Smets calls *conditional conjunctive independence* in his general Bayes theorem.

In fact, for binary spaces the factorisation (17.4) generalises to all subsets $A \subseteq \mathbb{X}_1 \times \cdots \times \mathbb{X}_n$ which are Cartesian products of subsets of $\mathbb{X}_1, \ldots, \mathbb{X}_n$, respectively: $A = A_1 \times \cdots \times A_n$, $A_i \subseteq \mathbb{X}_i$ for all i.

Corollary 24. *Whenever $A_i \subseteq \mathbb{X}_i$, $i = 1, \ldots, n$, and $\mathbb{X}_i = \mathbb{X} = \{T, F\}$ for all i, under conjunctive combination we have that*

$$Bel_{\mathbb{X}_1 \times \cdots \times \mathbb{X}_n}(A_1 \times \cdots \times A_n | \theta) = \prod_{i=1}^{n} Bel_{\mathbb{X}_i}(A_i | \theta). \tag{17.7}$$

Proof. As, by Lemma 21, all the focal elements of $Bel_{\mathbb{X}_1} \textcircled{\cap} \cdots \textcircled{\cap} Bel_{\mathbb{X}_n}$ are Cartesian products of the form $B = B_1 \times \cdots \times B_n$, $B_i \subseteq \mathbb{X}_i$,

$$Bel_{\mathbb{X}_1 \times \cdots \times \mathbb{X}_n}(A_1 \times \cdots \times A_n | \theta) = \sum_{\substack{B \subseteq A_1 \times \cdots \times A_n \\ B = B_1 \times \cdots \times B_n}} m_{\mathbb{X}_1}(B_1) \cdot \ldots \cdot m_{\mathbb{X}_n}(B_n).$$

But

$$\{B \subseteq A_1 \times \cdots \times A_n, B = B_1 \times \cdots \times B_n\} = \{B = B_1 \times \cdots \times B_n, B_i \subseteq A_i \forall i\},$$

since, if $B_j \not\subset A_j$ for some j, the resulting Cartesian product would not be contained within $A_1 \times \cdots \times A_n$. Therefore,

$$Bel_{\mathbb{X}_1 \times \cdots \times \mathbb{X}_n}(A_1 \times \cdots \times A_n | \theta) = \sum_{\substack{B = B_1 \times \cdots \times B_n \\ B_i \subseteq A_i \forall i}} m_{\mathbb{X}_1}(B_1) \cdot \ldots \cdot m_{\mathbb{X}_n}(B_n).$$

For all those A_i's, $i = i_1, \ldots, i_m$ that are singletons of $\mathbb{X}_i$, necessarily $B_i = A_i$ and we can write the above expression as

$$m(A_{i_1}) \cdot \ldots \cdot m(A_{i_m}) \sum_{B_j \subseteq A_j, j \neq i_1, \ldots, i_m} \prod_{j \neq i_1, \ldots, i_m} m_{\mathbb{X}_j}(B_j).$$

If the frames involved are binary, $\mathbb{X}_i = \{T, F\}$, those A_i's that are not singletons coincide with $\mathbb{X}_i$, so that we have

$$m(A_{i_1}) \cdot \ldots \cdot m(A_{i_m}) \sum_{B_j \subseteq \mathbb{X}_j, j \neq i_1, \ldots, i_m} \prod_j m_{\mathbb{X}_j}(B_j).$$

The quantity $\sum_{B_j \subseteq \mathbb{X}_j, j \neq i_1, \ldots, i_m} \prod_j m_{\mathbb{X}_j}(B_j)$ is, according to the definition of the conjunctive combination, the sum of the masses of all the possible intersections of (cylindrical extensions of) focal elements of $Bel_{\mathbb{X}_j}$, $j \neq i_1, \ldots, i_m$. Thus, it adds up to 1.

In conclusion (forgetting the conditioning on θ in the derivation for the sake of readability),

$$\begin{aligned} & Bel_{\mathbb{X}_1\times\cdots\times\mathbb{X}_n}(A_1\times\cdots\times A_n) \\ &= m(A_{i_1})\cdot\ldots\cdot m(A_{i_m})\cdot 1\cdot\ldots\cdot 1 \\ &= Bel_{\mathbb{X}_{i_1}}(A_{i_1})\cdot\ldots\cdot Bel_{\mathbb{X}_{i_m}}(A_{i_m})\cdot Bel_{\mathbb{X}_{j_1}}(\mathbb{X}_{j_1})\cdot\ldots\cdot Bel_{\mathbb{X}_{j_k}}(\mathbb{X}_{j_k}) \\ &= Bel_{\mathbb{X}_{i_1}}(A_{i_1})\cdot\ldots\cdot Bel_{\mathbb{X}_{i_m}}(A_{i_m})\cdot Bel_{\mathbb{X}_{j_1}}(A_{j_1})\cdot\ldots\cdot Bel_{\mathbb{X}_{j_k}}(A_{j_k}), \end{aligned}$$

and we have (17.7).

□

Corollary 24 states that conditional conjunctive independence always holds, for events that are Cartesian products, whenever the frames of discernment concerned are binary.

Binary trials: The disjunctive case Similar factorisation results hold when the (more cautious) disjunctive combination $\textcircled{\cup}$ is used. As in the conjunctive case, we first analyse the case $n = 2$.

Structure of the focal elements We seek the disjunctive combination $Bel_{\mathbb{X}_1}\textcircled{\cup}Bel_{\mathbb{X}_2}$, where each $Bel_{\mathbb{X}_i}$ has as focal elements $\{T\}$, $\{F\}$ and $\mathbb{X}_i$. Figure 17.1(b) is a diagram of all the unions of focal elements of the two input belief functions on their common refinement $\mathbb{X}_1\times\mathbb{X}_2$. There are $5 = 2^2+1$ distinct such unions, which correspond to the focal elements of $Bel_{\mathbb{X}_1}\textcircled{\cup}Bel_{\mathbb{X}_2}$, with mass values

$$\begin{aligned} m_{Bel_{\mathbb{X}_1}\oplus Bel_{\mathbb{X}_2}}(\{(x_i,x_j)\}^c) &= m_{\mathbb{X}_1}(\{x_i\}^c)\cdot m_{\mathbb{X}_2}(\{x_j\}^c), \\ m_{Bel_{\mathbb{X}_1}\oplus Bel_{\mathbb{X}_2}}(\mathbb{X}_1\times\mathbb{X}_2) &= 1-\sum_{i,j} m_{\mathbb{X}_1}(\{x_i\}^c)\cdot m_{\mathbb{X}_2}(\{x_j\}^c). \end{aligned}$$

We can now prove the following lemma.

Lemma 22. *The belief function $Bel_{\mathbb{X}_1}\textcircled{\cup}\cdots\textcircled{\cup}Bel_{\mathbb{X}_n}$, where $\mathbb{X}_i = \mathbb{X} = \{T,F\}$, has 2^n+1 focal elements, namely all the complements of the n-tuples $\mathbf{x} = (x_1,\ldots,x_n)$ of singleton elements $x_i\in\mathbb{X}_i$, with BPA*

$$m_{Bel_{\mathbb{X}_1}\textcircled{\cup}\cdots\textcircled{\cup}Bel_{\mathbb{X}_n}}(\{(x_1,\ldots,x_n)\}^c) = m_{\mathbb{X}_1}(\{x_1\}^c)\cdot\ldots\cdot m_{\mathbb{X}_n}(\{x_n\}^c), \quad (17.8)$$

plus the Cartesian product $\mathbb{X}_1\times\cdots\times\mathbb{X}_n$ itself, with its mass value given by normalisation.

Proof. The proof is by induction. The case $n = 2$ was proven above. In the induction step, we assume that the thesis is true for n, namely that the focal elements of $Bel_{\mathbb{X}_1}\textcircled{\cup}\cdots\textcircled{\cup}Bel_{\mathbb{X}_n}$ have the form

$$A = \Big\{(x_1,\ldots,x_n)\Big\}^c = \Big\{(x'_1,\ldots,x'_n)\Big|\exists i : \{x'_i\} = \{x_i\}^c\Big\}, \quad (17.9)$$

where $x_i\in\mathbb{X}_i = \mathbb{X}$. We need to prove it is true for $n+1$.

The vacuous extension of (17.9) has, trivially, the form

$$A' = \Big\{(x'_1, \ldots, x'_n, x_{n+1}) \Big| \exists i \in \{1, \ldots, n\} : \{x'_i\} = \{x_i\}^c, x_{n+1} \in \mathbb{X}\Big\}.$$

Note that only $2 = |\mathbb{X}|$ singletons of $\mathbb{X}_1 \times \cdots \times \mathbb{X}_{n+1}$ are *not* in A', for any given tuple $(x_1, \ldots, x_n)$. The vacuous extension to $\mathbb{X}_1 \times \cdots \times \mathbb{X}_{n+1}$ of a focal element $B = \{x_{n+1}\}$ of $Bel_{\mathbb{X}_{n+1}}$ is instead $B' = \{(y_1, \ldots, y_n, x_{n+1}), y_i \in \mathbb{X} \; \forall i = 1, \ldots, n\}$.

Now, all the elements of B', except for $(x_1, \ldots, x_n, x_{n+1})$, are also elements of A'. Hence, the union $A' \cup B'$ reduces to the union of A' and $(x_1, \ldots, x_n, x_{n+1})$. The only singleton element of $\mathbb{X}_1 \times \cdots \times \mathbb{X}_{n+1}$ not in $A' \cup B'$ is therefore $(x_1, \ldots, x_n, x'_{n+1})$, $\{x'_{n+1}\} = \{x_{n+1}\}^c$, for it is neither in A' nor in B'. All such unions are distinct, and therefore by the definition of the disjunctive combination their mass is $m(\{(x_1, \ldots, x_n)\}^c) \cdot m(\{x_{n+1}\}^c)$, which by the inductive hypothesis is equal to (17.8). Unions involving either $\mathbb{X}_{n+1}$ or $\mathbb{X}_1 \times \cdots \times \mathbb{X}_n$ are equal to $\mathbb{X}_1 \times \cdots \times \mathbb{X}_{n+1}$ by the property of the union operator.

□

Factorisation

Theorem 109. *In the hypotheses of Lemma 22, when the disjunctive combination* $\textcircled{\cup}$ *is used in the definition of the belief likelihood function, the following decomposition holds:*

$$Bel_{\mathbb{X}_1 \times \cdots \times \mathbb{X}_n}(\{(x_1, \ldots, x_n)\}^c | \theta) = \prod_{i=1}^n Bel_{\mathbb{X}_i}(\{x_i\}^c | \theta). \tag{17.10}$$

Proof. Since $\{(x_1, \ldots, x_n)\}^c$ contains only itself as a focal element,

$$Bel_{\mathbb{X}_1 \times \cdots \times \mathbb{X}_n}(\{(x_1, \ldots, x_n)\}^c | \theta) = m(\{(x_1, \ldots, x_n)\}^c | \theta).$$

By Lemma 22, the latter becomes

$$Bel_{\mathbb{X}_1 \times \cdots \times \mathbb{X}_n}(\{(x_1, \ldots, x_n)\}^c | \theta) = \prod_{i=1}^n m_{\mathbb{X}_i}(\{x_i\}^c | \theta) = \prod_{i=1}^n Bel_{\mathbb{X}_i}(\{x_i\}^c | \theta),$$

as $\{x_i\}^c$ is a singleton element of $\mathbb{X}_i$, and we have (17.10).

□

Note that $Pl_{\mathbb{X}_1 \times \cdots \times \mathbb{X}_n}(\{(x_1, \ldots, x_n)\}^c | \theta) = 1$ for all tuples $(x_1, \ldots, x_n)$, as the set $\{(x_1, \ldots, x_n)\}^c$ has non-empty intersection with all the focal elements of $Bel_{\mathbb{X}_1} \textcircled{\cup} \cdots \textcircled{\cup} Bel_{\mathbb{X}_n}$.

General factorisation results A look at the proof of Lemma 21 shows that the argument is in fact valid for the conjunctive combination of belief functions defined on an arbitrary collection $\mathbb{X}_1, \ldots, \mathbb{X}_n$ of finite spaces. Namely, we have the following theorem.

Theorem 110. *For any $n \in \mathbb{Z}$, the belief function $Bel_{\mathbb{X}_1} \oplus \cdots \oplus Bel_{\mathbb{X}_n}$, where $\mathbb{X}_1, \ldots, \mathbb{X}_n$ are finite spaces, has as focal elements all the Cartesian products $A_1 \times \cdots \times A_n$ of n focal elements $A_1 \subseteq \mathbb{X}_1, \ldots, A_n \subseteq \mathbb{X}_n$, with BPA*

$$m_{Bel_{\mathbb{X}_1} \oplus \cdots \oplus Bel_{\mathbb{X}_n}}(A_1 \times \cdots \times A_n) = \prod_{i=1}^{n} m_{\mathbb{X}_i}(A_i).$$

The proof is left to the reader. The corollary below follows.

Corollary 25. *When either $\textcircled{\cap}$ or $\oplus$ is used as a combination rule in the definition of the belief likelihood function, the following decomposition holds for tuples $(x_1, \ldots, x_n)$, $x_i \in \mathbb{X}_i$, which are the singleton elements of $\mathbb{X}_1 \times \cdots \times \mathbb{X}_n$, with $\mathbb{X}_1, \ldots, \mathbb{X}_n$ being arbitrary discrete frames of discernment:*

$$Bel_{\mathbb{X}_1 \times \cdots \times \mathbb{X}_n}(\{(x_1, \ldots, x_n)\}|\theta) = \prod_{i=1}^{n} Bel_{\mathbb{X}_i}(\{x_i\}|\theta). \tag{17.11}$$

Proof. For the singleton elements of $\mathbb{X}_1 \times \cdots \times \mathbb{X}_n$, since $\{(x_1, \ldots, x_n)\} = \{x_1\} \times \cdots \times \{x_n\}$, (17.11) becomes

$$\begin{aligned} & Bel_{\mathbb{X}_1 \times \cdots \times \mathbb{X}_n}(\{(x_1, \ldots, x_n)\}) \\ & = m_{Bel_{\mathbb{X}_1} \oplus \cdots \oplus Bel_{\mathbb{X}_n}}(\{(x_1, ..., x_n)\}) = \prod_{i=1}^{n} m_{\mathbb{X}_i}(\{x_i\}) = \prod_{i=1}^{n} Bel_{\mathbb{X}_i}(\{x_i\}), \end{aligned}$$

where the mass factorisation follows from Theorem 110, as mass and belief values coincide on singletons.

□

This new approach to inference with belief functions, based on the notion of the belief likelihood, opens up a number of interesting avenues: from a more systematic comparison with other inference approaches (see Section 4.1), to the possible generalisation of the factorisation results obtained above, and to the computation of belief and plausibility likelihoods for other major combination rules.

Lower and upper likelihoods of Bernoulli trials In the case of Bernoulli trials, where not only is there a single binary sample space $\mathbb{X}_i = \mathbb{X} = \{T, F\}$ and conditional independence holds, but also the random variables are assumed to be equally distributed, the conventional likelihood reads as $p^k(1-p)^{n-k}$, where $p = P(T)$, k is the number of successes (T) and n is as usual the total number of trials.

Let us then compute the lower likelihood function for a series of Bernoulli trials, under the similar assumption that all the belief functions $Bel_{\mathbb{X}_i} = Bel_{\mathbb{X}}$, $i = 1, \ldots, n$, coincide (the analogous of equidistribution), with $Bel_{\mathbb{X}}$ parameterised by $p = m(\{T\})$, $q = m(\{F\})$ (where, this time, $p + q \leq 1$).

Corollary 26. *Under the above assumptions, the lower and upper likelihoods of the sample $\mathbf{x} = (x_1, \ldots, x_n)$ are, respectively,*

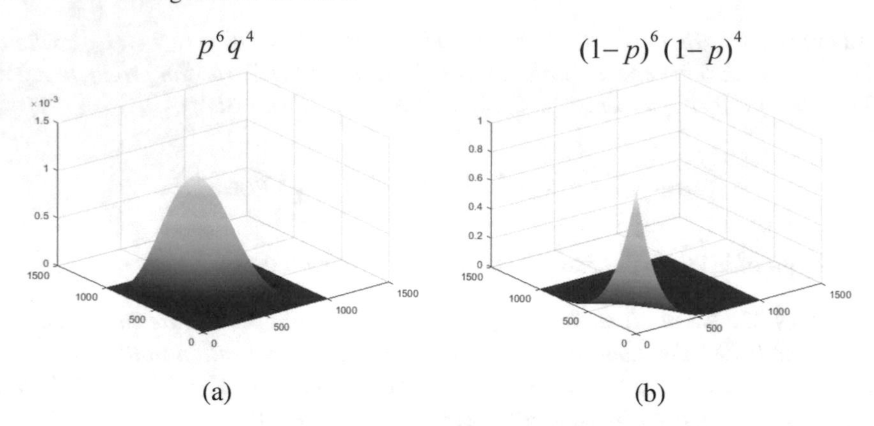

Fig. 17.2: Lower (a) and upper (b) likelihood functions plotted in the space of belief functions defined on the frame $\mathbb{X} = \{T, F\}$, parameterised by $p = m(T)$ (X axis) and $q = m(F)$ (Y axis), for the case of $k = 6$ successes in $n = 10$ trials.

$$
\begin{aligned}
\underline{L}(\mathbf{x}) &= \prod_{i=1}^{n} Bel_{\mathbb{X}}(\{x_i\}) = p^k q^{n-k}, \\
\overline{L}(\mathbf{x}) &= \prod_{i=1}^{n} Pl_{\mathbb{X}}(\{x_i\}) = (1-q)^k (1-p)^{n-k}.
\end{aligned}
\tag{17.12}
$$

The above decomposition for $\overline{L}(\mathbf{x})$, in particular, is valid under the assumption that Conjecture 1 holds, at least for Bernoulli trials, as the evidence seems to suggest. After normalisation, these lower and upper likelihoods can be seen as PDFs over the (belief) space $\mathcal{B}$ of all belief functions definable on $\mathbb{X}$ [334, 364].

Having observed a series of trials $\mathbf{x} = (x_1, \ldots, x_n)$, $x_i \in \{T, F\}$, one may then seek the belief function on $\mathbb{X} = \{T, F\}$ which best describes the observed sample, i.e., the optimal values of the two parameters p and q. In the next subsection we will address a similar problem in a generalised logistic regression setting.

Example 52: lower and upper likelihoods. *Figure 17.2 plots the lower and upper likelihoods (17.12) for the case of $k = 6$ successes in $n = 10$ trials. Both subsume the traditional likelihood $p^k(1-p)^{n-k}$ as their section for $p + q = 1$, although this is particularly visible for the lower likelihood (a). In particular, the maximum of the lower likelihood is the traditional maximum likelihood estimate $p = k/n$, $q = 1 - p$. This makes sense, for the lower likelihood is highest for the most committed belief functions (i.e., for probability measures). The upper likelihood (b) has a unique maximum in $p = q = 0$: this is the vacuous belief function on $\{T, F\}$, with $m(\{T, F\}) = 1$.*

The interval of belief functions joining $\max \overline{L}$ with $\max \underline{L}$ is the set of belief functions such that $\frac{p}{q} = \frac{k}{n-k}$, i.e., those which preserve the ratio between the observed empirical counts.

17.1.2 Generalised logistic regression

Based on Theorem 108 and Conjecture 1, we can also generalise logistic regression (Section 1.3.7) to a belief function setting by replacing the conditional probability $(\pi_i, 1-\pi_i)$ on $\mathbb{X} = \{T, F\}$ with a belief function ($p_i = m(\{T\})$, $q_i = m(\{F\})$, $p_i + q_i \leq 1$) on $2^{\mathbb{X}}$. Note that, just as in a traditional logistic regression setting (Section 1.3.7), the belief functions Bel_i associated with different input values x_i are not equally distributed.

Writing $T = 1, F = 0$, the lower and upper likelihoods can be computed as

$$\underline{L}(\beta|Y) = \prod_{i=1}^{n} p_i^{Y_i} q_i^{1-Y_i}, \quad \overline{L}(\beta|Y) = \prod_{i=1}^{n} (1-q_i)^{Y_i}(1-p_i)^{1-Y_i}.$$

The question becomes how to generalise the logit link between observations x and outputs y, in order to seek an analytical mapping between observations and belief functions over a binary frame. Just assuming

$$p_i = m(Y_i = 1|x_i) = \frac{1}{1 + e^{-(\beta_0 + \beta_1 x_i)}}, \tag{17.13}$$

as in the classical constraint (1.4), does not yield any analytical dependency for q_i. To address this issue, we can just (for instance) add a parameter β_2 such that the following relationship holds:

$$q_i = m(Y_i = 0|x_i) = \beta_2 \frac{e^{-(\beta_0 + \beta_1 x_i)}}{1 + e^{-(\beta_0 + \beta_1 x_i)}}. \tag{17.14}$$

We can then seek lower and upper optimal estimates for the parameter vector $\beta = [\beta_0, \beta_1, \beta_2]'$:

$$\arg\max_{\beta} \underline{L} \mapsto \underline{\beta}_0, \underline{\beta}_1, \underline{\beta}_2, \quad \arg\max_{\beta} \overline{L} \mapsto \overline{\beta}_0, \overline{\beta}_1, \overline{\beta}_2. \tag{17.15}$$

Plugging these optimal parameters into (17.13), (17.14) will then yield an upper and a lower family of conditional belief functions given x (i.e., an interval of belief functions),

$$Bel_{\mathbb{X}}(.|\underline{\beta}, x), \quad Bel_{\mathbb{X}}(.|\overline{\beta}, x).$$

Given any new test observation x', our generalised logistic regression method will then output a pair of lower and upper belief functions on $\mathbb{X} = \{T, F\} = \{1, 0\}$, as opposed to a sharp probability value as in the classical framework. As each belief function itself provides a lower and an upper probability value for each event, both the lower and the upper regressed BFs will provide an interval for the probability $P(T)$ of success, whose width reflects the uncertainty encoded by the training set of sample series.

Optimisation problem Both of the problems (17.15) are constrained optimisation problems (unlike the classical case, where $q_i = 1 - p_i$ and the optimisation problem is unconstrained). In fact, the parameter vector β must be such that

$$0 \leq p_i + q_i \leq 1 \quad \forall i = 1, \dots, n.$$

Fortunately, the number of constraints can be reduced by noticing that, under the analytical relations (17.13) and (17.14), $p_i + q_i \leq 1$ for all i whenever $\beta_2 \leq 1$.

The objective function can be simplified by taking the logarithm. In the lower-likelihood case, we get[80]

$$\log \underline{L}(\beta|Y) = \sum_{i=1}^{n} \Big\{ -\log(1 + e^{-(\beta_0+\beta_1 x_i)}) + (1 - Y_i)\big[\log \beta_2 - (\beta_0 + \beta_1 x_i)\big]\Big\}.$$

We then need to analyse the *Karush–Kuhn–Tucker* (KKT) necessary conditions for the optimality of the solution of a nonlinear optimisation problem subject to differentiable constraints.

Definition 137. *Suppose that the objective function $f : \mathbb{R}^n \to \mathbb{R}$ and the constraint functions $g_i : \mathbb{R}^n \to \mathbb{R}$ and $h_j : \mathbb{R}^n \to \mathbb{R}$ of a nonlinear optimisation problem*

$$\arg\max_{x} f(x),$$

subject to

$$g_i(x) \leq 0 \;\; i = 1, \dots, m, \quad h_j(x) = 0 \;\; j = 1, \dots, l,$$

are continuously differentiable at a point x^*. If x^* is a local optimum, under some regularity conditions there then exist constants μ_i ($i = 1, \dots, m$) and λ_j ($j = 1, \dots, l$), called KKT multipliers, *such that the following conditions hold:*

1. Stationarity*:* $\nabla f(x^*) = \sum_{i=1}^{m} \mu_i \nabla g_i(x^*) + \sum_{j=1}^{l} \lambda_j \nabla h_j(x^*)$.
2. Primal feasibility*:* $g_i(x^*) \leq 0$ *for all* $i = 1, \dots, m$, *and* $h_j(x^*) = 0$, *for all* $j = 1, \dots, l$.
3. Dual feasibility*:* $\mu_i \geq 0$ *for all* $i = 1, \dots, m$.
4. Complementary slackness*:* $\mu_i g_i(x^*) = 0$ *for all* $i = 1, \dots, m$.

For the optimisation problem considered here, we have only inequality constraints, namely

$$\begin{aligned} \beta_2 \leq 1 \quad & \equiv g_0 = \beta_2 - 1 \leq 0, \\ p_i + q_i \geq 0 & \equiv g_i = -\beta_2 - e^{\beta_0+\beta_1 x_i} \leq 0, \quad i = 1, \dots, n. \end{aligned}$$

The Lagrangian becomes

$$\Lambda(\beta) = \log \underline{L}(\beta) + \mu_0(\beta_2 - 1) - \sum_{i=1}^{n} \mu_i(\beta_2 + e^{\beta_0+\beta_1 x_i}).$$

[80] Derivations are omitted.

The stationarity conditions thus read as $\nabla \Lambda(\beta) = 0$, namely

$$\begin{cases} \displaystyle\sum_{i=1}^{n} \left[(1 - p_i) - (1 - Y_i) - \mu_i e^{\beta_0 + \beta_1 x_i} \right] = 0, \\ \displaystyle\sum_{i=1}^{n} \left[(1 - p_i) - (1 - Y_i) - \mu_i e^{\beta_0 + \beta_1 x_i} \right] x_i = 0, \\ \displaystyle\sum_{i=1}^{n} \left(\frac{1 - Y_i}{\beta_2} - \mu_i \right) + \mu_0 = 0, \end{cases} \qquad (17.16)$$

where, as usual, p_i is as in (17.13). The complementary slackness condition reads as follows:

$$\begin{cases} \mu_0(\beta_2 - 1) = 0, \\ \mu_i(\beta_2 + e^{\beta_0 + \beta_1 x_i}) = 0, \quad i = 1, \ldots, n. \end{cases} \qquad (17.17)$$

Gradient descent methods can then be applied to the above systems of equations to find the optimal parameters of our generalised logistic regressor.

Question 14. Is there an analytical solution to the generalised logistic regression inference problem, based on the above KKT necessity conditions?

Similar calculations hold for the upper-likelihood problem. A multi-objective optimisation setting in which the lower likelihood is minimised as the upper likelihood is maximised can also be envisaged, in order to generate the most cautious interval of estimates.

Research questions As far as generalised logistic regression is concerned, an analysis of the validity of such a direct extension of logit mapping and the exploration of alternative ways of generalising it are research questions that are potentially very interesting to pursue.

Question 15. What other parameterisations can be envisaged for the generalised logistic regression problem, in terms of the logistic links between the data and the belief functions Bel_i on $\{T, F\}$?

Comprehensive testing of the robustness of the generalised framework on standard estimation problems, including rare-event analysis, needs to be conducted to further validate this new procedure.

Note that the method, unlike traditional ones, can naturally cope with missing data (represented by vacuous observations $x_i = \mathbb{X}$), therefore providing a robust framework for logistic regression which can deal with incomplete series of observations.

17.1.3 The total probability theorem for random sets

Spies (Section 4.6.2) and others have posed the problem of generalising the law of total probability,

$$P(A) = \sum_{i=1}^{N} P(A|B_i)P(B_i),$$

where $\{B_1, \ldots, B_N\}$ is a disjoint partition of the sample space, to the case of belief functions. They mostly did so with the aim of generalising Jeffrey's combination rule – nevertheless, the question goes rather beyond their original intentions, as it involves understanding the space of solutions to the generalised total probability problem.

The law of total belief The problem of generalising the total probability theorem to belief functions can be posed as follows (Fig. 17.3).

Theorem 111. (Total belief theorem). *Suppose Θ and Ω are two frames of discernment, and $\rho : 2^{\Omega} \to 2^{\Theta}$ the unique refining between them. Let Bel_0 be a belief function defined over $\Omega = \{\omega_1, \ldots, \omega_{|\Omega|}\}$. Suppose there exists a collection of belief functions $Bel_i : 2^{\Pi_i} \to [0, 1]$, where $\Pi = \{\Pi_1, \ldots, \Pi_{|\Omega|}\}$, $\Pi_i = \rho(\{\omega_i\})$, is the partition of Θ induced by its coarsening Ω.*

Then, there exists a belief function $Bel : 2^{\Theta} \to [0, 1]$ such that:

- *(P1) Bel_0 is the marginal of Bel to Ω, $Bel_0 = Bel \upharpoonright_{\Omega}$ (2.13);*
- *(P2) $Bel \oplus Bel_{\Pi_i} = Bel_i$ for all $i = 1, \ldots, |\Omega|$, where Bel_{Π_i} is the categorical belief function with BPA $m_{\Pi_i}(\Pi_i) = 1$, $m_{\Pi_i}(B) = 0$ for all $B \neq \Pi_i$.*

Note that a disjoint partition $\Pi_1, \ldots, \Pi_{|\Omega|}$ of Θ defines a subalgebra $\mathbb{A}^{\rho}$ of 2^{Θ} as a Boolean algebra with set operations, which is isomorphic to the set algebra $\mathbb{A} = 2^{\Omega}$. We use the notation $Bel \upharpoonright_{\Omega}$ to denote the marginal of Bel to Ω.

Proof of existence Theorem 111's proof makes use of the two lemmas that follow. Here, basically, we adapt Smets's original proof of the generalised Bayesian theorem [1689] to the refinement framework presented here, which is more general than his multivariate setting, in which only Cartesian products of frames are considered. As usual, we denote by m_0 and m_i the mass functions of Bel_0 and Bel_i, respectively. Their sets of focal elements are denoted by $\mathcal{E}_{\Omega}$ and $\mathcal{E}_i$, respectively.

Suppose that $\Theta' \supseteq \Theta$ and m is a mass function over Θ. The mass function m can be identified with a mass function $\overrightarrow{m}_{\Theta'}$ over the larger frame Θ': for any $E' \subseteq \Theta'$, $\overrightarrow{m}_{\Theta'}(E') = m(E)$ if $E' = E \cup (\Theta' \setminus \Theta)$, and $\overrightarrow{m}_{\Theta'}(E') = 0$ otherwise. Such an $\overrightarrow{m}_{\Theta'}$ is called the *conditional embedding* of m into Θ'. When the context is clear, we can drop the subscript Θ'. It is easy to see that conditional embedding is the inverse of Dempster conditioning.

For the collection of belief functions $Bel_i : 2^{\Pi_i} \to [0, 1]$, let $\overrightarrow{Bel_i}$ be the conditional embedding of Bel_i into Θ and let $\overrightarrow{Bel}$ denote the Dempster combination of all $\overrightarrow{Bel_i}$, i.e., $\overrightarrow{Bel} = \overrightarrow{Bel_1} \oplus \cdots \oplus \overrightarrow{Bel_{|\Omega|}}$, with mass function $\overrightarrow{m}$.

Lemma 23. *The belief function $\overrightarrow{Bel}$ over Θ satisfies the following two properties: (1) each focal element $\overrightarrow{e}$ of $\overrightarrow{Bel}$ is the union of exactly one focal element e_i of each conditional belief function Bel_i; (2) the marginal $\overrightarrow{Bel} \upharpoonright_{\Omega}$ on Ω is the vacuous belief function over Ω.*

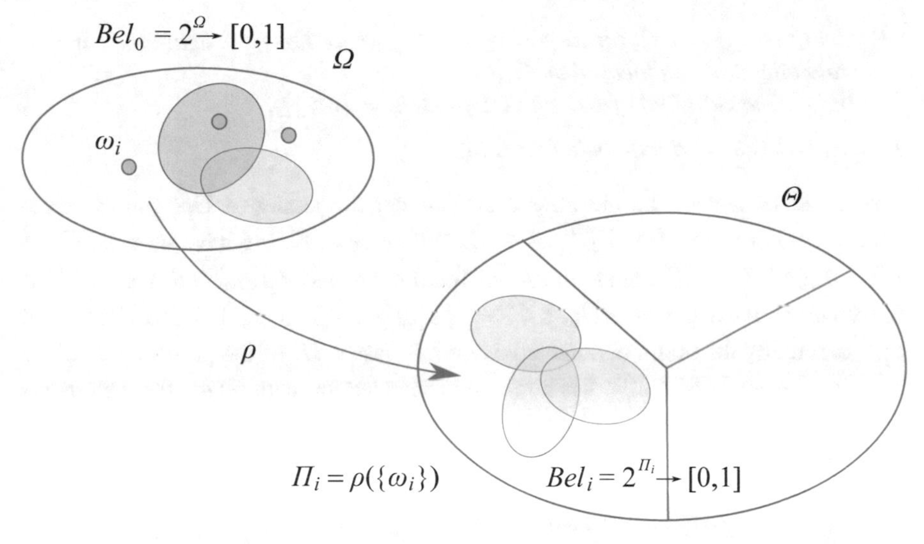

Fig. 17.3: Pictorial representation of the hypotheses of the total belief theorem (Theorem 111).

Proof. Each focal element $\overrightarrow{e_i}$ of $\overrightarrow{Bel_i}$ is of the form $(\bigcup_{j \neq i} \Pi_j) \cup e_i$, where e_i is some focal element of Bel_i. In other words, $\overrightarrow{e_i} = (\Theta \setminus \Pi_i) \cup e_i$. Since $\overrightarrow{Bel}$ is the Dempster combination of the $\overrightarrow{Bel_i}$s, it is easy to see that each focal element $\overrightarrow{e}$ of $\overrightarrow{Bel}$ is the union of exactly one focal element e_i from each conditional belief function Bel_i. In other words, $\overrightarrow{e} = \bigcup_{i=1}^{|\Omega|} e_i$, where $e_i \in \mathcal{E}_i$, and condition (1) is proven.

Let $\overrightarrow{\mathcal{E}}$ denote the set of all the focal elements of $\overrightarrow{Bel}$, namely

$$\overrightarrow{\mathcal{E}} = \left\{ \overrightarrow{e} \subseteq \Theta : \overrightarrow{e} = \bigcup_{i=1}^{|\Omega|} e_i \text{ where } e_i \text{ is a focal element of } Bel_i \right\}.$$

Note that e_i's coming from different conditional belief functions Bel_i's are disjoint. For each $\overrightarrow{e} \in \overrightarrow{\mathcal{E}}$, $\bar{\rho}(\overrightarrow{e}) = \Omega$ (where $\bar{\rho}$ denotes the outer reduction (2.12)). It follows from (2.13) that $\overrightarrow{m} \restriction_\Omega (\Omega) = 1$ – hence the marginal of $\overrightarrow{Bel}$ on Ω is the vacuous belief function there.

□

Let $Bel_0^{\uparrow\Theta}$ be the vacuous extension of Bel_0 from Ω to Θ. We define the desired total function Bel to be the Dempster combination of $Bel_0^{\uparrow\Theta}$ and $\overrightarrow{Bel}$, namely

$$Bel \doteq Bel_0^{\uparrow\Theta} \oplus \overrightarrow{Bel}. \tag{17.18}$$

Lemma 24. *The belief function Bel defined in (17.18) on the frame Θ satisfies the following two properties:*

1. *$Bel \oplus Bel_{\Pi_i} = Bel_i$ for all $i = 1, \ldots, |\Omega|$, where Bel_{Π_i} is again the categorical belief function focused on Π_i.*
2. *Bel_0 is the marginal of Bel on Ω, i.e., $Bel_0 = Bel \restriction_\Omega$.*

That is, Bel is a valid total belief function.

Proof. Let $\vec{m}$ and m_i be the mass functions corresponding to $\overrightarrow{Bel}$ and Bel_i, respectively. For each $\vec{e} = \bigcup_{i=1}^{|\Omega|} e_i \in \vec{\mathcal{E}}$, where $e_i \in \mathcal{E}_i$, we have that $\vec{m}(\vec{e}) = \prod_{i=1}^{|\Omega|} m_i(e_i)$. Let $\mathcal{E}_\Omega^{\uparrow\Theta}$ denote the set of focal elements of $Bel_0^{\uparrow\Theta}$. Since $Bel_0^{\uparrow\Theta}$ is the vacuous extension of Bel_0, $\mathcal{E}_\Omega^{\uparrow\Theta} = \{\rho(e_\Omega) : e_\Omega \in \mathcal{E}_\Omega\}$. Each element of $\mathcal{E}_\Omega^{\uparrow\Theta}$ is actually the union of some equivalence classes Π_i of the partition Π. Since each focal element of $Bel_0^{\uparrow\Theta}$ has non-empty intersection with all the focal elements $\vec{e} \in \vec{\mathcal{E}}$, it follows that

$$\sum_{e_\Omega \in \mathcal{E}_\Omega, \vec{e} \in \vec{\mathcal{E}}, \rho(e_\Omega) \cap \vec{e} \neq \emptyset} m_0^{\uparrow\Theta}(\rho(e_\Omega)) \vec{m}(\vec{e}) = 1. \tag{17.19}$$

Thus, the normalisation factor in the Dempster combination $Bel_0^{\uparrow\Theta} \oplus \overrightarrow{Bel}$ is equal to 1.

Now, let $\mathcal{E}$ denote the set of focal elements of the belief function $Bel = Bel_0^{\uparrow\Theta} \oplus \overrightarrow{Bel}$. By the Dempster sum (2.6), each element e of $\mathcal{E}$ is the union of focal elements of some conditional belief functions Bel_i, i.e., $e = e_{j_1} \cup e_{j_2} \cup \cdots \cup e_{j_K}$ for some K such that $\{j_1, \ldots, j_K\} \subseteq \{1, \ldots, |\Omega|\}$ and e_{j_l} is a focal element of Bel_{j_l}, $1 \leq l \leq K$. Let m denote the mass function for Bel. For each such $e \in \mathcal{E}$, $e = \rho(e_\Omega) \cap \vec{e}$ for some $e_\Omega \in \mathcal{E}_\Omega$ and $\vec{e} \in \vec{\mathcal{E}}$, so that $e_\Omega = \bar{\rho}(e)$. Thus we have

$$\begin{aligned}
&(m_0^{\uparrow\Theta} \oplus \vec{m})(e) \\
&= \sum_{e_\Omega \in \mathcal{E}_\Omega, \vec{e} \in \vec{\mathcal{E}}, \rho(e_\Omega) \cap \vec{e} = e} m_0^{\uparrow\Theta}(\rho(e_\Omega)) \vec{m}(\vec{e}) = \sum_{\substack{e_\Omega \in \mathcal{E}_\Omega, \vec{e} \in \vec{\mathcal{E}} \\ \rho(e_\Omega) \cap \vec{e} = e}} m_0(e_\Omega) \vec{m}(\vec{e}) \\
&= m_0(\bar{\rho}(e)) \sum_{\vec{e} \in \vec{\mathcal{E}}, \rho(\bar{\rho}(e)) \cap \vec{e} = e} \vec{m}(\vec{e}) \\
&= m_0(\bar{\rho}(e)) \cdot m_{j_1}(e_{j_1}) \cdots m_{j_K}(e_{j_K}) \prod_{j \notin \{j_1, \ldots, j_K\}} \sum_{e \in \mathcal{E}_j} m_j(e) \\
&= m_0(\bar{\rho}(e)) \prod_{k=1}^{K} m_{j_k}(e_{j_k}),
\end{aligned} \tag{17.20}$$

as $\vec{m}(\vec{e}) = \prod_{i=1}^{n} m_i(e_i)$ whenever $\vec{e} = \cup_{i=1}^{n} e_i$.

Without loss of generality, we will consider the conditional mass function $m(e_1|\Pi_1)$, where e_1 is a focal element of Bel_1 and Π_1 is the first partition class associated with the partition Π, and show that $m(e_1|\Pi_1) = m_1(e_1)$. In order to obtain $m(e_1|\Pi_1)$, which is equal to $(\sum_{e \in \mathcal{E}, e \cap \Pi_1 = e_1} m(e))/(Pl(\Pi_1))$, in the following we separately compute $\sum_{e \in \mathcal{E}, e \cap \Pi_1 = e_1} m(e)$ and $Pl(\Pi_1)$.

For any $e \in \mathcal{E}$, if $e \cap \Pi_1 \neq \emptyset$, $\bar{\rho}(e)$ is a subset of Ω including ω_1. Therefore,

$$
\begin{aligned}
& Pl(\Pi_1) \\
&= \sum_{e \in \mathcal{E}, e \cap \Pi_1 \neq \emptyset} m(e) = \sum_{\mathcal{C} \subseteq \{\Pi_2, \ldots, \Pi_{|\Omega|}\}} \left(\sum_{\rho(\bar{\rho}(e)) = \Pi_1 \cup (\bigcup_{E \in \mathcal{C}} E)} m(e) \right) \\
&= \sum_{\mathcal{C} \subseteq \{\Pi_2, \ldots, \Pi_{|\Omega|}\}} m_0^{\uparrow \Theta} \left(\Pi_1 \cup \bigcup_{E \in \mathcal{C}} E \right) \left(\sum_{e_1 \in \mathcal{E}_1} m_1(e_1) \prod_{\Pi_l \in \mathcal{C}} \sum_{e_l \in \mathcal{E}_l} m_l(e_l) \right) \\
&= \sum_{\mathcal{C} \subseteq \{\Pi_2, \ldots, \Pi_{|\Omega|}\}} m_0^{\uparrow \Theta} \left(\Pi_1 \cup \bigcup_{E \in \mathcal{C}} E \right) = \sum_{e_\Omega \in \mathcal{E}_\Omega, \omega_1 \in e_\Omega} m_0(e_\Omega) = Pl_0(\{\omega_1\}).
\end{aligned}
\tag{17.21}
$$

Similarly,

$$
\begin{aligned}
\sum_{\substack{e \in \mathcal{E} \\ e \cap \Pi_1 = e_1}} m(e) &= \sum_{\mathcal{C} \subseteq \{\Pi_2, \ldots, \Pi_{|\Omega|}\}} \sum_{\rho(\bar{\rho}(e)) = \bigcup_{E \in \mathcal{C}} E} m(e_1 \cup e) \\
&= m_1(e_1) \sum_{\mathcal{C} \subseteq \{\Pi_2, \ldots, \Pi_{|\Omega|}\}} m_0^{\uparrow \Theta} \left(\Pi_1 \cup \bigcup_{E \in \mathcal{C}} E \right) \prod_{\Pi_l \in \mathcal{C}} \sum_{e_l \in \mathcal{E}_l} m_l(e_l) \\
&= m_1(e_1) \sum_{\mathcal{C} \subseteq \{\Pi_2, \ldots, \Pi_{|\Omega|}\}} m_0^{\uparrow \Theta} \left(\Pi_1 \cup \bigcup_{E \in \mathcal{C}} E \right) \\
&= m_1(e_1) \sum_{e_\Omega \in \mathcal{E}_\Omega, \omega_1 \in e_\Omega} m_0(e_\Omega) = m_1(e_1) Pl_0(\{\omega_1\}).
\end{aligned}
\tag{17.22}
$$

From (17.21) and (17.22), it follows that

$$
m(e_1|\Pi_1) = \frac{\sum_{e \in \mathcal{E}, e \cap \Pi_1 = e_1} m(e)}{Pl(\Pi_1)} = m_1(e_1).
$$

This proves property 1. Proving 2 is much easier.

For any $e_\Omega \doteq \{\omega_{j_1}, \ldots, \omega_{j_K}\} \in \mathcal{E}_\Omega$,

$$
\begin{aligned}
m \restriction_\Omega (e_\Omega) &= \sum_{\bar{\rho}(e) = e_\Omega} m(e) = m_0^{\uparrow \Theta}(\rho(e_\Omega)) \prod_{l=1}^{K} \sum_{e \in \mathcal{E}_{j_l}} m_{j_l}(e) = m_0^{\uparrow \Theta}(\rho(e_\Omega)) \\
&= m_0(e_\Omega).
\end{aligned}
\tag{17.23}
$$

It follows that $Bel \restriction_\Omega = Bel_0$, and hence the thesis.

□

The proof of Theorem 111 immediately follows from Lemmas 23 and 24.

Example 53 *Suppose that the coarsening $\Omega := \{\omega_1, \omega_2, \omega_3\}$ considered induces a partition Π of Θ: $\{\Pi_1, \Pi_2, \Pi_3\}$. Suppose also that the conditional belief function Bel_1 considered, defined on Π_1, has two focal elements e_1^1 and e_1^2, that the conditional belief function Bel_2 defined on Π_2 has a single focal element e_2^1 and that Bel_3, defined on Π_3, has two focal elements e_3^1 and e_3^2 (see Fig. 17.4). According*

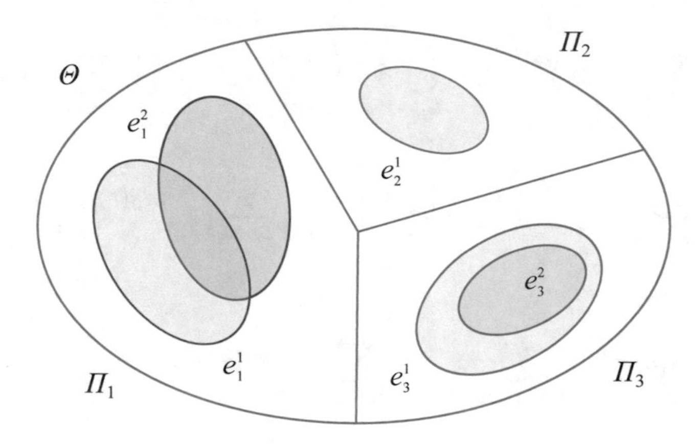

Fig. 17.4: The conditional belief functions considered in our case study. The set-theoretical relations between their focal elements are immaterial to the solution.

to Lemma 23, the Dempster combination $\overrightarrow{Bel}$ of the conditional embeddings of the Bel_i's has four focal elements, listed below:

$$\begin{array}{ll} e_1 = e_1^1 \cup e_2^1 \cup e_3^1, & e_2 = e_1^1 \cup e_2^1 \cup e_3^2, \\ e_3 = e_1^2 \cup e_2^1 \cup e_3^1, & e_4 = e_1^2 \cup e_2^1 \cup e_3^2. \end{array} \tag{17.24}$$

The four focal total elements can be represented as 'elastic bands' as in Fig. 17.5.

Without loss of generality, we assume that a prior Bel_0 on Ω has each subset of Ω as a focal element, i.e., $\mathcal{E}_\Omega = 2^\Omega$. It follows that each focal element e of the total belief function $Bel \doteq \overrightarrow{Bel} \oplus Bel_0^{\uparrow\Theta}$ is the union of some focal elements from different *conditional belief functions Bel_i. So, the set $\mathcal{E}$ of the focal elements of Bel is*

$$\mathcal{E} = \{e = \bigcup_{1 \le i \le I} e_i : 1 \le I \le 3, e_i \in \mathcal{E}_i\}$$

and is the union of the following three sets:

$$\begin{aligned} \mathcal{E}_{I=1} &:= \mathcal{E}_1 \cup \mathcal{E}_2 \cup \mathcal{E}_3, \\ \mathcal{E}_{I=2} &:= \{e \cup e' : (e, e') \in \mathcal{E}_i \times \mathcal{E}_j, 1 \le i, j \le 3, i \neq j\}, \\ \mathcal{E}_{I=3} &:= \{e_1 \cup e_2 \cup e_3 : e_i \in \mathcal{E}_i, 1 \le i \le 3\}. \end{aligned}$$

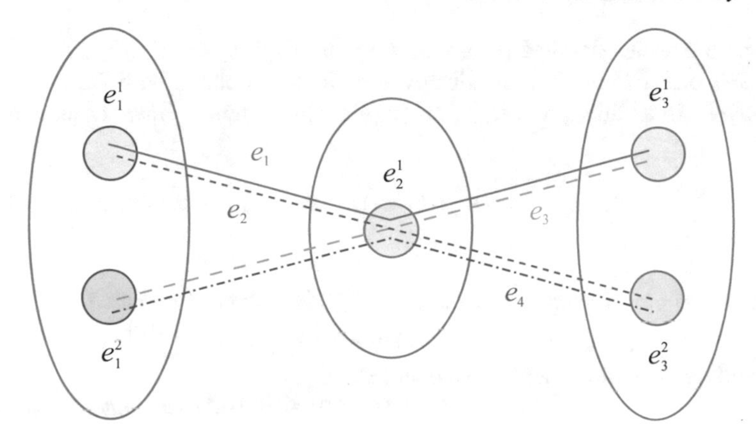

Fig. 17.5: Graphical representation of the four possible focal elements (17.24) of the total belief function (17.18) in our case study.

Hence $|\mathcal{E}| = 5 + 8 + 4 = 17$. *According to (17.20), it is very easy to compute the corresponding total mass function* m. *For example, for the two focal elements* $e_1^1 \cup e_3^2$ *and* $e_1^1 \cup e_2^1 \cup e_3^2$, *we have*

$$
\begin{aligned}
m(e_1^1 \cup e_3^2) &= m_0(\{w_1, w_3\}) m_1(e_1^1) m_3(e_3^2), \\
m(e_1^1 \cup e_2^1 \cup e_3^2) &= m_0(\Omega) m_1(e_1^1) m_2(e_2^1) m_3(e_3^2).
\end{aligned}
$$

Number of solutions The total belief function Bel obtained in Theorem 111 is not the only solution to the total belief problem (note that the Dempster combination (17.18) is itself unique, as the Dempster sum is unique).

Assume that Bel^* is a total belief function posssessing the two properties in Theorem 111. Let m^* and $\mathcal{E}^*$ denote its mass function and the set of its focal elements, respectively. Without loss of generality, we may still assume that the prior Bel_0 has every subset of Ω as its focal element, i.e., $\mathcal{E}_\Omega = 2^\Omega$. From property (P2), $Bel^* \oplus Bel_{\Pi_i} = Bel_i$, $1 \leq i \leq |\Omega|$, we derive that each focal element of Bel^* must be a union of focal elements of a number of conditional belief functions Bel_i. For, if e^* is a focal element of Bel^* and $e^* = e_l \cup e'$, where $\emptyset \neq e_l \subseteq \Pi_l$ and $e' \subseteq \Theta \setminus \Pi_l$ for some $1 \leq l \leq |\Omega|$, then $m_l(e_l) = (m^* \oplus m_{\Pi_l})(e_l) > 0$ and hence $e_l \in \mathcal{E}_l$. So we must have that $\mathcal{E}^* \subseteq \mathcal{E}$, where $\mathcal{E}$ is the set of focal elements of the total belief function Bel (17.18) obtained in Theorem 111,

$$
\mathcal{E} = \left\{ \bigcup_{j \in J} e_j : J \subseteq \{1, \ldots, |\Omega|\}, e_j \in \mathcal{E}_j \right\}.
$$

In order to find Bel^* (or m^*), we need to solve a group of linear equations which correspond to the constraints dictated by the two properties, in which the mass $m^*(e)$ of each focal element $e \in \mathcal{E}$ of the total solution (17.18) is treated as an unknown variable. There are $|\mathcal{E}|$ variables in the group.

From properties (P1) and (P2), we know that $Pl_0(\omega_i) = Pl^*(\Pi_i), 1 \leq i \leq |\Omega|$, where Pl_0 and Pl^* are the plausibility functions associated with Bel_0 and Bel^*, respectively. In addition, property (P1) implies the system of linear constraints

$$\left\{ \sum_{e \cap \Pi_i = e_i, e \in \mathcal{E}} m^*(e) = m_i(e_i) Pl_0(\omega_i), \quad \forall i = 1, \ldots, n, \ \forall e_i \in \mathcal{E}_i. \right. \tag{17.25}$$

The total number of such equations is $\sum_{j=1}^{|\Omega|} |\mathcal{E}_j|$. Since, for each $1 \leq i \leq |\Omega|$, $\sum_{e \in \mathcal{E}_i} m_i(e) = 1$, the system (17.25) includes a group of $\sum_{j=1}^{|\Omega|} |\mathcal{E}_j| - |\Omega|$ *independent* linear equations, which we denote as G_1.

From property (P2) (the marginal of Bel^* on Ω is Bel_0), we have the following constraints:

$$\left\{ \sum_{e \in \mathcal{E}, \overline{\rho}(e) = C} m^*(e) = m_0(C), \quad \forall \emptyset \neq C \subseteq \Omega. \right. \tag{17.26}$$

The total number of linear equations in (17.26) is the number of non-empty subsets of Ω. Since $\sum_{C \subseteq \Omega} m_0(C) = 1$, there is a subset of $|2^\Omega| - 2$ independent linear equations in (17.26), denoted by G_2.

As the groups of constraints G_1 and G_2 are independent, their union $G := G_1 \cup G_2$ completely specifies properties (P1) and (P2) in Theorem 111. Since $|G_1| = \sum_{j=1}^{|\Omega|} |\mathcal{E}_j| - |\Omega|$ and $|G_2| = |2^\Omega| - 2$, the cardinality of the union group G is $|G| = \sum_{j=1}^{|\Omega|} |\mathcal{E}_j| - |\Omega| + |2^\Omega| - 2$.

From Theorem 111, we know that the system of equations G is solvable and has at least one positive solution (in which each variable has a positive value). This implies that $|\mathcal{E}| \geq |G|$, i.e., the number of variables must be no less than that of the independent linear equations in G. If $|\mathcal{E}| > |G|$, in particular, we can apply the Fourier–Motzkin elimination method [963] to show that G has another distinct positive solution Bel^* (i.e., such that $m^*(e) \geq 0 \ \forall e \in \mathcal{E}$).

Example 54 *We employ Example 53 to illustrate the whole process of finding such an alternative total belief function Bel^*. We assume further that m_0 and m_i, $1 \leq i \leq 3$, take values*

$$m_1(e_1^1) = \frac{1}{2} = m_1(e_1^2), \quad m_2(e_2^1) = 1, \quad m_3(e_3^1) = \frac{1}{3}, \quad m_3(e_3^2) = \frac{2}{3},$$

$$m_0(\{\omega_1\}) = m_0(\{\omega_2\}) = m_0(\{\omega_3\}) = \frac{1}{16}, \quad m_0(\{\omega_1, \omega_2\}) = \frac{2}{16},$$

$$m_0(\{\omega_2, \omega_3\}) = \frac{4}{16}, \quad m_0(\{\omega_1, \omega_3\}) = \frac{3}{16}, \quad m_0(\Omega) = \frac{1}{4}.$$

If we follow the process prescribed above to translate the two properties into a group G of linear equations, we obtain 17 unknown variables $m^*(e)$, $e \in \mathcal{E}$ ($|\mathcal{E}| = 17$) and *eight independent linear equations ($|G| = 8$). From Theorem 111, we can construct a positive solution m defined according to (17.20). For this example,*

$$m(\{e_1^1, e_2^1, e_3^1\}) = m_0(\Omega)\, m_1(e_1^1)\, m_2(e_2^1)\, m_3(e_3^1) = \frac{1}{24},$$
$$m(\{e_1^2, e_2^1, e_3^2\}) = m_0(\Omega)\, m_1(e_1^2)\, m_2(e_2^1)\, m_3(e_3^2) = \frac{1}{12}.$$

When solving the equation group G via the Fourier–Motzkin elimination method, we choose $m^*(\{e_1^1, e_2^1, e_3^1\})$ and $m^*(\{e_1^2, e_2^1, e_3^2\})$ to be the last two variables to be eliminated. Moreover, there is a sufficiently small positive number ϵ such that $m^*(\{e_1^1, e_2^1, e_3^1\}) = \frac{1}{24} - \epsilon > 0$, $m^*(\{e_1^2, e_2^1, e_3^2\}) = \frac{1}{12} + \epsilon$, and all other variables also take positive values. It is easy to see that the m^* so obtained is different from the m obtained via Theorem 111.

Clearly, this new solution has the same focal elements as the original one; other solutions may exist with a different set of focal elements (as can indeed be empirically shown). We will discuss this issue further in a couple of pages.

Case of Bayesian prior Note that when the prior belief function Bel_0 is Bayesian, however, the total belief function obtained according to (17.20) is the unique one satisfying the two properties in Theorem 111.

Corollary 27. *For the belief function Bel_0 over Ω and the conditional belief functions Bel_i over Π_i in Theorem 111, if Bel_0 is Bayesian (i.e., a probability measure), then there exists a unique total belief function $Bel : 2^\Theta \to [0,1]$ such that (P1) and (P2) are satisfied. Moreover, the total mass function m of Bel is*

$$m(e) = \begin{cases} m_i(e)\, m_0(\omega_i) & \text{if } e \in \mathcal{E}_i \text{ for some } i, \\ 0 & \text{otherwise.} \end{cases}$$

Proof. It is easy to check that the total mass function m defined above satisfies the two properties. Now we need to show that it is unique. Since Bel_0 is Bayesian, $\mathcal{E}_\Omega \subseteq \{\{\omega_1\}, \ldots, \{\omega_{|\Omega|}\}\}$. In other words, all focal elements of Bel_0 are singletons. It follows that $\mathcal{E}_\Omega^{\uparrow\Theta} \subseteq \{\Pi_1, \ldots, \Pi_{|\Omega|}\}$. If e is a focal element of Bel, we obtain from (17.20) that $e \in \mathcal{E}_i$ for some $i \in \{1, \ldots, |\Omega|\}$ and $m(e) = m_0(\omega_i) m_i(e)$. This implies that, if $e \notin \bigcup_{i=1}^{|\Omega|} \mathcal{E}_i$, i.e., e is not a focal element of any conditional belief function Bel_i, $m(e) = 0$. So we have shown that the total mass function m is the unique one satisfying (P1) and (P2).

□

Generalisation If the first requirement (P1) is modified to include conditional constraints with respect to *unions* of equivalence classes, the approach used to prove Theorem 111 is no longer valid.

For each non-empty subset $\{i_1, \ldots, i_J\} \subseteq \{1, \ldots, |\Omega|\}$, let

$$Bel_{\bigcup_{j=1}^J \Pi_{i_j}}, \quad Bel_{i_1 \cdots i_J}$$

denote the categorical belief function with $\bigcup_{j=1}^{J} \Pi_{i_j}$ as its only focal element, and a conditional belief function on $\bigcup_{j=1}^{J} \Pi_{i_j}$, respectively.
We can then introduce a new requirement by generalising property (P1) as follows:

- $(P1') : Bel \oplus Bel_{\bigcup_{j=1}^{J} \Pi_{i_j}} = Bel_{i_1 \cdots i_J} \; \forall \emptyset \neq \{i_1, \ldots, i_J\} \subseteq \{1, \ldots, |\Omega|\}$.

Let $\overrightarrow{Bel}_{i_1 \cdots i_J}$ denote the conditional embeddings of $Bel_{i_1 \cdots i_J}$, and $\overrightarrow{Bel}_{\text{new}}$ the Dempster combination of all these conditional embeddings. Let $Bel_{\text{new}} = \overrightarrow{Bel}_{\text{new}} \oplus Bel_0^{\uparrow \Theta}$. It is easy to show that Bel_{new} satisfies neither $(P1')$ nor $(P2)$.

Relation to generalised Jeffrey's rules In spirit, our approach in this chapter is similar to Spies's Jeffrey's rule for belief functions in [1749]. His total belief function is also the Dempster combination of the prior on the subalgebra generated by the partition Π of Θ with the conditional belief functions on all the equivalence classes Π_i. Moreover, he showed that this total belief function satisfies the two properties in Theorem 111. However, his definition of the conditional belief function is different from the one used here, which is derived from Dempster's rule of combination. His definition falls within the framework of random sets, in which a conditional belief function is a *second-order* belief function whose focal elements are conditional events, defined as sets of subsets of the underlying frame of discernment. The biggest difference between Spies's approach and ours is thus that his framework depends on probabilities, whereas ours does not. It would indeed be interesting to explore the connection of our total belief theorem with his Jeffrey's rule for belief functions.

Smets [1690] also generalised Jeffrey's rule within the framework of models based on belief functions, without relying on probabilities. Recall that ρ is a refining mapping from 2^Ω to 2^Θ, and $\mathbb{A}^\rho$ is the Boolean algebra generated by the set of equivalence classes Π_i associated with ρ. Contrary to our total belief theorem, which assumes conditional constraints only with respect to the equivalence classes Π_i (the atoms of $\mathbb{A}^\rho$), Smets's generalised Jeffrey's rule considers constraints with respect to unions of equivalence classes, i.e., arbitrary elements of $\mathbb{A}^\rho$ (see the paragraph 'Generalisation' above).

Given two belief functions Bel_1 and Bel_2 over Θ, his general idea is to find a BF Bel_3 there such that:

- (Q1) its marginal on Ω is the same as that of Bel_1, i.e., $Bel_3 \restriction_\Omega = Bel_1 \restriction_\Omega$;
- (Q2) its conditional constraints with respect to elements of $\mathbb{A}^\rho$ are the same as those of Bel_2.

In more detail, let m be a mass function over Θ. Smets defines two kinds of conditioning for conditional constraints: for any $E \in \mathbb{A}^\rho$ and $e \subseteq E$ such that $\rho(\bar{\rho}(e)) = E$,

$$m^{\text{in}}(e|E) \doteq \frac{m(e)}{\sum_{\rho(\bar{\rho}(e'))=E} m(e')}, \quad m^{\text{out}}(e|E) \doteq \frac{m(e|E)}{\sum_{\rho(\bar{\rho}(e'))=E} m(e'|E)}.$$

The first one is the well-known geometric conditioning (Section 4.5.3), whereas the second is called *outer conditioning*. Both are distinct from Dempster's rule of conditioning. From these two conditioning rules, he obtains two different forms of generalised Jeffrey's rule, namely, for any $e \subseteq \Theta$,

- $m_3^{\text{in}}(e) = m_1^{\text{in}}(e|E)m_2(E)$, where $E = \rho(\bar{\rho}(e))$;
- $m_3^{\text{out}}(e) = m_1^{\text{out}}(e|E)m_2(E)$.

Both m_3^{in} and m_3^{out} satisfy (Q1). As for (Q2), m_3^{in} applies, whereas m_3^{out} does so only partially, since $(m_3^{\text{out}})^{\text{in}}(e|E) = m_1^{\text{out}}(e|E)$ [2117].

In [1227], Ma et al. defined a new Jeffrey's rule where the conditional constraints are indeed defined according to Dempster's rule of combination, and with respect to the whole power set of the frame instead of a subalgebra as in Smets's framework. In their rule, however, the conditional constraints are not preserved by their total belief functions.

Exploring the space of solutions Lemma 23 ensures that the solution (17.18) of the total belief theorem has focal elements which are unions of one focal element e_i of a subset of conditional belief functions Bel_i.

The following result prescribes that *any* solution to the total belief problem must have focal elements which obey the above structure (see [364] for a proof).

Proposition 67. *Each focal element e of a total belief function Bel meeting the requirements of Theorem 111 is the union of* exactly one *focal element for each of the conditional belief functions whose domain Π_i is a subset of $\rho(E)$, where E is the smallest focal element of the a priori belief function Bel_0 such that $e \subset \rho(E)$. Namely,*

$$e = \bigcup_{i:\Pi_i \subset \rho(E)} e_i^{j_i}, \qquad (17.27)$$

where $e_i^{j_i} \in \mathcal{E}_i \ \forall i$, and $\mathcal{E}_i$ denotes the list of focal elements of Bel_i.

The special total belief theorem If we require the a priori function Bel_0 to have only *disjoint* focal elements (i.e., we force Bel_0 to be the vacuous extension of a Bayesian function defined on some coarsening of Ω), we have what we call the *restricted* or *special* total belief theorem [364]. The situation generalises the case considered in Corollary 27.

In this special case, it suffices to solve separately the $|\mathcal{E}_\Omega|$ subproblems obtained by considering each focal element E of Bel_0, and then combine the resulting partial solutions by simply weighting the resulting basic probability assignments using the a priori mass $m_0(E)$, to obtain a fully normalised total belief function.

Candidate solutions as linear systems For each individual focal element of Bel_0, the task of finding a suitable solution translates into a linear algebra problem.

Let $N = |E|$ be the cardinality of E. A candidate solution to the subproblem of the restricted total belief problem associated with $E \in \mathcal{E}_\Omega$ is the solution to a linear system with $n_{\min} = \sum_{i=1,\ldots,N}(n_i - 1) + 1$ equations and $n_{\max} = \prod_i n_i$ unknowns,

$$A\mathbf{x} = \mathbf{b}, \tag{17.28}$$

where n_i is the number of focal elements of the conditional belief function Bel_i. Each column of A is associated with an admissible (i.e., satisfying the structure of Proposition 67) focal element e_j of the candidate total belief function with mass assignment m, $\mathbf{x} = [m(e_1), \ldots, m(e_n)]'$, and $n = n_{\min}$ is the number of equalities generated by the N conditional constraints.

Minimal solutions Since the rows of the solution system (17.28) are linearly independent, any system of equations obtained by selecting $n_{\min}$ columns from A has a unique solution. A *minimal* (i.e., with the minimum number of focal elements) solution to the special total belief problem is then uniquely determined by the solution of a system of equations obtained by selecting $n_{\min}$ columns from the $n_{\max}$ columns of A.

A class of linear transformations Let us represent each focal element of the candidate total function $e = \bigcup_{i:\Pi_i \subset \rho(E)} e_i^{j_i}$, where $e_i^{j_i} \in \mathcal{E}_i \ \forall i$ (see (17.27)), as the vector of indices of its constituent focal elements $e = [j_1, \ldots, j_N]'$.

Definition 138. *We define a class $\mathcal{T}$ of transformations acting on columns e of a candidate minimal solution system via the following formal sum:*

$$e \mapsto e' = -e + \sum_{i \in \mathcal{C}} e_i - \sum_{j \in \mathcal{S}} e_j, \tag{17.29}$$

where $\mathcal{C}$, $|\mathcal{C}| < N$, is a covering set of 'companions' of e (i.e., such that every component of e is present in at least one of them), and a number of 'selection' columns $\mathcal{S}$, $|\mathcal{S}| = |\mathcal{C}| - 2$, are employed to compensate for the side effects of $\mathcal{C}$ to yield an admissible column (i.e., a candidate focal element satisfying the structure of Proposition 67).

We call the elements of $\mathcal{T}$ *column substitutions*.

A sequence of column substitutions induces a discrete path in the solution space: the values $m(e_i)$ of the solution components associated with the columns e_i vary, and in a predictable way. If we denote by $s < 0$ the (negative) solution component associated with the old column e:

1. The new column e' has as its solution component $-s > 0$.
2. The solution component associated with each companion column is decreased by $|s|$.
3. The solution component associated with each selection is increased by $|s|$.
4. All other columns retain the old values of their solution components.

The proof is a direct consequence of the linear nature of the transformation (17.29).

Clearly, if we choose to substitute the column with the most negative solution component, the overall effect is that the most negative component is changed into a positive one, components associated with selection columns become more positive (or less negative), and, as for companion columns, while some of them may end up being assigned negative solution components, these will be smaller in absolute value than $|s|$ (since their initial value was positive). Hence we have the following proposition.

Proposition 68. *Column substitutions of the class $\mathcal{T}$ reduce the absolute value of the most negative solution component.*

An alternative existence proof We can then use Theorem 68 to prove that there always exists a selection of columns of A (the possible focal elements of the total belief function Bel) such that the resulting square linear system has a positive vector as a solution. This can be done in a constructive way, by applying a transformation of the type (17.29) recursively to the column associated with the most negative component, to obtain a path in the solution space leading to the desired solution.

Such an existence proof for the special total belief theorem, as an alternative to that provided earlier in this section, exploits the effects on solution components of column substitutions of type $\mathcal{T}$:

1. By Theorem 68, at each column substitution, the most negative solution component decreases.
2. If we keep replacing the most negative variable, we keep obtaining *distinct* linear systems, for at each step the transformed column is assigned a positive solution component and, therefore, if we follow the proposed procedure, *cannot be changed back to a negative one* by applying transformations of class $\mathcal{T}$.
3. This implies that there can be no cycles in the associated path in the solution space.
4. The number $\binom{n_{\max}}{n_{\min}}$ of solution systems is obviously finite, and hence the procedure must terminate.

Unfortunately, counter-examples show that there are 'transformable' columns (associated with negative solution components) which do not admit a transformation of the type (17.29). Although they do have companions on every partition Π_i, such counter-examples do not admit a complete collection of 'selection' columns.

Solution graphs The analysis of significant particular cases confirms that, unlike the classical law of total probability, the total belief theorem possesses more than one admissible minimal solution, even in the special case.

In [364], we noticed that all the candidate minimal solution systems related to a total belief problem of a given size $\{n_i, \ i = 1, \dots, N\}$ can be arranged into a *solution graph*, whose structure and symmetries can be studied to compute the number of minimal total belief functions for any given instance of the problem. We focused on the group of permutations of focal elements of the conditional belief functions Bel_i,

$$G = S_{n_1} \times \cdots \times S_{n_N}, \tag{17.30}$$

namely the product of the permutation groups S_{n_i} acting on the collections of focal elements of each individual conditional belief function Bel_i. The group G acts on the solution graph by generating orbits, i.e., the set of all candidate solution systems (nodes of the graph) obtained by some permutation of the focal elements within at least some of the partition elements Π_i. We hypothesised that such orbits are in 1–1 correspondence with the number of admissible solutions.

Future agenda A number of research directions remain open, based on the results obtained so far. The analysis of the special law of total belief needs to be completed, by finalising the alternative proof of existence and, through the latter, completing the analysis of solution graphs and the number of admissible solutions.

Further down the road, the full description of all minimal solutions needs to be extended to the general case of an arbitrary prior Bel_0, which our analysis earlier in this section sheds some light on but fails to provide a comprehensive understanding. Distinct versions of the law of total belief may arise from replacing Dempster conditioning with other accepted forms of conditioning for belief functions, such as credal [584], geometric [1788], conjunctive and disjunctive [1690] conditioning. As belief functions are a special type of coherent lower probabilities, which in turn can be seen as a special class of lower previsions (see [1874], Section 5.13), marginal extension [1294] can be applied to them to obtain a total lower prevision. The relationship between marginal extension and the law of total belief therefore needs to be understood.

Finally, fascinating relationships exist between the total belief problem and transversal matroids [1374], on the one hand, and the theory of positive linear systems [590], on the other, which we also plan to investigate in the near future.

17.1.4 Limit theorems for random sets

The law of total probability is only one important result of classical probability theory that needs to be generalised to the wider setting of random sets.

In order to properly define a Gaussian belief function, for instance, we would need to generalise the classical central limit theorem to random sets. The old proposal of Dempster and Liu merely transfers normal distributions on the real line by the Cartesian product with $\mathbb{R}^m$ (see Chapter 3, Section 3.3.4). In fact, both the central limit theorem and the law(s) of large numbers have already been generalised to imprecise probabilities [400, 314].[81] Here we review the most recent and relevant attempts at formulating similar laws for belief functions.

Central limit theorems Chareka [243] and later Terán [1806] have conducted interesting work on the central limit theorem for capacities. More recently, specific attention has been directed at central limit theorem results for belief measures [576, 1638].

[81] See http://onlinelibrary.wiley.com/book/10.1002/9781118763117.

Choquet's definition of belief measures Let Θ be a Polish space[82] and $\mathcal{B}(\Theta)$ be the Borel σ-algebra on Θ. Let us denote by $\mathcal{K}(\Theta)$ the collection of compact subsets of Θ, and by $\mathcal{P}(\Theta)$ the space of all probability measures on Θ. The set $\mathcal{K}(\Theta)$ will be endowed with the Hausdorff topology[83] generated by the topology of Θ.

Definition 139. *A belief measure on $(\Theta, \mathcal{B}(\Theta))$ is defined as a set function $Bel : \mathcal{B}(\Theta) \to [0,1]$ satisfying:*

- $Bel(\emptyset) = 0$ *and* $Bel(\Theta) = 1$;
- $Bel(A) \leq Bel(B)$ *for all Borel sets* $A \subset B$;
- $Bel(B_n) \downarrow Bel(B)$ *for all sequences of Borel sets* $B_n \downarrow B$;
- $Bel(G) = \sup\{Bel(K) : K \subset G, K \in \mathcal{K}(\Theta)\}$, *for all open sets* G;
- *Bel is totally monotone (or ∞-monotone): for all Borel sets* $B_1, \ldots, B_n$,

$$Bel\left(\bigcup_{i=1}^{n} B_i\right) \geq \sum_{\emptyset \neq I \subset \{1,\ldots,n\}} (-1)^{|I|+1} Bel\left(\bigcap_{i \in I} B_i\right).$$

By [1425], for all $A \in \mathcal{B}(\Theta)$ the collection $\{K \in \mathcal{K}(\Theta) : K \subset A\}$ is universally measurable.[84] Let us denote by $\mathcal{B}_u(\mathcal{K}(\Theta))$ the σ-algebra of all subsets of $\mathcal{K}(\Theta)$ which are universally measurable. The following result is due to Choquet [283].

Proposition 69. *The set function $Bel : \mathcal{B}(\Theta) \to [0,1]$ is a belief measure if and only if there exists a probability measure P_{Bel} on $(\mathcal{K}(\Theta), \mathcal{B}(\mathcal{K}(\Theta)))$ such that*

$$Bel(A) = P_{\text{Bel}}(\{K \in \mathcal{K}(\Theta) : K \subset A\}), \quad \forall A \in \mathcal{B}(\Theta).$$

Moreover, there exists a unique extension of P_{Bel} to $(\mathcal{K}(\Theta), \mathcal{B}_u(\mathcal{K}(\Theta)))$, which we can still denote by P_{Bel}.

By comparison with the usual random-set formulation (see Section 3.1.1), we can appreciate that Proposition 69 confirms that a belief measure is the total probability induced by a probability measure on a source space – in this case, the source space $\Omega = \mathcal{K}(\Theta)$ is the collection of all compact subspaces of Θ.

For any $A \in \mathcal{B}(\Theta^\infty)$, we define

$$Bel^\infty(A) \doteq P^\infty_{Bel}(\{K = K_1 \times K_2 \times \cdots \in (\mathcal{K}(\Theta))^\infty : K \subset A\}), \tag{17.31}$$

where P^∞_{Bel} is the i.i.d. product probability measure. Bel^∞ is the unique belief measure on $(\Theta^\infty, \mathcal{B}(\Theta^\infty)$ induced by P^∞_{Bel}.

[82] A separable completely metrisable topological space; that is, a space homeomorphic to a complete metric space that has a countable dense subset.

[83] A *Hausdorff space*, or 'separated' space, is a topological space in which distinct points have disjoint neighbourhoods. It implies the uniqueness of limits of sequences, nets and filters.

[84] A subset A of a Polish space X is called *universally measurable* if it is measurable with respect to every complete probability measure on X that measures all Borel subsets of X.

Bernoulli variables A central limit theorem for belief functions, in their Choquet formulation, was recently proposed by Epstein and Seo for the case of Bernoulli random variables [576].

Let[85] $\Theta = \{T, F\}$, and consider the set Θ^∞ of all infinite series of samples $\theta^\infty = (\theta_1, \theta_2, \ldots)$ extracted from Θ. Let $\Phi_n(\theta^\infty)$ be the empirical frequency of the outcome T in the first n experiments in sample θ^∞. Let Bel be a belief function on Θ induced by a measure P_{Bel}, and Bel^∞ the belief measure (17.31) on Θ^∞.

The law of large numbers asserts certainty that asymptotic empirical frequencies will lie in the interval $[Bel(T), 1 - Bel(F)]$, that is,

$$Bel^\infty \Big\{ \theta^\infty : [\liminf \Phi_n(\theta^\infty), \limsup \Phi_n(\theta^\infty)] \subset [Bel(T), 1 - Bel(F)] \Big\} = 1.$$

The authors of [576] proved the following proposition.

Proposition 70. *The following hold:*

$$\lim_{n \to \infty} Bel^\infty \left(\left\{ \theta^\infty : \sqrt{n} \frac{\Phi_n(\theta^\infty) - Pl(T)}{\sqrt{Bel(F)(1 - Bel(F))}} \leq \alpha \right\} \right) = \mathcal{N}(\alpha),$$

$$\lim_{n \to \infty} Bel^\infty \left(\left\{ \theta^\infty : \sqrt{n} \frac{\Phi_n(\theta^\infty) - Bel(T)}{\sqrt{Bel(T)(1 - Bel(T))}} \leq \alpha \right\} \right) = 1 - \mathcal{N}(\alpha),$$

where $\mathcal{N}$ denotes the usual normal distribution on real numbers.

The proof follows from showing that, for the events indicated, the minimising measures are i.i.d. As a result, classical limit theorems applied to these measures deliver corresponding limit theorems for the i.i.d. product Bel^∞.

The central result of [576], however, is the following.

Proposition 71. *Suppose that $G : \mathbb{R} \to \mathbb{R}$ is bounded, quasi-concave and upper-semicontinuous. Then*

$$\int G(\Phi_n(\theta^\infty)) \, \mathrm{d}Bel^\infty(\theta^\infty) = E[G(X'_{1n}), G(X'_{2n})] + O\left(\frac{1}{\sqrt{n}}\right),$$

where (X'_{1n}, X'_{2n}) is normally distributed with mean $(Bel(T), Pl(T))$ and covariance

$$\frac{1}{n} \begin{bmatrix} Bel(T)(1 - Bel(T)) & Bel(T)Bel(F) \\ Bel(T)Bel(F) & (1 - Bel(F))Bel(F) \end{bmatrix}.$$

That is,

$$\limsup_{n \to \infty} \sqrt{n} \left| \int G(\Phi_n(\theta^\infty)) \, \mathrm{d}Bel^\infty(\theta^\infty) - E[G(X'_{1n}), G(X'_{2n})] \right| \leq K$$

for some constant K.

[85] We adopt the notation used in this book, in place of Epstein and Seo's original one.

Generalisation Xiaomin Shi [1638] has recently (2015) generalised the result of Epstein and Seo [576] from Bernoulli random variables to general bounded random variables,[86] thanks to a theorem in a seminar paper [283] by Choquet.

Proposition 72. *Let Y_i, $i = 1, \ldots$, be equally distributed random variables on $(\Theta, \mathcal{B}(\Theta))$, bounded by a constant M, and let us define $X_i(\theta_1, \theta_2, ..) \doteq Y_i(\theta_i)$, $i \geq 1$. Then we have, for all $\alpha \in \mathbb{R}$,*

$$\lim_{n \to \infty} Bel^\infty \left(\frac{\sum_{i=1}^n X_i - n\underline{\mu}}{\sqrt{n}\underline{\sigma}} \geq \alpha \right) = 1 - \mathcal{N}(\alpha),$$

and

$$\lim_{n \to \infty} Bel^\infty \left(\frac{\sum_{i=1}^n X_i - n\overline{\mu}}{\sqrt{n}\overline{\sigma}} < \alpha \right) = 1 - \mathcal{N}(\alpha),$$

where $\underline{\mu} = E_{P_{Bel}^\infty}[\underline{Z}_i]$, $\overline{\mu} = E_{P_{Bel}^\infty}[\overline{Z}_i]$ and

$$\begin{array}{l} \underline{Z}_i(K_1 \times K_2 \times \cdots) = \inf_{\theta_i \in K_i} X_i(\theta_1, \theta_2, \ldots), \\ \overline{Z}_i(K_1 \times K_2 \times \cdots) = \sup_{\theta_i \in K_i} X_i(\theta_1, \theta_2, \ldots), \end{array}$$

for $K_i \in \mathcal{K}(\Theta)$, $i \geq 1$. The expressions for $\underline{\sigma}$ and $\overline{\sigma}$ are given in [1638], Theorem 3.1.

A central limit theorem for two-sided intervals was also provided ([1638], Theorem 4.1).

Gaussian random sets One can note, however, that the above results do not really address the question of whether there exists, among all (compact) random sets, a special class (playing the role of Gaussian probability measures) characterised by all sums of independent and equally distributed random sets converging to an object of that class (which it would make sense to call a proper 'Gaussian random set'). The issue remains open for future investigation.

Question 16. Does there exist a (parameterisable) class of random sets, playing a role similar to that which Gaussian distributions play in standard probability, such that sample averages of (the equivalent of) i.i.d. random sets converge in some sense to an object of such a class?

Laws of large numbers Several papers have been published on laws of large numbers for capacities [1230], monotone measures [12] and non-additive measures [1468, 1807], although not strictly for belief measures. Molchanov [1300], among others, studied the Glivenko–Cantelli theorem for capacities induced by random sets. A strong law of large numbers for set-valued random variables in a G_α space was recently produced by Guan Li [746, 1148], based on Taylor's result for single-valued random variables. The issue was also recently addressed by Kerkvliet in his PhD dissertation [957].

[86] `https://arxiv.org/pdf/1501.00771.pdf`.

Definition 140. *The* weak law of large numbers *states that the sample average converges in probability towards the expected value. Consider an infinite sequence of i.i.d. Lebesgue-integrable random variables $X_1, X_2, \ldots$ with expected value μ and sample average $\overline{X}_n = (X_1 + \cdots + X_n)/n$. Then, for any positive number ε,*

$$\lim_{n \to \infty} P\big(|\overline{X}_n - \mu| > \varepsilon \big) = 0. \tag{17.32}$$

The strong law of large numbers *states that the sample average converges almost surely to the expected value. Namely,*

$$P\Big(\lim_{n \to \infty} \overline{X}_n = \mu \Big) = 1. \tag{17.33}$$

When the probability measure is replaced by a set function ν assumed to be completely monotone, the empirical frequencies $f_n(A)$ of an event A over a sequence of Bernoulli trials satisfy

$$\nu \left(\nu(A) \leq \liminf_n f_n(A) \leq \limsup_n f_n(A) \leq 1 - \nu(A^c) \right) = 1.$$

In order to replace the event by a random variable X on a measurable space $(\Theta, \mathcal{B})$, one needs to calculate expectations with respect to ν and $\overline{\nu} = 1 - \nu(A^c)$ in such a way that $E_\nu[I_A] = \nu(A)$, $E_{\overline{\nu}}[I_A] = 1 - \nu(A^c)$, and both expectations equal the usual definition when ν is a probability measure. That is achieved by using the Choquet integral [283].

The resulting type of limit theorem has already been considered by Marinacci [1243] and Maccheroni and Marinacci [1230], and in more recent papers [270, 400, 314, 1468, 1469]. In those papers, the weakening of the axiomatic properties of probability are balanced by the incorporation of extra technical assumptions about the properties of Θ and/or the random variables, for instance that Θ is a compact or Polish [1230] topological space or that the random variables are continuous functions, are bounded, have sufficiently high moments, are independent in an unreasonably strong sense or satisfy ad hoc regularity requirements.

In [1807], it was shown that additivity can be replaced by complete monotonicity in the law of large numbers under the reasonable first-moment condition that both of the Choquet expectations $E_\nu[X]$ and $E_{\overline{\nu}}[X]$ are finite, with no additional assumptions. Under the assumption that the sample space is endowed with a topology, the continuity condition for monotone sequences can also be relaxed. The main result reads as follows ([1807], Theorem 1.1).

Proposition 73. *Let ν be a completely monotone set function on a measurable space $(\Theta, \mathcal{B})$. Let X be an integrable[87] random variable, and let $\{X_n, n\}$ be pairwise preindependent and identically distributed to X. If ν is a capacity, then for every $\epsilon > 0$,*

$$\nu \left(E_\nu[X] - \epsilon < \overline{X}_n < E_{\overline{\nu}}[X] + \epsilon \right) \to 1.$$

[87] That is, its lower and upper expectations $E_\nu[X]$ and $E_{\overline{\nu}}[X]$ are finite.

If $(\Theta, \mathcal{B})$ is a topological space with a Borel σ-algebra, ν is a topological capacity and the X_n are continuous functions on Θ, then

$$\nu\left(E_{\nu}[X] \leq \liminf_n \overline{X}_n \leq \limsup_n \overline{X}_n \leq E_{\overline{\nu}}[X] \right) = 1.$$

17.1.5 Frequentist inference with random sets

Random sets are mathematical objects detached from any specific interpretation. Just as probability measures are used by both Bayesians and frequentists for their analyses, random sets can also be employed in different ways according to the interpretation they are provided with.

In particular, it is natural to imagine a generalised frequentist framework in which random experiments are designed by assuming a specific random-set distribution, rather than a conventional one, in order to better cope with the ever-occurring set-valued observations (see the Introduction).

Parameterised families of random sets The first necessary step is to introduce parametric models based on random sets.

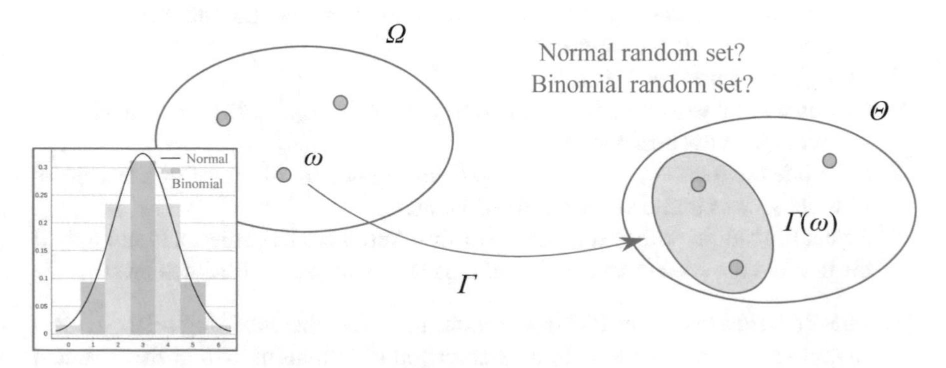

Fig. 17.6: Describing the family of random sets (right) induced by families of probability distributions in the source probability space (left) is a likely first step towards a generalisation of frequentist inference to random sets.

Recall Dempster's random-set interpretation (Fig. 17.6). Should the multivalued mapping Γ which defines a random set be 'designed', or derived from the problem? For instance, in the cloaked die example (Section 1.3.5) it is the occlusion which generates the multivalued mapping and we have no control over it. In other situations, however, it may make sense to impose a parameterised family of mappings

$$\Gamma(.|\pi) : \Omega \to 2^{\Theta}, \quad \pi \in \Pi,$$

which, given a (fixed) probability on the source space Ω, would yield as a result a parameterised family of random sets.

The alternative is to fix the multivalued mapping (e.g., when it is given by the problem), and model the source probability by a classical parametric model. A Gaussian or binomial family of source probabilities would then induce a family of 'Gaussian'[88] or 'binomial' random sets (see Fig. 17.6 again).

Question 17. What is the analytical form of random sets induced by say, Gaussian, binomial or exponential families of (source) probability distributions?

Hypothesis testing with random sets As we know, in hypothesis testing (Section 1.2.3), designing an experiment amounts to choosing a family of probability distributions which is assumed to generate the observed data. If parameterised families of random sets can be contructed, they can then be plugged into the frequentist inference machinery, after an obvious generalisation of some of the steps involved.

Hypothesis testing with random sets would then read as follows:

1. State the relevant null hypothesis H_0 and the alternative hypothesis.
2. State the assumptions about the form of the ~~distribution~~ *random set* (*mass assignment*) describing the observations.
3. State the relevant test statistic T (a quantity derived from the sample) – only this time the sample contains set-valued observations!
4. Derive the ~~distribution~~ *mass assignment* of the test statistic under the null hypothesis (from the assumptions).
5. Set a significance level (α).
6. Compute from the observations the observed value t_{obs} of the test statistic T – this will also now be set-valued.
7. Calculate the ~~p-value~~ *conditional belief value* under H_0 of sampling a test statistic at least as extreme as the observed value.
8. Reject the null hypothesis, in favour of the alternative hypothesis, if and only if ~~the p-value~~ the *conditional belief value* is less than the significance level.

Question 18. How would results from random-set hypothesis testing relate to classical hypothesis testing applied to parameterised distributions within the assumed random set?

17.1.6 Random-set random variables

We know that random sets are set-valued random variables: nevertheless, the question remains as to whether one can build random variables on top of random-set (belief) spaces, rather than the usual probability space. It would be natural to term such objects *random-set random variables*.

Just as in the classical case, we would need a mapping from Θ to a measurable space (e.g. the positive real half-line)

$$f : \Theta \to \mathbb{R}_+ = [0, +\infty],$$

[88] Note that the Gaussian random sets so defined would be entirely distinct from Gaussian random sets as advocated in the context of the generalised central limit theorem setting.

where this time Θ is itself the codomain of a multivalued mapping $\Gamma : \Omega \to 2^\Theta$, with source probability space Ω.

Generalising the Radon–Nikodym derivative For a classical continuous random variable X on a measurable space $(\Theta, \mathcal{F}(\Theta))$, we can compute its PDF as its *Radon–Nikodym derivative* (RND), namely the measurable function $p : \Theta \to [0, \infty)$ such that for any measurable subset $A \subset \Theta$,

$$P[X \in A] = \int_A p \, \mathrm{d}\mu,$$

where μ is a σ-finite measure on $(\Theta, \mathcal{F}(\Theta))$. We can pose ourselves the following research question.

Question 19. Can a (generalised) PDF be defined for a random-set random variable as defined above?

Answering Question 19 requires extending the notion of a Radon–Nikodym derivative to random sets. Graf [725] pioneered the study of Radon–Nikodym derivatives for the more general case of *capacities*. The extension of the RND to the even more general case of *set functions* was studied by Harding et al. in 1997 [793]. The problem was investigated more recently by Rebille in 2009 [1467]. The following summary of the problem is abstracted from Molchanov's Theory of Random Sets [1304].

Absolute continuity Let us assume that the two capacities μ, ν are monotone, subadditive and continuous from below.

Definition 141. *A capacity ν is* absolutely continuous *with respect to another capacity μ if, for every $A \in \mathcal{F}$, $\nu(A) = 0$ whenever $\mu(A) = 0$.*

This definition is the same as for standard measures. However, whereas for standard measures absolute continuity is equivalent to the integral relation $\mu = \int \nu$ (the existence of the corresponding RND), this is no longer true for general capacities. To understand this, consider the case of a finite Θ, $|\Theta| = n$. Then any measurable function $f : \Theta \to \mathbb{R}_+$ is determined by just n numbers, which do not suffice to uniquely define a capacity on 2^Θ (which has 2^n degrees of freedom) [1304].

Strong decomposition Nevertheless, a Radon–Nikodym theorem for capacities can be established after introducing the notion of *strong decomposition*.

Consider a pair of capacities (μ, ν) that are monotone, subadditive and continuous from below.

Definition 142. *The pair (μ, ν) is said to have the* strong decomposition *property if, $\forall \alpha \geq 0$, there exists a measurable set $A_\alpha \in \mathcal{F}$ such that*

$$\begin{array}{ll} \alpha(\nu(A) - \nu(B)) \leq \mu(A) - \mu(B) & \text{if } B \subset A \subset A_\alpha, \\ \alpha(\nu(A) - \nu(A \cap A_\alpha)) \geq \mu(A) - \mu(A \cap A_\alpha) \ \forall A. & \end{array}$$

Roughly speaking, the strong decomposition condition states that, for each bound α, the 'incremental ratio' of the two capacities is bounded by α in the sub-power set capped by some event A_α. Note that all standard measures possess the strong decomposition property.

A Radon–Nikodym theorem for capacities The following result was proved by Graf [725].

Theorem 112. *For every two capacities μ and ν, ν is an indefinite integral of μ if and only if the pair (μ, ν) has the strong decomposition property* and *ν is absolutely continuous with respect to μ.*

Nevertheless, a number of issues remain open. Most relevantly to our ambition to develop a statistical theory based on random sets, the conditions of Theorem 112 (which holds for general capacities) need to be elaborated for the case of completely alternating capacities (distributions of random closed sets). As a first step, Molchanov notes ([1304], page 75) that the strong decomposition property for two capacity functionals $\nu = T_X$ and $\mu = T_Y$ associated with the random sets X, Y implies that $\alpha T_X(\mathcal{C}_A^B) \leq T_Y(\mathcal{C}_A^B)$ if $B \subset A \subset A_\alpha$, and $\alpha T_X(\mathcal{C}_A^{A \cap A_\alpha}) \geq T_Y(\mathcal{C}_A^{A \cap A_\alpha})$ $\forall A$, where $\mathcal{C}_A^B = \{C \in \mathcal{C}, B \subset C \subset A\}$.

A specific result on the Radon–Nikodym derivative for random (closed) sets is still missing, and with it the possibility of generalising the notion of a PDF to these more complex objects.

17.2 Developing the geometric approach

The geometric approach to uncertainty measures, the main focus of this book, also has much room for further development. On the one hand, the geometric language needs to be extended to all aspects of the reasoning chain, including combination and conditioning, but also (potentially) inference.

Question 20. Can the inference problem can be posed in a geometric setting too, by representing both data and belief measures in a common geometric space?

An answer to such an intriguing question would require a geometric representation general enough to encode both the data driving the inference *and* the (belief) measures possibly resulting from the inference, in such a way that the inferred measure minimises some sort of distance from the empirical data. The question of what norm is the most appropriate to minimise for inference purposes would also arise.

On the other hand, while this book has mainly concerned itself with the geometric representation of finite belief measures, as we move on from belief fuctions on finite frames to random sets on arbitrary domains, the nature of the geometry of these continuous formulations poses new questions. The formalism needs to tackle (besides probability, possibility and belief measures) other important mathematical descriptions of uncertainty, first of all general monotone capacities and gambles (or variations thereof).

Finally, new, more sophisticated geometric representations of belief measures can be sought, in terms of either exterior algebras or areas of projections of convex bodies.

Some of these aspects are briefly considered in this section.

17.2.1 Geometry of general combination

The study of the geometry of the notion of evidence combination and belief updating, which we started with the analysis of Dempster's rule provided in Chapter 8, will need to be extended to the geometric behaviour of the other main combination operators. A comparative geometric analysis of combination rules would allow us to describe the 'cone' of possible future belief states under stronger or weaker assumptions about the reliability and independence of sources.

We can start by giving a general definition of a conditional subspace (see Chapter 8, Definition 122).

Definition 143. *Given a belief function $Bel \in \mathcal{B}$, we define the* conditional subspace *$\langle Bel \rangle_{\odot}$ as the set of all $\odot$ combinations of Bel with any other belief function on the same frame, where $\odot$ is an arbitrary combination rule, assuming such a combination exists. Namely,*

$$\langle Bel \rangle_{\odot} \doteq \left\{ Bel \odot Bel', \; Bel' \in \mathcal{B} \text{ s.t. } \exists \; Bel \odot Bel' \right\}. \tag{17.34}$$

Geometry of Yager's and Dubois's rules

Analysis on binary frames On binary frames $\Theta = \{x, y\}$, Yager's rule (4.23) and Dubois's rule (4.24) coincide (see Section 4.3.2), as the only conflicting focal elements there are $\{x\}$ and $\{y\}$, whose union is Θ itself. Namely,

$$\begin{array}{ll} m_{\circledY}(x) &= m_1(x)(1 - m_2(y)) + m_1(\Theta)m_2(x), \\ m_{\circledY}(y) &= m_1(y)(1 - m_2(x)) + m_1(\Theta)m_2(y), \\ m_{\circledY}(\Theta) &= m_1(x)m_2(y) + m_1(y)m_2(x) + m_1(\Theta)m_2(\Theta). \end{array} \tag{17.35}$$

Using (17.35), we can easily show that

$$\begin{array}{ll} Bel \circledY Bel_x &= [m(x) + m(\Theta), 0, m(y)]', \\ Bel \circledY Bel_y &= [0, m(y) + m(\Theta), m(x)]', \\ Bel \circledY Bel_\Theta &= Bel = [m(x), m(y), m(\Theta)], \end{array} \tag{17.36}$$

adopting the usual vector notation $Bel = [Bel(x), Bel(y), Bel(\Theta)]'$. The global behaviour of Yager's (and Dubois's) rule in the binary case is then pretty clear: the conditional subspace $\langle Bel \rangle_{\circledY}$ is the convex closure

$$\langle Bel \rangle_{\circledY} = Cl(Bel, Bel \circledY Bel_x, Bel \circledY Bel_y)$$

of the points (17.36): see Fig. 17.7.

Comparing (17.35) with (17.36), it is easy to see that

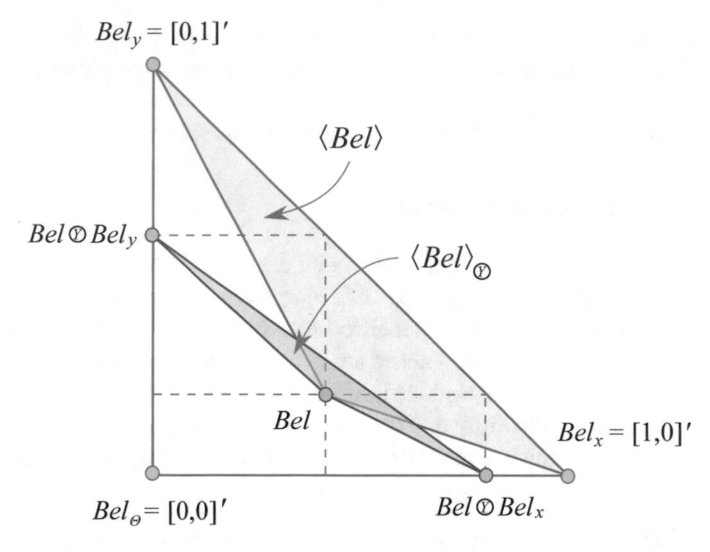

Fig. 17.7: Conditional subspace $\langle Bel \rangle_{\textcircled{Y}}$ for Yager's (and Dubois's) combination rule on a binary frame $\Theta = \{x, y\}$. Dempster's conditional subspace $\langle Bel \rangle$ is also shown for the sake of comparison.

$$Bel_1 \textcircled{Y} Bel_2 = m_2(x) Bel_1 \textcircled{Y} Bel_x + m_2(y) Bel_1 \textcircled{Y} Bel_y + m_2(\Theta) Bel_1 \textcircled{Y} Bel_\Theta,$$

i.e., the simplicial coordinates of Bel_2 in the binary belief space $\mathcal{B}_2$ and of the Yager combination $Bel_1 \textcircled{Y} Bel_2$ in the conditional subspace $\langle Bel_1 \rangle_{\textcircled{Y}}$ coincide.

We can then conjecture the following.

Conjecture 2. Yager combination and affine combination commute. Namely,

$$Bel \textcircled{Y} \left(\sum_i \alpha_i Bel_i \right) = \sum_i \alpha_i Bel \textcircled{Y} Bel_i.$$

As commutativity is the basis of our geometric analysis of Dempster's rule (see Chapter 8), this opens the way for a similar geometric construction for Yager's rule.

Geometric construction However, as shown in Fig. 17.8, the images of constant-mass loci under Yager's rule are parallel, so that the related conditional subspaces have no foci. Indeed, from (17.35),

$$\lim_{m_2(y) \to -\infty} \frac{m_{\textcircled{Y}}(y)}{m_{\textcircled{Y}}(x)} = \frac{m_1(y)(1 - m_2(x)) + m_1(\Theta) m_2(y)}{m_1(x)(1 - m_2(y)) + m_1(\Theta) m_2(x)} = -\frac{m_1(\Theta)}{m_1(x)},$$

and similarly for the loci with $m_2(y) = \text{const}$.

Nevertheless, the general principle of geometric construction of intersecting the linear spaces which are images of constant-mass loci still holds.

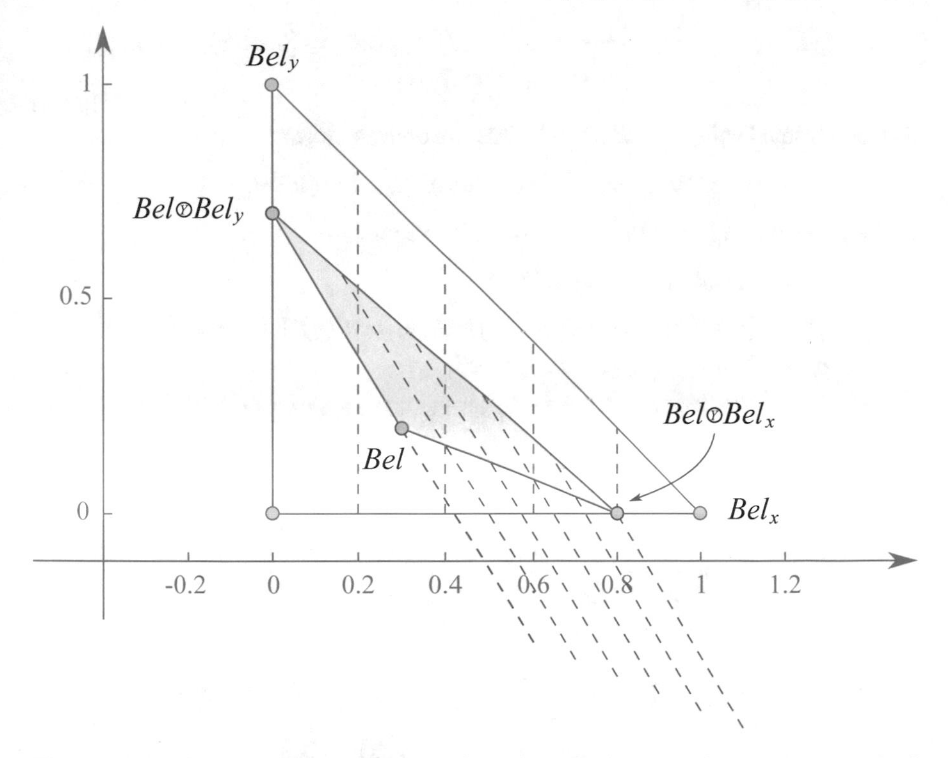

Fig. 17.8: In the case of the Yager combination, the images of constant-mass loci (dashed blue segments) do not converge to a focus, but are parallel lines (dashed purple lines; compare Fig 8.5, Chapter 8).

Geometry of disjunctive combination As we argued earlier in this book, disjunctive combination is the natural, cautious dual of conjunctive/Dempster combination. The operator follows from the assumption that the consensus between two sources of evidence is best represented by the union of the supported hypotheses, rather than by their intersection.

Again, here we will briefly analyse its geometry in the case of a binary frame, and make conjectures about its general behaviour. We will first consider normalised belief functions, to later extend our analysis to the unnormalised case.

Global behaviour Let us first understand the shape of the disjunctive conditional subspace. By definition,

$$\begin{array}{c} m_{\textcircled{\cup}}(x) = m_1(x)m_2(x), \quad m_{\textcircled{\cup}}(y) = m_1(y)m_2(y), \\ m_{\textcircled{\cup}}(\Theta) = 1 - m_1(x)m_2(x) - m_1(y)m_2(y), \end{array} \tag{17.37}$$

so that, adopting the usual vector notation $[Bel(x), Bel(y), Bel(\Theta)]'$,

$$Bel \textcircled{\cup} Bel_x = [m(x), 0, 1 - m(x)]', \quad Bel \textcircled{\cup} Bel_y = [0, m(y), 1 - m(y)]',$$
$$Bel \textcircled{\cup} Bel_\Theta = Bel_\Theta. \tag{17.38}$$

The conditional subspace $\langle Bel \rangle_{\textcircled{\cup}}$ is thus the convex closure

$$\langle Bel \rangle_{\textcircled{\cup}} = Cl(Bel, Bel \textcircled{\cup} Bel_x, Bel \textcircled{\cup} Bel_y)$$

of the points (17.38): see Fig. 17.9. As in the Yager case,

$$\begin{aligned} & Bel \textcircled{\cup} [\alpha Bel' + (1 - \alpha) Bel''] \\ &= \Big[m(x)(\alpha m'(x) + (1 - \alpha) m''(x)), m(y)(\alpha m'(y) + (1 - \alpha) m''(y)) \Big]' \\ &= \alpha Bel \textcircled{\cup} Bel' + (1 - \alpha) Bel \textcircled{\cup} Bel'', \end{aligned}$$

i.e., $\textcircled{\cup}$ *commutes with affine combination*, at least in the binary case.

Question 21. Does disjunctive combination commute with affine combination in general belief spaces?

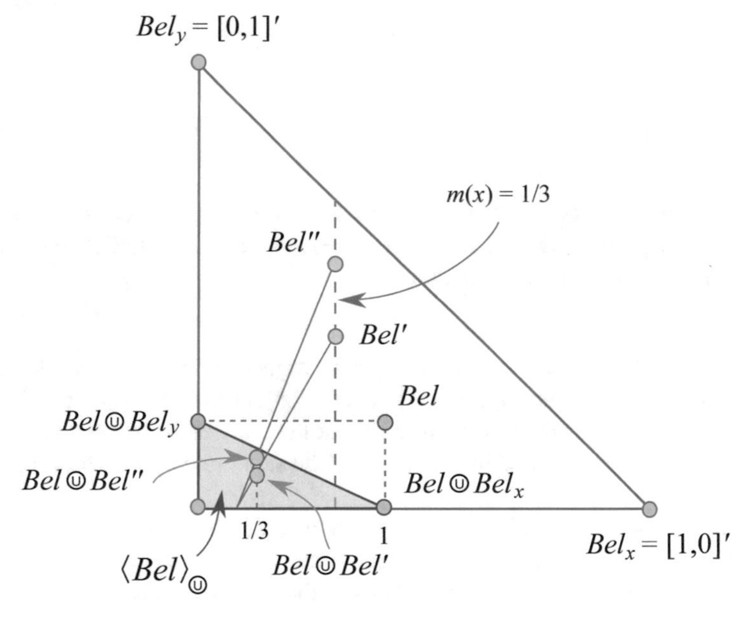

Fig. 17.9: Conditional subspace $\langle Bel \rangle_{\textcircled{\cup}}$ for disjunctive combination on a binary frame. The pointwise behaviour of $\textcircled{\cup}$ is also illustrated (see text).

Pointwise behaviour As in the Yager case, for disjunctive combination the images of constant-mass loci are parallel to each other. In fact, they are parallel to the corresponding constant-mass loci and the coordinate axes (observe, in Fig. 17.9, the locus $m(x) = \frac{1}{3}$ and its image in the conditional subspace $\langle Bel \rangle_{\textcircled{\cup}}$).

We can prove the following.

Theorem 113. *In the binary case $\Theta = \{x, y\}$, the lines joining Bel' and $Bel \circledcirc Bel'$ for any $Bel' \in \mathcal{B}$ intersect in the point*

$$\overline{m(x)} = m'(x)\frac{m(x) - m(y)}{1 - m(y)}, \quad \overline{m(y)} = 0. \tag{17.39}$$

Proof. Recalling the equation of the line joining two points (χ_1, υ_1) and (χ_2, υ_2) of $\mathbb{R}^2$, with coordinates (χ, υ),

$$\upsilon - \upsilon_1 = \frac{\upsilon_2 - \upsilon_1}{\chi_2 - \chi_1}(\chi - \chi_1),$$

we can identify the line joining Bel' and $Bel \circledcirc Bel'$ as

$$\upsilon - m'(y) = \frac{m(y)m'(y) - m'(y)}{m(x)m'(x) - m'(x)}(\chi - m'(x)).$$

Its intersection with $\upsilon = 0$ is the point (17.39), which does not depend on $m'(y)$ (i.e., on the vertical location of Bel' on the constant-mass loci).

□

A valid geometric construction for the disjunctive combination $Bel \circledcirc Bel'$ of two belief functions in $\mathcal{B}_2$ is provided by simple trigonometric arguments. Namely (see Fig. 17.10):

1. Starting from Bel', find its orthogonal projection onto the horizontal axis, with coordinate $m'(x)$ (point 1).
2. Draw the line with slope 45° passing through this projection, and intersect it with the vertical axis, at coordinate $\upsilon = m'(x)$ (point 2).
3. Finally, take the line l passing through Bel_y and the orthogonal projection of Bel onto the horizontal axis, and draw a parallel line l' through point 2 – its intersection with the horizontal axis (point 3) is the x coordinate $m(x)m'(x)$ of the desired combination.

A similar construction (shown in magenta) allows us to locate the y coordinate of the combination (as also shown in Fig. 17.10).

Geometry of combination of unnormalised belief functions In the case of unnormalised belief functions, Dempster's rule is replaced by the conjunctive combination. The disjunctive combination itself needs to be reassessed for UBFs as well.

In the unnormalised case, a distinction exists between the *belief* measure

$$Bel(A) = \sum_{\emptyset \neq B \subseteq A} m(B)$$

and the *believability* (in Smets's terminology) measure of an event A, denoted by $b(A)$,

$$b(A) = \sum_{\emptyset \subseteq B \subseteq A} m(B). \tag{17.40}$$

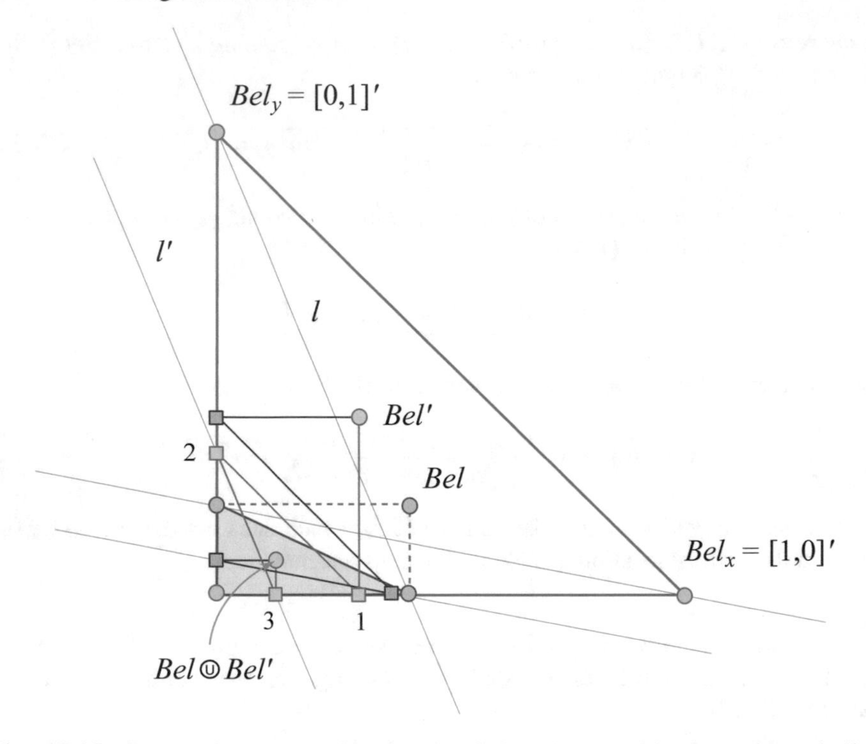

Fig. 17.10: Geometric construction for the disjunctive combination of two belief functions Bel, Bel' on a binary frame.

Here we analyse the geometric behaviour of the latter, in which case $\emptyset$ is not treated as an exception: the case of belief measures is left for future work. As $Bel(\Theta) = b(\Theta) = 1$, as usual, we neglect the related coordinate and represent believability functions as points of a Cartesian space of dimension $|2^\Theta| - 1$ (as $\emptyset$ cannot be ignored any more).

Conjunctive combination on a binary frame In the case of a binary frame, the conjunctive combination of two belief functions Bel_1 and Bel_2, with BPAs m_1, m_2, yields

$$\begin{aligned} m_{\textcircled{\cap}}(\emptyset) &= m_1(\emptyset) + m_2(\emptyset) - m_1(\emptyset)m_2(\emptyset) + m_1(x)m_2(y) + m_1(y)m_2(x), \\ m_{\textcircled{\cap}}(x) &= m_1(x)(m_2(x) + m_2(\Theta)) + m_1(\Theta)m_2(x), \\ m_{\textcircled{\cap}}(y) &= m_1(y)(m_2(y) + m_2(\Theta)) + m_1(\Theta)m_2(y), \\ m_{\textcircled{\cap}}(\Theta) &= m_1(\Theta)m_2(\Theta). \end{aligned} \tag{17.41}$$

Conditional subspace for conjunctive combination The global behaviour of $\textcircled{\cap}$ in the binary (unnormalised) case can then be understood in terms of its conditional subspace. We have

$$\begin{aligned}
b \textcircled{\cap} b_\emptyset &= b_\emptyset = [1,1,1]', \\
b \textcircled{\cap} b_x &= (m_1(\emptyset) + m_1(y)) b_\emptyset + (m_1(x) + m_1(\Theta)) b_x \\
&= [m_1(\emptyset) + m_1(y), 1, m_1(\emptyset) + m_1(y)]' \\
&= b_1(y) b_\emptyset + (1 - b_1(y)) b_x, \\
b \textcircled{\cap} b_y &= (m_1(\emptyset) + m_1(x)) b_\emptyset + (m_1(y) + m_1(\Theta)) b_y \\
&= [m_1(\emptyset) + m_1(x), m_1(\emptyset) + m_1(x), 1]' \\
&= b_1(x) b_\emptyset + (1 - b_1(x)) b_y, \\
b \textcircled{\cap} b_\Theta &= b,
\end{aligned} \tag{17.42}$$

as $b_x = [0,1,0]'$, $b_y = [0,0,1]'$, $b_\emptyset = [1,1,1]'$ and $b_\Theta = [0,0,0]'$.

From (17.42), we can note that the vertex $b \textcircled{\cap} b_x$ belongs to the line joining $b_\emptyset$ and b_x, with its coordinate given by the believability assigned by b to the other outcome y. Similarly, the vertex $b \textcircled{\cap} b_y$ belongs to the line joining $b_\emptyset$ and b_y, with its coordinate given by the believability assigned by b to the complementary outcome x (see Fig. 17.11).

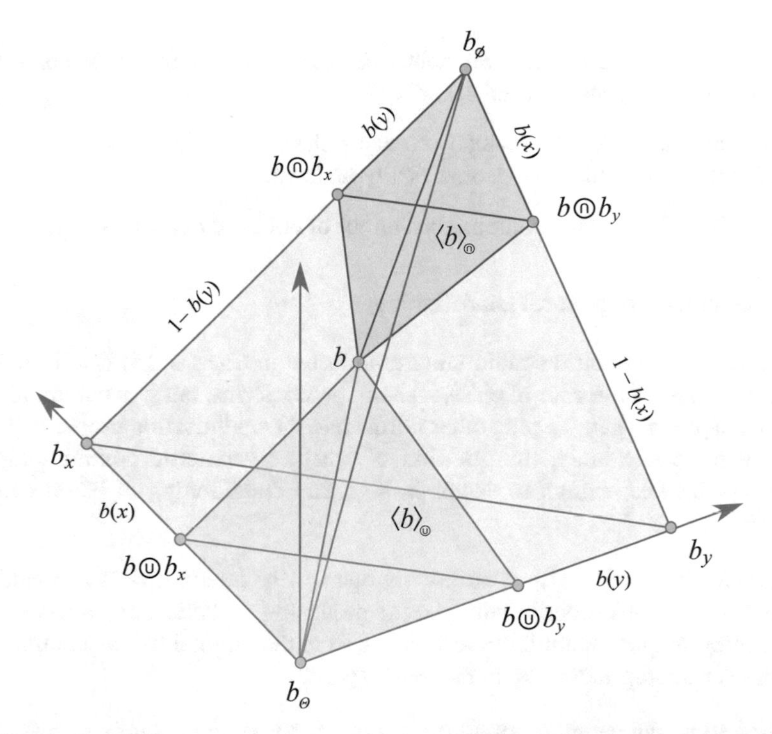

Fig. 17.11: Conditional subspaces induced by $\textcircled{\cap}$ and $\textcircled{\cup}$ in a binary frame, for the case of unnormalised belief functions.

Conditional subspace for disjunctive combination As for the disjunctive combination, it is easy to see that in the unnormalised case we get

$$b \circledcup b_\Theta = b_\Theta, \quad b \circledcup b_x = b(x) b_x + (1 - b(x)) b_\Theta,$$
$$b \circledcup b_\emptyset = b, \quad b \circledcup b_y = b(y) b_y + (1 - b(y)) b_\Theta,$$

so that the conditional subspace is as in Fig. 17.11. Note that, in the unnormalised case, there is a unit element with respect to $\circledcup$, namely $b_\emptyset$.

In the unnormalised case, we can observe a clear symmetry between the conditional subspaces induced by disjunctive and conjunctive combination.

Research questions A number of questions remain open after this preliminary geometric analysis of other combination rules on binary spaces, and its extension to the case of unnormalised belief functions.

Question 22. What is the general pointwise geometric behaviour of disjunctive combination, in both the normalised and the unnormalised case?

A dual question concerns the conjunctive rule, as the alter ego of Dempster's rule in the unnormalised case.

Question 23. What is the general pointwise geometric behaviour of conjunctive combination in the unnormalised case?

Bold and cautious rules, which are also inherently defined for unnormalised belief functions, are still to be geometrically understood.

Question 24. What is the geometric behaviour of bold and cautious rules?

17.2.2 Geometry of general conditioning

Our analysis of geometric conditioning (conducted in Chapter 15) is still in its infancy. We analysed the case of classical Minkowski norms, but it is natural to wonder what happens when we plug other norms into the optimisation problem (15.1).

Even more relevantly, the question of whether geometric conditioning is a framework flexible enough to encompass general conditioning in belief calculus arises.

Question 25. Can any major conditioning operator be interpreted as geometric (in the sense of this book) conditioning, i.e., as producing the belief function within the conditioning simplex at a minimum distance from the original belief function, with respect to an appropriate norm in the belief space?

In addition, the geometry of all the major conditioning operators that we listed in Section 4.5 (including Fagin's lower and upper envelopes (4.46), Suppes and Zanotti's 'geometric' conditioning (4.48), and Smets's unnormalised conditioning (4.49)) remains to be understood.

17.2.3 A true geometry of uncertainty

A true geometry of uncertainty will require the ability to manipulate in our geometric language any (or most) forms of uncertainty measures (compare the partial hierarchy reported in Chapter 6).

Probability and possibility measures are, as we know, special cases of belief functions: therefore, their geometric interpretation does not require any extension of the notion of a belief space (as we learned extensively in Parts II and III of this book). Most other uncertainty measures, however, are not special cases of belief functions – in fact, a number of them are more general than belief functions, such as probability intervals (2-monotone capacities), general monotone capacities and lower previsions.

Tackling these more general measures requires, therefore, extending the concept of a geometric belief space in order to encapsulate the most general such representation. In the medium term, the aim is to develop a general geometric theory of imprecise probabilities, encompassing capacities, random sets induced by capacity functionals, and sets of desirable gambles.

Geometry of capacities and random sets Developing a geometric theory of general monotone capacities appears to be rather challenging. We can gain some insight into this issue, however, from an analysis of 2-monotone capacities (systems of probability intervals) and, more specifically, when they are defined on binary frames.

Geometry of 2-monotone capacities Recall that a capacity μ is called *2-monotone* iff, for any A, B in the relevant σ-algebra (see (6.9)),

$$\mu(A \cup B) + \mu(A \cap B) \geq \mu(A) + \mu(B),$$

whereas it is termed *2-alternating* whenever

$$\mu(A \cup B) + \mu(A \cap B) \leq \mu(A) + \mu(B).$$

Belief measures are 2-monotone capacities, while their dual plausibility measures are 2-alternating ones. More precisely, belief functions are infinitely monotone capacities, as they satisfy (6.5) for every value of $k \in \mathbb{N}$.

A well-known lemma by Chateauneuf and Jaffray states the following.

Proposition 74. *A capacity μ on Θ is 2-monotone if and only if*

$$\sum_{\{x,y\} \subseteq E \subseteq A} m(E) \geq 0 \quad \forall x, y \in \Theta, \ \forall A \ni x, y, \tag{17.43}$$

where m is the Möbius transform (6.6) of μ.

We can use Proposition 74 to show that the following theorem is true.

Theorem 114. *The space of all 2-monotone capacities defined on a frame Θ, which we denote by $\mathcal{M}^2_\Theta$, is convex.*

Proof. Suppose that μ_1 and μ_2, with Möbius inverses m_1 and m_2, respectively, are 2-monotone capacities (therefore satisfying the constraint (17.43)). Suppose that a capacity μ is such that $\mu = \alpha\mu_1 + (1-\alpha)\mu_2, 0 \leq \alpha \leq 1$.

Since Möbius inversion is a linear operator, we have that

$$\begin{aligned} \sum_{\{x,y\} \subseteq E \subseteq A} m(E) &= \sum_{\{x,y\} \subseteq E \subseteq A} \alpha m_1 + (1-\alpha) m_2 \\ &= \alpha \sum_{\{x,y\} \subseteq E \subseteq A} m_1 + (1-\alpha) \sum_{\{x,y\} \subseteq E \subseteq A} m_2 \geq 0, \end{aligned}$$

since $\sum_{\{x,y\} \subseteq E \subseteq A} m_1 \geq 0$ and $\sum_{\{x,y\} \subseteq E \subseteq A} m_2 \geq 0$ for all $x, y \in \Theta$, $\forall A \ni x, y$.

□

Note that Theorem 114 states the convexity of the *space* of 2-monotone capacities (analogously to what we did for the belief space in Chapter 7): it is, instead, well known that every 2-monotone capacity corresponds to a convex set of probability measures in their simplex.

It is useful to get an intuition about the problem by considering the case of 2-monotone capacities defined on the ternary frame $\Theta = \{x, y, z\}$. There, the constraints (17.43) read as

$$\begin{cases} m(\{x,y\}) \geq 0, \\ m(\{x,y\}) + m(\Theta) \geq 0, \\ m(\{x,z\}) \geq 0, \\ m(\{x,z\}) + m(\Theta) \geq 0, \\ m(\{y,z\}) \geq 0, \\ m(\{y,z\}) + m(\Theta) \geq 0, \end{cases} \tag{17.44}$$

whereas for $\Theta = \{x, y, z, w\}$ they become

$$\begin{cases} m(\{x_i,x_j\}) \geq 0, \\ m(\{x_i,x_j\}) + m(\{x_i,x_j,x_k\}) \geq 0, \\ m(\{x_i,x_j\}) + m(\{x_i,x_j,x_l\}) \geq 0, \\ m(\{x_i,x_j\}) + m(\{x_i,x_j,x_k\}) + m(\{x_i,x_j,x_l\}) + m(\Theta) \geq 0, \end{cases}$$

for each possible choice of $x_i, x_j, x_k, x_l \in \Theta$.

The constraints (17.44) are clearly equivalent to

$$\begin{cases} m(\{x,y\}) \geq 0, \\ m(\{x,z\}) \geq 0, \\ m(\{y,z\}) \geq 0, \\ m(\Theta) \geq -\min\Big\{m(\{x,y\}), m(\{x,z\}), m(\{y,z\})\Big\}, \end{cases} \tag{17.45}$$

so that the space $\mathcal{M}^2$ of 2-monotone capacities on $\{x, y, z\}$ is the set of capacities

$$\mathcal{M}^2_{\{x,y,z\}} = \left\{ \mu : 2^\Theta \to [0,1] \,\middle|\, \sum_{A \subseteq \Theta} m(A) = 1, m(A) \geq 0 \; \forall A : |A| = 2, \right.$$
$$\left. m(\Theta) \geq -\min\left\{ m(\{x,y\}), m(\{x,z\}), m(\{y,z\}) \right\} \right\}.$$

We can begin to understand the structure of this convex set and its vertices by studying the shape of the convex subset of $\mathbb{R}^3$

$$z \geq -\min\{x,y\}, \quad x,y \geq 0, \quad x+y+z=1. \tag{17.46}$$

The latter is the intersection of the following loci:

$$V = \{x+y+z=1\}, \quad H_1 = \{x \geq 0\}, \quad H_2 = \{y \geq 0\},$$

$$H_3' = \{z \geq -x, x \leq y\}, \quad H_3'' = \{z \geq -y, y \leq x\},$$

the first of which is a linear space, whereas the others are half-spaces determined by a plane in $\mathbb{R}^3$. The resulting convex body is the convex closure

$$Cl([1,0,0],[0,1,0],[0,0,1],[1,1,-1])$$

depicted in Fig. 17.12. The four vertices are associated with all possible binary configurations of the variables x and y, with the z component determined by the other constraints. One can notice that vertices are allowed to have negative mass values on some component (focal elements of Θ). As a result, we can conjecture that the space of 2-monotone capacities on a frame Θ, while being embedded in a Cartesian space of the same dimension as the belief space, will contain the belief space by virtue of its extra vertices (e.g. $[1,1,-1]$ in Fig. 17.12).

As the geometry of 2-monotone capacities involves a hierarchical set of constraints involving events of increasing cardinality, however, the analysis of the toy problem (17.46) cannot be directly generalised to it.

Geometry of capacities and random sets The geometry of capacities is related to that of random sets, via the capacity functional T_X (recall Definition 74, Section 4.9.3). We will analyse this point further in the near future.

Geometry of functionals Any study of the geometry of gambles requires an analysis of the geometry of *functionals*, i.e., mappings from a space of functions to the real line. The same holds for other uncertainty representations which make use of functions on the real line, such as MV algebras (Section 4.9.5) and belief measures on fuzzy events (Section 6.5.2). For a very interesting recent publication (2016) on the geometry of MV algebras in terms of rational polyhedra, see [1324].

Geometric functional analysis [89] [833, 834] is a promising tool from this perspective. A central question of geometric functional analysis is: what do typical n-dimensional structures look like when n grows to infinity? This is exactly what happens when we try to generalise the notion of a belief space to continuous domains.

[89] https://link.springer.com/book/10.1007/978-1-4684-9369-6.

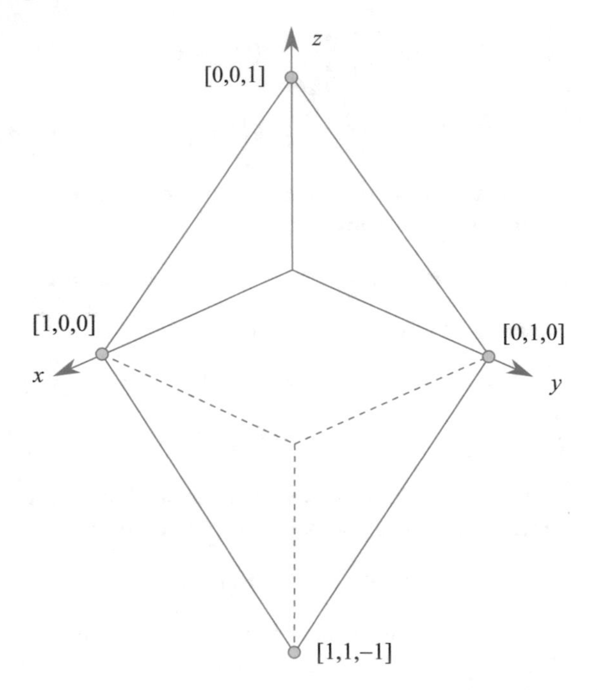

Fig. 17.12: The geometry of the convex body (17.46) in $\mathbb{R}^3$.

One of the main tools of geometric functional analysis is the theory of *concentration of measure*, which offers a geometric view of the limit theorems of probability theory. Geometric functional analysis thus bridges three areas: functional analysis, convex geometry and probability theory [1855].

In more detail, a *norm* on a vector space X is a function $\|\| : X \to \mathbb{R}$ that satisfies (i) non-negativity, (ii) homogeneity and (iii) the triangle inequality. A *Banach space* [581] is a complete normed space. A convex set K in $\mathbb{R}^n$ is called *symmetric* if $K = -K$ (i.e., $x \in K$ implies $-x \in K$).

Proposition 75. *(Banach spaces and symmetric convex bodies) Let $X = (\mathbb{R}^n, \|\cdot\|)$ be a Banach space. Then its unit ball B_X is a symmetric convex body in $\mathbb{R}^n$. Further, let K be a symmetric convex body in $\mathbb{R}^n$. Then K is the unit ball of some normed space with norm (its* Minkowski functional, *Fig. 17.13(a))*

$$\|x\|_K = \inf\left\{t > 0, \frac{x}{t} \in K\right\}.$$

The correspondence between unit balls in Banach spaces and convex bodies established in Proposition 75 allows arguments from convex geometry to be used in functional analysis and vice versa. Using Proposition 75, one can show that subspaces of Banach spaces correspond to *sections* of convex bodies, and quotient spaces correspond to *projections* of convex bodies.

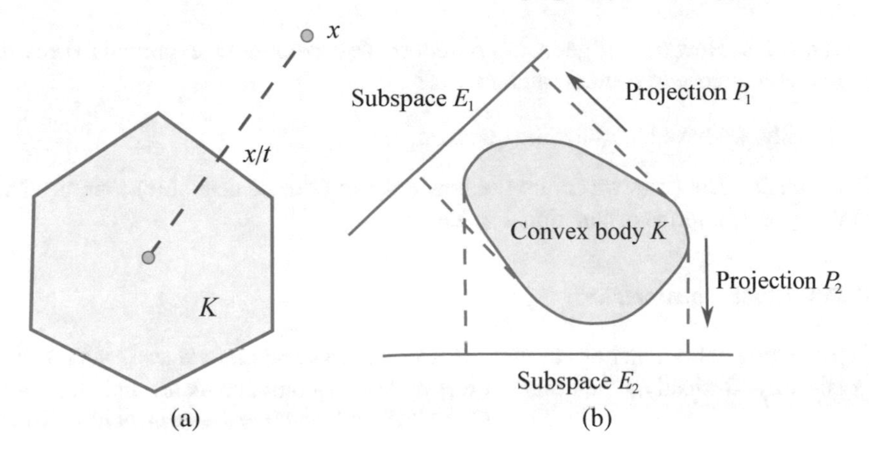

Fig. 17.13: (a) The Minkowski functional of a symmetric convex body in $\mathbb{R}^2$ [1855]. (b) Projections of a convex body.

The phenomenon of concentration of measure was the driving force for the early development of geometric functional analysis. It tells us that anti-intuitive phenomena occur in high dimensions. For example, the 'mass' of a high-dimensional ball is concentrated only in a thin band around any equator. This reflects the philosophy that metric and measure should be treated very differently: a set can have a large diameter but carry little mass [1855].

The shape of random projections of symmetric convex bodies onto k-dimensional subspaces (Fig. 17.13(b)) can also be studied within geometric functional analysis. It is easy to see that a projection captures only the extremal points of K. Hence, one quantity affected by projections is the diameter of the body.

Proposition 76. *(Diameters under random projections, [688]) Let K be a symmetric convex body in $\mathbb{R}^n$, and let P be a projection onto a random subspace of dimension k. Then, with probability at least $1 - e^{-k}$,*

$$diam(PK) \leq C \left(M^*(K) + \sqrt{\frac{k}{n}} diam(K) \right),$$

where $M^*(K)$ is the mean width of K.

Research questions As this is one of the most challenging topics for future research, most issues are still wide open, in particular the following.

Question 26. What is the most appropriate framework for describing the geometry of monotone capacities?

Question 27. Can we provide a representation in the theory of convex bodies of the set of 2-monotone capacities similar to that of the belief space, i.e., by providing an analytical expression for the vertices of this set?

Question 28. How can the geometry of monotone capacities be exploited to provide a geometric theory of general random sets?

Finally, we have the following question.

Question 29. How can we frame the geometry of (sets of desirable) gambles and MV algebras in terms of functional spaces?

17.2.4 Fancier geometries

Representing belief functions as mere vectors of mass or belief values is not entirely satisfactory. Basically, when this is done all vector components are indistinguishable, whereas they in fact correspond to values assigned to subsets of Θ of different cardinality.

Exterior algebras Other geometrical representations of belief functions on finite spaces can nevertheless be imagined, which take into account the qualitative difference between events of different cardinalities.

For instance, the exterior algebra [726] is a promising tool, as it allows us to encode focal elements of cardinality k with exterior powers (*k-vectors*) of the form $x_1 \wedge x_2 \wedge \cdots \wedge x_k$ and belief functions as linear combinations of decomposable k-vectors, namely

$$\mathbf{bel} = \sum_{A=\{x_1,\ldots,x_k\}\subset\Theta} m(A)\, x_1 \wedge x_2 \wedge \cdots \wedge x_k.$$

Under an exterior algebra representation, therefore, focal elements of different cardinalities are associated with qualitatively different objects (the k-vectors) which encode the geometry of volume elements of different dimensions (e.g. the area of parallelograms in $\mathbb{R}^2$).

The question remains of how Dempster's rule may fit into such a representation, and what its relationship with the exterior product operator $\wedge$ is.

Capacities as isoperimeters of convex bodies *Convex bodies*, as we have seen above, are also the subject of a fascinating field of study.

In particular, any convex body in $\mathbb{R}^n$ possesses 2^n distinct orthogonal projections onto the 2^n subspaces generated by all possible subsets of coordinate axes. It is easy to see that, given a convex body K in the Cartesian space $\mathbb{R}^n$, endowed with coordinates $x_1, \ldots, x_n$, the function ν that assigns to each subset of coordinates $S = \{x_{i_1}, \ldots, x_{i_k}\}$ the (hyper)volume $\nu(S)$ of the orthogonal projection $K|S$ of $\mathcal{K}$ onto the linear subspace generated by $S = \{x_{i_1}, \ldots, x_{i_k}\}$ is potentially a capacity. In order for that to happen, volumes in linear spaces of different dimensions need to be compared, and conditions need to be enforced in order for volumes of projections onto higher-dimensional subspaces to be greater than those of projections on any of their subspaces.

As we have just seen, symmetric convex bodies are in a 1–1 relationship with (balls in) Banach spaces, which can then be used as a tool to represent belief measures as projections of symmetric convex bodies.

The following research questions arise.

Question 30. Under what conditions is the above capacity monotone?

Question 31. Under what conditions is this capacity a belief function (i.e. an infinitely monotone capacity)?

17.2.5 Relation to integral and stochastic geometry

Integral geometry, or geometric probability This is the theory of invariant measures (with respect to continuous groups of transformations of a space onto itself) on sets consisting of submanifolds of the space (for example lines, planes, geodesics and convex surfaces; in other words, manifolds preserving their type under the transformations in question). Integral geometry has been constructed for various spaces, primarily Euclidean, projective and homogeneous spaces, and arose in connection with refinements of statements of problems in geometric probabilities.

In order to introduce an invariant measure, one needs to identify a function depending on the coordinates of the space under consideration whose integral over some region of the space is not changed under any continuous coordinate transformation belonging to a specified Lie group. This requires finding an integral invariant of the Lie group.

A classic problem of integral geometry is Buffon's *needle problem* (dating from 1733).

Proposition 77. *Some parallel lines on a wooden floor are a distance d apart from each other. A needle of length $l < d$ is randomly dropped onto the floor (Fig. 17.14, left). Then the probability that the needle will touch one of the lines is*

$$P = \frac{2l}{\pi d}.$$

Another classical problem is Firey's *colliding dice* problem (from 1974).

Proposition 78. *Suppose Ω_1 and Ω_2 are disjoint unit cubes in $\mathbb{R}^3$. In a random collision, the probability that the cubes collide edge-to-edge slightly exceeds the probability that the cubes collide corner-to-face. In fact,*

$$0.54 \approxeq P(\textit{edge-to-edge collision}) > P(\textit{corner-to-face collision}) \approxeq 0.46.$$

Roughly speaking, integral geometry is the problem of computing probabilities of events which involve geometric loci, as opposed to the geometric approach to uncertainty proposed in this book, which is about understanding the geometric properties of (uncertainty) measures. The main technical tool is Poincaré's integral-geometric formula.

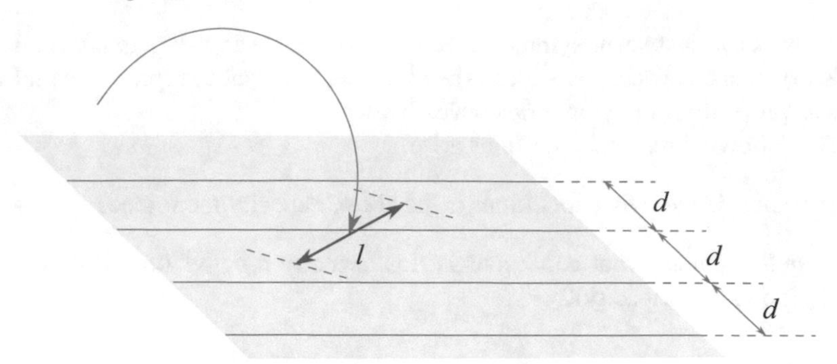

Fig. 17.14: Buffon's needle problem.

After representing lines in the plane using coordinates (p, θ), $\cos(\theta)x+\sin(\theta)y = p$, the *kinematic measure* for such lines is defined as $\mathrm{d}K = \mathrm{d}p \wedge \mathrm{d}\theta$, as the measure on a set of lines which is invariant under rigid motion, i.e., the Jacobian of the coordinate transformation is equal to 1. Now, let C be a piecewise $\mathcal{C}^1$ curve in the plane. Given a line L in the plane, let $n(L \cap C)$ be the number of intersection points. If C contains a linear segment and L agrees with that segment, $n(C \cap L) = \infty$.

Proposition 79. *(Poincaré formula for lines, 1896) Let C be a piecewise $\mathcal{C}^1$ curve in the plane. Then the measure of unoriented lines meeting C, counted with multiplicity, is given by*

$$2L(C) = \int_{L: L\cap C \neq \emptyset} n(C \cap L) \, \mathrm{d}K(L).$$

Integral-geometric formulae hold for convex sets Ω as well. Since $n(L \cap \partial\Omega)$, where $\partial\Omega$ is the boundary of Ω, is either zero or two for $\mathrm{d}K$-almost all lines L, the measure of unoriented lines that meet the convex set Ω is given by

$$L(\partial\Omega) = \int_{L: L\cap\Omega \neq \emptyset} \mathrm{d}K.$$

One can then prove the following.

Proposition 80. *(Sylvester's problem, 1889) Let $\Theta \subset \Omega$ be two bounded convex sets in the plane. Then the probability that a random line meets Θ given that it meets Ω is $P = L(\partial\Theta)/L(\partial\Omega)$.*

Corollary 28. *Let C be a piecewise $\mathcal{C}^1$ curve contained in a compact convex set Ω. The expected number of intersections with C of all random lines that meet Ω is*

$$\mathbb{E}(n) = \frac{2L(C)}{L(\partial\Omega)}.$$

Integral geometry potentially provides a powerful tool for designing new geometric representations of belief measures, on the one hand. On the other hand, we can foresee generalisations of integral geometry in terms of non-additive measures and capacities.

Question 32. How does geometric probability generalise when we replace standard probabilities with other uncertainty measures?

Stochastic geometry Stochastic geometry is a branch of probability theory which deals with set-valued random elements. It describes the behaviour of random configurations such as random graphs, random networks, random cluster processes, random unions of convex sets, random mosaics and many other random geometric structures [855]. The topic is very much intertwined with the theory of random sets, and Matheron's work in particular. The name appears to have been coined by David Kendall and Klaus Krickeberg [277].

Whereas geometric probability considers a fixed number of random objects of a fixed shape and studies their interaction when some of the objects move randomly, since the 1950s the focus has switched to models involving a random number of randomly chosen geometric objects. As a consequence, the notion of a *point process* started to play a prominent role in this field.

Definition 144. *A* point process η *is a measurable map from some probability space* $(\Omega, \mathcal{F}, P)$ *to the locally finite subsets of a Polish space* $\mathbb{X}$ *(endowed with a suitable* σ*-algebra), called the state space. The* intensity measure *of* η*, evaluated at a measurable set* $A \subset \mathbb{X}$*, is defined by* $\mu(A) = E[\eta(A)]$*, and equals the mean number of elements of* η *lying in* A*.*

Typically, $\mathbb{X}$ is either $\mathbb{R}^n$, the space of compact subsets of $\mathbb{R}^n$ or the set of all affine subspaces there (the Grassmannian). A point process can be written as $\eta = \sum_{i=1}^{\tau} \delta_{\zeta_i}$, where τ is a random variable taking values in $\mathbb{N}_0 \cup \{\infty\}$, and $\zeta_1, \zeta_2, \ldots$ is a sequence of random points in $\mathbb{X}$ (Fig. 17.15). 'Random point field' is sometimes considered a more appropriate terminology, as it avoids confusion with stochastic processes.

A point process is, in fact, a special case of a *random element*, a concept introduced by Maurice Fréchet [643] as a generalisation of the idea of a random variable to situations in which the outcome is a complex structure (e.g., a vector, a function, a series or a subset).

Definition 145. *Let* $(\Omega, \mathcal{F}, P)$ *be a probability space, and* $(E, \mathcal{E})$ *a measurable space. A* random element *with values in* E *is a function* $X : \Omega \to E$ *which is* $(\mathcal{F}, \mathcal{E})$*-measurable, that is, a function* X *such that for any* $B \in \mathcal{E}$*, the pre-image of* B *lies in* $\mathcal{F}$*.*

Random elements with values in E are sometimes called E-valued random variables (hence the connection with random sets).

Geometric processes are defined as point processes on manifolds that represent spaces of events. Thus, processes of straight lines in the plane are defined as point processes on a Möbius strip (as the latter represents the space of straight lines in $\mathbb{R}^2$).

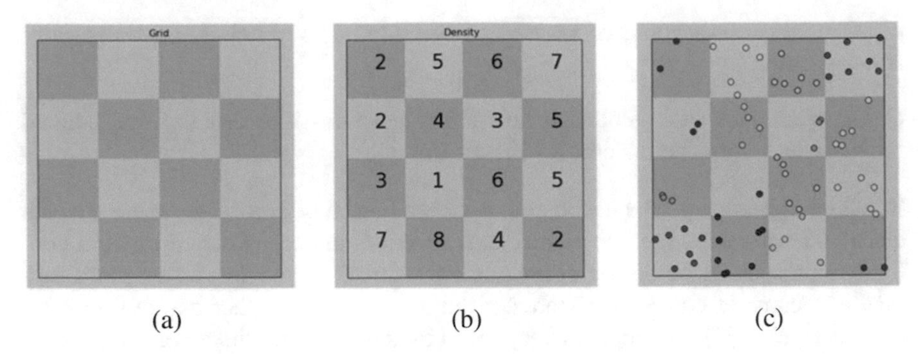

Fig. 17.15: Examples of point processes (from http://stats.stackexchange.com/questions/16282/how-to-generate-nd-point-process : (a) grid; (b) Poisson; (c) uniform.

Processes on manifolds form a more general concept; here, once again, stochastic geometry is linked with the theory of random sets [1268]. The 'intrinsic volume' of convex bodies and Minkowski combinations of them (*polyconvex* bodies) is a topic of stochastic geometry which relates to the questions we posed in Section 17.2.4.

A peculiarity which distinguishes stochastic geometry from the theory of random sets is the interest stochastic geometry has for geometric processes with distributions that are invariant relative to groups acting on the fundamental space $\mathbb{X}$. For instance, one may examine the class of those processes of straight lines on $\mathbb{R}^2$ which are invariant relative to the group of Euclidean motions of the plane.

17.3 High-impact developments

The theory of random sets is very well suited to tackling a number of high-profile applications, which range from the formulation of cautious climate change prediction models (Section 17.3.1) to the design of novel machine learning tool generalising classical mathematical frameworks (e.g. max-entropy classifiers and random forests, Section 17.3.2) and to the analysis of the very foundations of statistical machine learning (Section 17.3.3).

Exciting work has recently been conducted on Bayesian formulations of the foundations of quantum mechanics [237, 120]. Further analyses of the problem in the extended framework of imprecise probabilities are likely to follow.

17.3.1 Climate change

Climate change[90] is a paramount example of a problem which requires predictions to be made under heavy uncertainty, due to the imprecision associated with any climate model, the long-term nature of the predictions involved and the second-order uncertainty affecting the statistical parameters involved.

[90] https://link.springer.com/journal/10584.

A typical question a policymaker may ask a climate scientist is, for example [1507],

"*What is the probability that a doubling of atmospheric CO_2 from pre-industrial levels will raise the global mean temperature by at least* $2°C$*?*"

Rougier [1507] nicely outlined a Bayesian approach to climate modelling and prediction, in which a predictive distribution for a future climate is found by conditioning the future climate on observed values for historical and current climates. A number of challenges arise:

- in climate prediction, the collection of uncertain quantities for which the climate scientist must specify prior probabilities can be large;
- specifying a prior distribution over climate vectors is very challenging.

According to Rougier, considering that people are spending thousands of hours on collecting climate data and constructing climate models, it is surprising that little attention is being devoted to quantifying our judgements about how the two are related.

In this section, climate is represented as a vector of measurements y, collected at a given time. Its components include, for instance, the level of CO_2 concentration on the various points of a grid. More precisely, the *climate vector* $y = (y_h, y_f)$ collects together both historical and present (y_h) and future (y_f) climate values. A *measurement error* e is introduced to take into account errors due to, for instance, a seasick technician or atmospheric turbulence. The actual measurement vector is therefore

$$z \doteq y_h + e.$$

The Bayesian treatment of the problem makes use of a number of assumptions. For starters, we have the following axioms.

Axiom 3 *Climate and measurement error are independent:* $e \perp y$.

Axiom 4 *The measurement error is Gaussian distributed, with mean* $\mathbf{0}$ *and covariance* Σ^e: $e \sim \mathcal{N}(\mathbf{0}, \Sigma^e)$.

Thanks to these assumptions, the predictive distribution for the climate given the measured values $z = \tilde{z}$ is

$$p(y|z = \tilde{z}) \sim \mathcal{N}(\tilde{z} - y_h|\mathbf{0}, \Sigma^e)p(y), \tag{17.47}$$

which requires us to specify a prior distribution for the climate vector y itself.

Climate models The choice of such a prior $p(y)$ is extremely challenging, because y is such a large collection of quantities, and these component quantities are linked by complex interdependencies, such as those arising from the laws of nature. The role of the climate model is then to induce a distribution for the climate itself, and plays the role of a parametric model in statistical inference (Section 1.2.3).

Namely, a *climate model* is a deterministic mapping from a collection of parameters x (equation coefficients, initial conditions, forcing functions) to a vector of measurements (the 'climate'), namely

$$x \to y = g(x), \tag{17.48}$$

where g belongs to a predefined 'model space' $\mathcal{G}$. Climate scientists call an actual value of $g(x)$, computed for some specific set of parameter values x, a *model evaluation*. The reason is that the analytical mapping (17.48) is generally not known, and only images of specific input parameters can be computed or sampled (at a cost). A climate scientist considers, on a priori grounds based on their past experience, that some choices of x are better than others, i.e., that there exists a set of parameter values x^* such that

$$y = g(x^*) + \epsilon^*,$$

where ϵ^* is termed the 'model discrepancy'.

Prediction via a parametric model The difference between the climate vector and any model evaluation can be decomposed into two parts:

$$y - g(x) = g(x^*) - g(x) + \epsilon^*.$$

The first part is a contribution that may be reduced by a better choice of the model g; the second part is, in contrast, an irreducible contribution that arises from the model's own imperfections. Note that x^* is not just a statistical parameter, though, for it relates to physical quantities, so that climate scientists have a clear intuition about its effects. Consequently, scientists may be able to exploit their expertise to provide a prior $p(x^*)$ on the input parameters.

In this Bayesian framework for climate prediction, two more assumptions are needed.

Axiom 5 *The 'best' input, discrepancy and measurement error are mutually (statistically) independent:*

$$x^* \perp \epsilon^* \perp e.$$

Axiom 6 *The model discrepancy ϵ^* is Gaussian distributed, with mean* **0** *and covariance Σ^ϵ.*

Axioms 5 and 6 then allow us to compute the desired climate prior as

$$p(y) = \int \mathcal{N}(y - g(x^*)|\mathbf{0}, \Sigma^\epsilon) p(x^*) \, \mathrm{d}x^*, \tag{17.49}$$

which can be plugged into (17.47) to yield a Bayesian prediction of future climate values.

In practice, as we have said, the climate model function $g(.)$ is not known – we only possess a sample of model evaluations $\{g(x_1), \ldots, g(x_n)\}$. We call the process of tuning the covariances $\Sigma^\epsilon, \Sigma^e$ and checking the validity of the Gaussianity assumptions (Axioms 4 and 6) *model validation*. This can be done by using (17.47) to predict past/present climates $p(z)$, and applying some hypothesis testing to the result. If the observed value $\tilde{z}$ is in the tail of the distribution, the model parameters (if not the entire set of model assumptions) need to be corrected. As Rougier admits [1507], responding to bad validation results is not straightforward.

Model calibration Assuming that the model has been validated, it needs to be 'calibrated', i.e., we need to find the desired 'best' value x^* of the model's parameters.

Under Axioms 3–6, we can compute

$$p(x^*|z = \tilde{z}) = p(z = \tilde{z}|x^*) = \mathcal{N}(\tilde{z} = g(x^*)|\mathbf{0}, \Sigma^\epsilon + \Sigma^e)p(x^*).$$

As we know, maximum a posteriori estimation (see Section 1.2.5) could be applied to the above posterior distribution: however, the presence of multiple modes might make it ineffective.

As an alternative, we can apply full Bayesian inference to compute

$$p(y_f|z = \tilde{z}) = \int p(y_f|x^*, z = \tilde{z})p(x^*|z = \tilde{z}) \, \mathrm{d}x^*, \tag{17.50}$$

where $p(y_f|x^*, z = \tilde{z})$ is Gaussian with a mean which depends on $\tilde{z} - g(x)$. The posterior prediction (17.50) highlights two routes for climate data to impact on future climate predictions:

- by concentrating the distribution $p(x^*|z = \tilde{z})$ relative to the prior $p(x^*)$, depending on both the quantity and the quality of the climate data;
- by shifting the mean of $p(y_f|x^*, z = \tilde{z})$ away from $g(x)$, depending on the size of the difference $\tilde{z} - g(x)$.

Role of model evaluations Let us go back to the initial question: what is the probability that a doubling of atmospheric CO_2 will raise the global mean temperature by at least $2°C$ by 2100?

Let $Q \subset \mathcal{Y}$ be the set of climates y for which the global mean temperature is at least $2°C$ higher in 2100. The probability of the event of interest can then be computed by integration, as follows:

$$Pr(y_f \in Q|z = \tilde{z}) = \int f(x^*)p(x^*|z = \tilde{z}) \, \mathrm{d}x^*.$$

The quantity $f(x) \doteq \int_Q \mathcal{N}(y_f|\mu(x), \Sigma) \; dy_f$ can be computed directly, while the overall integral in dx^* requires numerical integration, for example via a naive Monte-Carlo approach,

$$\int \cong \frac{\sum_{i=1}^n f(x_i)}{n}, \quad x_i \sim p(x^*|z = \tilde{z}),$$

or by weighted sampling

$$\int \cong \frac{\sum_{i=1}^{n} w_i f(x_i)}{n}, \quad x_i \sim p(x^*|z = \tilde{z}),$$

weighted by the likelihood $w_i \propto p(z = \tilde{z}|x^* = x_i)$. A trade-off appears, in the sense that sophisticated models which take a long time to evaluate may not provide enough samples for the prediction to be statistically significant – however, they may make the prior $p(x^*)$ and covariance Σ^ϵ easier to specify.

Modelling climate with belief functions As we have seen, a number of issues arise in making climate inferences in the Bayesian framework.

Numerous assumptions are necessary to make calculations practical, which are only weakly related to the actual problem to be solved. Although the prior on *climates* is reduced to a prior $p(x^*)$ on the parameters of a climate *model*, there is no obvious way of picking the latter (confirming our criticism of Bayesian inference in the Introduction). In general, it is far easier to say what choices are definitively wrong (e.g. uniform priors) than to specify 'good' ones. Finally, significant parameter tuning is required (e.g. to set the values of Σ^ϵ, Σ^e and so on).

Formulating a climate prediction method based on the theory of belief functions will require quite a lot of work, but a few landmark features can already be identified, remembering our discussion of inference in Chapter 4:

- such a framework will have to avoid committing to priors $p(x^*)$ on the 'correct' climate model parameters;
- it will likely use a climate model as a parametric model to infer either a belief function on the space of climates $\mathcal{Y}$
- or a belief function on the space of parameters (e.g. covariances, etc.) of the distribution on $\mathcal{Y}$.

Question 33. Can a climate model allowing predictions about future climate be formulated in a belief function framework, in order to better describe the huge Knightian uncertainties involved?

17.3.2 Machine learning

Maximum entropy classifiers[91] are a widespread tool in machine learning. There, the Shannon entropy of a sought joint (or conditional) probability distribution $p(C_k|x)$ of observed data x and an associated class $C_k \in \mathcal{C} = \{C_1, \ldots, C_K\}$ is maximised, following the maximum entropy principle that the least informative distribution matching the available evidence should be chosen. Once one has picked a set of *feature functions*, chosen to efficiently encode the training information, the joint distribution is subject to the constraint that the empirical expectation of the feature functions equals the expectation associated with it.

Clearly, the two assumptions that (i) the training and test data come from the same probability distribution, and that (ii) the empirical expectation of the training

[91] http://www.kamalnigam.com/papers/maxent-ijcaiws99.pdf.

data is correct, and the model expectation should match it, are rather strong, and work against generalisation power. A more robust learning approach can instead be sought by generalising max-entropy classification in a belief function setting. We take the view here that a training set does not provide, in general, sufficient information to precisely estimate the joint probability distribution of the class and the data. We assume instead that a belief measure can be estimated, providing lower and upper bounds on the joint probability of the data and class.

As in the classical case, an appropriate measure of entropy for belief measures is maximised. In opposition to the classical case, however, the empirical expectation of the chosen feature functions is assumed to be merely 'compatible' with the lower and upper bounds associated with the sought belief measure. This leads to a constrained optimisation problem with inequality constraints, rather than equality ones, which needs to be solved by looking at the Karush-Kuhn-Tucker conditions.

Generalised max-entropy classifiers Given a training set in which each observation is attached to a class, namely

$$\mathcal{D} = \left\{ (x_i, y_i), i = 1, \ldots, N \middle| x_i \in X, y_i \in \mathcal{C} \right\}, \tag{17.51}$$

a set M of *feature maps* is designed,

$$\phi(x, C_k) = [\phi_1(x, C_k), \ldots, \phi_M(x, C_k)]', \tag{17.52}$$

whose values depend on both the object observed and its class. Each feature map $\phi_m : X \times \mathcal{C} \to \mathbb{R}$ is then a random variable, whose expectation is obviously

$$E[\phi_m] = \sum_{x,k} p(x, C_k)\phi_m(x, C_k). \tag{17.53}$$

In opposition, the *empirical* expectation of ϕ_m is

$$\hat{E}[\phi_m] = \sum_{x,k} \hat{p}(x, C_k)\phi_m(x, C_k), \tag{17.54}$$

where $\hat{p}$ is a histogram constructed by counting occurrences of the pair (x, C_k) in the training set:

$$\hat{p}(x, C_k) = \frac{1}{N} \sum_{(x_i, y_i) \in \mathcal{D}} \delta(x_i = x \wedge y_i = C_k).$$

The theoretical expectation (17.53) can be approximated by decomposing via Bayes' rule $p(x, C_k) = p(x)p(C_k|x)$, and approximating the (unknown) prior of the observations $p(x)$ by the empirical prior $\hat{p}$, i.e., the histogram of the observed values in the training set:

$$\tilde{E}[\phi_m] = \sum_{x,k} \hat{p}(x)p(C_k|x)\phi_m(x, C_k).$$

For values of x that are not observed in the training set, $\hat{p}(x)$ is obviously equal to zero.

Definition 146. *Given a training set (17.51) related to the problem of classifying $x \in X$ as belonging to one of the classes $\mathcal{C} = \{C_1, \ldots, C_K\}$, the* max-entropy classifier *is the conditional probability $p^*(C_k|x)$ which solves the following constrained optimisation problem:*

$$p^*(C_k|x) \doteq \arg \max_{p(C_k|x)} H_s(P), \tag{17.55}$$

where H_s is the traditional Shannon entropy, subject to

$$\tilde{E}_p[\phi_m] = \hat{E}[\phi_m] \quad \forall m = 1, \ldots, M. \tag{17.56}$$

The constraint (17.56) requires the classifier to be consistent with the empirical frequencies of the features in the training set, while seeking the least informative probability distribution that does so (17.55).

The solution of the maximum entropy classification problem (Definition 146) is the so-called *log-linear model*,

$$p^*(C_k|x) = \frac{1}{Z_\lambda(x)} e^{\sum_m \lambda_m \phi_m(x, C_k)}, \tag{17.57}$$

where $\lambda = [\lambda_1, \ldots, \lambda_M]'$ are the Lagrange multipliers associated with the linear constraints (17.56) in the above constrained optimisation problem, and $Z_\lambda(x)$ is a normalisation factor.

The classification function associated with (17.57) is

$$y(x) = \arg \max_k \sum_m \lambda_m \phi_m(x, C_k),$$

i.e., x is assigned to the class which maximises the linear combination of the feature functions with coefficients λ.

Generalised maximum entropy optimisation problem When generalising the maximum entropy optimisation problem (Definition 146) to the case of belief functions, we need to (i) choose an appropriate measure of entropy for belief functions as the objective function, and (ii) revisit the constraints that the (theoretical) expectations of the feature maps are equal to the empirical ones computed over the training set.

As for (i), in Chapter 4, Section 4.2 we reviewed the manifold generalisations of the Shannon entropy to the case of belief measures. As for (ii), it is sensible to require that the empirical expectation of the feature functions is bracketed by the lower and upper expectations associated with the sought belief function $Bel : 2^{X \times \mathcal{C}} \rightarrow [0, 1]$, namely

$$\sum_{(x, C_k) \in X \times \mathcal{C}} Bel(x, C_k) \phi_m(x, C_k) \leq \hat{E}[\phi_m] \leq \sum_{(x, C_k) \in X \times \mathcal{C}} Pl(x, C_k) \phi_m(x, C_k). \tag{17.58}$$

Note that this makes use only of the 2-monotonicity of belief functions, as only probability intervals on singleton elements of $X \times \mathcal{C}$ are considered.

Question 34. Should constraints of this form be enforced on all possible subsets $A \subset X \times \mathcal{C}$, rather than just singleton pairs (x, C_k)? This is linked to the question of what information a training set actually carries (see Section 17.3.3).

More general constraints would require extending the domain of feature functions to set values – we will investigate this idea in the near future.

We can thus pose the generalised maximum-entropy optimisation problem as follows [355].

Definition 147. *Given a training set (17.51) related to the problem of classifying* $x \in X$ *as belonging to one of the classes* $\mathcal{C} = \{C_1, \ldots, C_K\}$, *the* maximum random set entropy classifier *is the joint belief measure* $Bel^*(x, C_k) : 2^{X \times \mathcal{C}} \rightarrow [0, 1]$ *which solves the following constrained optimisation problem:*

$$Bel^*(x, C_k) \doteq \arg \max_{Bel(x,C_k)} H(Bel), \tag{17.59}$$

subject to the constraints

$$\sum_{x,k} Bel(x, C_k)\phi_m(x, C_k) \leq \hat{E}[\phi_m] \leq \sum_{x,k} Pl(x, C_k)\phi_m(x, C_k) \;\; \forall m = 1, \ldots, M, \tag{17.60}$$

where H is an appropriate measure of entropy for belief measures.

Question 35. An alternative course of action is to pursue the *least committed*, rather than the maximum-entropy, belief function satisfying the constraints (17.60). This is left for future work.

KKT conditions and sufficiency As the optimisation problem in Definition 147 involves inequality constraints (17.60), as opposed to the equality constraints of traditional max-entropy classifiers (17.56), we need to analyse the KKT (recall Definition 137) [949] necessary conditions for a belief function Bel to be an optimal solution to the problem.

Interestingly, the KKT conditions are also sufficient whenever the objective function f is concave, the inequality constraints g_i are continuously differentiable convex functions and the equality constraints h_j are affine.[92] This provides us with a way to tackle the generalised maximum entropy problem.

Concavity of the entropy objective function As for the objective function, in Section 4.2 we appreciated that several generalisations of the standard Shannon entropy to random sets are possible. It is a well-known fact that the Shannon entropy is a concave function of probability distributions, represented as vectors of probability

[92] More general sufficient conditions can be given in terms of *invexity* [109] requirements. A function $f : \mathbb{R}^n \rightarrow \mathbb{R}$ is 'invex' if there exists a vector-valued function g such that $f(x) - f(u) \geq g(x, u)\nabla f(u)$. Hanson [791] proved that if the objective function and constraints are invex with respect to the same function g, then the KKT conditions are sufficient for optimality.

values.[93] A number of other properties of concave functions are well known: any linear combination of concave functions is concave, a monotonic and concave function of a concave function is still concave, and the logarithm is a concave function. As we have also learned in Section 4.7.1, the transformations which map mass vectors to vectors of belief (and commonality) values are linear, as they can be expressed in the form of matrices. In particular, $\mathbf{bel} = BfrM\mathbf{m}$, where $BfrM$ is a matrix whose (A, B) entry is

$$BfrM(A,B) = \begin{cases} 1 & B \subseteq A, \\ 0 & \text{otherwise}. \end{cases}$$

The same can be said of the mapping $\mathbf{q} = QfrM\mathbf{m}$ between a mass vector and the associated commonality vector. As a consequence, belief, plausibility and commonality are all linear (and therefore concave) functions of a mass vector.

Using this matrix representation, it is easy to conclude that several of the entropies defined in Section 4.2 are indeed concave. In particular, Smets's specificity measure $H_t = \sum_A \log(1/Q(A))$ (4.13) is concave, as a linear combination of concave functions. Nguyen's entropy (4.9) $H_n = -\sum_A m(A)\log(m(A)) = H_s(m)$ is also concave as a function of m, as the Shannon entropy of a mass assignment. Dubois and Prade's measure $H_d = \sum_A m(A)\log(|A|)$ (4.12) is also concave with respect to m, as a linear combination of mass values. Straighforward applications of the Shannon entropy function to Bel and Pl,

$$H_{Bel}[m] = H_s[Bel] = \sum_{A \subseteq \Theta} Bel(A) \log\left(\frac{1}{Bel(A)}\right) \tag{17.61}$$

and

$$H_{Pl}[m] = H_s[Pl] = \sum_{A \subseteq \Theta} Pl(A) \log\left(\frac{1}{Pl(A)}\right),$$

are also trivially concave, owing to the concavity of the entropy function and to the linearity of the mapping from m to Bel, Pl.

Drawing conclusions about the other measures recalled in Section 4.2 is less immediate, as they involve products of concave functions (which are not, in general, guaranteed to be concave).

Convexity of the interval expectation constraints As for the constraints (17.60) of the generalised maximum entropy problem, we first note that (17.60) can be decomposed into the following pair of constraints:

$$g^1_m(m) \doteq \sum_{x,k} Bel(x, C_k)\phi_m(x, C_k) - \hat{E}[\phi_m] \leq 0,$$

$$g^2_m(m) = \sum_{x,k} \phi_m(x, C_k)[\hat{p}(x, C_k) - Pl(x, C_k)] \leq 0,$$

[93] `http://projecteuclid.org/euclid.lnms/1215465631`

for all $m = 1, \ldots, M$. The first inequality constraint is a linear combination of linear functions of the sought mass assignment $m^* : 2^{X \times \mathcal{C}} \to [0, 1]$ (since Bel^* results from applying a matrix transformation to m^*). As $\mathbf{pl} = 1 - J\mathbf{bel} = 1 - JBfrM\mathbf{m}$, the constraint on g_m^2 is also a linear combination of mass values. Hence, as linear combinations, the constraints on g_m^1 and g_m^2 are both concave and convex.

KKT conditions for the generalised maximum entropy problem We can conclude the following.

Theorem 115. *If either H_t, H_n, H_d, H_{Bel} or H_{Pl} is adopted as the measure of entropy, the generalised maximum entropy optimisation problem (Definition 147) has a concave objective function and convex constraints. Therefore, the KKT conditions are sufficient for the optimality of its solution(s).*

Let us then compute the KKT conditions (see Definition 137) for the Shannon-like entropy (17.61). Condition 1 (stationarity), applied to the sought optimal belief function $Bel^* : 2^{X \times \mathcal{C}} \to [0, 1]$ reads as

$$\nabla H_{Bel}(Bel^*) = \sum_{m=1}^{M} \mu_m^1 \nabla g_m^1(Bel^*) + \mu_m^2 \nabla g_m^2(Bel^*).$$

The components of $\nabla H_{Bel}(Bel)$ are the partial derivatives of the entropy with respect to the mass values $m(\overline{B})$, for all $\overline{B} \subseteq \Theta = X \times \mathcal{C}$. They read as

$$\begin{aligned}
\frac{\partial H_{Bel}}{\partial m(\overline{B})} &= \frac{\partial}{\partial m(\overline{B})} \sum_{A \supseteq \overline{B}} \left[- \left(\sum_{B \subseteq A} m(B) \right) \log \left(\sum_{B \subseteq A} m(B) \right) \right] \\
&= - \sum_{A \supseteq \overline{B}} \left[1 \cdot \log \left(\sum_{B \subseteq A} m(B) \right) + \left(\sum_{B \subseteq A} m(B) \right) \frac{1}{\left(\sum_{B \subseteq A} m(B) \right)} \right] \\
&= - \sum_{A \supseteq \overline{B}} \left[1 + \log \left(\sum_{B \subseteq A} m(B) \right) \right] = - \sum_{A \supseteq \overline{B}} [1 + \log Bel(A)].
\end{aligned}$$

For $\nabla g_m^1(Bel)$, we easily have

$$\begin{aligned}
\frac{\partial g_m^1}{\partial m(\overline{B})} &= \frac{\partial}{\partial m(\overline{B})} \sum_{(x, C_k) \in \Theta} Bel(x, C_k) \phi_m(x, C_k) - \hat{E}[\phi_m] \\
&= \frac{\partial}{\partial m(\overline{B})} \sum_{(x, C_k) \in \Theta} m(x, C_k) \phi_m(x, C_k) - \hat{E}[\phi_m] \qquad (17.62) \\
&= \begin{cases} \phi_m(x, C_k) & \overline{B} = \{(x, C_k)\}, \\ 0 & \text{otherwise.} \end{cases}
\end{aligned}$$

Note that, if we could define feature functions over non-singleton subsets $A \subseteq \Theta$, (17.62) would simply generalise to

$$\frac{\partial g^1_m}{\partial m(\overline{B})} = \phi(\overline{B}) \quad \forall \overline{B} \subseteq \Theta.$$

As for the second set of constraints,

$$\begin{aligned}
\frac{\partial g^2_m}{\partial m(\overline{B})} &= \frac{\partial}{\partial m(\overline{B})} \sum_{(x,C_k)\in\Theta} \phi_m(x,C_k)[\hat{p}(x,C_k) - Pl(x,C_k)] \\
&= \frac{\partial}{\partial m(\overline{B})} \sum_{(x,C_k)\in\Theta} \phi_m(x,C_k) \left[\hat{p}(x,C_k) - \sum_{B\cap\{(x,C_k)\}\neq\emptyset} m(B)\right] \\
&= \frac{\partial}{\partial m(\overline{B})} \left(- \sum_{(x,C_k)\in\Theta} \phi_m(x,C_k) \sum_{B\supseteq\{(x,C_k)\}} m(B) \right) \\
&= - \sum_{(x,C_k)\in\overline{B}} \phi_m(x,C_k).
\end{aligned}$$

Putting everything together, the KKT stationarity conditions for the generalised maximum entropy problem amount to the following system of equations:

$$\begin{cases}
- \sum_{A\supseteq\overline{B}} [1 + \log Bel(A)] = \sum_{m=1}^{M} \phi_m(\overline{x},\overline{C}_k)[\mu^1_m - \mu^2_m], \quad \overline{B} = \{(\overline{x},\overline{C}_k)\}, \\
- \sum_{A\supseteq\overline{B}} [1 + \log Bel(A)] = \sum_{m=1}^{M} \mu^2_m \sum_{(x,C_k)\in\overline{B}} \phi_m(x,C_k), \ \overline{B} \subset X \times \mathcal{C}, |\overline{B}| > 1.
\end{cases} \tag{17.63}$$

The other conditions are (17.60) (primal feasibility), $\mu^1_m, \mu^2_m \geq 0$ (dual feasibility) and complementary slackness,

$$\begin{aligned}
\mu^1_m \left[\sum_{(x,C_k)\in\Theta} Bel(\{(x,C_k)\})\phi_m(x,C_k) - \hat{E}[\phi_m] \right] &= 0, \\
\mu^2_m \left[\sum_{(x,C_k)\in\Theta} \phi_m(x,C_k) \left[\hat{p}(x,C_k) - Pl(\{(x,C_k)\})\right] \right] &= 0
\end{aligned} \tag{17.64}$$

In the near future we will address the analytical form of the solution of the above system of constraints, and its alternative versions associated with different entropy measures.

Question 36. Derive the analytical form of the solution to the above generalised maximum entropy classification problem.

Question 37. How does this analytical solution compare with the log-linear model solution to the traditional maximum entropy problem?

Question 38. Derive and compare the constraints and analytical solutions for the alternative generalised maximum entropy frameworks obtained by plugging in other concave generalised entropy functions.

17.3.3 Generalising statistical learning theory

Current machine learning paradigms focus on explaining the observable outputs in the training data ('overfitting'), which may, for instance, lead an autonomous driving system to perform well in validation training sessions but fail catastrophically when tested in the real world. Methods are often designed to handle a specific set of testing conditions, and thus are unable to predict how a system will behave in a radically new setting (e.g., how would a smart car [671] cope with driving in extreme weather conditions?). With the pervasive deployment of machine learning algorithms in 'mission-critical' artificial intelligence systems for which failure is not an option (e.g. smart cars navigating a complex, dynamic environment, robot surgical assistants capable of predicting the surgeon's needs even before they are expressed, or smart warehouses monitoring the quality of the work conducted: see Fig. 17.16), it is imperative to ensure that these algorithms behave predictably 'in the wild'.

Vapnik's classical *statistical learning theory* [1849, 1851, 1850] is effectively useless for model selection, as the bounds on generalisation errors that it predicts are too wide to be useful, and rely on the assumption that the training and testing data come from the same (unknown) distribution. As a result, practitioners regularly resort to cross-validation[94] to approximate the generalisation error and identify the parameters of the optimal model.

PAC learning Statistical learning theory [1849] considers the problem of predicting an output $y \in \mathcal{Y}$ given an input $x \in \mathcal{X}$, by means of a mapping $h : \mathcal{X} \to \mathcal{Y}, h \in \mathcal{H}$ called a *model* or *hypothesis*, which lives in a specific model space $\mathcal{H}$. The error committed by a model is measured by a *loss function* $l : (\mathcal{X} \times \mathcal{Y}) \times \mathcal{H} \to \mathbb{R}$, for instance the zero–one loss $l((x, y), h) = \mathbb{I}[y \neq h(x)]$. The input–output pairs are assumed to be generated a probability distribution p^*.

The *expected risk* of a model h,

$$L(h) \doteq \mathbb{E}_{(x,y)\sim p^*}[l((x, y), h)], \tag{17.65}$$

is measured as its expected loss l assuming that the pairs $(x_1, y_1), (x_2, y_2), \ldots$ are sampled i.i.d. from the probability distribution p^*. The *expected risk minimiser*,

[94] https://en.wikipedia.org/wiki/Cross-validation_(statistics).

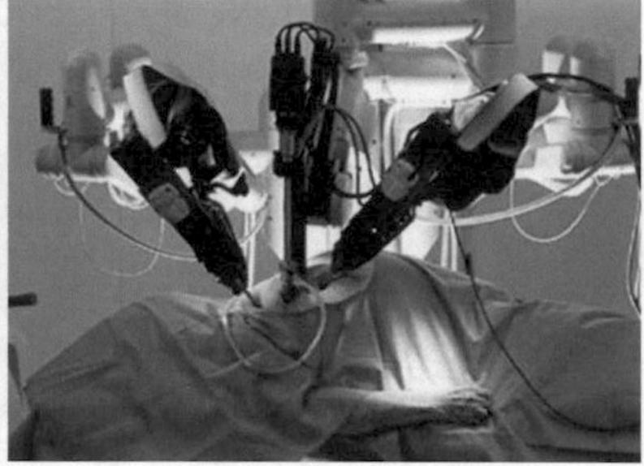

Fig. 17.16: New applications of machine learning to unconstrained environments (e.g. smart vehicles, left, or robotic surgical assistants, right) require new, robust foundations for statistical learning theory.

$$h^* \doteq \arg\min_{h \in \mathcal{H}} L(h), \tag{17.66}$$

is any hypothesis in the given model space that minimises the expected risk (not a random quantity at all). Given n training examples, also drawn i.i.d. from p^*, the *empirical risk* of a hypothesis h is the average loss over the available training set $\mathcal{D} = \{(x_1, y_1), \ldots, (x_n, y_n)\}$:

$$\hat{L}(h) \doteq \frac{1}{n} \sum_{i=1}^{n} l((x_i, y_i), h). \tag{17.67}$$

We can then define the *empirical risk minimiser* (ERM), namely

$$\hat{h} \doteq \arg\min_{h \in \mathcal{H}} \hat{L}(h), \tag{17.68}$$

which is instead a random variable which depends on the collection of training data.

Statistical learning theory is interested in the expected risk of the ERM (which is the only thing we can compute from the available training set). The core notion is that of *probably approximately correct* algorithms.

Definition 148. *A learning algorithm is* probably approximately correct *(PAC) if it finds with probability at least* $1-\delta$ *a model* $h \in \mathcal{H}$ *which is 'approximately correct', i.e., it makes a training error of no more than* ϵ.

According to this definition, PAC learning aims at providing bounds of the kind

$$P[L(\hat{h}) - L(h^*) > \epsilon] \leq \delta,$$

on the difference between the loss of the ERM and the minimum theoretical loss for that class of models. This is to account for the generalisation problem, due to the fact that the error we commit when training a model is different from the error one can expect in general cases in which data we have not yet observed are presented to the system.

As we will see in more detail later on, statistical learning theory makes predictions about the reliability of a training set, based on simple quantities such as the number of samples n. In particular, the main result of PAC learning, in the case of finite model spaces $\mathcal{H}$, is that we can relate the required size n of a training sample to the size of the model space $\mathcal{H}$, namely

$$\log|\mathcal{H}| \leq n\epsilon - \log\frac{1}{\delta},$$

so that the minimum number of training examples given ϵ, δ and $|\mathcal{H}|$ is

$$n \geq \frac{1}{\epsilon}\left(\log|\mathcal{H}| + \log\frac{1}{\delta}\right).$$

Vapnik–Chervonenkis dimension Obviously, for infinite-dimensional hypothesis spaces $\mathcal{H}$, the previous relation does not make sense. However, a similar constraint on the number of samples can be obtained after we introduce the following notion.

Definition 149. *The* Vapnik–Chervonenkis (VC) dimension[95] *of a model space $\mathcal{H}$ is the maximum number of points that can be successfully 'shattered' by a model $h \in \mathcal{H}$ (i.e., they can be correctly classified by some $h \in \mathcal{H}$ for all possible binary labellings of these points).*

Unfortunately, the VC dimension dramatically overestimates the number of training instances required. Nevertheless, arguments based on the VC dimension provide the only justification for max-margin linear SVMs, the dominant force in classification for the last two decades, before the advent of deep learning. As a matter of fact, for the space $\mathcal{H}_m$ of linear classifiers with margin m, the following expression holds:

$$VC_{\text{SVM}} = \min\left\{d, \frac{4R^2}{m^2}\right\} + 1,$$

where d is the dimensionality of the data and R is the radius of the smallest hypersphere enclosing all the training data. As the VC dimension of $\mathcal{H}_m$ decreases as m grows, it is then desirable to select the linear boundaries which have maximum margin, as in support vector machines.

Towards imprecise-probabilistic foundations for machine learning These issues with Vapnik's traditional statistical learning theory have recently been recognised by many scientists.

There is exciting research exploring a variety of options, for instance risk-sensitive reinforcement learning [1622]; approaches based on situational awareness, i.e., the ability to perceive the current status of an environment in terms of time and space, comprehend its meaning and behave with a corresponding risk attitude; robust optimisation using minimax learning [2118]; and budgeted adversary models [964]. The latter, rather than learning models that predict accurately on a target distribution, use minimax optimisation to learn models that are

[95] `http://www.svms.org/vc-dimension/`.

suitable for any target distribution within a 'safe' family. A portfolio of models have been proposed [231], including Satzilla [1998] and IBM Watson (`http://www.ibm.com/watson/`), which combines more than 100 different techniques for analysing natural language, identifying sources, finding and generating hypotheses, finding and scoring evidence, and merging and ranking hypotheses. By an ensemble of models, however, these methods mean sets of models of different nature (e.g. SVMs, random forests) or models of the same kind learned from different slices of data (e.g. boosting). Exciting recent progress in domain adaptation includes multiview learning [170], multifaceted understanding [174] and learning of transferable components [703, 1211].

Some consensus exists that robust approaches should provide worst-case guarantees, as it is not possible to rule out completely unexpected behaviours or catastrophic failures. The minimax approach, in particular, evokes the concept of imprecise probability, at least in its credal, robust-statistical incarnation. As we know, imprecise probabilities naturally arise whenever the data are insufficient to allow the estimation of a probability distribution. As we discussed in the Introduction to this book, training sets in virtually all applications of machine learning constitute a glaring example of data which is both:

- *insufficient in quantity* (think of a Google routine for object detection from images, trained on a few million images, compared with the thousands of billions of images out there), and
- *insufficient in quality* (as they are selected based on criteria such as cost, availability or mental attitudes, therefore biasing the whole learning process).

Uncertainty theory therefore has the potential to address model adaptation in new, original ways, so providing a new paradigm for robust statistical learning. A sensible programme for future research could then be based on the following developments (see Fig. 17.17):

1. Addressing the domain adaptation problem by allowing the test data to be sampled from a different probability distribution than the training data, under the weaker assumption that both belong to the same convex set of distributions (credal set) (Fig. 17.17(a)).
2. Extending further the statistical-learning-theory framework by moving away from the selection of a single model from a class of hypotheses to the identification of a convex set of models induced by the available training data (Fig. 17.17(b)).
3. Applying the resulting convex-theoretical learning theory (with respect to both the data-generating probabilities and the model space) to the functional spaces associated with convolutional neural networks (e.g., series of convolutions and max-pooling operations, Fig. 17.17(c)) in order to lay solid theoretical foundations for deep learning.

Bounds for realisable finite hypothesis classes To conclude the chapter, we consider the domain adaptation issue in the case of realisable finite hypothesis classes, and outline a credal solution for it.

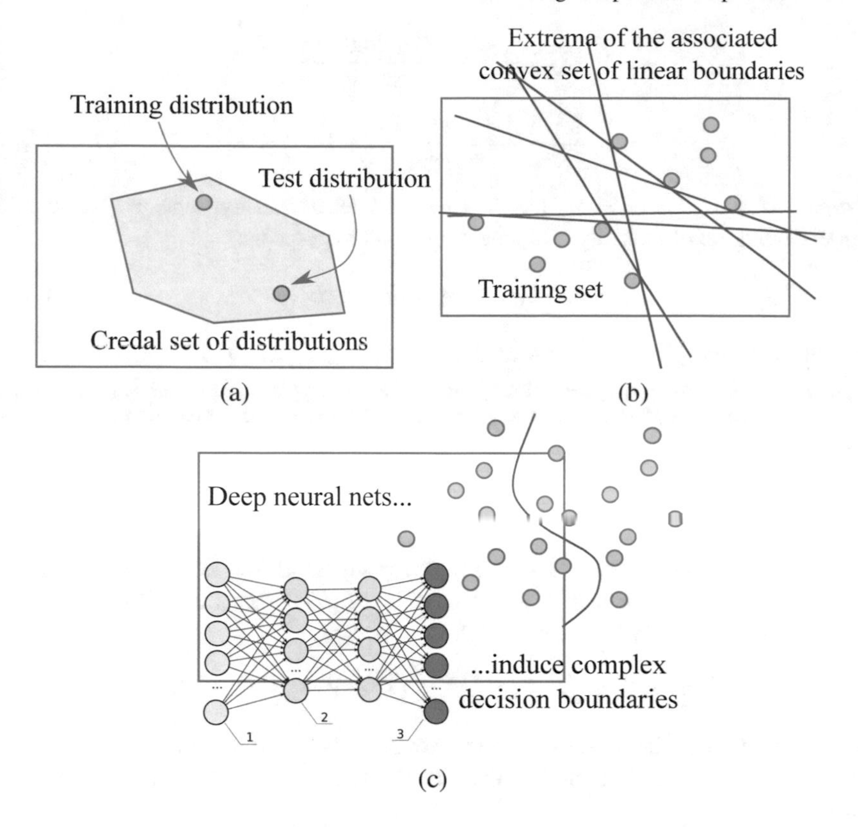

Fig. 17.17: (a) Training and test distributions from a common random set. (b) A training set of separable points determines only a convex set of admissible linear boundaries. (c) Going against conventional wisdom, deep neural models have very good generalisation power while generating, at the same time, complex decision boundaries.

Under the assumption that (1) the model space $\mathcal{H}$ is finite, and (2) there exists a hypothesis $h^* \in \mathcal{H}$ that obtains zero expected risk, that is,

$$L(h^*) = \mathbb{E}_{(x,y)\sim p^*}[l((x,y),h^*)] = 0, \tag{17.69}$$

a property called *realisability*, the following result holds[96].

Theorem 116. *Let $\mathcal{H}$ be a hypothesis class, where each hypothesis $h \in \mathcal{H}$ maps some $\mathcal{X}$ to $\mathcal{Y}$; let l be the zero–one loss, $l((x,y),h) = \mathbb{I}[y \neq h(x)]$; let p^* be any distribution over $\mathcal{X} \times \mathcal{Y}$; and let $\hat{h}$ be the empirical risk minimiser (17.68). Assume that (1) and (2) hold. Then the following two equivalent statements hold, with probability at least $1-\delta$:*

[96] https://web.stanford.edu/class/cs229t/notes.pdf.

$$L(\hat{h}) \le \frac{\log|\mathcal{H}| + \log(1/\delta)}{n}, \tag{17.70}$$

$$n \ge \frac{\log|\mathcal{H}| + \log(1/\delta)}{\epsilon} \Rightarrow L(\hat{h}) \le \epsilon. \tag{17.71}$$

Proof. Let $B = \{h \in \mathcal{H} : L(h) > \epsilon\}$ be the set of bad hypotheses. We wish to upper-bound the probability[97] of selecting a bad hypothesis:

$$P[L(\hat{h}) > \epsilon] = P[\hat{h} \in B]. \tag{17.72}$$

Recall that the empirical risk of the ERM is always zero, i.e., $\hat{L}(\hat{h}) = 0$, since at least $\hat{L}(h^*) = L(h^*) = 0$. So if we have selected a bad hypothesis ($\hat{h} \in B$), then some bad hypothesis must have zero empirical risk: $\hat{h} \in B$ implies $\exists h \in B : \hat{L}(h) = 0$, which is equivalent to saying that

$$P[\hat{h} \in B] \le P[\exists h \in B : \hat{L}(h) = 0]. \tag{17.73}$$

Firstly, we need to bound $P[\hat{L}(h) = 0]$ for any fixed $h \in B$. For each example, hypothesis h does not err, with probability $1 - L(h)$. Since the training examples are i.i.d. and since $L(h) > \epsilon$ for $h \in B$,

$$P[\hat{L}(h) = 0] = (1 - L(h))^n \le (1 - \epsilon)^n \le e^{-\epsilon n} \quad \forall h \in B,$$

where the first equality comes from the examples being i.i.d., and the last step follows since $1 - a \le e^{-a}$. Note that the probability $P[\hat{L}(h) = 0]$ decreases exponentially with n.

Secondly, we need to show that the above bound holds simultaneously for all $h \in B$. Recall that $P(A_1 \cup \cdots \cup A_K) \le P(\cup_{k=1}^{k} A_i)$. When applied to the (non-disjoint) events $A_h = \{\hat{L}(h) = 0\}$, the union bound yields

$$P[\exists h \in B : \hat{L}(h) = 0] \le \sum_{h \in B} P[\hat{L}(h) = 0].$$

Finally,

$$P[L(\hat{h}) \le \epsilon] = P[\hat{h} \in B] \le P[\exists h \in B : \hat{L}(h) = 0] \le \sum_{h \in B} P[\hat{L}(h) = 0]$$
$$\le |B| e^{-en} \le |\mathcal{H}| e^{-en} \doteq \delta,$$

and by rearranging we have $\epsilon = (\log|\mathcal{H}| + \log(1/\delta))/n$, i.e., (17.70).

The inequality (17.70) should be interpreted in the following way: with probability at least $1-\delta$, the expected loss of the empirical risk minimiser $\hat{h}$ (the particular model selected after training on n examples) is bounded by the ratio on the right-hand side, so that $L(\hat{h}) = O(\log|\mathcal{H}|/n)$. The inequality (17.70) amounts to saying

[97] Note that this is a probability measure on the space of models $\mathcal{H}$, completely distinct from the data-generating probability p^*.

that if I need to obtain an expected risk of at most ϵ with confidence at least $1-\delta$, I need to train the model on at least $(\log|\mathcal{H}| + \log(1/\delta))/\epsilon$ examples.

The result is *distribution-free*, as it is independent of the choice of $p^*(x,y)$. However, it does rely on the assumption that the training and test distributions are the same.

Bounds under credal generalisation A credal generalisation of Theorem 116 would thus read as follows.

Theorem 117. *Let $\mathcal{H}$ be a finite hypothesis class, let $\mathcal{P}$ be a credal set over $\mathcal{X} \times \mathcal{Y}$ and let a training set of samples $\mathcal{D} = \{(x_i, y_i), i = 1, \ldots, n\}$ be drawn from one of the distributions $\hat{p} \in \mathcal{P}$ in this credal set. Let*

$$\hat{h} \doteq \arg\min_{h \in \mathcal{H}} \hat{L}(h) = \arg\min_{h \in \mathcal{H}} \frac{1}{n} \sum_{i=1}^{n} l((x_i, y_i), h)$$

be the empirical risk minimiser, where l is the 0–1 loss. Assume that

$$\exists h^* \in \mathcal{H}, p^* \in \mathcal{P} : \mathbb{E}_{p^*}[l] = L_{p^*}(h^*) = 0 \tag{17.74}$$

(credal realisability) *holds.*

Then, with probability at least $1-\delta$, the following bound holds:

$$P\left[\max_{p \in \mathcal{P}} L_p(\hat{h}) > \epsilon\right] \leq \epsilon(\mathcal{H}, \mathcal{P}, \delta), \tag{17.75}$$

where ϵ is a function of the size of the model space $\mathcal{H}$, the size of the credal set $\mathcal{P}$ and δ.

How does the proof of Theorem 116 generalise to the credal case? We can note that, if we define

$$B \doteq \left\{h : \max_{p \in \mathcal{P}} L_p(h) > \epsilon\right\}$$

as the set of models which do not guarantee the bound *in the worst case*, we get, just as in the classical result (17.72),

$$P\left[\max_{p \in \mathcal{P}} L_p(\hat{h}) > \epsilon\right] = P[\hat{h} \in B].$$

However, whereas in the classical case $\hat{L}(\hat{h}) = 0$, in the credal case we can only say that

$$\hat{L}(h^*) = L_{\hat{p}}(h^*) \geq \min_{p \in \mathcal{P}} L_p(h^*) = 0,$$

by the credal realisability assumption (17.74). Thus, (17.73) does not hold, and the above argument which tries to bound $P[\hat{L}(h) = 0]$ does not apply here.

One option is to assume that $\hat{p} = p^* = \arg\min_p L_p(h^*)$. This would allow (17.73) to remain valid.

A different generalisation of realisability can also be proposed, in place of (17.74):

$$\forall p \in \mathcal{P}, \exists h^*_p \in \mathcal{H} : \mathbb{E}_p[l(h^*_p)] = L_p(h^*_p) = 0, \tag{17.76}$$

which we can call *uniform credal realisability*. Under the latter assumption, we would get

$$\hat{L}(\hat{h}) = \min_{h \in \mathcal{H}} \hat{L}(h) = \min_{h \in \mathcal{H}} L_{\hat{p}}(h) = L_{\hat{p}}(h^*_{\hat{p}}) = 0.$$

We will complete this analysis and extend it to the case of infinite, non-realisable model spaces in the near future.

References

1. Joaquín Abellán, *Combining nonspecificity measures in Dempster–Shafer theory of evidence*, International Journal of General Systems **40** (2011), no. 6, 611–622.
2. Joaquín Abellán and Andrés Masegosa, *Requirements for total uncertainty measures in DempsterShafer theory of evidence*, International Journal of General Systems **37** (2008), no. 6, 733–747.
3. Joaquín Abellán and Serafín Moral, *Completing a total uncertainty measure in the Dempster–Shafer theory*, International Journal of General Systems **28** (1999), no. 4-5, 299–314.
4. ______, *Building classification trees using the total uncertainty criterion*, International Journal of Intelligent Systems **18** (2003), no. 12, 1215–1225.
5. ______, *Corrigendum: A non-specificity measure for convex sets of probability distributions*, International Journal of Uncertaintainty, Fuzziness and Knowledge-Based Systems **13** (August 2005), no. 4, 467–467.
6. ______, *A non-specificity measure for convex sets of probability distributions*, International Journal of Uncertaintainty, Fuzziness and Knowledge-Based Systems **8** (June 2000), no. 3, 357–367.
7. ______, *Maximum of entropy for credal sets*, International Journal of Uncertainty, Fuzziness and Knowledge-Based Systems **11** (October 2003), no. 5, 587–597.
8. Joaqun Abellán and Manuel Gómez, *Measures of divergence on credal sets*, Fuzzy Sets and Systems **157** (2006), no. 11, 1514–1531.
9. Tomasz Adamek and Noel E. O'Connor, *Using Dempster–Shafer theory to fuse multiple information sources in region-based segmentation*, Proceedings of the IEEE International Conference on Image Processing (ICIP 2007), vol. 2, September 2007, pp. 269–272.
10. Ernest W. Adams, *Probability and the logic of conditionals*, Aspects of Inductive Logic (Jaakko Hintikka and Patrick Suppes, eds.), Studies in Logic and the Foundations of Mathematics, vol. 43, Elsevier, 1966, pp. 265–316.
11. Michael Aeberhard, Sascha Paul, Nico Kaempchen, and Torsten Bertram, *Object existence probability fusion using Dempster–Shafer theory in a high-level sensor data fusion architecture*, Proceedings of the IEEE Intelligent Vehicles Symposium (IV), June 2011, pp. 770–775.
12. Hamzeh Agahi, Adel Mohammadpour, Radko Mesiar, and Yao Ouyang, *On a strong law of large numbers for monotone measures*, Statistics and Probability Letters **83** (2013), no. 4, 1213–1218.
13. Mohammad R. Ahmadzadeh and Maria Petrou, *An expert system with uncertain rules based on Dempster–Shafer theory*, Proceedings of the IEEE International Geoscience and Remote Sensing Symposium (IGARSS '01), vol. 2, 2001, pp. 861–863.
14. ______, *Use of Dempster–Shafer theory to combine classifiers which use different class boundaries*, Pattern Analysis and Applications **6** (2003), no. 1, 41–46.

F. Cuzzolin, *The Geometry of Uncertainty*, Artificial Intelligence: Foundations, Theory, and Algorithms, https://doi.org/10.1007/978-3-030-63153-6

15. Mohammad R. Ahmadzadeh, Maria Petrou, and K. R. Sasikala, *The Dempster–Shafer combination rule as a tool to classifier combination*, Proceedings of the IEEE International Geoscience and Remote Sensing Symposium (IGARSS 2000), vol. 6, 2000, pp. 2429–2431.
16. Muhammad R. Ahmed, Xu Huang, and Dharmendra Sharma, *A novel misbehavior evaluation with Dempster–Shafer theory in wireless sensor networks*, Proceedings of the Thirteenth ACM International Symposium on Mobile Ad Hoc Networking and Computing (MobiHoc '12), 2012, pp. 259–260.
17. Mikel Aickin, *Connecting Dempster–Shafer belief functions with likelihood-based inference*, Synthese **123** (2000), no. 3, 347–364.
18. Martin Aigner, *Combinatorial theory*, Classics in Mathematics, Springer, New York, 1979.
19. John Aitchinson, *Discussion on professor Dempster's paper*, Journal of the Royal Statistical Society, Series B **30** (1968), 234–237.
20. Ahmed Al-Ani and Mohamed Deriche, *A Dempster–Shafer theory of evidence approach for combining trained neural networks*, Proceedings of the IEEE International Symposium on Circuits and Systems (ISCAS 2001), vol. 3, 2001, pp. 703–706.
21. ______, *A new technique for combining multiple classifiers using the Dempster–Shafer theory of evidence*, Journal of Artificial Intelligence Research **17** (2002), 333–361.
22. ______, *A new technique for combining multiple classifiers using the Dempster–Shafer theory of evidence*, ArXiv preprint arXiv:1107.0018, June 2011.
23. M. M. Al-Fadhala and Bilal M. Ayyub, *Decision analysis using belief functions for assessing existing structures*, Proceedings of the International Conference on Structural Safety and Reliability (ICCOSSAR'01) (Ross B. Corotis, Gerhart I. Schueeller, and Masanobu Shinozuka, eds.), June 2001.
24. Bilal Al Momani, Sally McClean, and Philip Morrow, *Using Dempster–Shafer to incorporate knowledge into satellite image classification*, Artificial Intelligence Review **25** (2006), no. 1-2, 161–178.
25. David W. Albrecht, Ingrid Zukerman, Ann E. Nicholson, and Ariel Bud, *Towards a Bayesian model for keyhole plan recognition in large domains*, User Modeling (A. Jameson, C. Paris, and C. Tasso, eds.), International Centre for Mechanical Sciences, vol. 383, Springer, 1997, pp. 365–376.
26. Carlos E. Alchourrón, Peter Gärdenfors, and David Makinson, *On the logic of theory change: Partial meet contraction and revision functions*, Readings in Formal Epistemology: Sourcebook (Horacio Arló-Costa, Vincent F. Hendricks, and Johan van Benthem, eds.), Springer International Publishing, 2016, pp. 195–217.
27. Tazid Ali, Palash Dutta, and Hrishikesh Boruah, *A new combination rule for conflict problem of Dempster–Shafer evidence theory*, International Journal of Energy, Information and Communications **3** (2012), 35–40.
28. Russell Almond, *Fusion and propagation in graphical belief models*, Research Report S-121, Harvard University, Department of Statistics, 1988.
29. ______, *Belief function models for simple series and parallel systems*, Tech. Report 207, Department of Statistics, University of Washington, 1991.
30. ______, *Fiducial inference and belief functions*, Tech. report, University of Washington, 1992.
31. Russell G. Almond, *Fusion and propagation of graphical belief models: an implementation and an example*, PhD dissertation, Department of Statistics, Harvard University, 1990.

32. ______, *Graphical belief modeling*, Chapman and Hall/CRC, 1995.
33. Hakan Altınçay, *A Dempster–Shafer theoretic framework for boosting based ensemble design*, Pattern Analysis and Applications **8** (2005), no. 3, 287–302.
34. ______, *On the independence requirement in Dempster–Shafer theory for combining classifiers providing statistical evidence*, Applied Intelligence **25** (2006), no. 1, 73–90.
35. Hakan Altncay and Mubeccel Demirekler, *Speaker identification by combining multiple classifiers using Dempster–Shafer theory of evidence*, Speech Communication **41** (2003), no. 4, 531–547.
36. Diego A. Alvarez, *On the calculation of the bounds of probability of events using infinite random sets*, International Journal of Approximate Reasoning **43** (2006), no. 3, 241–267.
37. Farzad Aminravan, Rehan Sadiq, Mina Hoorfar, Manuel J. Rodriguez, Alex Francisque, and Homayoun Najjaran, *Evidential reasoning using extended fuzzy Dempster–Shafer theory for handling various facets of information deficiency*, International Journal of Intelligent Systems **26** (2011), no. 8, 731–758.
38. P. An and Wooil M. Moon, *An evidential reasoning structure for integrating geophysical, geological and remote sensing data*, Proceedings of the IEEE International Geoscience and Remote Sensing Symposium (IGARSS '93), vol. 3, August 1993, pp. 1359–1361.
39. Zhi An, *Relative evidential support*, PhD dissertation, University of Ulster, 1991.
40. Zhi An, David A. Bell, and John G. Hughes, *R∊S – A relative method for evidential reasoning*, Uncertainty in Artificial Intelligence, Elsevier, 1992, pp. 1–8.
41. ______, *REL – A logic for relative evidential support*, International Journal of Approximate Reasoning **8** (1993), no. 3, 205–230.
42. Zhi An, David A. Bell, and John G. Hughes, *Relation-based evidential reasoning*, International Journal of Approximate Reasoning **8** (1993), no. 3, 231–251.
43. Kim Allan Andersen and John N. Hooker, *A linear programming framework for logics of uncertainty*, Decision Support Systems **16** (1996), no. 1, 39–53.
44. Bernhard Anrig, Rolf Haenni, and Norbert Lehmann, *ABEL - a new language for assumption-based evidential reasoning under uncertainty*, Tech. Report 97-01, Institute of Informatics, University of Fribourg, 1997.
45. Francis J. Anscombe, *Book reviews: The foundations of statistics*, Journal of the American Statistical Association **51** (1956), no. 276, 657–659.
46. Violaine Antoine, Benjamin Quost, Marie-Hélène Masson, and Thierry Denœux, *CECM: Constrained evidential C-means algorithm*, Computational Statistics and Data Analysis **56** (2012), no. 4, 894–914.
47. ______, *Evidential clustering with constraints and adaptive metric*, Proceedings of the Workshop on the Theory of Belief Functions (BELIEF 2010), April 2010.
48. Alessandro Antonucci and Fabio Cuzzolin, *Credal sets approximation by lower probabilities: Application to credal networks*, Computational Intelligence for Knowledge-Based Systems Design (Eyke Hllermeier, Rudolf Kruse, and Frank Hoffmann, eds.), Lecture Notes in Computer Science, vol. 6178, Springer, Berlin Heidelberg, 2010, pp. 716–725.
49. Alain Appriou, *Probabilités et incertitude en fusion de données multi-senseurs*, Revue Scientifique et Technique de la Défense **11** (1991), 27–40.
50. ______, *Formulation et traitement de l'incertain en analyse multi-senseurs*, 14 Colloque sur le traitement du signal et des images, 1993.
51. ______, *Uncertain data aggregation in classification and tracking processes*, Aggregation and Fusion of Imperfect Information (Bernadette Bouchon-Meunier, ed.), Physica-Verlag HD, Heidelberg, 1998, pp. 231–260.

52. Oya Aran, Thomas Burger, Alice Caplier, and Lale Akarun, *A belief-based sequential fusion approach for fusing manual and non-manual signs*, Pattern Recognition **42** (2009), no. 5, 812–822.
53. ———, *Sequential belief-based fusion of manual and non-manual information for recognizing isolated signs*, Proceedings of the International Gesture Workshop, Springer, 2009, pp. 134–144.
54. Astride Aregui and Thierry Denœux, *Novelty detection in the belief functions framework*, Proceedings of the International Conference on Information Processing and Management of Uncertainty in Knowledge-Based Systems (IPMU 2006), vol. 6, 2006, pp. 412–419.
55. Astride Aregui and Thierry Denœux, *Consonant belief function induced by a confidence set of pignistic probabilities*, Symbolic and Quantitative Approaches to Reasoning with Uncertainty: 9th European Conference, ECSQARU 2007, Hammamet, Tunisia, October 31 - November 2, 2007. Proceedings (Khaled Mellouli, ed.), Springer, Berlin Heidelberg, 2007, pp. 344–355.
56. Astride Aregui and Thierry Denœux, *Constructing predictive belief functions from continuous sample data using confidence bands*, Proceedings of the Fifth International Symposium on Imprecise Probability: Theories and Applications (ISIPTA'07), 2007, pp. 11–20.
57. ———, *Fusion of one-class classifiers in the belief function framework*, Proceedings of the 10th International Conference on Information Fusion (FUSION 2007), 2007, pp. 1–8.
58. ———, *Constructing consonant belief functions from sample data using confidence sets of pignistic probabilities*, International Journal of Approximate Reasoning **49** (2008), no. 3, 575–594.
59. M. Arif, T. Brouard, and N. Vincent, *A fusion methodology based on Dempster–Shafer evidence theory for two biometric applications*, Proceedings of the 18th International Conference on Pattern Recognition (ICPR 2006), vol. 4, 2006, pp. 590–593.
60. Stephan K. Asare and A. Wright, *Hypothesis revision strategies in conducting analytical procedures*, Accounting, Organizations and Society **22** (1997), no. 8, 737–755.
61. Yuksel Alp Aslandogan and Clement T. Yu, *Evaluating strategies and systems for content based indexing of person images on the web*, Proceedings of the Eighth ACM International Conference on Multimedia (MULTIMEDIA '00), 2000, pp. 313–321.
62. Krassimir T. Atanassov, *Intuitionistic fuzzy sets*, Fuzzy Sets and Systems **20** (1986), no. 1, 87–96.
63. Will D. Atkinson and Alexander Gammerman, *An application of expert systems technology to biological identification*, Taxon **36** (1987), no. 4, 705–714.
64. Nii Attoh-Okine, Yaw Adu-Gyamfi, and Stephen Mensah, *Potential application of hybrid belief functions and Hilbert-Huang transform in layered sensing*, IEEE Sensors Journal **11** (2011), no. 3, 530–535.
65. Nii O. Attoh-Okine, *Aggregating evidence in pavement management decision-making using belief functions and qualitative Markov tree*, IEEE Transactions on Systems, Man, and Cybernetics, Part C (Applications and Reviews) **32** (August 2002), no. 3, 243–251.
66. Thomas Augustin, *Modeling weak information with generalized basic probability assignments*, Data Analysis and Information Systems - Statistical and Conceptual Approaches (H. H. Bock and W. Polasek, eds.), Springer, 1996, pp. 101–113.
67. Thomas Augustin, *Expected utility within a generalized concept of probability – a comprehensive framework for decision making under ambiguity*, Statistical Papers **43** (2002), no. 1, 5–22.

68. Thomas Augustin, *Generalized basic probability assignments*, International Journal of General Systems **34** (2005), no. 4, 451–463.
69. Anjali Awasthi and Satyaveer S. Chauhan, *Using AHP and Dempster–Shafer theory for evaluating sustainable transport solutions*, Environmental Modelling and Software **26** (2011), no. 6, 787–796.
70. André Ayoun and Philippe Smets., *Data association in multi-target detection using the transferable belief model*, International Journal of Intelligent Systems **16** (2001), no. 10, 1167–1182.
71. Birsel Ayrulu, Billur Barshan, and Simukai W. Utete, *Target identification with multiple logical sonars using evidential reasoning and simple majority voting*, Proceedings of the IEEE International Conference on Robotics and Automation (ICRA'97), vol. 3, April 1997, pp. 2063–2068.
72. Ha-Rok Bae, Ramana V. Grandhi, and Robert A. Canfield, *An approximation approach for uncertainty quantification using evidence theory*, Reliability Engineering and System Safety **86** (2004), no. 3, 215–225.
73. Dennis Bahler and Laura Navarro, *Methods for combining heterogeneous sets of classifiers*, Proceedings of the 17th National Conf. on Artificial Intelligence (AAAI), Workshop on New Research Problems for Machine Learning, 2000.
74. Christopher A. Bail, *Lost in a random forest: Using Big Data to study rare events*, Big Data and Society **2** (2015), no. 2, 1–3.
75. Jayanth Balasubramanian and Ignacio E. Grossmann, *Scheduling optimization under uncertainty – an alternative approach*, Computers and Chemical Engineering **27** (2003), no. 4, 469–490.
76. James F. Baldwin, *Support logic programming*, Technical report ITRC 65, Information Technology Research Center, University of Bristol, 1985.
77. James F. Baldwin, *Evidential support logic programming*, Fuzzy Sets and Systems **24** (1987), no. 1, 1–26.
78. James F. Baldwin, *Combining evidences for evidential reasoning*, International Journal of Intelligent Systems **6** (1991), no. 6, 569–616.
79. James F. Baldwin, *Fuzzy, probabilistic and evidential reasoning in Fril*, Proceedings of the Second IEEE International Conference on Fuzzy Systems, vol. 1, March 1993, pp. 459–464.
80. James F. Baldwin, *A calculus for mass assignments in evidential reasoning*, Advances in the Dempster-Shafer Theory of Evidence (Ronald R. Yager, Janusz Kacprzyk, and Mario Fedrizzi, eds.), John Wiley and Sons, Inc., 1994, pp. 513–531.
81. ———, *A theory of mass assignments for artificial intelligence*, Proceedings of the IJCAI '91 Workshops on Fuzzy Logic and Fuzzy Control, Sydney, Australia, August 24, 1991 (Dimiter Driankov, Peter W. Eklund, and Anca L. Ralescu, eds.), Springer, Berlin Heidelberg, 1994, pp. 22–34.
82. James F. Baldwin, *Towards a general theory of evidential reasoning*, Proceedings of the 3rd International Conference on Information Processing and Management of Uncertainty in Knowledge-Based Systems (IPMU'90) (B. Bouchon-Meunier, R. R. Yager, and L. A. Zadeh, eds.), July 1990, pp. 360–369.
83. James F. Baldwin, James P. Martin, and B. W. Pilsworth, *Fril- fuzzy and evidential reasoning in artificial intelligence*, John Wiley and Sons, Inc., New York, NY, USA, 1995.
84. Mohua Banerjee and Didier Dubois, *A simple modal logic for reasoning about revealed beliefs*, Proceedings of the 10th European Conference on Symbolic and Quantitative Approaches to Reasoning with Uncertainty (ECSQARU 2009), Verona, Italy

(Claudio Sossai and Gaetano Chemello, eds.), Springer, Berlin, Heidelberg, July 2009, pp. 805–816.
85. G. Banon, *Distinction between several subsets of fuzzy measures*, Fuzzy Sets and Systems **5** (1981), no. 3, 291–305.
86. Piero Baraldi and Enrico Zio, *A comparison between probabilistic and Dempster–Shafer theory approaches to model uncertainty analysis in the performance assessment of radioactive waste repositories*, Risk Analysis **30** (2010), no. 7, 1139–1156.
87. George A. Barnard, *Review of Wald's 'Sequential analysis'*, Journal of the American Statistical Association **42** (1947), no. 240, 658–669.
88. Jeffrey A. Barnett, *Computational methods for a mathematical theory of evidence*, Proceedings of the 7th National Conference on Artificial Intelligence (AAAI-88), 1981, pp. 868–875.
89. ______, *Calculating Dempster–Shafer plausibility*, IEEE Transactions on Pattern Analysis and Machine Intelligence **13** (1991), no. 6, 599–602.
90. Jonathan Baron, *Second-order probabilities and belief functions*, Theory and Decision **23** (1987), no. 1, 25–36.
91. Pietro Baroni, *Extending consonant approximations to capacities*, Proceedings of Information Processing with Management of Uncertainty in Knowledge-based Systems (IPMU'04), 2004, pp. 1127–1134.
92. Pietro Baroni and Paolo Vicig, *Transformations from imprecise to precise probabilities*, Proceedings of the European Conference on Symbolic and Quantitative Approaches to Reasoning with Uncertainty (ECSQARU'03), 2003, pp. 37–49.
93. ______, *An uncertainty interchange format with imprecise probabilities*, International Journal of Approximate Reasoning **40** (2005), no. 3, 147–180.
94. Pietro Baroni and Paolo Vicig, *On the conceptual status of belief functions with respect to coherent lower probabilities*, Proceedings of the 6th European Conference on Symbolic and Quantitative Approaches to Reasoning with Uncertainty (ECSQARU 2001), Toulouse, France (Salem Benferhat and Philippe Besnard, eds.), Springer, Berlin, Heidelberg, September 2001, pp. 328–339.
95. Jean-Pierre Barthélemy, *Monotone functions on finite lattices: An ordinal approach to capacities, belief and necessity functions*, Preferences and Decisions under Incomplete Knowledge (János Fodor, Bernard De Baets, and Patrice Perny, eds.), Physica-Verlag HD, Heidelberg, 2000, pp. 195–208.
96. Otman Basir, Fakhri Karray, and Hongwei Zhu, *Connectionist-based Dempster–Shafer evidential reasoning for data fusion*, IEEE Transactions on Neural Networks **16** (November 2005), no. 6, 1513–1530.
97. Diderik Batens, Chris Mortensen, Graham Priest, and Jean-Paul Van Bendegem, *Frontiers of paraconsistent logic*, Studies in logic and computation, vol. 8, Research Studies Press, 2000.
98. Cédric Baudrit, Arnaud Hélias, and Nathalie Perrot, *Joint treatment of imprecision and variability in food engineering: Application to cheese mass loss during ripening*, Journal of Food Engineering **93** (2009), no. 3, 284–292.
99. Mathias Bauer, *A Dempster–Shafer approach to modeling agent preferences for plan recognition*, User Modeling and User-Adapted Interaction **5** (1995), no. 3–4, 317–348.
100. ______, *Approximation algorithms and decision making in the Dempster–Shafer theory of evidence – An empirical study*, International Journal of Approximate Reasoning **17** (1997), no. 2-3, 217–237.
101. ______, *Approximations for decision making in the Dempster–Shafer theory of evidence*, Proceedings of the Twelfth Conference on Uncertainty in Artificial Intelli-

gence (UAI'96) (E. Horvitz and F. Jensen, eds.), Portland, OR, USA, 1-4 August 1996, pp. 73–80.

102. Kristopher R. Beevers and Wesley H. Huang, *Loop closing in topological maps*, Proceedings of the 2005 IEEE International Conference on Robotics and Automation (ICRA 2005), April 2005, pp. 4367–4372.
103. David A. Bell, Yaxin Bi, and Jiwen Guan, *On combining classifier mass functions for text categorization*, IEEE Transactions on Knowledge and Data Engineering **17** (2005), no. 10, 1307–1319.
104. David A. Bell, Jiwen Guan, and Suk Kyoon Lee, *Generalized union and project operations for pooling uncertain and imprecise information*, Data and Knowledge Engineering **18** (1996), no. 2, 89–117.
105. David A. Bell, Jiwen Guan, and G. M. Shapcott, *Using the Dempster–Shafer orthogonal sum for reasoning which involves space*, Kybernetes **27** (1998), no. 5, 511–526.
106. Amandine Bellenger and Sylvain Gatepaille, *Uncertainty in ontologies: Dempster–Shafer theory for data fusion applications*, Workshop on Theory of Belief Functions, 2010.
107. Yakov Ben-Haim, *Info-gap decision theory*, second edition ed., Academic Press, Oxford, 2006.
108. Sarra Ben Hariz, Zied Elouedi, and Khaled Mellouli, *Clustering approach using belief function theory*, Artificial Intelligence: Methodology, Systems, and Applications: 12th International Conference, AIMSA 2006, Varna, Bulgaria, September 12-15, 2006. Proceedings (Jérôme Euzenat and John Domingue, eds.), Springer, Berlin, Heidelberg, 2006, pp. 162–171.
109. Adi Ben-Israel and Bertram Mond, *What is invexity?*, Journal of the Australian Mathematical Society. Series B. Applied Mathematics **28** (1986), 1–9.
110. Amel Ben Yaghlane, Thierry Denœux, and Khaled Mellouli, *Coarsening approximations of belief functions*, Proceedings of the 6th European Conference on Symbolic and Quantitative Approaches to Reasoning and Uncertainty (ECSQARU-2001) (S. Benferhat and P. Besnard, eds.), 2001, pp. 362–373.
111. ———, *Elicitation of expert opinions for constructing belief functions*, Uncertainty and intelligent information systems, World Scientific, 2008, pp. 75–89.
112. Boutheina Ben Yaghlane and Khaled Mellouli, *Belief function propagation in directed evidential networks*, Proceedings of Information Processing with Management of Uncertainty in Knowledge-based Systems (IPMU'06), vol. 2, 2006, pp. 1451–1458.
113. ———, *Inference in directed evidential networks based on the transferable belief model*, International Journal of Approximate Reasoning **48** (2008), no. 2, 399–418.
114. Boutheina Ben Yaghlane, Philippe Smets, and Khaled Mellouli, *Independence concepts for belief functions*, Proceedings of International Conference on Information Processing and Management of Uncertainty in Knowledge-Based Systems (IPMU'2000), 2000.
115. ———, *Independence concepts for belief functions*, Technologies for Constructing Intelligent Systems 2: Tools (Bernadette Bouchon-Meunier, Julio Gutiérrez-Ríos, Luis Magdalena, and Ronald R. Yager, eds.), Physica-Verlag HD, Heidelberg, 2002, pp. 45–58.
116. ———, *Directed evidential networks with conditional belief functions*, Symbolic and Quantitative Approaches to Reasoning with Uncertainty: 7th European Conference, ECSQARU 2003 Aalborg, Denmark, July 2-5, 2003 Proceedings (Thomas Dyhre Nielsen and Nevin Lianwen Zhang, eds.), Springer, Berlin, Heidelberg, 2003, pp. 291–305.

117. Alessio Benavoli, *Belief function and multivalued mapping robustness in statistical estimation*, International Journal of Approximate Reasoning **55** (January 2014), no. 1, 311–329.
118. Alessio Benavoli, Luigi Chisci, Alfonso Farina, and Branko Ristic, *Modelling uncertain implication rules in evidence theory*, Proceedings of the 11th International Conference on Information Fusion (FUSION 2008), June 2008, pp. 1–7.
119. Alessio Benavoli, Alessandro Facchini, and Marco Zaffalon, *Bayes + Hilbert = Quantum Mechanics*, ArXiv preprint arXiv:1605.08177, 2016.
120. Alessio Benavoli, Alessandro Facchini, and Marco Zaffalon, *Bayes + Hilbert = Quantum Mechanics*, Proceedings of the 14th International Conference on Quantum Physics and Logic (QPL 2017), 2017.
121. Kent Bendall, *Belief-theoretic formal semantics for first-order logic and probability*, Journal of Philosophical Logic **8** (1979), no. 1, 375–397.
122. Azzeddine Bendjebbour and Wojciech Pieczynski, *Unsupervised image segmentation using Dempster–Shafer fusion in a Markov fields context*, Proceedings of the International Conference on Multisource-Multisensor Information Fusion (FUSION'98) (R. Hamid, A. Zhu, and D. Zhu, eds.), vol. 2, Las Vegas, NV, USA, 6-9 July 1998, pp. 595–600.
123. Azzedine Bendjebbour, Yves Delignon, Laurent Fouque, Vincent Samson, and Wojciech Pieczynski, *Multisensor image segmentation using Dempster–Shafer fusion in Markov fields context*, IEEE Transactions on Geoscience and Remote Sensing **39** (2001), no. 8, 1789–1798.
124. Salem Benferhat, Alessandro Saffiotti, and Philippe Smets, *Belief functions and default reasoning*, Proceedings of the 11th International Joint Conference on Artificial intelligence (IJCAI-89), 1995, pp. 19–26.
125. ———, *Belief functions and default reasoning*, Tech. Report TR/IRIDIA/95-5, Université Libre de Bruxelles, 1995.
126. M. Bengtsson and J. Schubert, *Dempster–Shafer clustering using Potts spin mean field theory*, Soft Computing **5** (2001), no. 3, 215–228.
127. Layachi Bentabet, Yue Min Zhu, Olivier Dupuis, Valérie Kaftandjian, Daniel Babot, and Michèle Rombaut, *Use of fuzzy clustering for determining mass functions in Dempster–Shafer theory*, Proceedings of the 5th International Conference on Signal Processing (WCCC-ICSP 2000), vol. 3, 2000, pp. 1462–1470.
128. Rudolf J. Beran, *On distribution-free statistical inference with upper and lower probabilities*, Annals of Mathematical Statistics **42** (1971), no. 1, 157–168.
129. James O. Berger, *Robust Bayesian analysis: Sensitivity to the prior*, Journal of Statistical Planning and Inference **25** (1990), 303–328.
130. Ulla Bergsten and Johan Schubert, *Dempster's rule for evidence ordered in a complete directed acyclic graph*, International Journal of Approximate Reasoning **9** (1993), no. 1, 37–73.
131. Ulla Bergsten, Johan Schubert, and Per Svensson, *Applying data mining and machine learning techniques to submarine intelligence analysise*, Proceedings of the Third International Conference on Knowledge Discovery and Data Mining (KDD'97) (D. Heckerman, H. Mannila, D. Pregibon, and R. Uthurusamy, eds.), August 1997, pp. 127–130.
132. José M. Bernardo and Adrian F. M. Smith, *Bayesian theory*, Wiley, 1994.
133. Iacopo Bernetti, Christian Ciampi, Claudio Fagarazzi, and Sandro Sacchelli, *The evaluation of forest crop damages due to climate change. an application of Dempster–Shafer method*, Journal of Forest Economics **17** (2011), no. 3, 285–297.

134. Manfred Berres, *On a multiplication and a theory of integration for belief and plausibility functions*, Journal of Mathematical Analysis and Applications **121** (1987), no. 2, 487–505.
135. ———, *λ-additive measures on measure spaces*, Fuzzy Sets and Systems **27** (1988), no. 2, 159–169.
136. Roman Bertschy and Paul André Monney, *A generalization of the algorithm of Heidtmann to non-monotone formulas*, Journal of Computational and Applied Mathematics **76** (1996), no. 1-2, 55–76.
137. Philippe Besnard and Jürg Kohlas, *Evidence theory based on general consequence relations*, International Journal of Foundations of Computer Science **6** (1995), no. 2, 119–135.
138. Bernard Besserer, Stéphane Estable, and B. Ulmer, *Multiple knowledge sources and evidential reasoning for shape recognition*, Proceedings of the 4th International Conference on Computer Vision (ICCV'93), 1993, pp. 624–631.
139. Albrecht Beutelspacher and Ute Rosenbaum, *Projective geometry*, Cambridge University Press, Cambridge, 1998.
140. Truman F. Bewley, *Knightian decision theory. Part I*, Decisions in Economics and Finance **25** (2002), no. 2, 79–110.
141. Malcolm Beynon, *DS/AHP method: A mathematical analysis, including an understanding of uncertainty*, European Journal of Operational Research **140** (2002), no. 1, 148–164.
142. Malcolm Beynon, Darren Cosker, and David Marshall, *An expert system for multicriteria decision making using Dempster Shafer theory*, Expert Systems with Applications **20** (2001), no. 4, 357–367.
143. Malcolm Beynon, Bruce Curry, and Peter Morgan, *The Dempster–Shafer theory of evidence: an alternative approach to multicriteria decision modelling*, Omega **28** (2000), no. 1, 37–50.
144. Malcolm J. Beynon, *A method of aggregation in DS/AHP for group decision-making with the non-equivalent importance of individuals in the group*, Computers and Operations Research **32** (2005), no. 7, 1881–1896.
145. James C. Bezdek, *Pattern recognition with fuzzy objective function algorithms*, Kluwer Academic Publishers, 1981.
146. Deepika Bhalla, Raj Kumar Bansal, and Hari Om Gupta, *Integrating AI based DGA fault diagnosis using Dempster–Shafer theory*, International Journal of Electrical Power and Energy Systems **48** (2013), no. 0, 31–38.
147. Kamal K. Bharadwaj and G. C. Goel, *Hierarchical censored production rules (HCPRs) system employing the Dempster–Shafer uncertainty calculus*, Information and Software Technology **36** (1994), no. 3, 155–164.
148. Prabir Bhattacharya, *On the Dempster–Shafer evidence theory and non-hierarchical aggregation of belief structures*, IEEE Transactions on Systems, Man, and Cybernetics - Part A: Systems and Humans **30** (September 2000), no. 5, 526–536.
149. Anil Bhattacharyya, *On a measure of divergence between two multinomial populations*, Sankhy: The Indian Journal of Statistics (1933-1960) **7** (1946), no. 4, 401–406.
150. Yaxin Bi, *Combining multiple classifiers for text categorization using the Dempster–Shafer theory of evidence*, PhD dissertation, University of Ulster, 2004.
151. ———, *The impact of diversity on the accuracy of evidential classifier ensembles*, International Journal of Approximate Reasoning **53** (2012), no. 4, 584–607.
152. Yaxin Bi, Jiwen Guan, and David A. Bell, *The combination of multiple classifiers using an evidential reasoning approach*, Artificial Intelligence **172** (2008), no. 15, 1731–1751.

153. Loredana Biacino, *Fuzzy subsethood and belief functions of fuzzy events*, Fuzzy Sets and Systems **158** (January 2007), no. 1, 38–49.
154. E. Binaghi, L. Luzi, P. Madella, F. Pergalani, and A. Rampini, *Slope instability zonation: a comparison between certainty factor and fuzzy Dempster–Shafer approaches*, Natural Hazards **17** (1998), no. 1, 77–97.
155. Elisabetta Binaghi, Ignazio Gallo, and Paolo Madella, *A neural model for fuzzy Dempster–Shafer classifiers*, International Journal of Approximate Reasoning **25** (2000), no. 2, 89–121.
156. Elisabetta Binaghi and Paolo Madella, *Fuzzy Dempster–Shafer reasoning for rule-based classifiers*, International Journal of Intelligent Systems **14** (1999), no. 6, 559–583.
157. Elisabetta Binaghi, Paolo Madella, Ignazio Gallo, and Anna Rampini, *A neural refinement strategy for a fuzzy Dempster–Shafer classifier of multisource remote sensing images*, Proceedings of the SPIE - Image and Signal Processing for Remote Sensing IV, vol. 3500, Barcelona, Spain, 21-23 September 1998, pp. 214–224.
158. Thomas O. Binford and Tod S. Levitt, *Evidential reasoning for object recognition*, IEEE Transactions on Pattern Analysis and Machine Intelligence **25** (2003), no. 7, 837–851.
159. Garrett Birkhoff, *Abstract linear dependence and lattices*, American Journal of Mathematics **57** (1935), 800–804.
160. ______, *Lattice theory (3rd edition)*, vol. 25, American Mathematical Society Colloquium Publications, Providence, RI, 1967.
161. R. Bissig, Jürg Kohlas, and Norbert Lehmann, *Fast-division architecture for Dempster–Shafer belief functions*, Proceedings of the First International Joint Conference on Qualitative and Quantitative Practical Reasoning (ECSQARU–FAPR'97) (D. Gabbay, R. Kruse, A. Nonnengart, and H. J. Ohlbach, eds.), Lecture Notes in Artificial Intelligence, Springer, 1997, pp. 198–209.
162. Gautam Biswas and Tejwansh S. Anand, *Using the Dempster–Shafer scheme in a mixed-initiative expert system shell*, Uncertainty in Artificial Intelligence (L. N. Kanal, T. S. Levitt, and J .F. Lemmer, eds.), vol. 3, North-Holland, 1989, pp. 223–239.
163. Paul K. Black, *Is Shafer general Bayes?*, Proceedings of the Third Conference on Uncertainty in Artificial Intelligence (UAI'87), 1987, pp. 2–9.
164. ______, *An examination of belief functions and other monotone capacities*, PhD dissertation, Department of Statistics, Carnegie Mellon University, 1996.
165. ______, *Geometric structure of lower probabilities*, Random Sets: Theory and Applications (Goutsias, Malher, and Nguyen, eds.), Springer, 1997, pp. 361–383.
166. ______, *Is Shafer general Bayes?*, ArXiv preprint arXiv:1304.2711, 2013.
167. Paul K. Black and Kathryn B. Laskey, *Hierarchical evidence and belief functions*, Machine Intelligence and Pattern Recognition, vol. 9, Elsevier, 1990, pp. 207–215.
168. ______, *Hierarchical evidence and belief functions*, ArXiv preprint arXiv:1304.2342, 2013.
169. Samuel Blackman and Robert Popoli, *Design and analysis of modern tracking systems*, Artech House Publishers, 1999.
170. John Blitzer, Sham Kakade, and Dean P. Foster, *Domain adaptation with coupled subspaces*, Proceedings of the 14th International Conference on Artificial Intelligence and Statistics (AISTATS-11) (Geoffrey J. Gordon and David B. Dunson, eds.), vol. 15, 2011, pp. 173–181.
171. Isabelle Bloch, *Some aspects of Dempster–Shafer evidence theory for classification of multi-modality medical images taking partial volume effect into account*, Pattern Recognition Letters **17** (1996), no. 8, 905–919.

172. ______, *Information combination operators for data fusion: a comparative review with classification*, IEEE Transactions on Systems, Man, and Cybernetics - Part A: Systems and Humans **26** (January 1996), no. 1, 52–67.
173. Isabelle Bloch and H. Maitre, *Data fusion in 2D and 3D image processing: an overview*, Proceedings of the X Brazilian Symposium on Computer Graphics and Image Processing, October 1997, pp. 127–134.
174. Avrim Blum and Tom Mitchell, *Combining labeled and unlabeled data with co-training*, Proceedings of the 11th Annual Conference on Computational Learning Theory (COLT'98) (New York, NY, USA), ACM, 1998, pp. 92–100.
175. Daniel Bobrow, *Qualitative reasoning about physical systems*, Artificial intelligence **24** (1984), no. 1-3, 1–5.
176. Philip L. Bogler, *Shafer-Dempster reasoning with applications to multisensor target identification systems*, IEEE Transactions on Systems, Man and Cybernetics **17** (1987), no. 6, 968–977.
177. R. Bonnefon, Pierre Dhérété, and Jacky Desachy, *Geographic information system updating using remote sensing images*, Pattern Recognition Letters **23** (2002), no. 9, 1073–1083.
178. Robert F. Bordley, *Reformulating decision theory using fuzzy set theory and Shafer's theory of evidence*, Fuzzy Sets and Systems **139** (2003), no. 2, 243–266.
179. Hermann Borotschnig, Lucas Paletta, Manfred Prantl, and Axel Pinz, *A comparison of probabilistic, possibilistic and evidence theoretic fusion schemes for active object recognition*, Computing **62** (1999), no. 4, 293–319.
180. Michael Boshra and Hong Zhang, *Accommodating uncertainty in pixel-based verification of 3-D object hypotheses*, Pattern Recognition Letters **20** (1999), no. 7, 689–698.
181. Eloi Bossé and Jean Roy, *Fusion of identity declarations from dissimilar sources using the Dempster–Shafer theory*, Optical Engineering **36** (March 1997), no. 3, 648–657.
182. Eloi Bossé and Marc-Alain Simard, *Managing evidential reasoning for identity information fusion*, Optical Engineering **37** (1998), no. 2, 391–400.
183. J. Robert Boston, *A signal detection system based on Dempster–Shafer theory and comparison to fuzzy detection*, IEEE Transactions on Systems, Man, and Cybernetics - Part C: Applications and Reviews **30** (February 2000), no. 1, 45–51.
184. Mathieu Bouchard, Anne-Laure Jousselme, and Pierre-Emmanuel Doré, *A proof for the positive definiteness of the Jaccard index matrix*, International Journal of Approximate Reasoning **54** (2013), no. 5, 615–626.
185. Luke Boucher, Tony Simons, and Phil Green, *Evidential reasoning and the combination of knowledge and statistical techniques in syllable based speech recognition*, Speech recognition and understanding: Recent advances, trends and applications (P. Laface and R. De Mori, eds.), Springer Science and Business Media, July 1990, pp. 487–492.
186. Abdel-Ouahab Boudraa, Ayachi Bentabet, and Fabien Salzenstein, *Dempster–Shafer's basic probability assignment based on fuzzy membership functions*, Electronic Letters on Computer Vision and Image Analysis **4** (2004), no. 1, 1–10.
187. Mohamed Ayman Boujelben, Yves De Smet, Ahmed Frikha, and Habib Chabchoub, *Building a binary outranking relation in uncertain, imprecise and multi-experts contexts: The application of evidence theory*, International Journal of Approximate Reasoning **50** (2009), no. 8, 1259–1278.
188. Khaled Boukharouba, Laurent Bako, and S. Lecoeuche, *Identification of piecewise affine systems based on Dempster–Shafer theory*, IFAC Proceedings Volumes **42** (2009), no. 10, 1662–1667.

189. Imen Boukhris, Salem Benferhat, and Zied Elouedi, *Representing belief function knowledge with graphical models*, Knowledge Science, Engineering and Management (Hui Xiong and W. B. Lee, eds.), Lecture Notes in Computer Science, vol. 7091, Springer, Berlin Heidelberg, 2011, pp. 233–245.
190. Oumaima Boussarsar, Imen Boukhris, and Zied Elouedi, *Representing interventional knowledge in causal belief networks: Uncertain conditional distributions per cause*, Information Processing and Management of Uncertainty in Knowledge-Based Systems (Anne Laurent, Oliver Strauss, Bernadette Bouchon-Meunier, and Ronald R. Yager, eds.), Communications in Computer and Information Science, vol. 444, Springer International Publishing, 2014, pp. 223–232.
191. Boris R. Bracio, Wolfgang Horn, and Dietmar P. F. Moller, *Sensor fusion in biomedical systems*, Proceedings of the 19th Annual International Conference of the IEEE Engineering in Medicine and Biology Society, vol. 3, October 1997, pp. 1387–1390.
192. Jerome J. Braun, *Dempster–Shafer theory and Bayesian reasoning in multisensor data fusion*, Proceedings of SPIE - Sensor Fusion: Architectures, Algorithms, and Applications IV, vol. 4051, 2000, pp. 255–266.
193. Leo Breiman, *Bagging predictors*, Machine Learning **24** (August 1996), no. 2, 123–140.
194. Andrzej K. Brodzik and Robert H. Enders, *Semigroup structure of singleton Dempster–Shafer evidence accumulation*, IEEE Transactions on Information Theory **55** (2009), no. 11, 5241–5250.
195. Martin Bruning and Dieter Denneberg, *Max-min σ-additive representation of monotone measures*, Statistical Papers **34** (2002), no. 1, 23–35.
196. Randal E. Bryant, *Graph-based algorithms for Boolean function manipulation*, IIEEE Transactions on Computers **35** (August 1986), no. 8, 677–691.
197. Noel Bryson and Ayodele Mobolurin, *A qualitative discriminant approach for generating quantitative belief functions*, IEEE Transactions on Knowledge and Data Engineering **10** (1998), no. 2, 345–348.
198. ———, *A process for generating quantitative belief functions*, European Journal of Operational Research **115** (1999), no. 3, 624–633.
199. Noel (Kweku-Muata) Bryson and Anito Joseph, *Generating consensus priority interval vectors for group decision-making in the AHP*, Journal of Multi-Criteria Decision Analysis **9** (2000), no. 4, 127–137.
200. Bruce G. Buchanan, Edward Hance Shortliffe, et al., *Rule-based expert systems*, vol. 3, Addison-Wesley Reading, MA, 1984.
201. James J. Buckley, *Possibility and necessity in optimization*, Fuzzy Sets and Systems **25** (1988), no. 1, 1–13.
202. Dennis M. Buede and Paul Girardi, *A target identification comparison of Bayesian and Dempster-Shafer multisensor fusion*, IEEE Transactions on Systems, Man, and Cybernetics-Part A: Systems and Humans **27** (1997), no. 5, 569–577.
203. Dennis M. Buede and Paul Girardi, *Target identification comparison of Bayesian and Dempster–Shafer multisensor fusion*, IEEE Transactions on Systems, Man, and Cybernetics Part A: Systems and Humans. **27** (1997), no. 5, 569–577.
204. Dennis M. Buede and J. W. Martin, *Comparison of bayesian and Dempster–Shafer fusion*, Proceedings of the Tri-Service Data Fusion Symposium, vol. 1, 1989, pp. 81–101.
205. Alan Bundy, *Incidence calculus: A mechanism for probability reasoning*, Journal of Automated Reasoning **1** (1985), 263–283.

206. Thomas Burger, *Defining new approximations of belief function by means of Dempster's combination*, Proceedings of the Workshop on the Theory of Belief Functions (BELIEF 2010), 2010.
207. Thomas Burger, Oya Aran, and Alice Caplier, *Modeling hesitation and conflict: A belief-based approach for multi-class problems*, Proceedings of the 5th International Conference on Machine Learning and Applications (ICMLA '06), 2006, pp. 95–100.
208. Thomas Burger, Oya Aran, Alexandra Urankar, Alice Caplier, and Lale Akarun, *A Dempster–Shafer theory based combination of classifiers for hand gesture recognition*, Computer Vision and Computer Graphics. Theory and Applications (José Braz, Alpesh Ranchordas, HélderJ. Araújo, and João Madeiras Pereira, eds.), Communications in Computer and Information Science, vol. 21, Springer, 2008, pp. 137–150.
209. Thomas Burger and Alice Caplier, *A generalization of the pignistic transform for partial bet*, Proceedings of the 10th European Conference on Symbolic and Quantitative Approaches to Reasoning with Uncertainty (ECSQARU'09), Springer-Verlag, July 2009, pp. 252–263.
210. Thomas Burger and Fabio Cuzzolin, *The barycenters of the k-additive dominating belief functions and the pignistic k-additive belief functions*, Proceedings of the First International Workshop on the Theory of Belief Functions (BELIEF 2010), 2010.
211. Thomas Burger, Yousri Kessentini, and Thierry Paquet, *Dealing with precise and imprecise decisions with a Dempster–Shafer theory based algorithm in the context of handwritten word recognition*, Proceedings of the 12th International Conference on Frontiers in Handwriting Recognition, 2010, pp. 369–374.
212. Prabir Burman, *A comparative study of ordinary cross-validation, v-fold cross-validation and the repeated learning-testing methods*, Biometrika **76** (1989), no. 3, 503–514.
213. A. C. Butler, F. Sadeghi, S. S. Rao, and S. R. LeClair, *Computer-aided design/engineering of bearing systems using the Dempster–Shafer theory*, Artificial Intelligence for Engineering Design, Analysis and Manufacturing **9** (January 1995), no. 1, 1–11.
214. Richard Buxton, *Modelling uncertainty in expert systems*, International Journal of Man-machine Studies **31** (1989), no. 4, 415–476.
215. Anhui Cai, Toshio Fukuda, and Fumihito Arai, *Integration of distributed sensing information in DARS based on evidential reasoning*, Distributed Autonomous Robotic Systems 2 (Hajime Asama, Toshio Fukuda, Tamio Arai, and Isao Endo, eds.), Springer, 1996, pp. 268–279.
216. Colin Camerer and Martin Weber, *Recent developments in modeling preferences: Uncertainty and ambiguity*, Journal of Risk and Uncertainty **5** (1992), no. 4, 325–370.
217. Fabio Campos and Sérgio Cavalcante, *An extended approach for Dempster–Shafer theory*, Proceedings of the Fifth IEEE Workshop on Mobile Computing Systems and Applications, October 2003, pp. 338–344.
218. Fabio Campos and Fernando M. Campello de Souza, *Extending Dempster–Shafer theory to overcome counter intuitive results*, Proceedings of the IEEE International Conference on Natural Language Processing and Knowledge Engineering (NLP-KE '05), vol. 3, 2005, pp. 729–734.
219. José Cano, Miguel Delgado, and Serafín Moral, *An axiomatic framework for propagating uncertainty in directed acyclic networks*, International Journal of Approximate Reasoning **8** (1993), no. 4, 253–280.
220. Anne-Sophie Capelle-Laizé, Christine Fernandez-Maloigne, and Olivier Colot, *Introduction of spatial information within the context of evidence theory*, Proceedings of the

IEEE International Conference on Acoustics, Speech, and Signal Processing (ICASSP '03), vol. 2, April 2003, pp. II–785–8.

221. Andrea Capotorti and Barbara Vantaggi, *The consistency problem in belief and probability assessments*, Proceedings of the Sixth International Conference on Information Processing and Management of Uncertainty in Knowledge-Based Systems (IPMU 96), 1996, pp. 55–59.
222. ______, *The consistency problem in belief and probability assessments*, Proceedings of the 6th International Conference on Information Processing and Management of Uncertainty in Knowledge-Based Systems (IPMU'96), vol. 2, 1996, pp. 751–755.
223. Sandra Carberry, *Incorporating default inferences into plan recognition*, Proceedings of the Eighth National Conference on Artificial Intelligence (AAAI'90), vol. 1, 1990, pp. 471–478.
224. Jennifer Carlson and Robin R. Murphy, *Use of Dempster–Shafer conflict metric to adapt sensor allocation to unknown environments*, Tech. report, Safety Security Rescue Research Center, University of South Florida, 2005.
225. ______, *Use of Dempster–Shafer conflict metric to adapt sensor allocation to unknown environments.*, Proceedings of the FLAIRS Conference, 2006, pp. 866–867.
226. ______, *Use of Dempster–Shafer conflict metric to detect interpretation inconsistency*, Proceedings of the Twenty-First Conference on Uncertainty in Artificial Intelligence (UAI 2005), 2012, pp. 94–104.
227. Jennifer Carlson, Robin R. Murphy, Svetlana Christopher, and Jennifer Casper, *Conflict metric as a measure of sensing quality*, Proceedings of the 2005 IEEE International Conference on Robotics and Automation (ICRA 2005), April 2005, pp. 2032–2039.
228. Rudolf Carnap, *Logical foundations of probability*, University of Chicago Press, 1962.
229. François Caron, Saiedeh Navabzadeh Razavi, Jongchul Song, Philippe Vanheeghe, Emmanuel Duflos, Carlos Caldas, and Carl Haas, *Locating sensor nodes on construction projects*, Autonomous Robots **22** (2007), no. 3, 255–263.
230. François Caron, Branko Ristic, Emmanuel Duflos, and Philippe Vanheeghe, *Least committed basic belief density induced by a multivariate Gaussian: Formulation with applications*, International Journal of Approximate Reasoning **48** (2008), no. 2, 419–436.
231. Rich Caruana, Alexandru Niculescu-Mizil, Geoff Crew, and Alex Ksikes, *Ensemble selection from libraries of models*, Proceedings of the 21st International Conference on Machine Learning (ICML'04) (New York, NY, USA), ACM, 2004, pp. 18–.
232. Montserrat Casanovas and José M. Merigó, *Using fuzzy OWA operators in decision making with Dempster–Shafer belief structure*, Proceedings of the AEDEM International Conference, 2007, pp. 475–486.
233. ______, *Fuzzy aggregation operators in decision making with Dempster–Shafer belief structure*, Expert Systems with Applications **39** (2012), no. 8, 7138–7149.
234. William F. Caselton and Wuben Luo, *Decision making with imprecise probabilities: Dempster–Shafer theory and application*, Water Resources Research **28** (1992), no. 12, 3071–3083.
235. Cristiano Castelfranchi, *Mind as an anticipatory device: For a theory of expectations*, Proceedings of the First International Symposium on Brain, Vision, and Artificial Intelligence (BVAI 2005) (Massimo De Gregorio, Vito Di Maio, Maria Frucci, and Carlo Musio, eds.), Springer, Berlin Heidelberg, October 2005, pp. 258–276.
236. Marco E. G. V. Cattaneo, *Combining belief functions issued from dependent sources*, Proceedings of the Third International Symposium on Imprecise Probabilities and Their Applications (ISIPTA'03), 2003, pp. 133–147.

237. Carlton M. Caves, Christopher A. Fuchs, and Rüdiger Schack, *Quantum probabilities as Bayesian probabilities*, Physical review A **65** (January 2002), no. 2, 022305.
238. Luis Cayuela, Duncan J. Golicher, and Jose Maria Rey-Benayas, *The extent, distribution, and fragmentation of vanishing montane cloud forest in the highlands of Chiapas, Mexico*, Biotropica **38** (2006), no. 4, 544–554.
239. Subhash Challa and Don Koks, *Bayesian and Dempster–Shafer fusion*, Sadhana **29** (2004), no. 2, 145–174.
240. Robert G. Chambers and Tigran Melkonyan, *Degree of imprecision: Geometric and algorithmic approaches*, International Journal of Approximate Reasoning **10** (2007), no. 1, 106–122.
241. J. J. Chao, Elias Drakopoulos, and Chung-Chieh Lee, *An evidential reasoning approach to distributed multiple-hypothesis detection*, Proceedings of the 26th IEEE Conference on Decision and Control (CDC'87), vol. 26, December 1987, pp. 1826–1831.
242. Michel Chapron, *A color edge detector based on Dempster–Shafer theory*, Proceedings of the International Conference on Image Processing (ICIP 2000), vol. 2, September 2000, pp. 812–815.
243. Patrick Chareka, *The central limit theorem for capacities*, Statistics and Probability Letters **79** (2009), no. 12, 1456–1462.
244. Thierry Chateau, Michel Berducat, and Pierre Bonton, *An original correlation and data fusion based approach to detect a reap limit into a gray level image*, Proceedings of the 1997 IEEE/RSJ International Conference on Intelligent Robots and Systems (IROS'97), vol. 3, September 1997, pp. 1258–1263.
245. A. Chateauneuf and Jean-Yves Jaffray, *Some characterization of lower probabilities and other monotone capacities through the use of Möebius inversion*, Mathematical social sciences (1989), no. 3, 263–283.
246. Alain Chateauneuf, *On the use of capacities in modeling uncertainty aversion and risk aversion*, Journal of Mathematical Economics **20** (1991), no. 4, 343–369.
247. ______, *Combination of compatible belief functions and relations of specificity*, Papiers d'Economie Mathematique et Applications 92-59, Universit Panthon Sorbonne (Paris 1), 1992.
248. ______, *Decomposable capacities, distorted probabilities and concave capacities*, Mathematical Social Sciences **31** (1996), no. 1, 19–37.
249. Alain Chateauneuf and Jean-Yves Jaffray, *Local Möbius transforms of monotone capacities*, Symbolic and Quantitative Approaches to Reasoning and Uncertainty (Christine Froidevaux and Jürg Kohlas, eds.), Lecture Notes in Computer Science, vol. 946, Springer, Berlin Heidelberg, 1995, pp. 115–124.
250. Alain Chateauneuf and Jean-Christophe Vergnaud, *Ambiguity reduction through new statistical data*, International Journal of Approximate Reasoning **24** (2000), no. 2-3, 283–299.
251. C. W. R. Chau, Pawan Lingras, and S. K. Michael Wong, *Upper and lower entropies of belief functions using compatible probability functions*, Proceedings of the 7th International Symposium on Methodologies for Intelligent Systems (ISMIS'93) (J. Komorowski and Z. W. Ras, eds.), June 1993, pp. 306–315.
252. Ali Cheaito, Michael Lecours, and Eloi Bossé, *Modified Dempster–Shafer approach using an expected utility interval decision rule*, Proceedings of SPIE - Sensor Fusion: Architectures, Algorithms, and Applications III, vol. 3719, 1999, pp. 34–42.
253. ______, *A non-ad-hoc decision rule for the Dempster–Shafer method of evidential reasoning*, Proceedings of the SPIE - Sensor Fusion: Architectures, Algorithms, and Applications II, vol. 3376, Orlando, FL, USA, 16-17 April 1998, pp. 44–57.

254. ______, *Study of a modified Dempster–Shafer approach using an expected utility interval decision rule*, Proceedings of the SPIE - Sensor Fusion: Architectures, Algorithms, and Applications III, vol. 3719, Orlando, FL, USA, 7-9 April 1999, pp. 34–42.
255. Peter Cheeseman, *In defense of probability*, Proceedings of the 9th International Joint Conference on Artificial Intelligence (IJCAI'85), vol. 2, Morgan Kaufmann, 1985, pp. 1002–1009.
256. ______, *Probabilistic versus fuzzy reasoning*, Machine Intelligence and Pattern Recognition (L. Kanal and J. Lemmer, eds.), vol. 4, Elsevier, 1986, pp. 85–102.
257. ______, *A method of computing maximum entropy probability values for expert systems*, Maximum-entropy and Bayesian spectral analysis and estimation problems, Springer, 1987, pp. 229–240.
258. Li Chen and S. S. Rao, *A modified Dempster–Shafer theory for multicriteria optimization*, Engineering Optimization **30** (1998), no. 3-4, 177–201.
259. Liang-Hsuan Chen, *An extended rule-based inference for general decision-making problems*, Information Sciences **102** (1997), no. 1-4, 111–131.
260. Qi Chen and Uwe Aickelin, *Dempster–Shafer for anomaly detection*, Proceedings of the International Conference on Data Mining (DMIN 2006), 2006, pp. 232–238.
261. Shiuh-Yung Chen, Wei-Chung Lin, and Chin-Tu Chen, *Evidential reasoning based on Dempster–Shafer theory and its application to medical image analysis*, Proceedings of SPIE - Neural and Stochastic Methods in Image and Signal Processing II (Su-Shing Chen, ed.), vol. 2032, July 1993, pp. 35–46.
262. ______, *Spatial reasoning based on multivariate belief functions*, Proceedings of the IEEE Computer Society Conference on Computer Vision and Pattern Recognition (CVPR'92), June 1992, pp. 624–626.
263. Shiuh-Yung J. Chen, Wei-Chung Lin, and Chin-Tu Chen, *Medical image understanding system based on Dempster–Shafer reasoning*, Proceedings of SPIE - Medical Imaging V: Image Processing, vol. 1445, 1991, pp. 386–397.
264. Su-shing Chen, *Evidential logic and Dempster–Shafer theory*, Proceedings of the ACM SIGART International Symposium on Methodologies for Intelligent Systems (ISMIS '86), 1986, pp. 201–206.
265. Thomas M. Chen and Varadharajan Venkataramanan, *Dempster–Shafer theory for intrusion detection in ad hoc networks*, IEEE Internet Computing **9** (2005), no. 6, 35–41.
266. Yang Chen and Barbara Blyth, *An evidential reasoning approach to composite combat identification (CCID)*, Proceedings of the 2004 IEEE Aerospace Conference, vol. 3, March 2004, pp. 2042–2053.
267. Yi-lei Chen and Jun-jie Wang, *An improved method of D-S evidential reasoning*, Acta Simulata Systematica Sinica **1** (2004).
268. Yuan Yan Chen, *Statistical inference based on the possibility and belief measures*, Transactions of the American Mathematical Society **347** (1995), no. 5, 1855–1863.
269. Zengjing Chen, *Strong laws of large numbers for sub-linear expectations*, Science China Mathematics **59** (2016), no. 5, 945–954.
270. ______, *Strong law of large numbers for sub-linear expectations*, ArXiv preprint arXiv:1006.0749, June 2010.
271. Yizong Cheng and Rangasami L. Kashyap, *A study of associative evidential reasoning*, IEEE Transactions on Pattern Analysis and Machine Intelligence **11** (1989), no. 6, 623–631.
272. Kwai-Sang Chin, Kwong-Chi Lo, and Jendy P. F. Leung, *Development of user-satisfaction-based knowledge management performance measurement system with evidential reasoning approach*, Expert Systems with Applications **37** (2010), no. 1, 366–382.

273. Kwai-Sang Chin, Ying-Ming Wang, Gary Ka Kwai Poon, and Jian-Bo Yang, *Failure mode and effects analysis using a group-based evidential reasoning approach*, Computers and Operations Research **36** (2009), no. 6, 1768–1779.
274. Kwai-Sang Chin, Ying-Ming Wang, Jian-Bo Yang, and Ka Kwai Gary Poon, *An evidential reasoning based approach for quality function deployment under uncertainty*, Expert Systems with Applications **36** (2009), no. 3, 5684–5694.
275. Kwai-Sang Chin, Jian-Bo Yang, Min Guo, and James P.-K. Lam, *An evidential-reasoning-interval-based method for new product design assessment*, IEEE Transactions on Engineering Management **56** (2009), no. 1, 142–156.
276. Salim Chitroub, Amrane Houacine, and Boualem Sansal, *Evidential reasoning-based classification method for remotely sensed images*, Proceedings of SPIE - Image and Signal Processing for Remote Sensing VII, vol. 4541, 2002, pp. 340–351.
277. Sung Nok Chiu, Dietrich Stoyan, Wilfrid S. Kendall, and Joseph Mecke, *Stochastic geometry and its applications*, John Wiley & Sons, 2013.
278. Bassam A. Chokr and Vladik Kreinovich, *How far are we from the complete knowledge? Complexity of knowledge acquisition in the Dempster–Shafer approach*, Advances in the Dempster–Shafer Theory of Evidence (Ronald R. Yager, Janusz Kacprzyk, and Mario Fedrizzi, eds.), John Wiley and Sons, Inc., 1994, pp. 555–576.
279. Laurence Cholvy, *Applying theory of evidence in multisensor data fusion: a logical interpretation*, Proceedings of the Third International Conference on Information Fusion (FUSION 2000), vol. 1, 2000, pp. TUB4/17–TUB4/24.
280. ______, *Towards another logical interpretation of theory of evidence and a new combination rule*, Intelligent Systems for Information Processing, Elsevier, 2003, pp. 223–232.
281. Laurence Cholvy, *Using logic to understand relations between DSmT and Dempster–Shafer theory*, Symbolic and Quantitative Approaches to Reasoning with Uncertainty (Claudio Sossai and Gaetano Chemello, eds.), Lecture Notes in Computer Science, vol. 5590, Springer Berlin Heidelberg, 2009, pp. 264–274.
282. ______, *Using logic to understand relations between DSmT and Dempster–Shafer theory*, Symbolic and Quantitative Approaches to Reasoning with Uncertainty: 10th European Conference, ECSQARU 2009, Verona, Italy, July 1-3, 2009. Proceedings (Claudio Sossai and Gaetano Chemello, eds.), Springer, Berlin Heidelberg, 2009, pp. 264–274.
283. Gustave Choquet, *Theory of capacities*, Annales de l'Institut Fourier **5** (1953), 131–295.
284. Francis C. Chu and Joseph Y. Halpern, *Great expectations. Part II: Generalized expected utility as a universal decision rule*, Artificial Intelligence **159** (2004), no. 1, 207–229.
285. ______, *Great expectation. Part I: On the customizability of generalized expected utility*, Theory and Decision **64** (2008), no. 1, 1–36.
286. Jennifer Chu-Carroll and Michael K. Brown, *Tracking initiative in collaborative dialogue interactions*, Proceedings of the 35th Annual Meeting of the Association for Computational Linguistics and Eighth Conference of the European Chapter of the Association for Computational Linguistics (ACL '98), 1997, pp. 262–270.
287. Esma N. Cinicioglu and Prakash P. Shenoy, *On Walley's combination rule for statistical evidence*, Proceedings of the Eleventh International Conference on Information Processing and Management of Uncertainty in Knowledge-Based Systems (IPMU'06), 2006, pp. 386–394.

288. Mike Clarke and Nic Wilson, *Efficient algorithms for belief functions based on the relationship between belief and probability*, Proceedings of the European Conference on Symbolic and Quantitative Approaches to Uncertainty (ECSQARU'91) (R. Kruse and P. Siegel, eds.), October 1991, pp. 48–52.
289. Fritz Class, Alfred Kaltenmeier, and Peter Regel-Brietzmann, *Soft-decision vector quantization based on the Dempster/Shafer theory*, Proceedings of the International Conference on Acoustics, Speech, and Signal Processing (ICASSP-91), vol. 1, April 1991, pp. 665–668.
290. Johan Van Cleynenbreugel, S. A. Osinga, Freddy Fierens, Paul Suetens, and André Oosterlinck, *Road extraction from multitemporal satellite images by an evidential reasoning approach*, Pattern Recognition Letters **12** (June 1991), no. 6, 371–380.
291. Barry R. Cobb and Prakash P. Shenoy, *A comparison of Bayesian and belief function reasoning*, Information Systems Frontiers **5** (2003), no. 4, 345–358.
292. Barry R. Cobb and Prakash P. Shenoy, *On transforming belief function models to probability models*, Tech. report, School of Business, University of Kansas, July 2003.
293. Barry R. Cobb and Prakash P. Shenoy, *On the plausibility transformation method for translating belief function models to probability models*, International Journal of Approximate Reasoning **41** (2006), no. 3, 314–330.
294. ______, *On transforming belief function models to probability models*, Tech. report, University of Kansas, School of Business, Working Paper No. 293, February 2003.
295. ______, *A comparison of methods for transforming belief function models to probability models*, Proceedings of the 7th European Conference on Symbolic and Quantitative Approaches to Reasoning and Uncertainty (ECSQARU-2003), July 2003, pp. 255–266.
296. Paul R. Cohen and Milton R. Grinberg, *A framework for heuristic reasoning about uncertainty*, Proceedings of the Eighth International Joint Conference on Artificial Intelligence (IJCAI'83), vol. 1, Morgan Kaufmann, 1983, pp. 355–357.
297. ______, *A theory of heuristic reasoning about uncertainty*, Readings from the AI Magazine (Robert Engelmore, ed.), American Association for Artificial Intelligence, 1988, pp. 559–566.
298. Giulianella Coletti and Marcello Mastroleo, *Conditional belief functions: a comparison among different definitions*, Proceeding of 7th Workshop on Uncertainty Processing (WUPES), 2006, pp. 24–35.
299. Michael J. Collins, *Information fusion in sea ice remote sensing*, Microwave Remote Sensing of Sea Ice, American Geophysical Union, 2013, pp. 431–441.
300. Alexis J. Comber, Steve Carver, Steffen Fritz, Robert McMorran, Justin Washtell, and Peter Fisher, *Different methods, different wilds: Evaluating alternative mappings of wildness using fuzzy MCE and Dempster–Shafer MCE*, Computers, Environment and Urban Systems **34** (2010), no. 2, 142–152.
301. Alexis J. Comber, Alistair N. R. Law, and J. Rowland Lishman, *A comparison of Bayes, Dempster–Shafer and Endorsement theories for managing knowledge uncertainty in the context of land cover monitoring*, Computers, Environment and Urban Systems **28** (2004), no. 4, 311–327.
302. Etienne Côme, Laurent Bouillaut, Patrice Aknin, and Allou Samé, *Bayesian network for railway infrastructure diagnosis*, Proceedings of Information Processing with Management of Uncertainty in Knowledge-based Systems (IPMU'06), 2006.
303. Etienne Côme, Latifa Oukhellou, Thierry Denœux, and Patrice Aknin, *Mixture model estimation with soft labels*, Soft Methods for Handling Variability and Imprecision (Didier Dubois, M. Asunción Lubiano, Henri Prade, María Ángeles Gil, Przemysław

Grzegorzewski, and Olgierd Hryniewicz, eds.), Springer, Berlin Heidelberg, 2008, pp. 165–174.

304. Etienne Côme, Latifa Oukhellou, Thierry Denœux, and Patrice Aknin, *Learning from partially supervised data using mixture models and belief functions*, Pattern Recognition **42** (2009), no. 3, 334–348.
305. Roger Cooke and Philippe Smets, *Self-conditional probabilities and probabilistic interpretations of belief functions*, Annals of Mathematics and Artificial Intelligence **32** (2001), no. 1-4, 269–285.
306. Kimberly Coombs, Debra Freel, Douglas Lampert, and Steven J. Brahm, *Using Dempster–Shafer methods for object classification in the theater ballistic missile environment*, Proceedings of the SPIE - Sensor Fusion: Architectures, Algorithms, and Applications III, vol. 3719, Orlando, FL, USA, 7-9 April 1999, pp. 103–113.
307. Flavio S. Correa da Silva and Alan Bundy, *On some equivalence relations between incidence calculus and Dempster–Shafer theory of evidence*, Proceedings of the Sixth Conference on Uncertainty in Artificial Intelligence (UAI1990), 1990, pp. 378–383.
308. ______, *On some equivalence relations between incidence calculus and Dempster–Shafer theory of evidence*, ArXiv preprint arXiv:1304.1126, 2013.
309. Enrique Cortes-Rello and Forouzan Golshani, *Uncertain reasoning using the Dempster–Shafer method: an application in forecasting and marketing management*, Expert Systems **7** (1990), no. 1, 9–18.
310. Fabio Gagliardi Cozman, *Calculation of posterior bounds given convex sets of prior probability measures and likelihood functions*, Journal of Computational and Graphical Statistics **8** (1999), no. 4, 824–838.
311. ______, *Computing posterior upper expectations*, International Journal of Approximate Reasoning **24** (2000), no. 2-3, 191–205.
312. ______, *Credal networks*, Artificial Intelligence **120** (2000), no. 2, 199–233.
313. ______, *Graphical models for imprecise probabilities*, International Journal of Approximate Reasoning **39** (2005), no. 2, 167–184.
314. ______, *Concentration inequalities and laws of large numbers under epistemic and regular irrelevance*, International Journal of Approximate Reasoning **51** (2010), no. 9, 1069–1084.
315. Fabio Gagliardi Cozman and Serafín Moral, *Reasoning with imprecise probabilities*, International Journal of Approximate Reasoning **24** (2000), no. 2-3, 121–123.
316. Harald Cramer, *Mathematical methods of statistics (pms-9)*, vol. 9, Princeton University Press, 2016.
317. Henry H. Crapo and Gian-Carlo Rota, *On the foundations of combinatorial theory: Combinatorial geometries*, MIT Press, Cambridge, Mass., 1970.
318. Valerie Cross and Thomas Sudkamp, *Compatibility and aggregation in fuzzy evidential reasoning*, Proceedings of the IEEE International Conference on Systems, Man, and Cybernetics (SMC'91), vol. 3, October 1991, pp. 1901–1906.
319. Valerie Cross and Thomas Sudkamp, *Compatibility measures for fuzzy evidential reasoning*, Proceedings of the Fourth International Conference on Industrial and Engineering Applications of Artificial Intelligence and Expert Systems, June 1991, pp. 72–78.
320. Peter Cucka and Azriel Rosenfeld, *Evidence-based pattern-matching relaxation*, Pattern Recognition **26** (1993), no. 9, 1417–1427.
321. W. Cui and David I. Blockley, *Interval probability theory for evidential support*, International Journal of Intelligent Systems **5** (1990), no. 2, 183–192.

322. Alison C. Cullen and H. Christopher Frey, *Probabilistic techniques in exposure assessment: A handbook for dealing with variability and uncertainty in models and inputs*, Plenum Press: New York, 1999.
323. Shawn P. Curley and James I. Golden, *Using belief functions to represent degrees of belief*, Organizational Behavior and Human Decision Processes **58** (1994), no. 2, 271–303.
324. Shawn P. Curley and J. Frank Yates, *An empirical evaluation of descriptive models of ambiguity reactions in choice situations*, Journal of Mathematical Psychology **33** (1989), no. 4, 397–427.
325. Fabio Cuzzolin, *Lattice modularity and linear independence*, Proceedings of the 18th British Combinatorial Conference (BCC'01), 2001.
326. ______, *Visions of a generalized probability theory*, PhD dissertation, Università degli Studi di Padova, 19 February 2001.
327. ______, *Geometry of Dempster's rule of combination*, IEEE Transactions on Systems, Man and Cybernetics part B **34** (2004), no. 2, 961–977.
328. ______, *Simplicial complexes of finite fuzzy sets*, Proceedings of the 10th International Conference on Information Processing and Management of Uncertainty (IPMU'04), vol. 4, 2004, pp. 4–9.
329. ______, *Algebraic structure of the families of compatible frames of discernment*, Annals of Mathematics and Artificial Intelligence **45** (2005), no. 1-2, 241–274.
330. ______, *On the orthogonal projection of a belief function*, Proceedings of the International Conference on Symbolic and Quantitative Approaches to Reasoning with Uncertainty (ECSQARU'07), Lecture Notes in Computer Science, vol. 4724, Springer, Berlin / Heidelberg, 2007, pp. 356–367.
331. ______, *On the relationship between the notions of independence in matroids, lattices, and Boolean algebras*, Proceedings of the British Combinatorial Conference (BCC'07), 2007.
332. ______, *Relative plausibility, affine combination, and Dempster's rule*, Tech. report, INRIA Rhone-Alpes, 2007.
333. ______, *Two new Bayesian approximations of belief functions based on convex geometry*, IEEE Transactions on Systems, Man, and Cybernetics - Part B **37** (2007), no. 4, 993–1008.
334. ______, *A geometric approach to the theory of evidence*, IEEE Transactions on Systems, Man, and Cybernetics, Part C: Applications and Reviews **38** (2008), no. 4, 522–534.
335. ______, *Alternative formulations of the theory of evidence based on basic plausibility and commonality assignments*, Proceedings of the Pacific Rim International Conference on Artificial Intelligence (PRICAI'08), 2008, pp. 91–102.
336. ______, *Boolean and matroidal independence in uncertainty theory*, Proceedings of the International Symposium on Artificial Intelligence and Mathematics (ISAIM 2008), 2008.
337. ______, *Dual properties of the relative belief of singletons*, PRICAI 2008: Trends in Artificial Intelligence (Tu-Bao Ho and Zhi-Hua Zhou, eds.), vol. 5351, Springer, 2008, pp. 78–90.
338. ______, *An interpretation of consistent belief functions in terms of simplicial complexes*, Proceedings of the International Symposium on Artificial Intelligence and Mathematics (ISAIM 2008), 2008.
339. ______, *On the credal structure of consistent probabilities*, Logics in Artificial Intelligence (Steffen Hlldobler, Carsten Lutz, and Heinrich Wansing, eds.), Lecture Notes in Computer Science, vol. 5293, Springer, Berlin Heidelberg, 2008, pp. 126–139.

340. ______, *Semantics of the relative belief of singletons*, Interval/Probabilistic Uncertainty and Non-Classical Logics, Springer, 2008, pp. 201–213.
341. ______, *Semantics of the relative belief of singletons*, Proceedings of the International Workshop on Interval/Probabilistic Uncertainty and Non-Classical Logics (UncLog'08), 2008.
342. ______, *Complexes of outer consonant approximations*, Proceedings of the 10th European Conference on Symbolic and Quantitative Approaches to Reasoning with Uncertainty (ECSQARU'09), 2009, pp. 275–286.
343. ______, *The intersection probability and its properties*, Symbolic and Quantitative Approaches to Reasoning with Uncertainty (Claudio Sossai and Gaetano Chemello, eds.), Lecture Notes in Computer Science, vol. 5590, Springer, Berlin Heidelberg, 2009, pp. 287–298.
344. ______, *Credal semantics of Bayesian transformations in terms of probability intervals*, IEEE Transactions on Systems, Man, and Cybernetics, Part B: Cybernetics **40** (2010), no. 2, 421–432.
345. ______, *The geometry of consonant belief functions: simplicial complexes of necessity measures*, Fuzzy Sets and Systems **161** (2010), no. 10, 1459–1479.
346. ______, *Geometry of relative plausibility and relative belief of singletons*, Annals of Mathematics and Artificial Intelligence **59** (2010), no. 1, 47–79.
347. ______, *Three alternative combinatorial formulations of the theory of evidence*, Intelligent Data Analysis **14** (2010), no. 4, 439–464.
348. ______, *Geometric conditional belief functions in the belief space*, Proceedings of the 7th International Symposium on Imprecise Probabilities and Their Applications (ISIPTA'11), 2011.
349. ______, *On consistent approximations of belief functions in the mass space*, Symbolic and Quantitative Approaches to Reasoning with Uncertainty (Weiru Liu, ed.), Lecture Notes in Computer Science, vol. 6717, Springer, Berlin Heidelberg, 2011, pp. 287–298.
350. ______, *On the relative belief transform*, International Journal of Approximate Reasoning **53** (2012), no. 5, 786–804.
351. ______, *Chapter 12: An algebraic study of the notion of independence of frames*, Mathematics of Uncertainty Modeling in the Analysis of Engineering and Science Problems (S. Chakraverty, ed.), IGI Publishing, 2014.
352. Fabio Cuzzolin, *Lp consonant approximations of belief functions*, IEEE Transactions on Fuzzy Systems **22** (2014), no. 2, 420–436.
353. Fabio Cuzzolin, *Lp consonant approximations of belief functions*, IEEE Transactions on Fuzzy Systems **22** (2014), no. 2, 420–436.
354. ______, *On the fiber bundle structure of the space of belief functions*, Annals of Combinatorics **18** (2014), no. 2, 245–263.
355. Fabio Cuzzolin, *Generalised max entropy classifiers*, Belief Functions: Theory and Applications (Cham) (Sébastien Destercke, Thierry Denœux, Fabio Cuzzolin, and Arnaud Martin, eds.), Springer International Publishing, 2018, pp. 39–47.
356. Fabio Cuzzolin, *Geometric conditioning of belief functions*, Proceedings of the Workshop on the Theory of Belief Functions (BELIEF'10), April 2010.
357. ______, *Geometry of upper probabilities*, Proceedings of the 3rd Internation Symposium on Imprecise Probabilities and Their Applications (ISIPTA'03), July 2003.
358. ______, *The geometry of relative plausibilities*, Proceedings of the 11th International Conference on Information Processing and Management of Uncertainty (IPMU'06), special session on "Fuzzy measures and integrals, capacities and games" (Paris, France), July 2006.

359. ______, *Lp consonant approximations of belief functions in the mass space*, Proceedings of the 7th International Symposium on Imprecise Probability: Theory and Applications (ISIPTA'11), July 2011.
360. ______, *Consistent approximation of belief functions*, Proceedings of the 6th International Symposium on Imprecise Probability: Theory and Applications (ISIPTA'09), June 2009.
361. ______, *Geometry of Dempster's rule*, Proceedings of the 1st International Conference on Fuzzy Systems and Knowledge Discovery (FSKD'02), November 2002.
362. ______, *On the properties of relative plausibilities*, Proceedings of the International Conference of the IEEE Systems, Man, and Cybernetics Society (SMC'05), vol. 1, October 2005, pp. 594–599.
363. ______, *Families of compatible frames of discernment as semimodular lattices*, Proceedings of the International Conference of the Royal Statistical Society (RSS 2000), September 2000.
364. ______, *Visions of a generalized probability theory*, Lambert Academic Publishing, September 2014.
365. Fabio Cuzzolin and Ruggero Frezza, *An evidential reasoning framework for object tracking*, Proceedings of SPIE - Photonics East 99 - Telemanipulator and Telepresence Technologies VI (Matthew R. Stein, ed.), vol. 3840, 19-22 September 1999, pp. 13–24.
366. ______, *Evidential modeling for pose estimation*, Proceedings of the 4th Internation Symposium on Imprecise Probabilities and Their Applications (ISIPTA'05), July 2005.
367. ______, *Integrating feature spaces for object tracking*, Proceedings of the International Symposium on the Mathematical Theory of Networks and Systems (MTNS 2000), June 2000.
368. ______, *Geometric analysis of belief space and conditional subspaces*, Proceedings of the 2nd International Symposium on Imprecise Probabilities and their Applications (ISIPTA'01), June 2001.
369. ______, *Lattice structure of the families of compatible frames*, Proceedings of the 2nd International Symposium on Imprecise Probabilities and their Applications (ISIPTA'01), June 2001.
370. Fabio Cuzzolin and Wenjuan Gong, *Belief modeling regression for pose estimation*, Proceedings of the 16th International Conference on Information Fusion (FUSION 2013), 2013, pp. 1398–1405.
371. Wagner Texeira da Silva and Ruy Luiz Milidiu, *Algorithms for combining belief functions*, International Journal of Approximate Reasoning **7** (1992), no. 1-2, 73–94.
372. Bruce D'Ambrosio, *A hybrid approach to reasoning under uncertainty*, International Journal of Approximate Reasoning **2** (1988), no. 1, 29–45.
373. Milan Daniel, *Associativity and contradiction in combination of belief functions*, Proceedings of the 8th International Conference on Information Processing and Management Uncertainty in Knowledge-based Systems (IPMU 2000), 2000, pp. 133–140.
374. ______, *Distribution of contradictive belief masses in combination of belief functions*, Information, Uncertainty and Fusion (Bernadette Bouchon-Meunier, Ronald R. Yager, and Lotfi A. Zadeh, eds.), Springer US, Boston, MA, 2000, pp. 431–446.
375. ______, *Combination of belief functions and coarsening/refinement*, Proceedings of the International Conference on Information Processing and Management Uncertainty in Knowledge-based Systems (IPMU02), 2002, pp. 587–594.
376. ______, *Associativity in combination of belief functions; a derivation of minC combination*, Soft Computing **7** (2003), no. 5, 288–296.

377. ———, *Transformations of belief functions to probabilities*, Proceedings of the 6th Worskop on Uncertainty Processing (WUPES 2003) (2003).
378. ———, *Algebraic structures related to the combination of belief functions*, Scientiae Mathematicae Japonicae **60** (2004), no. 2, 245–256.
379. ———, *Consistency of probabilistic transformations of belief functions*, Proceedings of the International Conference on Information Processing and Management of Uncertainty in Knowledge-Based Systems (IPMU 2004), 2004, pp. 1135–1142.
380. ———, *Consistency of probabilistic transformations of belief functions*, Modern Information Processing, Elsevier, 2006, pp. 49–60.
381. ———, *MinC combination of belief functions: derivation and formulas*, Tech. Report 964, Academy of Science of the Czech Republic, 2006.
382. ———, *On transformations of belief functions to probabilities*, International Journal of Intelligent Systems **21** (2006), no. 3, 261–282.
383. ———, *Contribution of DSm approach to the belief function theory*, Proceedings of the 12th International Conference on Information Processing and Management of Uncertainty in Knowledge-Based Systems (IPMU 2008), vol. 8, 2008, pp. 417–424.
384. ———, *New approach to conflicts within and between belief functions*, Tech. Report 1062, Institute of Computer Science, Academy of Sciences of the Czech Republic, 2009.
385. ———, *A generalization of the classic combination rules to DSm hyper-power sets*, Information and Security Journal **20** (January 2006), 50–64.
386. ———, *Non-conflicting and conflicting parts of belief functions*, Proceedings of the 7th International Symposium on Imprecise Probability: Theories and Applications (ISIPTA'11), July 2011.
387. Milan Daniel, *Introduction to an algebra of belief functions on three-element frame of discernment — A quasi Bayesian case*, Proceedings of the 14th International Conference on Information Processing and Management of Uncertainty in Knowledge-Based Systems (IPMU 2012) (Berlin, Heidelberg) (Salvatore Greco, Bernadette Bouchon-Meunier, Giulianella Coletti, Mario Fedrizzi, Benedetto Matarazzo, and Ronald R. Yager, eds.), vol. III, Springer, July 2012, pp. 532–542.
388. ———, *Conflicts within and between belief functions*, Proceedings of the 13th International Conference on Information Processing and Management of Uncertainty (IPMU 2010) (Eyke Hüllermeier, Rudolf Kruse, and Frank Hoffmann, eds.), Springer, Berlin Heidelberg, June-July 2010, pp. 696–705.
389. Milan Daniel, *Algebraic structures related to Dempster–Shafer theory*, Proceedings of the 5th International Conference on Information Processing and Management of Uncertainty in Knowledge-Based Systems (IPMU'94) (B. Bouchon-Meunier, R. R. Yager, and L. A. Zadeh, eds.), Paris, France, 4-8 July 1994, pp. 51–61.
390. Mats Danielson and Love Ekenberg, *A framework for analysing decisions under risk*, European Journal of Operational Research **104** (1998), no. 3, 474–484.
391. Vladimir I. Danilov and Gleb A. Koshevoy, *Cores of cooperative games, superdifferentials of functions and the Minkovski difference of sets*, Journal of Mathematical Analysis Applications **247** (2000), no. 1, 1–14.
392. A. Philip Dawid and Vladimir G. Vovk, *Prequential probability: Principles and properties*, Bernoulli **5** (1999), no. 1, 125–162.
393. Luis M. de Campos, Juan F. Huete, and Serafín Moral, *Probability intervals: a tool for uncertain reasoning*, International Journal of Uncertainty, Fuzziness and Knowledge-Based Systems **2** (1994), no. 2, 167–196.
394. Gert de Cooman, *A gentle introduction to imprecise probability models and their behavioural interpretation*, SIPTA, 2003.

395. Gert de Cooman, *A behavioural model for vague probability assessments*, Fuzzy Sets and Systems **154** (2005), no. 3, 305–358.
396. Gert de Cooman and Dirk Aeyels, *A random set description of a possibility measure and its natural extension*, IEEE Transactions on Systems, Man, and Cybernetics-Part A: Systems and Humans **30** (2000), no. 2, 124–130.
397. Gert de Cooman and Filip Hermans, *Imprecise probability trees: Bridging two theories of imprecise probability*, Artificial Intelligence **172** (2008), no. 11, 1400–1427.
398. Gert de Cooman, Filip Hermans, and Erik Quaeghebeur, *Imprecise Markov chains and their limit behavior*, Tech. report, Universiteit Gent, 2009.
399. ______, *Imprecise markov chains and their limit behavior*, Probability in the Engineering and Informational Sciences **23** (2009), no. 4, 597–635.
400. Gert de Cooman and Enrique Miranda, *Weak and strong laws of large numbers for coherent lower previsions*, Journal of Statistical Planning and Inference **138** (2008), no. 8, 2409–2432.
401. Gert de Cooman and Marco Zaffalon, *Updating beliefs with incomplete observations*, Artificial Intelligence **159** (2004), no. 1-2, 75–125.
402. J. Kampé de Fériet, *Interpretation of membership functions of fuzzy sets in terms of plausibility and belief*, Fuzzy Information and Decision Processes (M. M. Gupta and E. Sanchez, eds.), vol. 1, North-Holland, Amsterdam, 1982, pp. 93–98.
403. Bruno de Finetti, *Theory of probability*, Wiley, London, 1974.
404. ______, *Foresight. Its logical laws, its subjective sources*, Breakthroughs in statistics, Springer, New York, NY, 1992, pp. 134–174.
405. Johan De Kleer, *An assumption-based TMS*, Artificial intelligence **28** (1986), no. 2, 127–162.
406. André de Korvin and Margaret F. Shipley, *A Dempster–Shafer-based approach to compromise decision making with multiattributes applied to product selection*, IEEE Transactions on Engineering Management **40** (1993), no. 1, 60–67.
407. Florence Dupin de Saint-Cyr, Jérôme Lang, and Thomas Schiex, *Penalty logic and its link with Dempster–Shafer theory*, Proceedings of the Tenth International Conference on Uncertainty in Artificial Intelligence (UAI'94), 1994, pp. 204–211.
408. Florence Dupin De Saint-Cyr, Jérôme Lang, and Thomas Schiex, *Penalty logic and its link with Dempster–Shafer theory*, Uncertainty Proceedings 1994, Elsevier, 1994, pp. 204–211.
409. François Delmotte and Laurent Dubois, *Multi-models and belief functions for decision support systems*, Proceedings of the IEEE International Conference on Systems, Man, and Cybernetics, vol. 1, October 1996, pp. 181–186.
410. François Delmotte and David Gacquer, *Detection of defective sources with belief functions*, Proceedings of the 12th International Conference Information Processing and Management of Uncertainty for Knowledge-Based Systems (IPMU 2008), 2008, pp. 337–344.
411. Sabrina Démotier, W. Schon, Thierry Denœux, and Khaled Odeh, *A new approach to assess risk in water treatment using the belief function framework*, Proceedings of the IEEE International Conference on Systems, Man and Cybernetics (SMC'03), vol. 2, October 2003, pp. 1792–1797.
412. Sabrina Démotier, Walter Schon, and Thierry Denœux, *Risk assessment based on weak information using belief functions: a case study in water treatment*, IEEE Transactions on Systems, Man and Cybernetics, Part C **36** (May 2006), no. 3, 382–396.
413. Arthur P. Dempster, *New methods for reasoning towards posterior distributions based on sample data*, The Annals of Mathematical Statistics **37** (1966), no. 2, 355–374.

414. ______, *Upper and lower probabilities induced by a multivalued mapping*, Annals of Mathematical Statistics **38** (1967), no. 2, 325–339.
415. ______, *Upper and lower probability inferences based on a sample from a finite univariate population*, Biometrika **54** (1967), no. 3-4, 515–528.
416. ______, *A generalization of Bayesian inference*, Journal of the Royal Statistical Society, Series B **30** (1968), no. 2, 205–247.
417. ______, *Upper and lower probabilities generated by a random closed interval*, Annals of Mathematical Statistics **39** (1968), no. 3, 957–966.
418. ______, *Upper and lower probabilities inferences for families of hypothesis with monotone density ratios*, Annals of Mathematical Statistics **40** (1969), no. 3, 953–969.
419. ______, *Bayes, Fischer, and belief fuctions*, Bayesian and Likelihood Methods in Statistics and Economics. Essays in Honor of George Barnard (S. Geisser, J. S. Hodges, S. J. Press, and A. Zellner, eds.), North-Holland, 1990, pp. 35–47.
420. ______, *Construction and local computation aspects of network belief functions*, Influence Diagrams, Belief Nets and Decision Analysis (R. M. Oliver and J. Q. Smith, eds.), Wiley, 1990.
421. ______, *Normal belief functions and the Kalman filter*, Tech. report, Department of Statistics, Harvard Univerisity, 1990.
422. ______, *Belief functions in the 21st century: A statistical perspective*, Proceedings of the Institute for Operations Research and Management Science Annual meeting (INFORMS), 2001.
423. ______, *Normal belief functions and the Kalman filter*, Data Analysis From Statistical Foundations (A. K. M. E. Saleh, ed.), Commack, NY: Nova, 2001, pp. 65–84.
424. ______, *A generalization of Bayesian inference*, Classic Works of the Dempster-Shafer Theory of Belief Functions, Springer, 2008, pp. 73–104.
425. ______, *A generalization of Bayesian inference*, Classic Works of the Dempster-Shafer Theory of Belief Functions (Roland R. Yager and Liping Liu, eds.), Studies in Fuzziness and Soft Computing, Springer, 2008, pp. 73–104.
426. ______, *Lindley's paradox: Comment*, Journal of the American Statistical Association **77** (June 1982), no. 378, 339–341.
427. Arthur P. Dempster and Wai Fung Chiu, *Dempster–Shafer models for object recognition and classification*, International Journal of Intelligent Systems **21** (2006), no. 3, 283–297.
428. Arthur P. Dempster and A. Kong, *Uncertain evidence and artificial analysis*, Tech. Report S-108, Department of Statistics, Harvard University, 1986.
429. Arthur P. Dempster and Augustine Kong, *Probabilistic expert systems in medicine: Practical issues in handling uncertainty: Comment*, Journal of Statistical Sciences **2** (February 1987), no. 1, 32–36.
430. Arthur P. Dempster, Nan M. Laird, and Donald B. Rubin, *Maximum likelihood from incomplete data via the EM algorithm*, Journal of the Royal Statistical Society. Series B (methodological) **39** (1977), no. 1, 1–22.
431. Eric den Breejen, Klamer Schutte, and Frank Cremer, *Sensor fusion for antipersonnel landmine detection: a case study*, Proceedings of SPIE - Detection and Remediation Technologies for Mines and Minelike Targets IV, vol. 3710, 1999, pp. 1235–1245.
432. Mingrong Deng, Weixuan Xu, and Jian-Bo Yang, *Estimating the attribute weights through evidential reasoning and mathematical programming*, International Journal of Information Technology and Decision Making **3** (2004), no. 3, 419–428.
433. Yong Deng and Felix T. S. Chan, *A new fuzzy Dempster MCDM method and its application in supplier selection*, Expert Systems with Applications **38** (2011), no. 8, 9854–9861.

434. Yong Deng, Wen Jiang, and Rehan Sadiq, *Modeling contaminant intrusion in water distribution networks: A new similarity-based DST method*, Expert Systems with Applications **38** (2011), no. 1, 571–578.
435. Yong Deng and Wen kang Shi, *A modified combination rule of evidence theory*, Journal-Shanghai Jiaotong University **37** (2003), no. 8, 1275–1278.
436. Yong Deng, Rehan Sadiq, Wen Jiang, and Solomon Tesfamariam, *Risk analysis in a linguistic environment: A fuzzy evidential reasoning-based approach*, Expert Systems with Applications **38** (2011), no. 12, 15438–15446.
437. Yong Deng, Dong Wang, and Qi Li, *An improved combination rule in fault diagnosis based on Dempster Shafer theory*, Proceedings of the International Conference on Machine Learning and Cybernetics, vol. 1, July 2008, pp. 212–216.
438. ______, *An improved combination rule in fault diagnosis based on Dempster Shafer theory*, Proceedings of the 2008 International Conference on Machine Learning and Cybernetics, vol. 1, July 2008, pp. 212–216.
439. Dieter Denneberg, *Conditioning (updating) non-additive measures*, Annals of Operations Research **52** (1994), no. 1, 21–42.
440. ______, *Representation of the Choquet integral with the 6-additive Möbius transform*, Fuzzy Sets and Systems **92** (1997), no. 2, 139–156.
441. ______, *Totally monotone core and products of monotone measures*, International Journal of Approximate Reasoning **24** (2000), no. 2-3, 273–281.
442. Dieter Denneberg and Michel Grabisch, *Interaction transform of set functions over a finite set*, Information Sciences **121** (1999), no. 1-2, 149–170.
443. Thierry Denœux, *A k-nearest neighbor classification rule based on Dempster–Shafer theory*, IEEE Transactions on Systems, Man, and Cybernetics **25** (1995), no. 5, 804–813.
444. ______, *Analysis of evidence-theoretic decision rules for pattern classification*, Pattern Recognition **30** (1997), no. 7, 1095–1107.
445. ______, *Reasoning with imprecise belief structures*, International Journal of Approximate Reasoning **20** (1999), no. 1, 79–111.
446. Thierry Denœux, *Allowing imprecision in belief representation using fuzzy-valued belief structures*, Information, Uncertainty and Fusion (Bernadette Bouchon-Meunier, Ronald R. Yager, and Lotfi A. Zadeh, eds.), Springer US, Boston, MA, 2000, pp. 269–281.
447. ______, *Modeling vague beliefs using fuzzy-valued belief structures*, Fuzzy sets and systems **116** (2000), no. 2, 167–199.
448. Thierry Denœux, *A neural network classifier based on Dempster–Shafer theory*, IEEE Transactions on Systems, Man, and Cybernetics - Part A: Systems and Humans **30** (2000), no. 2, 131–150.
449. ______, *Inner and outer approximation of belief structures using a hierarchical clustering approach*, International Journal of Uncertainty, Fuzziness and Knowledge-Based Systems **9** (2001), no. 4, 437–460.
450. ______, *Inner and outer approximation of belief structures using a hierarchical clustering approach*, International Journal of Uncertainty, Fuzziness and Knowledge-Based Systems **9** (2001), no. 4, 437–460.
451. ______, *The cautious rule of combination for belief functions and some extensions*, Proceedings of the 9th International Conference on Information Fusion (FUSION 2006) (S. Coraluppi, A. Baldacci, C. Carthel, P. Willett, R. Lynch, S. Marano, and A. Farina, eds.), 2006, pp. 1–8.
452. ______, *Constructing belief functions from sample data using multinomial confidence regions*, International Journal of Approximate Reasoning **42** (2006), no. 3, 228–252.

453. ______, *Construction of predictive belief functions using a frequentist approach*, Proceedings of Information Processing with Management of Uncertainty in Knowledge-based Systems (IPMU'06), 2006, pp. 1412–1419.
454. ______, *Conjunctive and disjunctive combination of belief functions induced by non distinct bodies of evidence*, Artificial Intelligence (2008), no. 2-3, 234–264.
455. ______, *Conjunctive and disjunctive combination of belief functions induced by nondistinct bodies of evidence*, Artificial Intelligence **172** (2008), no. 2, 234–264.
456. Thierry Denœux, *A k-nearest neighbor classification rule based on Dempster–Shafer theory*, Classic Works of the Dempster-Shafer Theory of Belief Functions (Roland R. Yager and Liping Liu, eds.), Studies in Fuzziness and Soft Computing, vol. 219, Springer, 2008, pp. 737–760.
457. Thierry Denœux, *Extending stochastic ordering to belief functions on the real line*, Information Sciences **179** (2009), no. 9, 1362–1376.
458. Thierry Denœux, *Maximum likelihood from evidential data: An extension of the EM algorithm*, Combining Soft Computing and Statistical Methods in Data Analysis (Christian Borgelt, Gil González-Rodríguez, Wolfgang Trutschnig, María Asunción Lubiano, María Ángeles Gil, Przemysław Grzegorzewski, and Olgierd Hryniewicz, eds.), Springer, Berlin, Heidelberg, 2010, pp. 181–188.
459. Thierry Denœux, *Maximum likelihood estimation from uncertain data in the belief function framework*, IEEE Transactions on Knowledge and Data Engineering **25** (2013), no. 1, 119–130.
460. ______, *Likelihood-based belief function: Justification and some extensions to low-quality data*, International Journal of Approximate Reasoning **55** (2014), no. 7, 1535–1547.
461. ______, *Rejoinder on 'Likelihood-based belief function: Justification and some extensions to low-quality data'*, International Journal of Approximate Reasoning **55** (2014), no. 7, 1614–1617.
462. ______, *Decision-making with belief functions*, 2015.
463. ______, *Allowing imprecision in belief representation using fuzzy-valued belief structures*, Proceedings of the International Conference on Information Processing and Management of Uncertainty in Knowledge-Based Systems (IPMU'98), vol. 1, July 1998, pp. 48–55.
464. ______, *Function approximation in the framework of evidence theory: A connectionist approach*, Proceedings of the International Conference on Neural Networks (ICNN'97), vol. 1, June 1997, pp. 199–203.
465. ______, *An evidence-theoretic neural network classifier*, Proceedings of the IEEE International Conference on Systems, Man, and Cybernetics (SMC'95), vol. 3, October 1995, pp. 712–717.
466. Thierry Denœux and Amel Ben Yaghlane, *Approximating the combination of belief functions using the fast Möbius transform in a coarsened frame*, International Journal of Approximate Reasoning **31** (October 2002), no. 1–2, 77–101.
467. Thierry Denœux and M. Skarstein Bjanger, *Induction of decision trees from partially classified data using belief functions*, Proceedings of the IEEE International Conference on Systems, Man, and Cybernetics (SMC 2000), vol. 4, 2000, pp. 2923–2928.
468. Thierry Denœux and Arthur P. Dempster, *The Dempster–Shafer calculus for statisticians*, International Journal of Approximate Reasoning **48** (2008), no. 2, 365–377.
469. Thierry Denœux and Gérard Govaert, *Combined supervised and unsupervised learning for system diagnosis using Dempster–Shafer theory*, Proceedings of the International Conference on Computational Engineering in Systems Applications, Sympo-

sium on Control, Optimization and Supervision, CESA '96 IMACS Multiconference, vol. 1, Lille, France, 9-12 July 1996, pp. 104–109.
470. Thierry Denœux, Orakanya Kanjanatarakul, and Songsak Sriboonchitta, *EK-NNclus: A clustering procedure based on the evidential K-nearest neighbor rule*, Knowledge-Based Systems **88** (2015), 57–69.
471. Thierry Denœux, Jürg Kohlas, and Paul-André Monney, *An algebraic theory for statistical information based on the theory of hints*, International Journal of Approximate Reasoning **48** (2008), no. 2, 378–398.
472. Thierry Denœux and Marie-Hélène Masson, *Evidential reasoning in large partially ordered sets*, Annals of Operations Research **195** (2012), no. 1, 135–161.
473. Thierry Denœux and Marie-Hélène Masson, *EVCLUS: evidential clustering of proximity data*, IEEE Transactions on Systems, Man, and Cybernetics, Part B (Cybernetics) **34** (February 2004), no. 1, 95–109.
474. Thierry Denœux and Marie-Hélène Masson, *Clustering of proximity data using belief functions*, Proceedings of the Ninth International Conference on Information Processing and Management of Uncertainty in Knowledge-based Systems (IPMU'2002), July 2002, pp. 609–616.
475. Thierry Denœux and Philippe Smets, *Classification using belief functions: Relationship between case-based and model-based approaches*, IEEE Transactions on Systems, Man, and Cybernetics Part B **36** (December 2006), no. 6, 1395–1406.
476. Thierry Denœux, Zoulficar Younes, and Fahed Abdallah, *Representing uncertainty on set-valued variables using belief functions*, Artificial Intelligence **174** (2010), no. 7, 479–499.
477. Thierry Denœux and Lalla M. Zouhal, *Handling possibilistic labels in pattern classification using evidential reasoning*, Fuzzy Sets and Systems **122** (2001), no. 3, 409–424.
478. Michel C. Desmarais and Jiming Liu, *Exploring the applications of user-expertise assessment for intelligent interfaces*, Proceedings of the INTERACT '93 and CHI '93 Conference on Human Factors in Computing Systems (CHI '93), 1993, pp. 308–313.
479. Michel C. Desmarais and Jiming Liu, *Experimental results on user knowledge assessment with an evidential reasoning methodology*, Proceedings of the 1993 International Workshop on Intelligent User Interfaces (W. D. Gray, W. E. Hefley, and D. Murray, eds.), January 1993, pp. 223–225.
480. Sébastien Destercke, *Independence concepts in evidence theory: some results about epistemic irrelevance and imprecise belief functions*, Proceedings of the First Workshop on the theory of belief functions (BELIEF 2010), 2010.
481. Sébastien Destercke and Thomas Burger, *Revisiting the notion of conflicting belief functions*, Belief Functions: Theory and Applications: Proceedings of the 2nd International Conference on Belief Functions, Compiègne, France 9-11 May 2012 (Thierry Denœux and Marie-Hélène Masson, eds.), Springer, Berlin, Heidelberg, 2012, pp. 153–160.
482. Sébastien Destercke and Thomas Burger, *Toward an axiomatic definition of conflict between belief functions*, IEEE Transactions on Cybernetics **43** (2013), no. 2, 585–596.
483. Sébastien Destercke and Didier Dubois, *Idempotent conjunctive combination of belief functions: Extending the minimum rule of possibility theory*, Information Sciences **181** (2011), no. 18, 3925–3945.
484. Sébastien Destercke and Didier Dubois, *Can the minimum rule of possibility theory be extended to belief functions?*, Proceedings of the 10th European Conference on Symbolic and Quantitative Approaches to Reasoning with Uncertainty (ECSQARU

2009) (Claudio Sossai and Gaetano Chemello, eds.), Springer, Berlin, Heidelberg, July 2009, pp. 299–310.
485. Sébastien Destercke, Didier Dubois, and Eric Chojnacki, *Unifying practical uncertainty representations – I: Generalized p-boxes*, International Journal of Approximate Reasoning **49** (2008), no. 3, 649–663.
486. Sébastien Destercke, Didier Dubois, and Eric Chojnacki, *Unifying Practical Uncertainty Representations: I. Generalized P-Boxes*, ArXiv preprint arXiv:0808.2747, 2008.
487. Sébastien Destercke, Didier Dubois, and Eric Chojnacki, *Cautious conjunctive merging of belief functions*, Proceedings of the 9th European Conference on Symbolic and Quantitative Approaches to Reasoning with Uncertainty (ECSQARU 2007) (Khaled Mellouli, ed.), Springer, Berlin, Heidelberg, October-November 2007, pp. 332–343.
488. Mary Deutsch-McLeish, *A study of probabilities and belief functions under conflicting evidence: comparisons and new method*, Proceedings of the 3rd International Conference on Information Processing and Management of Uncertainty in Knowledge-Based Systems (IPMU'90) (B. Bouchon-Meunier, R. R. Yager, and L. A. Zadeh, eds.), July 1990, pp. 41–49.
489. Mary Deutsch-McLeish, Paulyn Yao, and Fei Song, *Knowledge acquisition methods for finding belief functions with an application to medical decision making*, Proceedings of the International Symposium on Artificial Intelligence Applications in Informatics (F. J. Cantu-Ortiz and H. Terashima-Marin, eds.), November 1991, pp. 231–237.
490. Jean Dezert, *Foundations for a new theory of plausible and paradoxical reasoning*, Information and Security (Tzv. Semerdjiev, ed.), vol. 9, Bulgarian Academy of Sciences, 2002, pp. 13–57.
491. Jean Dezert, *An introduction to the theory of plausible and paradoxical reasoning*, Numerical Methods and Applications: 5th International Conference, NMA 2002 Borovets, Bulgaria, August 20–24, 2002 Revised Papers (Ivan Dimov, Ivan Lirkov, Svetozar Margenov, and Zahari Zlatev, eds.), Springer, Berlin, Heidelberg, 2003, pp. 12–23.
492. Jean Dezert and Florentin Smarandache, *On the generation of hyper-powersets for the DSmT*, Proceedings of the International Conference on Information Fusion (FUSION 2003), 2003, pp. 8–11.
493. Jean Dezert and Florentin Smarandache, *A new probabilistic transformation of belief mass assignment*, Proceedings of the 11th International Conference on Information Fusion (FUSION 2008), 2008, pp. 1–8.
494. Jean Dezert and Florentin Smarandache, *Partial ordering of hyper-powersets and matrix representation of belief functions within DSmT*, Proceedings of the Sixth International Conference of Information Fusion (FUSION 2003), vol. 2, July 2003, pp. 1230–1238.
495. Jean Dezert, Florentin Smarandache, and Milan Daniel, *The generalized pignistic transformation*, Proceedings of the Seventh International Conference on Information Fusion (FUSION 2004), June-July 2004, pp. 384–391.
496. Jean Dezert, Jean-Marc Tacnet, Mireille Batton-Hubert, and Florentin Smarandache, *Multi-criteria decision making based on DSmT-AHP*, Proceedings of the International Workshop on Belief Functions (BELIEF 2010), April 2010.
497. Jean Dezert, Albena Tchamova, Florentin Smarandache, and Pavlina Konstantinova, *Target type tracking with PCR5 and Dempster's rules: A comparative analysis*, Proceedings of the 9th International Conference on Information Fusion (FUSION 2006), July 2006, pp. 1–8.

498. Jean Dezert, Pei Wang, and Albena Tchamova, *On the validity of Dempster–Shafer theory*, Proceedings of the 15th International Conference on Information Fusion (FUSION 2012), July 2012, pp. 655–660.
499. Persi Diaconis, *Review of 'a mathematical theory of evidence'*, Journal of American Statistical Society **73** (1978), no. 363, 677–678.
500. Persi Diaconis and Sandy L. Zabell, *Updating subjective probability*, Journal of the American Statistical Association **77** (1982), no. 380, 822–830.
501. Javier Diaz, Maria Rifqi, and Bernadette Bouchon-Meunier, *A similarity measure between basic belief assignments*, Proceedings of the 9th International Conference on Information Fusion (FUSION'06), 2006, pp. 1–6.
502. L. Díaz-Más, R. Muñoz Salinas, F. J. Madrid-Cuevas, and R. Medina-Carnicer, *Shape from silhouette using Dempster–Shafer theory*, Pattern Recognition **43** (June 2010), no. 6, 2119–2131.
503. Robert P. Dilworth, *Dependence relations in a semimodular lattice*, Duke Mathematical Journal **11** (1944), no. 3, 575–587.
504. Shuai Ding, Shan-Lin Yang, and Chao Fu, *A novel evidential reasoning based method for software trustworthiness evaluation under the uncertain and unreliable environment*, Expert Systems with Applications **39** (2012), no. 3, 2700–2709.
505. Zoltan Domotor, *Higher order probabilities*, Philosophical Studies **40** (1981), 31–46.
506. Fei Dong, Sol M. Shatz, and Haiping Xu, *Inference of online auction shills using Dempster–Shafer theory*, Proceedings of the Sixth International Conference on Information Technology: New Generations (ITNG '09), April 2009, pp. 908–914.
507. Qu Dongcai, Meng Xiangwei, Huang Juan, and He You, *Research of artificial neural network intelligent recognition technology assisted by Dempster–Shafer evidence combination theory*, Proceedings of the 7th International Conference on Signal Processing (ICSP '04), vol. 1, August 2004, pp. 46–49.
508. Pierre-Emmanuel Doré, Anthony Fiche, and Arnaud Martin, *Models of belief functions – Impacts for patterns recognitions*, Proceedings of the 13th International Conference on Information Fusion (FUSION 2010), 2010, pp. 1–6.
509. Pierre-Emmanuel Doré, Arnaud Martin, Irène Abi-Zeid, Anne-Laure Jousselme, and Patrick Maupin, *Theory of belief functions for information combination and update in search and rescue operations*, Proceedings of the 12th International Conference on Information Fusion (FUSION'09), 2009, pp. 514–521.
510. Doré, Pierre-Emmanuel, Martin, Arnaud, Abi-Zeid, Irène, Jousselme, Anne-Laure, and Maupin, Patrick, *Belief functions induced by multimodal probability density functions, an application to the search and rescue problem*, RAIRO-Operations Research - Recherche Opérationnelle **44** (2010), no. 4, 323–343.
511. Arnaud Doucet and Adam M. Johansen, *A tutorial on particle filtering and smoothing: Fifteen years later*, Handbook of nonlinear filtering **12** (2009), no. 656–704, 3.
512. James Dow and Sergio Ribeiro da Costa Werlang, *Excess volatility of stock prices and knightian uncertainty*, European Economic Review **36** (1992), no. 2-3, 631–638.
513. James Dow and Sergio Werlang, *Uncertainty aversion, risk aversion, and the optimal choice of portfolio*, Econometrica **60** (1992), no. 1, 197–204.
514. Aldo Franco Dragoni, Paolo Giorgini, and Alessandro Bolognini, *Distributed knowledge elicitation through the Dempster–Shafer theory of evidence: a simulation study*, Proceedings of the Second International Conference on Multi-Agent Systems (ICMAS'96), December 1996, p. 433.
515. Cyril Drocourt, Laurent Delahoche, Claude Pegard, and Arnaud Clerentin, *Mobile robot localization based on an omnidirectional stereoscopic vision perception sys-*

tem, Proceedings of the IEEE International Conference on Robotics and Automation (ICRA'99), vol. 2, 1999, pp. 1329–1334.

516. Feng Du, Wen kang Shi, and Yong Deng, *Feature extraction of evidence and its application in modification of evidence theory*, Journal of Shanghai Jiaotong University **38** (2004), 164–168.
517. Werner Dubitzky, Alex G. Bochner, John G. Hughes, and David A. Bell, *Towards concept-oriented databases*, Data and Knowledge Engineering **30** (1999), no. 1, 23–55.
518. Didier Dubois and Thierry Denœux, *Statistical inference with belief functions and possibility measures: A discussion of basic assumptions*, Combining Soft Computing and Statistical Methods in Data Analysis (Christian Borgelt, Gil González-Rodríguez, Wolfgang Trutschnig, María Asunción Lubiano, María Ángeles Gil, Przemysław Grzegorzewski, and Olgierd Hryniewicz, eds.), Springer, Berlin Heidelberg, 2010, pp. 217–225.
519. Didier Dubois, Hélène Fargie, and Henri Prade, *Comparative uncertainty, belief functions and accepted beliefs*, Proceedings of the Fourteenth Conference on Uncertainty in Artificial Intelligence (UAI'98), Morgan Kaufmann, 1998, pp. 113–120.
520. Didier Dubois, Michel Grabisch, François Modave, and Henri Prade, *Relating decision under uncertainty and MCDM models*, Proceedings of the AAAI Fall Symposium on Frontiers in Soft Computing and Decisions Systems, 1997, pp. 6–15.
521. Didier Dubois, Michel Grabisch, Henri Prade, and Philippe Smets, *Assessing the value of a candidate: Comparing belief function and possibility theories*, Proceedings of the Fifteenth Conference on Uncertainty in Artificial Intelligence (UAI'99), Morgan Kaufmann, 1999, pp. 170–177.
522. Didier Dubois, Michel Grabisch, Henri Prade, and Philippe Smets, *Using the transferable belief model and a qualitative possibility theory approach on an illustrative example: the assessment of the value of a candidate*, International journal of intelligent systems **16** (2001), no. 11, 1245–1272.
523. Didier Dubois and Henri Prade, *On several representations of an uncertain body of evidence*, Fuzzy Information and Decision Processes (M. M. Gupta and E. Sanchez, eds.), North Holland, 1982, pp. 167–181.
524. ______, *Unfair coins and necessity measures: Towards a possibilistic interpretation of histograms*, Fuzzy Sets and Systems **10** (1983), no. 1, 15–20.
525. Didier Dubois and Henri Prade, *Combination and propagation of uncertainty with belief functions: A reexamination*, Proceedings of the 9th International Joint Conference on Artificial Intelligence (IJCAI'85), vol. 1, Morgan Kaufmann, 1985, pp. 111–113.
526. ______, *On the unicity of Dempster rule of combination*, International Journal of Intelligent Systems **1** (1986), no. 2, 133–142.
527. Didier Dubois and Henri Prade, *A set-theoretic view of belief functions Logical operations and approximations by fuzzy sets*, International Journal of General Systems **12** (1986), no. 3, 193–226.
528. ______, *The mean value of a fuzzy number*, Fuzzy Sets and Systems **24** (1987), no. 3, 279–300.
529. ______, *The principle of minimum specificity as a basis for evidential reasoning*, Uncertainty in Knowledge-Based Systems (B. Bouchon and R. R. Yager, eds.), Springer-Verlag, 1987, pp. 75–84.
530. ______, *Properties of measures of information in evidence and possibility theories*, Fuzzy Sets and Systems **24** (1987), no. 2, 161–182.
531. ______, *Possibility theory*, Plenum Press, New York, 1988.

532. ______, *Representation and combination of uncertainty with belief functions and possibility measures*, Computational Intelligence **4** (1988), no. 3, 244–264.
533. ______, *Consonant approximations of belief functions*, International Journal of Approximate Reasoning **4** (1990), 419–449.
534. ______, *Modeling uncertain and vague knowledge in possibility and evidence theories*, Uncertainty in Artificial Intelligence, volume 4 (R. D. Shachter, T. S. Levitt, L. N. Kanal, and J. F. Lemmer, eds.), North-Holland, 1990, pp. 303–318.
535. ______, *Epistemic entrenchment and possibilistic logic*, Artificial Intelligence **50** (1991), no. 2, 223–239.
536. ______, *Focusing versus updating in belief function theory*, Tech. Report IRIT/91-94/R, IRIT, Universite P. Sabatier, Toulouse, 1991.
537. ______, *Random sets and fuzzy interval analysis*, Fuzzy Sets and Systems **42** (1991), no. 1, 87–101.
538. ______, *Evidence, knowledge, and belief functions*, International Journal of Approximate Reasoning **6** (1992), no. 3, 295–319.
539. ______, *On the combination of evidence in various mathematical frameworks*, Reliability Data Collection and Analysis (J. Flamm and T. Luisi, eds.), Springer, Dordrecht, 1992, pp. 213–241.
540. ______, *On the relevance of non-standard theories of uncertainty in modeling and pooling expert opinions*, Reliability Engineering and System Safety **36** (1992), no. 2, 95–107.
541. ______, *A survey of belief revision and updating rules in various uncertainty models*, International Journal of Intelligent Systems **9** (1994), no. 1, 61–100.
542. ______, *Bayesian conditioning in possibility theory*, Fuzzy Sets and Systems **92** (1997), no. 2, 223–240.
543. ______, *Properties of measures of information in evidence and possibility theories*, Fuzzy Sets and Systems **100** (1999), 35–49.
544. Didier Dubois and Henri Prade, *Possibility theory: an approach to computerized processing of uncertainty*, Springer Science & Business Media, 2012.
545. ______, *A tentative comparison of numerical approximate reasoning methodologies*, International Journal of Man-Machine Studies **27** (November 1987), no. 5-6, 717–728.
546. Didier Dubois, Henri Prade, and Sandra Sandri, *On possibility/probability transformations*, Fuzzy Logic: State of the Art (R. Lowen and M. Lowen, eds.), Kluwer Academic Publisher, 1993, pp. 103–112.
547. Didier Dubois, Henri Prade, and Philippe Smets, *New semantics for quantitative possibility theory*, Proceedings of the 6th European Conference on Symbolic and Quantitative Approaches to Reasoning and Uncertainty (ECSQARU 2001) (S. Benferhat and Ph. Besnard, eds.), Springer-Verlag, 2001, pp. 410–421.
548. ______, *A definition of subjective possibility*, International Journal of Approximate Reasoning **48** (2008), no. 2, 352–364.
549. Didier Dubois and Arthur Ramer, *Extremal properties of belief measures in the theory of evidence*, International Journal of Uncertainty, Fuzziness and Knowledge-Based Systems **1** (1993), no. 1, 57–68.
550. Didier Dubois and Ronald R. Yager, *Fuzzy set connectives as combinations of belief structures*, Information Sciences **66** (1992), no. 3, 245–276.
551. Boris A. Dubrovin, Sergei P. Novikov, and Anatolij T. Fomenko, *Sovremennaja geometrija. metody i prilozenija*, Nauka, Moscow, 1986.
552. ______, *Geometria contemporanea 3*, Editori Riuniti, 1989.

553. Jacques Dubucs and Glenn Shafer, *A betting interpretation for probabilities and Dempster–Shafer degrees of belief*, International Journal of Approximate Reasoning **52** (2011), no. 2, 127–136.
554. Richard O. Duda, Peter E. Hart, and Nils J. Nilsson, *Subjective Bayesian methods for rule-based inference systems*, Proceedings of the National Computer Conference and Exposition (AFIPS '76), June 1976, pp. 1075–1082.
555. Vincent Dugat and Sandra Sandri, *Complexity of hierarchical trees in evidence theory*, ORSA Journal of Computing **6** (1994), no. 1, 37–49.
556. Stephen D. Durham, Jeffery S. Smolka, and Marco Valtorta, *Statistical consistency with Dempster's rule on diagnostic trees having uncertain performance parameters*, International Journal of Approximate Reasoning **6** (1992), no. 1, 67–81.
557. Amitava Dutta, *Reasoning with imprecise knowledge in expert systems*, Information Sciences **37** (1985), no. 1-3, 3–24.
558. Ludmila Dymova, Pavel Sevastianov, and Pavel Bartosiewicz, *A new approach to the rule-base evidential reasoning: Stock trading expert system application* , Expert Systems with Applications **37** (2010), no. 8, 5564–5576.
559. Ludmila Dymova, Pavel Sevastianov, and Krzysztof Kaczmarek, *A stock trading expert system based on the rule-base evidential reasoning using Level 2 Quotes*, Expert Systems with Applications **39** (2012), no. 8, 7150–7157.
560. Ludmila Dymova and Pavel Sevastjanov, *An interpretation of intuitionistic fuzzy sets in terms of evidence theory: Decision making aspect*, Knowledge-Based Systems **23** (2010), no. 8, 772–782.
561. William F. Eddy, *Approximating belief functions in a rule-based system*, American Journal of Mathematical and Management Sciences **9** (1989), no. 3-4, 211–228.
562. William F. Eddy and Gabriel P. Pei, *Structures of rule-based belief functions*, IBM journal of research and development **30** (1986), no. 1, 93–101.
563. Paul T. Edlefsen, Chuanhai Liu, and Arthur P. Dempster, *Estimating limits from Poisson counting data using Dempster–Shafer analysis*, The Annals of Applied Statistics **3** (2009), no. 2, 764–790.
564. Anthony W. F. Edwards, *The history of likelihood*, 1974, pp. 9–15.
565. Jürgen Eichberger and David Kelsey, *E-capacities and the Ellsberg paradox*, Theory and Decision **46** (1999), no. 2, 107–138.
566. Hillel J. Einhorn and Robin M. Hogarth, *Decision making under ambiguity*, Journal of Business **59** (1986), S225–S250.
567. Siegfried Eisinger and Uwe K. Rakowsky, *Modeling of uncertainties in reliability centered maintenance – a probabilistic approach*, Reliability Engineering and System Safety **71** (2001), no. 2, 159–164.
568. Nour-Eddin El Faouzi, Henry Leung, and Ajeesh Kurian, *Data fusion in intelligent transportation systems: Progress and challenges – A survey*, Information Fusion **12** (2011), no. 1, 4–10, Special Issue on Intelligent Transportation Systems.
569. Daniel Ellsberg, *Risk, ambiguity, and the Savage axioms*, The Quarterly Journal of Economics (1961), 643–669.
570. Zied Elouedi, Khaled Mellouli, and Philippe Smets, *Classification with belief decision trees*, Proceedings of the Ninth International Conference on Artificial Intelligence: Methodology, Systems, Architectures (AIMSA 2000), 2000, pp. 80–90.
571. ———, *Belief decision trees: theoretical foundations*, International Journal of Approximate Reasoning **28** (2001), no. 2, 91–124.
572. ———, *Assessing sensor reliability for multisensor data fusion within the transferable belief model*, IEEE Transactions on Systems, Man, and Cybernetics, Part B (Cybernetics) **34** (February 2004), no. 1, 782–787.

573. ______, *Decision trees using the belief function theory*, Proceedings of the Eighth International Conference on Information Processing and Management of Uncertainty in Knowledge-based Systems (IPMU 2000), vol. 1, Madrid, 2000, pp. 141–148.
574. Kurt J. Engemann, Holmes E. Miller, and Ronald R. Yager, *Decision making with belief structures: an application in risk management*, International Journal of Uncertainty, Fuzziness and Knowledge-Based Systems **4** (1996), no. 1, 1–25.
575. Amel Ennaceur, Zied Elouedi, and Eric Lefévre, *Introducing incomparability in modeling qualitative belief functions*, Modeling Decisions for Artificial Intelligence: 9th International Conference, MDAI 2012, Girona, Catalonia, Spain, November 21-23, 2012. Proceedings (Vicenç Torra, Yasuo Narukawa, Beatriz López, and Mateu Villaret, eds.), Springer, Berlin, Heidelberg, 2012, pp. 382–393.
576. Larry G. Epstein and Kyoungwon Seo, *A central limit theorem for belief functions*, 2011.
577. Aydan M. Erkmen and Harry E. Stephanou, *Information fractals for evidential pattern classification*, IEEE Transactions on Systems, Man, and Cybernetics **20** (September 1990), no. 5, 1103–1114.
578. Mansour Esmaili, Reihaneh Safavi-Naini, and Josef Pieprzyk, *Evidential reasoning in network intrusion detection systems*, Proceedings of the First Australasian Conference on Information Security and Privacy (ACISP '96), Springer-Verlag, 1996, pp. 253–265.
579. Francesc Esteva and Lluís Godo, *On uncertainty and Kripke modalities in t-norm fuzzy logics*, Witnessed Years - Essays in Honour of Petr Hájek (P. Cintula, Z. Haniková, and V. Svejdar, eds.), no. 10, College Publications, 2009, pp. 173–192.
580. Jacob Marschak et al., *Personal probabilities of probabilities*, Theory and Decision **6** (1975), no. 2, 121–153.
581. Marián Fabian, Petr Habala, Petr Hájek, Vicente Montesinos Santalucía, Jan Pelant, and Václav Zizler, *Functional analysis and infinite-dimensional geometry*, Springer Science and Business Media, 2013.
582. Ronald Fagin and Joseph Y. Halpern, *Uncertainty, belief, and probability*, Proceedings of the AAAI Conference on Artificial Intelligence (AAAI'89), 1989, pp. 1161–1167.
583. ______, *Uncertainty, belief and probability*, Proceedings of the 11th International Joint Conference on Artificial intelligence (IJCAI-89), 1989, pp. 1161–1167.
584. Ronald Fagin and Joseph Y. Halpern, *A new approach to updating beliefs*, Proceedings of the Sixth Annual Conference on Uncertainty in Artificial Intelligence (UAI'90), 1990, pp. 347–374.
585. Ronald Fagin and Joseph Y Halpern, *Uncertainty, belief, and probability*, Computational Intelligence **7** (1991), no. 3, 160–173.
586. Ronald Fagin, Joseph Y. Halpern, and Nimrod Megiddo, *A logic for reasoning about probabilities*, Information and computation **87** (July/August 1990), no. 1-2, 78–128.
587. Michael Falk, Jürg Hüsler, and Rolf-Dieter Reiss, *Laws of small numbers: extremes and rare events*, Springer Science and Business Media, 2010.
588. Thomas C. Fall, *Evidential reasoning with temporal aspects*, Proceedings of the Fifth National Conference on Artificial Intelligence (AAAI'86), 1986, pp. 891–895.
589. Xianfeng Fan and Ming J. Zuo, *Fault diagnosis of machines based on D–S evidence theory. Part 1: D–S evidence theory and its improvement*, Pattern Recognition Letters **27** (2006), no. 5, 366–376.
590. Lorenzo Farina and Sergio Rinaldi, *Positive linear systems: theory and applications*, vol. 50, John Wiley & Sons, 2011.

591. Olivier D. Faugeras, *Relaxation labeling and evidence gathering*, Proceedings of the 6th IEEE Computer Society International Conference on Pattern Recognition (ICPR), 1982, pp. 405–412.
592. Francis Faux and Franck Luthon, *Robust face tracking using colour Dempster–Shafer fusion and particle filter*, Proceedings of the 9th International Conference on Information Fusion (FUSION 2006), July 2006, pp. 1–7.
593. Rebecca Fay, Friedhelm Schwenker, Christian Thiel, and Gunther Palm, *Hierarchical neural networks utilising Dempster–Shafer evidence theory*, Artificial Neural Networks in Pattern Recognition (Friedhelm Schwenker and Simone Marinai, eds.), Lecture Notes in Computer Science, vol. 4087, Springer, Berlin Heidelberg, 2006, pp. 198–209.
594. Liguo Fei, Yong Deng, and Sankaran Mahadevan, *Which is the best belief entropy?*, 2015.
595. Bakhtiar Feizizadeh, Thomas Blaschke, and Hossein Nazmfar, *GIS-based ordered weighted averaging and Dempster–Shafer methods for landslide susceptibility mapping in the Urmia Lake Basin, Iran*, International Journal of Digital Earth **7** (2014), no. 8, 688–708.
596. Huamin Feng, Rui Shi, and Tat-Seng Chua, *A bootstrapping framework for annotating and retrieving WWW images*, Proceedings of the 12th Annual ACM International Conference on Multimedia (MULTIMEDIA '04), 2004, pp. 960–967.
597. Ruijia Feng, Guangyuan Zhang, and Bo Cheng, *An on-board system for detecting driver drowsiness based on multi-sensor data fusion using Dempster–Shafer theory*, Proceedings of the International Conference on Networking, Sensing and Control (ICNSC '09), March 2009, pp. 897–902.
598. Tao Feng, Shao-Pu Zhang, and Ju-Sheng Mi, *The reduction and fusion of fuzzy covering systems based on the evidence theory*, International Journal of Approximate Reasoning **53** (2012), no. 1, 87–103.
599. Norman Fenton, Bev Littlewood, Martin Neil, Lorenzo Strigini, Alistair Sutcliffe, and David Wright, *Assessing dependability of safety critical systems using diverse evidence*, IEE Proceedings – Software **145** (1998), no. 1, 35–39.
600. Juan M. Fernndez-Luna, Juan F. Huete, Benjamin Piwowarski, Abdelaziz Kallel, and Sylvie Le Hégarat-Mascle, *Combination of partially non-distinct beliefs: The cautious-adaptive rule*, International Journal of Approximate Reasoning **50** (2009), no. 7, 1000–1021.
601. Carlo Ferrari and Gaetano Chemello, *Coupling fuzzy logic techniques with evidential reasoning for sensor data interpretation*, Proceedings of Intelligent Autonomous Systems (T. Kanade, F. C. A. Groen, and L. O. Hertzberger, eds.), vol. 2, December 1989, pp. 965–971.
602. Scott Ferson, Janos Hajagos, David S. Myers, and W. Troy Tucker, *Constructor: synthesizing information about uncertain variables.*, Tech. Report SAN D2005-3769, Sandia National Laboratories, 2005.
603. Scott Ferson, Vladik Kreinovich, Lev Ginzburg, Davis S. Myers, and Kari Sentz, *Constructing probability boxes and Dempster–Shafer structures*, Tech. Report SAND-2015-4166J, Sandia National Laboratories, January 2003.
604. ———, *Constructing probability boxes and Dempster–Shafer structures*, Tech. Report SAND2002-4015, Sandia National Laboratories, 2003.
605. Scott Ferson, Roger B. Nelsen, Janos Hajagos, Daniel J. Berleant, and Jianzhong Zhang, *Dependence in probabilistic modeling, Dempster–Shafer theory, and probability bounds analysis*, Tech. Report SAND2004-3072, Sandia National Laboratories, October October 2004.

606. Thomas Fetz, *Sets of joint probability measures generated by weighted marginal focal sets*, Proceedings of the 2st International Symposium on Imprecise Probabilities and Their Applications (ISIPTA'01), 2001.
607. ______, *Modelling uncertainties in limit state functions*, International Journal of Approximate Reasoning **53** (2012), no. 1, 1–23.
608. Anthony Fiche, Jean-Christophe Cexus, Arnaud Martin, and Ali Khenchaf, *Features modeling with an α-stable distribution: Application to pattern recognition based on continuous belief functions*, Information Fusion **14** (2013), no. 4, 504–520.
609. ______, *Features modeling with an α-stable distribution: application to pattern recognition based on continuous belief functions*, ArXiv preprint arXiv:1501.05612, January 2015.
610. Anthony Fiche and Arnaud Martin, *Bayesian approach and continuous belief functions for classification*, Proceedings of the Rencontre francophone sur la Logique Floue et ses Applications, 2009, pp. 5–6.
611. Anthony Fiche, Arnaud Martin, Jean-Christophe Cexus, and Ali Khenchaf, *Continuous belief functions and α-stable distributions*, Proceedings of the 13th Conference on Information Fusion (FUSION 2010), 2010, pp. 1–7.
612. Pablo Ignacio Fierens and Terrence L. Fine, *Towards a frequentist interpretation of sets of measures*, Proceedings of the 2nd International Symposium on Imprecise Probabilities and Their Applications (ISIPTA'01), vol. 1, June 2001, pp. 179–187.
613. Arthur Filippidis, *A comparison of fuzzy and Dempster–Shafer evidential reasoning fusion methods for deriving course of action from surveillance observations*, International Journal of Knowledge-Based Intelligent Engineering Systems **3** (October 1999), no. 4, 215–222.
614. ______, *Fuzzy and Dempster–Shafer evidential reasoning fusion methods for deriving action from surveillance observations*, Proceedings of the Third International Conference on Knowledge-Based Intelligent Information Engineering Systems, September 1999, pp. 121–124.
615. Terrence L. Fine, *II - Axiomatic comparative probability*, Theories of Probability (Terrence L. Fine, ed.), Academic Press, 1973, pp. 15–57.
616. ______, *Book reviews: A mathematical theory of evidence*, Bulletin of the American Mathematical Society **83** (1977), no. 4, 667–672.
617. ______, *Lower probability models for uncertainty and nondeterministic processes*, Journal of Statistical Planning and Inference **20** (1988), no. 3, 389–411.
618. Guido Fioretti, *A mathematical theory of evidence for G.L.S. Shackle*, Mind and Society **2** (2001), no. 1, 77–98.
619. Peter C. Fishburn, *Decision and value theory*, Wiley, New York, 1964.
620. Ronald A. Fisher, *On the mathematical foundations of theoretical statistics*, Philosophical Transactions of the Royal Society of London A: Mathematical, Physical and Engineering Sciences **222** (1922), no. 594-604, 309–368.
621. ______, *The fiducial argument in statistical inference*, Annals of Human Genetics **6** (1935), no. 4, 391–398.
622. T. Fister and R. Mitchell, *Modified Dempster–Shafer with entropy based belief body compression*, Proceedings of the 1994 Joint Service Combat Identification Systems Conference (CISC-94), 1994, pp. 281–310.
623. Dale Fixen and Ronald P. S. Mahler, *The modified Dempster–Shafer approach to classification*, IEEE Transactions on Systems, Man, and Cybernetics - Part A: Systems and Humans **27** (January 1997), no. 1, 96–104.

624. Dale Fixsen and Ronald P. S. Mahler, *A Dempster–Shafer approach to Bayesian classification*, Proceedings of the Fifth International Symposium on Sensor Fusion, vol. 1, 1992, pp. 213–231.
625. Tommaso Flaminio and Lluís Godo, *A betting metaphor for belief functions on MV-algebras and fuzzy epistemic states*, Proceedings of the International Workshops on the Logical and Algebraic Aspects of Many-valued Reasoning (MANYVAL 2013) (Prague, Czech Republic) (Tomas Kroupa, ed.), September 2013, pp. 30–32.
626. Tommaso Flaminio, Lluís Godo, and Tomas Kroupa, *Combination and soft-normalization of belief functions on MV-algebras*, Modeling Decisions for Artificial Intelligence: 9th International Conference, MDAI 2012, Girona, Catalonia, Spain, November 21-23, 2012. Proceedings (Vicenc Torra, Yasuo Narukawa, Beatriz Lopez, and Mateu Villaret, eds.), Springer, Berlin Heidelberg, 2012, pp. 23–34.
627. Tommaso Flaminio, Lluís Godo, and Enrico Marchioni, *Belief functions on MV-algebras of fuzzy events based on fuzzy evidence*, Symbolic and Quantitative Approaches to Reasoning with Uncertainty: 11th European Conference, ECSQARU 2011, Belfast, UK, June 29–July 1, 2011. Proceedings (Weiru Liu, ed.), Springer, Berlin Heidelberg, 2011, pp. 628–639.
628. Tommaso Flaminio, Lluís Godo, and Enrico Marchioni, *Logics for belief functions on MV-algebras*, International Journal of Approximate Reasoning **54** (2013), no. 4, 491–512.
629. Mihai C. Florea and Eloi Bossé, *Crisis management using Dempster Shafer theory: Using dissimilarity measures to characterize sources' reliability*, C3I for Crisis, Emergency and Consequence Management, RTO-MP-IST-086, 2009.
630. Mihai C. Florea, Eloi Bossé, and Anne-Laure Jousselme, *Metrics, distances and dissimilarity measures within Dempster–Shafer theory to characterize sources' reliability*, Proceedings of the Cognitive Systems with Interactive Sensors Conference (COGIS09), 2009.
631. Mihai C. Florea, Jean Dezert, Pierre Valin, Florentin Smarandache, and Anne-Laure Jousselme, *Adaptative combination rule and proportional conflict redistribution rule for information fusion*, Proceedings of the Cogis '06 Conference, March 2006.
632. Mihai C. Florea, Anne-Laure Jouselme, Dominic Grenier, and Eloi Bossé, *Combining belief functions and fuzzy membership functions*, Proceedings of SPIE - Sensor Fusion : Architectures, Algorithms, and Applications VII, vol. 5099, 2003, pp. 113–122.
633. Mihai C. Florea, Anne-Laure Jousselme, Dominic Grenier, and Eloi Bossé, *Approximation techniques for the transformation of fuzzy sets into random sets*, Fuzzy Sets and Systems **159** (2008), no. 3, 270–288.
634. Marco Fontani, Tiziano Bianchi, Alessia De Rosa, Alessandro Piva, and Mauro Barni, *A Dempster–Shafer framework for decision fusion in image forensics*, Proceedings of the IEEE International Workshop on Information Forensics and Security (WIFS 2011), November 2011, pp. 1–6.
635. Malcolm R. Forster, *Counterexamples to a likelihood theory of evidence*, Minds and Machines **16** (2006), no. 3, 319–338.
636. Philippe Fortemps, *Jobshop scheduling with imprecise durations: A fuzzy approach*, IEEE Transactions on Fuzzy Systems **5** (1997), no. 4, 557–569.
637. Samuel Foucher, J.-M. Boucher, and Goze B. Benie, *Multiscale and multisource classification using Dempster–Shafer theory*, Proceedings of the International Conference on Image Processing (ICIP'99), vol. 1, 1999, pp. 124–128.
638. Samuel Foucher, Mickael Germain, J. M. Boucher, and Goze B. Benie, *Multisource classification using ICM and Dempster–Shafer theory*, IEEE Transactions on Instrumentation and Measurement **51** (2002), no. 2, 277–281.

639. Samuel Foucher, France Laliberte, Gilles Boulianne, and Langis Gagnon, *A Dempster–Shafer based fusion approach for audio-visual speech recognition with application to large vocabulary French speech*, Proceedings of the 2006 IEEE International Conference on Acoustics, Speech and Signal Processing (ICASSP 2006), vol. 1, May 2006, pp. I–I.
640. Laurent Fouque, Alain Appriou, and Wojciech Pieczynski, *An evidential Markovian model for data fusion and unsupervised image classification*, Proceedings of the Third International Conference on Information Fusion (FUSION 2000), vol. 1, July 2000, pp. TUB4/25–TUB4/32.
641. Jérémie François, Yves Grandvalet, Thierry Denœux, and Jean-Michel Roger, *Resample and combine: an approach to improving uncertainty representation in evidential pattern classification*, Information Fusion **4** (2003), no. 2, 75–85.
642. Jürgen Franke and Eberhard Mandler, *A comparison of two approaches for combining the votes of cooperating classifiers*, Pattern Recognition, 1992. Vol.II. Conference B: Pattern Recognition Methodology and Systems, Proceedings., 11th IAPR International Conference on, 1992, pp. 611–614.
643. Maurice Fréchet, *Les éléments aléatoires de nature quelconque dans un espace distancié*, Annales de l'institut Henri Poincaré, vol. 10, 1948, pp. 215–310.
644. David A. Freedman, *On the asymptotic behavior of Bayes' estimates in the discrete case*, The Annals of Mathematical Statistics (1963), 1386–1403.
645. Chao Fu, Michael Huhns, and Shan-Lin Yang, *A consensus framework for multiple attribute group decision analysis in an evidential reasoning context*, Information Fusion **17** (2014), 22–35, Special Issue: Information fusion in consensus and decision making.
646. Chao Fu and Shan-Lin Yang, *The group consensus based evidential reasoning approach for multiple attributive group decision analysis*, European Journal of Operational Research **206** (2010), no. 3, 601–608.
647. ______, *An evidential reasoning based consensus model for multiple attribute group decision analysis problems with interval-valued group consensus requirements*, European Journal of Operational Research **223** (2012), no. 1, 167–176.
648. ______, *Group consensus based on evidential reasoning approach using interval-valued belief structures*, Knowledge-Based Systems **35** (2012), no. 0, 201–210.
649. ______, *Conjunctive combination of belief functions from dependent sources using positive and negative weight functions*, Expert Systems with Applications **41** (2014), no. 4, Part 2, 1964–1972.
650. Chao Fu and Shanlin Yang, *The combination of dependence-based interval-valued evidential reasoning approach with balanced scorecard for performance assessment*, Expert Systems with Applications **39** (February 2012), no. 3, 3717–3730.
651. Cong fu Xu, Wei-Dong Geng, and Yun-He Pan, *Review of Dempster–Shafer method for data fusion*, Chinese Journal of Electronics **29** (2001), no. 3, 393–396.
652. Pascal Fua, *Using probability-density functions in the framework of evidential reasoning*, Uncertainty in Knowledge-Based Systems (B. Bouchon and R. R. Yager, eds.), Lecture Notes in Computer Science, vol. 286, Springer, Berlin Heidelberg, 1987, pp. 103–110.
653. Robert M. Fung and Chee Yee Chong, *Metaprobability and dempster-shafer in evidential reasoning*, Machine Intelligence and Pattern Recognition, vol. 4, Elsevier, 1986, pp. 295–302.
654. ______, *Metaprobability and Dempster–Shafer in evidential reasoning*, ArXiv preprint arXiv:1304.3427, 2013.

655. Zhun ga Liu, Yongmei Cheng, Quan Pan, and Zhuang Miao, *Combination of weighted belief functions based on evidence distance and conflicting belief*, Control Theory and Applications **26** (2009), no. 12, 1439–1442.
656. ______, *Weight evidence combination for multi-sensor conflict information*, Chinese Journal of Sensors and Actuators **3** (2009), 366–370.
657. Zhun ga Liu, Jean Dezert, Grégoire Mercier, and Quan Pan, *Belief C-means: An extension of fuzzy C-means algorithm in belief functions framework*, Pattern Recognition Letters **33** (2012), no. 3, 291–300.
658. Zhun ga Liu, Jean Dezert, Quan Pan, and Grégoire Mercier, *Combination of sources of evidence with different discounting factors based on a new dissimilarity measure*, Decision Support Systems **52** (2011), no. 1, 133–141.
659. Zhun ga Liu, Quan Pan, Jean Dezert, and Grégoire Mercier, *Credal C-means clustering method based on belief functions*, Knowledge-Based Systems **74** (2015), 119–132.
660. Haim Gaifman, *A theory of higher order probabilities*, Causation, Chance and Credence: Proceedings of the Irvine Conference on Probability and Causation (Brian Skyrms and William L. Harper, eds.), vol. 1, Springer Netherlands, Dordrecht, 1988, pp. 191–219.
661. Thibault Gajdos, Takashi Hayashi, Jean-Marc Tallon, and J.-C. Vergnaud, *Attitude toward imprecise information*, Journal of Economic Theory **140** (2008), no. 1, 27–65.
662. Fabio Gambino, Giuseppe Oriolo, and Giovanni Ulivi, *Comparison of three uncertainty calculus techniques for ultrasonic map building*, Proceedings of SPIE - Applications of Fuzzy Logic Technology III, vol. 2761, 1996, pp. 249–260.
663. Fabio Gambino, Giovanni Ulivi, and Marilena Vendittelli, *The transferable belief model in ultrasonic map building*, Proceedings of 6th International Conference on Fuzzy Systems (FUZZ-IEEE), 1997, pp. 601–608.
664. H. Garcia-Compeán, J. M. López-Romero, M. A. Rodriguez-Segura, and M. Socolovsky, *Principal bundles, connections and BRST cohomology*, Tech. report, Los Alamos National Laboratory, hep-th/9408003, July 1994.
665. Peter Gardenfors, *Knowledge in flux: Modeling the dynamics of epistemic states*, MIT Press, Cambridge, MA, 1988.
666. Peter Gardenfors, Bengt Hansson, and Nils-Eric Sahlin, *Evidentiary value: philosophical, judicial and psychological aspects of a theory*, CWK Gleerups, 1983.
667. Thomas D. Garvey, *Evidential reasoning for geographic evaluation for helicopter route planning*, IEEE Transactions on Geoscience and Remote Sensing **GE-25** (1987), no. 3, 294–304.
668. Thomas D. Garvey, John D. Lowrance, and Martin A. Fischler, *An inference technique for integrating knowledge from disparate sources*, Proceedings of the 7th International Joint Conference on Artificial Intelligence (IJCAI'81), vol. 1, Morgan Kaufmann, 1981, pp. 319–325.
669. Wen-Lung Gau and Daniel J. Buehrer, *Vague sets*, IEEE Transactions on Systems, Man, and Cybernetics **23** (March 1993), no. 2, 610–614.
670. Jorg Gebhardt and Rudolf Kruse, *The context model: An integrating view of vagueness and uncertainty*, International Journal of Approximate Reasoning **9** (1993), no. 3, 283–314.
671. Andreas Geiger, Philip Lenz, and Raquel Urtasun, *Are we ready for autonomous driving? The KITTI vision benchmark suite*, Proceedings of the 2012 IEEE Conference on Computer Vision and Pattern Recognition (CVPR 2012), 2012, pp. 3354–3361.
672. Andrew Gelman, *The boxer, the wrestler, and the coin flip: A paradox of robust Bayesian inference and belief functions*, The American Statistician **60** (2006), no. 2, 146–150.

673. Andrew Gelman, Nate Silver, and Aaron Edlin, *What is the probability your vote will make a difference?*, Economic Inquiry **50** (2012), no. 2, 321–326.
674. Giambattista Gennari, Alessandro Chiuso, Fabio Cuzzolin, and Ruggero Frezza, *Integrating shape and dynamic probabilistic models for data association and tracking*, Proceedings of the 41st IEEE Conference on Decision and Control (CDC'02), vol. 3, December 2002, pp. 2409–2414.
675. Wu Genxiu, *Belief function combination and local conflict management*, Computer Engineering and Applications **40** (2004), no. 34, 81–84.
676. See Ng Geok and Singh Harcharan, *Data equalisation with evidence combination for pattern recognition*, Pattern Recognition Letters **19** (1998), no. 3-4, 227–235.
677. Thomas George and Nikhil R. Pal, *Quantification of conflict in Dempster–Shafer framework: a new approach*, International Journal Of General Systems **24** (1996), no. 4, 407–423.
678. Mickaël Germain, Jean-Marc Boucher, and Goze B. Benie, *A new mass functions assignment in the Dempster–Shafer theory : the fuzzy statistical approach*, Proceedings of the IEEE Instrumentation and Measurement Technology Conference (IMTC 2008), May 2008, pp. 825–829.
679. Janos J. Gertler and Kenneth C. Anderson, *An evidential reasoning extension to quantitative model-based failure diagnosis*, IEEE Transactions on Systems, Man and Cybernetics **22** (1992), no. 2, 275–289.
680. Jamal Ghasemi, Reza Ghaderi, M. R. Karami Mollaei, and S. A. Hojjatoleslami, *A novel fuzzy Dempster–Shafer inference system for brain {MRI} segmentation*, Information Sciences **223** (2013), no. 0, 205–220.
681. Jamal Ghasemi, Mohammad Reza Karami Mollaei, Reza Ghaderi, and Ali Hojjatoleslami, *Brain tissue segmentation based on spatial information fusion by Dempster–Shafer theory*, Journal of Zhejiang University SCIENCE C **13** (2012), no. 7, 520–533.
682. Paolo Ghirardato and Michel Le Breton, *Choquet rationality*, Journal of Economic Theory **90** (2000), no. 2, 277–285.
683. Giorgio Giacinto, R. Paolucci, and Fabio Roli, *Application of neural networks and statistical pattern recognition algorithms to earthquake risk evaluation*, Pattern Recognition Letters **18** (1997), no. 11-13, 1353–1362.
684. Giorgio Giacinto and Fabio Roli, *Design of effective neural network ensembles for image classification purposes*, Image and Vision Computing **19** (2001), no. 9-10, 699–707.
685. Phan H. Giang, *Decision with Dempster–Shafer belief functions: Decision under ignorance and sequential consistency*, International Journal of Approximate Reasoning **53** (2012), no. 1, 38–53.
686. Phan H. Giang and Prakash P. Shenoy, *Decision making with partially consonant belief functions*, Proceedings of the Nineteenth Conference on Uncertainty in Artificial Intelligence (UAI'03), Morgan Kaufmann, 2003, pp. 272–280.
687. ———, *A decision theory for partially consonant belief functions*, International Journal of Approximate Reasoning **52** (2011), no. 3, 375–394.
688. Apostolos A. Giannopoulos and V. D. Milman, *Asymptotic convex geometry short overview*, Different Faces of Geometry (Simon Donaldson, Yakov Eliashberg, and Mikhael Gromov, eds.), Springer US, Boston, MA, 2004, pp. 87–162.
689. Itzhak Gilboa, *Expected utility with purely subjective non-additive probabilities*, Journal of Mathematical Economics **16** (1987), no. 1, 65–88.
690. Itzhak Gilboa and David Schmeidler, *Maxmin expected utility with non-unique prior*, Journal of Mathematical Economics **18** (1989), no. 2, 141–153.

691. ______, *Updating ambiguous beliefs*, Journal of Economic Theory **59** (1993), no. 1, 33–49.
692. ______, *Additive representations of non-additive measures and the Choquet integral*, Annals of Operations Research **52** (1994), no. 1, 43–65.
693. ______, *Additive representations of non-additive measures and the Choquet integral*, Annals of Operations Research **52** (1994), no. 1, 43–65.
694. Robin Giles, *Foundations for a theory of possibility*, Fuzzy Information and Decision Processes (M. M. Gupta and E. Sanchez, eds.), North-Holland, 1982, pp. 183–195.
695. Peter R. Gillett, *Monetary unit sampling: a belief-function implementation for audit and accounting applications*, International Journal of Approximate Reasoning **25** (2000), no. 1, 43–70.
696. Matthew L. Ginsberg, *Non-monotonic reasoning using Dempster's rule*, Proceedings of the 3rd National Conference on Artificial Intelligence (AAAI-84), 1984, pp. 126–129.
697. Michele Giry, *A categorical approach to probability theory*, Categorical aspects of topology and analysis, Springer, 1982, pp. 68–85.
698. Lluís Godo, Petr Hájek, and Francesc Esteva, *A fuzzy modal logic for belief functions*, Fundamenta Informaticae **57** (February 2003), no. 2-4, 127–146.
699. J. A. Goguen, *Categories of v-sets*, Bulletin of the American Mathematical Society **75** (1969), no. 3, 622–624.
700. M. Goldszmidt and Judea Pearl, *Default ranking: A practical framework for evidential reasoning, belief revision and update*, Proceedings of the 3rd International Conference on Knowledge Representation and Reasoning, Morgan Kaufmann, 1992, pp. 661–672.
701. Moisés Goldszmidt and Judea Pearl, *System-Z+: A formalism for reasoning with variable-strength defaults*, Proceedings of the Ninth National Conference on Artificial Intelligence (AAAI'91), vol. 1, AAAI Press, 1991, pp. 399–404.
702. Forouzan Golshani, Enrique Cortes-Rello, and Thomas H. Howell, *Dynamic route planning with uncertain information*, Knowledge-based Systems **9** (1996), no. 4, 223–232.
703. Mingming Gong, Kun Zhang, Tongliang Liu, Dacheng Tao, Clark Glymour, and Bernhard Schölkopf, *Domain adaptation with conditional transferable components*, Proceedings of the 33nd International Conference on Machine Learning (ICML 2016), JMLR Workshop and Conference Proceedings, vol. 48, 2016, pp. 2839–2848.
704. Wenjuan Gong and Fabio Cuzzolin, *A belief-theoretical approach to example-based pose estimation*, IEEE Transactions on Fuzzy Systems **26** (2017), no. 2, 598–611.
705. Irving J. Good, *Symposium on current views of subjective probability: Subjective probability as the measure of a non-measurable set*, Logic, Methodology and Philosophy of Science (Ernest Nagel, Patrick Suppes, and Alfred Tarski, eds.), vol. 44, Elsevier, 1966, pp. 319–329.
706. ______, *Lindley's paradox: Comment*, Journal of the American Statistical Association **77** (1982), no. 378, 342–344.
707. Irwin R. Goodman, *Fuzzy sets as equivalence classes of random sets*, Recent Developments in Fuzzy Sets and Possibility Theory (Ronald R. et al. Yager, ed.), Pergamon Press, 1982, pp. 327–343.
708. Irwin R. Goodman and Hung T. Nguyen, *Uncertainty models for knowledge-based systems; a unified approach to the measurement of uncertainty*, Elsevier Science, New York, NY, USA, 1985.
709. ______, *A theory of conditional information for probabilistic inference in intelligent systems: Ii. product space approach*, Information Sciences **76** (1994), no. 1-2, 13–42.

710. Jean Gordon and Edward H. Shortliffe, *A method for managing evidential reasoning in a hierarchical hypothesis space*, Artificial Intelligence **26** (1985), no. 3, 323–357.
711. Jean Gordon and Edward H. Shortliffe, *The Dempster–Shafer theory of evidence*, Readings in Uncertain Reasoning (Glenn Shafer and Judea Pearl, eds.), Morgan Kaufmann, 1990, pp. 529–539.
712. Jean Gordon and Edward H. Shortliffe, *A method for managing evidential reasoning in a hierarchical hypothesis space: a retrospective*, Artificial Intelligence **59** (February 1993), no. 1-2, 43–47.
713. Pece V. Gorsevski, Piotr Jankowski, and Paul E. Gessler, *Spatial prediction of landslide hazard using fuzzy k-means and Dempster–Shafer theory*, Transactions in GIS **9** (2005), no. 4, 455–474.
714. Jean Goubault-Larrecq, *Continuous capacities on continuous state spaces*, Automata, Languages and Programming: 34th International Colloquium, ICALP 2007, Wrocław, Poland, July 9-13, 2007. Proceedings (Lars Arge, Christian Cachin, Tomasz Jurdziński, and Andrzej Tarlecki, eds.), Springer, Berlin, Heidelberg, 2007, pp. 764–776.
715. Henry W. Gould, *Combinatorial identities*, Morgantown, W. Va., 1972.
716. John Goutsias, *Modeling random shapes: an introduction to random closed set theory*, Tech. report, Department of Electrical and Computer Engineering, John Hopkins University, Baltimore, JHU/ECE 90-12, April 1998.
717. John Goutsias, Ronald P. S. Mahler, and Hung T. Nguyen, *Random sets: theory and applications*, IMA Volumes in Mathematics and Its Applications, vol. 97, Springer-Verlag, December 1997.
718. Michel Grabisch, *K-order additive discrete fuzzy measures and their representation*, Fuzzy sets and systems **92** (1997), no. 2, 167–189.
719. Michel Grabisch, *On lower and upper approximation of fuzzy measures by k-order additive measures*, Information, Uncertainty and Fusion (Bernadette Bouchon-Meunier, Ronald R. Yager, and Lotfi A. Zadeh, eds.), Springer US, Boston, MA, 2000, pp. 105–118.
720. Michel Grabisch, *The Möbius transform on symmetric ordered structures and its application to capacities on finite sets*, Discrete Mathematics **287** (2004), no. 1–3, 17–34.
721. ______, *Belief functions on lattices*, International Journal of Intelligent Systems **24** (2009), no. 1, 76–95.
722. ______, *Upper approximation of non-additive measures by k-additive measures - the case of belief functions*, Proceedings of the 1st International Symposium on Imprecise Probabilities and Their Applications (ISIPTA'99), June 1999.
723. Michel Grabisch, Hung T. Nguyen, and Elbert A. Walker, *Fundamentals of uncertainty calculi with applications to fuzzy inference*, vol. 30, Springer Science & Business Media, 2013.
724. Michel Grabisch, Michio Sugeno, and Toshiaki Murofushi, *Fuzzy measures and integrals: theory and applications*, New York: Springer, 2000.
725. Siegfried Graf, *A Radon-Nikodym theorem for capacities*, Journal fr die reine und angewandte Mathematik **320** (1980), 192–214.
726. Hermann Grassmann, *Die Ausdehnungslehre von 1844 oder die lineale Ausdehnungslehre* , Wigand, 1878.
727. Frank J. Groen and Ali Mosleh, *Foundations of probabilistic inference with uncertain evidence*, International Journal of Approximate Reasoning **39** (2005), no. 1, 49–83.
728. X. E. Gros, J. Bousigue, and K. Takahashi, {*NDT*} *data fusion at pixel level*, {NDT} & E International **32** (1999), no. 5, 283–292.

729. Emmanuéle Grosicki, Matthieu Carre, Jean-Marie Brodin, and Edouard Geoffrois, *Results of the RIMES evaluation campaign for handwritten mail processing*, Proceedings of the International Conference on Document Analysis and Recognition, 2009, pp. 941–945.
730. Benjamin N. Grosof, *An inequality paradigm for probabilistic knowledge*, Proceedings of the First Conference on Uncertainty in Artificial Intelligence (UAI1985), 1985, pp. 1–8.
731. ______, *Evidential confirmation as transformed probability: On the duality of priors and updates*, Machine Intelligence and Pattern Recognition, vol. 4, Elsevier, 1986, pp. 153–166.
732. ______, *Evidential confirmation as transformed probability*, ArXiv preprint arXiv:1304.3439, 2013.
733. Peter Grünwald, *The minimum description length principle and reasoning under uncertainty*, PhD dissertation, Universiteit van Amsterdam, 1998.
734. Jason Gu, Max Meng, Al Cook, and Peter X. Liu, *Sensor fusion in mobile robot: some perspectives*, Proceedings of the 4th World Congress on Intelligent Control and Automation, vol. 2, 2002, pp. 1194–1199.
735. Jiwen Guan and David A. Bell, *A generalization of the Dempster–Shafer theory*, Proceedings of the 13th International Joint Conference on Artifical Intelligence (IJCAI'93), vol. 1, Morgan Kaufmann, 1993, pp. 592–597.
736. ______, *Approximate reasoning and evidence theory*, Information Sciences **96** (1997), no. 3, 207–235.
737. ______, *Generalizing the Dempster–Shafer rule of combination to Boolean algebras*, Proceedings of the IEEE International Conference on Developing and Managing Intelligent System Projects, March 1993, pp. 229–236.
738. ______, *The Dempster–Shafer theory on Boolean algebras*, Chinese Journal of Advanced Software Research **3** (November 1996), no. 4, 313–343.
739. Jiwen Guan, David A. Bell, and Victor R. Lesser, *Evidential reasoning and rule strengths in expert systems*, AI and Cognitive Science 90 (Michael F. McTear and Norman Creaney, eds.), Workshops in Computing, Springer, 1991, pp. 378–390.
740. Jiwen Guan, David A. Bell, and Victor R. Lesser, *Evidential reasoning and rule strengths in expert systems*, Proceedings of AI and Cognitive Science '90 (M. F. McTear and N. Creaney, eds.), September 1990, pp. 378–390.
741. Jiwen Guan, Jasmina Pavlin, and Victor R. Lesser, *Combining evidence in the extended Dempster–Shafer theory*, AI and Cognitive Science 89 (Alan F. Smeaton and Gabriel McDermott, eds.), Workshops in Computing, Springer, 1990, pp. 163–178.
742. Jiwen W. Guan and David A. Bell, *Discounting and combination operations in evidential reasoning*, Proceedings of the Ninth Conference on Uncertainty in Artificial Intelligence (UAI'93) (D. Heckerman and A. Mamdani, eds.), July 1993, pp. 477–484.
743. ______, *Evidential reasoning in intelligent system technologies*, Proceedings of the Second Singapore International Conference on Intelligent Systems (SPICIS'94), vol. 1, November 1994, pp. 262–267.
744. ______, *A linear time algorithm for evidential reasoning in knowledge base systems*, Proceedings of the Third International Conference on Automation, Robotics and Computer Vision (ICARCV'94), vol. 2, November 1994, pp. 836–840.
745. Jiwen W. Guan, David A. Bell, and Zhong Guan, *Evidential reasoning in expert systems: computational methods*, Proceedings of the Seventh International Conference on Industrial and Engineering Applications of Artificial Intelligence and Expert Systems (IEA/AIE-94) (F. D. Anger, R. V. Rodriguez, and M. Ali, eds.), May-June 1994, pp. 657–666.

746. Li Guan and Shou mei Li, *A strong law of large numbers for set-valued random variables in rademacher type P Banach space*, Proceedings of the International Conference on Machine Learning and Cybernetics, 2006, pp. 1768–1773.
747. Xin Guan, Xiao Yi, and You He, *An improved Dempster–Shafer algorithm for resolving the conflicting evidences*, International Journal of Information Technology **11** (2005), no. 12, 68–75.
748. Xin Guan, Xiao Yi, Xiaoming Sun, and You He, *Efficient fusion approach for conflicting evidence*, Journal of Tsinghua University (Science and Technology) **1** (2009).
749. Zhang Guang-Quan, *Semi-lattice structure of all extensions of the possibility measure and the consonant belief function on the fuzzy set*, Fuzzy Sets and Systems **43** (1991), no. 2, 183–188.
750. Mickael Guironnet, Denis Pellerin, and Michéle Rombaut, *Camera motion classification based on the transferable belief model*, Proceedings of the 14th European Signal Processing Conference (EUSIPCO'06), 2006, pp. 1–5.
751. H. Y. Guo, *Structural damage detection using information fusion technique*, Mechanical Systems and Signal Processing **20** (2006), no. 5, 1173–1188.
752. Lan Guo, *Software quality and reliability prediction using Dempster–Shafer theory*, Ph.D. thesis, West Virginia University, 2004.
753. Lan Guo, Bojan Cukic, and Harshinder Singh, *Predicting fault prone modules by the Dempster–Shafer belief networks*, Proceedings of the 18th IEEE International Conference on Automated Software Engineering, October 2003, pp. 249–252.
754. Min Guo, Jian-Bo Yang, Kwai-Sang Chin, Hong-Wei Wang, and Xin bao Liu, *Evidential reasoning approach for multiattribute decision analysis under both fuzzy and interval uncertainty*, IEEE Transactions on Fuzzy Systems **17** (2009), no. 3, 683–697.
755. Min Guo, Jian-Bo Yang, Kwai-Sang Chin, and Hongwei Wang, *Evidential reasoning based preference programming for multiple attribute decision analysis under uncertainty*, European Journal of Operational Research **182** (2007), no. 3, 1294–1312.
756. Michael A. S. Guth, *Uncertainty analysis of rule-based expert systems with Dempster–Shafer mass assignments*, International Journal of Intelligent Systems **3** (1988), no. 2, 123–139.
757. Vu Ha, AnHai Doan, Van H. Vu, and Peter Haddawy, *Geometric foundations for interval-based probabilities*, Annals of Mathematics and Arti
cal Inteligence **24** (1998), no. 1-4, 1–21.
758. Vu Ha and Peter Haddawy, *Theoretical foundations for abstraction-based probabilistic planning*, Proceedings of the 12th International Conference on Uncertainty in Artificial Intelligence (UAI'96), August 1996, pp. 291–298.
759. Minh Ha-Duong, *Hierarchical fusion of expert opinion in the Transferable Belief Model, application on climate sensivity*, Working paper halshs-00112129-v3, HAL, 2006.
760. ———, *Hierarchical fusion of expert opinions in the Transferable Belief Model, application to climate sensitivity*, International Journal of Approximate Reasoning **49** (2008), no. 3, 555–574.
761. Jay K. Hackett and Mubarak Shah, *Multi-sensor fusion: a perspective*, Proceedings of the IEEE International Conference on Robotics and Automation (ICRA 1990), vol. 2, May 1990, pp. 1324–1330.
762. Peter Haddawy, *A variable precision logic inference system employing the Dempster–Shafer uncertainty calculus*, PhD dissertation, University of Illinois at Urbana-Champaign, 1987.

763. Rolf Haenni, *Are alternatives to Dempster's rule of combination real alternatives?: Comments on About the belief function combination and the conflict management"*, Information Fusion **3** (2002), no. 3, 237–239.
764. ———, *Ordered valuation algebras: a generic framework for approximating inference*, International Journal of Approximate Reasoning **37** (2004), no. 1, 1–41.
765. ———, *Aggregating referee scores: an algebraic approach*, Proceedings of the 2nd International Workshop on Computational Social Choice (COMSOC'08) (U. Endriss and W. Goldberg, eds.), 2008, pp. 277–288.
766. ———, *Shedding new light on Zadeh's criticism of Dempster's rule of combination*, Proceedings of the 7th International Conference on Information Fusion (FUSION 2005), vol. 2, July 2005, pp. 6–.
767. ———, *Towards a unifying theory of logical and probabilistic reasoning*, Proceedings of the 4th International Symposium on Imprecise Probabilities and Their Applications (ISIPTA'05), vol. 5, July 2005, pp. 193–202.
768. Rolf Haenni and Stephan Hartmann, *Modeling partially reliable information sources: A general approach based on Dempster–Shafer theory*, Information Fusion **7** (2006), no. 4, 361–379, Special Issue on the Seventh International Conference on Information Fusion-Part I Seventh International Conference on Information Fusion.
769. Rolf Haenni and Norbert Lehmann, *Resource bounded and anytime approximation of belief function computations*, International Journal of Approximate Reasoning **31** (2002), no. 1, 103–154.
770. Rolf Haenni and Norbert Lehmann, *Implementing belief function computations*, International Journal of Intelligent Systems **18** (2003), no. 1, 31–49.
771. ———, *Probabilistic argumentation systems: a new perspective on the Dempster–Shafer theory*, International Journal of Intelligent Systems **18** (2003), no. 1, 93–106.
772. Rolf Haenni, Jan-Willem Romeijn, Gregory Wheeler, and Jon Williamson, *Possible semantics for a common framework of probabilistic logics*, Proceedings of the International Workshop on Interval/Probabilistic Uncertainty and Non-Classical Logics (UncLog'08) (Ishikawa, Japan) (V.-N. Huynh, Y. Nakamori, H. Ono, J. Lawry, V. Kreinovich, and Hung T. Nguyen, eds.), Advances in Soft Computing, no. 46, 2008, pp. 268–279.
773. Petr Hájek, *Deriving Dempster's rule*, Proceeding of the International Conference on Information Processing and Management of Uncertainty in Knowledge-Based Systems (IPMU'92), 1992, pp. 73–75.
774. Petr Hájek, *Systems of conditional beliefs in Dempster–Shafer theory and expert systems*, International Journal of General Systems **22** (1993), no. 2, 113–124.
775. ———, *On logics of approximate reasoning*, Knowledge Representation and Reasoning Under Uncertainty: Logic at Work (Michael Masuch and László Pólos, eds.), Springer, Berlin Heidelberg, 1994, pp. 17–29.
776. ———, *On logics of approximate reasoning II*, Proceedings of the ISSEK94 Workshop on Mathematical and Statistical Methods in Artificial Intelligence (G. Della Riccia, R. Kruse, and R. Viertl, eds.), Springer, 1995, pp. 147–155.
777. ———, *Getting belief functions from Kripke models*, International Journal of General Systems **24** (1996), no. 3, 325–327.
778. Petr Hájek, *A note on belief functions in MYCIN-like systems*, Proceedings of Aplikace Umele Inteligence AI '90, March 1990, pp. 19–26.
779. Petr Hájek and David Harmanec, *An exercise in Dempster–Shafer theory*, International Journal of General Systems **20** (1992), no. 2, 137–142.
780. ———, *On belief functions (the present state of Dempster–Shafer theory)*, Advanced topics in AI, Springer-Verlag, 1992.

781. Jim W. Hall and Jonathan Lawry, *Generation, combination and extension of random set approximations to coherent lower and upper probabilities*, Reliability Engineering and System Safety **85** (2004), no. 1-3, 89–101.
782. Jim W. Hall, Robert J. Lempert, Klaus Keller, Andrew Hackbarth, Christophe Mijere, and David J. McInerney, *Robust climate policies under uncertainty: A comparison of robust decision making and info-gap methods*, Risk Analysis **32** (2012), no. 10, 1657–1672.
783. Paul R. Halmos, *Measure theory*, vol. 18, Springer, 2013.
784. Joseph Y. Halpern, *Reasoning about uncertainty*, MIT Press, 2017.
785. Joseph Y. Halpern and Ronald Fagin, *Two views of belief: Belief as generalized probability and belief as evidence*, Artificial Intelligence **54** (1992), no. 3, 275–317.
786. Joseph Y. Halpern and Riccardo Pucella, *A logic for reasoning about evidence*, Proceedings of the Nineteenth Conference on Uncertainty in Artificial Intelligence (UAI'03), Morgan Kaufmann, 2003, pp. 297–304.
787. Joseph Y. Halpern and Riccardo Pucella, *Reasoning about expectation*, Proceedings of the Eighteenth Conference on Uncertainty in Artificial Intelligence (UAI2002), 2014, pp. 207–215.
788. Deqiang Han, Jean Dezert, Chongzhao Han, and Yi Yang, *Is entropy enough to evaluate the probability transformation approach of belief function?*, Advances and Applications of DSmT for Information Fusion, vol. 4, 2015, pp. 37–43.
789. ______, *Is entropy enough to evaluate the probability transformation approach of belief function?*, Proceedings of the 13th Conference on Information Fusion (FUSION 2010), July 2010, pp. 1–7.
790. Deqiang Han, Chongzhao Han, and Yi Yang, *A modified evidence combination approach based on ambiguity measure*, Proceedings of the 11th International Conference on Information Fusion (FUSION 2008), 2008, pp. 1–6.
791. Morgan A. Hanson, *On sufficiency of the Kuhn-Tucker conditions*, Journal of Mathematical Analysis and Applications **80** (1981), no. 2, 545–550.
792. Frank Harary and Dominic Welsh, *Matroids versus graphs*, The many facets of graph theory, Lecture Notes in Math., vol. 110, Springer-Verlag, Berlin, 1969, pp. 155–170.
793. John Harding, Massimo Marinacci, Nhu T. Nguyen, and Tonghui Wang, *Local Radon-Nikodym derivatives of set functions*, International Journal of Uncertainty, Fuzziness and Knowledge-Based Systems **5** (1997), no. 3, 379–394.
794. David Harmanec, *Toward a characterization of uncertainty measure for the Dempster–Shafer theory*, Proceedings of the Eleventh Conference on Uncertainty in Artificial Intelligence (UAI'95) (San Francisco, CA, USA), Morgan Kaufmann Publishers Inc., 1995, pp. 255–261.
795. ______, *Uncertainty in Dempster–Shafer theory*, Ph.D. thesis, State University of New York at Binghamton, 1997.
796. David Harmanec, *Faithful approximations of belief functions*, Proceedings of the Fifteenth Conference on Uncertainty in Artificial Intelligence (UAI1999), 1999, pp. 271–278.
797. ______, *Faithful approximations of belief functions*, ArXiv preprint arXiv:1301.6703, 2013.
798. ______, *Toward a characterisation of uncertainty measure for the Dempster–Shafer theory*, Proceedings of the Eleventh Conference on Uncertainty in Artificial Intelligence (UAI'95) (P. Besnard and S. Hanks, eds.), Montreal, Quebec, Canada, 18-20 August 1995, pp. 255–261.
799. David Harmanec and Petr Hájek, *A qualitative belief logic*, International Journal of Uncertainty, Fuzziness and Knowledge-Based Systems **2** (1994), no. 2, 227–236.

800. David Harmanec, George Klir, and Germano Resconi, *On modal logic interpretation of Dempster–Shafer theory*, International Journal of Intelligent Systems **9** (1994), no. 10, 941–951.
801. David Harmanec and George J. Klir, *Measuring total uncertainty in Dempster–Shafer theory: A novel approach*, International Journal of General Systems **22** (1994), no. 4, 405–419.
802. ______, *On information-preserving transformations*, International Journal of General Systems **26** (1997), no. 3, 265–290.
803. David Harmanec, George J. Klir, and Germano Resconi, *On modal logic interpretation of Dempster–Shafer theory of evidence*, International Journal of Intelligent Systems **9** (1994), no. 10, 941–951.
804. David Harmanec, George J. Klir, and Zhenyuan Wang, *Modal logic interpretation of Dempster–Shafer theory: An infinite case*, International Journal of Approximate Reasoning **14** (1996), no. 2-3, 81–93.
805. David Harmanec, Germano Resconi, George J. Klir, and Yin Pan, *On the computation of uncertainty measure in Dempster–Shafer theory*, International Journal Of General System **25** (1996), no. 2, 153–163.
806. William L. Harper, *Rational belief change, Popper functions and counterfactuals*, Synthese **30** (1975), no. 1-2, 221–262.
807. Ralph V. L. Hartley, *Transmission of information*, Bell System Technical Journal **7** (1928), no. 3, 535–563.
808. Hai-Yen Hau and Rangasami L. Kashyap, *Belief combination and propagation in a lattice-structured interference network*, IEEE Transactions on Systems, Man, and Cybernetics **20** (1990), no. 1, 45–57.
809. Hai-Yen Hau and Rangasami L. Kashyap, *On the robustness of Dempster's rule of combination*, Proceedings of the IEEE International Workshop on Tools for Artificial Intelligence, Architectures, Languages and Algorithms, October 1989, pp. 578–582.
810. Daqing He, Ayse Göker, and David J. Harper, *Combining evidence for automatic web session identification*, Information Processing and Management **38** (2002), no. 5, 727–742.
811. Wei He, Nicholas Williard, Michael Osterman, and Michael Pecht, *Prognostics of lithium-ion batteries based on Dempster–Shafer theory and the Bayesian Monte Carlo method*, Journal of Power Sources **196** (2011), no. 23, 10314–10321.
812. David Heckerman, *Probabilistic interpretations for MYCIN's certainty factors*, Machine Intelligence and Pattern Recognition, vol. 4, Elsevier, 1986, pp. 167–196.
813. David Heckerman, *An empirical comparison of three inference methods*, Machine Intelligence and Pattern Recognition, vol. 9, Elsevier, 1990, pp. 283–302.
814. ______, *An empirical comparison of three inference methods*, ArXiv preprint arXiv:1304.2357, 2013.
815. Sylvie Le Hégarat-Mascle, Isabelle Bloch, and Daniel Vidal-Madjar, *Introduction of neighborhood information in evidence theory and application to data fusion of radar and optical images with partial cloud cover*, Pattern Recognition **31** (1998), no. 11, 1811–1823.
816. Stanisaw Heilpern, *Representation and application of fuzzy numbers*, Fuzzy Sets and Systems **91** (1997), no. 2, 259–268.
817. Lothar Heinrich and Ilya Molchanov, *Central limit theorem for a class of random measures associated with germ-grain models*, Advances in Applied Probability **31** (1999), no. 2, 283–314.

818. Jon C. Helton, J. D. Johnson, William L. Oberkampf, and Curtis B. Storlie, *A sampling-based computational strategy for the representation of epistemic uncertainty in model predictions with evidence theory*, Computer Methods in Applied Mechanics and Engineering **196** (2007), no. 37-40, 3980–3998.
819. Jon C. Helton, Jay D. Johnson, and William L. Oberkampf, *An exploration of alternative approaches to the representation of uncertainty in model predictions*, Reliability Engineering and System Safety **85** (2004), no. 1-3, 39–71.
820. Ebbe Hendon, Hans JØrgen Jacobsen, Birgitte Sloth, and Torben TranÆs, *Expected utility with lower probabilities*, Journal of Risk and Uncertainty **8** (1994), no. 2, 197–216.
821. Ebbe Hendon, Hans Jorgen Jacobsen, Birgitte Sloth, and Torben Tranaes, *The product of capacities and belief functions*, Mathematical Social Sciences **32** (1996), no. 2, 95–108.
822. Steven J. Henkind and Malcolm C. Harrison, *An analysis of four uncertainty calculi*, IEEE Transactions on Systems, Man, and Cybernetics **18** (1988), no. 5, 700–714.
823. Enric Hernandez and Jordi Recasens, *On possibilistic and probabilistic approximations of unrestricted belief functions based on the concept of fuzzy t-preorder*, International Journal of Uncertainty, Fuzziness and Knowledge-Based Systems **10** (2002), no. 2, 185–200.
824. ———, *Indistinguishability relations in Dempster–Shafer theory of evidence*, International Journal of Approximate Reasoning **37** (2004), no. 3, 145–187.
825. Timothy Herron, Teddy Seidenfeld, and Larry Wasserman, *Divisive conditioning: further results on dilation*, Philosophy of Science **64** (1997), 411–444.
826. H. T. Hestir, Hung T. Nguyen, and G. S. Rogers, *A random set formalism for evidential reasoning*, Conditional Logic in Expert Systems, North Holland, 1991, pp. 309–344.
827. Cokky Hilhorst, Piet Ribbers, Eric van Heck, and Martin Smits, *Using Dempster–Shafer theory and real options theory to assess competing strategies for implementing IT infrastructures: A case study*, Decision Support Systems **46** (2008), no. 1, 344–355.
828. Bruce M. Hill, *Lindley's paradox: Comment*, Journal of the American Statistical Association **77** (1982), no. 378, 344–347.
829. J. Hodges, S. Bridges, C. Sparrow, B. Wooley, B. Tang, and C. Jun, *The development of an expert system for the characterization of containers of contaminated waste*, Expert Systems with Applications **17** (1999), no. 3, 167–181.
830. James C. Hoffman and Robin R. Murphy, *Comparison of Bayesian and Dempster–Shafer theory for sensing: A practitioner's approach*, Proceedings of SPIE - Neural and Stochastic Methods in Image and Signal Processing II, vol. 2032, 1993, pp. 266–279.
831. Owen Hoffman and Jana S. Hammonds, *Propagation of uncertainty in risk assessments: The need to distinguish between uncertainty due to lack of knowledge and uncertainty due to variability*, Risk Analysis **14** (1994), no. 5, 707–712.
832. Ulrich Hohle, *Entropy with respect to plausibility measures*, Proceedings of the 12th IEEE Symposium on Multiple-Valued Logic, 1982, pp. 167–169.
833. Richard B. Holmes, *Principles of Banach Spaces*, Geometric Functional Analysis and its Applications, Springer New York, New York, NY, 1975, pp. 119–201.
834. ———, *Geometric functional analysis and its applications*, vol. 24, Springer Science and Business Media, 2012.
835. Aoi Honda and Michel Grabisch, *Entropy of capacities on lattices and set systems*, Information Sciences **176** (2006), no. 23, 3472–3489.

836. Lang Hong, *Recursive algorithms for information fusion using belief functions with applications to target identification*, Proceedings of the First IEEE Conference on Control Applications (CCTA 1992), 1992, pp. 1052–1057.
837. Lang Hong and Andrew Lynch, *Recursive temporal-spatial information fusion with applications to target identification*, IEEE Transactions on Aerospace and Electronic Systems **29** (1993), no. 2, 435–445.
838. Xin Hong, Chris Nugent, Maurice Mulvenna, Sally McClean, Bryan Scotney, and Steven Devlin, *Evidential fusion of sensor data for activity recognition in smart homes*, Pervasive and Mobile Computing **5** (2009), no. 3, 236–252.
839. Takahiko Horiuchi, *A new theory of evidence for non-exclusive elementary propositions*, International Journal of Systems Science **27** (1996), no. 10, 989–994.
840. ______, *Decision rule for pattern classification by integrating interval feature values*, IEEE Transactions on Pattern Analysis and Machine Intelligence **20** (April 1998), no. 4, 440–448.
841. Eric J. Horvitz, David E. Heckerman, and Curtis P. Langlotz, *A framework for comparing alternative formalisms for plausible reasoning*, Proceedings of the Fifth AAAI National Conference on Artificial Intelligence (AAAI'86), 1986, pp. 210–214.
842. Reza HoseinNezhad, Behzad Moshiri, and Mohammad R. Asharif, *Sensor fusion for ultrasonic and laser arrays in mobile robotics: a comparative study of fuzzy, Dempster and Bayesian approaches*, Proceedings of the First IEEE International Conference on Sensors (Sensors 2002), vol. 2, 2002, pp. 1682–1689.
843. Liqun Hou and Neil W. Bergmann, *Induction motor fault diagnosis using industrial wireless sensor networks and Dempster–Shafer classifier fusion*, Proceedings of the 37th Annual Conference of the IEEE Industrial Electronics Society (IECON 2011), Nov 2011, pp. 2992–2997.
844. Andrew Howard and Les Kitchen, *Generating sonar maps in highly specular environments*, Proceedings of the Fourth International Conference on Control Automation Robotics and Vision, 1996, pp. 1870–1874.
845. Yen-Teh Hsia, *A belief function semantics for cautious non-monotonicity*, Tech. Report TR/IRIDIA/91-3, Université Libre de Bruxelles, 1991.
846. Yen-Teh Hsia, *Characterizing belief with minimum commitment*, Proceedings of the 12th International Joint Conference on Artificial Intelligence (IJCAI'91), vol. 2, Morgan Kaufmann, 1991, pp. 1184–1189.
847. Yen-Teh Hsia and Prakash P. Shenoy, *An evidential language for expert systems*, Methodologies for Intelligent Systems (Z. Ras, ed.), North Holland, 1989, pp. 9–16.
848. ______, *MacEvidence: A visual evidential language for knowledge-based systems*, Tech. Report 211, School of Business, University of Kansas, 1989.
849. Yen-Teh Hsia and Philippe Smets, *Belief functions and non-monotonic reasoning*, Tech. Report IRIDIA/TR/1990/3, Université Libre de Bruxelles, 1990.
850. Chang-Hua Hu, Xiao-Sheng Si, and Jian-Bo Yang, *System reliability prediction model based on evidential reasoning algorithm with nonlinear optimization*, Expert Systems with Applications **37** (2010), no. 3, 2550–2562.
851. Chang-hua Hu, Xiao-sheng Si, Zhi-jie Zhou, and Peng Wang, *An improved D-S algorithm under the new measure criteria of evidence conflict*, Chinese Journal of Electronics **37** (2009), no. 7, 1578.
852. Lifang Hu, Xin Guan, Yong Deng, Deqiang Han, and You He, *Measuring conflict functions in generalized power space*, Chinese Journal of Aeronautics **24** (2011), no. 1, 65–73.

853. Wei Hu, Jianhua Li, and Qiang Gao, *Intrusion detection engine based on Dempster–Shafer's theory of evidence*, Proceedings of the International Conference on Communications, Circuits and Systems, vol. 3, June 2006, pp. 1627–1631.
854. Zhongsheng Hua, Bengang Gong, and Xiaoyan Xu, *A DS–AHP approach for multi-attribute decision making problem with incomplete information* , Expert Systems with Applications **34** (2008), no. 3, 2221–2227.
855. Daniel Hug and Matthias Reitzner, *Introduction to stochastic geometry*, Stochastic Analysis for Poisson Point Processes: Malliavin Calculus, Wiener-Itô Chaos Expansions and Stochastic Geometry (Giovanni Peccati and Matthias Reitzner, eds.), Springer International Publishing, 2016, pp. 145–184.
856. Kenneth F. Hughes and Robin R. Murphy, *Ultrasonic robot localization using Dempster–Shafer theory*, Proceedings of SPIE - Neural and Stochastic Methods in Image and Signal Processing, vol. 1766, 1992, pp. 2–11.
857. Eyke Hullermeier, *Similarity-based inference as evidential reasoning*, International Journal of Approximate Reasoning **26** (2001), no. 2, 67–100.
858. Robert A. Hummel and Michael S. Landy, *A statistical viewpoint on the theory of evidence*, IEEE Transactions on Pattern Analysis and Machine Intelligence **10** (1988), no. 2, 235–247.
859. Robert A. Hummel and Larry M. Manevitz, *Combining bodies of dependent information*, Proceedings of the International Joint Conference on Artificial Intelligence (IJCAI'87), 1987, pp. 1015–1017.
860. Anthony Hunter and Weiru Liu, *Fusion rules for merging uncertain information*, Information Fusion **7** (2006), no. 1, 97–134.
861. Daniel Hunter, *Dempster–Shafer versus probabilistic logic*, Proceedings of the Third Conference on Uncertainty in Artificial Intelligence (UAI'87), 1987, pp. 22–29.
862. ______, *Dempster–Shafer vs. probabilistic logic*, Proceedings of the Third Conference on Uncertainty in Artificial Intelligence (UAI'97), 1987, pp. 22–29.
863. V.-N. Huynh, Y. Nakamori, H. Ono, J. Lawry, V. Kreinovich, and Hung T. Nguyen (eds.), *Interval / probabilistic uncertainty and non-classical logics*, Springer, 2008.
864. Van-Nam Huynh, Yoshiteru Nakamori, and Tu-Bao Ho, *Assessment aggregation in the evidential reasoning approach to MADM under uncertainty: Orthogonal versus weighted sum*, Advances in Computer Science - ASIAN 2004. Higher-Level Decision Making: 9th Asian Computing Science Conference. Dedicated to Jean-Louis Lassez on the Occasion of His 5th Birthday. Chiang Mai, Thailand, December 8-10, 2004. Proceedings (Michael J. Maher, ed.), Springer, Berlin Heidelberg, 2005, pp. 109–127.
865. Van-Nam Huynh, Yoshiteru Nakamori, Tu-Bao Ho, and Tetsuya Murai, *Multiple-attribute decision making under uncertainty: the evidential reasoning approach revisited*, IEEE Transactions on Systems, Man and Cybernetics, Part A: Systems and Humans **36** (2006), no. 4, 804–822.
866. Van-Nam Huynh, Tri Thanh Nguyen, and Cuong Anh Le, *Adaptively entropy-based weighting classifiers in combination using Dempster–Shafer theory for word sense disambiguation*, Computer Speech and Language **24** (2010), no. 3, 461–473.
867. Ion Iancu, *Prosum-prolog system for uncertainty management*, International Journal of Intelligent Systems **12** (1997), no. 9, 615–627.
868. Laurie Webster II, Jen-Gwo Chen, Simon S. Tan, Carolyn Watson, and André de Korvin, *Vadidation of authentic reasoning expert systems*, Information Sciences **117** (1999), no. 1-2, 19–46.
869. Horace H. S. Ip and Richard C. K. Chiu, *Evidential reasoning for facial gestures recognition from cartoon images*, Proceedings of the Australian New Zealand Intelligent Information Systems Conference (ANZIIS'94), 1994, pp. 397–401.

870. Horace H. S. Ip and Hon-Ming Wong, *Evidential reasoning in foreign exchange rates forecasting*, Proceedings of the First International Conference on Artificial Intelligence Applications on Wall Street, 1991, pp. 152–159.
871. Mitsuru Ishizuka, *Inference methods based on extended Dempster and Shafer's theory for problems with uncertainty/fuzziness*, New Generation Computing **1** (1983), no. 2, 159–168.
872. Mitsuru Ishizuka, King Sun Fu, and James T. P. Yao, *Inference procedures under uncertainty for the problem-reduction method*, Information Sciences **28** (1982), no. 3, 179–206.
873. Makoto Itoh and Toshiyuki Inagaki, *A new conditioning rule for belief updating in the Dempster–Shafer theory of evidence*, Transactions of the Society of Instrument and Control Engineers **31** (1995), no. 12, 2011–2017.
874. Nathan Jacobson, *Basic algebra I*, Freeman and Company, New York, 1985.
875. Jean-Yves Jaffray, *Application of linear utility theory to belief functions*, Uncertainty and Intelligent Systems, Springer-Verlag, 1988, pp. 1–8.
876. ______, *Coherent bets under partially resolving uncertainty and belief functions*, Theory and Decision **26** (1989), no. 2, 99–105.
877. ______, *Linear utility theory for belief functions*, Operation Research Letters **8** (1989), no. 2, 107–112.
878. Jean-Yves Jaffray, *Belief functions, convex capacities, and decision making*, Mathematical Psychology: Current Developments (Jean-Paul Doignon and Jean-Claude Falmagne, eds.), Springer, New York, 1991, pp. 127–134.
879. Jean-Yves Jaffray, *Linear utility theory and belief functions: A discussion*, Progress in Decision, Utility and Risk Theory (Attila Chikán, ed.), Springer Netherlands, 1991, pp. 221–229.
880. ______, *Bayesian updating and belief functions*, IEEE Transactions on Systems, Man and Cybernetics **22** (1992), no. 5, 1144–1152.
881. ______, *Dynamic decision making with belief functions*, Advances in the Dempster-Shafer Theory of Evidence (R. R. Yager, M. Fedrizzi, and J. Kacprzyk, eds.), Wiley, 1994, pp. 331–352.
882. ______, *On the maximum-entropy probability which is consistent with a convex capacity*, International Journal of Uncertainty, Fuzziness and Knowledge-Based Systems **3** (1995), no. 1, 27–33.
883. Jean-Yves Jaffray and Peter P. Wakker, *Decision making with belief functions: Compatibility and incompatibility with the sure-thing principle*, Journal of Risk and Uncertainty **8** (1994), no. 3, 255–271.
884. Fabrice Janez, *Fusion de sources d'information definies sur des referentiels non exhaustifs differents. solutions proposees sous le formalisme de la theorie de l'evidence*, PhD dissertation, University of Angers, France, 1996.
885. Fabrice Janez and Alain Appriou, *Theory of evidence and non-exhaustive frames of discernment: Plausibilities correction methods*, International Journal of Approximate Reasoning **18** (1998), no. 1, 1–19.
886. Fabrice Janez, Olivier Goretta, and Alain Michel, *Automatic map updating by fusion of multispectral images in the Dempster–Shafer framework*, Proceedings of SPIE - Applications of Digital Image Processing XXIII, vol. 4115, 2000, pp. 245–255.
887. Iman Jarkass and Michèle Rombaut, *Dealing with uncertainty on the initial state of a Petri net*, Proceedings of the Fourteenth Conference on Uncertainty in Artificial Intelligence (UAI'98), Morgan Kaufmann, 1998, pp. 289–295.
888. Edwin T. Jaynes, *Information theory and statistical mechanics*, Physical review **106** (1957), no. 4, 620.

889. Richard C. Jeffrey, *The logic of decision*, Mc Graw - Hill, 1965.
890. Richard C. Jeffrey, *Dracula meets wolfman: Acceptance vs. partial belief*, Induction, Acceptance and Rational Belief (Marshall Swain, ed.), Springer Netherlands, Dordrecht, 1970, pp. 157–185.
891. Richard C. Jeffrey, *Preference among preferences*, Journal of Philosophy **71** (1974), 377–391.
892. ______, *Conditioning, kinematics, and exchangeability*, Causation, chance, and credence (Brian Skyrms and William L. Harper, eds.), vol. 41, Kluwer Academic Publishers, 1988, pp. 221–255.
893. ______, *The logic of decision*, University of Chicago Press, 1990.
894. Harold Jeffreys, *Some tests of significance, treated by the theory of probability*, Mathematical Proceedings of the Cambridge Philosophical Society, vol. 31, Cambridge University Press, 1935, pp. 203–222.
895. ______, *An invariant form for the prior probability in estimation problems*, Proceedings of the Royal Society of London. Series A, Mathematical and Physical Sciences **186** (1946), no. 1007, 453–461.
896. Frank Jensen, Finn V. Jensen, and Søren L. Dittmer, *From influence diagrams to junction trees*, Proceedings of the Tenth International Conference on Uncertainty in Artificial Intelligence (UAI'94), Morgan Kaufmann, 1994, pp. 367–373.
897. Qiang Ji and Michael M. Marefat, *A Dempster–Shafer approach for recognizing machine features from {CAD} models*, Pattern Recognition **36** (2003), no. 6, 1355–1368.
898. Qiang Ji, Michael M. Marefat, and Paul J. Lever, *An evidential reasoning approach for recognizing shape features*, Proceedings of the 11th Conference on Artificial Intelligence for Applications, February 1995, pp. 162–168.
899. Jiang Jiang, Xuan Li, Zhi jie Zhou, Dong ling Xu, and Ying wu Chen, *Weapon system capability assessment under uncertainty based on the evidential reasoning approach*, Expert Systems with Applications **38** (2011), no. 11, 13773–13784.
900. Wen Jiang, Yong Deng, and Jinye Peng, *A new method to determine BPA in evidence theory*, Journal of Computers **6** (2011), no. 6, 1162–1167.
901. Wen Jiang, Dejie Duanmu, Xin Fan, and Yong Deng, *A new method to determine basic probability assignment under fuzzy environment*, Proceedings of the International Conference on Systems and Informatics (ICSAI 2012), May 2012, pp. 758–762.
902. Wen Jiang, An Zhang, and Qi Yang, *A new method to determine evidence discounting coefficient*, Proceedings of the International Conference on Intelligent Computing, vol. 5226/2008, 2008, pp. 882–887.
903. Radim Jirousek, *Composition of probability measures on finite spaces*, Proceedings of the Thirteenth Conference on Uncertainty in Artificial Intelligence (UAI'97), Morgan Kaufmann, 1997, pp. 274–281.
904. ______, *An attempt to define graphical models in Dempster–Shafer theory of evidence*, Combining Soft Computing and Statistical Methods in Data Analysis (Christian Borgelt, Gil González-Rodríguez, Wolfgang Trutschnig, María Asunción Lubiano, María Ángeles Gil, Przemysław Grzegorzewski, and Olgierd Hryniewicz, eds.), Springer, Berlin, Heidelberg, 2010, pp. 361–368.
905. ______, *Factorization and decomposable models in Dempster–Shafer theory of evidence*, Proceedings of the Workshop on Theory of Belief Functions, 2010.
906. ______, *Local computations in Dempster–Shafer theory of evidence*, International Journal of Approximate Reasoning **53** (November 2012), no. 8, 1155–1167.
907. Radim Jirousek and Prakash P. Shenoy, *A note on factorization of belief functions*, Proceedings of the 14th Czech-Japan Seminar on Data Analysis and Decision Making under Uncertainty, 2011.

908. Radim Jirousek and Prakash P. Shenoy, *Entropy of belief functions in the Dempster–Shafer theory: A new perspective*, Belief Functions: Theory and Applications: 4th International Conference, BELIEF 2016, Prague, Czech Republic, September 21-23, 2016, Proceedings (Jirina Vejnarova and Vaclav Kratochvil, eds.), Springer International Publishing, 2016, pp. 3–13.
909. Radim Jirousek and Prakash P. Shenoy, *A new definition of entropy of belief functions in the Dempster–Shafer theory*, Working Paper 330, Kansas University School of Business, 2016.
910. Radim Jirousek and Jirina Vejnarova, *Compositional models and conditional independence in evidence theory*, International Journal of Approximate Reasoning **52** (2011), no. 3, 316–334.
911. Radim Jirousek, Jirina Vejnarova, and Milan Daniel, *Compositional models for belief functions*, Proceedings of the Fifth International Symposium on Imprecise Probabilities and Their Applications (ISIPTA'07), vol. 7, 2007, pp. 243–252.
912. Pontus Johnson, Lars Nordström, and Robert Lagerström, *Formalizing analysis of enterprise architecture*, Enterprise Interoperability (Guy Doumeingts, Jörg Müller, Gérard Morel, and Bruno Vallespir, eds.), Springer, 2007, pp. 35–44.
913. Lianne Jones, Malcolm J. Beynon, Catherine A. Holt, and Stuart Roy, *An application of the Dempster–Shafer theory of evidence to the classification of knee function and detection of improvement due to total knee replacement surgery*, Journal of Biomechanics **39** (2006), no. 13, 2512–2520.
914. Audun Jøsang, *Artificial reasoning with subjective logic*, Proceedings of the Second Australian Workshop on Commonsense Reasoning, vol. 48, 1997, p. 34.
915. ———, *The consensus operator for combining beliefs*, Artificial Intelligence **141** (2002), no. 1, 157–170.
916. ———, *Subjective evidential reasoning*, Proceedings of the International Conference on Information Processing and Management of Uncertainty (IPMU2002), 2002.
917. ———, *Subjective logic – a formalism for reasoning under uncertainty*, Artificial Intelligence: Foundations, Theory, and Algorithms, Springer, 2016.
918. ———, *A logic for uncertain probabilities*, International Journal of Uncertainty, Fuzziness and Knowledge-Based Systems **9** (June 2001), no. 3, 279–311.
919. Audun Jøsang, Milan Daniel, and Patrick Vannoorenberghe, *Strategies for combining conflicting dogmatic beliefs*, Proceedings of the Sixth International Conference on Information Fusion (FUSION 2003), vol. 2, July 2003, pp. 1133–1140.
920. Audun Jøsang, Javier Diaz, and Maria Rifqi, *Cumulative and averaging fusion of beliefs*, Information Fusion **11** (2010), no. 2, 192–200.
921. Audun Jøsang and Zied Elouedi, *Interpreting belief functions as Dirichlet distributions*, Proceedings of the 9th European Conference on Symbolic and Quantitative Approaches to Reasoning with Uncertainty (ECSQARU 2007), Hammamet, Tunisia (Khaled Mellouli, ed.), Springer, Berlin, Heidelberg, October-November 2007, pp. 393–404.
922. Audun Jøsang and David McAnally, *Multiplication and comultiplication of beliefs*, International Journal of Approximate Reasoning **38** (2005), no. 1, 19–51.
923. Audun Jøsang and Simon Pope, *Dempster's rule as seen by little colored balls*, Computational Intelligence **28** (November 2012), no. 4, 453–474.
924. Audun Jøsang, Simon Pope, and David McAnally, *Normalising the consensus operator for belief fusion*, Proceedings of Information Processing with Management of Uncertainty in Knowledge-based Systems (IPMU'06), 2006, pp. 1420–1427.

925. A. V. Joshi, S. C. Sahasrabudhe, and K. Shankar, *Sensitivity of combination schemes under conflicting conditions and a new method*, Advances in Artificial Intelligence: 12th Brazilian Symposium on Artificial Intelligence SBIA '95, Campinas, Brazil, October 10–12, 1995 Proceedings (Jacques Wainer and Ariadne Carvalho, eds.), Springer, Berlin Heidelberg, 1995, pp. 39–48.
926. Cliff Joslyn, *Towards an empirical semantics of possibility through maximum uncertainty*, Proceedings of the International Fuzzy Systems Association Conference (IFSA 1991) (R. Lowen and M. Roubens, eds.), vol. A, 1991, pp. 86–89.
927. ———, *Possibilistic normalization of inconsistent random intervals*, Advances in Systems Science and Applications (1997), 44–51.
928. Cliff Joslyn and Scott Ferson, *Approximate representations of random intervals for hybrid uncertain quantification in engineering modeling*, Proceedings of the 4th International Conference on Sensitivity Analysis of Model Output (SAMO 2004) (K. M. Hanson and F. M. Hemez, eds.), 2004, pp. 453–469.
929. Cliff Joslyn and Jon C. Helton, *Bounds on belief and plausibility of functionally propagated random sets*, Proceedings of the Annual Meeting of the North American Fuzzy Information Processing Society (NAFIPS-FLINT 2002), 2002, pp. 412–417.
930. Cliff Joslyn and George Klir, *Minimal information loss possibilistic approximations of random sets*, Proceedings of the IEEE International Conference on Fuzzy Systems (FUZZ-IEEE) (Jim Bezdek, ed.), 1992, pp. 1081–1088.
931. Cliff Joslyn and Luis Rocha, *Towards a formal taxonomy of hybrid uncertainty representations*, Information Sciences **110** (1998), no. 3-4, 255–277.
932. Alexandre Jouan, Langis Gagnon, Elisa Shahbazian, and Pierre Valin, *Fusion of imagery attributes with non-imaging sensor reports by truncated Dempster–Shafer evidential reasoning*, Proceedings of the International Conference on Multisource-Multisensor Information Fusion (FUSION'98) (R. Hamid, A. Zhu, and D. Zhu, eds.), vol. 2, Las Vegas, NV, USA, 6-9 July 1998, pp. 549–556.
933. Anne-Laure Jousselme, Dominic Grenier, and Eloi Bossé, *A new distance between two bodies of evidence*, Information Fusion **2** (2001), no. 2, 91–101.
934. Anne-Laure Jousselme, Dominic Grenier, and Eloi Bossé, *Analyzing approximation algorithms in the theory of evidence*, Proceedings of SPIE - Sensor Fusion: Architectures, Algorithms, and Applications VI (B. V. Dasarathy, ed.), vol. 4731, March 2002, pp. 65–74.
935. Anne-Laure Jousselme, Dominic Grenier, and Eloi Bossé, *Analyzing approximation algorithms in the theory of evidence*, Proceedingsg of SPIE - Sensor Fusion: Architectures, Algorithms, and Applications VI, vol. 4731, March 2002, pp. 65–74.
936. Anne-Laure Jousselme, Chunsheng Liu, Dominic Grenier, and Eloi Bossé, *Measuring ambiguity in the evidence theory*, IEEE Transactions on Systems, Man, and Cybernetics - Part A: Systems and Humans **36** (2006), no. 5, 890–903.
937. Anne-Laure Jousselme and P. Maupin, *On some properties of distances in evidence theory*, Proceedings of the Workshop on the Theory of Belief Functions (BELIEF'10), 2010, pp. 1–6.
938. Anne-Laure Jousselme and Patrick Maupin, *Distances in evidence theory: Comprehensive survey and generalizations*, International Journal of Approximate Reasoning **53** (2012), no. 2, 118–145.
939. Henry E. Kyburg Jr., *Interval-valued probabilities*, Tech. report, Imprecise Probabilities Project, 1998.
940. Imene Jraidi and Zied Elouedi, *Belief classification approach based on generalized credal EM*, Symbolic and Quantitative Approaches to Reasoning with Uncertainty: 9th

European Conference, ECSQARU 2007, Hammamet, Tunisia, October 31 - November 2, 2007. Proceedings (Khaled Mellouli, ed.), Springer, Berlin Heidelberg, 2007, pp. 524–535.

941. Khaled Kaaniche, Benjamin Champion, Claude Pégard, and Pascal Vasseur, *A vision algorithm for dynamic detection of moving vehicles with a UAV*, Proceedings of the 2005 IEEE International Conference on Robotics and Automation (ICRA 2005), April 2005, pp. 1878–1883.
942. Gregory J. Kacprzynski, Michael J. Roemer, Girish Modgil, Andrea Palladino, and Kenneth Maynard, *Enhancement of physics-of-failure prognostic models with system level features*, Proceedings of the 2002 IEEE Aerospace Conference, vol. 6, 2002, pp. 2919–2925.
943. Daniel Kahneman and Amos Tversky, *Variants of uncertainty*, Cognition **11** (1982), no. 2, 143–157.
944. Rudolph Emil Kalman et al., *A new approach to linear filtering and prediction problems*, Journal of basic Engineering **82** (1960), no. 1, 35–45.
945. Tin Kam Ho, *Data complexity analysis for classifier combination*, Multiple Classifier Systems (Josef Kittler and Fabio Roli, eds.), Lecture Notes in Computer Science, vol. 2096, Springer, Berlin Heidelberg, 2001, pp. 53–67.
946. H. Kang, J. Cheng, I. Kim, and G. Vachtsevanos, *An application of fuzzy logic and Dempster–Shafer theory to failure detection and identification*, Proceedings of the 30th IEEE Conference on Decision and Control (CDC'91), vol. 2, December 1991, pp. 1555–1560.
947. Fatma Karem, Mounir Dhibi, and Arnaud Martin, *Combination of supervised and unsupervised classification using the theory of belief functions*, Belief Functions: Theory and Applications: Proceedings of the 2nd International Conference on Belief Functions, Compiègne, France 9-11 May 2012 (Thierry Denœux and Marie-Hélène Masson, eds.), Springer, Berlin Heidelberg, 2012, pp. 85–92.
948. Iman Karimi and Eyke Hüllermeier, *Risk assessment system of natural hazards: A new approach based on fuzzy probability*, Fuzzy Sets and Systems **158** (2007), no. 9, 987–999.
949. William Karush, *Minima of functions of several variables with inequalities as side constraints*, MSc dissertation, Department of Mathematics, University of Chicago, Chicago, Illinois, 1939.
950. David G. Kendall, *Foundations of a theory of random sets*, Stochastic Geometry (E. F. Harding and D. G. Kendall, eds.), Wiley, London, 1974, pp. 322–376.
951. Robert Kennes, *Evidential reasoning in a categorial perspective: conjunction and disjunction on belief functions*, Uncertainty in Artificial Intelligence 6 (B. D'Ambrosio, Ph. Smets, and P. P. Bonissone, eds.), Morgan Kaufann, 1991, pp. 174–181.
952. ———, *Computational aspects of the Möbius transformation of graphs*, IEEE Transactions on Systems, Man, and Cybernetics **22** (1992), no. 2, 201–223.
953. Robert Kennes and Philippe Smets, *Computational aspects of the Möbius transform*, Proceedings of the 6th Conference on Uncertainty in Artificial Intelligence (UAI'90), 1990, pp. 344–351.
954. ———, *Fast algorithms for Dempster–Shafer theory*, Uncertainty in Knowledge Bases (B. Bouchon-Meunier, R. R. Yager, and L. A. Zadeh, eds.), Lecture Notes in Computer Science, vol. 521, Springer-Verlag, 1991, pp. 14–23.
955. ———, *Computational aspects of the Möbius transform*, ArXiv preprint arXiv:1304.1122, 2013.

956. Jeroen Keppens, *Conceptions of vagueness in subjective probability for evidential reasoning*, Proceedings of the Twenty-Second Annual Conference on Legal Knowledge and Information Systems (JURIX 2009), 2009, pp. 89–99.
957. Timber Kerkvliet, *Uniform probability measures and epistemic probability*, PhD dissertation, Vrije Universiteit, 2017.
958. Gabriele Kern-Isberner, *Conditionals in nonmonotonic reasoning and belief revision: Considering conditionals as agents*, vol. 2087, Springer Science and Business Media, 2001.
959. Yousri Kessentini, Thomas Burger, and Thierry Paquet, *Evidential ensemble HMM classifier for handwriting recognition*, Proceedings of the International Conference on Information Processing and Management of Uncertainty in Knowledge-Based Systems (IPMU 2010), 2010, pp. 445–454.
960. Yousri Kessentini, Thierry Paquet, and AbdelMajid Ben Hamadou, *Off-line handwritten word recognition using multi-stream hidden Markov models*, Pattern Recognition Letters **30** (2010), no. 1, 60–70.
961. John Maynard Keynes, *A treatise on probability*, Macmillan and Company, 1921.
962. John Maynard Keynes, *A treatise on probability, ch. 4: Fundamental ideas*, Macmillan, 1921.
963. Leonid Khachiyan, *Fourier–Motzkin elimination method*, Encyclopedia of Optimization (Christodoulos A. Floudas and Panos M. Pardalos, eds.), Springer US, Boston, MA, 2009, pp. 1074–1077.
964. Fereshte Khani, Martin Rinard, and Percy Liang, *Unanimous prediction for 100% precision with application to learning semantic mappings*, Proceedings of the 54th Annual Meeting of the Association for Computational Linguistics (Volume 1: Long Papers), 2016, pp. 952–962.
965. Vahid Khatibi and G. A. Montazer, *A new evidential distance measure based on belief intervals*, Scientia Iranica - Transactions D: Computer Science and Engineering and Electrical Engineering **17** (2010), no. 2, 119–132.
966. Rashid Hafeez Khokhar, David A. Bell, Jiwen Guan, and Qing Xiang Wu, *Risk assessment of e-commerce projects using evidential reasoning*, Fuzzy Systems and Knowledge Discovery (Lipo Wang, Licheng Jiao, Guanming Shi, Xue Li, and Jing Liu, eds.), Lecture Notes in Computer Science, vol. 4223, Springer, Berlin Heidelberg, 2006, pp. 621–630.
967. Kourosh Khoshelham, Carla Nardinocchi, Emanuele Frontoni, Adriano Mancini, and Primo Zingaretti, *Performance evaluation of automated approaches to building detection in multi-source aerial data*, ISPRS Journal of Photogrammetry and Remote Sensing **65** (2010), no. 1, 123–133.
968. Hakil Kim and Philip H. Swain, *Evidential reasoning approach to multisource-data classification in remote sensing*, IEEE Transactions on Systems, Man, and Cybernetics **25** (1995), no. 8, 1257–1265.
969. Jin H. Kim and Judea Pearl, *A computational model for causal and diagnostic reasoning in inference systems*, Proceedings of the Eighth International Joint Conference on Artificial Intelligence (IJCAI'83), vol. 1, 1983, pp. 190–193.
970. Gary King and Langche Zeng, *Logistic regression in rare events data*, Political Analysis **9** (2001), no. 2, 137–163.
971. Josef Kittler, Mohamad Hatef, Robert P. W. Duin, and Jiri Matas, *On combining classifiers*, IEEE Transactions on Pattern Analysis and Machine Intelligence **20** (1998), no. 3, 226–239.
972. Daniel A. Klain and Gian-Carlo Rota, *Introduction to geometric probability*, Cambridge University Press, 1997.

973. D. Klaua, *Über einen Ansatz zur mehrwertigen Mengenlehre*, Monatsb. Deutsch. Akad. Wiss. Berlin **7** (1965), 859–876.
974. Frank Klawonn and Erhard Schwecke, *On the axiomatic justification of Dempster's rule of combination*, International Journal of Intelligent Systems **7** (1992), no. 5, 469–478.
975. Frank Klawonn and Philippe Smets, *The dynamic of belief in the transferable belief model and specialization-generalization matrices*, Proceedings of the Eighth International Conference on Uncertainty in Artificial Intelligence (UAI'92), Morgan Kaufmann, 1992, pp. 130–137.
976. John Klein and Olivier Colot, *Automatic discounting rate computation using a dissent criterion*, Proceedings of the Workshop on the theory of belief functions (BELIEF 2010), 2010, pp. 1–6.
977. John Klein, Christéle Lecomte, and Pierre Miché, *Hierarchical and conditional combination of belief functions induced by visual tracking*, International Journal of Approximate Reasoning **51** (2010), no. 4, 410–428.
978. Lawrence A. Klein, *Sensor and data fusion: A tool for information assessment and decision making*, vol. 138, SPIE Digital Library, 2004.
979. Robert Kleyle and André de Korvin, *A belief function approach to information utilization in decision making*, Journal of the American Society for Information Science **41** (1990), no. 8, 569–580.
980. George J. Klir, *Where do we stand on measures of uncertainty, ambiguity, fuzziness, and the like?*, Fuzzy Sets and Systems **24** (1987), no. 2, 141–160.
981. ———, *Generalized information theory*, Fuzzy Sets and Systems **40** (1991), no. 1, 127–142.
982. ———, *Dynamic decision making with belief functions*, Measures of uncertainty in the Dempster–Shafer theory of evidence (R. R. Yager, M. Fedrizzi, and J. Kacprzyk, eds.), Wiley, 1994, pp. 35–49.
983. George J. Klir, *Measures of uncertainty in the Dempster–Shafer theory of evidence*, Advances in the Dempster-Shafer Theory of Evidence (Ronald R. Yager, Janusz Kacprzyk, and Mario Fedrizzi, eds.), John Wiley and Sons, Inc., 1994, pp. 35–49.
984. George J. Klir, *Principles of uncertainty: What are they? Why do we need them?*, Fuzzy Sets and Systems **74** (1995), no. 1, 15–31.
985. ———, *Basic issues of computing with granular probabilities*, Proceedings of the IEEE International Conference on Fuzzy Systems, 1998, pp. 101–105.
986. ———, *On fuzzy-set interpretation of possibility theory*, Fuzzy Sets and Systems **108** (1999), no. 3, 263–273.
987. ———, *Generalized information theory: aims, results, and open problems*, Reliability Engineering and System Safety **85** (2004), no. 1-3, 21–38, Alternative Representations of Epistemic Uncertainty.
988. ———, *Statistical modeling with imprecise probabilities*, Tech. Report AFRL-IF-RS-TR-1998-166, SUNY Binghamton, August 1998.
989. George J. Klir, *Probabilistic versus possibilistic conceptualization of uncertainty*, Proceedings of the First International Symposium on Uncertainty Modeling and Analysis, December 1990, pp. 38–41.
990. George J. Klir and Tina A. Folger, *Fuzzy sets, uncertainty and information*, Prentice Hall, Englewood Cliffs (NJ), 1988.
991. George J. Klir and David Harmanec, *Generalized information theory*, Kybernetes **25** (1996), no. 7-8, 50–67.

992. George J. Klir and H. W. Lewis, *Remarks on "Measuring ambiguity in the evidence theory"*, IEEE Transactions on Systems, Man, and Cybernetics - Part A: Systems and Humans **38** (2008), no. 4, 995–999.
993. George J. Klir and H. W. III Lewis, *Remarks on measuring ambiguity in the evidence theory*, IEEE Transactions on Systems, Man and Cybernetics, Part A: Systems and Humans **38** (July 2008), no. 4, 995–999.
994. George J. Klir and Matthew Mariano, *On the uniqueness of possibilistic measure of uncertainty and information*, Fuzzy Sets and Systems **24** (November 1987), no. 2, 197–219.
995. George J. Klir and Behzad Parviz, *A note on the measure of discord*, Proceedings of the Eighth Conference on Uncertainty in Artificial Intelligence (UAI '92), Morgan Kaufmann, 1992, pp. 138–141.
996. George J. Klir and Arthur Ramer, *Uncertainty in the Dempster–Shafer theory: a critical re-examination*, International Journal of General Systems **18** (1990), no. 2, 155–166.
997. George J. Klir, Zhenyuan Wang, and David Harmanec, *Constructing fuzzy measures in expert systems*, Fuzzy Sets and Systems **92** (1997), no. 2, 251–264.
998. George J. Klir and Mark Wierman, *Uncertainty-based information: elements of generalized information theory*, vol. 15, Springer Science and Business Media, 1999.
999. George J. Klir and Bo Yuan, *Fuzzy sets and fuzzy logic: theory and applications*, Prentice Hall PTR, Upper Saddle River, NJ, 1995.
1000. George J Klir and Bo Yuan, *Fuzzy sets and fuzzy logic: theory and applications*, Possibility Theory versus Probab. Theory **32** (1996), no. 2, 207–208.
1001. Mieczyslaw A. Klopotek, *Identification of belief structure in Dempster–Shafer theory*, Foundations of Computing and Decison Sciences **21** (1996), no. 1, 35–54.
1002. ______, *Methods of identification and interpretations of belief distributions in the Dempster–Shafer theory*, PhD dissertation, Institute of Computer Science of the Polish Academy of Sciences, 1998.
1003. Mieczyslaw A. Klopotek, Andrzej Matuszewski, and Sawomir T. Wierzchon, *Overcoming negative-valued conditional belief functions when adapting traditional knowledge acquisition tools to Dempster–Shafer theory*, Proceedings of the International Conference on Computational Engineering in Systems Applications, Symposium on Modelling, Analysis and Simulation, CESA '96 IMACS Multiconference, vol. 2, Lille, France, 9-12 July 1996, pp. 948–953.
1004. Mieczyslaw A. Klopotek and Sawomir T. Wierzchon, *An interpretation for the conditional belief function in the theory of evidence*, Proceedings of the International Symposium on Methodologies for Intelligent Systems, Springer, 1999, pp. 494–502.
1005. Mieczysław A. Kłopotek and Sławomir T. Wierzchoń, *A new qualitative rough-set approach to modeling belief functions*, Rough Sets and Current Trends in Computing: First International Conference, RSCTC'98 Warsaw, Poland, June 22–26, 1998 Proceedings (Lech Polkowski and Andrzej Skowron, eds.), Springer, Berlin Heidelberg, 1998, pp. 346–354.
1006. Mieczysław Alojzy Kłopotek and Sławomir Tadeusz Wierzchoń, *Empirical models for the Dempster–Shafer-theory*, Belief Functions in Business Decisions (Rajendra P. Srivastava and Theodore J. Mock, eds.), Physica-Verlag HD, Heidelberg, 2002, pp. 62–112.
1007. Frank H. Knight, *Risk, uncertainty and profit*, Courier Corporation, 2012.
1008. Waldemar W. Koczkodaj, *A new definition of consistency of pairwise comparisons*, Mathematical and Computer Modelling **18** (1993), no. 7, 79–84.

1009. Earl T. Kofler and Cornelius T. Leondes, *Algorithmic modifications to the theory of evidential reasoning*, Journal of Algorithms **17** (September 1994), no. 2, 269–279.
1010. Jürg Kohlas, *Modeling uncertainty for plausible reasoning with belief*, Tech. Report 116, Institute for Automation and Operations Research, University of Fribourg, 1986.
1011. ———, *The logic of uncertainty. potential and limits of probability. theory for managing uncertainty in expert systems*, Tech. Report 142, Institute for Automation and Operations Research, University of Fribourg, 1987.
1012. ———, *Conditional belief structures*, Probability in Engineering and Information Science **2** (1988), no. 4, 415–433.
1013. ———, *Modeling uncertainty with belief functions in numerical models*, European Journal of Operational Research **40** (1989), no. 3, 377–388.
1014. ———, *Evidential reasoning about parametric models*, Tech. Report 194, Institute for Automation and Operations Research, University Fribourg, 1992.
1015. ———, *Support and plausibility functions induced by filter-valued mappings*, International Journal of General Systems **21** (1993), no. 4, 343–363.
1016. ———, *Mathematical foundations of evidence theory*, Mathematical Models for Handling Partial Knowledge in Artificial Intelligence (Giulianella Coletti, Didier Dubois, and Romano Scozzafava, eds.), Plenum Press, 1995, pp. 31–64.
1017. ———, *The mathematical theory of evidence – A short introduction*, System Modelling and Optimization (J. Dolezal and J. Fidler, eds.), IFIP Advances in Information and Communication Technology, Chapman and Hall, 1995, pp. 37–53.
1018. ———, *Allocation of arguments and evidence theory*, Theoretical Computer Science **171** (1997), no. 1–2, 221–246.
1019. ———, *Uncertain information: Random variables in graded semilattices*, International Journal of Approximate Reasoning **46** (2007), no. 1, 17–34, Special Section: Random Sets and Imprecise Probabilities (Issues in Imprecise Probability).
1020. Jürg Kohlas and Philippe Besnard, *An algebraic study of argumentation systems and evidence theory*, Tech. report, Institute of Informatics, University of Fribourg, 1995.
1021. Jürg Kohlas and Hans Wolfgang Brachinger, *Argumentation systems and evidence theory*, Proceedings of the International Conference on Information Processing and Management of Uncertainty in Knowledge-Based Systems (IPMU'94) (B. Bouchon-Meunier, R. R. Yager, and L. A. Zadeh, eds.), Advances in Intelligent Computing, Springer-Verlag, 1994, pp. 41–50.
1022. Jürg Kohlas and Christian Eichenberger, *Uncertain information*, Formal Theories of Information (Giovanni Sommaruga, ed.), Lecture Notes in Computer Science, vol. 5363, Springer-Verlag, Berlin Heidelberg, 2009, pp. 128–160.
1023. Jürg Kohlas and Paul-André Monney, *Modeling and reasoning with hints*, Tech. Report 174, Institute for Automation and Operations Research, University of Fribourg, 1990.
1024. ———, *Propagating belief functions through constraint systems*, International Journal of Approximate Reasoning **5** (1991), no. 5, 433–461.
1025. ———, *Representation of evidence by hints*, Advances in the Dempster-Shafer Theory of Evidence (R. R. Yager, J. Kacprzyk, and M. Fedrizzi, eds.), John Wiley, New York, 1994, pp. 473–492.
1026. ———, *Theory of evidence - A survey of its mathematical foundations, applications and computational anaylsis*, Zeitschrift fr Operations Researc - Mathematical Methods of Operations Research **39** (1994), no. 1, 35–68.
1027. ———, *A mathematical theory of hints: An approach to Dempster–Shafer theory of evidence*, Lecture Notes in Economics and Mathematical Systems, vol. 425, Springer-Verlag, 1995.

1028. Jürg Kohlas, Paul-André Monney, Rolf Haenni, and Norbert Lehmann, *Model-based diagnostics using hints*, Proceedings of the European Conference on Symbolic and Quantitative Approaches to Uncertainty (ECSQARU'95) (Ch. Fridevaux and Jürg Kohlas, eds.), Lecture Notes in Computer Science, vol. 946, Springer, 1995, pp. 259–266.
1029. Don Koks and Subhash Challa, *An introduction to Bayesian and Dempster–Shafer data fusion*, Tech. Report DSTO-TR-1436, Defence Science and Technology Organisation, Salisbury, Australia, 2003.
1030. Andrei N. Kolmogorov, *Grundbegriffe der wahrscheinlichkeitsrechnung*, Springer, Berlin, 1933.
1031. ______, *Foundations of the theory of probability*, Chealsea Pub. Co., Oxford, 1950.
1032. Andrei N. Kolmogorov and Albert T. Bharucha-Reid, *Foundations of the theory of probability: Second english edition*, Courier Dover Publications, 2018.
1033. Augustine Kong, *Multivariate belief functions and graphical models*, PhD dissertation, Harvard University, Department of Statistics, 1986.
1034. C. T. A. Kong, *A belief function generalization of Gibbs ensemble*, Tech. Report S-122, Harvard University, 1988.
1035. Jerzy Konorski and Rafal Orlikowski, *Data-centric Dempster–Shafer theory-based selfishness thwarting via trust evaluation in MANETs and WSNs*, Proceedings of the 3rd International Conference on New Technologies, Mobility and Security (NTMS), December 2009, pp. 1–5.
1036. Bernard O. Koopman, *The axioms and algebra of intuitive probability*, Annals of mathematics **41** (1940), 269–292.
1037. ______, *The bases of probability*, Bulletin of the American Mathematical Society **46** (1940), no. 10, 763–774.
1038. Mieczysaw A. Kopotek and Sawomir T. Wierzchon, *Empirical models for the Dempster–Shafer-theory*, Belief Functions in Business Decisions (Rajendra P. Srivastava and Theodore J. Mock, eds.), Studies in Fuzziness and Soft Computing, vol. 88, Physica-Verlag HD, 2002, pp. 62–112.
1039. Petri Korpisaari and J. Saarinen, *Dempster–Shafer belief propagation in attribute fusion*, Proceedings of the Second International Conference on Information Fusion (FUSION'99), vol. 2, Sunnyvale, CA, USA, 6-8 July 1999, pp. 1285–1291.
1040. Gleb A. Koshevoy, *Distributive lattices and products of capacities*, Journal of Mathematical Analysis Applications **219** (1998), no. 2, 427–441.
1041. Guy Kouemou, Christoph Neumann, and Felix Opitz, *Exploitation of track accuracy information in fusion technologies for radar target classification using Dempster–Shafer rules*, Proceedings of the 12th International Conference on Information Fusion (FUSION '09), July 2009, pp. 217–223.
1042. Volker Kraetschmer, *Constraints on belief functions imposed by fuzzy random variables: Some technical remarks on Römer-Kandel*, IEEE Transactions on Systems, Man, and Cybernetics, Part B: Cybernetics **28** (1998), no. 6, 881–883.
1043. Ivan Kramosil, *Expert systems with non-numerical belief functions*, Problems of Control and Information Theory **17** (1988), no. 5, 285–295.
1044. ______, *Possibilistic belief functions generated by direct products of single possibilistic measures*, Neural Network World **9** (1994), no. 6, 517–525.
1045. ______, *Approximations of believeability functions under incomplete identification of sets of compatible states*, Kybernetika **31** (1995), no. 5, 425–450.
1046. ______, *Dempster–Shafer theory with indiscernible states and observations*, International Journal of General Systems **25** (1996), no. 2, 147–152.

1047. ———, *Expert systems with non-numerical belief functions*, Problems of control and information theory **16** (1996), no. 1, 39–53.
1048. ———, *Belief functions generated by signed measures*, Fuzzy Sets and Systems **92** (1997), no. 2, 157–166.
1049. ———, *Probabilistic analysis of Dempster–Shafer theory. Part one*, Tech. Report 716, Academy of Science of the Czech Republic, 1997.
1050. ———, *Probabilistic analysis of Dempster–Shafer theory. Part three.*, Technical Report 749, Academy of Science of the Czech Republic, 1998.
1051. ———, *Probabilistic analysis of Dempster–Shafer theory. Part two.*, Tech. Report 749, Academy of Science of the Czech Republic, 1998.
1052. ———, *Measure-theoretic approach to the inversion problem for belief functions*, Fuzzy Sets and Systems **102** (1999), no. 3, 363–369.
1053. ———, *Nonspecificity degrees of basic probability assignments in Dempster–Shafer theory*, Computing and Informatics **18** (1999), no. 6, 559–574.
1054. ———, *Dempster combination rule with Boolean-like processed belief functions*, International Journal of Uncertainty, Fuzziness and Knowledge-Based Systems **9** (2001), no. 1, 105–121.
1055. ———, *Probabilistic analysis of dempster combination rule*, Probabilistic Analysis of Belief Functions, Springer, 2001, pp. 57–67.
1056. ———, *Probabilistic analysis of belief functions*, ISFR International Series on Systems Science and Engineering, vol. 16, Kluwer Academic / Plenum Publishers, 2002.
1057. ———, *Belief functions generated by fuzzy and randomized compatibility relations*, Fuzzy Sets and Systems **135** (2003), no. 3, 341–366.
1058. ———, *Belief functions with nonstandard values*, Qualitative and Quantitative Practical Reasoning (Dav Gabbay, Rudolf Kruse, Andreas Nonnengart, and Hans Jurgen Ohlbach, eds.), Lecture Notes in Computer Science, vol. 1244, Springer, Berlin Heidelberg, Bonn, June 1997, pp. 380–391.
1059. ———, *Dempster combination rule for signed belief functions*, International Journal of Uncertainty, Fuzziness and Knowledge-Based Systems **6** (February 1998), no. 1, 79–102.
1060. ———, *Monte-Carlo estimations for belief functions*, Proceedings of the Fourth International Conference on Fuzzy Sets Theory and Its Applications, vol. 16, February 1998, pp. 339–357.
1061. ———, *Definability of belief functions over countable sets by real-valued random variables*, Proceedings of the International Conference on Information Processing and Management of Uncertainty in Knowledge-Based Systems (IPMU'94) (Svoboda V., ed.), vol. 3, July 1994, pp. 49–50.
1062. ———, *Jordan decomposition of signed belief functions*, Proceedings of the International Conference on Information Processing and Management of Uncertainty in Knowledge-Based Systems (IPMU'96), July 1996, pp. 431–434.
1063. ———, *Toward a Boolean-valued Dempster–Shafer theory*, LOGICA '92 (Svoboda V., ed.), Prague, 1993, pp. 110–131.
1064. ———, *Measure-theoretic approach to the inversion problem for belief functions*, Proceedings of the Seventh International Fuzzy Systems Association World Congress (IFSA'97), vol. 1, Prague, June 1997, pp. 454–459.
1065. ———, *Strong law of large numbers for set-valued random variables*, Proceedings of the 3rd Workshop on Uncertainty Processing in Expert Systems, September 1994, pp. 122–142.
1066. David H. Krantz and John Miyamoto, *Priors and likelihood ratios as evidence*, Journal of the American Statistical Association **78** (June 1983), no. 382, 418–423.

1067. Paul Krause and Dominic Clark, *Representing uncertain knowledge: An artificial intelligence approach*, Kluwer Academic Publishers, Norwell, MA, USA, 1993.
1068. ———, *Representing uncertain knowledge: an artificial intelligence approach*, Springer Science and Business Media, 2012.
1069. Vladik Kreinovich, Andrew Bernat, Walter Borrett, Yvonne Mariscal, and Elsa Villa, *Monte–Carlo methods make Dempster–Shafer formalism feasible*, Technical Report NASA-CR-192945, NASA, 1991.
1070. Vladik Kreinovich, Andrew Bernat, Walter Borrett, Yvonne Mariscal, and Elsa Villa, *Monte-Carlo methods make Dempster–Shafer formalism feasible*, Advances in the Dempster–Shafer theory of evidence, John Wiley and Sons, Inc., 1994, pp. 175–191.
1071. Vladik Kreinovich, Claude Langrand, and Hung T. Nguyen, *Combining fuzzy and probabilistic knowledge using belief functions*, Departmental Technical Reports (CS) 414, University of Texas at El Paso, 2001.
1072. Vladik Kreinovich, Gang Xiang, and Scott Ferson, *Computing mean and variance under Dempster–Shafer uncertainty: Towards faster algorithms*, International Journal of Approximate Reasoning **42** (2006), no. 3, 212–227.
1073. Elmar Kriegler, *Imprecise probability analysis for integrated assessment of climate change*, Ph.D. thesis, University of Potsdam, 2005.
1074. Ganesh Krishnamoorthy, Theodore J. Mock, and Mary T. Washington, *A comparative evaluation of belief revision models in auditing*, AUDITING: A Journal of Practice and Theory **18** (1999), no. 2, 105–127.
1075. Raghu Krishnapuram and James M. Keller, *Fuzzy set theoretic approach to computer vision: An overview*, Proceedings of the IEEE International Conference on Fuzzy Systems, March 1992, pp. 135–142.
1076. Tomas Kroupa, *Belief functions on formulas in Lukasiewicz logic*, Proceedings of the 8th Workshop on Uncertainty Processing (WUPES) (Tomas Kroupa and Jirina Vejnarov, eds.), 2009, p. 156.
1077. Tomas Kroupa, *From probabilities to belief functions on MV-algebras*, Combining Soft Computing and Statistical Methods in Data Analysis (Christian Borgelt, Gil Gonzalez-Rodriguez, Wolfgang Trutschnig, Maria Asuncion Lubiano, Maria Angeles Gil, Przemyslaw Grzegorzewski, and Olgierd Hryniewicz, eds.), Springer, Berlin Heidelberg, 2010, pp. 387–394.
1078. ———, *Extension of belief functions to infinite-valued events*, Soft Computing **16** (2012), no. 11, 1851–1861.
1079. Rudolf Kruse, Detlef Nauck, and Frank Klawonn, *Reasoning with mass distributions*, Uncertainty in Artificial Intelligence (B. D. D'Ambrosio, Ph. Smets, and P. P. Bonissone, eds.), Morgan Kaufmann, 1991, pp. 182–187.
1080. Rudolf Kruse and Erhard Schwecke, *Specialization – A new concept for uncertainty handling with belief functions*, International Journal of General Systems **18** (1990), no. 1, 49–60.
1081. Rudolf Kruse, Erhard Schwecke, and Frank Klawonn, *On a tool for reasoning with mass distribution*, Proceedings of the 12th International Joint Conference on Artificial Intelligence (IJCAI'91), vol. 2, 1991, pp. 1190–1195.
1082. Rudolf Kruse, Erhard Schwecke, and Frank Klawonn, *On a tool for reasoning with mass distributions.*, Proceedings of the International Joint Conference on Artificial Intelligence (IJCAI'91), 1991, pp. 1190–1195.
1083. Jan Kühr and Daniele Mundici, *De Finetti theorem and Borel states in* $[0,1]$*-valued algebraic logic*, International Journal of Approximate Reasoning **46** (2007), no. 3, 605–616.

1084. Masayuki Kumon, Akimichi Takemura, and Kei Takeuchi, *Game-theoretic versions of strong law of large numbers for unbounded variables*, Stochastics - An International Journal of Probability and Stochastic Processes **79** (2007), no. 5, 449–468.
1085. Ludmila I. Kuncheva, James C. Bezdek, and Robert P.W. Duin, *Decision templates for multiple classifier fusion: An experimental comparison*, Pattern Recognition **34** (2001), no. 2, 299–314.
1086. Henry E. Kyburg, *Bayesian and non-Bayesian evidential updating*, Artificial Intelligence **31** (1987), no. 3, 271–294.
1087. ———, *Higher order probabilities*, Tech. report, University of Rochester, 1988.
1088. Henry E. Kyburg, *Getting fancy with probability*, Synthese **90** (1992), no. 2, 189–203.
1089. Hicham Laanaya, Arnaud Martin, Driss Aboutajdine, and Ali Khenchaf, *Support vector regression of membership functions and belief functions – Application for pattern recognition*, Information Fusion **11** (2010), no. 4, 338–350.
1090. Maria T. Lamata and Serafín Moral, *Measures of entropy in the theory of evidence*, International Journal of General Systems **14** (1988), no. 4, 297–305.
1091. ———, *Classification of fuzzy measures*, Fuzzy Sets and Systems **33** (1989), no. 2, 243–253.
1092. ———, *Calculus with linguistic probabilites and belief*, Advances in the Dempster-Shafer Theory of Evidence (R. R. Yager, M. Fedrizzi, and J. Kacprzyk, eds.), Wiley, 1994, pp. 133–152.
1093. Pierre Lanchantin and Wojciech Pieczynski, *Unsupervised restoration of hidden non-stationary Markov chains using evidential priors*, IEEE Transactions on Signal Processing **53** (2005), no. 8, 3091–3098.
1094. Saunders Mac Lane, *A lattice formulation for transcendence degrees and p-bases*, Duke Mathematical Journal **4** (1938), no. 3, 455–468.
1095. Andrew T. Langewisch and F. Fred Choobineh, *Mean and variance bounds and propagation for ill-specified random variables*, IEEE Transactions on Systems, Man, and Cybernetics - Part A: Systems and Humans **34** (July 2004), no. 4, 494–506.
1096. Pierre Simon Laplace, *Essai philosophique sur les Probabilités*, H. Remy, Bruxelles, 1829.
1097. Kathryn B. Laskey, *Belief in belief functions: An examination of Shafer's canonical examples*, ArXiv preprint arXiv:1304.2715, 2013.
1098. ———, *Beliefs in belief functions: an examination of Shafer's canonical examples*, Proceedings of the Third Conference on Uncertainty in Artificial Intelligence (UAI'87), Seattle, 1987, pp. 39–46.
1099. Kathryn B. Laskey and Paul E. Lehner, *Belief manteinance: an integrated approach to uncertainty management*, Proceeding of the Seventh National Conference on Artificial Intelligence (AAAI-88), vol. 1, 1988, pp. 210–214.
1100. ———, *Assumptions, beliefs and probabilities*, Artificial Intelligence **41** (1989), no. 1, 65–77.
1101. Steffen L. Lauritzen, *Graphical models*, vol. 17, Clarendon Press, 1996.
1102. Steffen L. Lauritzen and Finn V. Jensen, *Local computation with valuations from a commutative semigroup*, Annals of Mathematics and Artificial Intelligence **21** (1997), no. 1, 51–69.
1103. Steffen L. Lauritzen and David J. Spiegelhalter, *Local computations with probabilities on graphical structures and their application to expert systems*, Readings in Uncertain Reasoning (Glenn Shafer and Judea Pearl, eds.), Morgan Kaufmann, 1990, pp. 415–448.

1104. CuongAnh Le, Van-Nam Huynh, and Akira Shimazu, *An evidential reasoning approach to weighted combination of classifiers for word sense disambiguation*, Machine Learning and Data Mining in Pattern Recognition (Petra Perner and Atsushi Imiya, eds.), Lecture Notes in Computer Science, vol. 3587, Springer Science and Business Media, 2005, pp. 516–525.
1105. Sylvie Le Hégarat-Mascle, Isabelle Bloch, and D. Vidal-Madjar, *Unsupervised multisource remote sensing classification using Dempster–Shafer evidence theory*, Proceedings of SPIE - Synthetic Aperture Radar and Passive Microwave Sensing, vol. 2584, 1995, pp. 200–211.
1106. Sylvie Le Hegarat-Mascle, Isabelle Bloch, and Daniel Vidal-Madjar, *Application of Dempster–Shafer evidence theory to unsupervised classification in multisource remote sensing*, IEEE Transactions on Geoscience and Remote Sensing **35** (1997), no. 4, 1018–1031.
1107. Sylvie Le Hegarat-Mascle, D. Richard, and C. Ottle, *Multi-scale data fusion using Dempster–Shafer evidence theory*, Proceedings of the IEEE International Geoscience and Remote Sensing Symposium (IGARSS '02), vol. 2, 2002, pp. 911–913.
1108. Duncan Ermini Leaf and Chuanhai Liu, *Inference about constrained parameters using the elastic belief method*, International Journal of Approximate Reasoning **53** (2012), no. 5, 709–727.
1109. Yann LeCun, Yoshua Bengio, and Geoffrey Hinton, *Deep learning*, Nature **521** (2015), no. 7553, 436–444.
1110. François Leduc, Basel Solaiman, and François Cavayas, *Combination of fuzzy sets and Dempster–Shafer theories in forest map updating using multispectral data*, Proceedings of SPIE - Sensor Fusion: Architectures, Algorithms, and Applications V, vol. 4385, 2001, pp. 323–334.
1111. Chia-Hoang Lee, *A comparison of two evidential reasoning schemes*, Artificial Intelligence **35** (1988), no. 1, 127–134.
1112. E. Stanley Lee and Qing Zhu, *An interval Dempster–Shafer approach*, Computers and Mathematics with Applications **24** (1992), no. 7, 89–95.
1113. E. Stanley Lee and Qing Zhu, *Fuzzy and evidential reasoning*, Physica-Verlag, Heidelberg, 1995.
1114. Rae H. Lee and Richard M. Leahy, *Multispectral tissue classification of MR images using sensor fusion approaches*, Proceedings of SPIE - Medical Imaging IV: Image Processing, vol. 1233, 1990, pp. 149–157.
1115. Seung-Jae Lee, Sang-Hee Kang, Myeon-Song Choi, Sang-Tae Kim, and Choong-Koo Chang, *Protection level evaluation of distribution systems based on Dempster–Shafer theory of evidence*, Proceedings of the IEEE Power Engineering Society Winter Meeting, vol. 3, Singapore, 23-27 January 2000, pp. 1894–1899.
1116. Suk Kyoon Lee, *Imprecise and uncertain information in databases: An evidential approach*, Proceedings of the Eighth International Conference on Data Engineering, February 1992, pp. 614–621.
1117. Eric Lefévre and Olivier Colot, *A classification method based on the Dempster–Shafers theory and information criteria*, Proceeding of the 2nd International Conference on Information Fusion (FUSION'99), 1999, pp. 1179–1184.
1118. Eric Lefévre, Olivier Colot, and Patrick Vannoorenberghe, *Belief function combination and conflict management*, Information Fusion **3** (2002), no. 2, 149–162.
1119. ______, *Reply to the Comments of R. Haenni on the paper "Belief functions combination and conflict management*, Information Fusion **4** (2003), no. 1, 63–65.

1120. Eric Lefévre, Olivier Colot, Patrick Vannoorenberghe, and D. de Brucq, *A generic framework for resolving the conflict in the combination of belief structures*, Proceedings of the Third International Conference on Information Fusion (FUSION 2000), vol. 1, July 2000, pp. MOD4/11–MOD4/18.
1121. Eric Lefévre, Olivier Colot, Patrick Vannoorenberghe, and Denis de Brucq, *Knowledge modeling methods in the framework of Evidence Theory: An experimental comparison for melanoma detection*, Proceedings of the IEEE International Conference on Systems, Man, and Cybernetics (SMC 2000), vol. 4, 2000, pp. 2806–2811.
1122. Eric Lefévre and Zied Elouedi, *How to preserve the conflict as an alarm in the combination of belief functions?*, Decision Support Systems **56** (2013), 326–333.
1123. Éric Lefèvre, Zied Elouedi, and David Mercier, *Towards an alarm for opposition conflict in a conjunctive combination of belief functions*, Symbolic and Quantitative Approaches to Reasoning with Uncertainty: 11th European Conference, ECSQARU 2011, Belfast, UK, June 29–July 1, 2011. Proceedings (Weiru Liu, ed.), Springer, Berlin, Heidelberg, 2011, pp. 314–325.
1124. Eric Lefévre, Patrick Vannoorenberghe, and Olivier Colot, *Using information criteria in Dempster–Shafer's basic belief assignment*, Proceedings of the IEEE International Fuzzy Systems Conference (FUZZ-IEEE '99), vol. 1, August 1999, pp. 173–178.
1125. Norbert Lehmann, *Argumentation systems and belief functions*, PhD dissertation, Université de Fribourg, 2001.
1126. Norbert Lehmann and Rolf Haenni, *Belief function propagation based on Shenoy's fusion algorithm*, Proceedings of the 8th International Conference on Information Processing and Management of Uncertainty in Knowledge-Based Systems (IPMU00), 2000, pp. 1928–1931.
1127. ______, *An alternative to outward propagation for Dempster–Shafer belief functions*, Proceedings of The Fifth European Conference on Symbolic and Quantitative Approaches to Reasoning with Uncertainty (ECSQARU'99), Lecture Notes in Computer Science, Springer, July 1999, pp. 256–267.
1128. Ehud Lehrer, *Updating non-additive probabilities - a geometric approach*, Games and Economic Behavior **50** (2005), 42–57.
1129. Hui Lei and G. C. S. Shoja, *A distributed trust model for e-commerce applications*, Proceedings of the 2005 IEEE International Conference on e-Technology, e-Commerce and e-Service (EEE '05), March 2005, pp. 290–293.
1130. Benoit Lelandais, Isabelle Gardin, Laurent Mouchard, Pierre Vera, and Su Ruan, *Using belief function theory to deal with uncertainties and imprecisions in image processing*, Belief Functions: Theory and Applications (Thierry Denœux and Marie-Hélène Masson, eds.), Advances in Intelligent and Soft Computing, vol. 164, Springer, Berlin Heidelberg, 2012, pp. 197–204.
1131. Benot Lelandais, Su Ruan, Thierry Denœux, Pierre Vera, and Isabelle Gardin, *Fusion of multi-tracer PET images for dose painting*, Medical Image Analysis **18** (2014), no. 7, 1247–1259.
1132. John F. Lemmer, *Confidence factors, empiricism, and the Dempster–Shafer theory of evidence*, Uncertainty in Artificial Intelligence (L. N. Kanal and J. F. Lemmers, eds.), North Holland, 1986, pp. 167–196.
1133. ______, *Confidence factors, empiricism and the Dempster–Shafer theory of evidence*, 2013.
1134. John F. Lemmer and Henry E. Kyburg, *Conditions for the existence of belief functions corresponding to intervals of belief*, Proceedings of the Ninth National Conference on Artificial Intelligence, (AAAI-91), July 1991, pp. 488–493.

1135. Vasilica Lepar and Prakash P. Shenoy, *A comparison of Lauritzen–Spiegelhalter, Hugin, and Shenoy–Shafer architectures for computing marginals of probability distributions*, Proceedings of the International Conference on Uncertainty in Artificial Intelligence (UAI'98), 1998, pp. 328–337.
1136. Stephen A. Lesh, *An evidential theory approach to judgement-based decision making*, PhD dissertation, Department of Forestry and Environmental Studies, Duke University, December 1986.
1137. Henry Leung, Yifeng Li, Eloi Bossé, Martin Blanchette, and Keith C. C. Chan, *Improved multiple target tracking using Dempster–Shafer identification*, Proceedings of the SPIE - Signal Processing, Sensor Fusion, and Target Recognition VI, vol. 3068, Orlando, FL, USA, 21-24 April 1997, pp. 218–227.
1138. Henry Leung and Jiangfeng Wu, *Bayesian and Dempster–Shafer target identification for radar surveillance*, IEEE Transactions on Aerospace and Electronic Systems **36** (April 2000), no. 2, 432–447.
1139. Kwong-Sak Leung and Wai Lam, *Fuzzy concepts in expert systems*, Computer **21** (1988), no. 9, 43–56.
1140. Yee Leung, Nan-Nan Ji, and Jiang-Hong Ma, *An integrated information fusion approach based on the theory of evidence and group decision-making*, Information Fusion **14** (2013), no. 4, 410–422.
1141. Isaac Levi, *The enterprise of knowledge: An essay on knowledge, credal probability, and chance*, The MIT Press, Cambridge, Massachusetts, 1980.
1142. ______, *Consonance, dissonance and evidentiary mechanism*, Festschrift for Soren Hallden, Theoria, 1983, pp. 27–42.
1143. David Lewis, *Probabilities of conditionals and conditional probabilities*, Philosophical Review **85** (July 1976), 297–315.
1144. David K. Lewis, *Completeness and decidability of three logics of counterfactual conditionals*, Theoria **37** (1971), no. 1, 74–85.
1145. ______, *Counterfactuals*, John Wiley and Sons, 2013.
1146. Baohua Li, Yunmin Zhu, and X. Rong Li, *Fault-tolerant interval estimation fusion by Dempster–Shafer theory*, Proceedings of the Fifth International Conference on Information Fusion (FUSION 2002), vol. 2, July 2002, pp. 1605–1612.
1147. Bicheng Li, Bo Wang, Jun Wei, Yuqi Huang, and Zhigang Guo, *Efficient combination rule of evidence theory*, Proceedings of SPIE - Object Detection, Classification, and Tracking Technologies, vol. 4554, 2001, pp. 237–240.
1148. Guan Li, *A strong law of large numbers for set-valued random variables in $g\alpha$ space*, Journal of Applied Mathematics and Physics **3** (2015), no. 7, 797–801.
1149. Jiaming Li, Suhuai Luo, and Jesse S. Jin, *Sensor data fusion for accurate cloud presence prediction using Dempster–Shafer evidence theory*, Sensors **10** (2010), no. 10, 9384–9396.
1150. Jing Li, Jian Liu, and Keping Long, *Reliable cooperative spectrum sensing algorithm based on Dempster–Shafer theory*, Proceedings of the IEEE Global Telecommunications Conference (GLOBECOM 2010), December 2010, pp. 1–5.
1151. Jinping Li, Qingbo Yang, and Bo Yang, *Dempster–Shafer theory is a special case of vague sets theory*, Proceedings of the 2004 International Conference on Information Acquisition, June 2004, pp. 50–53.
1152. Peng Li and Si-Feng Liu, *Interval-valued intuitionistic fuzzy numbers decision-making method based on grey incidence analysis and DS theory of evidence*, Acta Automatica Sinica **37** (2011), no. 8, 993–998.

1153. Wenjia Li and Anupam Joshi, *Outlier detection in ad hoc networks using Dempster–Shafer theory*, Proceedings of the Tenth International Conference on Mobile Data Management: Systems, Services and Middleware (MDM'09), May 2009, pp. 112–121.
1154. Winston Li, Henry Leung, Chiman Kwan, and Bruce R. Linnell, *E-nose vapor identification based on Dempster–Shafer fusion of multiple classifiers*, IEEE Transactions on Instrumentation and Measurement **57** (2008), no. 10, 2273–2282.
1155. Xinde Li, Xianzhong Dai, Jean Dezert, and Florentin Smarandache, *Fusion of imprecise qualitative information*, Applied Intelligence **33** (2010), no. 3, 340–351.
1156. Ze-Nian Li and Leonard Uhr, *Evidential reasoning in a computer vision system*, Proceedings of the 2nd International Conference on Uncertainty in Artificial Intelligence (UAI'86), Elsevier Science, 1986, p. 403.
1157. ______, *Evidential reasoning in a computer vision system*, Machine Intelligence and Pattern Recognition, vol. 5, Elsevier, 1988, pp. 403–412.
1158. ______, *Evidential reasoning in a computer vision system*, Machine Intelligence and Pattern Recognition, vol. 5, Elsevier, 1988, pp. 403–412.
1159. Chen Liang-zhou, Shi Wen-kang, Deng Yong, and Zhu Zhen-fu, *A new fusion approach based on distance of evidences*, Journal of Zhejiang University SCIENCE A **6** (2005), no. 5, 476–482.
1160. Jing Liao, Yaxin Bi, and Chris Nugent, *Using the Dempster–Shafer theory of evidence with a revised lattice structure for activity recognition*, IEEE Transactions on Information Technology in Biomedicine **15** (2011), no. 1, 74–82.
1161. Ruijin Liao, Hanbo Zheng, S. Grzybowski, Lijun Yang, Yiyi Zhang, and Yuxiang Liao, *An integrated decision-making model for condition assessment of power transformers using fuzzy approach and evidential reasoning*, IEEE Transactions on Power Delivery **26** (2011), no. 2, 1111–1118.
1162. Othniel K. Likkason, Elisha M. Shemang, and Cheo E. Suh, *The application of evidential belief function in the integration of regional geochemical and geological data over the Ife-Ilesha goldfield, Nigeria*, Journal of African Earth Sciences **25** (1997), no. 3, 491–501.
1163. Ee-Peng Lim, Jaideep Srivastava, and Shashi Shekhar, *An evidential reasoning approach to attribute value conflict resolution in database integration*, IEEE Transactions on Knowledge and Data Engineering **8** (1996), no. 5, 707–723.
1164. ______, *Resolving attribute incompatibility in database integration: An evidential reasoning approach*, Proceedings of the 10th International Conference on Data Engineering, February 1994, pp. 154–163.
1165. Philipp Limbourg, *Multi-objective optimization of problems with epistemic uncertainty*, Evolutionary Multi-Criterion Optimization: Third International Conference, EMO 2005, Guanajuato, Mexico, March 9-11, 2005. Proceedings (Carlos A. Coello, Arturo Hernández Aguirre, and Eckart Zitzler, eds.), Springer, Berlin Heidelberg, 2005, pp. 413–427.
1166. Tsau Young Lin, *Granular computing on binary relations II: Rough set representations and belief functions*, Rough Sets In Knowledge Discovery, PhysicaVerlag, 1998, pp. 121–140.
1167. ______, *Measure theory on granular fuzzy sets*, Proceedings of the 18th International Conference of the North American Fuzzy Information Processing Society (NAFIPS), 1999, pp. 809–813.
1168. ______, *Fuzzy partitions II: Belief functions A probabilistic view*, Rough Sets and Current Trends in Computing: First International Conference, RSCTC'98 Warsaw,

Poland, June 22–26, 1998 Proceedings (Lech Polkowski and Andrzej Skowron, eds.), Springer, Berlin Heidelberg, February 1999, pp. 381–386.

1169. Tsau Young Lin and Lin Churn-Jung Liau, *Belief functions based on probabilistic multivalued random variables*, Proceedings of the Joint Conference of Information Sciences, 1997, pp. 269–272.

1170. Tsau Young Lin and Yiyu (Y. Y.) Yao, *Neighborhoods systems: measure, probability and belief functions*, Proceedings of the Fourth International Workshop on Rough Sets, Fuzzy Sets and Machine Discovery, 1996, pp. 202–207.

1171. Tzu-Chao Lin, *Partition belief median filter based on Dempster–Shafer theory for image processing*, Pattern Recognition **41** (2008), no. 1, 139–151.

1172. Tzu-Chao Lin and Pao-Ta Yu, *Thresholding noise-free ordered mean filter based on Dempster–Shafer theory for image restoration*, IEEE Transactions on Circuits and Systems I: Regular Papers **53** (2006), no. 5, 1057–1064.

1173. Zuo-Quan Lin, Ke-Dian Mu, and Qing Han, *An approach to combination of conflicting evidences by disturbance of ignorance*, Journal of Software **8** (2004), 5.

1174. Dennis V. Lindley, *The probability approach to the treatment of uncertainty in artificial intelligence and expert systems*, Statistical Science **2** (1987), no. 1, 17–24.

1175. Dennis V. Lindley, *Uncertainty*, Understanding Uncertainty, John Wiley and Sons, Inc., 2006, pp. 1–14.

1176. Pawan Lingras and S. K. Michael Wong, *Two perspectives of the Dempster–Shafer theory of belief functions*, International Journal of Man-Machine Studies **33** (1990), no. 4, 467–487.

1177. Baoding Liu, *Uncertainty theory*, Springer-Verlag, 2004.

1178. ______, *Some research problems in uncertainty theory*, Journal of Uncertain Systems **3** (2009), no. 1, 3–10.

1179. Baoding Liu and Yian-Kui Liu, *Expected value of fuzzy variable and fuzzy expected value models*, IEEE Transactions on Fuzzy Systems **10** (2002), no. 4, 445–450.

1180. Da You Liu, Ji Hong Ouyang, Hai Ying, Chen Jian Zhong, and Yu Qiang Yuan, *Research on a simplified evidence theory model*, Journal of Computer Research and Development **2** (1999).

1181. Dayou Liu and Yuefeng Li, *The interpretation of generalized evidence theory*, Chinese Journal of Computers **20** (1997), no. 2, 158–164.

1182. Guilong Liu, *Rough set theory based on two universal sets and its applications*, Knowledge-Based Systems **23** (2010), no. 2, 110–115.

1183. Hu-Chen Liu, Long Liu, Qi-Hao Bian, Qin-Lian Lin, Na Dong, and Peng-Cheng Xu, *Failure mode and effects analysis using fuzzy evidential reasoning approach and grey theory*, Expert Systems with Applications **38** (2011), no. 4, 4403–4415.

1184. Hu-Chen Liu, Long Liu, and Qing-Lian Lin, *Fuzzy failure mode and effects analysis using fuzzy evidential reasoning and belief rule-based methodology*, IEEE Transactions on Reliability **62** (March 2013), no. 1, 23–36.

1185. Jian Liu, Jing Li, and Keping Long, *Enhanced asynchronous cooperative spectrum sensing based on Dempster–Shafer theory*, Proceedings of the IEEE Global Telecommunications Conference (GLOBECOM 2011), December 2011, pp. 1–6.

1186. Jiming Liu and Michel C. Desmarais, *Method of learning implication networks from empirical data: algorithm and Monte-Carlo simulation-based validation*, IEEE Transactions on Knowledge and Data Engineering **9** (1997), no. 6, 990–1004.

1187. Jun Liu, Jian-Bo Yang, Jin Wang, and How-Sing Sii, *Engineering system safety analysis and synthesis using the fuzzy rule-based evidential reasoning approach*, Quality and Reliability Engineering International **21** (2005), no. 4, 387–411.

1188. Lei Jian Liu, Jing Yu Yang, and Jian Feng Lu, *Data fusion for detection of early stage lung cancer cells using evidential reasoning*, Proceedings of SPIE - Sensor Fusion VI (Paul S. Schenker, ed.), vol. 2059, September 1993, pp. 202–212.
1189. Liping Liu, *Model combination using Gaussian belief functions*, Tech. report, School of Business, University of Kansas, 1995.
1190. ______, *Propagation of Gaussian belief functions*, Learning Models from Data: AI and Statistics (D. Fisher and H. J. Lenz, eds.), Springer, 1996, pp. 79–88.
1191. ______, *A theory of Gaussian belief functions*, International Journal of Approximate Reasoning **14** (1996), no. 2-3, 95–126.
1192. ______, *Local computation of Gaussian belief functions*, International Journal of Approximate Reasoning **22** (1999), no. 3, 217–248.
1193. Liping Liu, *A relational representation of belief functions*, Belief Functions: Theory and Applications: Third International Conference, BELIEF 2014, Oxford, UK, September 26-28, 2014. Proceedings (Fabio Cuzzolin, ed.), Springer International Publishing, 2014, pp. 161–170.
1194. ______, *A new matrix addition rule for combining linear belief functions*, Belief Functions: Theory and Applications: 4th International Conference, BELIEF 2016, Prague, Czech Republic, September 21-23, 2016, Proceedings (Jivrina Vejnarová and Václav Kratochvíl, eds.), Springer International Publishing, 2016, pp. 14–24.
1195. Liping Liu, Catherine Shenoy, and Prakash P. Shenoy, *Knowledge representation and integration for portfolio evaluation using linear belief functions*, IEEE Transactions on Systems, Man, and Cybernetics - Part A: Systems and Humans **36** (July 2006), no. 4, 774–785.
1196. Weiru Liu, *Analyzing the degree of conflict among belief functions*, Artificial Intelligence **170** (2006), no. 11, 909–924.
1197. ______, *Measuring conflict between possibilistic uncertain information through belief function theory*, Knowledge Science, Engineering and Management (Jorg Siekmann, ed.), Lecture Notes in Computer Science, vol. 4092, Springer, 2006, pp. 265–277.
1198. Weiru Liu, *Propositional, probabilistic and evidential reasoning: Integrating numerical and symbolic approaches*, 1st ed., Physica-Verlag, 2010.
1199. Weiru Liu and Alan Bundy, *The combination of different pieces of evidence using incidence calculus*, Technical report 599, Department of Artificial Intelligence, University of Edinburgh, 1992.
1200. Weiru Liu and Alan Bundy, *A comprehensive comparison between generalized incidence calculus and the Dempster–Shafer theory of evidence*, International Journal of Human-Computer Studies **40** (June 1994), no. 6, 1009–1032.
1201. Weiru Liu and Jun Hong, *Reinvestigating Dempster's idea on evidence combination*, Knowledge and Information Systems **2** (2000), no. 2, 223–241.
1202. Weiru Liu, Jun Hong, Michael F. McTear, and John G. Hughes, *An extended framework for evidential reasoning system*, International Journal of Pattern Recognition and Artificial Intelligence **7** (June 1993), no. 3, 441–457.
1203. Weiru Liu, Jun Hong, and Micheal F. McTear, *An extended framework for evidential reasoning systems*, Proceedings of the 2nd International Conference on Tools for Artificial Intelligence (ICTAI), 1990, pp. 731–737.
1204. Weiru Liu, John G. Hughes, and Michael F. McTear, *Representing heuristic knowledge and propagating beliefs in Dempster–Shafer theory of evidence*, Advances in the Dempster-Shafer theory of evidence, John Wiley and Sons, 1992, pp. 441–471.
1205. Weiru Liu, David McBryan, and Alan Bundy, *The method of assigning incidences*, Applied Intelligence **9** (1998), no. 2, 139–161.

1206. Xue-Hua Liu, Andrew K. Skidmore, and H. Van Oosten, *Integration of classification methods for improvement of land-cover map accuracy*, ISPRS Journal of Photogrammetry and Remote Sensing **56** (2002), no. 4, 257–268.
1207. Yi Liu and Yuan F. Zheng, *One-against-all multi-class SVM classification using reliability measures*, Proceedings of the 2005 IEEE International Joint Conference on Neural Networks (IJCNN), vol. 2, July 2005, pp. 849–854.
1208. Zheng Liu, David S. Forsyth, Mir-Saeed Safizadeh, and Abbas Fahr, *A data-fusion scheme for quantitative image analysis by using locally weighted regression and Dempster–Shafer theory*, IEEE Transactions on Instrumentation and Measurement **57** (2008), no. 11, 2554–2560.
1209. Kin-Chung Lo, *Agreement and stochastic independence of belief functions*, Mathematical Social Sciences **51** (2006), no. 1, 1–22.
1210. Gabriele Lohmann, *An evidential reasoning approach to the classification of satellite images*, Symbolic and Qualitative Approaches to Uncertainty (R. Kruse and P. Siegel, eds.), Springer-Verlag, 1991, pp. 227–231.
1211. Mingsheng Long, Jianmin Wang, and Michael I. Jordan, *Deep transfer learning with joint adaptation networks*, Proceedings of the International Conference on Machine Learning (ICML 2017), 2017, pp. 2208–2217.
1212. Pierre Loonis, El-Hadi Zahzah, and Jean-Pierre Bonnefoy, *Multi-classifiers neural network fusion versus Dempster–Shafer's orthogonal rule*, Proceedings of the International Conference on Neural Networks (ICNN'95), vol. 4, 1995, pp. 2162–2165.
1213. John D. Lowrance, *Evidential reasoning with Gister: A manual*, Tech. report, Artificial Intelligence Center, SRI International, 1987.
1214. ______, *Automated argument construction*, Journal of Statistical Planning Inference **20** (1988), no. 3, 369–387.
1215. ______, *Evidential reasoning with Gister-CL: A manual*, Tech. report, Artificial Intelligence Center, SRI International, 1994.
1216. John D. Lowrance and Thomas D. Garvey, *Evidential reasoning: A developing concept*, Proceedings of the Internation Conference on Cybernetics and Society (Institute of Electrical and Electronical Engineers, eds.), 1982, pp. 6–9.
1217. ______, *Evidential reasoning: an implementation for multisensor integration*, Tech. Report 307, SRI International, Menlo Park, CA, 1983.
1218. John D. Lowrance, Thomas D. Garvey, and Thomas M. Strat, *A framework for evidential-reasoning systems*, Proceedings of the National Conference on Artificial Intelligence (AAAI-86) (American Association for Artificial Intelligence, ed.), 1986, pp. 896–903.
1219. ______, *A framework for evidential reasoning systems*, Readings in uncertain reasoning (Glenn Shafer and Judea Pearl, eds.), Morgan Kaufman, 1990, pp. 611–618.
1220. Shin-Yee Lu and Harry E. Stephanou, *A set-theoretic framework for the processing of uncertain knowledge*, Proceedings of the National Conference on Artificial Intelligence (AAAI'84), 1984, pp. 216–221.
1221. Yi Lu, *Knowledge integration in a multiple classifier system*, Applied Intelligence **6** (1996), no. 2, 75–86.
1222. Yi Hui Lu, John Tinder, and Kurt Kubik, *Automatic building detection using the Dempster–Shafer algorithm*, Photogrammetric Engineering And Remote Sensing **72** (2006), no. 4, 395–403.
1223. Caro Lucas and Babak Nadjar Araabi, *Generalization of the Dempster–Shafer theory: a fuzzy-valued measure*, IEEE Transactions on Fuzzy Systems **7** (1999), no. 3, 255–270.

1224. Caro Lucas and Babak Nadjar Araabi, *Generalization of the Dempster–Shafer theory: a fuzzy-valued measure*, IEEE Transactions on Fuzzy Systems **7** (1999), no. 3, 255–270.
1225. Wuben B. Luo and Bill Caselton, *Using Dempster–Shafer theory to represent climate change uncertainties*, Journal of Environmental Management **49** (1997), no. 1, 73–93.
1226. Jianbing Ma, Weiru Liu, Didier Dubois, and Henri Prade, *Revision rules in the theory of evidence*, Proceedings of the 22nd IEEE International Conference on Tools with Artificial Intelligence (ICTAI 2010), vol. 1, 2010, pp. 295–302.
1227. ______, *Bridging Jeffrey's rule, AGM revision and Dempster conditioning in the theory of evidence*, International Journal on Artificial Intelligence Tools **20** (2011), no. 4, 691–720.
1228. Yong Ma and David C. Wilkins, *Induction of uncertain rules and the sociopathicity property in Dempster–Shafer theory*, Symbolic and Quantitative Approaches to Uncertainty (Rudolf Kruse and Pierre Siegel, eds.), Lecture Notes in Computer Science, vol. 548, Springer, Berlin Heidelberg, 1991, pp. 238–245.
1229. Sebastian Maass, *A philosophical foundation of non-additive measure and probability*, Theory and decision **60** (2006), no. 2-3, 175–191.
1230. Fabio Maccheroni and Massimo Marinacci, *A strong law of large numbers for capacities*, The Annals of Probability **33** (2005), no. 3, 1171–1178.
1231. Yutaka Maeda and Hidetomo Ichihashi, *An uncertainty measure with monotonicity under the random set inclusion*, International Journal of General Systems **21** (1993), no. 4, 379–392.
1232. Ronald P. S. Mahler, *Using a priori evidence to customize Dempster–Shafer theory*, Proceedings of the 6th National Symposium on Sensors and Sensor Fusion, vol. 1, 1993, pp. 331–345.
1233. Ronald P. S. Mahler, *Classification when a priori evidence is ambiguous*, Proceedings of SPIE - Automatic Object Recognition IV, vol. 2234, 1994, pp. 296–304.
1234. ______, *Random-set approach to data fusion*, Proceedings of SPIE - Automatic Object Recognition IV, vol. 2234, 1994, pp. 287–295.
1235. Ronald P. S. Mahler, *Combining ambiguous evidence with respect to ambiguous a priori knowledge. Part II: Fuzzy logic*, Fuzzy Sets and Systems **75** (1995), no. 3, 319–354.
1236. ______, *Statistical multisource-multitarget information fusion*, Artech House, Inc., Norwood, MA, USA, 2007.
1237. ______, *"Statistics 101" for multisensor, multitarget data fusion*, IEEE Aerospace and Electronic Systems Magazine **19** (January 2004), no. 1, 53–64.
1238. ______, *Can the Bayesian and Dempster–Shafer approaches be reconciled? Yes*, Proceedings of the 7th International Conference on Information Fusion (FUSION 2005), vol. 2, June-July 2005, p. 8.
1239. José A. Malpica, Maria C. Alonso, and Maria A. Sanz, *Dempster–Shafer theory in geographic information systems: A survey*, Expert Systems with Applications **32** (2007), no. 1, 47–55.
1240. David A. Maluf, *Monotonicity of entropy computations in belief functions*, Intelligent Data Analysis **1** (1997), no. 1, 207–213.
1241. Eberhard Mandler and Jürgen Schmann, *Combining the classification results of independent classifiers based on the Dempster/Shafer theory of evidence*, Machine Intelligence and Pattern Recognition, vol. 7, Elsevier, 1988, pp. 381–393.
1242. Jean-Luc Marichal, *Entropy of discrete Choquet capacities*, European Journal of Operational Research **137** (2002), no. 3, 612–624.

1243. Massimo Marinacci, *Limit laws for non-additive probabilities and their frequentist interpretation*, Journal of Economic Theory **84** (1999), no. 2, 145–195.
1244. George Markakis, *A boolean generalization of the Dempster-Shafer construction of belief and plausibility functions*, Tatra Mountains Mathematical Publications **16** (1999), no. 117, 117–125.
1245. George Markakis, *A boolean generalization of the Dempster–Shafer construction of belief and plausibility functions*, Proceedings of the Fourth International Conference on Fuzzy Sets Theory and Its Applications, Liptovsky Jan, Slovakia, 2-6 Feb. 1998, pp. 117–125.
1246. Ivan Marsic, *Evidential reasoning in visual recognition*, Proceedings of the Artificial Neural Networks in Engineering Conference (ANNIE'94) (C. H. Dagli, B. R. Fernandez, J. Ghosh, and R. T. S. Kumara, eds.), vol. 4, November 1994, pp. 511–516.
1247. Arnaud Martin, *Implementing general belief function framework with a practical codification for low complexity*, ArXiv preprint arXiv:0807.3483, 2008.
1248. ______, *Implementing general belief function framework with a practical codification for low complexity*, Algorithms **18** (2008), 33.
1249. Arnaud Martin, *About conflict in the theory of belief functions*, Belief Functions: Theory and Applications: Proceedings of the 2nd International Conference on Belief Functions, Compiègne, France 9-11 May 2012 (Thierry Denœux and Marie-Hélène Masson, eds.), Springer, Berlin Heidelberg, 2012, pp. 161–168.
1250. Arnaud Martin, *Reliability and combination rule in the theory of belief functions*, Proceedings of the 12th International Conference on Information Fusion (FUSION 2009), July 2009, pp. 529–536.
1251. Arnaud Martin, Anne-Laure Jousselme, and Christophe Osswald, *Conflict measure for the discounting operation on belief functions*, Proceedings of the 11th International Conference on Information Fusion (FUSION 2008), June 2008, pp. 1–8.
1252. Arnaud Martin, Marie-Hélène Masson, Frédéric Pichon, Didier Dubois, and Thierry Denœux, *Relevance and truthfulness in information correction and fusion*, International Journal of Approximate Reasoning **53** (2012), no. 2, 159–175.
1253. Arnaud Martin and Christophe Osswald, *A new generalization of the proportional conflict redistribution rule stable in terms of decision*, Advances and Applications of DSmT for Information Fusion (Florentin Smarandache and Jean Dezert, eds.), Infinite Study, 2006, pp. 69–88.
1254. Arnaud Martin and Christophe Osswald, *Toward a combination rule to deal with partial conflict and specificity in belief functions theory*, Proceedings of the 10th International Conference on Information Fusion (FUSION 2007), 2007, pp. 1–8.
1255. Arnaud Martin and Isabelle Quidu, *Decision support with belief functions theory for seabed characterization*, ArXiv preprint arXiv:0805.3939, 2008.
1256. ______, *Decision support with belief functions theory for seabed characterization*, Proceedings of the 11th International Conference on Information Fusion (FUSION 2008), 2008, pp. 1–8.
1257. Ryan Martin and Chuanhai Liu, *Inferential models: A framework for prior-free posterior probabilistic inference*, Journal of the American Statistical Association **108** (2013), no. 501, 301–313.
1258. ______, *Inferential models: A framework for prior-free posterior probabilistic inference*, ArXiv preprint arXiv:1206.4091, June 2012.
1259. Ryan Martin, Jianchun Zhang, and Chuanhai Liu, *Dempster-Shafer theory and statistical inference with weak beliefs*, Statistical Science **25** (2010), no. 1, 72–87.

1260. Francis Martinerie and Philippe Foster, *Data association and tracking from distributed sensors using hidden Markov models and evidential reasoning*, Proceedings of 31st Conference on Decision and Control (CDC'92), December 1992, pp. 3803–3804.
1261. Marie-Hélène Masson and Thierry Denœux, *Clustering interval-valued proximity data using belief functions*, Pattern Recognition Letters **25** (2004), no. 2, 163–171.
1262. ———, *ECM: An evidential version of the fuzzy c-means algorithm*, Pattern Recognition **41** (2008), no. 4, 1384–1397.
1263. ———, *RECM: Relational Evidential c-means algorithm*, Pattern Recognition Letters **30** (2009), no. 11, 1015–1026.
1264. ———, *Ensemble clustering in the belief functions framework*, International Journal of Approximate Reasoning **52** (2011), no. 1, 92–109.
1265. ———, *Ranking from pairwise comparisons in the belief functions framework*, Advances in Soft Computing – 2nd International Conference on Belief Functions (BELIEF 2012), May 2012, pp. 311–318.
1266. Marie-Hélène Masson and Thierry Denœux, *Belief functions and cluster ensembles*, Proceedings of the 10th European Conference on Symbolic and Quantitative Approaches to Reasoning with Uncertainty (ECSQARU 2009) (Berlin, Heidelberg) (Claudio Sossai and Gaetano Chemello, eds.), Springer, Verona, Italy, July 2009, pp. 323–334.
1267. Benson Mates, *Elementary logic*, Oxford University Press, 1972.
1268. Georges Matheron, *Random sets and integral geometry*, Wiley Series in Probability and Mathematical Statistics, New York, 1975.
1269. Bree R. Mathon, Metin M. Ozbek, and George F. Pinder, *Dempster–Shafer theory applied to uncertainty surrounding permeability*, Mathematical Geosciences **42** (2010), no. 3, 293–307.
1270. Takashi Matsuyama, *Belief formation from observation and belief integration using virtual belief space in Dempster–Shafer probability model*, Proceedings of the IEEE International Conference on Multisensor Fusion and Integration for Intelligent Systems (MFI '94), October 1994, pp. 379–386.
1271. Takashi Matsuyama and Mitsutaka Kurita, *Pattern classification based on Dempster–Shafer probability model-belief formation from observation and belief integration using virtual belief space*, IEICE Transactions **76** (1993), no. 4, 843–853.
1272. Sally McClean and Bryan Scotney, *Using evidence theory for the integration of distributed databases*, International Journal of Intelligent Systems **12** (1997), no. 10, 763–776.
1273. Sally McClean, Bryan Scotney, and Mary Shapcott, *Using background knowledge in the aggregation of imprecise evidence in databases*, Data and Knowledge Engineering **32** (2000), no. 2, 131–143.
1274. Mary McLeish, *Nilson's probabilistic entailment extended to Dempster–Shafer theory*, International Journal of Approximate Reasoning **2** (1988), no. 3, 339–340.
1275. ———, *A model for non-monotonic reasoning using Dempster's rule*, Uncertainty in Artificial Intelligence 6 (P. P. Bonissone, M. Henrion, L. N. Kanal, and J. F. Lemmer, eds.), Elsevier Science Publishers, 1991, pp. 481–494.
1276. Mary McLeish, Paulyn Yao, Matthew Cecile, and T. Stirtzinger, *Experiments using belief functions and weights of evidence incorporating statistical data and expert opinions*, Proceedings of the Fifth Conference on Uncertainty in Artificial Intelligence (UAI1989), 1989, pp. 253–264.
1277. George Meghabghab and D. Meghabghab, *Multiversion information retrieval: performance evaluation of neural networks vs. Dempster–Shafer model*, Proceedings of the

Third Golden West International Conference on Intelligent Systems (E. A. Yfantis, ed.), Kluwer Academic, June 1994, pp. 537–545.

1278. Khaled Mellouli, *On the propagation of beliefs in networks using the Dempster–Shafer theory of evidence*, PhD dissertation, University of Kansas, School of Business, 1986.

1279. Khaled Mellouli and Zied Elouedi, *Pooling experts opinion using Dempster–Shafer theory of evidence*, Proceedings of the International Conference on Systems, Man, and Cybernetics (SMC'97), 1997, pp. 1900–1905.

1280. Khaled Mellouli, Glenn Shafer, and Prakash P. Shenoy, *Qualitative Markov networks*, Uncertainty in Knowledge-Based Systems: International Conference on Information Processing and Management of Uncertainty in Knowledge-Based Systems Paris, France, June 30 – July 4, 1986 Selected and Extended Contributions (B. Bouchon and R. R. Yager, eds.), Springer, Berlin Heidelberg, 1987, pp. 67–74.

1281. David Mercier, Genevieve Cron, Thierry Denœux, and Marie-Hélène Masson, *Fusion of multi-level decision systems using the transferable belief model*, Proceedings of the 7th International Conference on Information Fusion (FUSION 2005), vol. 2, 2005, pp. 8 pp.–.

1282. David Mercier, Thierry Denœux, and Marie-Hélène Masson, *General correction mechanisms for weakening or reinforcing belief functions*, Proceedings of the 9th International Conference on Information Fusion (FUSION 2006), 2006, pp. 1–7.

1283. ______, *Refined sensor tuning in the belief function framework using contextual discounting*, Proceedings of Information Processing with Management of Uncertainty in Knowledge-based Systems (IPMU'06), vol. 2, 2006, pp. 1443–1450.

1284. David Mercier, Thierry Denœux, and Marie-Hélène Masson, *Belief function correction mechanisms*, Foundations of Reasoning under Uncertainty (Bernadette Bouchon-Meunier, Luis Magdalena, Manuel Ojeda-Aciego, José-Luis Verdegay, and Ronald R. Yager, eds.), Springer, Berlin Heidelberg, 2010, pp. 203–222.

1285. David Mercier, Eric Lefévre, and François Delmotte, *Belief functions contextual discounting and canonical decompositions*, International Journal of Approximate Reasoning **53** (2012), no. 2, 146–158.

1286. David Mercier, Benjamin Quost, and Thierry Denœux, *Contextual discounting of belief functions*, Symbolic and Quantitative Approaches to Reasoning with Uncertainty: 8th European Conference, ECSQARU 2005, Barcelona, Spain, July 6-8, 2005. Proceedings (Lluís Godo, ed.), Springer, Berlin Heidelberg, 2005, pp. 552–562.

1287. David Mercier, Benjamin Quost, and Thierry Denœux, *Refined modeling of sensor reliability in the belief function framework using contextual discounting*, Information Fusion **9** (2008), no. 2, 246–258.

1288. José M. Merigó and Montserrat Casanovas, *Induced aggregation operators in decision making with the Dempster–Shafer belief structure*, International Journal of Intelligent Systems **24** (2009), no. 8, 934–954.

1289. José M. Merigó, Montserrat Casanovas, and Luis Martínez, *Linguistic aggregation operators for linguistic decision making based on the Dempster–Shafer theory of evidence*, International Journal of Uncertainty, Fuzziness and Knowledge-Based Systems **18** (2010), no. 3, 287–304.

1290. Aaron Meyerowitz, Fred Richman, and Elbert Walker, *Calculating maximum-entropy probability densities for belief functions*, International Journal of Uncertainty, Fuzziness and Knowledge-Based Systems **2** (1994), no. 4, 377–389.

1291. Nada Milisavljevic, Isabelle Bloch, and Marc P. J. Acheroy, *Modeling, combining, and discounting mine detection sensors within the Dempster–Shafer framework*, Proceedings of SPIE - Detection and Remediation Technologies for Mines and Minelike Targets V, vol. 4038, 2000, pp. 1461–1472.

1292. Enrique Miranda, *A survey of the theory of coherent lower previsions*, International Journal of Approximate Reasoning **48** (2008), no. 2, 628–658.
1293. Enrique Miranda, Inés Couso, and Pedro Gil, *Extreme points of credal sets generated by 2-alternating capacities*, International Journal of Approximate Reasoning **33** (2003), no. 1, 95–115.
1294. Enrique Miranda and Gert de Cooman, *Marginal extension in the theory of coherent lower previsions*, International Journal of Approximate Reasoning **46** (2007), no. 1, 188–225.
1295. Enrique Miranda, Gert de Cooman, and Inés Couso, *Lower previsions induced by multi-valued mappings*, Journal of Statistical Planning and Inference **133** (2005), no. 1, 173–197.
1296. Pedro Miranda, Michel Grabisch, and Pedro Gil, *On some results of the set of dominating k-additive belief functions*, Proceedings of Information Processing with Management of Uncertainty in Knowledge-based Systems (IPMU'04), 2004, pp. 625–632.
1297. ______, *Dominance of capacities by k-additive belief functions*, European Journal of Operational Research **175** (2006), no. 2, 912–930.
1298. Stefano Mizzaro, *Relevance: The whole history*, Journal of the American Society for Information Science **48** (1997), no. 9, 810–832.
1299. Kambiz Mokhtari, Jun Ren, Charles Roberts, and Jin Wang, *Decision support framework for risk management on sea ports and terminals using fuzzy set theory and evidential reasoning approach*, Expert Systems with Applications **39** (2012), no. 5, 5087–5103.
1300. Ilya Molchanov, *Uniform laws of large numbers for empirical associated functionals of random closed sets*, Theory of Probability and Its Applications **32** (1988), no. 3, 556–559.
1301. ______, *Limit theorems for unions of random closed sets*, Springer, 1993.
1302. ______, *Statistical problems for random sets*, Random Sets, Springer, 1997, pp. 27–45.
1303. ______, *On strong laws of large numbers for random upper semicontinuous functions*, Journal of mathematical analysis and applications **235** (1999), no. 1, 349–355.
1304. ______, *Theory of random sets*, Springer-Verlag, 2005.
1305. Paul-André Monney, *A mathematical theory of arguments for statistical evidence*, Physica, 19 November 2002.
1306. ______, *Planar geometric reasoning with the thoery of hints*, Computational Geometry. Methods, Algorithms and Applications (H. Bieri and H. Noltemeier, eds.), Lecture Notes in Computer Science, vol. 553, Springer Science and Business Media, 1991, pp. 141–159.
1307. ______, *Analyzing linear regression models with hints and the Dempster–Shafer theory*, International Journal of Intelligent Systems **18** (2003), no. 1, 5–29.
1308. Paul-André Monney, *Dempster specialization matrices and the combination of belief functions*, Proceedings of the 6th European Conference on Symbolic and Quantitative Approaches to Reasoning with Uncertainty (ECSQARU 2001) (Salem Benferhat and Philippe Besnard, eds.), Springer, Berlin, Heidelberg, September 2001, pp. 316–327.
1309. Paul-André Monney, Moses W. Chan, Enrique H. Ruspini, and Marco E. G. V. Cattaneo, *Belief functions combination without the assumption of independence of the information sources*, International Journal of Approximate Reasoning **52** (2011), no. 3, 299–315.
1310. Paul-André Monney, Moses W. Chan, Enrique H. Ruspini, and Johan Schubert, *Dependence issues in knowledge-based systems conflict management in Dempster–*

Shafer theory using the degree of falsity, International Journal of Approximate Reasoning **52** (2011), no. 3, 449–460.

1311. Wooil M. Moon, *Integration of remote sensing and geophysical/geological data using Dempster–Shafer approach*, Proceedings of the 12th Canadian Symposium on Remote Sensing Geoscience and Remote Sensing Symposium (IGARSS'89), vol. 2, July 1989, pp. 838–841.
1312. Brice Mora, Richard A. Fournier, and Samuel Foucher, *Application of evidential reasoning to improve the mapping of regenerating forest stands*, International Journal of Applied Earth Observation and Geoinformation **13** (2011), no. 3, 458–467.
1313. Serafín Moral, *Comments on 'Likelihood-based belief function: Justification and some extensions to low-quality data' by thierry denœux*, International Journal of Approximate Reasoning **55** (2014), no. 7, 1591–1593.
1314. Serafín Moral and Luis M. de Campos, *Partially specified belief functions*, ArXiv preprint arXiv:1303.1513, 2013.
1315. Serafín Moral and Luis M. de Campos, *Partially specified belief functions*, Proceedings of the Ninth Conference on Uncertainty in Artificial Intelligence (UAI'93) (D. Heckerman and A. Mamdani, eds.), July 1993, pp. 492–499.
1316. Serafín Moral and Antonio Salmeron, *A Monte Carlo algorithm for combining Dempster–Shafer belief based on approximate pre-computation*, Proceedings of the European Conference on Symbolic and Quantitative Approaches to Reasoning and Uncertainty (ECSQARU '95), 1999, pp. 305–315.
1317. Serafín Moral and Nic Wilson, *Markov Chain Monte-Carlo algorithms for the calculation of Dempster–Shafer belief*, Proceedings of the Twelfth National Conference on Artificial Intelligence (AAAI'94), vol. 1, 1994, pp. 269–274.
1318. Serafín Moral and Nic Wilson, *Importance sampling Monte–Carlo algorithms for the calculation of Dempster–Shafer belief*, Proceedings of the 6th International Conference on Information Processing and Management of Uncertainty in Knowledge-Based Systems (IPMU'96), July 1996, pp. 1337–1344.
1319. Ekaterini Moutogianni and Mounia Lalmas, *A Dempster–Shafer indexing for structured document retrieval: implementation and experiments on a Web museum collection*, IEE Colloquium on Microengineering in Optics and Optoelectronics (Ref. No. 1999/187), Glasgow, UK, 11-12 Nov. 1999, pp. 20–21.
1320. Kedian Mu, Zhi Jin, Ruqian Lu, and Weiru Liu, *Measuring inconsistency in requirements specifications*, Symbolic and Quantitative Approaches to Reasoning with Uncertainty (Lluís Godo, ed.), Lecture Notes in Computer Science, vol. 3571, Springer-Verlag, Berlin/Heidelberg, 2005, pp. 440–451.
1321. Chantal Muller, Michéle Rombaut, and Marc Janier, *Dempster Shafer approach for high level data fusion applied to the assessment of myocardial viability*, Functional Imaging and Modeling of the Heart (Toivo Katila, Jukka Nenonen, Isabelle E. Magnin, Patrick Clarysse, and Johan Montagnat, eds.), Lecture Notes in Computer Science, vol. 2230, Springer, Berlin Heidelberg, 2001, pp. 104–112.
1322. Juliane Müller and Robert Piché, *Mixture surrogate models based on Dempster–Shafer theory for global optimization problems*, Journal of Global Optimization **51** (2011), no. 1, 79–104.
1323. Juliane Müller and Robert Piché, *Mixture surrogate models based on Dempster–Shafer theory for global optimization problems*, Journal of Global Optimization **51** (2011), no. 1, 79–104.
1324. Daniele Mundici, *A geometric approach to MV-algebras*, On Logical, Algebraic, and Probabilistic Aspects of Fuzzy Set Theory (Susanne Saminger-Platz and Radko Mesiar, eds.), Springer International Publishing, 2016, pp. 57–70.

1325. Tetsuya Murai, Yasuo Kudo, and Yoshiharu Sato, *Association rules and Dempster–Shafer theory of evidence*, International Conference on Discovery Science, Springer, 2003, pp. 377–384.
1326. Tetsuya Murai, Masaaki Miyakoshi, and Masaru Shimbo, *Soundness and completeness theorems between the Dempster–Shafer theory and logic of belief*, Proceedings of the IEEE World Congress on Computational Intelligence - Proceedings of the Third IEEE Conference on Fuzzy Systems, vol. 2, 1994, pp. 855–858.
1327. Toshiaki Murofushi and Michio Sugeno, *Some quantities represented by the Choquet integral*, Fuzzy Sets and Systems **56** (1993), no. 2, 229–235.
1328. Catherine K. Murphy, *Combining belief functions when evidence conflicts*, Decision Support Systems **29** (2000), no. 1, 1–9.
1329. Robin R. Murphy, *Application of Dempster–Shafer theory to a novel control scheme for sensor fusion*, Proceedings of SPIE - Stochastic and Neural Methods in Signal Processing, Image Processing, and Computer Vision, vol. 1569, 1991, pp. 55–68.
1330. ______, *Adaptive rule of combination for observations over time*, Proceedings of the IEEE/SICE/RSJ International Conference on Multisensor Fusion and Integration for Intelligent Systems, December 1996, pp. 125–131.
1331. ______, *Dempster–Shafer theory for sensor fusion in autonomous mobile robots*, IEEE Transactions on Robotics and Automation **14** (1998), no. 2, 197–206.
1332. Robin R. Murphy and Ronald C. Arkin, *SFX: An architecture for action-oriented sensor fusion*, Proceedings of the 1992 IEEE/RSJ International Conference on Intelligent Robots and Systems, vol. 2, July 1992, pp. 1079–1086.
1333. Sohail Nadimi and Bir Bhanu, *Multistrategy fusion using mixture model for moving object detection*, Proceedings of the International Conference on Multisensor Fusion and Integration for Intelligent Systems (MFI 2001), 2001, pp. 317–322.
1334. Homayoun Najjaran, Rehan Sadiq, and Balvant Rajani, *Condition assessment of water mains using fuzzy evidential reasoning*, Proceedings of the IEEE International Conference on Systems, Man and Cybernetics (SMC'05), vol. 4, October 2005, pp. 3466–3471.
1335. Kazuhiro Nakadai, Hiroshi G. Okuno, and Hiroaki Kitano, *Real-time sound source localization and separation for robot audition*, Proceedings of the IEEE International Conference on Spoken Language Processing, 2002, 2002, pp. 193–196.
1336. Eduardo F. Nakamura, Fabiola G. Nakamura, Carlos M. Figueiredo, and Antonio A. Loureiro, *Using information fusion to assist data dissemination in wireless sensor networks*, Telecommunication Systems **30** (2005), no. 1-3, 237–254.
1337. Louis Narens, *Theories in probability: An examination of logical and qualitative foundations*, World Scientific Publishing, 2007.
1338. Ghalia Nassreddine, Fahed Abdallah, and Thierry Denœux, *A state estimation method for multiple model systems using belief function theory*, Proceedings of the 12th International Conference on Information Fusion (FUSION '09), 2009, pp. 506–513.
1339. ______, *State estimation using interval analysis and belief-function theory: Application to dynamic vehicle localization*, IEEE Transactions on Systems, Man, and Cybernetics, Part B (Cybernetics) **40** (2010), no. 5, 1205–1218.
1340. Richard E. Neapolitan, *The interpretation and application of belief functions*, Applied Artificial Intelligence **7** (1993), no. 2, 195–204.
1341. Constantin V. Negoita, *Expert systems and fuzzy systems*, Benjamin-Cummings Pub. Co., 1984.
1342. Geok See Ng and Harcharan Singh, *Data equalisation with evidence combination for pattern recognition*, Pattern Recognition Letters **19** (1998), no. 3-4, 227–235.

1343. Raymond T. Ng and V. S. Subrahmanian, *Relating Dempster–Shafer theory to stable semantics*, Tech. Report UMIACS-TR-91-49, University of Maryland at College Park, College Park, MD, USA, 1991.
1344. Hung T. Nguyen, *On random sets and belief functions*, Journal of Mathematical Analysis and Applications **65** (1978), 531–542.
1345. ______, *On modeling of linguistic information using random sets*, Information Sciences **34** (1984), no. 3, 265–274.
1346. ______, *On entropy of random sets and possibility distributions*, The Analysis of Fuzzy Information (J. C. Bezdek, ed.), CRC Press, 1985, pp. 145–156.
1347. ______, *Fuzzy and random sets*, Fuzzy Sets and Systems **156** (2005), no. 3, 349–356.
1348. ______, *An introduction to random sets*, Taylor and Francis, 2006.
1349. ______, *On random sets and belief functions*, Classic Works of the Dempster-Shafer Theory of Belief Functions (Ronald R. Yager and Liping Liu, eds.), Springer, 2008, pp. 105–116.
1350. Hung T. Nguyen, *On belief functions and random sets*, Advances in Intelligent and Soft Computing (Thierry Denœux and Marie-Hélène Masson, eds.), vol. 164, Springer, Berlin Heidelberg, 2012, pp. 1–19.
1351. Hung T. Nguyen and Philippe Smets, *On dynamics of cautious belief and conditional objects*, International Journal of Approximate Reasoning **8** (1993), no. 2, 89–104.
1352. Hung T. Nguyen and Elbert A. Walker, *On decision making using belief functions*, Advances in the Dempster-Shafer Theory of Evidence (Ronald R. Yager, Janusz Kacprzyk, and Mario Fedrizzi, eds.), John Wiley and Sons, Inc., 1994, pp. 311–330.
1353. Hung T. Nguyen and Tonghui Wang, *Belief functions and random sets*, Applications and Theory of Random Sets, The IMA Volumes in Mathematics and its Applications, Vol. 97, Springer, 1997, pp. 243–255.
1354. Nhan Nguyen-Thanh and Insoo Koo, *An enhanced cooperative spectrum sensing scheme based on evidence theory and reliability source evaluation in cognitive radio context*, IEEE Communications Letters **13** (July 2009), no. 7, 492–494.
1355. Ojelanki K. Ngwenyama and Noel Bryson, *Generating belief functions from qualitative preferences: An approach to eliciting expert judgments and deriving probability functions*, Data and Knowledge Engineering **28** (1998), no. 2, 145–159.
1356. Nguyen-Thanh Nhan, Xuan Thuc Kieu, and Insoo Koo, *Cooperative spectrum sensing using enhanced Dempster–Shafer theory of evidence in cognitive radio*, Emerging Intelligent Computing Technology and Applications. With Aspects of Artificial Intelligence (De-Shuang Huang, Kang-Hyun Jo, Hong-Hee Lee, Hee-Jun Kang, and Vitoantonio Bevilacqua, eds.), Lecture Notes in Computer Science, vol. 5755, Springer, 2009, pp. 688–697.
1357. Dennis Nienhuser, Thomas Gumpp, and J. Marius Zollner, *A situation context aware Dempster–Shafer fusion of digital maps and a road sign recognition system*, Proceedings of the IEEE Intelligent Vehicles Symposium, June 2009, pp. 1401–1406.
1358. Nils J. Nilsson, *Probabilistic logic*, Artificial Intelligence **28** (February 1986), no. 1, 71–88.
1359. Gang Niu and Bo-Suk Yang, *Dempster–Shafer regression for multi-step-ahead time-series prediction towards data-driven machinery prognosis*, Mechanical Systems and Signal Processing **23** (2009), no. 3, 740–751.
1360. Rafael Mu noz Salinas, Rafael Medina-Carnicer, Francisco J. Madrid-Cuevas, and Angel Carmona-Poyato, *Multi-camera people tracking using evidential filters*, International Journal of Approximate Reasoning **50** (2009), no. 5, 732–749.

1361. Donald Nute and William Mitcheltree, *Reviewed work: The logic of conditionals: An application of probability to deductive logic. by Ernest W. Adams*, Noûs **15** (1981), no. 3, 432–436.
1362. Michael Oberguggenberger and Wolfgang Fellin, *Reliability bounds through random sets: Non-parametric methods and geotechnical applications*, Computers and Structures **86** (2008), no. 10, 1093–1101.
1363. William Oberkampf, Jon Helton, and Kari Sentz, *Mathematical representation of uncertainty*, Proceedings of the 19th AIAA Applied Aerodynamics Conference, Fluid Dynamics and Co-located Conferences, 2001, p. 1645.
1364. H. Ogawa, King Sun Fu, and James Tsu Ping Yao, *An inexact inference for damage assessment of existing structures*, International Journal of Man-Machine Studies **22** (1985), no. 3, 295–306.
1365. C. Ordonez and E. Omiecinski, *FREM: Fast and robust EM clustering for large data sets*, Proceedings of the ACM CIKM Conference, 2002.
1366. Pekka Orponen, *Dempster's rule of combination is #p-complete*, Artificial Intelligence **44** (1990), no. 1, 245–253.
1367. Farhad Orumchian, Babak Nadjar Araabi, and Elham Ashoori, *Using plausible inferences and Dempster–Shafer theory of evidence for adaptive information filtering*, Proceedings of the 4th International Conference on Recent Advances in Soft Computing, 2002, pp. 248–254.
1368. Roberto A. Osegueda, Seetharami R. Seelam, Ana C. Holguin, Vladik Kreinovich, Chin-Wang Tao, and Hung T. Nguyen, *Statistical and Dempster-Shafer techniques in testing structural integrity of aerospace structures*, International Journal of Uncertainty, Fuzziness and Knowledge-Based Systems **9** (2001), no. 6, 749–758.
1369. Roberto A. Osegueda, Seetharami R. Seelam, Bharat Mulupuru, and Vladik Kreinovich, *Statistical and Dempster–Shafer techniques in testing structural integrity of aerospace structures*, Proceedings of SPIE - Smart nondestructive evaluation and health monitoring of structural and biological systems II, vol. 5047, 2003, pp. 140–151.
1370. Kweku-Muata Osei-Bryson, *Supporting knowledge elicitation and consensus building for Dempster–Shafer decision models*, International Journal of Intelligent Systems **18** (2003), no. 1, 129–148.
1371. Ahmad Osman, Valérie Kaftandjian, and Ulf Hassler, *Improvement of X-ray castings inspection reliability by using Dempster–Shafer data fusion theory*, Pattern Recognition Letters **32** (2011), no. 2, 168–180.
1372. Christophe Osswald and Arnaud Martin, *Understanding the large family of Dempster–Shafer theory's fusion operators - a decision-based measure*, Proceedings of the 9th International Conference on Information Fusion (FUSION 2006), July 2006, pp. 1–7.
1373. Latifa Oukhellou, Alexandra Debiolles, Thierry Denœux, and Patrice Aknin, *Fault diagnosis in railway track circuits using Dempster–Shafer classifier fusion*, Engineering Applications of Artificial Intelligence **23** (2010), no. 1, 117–128.
1374. James G. Oxley, *Matroid theory*, Oxford University Press, 1992.
1375. Amir Padovitz, Arkady Zaslavsky, and Seng W. Loke, *A unifying model for representing and reasoning about context under uncertainty*, Proceedings of the 11th International Conference on Information Processing and Management of Uncertainty in Knowledge-Based Systems (IPMU'06, 2006, pp. 1982–1989.
1376. Daniel Pagac, Eduardo M. Nebot, and Hugh Durrant-Whyte, *An evidential approach to map-building for autonomous vehicles*, IEEE Transactions on Robotics and Automation **14** (August 1998), no. 4, 623–629.

1377. Nikhil R. Pal, *On quantification of different facets of uncertainty*, Fuzzy Sets and Systems **107** (1999), no. 1, 81–91.
1378. Nikhil R. Pal, James C. Bezdek, and Rohan Hemasinha, *Uncertainty measures for evidential reasoning I: A review*, International Journal of Approximate Reasoning **7** (1992), no. 3-4, 165–183.
1379. ______, *Uncertainty measures for evidential reasoning II: A new measure of total uncertainty*, International Journal of Approximate Reasoning **8** (1993), no. 1, 1–16.
1380. Nikhil R. Pal and Susmita Ghosh, *Some classification algorithms integrating Dempster–Shafer theory of evidence with the rank nearest neighbor rules*, IEEE Transactions on Systems, Man, and Cybernetics - Part A: Systems and Humans **31** (January 2001), no. 1, 59–66.
1381. Prasad Palacharla and Peter Nelson, *Understanding relations between fuzzy logic and evidential reasoning methods*, Proceedings of Third IEEE International Conference on Fuzzy Systems, vol. 1, 1994, pp. 1933–1938.
1382. Prasad Palacharla and Peter C. Nelson, *Evidential reasoning in uncertainty for data fusion*, Proceedings of the Fifth International Conference on Information Processing and Management of Uncertainty in Knowledge-Based Systems (IPMU'94), vol. 1, 1994, pp. 715–720.
1383. Wei Pan, Yangsheng Wang, and Hongji Yang, *Decision rule analysis of Dempster–Shafer theory of evidence*, Computer Engineering and Application **5** (2004), 17–20.
1384. Wei Pan and Hongji Yang, *New methods of transforming belief functions to pignistic probability functions in evidence theory*, Proceedings of the International Workshop on Intelligent Systems and Applications, 2009, pp. 1–5.
1385. Yin Pan and George J. Klir, *Bayesian inference based on interval probabilities*, Journal of Intelligent and Fuzzy Systems **5** (1997), no. 3, 193–203.
1386. Suvasini Panigrahi, Amlan Kundu, Shamik Sural, and A. K. Majumdar, *Use of Dempster–Shafer theory and Bayesian inferencing for fraud detection in mobile communication networks*, Information Security and Privacy (Josef Pieprzyk, Hossein Ghodosi, and Ed Dawson, eds.), Lecture Notes in Computer Science, vol. 4586, Springer, Berlin Heidelberg, 2007, pp. 446–460.
1387. Suvasini Panigrahi, Amlan Kundu, Shamik Sural, and Arun K. Majumdar, *Credit card fraud detection: A fusion approach using Dempster–Shafer theory and Bayesian learning*, Information Fusion **10** (2009), no. 4, 354–363.
1388. Jeff B. Paris, *A note on the Dutch Book method*, Proceedings of the Second International Symposium on Imprecise Probabilities and Their applications (ISIPTA'01), 2001.
1389. Jeff B. Paris, David Picado-Muino, and Michael Rosefield, *Information from inconsistent knowledge: A probability logic approach*, Interval / Probabilistic Uncertainty and Non-classical Logics, Advances in Soft Computing (Van-Nam Huynh, Y. Nakamori, H. Ono, J. Lawry, V. Kreinovich, and Hung T. Nguyen, eds.), vol. 46, Springer-Verlag, Berlin - Heidelberg, 2008.
1390. Simon Parsons and Ebrahim H. Mamdani, *Qualitative Dempster–Shafer theory*, Proceedings of the III Imacs International Workshop on Qualitative Reasoning and Decision Technologies, June 1993.
1391. Simon Parsons and Alessandro Saffiotti, *A case study in the qualitative verification and debugging of numerical uncertainty*, International Journal of Approximate Reasoning **14** (1996), no. 2-3, 187–216.
1392. Saeid Pashazadeh and Mohsen Sharifi, *Reliability assessment under uncertainty using Dempster–Shafer and vague set theories*, Proceedings of the IEEE International Con-

ference on Computational Intelligence for Measurement Systems and Applications (CIMSA 2008), July 2008, pp. 131–136.
1393. Zdzislaw Pawlak, *Rough sets*, International Journal of Computer and Information Sciences **11** (1982), no. 5, 341–356.
1394. ______, *Rough sets: Theoretical aspects of reasoning about data*, Kluwer Academic Publishers, 1992.
1395. ______, *Vagueness and uncertainty: a rough set perspective*, Computational Intelligence **11** (1995), no. 2, 227–232.
1396. ______, *Rough set theory and its applications to data analysis*, Cybernetics and Systems **29** (1998), no. 7, 661–688.
1397. Zdzislaw Pawlak, Jerzy Grzymala-Busse, Roman Slowinski, and Wojciech Ziarko, *Rough sets*, Communications of the ACM **38** (1995), no. 11, 88–95.
1398. Judea Pearl, *Fusion, propagation, and structuring in belief networks*, Artificial intelligence **29** (1986), no. 3, 241–288.
1399. ______, *On evidential reasoning in a hierarchy of hypotheses*, Artificial Intelligence **28** (1986), no. 1, 9–15.
1400. ______, *Evidential reasoning using stochastic simulation of causal models*, Artificial Intelligence **32** (1987), no. 2, 245–257.
1401. ______, *Evidential reasoning under uncertainty*, Tech. Report R-107-S, Computer Science Department, University of California at Los Angeles, 1988.
1402. ______, *Evidential reasoning under uncertainty*, Exploring Artificial Intelligence (H. E. Shrobe, ed.), Morgan Kaufmann, 1988, pp. 381–418.
1403. ______, *On probability intervals*, International Journal of Approximate Reasoning **2** (1988), no. 3, 211–216.
1404. ______, *On probability intervals*, Tech. Report R-105, Computer Science Department, University of California at Los Angeles, 1988.
1405. ______, *Probabilistic reasoning in intelligent systems: Networks of plausible inference*, Morgan Kaufmann, 1988.
1406. ______, *Reasoning with belief functions: a critical assessment*, Tech. Report R-136, University of California at Los Angeles, 1989.
1407. ______, *Bayesian and belief-functions formalisms for evidential reasoning: A conceptual analysis*, Readings in Uncertain Reasoning (Glenn Shafer and Judea Pearl, eds.), Morgan Kaufmann, 1990, pp. 540–574.
1408. ______, *Reasoning with belief functions: an analysis of compatibility*, International Journal of Approximate Reasoning **4** (1990), no. 5-6, 363–389.
1409. ______, *Rejoinder to comments on 'Reasoning with belief functions: an analysis of compatibility*, International Journal of Approximate Reasoning **6** (1992), no. 3, 425–443.
1410. ______, *Probabilistic reasoning in intelligent systems: networks of plausible inference*, Morgan Kaufmann, 2014.
1411. ______, *The sure-thing principle*, Tech. Report R-466, University of California at Los Angeles, Computer Science Department, February 2016.
1412. Mario Pechwitz, S. Snoussi Maddouri, Volker Maergner, Noureddine Ellouze, and Hamid Amiri, *IFN/ENIT - database of handwritten arabic words*, Proceedings of Colloque International Francophone sur l'Ecrit et le Doucement (CIFED), vol. 2, October 2002, pp. 127–136.
1413. Derek R. Peddle, *Knowledge formulation for supervised evidential classification*, Photogrammetric Engineering and Remote Sensing **61** (1995), no. 4, 409–417.

1414. Derek R. Peddle and David T. Ferguson, *Optimisation of multisource data analysis: an example using evidential reasoning for {GIS} data classification*, Computers and Geosciences **28** (2002), no. 1, 45–52.
1415. Charles Sanders Peirce, *Collected papers of Charles Sanders Peirce*, vol. 2, Harvard University Press, Cambridge, MA, 1960.
1416. Liu Peizhi and Zhang Jian, *A context-aware application infrastructure with reasoning mechanism based on Dempster–Shafer evidence theory*, Proceedings of the IEEE Vehicular Technology Conference (VTC Spring), May 2008, pp. 2834–2838.
1417. Andrés Perea, *A model of minimal probabilistic belief revision*, Theory and Decision **67** (2009), no. 2, 163–222.
1418. Joseph S. J. Peri, *Dempster–Shafer theory, Bayesian theory, and measure theory*, Proceedings of SPIE - Signal Processing, Sensor Fusion, and Target Recognition XIV, vol. 5809, 2005, pp. 378–389.
1419. C. Perneel, H. Van De Velde, and M. Acheroy, *A heuristic search algorithm based on belief functions*, Proceedings of the Fourteenth International Avignon Conference (AFIA -AI'94), vol. 1, June 1994, pp. 99–107.
1420. Laurent Perrussel, Luis Enrique Sucar, and Michael Scott Balch, *Mathematical foundations for a theory of confidence structures*, International Journal of Approximate Reasoning **53** (2012), no. 7, 1003–1019.
1421. Walter L. Perry and Harry E. Stephanou, *Belief function divergence as a classifier*, Proceedings of the 1991 IEEE International Symposium on Intelligent Control, August 1991, pp. 280–285.
1422. Chariya Peterson, *Local Dempster Shafer theory*, Tech. Report CSC-AMTAS-98001, Computer Sciences Corporation, 1998.
1423. Simon Petit-Renaud and Thierry Denœux, *Nonparametric regression analysis of uncertain and imprecise data using belief functions*, International Journal of Approximate Reasoning **35** (2004), no. 1, 1–28.
1424. ______, *Handling different forms of uncertainty in regression analysis: a fuzzy belief structure approach*, Proceedings of the Fifth European Conference on Symbolic and Quantitative Approaches to Reasoning with Uncertainty (ECSQARU'99), Springer, Berlin, Heidelberg, London, 5-9 July 1999, pp. 340–351.
1425. Fabrice Philippe, Gabriel Debs, and Jean-Yves Jaffray, *Decision making with monotone lower probabilities of infinite order*, Mathematics of Operations Research **24** (1999), no. 3, 767–784.
1426. Frédéric Pichon, *Belief functions: Canonical decompositions and combination rules*, 2009.
1427. Frédéric Pichon and Thierry Denœux, *A new justification of the unnormalized Dempster's rule of combination from the Least Commitment Principle*, Proceedings of the 21st FLAIRS Conference (FLAIRS'08), Special Track on Uncertain Reasoning, 2008, pp. 666–671.
1428. Frédéric Pichon and Thierry Denœux, *T-norm and uninorm-based combination of belief functions*, Proceedings of the Annual Meeting of the North American Fuzzy Information Processing Society (NAFIPS 2008), 2008, pp. 1–6.
1429. ______, *Interpretation and computation of alpha-junctions for combining belief functions*, Proceedings of the 6th International Symposium on Imprecise Probability: Theories and Applications (ISIPTA'09), 2009.
1430. Wojciech Pieczynski, *Arbres de Markov triplet et fusion de Dempster–Shafer*, Comptes Rendus Mathematique **336** (2003), no. 10, 869–872.

1431. ———, *Unsupervised Dempster–Shafer fusion of dependent sensors*, Proceedings of the 4th IEEE Southwest Symposium on Image Analysis and Interpretation, Austin, TX, USA, 2-4 April 2000, pp. 247–251.
1432. Luc Pigeon, Basel Solaiman, Thierry Toutin, and Keith P. B. Thomson, *Dempster–Shafer theory for multi-satellite remotely sensed observations*, Proceedings of SPIE - Sensor Fusion: Architectures, Algorithms, and Applications IV, vol. 4051, 2000, pp. 228–236.
1433. Wang Ping and Genqing Yang, *Improvement method for the combining rule of Dempster–Shafer evidence theory based on reliability*, Journal of Systems Engineering and Electronics **16** (June 2005), no. 2, 471–474.
1434. Gadi Pinkas, *Propositional non-monotonic reasoning and inconsistency in symmetric neural networks*, Proceedings of the 12th International Joint Conference on Artificial Intelligence (IJCAI'91), vol. 1, Morgan Kaufmann, 1991, pp. 525–530.
1435. Axel Pinz, Manfred Prantl, Harald Ganster, and Hermann Kopp-Borotschnig, *Active fusion - A new method applied to remote sensing image interpretation*, Pattern Recognition Letters **17** (1996), no. 13, 1349–1359.
1436. Tomaso Poggio and Federico Girosi, *A theory of networks for approximation and learning*, Tech. report, Massachusetts Institute of Technology, Cambridge, MA, USA, 1989.
1437. L. Polkowski and A. Skowron, *Rough mereology: A new paradigm for approximate reasoning*, International Journal of Approximate Reasoning **15** (1996), no. 4, 333–365.
1438. Peter Pong and Subhash Challa, *Empirical analysis of generalised uncertainty measures with Dempster Shafer fusion*, Proceedings of the 10th International Conference on Information Fusion (FUSION 2007), July 2007, pp. 1–9.
1439. Karl R. Popper, *The propensity interpretation of probability*, The British Journal for the Philosophy of Science **10** (1959), no. 37, 25–42.
1440. Graham Priest, Richard Routley, and Jean Norman, *Paraconsistent Logic: Essays on the Inconsistent*, Philosophia Verlag, 1989.
1441. Gregory Provan, *An analysis of exact and approximation algorithms for Dempster–Shafer theory*, Tech. Report 90-15, Department of Computer Science, University of British Columbia, 1990.
1442. ———, *A logic-based analysis of Dempster–Shafer theory*, International Journal of Approximate Reasoning **4** (1990), no. 5-6, 451–495.
1443. ———, *The validity of Dempster–Shafer belief functions*, International Journal of Approximate Reasoning **6** (1992), no. 3, 389–399.
1444. Gregory M. Provan, *An analysis of ATMS-based techniques for computing Dempster–Shafer belief functions*, Proceedings of the 11th International Joint Conference on Artificial Intelligence (IJCAI'89), vol. 2, Morgan Kaufmann, 1989, pp. 1115–1120.
1445. ———, *The application of Dempster Shafer theory to a logic-based visual recognition system*, Proceedings of the Fifth Annual Conference on Uncertainty in Artificial Intelligence (UAI'89), North-Holland, 1990, pp. 389–406.
1446. Ronald L. Pryor, *Principles of nonspecificity*, PhD dissertation, State University of New York at Binghamton, 2007.
1447. Peng Qihang, Zeng Kun, Wang Jun, and Li Shaoqian, *A distributed spectrum sensing scheme based on credibility and evidence theory in cognitive radio context*, Proceedings of the 17th IEEE International Symposium on Personal, Indoor and Mobile Radio Communications, September 2006, pp. 1–5.
1448. Du Qingdong, Xu Lingyu, and Zhao Hai, *D-S evidence theory applied to fault diagnosis of generator based on embedded sensors*, Proceedings of the Third International

Conference on Information Fusion (FUSION 2000), vol. 1, July 2000, pp. TUD5/3–TUD5/8.

1449. Philippe Quinio and Takashi Matsuyama, *Random closed sets: A unified approach to the representation of imprecision and uncertainty*, Symbolic and Quantitative Approaches to Uncertainty: European Conference ECSQARU Marseille, France, October 15–17, 1991 Proceedings (Rudolf Kruse and Pierre Siegel, eds.), Springer, Berlin Heidelberg, 1991, pp. 282–286.
1450. John R. Quinlan, *Inferno: a cautious approach to uncertain inference*, The Computer Journal **26** (1983), no. 3, 255–269.
1451. Benjamin Quost, Thierry Denœux, and Marie-Hélène Masson, *One-against-all classifier combination in the framework of belief functions*, Proceedings of Information Processing with Management of Uncertainty in Knowledge-based Systems (IPMU'06), 2006, pp. 356–363.
1452. Benjamin Quost, Marie-Hélène Masson, and Thierry Denœux, *Refined classifier combination using belief functions*, Proceedings of the 11th International Conference on Information Fusion (FUSION 2008), 2008, pp. 1–7.
1453. ______, *Classifier fusion in the Dempster–Shafer framework using optimized t-norm based combination rules*, International Journal of Approximate Reasoning **52** (2011), no. 3, 353–374.
1454. Deepashree Raje and Pradeep P. Mujumdar, *Hydrologic drought prediction under climate change: Uncertainty modeling with Dempster–Shafer and Bayesian approaches*, Advances in Water Resources **33** (2010), no. 9, 1176–1186.
1455. Andrej Rakar, Dani Juricic, and Peter Ballé, *Transferable belief model in fault diagnosis*, Engineering Applications of Artificial Intelligence **12** (1999), no. 5, 555–567.
1456. Emmanuel Ramasso, Michéle Rombaut, and Denis Pellerin, *Forward-Backward-Viterbi procedures in the Transferable Belief Model for state sequence analysis using belief functions*, Symbolic and Quantitative Approaches to Reasoning with Uncertainty, Springer, 2007, pp. 405–417.
1457. Emmanuel Ramasso, Michèle Rombaut, and Denis Pellerin, *State filtering and change detection using TBM conflict application to human action recognition in athletics videos*, IEEE Transactions on Circuits and Systems for Video Technology **17** (July 2007), no. 7, 944–949.
1458. Arthur Ramer, *Uniqueness of information measure in the theory of evidence*, Random Sets and Systems **24** (1987), no. 2, 183–196.
1459. ______, *Texts on evidence theory: Comparative review*, International Journal of Approximate Reasoning **14** (1996), no. 2-3, 217–220.
1460. Arthur Ramer and George J. Klir, *Measures of discord in the Dempster–Shafer theory*, Information Sciences **67** (1993), no. 1-2, 35–50.
1461. Youhua Ran, Xin Li, Ling Lu, and Z. Y. Li, *Large-scale land cover mapping with the integration of multi-source information based on the Dempster–Shafer theory*, International Journal of Geographical Information Science **26** (2012), no. 1, 169–191.
1462. S. S. Rao and Kiran K. Annamdas, *Dempster–Shafer theory in the analysis and design of uncertain engineering systems*, Product Research (N. R. Srinivasa Raghavan and John A. Cafeo, eds.), Springer, 2009, pp. 135–160.
1463. Ali Rashidi and Hassan Ghassemian, *Extended Dempster–Shafer theory for multi-system/sensor decision fusion*, 2003, pp. 31–37.
1464. Carl E. Rasmussen and Christopher K. I. Williams, *Gaussian processes for machine learning*, MIT Press, 2006.
1465. Bonnie K. Ray and David H. Krantz, *Foundations of the theory of evidence: Resolving conflict among schemata*, Theory and Decision **40** (1996), no. 3, 215–234.

1466. Yann Rébillé, *A Yosida–Hewitt decomposition for totally monotone set functions on locally compact σ-compact topological spaces*, International Journal of Approximate Reasoning **48** (2008), no. 3, 676–685, Special Section on Choquet Integration in honor of Gustave Choquet (19152006) and Special Section on Nonmonotonic and Uncertain Reasoning.
1467. ______, *A Radon-Nikodym derivative for almost subadditive set functions*, Working Paper hal-00441923, HAL, 2009.
1468. ______, *Law of large numbers for non-additive measures*, Journal of Mathematical Analysis and Applications **352** (2009), no. 2, 872–879.
1469. ______, *Laws of large numbers for continuous belief measures on compact spaces*, International Journal of Uncertainty, Fuzziness and Knowledge-Based Systems **17** (2009), no. 5, 685–704.
1470. B. Srinath Reddy and Otman A. Basir, *Concept-based evidential reasoning for multimodal fusion in human-computer interaction*, Applied Soft Computing **10** (2010), no. 2, 567–577.
1471. D. Krishna Sandeep Reddy and Arun K. Pujari, *N-gram analysis for computer virus detection*, Journal in Computer Virology **2** (2006), no. 3, 231–239.
1472. Steven Reece, *Qualitative model-based multisensor data fusion and parameter estimation using ∞-norm Dempster–Shafer evidential reasoning*, Proceedings of the SPIE - Signal Processing, Sensor Fusion, and Target Recognition VI (D. Heckerman and A. Mamdani, eds.), vol. 3068, April 1997, pp. 52–63.
1473. Scott Reed, Yvan Petillot, and J. Bell, *Automated approach to classification of mine-like objects in sidescan sonar using highlight and shadow information*, IEE Proceedings - Radar, Sonar and Navigation **151** (2004), no. 1, 48–56.
1474. Marek Reformat, Michael R. Berthold, Hatem Masri, and Fouad Ben Abdelaziz, *Belief linear programming*, International Journal of Approximate Reasoning **51** (2010), no. 8, 973–983.
1475. Marek Reformat and Ronald R. Yager, *Building ensemble classifiers using belief functions and OWA operators*, Soft Computing **12** (2008), no. 6, 543–558.
1476. Helen M. Regan, Scott Ferson, and Daniel Berleant, *Equivalence of methods for uncertainty propagation of real-valued random variables*, International Journal of Approximate Reasoning **36** (2004), no. 1, 1–30.
1477. Giuliana Regoli, *Rational comparisons and numerical representations*, Decision Theory and Decision Analysis: Trends and Challenges (Sixto Ríos, ed.), Springer Netherlands, Dordrecht, 1994, pp. 113–126.
1478. Thomas Reineking, *Particle filtering in the Dempster–Shafer theory*, International Journal of Approximate Reasoning **52** (2011), no. 8, 1124–1135.
1479. ______, *Belief functions: Theory and algorithms*, PhD dissertation, Universität Bremen, 2014.
1480. Raymond Reiter, *A logic for default reasoning*, Artificial intelligence **13** (1980), no. 1-2, 81–132.
1481. ______, *A logic for default reasoning*, Readings in Nonmonotonic Reasoning (Matthew L. Ginsberg, ed.), Morgan Kaufmann, 1987, pp. 68–93.
1482. Xiaohui Ren, Jinfeng Yang, Henghui Li, and Renbiao Wu, *Multi-fingerprint information fusion for personal identification based on improved Dempster–Shafer evidence theory*, Proceedings of the International Conference on Electronic Computer Technology, February 2009, pp. 281–285.
1483. Germano Resconi, George J. Klir, and Ute St Clair, *Hierarchical uncertainty metatheory based upon modal logic*, International Journal of General Systems **21** (1992), no. 1, 23–50.

1484. Germano Resconi, George J. Klir, Ute St Clair, and David Harmanec, *On the integration of uncertainty theories*, International Journal of Uncertainty, Fuzziness and Knowledge-Based Systems **1** (1993), no. 1, 1–18.
1485. Germano Resconi, George J Klir, David Harmanec, and Ute St Clair, *Interpretations of various uncertainty theories using models of modal logic: a summary*, Fuzzy Sets and Systems **80** (1996), no. 1, 7–14.
1486. Germano Resconi, A. J. van der Wal, and D. Ruan, *Speed-up of the Monte Carlo method by using a physical model of the Dempster–Shafer theory*, International Journal of Intelligent Systems **13** (1998), no. 2-3, 221–242.
1487. M. Rey, James K. E. Tunaley, and Timothy Sibbald, *Use of the Dempster–Shafer algorithm for the detection of SAR ship wakes*, IEEE Transactions on Geoscience and Remote Sensing **31** (September 1993), no. 5, 1114–1118.
1488. John A. Richards and Xiuping Jia, *A Dempster–Shafer relaxation approach to context classification*, IEEE Transactions on Geoscience and Remote Sensing **45** (2007), no. 5, 1422–1431.
1489. Jorma Rissanen, *Modeling by shortest data description*, Automatica **14** (1978), no. 5, 465–471.
1490. Branko Ristic and Philippe Smets, *Belief function theory on the continuous space with an application to model based classification*, Modern Information Processing, 2006, pp. 11–24.
1491. ———, *The TBM global distance measure for the association of uncertain combat ID declarations*, Information Fusion **7** (2006), no. 3, 276–284.
1492. Michael J. Roemer, Gregory J. Kacprzynski, and Rolf F. Orsagh, *Assessment of data and knowledge fusion strategies for prognostics and health management*, Proceedings of the IEEE Aerospace Conference, vol. 6, March 2001, pp. 2979–2988.
1493. Michael J. Roemer, Gregory J. Kacprzynski, and Michael H. Schoeller, *Improved diagnostic and prognostic assessments using health management information fusion*, Proceedings of the IEEE Systems Readiness Technology Conference (AUTOTESTCON), August 2001, pp. 365–377.
1494. Christopher Roesmer, *Nonstandard analysis and Dempster-shafer theory*, International Journal of Intelligent Systems **15** (2000), no. 2, 117–127.
1495. Galina Rogova, *Combining the results of several neural network classifiers*, Neural Networks **7** (1994), no. 5, 777–781.
1496. Laurent Romary and Jean-Marie Pierrel, *The use of the Dempster–Shafer rule in the lexical component of a man-machine oral dialogue system*, Speech Communication **8** (1989), no. 2, 159–176.
1497. Michèle Rombaut, Iman Jarkass, and Thierry Denœux, *State recognition in discrete dynamical systems using Petri nets and evidence theory*, Symbolic and Quantitative Approaches to Reasoning and Uncertainty (Anthony Hunter and Simon Parsons, eds.), Springer, Berlin Heidelberg, 1999, pp. 352–361.
1498. Christoph Römer and Abraham Kandel, *Applicability analysis of fuzzy inference by means of generalized Dempster–Shafer theory*, IEEE Transactions on Fuzzy Systems **3** (1995), no. 4, 448–453.
1499. ———, *Constraints on belief functions imposed by fuzzy random variables*, IEEE Transactions on Systems, Man, and Cybernetics, Part B: Cybernetics **25** (1995), no. 1, 86–99.
1500. ———, *Applicability analysis of fuzzy inference by means of generalized Dempster–Shafer theory*, IEEE Transactions on Fuzzy Systems **4** (November 1995), no. 4, 448–453.

1501. Kimmo I. Rosenthal, *Quantales and their applications*, Longman scientific and technical, Longman house, Burnt Mill, Harlow, Essex, UK, 1990.
1502. David Ross, *Random sets without separability*, Annals of Probability **14** (July 1986), no. 3, 1064–1069.
1503. Dan Roth, *On the hardness of approximate reasoning*, Artificial Intelligence **82** (1996), no. 1-2, 273–302.
1504. Franz Rottensteiner, John Trinder, Simon Clode, and Kurt Kubik, *Using the Dempster–Shafer method for the fusion of {LIDAR} data and multi-spectral images for building detection*, Information Fusion **6** (2005), no. 4, 283–300.
1505. ———, *Building detection by fusion of airborne laser scanner data and multi-spectral images: Performance evaluation and sensitivity analysis*, ISPRS Journal of Photogrammetry and Remote Sensing **62** (2007), no. 2, 135–149.
1506. Franz Rottensteiner, John Trinder, Simon Clode, Kurt Kubik, and Brian Lovell, *Building detection by Dempster–Shafer fusion of LIDAR data and multispectral aerial imagery*, Proceedings of the Seventeenth International Conference on Pattern Recognition (ICPR'04), 2004, pp. 339–342.
1507. Jonathan Rougier, *Probabilistic inference for future climate using an ensemble of climate model evaluations*, Climatic Change **81** (April 2007), no. 3, 247–264.
1508. Cyril Royere, Dominique Gruyer, and Veronique Cherfaoui, *Data association with believe theory*, Proceedings of the Third International Conference on Information Fusion (FUSION 2000), vol. 1, July 2000, pp. 3–9.
1509. Enrique H. Ruspini, *Approximate deduction in single evidential bodies*, Proceedings of the Second Conference on Uncertainty in Artificial Intelligence (UAI'86), 1986, pp. 215–222.
1510. ———, *The logical foundations of evidential reasoning*, Tech. Report 408, SRI International, Menlo Park, CA, 1986.
1511. ———, *Epistemic logics, probability and the calculus of evidence*, Proceedings of the 10th International Joint Conference on Artificial intelligence (IJCAI-87), vol. 2, 1987, pp. 924–931.
1512. ———, *Approximate reasoning: Past, present, future*, Information Sciences **57** (1991), 297–317.
1513. Enrique H. Ruspini, *Epistemic logics, probability, and the calculus of evidence*, Classic Works of the Dempster-Shafer Theory of Belief Functions (Roland R. Yager and Liping Liu, eds.), Springer, Berlin, Heidelberg, 2008, pp. 435–448.
1514. Enrique H. Ruspini, *Approximate deduction in single evidential bodies*, ArXiv preprint arXiv:1304.3104, 2011.
1515. Enrique H. Ruspini, John D. Lowrance, and Thomas M. Strat, *Understanding evidential reasoning*, International Journal of Approximate Reasoning **6** (1992), no. 3, 401–424.
1516. Matthew Ryan, *Violations of belief persistence in Dempster–Shafer equilibrium*, Games and Economic Behavior **39** (2002), no. 1, 167–174.
1517. Matthew Ryan, Rhema Vaithianathan, and Luca Rigotti, *Throwing good money after bad*, Working Paper 2014-01, Auckland University of Technology, Department of Economics, 2014.
1518. Régis Sabbadin, *Decision as abduction*, Proceedings of the 13th European Conference on Arti
cial Intelligence (ECAI 98) (Henri Prade, ed.), Wiley, 1998, pp. 600–604.
1519. Michael Sabourin, Amar Mitiche, Danny Thomas, and George Nagy, *Classifier combination for hand-printed digit recognition*, Proceedings of the Second International Conference on Document Analysis and Recognition, October 1993, pp. 163–166.

1520. Rehan Sadiq, Yehuda Kleiner, and Balvant Rajani, *Estimating risk of contaminant intrusion in water distribution networks using Dempster–Shafer theory of evidence*, Civil Engineering and Environmental Systems **23** (2006), no. 3, 129–141.

1521. Rehan Sadiq, Homayoun Najjaran, and Yehuda Kleiner, *Investigating evidential reasoning for the interpretation of microbial water quality in a distribution network*, Stochastic Environmental Research and Risk Assessment **21** (2006), no. 1, 63–73.

1522. A Saffiotti and E Umkehrer, *Pulcinella a general tool for propagating uncertainty in valuation networks*, Proceedings of the 7th Conference on Uncertainty in Artificial Intelligence (UAI'91) (B. D'Ambrosio, Ph. Smets, and P. P. Bonisonne, eds.), Morgan Kaufmann, 1991.

1523. Alessandro Saffiotti, *A hybrid framework for representing uncertain knowledge*, Proceedings of the Eighth National conference on Artificial intelligence (AAAI), vol. 1, 1990, pp. 653–658.

1524. ______, *A hybrid belief system for doubtful agents*, Uncertainty in Knowledge Bases, Lecture Notes in Computer Science, vol. 251, Springer-Verlag, 1991, pp. 393–402.

1525. ______, *Using Dempster–Shafer theory in knowledge representation*, Uncertainty in Artificial Intelligence 6 (B. D'Ambrosio, Philippe Smets, and P. P. Bonissone, eds.), Morgan Kaufann, 1991, pp. 417–431.

1526. ______, *A belief-function logic*, Proceedings of the 10th National Conference on Artificial Intelligence (AAAI'92), 1992, pp. 642–647.

1527. ______, *Issues of knowledge representation in Dempster–Shafer's theory*, Advances in the Dempster-Shafer theory of evidence (R. R. Yager, M. Fedrizzi, and J. Kacprzyk, eds.), Wiley, 1994, pp. 415–440.

1528. ______, *Using Dempster–Shafer theory in knowledge representation*, ArXiv preprint arXiv:1304.1123, March 2013.

1529. Alessandro Saffiotti and E. Umkehrer, *PULCINELLA: A general tool for propagation uncertainty in valuation networks*, Tech. report, IRIDIA, Libre Universite de Bruxelles, 1991.

1530. Alessandro Saffiotti, Elisabeth Umkehrer, and Simon Parsons, *Comparing uncertainty management techniques*, Microcomputers in Civil Engineering **9** (1994), no. 5, 367–383.

1531. Antonio Sanfilippo, Bob Baddeley, Christian Posse, and Paul Whitney, *A layered Dempster–Shafer approach to scenario construction and analysis*, Proceedings of the IEEE International Conference on Intelligence and Security Informatics (ISI 2007), May 2007, pp. 95–102.

1532. Davy Sannen, Hendrik Van Brussel, and Marnix Nuttin, *Classifier fusing using discounted Dempster–Shafer combination*, Proceedings of the International Conference on Machine Learning and Data Mining in Pattern Recognition, July 2007, pp. 216–230.

1533. Anjan Sarkar, Anjan Banerjee, Nilanjan Banerjee, Siddhartha Brahma, B. Kartikeyan, Manab Chakraborty, and Kantilal L. Majumder, *Landcover classification in MRF context using Dempster–Shafer fusion for multisensor imagery*, IEEE Transactions on Image Processing **14** (2005), no. 5, 634–645.

1534. Manish Sarkar, *Modular pattern classifiers: A brief survey*, Proceedings of the IEEE International Conference on Systems, Man, and Cybernetics (SMC'00), vol. 4, 2000, pp. 2878–2883.

1535. V. V. S. Sarma and Savithri Raju, *Multisensor data fusion and decision support for airborne target identification*, IEEE Transactions on Systems, Man and Cybernetics **21** (1991), no. 5, 1224–1230.

1536. David Saunders, *Representing uncertain knowledge: An artificial intelligence aproach*, AI Communications **7** (1994), no. 2, 130–131.
1537. Leonard J. Savage, *The foundations of statistics*, John Wiley and Sons, Inc., 1954.
1538. Björn Scheuermann and Bodo Rosenhahn, *Feature quarrels: The Dempster–Shafer evidence theory for image segmentation using a variational framework*, Computer Vision – ACCV 2010 (Ron Kimmel, Reinhard Klette, and Akihiro Sugimoto, eds.), Lecture Notes in Computer Science, vol. 6493, Springer Science and Business Media, 2011, pp. 426–439.
1539. David Schmeidler, *Integral representation without additivity*, Proceedings of the American Mathematical Society **97** (1986), no. 2, 255–261.
1540. ______, *Subjective probability and expected utility without additivity*, Econometrica **57** (1989), no. 3, 571–587.
1541. Kristin Schneider, *Dempster-Shafer analysis for species presence prediction of the winter wren (Troglodytes troglodytes)*, Proceedings of the 1st International Conference on GeoComputation (R. J. Abrahart, ed.), vol. 2, Leeds, UK, 17-19 September 1996, p. 738.
1542. Shimon Schocken and Robert A. Hummel, *On the use of the Dempster Shafer model in information indexing and retrieval applications*, International Journal of Man-Machine Studies **39** (1993), no. 5, 843–879.
1543. Johan Schubert, *On nonspecific evidence*, International Journal of Intelligent Systems **8** (1993), no. 6, 711–725.
1544. Johan Schubert, *Cluster-based specification techniques in Dempster–Shafer theory for an evidential intelligence analysis of multipletarget tracks*, PhD dissertation, Royal Institute of Technology, Sweden, 1994.
1545. ______, *Cluster-based specification techniques in Dempster–Shafer theory*, Symbolic and Quantitative Approaches to Reasoning and Uncertainty (C. Froidevaux and J. Kohlas, eds.), Springer Science and Business Media, 1995, pp. 395–404.
1546. ______, *Cluster-based specification techniques in Dempster–Shafer theory for an evidential intelligence analysis of multipletarget tracks*, AI Communications **8** (1995), no. 2, 107–110.
1547. ______, *Finding a posterior domain probability distribution by specifying nonspecific evidence*, International Journal of Uncertainty, Fuzziness and Knowledge-Based Systems **3** (1995), no. 2, 163–185.
1548. ______, *On ϱ in a decision-theoretic apparatus of Dempster–Shafer theory*, International Journal of Approximate Reasoning **13** (1995), no. 3, 185–200.
1549. ______, *Specifying nonspecific evidence*, International Journal of Intelligent Systems **11** (1996), no. 8, 525–563.
1550. Johan Schubert, *Creating prototypes for fast classification in Dempster–Shafer clustering*, Qualitative and Quantitative Practical Reasoning (Dov M. Gabbay, Rudolf Kruse, Andreas Nonnengart, and Hans Jürgen Ohlbach, eds.), Lecture Notes in Computer Science, vol. 1244, Springer, Berlin Heidelberg, 1997, pp. 525–535.
1551. ______, *Fast Dempster–Shafer clustering using a neural network structure*, Information, Uncertainty and Fusion (Bernadette Bouchon-Meunier, Ronald R. Yager, and Lotfi A. Zadeh, eds.), The Springer International Series in Engineering and Computer Science, vol. 516, Springer, 2000, pp. 419–430.
1552. ______, *Managing inconsistent intelligence*, Proceedings of the Third International Conference on Information Fusion (FUSION 2000), vol. 1, July 2000, pp. TUB4/10–TUB4/16.
1553. Johan Schubert, *Creating prototypes for fast classification in Dempster–Shafer clustering*, ArXiv preprint arXiv:cs/0305021, 2003.

1554. ———, *Fast Dempster–Shafer clustering using a neural network structure*, ArXiv preprint arXiv:cs/0305026, 2003.
1555. ———, *Clustering belief functions based on attracting and conflicting metalevel evidence using Potts spin mean field theory*, Information Fusion **5** (2004), no. 4, 309–318.
1556. ———, *Clustering decomposed belief functions using generalized weights of conflict*, International Journal of Approximate Reasoning **48** (2008), no. 2, 466–480.
1557. ———, *Conflict management in Dempster–Shafer theory by sequential discounting using the degree of falsity*, Proceedings of the 12th International Conference Information Processing and Management of Uncertainty for Knowledge-Based Systems (IPMU 2008), vol. 8, 2008, p. 299.
1558. ———, *Managing decomposed belief functions*, Uncertainty and Intelligent Information Systems, World Scientific, 2008, pp. 91–103.
1559. ———, *Conflict management in Dempster–Shafer theory using the degree of falsity*, International Journal of Approximate Reasoning **52** (2011), no. 3, 449–460.
1560. ———, *Constructing and evaluating alternative frames of discernment*, International Journal of Approximate Reasoning **53** (2012), no. 2, 176–189.
1561. Johan Schubert, *The internal conflict of a belief function*, Belief Functions: Theory and Applications: Proceedings of the 2nd International Conference on Belief Functions, Compiègne, France 9-11 May 2012 (Thierry Denœux and Marie-Hélène Masson, eds.), Springer, Berlin, Heidelberg, 2012, pp. 169–177.
1562. Johan Schubert, *Reliable force aggregation using a refined evidence specification from Dempster–Shafer clustering*, Proceedings of the Fourth Annual Conference on Information Fusion (FUSION 2001), August 2001, pp. TuB3/15–22.
1563. ———, *Simultaneous Dempster–Shafer clustering and gradual determination of number of clusters using a neural network structure*, Proceedings of the Information, Decision and Control Conference (IDC'99), February 1999, pp. 401–406.
1564. Johan Schubert, *Simultaneous Dempster–Shafer clustering and gradual determination of number of clusters using a neural network structure*, Proceedings of Information, Decision and Control (IDC-99), February 1999, pp. 401–406.
1565. Johan Schubert, *Evidential force aggregation*, Proceedings of the Sixth International Conference on Information Fusion (FUSION 2003), July 2003, pp. 1223–1229.
1566. ———, *A neural network and iterative optimization hybrid for Dempster–Shafer clustering*, ArXiv preprint arXiv:cs/0305024, May 2003.
1567. ———, *Reliable force aggregation using a refined evidence specification from Dempster–Shafer clustering*, ArXiv preprint arXiv:cs/0305030, May 2003.
1568. ———, *A neural network and iterative optimization hybrid for Dempster–Shafer clustering*, Proceedings of the EuroFusion98 International Conference on Data Fusion (EF'98) (M. Bedworth and J. O'Brien, eds.), October 1998, pp. 29–36.
1569. Alan C. Schultz and William Adams, *Continuous localization using evidence grids*, Proceedings of the IEEE International Conference on Robotics and Automation (ICRA'98), vol. 4, May 1998, pp. 2833–2839.
1570. Romano Scozzafava, *Subjective probability versus belief functions in artificial intelligence*, International Journal of General Systems **22** (1994), no. 2, 197–206.
1571. Teddy Seidenfeld, *Statistical evidence and belief functions*, Proceedings of the Biennial Meeting of the Philosophy of Science Association, 1978, pp. 478–489.
1572. ———, *Some static and dynamic aspects of rubust Bayesian theory*, Random Sets: Theory and Applications (Goutsias, Malher, and Nguyen, eds.), Springer, 1997, pp. 385–406.

1573. Teddy Seidenfeld, Mark Schervish, and Joseph Kadane, *Coherent choice functions under uncertainty*, Proceedings of the Fifth International Symposium on Imprecise Probabilities and Their Applications (ISIPTA'07), 2007.
1574. ______, *Coherent choice functions under uncertainty*, Synthese **172** (2010), no. 1, 157.
1575. Teddy Seidenfeld and Larry Wasserman, *Dilation for convex sets of probabilities*, Annals of Statistics **21** (1993), 1139–1154.
1576. Kari Sentz and Scott Ferson, *Combination of evidence in Dempster–Shafer theory*, Tech. Report SAND2002-0835, SANDIA, April 2002.
1577. Lisa Serir, Emmanuel Ramasso, and Noureddine Zerhouni, *Evidential evolving Gustafson–Kessel algorithm for online data streams partitioning using belief function theory*, International Journal of Approximate Reasoning **53** (2012), no. 5, 747–768.
1578. Lisa Serir, Emmanuel Ramasso, and Noureddine Zerhouni, *E2GK: Evidential evolving Gustafsson–Kessel algorithm for data streams partitioning using belief functions*, Proceedings of the 11th European Conference on Symbolic and Quantitative Approaches to Reasoning with Uncertainty (ECSQARU 2011), Belfast, UK (Weiru Liu, ed.), Springer, Berlin, Heidelberg, June-July 2011, pp. 326–337.
1579. Ronald Setia and Gary S. May, *Run-to-run failure detection and diagnosis using neural networks and Dempster–Shafer theory: an application to excimer laser ablation*, IEEE Transactions on Electronics Packaging Manufacturing **29** (2006), no. 1, 42–49.
1580. Pavel Sevastianov, *Numerical methods for interval and fuzzy number comparison based on the probabilistic approach and Dempster–Shafer theory*, Information Sciences **177** (2007), no. 21, 4645–4661.
1581. Pavel Sevastianov and Ludmila Dymova, *Synthesis of fuzzy logic and Dempster–Shafer theory for the simulation of the decision-making process in stock trading systems*, Mathematics and Computers in Simulation **80** (2009), no. 3, 506–521.
1582. Pavel Sevastianov, Ludmila Dymova, and Pavel Bartosiewicz, *A framework for rule-base evidential reasoning in the interval setting applied to diagnosing type 2 diabetes*, Expert Systems with Applications **39** (2012), no. 4, 4190–4200.
1583. Glenn Shafer, *A mathematical theory of evidence*, Princeton University Press, 1976.
1584. ______, *A theory of statistical evidence*, Foundations of Probability Theory, Statistical Inference, and Statistical Theories of Science (W. L. Harper and C. A. Hooker, eds.), vol. 2, Reidel, Dordrecht, 1976, pp. 365–436.
1585. ______, *Nonadditive probabilites in the work of Bernoulli and Lambert*, Archive for History of Exact Sciences **19** (1978), 309–370.
1586. Glenn Shafer, *Two theories of probability*, PSA: Proceedings of the Biennial Meeting of the Philosophy of Science Association, vol. 1978, Philosophy of Science Association, 1978, pp. 441–465.
1587. Glenn Shafer, *Allocations of probability*, Annals of Probability **7** (1979), no. 5, 827–839.
1588. ______, *Jeffrey's rule of conditioning*, Philosophy of Sciences **48** (1981), 337–362.
1589. ______, *Belief functions and parametric models*, Journal of the Royal Statistical Society: Series B (Methodological) **44** (1982), no. 3, 322–339.
1590. ______, *Lindley's paradox*, Journal of the American Statistical Association **77** (1982), no. 378, 325–334.
1591. ______, *The combination of evidence*, Tech. Report 162, School of Business, University of Kansas, 1984.
1592. ______, *Conditional probability*, International Statistical Review **53** (1985), no. 3, 261–277.

1593. ———, *Nonadditive probability*, Encyclopedia of Statistical Sciences (Kotz and Johnson, eds.), vol. 6, Wiley, 1985, pp. 271–276.
1594. ———, *Probability judgement in artificial intelligence*, Proceedings of the First Conference on Uncertainty in Artificial Intelligence (UAI1985), 1985, pp. 91–98.
1595. ———, *The combination of evidence*, International Journal of Intelligent Systems **1** (1986), no. 3, 155–179.
1596. ———, *Belief functions and possibility measures*, The analysis of fuzzy information, vol. 1, CRC Press, 1987, pp. 51–84.
1597. ———, *Probability judgment in artificial intelligence and expert systems*, Statistical Science **2** (1987), no. 1, 3–44.
1598. ———, *Perspectives on the theory and practice of belief functions*, International Journal of Approximate Reasoning **4** (1990), no. 5, 323–362.
1599. ———, *A note on Dempster's Gaussian belief functions*, Tech. report, School of Business, University of Kansas, 1992.
1600. ———, *Rejoinders to comments on 'Perspectives on the theory and practice of belief functions'*, International Journal of Approximate Reasoning **6** (1992), no. 3, 445–480.
1601. ———, *Response to the discussion on belief functions*, Working paper 231, School of Business, University of Kansas, 1992.
1602. ———, *Probabilistic expert systems*, SIAM (Society for Industrial and Applied Mathematics), Philadelphia, 1996.
1603. ———, *Comments on 'Constructing a logic of plausible inference: a guide to Cox's Theorem', by Kevin S. Van Horn*, International Journal of Approximate Reasoning **35** (2004), no. 1, 97–105.
1604. ———, *Game-theoretic probability: Theory and applications*, Proceedings of the Fifth International Symposium on Imprecise Probabilities and Their Applications (ISIPTA'07), 2007.
1605. Glenn Shafer, *A betting interpretation for probabilities and Dempster–Shafer degrees of belief*, International Journal of Approximate Reasoning **52** (2011), no. 2, 127–136.
1606. Glenn Shafer, *Bayes's two arguments for the rule of conditioning*, Annals of Statistics **10** (December 1982), no. 4, 1075–1089.
1607. ———, *Constructive probability*, Synthese **48** (July 1981), no. 1, 1–60.
1608. Glenn Shafer and Roger Logan, *Implementing Dempster's rule for hierarchical evidence*, Artificial Intelligence **33** (1987), no. 3, 271–298.
1609. Glenn Shafer and Prakash P. Shenoy, *Local computation on hypertrees*, Working paper 201, School of Business, University of Kansas, 1988.
1610. ———, *Probability propagation*, Preliminary Papers of the Second International Workshop on Artificial Intelligence and Statistics, 1989, pp. 1–10.
1611. Glenn Shafer, Prakash P. Shenoy, and Khaled Mellouli, *Propagating belief functions in qualitative Markov trees*, International Journal of Approximate Reasoning **1** (1987), no. 4, 349–400.
1612. Glenn Shafer and Rajendra Srivastava, *The Bayesian and belief-function formalism: A general perspective for auditing*, Auditing: A Journal of Practice and Theory **9** (1990), no. Supplement, 110–148.
1613. Glenn Shafer and Amos Tversky, *Languages and designs for probability judgment*, Classic Works of the Dempster-Shafer Theory of Belief Functions (Roland R. Yager and Liping Liu, eds.), Springer, Berlin, Heidelberg, 2008, pp. 345–374.
1614. Glenn Shafer and Amos Tversky, *Weighing evidence: The design and comparison of probability thought experiments*, Technical report AD-A131475, Stanford University, Department of Psychology, June 1983.

1615. Glenn Shafer and Vladimir Vovk, *Probability and finance: It's only a game!*, Wiley, New York, 2001.
1616. Claude E. Shannon, *A mathematical theory of communication*, Bell System Technical Journal **27** (1948), no. 3, 379–423.
1617. Lloyd S. Shapley, *A value for n-person games*, Contributions to the Theory of Games (H. W. Kuhn and A. W. Tucker, eds.), vol. 2, 1953, pp. 307–317.
1618. ———, *Cores of convex games*, International Journal of Game Theory **1** (1971), no. 1, 11–26.
1619. Harold I. Sharlin, *William Whewell's theory of scientific method*, Science **168** (1970), no. 3936, 1195–1196.
1620. Lokendra Shastri, *Evidential reasoning in semantic networks: A formal theory and its parallel implementation (inheritance, categorization, connectionism, knowledge representation)*, Ph.D. thesis, The University of Rochester, 1985.
1621. Lokendra Shastri and Jerome A. Feldman, *Evidential reasoning in semantic networks: A formal theory*, Proceedings of the 9th International Joint Conference on Artificial Intelligence (IJCAI'85), vol. 1, Morgan Kaufmann, 1985, pp. 465–474.
1622. Yun Shen, Michael J. Tobia, Tobias Sommer, and Klaus Obermayer, *Risk-sensitive reinforcement learning*, Machine learning **49** (2002), no. 2-3, 267–290.
1623. Prakash P. Shenoy, *A valuation-based language for expert systems*, International Journal of Approximate Reasoning **3** (1989), no. 5, 383–411.
1624. ———, *On Spohn's rule for revision of beliefs*, International Journal of Approximate Reasoning **5** (1991), no. 2, 149–181.
1625. Prakash P. Shenoy, *On Spohn's theory of epistemic beliefs*, Uncertainty in Knowledge Bases: 3rd International Conference on Information Processing and Management of Uncertainty in Knowledge-Based Systems, IPMU '90 Paris, France, July 2–6, 1990 Proceedings (Bernadette Bouchon-Meunier, Ronald R. Yager, and Lotfi A. Zadeh, eds.), Springer, Berlin, Heidelberg, 1991, pp. 1–13.
1626. ———, *Valuation networks and conditional independence*, Proceedings of the Ninth International Conference on Uncertainty in Artificial Intelligence (UAI'93), Morgan Kaufmann, 1993, pp. 191–199.
1627. Prakash P. Shenoy, *Using Dempster–Shafer's belief function theory in expert systems*, Advances in the Dempster-Shafer Theory of Evidence (R. R. Yager, M. Fedrizzi, and J. Kacprzyk, eds.), Wiley, 1994, pp. 395–414.
1628. ———, *Binary join trees for computing marginals in the Shenoy-Shafer architecture*, International Journal of Approximate Reasoning **17** (1997), no. 2, 239–263.
1629. ———, *No double counting semantics for conditional independence*, Working paper 307, School of Business, University of Kansas, 2005.
1630. Prakash P. Shenoy and Glenn Shafer, *Propagating belief functions with local computations*, IEEE Expert **1** (1986), no. 3, 43–52.
1631. ———, *An axiomatic framework for Bayesian and belief function propagation*, Proceedings of the Fourth Conference on Uncertainty in Artificial Intelligence (UAI'88), 1988, pp. 307–314.
1632. ———, *Axioms for probability and belief functions propagation*, Uncertainty in Artificial Intelligence, 4 (R. D. Shachter, T. S. Lewitt, L. N. Kanal, and J. F. Lemmer, eds.), North Holland, 1990, pp. 159–198.
1633. Prakash P. Shenoy, Glenn Shafer, and Khaled Mellouli, *Propagation of belief functions: a distributed approach*, Proceedings of the Second Conference on Uncertainty in Artificial Intelligence (UAI'86), 1986, pp. 249–260.

1634. Prakash P Shenoy, Glenn Shafer, and Khaled Mellouli, *Propagation of belief functions: A distributed approach*, Machine Intelligence and Pattern Recognition, vol. 5, Elsevier, 1988, pp. 325–335.
1635. F. K. J. Sheridan, *A survey of techniques for inference under uncertainty*, Artificial Intelligence Review **5** (1991), no. 1-2, 89–119.
1636. Chao Shi, Yongmei Cheng, Quan Pan, and Yating Lu, *A new method to determine evidence distance*, Proceedings of the 2010 International Conference on Computational Intelligence and Software Engineering (CiSE), 2010, pp. 1–4.
1637. Lixin Shi, Jian-Yun Nie, and Guihong Cao, *Relating dependent indexes using Dempster–Shafer theory*, Proceedings of the 17th ACM Conference on Information and Knowledge Management (CIKM '08), 2008, pp. 429–438.
1638. Xiaomin Shi, *Central limit theorems for bounded random variables under belief measures*, ArXiv preprint arxiv:1501.00771v1, 2015.
1639. ______, *Central limit theorems for bounded random variables under belief measures*, Journal of Mathematical Analysis and Applications **460** (2018), no. 2, 546–560.
1640. Haruhisa Shimoda, Sun-pyo Hong, Kiyonari Fukue, and Toshibumi Sakata, *A multitemporal classification method using Dempster–Shafer model*, Proceedings of the Geoscience and Remote Sensing Symposium (IGARSS '91), vol. 3, June 1991, pp. 1831–1834.
1641. Almas Shintemirov, Wenhu Tang, and Qinghua Wu, *Transformer winding condition assessment using frequency response analysis and evidential reasoning*, IET Electric Power Applications **4** (2010), no. 3, 198–212.
1642. Margaret F. Shipley and André de Korvin, *Rough set theory fuzzy belief functions related to statistical confidence: application and evaluation for golf course closing*, Stochastic Analysis and Applications **13** (1995), no. 4, 487–502.
1643. Margaret F. Shipley, Charlene A. Dykman, and André de Korvin, *Project management: using fuzzy logic and the Dempster–Shafer theory of evidence to select team members for the project duration*, Proceedings of the 18th International Conference of the North American Fuzzy Information Processing Society (NAFIPS 1999), 1999, pp. 640–644.
1644. Seymour Shlien, *Multiple binary decision tree classifiers*, Pattern Recognition **23** (1990), no. 7, 757–763.
1645. Abel Shoshana, *The sum-and-lattice points method based on an evidential reasoning system applied to the real-time vehicle guidance problem*, Uncertainty in Artificial Intelligence 2 (J. F. Lemmer and L. N. Kanal, eds.), North Holland, 1988, pp. 365–370.
1646. Xiao-Sheng Si, Chang-Hua Hu, Jian-Bo Yang, and Qi Zhang, *On the dynamic evidential reasoning algorithm for fault prediction*, Expert Systems with Applications **38** (2011), no. 5, 5061–5080.
1647. Xiao-Sheng Si, Chang-Hua Hu, and Zhi-Jie Zhou, *Fault prediction model based on evidential reasoning approach*, Science China Information Sciences **53** (2010), no. 10, 2032–2046.
1648. Christos Siaterlis and Basil Maglaris, *Towards multisensor data fusion for dos detection*, Proceedings of the 2004 ACM Symposium on Applied Computing, 2004, pp. 439–446.
1649. H. S. Sii, T. Ruxton, and J. Wang, *Synthesis using fuzzy set theory and a Dempster–Shafer-based approach to compromise decision-making with multiple-attributes applied to risk control options selection*, Proceedings of the Institution of Mechanical Engineers, Part E: Journal of Process Mechanical Engineering **216** (2002), no. 1, 15–29.

1650. Roman Sikorski, *Boolean algebras*, Springer Verlag, 1964.
1651. Marc-Alain Simard, Jean Couture, and Eloi Bossé, *Data fusion of multiple sensors attribute information for target identity estimation using a Dempster–Shafer evidential combination algorithm*, Proceedings of the SPIE - Signal and Data Processing of Small Targets (P. G. Anderson and K. Warwick, eds.), vol. 2759, Orlando, FL, USA, 9-11 April 1996, pp. 577–588.
1652. Christophe Simon and Philippe Weber, *Bayesian networks implementation of the Dempster Shafer theory to model reliability uncertainty*, Proceedings of the First International Conference on Availability, Reliability and Security (ARES'06), 2006, pp. 6–pp.
1653. Christophe Simon, Philippe Weber, and Alexandre Evsukoff, *Bayesian networks inference algorithm to implement Dempster Shafer theory in reliability analysis*, Reliability Engineering and System Safety **93** (2008), no. 7, 950–963.
1654. William R. Simpson and John W. Sheppard, *The application of evidential reasoning in a portable maintenance aid*, Proceedings of the IEEE Systems Readiness Technology Conference, 'Advancing Mission Accomplishment' (AUTOTESTCON 90), September 1990, pp. 211–214.
1655. Richa Singh, Mayank Vatsa, Afzel Noore, and Sanjay K. Singh, *Dempster–Shafer theory based classifier fusion for improved fingerprint verification performance*, Computer Vision, Graphics and Image Processing (Prem K. Kalra and Shmuel Peleg, eds.), Lecture Notes in Computer Science, vol. 4338, Springer, Berlin Heidelberg, 2006, pp. 941–949.
1656. T. Sitamahalakshmi, A. Vinay Babu, and M. Jagadeesh, *Character recognition using Dempster-Shafer theory combining different distance measurement methods*, International Journal of Engineering Science and Technology **2** (2010), no. 5, 1177–1184.
1657. Andrzej Skowron, *Boolean reasoning for decision rules generation*, Methodologies for Intelligent Systems: 7th International Symposium, ISMIS'93 Trondheim, Norway, June 15–18, 1993 Proceedings (Jan Komorowski and Zbigniew W. Raś, eds.), Springer, Berlin Heidelberg, 1993, pp. 295–305.
1658. Andrzej Skowron and Jerzy Grzymalla-Busse, *From rough set theory to evidence theory*, Advances in the Dempster-Shafer Theory of Evidence (Ronald R. Yager, Janusz Kacprzyk, and Mario Fedrizzi, eds.), John Wiley and Sons, Inc., 1994, pp. 193–236.
1659. Andrzej Skowron and Cecylia Rauszer, *The discernibility matrices and functions in information systems*, Intelligent Decision Support: Handbook of Applications and Advances of the Rough Sets Theory (Roman Słowiński, ed.), Springer Netherlands, Dordrecht, 1992, pp. 331–362.
1660. Brian Skyrms, *Higher order degrees of belief*, Prospects for Pragmatism (D. H. Mellor, ed.), Cambridge University Press, 1980, pp. 109–137.
1661. Anna Slobodova, *Conditional belief functions and valuation-based systems*, Tech. report, Slovak Academy of Sciences, 1994.
1662. ______, *Multivalued extension of conditional belief functions*, Proceedings of the International Joint Conference on Qualitative and Quantitative Practical Reasoning (ECSQARU / FAPR '97), June 1997, pp. 568–573.
1663. ______, *A comment on conditioning in the Dempster–Shafer theory*, Proceedings of the International ICSC Symposia on Intelligent Industrial Automation and Soft Computing (P. G. Anderson and K. Warwick, eds.), Reading, UK, 26-28 March 1996, pp. 27–31.
1664. Florentin Smarandache and Jean Dezert, *An introduction to the DSm theory for the combination of paradoxical, uncertain and imprecise sources of information*, Proceedings of the 13th International Congress of Cybernetics and Systems, 2005, pp. 6–10.

1665. ______, *An introduction to dsm theory of plausible, paradoxist, uncertain, and imprecise reasoning for information fusion*, Octogon Mathematical Magazine **15** (2007), no. 2, 681–722.
1666. Florentin Smarandache, Jean Dezert, and Jean-Marc Tacnet, *Fusion of sources of evidence with different importances and reliabilities*, Proceedings of the 13th International Conference on Information Fusion (FUSION 2010), July 2010, pp. 1–8.
1667. Florentin Smarandache, Arnaud Martin, and Christophe Osswald, *Contradiction measures and specificity degrees of basic belief assignments*, Proceedings of the 14th International Conference on Information Fusion (FUSION 2011), 2011, pp. 1–8.
1668. Philippe Smets, *Theory of evidence and medical diagnostic*, Medical Informatics Europe **78** (1978), 285–291.
1669. ______, *Medical diagnosis : Fuzzy sets and degree of belief*, MIC 79 (J. Willems, ed.), Wiley, 1979, pp. 185–189.
1670. ______, *The degree of belief in a fuzzy event*, Information Sciences **25** (1981), 1–19.
1671. ______, *Medical diagnosis : Fuzzy sets and degrees of belief*, Fuzzy sets and Systems **5** (1981), no. 3, 259–266.
1672. ______, *Information content of an evidence*, International Journal of Man Machine Studies **19** (1983), no. 1, 33–43.
1673. ______, *Upper and lower probability functions versus belief functions*, Proceedings of the International Symposium on Fuzzy Systems and Knowledge Engineering, 1987, pp. 10–16.
1674. ______, *Belief functions*, Non-Standard Logics for Automated Reasoning (Philippe Smets, Abe Mamdani, Didier Dubois, and Henri Prade, eds.), Academic Press, London, 1988, pp. 253–286.
1675. ______, *Belief functions versus probability functions*, Proceedings of the International Conference on Information Processing and Management of Uncertainty in Knowledge-Based Systems (IPMU'88) (B. Bouchon, L. Saitta, and R. R. Yager, eds.), Springer Verlag, 1988, pp. 17–24.
1676. ______, *Transferable belief model versus Bayesian model*, Proceedings of the 8th European Conference on Artificial Intelligence (ECAI 88) (Y. Kodratoff and B. Ueberreiter, eds.), Pitman, 1988, pp. 495–500.
1677. ______, *The combination of evidence in the transferable belief model*, IEEE Transactions on Pattern Analysis and Machine Intelligence **12** (1990), no. 5, 447–458.
1678. Philippe Smets, *Constructing the pignistic probability function in a context of uncertainty*, Proceedings of the Fifth Annual Conference on Uncertainty in Artificial Intelligence (UAI '89), North-Holland, 1990, pp. 29–40.
1679. Philippe Smets, *The transferable belief model and possibility theory*, Proceedings of the Annual Conference of the North American Fuzzy Information processing Society (NAFIPS-90) (I. Burhan Turksen, ed.), 1990, pp. 215–218.
1680. ______, *About updating*, Proceedings of the 7th International Conference on Uncertainty in Artificial Intelligence (UAI'91) (B. D'Ambrosio, Ph. Smets, and P. P. Bonissone, eds.), 1991, pp. 378–385.
1681. ______, *Patterns of reasoning with belief functions*, Journal of Applied Non-Classical Logic **1** (1991), no. 2, 166–170.
1682. ______, *Probability of provability and belief functions*, Logique et Analyse **133-134** (1991), 177–195.
1683. ______, *The transferable belief model and other interpretations of Dempster–Shafer's model*, Uncertainty in Artificial Intelligence (P. P. Bonissone, M. Henrion, L. N. Kanal, and J. F. Lemmer, eds.), vol. 6, North-Holland, Amsterdam, 1991, pp. 375–383.

1684. ______, *Varieties of ignorance and the need for well-founded theories*, Information Sciences **57-58** (1991), 135–144.
1685. ______, *The nature of the unnormalized beliefs encountered in the transferable belief model*, Proceedings of the 8th Annual Conference on Uncertainty in Artificial Intelligence (UAI-92) (San Mateo, CA), Morgan Kaufmann, 1992, pp. 292–29.
1686. ______, *Resolving misunderstandings about belief functions*, International Journal of Approximate Reasoning **6** (1992), no. 3, 321–344.
1687. ______, *The transferable belief model and random sets*, International Journal of Intelligent Systems **7** (1992), no. 1, 37–46.
1688. ______, *The transferable belief model for expert judgments and reliability problems*, Reliability Engineering and System Safety **38** (1992), no. 1-2, 59–66.
1689. ______, *Belief functions : the disjunctive rule of combination and the generalized Bayesian theorem*, International Journal of Approximate Reasoning **9** (1993), no. 1, 1–35.
1690. ______, *Jeffrey's rule of conditioning generalized to belief functions*, Proceedings of the Ninth International Conference on Uncertainty in Artificial Intelligence (UAI'93), Morgan Kaufmann, 1993, pp. 500–505.
1691. ______, *No Dutch Book can be built against the TBM even though update is not obtained by Bayes rule of conditioning*, Workshop on probabilistic expert systems, Societa Italiana di Statistica (R. Scozzafava, ed.), 1993, pp. 181–204.
1692. ______, *No Dutch book can be built against the TBM even though update is not obtained by Bayes rule of conditioning*, Workshop on probabilistic expert systems, Societá Italiana di Statistica, Roma, 1993, pp. 181–204.
1693. ______, *Quantifying beliefs by belief functions: An axiomatic justification*, Proceedings of the 13th International Joint Conference on Artificial Intelligence (IJCAI'93), 1993, pp. 598–603.
1694. ______, *Belief induced by the knowledge of some probabilities*, Proceedings of the 10th International Conference on Uncertainty in Artificial Intelligence (UAI'94) (D. Heckerman, D. Poole, and R. Lopez de Mantaras, eds.), 1994, pp. 523–530.
1695. Philippe Smets, *Belief induced by the partial knowledge of the probabilities*, Uncertainty Proceedings, Elsevier, 1994, pp. 523–530.
1696. Philippe Smets, *What is Dempster–Shafer's model ?*, Advances in the Dempster–Shafer Theory of Evidence (R. R. Yager, M. Fedrizzi, and J. Kacprzyk, eds.), Wiley, 1994, pp. 5–34.
1697. ______, *The axiomatic justification of the transferable belief model*, Tech. Report TR/IRIDIA/1995-8.1, Université Libre de Bruxelles, 1995.
1698. ______, *The canonical decomposition of a weighted belief*, Proceedings of the International Joint Conference on Artificial Intelligence (IJCAI-95), 1995, pp. 1896–1901.
1699. ______, *Non standard probabilistic and non probabilistic representations of uncertainty*, Advances in Fuzzy Sets Theory and Technology (P.P. Wang, ed.), vol. 3, Duke University, 1995, pp. 125–154.
1700. ______, *Probability, possibility, belief : which for what ?*, Foundations and Applications of Possibility Theory (G. De Cooman, D. Ruan, and E. E. Kerre, eds.), World Scientific, 1995, pp. 20–40.
1701. ______, *The transferable belief model for uncertainty representation*, Proceedings of the Third International Symposium on Conceptual Tools for Understanding Nature, World Scientific, 1995, pp. 135–152.
1702. ______, *Imperfect information: Imprecision and uncertainty*, Uncertainty Management in Information Systems: From Needs to Solutions (Amihai Motro and Philippe Smets, eds.), Springer, 1997, pp. 225–254.

1703. ______, *The normative representation of quantified beliefs by belief functions*, Artificial Intelligence **92** (1997), no. 1-2, 229–242.
1704. ______, *The application of the transferable belief model to diagnostic problems*, International Journal of Intelligent Systems **13** (1998), no. 2-3, 127–158.
1705. Philippe Smets, *Numerical representation of uncertainty*, Belief change, Springer, 1998, pp. 265–309.
1706. Philippe Smets, *Probability, possibility, belief: Which and where ?*, Handbook of Defeasible Reasoning and Uncertainty Management Systems, Vol. 1: Quantified Representation of Uncertainty and Imprecision (D. Gabbay and Ph. Smets, eds.), Kluwer, 1998, pp. 1–24.
1707. ______, *The transferable belief model for quantified belief representation*, Handbook of Defeasible Reasoning and Uncertainty Management Systems, Vol. 1: Quantified Representation of Uncertainty and Imprecision (D. Gabbay and Ph. Smets, eds.), Kluwer, 1998, pp. 267–301.
1708. Philippe Smets, *Practical uses of belief functions*, Proceedings of the Fifteenth Conference on Uncertainty in Artificial Intelligence (UAI'99), Morgan Kaufmann, 1999, pp. 612–621.
1709. Philippe Smets, *Data fusion in the transferable belief model*, Proceedings of the Third International Conference on Information Fusion (FUSION 2000), vol. 1, 2000, pp. PS21–PS33.
1710. ______, *Quantified epistemic possibility theory seen as an hyper cautious Transferable Belief Model*, Rencontres Francophones sur la Logique Floue et ses Applications (LFA 2000), 2000.
1711. ______, *Decision making in a context where uncertainty is represented by belief functions*, Belief Functions in Business Decisions (R. Srivastava, ed.), Physica-Verlag, 2001, pp. 495–504.
1712. ______, *Decision making in a context where uncertainty is represented by belief functions*, Belief functions in business decisions, Springer, 2002, pp. 17–61.
1713. ______, *Showing why measures of quantified beliefs are belief functions*, Intelligent Systems for Information Processing: From representations to Applications, Elsevier, 2002, pp. 265–276.
1714. ______, *Belief functions on real numbers*, International Journal of Approximate Reasoning **40** (2005), no. 3, 181–223.
1715. ______, *Decision making in the TBM: the necessity of the pignistic transformation*, International Journal of Approximate Reasoning **38** (2005), no. 2, 133–147.
1716. ______, *Analyzing the combination of conflicting belief functions*, Information Fusion **8** (2007), no. 4, 387–412.
1717. ______, *Decision making in the TBM: the necessity of the pignistic transformation*, International Journal of Approximate Reasoning **38** (February 2005), no. 2, 133–147.
1718. ______, *Bayes' theorem generalized for belief functions*, Proceedings of the 7th European Conference on Artificial Intelligence (ECAI-86), vol. 2, July 1986, pp. 169–171.
1719. ______, *The concept of distinct evidence*, Proceedings of the 4th Conference on Information Processing and Management of Uncertainty in Knowledge-Based Systems (IPMU 92), july 92, pp. 789–794.
1720. Philippe Smets, *The α-junctions: Combination operators applicable to belief functions*, Proceedings of the First International Joint Conference on Qualitative and Quantitative Practical Reasoning (ECSQARU-FAPR'97) (Dov M. Gabbay, Rudolf Kruse, Andreas Nonnengart, and Hans Jürgen Ohlbach, eds.), Springer, Berlin, Heidelberg, June 1997, pp. 131–153.

1721. Philippe Smets, *Probability of deductibility and belief functions*, Proceedings of the European Conference on Symbolic and Quantitative Approaches to Reasoning and Uncertainty (ECSQARU'93) (M. Clark, R. Kruse, and Serafín Moral, eds.), November 1993, pp. 332–340.
1722. ———, *The application of the matrix calculus to belief functions*, International Journal of Approximate Reasoning **31** (October 2002), no. 1-2, 1–30.
1723. ———, *Data fusion in the transferable belief model*, Proceedings of the Third International Conference on Information Fusion (ICIF), vol. 1, Paris, France 2000, pp. 21–33.
1724. ———, *Applying the transferable belief model to diagnostic problems*, Proceedings of 2nd International Workshop on Intelligent Systems and Soft Computing for Nuclear Science and Industry (D. Ruan, P. D'hondt, P. Govaerts, and E. E. Kerre, eds.), September 1996, pp. 285–292.
1725. ———, *Belief functions and generalized Bayes theorem*, Proceedings of the Second IFSA Congress, Tokyo, Japan, 1987, pp. 404–407.
1726. Philippe Smets and Roger Cooke, *How to derive belief functions within probabilistic frameworks?*, Proceedings of the International Joint Conference on Qualitative and Quantitative Practical Reasoning (ECSQARU / FAPR '97), June 1997.
1727. Philippe Smets and Yen-Teh Hsia, *Defeasible reasoning with belief functions*, Tech. Report TR/IRIDIA/90-9, Universite' Libre de Bruxelles, 1990.
1728. Philippe Smets and Yen-Teh Hsia, *Default reasoning and the transferable belief model*, Proceedings of the Sixth Annual Conference on Uncertainty in Artificial Intelligence (UAI '90), Elsevier Science, 1991, pp. 495–504.
1729. Philippe Smets, Yen-Teh Hsia, Alessandro Saffiotti, Robert Kennes, Hong Xu, and E. Umkehren, *The transferable belief model*, Proceedings of the European Conference on Symbolic and Quantitative Approaches to Reasoning and Uncertainty (ECSQARU'91) (R. Kruse and P. Siegel, eds.), Lecture Notes in Computer Science, vol. 458, Springer Verlag, Berlin, 1991, pp. 91–96.
1730. Philippe Smets and Robert Kennes, *The Transferable Belief Model*, Artificial Intelligence **66** (1994), no. 2, 191–234.
1731. Philippe Smets and Robert Kennes, *The transferable belief model*, Artificial intelligence **66** (1994), no. 2, 191–234.
1732. Philippe Smets and Rudolf Kruse, *The transferable belief model for belief representation*, Uncertainty Management in information systems: from needs to solutions (A. Motro and Ph. Smets, eds.), Kluwer, 1997, pp. 343–368.
1733. Philippe Smets and Branko Ristic, *Kalman filter and joint tracking and classification in the TBM framework*, Proceedings of the Seventh International Conference on Information Fusion (FUSION 2004), vol. 1, 2004, pp. 46–53.
1734. Cedric A. B. Smith, *Consistency in statistical inference and decision*, Journal of the Royal Statistical Society, Series B **23** (1961), 1–37.
1735. ———, *Personal probability and statistical analysis*, Journal of the Royal Statistical Society, Series A **128** (1965), 469–489.
1736. Edward E. Smith, *Triplanetary*, Fantasy Press, 1948.
1737. Peter Smith and O. R. Jones, *The philosophy of mind: An introduction*, Cambridge University Press, 1986.
1738. Michael J. Smithson, *Ignorance and uncertainty: Emerging paradigms*, Springer Science & Business Media, 2012.
1739. Ronald D. Snee, *Statistical thinking and its contribution to total quality*, The American Statistician **44** (1990), no. 2, 116–121.

1740. Paul Snow, *The vulnerability of the transferable belief model to Dutch books*, Artificial Intelligence **105** (1998), no. 1-2, 345–354.
1741. Leen-Kiat Soh, *A mutliagent framework for collaborative conceptual learning using a Dempster–Shafer belief system*, Working Notes of the AAAI Spring Symposium on Collaborative Learning Agents, 2002, pp. 25–27.
1742. Leen-Kiat Soh, Costas Tsatsoulis, Todd Bowers, and Andrew Williams, *Representing sea ice knowledge in a Dempster–Shafer belief system*, Proceedings of the IEEE International Geoscience and Remote Sensing Symposium (IGARSS '98), vol. 4, July 1998, pp. 2234–2236.
1743. So Young Sohn and Sung Ho Lee, *Data fusion, ensemble and clustering to improve the classification accuracy for the severity of road traffic accidents in Korea*, Safety Science **41** (2003), no. 1, 1–14.
1744. Bassel Solaiman, Raphael K. Koffi, M.-C. Mouchot, and Alain Hillion, *An information fusion method for multispectral image classification postprocessing*, IEEE Transactions on Geoscience and Remote Sensing **36** (1998), no. 2, 395–406.
1745. Anne H. S. Solberg, Anil K. Jain, and Torfinn Taxt, *Multisource classification of remotely sensed data: fusion of Landsat TM and SAR images*, IEEE Transactions on Geoscience and Remote Sensing **32** (1994), no. 4, 768–778.
1746. Zenon A. Sosnowski and Jarosaw S. Walijewski, *Generating fuzzy decision rules with the use of Dempster–Shafer theory*, Proceedings of the 13th European Simulation Multiconference (H. Szczerbicka, ed.), vol. 2, Warsaw, Poland, 1-4 June 1999, pp. 419–426.
1747. Zenon A. Sosnowski and Jaroslaw S. Walijewski, *Genetic-based tuning of fuzzy Dempster–Shafer model*, Computational Intelligence, Theory and Applications (Bernd Reusch, ed.), Advances in Soft Computing, vol. 33, Springer Science and Business Media, 2006, pp. 747–755.
1748. C. Soyer, H. I. Bozma, and Y. Istefanopulos, *Attentional sequence-based recognition: Markovian and evidential reasoning*, IEEE Transactions on Systems, Man, and Cybernetics, Part B: Cybernetics **33** (2003), no. 6, 937–950.
1749. Marcus Spies, *Conditional events, conditioning, and random sets*, IEEE Transactions on Systems, Man, and Cybernetics **24** (1994), no. 12, 1755–1763.
1750. ———, *Evidential reasoning with conditional events*, Advances in the Dempster-Shafer Theory of Evidence (Ronald R. Yager, Janusz Kacprzyk, and Mario Fedrizzi, eds.), John Wiley and Sons, Inc., 1994, pp. 493–511.
1751. Richard Spillman, *Managing uncertainty with belief functions*, AI Expert **5** (May 1990), no. 5, 44–49.
1752. Wolfgang Spohn, *Ordinal conditional functions: A dynamic theory of epistemic states*, Causation in Decision, Belief Change, and Statistics: Proceedings of the Irvine Conference on Probability and Causation (William L. Harper and Brian Skyrms, eds.), Springer Netherlands, Dordrecht, 1988, pp. 105–134.
1753. ———, *A general non-probabilistic theory of inductive reasoning*, Proceedings of the Fourth Annual Conference on Uncertainty in Artificial Intelligence (UAI'88), North-Holland, 1990, pp. 149–158.
1754. K. Spurgeon, W. H. Tang, Q. H. Wu, Z. J. Richardson, and G. Moss, *Dissolved gas analysis using evidential reasoning*, IEE Proceedings - Science, Measurement and Technology **152** (2005), no. 3, 110–117.
1755. A. Srinivasan and J. A. Richards, *Knowledge-based techniques for multi-source classification*, International Journal of Remote Sensing **11** (1990), no. 3, 505–525.
1756. Rajendra P. Srivastava, *Decision making under ambiguity: A belief-function perspective*, Archives of Control Sciences **6** (1997), 5–28.

1757. ______, *Alternative form of Dempster's rule for binary variables*, International Journal of Intelligent Systems **20** (2005), no. 8, 789–797.
1758. ______, *An introduction to evidential reasoning for decision making under uncertainty: Bayesian and belief function perspectives*, International Journal of Accounting Information Systems **12** (2011), no. 2, 126–135, Special Issue on Methodologies in {AIS} Research.
1759. Rajendra P. Srivastava and Glenn Shafer, *Integrating statistical and nonstatistical audit evidence using belief functions: a case of variable sampling*, International Journal of Intelligent Systems **9** (June 1994), no. 6, 519–539.
1760. Rajendra P. Srivastava, Prakash P. Shenoy, and Glenn Shafer, *Propagating belief functions in AND-trees*, International Journal of Intelligent Systems **10** (1995), no. 7, 647–664.
1761. Roger Stein, *The Dempster–Shafer theory of evidential reasoning*, AI Expert **8** (August 1993), 26–31.
1762. Manfred Stern, *Semimodular lattices*, Cambridge University Press, 1999.
1763. Per R. Stokke, Thomas A. Boyce, John D. Lowrance, and William K. Ralston, *Evidential reasoning and project early warning systems*, Research-Technology Management, 1994.
1764. Per R. Stokke, Thomas A. Boyce, John D. Lowrance, and William K. Ralston, *Industrial project monitoring with evidential reasoning*, Nordic Advanced Information Technology Magazine **8** (July 1994), no. 1, 18–27.
1765. Ewa Straszecka, *Medical knowledge representation in terms of IF-THEN rules and the Dempster–Shafer theory*, Artificial Intelligence and Soft Computing - ICAISC 2004 (Leszek Rutkowski, Jörg H. Siekmann, Ryszard Tadeusiewicz, and Lotfi A. Zadeh, eds.), Lecture Notes in Computer Science, vol. 3070, Springer, Berlin Heidelberg, 2004, pp. 1056–1061.
1766. Ewa Straszecka, *On an application of Dempster–Shafer theory to medical diagnosis support*, Proceedings of the 6th European Congress on Intelligent Techniques and Soft Computing (EUFIT'98), vol. 3, Aachen, Germany, 1998, pp. 1848–1852.
1767. Thomas M. Strat, *The generation of explanations within evidential reasoning systems*, Proceedings of the Tenth International Joint Conference on Artificial Intelligence (IJCAI-87), 1987, pp. 1097–1104.
1768. ______, *Decision analysis using belief functions*, International Journal of Approximate Reasoning **4** (1990), no. 5, 391–417.
1769. ______, *Making decisions with belief functions*, Uncertainty in Artificial Intelligence, 5 (M. Henrion, R. D. Schachter, L. N. Kanal, and J. F. Lemmers, eds.), North Holland, 1990, pp. 351–360.
1770. ______, *Decision analysis using belief functions*, Advances in the Dempster-Shafer Theory of Evidence (R. R. Yager, M. Fedrizzi, and J. Kacprzyk, eds.), Wiley, 1994, pp. 275–309.
1771. ______, *Making decisions with belief functions*, ArXiv preprint arXiv:1304.1531, 2013.
1772. ______, *Continuous belief functions for evidential reasoning*, Proceedings of the Fourth National Conference on Artificial Intelligence (AAAI-84), August 1984, pp. 308–313.
1773. Thomas M. Strat and John D. Lowrance, *Explaining evidential analyses*, International Journal of Approximate Reasoning **3** (1989), no. 4, 299–353.
1774. Milan Studený, *On stochastic conditional independence: the problems of characterization and description*, Annals of Mathematics and Artificial Intelligence **35** (2002), no. 1, 323–341.

1775. Xiaoyan Su, Sankaran Mahadevan, Wenhua Han, and Yong Deng, *Combining dependent bodies of evidence*, Applied Intelligence **44** (2016), no. 3, 634–644.
1776. John J. Sudano, *Pignistic probability transforms for mixes of low- and high-probability events*, Proceedings of the Fourth International Conference on Information Fusion (FUSION 2001), 2001, pp. 23–27.
1777. ———, *Inverse pignistic probability transforms*, Proceedings of the Fifth International Conference on Information Fusion (FUSION 2002), vol. 2, 2002, pp. 763–768.
1778. John J. Sudano, *Equivalence between belief theories and naive Bayesian fusion for systems with independent evidential data: part I, the theory*, Proceedings of the Sixth International Conference on Information Fusion (FUSION 2003), vol. 2, July 2003, pp. 1239–1243.
1779. John J. Sudano, *Yet another paradigm illustrating evidence fusion (YAPIEF)*, Proceedings of the 9th International Conference on Information Fusion (FUSION'06), 2006, pp. 1–7.
1780. Thomas Sudkamp, *The consistency of Dempster–Shafer updating*, International Journal of Approximate Reasoning **7** (1992), no. 1-2, 19–44.
1781. Michio Sugeno, *Theory of fuzzy integrals and its applications*, PhD dissertation, Tokyo Institute of Technology, 1974, Tokyo, Japan.
1782. ———, *Fuzzy measures and fuzzy integrals: A survey*, Fuzzy Automata and Decision Processes (M. M Gupta, G. N. Saridis, and B. R. Gaines, eds.), North Holland, Amsterdam, 1977, pp. 89–102.
1783. Michio Sugeno, Yasuo Narukawa, and Toshiaki Murofushi, *Choquet integral and fuzzy measures on locally compact space*, Fuzzy Sets and Systems **99** (1998), no. 2, 205–211.
1784. Yoshiaki Sugie and Tetsunori Kobayashi, *Media-integrated biometric person recognition based on the Dempster–Shafer theory*, Proceedings of the 16th International Conference on Pattern Recognition (ICPR'02), vol. 4, August 2002, pp. 381–384.
1785. Doug Y. Suh, Russell M. Mersereau, Robert L. Eisner, and Roderic I. Pettigrew, *Automatic boundary detection on cardiac magnetic resonance image sequences for four dimensional visualization of the left ventricle*, Proceedings of the First Conference on Visualization in Biomedical Computing, May 1990, pp. 149–156.
1786. Hongyan Sun and Mohamad Farooq, *Conjunctive and disjunctive combination rules of evidence*, Proceedings of SPIE - Signal Processing, Sensor Fusion, and Target Recognition XIII (Ivan Kadar, ed.), vol. 5429, August 2004, pp. 392–401.
1787. Wanxiao Sun, Shunlin Liang, Gang Xu, Hongliang Fang, and Robert Dickinson, *Mapping plant functional types from MODIS data using multisource evidential reasoning*, Remote Sensing of Environment **112** (2008), no. 3, 1010–1024.
1788. Patrick Suppes and Mario Zanotti, *On using random relations to generate upper and lower probabilities*, Synthese **36** (1977), no. 4, 427–440.
1789. Nicolas Sutton-Charani, Sébastien Destercke, and Thierry Denœux, *Classification trees based on belief functions*, Belief Functions: Theory and Applications: Proceedings of the 2nd International Conference on Belief Functions, Compiègne, France 9-11 May 2012 (Thierry Denœux and Marie-Hélène Masson, eds.), Springer, Berlin Heidelberg, 2012, pp. 77–84.
1790. Johan A. K. Suykens and Joos Vandewalle, *Least squares support vector machine classifiers*, Neural processing letters **9** (1999), no. 3, 293–300.
1791. Laura P. Swiler, Thomas L. Paez, and Randall L. Mayes, *Epistemic uncertainty quantification tutorial*, Proceedings of the 27th International Modal Analysis Conference (IMAC-XXVII), February 2009.

1792. Gabor Szucs and Gyula Sallai, *Route planning with uncertain information using Dempster–Shafer theory*, Proceedings of the International Conference on Management and Service Science (MASS '09), September 2009, pp. 1–4.
1793. Mahdi Tabassian, Reza Ghaderi, and Reza Ebrahimpour, *Combination of multiple diverse classifiers using belief functions for handling data with imperfect labels*, Expert Systems with Applications **39** (2012), no. 2, 1698–1707.
1794. ———, *Combining complementary information sources in the Dempster–Shafer framework for solving classification problems with imperfect labels*, Knowledge-Based Systems **27** (2012), 92–102.
1795. Jean-Marc Tacnet and Jean Dezert, *Cautious OWA and evidential reasoning for decision making under uncertainty*, Proceedings of the 14th International Conference on Information Fusion (FUSION 2011), July 2011, pp. 1–8.
1796. Hideo Tanaka and Hisao Ishibuchi, *Evidence theory of exponential possibility distributions*, International Journal of Approximate Reasoning **8** (1993), no. 2, 123–140.
1797. Hideo Tanaka, Kazutomi Sugihara, and Yutaka Maeda, *Non-additive measures by interval probability functions*, Information Sciences **164** (August 2004), no. 1-4, 209–227.
1798. K. Tanaka and George J. Klir, *A design condition for incorporating human judgement into monitoring systems*, Reliability Engineering and System Safety **65** (1999), no. 3, 251–258.
1799. Wenhu Tang, K. Spurgeon, Q. H. Wu, and Z. J. Richardson, *An evidential reasoning approach to transformer condition assessments*, IEEE Transactions on Power Delivery **19** (2004), no. 4, 1696–1703.
1800. Yongchuan Tang and Jiacheng Zheng, *Dempster conditioning and conditional independence in evidence theory*, Proceedings of the Australasian Joint Conference on Artificial Intelligence, 2005, pp. 822–825.
1801. Majid H. Tangestani and Farid Moore, *The use of Dempster–Shafer model and GIS in integration of geoscientific data for porphyry copper potential mapping, north of Shahr-e-Babak, Iran*, International Journal of Applied Earth Observation and Geoinformation **4** (2002), no. 1, 65–74.
1802. Albena Tchamova and Jean Dezert, *On the behavior of Dempster's rule of combination and the foundations of Dempster–Shafer theory*, Proceedings of the 6th IEEE International Conference on Intelligent Systems, September 2012, pp. 108–113.
1803. Oswald Teichmuller, *p-algebren*, Deutsche Math. **1** (1936), 362–388.
1804. Paul Teller, *Conditionalization and observation*, Synthese **26** (1973), no. 2, 218–258.
1805. Abdelkader Telmoudi and Salem Chakhar, *Data fusion application from evidential databases as a support for decision making*, Information and Software Technology **46** (2004), no. 8, 547–555.
1806. Pedro Terán, *Counterexamples to a Central Limit Theorem and a Weak Law of Large Numbers for capacities*, Statistics and Probability Letters **96** (2015), 185–189.
1807. ———, *Laws of large numbers without additivity*, Transactions of the American Mathematical Society **366** (October 2014), no. 10, 5431–5451.
1808. Bjørnar Tessem, *Interval probability propagation*, International Journal of Approximate Reasoning **7** (1992), no. 3-4, 95–120.
1809. ———, *Interval probability propagation*, International Journal of Approximate Reasoning **7** (1992), no. 3, 95–120.
1810. ———, *Approximations for efficient computation in the theory of evidence*, Artificial Intelligence **61** (1993), no. 2, 315–329.

1811. Christian Thiel, Friedhelm Schwenker, and Günther Palm, *Using Dempster–Shafer theory in MCF systems to reject samples*, Multiple Classifier Systems (Nikunj C. Oza, Robi Polikar, Josef Kittler, and Fabio Roli, eds.), Lecture Notes in Computer Science, vol. 3541, Springer, Berlin Heidelberg, 2005, pp. 118–127.
1812. H. Mathis Thoma, *Belief function computations*, Conditional Logic in Expert Systems (Irwin R. Goodman, ed.), North Holland, 1991, pp. 269–308.
1813. Hans Mathis Thoma, *Factorization of belief functions*, Phd dissertation, Harvard University, Cambridge, MA, USA, 1989.
1814. Omoju Thomas and David J. Russomanno, *Applying the semantic Web expert system shell to sensor fusion using Dempster–Shafer theory*, Proceedings of the Thirty-Seventh Southeastern Symposium on System Theory (SSST '05), March 2005, pp. 11–14.
1815. Stelios C. Thomopoulos, *Theories in distributed decision fusion: comparison and generalization*, Proceedings of SPIE - Sensor Fusion III: 3D Perception and Recognition, vol. 1383, 1991, pp. 623–634.
1816. Terence R. Thompson, *Parallel formulation of evidential-reasoning theories*, Proceedings of the 9th International Joint Conference on Artificial Intelligence (IJCAI'85), vol. 1, Morgan Kaufmann, 1985, pp. 321–327.
1817. Sebastian Thrun, Wolfgang Burgard, and Dieter Fox, *A probabilistic approach to concurrent mapping and localization for mobile robots*, Autonomous Robots **5** (1998), no. 3-4, 253–271.
1818. Michael E. Tipping, *Sparse Bayesian learning and the relevance vector machine*, Journal of Machine Learning Research **1** (2001), 211–244.
1819. Arun P. Tirumalai, Brian G. Schunck, and Ramesh C. Jain, *Evidential reasoning for building environment maps*, IEEE Transactions on Systems, Man and Cybernetics **25** (1995), no. 1, 10–20.
1820. Bruce E. Tonn, *An algorithmic approach to combining belief functions*, International Journal of Intelligent Systems **11** (1996), no. 7, 463–476.
1821. Vicenç Torra, *A new combination function in evidence theory*, International Journal of Intelligent Systems **10** (1995), no. 12, 1021–1033.
1822. Asma Trabelsi, Zied Elouedi, and Eric Lefévre, *Handling uncertain attribute values in decision tree classifier using the belief function theory*, Artificial Intelligence: Methodology, Systems, and Applications: 17th International Conference, AIMSA 2016, Varna, Bulgaria, September 7-10, 2016, Proceedings (Christo Dichev and Gennady Agre, eds.), Springer International Publishing, 2016, pp. 26–35.
1823. Salsabil Trabelsi, Zied Elouedi, and Pawan Lingras, *Belief rough set classifier*, Advances in Artificial Intelligence: 22nd Canadian Conference on Artificial Intelligence, Canadian AI 2009 Kelowna, Canada, May 25-27, 2009 Proceedings (Yong Gao and Nathalie Japkowicz, eds.), Springer, Berlin Heidelberg, 2009, pp. 257–261.
1824. Salsabil Trabelsi, Zied Elouedi, and Pawan Lingras, *Classification systems based on rough sets under the belief function framework*, International Journal of Approximate Reasoning **52** (2011), no. 9, 1409–1432.
1825. Salsabil Trabelsi, Zied Elouedi, and Khaled Mellouli, *Pruning belief decision tree methods in averaging and conjunctive approaches*, International Journal of Approximate Reasoning **46** (2007), no. 3, 568–595.
1826. David Tritchler and Gina Lockwood, *Modelling the reliability of paired comparisons*, Journal of Mathematical Psychology **35** (1991), no. 3, 277–293.
1827. Matthias Troffaes, *Decision making under uncertainty using imprecise probabilities*, International Journal of Approximate Reasoning **45** (2007), no. 1, 17–29.

1828. C. Andy Tsao, *A note on Lindley's paradox*, TEST **15** (2006), no. 1, 125–139.
1829. Elena Tsiporkova, Bernard De Baets, and Veselka Boeva, *Dempster's rule of conditioning traslated into modal logic*, Fuzzy Sets and Systems **102** (1999), no. 3, 317–383.
1830. ______, *Evidence theory in multivalued models of modal logic*, Journal of Applied Non-Classical Logics **10** (2000), no. 1, 55–81.
1831. Elena Tsiporkova, Veselka Boeva, and Bernard De Baets, *Dempster–Shafer theory framed in modal logic*, International Journal of Approximate Reasoning **21** (1999), no. 2, 157–175.
1832. William T. Tutte, *Matroids and graphs*, Transactions of the American Mathematical Society **90** (1959), no. 3, 527–552.
1833. Simukai W. Utete, Billur Barshan, and Birsel Ayrulu, *Voting as validation in robot programming*, International Journal of Robotics Research **18** (1999), no. 4, 401–413.
1834. Lev V. Utkin and Thomas Augustin, *Decision making under incomplete data using the imprecise Dirichlet model*, International Journal of Approximate Reasoning **44** (2007), no. 3, 322–338.
1835. Mohammad A. T. Vakili, *Approximation of hints*, Technical report 209, Institute for Automation and Operation Research, University of Fribourg, 1993.
1836. Fabio Valente, *Multi-stream speech recognition based on Dempster–Shafer combination rule*, Speech Communication **52** (2010), no. 3, 213–222.
1837. Fabio Valente and Hynek Hermansky, *Combination of acoustic classifiers based on Dempster–Shafer theory of evidence*, Proceedings of the International Conference on Acoustics, Speech, and Signal Processing (ICASSP 2007), vol. 4, April 2007, pp. 1129–1132.
1838. Pierre Valin and David Boily, *Truncated Dempster–Shafer optimization and benchmarking*, Proceedings of SPIE - Sensor Fusion: Architectures, Algorithms, and Applications IV, vol. 4051, 2000, pp. 237–246.
1839. Pierre Valin, Pascal Djiknavorian, and Eloi Bossé, *A pragmatic approach for the use of Dempster–Shafer theory in fusing realistic sensor data.*, Journal of Advances in Information Fusion **5** (2010), no. 1, 32–40.
1840. Carine Van den Acker, *Belief function representation of statistical audit evidence*, International Journal of Intelligent Systems **15** (2000), no. 4, 277–290.
1841. A. W. van der Vaart, *10.2 bernstein–von mises theorem*, Asymptotic Statistics, Cambridge University Press, 1998.
1842. Kevin S. Van Horn, *Constructing a logic of plausible inference: a guide to Coxs theorem*, International Journal of Approximate Reasoning **34** (2003), no. 1, 3–24.
1843. Jean-Marc Vannobel, *Continuous belief functions: singletons plausibility function in conjunctive and disjunctive combination operations of consonant bbds*, Proceedings of the Workshop on the Theory of Belief Functions, 2010.
1844. Jean-Marc Vannobel, *Continuous belief functions: Focal intervals properties*, Belief Functions: Theory and Applications: Proceedings of the 2nd International Conference on Belief Functions, Compiègne, France 9-11 May 2012 (Thierry Denœux and Marie-Hélène Masson, eds.), Springer, Berlin Heidelberg, 2012, pp. 93–100.
1845. Patrick Vannoorenberghe, *On aggregating belief decision trees*, Information Fusion **5** (2004), no. 3, 179–188.
1846. Patrick Vannoorenberghe and Thierry Denœux, *Likelihood-based vs. distance-based evidential classifiers*, Proceedings of the 10th IEEE International Conference on Fuzzy Systems, vol. 1, 2001, pp. 320–323.

1847. ______, *Handling uncertain labels in multiclass problems using belief decision trees*, Proceedings of the Ninth International Conference on Information Processing and Management of Uncertainty in Knowledge-based Systems (IPMU 2002), vol. 3, 2002, pp. 1919–1926.
1848. Patrick Vannoorenberghe and Philippe Smets, *Partially supervised learning by a credal EM approach*, Symbolic and Quantitative Approaches to Reasoning with Uncertainty: 8th European Conference, ECSQARU 2005, Barcelona, Spain, July 6-8, 2005. Proceedings (Lluís Godo, ed.), Springer, Berlin Heidelberg, 2005, pp. 956–967.
1849. Vladimir N. Vapnik, *Statistical learning theory*, vol. 1, Wiley New York, 1998.
1850. ______, *An overview of statistical learning theory*, IEEE Transactions on Neural Networks **10** (1999), no. 5, 988–999.
1851. ______, *The nature of statistical learning theory*, Springer science & business media, 2013.
1852. Pascal Vasseur, Claude Pegard, El M. Mouaddib, and Laurent Delahoche, *Perceptual organization approach based on Dempster–Shafer theory*, Pattern Recognition **32** (1999), no. 8, 1449–1462.
1853. Jirina Vejnarova, *A few remarks on measures of uncertainty in Dempster–Shafer theory*, International Journal of General Systems **22** (1993), no. 2, 233–243.
1854. Jirina Vejnarova and George J. Klir, *Measure of strife in Dempster–shafer theory*, International Journal of General Systems **22** (1993), no. 1, 25–42.
1855. Roman Vershynin, *Lectures in geometric functional analysis*, 2011.
1856. Matti Vihola, *Random sets for multitarget tracking and data fusion*, PhD dissertation, Tampere University of Technology, Department of Information Technology, 2004.
1857. Ba-Ngu Vo, Sumeetpal S. Singh, and Arnaud Doucet, *Sequential Monte Carlo implementation of the PHD filter for multi-target tracking*, Proceedings of the Sixth International Conference on Information Fusion (FUSION 2003), vol. 2, 2003, pp. 792–799.
1858. John von Neumann and Oskar Morgenstern, *Theory of games and economic behavior*, Princeton University Press, 1944.
1859. F. Voorbraak, *A computationally efficient approximation of Dempster–Shafer theory*, International Journal on Man-Machine Studies **30** (1989), no. 5, 525–536.
1860. ______, *On the justification of Dempster's rule of combination*, Artificial Intelligence **48** (1991), no. 2, 171–197.
1861. Vladimir Vovk and Glenn Shafer, *Good randomized sequential probability forecasting is always possible*, Journal of the Royal Statistical Society: Series B (Statistical Methodology) **67** (2005), no. 5, 747–763.
1862. ______, *The game-theoretic capital asset pricing model*, International Journal of Approximate Reasoning **49** (2008), no. 1, 175–197.
1863. ______, *A game-theoretic explanation of the* $\sqrt{dt}$ *effect*, Working Paper 5, The Game-Theoretic Probability and Finance project, January 2003.
1864. Vladimir Vovk, Akimichi Takemura, and Glenn Shafer, *Defensive forecasting*, ArXiv preprint cs/0505083, 2005.
1865. Richard A. Wadsworth and Jane R. Hall, *Setting site specific critical loads: An approach using endorsement theory and Dempster–Shafer*, Water, Air, and Soil Pollution: Focus **7** (2007), no. 1-3, 399–405.
1866. Carl G. Wagner, *Consensus for belief functions and related uncertainty measures*, Theory and Decision **26** (1989), no. 3, 295–304.
1867. Peter P. Wakker, *Dempster-belief functions are based on the principle of complete ignorance*, Proceedings of the 1st International Sysmposium on Imprecise Probabilites and Their Applications (ISIPTA'99), June - July 1999, pp. 535–542.

1868. Peter P. Wakker, *Dempster belief functions are based on the principle of complete ignorance*, International Journal of Uncertainty, Fuzziness and Knowledge-Based Systems **8** (June 2000), no. 3, 271–284.
1869. Abraham Wald, *Statistical decision functions which minimize the maximum risk*, Annals of Mathematics **46** (1945), no. 2, 265–280.
1870. ______, *Statistical decision functions*, John Wiley and Sons, New York, 1950.
1871. Peter Walley, *Coherent lower (and upper) probabilities*, Tech. Report Statistics Research Report 22, University of Warwick, Coventry, UK, 1981.
1872. ______, *The elicitation and aggregation of beliefs*, Tech. Report Statistics Research Report 23, University of Warwick, Coventry, UK, 1982.
1873. ______, *Belief function representations of statistical evidence*, The Annals of Statistics **15** (1987), no. 4, 1439–1465.
1874. ______, *Statistical reasoning with imprecise probabilities*, Chapman and Hall, New York, 1991.
1875. ______, *Measures of uncertainty in expert systems*, Artificial Intelligence **83** (1996), no. 1, 1–58.
1876. ______, *Imprecise probabilities*, The Encyclopedia of Statistical Sciences (C. B. Read, D. L. Banks, and S. Kotz, eds.), Wiley, New York, NY, USA, 1997.
1877. ______, *Towards a unified theory of imprecise probability*, International Journal of Approximate Reasoning **24** (2000), no. 2-3, 125–148.
1878. Peter Walley and Terrence L. Fine, *Towards a frequentist theory of upper and lower probability*, The Annals of Statistics **10** (1982), no. 3, 741–761.
1879. Anton Wallner, *Maximal number of vertices of polytopes defined by f-probabilities*, Proceedings of the Fourth International Symposium on Imprecise Probabilities and Their Applications (ISIPTA 2005) (Fabio Gagliardi Cozman, R. Nau, and Teddy Seidenfeld, eds.), 2005, pp. 126–139.
1880. Nayer M. Wanas and Mohamed S. Kamel, *Decision fusion in neural network ensembles*, Proceedings of the International Joint Conference on Neural Networks (IJCNN '01), vol. 4, 2001, pp. 2952–2957.
1881. Chua-Chin Wang and Hon-Son Don, *Evidential reasoning using neural networks*, Proceedings of the IEEE International Joint Conference on Neural Networks (IJCNN'91), 1991, pp. 497–502.
1882. ______, *Evidential reasoning using neural networks*, Proceedings of the 1991 IEEE International Joint Conference on Neural Networks (IJCNN'91), vol. 1, November 1991, pp. 497–502.
1883. ______, *A geometrical approach to evidential reasoning*, Proceedings of the IEEE International Conference on Systems, Man, and Cybernetics (SMC'91), vol. 3, 1991, pp. 1847–1852.
1884. ______, *The majority theorem of centralized multiple BAMs networks*, Information Sciences **110** (1998), no. 3-4, 179–193.
1885. ______, *A robust continuous model for evidential reasoning*, Journal of Intelligent and Robotic Systems: Theory and Applications **10** (June 1994), no. 2, 147–171.
1886. ______, *A polar model for evidential reasoning*, Information Sciences **77** (March 1994), no. 3–4, 195–226.
1887. ______, *A continuous belief function model for evidential reasoning*, Proceedings of the Ninth Biennial Conference of the Canadian Society for Computational Studies of Intelligence (J. Glasgow and R. F. Hadley, eds.), May 1992, pp. 113–120.
1888. ______, *Evidential reasoning using neural networks*, Proceedings of the IEEE International Joint Conference on Neural Networks (IJCNN'91), vol. 1, November 1991, pp. 497–502.

1889. Jin Wang, Jian-Bo Yang, and Pratyush Sen, *Multi-person and multi-attribute design evaluations using evidential reasoning based on subjective safety and cost analyses*, Reliability Engineering and System Safety **52** (1996), no. 2, 113–128.

1890. P. Wang, N. Propes, Noppadon Khiripet, Y. Li, and George Vachtsevanos, *An integrated approach to machine fault diagnosis*, Proceedings of the IEEE Annual Textile, Fiber and Film Industry Technical Conference, 1999, pp. 7–pp.

1891. Pei Wang, *A defect in Dempster–Shafer theory*, Uncertainty Proceedings, Elsevier, 1994, pp. 560–566.

1892. ______, *Heuristics and normative models of judgment under uncertainty*, International Journal of Approximate Reasoning **14** (1996), no. 4, 221–235.

1893. ______, *The limitation of Bayesianism*, Artificial Intelligence **158** (2004), no. 1, 97–106.

1894. ______, *The limitation of Bayesianism*, Artificial Intelligence **158** (2004), no. 1, 97–106.

1895. Ping Wang, *The reliable combination rule of evidence in Dempster–Shafer theory*, Proceedings of the 2008 Congress on Image and Signal Processing (CISP '08), vol. 2, May 2008, pp. 166–170.

1896. ______, *The reliable combination rule of evidence in Dempster–Shafer theory*, Proceedings of the Congress on Image and Signal Processing (CISP '08), vol. 2, May 2008, pp. 166–170.

1897. Shijie Wang and Marco Valtorta, *On the exponential growth rate of Dempster–Shafer belief functions*, Proceedings of the SPIE - Applications of Artificial Intelligence X: Knowledge-Based Systems (Gautam Biswas, ed.), vol. 1707, April 1992, pp. 15–24.

1898. Weize Wang and Xinwang Liu, *Intuitionistic fuzzy information aggregation using Einstein operations*, IEEE Transactions on Fuzzy Systems **20** (2012), no. 5, 923–938.

1899. Ying-Ming Wang and Taha M. S. Elhag, *A comparison of neural network, evidential reasoning and multiple regression analysis in modelling bridge risks*, Expert Systems with Applications **32** (2007), no. 2, 336–348.

1900. Ying-Ming Wang and Taha M. S. Elhag, *Evidential reasoning approach for bridge condition assessment*, Expert Systems with Applications **34** (2008), no. 1, 689–699.

1901. Ying-Ming Wang, Jian-Bo Yang, and Dong-Ling Xu, *Environmental impact assessment using the evidential reasoning approach*, European Journal of Operational Research **174** (2006), no. 3, 1885–1913.

1902. Ying-Ming Wang, Jian-Bo Yang, Dong-Ling Xu, and Kwai-Sang Chin, *The evidential reasoning approach for multiple attribute decision analysis using interval belief degrees*, European Journal of Operational Research **175** (2006), no. 1, 35–66.

1903. ______, *On the combination and normalization of interval-valued belief structures*, Information Sciences **177** (2007), no. 5, 1230–1247, Including: The 3rd International Workshop on Computational Intelligence in Economics and Finance (CIEF2003).

1904. ______, *Consumer preference prediction by using a hybrid evidential reasoning and belief rule-based methodology*, Expert Systems with Applications **36** (2009), no. 4, 8421–8430.

1905. Yong Wang, Huihua Yang, Xingyu Wang, and Ruixia Zhang, *Distributed intrusion detection system based on data fusion method*, Proceedings of the Fifth World Congress on Intelligent Control and Automation (WCICA 2004), vol. 5, June 2004, pp. 4331–4334.

1906. Yonghong Wang and Munindar P. Singh, *Formal trust model for multiagent systems*, Proceedings of the 20th International Joint Conference on Artifical Intelligence (IJCAI'07), 2007, pp. 1551–1556.

1907. Yuan Wang, Yunhong Wang, and Tieniu Tan, *Combining fingerprint and voiceprint biometrics for identity verification: an experimental comparison*, Biometric Authentication (David Zhang and Anil K. Jain, eds.), Lecture Notes in Computer Science, vol. 3072, Springer, Berlin Heidelberg, 2004, pp. 663–670.
1908. Yuhong Wang and Yaoguo Dang, *Approach to interval numbers investment decision-making based on grey incidence coefficients and DS theory of evidence*, Systems Engineering - Theory and Practice **29** (2009), no. 11, 128–134.
1909. Zhenyuan Wang, *Semi-lattice structure of all extensions of possibility measure and consonant belief function*, Fuzzy Mathematics in Earthquake Research (D. Feng and X. Liu, eds.), Seismological Press, Beijing, 1985, pp. 332–336.
1910. Zhenyuan Wang and George J. Klir, *Fuzzy measure theory*, New York: Plenum Press, 1992.
1911. ———, *Choquet integrals and natural extensions of lower probabilities*, International Journal of Approximate Reasoning **16** (1997), no. 2, 137–147.
1912. Zhenyuan Wang and Wei Wang, *Extension of lower probabilities and coherence of belief measures*, Advances in Intelligent Computing — IPMU '94: 5th International Conference on Information Processing and Management of Uncertainty in Knowledge-Based Systems Paris, France, July 4–8, 1994 Selected Papers (Bernadette Bouchon-Meunier, Ronald R. Yager, and Lotfi A. Zadeh, eds.), Springer, Berlin Heidelberg, 1995, pp. 62–69.
1913. Larry Wasserman, *Some applications of belief functions to statistical inference*, PhD dissertation, University of Toronto, 1987.
1914. Larry A. Wasserman, *Belief functions and statistical inference*, Canadian Journal of Statistics **18** (1990), no. 3, 183–196.
1915. ———, *Prior envelopes based on belief functions*, Annals of Statistics **18** (1990), no. 1, 454–464.
1916. ———, *Comments on Shafer's 'Perspectives on the theory and practice of belief functions'*, International Journal of Approximate Reasoning **6** (1992), no. 3, 367–375.
1917. Junzo Watada, Yuji Kubo, and K. Kuroda, *Logical approach to evidential reasoning under a hierarchical structure*, Proceedings of the International Conference on Data and Knowledge Systems for Manufacturing and Engineering, vol. 1, May 1994, pp. 285–290.
1918. Philippe Weber and Christophe Simon, *Dynamic evidential networks in system reliability analysis: A Dempster Shafer approach*, Proceedings of the 16th Mediterranean Conference on Control and Automation, June 2008, pp. 603–608.
1919. Hua wei Guo, Wen kang Shi, Yong Deng, and Zhi jun Chen, *Evidential conflict and its 3D strategy: discard, discover and disassemble?*, Systems Engineering and Electronics **6** (2007), 890–898.
1920. Hua wei Guo, Wen kang Shi, Qing kun Liu, and Yong Deng, *A new combination rule of evidence*, Journal of Shanghai Jiaotong University - Chinese Edition **40** (2006), no. 11, 1895–1900.
1921. Kurt Weichselberger, *Interval probability on finite sample spaces*, Robust Statistics, Data Analysis, and Computer Intensive Methods: In Honor of Peter Huber's 60th Birthday (Helmut Rieder, ed.), Springer, New York, 1996, pp. 391–409.
1922. ———, *The theory of interval-probability as a unifying concept for uncertainty*, International Journal of Approximate Reasoning **24** (2000), no. 2, 149–170.
1923. Kurt Weichselberger and Sigrid Pohlmann, *A methodology for uncertainty in knowledge-based systems*, Lecture Notes in Artificial Intelligence, vol. 419, Springer, Berlin, 1990.

1924. Thomas Weiler, *Approximation of belief functions*, International Journal of Uncertainty, Fuzziness and Knowledge-Based Systems **11** (2003), no. 6, 749–777.
1925. Chenglin Wen, Xiaobin Xu, and Zhiliang Li, *Research on unified description and extension of combination rules of evidence based on random set theory*, The Chinese Journal of Electronics **17** (2008), no. 2, 279–284.
1926. Lv Wenhong, *Decision-making rules based on belief interval with D-S evidence theory*, Proceedings of the Second International Conference of Fuzzy Information and Engineering (ICFIE) (Bing-Yuan Cao, ed.), Springer, Berlin, Heidelberg, 2007, pp. 619–627.
1927. Leonard P. Wesley, *Evidential knowledge-based computer vision*, Optical Engineering **25** (1986), no. 3, 363–379.
1928. ______, *Autonomous locative reasoning: An evidential approach*, Proceedings of the IEEE International Conference on Robotics and Automation (ICRA), 1993, pp. 700–707.
1929. Joe Whittaker, Simon Garside, and Karel Lindveld, *Tracking and predicting a network traffic process*, International Journal of Forecasting **13** (1997), no. 1, 51–61.
1930. Thanuka L. Wickramarathne, Kamal Premaratne, and Manohar N. Murthi, *Focal elements generated by the Dempster–Shafer theoretic conditionals: A complete characterization*, Proceedings of the 13th Conference on Information Fusion (FUSION 2010), July 2010, pp. 1–8.
1931. Mark J. Wierman, *Measuring conflict in evidence theory*, Proceedings of the Joint 9th IFSA World Congress and 20th NAFIPS International Conference, vol. 3, 2001, pp. 1741–1745.
1932. ______, *Dempster, Shafer, and aggregate uncertainty*, Proceedings of the IEEE Conference on Norbert Wiener in the 21st Century (21CW), 2014, pp. 1–7.
1933. Slawomir T. Wierzchon and Mieczyslaw A. Klopotek, *Modified component valuations in valuation based systems as a way to optimize query processing*, Journal of Intelligent Information Systems **9** (1997), no. 2, 157–180.
1934. Slawomir T. Wierzchon, A. Pacan, and Mieczyslaw A. Klopotek, *An object-oriented representation framework for hierarchical evidential reasoning*, Proceedings of the Fourth International Conference on Artificial Intelligence: Methodology, Systems, Applications (AIMSA '90) (P. Jorrand and V. Sgurev, eds.), North Holland, September 1990, pp. 239–248.
1935. Elwood Wilkins and Simon H. Lavington, *Belief functions and the possible worlds paradigm*, Journal of Logic and Computation **12** (2002), no. 3, 475–495.
1936. Graeme G. Wilkinson and Jacques Megier, *Evidential reasoning in a pixel classification hierarchy-a potential method for integrating image classifiers and expert system rules based on geographic context*, International Journal of Remote Sensing **11** (October 1990), no. 10, 1963–1968.
1937. P. M. Williams, *Discussion of Shafer's paper*, Journal of the Royal Statistical Society B **44** (1982), 322–352.
1938. Peter M. Williams, *On a new theory of epistemic probability*, British Journal for the Philosophy of Science **29** (1978), 375–387.
1939. Nic Wilson, *Rules, belief functions and default logic*, Proceedings of the Sixth Conference on Uncertainty in Artificial Intelligence (UAI1990), 1990, pp. 443–449.
1940. ______, *A Monte-Carlo algorithm for Dempster–Shafer belief*, Proceedings of the Seventh Conference on Uncertainty in Artificial Intelligence (UAI'91), Morgan Kaufmann, 1991, pp. 414–417.
1941. ______, *The combination of belief: when and how fast?*, International Journal of Approximate Reasoning **6** (1992), no. 3, 377–388.

1942. ______, *How much do you believe?*, International Journal of Approximate Reasoning **6** (1992), no. 3, 345–365.
1943. ______, *The assumptions behind Dempster's rule*, Uncertainty in Artificial Intelligence, Morgan Kaufmann, 1993, pp. 527–534.
1944. ______, *Decision-making with belief functions and pignistic probabilities*, Proceedings of the European Conference on Symbolic and Quantitative Approaches to Reasoning and Uncertainty (ECSQARU '93), Springer-Verlag, 1993, pp. 364–371.
1945. Nic Wilson, *Default logic and Dempster–Shafer theory*, Symbolic and Quantitative Approaches to Reasoning and Uncertainty (Michael Clarke, Rudolf Kruse, and Serafín Moral, eds.), Lecture Notes in Computer Science, vol. 747, Springer, Berlin Heidelberg, 1993, pp. 372–379.
1946. ______, *Algorithms for Dempster–Shafer theory*, Handbook of Defeasible Reasoning and Uncertainty Management Systems (Jürg Kohlas and Serafín Moral, eds.), vol. 5, Springer Netherlands, 2000, pp. 421–475.
1947. Nic Wilson, *Algorithms for dempster-shafer theory*, Algorithms for Uncertainty and Defeasible Reasoning (Dov M. Gabbay and Philippe Smets, eds.), Handbook of Defeasible Reasoning and Uncertainty Management Systems, Springer, 2001, pp. 421–476.
1948. ______, *Justification, computational efficiency and generalisation of the Dempster–Shafer theory*, Research Report 15, Department of Computing and Mathematical Sciences, Oxford Polytechnic, June 1989.
1949. ______, *The representation of prior knowledge in a Dempster–Shafer approach*, Proceedings of the DRUMS workshop on Integration of Uncertainty Formalisms, June 1991.
1950. Nic Wilson and Serafín Moral, *Fast Markov chain algorithms for calculating Dempster–Shafer belief*, Proceedings of the 12th European Conference on Artificial Intelligence (ECAI'96), John Wiley and Sons, August 1996, pp. 672–676.
1951. Peter Nicholas Wilson, *Some theoretical aspects of the Dempster–Shafer theory*, PhD dissertation, Oxford Polytechnic, 1992.
1952. S. K. Michael Wong and Pawan Lingas, *Generation of belief functions from qualitative preference relations*, Proceedings of the Third International Conference on Information Processing and Management of Uncertainty in Knowledge-Based Systems (IPMU'90), 1990, pp. 427–429.
1953. S. K. Michael Wong and Pawan Lingras, *Representation of qualitative user preference by quantitative belief functions*, IEEE Transactions on Knowledge and Data Engineering **6** (February 1994), no. 1, 72–78.
1954. S. K. Michael Wong, Pawan Lingras, and Yiyu (Y. Y.) Yao, *Propagation of preference relations in qualitative inference networks*, Proceedings of the 12th International Joint Conference on Artificial Intelligence (IJCAI'91), vol. 2, Morgan Kaufmann, 1991, pp. 1204–1209.
1955. S. K. Michael Wong, Lusheng Wang, and Yiyu (Y. Y.) Yao, *Interval structure: A framework for representing uncertain information*, Proceedings of the Eighth International Conference on Uncertainty in Artificial Intelligence (UAI'92), Morgan Kaufmann, 1992, pp. 336–343.
1956. ______, *Nonnumeric belief structures*, Proceedings of the Fourth International Conference on Computing and Information (ICCI '92), 1992, pp. 274–275.
1957. S. K. Michael Wong, Yiyu (Y. Y.) Yao, and Peter Bollmann, *Characterization of comparative belief structures*, International Journal of Man-Machine Studies **37** (1992), no. 1, 123–133.

1958. S. K. Michael Wong, Yiyu (Y. Y.) Yao, Peter Bollmann, and H. C. Burger, *Axiomatization of qualitative belief structure*, IEEE Transactions on Systems, Man, and Cybernetics **21** (1990), no. 4, 726–734.
1959. Chong Wu and David Barnes, *Formulating partner selection criteria for agile supply chains: A Dempster–Shafer belief acceptability optimisation approach*, International Journal of Production Economics **125** (2010), no. 2, 284–293.
1960. Fa-Yueh Wu, *The Potts model*, Reviews of Modern Physics **54** (January 1982), no. 1, 235–268.
1961. Huadong Wu, *Sensor data fusion for context-aware computing using Dempster–Shafer theory*, PhD dissertation, Carnegie Mellon University, 2003.
1962. Huadong Wu, Mel Siegel, and Sevim Ablay, *Sensor fusion for context understanding*, Proceedings of the 19th IEEE Instrumentation and Measurement Technology Conference (IMTC/2002), vol. 1, 2002, pp. 13–17.
1963. Huadong Wu, Mel Siegel, Rainer Stiefelhagen, and Jie Yang, *Sensor fusion using Dempster–Shafer theory for context-aware HCI*, Proceedings of the 19th IEEE Instrumentation and Measurement Technology Conference (IMTC/2002), vol. 1, 2002, pp. 7–12.
1964. Wei-Zhi Wu, *Attribute reduction based on evidence theory in incomplete decision systems*, Information Sciences **178** (2008), no. 5, 1355–1371.
1965. Wei-Zhi Wu, Yee Leung, and Ju-Sheng Mi, *On generalized fuzzy belief functions in infinite spaces*, IEEE Transactions on Fuzzy Systems **17** (2009), no. 2, 385–397.
1966. Wei-Zhi Wu, Yee Leung, and Wen-Xiu Zhang, *Connections between rough set theory and Dempster–Shafer theory of evidence*, International Journal of General Systems **31** (2002), no. 4, 405–430.
1967. Wei-Zhi Wu and Ju-Sheng Mi, *Knowledge reduction in incomplete information systems based on Dempster–Shafer theory of evidence*, Rough Sets and Knowledge Technology: First International Conference, RSKT 2006, Chongquing, China, July 24-26, 2006. Proceedings (Guo-Ying Wang, James F. Peters, Andrzej Skowron, and Yiyu (Y. Y.) Yao, eds.), Springer, Berlin Heidelberg, 2006, pp. 254–261.
1968. Wei-Zhi Wu, Mei Zhang, Huai-Zu Li, and Ju-Sheng Mi, *Knowledge reduction in random information systems via Dempster–Shafer theory of evidence*, Information Sciences **174** (2005), no. 3-4, 143–164.
1969. Xiaoping Wu, Qing Ye, and Lingyan Liu, *Dempster–Shafer theory of evidence based on improved BP and its application*, Journal of Wuhan University of Technology **29** (2007), 158–161.
1970. Yong-Ge Wu, Jing-Yu Yang, Ke-Liu, and Lei-Jian Liu, *On the evidence inference theory*, Information Sciences **89** (1996), no. 3, 245–260.
1971. Youfeng Wu and James R. Larus, *Static branch frequency and program profile analysis*, Proceedings of the 27th Annual International Symposium on Microarchitecture (MICRO 27), 1994, pp. 1–11.
1972. Yan Xia, S. Sitharama Iyengar, and Nathan E. Brener, *An event driven integration reasoning scheme for handling dynamic threats in an unstructured environment*, Artificial Intelligence **95** (1997), no. 1, 169–186.
1973. Zhi Xiao, Xianglei Yang, Ying Pang, and Xin Dang, *The prediction for listed companies' financial distress by using multiple prediction methods with rough set and Dempster–Shafer evidence theory*, Knowledge-Based Systems **26** (2012), no. 0, 196–206.
1974. Yan Ming Xiong and Zhan Ping Yang, *A novel combination method of conflict evidence in multi-sensor target recognition*, Advanced Materials Research, vol. 143, Trans Tech Publications, 2011, pp. 920–924.

1975. D. L. Xu and J. B. Yang, *Intelligent decision system based on the evidential reasoning approach and its applications*, Journal of Telecommunications and Information Technology **3** (2005), 73–80.
1976. Dong-Ling Xu, *An introduction and survey of the evidential reasoning approach for multiple criteria decision analysis*, Annals of Operations Research **195** (2012), no. 1, 163–187.
1977. Dong-Ling Xu and Jian-Bo Yang, *Intelligent decision system for self-assessment*, Journal of Multi-Criteria Decision Analysis **12** (2003), no. 1, 43–60.
1978. Dong-Ling Xu and Jian-Bo Yang, *Intelligent decision system based on the evidential reasoning approach and its applications*, Journal of Telecommunications and Information Technology **3** (2005), 73–80.
1979. Dong-Ling Xu, Jian-Bo Yang, and Ying-Ming Wang, *The evidential reasoning approach for multi-attribute decision analysis under interval uncertainty*, European Journal of Operational Research **174** (2006), no. 3, 1914–1943.
1980. Guangquan Xu, Zhiyong Feng, Huabei Wu, and Dexin Zhao, *Swift trust in a virtual temporary system: A model based on the Dempster–Shafer theory of belief functions*, International Journal of Electronic Commerce **12** (2007), no. 1, 93–126.
1981. Guoping Xu, Weifeng Tian, Li Qian, and Xiangfen Zhang, *A novel conflict reassignment method based on grey relational analysis (gra)*, Pattern Recognition Letters **28** (2007), no. 15, 2080–2087.
1982. Hong Xu, *An efficient implementation of the belief function propagation*, Proceedings of the 7th International Conference on Uncertainty in Artificial Intelligence (UAI'91) (B. D. DÁmbrosio, Ph. Smets, and P. P. Bonissone, eds.), Morgan Kaufmann, 1991, pp. 425–432.
1983. ———, *A decision calculus for belief functions in valuation-based systems*, Proceedings of the 8th International Conference on Uncertainty in Artificial Intelligence (UAI'92) (D. Dubois, M. P. Wellman, B. D'Ambrosio, and Ph. Smets, eds.), 1992, pp. 352–359.
1984. ———, *An efficient tool for reasoning with belief functions*, Proceedings of the 4th International Conference on Information Proceeding and Management of Uncertainty in Knowledge-Based Systems (IPMU'92), 1992, pp. 65–68.
1985. ———, *An efficient tool for reasoning with belief functions uncertainty in intelligent systems*, Advances in the Dempster-Shafer Theory of Evidence (B. Bouchon-Meunier, L. Valverde, and R. R. Yager, eds.), North-Holland: Elsevier Science, 1993, pp. 215–224.
1986. ———, *Computing marginals from the marginal representation in Markov trees*, Proceedings of the 5th International Conference on Information Proceeding and Management of Uncertainty in Knowledge-Based Systems (IPMU'94), Springer, 1994, pp. 108–116.
1987. ———, *Computing marginals for arbitrary subsets from marginal representation in Markov trees*, Artificial Intelligence **74** (1995), no. 1, 177–189.
1988. ———, *Valuation-based systems for decision analysis using belief functions*, Decision Support Systems **20** (1997), no. 2, 165–184.
1989. ———, *A decision calculus for belief functions in valuation-based systems*, ArXiv preprint arXiv:1303.5439, 2013.
1990. Hong Xu, Yen-Teh Hsia, and Philippe Smets, *A belief-function based decision support system*, Proceedings of the 9th International Conference on Uncertainty in Artificial Intelligence (UAI'93) (D. Heckerman and A Mamdani, eds.), 1993, pp. 535–542.
1991. ———, *Transferable belief model for decision making in valuation based systems*, IEEE Transactions on Systems, Man, and Cybernetics **26** (1996), no. 6, 698–707.

1992. Hong Xu and Robert Kennes, *Steps towards an efficient implementation of Dempster–Shafer theory*, Advances in the Dempster-Shafer Theory of Evidence (Ronald R. Yager, Mario Fedrizzi, and Janusz Kacprzyk, eds.), John Wiley and Sons, Inc., 1994, pp. 153–174.
1993. Hong Xu and Philippe Smets, *Evidential reasoning with conditional belief functions*, Proceedings of the 10th International Conference on Uncertainty in Artificial Intelligence (UAI'94) (R. Lopez de Mantaras and D. Poole, eds.), 1994, pp. 598–605.
1994. ______, *Generating explanations for evidential reasoning*, Proceedings of the 11th International Conference on Uncertainty in Artificial Intelligence (UAI'95) (Ph. Besnard and S. Hanks, eds.), 1995, pp. 574–581.
1995. ______, *Reasoning in evidential networks with conditional belief functions*, International Journal of Approximate Reasoning **14** (1996), no. 2-3, 155–185.
1996. ______, *Some strategies for explanations in evidential reasoning*, IEEE Transactions on Systems, Man and Cybernetics **26** (1996), no. 5, 599–607.
1997. Lei Xu, Adam Krzyzak, and Ching Y. Suen, *Methods of combining multiple classifiers and their applications to handwriting recognition*, IEEE Transactions on Systems, Man, and Cybernetics **22** (1992), no. 3, 418–435.
1998. Lin Xu, Frank Hutter, Holger H. Hoos, and Kevin Leyton-Brown, *SATzilla: Portfolio-based Algorithm Selection for SAT*, Journal of Artificial Intelligence Research **32** (June 2008), no. 1, 565–606.
1999. Ronald R. Yager, *On a general class of fuzzy connectives*, Fuzzy Sets and Systems **4** (1980), no. 3, 235–242.
2000. ______, *Generalized probabilities of fuzzy events from fuzzy belief structures*, Information Sciences **28** (1982), no. 1, 45–62.
2001. ______, *Entropy and specificity in a mathematical theory of evidence*, International Journal of General Systems **9** (1983), no. 4, 249–260.
2002. ______, *Hedging in the combination of evidence*, Journal of Information and Optimization Sciences **4** (1983), no. 1, 73–81.
2003. ______, *On the relationship of methods of aggregating evidence in expert systems*, Cybernetics and Systems **16** (1985), no. 1, 1–21.
2004. Ronald R. Yager, *Reasoning with uncertainty for expert systems*, Proceedings of the 9th International Joint Conference on Artificial Intelligence (IJCAI'85), vol. 2, 1985, pp. 1295–1297.
2005. Ronald R. Yager, *Arithmetic and other operations on Dempster–Shafer structures*, International Journal of Man-Machine Studies **25** (1986), no. 4, 357–366.
2006. ______, *The entailment principle for Dempster–Shafer granules*, International Journal of Intelligent Systems **1** (1986), no. 4, 247–262.
2007. Ronald R. Yager, *Toward a general theory of reasoning with uncertainty. I: Nonspecificity and fuzziness*, International Journal of Intelligent Systems **1** (1986), no. 1, 45–67.
2008. Ronald R. Yager, *Toward a general theory of reasoning with uncertainty. Part II: Probability*, International Journal of Man-Machine Studies **25** (1986), no. 6, 613–631.
2009. ______, *On the Dempster–Shafer framework and new combination rules*, Information Sciences **41** (1987), no. 2, 93–138.
2010. ______, *Quasiassociative operations in the combination of evidence*, Kybernetes **16** (1987), no. 1, 37–41.
2011. ______, *Non-monotonic compatibility relations in the theory of evidence*, Machine Learning and Uncertain Reasoning (B. R. Gaines and J. H. Boose, eds.), Academic Press Ltd., 1990, pp. 291–311.

2012. Ronald R. Yager, *Credibility discounting in the theory of approximate reasoning*, Proceedings of the Sixth Annual Conference on Uncertainty in Artificial Intelligence (UAI'90), Elsevier Science, 1991, pp. 299–310.
2013. Ronald R. Yager, *Decision making under Dempster–Shafer uncertainties*, Tech. Report MII-915, Machine Intelligence Institute, Iona College, 1991.
2014. ______, *Decision making under Dempster–Shafer uncertainties*, International Journal of General Systems **20** (1992), no. 3, 233–245.
2015. ______, *On considerations of credibility of evidence*, International Journal of Approximate Reasoning **7** (1992), no. 1, 45–72.
2016. ______, *On the aggregation of prioritized belief structures*, IEEE Transactions on Systems, Man, and Cybernetics - Part A: Systems and Humans **26** (1996), no. 6, 708–717.
2017. ______, *On the normalization of fuzzy belief structures*, International Journal of Approximate Reasoning **14** (1996), no. 2-3, 127–153.
2018. Ronald R. Yager, *A class of fuzzy measures generated from a Dempster–Shafer belief structure*, International Journal of Intelligent Systems **14** (1999), no. 12, 1239–1247.
2019. Ronald R. Yager, *Modeling uncertainty using partial information*, Information Sciences **121** (1999), no. 3-4, 271–294.
2020. ______, *Dempster–Shafer belief structures with interval valued focal weights*, International Journal of Intelligent Systems **16** (2001), no. 4, 497–512.
2021. ______, *Nonmonotonicity and compatibility relations in belief structures*, Annals of Mathematics and Artificial Intelligence **34** (2002), no. 1-3, 161–176.
2022. ______, *Decision making using minimization of regret*, International Journal of Approximate Reasoning **36** (2004), no. 2, 109–128.
2023. ______, *Aggregating non-independent Dempster–Shafer belief structures*, Proceedings of the 12th International Conference on Information Processing and Management of Uncertainty in Knowledge-Based Systems (IPMU 2008, 2008, pp. 289–297.
2024. ______, *Decision making under Dempster–Shafer uncertainties*, Classic Works of the Dempster–Shafer Theory of Belief Functions (Roland R. Yager and Liping Liu, eds.), Studies in Fuzziness and Soft Computing, vol. 219, Springer, Berlin Heidelberg, 2008, pp. 619–632.
2025. ______, *Human behavioral modeling using fuzzy and Dempster–Shafer theory*, Social Computing, Behavioral Modeling, and Prediction (Huan Liu, John J. Salerno., and Michael J. Young, eds.), Springer, 2008, pp. 89–99.
2026. ______, *Comparing approximate reasoning and probabilistic reasoning using the Dempster–Shafer framework*, International Journal of Approximate Reasoning **50** (2009), no. 5, 812–821.
2027. Ronald R. Yager, *Joint cumulative distribution functions for Dempster–Shafer belief structures using copulas*, Fuzzy Optimization and Decision Making **12** (2013), no. 4, 393–414.
2028. Ronald R. Yager, *Minimization of regret decision making with Dempster–Shafer uncertainty*, Proceedings of the IEEE International Conference on Fuzzy Systems, vol. 1, July 2004, pp. 511–515.
2029. ______, *Hierarchical aggregation functions generated from belief structures*, IEEE Transactions on Fuzzy Systems **8** (October 2000), no. 5, 481–490.
2030. ______, *Cumulative distribution functions from Dempster–Shafer belief structures*, IEEE Transactions on Systems, Man, and Cybernetics, Part B: Cybernetics **34** (October 2004), no. 5, 2080–2087.

2031. Ronald R. Yager and Dimitar P. Filev, *Including probabilistic uncertainty in fuzzy logic controller modeling using Dempster–Shafer theory*, IEEE Transactions on Systems, Man, and Cybernetics **25** (1995), no. 8, 1221–1230.
2032. Ronald R. Yager and Antoine Kelman, *Decision making under various types of uncertainties*, Journal of Intelligent and Fuzzy Systems **3** (July 1995), no. 4, 317–323.
2033. Ronald R. Yager and Liping Liu, *Classic works of the Dempster–Shafer theory of belief functions*, 1st ed., Springer Publishing Company, 2010.
2034. Koichi Yamada, *A new combination of evidence based on compromise*, Fuzzy Sets and Systems **159** (2008), no. 13, 1689–1708.
2035. Yang Yan, Jing Zhanrong, Gao Tian, and Wang Huilong, *Multi-sources information fusion algorithm in airborne detection systems*, Journal of Systems Engineering and Electronics **18** (2007), no. 1, 171–176.
2036. J. B. Yang and P. Sen, *Evidential reasoning based hierarchical analysis for design selection of ship retro-fit options*, Artificial Intelligence in Design 94 (John S. Gero and Fay Sudweeks, eds.), Springer Netherlands, 1994, pp. 327–344.
2037. Jang-Bo Yang, L. Xu, and Mingrong Deng, *Nonlinear Regression to Estimate Both Weights and Utilities Via Evidential Reasoning for MADM*, Proceedings of the 5th International Conference on Optimisation: Techniques and Applications, 2001, pp. 15–17.
2038. Jian-Bo Yang, *Rule and utility based evidential reasoning approach for multiattribute decision analysis under uncertainties*, European Journal of Operational Research **131** (2001), no. 1, 31–61.
2039. Jian-Bo Yang, Jun Liu, Jin Wang, How-Sing Sii, and Hong-Wei Wang, *Belief rule-base inference methodology using the evidential reasoning Approach – RIMER*, IEEE Transactions on Systems, Man and Cybernetics, Part A: Systems and Humans **36** (2006), no. 2, 266–285.
2040. ______, *Belief rule-base inference methodology using the evidential reasoning approach – RIMER*, IEEE Transactions on Systems, Man, and Cybernetics A: Systems and Humans **36** (2006), no. 2, 266–285.
2041. Jian-Bo Yang and Madan G. Singh, *An evidential reasoning approach for multiple-attribute decision making with uncertainty*, IEEE Transactions on Systems, Man, and Cybernetics **24** (January 1994), no. 1, 1–18.
2042. Jian-Bo Yang, Ying-Ming Wang, Dong-Ling Xu, and Kwai-Sang Chin, *The evidential reasoning approach for MADA under both probabilistic and fuzzy uncertainties*, European Journal of Operational Research **171** (2006), no. 1, 309–343.
2043. Jian-Bo Yang and Dong-Ling Xu, *On the evidential reasoning algorithm for multiple attribute decision analysis under uncertainty*, IEEE Transactions on Systems, Man, and Cybernetics - Part A: Systems and Humans **32** (2002), no. 3, 289–304.
2044. ______, *Evidential reasoning rule for evidence combination*, Artificial Intelligence **205** (2013), 1–29.
2045. ______, *Nonlinear information aggregation via evidential reasoning in multiattribute decision analysis under uncertainty*, IEEE Transactions on Systems, Man and Cybernetics, Part A: Systems and Humans **32** (May 2002), no. 3, 376–393.
2046. Jianping Yang, Hong-Zhong Huang, Qiang Miao, and Rui Sun, *A novel information fusion method based on Dempster–Shafer evidence theory for conflict resolution*, Intelligent Data Analysis **15** (2011), no. 3, 399–411.
2047. Miin-Shen Yang, Tsang-Chih Chen, and Kuo-Lung Wu, *Generalized belief function, plausibility function, and Dempster's combinational rule to fuzzy sets*, International Journal of Intelligent Systems **18** (2003), no. 8, 925–937.

2048. Shan-Lin Yang and Chao Fu, *Constructing confidence belief functions from one expert*, Expert Systems with Applications **36** (2009), no. 4, 8537–8548.
2049. Yanli Yang, Ali A. Minai, and Marios M. Polycarpou, *Evidential map-building approaches for multi-UAV cooperative search*, Proceedings of the 2005 American Control Conference, June 2005, pp. 116–121.
2050. łZai-Li Yang, Jiangping Wang, Steve Bonsall, and Quan-Gen Fang, *Use of fuzzy evidential reasoning in maritime security assessment*, Risk Analysis **29** (2009), no. 1, 95–120.
2051. Shuang Yao and Wei-Qiang Huang, *Induced ordered weighted evidential reasoning approach for multiple attribute decision analysis with uncertainty*, International Journal of Intelligent Systems **29** (2014), no. 10, 906–925.
2052. Yan-Qing Yao, Ju-Sheng Mi, and Zhou-Jun Li, *Attribute reduction based on generalized fuzzy evidence theory in fuzzy decision systems*, Fuzzy Sets and Systems **170** (2011), no. 1, 64–75.
2053. Yiyu (Y. Y.) Yao, *Two views of the theory of rough sets in finite universes*, International Journal of Approximate Reasoning **15** (1996), no. 4, 291–317.
2054. ———, *A comparative study of fuzzy sets and rough sets*, Information Sciences **109** (1998), no. 1-4, 227–242.
2055. Yiyu (Y. Y.) Yao, *Granular computing: basic issues and possible solutions*, Proceedings of the 5th Joint Conference on Information Sciences, 2000, pp. 186–189.
2056. Yiyu (Y. Y.) Yao, Churn-Jung Liau, and Ning Zhong, *Granular computing based on rough sets, quotient space theory, and belief functions*, Foundations of Intelligent Systems: 14th International Symposium, ISMIS 2003, Maebashi City, Japan, October 28-31, 2003. Proceedings (Ning Zhong, Zbigniew W. Raś, Shusaku Tsumoto, and Einoshin Suzuki, eds.), Springer, Berlin Heidelberg, 2003, pp. 152–159.
2057. Yiyu (Y. Y.) Yao and Pawan Lingras, *Interpretations of belief functions in the theory of rough sets*, Information Sciences **104** (1998), no. 1-2, 81–106.
2058. Yiyu (Y. Y.) Yao and Xin Wang, *Interval based uncertain reasoning using fuzzy and rough sets*, Advances in Machine Intelligence and Soft-Computing (Paul P. Wang, ed.), vol. IV, Duke University, 1997, pp. 196–215.
2059. Qing Ye, Xiao ping Wu, and Ye xin Song, *An evidence combination method of introducing weight factors*, Fire Control and Command Control **32** (2007), no. 6, 21–24.
2060. John Yen, *A reasoning model based on an extended Dempster–Shafer theory*, Proceedings of the Fifth AAAI National Conference on Artificial Intelligence (AAAI'86), 1986, pp. 125–131.
2061. John Yen, *Implementing evidential reasoning in expert systems*, Proceedings of the Third Conference on Uncertainty in Artificial Intelligence (UAI'87), 1987, pp. 180–189.
2062. ———, *Generalizing the Dempster–Shafer theory to fuzzy sets*, IEEE Transactions on Systems, Man, and Cybernetics **20** (1990), no. 3, 559–569.
2063. ———, *Computing generalized belief functions for continuous fuzzy sets*, International Journal of Approximate Reasoning **6** (1992), no. 1, 1–31.
2064. ———, *Generalizing the Dempster–Shafer theory to fuzzy sets*, Classic Works of the Dempster-Shafer Theory of Belief Functions (Roland R. Yager and Liping Liu, eds.), Studies in Fuzziness and Soft Computing, vol. 219, Springer, 2008, pp. 529–554.
2065. ———, *Can evidence be combined in the Dempster–Shafer theory*, Proceedings of the Third Conference on Uncertainty in Artificial Intelligence (UAI'87), 2013, pp. 70–76.
2066. ———, *Implementing evidential reasoning in expert systems*, ArXiv preprint arXiv:1304.2731, 2013.

2067. John Yen, *GERTIS: A Dempster–Shafer approach to diagnosing hierarchical hypotheses*, Communications of the ACM **32** (May 1989), no. 5, 573–585.
2068. Lu Yi, *Evidential reasoning in a multiple classifier system*, Proceedings of the Sixth International Conference on Industrial and Engineering Applications of Artificial Intelligence and Expert Systems (IEA/AIE 93) (P. W. H. Chung, G. Lovegrove, and M. Ali, eds.), June 1993, pp. 476–479.
2069. Zou Yi, Yeong K. Ho, Chin Seng Chua, and Xiao Wei Zhou, *Multi-ultrasonic sensor fusion for autonomous mobile robots*, Proceedings of SPIE - Sensor Fusion: Architectures, Algorithms, and Applications IV, vol. 4051, 2000, pp. 314–321.
2070. Zou Yi, Ho Yeong Khing, Chua Chin Seng, and Zhou Xiao Wei, *Multi-ultrasonic sensor fusion for mobile robots*, Proceedings of the IEEE Intelligent Vehicles Symposium, October 2000, pp. 387–391.
2071. Deng Yong, Shi WenKang, Zhu ZhenFu, and Liu Qi, *Combining belief functions based on distance of evidence*, Decision Support Systems **38** (2004), no. 3, 489–493.
2072. Chang yong Liang, Zeng ming Chen, Yong qing Huang, and Jian jun Tong, *A method of dispelling the absurdities of Dempster–Shafers rule of combination*, Systems Engineering-theory and Practice **25** (2005), no. 3, 7–12.
2073. Jia Yonghong and Li Deren, *Feature fusion based on Dempster–Shafer's evidential reasoning for image texture classification*, 2004.
2074. Virginia R. Young, *Families of update rules for non-additive measures: Applications in pricing risks*, Insurance: Mathematics and Economics **23** (1998), no. 1, 1–14.
2075. Virginia R. Young and Shaun S. Wang, *Updating non-additive measures with fuzzy information*, Fuzzy Sets and Systems **94** (1998), no. 3, 355–366.
2076. Bin Yu, Joseph Giampapa, Sean Owens, and Katia Sycara, *An evidential model of multisensor decision fusion for force aggregation and classification*, Proceedings of the 8th International Conference on Information Fusion (FUSION 2005), vol. 2, July 2005, p. 8.
2077. Bin Yu and Munindar P. Singh, *Distributed reputation management for electronic commerce*, Computational Intelligence **18** (2002), no. 4, 535–549.
2078. Bin Yu and Munindar P. Singh, *An evidential model of distributed reputation management*, Proceedings of First International Conference on Autonomous Agents and MAS, July 2002, pp. 294–301.
2079. Bo Yu and Guoray Cai, *A query-aware document ranking method for geographic information retrieval*, Proceedings of the 4th ACM Workshop on Geographical Information Retrieval (GIR '07), 2007, pp. 49–54.
2080. Chunhai Yu and Fahard Arasta, *On conditional belief functions*, International Journal of Approxiomate Reasoning **10** (1994), no. 2, 155–172.
2081. Dong Yu and Deborah Frincke, *Alert confidence fusion in intrusion detection systems with extended Dempster–Shafer theory*, Proceedings of the 43rd Annual Southeast Regional Conference, vol. 2, 2005, pp. 142–147.
2082. Cao Yujun, *The evidence aggregation method in the theory of evidence*, Journal Of Xi'an Jiaotong University **31** (1997), no. 6, 106–110.
2083. Lotfi A. Zadeh, *Fuzzy sets*, Information and Control **8** (1965), no. 3, 338–353.
2084. ______, *Fuzzy sets as a basis for a theory of possibility*, Fuzzy Sets and Systems **1** (1978), 3–28.
2085. ______, *On the validity of Dempster's rule of combination of evidence*, Tech. Report ERL M79/24, University of California, Berkeley, 1979.
2086. ______, *A mathematical theory of evidence (book review)*, AI Magazine **5** (1984), no. 3, 81–83.

2087. ———, *Syllogistic reasoning as a basis for combination of evidence in expert systems*, Proceedings of the 9th International Joint Conference on Artificial Intelligence (IJCAI'85), vol. 1, Morgan Kaufmann, 1985, pp. 417–419.
2088. ———, *Is probability theory sufficient for dealing with uncertainty in AI: A negative view*, Uncertainty in Artificial Intelligence (L. N. Kanal and J. F. Lemmer, eds.), vol. 2, North-Holland, Amsterdam, 1986, pp. 103–116.
2089. ———, *A simple view of the Dempster–Shafer theory of evidence and its implication for the rule of combination*, AI Magazine **7** (1986), no. 2, 85–90.
2090. ———, *Fuzzy sets and information granularity*, Fuzzy Sets, Fuzzy Logic, and Fuzzy Systems (George J. Klir and Bo Yuan, eds.), World Scientific Publishing, 1996, pp. 433–448.
2091. ———, *Toward a perception-based theory of probabilistic reasoning with imprecise probabilities*, Soft Methods in Probability, Statistics and Data Analysis (Przemysław Grzegorzewski, Olgierd Hryniewicz, and María Ángeles Gil, eds.), Physica-Verlag HD, Heidelberg, 2002, pp. 27–61.
2092. ———, *Toward a generalized theory of uncertainty (GTU) – an outline*, Information Sciences **172** (2005), no. 1-2, 1–40.
2093. ———, *Generalized theory of uncertainty (GTU) - principal concepts and ideas*, Computational Statistics and Data Analysis **51** (2006), no. 1, 15–46.
2094. Marco Zaffalon and Enrico Fagiuoli, *Tree-based credal networks for classification*, Reliable computing **9** (2003), no. 6, 487–509.
2095. Debra K. Zarley, *An evidential reasoning system*, Tech. Report 206, University of Kansas, 1988.
2096. Debra K. Zarley, Yen-Teh Hsia, and Glenn Shafer, *Evidential reasoning using DELIEF*, Proceedings of the Seventh National Conference on Artificial Intelligence (AAAI-88), vol. 1, 1988, pp. 205–209.
2097. Bernard P. Zeigler, *Some properties of modified Dempster–Shafer operators in rule based inference systems*, International Journal of General Systems **14** (1988), no. 4, 345–356.
2098. John Zeleznikow and James R. Nolan, *Using soft computing to build real world intelligent decision support systems in uncertain domains*, Decision Support Systems **31** (2001), no. 2, 263–285.
2099. Chen Zewang, Sun Yongrong, and Yuan Xing, *Development of an algorithm for car navigation system based on Dempster–Shafer evidence reasoning*, Proceedings of the IEEE 5th International Conference on Intelligent Transportation Systems, September 2002, pp. 534–537.
2100. Lian-Yin Zhai, Li-Pheng Khoo, and Sai-Cheong Fok, *Feature extraction using rough set theory and genetic algorithms – an application for the simplification of product quality evaluation*, Computers and Industrial Engineering **43** (2002), no. 4, 661–676.
2101. Yanmei Zhan, H. Leung, Keun-Chang Kwak, and Hosub Yoon, *Automated speaker recognition for home service robots using genetic algorithm and Dempster–Shafer fusion technique*, IEEE Transactions on Instrumentation and Measurement **58** (2009), no. 9, 3058–3068.
2102. Bin Zhang and Sargur N. Srihari, *Class-wise multi-classifier combination based on Dempster–Shafer theory*, Proceedings of the 7th International Conference on Control, Automation, Robotics and Vision (ICARCV 2002), vol. 2, December 2002, pp. 698–703.
2103. Daqiang Zhang, Jiannong Cao, Jingyu Zhou, and Minyi Guo, *Extended Dempster–Shafer theory in context reasoning for ubiquitous computing environments*, Proceed-

ings of the International Conference on Computational Science and Engineering (CSE'09), vol. 2, August 2009, pp. 205–212.
2104. Jianchun Zhang and Chuanhai Liu, *Dempster–Shafer inference with weak beliefs*, Statistica Sinica **21** (2011), 475–494.
2105. Jiankang Zhang, *Subjective ambiguity, expected utility and Choquet expected utility*, Economic Theory **20** (2002), no. 1, 159–181.
2106. Lianwen Zhang, *Representation, independence and combination of evidence in the Dempster–Shafer theory*, Advances in Dempster-Shafer theory of evidence (R. R. Yager, M. Fedrizzi, and J. Kacprzyk, eds.), John Wiley and Sons, New York, 1994, pp. 51–69.
2107. Mei Zhang, Li Da Xu, Wen-Xiu Zhang, and Huai-Zu Li, *A rough set approach to knowledge reduction based on inclusion degree and evidence reasoning theory*, Expert Systems **20** (2003), no. 5, 298–304.
2108. Nevin L. Zhang, *Weights of evidence and internal conflict for support functions*, Classic Works of the Dempster-Shafer Theory of Belief Functions (Roland R. Yager and Liping Liu, eds.), Springer, Berlin Heidelberg, 2008, pp. 411–418.
2109. Shan-Ying Zhang, Quan Pan, and Hong-Cai Zhang, *Conflict problem of Dempster–Shafer evidence theory*, Acta Aeronautica Et Astronautica Sinica **4** (2001).
2110. Shanying Zhang, Pan Quan, and Hongcai Zhang, *A new kind of combination rule of evidence theory*, Control and Decision **15** (2000), no. 5, 540–544.
2111. Xin-Man Zhang, Jiu-Qiang Han, and Xue-Bin Xu, *Dempster–Shafer reasoning with application multisensor object recognition system*, Proceedings of the 2004 International Conference on Machine Learning and Cybernetics, vol. 2, August 2004, pp. 975–977.
2112. Wentao Zhao, Tao Fang, and Yan Jiang, *Data fusion using improved Dempster–Shafer evidence theory for vehicle detection*, Fourth International Conference on Fuzzy Systems and Knowledge Discovery (FSKD 2007), vol. 1, August 2007, pp. 487–491.
2113. Xueqiang Zheng, Jinlong Wang, Qihui Wu, and Juan Chen, *Cooperative spectrum sensing algorithm based on Dempster–Shafer theory*, Proceedings of the 11th IEEE Singapore International Conference on Communication Systems (ICCS 2008), November 2008, pp. 218–221.
2114. Chunlai Zhou, *Belief functions on distributive lattices*, Proceedings of the National Conference on Artificial Intelligence (AAAI 2012), 2012, pp. 1968–1974.
2115. ______, *Belief functions on distributive lattices*, Artificial Intelligence **201** (2013), 1–31.
2116. ______, *Logical foundations of evidential reasoning with contradictory information*, J. Michael Dunn on Information Based Logics (Katalin Bimbó, ed.), Springer International Publishing, 2016, pp. 213–246.
2117. Chunlai Zhou, Mingyue Wang, and Biao Qin, *Belief-kinematics Jeffrey-s rules in the theory of evidence*, Proceedings of the Thirtieth Conference on Uncertainty in Artificial Intelligence (UAI 2014), July 2014, pp. 917–926.
2118. Dengyong Zhou, John C. Platt, Sumit Basu, and Yi Mao, *Learning from the wisdom of crowds by minimax entropy*, Proceedings of the 25th International Conference on Neural Information Processing Systems (NIPS'12), 2012, pp. 2195–2203.
2119. Jun Zhou, Zissimos P. Mourelatos, and Clifton Ellis, *Design under uncertainty using a combination of evidence theory and a Bayesian approach*, SAE International Journal of Materials and Manufacturing **1** (2009), no. 1, 122–135.
2120. Kuang Zhou, Arnaud Martin, Quan Pan, and Zhun ga Liu, *Median evidential c-means algorithm and its application to community detection*, ArXiv preprint 1501.01460, 2015.

2121. ______, *Median evidential c-means algorithm and its application to community detection*, Knowledge-Based Systems **74** (2015), 69–88.
2122. Mi Zhou, Xin-Bao Liu, and Jian-Bo Yang, *Evidential reasoning-based nonlinear programming model for MCDA under fuzzy weights and utilities*, International Journal of Intelligent Systems **25** (2010), no. 1, 31–58.
2123. Hongwei Zhu and Otman Basir, *Extended discounting scheme for evidential reasoning as applied to MS lesion detection*, Proceedings of the 7th International Conference on Information Fusion (FUSION 2004) (Per Svensson and Johan Schubert, eds.), 2004, pp. 280–287.
2124. ______, *Automated brain tissue segmentation and MS lesion detection using fuzzy and evidential reasoning*, Proceedings of the 10th IEEE International Conference on Electronics, Circuits and Systems, vol. 3, December 2003, pp. 1070–1073.
2125. ______, *A K-NN associated fuzzy evidential reasoning classifier with adaptive neighbor selection*, Proceedings of the Third IEEE International Conference on Data Mining (ICDM 2003), November 2003, pp. 709–712.
2126. Hongwei Zhu, Otman Basir, and Fakhreddine Karray, *Data fusion for pattern classification via the Dempster–Shafer evidence theory*, Proceedings of the IEEE International Conference on Systems, Man and Cybernetics (SMC'02), vol. 7, October 2002, p. 2.
2127. Hongwei Zhu and Otmar Basir, *A scheme for constructing evidence structures in Dempster–Shafer evidence theory for data fusion*, Proceedings of the IEEE International Symposium on Computational Intelligence in Robotics and Automation, vol. 2, July 2003, pp. 960–965.
2128. Qing Zhu and E. S. Lee, *Dempster–Shafer approach in propositional logic*, International Journal of Intelligent Systems **8** (1993), no. 3, 341–349.
2129. Yue Min Zhu, Layachi Bentabet, Olivier Dupuis, Valérie Kaftandjian, Daniel Babot, and Michele Rombaut, *Automatic determination of mass functions in Dempster–Shafer theory using fuzzy C-means and spatial neighborhood information for image segmentation*, Optical Engineering **41** (April 2002), 760–770.
2130. Yunmin Zhu and X. Rong Li, *Extended Dempster–Shafer combination rules based on random set theory*, Proceedings of SPIE - Multisensor, Multisource Information Fusion: Architectures, Algorithms, and Applications, vol. 5434, 2004, pp. 112–120.
2131. Stephen T. Ziliak and Deirdre N. McCloskey, *The cult of statistical significance: How the standard error costs us jobs, justice, and lives*, University of Michigan Press, 2008.
2132. Hans-Jürgen Zimmermann, *An application-oriented view of modeling uncertainty*, European Journal of Operational Research **122** (2000), no. 2, 190–198.
2133. Loai Zomlot, Sathya Chandran Sundaramurthy, Kui Luo, Xinming Ou, and S. Raj Rajagopalan, *Prioritizing intrusion analysis using Dempster–Shafer theory*, Proceedings of the 4th ACM Workshop on Security and Artificial Intelligence (AISec '11), 2011, pp. 59–70.
2134. Lalla M. Zouhal and Thierry Denœux, *An adaptive k-nn rule based on Dempster–Shafer theory*, Proceedings of the 6th International Conference on Computer Analysis of Images and Patterns (CAIP'95) (V. Hlavac and R. Sara, eds.), Prague, Czech Republic, 6-8 September 1995, pp. 310–317.
2135. Lalla Meriem Zouhal and Thierry Denœux, *An evidence-theoretic k-NN rule with parameter optimization*, IEEE Transactions on Systems, Man and Cybernetics Part C: Applications and Reviews **28** (1998), no. 2, 263–271.
2136. ______, *Generalizing the evidence-theoretic k-NN rule to fuzzy pattern recognition*, Proceedings of the Second International Symposium on Fuzzy Logic and Applications (ISFL'97), ICSC Academic Press, February 1997, pp. 294–300.

2137. Mourad Zribi and Mohammed Benjelloun, *Parametric estimation of Dempster–Shafer belief functions*, Proceedings of the Sixth International Conference on Information Fusion (FUSION 2003), vol. 1, July 2003, pp. 485–491.

Printed in the United States
by Baker & Taylor Publisher Services